RIBOSOMES

Proceedings of the 9th Steenbock Symposium held at the University of Wisconsin-Madison, July 5–8, 1979

RIBOSOMES
Structure, Function, and Genetics

Edited by

G. Chambliss, Ph.D.
Department of Bacteriology

G. R. Craven, Ph.D.
Laboratory of Molecular Biology and
Department of Genetics

J. Davies, Ph.D.
Department of Biochemistry

K. Davis
Department of Biochemistry

L. Kahan, Ph.D.
Department of Physiological Chemistry

M. Nomura, Ph.D.
Institute for Enzyme Research

University of Wisconsin
Madison, Wisconsin

University Park Press
Baltimore

UNIVERSITY PARK PRESS
International Publishers in Science, Medicine, and Education
233 East Redwood Street
Baltimore, Maryland 21202

Manufactured in the United States of America by
The Maple Press Company.

ISBN 0-8391-1569-5

Library of Congress Cataloging in Publication Data

Steenbock Symposium on Ribosomes, 9th, University of Wisconsin—Madison, 1979.
Ribosomes, structure, function, and genetics.

Includes index.
1. Ribosomes—Congresses. I. Chambliss, Glen. II. Title.
QH603.R5S73 1979 574.87'34 79-27158
ISBN 0-8391-1569-5

Contents

Participants

Arlette Adoutte
Department of Biochemistry
University of Wisconsin-Madison
Madison, WI 53706

Paul Ahlquist
Biophysics Laboratory
University of Wisconsin-Madison
Madison, WI 53706

Jean-Hervé Alix
Inst. Biol. Physico-Chimique
Fond. Edmond de Rothschild
13, Rue Pierre et Marie Curie
75005 Paris, FRANCE

Ricardo Amils
Inst. Bioquimica Macromol.
Centro Biol. Mol., CSIC & UAM
Canto Blanco, Madrid-34, SPAIN

Philip E. Auron
The Pennsylvania State Univ.
201 Althouse Laboratory
University Park, PA 16802

Gail A. Baughman
Institute for Enzyme Research
University of Wisconsin-Madison
Madison, WI 53706

Joël Begueret
Lab. de Genetique
Faculte des Sciences
33405-Talence, FRANCE

Donald P. Blair
Department of Chemistry
University of Montana
Missoula, MT 59812

Robert M. Bock
Dean, The Graduate School
Laboratory of Molecular Biology
University of Wisconsin-Madison
Madison, WI 53706

Alexel A. Bogdanov
A. N. Belozersky Lab. Mol. Biol.
& Bioorgan. Chem., Bldg. A
Moscow State University
Moscow, U.S.S.R.

Guy Boileau
Department of Biological Chemistry
University of California-Davis
Davis, CA 95616

Alex Bollen
Fac. Sci., Lab. Genetique
Univ. Libre de Bruxelles
Rue des Chevaux 67-1640
Rhode-St-Genese, BELGIUM

Leendert Bosch
Biochemisch Laboratory
Rijksuniversiteit Leiden
Wassenaarseweg 64
2333 Al Leiden,
THE NETHERLANDS

Miroslav Boublik
Roche Institute of Molecular Biology
Nutley, NJ 07110

John E. Boynton
Department of Zoology
Duke University
Durham, NC 27706

Léa Brakier-Gingras
Dept. de Biochemistry
Universite de Montreal
H3C-3J7, Montreal, Quebec,
CANADA

Christiane Branlant
IBMC, CNRS
15 Rue Descartes
67000 Strasbourg, FRANCE

Jürgen Brosius
Div. Natural Sciences
Thimann Laboratories
Univ. California, Santa Cruz
Santa Cruz, CA 95064

Richard R. Burgess
McArdle Laboratory for
Cancer Research
University of Wisconsin-Madison
Madison, WI 53706

Michael Cannon
Dept. Biochemistry
Univ. London King's College
Strand, London,
ENGLAND WC2R 2LS

Charles R. Cantor
Department of Chemistry
Columbia University
Havemeyer Hall
New York, NY 10027

Jonathan B. Chaires
Biochemistry and Biophysics Section
The University of Connecticut, U-125
Storrs, CT 06268

Kaldana Chakraburtty
Department of Biochemistry
The Medical College of Wisconsin
8701 Watertown Plank Road
Milwaukee, WI 53226

Glenn Chambliss
Department of Bacteriology
University of Wisconsin-Madison
Madison, WI 53706

W. Scott Champney
Department of Biochemistry
The University of Georgia
Athens, GA 30602

Li-Ming Changchien
Laboratory of Molecular Biology
University of Wisconsin-Madison
Madison, WI 53706

Brian F. C. Clark
Aarhus University
Department of Chemistry
Kemisk Institut
Langelandsgade 140
DK-8000 Aarhus C, DENMARK

Jeffrey A. Cohlberg
Chemistry Department
California State University
Long Beach, CA 90840

Richard C. Conrad
Laboratory of Molecular Biology
University of Wisconsin-Madison
Madison, WI 53706

Barry S. Cooperman
Department of Chemistry
University of Pennsylvania
Philadelphia, PA 19174

C. Cowgill de Narvaez
Dept. Biochemistry, Biophysics
Oregon State University
Corvallis, OR 97331

Gary R. Craven
Department of Molecular Biology and
Biophysics
University of Wisconsin-Madison
Madison, WI 53706

Eric Cundliffe
Department of Biochemistry
University of Leicester
Adrian Building, University Road
Leicester LEI 7RH, ENGLAND

Albert Dahlberg
Department of Physiological Chemistry
589 Medical Sciences Building
University of Wisconsin-Madison
Madison, WI 53706

James Dahlberg
Department of Physiological Chemistry
589 Medical Sciences Building
University of Wisconsin-Madison
Madison, WI 53706

Pat P. Dennis
Department of Biochemistry
University of British Columbia
Vancouver, B.C., CANADA

Dipak B. Datta
Laboratory of Molecular Biology
University of Wisconsin-Madison
Madison, WI 53706

Julian E. Davies
Department of Biochemistry
University of Wisconsin-Madison
Madison, WI 53706

Dennis R. Dean
Institute for Enzyme Research
University of Wisconsin-Madison
Madison, WI 53706

Dr. Herman A. Deboer
Institute for Enzyme Research
University of Wisconsin-Madison
Madison, WI 53706

Jan Dijk
Max-Planck-Institut for Molekulare
Genetik
1000 Berlin 33 (Dahlem)
Ihnestrasse 63-73, GERMANY

Ferdinand Dohme
Thimann Laboratory
University of California-Santa Cruz
Santa Cruz, CA 95064

David E. Draper
Dept. Mol. Cell & Dev. Biol.
Univ. Colorado
Boulder, CO 80309

Jean Pierre Ebel
IBMC, CNRS
14 rue Descartes
67000 Strasbourg, FRANCE

Chantal Ehresmann
IBMC, CNRS
15 rue Descartes
67000 Strasbourg, FRANCE

Lea Eisenbach
Institute for Enzyme Research
University of Wisconsin-Madison
Madison, WI 53706

David Elson
The Weizmann Institute of Science
Biochemistry Department
Rehovot, ISRAEL

Volker A. Erdmann
Max-Planck Institute fur
Molekulare Genetik
1000 Berlin 33 (Dahlem)
Ilmestrasse 63-73, GERMANY

Alain Expert-Bezancon
Inst. Biol. Physico-Chimique
Fond. Edmond de Rothschild
13 rue Paerre et Marie Curie
75005 Paris, FRANCE

Stephen R. Fahnestock
Department of Biochemistry
& Biophysics
Pennsylvania State University
University Park, PA 16802

James D. Friesen
Department of Biology
York University
4700 Keele Street
Downsview, Ontario,
CANADA M3J IP3

Jonathan A. Gallant
Department of Genetics
University of Washington
Seattle, WA 98195

Manuel Garcia-Patrone
Department of Biological Chemistry
School of Medicine
University of California-Davis
Davis, CA 95616

Hans Günter Gassen
Tech. Hochschule Darmstadt
Inst. Organische Chemie
61 Darmstadt, Petersenstrasse 22
GERMANY

Kirsten Gausing
Department of Molecular Biology
University of Aarhus
DK-8000 Arhus, DENMARK

Susan Gerbi
Biomedical Division
Brown University
Box G
Providence, RI 02912

Nicholas W. Gillham
Department of Zoology
Duke University
Durham, NC 27706

Lary Gold
Dept. Mol., Cell. & Dev. Biol.
University of Colorado
Boulder, CO 80309

Julian Gordon
Friedrich Miescher-Institut
Postfach 273
CH-4002 Basel, SWITZERLAND

Madge Yang Graham
Department of Microbiology and
Immunology
Washington University School of
Medicine
St. Louis, MO 63110

Linda R. Gritz
Department of Biochemistry
University of Wisconsin-Madison
Madison, WI 53706

Carol A. Gross
McArdle Laboratory for Cancer
Research
University of Wisconsin-Madison
Madison, WI 53706

Marianne Grunberg-Manago
Inst. Biol. Physico-Chimique
Fond. Edmond de Rothschild
13, rue Pierre et Marie Curie
75005 Paris, FRANCE

Richard Hallberg
Plant Science Building
Cornell University
Ithaca, NY 14853

Marc Hansen
Department of Bacteriology
University of Wisconsin-Madison
Madison, WI 53706

Boyd Hardesty
Department of Chemistry
University of Texas at Austin
Austin, TX 78712

Donal H. Hayes
Inst. Biol. Physico-Chimique
Fond. Edmond de Rothschild
13 rue Pierre et Marie Curie
75005 Paris, FRANCE

Tina M. Henkin
Department of Bacteriology
University of Wisconsin-Madison
Madison, WI 53706

James R. Hildebrandt
Department of Chemistry
Columbia University
New York, NY 10027

Heinz-Kurt
Friedrich Miescher-Inst.
P. O. Box 273
CH-4002 Basel, SWITZERLAND

John J. Hopfield
Department of Physics
Princeton University
Princeton, NJ 08540

Jack Horowitz
Biochemistry and Biophysics Dept.
Iowa State University
Ames, IA 50011

Jean-Francois Houssais
Institut Curie, Bat. 110
Universite Paris-Sud
91405 Orsay, FRANCE

Leif Isaksson
The Wallenberg Laboratory
University of Uppsala
Uppsala, SWEDEN

Katsumi I. Isono
Max-Planck-Institut fur
Molekulare Genetik
1 Berlin 33 (Dahlem)
Ihnestrasse 63-73, GERMANY

Rosemary Jagus
Lab. Molecular Hematology
NIH, Bldg. 10, Rm. 7012
Bethesda, MD 20205

Sue Jinks
Institute for Enzyme Research
University of Wisconsin-Madison
Madison, WI 53706

Arthur E. Johnson
Department of Chemistry
University of Oklahoma
Norman, OK 73019

Paul Kaesberg
Depts. Biochemistry & Biophysics
University of Wisconsin-Madison
Madison, WI 53706

Sarah Kagan
Department of Biochemistry
University of Wisconsin-Madison
Madison, WI 53706

Lawrence Kahan
Department of Physiological
Chemistry
University of Wisconsin-Madison
Madison, WI 53706

Andrea Kaminski
Department of Bacteriology
1550 Linden Drive
Madison, WI 53706

Terrence Kayes
Department of Food Science
103 Babcock Hall
University of Wisconsin-Madison
Madison, WI 53706

Marian E. Kenerley
Institute for Enzyme Research
University of Wisconsin-Madison
Madison, WI 53706

Ken-ichi Komatsu
Department of Biochemistry
University of Wisconsin-Madison
Madison, WI 53706

Charles G. Kurland
The Wallenberg Laboratory
University of Uppsala
Uppsala, SWEDEN

Thomas King
Department of Microbiology
Washington Univ. Sch. Med.
St. Louis, MO 63110

Gisela Kramer
Department of Chemistry
University of Texas at Austin
Austin, TX 78712

Judy Krueger
Institute for Enzyme Research
University of Wisconsin-Madison
Madison, WI 53706

Ernst Kuechler
Inst. fur Biochem. Univ. Wien
A-1090 Wien IX
Wahringerstrasse 17, AUSTRIA

Werner Kühlbrandt
MRC Lab. of Mol. Biol.
Univ. of Cambridge
Hills Road
Cambridge, ENGLAND CB2 2QH

James Lake
Molecular Biology Institute
University of California, Los Angeles
Los Angeles, CA 90024

John M. Lambert
Department of Biological Chemistry
Univ. California-Davis
School of Medicine
Davis, CA 95616

Alan Lambowitz
Department of Microbiology
St. Louis Univ. Med. Ctr.
St. Louis, MO 63104

Jerome A. Langer
Department of Biology
Univ. California-Los Angeles
Los Angeles, CA 90024

David P. Leader
Department of Biochemistry
The University of Glasgow
Glasgow, SCOTLAND G12 8QQ

Jean Leboul
Department of Biochemistry
University of Wisconsin-Madison
Madison, WI 53706

John C. Lee
Department of Biochemistry
Univ. Texas Health Sciences Ctr.
San Antonio, TX 78284

Judy Leventhal
Department of Bacteriology
University of Wisconsin-Madison
Madison, WI 53706

Hope Liebke
Biological Sciences Group
Genetics and Cell Biology
University of Connecticut
Storrs, CT 06066

Terrance Leighton
Department of Bacteriology &
Immunology
University of California-Berkeley
Berkeley, CA 94720

Lasse Lindahl
Department of Biology
University of Rochester
Rochester, NY 14627

Thomas Linn
Laboratory of Molecular Biology
University of Wisconsin-Madison
Madison, WI 53706

Jennifer Littlechild
Max-Planck-Institut fur
Molekulare Genetik
1000 Berlin 33 (Dahlem)
Ihnestrasse 63-73, GERMANY

Barbara Littlewood
Department of Biochemistry
University of Wisconsin-Madison
Madison, WI 53706

Jean Lucas-Lenard
Biochemistry and Biophysics Section
University of Connecticut
Storrs, CT 06268

Kenneth Luehrsen
Department of Biophysical Sciences
University of Houston
Houston, TX 77004

Reinhard Lührmann
Max-Planck-Institut fur Molekulare Genetik
1000 Berlin 33 (Dahlem)
Ihnestrasse 63-73, GERMANY

Elsebet Lund
Department of Physiological Chemistry
University of Wisconsin-Madison
Madison, WI 53706

George A. Mackie
Department of Biochemistry
University of Western Ontario
London, Ontario, CANADA

Jean-Jacques Madjar
Department of Biological Chemistry
University of California-Davis
Davis, CA 95616

David Marquis
The Pennsylvania State University
108 Althouse Laboratory
University Park, PA 16802

Al T. Matheson
Department of Biochemistry
University of Victoria
Vancouver Island, CANADA

John R. Menninger
Zoology Department
University of Iowa
Iowa City, IA 52242

Valerie A. Mihelc
Department of Bacteriology
University of Wisconsin-Madison
Madison, WI 53706

Juan Modolell
Centro Biologia Molecular
Univ. Autonoma de Madrid
Fac. de Ciencias, Canto Blanco
Madrid 34, SPAIN

Terry E. Moen
Department of Bacteriology
University of Wisconsin-Madison
Madison, WI 53706

Wim Moller
Laboratorium Voor Fysiologische Scheikunde
Faculteit Geneeskunde
Rijksuniversiteit Leiden
Wassenaarseweg 72
Leiden, THE NETHERLANDS

Peter B. Moore
Department of Chemistry
Yale University
New Haven, CT 06510

Knud Nierhaus
Max-Planck-Institut fur Molkulare Genetik
1000 Berlin 33 (Dahlem)
Ihnestrasse 63-73, GERMANY

Franz Noll
Akad. Wissenschaften der GDR
Zentralinst. Molekularbiol.
GDR 1115 Berlin-Buch, GERMANY

Harry F. Noller
Thimann Laboratories
Division of Natural Sciences
University of California
Santa Cruz, CA 95064

Masayasu Nomura
The Institute for Enzyme Research
University of Wisconsin-Madison
Madison, WI 53706

Volker Nowotny
Max-Planck-Institut fur Molekulare Genetik
Ihnestrasse 63-73
D-1000 Berlin 33, GERMANY

Thomas W. O'Brien
Department of Biochemistry and Molecular Biology
University of Florida
Gainesville, FL 32610

James Ofengand
Roche Institute of Molecular Biology
Nutley, NJ 07110

Peter Olins
The Institute for Enzyme Research
University of Wisconsin-Madison
Madison, WI 53706

Syozo Osawa
Research Institute for Nuclear
Medicine and Biology
Hiroshima University
Kasumi 1-2-3, Hiroshima, JAPAN

Christian Oste
Department of Biochemistry
The Weizmann Institute of Science
Rehovot, ISRAEL

Andrea Parmeggiani
Ecole Polytechnique
Lab. de Biochimie
91128 Palaiseau, FRANCE

Nancy J. Pearson
Rosenstiel Basic Medical Sciences
Research Center
Brandeis University
Waltham, MA 02154

Hans U. Petersen
Kemisk Institute
Aarhus University
Langelandsgade 140
DK-8000 Aarhus C, DENMARK

Marguerite Picard-Bennoun
Lab. de Genetique de l'Univ. Paris-Suc
Batiment 400
91405 Orsay-Cedex, FRANCE

Wolfgang Piepersberg
Lehrstuhl fur Mikrobiologie
Universitat Regensburg
Postfach D-8400 Rengensburg
FRG, GERMANY

Rudi J. Planta
Biochemisch Laboratorium
Subfaculteit der Scheikunde
Vrije Universiteit
De Boelelaan 1085
Amsterdam (Buitenveldert),
THE NETHERLANDS

Hubert Pöche
Max-Planck-Institut fur Molekulare
Genetik
1000 Berlin 33 (Dahlem)
Ihnestrasse 63-73, GERMANY

Leonard Post
The Institute for Enzyme Research
University of Wisconsin-Madison
Madison, WI 53706

Gilda Quan
Department of Bacteriology
University of Wisconsin-Madison
Madison, WI 53706

Bimal Ray
Department of Microbiology and
Immunology
Washington University School of
Medicine
St. Louis, MO 63110

Jean-Paul Reboud
Univ. Lyon 1, Lab. Biochem.,
Medicale, Unite Lyon Nord
43 Boulevard
69 Villeurbanne, FRANCE

Tom Record
Department of Chemistry
University of Wisconsin-Madison
Madison, WI 53706

Vernon E. Reichenbecher, Jr.
Department of Genetics
Baylor College of Medicine
Houston, TX 77030

Fritz Reusser
The Upjohn Company
Kalamazoo, MI 49001

Alexander Rich
Department of Biology
Massachusetts Institute of Technology
Cambridge, MA 02139

Dietmar Richter
Institut fur Physiologische Chemie
Universitat Hamburg
Abt. Zellbiochemie
2 Hamburg 20, Martinstr. 52,
GERMANY

Jutta Rinke
Max-Planck-Institut fur Molekulare
Genetik
1000 Berlin 33 (Dahlem)
Ihnestrasse 63-73, GERMANY

Peter J. Russell
Biology Department
Reed College
Portland, OR 97202

Melvin Santer
Biology Department
Haverford College
Haverford, PA 19041

Henry W. Schaup
Department of Molecular Biology
University of Wisconsin-Madison
Madison, WI 53706

David Schlessinger
Department of Microbiology and Immunology
Washington University School of Medicine
St. Louis, MO 63110

Ira Schwartz
Department of Biochemistry
The Commonwealth of Massachusetts
University of Massachusetts
Amherst, MA 01003

Jean Schwarzbauer
Laboratory of Molecular Biology
University of Wisconsin-Madison
Madison, WI 53706

Paul F. G. Sims
Department of Biochemistry
University of Oxford
Oxford, UNITED KINGDOM OX1 3QU

Arjun Singh
Department of Biochemistry
University of Wisconsin-Madison
Madison, WI 53706

Pnina Spitnik-Elson
The Weizmann Inst. of Science
Biochemistry Department
Rehovot, ISRAEL

Mathias Springer
Inst. Biol. Physico-Chimique
Fond. Edmond de Rothschild
13 rue Pierre et Marie Curie
Paris Ve, FRANCE

Joan Steitz
Department of Molecular Biophysics and Biochemistry
Yale University
New Haven, CT 06510

Scott Stibitz
Department of Biochemistry
University of Wisconsin-Madison
Madison, WI 53706

William A. Strycharz
The Institute for Enzyme Research
University of Wisconsin-Madison
Madison, WI 53706

Muttaiya Sundaralingam
Department of Biochemistry
University of Wisconsin-Madison
Madison, WI 53706

Claudia Sutton
Institute for Enzyme Research
University of Wisconsin-Madison
Madison, WI 53706

Jill R. Thompson
Department of Biochemistry
Univ. Leicester
Adrian Bldg., University Rd.
Leicester LE1 7RH, ENGLAND

David L. Thurlow
Department of Biochemistry
University of Massachusetts
Amherst, MA 01003

Robert R. Traut
Department of Biological Chemistry
University of California, Davis
Davis, CA 95616

Elio F. Vanin
Genetics Department
University of Wisconsin-Madison
Madison, WI 53706

Charles A. Vaslet
Rosenstiel Basic Medical Sciences Research Center
Brandeis University
Waltham, MA 02154

David Vázquez
Instituto de Bioquimica de Macromoleculas
CSIC and UAM
Canto Blanco, Madrid-34, SPAIN

Jonathan R. Warner
Department of Biochemistry
Albert Einstein College of Medicine of Yeshiva University
1300 Morris Park Avenue, Bronx
New York, NY 10461

James D. Watson
Cold Spring Harbor Laboratory
Cold Spring Harbor, Long Island, NY 11724

Herbert Weissbach
Roche Institute of Molecular Biology
Nutley, NJ 07110

Peter Wejksnora
Department of Biochemistry
Albert Einstein College of Medicine
1300 Morris Park Ave.
Bronx, NY 10461

Donald Winkelmann
Physiological Chemistry
589 Medical Sciences Bldg.
University of Wisconsin-Madison
Madison, WI 53706

Heinz-Gunter Wittmann
Max-Planck-Institut fur Molekulare Genetik
1000 Berlin 33 (Dahlem)
Ihnestrasse 63-73, GERMANY

Carl Woese
Department of Microbiology
University of Illinois
Urbana, IL 61801

Paul L. Wollenzien
Department of Chemistry
Columbia University
New York, NY 10027

Ira G. Wool
Department of Biochemistry
The University of Chicago
920 East 58th Street
Chicago, IL 60637

John Woolford, Jr.
Rosenstiel Basic Medical Sciences Center
Brandeis University
Waltham, MA 02154

Paul Wrede
Department of Biology
Massachusetts Institute of Technology
Cambridge, MA 02139

Peter Wurmbach
Max-Planck-Institut fur Molekulare Genetik
100 Berlin
Ihnestrasse 63-73, GERMANY

Makoto Yaguchi
Division of Biological Sciences
National Research Council
Ottawa, Ontario, CANADA K1A OR6

John Yates
The Institute for Enzyme Research
University of Wisconsin-Madison
Madison, WI 53706

Janice Zengel
Department of Biology
University of Rochester
Rochester, NY 14627

Robert A. Zimmermann
Commonwealth of Massachusetts
University of Massachusetts
Department of Biochemistry
Amherst, MA 01003

Preface

The 1974 Cold Spring Harbor Symposium on Ribosomes, organized by Peter Lengyel, Masayasu Nomura, and Alfred Tissieres, was significant in that it provided the first text that summarized diverse studies on ribosomes. Since that time, although there have been excellent EMBO Ribosome Workshops to provide a medium for information exchange between workers in the field, there has been no general summary of the state-of-the-art for scientists whose research interests were not directly involved with this magic organelle.

It is for this reason that the Steenbock Symposium was planned, and this book is the outcome. During the intervening five years, the ribosome field has taken creative advantage of new physical and chemical techniques to provide a wealth of new information on ribosome structure and biosynthesis and its function in protein synthesis. These advances have been obtained by imaginative use of many techniques, including nucleic acid and protein sequencing, immunoelectron microscopy, neutron scattering, chemical and photochemical modification, and, of course, genetics to provide insights into protein and RNA interactions in the ribosome structure and the roles that they play in ribosome function. There has been renewed interest in such subjects as the fidelity of translation, and considerable success in the application of these various methods to the study of eukaryotic ribosomes.

The editors have used the Cold Spring Harbor book as a basis and have tried to build on this text to bring the reader up-to-date. Although the book presents very substantial progress, it should be patently obvious that many puzzles remain to be unraveled. In spite of the large volume of work, we still lack details of the chemistry (or physics) of protein synthesis, and how the ribosome is concerned with this process. The search for the ethereal peptidyltransferase will be with us for some time! However, we are confident that, as structural and functional details become collated, the pieces of the puzzle will surely fall into place.

Notwithstanding the success of the symposium, we all recognized the sad absence of Luigi Gorini. Luigi contributed much originality and industry to the ribosome field, and the benefits of his work will be apparent for years to come; Luigi enriched our lives with his humor, energy, and enthusiasm.

The organizers of the symposium wish to thank Mrs. Evelyn Steenbock for the support of the proceedings and University Park Press for their helpful encouragement. The participants in the symposium and the authors of chapters were tolerant of our organization and of our regulations for manuscripts. We thank them for their many contributions to a rewarding and enjoyable week in Madison in July 1979.

G. Chambliss
G. Craven
J. Davies
K. Davis
L. Kahan
M. Nomura

AMERICA'S DAIRYLAND
RIBOSM
JAN WISCONSIN

AMERICA'S DAIRYLAND
RIBOSM
JAN WISCONSIN

TOPOGRAPHY OF
A.A.Bogdanov

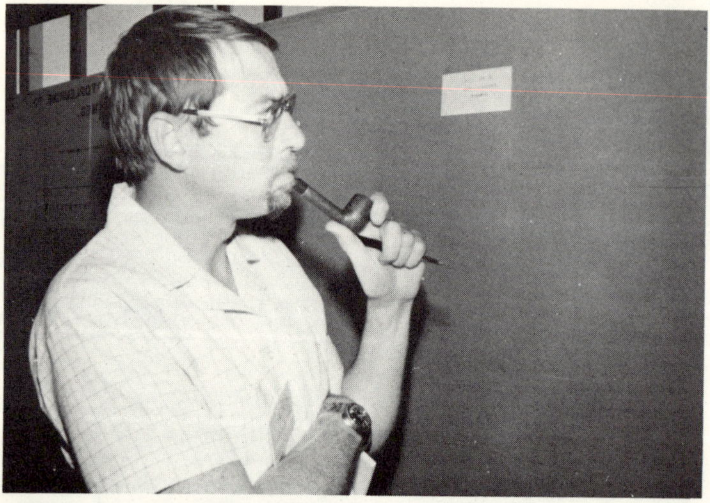

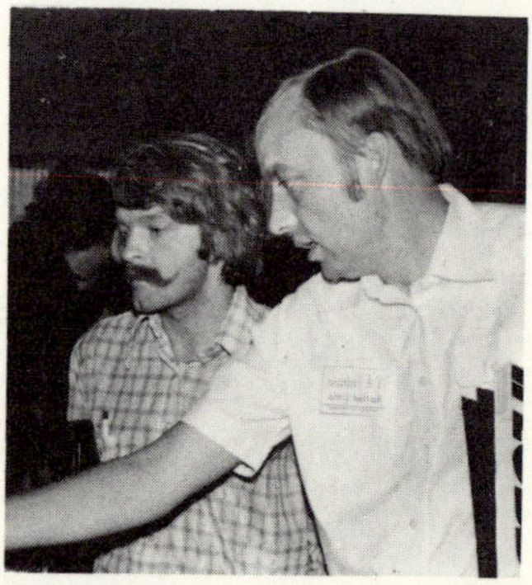

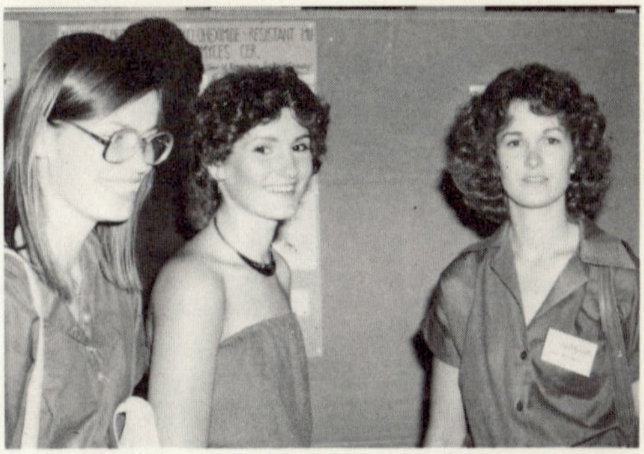

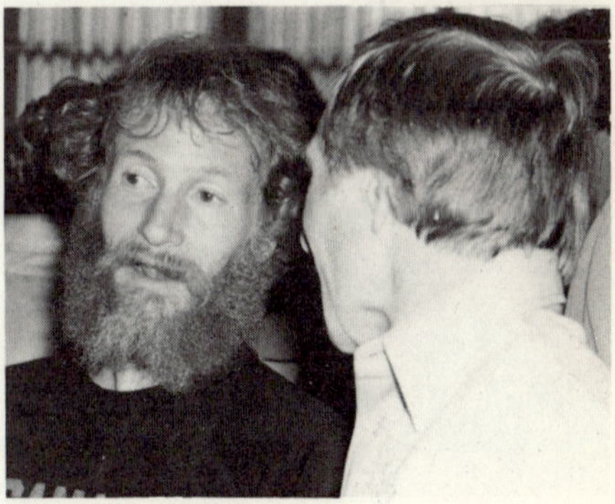

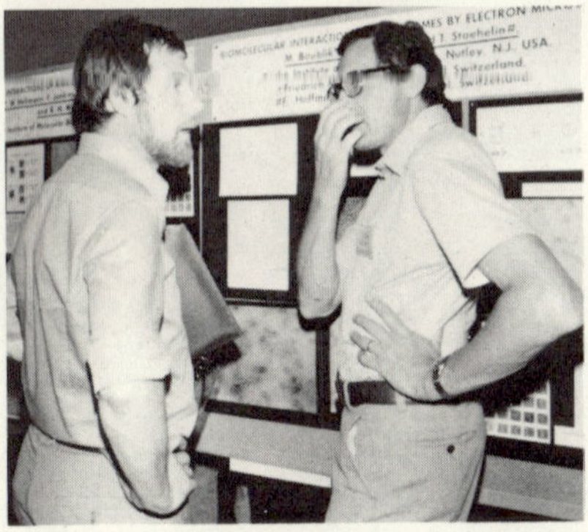
BY ELECTRON
Staehelin#
Nutley N.J. USA
Switzerland

PART 1: STRUCTURE AND IN VITRO ASSEMBLY

STRUCTURE AND TOPOGRAPHY OF RIBOSOMAL RNA

HARRY F. NOLLER

Thimann Laboratories,
University of California,
Santa Cruz, CA 95064

INTRODUCTION

A striking feature of the ribosome is that it is constructed from both protein and nucleic acid components. It is not yet clear why the ribosome, but not other macromolecular assemblies of the genetic apparatus such as RNA polymerase, should have RNA as a major part of its structure. A reasonable guess is that it emerged very early in molecular evolution, and that ribosomal RNA played some crucial role, replacement or alteration of which would have been lethal. Very likely, rRNA continues to play such a role in modern ribosomes, and so its elucidation could provide clues to the understanding of early events in molecular evolution (see Woese, this volume).

As recently as five years ago, an accepted answer to these questions was that the role of rRNA is to provide a scaffold for the assembly of ribosomal proteins (1). This view was to some extent a reflection of the great advances in ribosomal protein assembly, structure and function that had been made at that time. It is now well established that rRNA also participates directly in the functioning of the ribosome, in addition to its undisputed contributions to structural organization. Among the functions for which there is evidence of rRNA participation are 1) mRNA selection (2-4); 2) tRNA binding (5,6); 3) antibiotic sensitivity/resistance (7-9); and 4) subunit association (10-12).

New knowledge of the nature of ribosomal RNA should, then, illuminate our perception of the evolution, assembly, structure and function of ribosomes. I present here a summary of recent progress on the structure and topography of rRNA, and explore some of the implications that emerge.

Primary Structure

5S RNA

E. coli 5S RNA was one of the first nucleic acids to be sequenced (13). Many others have now been completed, and are summarized by Erdmann (14) and Hori (15).

16S RNA

The pioneering work by the Strasbourg group using conventional RNA fingerprinting methods resulted in a partially complete sequence (16). The complete 16S RNA sequence was obtained by DNA sequencing using the cloned rrnB gene (17) and by rapid RNA methods (18). The sequence of the rrnB 16S rRNA is 1541 nucleotides long, and is shown in Fig. 1. There are eight discrepancies between the two complete sequences. One of these is probably due to an idiosyncratic sequence in the rrnB cistron around position 90. About half of another 16S RNA gene has been sequenced by a third group (53). Their findings are in agreement with those of Brosius et al. (17) with respect to the remaining controversial nucleotides.

23S RNA

Early studies on the primary structure of 23S RNA, like 16S RNA, were based on classical RNA sequence methods (19, 20). A complete sequence for the E. coli rrnB 23S RNA has recently been deduced (J. Brosius, T. Dull and H. Noller, unpublished), and is shown in Figure 2. It contains 2904 nucleotides, in good agreement with previous estimates. This sequence was obtained using the Maxam-Gilbert DNA sequence method on cloned restriction fragments of the rrnB gene. Another group has reported the sequence of about 630 nucleotides of a 23S RNA gene by DNA methods (53). There are numerous discrepancies between their sequence and that shown in Figure 2.

5S RNA Topography

Unlike proteins, nucleic acids have simple rules for base pairing which, in principle, make possible the deduction of secondary structure from primary structure. The early success of the cloverleaf model for tRNA (21) was followed, however, by an ominous vacuum. The secondary structure of the slightly more complex 5S RNA has been a matter of active controversy for more than a decade since the completion of its primary

pAAAUUGAAGAGUUUGAUCAUGGCUCAGAUUGAACGCUGGCGGCAGGCCUA 50
ACACAUGCAAGUCGAACGGUAACAGGAAGAAGCUUGCUCUUUGCUGACGA 100
GUGGCGGACGGGUGAGUAAUGUCUGGGAAACUGCCUGAUGGAGGGGGAUA 150
ACUACUGGAAACGGUAGCUAAUACCGCAUAACGUCGCAAGACCAAAGAGG 200
GGGACCUUCGGGCCUCUUGCCAUCGGAUGUGCCCAGAUGGGAUUAGCUAG 250
UAGGUGGGGUAACGGCUCACCUAGGCGACGAUCCCUAGCUGGUCUGAGAG 300
GAUGACCAGCCACACUGGAACUGAGACACGGUCCAGACUCCUACGGGAGG 350
CAGCAGUGGGGAAUAUUGCACAAUGGGCGCAAGCCUGAUGCAGCCAUGCC 400
GCGUGUAUGAAGAAGGCCUUCGGGUUGUAAAGUACUUUCAGCGGGGAGGA 450
AGGGAGUAAAGUUAAUACCUUUGCUCAUUGACGUUACCCGCAGAAGAAGC 500
ACCGGCUAACUCCGUGCCAGCAGCCGCGGUAAUACGGAGGGUGCAAGCGU 550
UAAUCGGAAUUACUGGGCGUAAAGCGCACGCAGGCGGUUUGUUAAGUCAG 600
AUGUGAAAUCCCCGGGCUCAACCUGGGAACUGCAUCUGAUACUGGCAAGC 650
UUGAGUCUCGUAGAGGGGGGUAGAAUUCCAGGUGUAGCGGUGAAAUGCGU 700
AGAGAUCUGGAGGAAUACCGGUGGCGAAGGCGGCCCCCUGGACGAAGACU 750
GACGCUCAGGUGCGAAAGCGUGGGGAGCAAACAGGAUUAGAUACCCUGGU 800
AGUCCACGCCGUAAACGAUGUCGACUUGGAGGUUGUGCCCUUGAGGCGUG 850
GCUUCCGGAGCUAACGCGUUAAGUCGACCGCCUGGGGAGUACGGCCGCAA 900
GGUUAAAACUCAAAUGAAUUGACGGGGGCCCGCACAAGCGGUGGAGCAUG 950
UGGUUUAAUUCGAUGCAACGCGAAGAACCUUACCUGGUCUUGACAUCCAC 1000
GGAAGUUUUCAGAGAUGAGAAUGUGCCUUCGGGAACCGUGAGACAGGUGC 1050
UGCAUGGCUGUCGUCAGCUCGUGUUGUGAAAUGUUGGGUUAAGUCCCGCA 1100
ACGAGCGCAACCCUUAUCCUUUGUUGCCAGCGGUCCGGCCGGGAACUCAA 1150
AGGAGACUGCCAGUGAUAAACUGGAGGAAGGUGGGGAUGACGUCAAGUCA 1200
UCAUGGCCCUUACGACCAGGGCUACACACGUGCUACAAUGGCGCAUACAA 1250
AGAGAAGCGACCUCGCGAGAGCAAGCGGACCUCAUAAAGUGCGUCGUAGU 1300
CCGGAUUGGAGUCUGCAACUCGACUCCAUGAAGUCGGAAUCGCUAGUAAU 1350
CGUGGAUCAGAAUGCCACGGUGAAUACGUUCCCGGGCCUUGUACACACCG 1400
CCCGUCACACCAUGGGAGUGGGUUGCAAAAGAAGUAGGUAGCUUAACCUU 1450
CGGGAGGGCGCUUACCACUUUGUGAUUCAUGACUGGGGUGAAGUCGUAAC 1500
AAGGUAACCGUAGGGGAACCUGCGGUUGGAUCACCUCCUUA$_{OH}$ 1541

Fig. 1. Primary structure of E. coli 16S RNA, from the DNA sequence of the rrn B gene (17).

```
pGGUUAAGCGACUAAGCGUACACGGUGGAUGCCCUGGCAGUCAGAGGCGAU     50
                                           K
 GAAGGACGUGCUAAUCUGCGAUAAGCGUCGGUAAGGUGAUAUGAACCGUU    100
                                         K
 AUAACCGGCGAUUUCCGAAUGGGGAAACCCAGUGUGUUUCGACACACUAU    150
 CAUUAACUGAAUCCAUAGGUUAAUGAGGCGAACCGGGGGAACUGAAACAU    200
 CUAAGUACCCCGAGGAAAAGAAAUCAACCGAGAUUCCCCCAGUAGCGGCG    250
                           K
 AGCGAACGGGGAGCAGCCCAGAGCCUGAAUCAGUGUGUGUGUUAGUGGAA    300
 GCGUCUGGAAAGGCGCGCGAUACAGGGUGACAGCCCCGUACACAAAAAUG    350
                                       K
 CACAUGCUGUGAGCUCGAUGAGUAGGGCGGGACACGUGGUAUCCUGUCUG    400
 AAUAUGGGGGGACCAUCCUCCAAGGCUAAAUACUCCUGACUGACCGAUAG    450
 UGAACCAGUACCGUGAGGGAAAGGCGAAAAGAACCCCGGCGAGGGGAGUG    500
 AAAAAGAACCUGAAACCGUGUACGUACAAGCAGUGGGAGCACGCUUAGGC    550
 GUGUGACUGCGUACCUUUUGUAUAAUGGGUCAGCGACUUAUAUUCUGUAG    600
            K
 CAAGGUUAACCGAAUAGGGGAGCCGAAGGGAAACCGAGUCUUAACUGGGC    650
 GUUAAGUUGCAGGGUAUAGACCCGAAACCCGGUGAUCUAGCCAUGGGCAG    700
                                              *
 GUUGAAGGUUGGGUAACACUAACUGGAGGACCGAACCGACUAAUGΨTGAA    750
 AAAUUAGCGGAUGACUUGUGGCUGGGGGUGAAAGGCCAAUCAAACCGGGA    800
 GAUAGCUGGUUCUCCCCGAAAGCUAUUUAGGUAGCGCCUCGUGAAUUCAU    850
 CUCCGGGGGUAGAGCACUGUUUCGGCAAGGGGGUCAUCCCGACUUACCAA    900
       K
 CCCGAUGCAAACUGCGAAUACCGGAGAAUGUUAUCACGGGAGACACACGG    950
 CGGGUGCUAACGUCCGUCGUGAAGAGGGAAACAACCCAGACCGCCAGCUA   1000
 AGGUCCCAAAGUCAUGGUUAAGUGGGAAACGAUGUGGGAAGGCCCAGACA   1050
                                           K
 GCCAGGAUGUUGGCUUAGAAGCAGCCAUCAUUUAAAGAAAGCGUAAUAGC   1100
 UCACUGGUCGAGUCGGCCUGCGCGGAAGAUGUAACGGGGCUAAACCAUGC   1150
 ACCGAAGCUGCGGCAGCGACGCUUAUGCGUUGUUGGGUAGGGGAGCGUUC   1200
 UGUAAGCCUGCGAAGGUGUGCUGUGAGGCAUGCUGGAGGUAUCAGAAGUG   1250
 CGAAUGCUGACAUAAGUAACGAUAAAGCGGGUGAAAAGCCCGCUCGCCGG   1300
 AAGACCAAGGGUUCCUGUCCAACGUUAAUCGGGGCAGGGUGAGUCGACCC   1350
 CUAAGGCGAGGCCGAAAGGCGUAGUCGAUGGGAAACAGGUUAAUAUUCCU   1400
 GUACUUGGUGUUACUGCGAAGGGGGGACGGAGAAGGCUAUGUUGGCCGGG   1450
 CGACGGUUGUCCCGGUUUAAGCGUGUAGGCUGGUUUUCCAGGCAAAUCCG   1500
 GAAAAUCAAGGCUGAGGCGUGAUGACGAGGCACUACGGUGCUGAAGCAAC   1550
                       K
 AAAUGCCCUGCUUCCAGGAAAAGCCUCUAAGCAUCAGGUAACAUCAAAUC   1600
```

```
                  *
GUACCCCAAACCGACACAGGUGGUCAGGUAGAGAAUACCAAGGCGCUUGA    1650

GAGAACUCGGGUGAAGGAACUAGGCAAAAUGGUGCCGUAACUUCGGGAGA    1700

AGGCACGCUGAUAUGUAGGUGAGGUCCCUCGCGGAUGGAGCUGAAAUCAG    1750

UCGAAGAUACCAGCUGGCUGCAACUGUUUAUUAAAAACACAGCACUGUGC    1800

AAACACGAAAGUGGACGUAUACGGUGUGACGCCUGCCCGGUGCCGGAAGG    1850

UUAAUUGAUGGGGUUAGCGCAAGCGAAGCUCUUGAUCGAAGCCCCGGUAA    1900
               *
ACGGCGGCCGΨAACUAΨAACGGUCCUAAGGUAGCGAAATUCCUUGUCGGG    1950
   K
UAAGUUCCGACCUGCACGAAUGGCGUAAUGAUGGCCAGGCUGUCUCCACC    2000

CGAGACUCAGUGAAAUUGAACUCGCUGUGAAGAUGCAGUGUACCCGCGGC    2050
                   *                      K
AAGACGGAAAGACCCCGUGAACCUUUACUAUAGCUUGACACUGAACAUUG    2100
 K             K
AGCCUUGAUGUGUAGGAUAGGUGGGAGGCUUUGAAGUGUGGACGCCAGUC    2150

UGCAUGGAGCCGACCUUGAAAUACCACCCUUUAAUGUUUGAUGUUCUAAC    2200

GUUGACCCGUAAUCCGGGUUGCGGACAGUGUCUGGUGGGUAGUUUGACUG    2250
*
GGGCGGUCUCCUCCUAAAGAGUAACGGAGGAGCACGAAGGUUGGCUAAUC    2300
      (K)(K)
CUGGUCGGACAUCAGGAGGUUAGUGCAAUGGCAUAAGCCAGCUUGACUGC    2350
                                         K  *
GAGCGUGACGGCGCGAGCAGGUGCGAAAGCAGGUCAUAGUGAUCCGGUGG    2400
                                                *
UUCUGAAUGGAAGGGCCAUCGCUCAACGGAUAAAAGGUACUCCGGGGAUA    2450
       (K)
ACAGGCUGAUACCGCCCAAGAGUUCAUAUCGACGGCGGUGUUUGGCACCU    2500

CGAUGUCGGCUCAUCACAUCCUGGGGCUGAAGUAGGUCCCAAGGGUAUGG    2550

CUGUUCGCCAUUUAAAGUGGUACGCGAGCUGGGUUUAGAACGUCGUGAGA    2600

CAGUUCGGUCCCUAUCUGCCGUGGGCGCUGGAGAACUGAGGGGGGCUGCU    2650

CCUAGUACGAGAGGACCGGAGUGGACGCAUCACUGGUGUUCGGGUUGUCA    2700

UGCCAAUGGCACUGCCCGGUAGCUAAAUGCGGAAGAGAUAAGUGCUGAAA    2750
(K)
GCAUCUAAGCACGAAACUUGCCCCGAGAUGAGUUCUCCCUGACCCUUUAA    2800

GGGUCCUGAAGGAACGUUGAAGACGACGACGUUGAUAGGCCGGGUGUGUA    2850

AGCGCAGCGAUGCGUUGAGCUAACCGGUACUAAUGAACCGUGAGGCUUAA    2900

CCUU(OH)                                              2904
```

Fig. 2. Primary structure of *E. coli* 23S RNA, from the DNA sequence of the *rrnB* gene (J. Brosius, T. Dull and H. Noller, unpublished). An asterisk denotes unidentified post-transcriptional modifications. Positions of many modified nucleotides remain undetermined. Guanine residues which are modified by Kethoxal in 50S subunits are marked by K. Circled K's indicate sites which are protected from modification by 30S subunits.

structure. Serious contemplation of the secondary structures of the large rRNA's invokes the need for a set of rigorous criteria to help discriminate between the overwhelmingly numerous possibilities. The ultimate test of x-ray crystallographic analysis appears to be out of reach for the foreseeable future.

The most appealing secondary structure model for the A form of 5S RNA was deduced by Fox and Woese (22). Their model, with minor modifications, is shown in Figure 3.

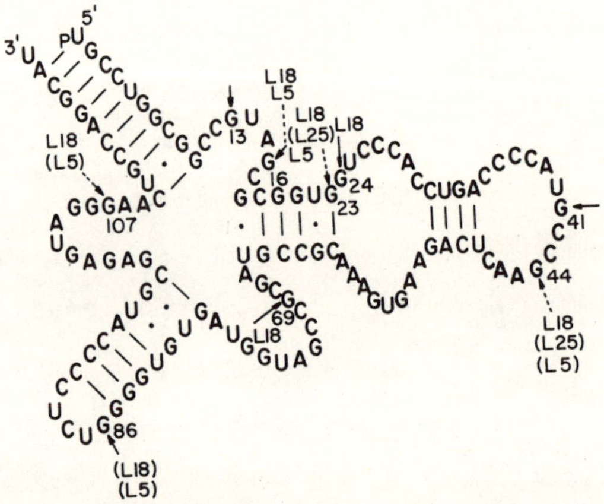

Fig. 3. A secondary structure model for 5S RNA based on that of Fox and Woese (22). Arrows show sites of kethoxal modification (23); dashed arrows indicate weak modification. Protection of sites from kethoxal by proteins L5, L18 and L25 is shown (33).

Positions of nuclease attack and chemical modification are in good agreement with the proposed structure. Most significantly, all known procaryotic 5S RNA sequences can be fitted to this model. The lack of accessibility of some unpaired regions indicates that there is additional structure not explained by the secondary structural scheme in Figure 3. Most notably, the GAAC sequence at G_{44} is resistant to modification by a variety of single strand-specific reagents (23, 24). This sequence has often been suggested to pair with the invariant GTΨC sequence in tRNA (25), which is itself involved in a very stable tertiary structure (26, 27) (see Ofengand, this volume).

A second stable conformer of 5S RNA, usually referred to as the B form, has a secondary structure quite distinct from that of the A form, from chemical, enzymatic and physical studies.

The inaccessibility of G_{41}, which is the most reactive G in the A form, and the high reactivity of G_{44} are examples in point (23). Very likely, all the helical sections in Figure 3, except for the 10 base pair stem involving the 5' and 3' termini, are disrupted in the B form, and at least one new helix involving pairing of residues 33-42 with 79-88, as suggested by Weidner et al. (28) is formed. The biological significance of the B form has not been established. It does not bind the 5S RNA binding proteins L5, L18 and L25 (29). Furthermore, the conformation of 5S RNA in the ribosome more closely resembles that of the A form (30). Whether or not a transition between the A and B conformers occurs in vivo, as suggested by Weidner et al (28) is unresolved at the present time.

Binding of proteins L5, L18 and L25 produces changes in the structure and accessibility of 5S RNA, most of which can be attributed to binding of L18. A 20 per cent increase in ellipticity at 263 nm is observed by circular dichroism upon binding of L18, interpreted either as an increase in helicity (31) or an alteration in the tilting of bases (32) in 5S RNA. Residues G_{24} and G_{69} become protected from reaction with kethoxal (see also below) by binding L18, and modification of these two sites parallels inactivation of binding, suggesting that they may be important for protein recognition (33).

With the proteins bound, the kethoxal reactivity pattern of 5S RNA resembles its reactivity in the ribosome (33), where only G_{13} and G_{41} are exposed (30). The only additional reactive site in the 5S RNA-protein complex is G_{86}, which becomes masked in the ribosome.

Österberg and coworkers (34) have proposed a model for the overall shape of 5S RNA, in the form of an elongated "Y" with overall dimensions 40x80x125Å. They proposed a detailed model which exhibits some discrepancies with the available criteria. The model presented in Figure 3 is arranged to conform with the general shape predicted by their studies.

16S RNA Topography

With the solution of the primary structure of 16S RNA, we can now contemplate questions such as 1) What is its base pairing arrangement? 2) What is its overall folding scheme? 3) How is it oriented with respect to the positions of proteins in the 30S subunit as determined by immunoelectron microscopy and other physical methods? 4) Does the folding pattern of the 16S RNA chain help us to understand its role in assembly?

And, 5) are clues to functional mechanisms revealed in its structure?

Electron microscopy has been used more and more successfully in recent studies on 16S RNA topography. Sloof et al. (35) and Cole et al. (36) have localized protein binding sites with this approach. Protein - RNA complexes are fixed, spread under denaturing conditions, stained and shadowed for microscopy. The RNA appears as an unfolded chain punctuated by features generated by the bound proteins. The latter authors have obtained electron micrographs of complexes of S4 and S8 with 16S RNA (36). The S4 complexes show that three rather widely separated sites, located roughly at positions 1-50, 250-550 and 750-900 on the 16S RNA chain, are held in juxtaposition by protein S4. Protein S8 stabilizes a hairpin structure estimated to contain about 65 nucleotides around position 650.

Exciting implications for the role of 16S RNA in formation of the structure of the 30S subunit come from the studies of Vasiliev. Electron micrographs of naked 16S RNA prepared under reconstitution conditions show particles whose shapes bear a strong resemblance to complete 30S subunits, and have approximately the same dimensions (37). This finding provides strong evidence that the information that determines the overall shape of the 30S subunit is contained in the 16S RNA. Thus, one can hope to understand key steps in the mechanism of 30S subunit assembly by studying the intramolecular interactions of 16S RNA.

Hochkeppel and Craven (71) have shown that 16S RNA prepared by acetic acid - urea extraction has significantly different properties from standard phenol - extracted RNA. Their 16S RNA preparation binds several additional 30S proteins, and shows a slower electrophoretic mobility in agarose gels. In addition, they have observed interconversion of the two electrophoretic forms by binding specific ribosomal proteins (72). These studies suggest that subtle changes in the 16S RNA structure may have a profound effect on our perception of ribosome assembly, and that protein-induced conformation changes may be an important part of the assembly process.

Fragmentation of the 30S subunit by nuclease digestion has led to isolation of subparticles containing sections of 16S RNA bound to a set of ribosomal proteins. A well-characterized example is the subfragment III of Brimacombe and coworkers (40), which contains a fragment of 16S RNA containing approximately residues 1094-1390. The 5' two-thirds of the RNA structure may be organized in the 30S subunit roughly into a 5' two-thirds/3' one-third division.

Partial nuclease digestion of 16S RNA by Ehresmann et al. (41, 42) in the course of RNA sequence studies has provided more detailed information about its structural organization. Several groups have also obtained extensive evidence of this kind as a result of protein binding site studies (see Zimmermann, this volume). These studies also provide a test of secondary structure, since the nucleases used are virtually single strand specific. Studies of this kind have led to the proposal by Garrett and coworkers (43, 44) of the existence of an RNA "core" containing the 5'-terminal 500 nucleotides of 16S RNA, which they argue acts as a nucleus for the assembly of the 30S subunit.

Another approach is the use of single-strand specific chemical modification. A particularly well suited reagent is kethoxal (α-keto-β-ethoxybutyraldehyde), which is guanine-specific, reacts under mild conditions, and can be removed at will (45). Studies with $tRNA^{phe}_{yeast}$ (45, 46) have shown that it is a sensitive and reliable probe of RNA conformation. In 30S subunits, it reacts at about 25 sites in 16S RNA, most of which are distributed from the middle to the 3' end (47). In 70S ribosomes, sites near the middle (positions 673, 702, 704, 790, 802 and 817) and the 3' terminus (positions 1165, 1404, 1496, 1504, 1515 and 1516) are strongly protected from modification (11). Modification of the latter sites causes loss of 50S subunit binding activity (12). These results argue that these sites are involved in important contacts with the 50S subunit. A group of sites at positions 1052, 1063 and 1067 become more reactive in 70S ribosomes. This indicates a 50S induced conformational change in the 30S subunit affecting this region of 16S RNA. Santer and Shane (10) have studied the effect of 50S subunits on 16S RNA topography, using RNAse T_1 as a probe, and have also found protection of sites in the middle and at the 3' terminus. In addition, they reported enhanced cleavage of sites near the 3' terminus of 16S RNA in 70S ribosomes.

Kethoxal modification of inactive 30S subunits, as defined by Zamir and coworkers (48), shows many more reactive guanines, and much higher levels of reaction (49). A conformational change in the 30S subunit affecting 16S RNA topography is therefore brought about by the inactivation process, in addition to changes in protein topography (48). Modification of ribonucleoprotein complexes containing proteins S4 (50), S8 and S15 (51) provide further detailed information about guanine accessibility in 16S RNA and help to establish the structure of the smaller RNA fragments obtained in these complexes.

Chemical modification of naked 16S RNA has been extensively explored by Woese and his collaborators (52). Use of bisulfite (C-specific), m-chloroperbenzoate (A-specific) and glyoxal (G-specific) reactions has provided a large body of detailed information that can be used to test potential secondary structure models.

Application of these criteria to previously proposed secondary structure models (1, 42) shows numerous discrepancies. This is not unexpected, since extensive revisions in the primary structure have been made since that time. It is appropriate to ask, how good a secondary structure can now be deduced, given the availability of the complete sequence and extensive data with which to test such a model? I therefore present a preliminary secondary structure model for 16S RNA which survives nearly all the critical tests that can be applied at this time (54). Undoubtedly, this model will subsequently be revised with the availability of more information, but for the present should be helpful to workers in this area, and provide a focus for discussions.

A Preliminary Secondary Structure Model For 16S RNA

The long range base pairing interactions are shown in the schematic diagram (Fig. 4) and the detailed model is given in Fig. 5. Although much data is available on the naked RNA, our intention is to reflect the

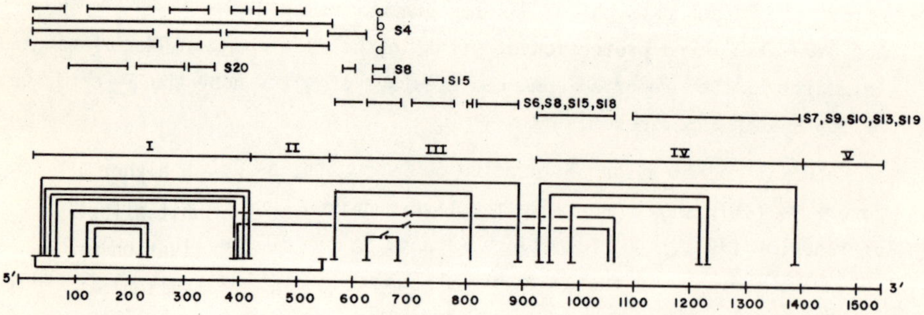

Figure 4. Schematic diagram of proposed long range base paired interacting in 16S RNA (54). RNA fragments obtained in protein binding site studies (40, 61, 73-76) are shown. Domains I-V are discussed in the text. (. .) indicates a potential "switch" mechanism.

the structure as it exists in the 30S ribosomal subunit. We have assumed that most features of the structure are the same in the RNA and the subunit, although there is some hazard in this assumption at present. Based on this, we have freely used both kinds of data, as well as that obtained from studies on ribonucleoprotein fragments, to test the model. In case of conflicting predictions, we have favored the 30S subunit data.

Base pairing possibilities were obtained for the rrnB sequence by computer generated arrays (N. Crawford and H. Noller, unpublished). No attempt was made to censor the computer output except to limit the run of base pairs to four or more, and to set a rough upper limit for the free energy of each helix by a simple arbitrary formula (A-U + 2G-C-4).

Potential helices were then tested by the following criteria:

1) Maximum predicted thermodynamic stability (55)
2) Resistance to cleavage by single strand specific nucleases (41, 42 and 56)
3) Unreactivity of G's toward kethoxal in
   a) active 30S or 70S ribosomes (11, 47)
   b) inactive 30S ribosomes (49)
   c) nucleoprotein fragments (50, 51)
4) Unreactivity in naked 16S RNA (52) toward
   a) bisulfite (C and U specific)
   b) m-chloroperbenzoate (A specific)
   c) glyoxal (G specific)
5) Existence of analogous structures in other procaryotes. Where possible, T_1 oligonucleotide catalogues (57) of procaryotic 16S RNA were used to identify large characteristic T oligomers containing altered sequences with respect to E. coli. In other cases, the partially completed primary structure of Bacillus brevis 16S RNA was used (58).
6) Isolation of base paired fragments (42, 58).

The structure in Figure 5 is presented in such a way that the relative certainty of any of its features can readily be assessed. The results of criteria 2-5 are shown as described in the legend to Figure 5. Helices for which there is phylogenetic evidence (criteria 5) as well as chemical and enzymatic evidence are very likely to exist. These are shaded in Figure 5. Another particularly strong test is criterion 6. The helix pairing residues 925-932/1383-1390 is supported by the isolation by Ehresmann et al. (42) of fragments (O' and A) containing these sequences

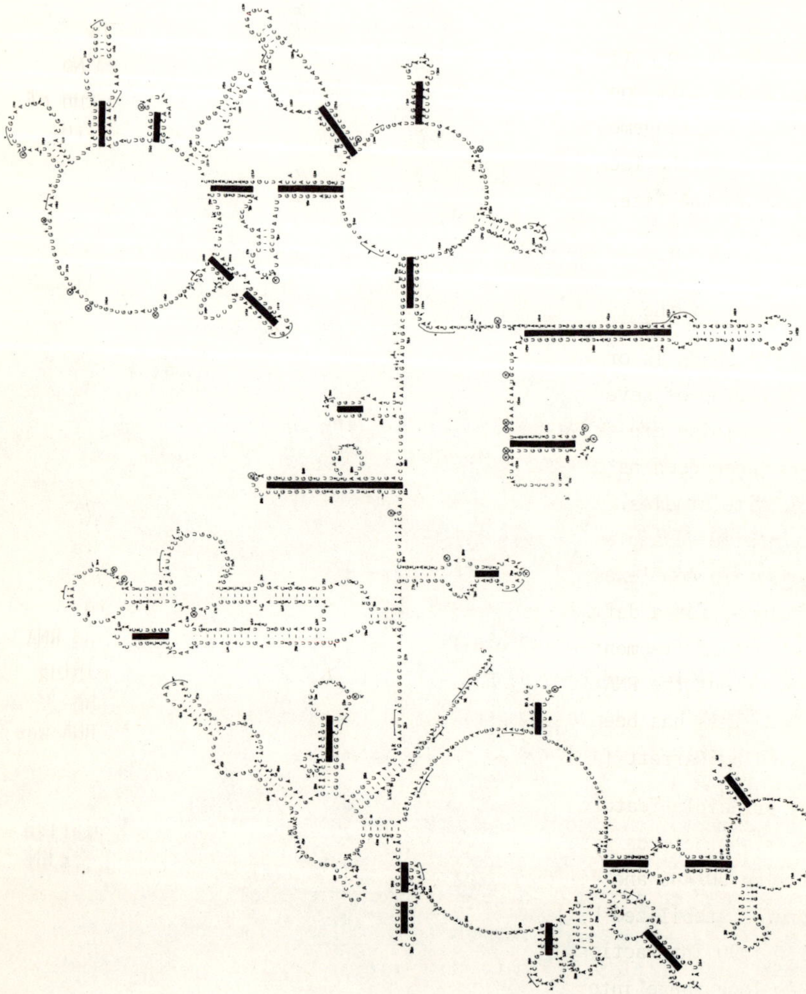

Figure 5. Preliminary secondary structure model for E. coli 16S RNA (54). ((K)), kethoxal reactive in active 30S subunits (47); (K), in inactive 30S (49); k, in ribonucleoprotein fragments (50, 51). Shaded circles indicate kethoxal sites that are important for subunit association (11, 12). Arrows indicate sites of cleavage by T, (41, 42), pancreatic (41, 42), or U_2 (52) nucleases. Nucleotides chemically reactive toward bisulfite (C), m-chloroperbenzoate (A) or glyoxal (G) in naked 16S RNA are indicated by small solid circles over the base; unreactive nucleotides by open circles (52). The boxed sequences base pair according to the scheme in Fig. 4. The domains I-V are discussed in the text. Shaded helices are those for which there is phylogenetic evidence (54).

held together noncovalently. The stem involving positions 1408-1490 is substantiated by isolation of several paired fragments from this region of B. brevis 16S RNA.

There is close agreement between the model and the chemical evidence. Some nuclease cuts conflict with the structure, but these might be due to structural rearrangements following the initial cuts. Also, there is uncertainty in the location of some of the nuclease cleavage sites as described in the literature. Therefore, we would weight the chemical data more heavily.

Implications of the 16S RNA Secondary Structure Model

The RNA chain is organized roughly into 5 structural domains (numbered I-V) by virtue of several long range base pairing interactions. The limits of these domains are shown in Figures 4 and 5. There is a close correlation between these domains and the regions of 16S RNA identified in protein binding site studies. This can most easily be seen in Figure 4, where representative examples of nuclease-resistant RNA fragments obtained in nucleoprotein complexes by several groups is summarized (see Zimmermann, this volume, for a definitive discussion). This correlation suggests that the pattern of fragments obtained in protein binding site studies reflects to some extent the RNA structure itself, as well as the effect of protein binding. This has been noted previously in the case of the S4 RNA, as discussed by Garrett (43).

The principle features of the model are established by long range interactions between positions 50 and 400, 40 and 900, and 930 and 1390. The 50-400 and 40-900 interactions agree well with the loop structures stabilized by protein S4 as observed by Cole et al. (36). The 930-1390 interaction has been directly observed (42, see above). Another long range interaction in domain IV, 985-1215, is supported by comparative phylogenetic evidence (54). In addition, we propose reversible interactions, or "switches" between 100 and 1060, and 400 and 1050. These switches are supported by the observation of enhanced kethoxal reactivity at positions 1052, 1063 and 1067 upon association with 50S subunits (11). Therefore, the switches would be "closed" in 30S subunits, and "open" in 70S ribosomes. As suggested previously, such a mechanism could provide access to a structure which is operational only in 70S ribosomes, such as an aminoacyl tRNA binding site (12). Note that closing this switch would bring together domains I and IV, which could result in a major change in the overall shape of the 30S particle.

The remaining long range interactions tend to reinforce the core interactions. The rest of the model consists of various versions of hairpin structures. So far, 19 of these have been strongly substantiated by phylogenetic evidence, and all are in good agreement with chemical and enzymatic tests. One helical stem, containing residues 586-651, clearly corresponds to the 65 nucleotide hairpin structure stabilized by protein S8 observed by electron microscopy (36). The two large helical features at positions 586-750 are supported independently by topographical studies on the S8/S15 nucleoprotein complex (51).

Correlation of Assembly, Structure and Function

This preliminary secondary structure model can be used as a framework for discussions of assembly, structure and function. Presumably, these are interdependent in the ribosomal structure, and so one should expect to find evidence of this in a model which accurately reflects aspects of the true structure.

Domains I (positions 25-420) and II (420-560) are relatively nuclease resistant (43) and have only two kethoxal reactive sites in active 30S subunits (47). The binding site for S20 is contained within domain I, and the S4 binding site covers domains I and II. This region corresponds to the proposed structural "core" (43) of the 30S subunit, and its binding proteins are crucial for assembly (59). Thus, this region is very likely involved in overall structural organization and assembly. The proposed long range interactions bridging this region with the rest of the 16S RNA may have important consequences for assembly. A possible functional role for this region has been suggested by affinity labeling of residues 461 and 473 with a codon analogue (60). Domains I and II are possibly located in the "body" of the electron microscope model of the 30S subunit, as suggested by Garrett et al. (43).

Domain III (residues 560-890) corresponds to the binding regions for proteins S6, S8, S15 and S18 (61). The kethoxal-reactive guanines at positions 673, 702, 704, 790, 802 and 817 are all strongly protected by association of 30S and 50S subunits, suggesting that this region of 16S RNA makes contact with the 50S subunit (11). This has been substantiated by experiments showing that these sites are also crucial for the association process (12). A possible "switch" mechanism is also suggested for this region. If the two long helices in the 586-750 region are aligned parallel to each other, as in Figure 5, it is evident that an alternate helix can be formed between residues 621-628/676-683 with little or no adjustment of the rest of the structure. Formation of the alternate helix liberates the sequence in the region 610-620 and 700-720. Interaction of this domain with the 50S subunit might be facilitated by such a mechanism. There is no direct evidence to support such a switch, however. Furthermore, a compensating decrease in free energy is needed to make up for the net loss of one helix during the proposed transition. Domain III is probably in the "platform" of the 30S subunit as judged from the protein antigenic sites located there (62, 63).

Domain IV (positions 940-1400) corresponds closely to the binding site fragment for protein S7 (61), to the RNA associated with a subparticle containing proteins S7, S9, S10, S13 and S19 (40), and to the RNA binding site fragment for the protein pair S13/S19 (64). The structure of this region is defined by the three long range interactions 930-1390, 950-1230 and 985-1215, two of which are solidly substantiated (see above). The isolation of this section of the molecule virtually intact, in association with various proteins, is further evidence for the correctness of this region of the model. There are several kethoxal-reactive sites in this domain, some of which show moderately decreased reactivity in 70S ribosomes, suggesting that they may lie at the subunit interface, but not in direct contact with the 50S subunit (11, 12). Recently, Ofengand and his collaborators have affinity-labeled the 5' - terminal base of the anticodon of P site-bound tRNA to a site somewhere in this region of 16S RNA (65). Domain IV therefore contains elements of the peptidyl tRNA binding site. As mentioned above, this site has been mapped on the "head" of the e.m. model of the 30S subunit (39). This is in agreement with the placement of the proteins which bind to this domain in the head region (62, 63). The conformational change noted earlier, associated with sites in the 1050 region (11), may be of functional importance in the transition from 30S to 70S ribosomes.

The 3' terminal region, domain V (residues 1400-1541), is made up of two main helical stems. The larger of these two, including positions 1408-1490, is strongly supported by direct isolation of several paired fragments from this region of B. brevis 16S RNA, using 2-d gel electrophoresis (58). The base pairing scheme shown is closely analogous to that which can be drawn for B. brevis in this region, although there are many base changes in the stems. In both cases gaps in the base pairing involving juxtaposition of A and G are found, at exactly the same position of the main stem. We suggest that these bases may form purine-purine base pairs, perhaps facilitated by ribosomal proteins. A precendent for A-G base pairing is in $tRNA^{phe}_{yeast}$ in which $^{2}_{m}G_{26}$ pairs with A_{44} (26, 27). Some alteration in the geometry of the backbone would be required in order to accomodate the greater glycosyl-glycosyl distance.

The position of at least one feature of domain V, the dimethyladenosine residues at positions 1517 and 1518, have been placed by immunoelectron microscopy in the 'platform' structure in the Lake model (62). In the Stöffler model (63), the platform is in two sections, and it is conceivable that domains III and V could correspond to these two features. This interpretation is perhaps beyond the resolving power of the available data, however.

The functional role of domain V RNA is at least two-fold. The participation of the 3' - terminal polypyrimidine sequence in mRNA selection seems firmly established (2-4). In addition, kethoxal modification studies show that guanines 1404, 1496, 1504, 1515 and 1516 are important for 50S subunit association (11, 12). Other indications of the functional importance of domain V are the loss of ribosome function caused by cleavage after position 1493 by colicin E3 (66, 67), resistance to kasugamycin when residues 1517 and 1518 are unmethylated (7), and the apparent proximity of protein S1 and initiation factors to the 3' terminus (68-70). These correlations are summarized in Table I.

Table I

Structural and Functional Organization of 16S RNA

Domain	Position	Proteins	Location	Function
I	25-420	S4, S20	body	assembly
II	420-560	S4	body	assembly
III	560-890	S6, S8, S15 and S18	platform	subunit association
IV	920-1400	S7, S9, S10, S13 and S19	head	tRNA binding
V	1400-1541	S1 and S21	platform	subunit association, initiation

CONCLUSION

We now appear to be somewhat closer to understanding the functional role of 16S RNA and its structural organization. It is still difficult to explain why RNA is uniquely suited to this task. The reason may lie in the ability of RNA to efficiently recognize other RNA sequences, and to do so with precise geometry. In addition to potential intermolecular interactions with mRNA, tRNA and 23S RNA in this way, the possible existence of intramolecular "switches", in which helices break and reform in the course of protein synthesis, has been raised. This idea has earlier been proposed in connection with 5S by others (22, 28). The large number of potential competing helices of similar energy in the large rRNA's suggest a rich variety of possibilities. The first step toward answering this fundamental question may be the proof of such a mechanism.

ACKNOWLEDGEMENTS

I wish to thank Drs. Carl Woese and Roger Garrett for stimulating discussions, in the course of which many of these ideas have emerged, and for the enjoyment of collaborating with them. I am indebted to my coworkers Dr. Jürgen Brosius, Nigel Crawford, Thomas Dull, Robin Gutell, Winship Herr, James Hogan, K. Katze, Jo Ann Kop, Dr. Samuel Politz and Mary Alice Raker for countless discussions and excellent science. This work was supported by NIH grant No. GM-17129.

REFERENCES

1. Fellner, P. (1924) in Ribosomes (pp. 169-191) (ed. Nomura, M., Tissieres, A. and Lengyel, P.) Cold Spring Harbor, New York.

2. Shine, J. and Dalgarno, L. (1974) Proc. Nat. Acad. Sci. USA 72, 4734-4738.

3. Steitz, J.A. and Jakes, K. (1975) Proc. Nat. Acad. Sci. USA 72, 4734-4738

4. Dunn, J.J., Buzash-Pollert, E. and Studier, W.F. (1978) Proc. Nat. Acad. Sci. USA 75, 2741-2745

5. Noller, H.F. and Chaires, J.B. (1972) Proc. Nat. Acad. Sci. USA 69, 3115-3118.

6. Ofengand, J., Liou, R., Schwartz, I, and Zimmermann, R.A. (1979) Biochemistry, in press.

7. Helser, T.L., Davies, J.E. and Dahlberg, J.E. (1972) Nature New Biol. 235, 6-8.

8. Lai, C.J., Dahlberg, J.E. and Weisblum, B. (1973) Biochemistry 12, 457-459.

9. Yamada, T., Mizugichi, Y., Nierhaus, K.H. and Wittmann, H.G. (1978) Nature 275, 460-461.

10. Santer, M. and Shane, S. (1977) J. Bacteriol. 130, 900-910.

11. Chapman, N.M. and Noller, H.F. (1977) J. Mol. Biol. 109, 131-149.

12. Herr, W., Chapman, N.M. and Noller, H.F. (1979) J. Mol. Biol. 130, 433-449.

13. Brownlee, G.G., Sanger, F. and Barrell, B.G. (1967) Nature 215, 735-736.

14. Erdmann, V.A. (1979) Nucleic Acids Res. 6, r29-r44.

15. Hori, H. and Osawa, S. (1979) Proc. Nat. Acad. Sci. USA 76, 381-385.

16. Ehresmann, C., Stiegler P., Carbon, P. and Ebel, J.P. (1977) FEBS Lett. 84, 337-341.

17. Brosius, J., Palmer, M.L., Kennedy, P.J. and Noller, H.F. (1978) Proc. Nat. Acad. Sci. USA 75, 4801-4805.

18. Carbon, P., Ehresmann, C., Ehresmann, B. and Ebel, J.P. (1978) FEBS Lett. 94, 152-156.

19. Branlant, C., Korobko, V. and Ebel, J.P. (1976) Eur. J. Biochem 70, 471-482.

20. Branlant, C., Sri Widada, J., Krol, A., Fellner, P. and Ebel, J.P. (1975) Biochimie 57, 175-225.

21. Holley, R.W., Apgar, J., Everett, G.A., Madison, J.T., Marquisee, M., Merrill, S.H., Penwick, J.R. and Zamir, A. (1965) Science 147, 1462-1464

22. Fox, G. and Woese, C.R. (1975) Nature 256, 505-506.

23. Noller, H.F. and Garrett, R.A. (1979) J. Mol. Biol., in press.

24. Peattie, D.A., Garrett, R.A. and Noller, H.F., unpublished.

25. Erdmann, V.A. (1976) Progr. Nucleic Acids Res. and Mol. Biol. 18, 45-90.

26. Rich, A. and Rajbhandary, U.L. (1976) Ann. Rev. Biochem. 45, 805-860.

27. Jack, A., Ladner, J.E. and Klug, A. (1976) J. Mol. Biol. 108, 619-649.

28. Weidner, H., Yuan, R. and Crothers, D.M. (1977) Nature 266, 193-194.

29. Bellemare, G., Jordan, B.R., Rocca-Serra, J. and Monier, R. (1972) Biochimie 54, 1453-1466.

30. Noller, H.F. and Herr, W. (1974) J. Mol. Biol. 90, 181-184.

31. Bear, D.G., Schleich, T., Noller, H.F. and Garrett, R.A. (1977) Nucleic Acids Res. 4, 2511-2526.

32. Spierer, P., Bogdanov, A. and Zimmermann, R.A. (1978) Biochemistry 17, 5394-5398.

33. Garrett, R.A. and Noller, H.F. (1979) J. Mol. Biol., in press.

34. Österberg, R., Sjöberg, B. and Garrett, R.A. (1976) Molec. Gen. Genet. 147, 129-138.

35. Sloof, P., Garrett, R.A. and Nanninga, N. (1976) Molec. Gen. Genet. 147, 129-138.

36. Cole, M.D., Beer, M., Koller, T., Strycharz, W.A. and Nomura, M. (1978) Proc. Nat. Acad. Sci. USA 75, 270-274.

37. Vasiliev, V.D., Selivanova, O.M. and Koteliansky, V.E. (1978) FEBS Lett. 95, 273-276.

38. Politz, S.M. and Glitz, D.G. (1977) Proc. Nat. Acad. Sci. USA 74, 1468-1472.

39. Keren-Zur, M., Boublik, M. and Ofengand, J. (1979) Proc. Nat. Acad. Sci. USA 76, 1054-1058.

40. Yuki, A. and Brimacombe, R. (1975) Eur. J. Biochem. 56, 23-34.

41. Ehresmann, C., Stiegler, P., Fellner, P. and Ebel, J.P. (1972) Biochimie 54, 901-967.

42. Ehresmann, C., Stiegler, P., Fellner, P. and Ebel, J.P. (1975) Biochimie 57, 711-748.

43. Garrett, R.A., Ungewickell, E., Newberry, V., Hunter, J. and Wagner, R. (1977) Cell Biol. Int. Reports 1, 487-502.

44. Ehresmann, C., Stiegler, P., Carbon, P., Ungewickell, E. and Garrett, R.A. (1977) FEBS Lett. 81, 188-192.

45. Litt, M. (1969) Biochemistry 8, 3249-3253.

46. Rhodes, D. (1975) J. Mol. Biol. 94, 449-460.

47. Noller, H.F. (1974) Biochemistry 13, 4694-4703.

48. Ginzberg, I. and Zamir, A. (1975) J. Mol. Biol. 93. 465-476.

49. Hogan, J.J. and Noller, H.F. (1978) Biochemistry 17, 587-593.

50. Noller, H.F. and Garrett, R.A. (1979) unpublished.

51. Müller, R., Garrett, R.A. and Noller, H.F. (1979) J. Biol. Chem. 254, 3873-3878.

52. Magrum, L., Stahl, D. and Woese, C.R. Unpublished results.

53. Sekiya, T. and Nishimura, S. (1979) Nucleic Acids Res. 6, 575-592.

54. Noller, H.F., Kop, J., Crawford, N., Brosius, J., Gupta, R., Fox, G.E. and Woese, C.R. Unpublished results.

55. Tinoco, I., Borer, P.N., Dengler, B., Levine, M.D., Uhlenbeck, D.C., Crothers, D.M. and Gralla, J. (1973) Nature New Biol. 246, 40-41.

56. Santer, M. and Santer, U. (1973) J. Bacteriol. 116, 1304-1313.

57. Woese, C.R., Fox, G. E., Zablen, L., Uchida, T., Bonen, L., Pechman, K., Lewis, B. and Stahl, D. (1975) Nature 254, 83-86.

58. Kop, J., Noller, H.F., Siegel, R., Gupta, R. and Woese, C.R. Unpublished.

59. Nomura, M. and Held, W.A. (1974) in Ribosomes (Nomura, M., Tissières, A. and Lengyel, P., eds.) pp. 193-223, Cold Spring Harbor, N.Y.

60. Wagner, R., Gassen, H.B., Ehresmann, C., Stiegler, P. and Ebel, J.P. (1976) FEBS Lett. 67, 312-315.

61. Zimmermann, R.A. (1974) in Ribosomes (Nomura, M., Tissières, A. and Lengyel, P., eds.) Cold Spring Harbor, N.Y.

62. Lake, J.A. (1976) J. Mol. Biol. 105, 131-149.

63. Stöffler, G. and Wittmann, H.G. (1977) in Molecular Mechanisms of Protein Biosynthesis (Weissbach, H. and Pestka, S., eds.) pp. 117-202, Academic Press, N.Y.

64. Müller, R. and Garrett, R.A. (1979) Unpublished.

65. Zimmermann, R.A., Gates, S.M., Schwartz, I. and Ofengand, J. (1979) Biochemistry, in press.

66. Bowman, C.M., Dahlberg, J.E., Ikemura, T., Konisky, J. and Nomura, M. (1971) Proc. Natl. Acad. Sci. USA 68, 964-968.

67. Senior, B.W. and Holland, I.B. (1971) Proc. Natl. Acad. Sci. USA 68, 959-963.

68. Kenner, R. A. (1973) Biochem. Biophys. Res. Commun. 51, 932-938.

69. Czernilofsky, A.P., Kurland, C. G. and Stöffler, G. (1975) FEBS Lett. 58, 281-283.

70. Van Duin, J., Kurland, C.G., Dondon, J. and Grunberg-Manago, M. (1975) FEBS Lett. 59, 287-290.

71. Hochkeppel, H.K. and Craven, G. R. (1976) Nucleic Acids Res. 3, 1883-1902.

72. Hochkeppel, H.K. and Craven, G.R. (1977) J. Mol. Biol. 113, 623-634.

73. Schaup, H.W., Sogin, M., Woese, C. and Kurland, C.G. (1971) Mol. Gen. Genet. 114, 1-8.

74. Mackie, G.A. and Zimmermann, R.A. (1975) J. Biol. Chem. 250, 4100-4112.

75. Ungewickell, E., Ehresmann, C., Stiegler, P. and Garrett, R.A. (1975) Nucleic Acids Res. 2, 1867-1888.

76. Zimmermann, R.A., Mackie, G.A., Muto, A., Garrett, R.A., Ungewickell, E., Ehresmann, C., Stiegler, P., Ebel, J.P. and Fellner, P. (1975) Nucleic Acids Res. 2, 279-302.

PHYSICAL AND CHEMICAL TECHNIQUES FOR THE STUDY OF RNA STRUCTURE ON THE RIBOSOME

CHARLES R. CANTOR

Barth Laboratory,
Departments of Chemistry and Biological Sciences,
Columbia University,
New York, NY 10027

INTRODUCTION

Most of the more powerful structural techniques that have been brought to bear on the ribosome have focused on the proteins. This is easy to understand - all of these techniques require the ability to label or analyze one specific portion of the structure at a time. The large number of single copy ribosomal proteins greatly facilititates immunological approaches necessary in electron microscopy crosslinking and affinity labeling. It also allows the specific probes necessary in neutron scattering and fluorescence. Unfortunately there is no known way to reversibly break up large ribosomal RNAs into convenient size pieces. Indeed there is not even a reasonable way to do this irreversibly.

Compared to proteins, the state of the art of RNA chemical modification is rather primative. With a few striking exceptions RNA immunology is all but unknown. It is not surprising then that progress in our understanding of the structure of RNAs on the ribosome has lagged far behind that of proteins. However a number of recent advances in the technology of dealing with nucleic acid structures will change this picture radically. These include rapid sequencing techniques which make it far easier to obtain RNA primary structure information than protein sequences. Such techniques will surely be adaptable for locating the sites of chemical modification (or enzymatic cleavage). The linear structure of RNA which greatly complicates many other techniques, is a powerful asset when electron microscopy is used to locate crosslinks within the RNA. Recombinant DNA techniques allow the isolation and manipulation of almost any desired fragments of a nucleic acid sequence. New techniques for the chemical modification and crosslinking of RNA now exist and these appear at least as useful as the corresponding techniques for proteins.

In most cases, except for the rapid appearance of rRNA sequence data, and the identification of a number of intra-RNA crosslinks, there has not yet been time for the bulk of the new technology to make its full impact on our understanding of rRNA. In other words, this article is premature. Rather than stress what is already known, I will first outline just how little we know. Then I will sketch a few of the areas that are somewhat developed and comment briefly on areas which are probably ripe for rapid development in the near future.

INFORMATION SOUGHT

At least six RNAs can occupy the prokaryotic ribosome simultaneously: the intrinsic structural 16S, 23S, and 5S rRNAs, the bound mRNA, and tRNAs bound at the peptidyl (P) and amino acyl (A) sites. Among the questions of interest are the structures of these individual RNAs and the nature and extent of any interactions among them. We shall focus all these questions almost exclusively on E. coli ribosomes in this article. However it will be most important in the future to see, once answers to these questions have been arrived at, whether they can be generalized to other prokaryotes and to the ribosomes from eukaryotic cells and organelles.

Structure of Individual RNAs

We know the high resolution crystal structure of one tRNA species (1), and there is lower resolution structural information on several others. The vast majority of available data suggests that all tRNAS that participate in protein synthesis will have virtually identical tertiary structures, and that the structure in the crystal and in solution are also nearly identical. In contrast there is virtually no high resolution structural information available on isolated rRNAs or functionally significant regions of mRNAs. Even with the simplest rRNA, the 5S rRNA, the body of data currently in hand allows less definitive conclusions than were available with just the first few tRNA sequences. We clearly have a long way to go.

Ulimately, one would like to know the structure of each RNA separately, and in various functional states of the ribosome complex. Then one will be able to ask definitively whether any RNA structural changes are important for ribosome function or self-assembly. Even in the absence of complete structures, however, it is possible to start to examine whether any large RNA structure changes accompany assembly or function. Transfer RNA is a particularly attractive target for such studies since one can exploit the known crystal structure to plan decisive experiments. In sum, the slim amount of data currently available supports the notion of local conformational changes in RNA regions which may be important for function or assembly. However there is every reason to believe that free RNAs are by and large good models for the structure of RNAs on the ribosome, except in the case of mRNA. Here, although there is very little data, it is hard to see how an RNA will be used as a message unless the extensive secondary structure demanded by thermodynamic considerations, is systematically disrupted during protein synthesis.

RNA-RNA Interactions

There are fifteen possible binary interactions among the six RNAs on the ribosome. There is no reason in principle why all might not occur simultaneously. At various times, almost all have been invoked as having particular functional significance. Yet the amount of hard data on this subject is surprisingly slim. Among the questions to be answered are which interactions actually occur and at which stages in protein synthesis. For each interaction it will be important to know the extent and the nature. For example, how many bases are involved? Is the interaction a normal double strand, some sort of non-Watson-Crick tertiary base pairing or triple strand formation, or intercalation? If the interactions are extensive it will also be necessary to examine the topological implications or restrictions involved. We will deal with this in detail later.

Very few of the fifteen potential RNA-RNA interactions are currently firmly established facts. Direct contact between mRNA and tRNA in the A site seems inescapable and the evidence is pretty strong for P site bound tRNA too (2-4). Nagging doubts remain as to whether both sets of contacts remain at all stages in protein synthesis and the detailed structure of the contacts (e.g. are all three base pairs made in the P site, are the anticodons 3'-stacked or 5'-stacked) is unknown. Indirect evidence exists for contacts between the two ribosome-bound tRNAs (5,6). Good evidence also exists for mRNA-16S rRNA contact at some state during initiation (7) but it does not yet seem clear at what stage this contact is broken. Results from affinity labeling point strongly towards contacts between P site bound tRNA and the 16S rRNA, via the anticodon (8,9) and the 23S rRNA, via the CCA terminus (6,10-12) but the functional significance of these contacts remain unproven, and nothing is know about their extent. The first experimental hints about interactions between the 16S and 23S rRNAs are just emerging. In spite of tremendous effort and many imaginative suggestions, the role of 5S rRNA interactions still remains somewhat elusive.

RNA SIZE AND SHAPE

Free RNAs

Information on the size and shape of free rRNA in solution is accessible from a variety of hydrodynamic and scattering techniques. What these generally show is that the apparent volume occupied by rRNAs varies quite a bit depending on the salt, pH, temperature and other conditions used for the measurement. Such variability has been found for other large RNAs. It presumably reflects a delicate balance between stronge base pairing interactions and stacking interactions which tend to keep much of the molecule collapsed into domains, and strong electrostatic repulsion which in the absence of sufficient divalent ions and supporting electrolyte will cause an inevitable expansion. For example it has been reported (13) that, at neutral pH variation of the ionic strength and added divalent counterions shifts the sedimentation constant of the 23S rRNA from 15 to 28S. After rigorous removal of divalent ions the range increases from 4 to 32S, while by lowering the pH to 3.0 one can raise the sedimentation constant to 50S. These are enormous changes which reflect a variation in the effective volume of the RNA of more than an order of magnitude.

In the face of such variability, it is fair to ask whether a large rRNA can have an unique stable folded conformation in solution under any set of environmental circumstances. The answer to this question is unknown, but one type of result argues in favor of specific conformations. Conditions can be found, not too far from physiological, where rRNAs have quite compact conformations. For example (14), in the typical buffers used for 30S ribosome reconstitution, the 16S rRNA has a sedimentation constant of 21.1S. This corresponds to nearly the same hydrodynamic radius as the intact 30S particle. In other words, the free rRNA under these circumstances has a size and shape which may be very similar to the structure it adopts in the intact particle. Since the 30S ribosome is half RNA by volume, and since there is much exposed protein, it is possible that many of the RNA-RNA contacts observed in the ribosome can be maintained in the free RNA under the right circumstances.

Most solution hydrodynamic techniques and straight-forward scattering studies provide only a spherically averaged view of the structure. Thus, except for some information about axial ratios of equivalent ellipsoids, most detailed shape information is lost. Ultimately it should be possible, by contrast variation methods to gain at least a low resolution picture of the shape of isolated ribosomal RNAs in solution, much as has been reported for intact particles (15). At present, the only detailed information about the shape of isolated rRNA comes from electron microscopy. Vasiliev and coworkers have recently reported studies (16) on the 16S rRNA in 6 mM Mg^{2+}, 30 mM NH_4^+, 1 M EtOH. In solution under such conditions, the RNA has a sedimentation constant of 21.5S and thus is compacted to ribosome size. When molecules are lyophilized from such a solution they appear as compact Y or V shaped structures with overall dimensions that allow them to fit within the 30S particle. From these results, it is possible that the free RNA can attain the same specific folded structure as it has in the 30S particle. But available data is probably not yet sufficient to prove this assertion.

RNAs on the Ribosome

We have discussed free 16S and 23S rRNA and considerable information is available about the shape of 5S rRNA, tRNA, and even some mRNAs in solution. One would also like to know the shapes of these RNAs on the ribosome. This type of study is in its infancy. Contrast variation neutron or x-ray scattering experiments have already revealed considerable infomation about the average location of the 16S and 23S rRNAs within their respective subunits. By more detailed analyses of such experiments it should be possible to get some information about the shapes of the RNAs. In principle it should also be possible to examine the low resolution shape of other ribosome-bound RNAs by neutron scattering using specifically deuterated components just in the same way as is now being done for proteins (17). For 5S rRNA and P-site bound tRNA, samples ought to be stable enough for the long exposure times required in these measurements. By examining the scattering interference terms between two deuterated components, one learns about the projection of their shapes along the intercomponent vector. Such studies are proving useful for ribosomal proteins and they ought to be applicable in some cases to RNAs also.

Indirect information about RNA shape on the ribosome is available through studies of ribosomal proteins. Much is known about the location of individual 30S and 50S proteins, and quite a bit is known about regions of the RNAs contacted by these proteins. By combining these two data sets one can infer the location of particular regions of the RNA. At present, results are too incomplete to make any kind of general summary useful. However they clearly point to a rather complex arrangement of the RNA chain. In both the 30S and 50S subunits there appear to be substantial regions where few or no proteins are found (18). If this result bears up under continued experimentation it almost certainly implies protein-free RNA domains which presumably have the same structure as in the free RNA.

ACCESSIBILITY TO SOLUTION REAGENTS

Limitations of the Approach

In principle, one of the simplest questions one can ask about a macromolecular structure is which residues are on the outside surface. One argues that such residues should show the same access to solution reagents as those in equivalent unstructured model compounds. Residues showing less accessibility are deemed to be progressively less exposed. A variety of such studies have been carried out on RNAs free and in the ribosome. Before dealing with the results of these studies it would be useful to comment on their limitations.

The reagents used to gauge accessibility vary enormously in size. The largest are immunoglobulin G molecules. If simple binding is measured these require about a 30 Å cavity around the target RNA hapten. If dimer or precipitate formation is measured even a larger access zone may be needed. The enzymes used to measure accessibility are mostly nucleases which are relatively small proteins. But these still require a substantial exposed area. Among the small covalent RNA reagents, the most popular for accessibility studies has been kethoxal which reacts with single stranded guanosines. $[^3H]$-H_2O is an even smaller reagent useful for accessibility studies. The most effective application is to observe tritium exchange into the C_8-position of purine bases. Because of the different sizes of the reagents involved it is understandable that these approaches probe somewhat different aspects of structure.

All surface accessibility studies suffer from two basic complications. They are all a convolution of two effects: the physical accessibility of the site to the reagent, implying whether or not the site is at the surface, and the intrinsic reactivity of the site, which is generally a function of all sorts of unknown factors about the microscopic environment. Usually it is safe to conclude that a reactive site is on the surface but it is not safe to conclude that an unreactive site is buried. The second complication arises from fluctuations or flexibility in molecular structure. Since chemical modification or exchange studies are carried out on a very long time scale, they can pick up configurations which may be rare equilibrium species, but very reactive when present. Certain dynamic accessibility probes like collisional quenching of a fluorescent chromophore operate on a fast enough time scale to eliminate such problems, but because such techniques require the introduction of a probe molecule, they raise problems of their own. Chemical and enzymatic modifications always entail the risk of progressive perturbation of the structure. Thus, studies of the first guanine residues to be kethoxalated, or the first phosphodiester bonds to be cleaved be a nuclease are quite reliable. In contrast, studies at very high degrees of reaction or digestion have to be viewed with great caution unless it can be shown by such criterea as biological activity, that the structure has not been too seriously perturbed. Exchange techniques are immune from such problems.

Covalent Modifications

Noller and his group have carried out an extensive series of studies on the accessibility of rRNA in ribosomes to reaction with kethoxal. These studies are summarized in detail elsewhere in this volume. Here we mention just a few highlights. In the inactive 30S subunit there are 70 to 80 reactive guanines. This is about 1/2 of those expected to be unpaired in the 16S rRNA and less than 1/6 of the total. About half of the reactive residues become inaccessible in the active 30S confiormation (19), and additional residues become protected when the 50S subunit is bound (20). One reactive site on the 16S rRNA is actually more accessible in the 70S particle than in the free subunit, implying possible changes in RNA structure or RNA-protein contacts upon formation of the

70S particle. Assignment of each of the kethoxal-reactive sites produces a detailed and complex picture of regions of accessibility. Because kethoxal is single-strand specific it also reveals some infomation about the base-pairing pattern of the 16S rRNA within the 30S particle. When similar studies have been performed on the free 16S rRNA, these should yield much useful secondary structure information.

A number of different chemical modification and other accessibility probes have focused on the 3' end of the 16S rRNA. This region is of obvious interest in view of indications that it plays a functional role in initiation. The reagents employed include T1 RNase digestion (21), colicin E3 cleavage (22), kethoxalation (19), semicarbazone formation after periodate oxidation (23), dynamic quenching of fluorescent dyes attached to the 3' end (23), and antibodies specific for $M^2{}_2A$-$M^2{}_2A$ sequence not far from the 3' end (24). Most of these studies indicate substantial differences in accessibility when the inactive and active conformations of the 30S particle are compared. There are also differences between the 30S subunit and the 70S particle. However the overall pattern of results is surprisingly complicated with some regions becoming less accessible and others more accessible upon increasing assembly. The simplest interpretation is a concerted pattern of secondary structure within the 16S rRNA itself, but altered contacts with proteins or the 23S rRNA cannot be excluded. The important thing is that even a limited region of the 16S rRNA shows a pattern of accessibility changes that is really quite complex to analyze.

Tritium Exchange

Among all the techniques for accessibility measurements, one that appears potentially especially informative is tritium exchange at the C_8-purine positions, henceforth called slow tritium exchange. An extensive series of model compound studies by Schimmel and his collaborators has calibrated the exchange rates expected for single and double stranded purines as a function of temperature (25). The activation energy for this exchange reaction is very high. Thus at 4°C exchange rates are negligble if extremes of pH are avoided. The pattern of tritium exchange can be monitored quantitatively by oligonucleotide fragmentation and standard sequencing techniques. These are greatly facilitated by the fact that no actual chemical modification has been suffered by the RNA. At 37°C rates are fast enough so that 4% exchange can be observed in an unrestricted single strand guanine in 24 hours. This is sufficient to allow quantitative monitoring if $[^3H]$-H_2O of high specific activity is used. At higher temperatures the exchange rates are vastly more rapid-however since these entail great risk of structure perturbation they are best avoided.

Neal Farber, in my laboratory has been using slow tritium exchange to compare the properties of free and ribosome-bound RNAs. Since exchange rates are markedly slower in double strands, studies on free RNAs yield information about secondary structure. Then studies on the ribosome allow inferences about protected regions. For example the overall exchange rates of 16S rRNA, free and ribosome-bound are shown in Table 1.

TABLE 1. Tritium Exchange into the C_8- Purine Positions of Ribosomal RNAs.[a]

	Exchange rate ratio[b]	
Sample	16S rRNA	5S rRNA
Free rRNA	6.1	5.2
Ribosomal subunit	9.2	23.5
70S Ribosome	25.1	23.5

[a]All measurements were made at 37°C at pH 7.0. The magnesium concentration was 10mM. All data is taken from reference 26.

[b]This is the ratio of the exchange rate of free purine mononucleotides to that observed in the intact RNA after correction for purine composition.

It is clear that the bulk of those residues free to exchange in the 16S rRNA are protected in the 70S particle. The surprising result is that there is relatively little protection in the 30S particle (26). Some of this must be due to a significant amount of exposed RNA surface. However, the magnitude of the effect suggests that the 30S subunit may be much looser and experience much more breathing motions than the intact 70S particle.

More detailed information is available on the 5S rRNA. Here the positions of individual exchanging protons have been identified by sequencing. The free RNA shows many exchangeable purine protons at 37°C. In contrast the exchange rates of all but a few protons in the 5S RNA component of the 50S or 70S ribosome are slowed significantly. This is clear even in the averaged results shown in Table 1. Apparently most single and double strand regions of the 5S rRNA are not accessible in the ribosome. We cannot say whether the extensive protection in the particle is due to direct contact with proteins or the 23S rRNA or is an indirect effect due to tightening of the 5S structure by constraining it to lie within the particle. However the dramatic decrease in exchange rates is so large compared to protection seen in the tRNA tertiary structure that it implies a 5S RNA largely buried in the ribosome. No difference is observed between the 50S and 70S particle.

A few positions on the 5S rRNA remain almost as exposed for tritium exchange in the free RNA. These include residues G_{13}, G_{41}, A_{108} and/or A_{109}. Since the pattern of guanine accessibility 5S rRNA has also been studied by kethoxalation (27), an unique opportinity for comparison of the two techniques in a nucleoprotein is available. Happily the agreement is excellent.

Slow tritium exchange has also been used (26), to provide some of the first information on the structure of ribosome-bound tRNA. Here the experimental strategy is to work under conditions where the bulk of the tRNA in the solution is ribosome bound. We used 20mM Mg^{2+}, 37°C to examine deacylated yeast $tRNA^{Phe}$ bound at the P site. An extensive series of studies by Schimmel and coworkers had already shown that the exchange pattern of free tRNA is in excellent agreement with the known crystal structure (28). When free and ribosome-bound tRNA are compared it is clear that there are substantial differences in exchange rates. Protons at both the anticodon loop and adjacent stem region, and at the CCA end and adjacent stem show markedly slower exchange in the ribosomal P site. From this we infer that these regions of the tRNA are in contact with the ribosome. In contrast, some protons near the contact region between the pseudouridine and dihydrouridine loops actually exchange more rapidly on the ribosome than off. This indicates the some of the tertiary contacts in this region may be disrupted upon ribosome binding. The overall pattern of tritium exchange of P-site bound tRNA is unaffected by the presence of message. This suggests that the structure and location of the bound tRNA depends only on the ribosome.

BASE-PAIRING PATTERN

Spectroscopic Studies

We are geared to think of RNA structure in terms of intramolecular base pairing, and even in the absence of any truly direct evidence, there is almost universal agreement that a majority of the bases in ribosomal RNAs are participants in Watson-Crick double strands. The best evidence for this comes from spectroscopic studies. In a small RNA like 5S rRNA, techniques such as NMR, raman spectroscopy, circular dichroism and UV absorbance are capable, in principle, not only in assessing the average extent of double strand, but also in identifying particular residues involved in base pairing. Tritium exchange is also very useful for this purpose. In practice these various methods do not yet agree on the extent of base pairing in 5S rRNA in solution nor on the assignment of particular double strands except for a few regions. None of these techniques is useful for examining the 5S rRNA on the ribosome except tritium exchange, and here interpretation is complicated by questions of accessibility. However all methods agree the 5S rRNA does have some double strands.

Mainly UV optical techniques have been used thus far to examine aspects of the structure of the larger rRNAs. Since RNA absorbance and CD on a per residue basis completely overwhelm protein contributions at wavelengths above 240nm it is possible to look at RNA within the ribosome and compare it with free molecules. The results show that the free RNAs have extensive amounts of base pairing- 65% is a typical estimate- and there are at most rather tiny net changes in secondary structure upon subunit assembly or 70S complex formation (29). However several caveats apply: large rearrangements in secondary structure would be largely invisible in optical spectra so long as the net fraction of base paired residues remained unaltered. In addition, it is really not possible to conclude with certainty from CD or hypochromism what fraction of base pairing is Watson-Crick as opposed to other tertiary types of pairing.

Psoralen Crosslinking Studies

Double-strand specific crosslinking reagents are a natural way to examine the pattern of secondary structure in a free RNA or a nucleoprotein. A number of reagents are potentially available including mustards (30), azide derivitives of intercalators (31), bis-glyoxal derivatives of intercalators (32), psoralens (33), and direct photoirradiation (30). It is not certain, however, that any of these reagents are totally specific for double strands. The most promising reagents, from both the points of view of specificity and yield are psoralens. The structures of two of these three ring intercalators are shown in Figure 1.

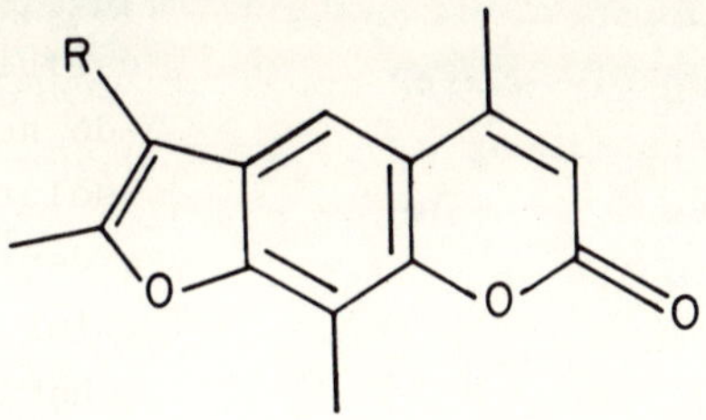

Compound, R =	Abbreviation
$HO-CH_2-$	HMT
$NH_3^+-CH_2-$	AMT

Figure 1. Structure of two of the psoralen crosslinkers useful for rRNA structural studies.

Upon irradiation at 250-260 nm, these reagents are capable of two successive cycloaddition reactions with pyrimidines on adjacent strands to form a double strand crosslink. A reagent intercalated into a homopurine-homo pyrimidine stretch will presumably only react with one strand. Thus, statistically, about half of all covalently attached psoralens should generate crosslinks.

In principle one could use standard oligonucleotide mapping techniques to find the sequences crosslinked by a psoralen. In practice this would be an extremely laborious task even for high yield products and a nearly impossible one for low yield products. Ultimately it should be possible to learn how to break psoralen crosslinks to assist sequencing. Even more desirable would be ways to adapt modern ladder sequencing techniques for the location of attached psoralens. Until such techniques are developed, the most convenient strategy is to use electron microscopy to determine the approximate location of psoralen crosslinks (33).

After rigorous formamide denaturation, rendered irreversible by formaldehyde treatment, and appropriate spreading in formamide, rRNA molecules examined by Kleinschmidt-type microscopic techniques appears as an almost uniform collection of somewhat flexible rod-like molecules. The presence of an intrastrand crosslink is potentially visualizable. It will generate a small bulge, for a crosslink between residues near by in the chain, or a loop, for a crosslink between distant residues. Thus if a double-strand specific crosslinker like a psoralen is used, the location of such loops in the electron microscope yields the location of double strand regions. When the 16S rRNA in aqueous solution is treated with the psoralen HMT there are numerous sites of covalent attachment. In collaboration with Paul Wollenzien, Pallaiah Thammana, and John Hearst, molecules containing typically 3 to 30 attached HMT's have been examined in the electron microscope. These show two striking effects of the psoralen treatment (34). The average molecular contour length is shortened by about 15% implying the presence of hairpin loops too small to show up as distinct bulges in the elctron microscope. However no distinct larger hairpin bulges are seen. Instead the average molecule contains one or more large loops. These range in size from 200 bases to essentially full circle molecules.

By quantitiative length measurements of many molecules we have been able to resolve the crosslinks seen into 11 distinct classes. The absolute orientation of several crosslinked features could be determined by R-loop mapping crosslinked molecules onto a plasmid carrying part of the 16S rRNA in a known orientation (35). Then molecules containing two or more loops simultaneously were used to assign the orientation of the remaining features (34). The accuracy of the microscope techniques used limits the resolution of each crosslink position to between ± 25 and ± 50 bases. While this leaves much to be desired, the overall crosslinking pattern is sufficiently unexpected and informative to allow a number of interesting conclusions.

The 16S rRNA crosslinks seen are summarized in Table 2.

TABLE 2. Location of the Major Psoralen Crosslinks Observed in 16S rRNA After Treatment with HMT.[a]

Loop Size	Crosslinked Residues[b]
1490	30 - 1520
1000	530 - 1530
680	0 - 680
630	740 - 1370
560	970 - 1530
500	140 - 640
370	150 - 520
300	650 - 950
240	0 - 240
210	280 - 490
200	1240 - 1440

[a]For further details see references 34.

[b]Residues are numbered so that the 5' terminus is 1, the 3' terminus is 1541. The actual locations given are uncertain to within ± 25 to 50 residues.

One of the first things to notice is that both the 3' and 5' ends seem to be involved in contacts with several other regions spaced throughout the molecule. It is possible that such a reaction pattern is simply a reflection of heterogeneity in the structure of the 16S rRNA molecule. However several pieces of evidence argue strongly against heterogeneity. The crosslinking pattern seen is by and large quite insensitive to substantial variation of environmental conditions (34). It would be hard to expect heterogeneous classes of molecules to all have comparable stability. The statistics of appearance of molecules with pairs of crosslinked features is explained by simply multiplying the observed frequencies for each separate feature. This is consistant with the crosslinks representing random sampling of molecules containing all features simultaneously whereas it would be difficult to explain by random crosslinking distinct mixtures of several structures. Finally at high degrees of psoralen reaction, molecules can be seen which have accumulated so many crosslinks that they do not spread well on the grid. Unfortunately one cannot, therefore, assign the locations of individual crosslinks within such molecules.

Probably the strongest evidence for a fairly definite secondary structure pattern in the free 16S rRNA comes from studies on the 30S ribosome (36). The reactivity of such particles towards psoralens is much less than that of the free rRNA, particularly when large amounts of magnesium are present. This is presumably because a tight structure is relatively resistant to the binding of intercalators, since such binding requires considerable lengthening and unwinding of the double helix adjacent to the intercalator. However the pattern of psoralen reactivity- that is the acutal crosslinked features seen, and in a number of cases their relative yields, is strikingly similar to that of the free RNA. All of the six features seen most frequently in the 16S rRNA are also prominent in the 30S particle. No features are seen in the 30S particle which are not also seen in the free RNA. The only significant difference observed is a substantial drop in yield of all those crosslinks which involve regions near the 3' end of the rRNA. This drop is seen only in activated 30S particles, not in the inactivated conformation. Thus psoralen crosslinking confirms the notion of structural changes in the 3' end of the 16S rRNA upon 30S activation.

The great overall similarity in the RNA crosslinking pattern of the free 16S rRNA and 30S ribosomes argues strongly that the free RNA has a definite folded structure. In fact this structure must resemble in many ways the structure of RNA within the assembled particle. The pattern of crosslinking seen in the 16S rRNA translates into a pattern of secondary structure quite unlike what has been proposed for most RNAs in the past. It implies that a substantial fraction of the base pairs are formed between residues that are far apart in the primary structure. When all eleven crosslinks are placed simultaneougly in the structure and assigned to plausible duplex lengths consistant with the known sequence, the RNA chain becomes constrained to a volume only two to three times the size of the ribosomal subunit. In otherwords, this is an RNA which has the capacity to fold up into a globular unit in the same way that protein chains do. Crosslinking results are fully consistant with hydrodynamic evidence described earlier.

Does the Sequence Help?

Much is known about the thermodynamics of individual single and double strand interactions which contribute to the overall secondary structure of an RNA (37). Thus it is possible to predict, with some confidence the most stable pattern of duplex for a short nucleotide sequence. Unfortunately for large RNAs, the complexity of the conformational energy surface is still very great and this would appear to preclude any kind of search for a global energy minimum. When the 16S rRNA sequence is scanned by a computer program a dense array of locally stable secondary structure regions can be found (38). But there is no way to select among a large number of mutually inconsistant structures to produce a small number of "best" structures. It is striking that there are almost no long stretches of continuous Watson-Crick base pairing possible. But there are a very large number of short stable duplexes. Thus the difficulty is that the 16S rRNA, as any random RNA sequence, is extraordinarily rich in rather mediocre duplex regions.

The electron microscopic assignment of crosslinks is not sufficiently accurate to assign particular sequence regions directly to duplexes. However it does place severe constraints on possible secondary structure patterns. Therefore we asked whether one could assign reasonable sections of sequences to each double helical region implied by the crosslinking (39). This was fairly easy to accomplish given the great density of potential secondary structure regions. The assignment of all 11 crosslinks was possible without reusing the same residues.

There is no reason to believe that the particular pattern of structure which results is the most stable or is the best fit to the crosslinking data. However the general character of the molecular structure which results is probably a reasonable representation of the true molecule. The important fact is that, even knowing approximate crosslinking locations, and using these to narrow the search, we were unable to find unique stable nearly-continuous stretches of double helix. The best that could be done was to match each crosslink with a long stretch of duplex containing short helical regions interspersed with frequent loops.

Future Directions

To evaluate models of the 16S rRNA structure we have to know not only about double strand regions, but also we have to have information about the proximity of single strand regions. We are just starting such studies in collaboration with Alain Expert-Bezancon and Donal Hayes. Using a thiol-containing glyoxal derivative maleimide and subsequent bridging reactions one can form single strand - single strand crosslinks and then map their positions by electron microscopy just in the same way as was done for psoralens. Although it is too early to draw quantitative conclusions, the overall appearance of crosslinks seen is quite similar to psoralen results (40). If this bears up to further scrutiny it means that most close approaches of single strands arise from double strand bulge loops rather than by contact between hairpins and bulge loops or other hairpins.

From our experinece to date, crosslinking monitored by electron microscopy seems to be a very effective strategy for examining rRNA structure. Further progress, however, requires several technical advances. While some improvement in electron microscope technique may be possible, it seems unrealistic at present to assume that a resolution will be reached where the crosslinked residues can by uniquely identified. Thus improved sequencing techniques for locating crosslinks are badly needed. Obviously 11 psoralen detected interactions represent only a tiny fraction of the total secondary structure regions. The simple strategy of applying more and more vigorous psoralen crosslinking conditions will not work. Even if the risks of structure perturbation and trapping could be avoided, such heavily crosslinked molecules will not spread well enough on grids to be mapped. Therefore what is needed are new crosslinking reagents or ways of modulating the specificity of existing reagents.

All the results discussed thus far have involved the 16S rRNA. Some preliminary experiments have also been carried out on free 23S rRNA and on 70S particles. With free 23S rRNA extradordinarily rich pattern of crosslinks is seen by electron microscopy of HMT treated molecules (41). Both ends of the RNA show a complex pattern of large loops, much like the results described for the 16S rRNA. The center third of the 23S rRNA is different in that it displays a number of prominent large hairpins. Beacuse of the large number of different features seen, it will take much quantitative mapping of individual molecules before features can be assigned with confidence. Such work is now in progress.

The intact 70S ribosome is rather resistant to HMT crosslinking. This is probably because the high Mg^{2+} concentrations required to maintain the 70S complex also tightens the structure of the RNA in the subunits considerably, as indicated by tritium exchange experiments (26). Thus to obtain reasonable numbers of covalently attached psoralens, we had to use the more reactive derivitive AMT (see Figure 1). Here at least 7 psoralens can be attached (42), and rRNAs isolated from such particles show large crosslinked loops. A question of longstanding interest in the ribosome field is whether any direct contact between the 16S and 23S rRNAs occurs in the 70S complex. To examine this Gloria Chu in my laboratory looked for higher molecular weight species by gel electrophoresis, and in some preparations, at very low yield, a species consistant with a covalent complex of the two rRNAs was seen. Because of the low yield we were concerned that such a species might be an artifact resulting from random inter-particle crosslinking. However in the electron microscope, a consistant Y-shaped sturcture was seen for the crosslinked molecules. Length measurements indicate that one end of the 16S rRNA can be crosslinked to a point on the 23S rRNA about 40% of the contour length from one end. However the absolute polarity of each of the molecules in the crosslinked pair is not yet known.

TOPOGRAPHICAL ASPECTS OF rRNA STRUCTURE

The pattern of crosslinks seen in the 16S rRNA implies an extent of base pairing between distant regions that had not previously been seriously considered for large RNAs. There is no thermodynamic difficulty with such interactions since the entropy cost of loop closure increases very slowly with loop size, and indeed can be conpensated for by just a few extra base pairs. In addition secondary structure interactions among residues within the loop will greatly diminish the cost of closure. However the type of base pairing pattern indicated by the results in Table 2 raises a series of complex topographical issues that affect both rRNA structure and assembly, and the nature of interactions between two RNA molecules.

The topographical problems which accompany base pairing between distant residues are best introduced by the example in Figure 2a. Here is a staggered arrangement of two base paired regions which corresponds to many of the pairs of crosslinks in Table 2. If each basepaired region is actually a full helix turn (11 base pairs for RNA) or more the nucleic acid strand is actually a knot. This is tested by taking the twisted string representation shown in the figure and pulling both ends simultaneously (43). A molecule without knots will pull out into a straight line whereas the knotted structure in the figure will leave a tight bolus. Using this critereon

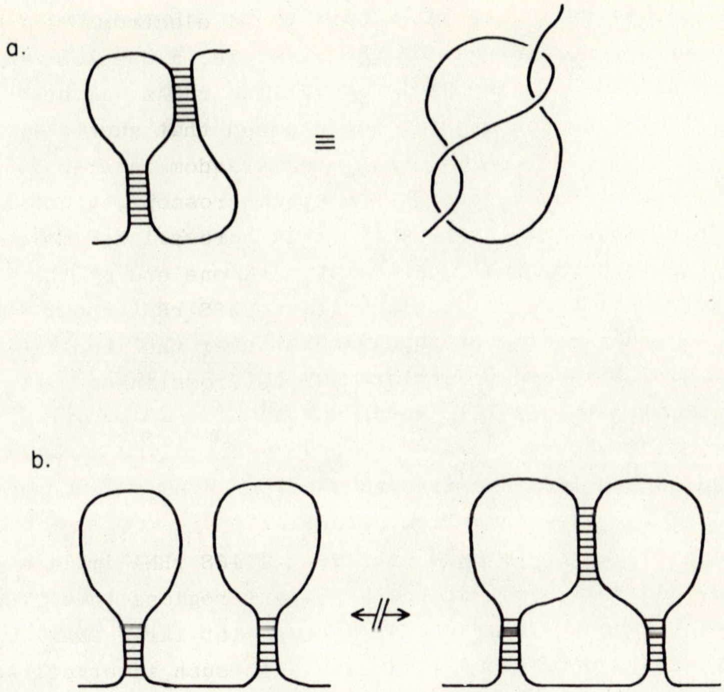

Figure 2. Schematic of secondary features that may correspond to knotted topologies. (a) A simple base pairing scheme which can produce a knot. (b) The order of assembly must be controlled to form this more complex knotted base pairing pattern.

a molecule containing all the duplex regions shown in Table 2 has the potential for forming eight to eleven knots depending on the exact assignment of some of the interactions.

The difficulties introduced by knotted topographies are illustrated by the example in Figure 2b. To form the structure containing the three duplex regions it is necessary to control the order of assembly of the three regions. If the two outer ones are formed first, the inner one cannot be formed. Thus, in general, the assembly of knotted structures will require a fine orchestration among possible assembly pathways so that the order of assembly can be regulated. The example in Figure 2b also shows the problems of base pairing between two different RNA molecules. Topographically a hairpin loop is effectively a circle. It is not possible to make extensive double helix between two single stranded circles without cutting one of the strands. Small bits of helix can be made so long as intervening regions are left-hand twisted to compensate for the righthand twist of the duplex.

In the example in Figure 2b the actual lenghts of the outer duplex regions are irrelevant. Even a single base pair sets up the topographical constraint. Thus even the presence of a single full turn of double helix in a molecule like 16S rRNA raises the spector of topographical restraints. It is worth pausing a second to explore why this is a particularly serious problem for nucleic acids but not for proteins. The protein analog of the types of secondary structure we have uncovered in the 16S rRNA is beta sheets. Here two or more sections of peptide chain from potentially distant regions are brought together to make helices. However the beta sheet is a two-fold helix so that the strands do not actually twist around each other. In practice, there is a slight twist to the most stable beta sheet structures but this is so gradual that even in sheets which stretch across an entire protein subunit, there is not a full turn of twist (44). As a result the peptide chains never become linked into knots. In contrast nucleic acid double helices (or triple helices for that matter) have a very high density of twist. There is a very great chance of forming topographical links resulting in knotted structures.

How can one tell if there are actually knots in an RNA chain? The test described earlier- grab the ends and pull- must be applied on the actual molecule. This can be done by placing covalent crosslinks in the molecule and then subjecting it to strongly denaturing conditions. A crosslink renders all of the chain between the two linked residues into a topographical circle and thus any knots contained within the circle become trapped. They can be visualized within the circle as complex loops or boluses depending on how tight the strand is stretched. Fortunately one of the psoralen crosslinks within the 16S rRNA molecule is essentially end to end. Electron microscopic examination of these molecules reveal loops or boluses with some frequency (39). A quantitative analysis is not yet complete, and we cannot yet say whether the loops seen arise from knots or from additional crosslinks. However the results already indicate that even if the 16S rRNA contains a knot or two, it contains far fewer knots than would be potentially introduced by the pattern of secondary structure implied by Table 2.

Two mechanisms may be operative in limiting the number of knots in the 16S rRNA and other large RNAs. The rather mediocre base pairing potential of the 16S rRNA has been alluded to previously. This means that almost no individual duplex regions long enough to form a link exist. The danger of links arises by the cumulative twist of successive duplexes. However single strand bulge loops between such duplexes may be able to twist in a left hand fashing so as to reduce or eliminate the total number of links (45). A second mechanism is suggested by the structure of rRNA precursors which place the 16S rRNA sequence within a loop closed by a very stable duplex region in the precursor sequence (4C). Suppose such a structure forms before the 16S rRNA structural region has had a chance to fold. As long as the ends fo the precursor are kept together and intact, the 16S rRNA is topographically a circle, and the resulting topographical constraints prevent any knots from forming.

DOES THE RIBOSOME HAVE AN RNA CORE?

The experimental results we have outlined indicate that the free 16S rRNA (and presumablv also the free 23S rRNA) is folded into a globular structure not too different in size from the 30S ribosome itself. At least some of the secondary structure features which are present in this structure are extremely stable and appear to be retained intact in the assembled ribosomal subunit. One way to summarize these results is to postulate that the rRNA is a structural core- that is- it plays the fundamental role in directing the binding and assembly of proteins.

Although a few specific complexes between isolated ribosomal proteins are known, by and large these appear quite rare. It is also not clear whether isolated ribosomal proteins have the same structures that they do when part of the ribosome. In contrast the bulk of the available evidence suggests that the RNA can have the same structure, isolated as assembled. This would make the RNA the core of the structure. However there is a catch. The source of virtually all mature rRNA we study is isolated from assembled ribosomes. Since RNA secondary structures can be very stable, there is no way at present to exclude the possibility that the RNA we study retains a memory of the structure it had in the ribosome. Exactly this same sort of problem has plagued arguments about the folding mechanisms of individual protein chains.

To argue against an RNA memory one would have to show that one could reversibly unfold and refold RNA structure. Unfortunately there is no real assay of total RNA structure. For example one could ask whether RNA was capable of reassembly into ribosomes, but to do this one has to add proteins. Thus one could always argue these proteins contain the information for proper assembly. In support of the idea of an RNA structure memory there is considerable evidence that the assembly pathways followed by both 16S rRNA and 23S rRNA depend on the method of isolation (47,48). Furthermore a number of physical properties of ribosomal RNAs are preparation-dependent (49,50). Unfortunately we really do not know the nature or extent of the structural differences that give rise to these striking differences in behavior. There is no way to know *a priori* what type of isolation procedure will strip the proteins off an RNA with the least perturbation in RNA structure.

Given the high density of secondary structures available to the 16S rRNA and topographical assembly constraints (and even worse potential problems dealing with topographical isomers) it is not surprising that the molecule can be prepared in various metastable structures. Indeed very early studies showed that denaturation and renaturation of ribosomal RNAs involved massive hysteresis (51). During *in vitro* assembly proteins surely help to accelerate the return of the RNA to structures capable of assembly (52). What happens *in vivo* remains largely the domain of future studies (53).

ACKNOWLEDGEMENTS

This work was supported by grants from the U.S. Public Health Service (GM 19843 and GM 14825) and the National Science Foundation (PCM75-14329). The author is grateful to a large number of coworkers and collaborators, not only for helpful discussions, but also for permission to cite work unpublished or in press. Included are John Hearst, Paul Wollenzien, Pallaiah Thammana, and Gloria Chu, Alain Bezancon and Donal Hayes for electron microscopic work; Neal Farber for tritium exchange, and Catherine and Craig Squires for analysis of 16S rRNA sequence data.

REFERENCES

1. Kim, S.H. (1978) in Transfer RNA (ed. by S. Altman) pp 248-293, MIT Press, Cambridge.

2. Peters, M., and Yarus, M. (1979) J. Mol. Biol. in press.

3. Fairclough, R.H., Cantor, C.R., Wintermeyer, W., and Zachau, H.G. (1979) J. Mol. Biol. in press.

4. Keuchler, E. (1979) in Transfer RNA Part 1: Structure, Properties, and Recognition (ed. P. Schimmel, D. Soll, and J. Abelson) Cold Spring Harbor, New York, in press.

5. Fairclough, R.H., and Cantor, C.R. (1979) J. Mol. Biol. in press.

6. Johnson, A.E., and Cantor, C.R. (1979) in preparation.

7. Steitz, J.A., and Jakes, K. (1975) Proc. Natl. Acad. Sci. 72, 4734-4738.

8. Schwartz, I., and Ofengand, J. (1978) Biochemistry 17, 2524-2530.

9. Prince, J.B., Hixson, S.S. and Zimmermann, R.A. (1979) J. Biol. Chem. 254, 4745-4749.

10. Yukioka, M., Hatayama, T., and Omori, K. (1977) Eur. J. Biochem. 73, 449-459.

11. Eckermann, D.J. and Symons, R.H. (1978) Eur. J. Biochem. 82, 225-234.

12. Sonenberg, N., Wilchek, M. and Zamir, A. (1975) Proc. Natl. Acad. Sci. 71, 4332-4336.

13. Nisbet, J.H. and Slayter, H.S. (1975) Biochemistry 14, 4003-4010.

14. Allen, S.H. and Wong, K.P. (1978) J. Biol. Chem. 253, 8759-8799.

15. Stuhrmann, H.B., Koch, M.H.J., Parfait, R., Haas, J., Ibel, K., and Crichton, R.R. (1977) Proc. Natl. Acad. Sci. 74, 2316-2320.

16. Vasiliev, V.D., Selinvanova, O.M., and Koteliansky, V.E. (1978) FEBS Lett. 95, 273-276.

17. Moore, P.B., Langer, J.A., Schoenborn, B.P., and Engleman D.M. (1977) J. Mol. Biol. 112, 199-227.

18. Brimacombe, R., Stöffler, G., and Wittmann, H.G. (1978) Ann. Rev. Biochem. 47, 217-249.

19. Hogan, J.J., and Noller, H.F. (1978) Biochemistry 17, 587-593.

20. Chapman, N.M. and Noller, H.F. (1977) J. Mol. Biol. 109, 131-149.

21. Santer, M. and Shane, S. (1977) J. Bacteriology 130, 900-910.

22. Nomura, M., Sidikaro, J., Jakes, K. and Zinder, N. (1974) in Ribosomes (ed. by M. Nomura, A. Tissieres and P. Lengyel) pp 805-814, Cold Spring Harbor, New York.

23. Schreiber, J.P., Hsiung, N., and Cantor, C.R. (1979) Nucleic Acids Res. 6, 181-193.

24. Thammana, R., and Cantor, C.R. (1978) Nucleic Acids Res. 5, 805-823.

25. Gamble. R., Schoemaker, H.J.P., Jekowsky, E., Schimmel, P., (1976) Biochemistry 15, 2791-2794.

26. Farber, N. and Cantor, C.R. (1979) manuscript in preparation.

27. Noller, H.F. and Herr, W. (1974) J. Mol. Biol. 90, 181-184.

28. Schoemaker, M.J.P., Gamble, R.C., Budzik, G.P., and Schimmel, P.R. (1976) Biochemistry, 15, 2800-2803.

29. Dunn, J.M. and Wong, K.P. (1979) J. Biol. Chem. in press

30. Zweib, C., Ross, A., Rinke, J., Meinke, M., and Brimacombe R. (1978). Nucleic Acids Res. 5, 2705-2720.

31. White, J.R., Yielding, W.E. and Yielding, K.L. (1977) in Methods in Enzymology, 46 (ed. by W. Jakoby and M. Wilchek) pp 644-649, Academic Press, New York.

32. Wagner, R. and Garrett, R.A. (1978) Nucleic Acids Res. 5, 4065-4075.

33. Wollenzien, P.L., Youvan, D.L. and Hearst, J.E. (1978) Proc. Natl. Acad. Sci. 75, 1642-1646.

34. Wollenzien, P.L., Hearst, J.E., Thammana, P., and Cantor, C.R. (1979) J. Mol. Biol. in press.

35. Wollenzien, P.L., Hearst, J.E., Squires, C.L., and Squires, C.L. (1979) submitted for publication.

36. Thammana, P., Cantor, C.R., Wollenzien, P.L., and Hearst, J.E., (1979) J. Mol. Biol. in press.

37. Tinoco, I. Jr., Borer, P.N., Dengler, B., Levine, H.D., Uhlenbeck, O.D., Crothers, D.M., and Gralla, J. (1973) Nature, New Biol. 246, 40-41.

38. Squires, C.L. and Squires, C.L. (1979) unpublished results.

39. Cantor, C.R., Wollenzien, P.R. and Hearst, J.E. (1979) m.s. in preparation.

40. Expert-Bezancon, A., Hayes, D. Wollenzien, P.L. and Cantor, C.R. (1979) unpublished results.

41. Wollenzien, P.L., Hearst, J.E., Thammana, P. and Cantor, C.R. (1979) unpublished results.

42. Chu, G.Y. and Cantor, C.R. (1979) unpublished results.

43. Neuwirth, L. (1979) Scientific American 240 (no. 6) 110-124.

44. Schultz, G.E. and Schirmer, R.H. (1979) Principles of Protein Structures, pp. 77-78, Springer Verlag, New York.

45. Squires, C.L. (1979) personal communication.

46. Young, R.A., and Steitz, J.A. (1978) Proc. Natl. Acad. Sci. 75, 3593-3597.

47. Hochkeppel, H.K., Spicer, E., and Craven, G.R. (1976) J. Mol. Biol. 101, 155-170.

48. Nowotny, V., Amils, R., Rheinberger, M.J. and Nierhaus, K.N. (1979) m.s. in preparation.

49. Hochkeppel, H.K., Gordon, J., and Brack, C. (1977) FEBS Lett. 77, 277-280.

50. Ungewickell, E., Garrett, R.A. and Le Bret, M. (1977) FEBS Lett. 84, 37-42.

51. Revzin, A., Neumann, E., and Katchalsky, A. (1973) J. Mol. Biol. 79, 95-114.

52. Bogdanov, A.A., Zimmermann, R.A., Wang, C.C., and Ford, N.C.Jr. (1978) Science 202, 999-1001.

53. de Narvaez, C.G., and Schaup, H.W. (1979) J. Mol. Biol. in press.

STRUCTURE OF RIBOSOMAL PROTEINS

H. G. WITTMANN, J. A. LITTLECHILD
and B. WITTMANN-LIEBOLD

Max-Planck-Institut für Molekulare Genetik
D-1000 Berlin-Dahlem, Germany

INTRODUCTION

To understand the function of the ribosomal machinery at a molecular level it is essential to have a detailed knowledge of the structure of its individual components. This is especially true for the numerous proteins and for the RNA strands present in the ribosome but it also applies to the protein factors and tRNAs interacting temporarily with the ribosomal particle.

In this chapter we summarize our present knowledge on the primary, secondary and tertiary structures of the ribosomal proteins and the factors involved in protein biosynthesis. Most of the information to date is derived from the ribosomes of *E. coli*. Therefore,we will deal mainly with the ribosomal proteins of this organism. This restriction should also minimize overlaps with other chapters of this book.

Number of Ribosomal Proteins

The number of ribosomal proteins is found to range between 50-60 in bacterial ribosomes and 70-80 in cytoplasmic ribosomes of eukaryotes. When the protein moiety extracted from *E. coli* ribosomes is separated by two-dimensional polyacrylamide gel electrophoresis, 21 spots (S1-S21) from the small and 34 spots (L1-L34) from the large ribosomal subunit are detected (1). In the small subunit the number of spots corresponds to the number of individual proteins (2) whereas in the large subunit the spot L8 was shown to consist of a complex of three proteins: L7/L12 and L10 (3). Furthermore, it was revealed by protein-chemical and immunological studies (2,4) that protein L26 from the large subunit is identical to protein S20 from the small subunit. Since the stoichiometry of protein S20 (0.8) and of protein L26 (0.2) corresponds to a total of one copy per 70S ribosome (5) it is likely that this protein is distributed between the two subunits when the 70S ribosome dissociates into its two subunits. Therefore, there are 32 individual proteins in the large subunit of the *E. coli* ribosome. Although one of them, L7, differs only in the presence of an acetyl group at its N-terminus from L12 (6) we continue to regard L7 and L12 as two individual proteins.

Protein Isolation

All *E. coli* ribosomal proteins have been isolated and characterized by chemical, physical and immunological techniques. The first method of isolation was to extract the proteins with 67% acetic acid from the ribosome and then to separate them by column chromatography on cellulose exchangers and by gel filtration in the presence of 6 M urea (7). The proteins isolated in this way have been used mainly for molecular weight determinations and for immunological and protein-chemical studies (2,4).

For physical studies, e.g. CD, NMR, shape determination and crystallization, which require the conservation of the native structure as far as possible, a procedure has been developed which completely avoids the use of acetic acid, urea and lyophilization (8). By washing of ribosomal subunits with increasing concentrations of salts, e.g. LiCl or NaCl, groups of proteins were removed and further fractionated by CM-Sephadex chromatography and gel filtration on Sephadex G75 in buffers of high ion concentrations but without urea. Many of the physical studies described below have been performed with proteins isolated by this gentler procedure. Furthermore, comparative studies have also been made with proteins isolated by the different methods.

The isolation of ribosomal proteins which are active in the reconstitution assay and in functional tests has recently been described (9). The isolation procedure consists of three steps: first, prefractionation by washing of 50S subunits with high concentrations of salt; second, gel filtration on Sephadex; and third, column chromatography on carboxymethyl cellulose in presence of 6 M urea.

PRIMARY STRUCTURE

Sequence Determination of *E. coli* Wild Type Ribosomal Proteins

The protein-chemical studies that were carried out on ribosomal proteins up until 1974 have been reviewed in some detail (2). They have included determination of amino acid composition, end group analysis, peptide mapping, isolation and analysis of tryptic peptides as well as determination of the amino acid sequence within the N-terminal regions by means of a liquid phase sequenator. Until then only the proteins S4 and L7/L12 had been sequenced completely.

The primary structures of 19 proteins from the small and of 29 proteins from the large subunit of the *E. coli* ribosome have now been determined (Table 1). Eludication of the remaining six proteins, namely S1 and S2 from the small and L2, L4, L9 and L17 from the large subunit, is in progress. Determination of their sequences, with the possible exception of protein S1, can be expected in the very near future.

In addition to the 48 *E. coli* ribosomal proteins the complete primary structure has been determined for the initiation factors IF-1 and IF-3 as well as for two proteins, NS1 and NS2, which tightly bind to native 30S subunits (Table 1). The latter two proteins are indistinguishable (76) from proteins HU (77) and HD (78) which have been identified as DNA binding proteins.

This rapid progress in the elucidation of the primary structure of ribosomal proteins was made possible by several improvements in the techniques for amino acid sequence analysis. Several major advances contributed towards the goal to sequence proteins and peptides more quickly, more reliably and with less material. These were first technical improvements to the Beckman liquid phase sequencer (79-80) including the construction of an automated conversion device for PTH-amino acids (81); second, improvements in the solid phase sequencing technique (82,83); third, availability of more sensitive amino acid analyzers; fourth, isolation of peptides by thin layer techniques and micro column procedures (13,84); fifth, introduction of a new reagent, namely dimethylaminoazobenzene isothiocyanate (85), for the Edman degradation reaction. Using the latter method an amount of 2-8

TABLE 1. List of *E. coli* ribosomal (and related) proteins whose primary structures have been determined

Protein	Amino acids	Mol. wt.	References: Wild type	References: Mutants	Protein	Amino acids	Mol. wt.	References: Wild type	References: Mutants
S3	232	25 852	10		L12	120	12 164	6	
S4	203	23 137	11	2,60	L13	142	16 019	36	
S5	166	17 515	12	61-68	L14	120	13 227	37	
S6	135	15 704	13	13	L15	144	14 981	38	
S7K	177	19 732	14		L16	136	15 296	39	
S7B	153	17 131	14		L18	117	12 770	40	
S8	129	13 996	15	70	L19	114	13 002	41	
S9	128	14 569	16		L20	117	13 366	42	
S10	(105)	(12 290)	17		L21	103	11 565	43	
S11	128	13 728	18		L22	109	12 111	44	
S12	123	13 606	19	60,63,66 71,72,75	L23	99	11 013	45	
S13	117	12 968	20		L24	103	11 185	46	74
S14	(98)	(11 162)	21		L25	94	10 694	47,48	
S15	87	10 001	22		L26	86	9 553	=S20	
S16	82	9 191	23		L27	84	8 993	49	
S17	83	9 573	24	69,73	L28	77	8 875	50	
S18	74	8 896	25		L29	63	7 273	51	
S19	91	10 299	26		L30	58	6 411	52	
S20	86	9 553	27		L31	62	6 971	53	
S21	70	8 369	28		L32	56	6 315	54	
L1	233	24 599	29		L33	54	6 255	55	
L3	209	22 258	30		L34	46	5 381	56	
L5	178	20 171	31		IF-1	71	8 119	57	
L6	176	18 832	32		IF-3	181	20 693	58	
L7	120	12 206	6		NS1	90	9 225	59	
L10	165	17 736	33,34		NS2	90	9 535	59	
L11	141	14 874	35						

nmol of peptide or protein is enough to sequence 20-30 amino acid residues. This results in a 10-20 fold higher sensitivity when compared to previous techniques. Whereas, for proteins sequenced about 5-10 years ago elucidation of the primary structure required 50-200 mg of material this could now be achieved with several milligrams only.

The application of the improved techniques has also accelerated the determination of the primary structure of the ribosomal proteins from other organisms, e.g. yeast and rat, and has also facilitated studies for which efficient and sensitive methods for sequence analysis of peptides or proteins are essential: e.g. determination of the amino acid residues involved in protein-protein or protein-RNA cross-links; elucidation of the amino acid alterations in mutant proteins; isolation of large protein fragments which were injected into rabbits for antibody production, which were used for investigations on protein-RNA interactions, or which were studied by NMR or CD technique. Examples of such studies are given below.

The amino acid sequence analysis of the ribosomal proteins has been rendered more difficult by the occurrence of a number of modified amino acids in several of these proteins. Acetylation of the N-terminus was identified in proteins S5 (12), S18 (25) and L7 (6). Because of the blocked N-termini the application of the sequenator was not possible for studies on the N-terminal regions of these proteins and hence slowed down the elucidation of their primary structure. The terminal amino acids of other proteins, namely S11, L11, L16, L33 and IF-3, were found to be modified by methylation (Table 2).

TABLE 2. Occurrence of modified amino acids in *E. coli* ribosomal proteins

Protein	Position	Modified amino acid	Ref.	Protein	Position	Modified amino acid	Ref.
S5	1	N-acetyl alanine	12	L11	1	N-trimethyl alanine	35
S11	1	N-monomethyl alanine	18,86	L11	3	N_ε-trimethyl lysine	35
S12	88	derivative of Asp	19,see text	L11	39	N_ε-trimethyl-lysine	35
S18	1	N-acetyl alanine	25	L16	1	N-monomethyl-methionine	39,86
L3	150	N^5-monomethyl glutamine	30	L16	81	derivative of Arg	39
L7	1	N-acetyl serine	6	L33	1	N-monomethyl alanine	55,86
L7/L12	81	N_ε-monomethyl lysine	6	IF-3	1	N-monomethyl methionine	58

Protein L11 is the most heavily methylated *E. coli* ribosomal protein. It contains nine methyl groups, namely three at the N-terminal amino acid, trimethyl-alanine, and six at the two internal trimethyl-lysines in positions 3 and 39, respectively (35). The occurrence of modified amino acids not at the N-terminus but in the internal regions of the protein has, besides in L11, also been demonstrated in several other proteins, namely S12, L3, L7/L12 and L16 (Table 2). The modified amino acid in position 88 of S12 (19) is likely to be a derivative of aspartic acid since the product of the Edman degradation step (85) gave a mass ion identical to that of aspartic acid although it does not comigrate with this component in thin-layer chromatography. Furthermore, determination of the DNA sequence of this region revealed a triplet corresponding to aspartic acid (87). In some of the S12 mutants an aspartic acid residue has been found in position 88 (19,63,75). The enzyme system modifying the aspartic acid residue in position 88 of the wild-type S12 protein has apparently become inactive in these mutants.

Since the DNA sequence has recently been determined for several genes of *E. coli* ribosomal proteins (87-89) it now is possible to compare the nucleotide sequences with the previously published amino acid sequences. There is complete agreement for most of the studied proteins, namely L1, L7/L12, L11, L14, S11 and S12, whereas a few discrepancies remain to be solved for proteins S17 and L10. The sequence Arg-Pro-Cys in positions 63-65 of S17 differs from Cys-Arg-Pro which was deduced from the DNA sequence. The sequence proposed for protein L10 is in agreement with that given in ref. 34, except that the glutamine in position 116 of L10 was postulated to be glutamic acid according to the DNA sequence, and the arginine residue in position 84 of L10 which has independently been determined by two protein sequencing groups (33,34) has not been found to correspond to a triplet in the DNA. Studies to resolve these discrepancies are in progress.

The knowledge of the primary structures of almost all of the proteins present in the *E. coli* ribosome (Table 1) allows an answer to the question as to whether there are significant homologies among this set of proteins. If so, this would indicate that the ribosomal proteins derive from a common ancestor. Therefore, the amino acid sequences of the *E. coli* ribosomal proteins (Table 1) were compared and searched for identical regions using computer programmes. The numbers of identical tri-, tetra-, penta-, or hexapeptides obtained in this way were compared to the numbers obtained by performing the same comparison with completely unrelated proteins whose primary structures were taken from the literature, and also with artificial proteins which have the same amino acid compositions and lengths as the *E. coli* ribosomal proteins but whose amino acid sequences were randomly generated by a computer program. The frequency pattern of the occurrence of identical peptides was similar in all three cases, the native ribosomal proteins, their artificially generated isomers and the unrelated proteins from the literature. Therefore, one can conclude that, with the exception of L7/L12 and S20/L26, there is no indication of strong structural homologies among the studied *E. coli* ribosomal proteins (90) although in a few cases there might exist very weak homologies (23,27,34,91). This result is in good agreement with the conclusions drawn from immunological studies with antibodies against the individual *E. coli* ribosomal proteins where no significant cross-reaction between any of the proteins was detected (92,93).

Sequence Analysis of *E. coli* Mutant Ribosomal Proteins

Many mutants of *E. coli* have been isolated which have altered ribosomal proteins as revealed by two-dimensional gel electrophoresis. As reviewed in reference (94) mutants are now available for almost all of the *E. coli* ribosomal proteins. Most of these mutants have been isolated by their temperature-sensitivity (95,96) or by their altered behavior towards antibiotics. A very efficient system for isolating ribosomal mutants used a streptomycin dependent strain which gives rise to mutations in a very large number of ribosomal proteins (97). In this way mutants with alterations in each of 50 proteins, i.e. 95% of those present in the *E. coli* ribosomes, have been isolated (98). There are mutants which harbor up to eight altered proteins in the same ribosome.

As listed in Table 1 and reviewed in reference (2) protein-chemical analyses have been carried out on many mutants. Most of these studies were performed with the mutationally altered proteins S5 and S12. Figure 1 summarizes the amino acid replacements revealed so far in protein S5. The amino acid exchanges in various spectinomycin resistant mutants are clustered in positions 19-21 of this protein (61,62). Clustering of amino acid replacements has also been found in neamine-resistant mutants altered in protein S5 (69) and in erythromycin-resistant mutants altered in protein L4 (99).

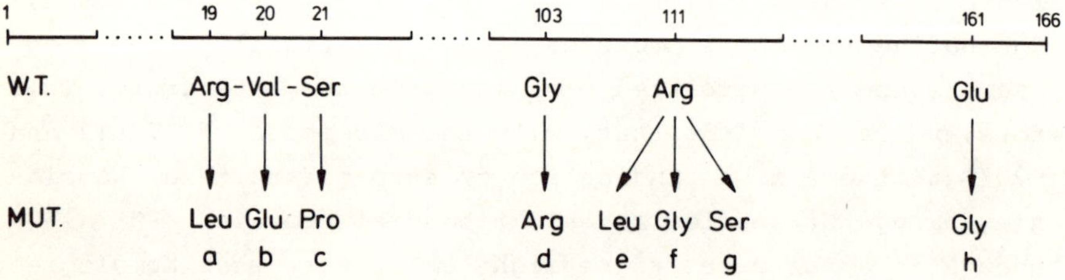

Figure 1. Amino acid replacements in protein S5 of the following E.coli mutants (with reference number). a: spc^r-9 (64); b: spc^r-7 (61); c: spc^r-13 (61), spc^r-15 (62); d: N660 (63); e: d1023 (63), N421 (65); f: nea-314 (69); g: nea-319 (69); h: 0-1, the mutation leads to a shorter protein chain (65).

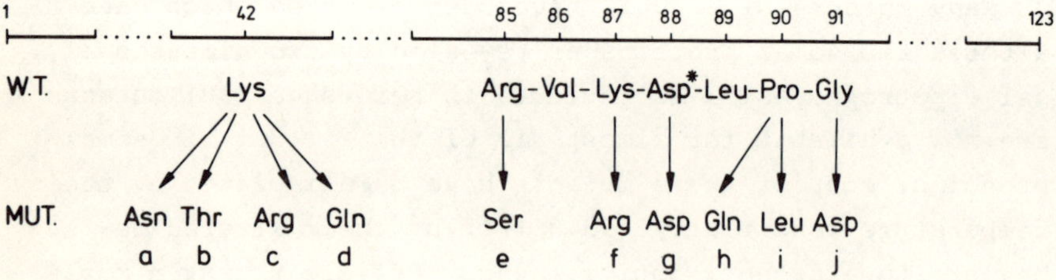

Figure 2. Amino acid replacements in protein S12 of the following E.coli mutants (with reference number). a: SM-10 (72), L44-1 (71), L1-401 (71); b: L44-2 (71); c: SM-5 (72), L11-His 1 (71); d: N421 (71); e: d1023 (63); f: SM-1 (72), L44-4 (71), L1-431 (71); g: d1023 (63), VT (75); h: nea-319 (69); i: T776/39:3 (60), T776/40:4 (60), T779/25:3 (60), nea-314 (69); j: N660 (63). Asp in position 88 is a derivative of aspartic acid.*

Fig. 2 illustrates the amino acid exchanges in protein S12. The following two points can be concluded from our protein-chemical studies: 1) All replacements are clustered in only two short regions of protein S12, namely in position 42 and within the region from positions 85-91. No amino acid exchanges in any of the many mutants analyzed so far have been detected outside of these two short regions. 2) The replacement of a single amino acid in protein S12 suffices to make the ribosome streptomycin-resistant. Since it has been shown by reconstitution tests (100) that protein S12 alone confers the altered streptomycin resistance it follows from the protein-chemical results that the replacement of only one out of the 8000 amino acids present in the *E. coli* ribosome induces the altered ribosomal reaction towards the antibiotic.

Comparisons between the Amino Acid Sequences of Ribosomal Proteins from *E. coli* and Other Organisms

Compared with the approximately 50 completed sequences of ribosomal proteins from *E. coli* little is known so far about the structure of ribosomal proteins from other organisms. The amino acid sequences of four such proteins, namely one from *Bacillus subtilis* (101), two from yeast (102,103) and one from the brine shrimp *Artemia salina* (104) have been reported. Furthermore, partial sequences of ribosomal proteins are available from several bacterial species (105-109), from yeast (110), from wheat germ (109), and from rat liver (111, 112).

Without going into the details which will be discussed elsewhere in this book (113), only three points are mentioned which derive from the protein-chemical studies on these proteins: 1) The amino acid sequences of ribosomal proteins from bacterial species, e.g. *E. coli* and *Bacillus subtilis*, belonging to different families show significant structural homologies. This is illustrated by the comparison of the complete primary structures of protein B-L9 from *B. subtilis* and proteins L7/L12 from *E. coli* (Fig. 3). More than 50% of the amino acid residues are identical in both proteins (101). 2) The amino acid sequence of a ribosomal protein, HL20, from the halophilic *Halobacter cutirubrum* resembles in different regions that of *E. coli* proteins L7/L12 and that of *Artemia salina* protein eL12 (104,108) whereas there is little homology between the *E. coli* and the *Artemia* protein (102,104). Thus the halophile *Halobacter* has an interesting phylogenetic position with properties of both pro- and eukaryotes. 3) The amino acid sequence of the few yeast ribosomal proteins which have so far been studied exhibits some homologies with that of other eukaryotic ribosomal proteins, e.g. from *Artemia salina* (104,110) or from rat liver (102,112). However, only a very low degree of sequence homology, if any, has so far been detected between ribosomal proteins from *E. coli* on the one hand and those from yeast, *Artemia* or rat on the other hand (102,104,109,112). It is possible that knowledge of the spatial structure of these proteins would show a higher degree of homology than the primary structure. Interestingly, the phylogenetic relationship among various eukaryotic species, e.g. yeast and rat, is (at least as far as the primary structure of ribosomal proteins is concerned) greater than between a unicellular eukaryote (yeast) and a unicellular prokaryote (*E. coli*).

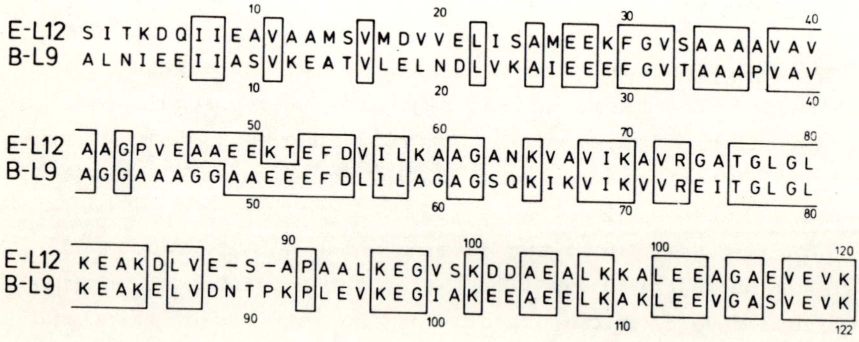

Figure 3. Comparison of the primary structures of E. coli protein E-L12 and B. subtilis protein B-L9 (101). Abbreviations for amino acids are according to the one-letter code.

SECONDARY STRUCTURE

Information about the secondary structure of ribosomal proteins has been derived from two sources: the prediction based on the amino acid sequence and the experimental results from circular dichroism studies.

Table 3. Prediction of secondary structure based on the amino acid sequence

Protein	Helix %	Extended structure %	Turn %	Protein	Helix %	Extended structure %	Turn %
S3	25	12	17	L13	27	4	22
S4	28	4	18	L14	15	15	25
S5	34	8	18	L15	26	6	28
S6	36	4	10	L16	27	9	24
S7	38	5	13	L18	44	9	15
S8	36	5	14	L19	25	9	11
S9	18	6	20	L20	49	2	19
S10	19	5	12	L21	20	n.c.	23
S11	16	9	26	L22	28	6	11
S12	7	13	16	L23	35	8	14
S13	21	n.c.	19	L24	9	8	21
S14	34	n.c.	18	L25	42	10	10
S15	30	n.c.	14	L26	64	n.c.	5
S16	22	5	15	L27	10	6	30
S17	13	8	16	L28	22	14	16
S18	15	5	15	L29	67	n.c.	5
S19	15	6	25	L30	n.c.	9	2
S20	64	n.c.	5	L31	n.c.	5	24
S21	57	n.c.	4	L32	16	2	34
L1	29	4	13	L33	33	6	15
L3	18	12	22	L34	17	n.c.	24
L5	26	6	15	IF-1	11	7	17
L6	24	10	17	IF-3	34-37	6-8	15-17
L7/L12	64	n.c.	3	NS1	43	n.c.	14
L10	47	3	9	NS2	43	1	16
L11	33	6	17				

n.c.: not calculated

Prediction of Secondary Structure

The secondary structures of ribosomal proteins have been predicted (114-116) for those ribosomal proteins whose primary structures are known. Four different predictive methods were used to calculate the conformational states such as α-helix, extended structure, random coil and turn or bend (Table 3). Illustrations were drawn for each of the proteins to summarize the information obtained and to show in which region of the amino acid sequence the different states are predicted by all four methods. An example is shown in Figure 4. When at least three of the four prediction methods agreed about the conformational state of the amino acid residues, this was taken to indicate a high probability of this particular structure occurring in this region of the sequence. It should be pointed out that by applying these strict criteria the numbers given in Table 3 which indicate how many percent of the amino acid residues form helix, turns etc., are minimal values.

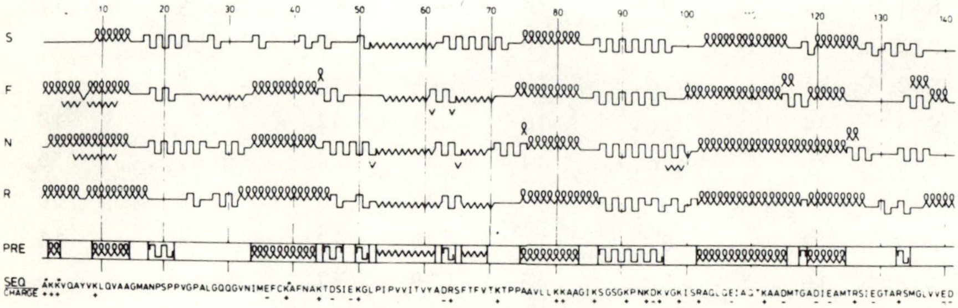

Figure 4. Secondary structure of protein L11 as predicted according to four different methods. The symbols represent residues in helical ( ), turn or bend ( ), extended ( ) and coil (———) conformational states, respectively. The line "PRE" summarizes the secondary structure obtained when at least three out of the four predictions are in agreement. The %-values for helix, turn etc. (illustrated in line PRE) are listed in Table 3 for the proteins sequenced so far. The amino acid sequence of protein L11 is shown in the bottom line in the one-letter code. For details see reference (115).

The results on the prediction of the secondary structures show that each of the ribosomal proteins studied has a unique secondary structure which differs from that of all other proteins. The α-helix contents for many proteins range between 20-40%. Several proteins, e.g. S20, L7/L12 and L29, have a high and others, e.g. S12 and L27, a low α-helix content. The values predicted for extended structure are low, i.e. less than 20%, for all proteins whereas those for turns range between 10-30% for most proteins. A comparison of these predictive values with those experimentally determined is given below.

It is possible to construct models of the ribosomal proteins based on these prediction methods. This can be facilitated by the use of computer graphics (117). It is clear that more information is needed to construct a feasible 3D model. This information can be obtained in the absence of crystallographic data from proton magnetic resonance results which could provide information as to the folding of the protein chain and from intramolecular cross-linking studies.

Circular Dichroism

Many studies have been made on the optical rotary dispersion (ORD) or circular dichroism (CD) of ribosomal proteins. Early studies (118-120) were made on a mixture of proteins and the general conclusion was that both in the ribosome and in the isolated state (usually by acetic acid extraction) the proteins contained ∿ 25% α-helix together with some ß-pleated sheet and random coil conformation. In an early study on individual ribosomal proteins (121) the CD spectrum was measured in dilute HCl (conditions under which proteins were soluble) of several purified protein fractions. An acidic protein, L7/L12, was found to have 50% α-helix and the remainder of the isolated proteins α-helix values varying from 19-49% with negligible amounts of ß-structure. Other measurements on the proteins L7/L12 have resulted in values of 60 ± 5% α-helix, 20% ß-structure (122); 45-50% α-helix (123) and 50-55% α-helix (124).

The CD spectra of proteins S3, S4, S6, S7 and S8 gave α-helix values varying from 18% for S6 to 50% for S7 and an average of 18% ß-structure. A change resulting in loss of structure was observed upon heating to 42°C (125). The protein S4 has also been studied by other groups, and values of 26-32% α-helix, little or no ß-structure (126) and 32% α-helix, 14% ß-structure (127) have been quoted.

It has been suggested (120,128) that the CD spectra of ribosomal proteins free in solution represent a unique conformation which is different from their conformation in the ribosome. This information was obtained by subtraction of the CD spectrum of the rRNA from the spectrum of the total ribosome. The ribosomal protein S1 has been estimated to contain 8-17%

α-helix, 32-41% ß-structure and 44-50% random coil (129).

In all of these studies there is a great deal of variation in the method of preparation of the ribosomal proteins or in the buffer conditions under which the measurements were performed. Until recently all methods used for the preparation of proteins from both the 30S or 50S subunit involved treatment with acetic acid, or urea and lyophilization (reviewed in 7) with the exception of proteins L7/L12 and protein S1 which can be washed free from the ribosome by other techniques (130,131). These procedures result in the denaturation of the protein during the isolation procedure. We have now developed methods to isolate both the 30S (132) and the 50S proteins (133) from the ribosome avoiding these harsh conditions (see above). These proteins show a reversed solubility to those isolated by the harsher conditions (134) and are soluble at high salt concentrations and less soluble at low salt concentrations. This enables experiments to be performed under reconstitution conditions (0.35 M salt) without the problems of protein precipitation.

When proteins prepared with acetic acid and urea (134) were used for CD studies (135), it was found that most of these proteins contained relatively low levels of ordered structure. In this case it was not possible to use either the method of Rosenkranz and Scholtan (136) or Chen et al. (137) in order to calculate values of α-helix, ß-pleated sheet and random coil.

The proteins prepared by the salt extraction method, from their CD spectra, show in all cases an increased amount of α-helix when they are compared to the urea-acetic acid extracted proteins described above (138,139). Figure 5 shows a comparison of the CD spectra of the ribosomal proteins S4. The spectra of several of the 30S and 50S proteins prepared by the gentler conditions are given in Figure 6. These proteins do not show significant differences in secondary structure when changing the salt concentration from 5 → 300 mM salt (KF). This result is in contrast to salt induced changes observed with the histone proteins (reviewed in reference (140)).When the salt extracted ribosomal proteins are dialysed into 6 M urea/methylamine pH 7.0 and back into a phosphate buffer containing 0.3 M potassium fluoride, the same CD spectra as seen

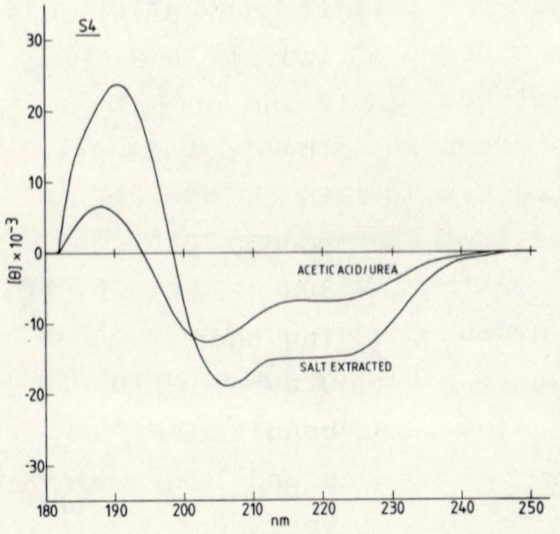

Figure 5. Comparison of the CD spectra of urea-acetic acid (134) prepared S4 measured in 5 mM potassium fluoride, 5 mM potassium phosphate pH 7.0, with salt extracted S4 in 300 mM potassium fluoride, 5 mM potassium phosphate pH 7.0.

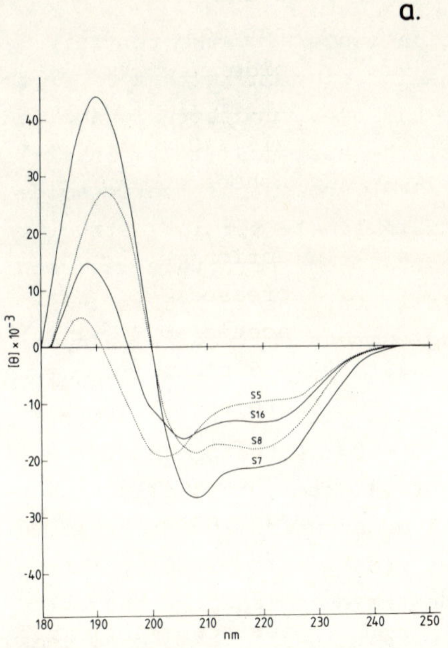

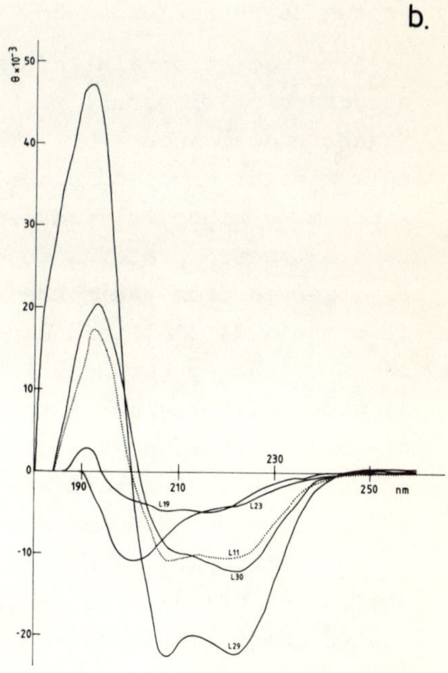

Figure 6. CD spectra of several salt extracted 30S a) and 50S b) proteins measured in 300 mM potassium fluoride, 5 mM potassium phosphate pH 7.0.

before this treatment are obtained for several of the ribosomal small subunit proteins. This implies that in these cases exposure to urea at pH 7.0 will unfold the protein and allow it to refold into the same secondary structure when it is dialyzed back into a "reconstitution type" buffer (141). It is clear from these studies that the proteins isolated by the gentler procedure contain more ordered structure than the proteins prepared by harsher methods.

When a comparison is made between the amount of α-helix predicted from the primary structure (described earlier in this review) with that obtained from CD measurements, good agreement is found for several of the proteins, i.e. S4, S8, L11, L27. However, there are several discrepancies such as the case of S20, S21 and L25 where a high α-helix value is predicted and a lower α-helix value is found experimentally from the CD results.

TERTIARY STRUCTURE

Proton Magnetic Resonance Techniques

A method to obtain information concerning the tertiary structure of ribosomal proteins in solution is that of Proton Magnetic Resonance (PMR). This has been successfully applied to a similar group of proteins, the histones (140). Initial studies on ribosomal proteins (142) showed the acetic acid-urea extracted proteins to contain little tertiary structure as observed from their PMR spectra. Some structure was seen in protein S4 (143) and especially in protein S16 as indicated by the appearance of ring current shifted resonances in the apolar region of the spectrum. These are the result of the interaction of apolar methyl groups with aromatic amino acids in the tertiary structure of the protein. These PMR spectra were recorded either in water or in dilute phosphate buffer pH 7.0, conditions under which the proteins were soluble. A comparative study of proteins isolated by the acetic acid-urea technique with those isolated by the salt extraction procedure (144) show that in all cases examined more shifted resonances, i.e. more tertiary structure is present in the second group of proteins.

We have examined the PMR spectra of most of the 30S and 50S proteins extracted by the gentler techniques. Most of the proteins appear to have a unique tertiary structure as observed from the width of the resonance peaks and the appearance of ring current shifted resonances (145).

These studies have been extended to look at proteolytic resistant "cores" of some of the proteins. In the case of the protein S4, a C-terminal fragment can be prepared similar to that described by other workers (146,147) that contains most of the structure present in the intact protein (148) (Fig. 7a). This fragment still retains its ability to bind specifically to the 16S RNA. A N-terminal fragment can be isolated from the protein L11 which appears very structured, whereas the remaining C-terminal fragment appears to resemble a random coil (Fig. 7b) from the PMR spectra (149).

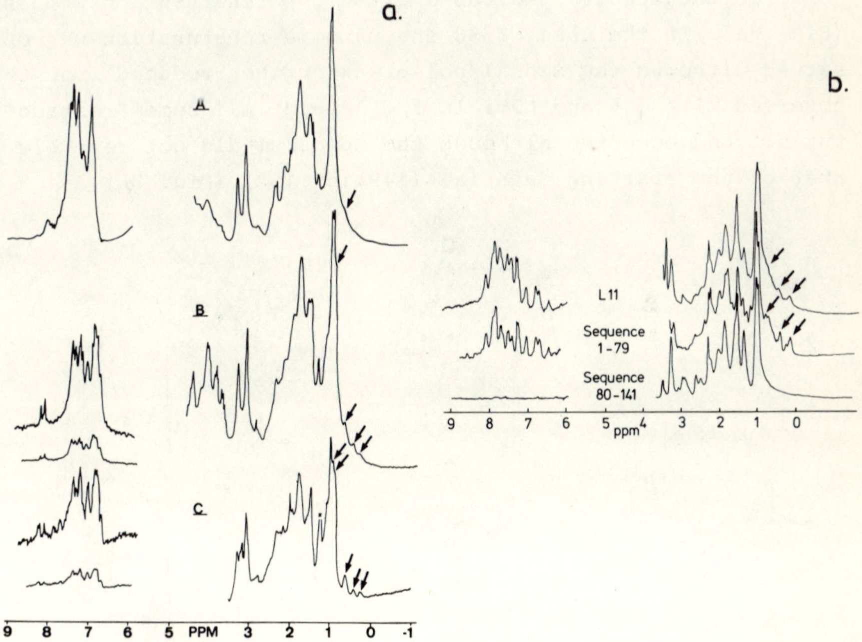

Figure 7. a) PMR profiles of protein S4, A acetic acid - urea extracted, B salt extracted, C C-terminal sequence 47-203 salt extracted. b) PMR profiles of protein L11, N-terminal sequence residues 1-79 and C-terminal sequence residues 80-141. Spectra measured in 350 mM potassium chloride, 50 mM potassium phosphate pD 7.0 except acetic acid-urea S4 which was measured in 5 mM potassium phosphate pD 7.0 in D_2O. Arrows point to ring current shifted resonances in the apolar region of the PMR spectrum revealing tertiary interactions.

In the case of proteins S4, S16 and L30 it has been possible to follow the titration of the histidine residues in these proteins. This provides information as to their environment in the tertiary structure of the protein (150,151).

The salt concentration was found to have little effect on the tertiary structure of the ribosomal proteins as observed from the PMR spectra (150,151). This agrees with the results on secondary structure obtained from comparison of CD spectra at different salt concentrations. When the effect of urea denaturation and renaturation was observed from the PMR spectra of the ribosomal proteins S4, S8, S16, L11 and L30, the results varied from one protein to another. In the case of S16 and L30 renaturation of the protein could be obtained but the results were not completely quantitative since the signal/noise of the spectra decreased after the renaturation process (Fig. 8a). In the case of S8 and S4 some renaturation was observed although the signal/noise was further reduced from that observed with S16 and L30. In the case of L11 some "restructuring" had occurred although the spectrum did not resemble that of the starting material (149,151,152) (Fig. 8b).

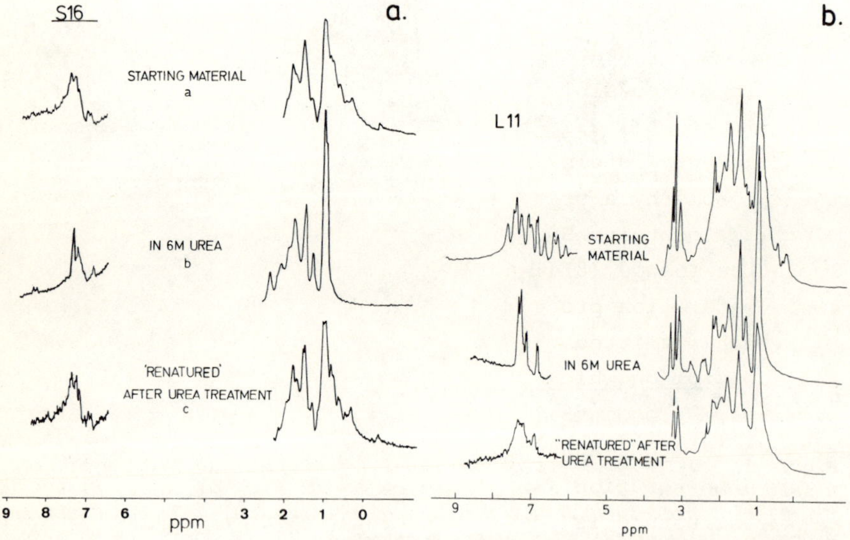

Figure 8. a) PMR spectra of protein S16, starting material, protein in 6 M urea, after removal of urea by dialysis. b) PMR spectra of protein L11 as for protein S16. All spectra measured in 350 mM potassium chloride, 50 mM potassium phosphate pD 7.0 in D_2O.

A PMR study on the acidic protein L7 has been reported (124). The two phenylalanine residues were assigned in the monomer protein. Phenylalanine residue 54 is found to be involved in the structure at the C-terminal region of the protein whereas phenylalanine residue 30 is involved in the dimerization of the protein pair. Shifted resonances are seen in the apolar region of the spectrum indicating tertiary structure in this protein.

The PMR spectrum of protein S1 has been reported (153). It was found that the protein has considerable tertiary structure in physiological buffer but more flexibility than normal for globular proteins. No difference was observed when the protein was prepared in the presence or absence of urea at neutral pH. The spectra obtained in this study resemble those obtained with salt extracted S1 (132,145).

Microcalorimetry

The technique of microcalorimetry has been applied to the ribosomal proteins in order to obtain information regarding the tertiary folding of these molecules (154,155). This technique looks at conformational transitions of the "order-disorder" type which are observed as heat absorption peaks during the process of heat denaturation of the protein. In this study it was possible to show that the proteins S4, S7, S8, S15, S16 and L11 displayed a cooperative tertiary structure in solution. Proteins S4, S7, S15 and S18 were extracted from the ribosome by a urea-LiCl technique followed by renaturation. The proteins S8, S16 and L11 were prepared by the mild isolation method (8) described above. A high reversibility of the denaturation process was observed for proteins S15 and S16 which permitted the calculation of the enthalpy value of heat absorption which was 60 kcal/mol and 50 kcal/mol respectively. Protein S18 showed no conformational transitions between 10-100°C and was therefore considered to be disordered in solution. In this context it is interesting that little structure is predicted for protein S18 (115).

A more detailed study with protein L7 and fragments of this protein has been described (124). The process of thermal denaturation was analyzed by comparing the values of calorimetry and effective enthalpies. The fragment of L7 from residues 1-26 was found to not affect the denaturation enthalpy of the monomer; hence the melting occurs in the C-terminal region after residue 30. When this result was compared with that obtained from the L7 dimer, an additional stage in the denaturation was observed in the latter sample. An ordered structure for the N-terminal region of the protein dependent on dimer formation was proposed.

Fluorimetric Studies

When the fluorescence properties of S4 and S7 were examined (156) it was found that the emissions observed were due to tyrosine and tryptophan for protein S4 and to tryptophan only for protein S7. The single tryptophan residue in S4 was found to be not exposed but could be released into the aqueous medium after heating to 35°C and above. In protein S7 the tryptophan residues can be split into two groups: exposed and partially buried. The spectral characteristics of the tryptophan residue in S4 and of the partially buried one of S7 suggest that these residues are located in α-helical regions of the proteins. The observation that the energy transfer observed between Tyr → Trp in protein S4 is very inefficient, implies an extended conformation for this protein where the N-terminal region of the protein containing 6 out of 8 tyrosines is far from the tryptophan in position 167. Another report specifically on the protein S4 (157) showed that the protein was present at room temperature in two conformations, and at 42°C under ribosome reconstitution conditions only one conformation was predominant. This effect was monitored by the fluorescence decay of the single tryptophan residue in the amino acid sequence. Both of these studies used ribosomal proteins which had been extracted with acetic acid and then treated with 6 M guanidium chloride before being dialysed into optimal conditions for 30S reconstitution. The presence of two conformations of protein S4 suggest a non-homogeneous preparation of the protein (see PMR results).

Static and nanosecond fluorescence polarization techniques have recently been applied to S1 reacted with AEDANS (158). This result suggests that S1 is a flexible protein with at least two domains that can rotate independently. A two domain structure is in agreement with low-angle X-ray scattering studies (159,160).

Many other studies have been described where a fluorescent probe has been used to specifically label a protein in the ribosome to obtain information about the distance between this protein and other ribosomal components. These studies have been reviewed (94) and will not be discussed here.

Proteolytic Fragments

Ribosomal proteins prepared by the new mild isolation procedure (132,133) have been used to look for their ability to produce resistant "core" particles after short treatment with proteolytic enzymes. Many of the proteins were found to produce resistant fragments (Fig. 9) (145,149,161). When proteins prepared by the acetic acid-urea procedure (134) were used in the same experiments they produced lower yields of the same fragments, different fragments or no fragments at all.

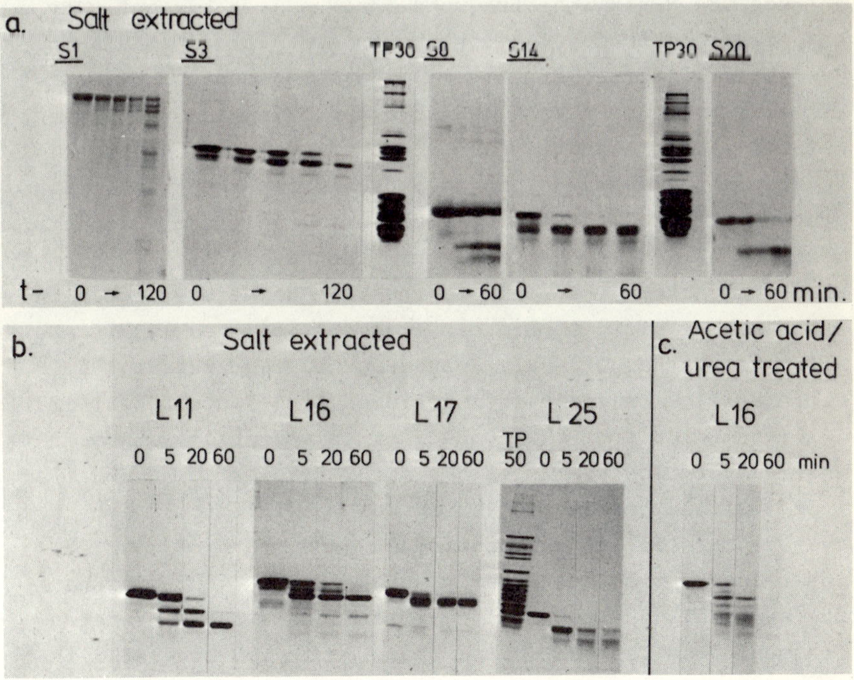

Figure 9. Salt extracted a) 30S and b) 50S proteins subjected to mild proteolytic digestion with chymotrypsin (30S) trypsin (50S); enzyme: protein ratio of 50-100:1, 0°C for time indicated in 0.35 M salt, 0.05 M potassium phosphate or HEPES buffer pH 7.0. c) Acetic acid-urea extracted L16 subjected to trypsin digestion under conditions as above.

This could explain the problems in preparing protein fragments from isolated acetic acid-urea proteins (162).

Proteins produced from the mild isolation method can be classified into three groups: 1) resistant under the conditions tested, 2) production of a resistant fragment, 3) totally degraded. Some proteins were found to be resistant: S15, S16, S17, L30. Proteins L7/L12 were resistant to trypsin, chymotrypsin and subtilisin, L16 to subtilisin, and L17 and L25 to chymotrypsin. Other proteins were completely degraded under these conditions: S2, L2, L3, L10, L29, and L33.

Several of these fragments obtained have been tested for biological activity. It has been found for the case of S8 and S20 that the fragments produced lose their ability to bind to rRNA (161). For the case of protein L11, a C-terminal fragment from residue 80-141 was found to have no activity to stimulate the binding of L16 to a 50S core particle (163) whereas the N-terminal fragment residue 1-79 had activity but less than that found for the intact protein (149). The C-terminal fragment of L16 (17-136) was found to have lost its ability to bind to a 50S core particle, and therefore the N-terminal residues 1-16 are involved in this activity (149).

Fragment production has been previously reported for the protein S4 (146,147). A low yield of a resistant fragment from residues 47-203 was isolated from the free protein, and only when the protein was bound to the 16S RNA before digestion was a higher yield of protein fragment obtained (147). A similar result has recently been reported for protein L18 where a large fragment from residues 18-117 could be isolated in high yields from the complex of this protein with 5S rRNA, whereas very low yields of a similar fragment were obtained from the free protein (164). It was possible to isolate a cyanogen bromide digestion product from protein S8 which interacted stoichiometrically with 16S RNA (165). On the basis of this and the ability to produce a small protein fragment from the RNA:protein complex it was suggested that the amino acid residues 25-47 are important for RNA binding.

The production of a resistant fragment from the specific protein:RNA complex presumably selects for the protein in the sample that is in a more "native" state, since the denatured material does not bind to rRNA. This is reflected by the efficiencies of protein prepared by two different methods to bind to rRNA. Three or four times more acetic acid-urea extracted protein being required for saturation than protein prepared by the gentler method. The possibility exists that some protein could renature under the conditions of the binding assay, i.e. incubation at 42°C, high salt, presence of rRNA.

A recent study on the ribosomal protein S1 (166) describes the production of a fragment from this protein which binds to 30S ribosomal subunits and to polyuridylic acid but is unable to support protein synthesis. This C-terminal fragment contains the two cysteine residues and represents 65% of the intact molecule (167). The protein was prepared by an improved LiCl/urea method (160). A similar fragment can be isolated from salt extracted S1 (145). The organisation of the tertiary structure of this protein into distinct subdomains agrees with the results found from X-ray scattering experiments (159,160). A mutant of S1 (168) which represents 75% of the intact molecule with the same N-terminus appears to be cleaved at a similar place to the normal S1 protein (169).

Such proteolytically resistant fragments are useful for several studies: 1) to provide information about the function of the ribosomal proteins, 2) to be used in structural studies such as CD and NMR, 3) to be used for crystallization experiments.

Ribosomal proteins have also been isolated from the thermophilic bacteria, *Bacillus stearothermophilus*, by a modification of the salt-extraction procedures (170,171). These proteins, when tested for their ability to produce proteolytically resistant fragments, appear in most cases to be more resistant than their *E. coli* counterparts, and they were either resistant or produced a resistant fragment. None of the thermophilic proteins tested under the conditions described were totally degraded. It would appear that these proteins have a "tighter" structure. The protein BL17 which has recently been crystallized (172) was totally resistant to protease attack under the conditions tested.

Tertiary Structure Prediction

A method has been proposed (173) to predict the tertiary structure of proteins. The main assumption of this mechanism is that the interaction of fluctuating α-helices is the physical principle which dictates a 3D distribution of the polypeptide chain material in the protein globule. At first local interactions form the fluctuating α-helices (174) and then long range interactions fold the protein chain into a highly helical globular intermediate. This is finally transformed into the native structure by unwinding of excess α-helices.

This approach has been used to predict the tertiary and quaternary structure of proteins L7/L12 (175). A globular structure for the C-terminal region of protein L7 was constructed which was formed by two adhering hydrophobic surfaces formed by α-helices from residues 51-59 and 65-73, and from residues 93-101 and 105-113. The dimer model proposed arranges the proteins in an antiparallel fashion where the hydrophobic surface of an α-helix from residues 4-11 of one molecule fits into the hydrophobic cavity (helices 51-59 and 93-101) on the surface of the globule from the other molecule and vice-versa.

Crystallography of Ribosomal Proteins

It has been reported (176) that crystals can be obtained from the proteins L7/L12. They were thought to be from the intact protein crystallizing in the space group $P6_122$ (or $P6_522$) but were later discovered to be a N-terminal fragment of this protein (177) from residues 1-36 in the amino acid sequence. The space group was $P6_1$ (or $P6_5$) and the resolution obtained in the diffraction pattern was to 4 Å. A crystallizable fragment has also been obtained from the C-terminal region of this protein, from residues 49-118. This produces crystals which resemble tetragonal bipyramids in the space group $P4_1\ 2_12$ ($P4_3\ 2_12$) and diffract to a resolution of 2.5 Å.

Recently it has been possible to obtain crystals from the first intact ribosomal protein, BL17 (172). This produces orthorhombic crystals with a space group $P_{2_12_12}$ which diffracts to a resolution of 3.5 Å. This *Bacillus stearothermophilus* ribosomal protein has not yet been correlated to its counterpart in the *E. coli* ribosome. It should be noted that this protein was prepared under mild conditions by a method based on that described for *E. coli* ribosomal proteins (8).

TABLE 4. Low-angle X-ray scattering studies of ribosomal proteins.

Protein	Preparation method *	Rg (Å)	Dimensions (Å)†	Ref.
S1	a	58($\pm$3.5)	Prolate ellipsoid A = 220 (190-260) B = 25-30	(129)
	b	72	'V'shape, 2 elliptic cylinders H = 220, A = 20, B = 8	(159)
	a	80	2 domain model largest dimension 280	(160)
S3	a	31	Elongated. Largest dimension 110	(178)
S4	a	33.6	Rod A=140, B=C=10	(179)
	b	42	Ellipsoid A=180, B=50, C=8 or rod A=180x21x21	(180, 181)
	a	26	Elongated. Largest dimension 100	(178)
S7	a	27	Elongated. Largest dimension 100	(178)
S8	b	23	Prolate ellipsoid A=96, B=C=28	(159)
S15	b	26.5	Prolate ellipsoid A=116, B=C=21	(159)
S16	b	21	Prolate ellipsoid A=86, B=C=24	(159)
S20	a	27	Elongated. Largest dimension 100	(178)
L11	b	34	Prolate ellipsoid. Largest dimension 150. Axial ratio 5-6:1, 2 ellipsoids joined by long axes	(182)
L7/L12	b*	36.5	Rod A=125.5, B=C=17.3	(183)
	b*	41$\pm$2	Ellipsoid A=180, B=32, C=12	(184)
L7/L12/L10	b*	45$\pm$2	A=180, B=90, C=12	(185)
L18	a	26$\pm$5	Prolate ellipsoid. Axial ratio 6:1 Longest dimension 113	(186)
L25	a	24$\pm$5	Prolate ellipsoid Axial ratio 5.7:1 Longest dimension 104	(186)

Preparation Method: a - methods involving the use of acetic acid or urea (this group is composed of many different methods and the reader is suggested to refer to the original paper for details. Some of these methods may be more denaturing to the proteins than others). b - mild method of salt extraction (8), b-method of NH_4Cl/ethanol wash (130). † A, B, C are full axes of protein.

SHAPE OF RIBOSOMAL PROTEINS

Low Angle X-Ray Scattering

Low-angle X-ray scattering can give some information as to the structure of ribosomal proteins in solution. The scattering curve so obtained can be fitted to a model for the protein shape. Many of the proteins so examined can be represented from their scattering curves as having elongated shapes. A summary of these results is presented in Table 4. Difficulties have been experienced by some groups due to the insolubility of many of the ribosomal proteins which only allows the experiments to be performed in water or at low ionic strength and at dilute protein concentration. The more gentle purification procedure (8) improves these problems of solubility although many of these proteins are still only soluble at concentrations between 5-10 mg/ml.

Neutron Scattering

Recently a globular conformation was reported (187) for several of the ribosomal proteins in solution. These proteins were S4, S7, S8, S15 and S16. The protein preparation procedure involves a renaturation of the proteins after urea treatment. The difference between the structure of these proteins and of those reported from other groups which have been characterised by other methods remains to be resolved. When protein S4 isolated by the mild methods described above was examined by the neutron scattering technique, a scattering curve almost identical to that previously obtained from X-ray scattering (180) was observed, indicating agreement between the two methods thus confirming an elongated shape for protein S4 (188).

Sedimentation, Viscosity, Diffusion

Many such studies have been reported that give information as to the shape of ribosomal proteins in solution. The

TABLE 5. Sedimentation, viscosity and diffusion studies of ribosomal proteins

Protein	Axial Ratio	f/fo Frictional coeff.	Method of preparation*	Reference
S1	10:1	1.7	a b[T]	(129)
	7.5-11:1	1.7, 1.55[a]	a	(189)
S2	5-7.5 :1	1.30[a]	b	(190)
S3	-	1.1	a*	(191)
	5-6:1	1.4[a]	a	(192)
S4	-	1.6	a*	(191)
	-	1.69[a]	a	(179)
S5	-	1.5	a*	(191)
	6-8:1	1.32[a]) high salt	a	(193)
	8-11:1	1.44[a]) low salt		
S6	-	1.3	a*	(191,195)
	3.5-4:1	1.23, 1.10[a]	a	(194)
S7	-	1.5	a*	(191)
S8	2.3:1	1.02[a]	a	(196)
	2.3:1	1.06[a]	b	(196)
S13	2.0:1	1.05[a]	a	(197)
S15	5-6:1	1.26	a	(198)
S17	3.5-4:1	1.24	a	(192)
S18	8-12:1	1.45	a	(198)
	-	1.40	a*	(195)
S20	-	1.80	a*	(191)
	5.5-7.2:1	1.47,1.33[a]	a	(194)
S21	8-9.4 :1	1.44[a]	a	(197)
L1	5-6:1	1.28	b	(199)
L3	5.5-6:1	1.46	b	(200)
L6	5:1	1.34[a]	a	(196)
	4.2:1	1.11[a]	b	(196)
L9	4-6.0:1	1.2	b	(199)
L7/L12	19.2-20.1:1	1.97-2.01	b*	(183)
L11	4-5:1	1.23[a]	b	(182)
L24	3-3.5:1	1.15	b	(200)
L25	3.5-4.5:1	1.2	b	(199)
L30	2.5-4:1	1.17	b	(199)

Preparation Method. a - methods involving the use of acetic acid or urea; a as method a, but involving a renaturation step with guanidinium hydrochloride followed by dialysis into reconstitution buffer.
b - mild method of salt extraction (8), b* - NH_4Cl/ethanol method (130), b^T - mild low salt wash of ribosomes (130).
[a]Frictional ratios corrected for hydration and asymmetry taken from M.W., $S_{20,w}$, or viscosity measurements.

results are summarised in Table 5. Several of the proteins appear to have an elongated shape in solution whereas others have a more globular shape. These results are in agreement with those reported from low angle X-ray scattering experiments.

PROTEIN-PROTEIN COMPLEXES

Protein:protein interaction must play an important part in the structure of the ribosome, and little is known about this to date. Chemical cross-linking studies (reviewed in (201) and this volume) give information as to proteins which are close to each other within the ribosomal particle. The complex L7/L12/L10 can be isolated from the ribosome by an NH_4Cl/ethanol washing procedure (130) and appears to be the most stable complex in the ribosome appearing as the 'L8' spot on the two-dimensional electrophoresis system (3) which is run in 4 M urea. By using the gentler methods of ribosomal protein isolation it is possible to isolate several potential protein:protein complexes (8). Two of these, S13/S19 and L7/L12/L10, have been found to interact specifically with the rRNA as complexes (202) whereas their component proteins bind weakly (203) or not at all (202). The binding of the latter complex to rRNA appears to be stimulated by yet another protein, L11, and vice versa (204). Clearly, by further study of these complexes we will be able to build up small substructures of the ribosomal subunits to aid our understanding of the complex interactions involved in the ribosome. The question still remains as to whether the structures of proteins isolated from the ribosome by however gentle a procedure, represent the same structures as in the intact particle. Studies on isolated protein:protein and protein:protein:RNA complexes should help to answer these questions.

CONCLUSION

An essential prerequisite for a thorough study on the structure of the ribosomal particle is the knowledge of the primary, secondary and tertiary structure as well as the shape of the individual ribosomal components. The determination of the primary structure of almost all ribosomal proteins (Table 1) and of the ribosomal RNAs from *E. coli* has made rapid progress. It has developed to a point where the complete elucidation of the protein and RNA primary structures can be expected in the very near future.

Information about the secondary structure of the ribosomal proteins derives from prediction based on the primary structure and from experimental investigations, mainly CD-studies. All of the ribosomal proteins appear to have a discrete secondary structure, most having an α-helix content 25 - 70%, some ß-structure, and random coil.

Valuable information is also appearing from results on the tertiary structure of the proteins in solution, using techniques such as PMR, microcalorimetry and limited proteolytic attack. Results should soon become available on the analysis of the tertiary structure of protein BL17 and of the crystallizable fragments from proteins L7/L12. It is hoped that other ribosomal proteins will crystallize in a form suitable for structural analysis. The techniques of low angle X-ray scattering, and sedimentation, viscosity, diffusion measurements are providing information as to the shape of the individual proteins. By the use of neutron -scattering studies, complexes of proteins and RNA can be examined in order to determine whether the proteins once complexed with RNA and other proteins have the same shape as in solution.

Clearly our knowledge of the structure of the bacterial ribosomal proteins has proceeded rapidly in the last few years. This knowledge will help to unravel the complex organization of the ribosome and its mechanism in protein biosynthesis.

REFERENCES

1. Kaltschmidt, E., and Wittmann, H.G. (1970) Proc. Nat. Acad. Sci. USA 67, 1276-1282.

2. Wittmann, H.G., and Wittmann-Liebold, B. (1974) Ribosomes pp. 115-140, Cold Spring Harbor Lab., Long Island, N.Y.

3. Pettersson, I., Hardy, S.J.S., and Liljas, A. (1976) FEBS Lett. 64, 135-138.

4. Stöffler, G. (1974) Ribosomes pp. 615-667, Cold Spring Harbor Lab., Long Island, N.Y.

5. Weber, H.J. (1972) Mol. Gen. Genet. 119, 233-248.

6. Terhorst, C., Möller, W., Laursen, R., and Wittmann-Liebold, B. (1973) Eur. J. Biochem. 34, 138-152.

7. Wittmann, H.G. (1974) Ribosomes pp. 93-114, Cold Spring Harbor Lab., Long Island, N.Y.

8. Dijk, J., and Littlechild, J. (1979) Meth. Enzym. 59, 481-502.

9. Wystup, G., Teraoka, H., Schulze, H., Hampl, H. and Nierhaus, K.H. (1979) manuscript submitted.

10. Brauer, D. and Röming, R. (1979) FEBS Lett., in press.

11. Schiltz, E., and Reinbolt, J. (1975) Eur. J. Biochem. 56, 467-481.

12. Wittmann-Liebold, B., and Greuer, B. (1978) FEBS Lett. 95, 91-98.

13. Hitz, H., Schäfer, D., and Wittmann-Liebold (1977) Eur. J. Biochem. 75, 497-512.

14. Reinbolt, J., Tritsch, D., and Wittmann-Liebold (1978) FEBS Lett. 91, 297-301.

15. Allen, G.,and Wittmann-Liebold (1978) Hoppe-Seyler's Z. Physiol. Chem. 359, 1509-1525.

16. Chen, R. (1977) Hoppe-Seyler's Z. Physiol. Chem. 358, 1415-1430.

17. Yaguchi, M., and Wittmann, H.G. (1979) unpublished results.

18. Chen, R., Kamp, R., and Wittmann-Liebold, B. (1979) manuscript in preparation.

19. Funatsu, G., Yaguchi, M., and Wittmann-Liebold, B. (1977) FEBS Lett. 73, 12-17.

20. Lindemann, H., and Wittmann-Liebold, B. (1977) Hoppe-Seyler's Z. Physiol. Chem. 358, 843-863.

21. Yaguchi, M., and Wittmann, H.G. (1979) manuscript in preparation.

22. Morinaga, T., Funatsu, G., Funatsu, M., and Wittmann, H.G. (1976) FEBS Lett. 64, 307-309.

23. Vandekerckhove, J., Rombauts, W., and Wittmann-Liebold, B. (1977) Hoppe-Seyler's Z. Physiol. Chem. 358, 989-1002.

24. Yaguchi, M., and Wittmann, H.G. (1978) FEBS Lett. 87, 37-40.

25. Yaguchi, M. (1977) FEBS Lett. 59, 217-220.

26. Yaguchi, M., and Wittmann, H.G. (1978) FEBS Lett. 88, 227-230.

27. Wittmann-Liebold, B., Marzinzig, E., and Lehmann, A. (1976) FEBS Lett. 68, 110-114.

28. Vandekerckhove, J., Rombauts, W., Peeters, P.,and Wittmann-Liebold, B. (1975) Hoppe-Seyler's Z. Physiol. Chem. 356, 1955-1976.

29. Brauer, D., and Öchsner, I. (1978) FEBS Lett. 96, 317-321.

30. Muranova, T.A., Muranov, A.V., Markova, L.F., and Ovchinnikov, Y.A. (1978) FEBS Lett. 96, 301-305.

31. Chen, R., and Ehrke, G. (1976) FEBS Lett. 69, 240-245.

32. Chen, R., Arfsten, U., and Chen-Schmeisser (1977) Hoppe-Seyler's Z. Physiol. Chem. 358, 531-535.

33. Dovgas, N.U., Vinukurov, L.M., Velmoga, I.S., Alakhov, Y. B., and Ovchinnikov, Y.A. (1976) FEBS Lett. 67, 58-61.

34. Heiland, I., Brauer, D., and Wittmann-Liebold, B. (1976) Hoppe-Seyler's Z. Physiol. Chem. 357, 1751-1770.

35. Dognin, M.J., and Wittmann-Liebold, B. (1977) FEBS Lett. 84, 342-346.

36. Mende, L. (1978) FEBS Lett. 96, 313-316.

37. Morinaga, T., Funatsu, G., Funatsu, M., Wittmann-Liebold, B., and Wittmann, H.G. (1978) FEBS Lett. 91, 74-77.

38. Giorginis, S., and Chen, R. (1977) FEBS Lett. 84, 347-350.

39. Brosius, J., and Chen, R. (1976) FEBS Lett. 68, 105-109.

40. Brosius, J., Schiltz, E., and Chen, R. (1975) FEBS Lett. 56, 359-361.

41. Brosius, J., and Arfsten, U. (1978) Biochemistry 17, 508-516.

42. Wittmann-Liebold, B., and Seib, C. (1979) FEBS Lett., in press.

43. Heiland, I., and Wittmann-Liebold, B. (1979) manuscript submitted.

44. Wittmann-Liebold, B. (1979) manuscript submitted.

45. Wittmann-Liebold, B., and Greuer, B. (1979) manuscript submitted.

46. Wittmann-Liebold, B. (1979) FEBS Lett., in press.

47. Dovgas, N.U., Markova, L.F., Mednikova, T.A., Vinokurov, L.M., Alakov, Y.B. and Ovchinnikov, Y.A. (1975) FEBS Lett. 53, 351-354.

48. Bitar, K.G., and Wittmann-Liebold, B. (1975) Hoppe-Seyler's Z. Physiol. Chem. 356, 1343-1352.

49. Chen, R., Mende, L., and Arfsten, U. (1975) FEBS Lett. 59, 96-99.

50. Wittmann-Liebold, B., and Marzinzig, E. (1977) FEBS Lett. 81, 214-217.

52. Bitar, K.G. (1975) Biochim. Biophys. Acta 386, 99-106.

52. Ritter, E., and Wittmann-Liebold, B. (1975) 60, 153-155.

53. Brosius, J. (1978) Biochemistry 17, 501-508.

54. Wittmann-Liebold, B., Greuer, B., and Pannenbecker, R. (1975) Hoppe-Seyler's Z. Physiol. Chem. 356, 1977-1979.

55. Wittmann-Liebold, B., and Pannenbecker, R. (1976) FEBS Lett. 68, 115-118.

56. Chen, R. (1976) Hoppe-Seyler's Z. Physiol. Chem. 357, 873-886.

57. Pon, C., Wittmann-Liebold, B., and Gualerzi, C. (1979)

FEBS Lett. 101, 157-160.

58. Brauer, D., and Wittmann-Liebold, B. (1977) FEBS Lett. 79, 269-275.

59. Mende, L., Timm, B., and Subramanian, A.R. (1978) FEBS Lett. 96, 395-398.

60. van Acken, U. (1975) Mol. Gen. Genet. 140, 61-68.

61. Funatsu, G., Schiltz, E., and Wittmann, H.G. (1971) Mol. Gen. Genet. 114, 106-111.

62. Funatsu, G., Nierhaus, K.H., and Wittmann-Liebold, B. (1972) J. Mol. Biol. 64, 201-209.

63. Itoh, T., and Wittmann, H.G. (1973) Mol. Gen. Genet. 127, 19-32.

64. DeWilde, M. and Wittmann-Liebold, B. (1973) Mol. Gen. Genet. 127, 273-276.

65. Piepersberg, W., Böck, A., Yaguchi, M., and Wittmann, H.G. (1975) Mol. Gen. Genet. 143, 43-52.

66. Yaguchi, M., Wittmann, H.G., Cabezon, T., DeWilde, M., Villarroel, R., Herzog, A., and Bollen, A. (1975) Mol. Gen. Genet. 142, 35-43.

67. Wittmann-Liebold, B., and Wittmann, H.G. (1971) Biochim. Biophys. Acta 251, 44-53.

68. Wittmann, H.G., Yaguchi, M., Piepersberg, W., and Böck, A. (1975) J. Mol. Biol. 98, 827-829.

69. Yaguchi, M., Wittmann, H.G., Cabezon, T., DeWilde, M., Villarroel, R., Herzog, A., and Bollen, A. (1976) J. Mol. Biol. 104, 617-620.

70. Zubke, G., Stadler, H., Ehrlich, R., Stöffler, G., Wittmann, H.G., and Apirion, D. (1977) Mol. Gen. Genet. 158, 129-139.

71. Funatsu, G., and Wittmann, H.G. (1972) J. Mol. Biol. 68, 547-550.

72. Funatsu, G., Nierhaus, K.H., and Wittmann, H.G. (1972) Biochim. Biophys. Acta 287, 282-291.

73. Herzog, A., Yaguchi, M., Cabezon, T., Corchuelo, M.C., Petre, J., and Bollen, A. (1979) Mol. Gen. Genet. 171, 15-22.

74. Cabezon, T., Herzog, A., Petre, J., Yaguchi, M., and Bollen, A.(1977) J. Mol. Biol. 116, 361-374.

75. Dabbs, E., and Wittmann-Liebold, B., unpublished results.

76. Suryanarayana, T., and Subramanian, A.R. (1978) Biochim. Biophys. Acta 520, 342-357.

77. Rouviere-Yaniv, J., and Gros, F. (1975) Proc. Nat. Acad. Sci. USA 72, 3428-3432.

78. Berthold, V., and Geider, K. (1976) Eur. J. Biochem. 71, 443-449.

79. Wittmann-Liebold, B. (1973) Hoppe-Seyler's Z. Physiol. Chem. 354, 1415-1431.

80. Wittmann-Liebold, B., Geissler, A.W. and Marzinzig, E. (1975) J. Supramol. Struct. 3, 426-447.

81. Wittmann-Liebold, B., Graffunder, H., and Kohls, H. (1976) Analyt. Biochem. 75, 621-633.

82. Wittmann-Liebold, B., and Lehmann, A. (1975) Solid phase methods in protein sequence analysis, pp. 81-90. Pierce Chemical Comp., Rockford, Illinois.

83. Wittmann-Liebold, B., Brauer, D., and Dognin, J. (1977) Solid phase methods in protein sequence analysis, pp. 219-232, Elsevier/North Holland Biomed. Press, Amsterdam.

84. Chen, R., and Wittmann-Liebold, B. (1975) FEBS Lett. 52, 139-140.

85. Chang, J.Y., Brauer, D., and Wittmann-Liebold, B. (1978) FEBS Lett. 93, 205-214.

86. Chen, R., Brosius, J., Wittmann-Liebold, B., and Schäfer, D. (1977) J. Mol. Biol. 111, 173-181.

87. Nomura, M. (1979) personal communication.

88. Post, L.E., Arfsten, A.E., Reusser, F., and Nomura, M. (1978) Cell 15, 215-229.

89. Post, L.E., Strycharz, G.D., Nomura, M., Lewis, H., and Dennis, P.P. (1979) Proc. Nat. Acad. Sci. USA 76, 1967-1701.

90. Wittmann-Liebold, B., and Dzionara, M. (1976) FEBS Lett. 65, 281-283.

91. Wittmann-Liebold, B., and Dzionara, M. (1976) FEBS Lett. 61, 14-19.

92. Stöffler, G., and Wittmann, H.G. (1971) J. Mol. Biol. 62, 407-409.

93. Stöffler, G., and Wittmann, H.G. (1971) Proc. Nat. Acad. Sci. USA 68, 2283-2287.

94. Brimacombe, R., Stöffler, G., and Wittmann, H.G. (1978) Ann. Rev. Biochem. 47, 217-249.

95. Isono, K., Krauss, J., and Hirota, Y. (1976) Mol. Gen. Genet. 149, 297-302.

96. Isono, K., Cumberlidge, A.G., Isono, S., and Hirota, Y. (1977) Mol. Gen. Genet. 152, 239-243.

97. Dabbs, E., and Wittmann, H.G. (1976) Mol. Gen. Genet. 149, 303-309.

98. Dabbs, E. (1978) Mol. Gen. Genet. 165, 73-78.

99. Otaka, E., Itoh, T., Osawa, S., Tanaka, K. and Tamaki, M. (1971) Mol. Gen. Genet. 114, 14-22.

100. Ozaki, M., Mizushima, S. and Nomura, M. (1970) Nature 222, 333-339.

101. Itoh, T., and Wittmann-Liebold, B. (1978) FEBS Lett. 96, 392-394.

102. Itoh, T., and Wittmann-Liebold, B. (1978) FEBS Lett. 96, 399-402 and J. Biochem., submitted.

103. Itoh, T. and Wittmann-Liebold, B. (1979) unpublished results.

104. Amons, R., Pluijms, W. and Möller, W. (1979) FEBS Lett., in press.

105. Yaguchi, M., Roy, C., Matheson, A.T., Visentin, L.P. (1973) Canad. J. Biochem. 51, 1215-1217.

106. Isono, K., Isono, S., Stöffler, G., Visentin, L.P., Yaguchi, M. and Matheson, A.T. (1973) Mol. Gen. Genet. 127, 191-195.

107. Higo, K.,and Loertscher, K. (1974) J. Bact. 118, 180-186.

108. Matheson, A.T., Yaguchi, M., Nazar, R.N., Visentin, L.P., and Willick, G.E. (1978) Energetics and structure of halophilic microorganisms, pp. 481-501, Elsevier/North Holland Biomed. Press, Amsterdam.

109. Visentin, L.P., Yaguchi, M., and Matheson, A.T. (1979) Canad. J. Biochem., in press.

110. Amons, R., van Agthoven, A., Pluijms, W., Möller, W., Higo, K., Itoh, T., and Osawa, S. (1977) FEBS Lett. 81, 308-310.

111. Amons, R., van Agthoven, A., Pluijms, W. and Möller, W. (1978) FEBS Lett. 86, 282-284.

112. Wittmann-Liebold, B., Geissler, A.W., Lin, A., and Wool, I.G. (1979) manuscript submitted.

113. Matheson, A.T., and Möller, W. (1979) this volume.

114. Wittmann-Liebold, B., Robinson, S.M.L., and Dzionara, M. (1977) FEBS Lett. 77, 301-307 and 81, 204-213.

115. Dzionara, M., Robinson, S.M.L., and Wittmann-Liebold (1977) Hoppe-Seyler's Z. Physiol. Chem. 358, 1003-1019.

116. Dzionara, M., Robinson, S.M.L., and Wittmann-Liebold (1977) J. Supramol. Struct. 7, 191-204.

117. Bond, P.J., Littlechild, J.A. and Wittmann-Liebold, B. (1979) unpublished data.

118. McPhie, P., and Gratzer, W.B. (1966) Biochemistry 5, 1310-1315.

119. Sarkar, P.K., Yang, J.T., Doty, P. (1967) Biopolymers 5, 1-4.

120. Cotter, R.I., and Gratzer, W.B. (1969) Eur. J. Biochem. 8, 352-356.

121. Dzionara, M. (1970) FEBS Lett. 8, 197-200.

122. Möller, W., Castleman, H., and Terhorst, C.P. (1970) FEBS Lett. 8, 192-196.

123. Boublik, M., Brot, N., and Weissbach, H. (1973) Biopolymers 12, 2083-2092.

124. Gudkov, A.T., Khechinashvili, N.N., Bushuev, V.N. (1978) Eur. J. Biochem. 90, 313-318.

125. Lemieux, G., Lefevre, J.-F., Daune, M. (1974) Eur. J. Biochem. 49, 185-194.

126. Morrison, C.A., Garrett, R.A., and Bradbury, E.M. (1977) Eur. J. Biochem. 78, 153-159.

127. Allen, S.H., and Wong, K.-P. (1978) Biophysical Journal 23, 359-373.

128. Allen, S.H., and Wong, K.-P. (1978) Biochemistry 17, 3971-3978.

129. Laughrea, M., and Moore, P.B. (1977) J. Mol. Biol. 112, 399-421.

130. Hamel, E., Koka, M., and Nakamoto, T.(1972) J.Biol.Chem. 247, 805-814.

131. Tal, M., Airam, M., Kanarek, A.,and Weiss, A. (1972) Biochim. Biophys. Acta 281, 381-392.

132. Littlechild, J.A., and Malcolm, A.L. (1978) Biochemistry 17, 3363-3369.

133. Dijk, J., and Ackermann,I.(1979)J. Biol. Chem.,submitted.

134. Hindennach, I., Stöffler, G., and Wittmann, H.G. (1971) Eur. J. Biochem. 23, 7-11.

135. Heiland, I., and Snatzke, G. (1979) Hoppe-Seyler'S Z. Physiol. Chem. 360, 279-280.

136. Rosenkranz, H., and Scholtan, W. (1971) Hoppe-Seyler's Z. Physiol. Chem. 352, 896-904.

137. Chen, J.-H., Yang, J.T., and Martinez, H.M. (1972) Biochemistry 11, 4120-4131.

138. Littlechild, J.A., Freund, A.M., Peyot, J., and Daune, M. (1979) manuscript in preparation.

139. Dijk, J., Freund, A.M., Peyot, J., and Daune, M. (1979) manuscript in preparation.

140. Bradbury, E.M., Hjelm, R.P., Carpenter, B.G., Baldwin, J.P., and Kneale, G.G. (1977) Nucleic Acid and Protein Recognition. Acad. Press pp. 117-137.

141. Littlechild, J.A., Freund, A.M. (1979) manuscript in preparation.

142. Morrison, C.A., Bradbury, E.M., and Garrett, R.A. (1977) FEBS Lett. 81, 435-439.

143. Morrison, C.A., Garrett, R.A., and Bradbury, E.M. (1977) Eur. J. Biochem. 78, 153-159.

144. Morrison, C.A., Bradbury, E.M., Littlechild, J., and Dijk, J. (1977) FEBS Lett. 83, 348-352.

145. Littlechild, J., Malcolm, A., Paterakis, K., Ackermann, I., Dijk, J. (1979) Biochemistry submitted.

146. Changchien,L.-M., and Craven, G.R. (1976) J. Mol. Biol. 108, 381-401.

147. Newberry, V., Yaguchi, M., and Garrett, R.A. (1977) Eur. J. Biochem. 76, 51-61.

148. Littlechild, J., Morrison, C.A., and Bradbury, E.M. (1979) FEBS Lett. in press.

149. Dijk, J. (1979) manuscript in preparation.

150. Littlechild, J., Cary, P., and Bradbury, E.M. (1979) manuscript in preparation.

151. Dijk, J., and Ackermann, I. (1979) FEBS Lett. submitted.

152. Littlechild, J. (1979) FEBS Lett. manuscript in preparation.

153. Moore, P.B., and Laughrea, M. (1979) Nucl. Acid Res. 6, 2355-2361.

154. Privalov, P.L., and Khechinashvili, N.N. (1974) J. Mol. Biol. 86, 665-684.

155. Khechinashvili, N.N., Koteliansky, V.E., Gogia, Z.V., Littlechild, J., and Dijk, J. (1978) FEBS Lett. 95, 270-272.

156. Gérard, D., Lemieux, G., and Laustriat, G.(1975) Photochemistry and Photobiology 22, 89-95.

157. Brochon, J.-C., Wahl, P., Vachette, P., and Daune, M.P. (1976) Eur. J. Biochem. 65, 35-39.

158. Chu Gloria, Y., and Cantor, C.R. (1979) Nucl. Acid Res. 6, 2363-2379.

159. Österberg, R., Sjöberg, B., and Littlechild, J. (1978) FEBS Lett. 93, 115-119.

160. Labischinski, M., and Subramanian, A.R. (1979) Eur. J. Biochem. 95, 359-366.

161. Paterakis, K., and Littlechild, J. (1979) manuscript in preparation.

162. Changchien, L.-M., Craven, G.R. (1979) Biochemistry 18, 1275-1281.

163. Schulze, H., and Nierhaus, K.H. (1979) manuscript in preparation.

164. Newberry, V., Brosius, J., and Garrett, R. (1978) Nucl. Acid Res. 5, 1753-1766.

165. Bruce, J., Firpo, E.J., and Schaup, H.W. (1977) Nucl. Acid Res. 4, 3327-3340.

166. Suryanarayana, T., and Subramanian, A.R. (1979) J. Mol. Biol. 127, 41-54.

167. Subramanian, A.R., unpublished results.

168. Subramanian, A.R., Mizushima, S. (1979) J. Biol. Chem. 254, 4309-4312.

169. Subramanian, A.R. (1979) manuscript in preparation.

170. Littlechild, J.A., and Malcolm, A.L. (1979) manuscript in preparation.

171. Appelt, K., Ackermann, I., and Dijk, J. (1979) manuscript in preparation.

172. Appelt, K., Dijk, J., and Epp, O. (1979) FEBS Lett. in press.

173. Lim, V.I. (1978) FEBS Lett. 89, 10-14.

174. Finkelstein, A.V., and Ptitsyn, O.B. (1976) J. Mol. Biol. 103, 15-24.

175. Gudkov, A.T., Behlke, J. Vtiurin, N.N., and Lim, V.I. (1977) FEBS Lett. 82, 125-129.

176. Liljas, A., and Kurland, C.G. (1976) FEBS Lett. 71, 130-132.

177. Liljas, A., Eriksson, S., Donner, D., and Kurland, C.G. (1978) FEBS Lett. 88, 300-304.

178. Gulik, A., Freund, A.M., and Vachette, P. (1978) J. Mol. Biol. 119, 391-397.

179. Paradies, H.H., and Franz, A. (1976) Eur. J. Biochem. 67, 23-29.

180. Österberg, R., Sjöberg, B., Garrett, R., and Littlechild, J. (1977) FEBS Lett. 73, 25-28.

181. Österberg, R., Sjöberg, B., Garrett, R.A., Littlechild, J. (1978) Nucl. Acid Res. 5, 3579-3587.

182. Giri, L., Dijk, J., Labischinski, H., Bradaczek, H. (1978) Biochemistry 17, 745-749.

183. Wong, K.-P., and Paradies, H.H. (1974) Biochem. Biophys. Res. Comm. 61, 178-184.

184. Österberg, R., Sjöberg, B., Liljas, A., and Pettersson, I. (1976) FEBS Lett. 66, 48-51.

185. Österberg, R., Sjöberg, B., Pettersson, I., Liljas, A., and Kurland, C.G. (1977) FEBS Lett. 73, 22-24.

186. Österberg, R., Sjöberg, B., Garrett, R.A. (1976) FEBS Lett. 65, 73-76.

187. Serdyuk, I.N., Zaccai, G., and Spirin, A.S. (1978) FEBS Lett. 94, 349-352.

188. Österberg, R., Sjöberg, B., Littlechild, J. (1979) unpublished results.

189. Giri, L., and Subramanian, A.R. (1977) FEBS Lett. 81, 199-203.

190. Georgalis, Y., Littlechild, J.A., and Giri, L. (1979) Biochemistry, submitted.

191. Rohde, M.F., O'Brien, S., Cooper, S., and Aune, K.C. (1975) Biochemistry 14, 1079-1087.

192. Franz, A., Georgalis, Y., and Giri, L. (1979) Biochim. Biophys. Acta, in press.

193. Georgalis, Y., and Giri, L. (1978) FEBS Lett. 95, 99-102.

194. Georgalis, Y., and Giri, L. (1979) manuscript in preparation.

195. Prakash, V., and Aune, K.C. (1978) Arch. Biochem. Biophys. 187, 399-405.

196. Giri, L., Littlechild, J., and Dijk, J. (1977) FEBS Lett. 79, 238-244.

197. Georgalis, Y. (1979) manuscript in preparation.

198. Giri, L., and Franz, A. (1978) FEBS Lett. 87, 31-36.

199. Giri, L., Franz, A., and Dijk, J. (1979) Biochemistry, in press.

200. Giri, L., and Dijk, J. (1979) Arch. Biochem. Biophys. 193, 122-129.

201. Traut, R.R., Heimark, R.L., Sun, T.T., Hershey, J.W.B., and Bollen, A. (1974) In Ribosomes pp. 271-208, Cold Spring Harbor Lab., Cold Spring Harbor, N.Y.

202. Dijk, J., Littlechild, J., and Garrett, R. (1977) FEBS Lett. 77, 295-300.

203. Marquardt, O., Roth, H.E., Wystup, G., and Nierhaus, K.H. (1979) manuscript submitted.

204. Dijk, J., Garrett, R.A., and Müller, R. (1979) Nucl. Acid Res., in press.

PROTEIN TOPOGRAPHY OF *ESCHERICHIA COLI* RIBOSOMAL SUBUNITS AS INFERRED FROM PROTEIN CROSSLINKING

R. R. TRAUT, J. M. LAMBERT, G. BOILEAU, AND J. W. KENNY

Department of Biological Chemistry,
School of Medicine
University of California,
Davis, CA 95616

INTRODUCTION

The results of intensive investigations of the ribosome of Escherichia coli during the 1960's led to the conclusion that knowledge of the spatial arrangement or topography of its protein constituents would be invaluable in understanding the function of the ribosome in protein synthesis. The ribosome was shown to contain fifty-three distinct proteins, most of which were present in quantities equimolar with the ribosome particle. The smaller 30S ribosomal subunit was reconstituted from purified proteins and ribosomal RNA; omission of single proteins in such reconstitution experiments indicated that ribosomal proteins acted cooperatively to determine functional properties of the ribosome; genetic studies provided independent evidence for cooperative interactions among ribosomal proteins. It was also shown that the 50S ribosomal subunit itself catalyzed peptide bond formation and hence could be considered as a complex enzyme containing thirty-three polypeptides and two RNA components (1). Several proteins as well as the RNA have been implicated at the peptidyl transferase center by affinity labeling studies (2-4).

The accumulated evidence strongly suggested that groups of ribosomal proteins were required to form active sites and that there were cooperative interactions among them. In no case has it been shown that a single protein is uniquely associated with a single functional property. Thus many laboratories undertook investigations of ribosomal protein topography. Recent evidence for an active role for RNA in ribosome structure and function does not diminish the importance of continuing the investigation of protein topography.

Chemical crosslinking of biological structures containing several polypeptide chains can provide useful information on the topography of the protein constituents. Crosslinking reagents based on bis(imido esters), such as dimethyl suberimidate (5), have been particularly useful since they react under mild conditions specifically with amino groups of proteins to yield amidine derivatives that retains the positive charge. Crosslinking studies using bis(imido esters) have been prominent in providing evidence on the spatial arrangement of ribosomal proteins (6-10). Such reagents have also provided useful information about the location of binding sites of initiation factors to the ribosome (11-14). The products of crosslinking reactions have been principally analyzed using polyacrylamide/dodecyl sulfate gel electrophoresis in which crosslinked species were evident from their increased molecular weights. Identification of the protein components of crosslinked species can be made using immunological techniques (7,12) or by cleavage of the crosslink (6,9).

Since the aim of such research efforts was to construct models placing each protein relative to the others, the need for techniques capable of detecting and identifying large numbers of crosslinked proteins became apparent. Analysis of crosslinked protein mixtures has been greatly facilitated by the development of reagents that contain within them an easily cleavable linkage (15,16). Bifunctional imido esters that introduce a disulfide bond as the crosslink between different proteins have been found to be particularly useful compounds of this kind. The analysis of the composition of such crosslinked products would be immensely simplified by virtue of the fact that the disulfide bond linking two proteins could be readily cleaved by reduction to regenerate monomeric proteins which would have the same electrophoretic mobilities as the native proteins. The reversibility of the crosslinking procedure suggested the use of two-dimensional polyacrylamide/dodecyl sulfate gel electrophoresis to simplify further the identification of crosslinked proteins. The disulfide crosslinking reagent, 2-iminothiolane (Fig. 1), formerly called methyl 4-mercaptobutyrimidate, has been especially productive in our investigation of ribosome protein topography of both the 30S (17-21) and 50S (22-24) ribosomal subunits.

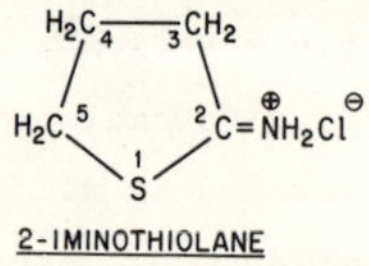

Figure 1. Chemical structure of 2-iminothiolane.

This article is divided into five parts: 1. a brief outline of the methodology employed in using 2-iminothiolane as a crosslinking reagent and in identifying crosslinked proteins. 2. preliminary evidence of proteins at the subunit interface. 3. evidence on the binding site for S1. 4. a summary of crosslinking evidence on the protein topography of the 30S subunit. 5. a summary of 50S crosslinks and a correlation with functional sites.

CROSSLINKING PROCEDURE

The Two-Step Crosslinking Procedure with 2-Iminothiolane

Figure 2 outlines the two-step crosslinking procedure: modification of ribosomal subunits with 2-iminothiolane followed by oxidation. In the first step, ribosomal subunits are incubated with 2-iminothiolane in triethanolamine, pH 8.0, under reducing conditions to permit formation of amidine linkages between lysine amino groups and imidate group of the reagent. After incubation with 2-iminothiolane approximately 42 sulfhydryl groups are added per 30S ribosomal subunit (25) and 97 per 50S subunit. Under the conditions employed two to three sulfhydryl groups are added per ribosomal protein on the average.

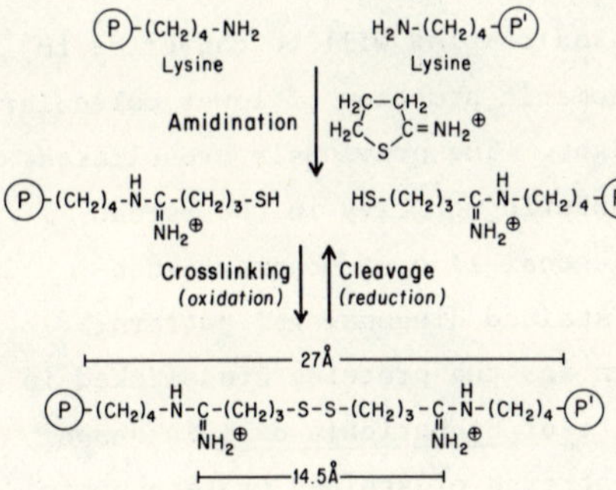

Figure 2. Modification of proteins with 2-iminothiolane and reversible crosslinking by disulfide bond formation.

After incubation with 2-iminothiolane and dialysis to remove sulfhydryl reagents, the modified ribosomal subunits were incubated for 30 min at 0° with 40 mM H_2O_2 to promote the formation of disulfide bonds. As a result of oxidation, it can be predicted that disulfide bonds of three types are formed: intermolecular protein:protein, intramolecular, and protein:mercaptan (21,25). Only the first are of interest in the studies described here and they can be detected by molecular weight analysis upon electrophoresis in polyacrylamide/dodecyl sulfate gels.

Analysis of Crosslinks by Two-Dimensional Polyacrylamide/Dodecyl Sulfate Gel Electrophoresis

The use of intermolecular disulfide bonds to produce crosslinks between proteins suggested two-dimensional polyacrylamide/dodecyl sulfate gel (diagonal gel) electrophoresis for the separation of crosslinked from non-crosslinked proteins. The method is shown schematically in Figure 3. In the first dimension, carried out in a tube gel, the oxidized ribosomal proteins are subjected to electrophoresis under non-reducing conditions. A second electrophoresis, with the tube gel embedded in a slab gel, run under the same non-reducing conditions would result in a pattern in which both monomeric and crosslinked proteins would fall on a diagonal line: the relative mobility of each component in the mixture would be the same in both dimensions. On the other hand, if the protein mixture contains disulfide linked complexes and if the first dimension tube gel is soaked in reducing agent prior to the second electrophoresis, then each crosslinked complex will be converted to monomeric proteins of lower molecular weight. The previously crosslinked proteins will have a greater electrophoretic mobility in the second dimension and will appear <u>below</u> the diagonal line of monomeric non-crosslinked ribosomal proteins on the stained diagonal gel pattern. An illustration of the diagonal pattern for the proteins crosslinked in the intact 50S or 30S ribosomal subunits of *Escherichia coli* is shown in Figure 4 or 7, respectively. The patterns of stained protein spots beneath the diagonal line represents a characteristic fingerprint of protein:protein interactions in the native ribosomal particle (18).

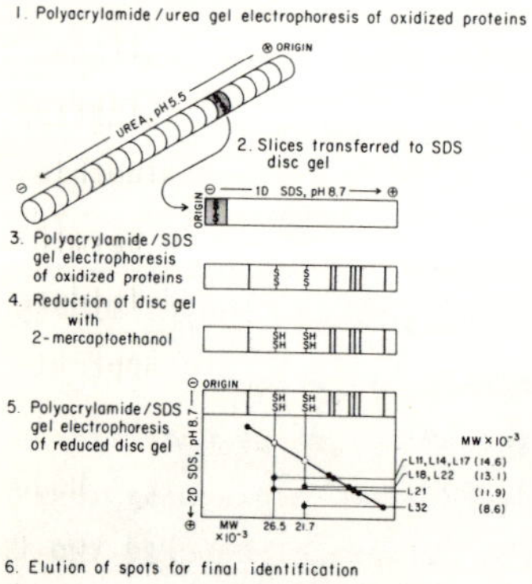

Figure 3. Electrophoretic purification of crosslinks and diagonal gel electrophoresis.

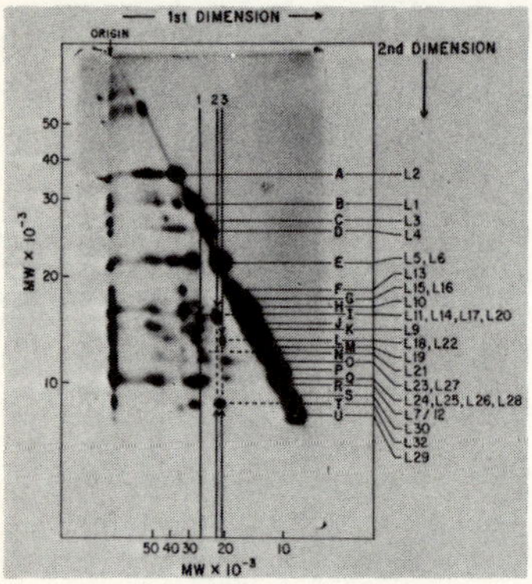

Figure 4. Diagonal gel electrophoresis of total 50S proteins extracted with LiCl/urea from intact 50S ribosomal subunits crosslinked with 2-iminothiolane. The 3 vertical axes indicate crosslinked pairs, which are analyzed in Figure 5.

Two criteria are employed to determine those individual proteins which had been members of specific dimers crosslinked by disulfide bonds: first, any monomeric components of an originally crosslinked species must necessarily fall on the same vertical line below the diagonal; second, the sum of the apparent molecular weights of the monomeric proteins arising by reduction of the disulfide crosslinked species must equal that of the crosslinked complex. The first criterion is obvious; the second depends upon the finding that crosslinked ribosomal protein dimers have apparent molecular weights (19), determined by electrophoretic mobilities in polyacrylamide/dodecyl sulfate gels, which are approximately the same as the molecular weight obtained by summing those of the monomeric protein members of the crosslinked complex. It was shown that crosslinked protein dimers and monomeric proteins fall on the same straight line when plotted on a standard mobility vs. log molecular weight graph (19). The vertical lines and arrows in Figure 4 indicate pairs of proteins which meet the criterion of molecular weight additivity and hence are concluded to have been crosslinked by disulfide bonds.

Identification of Crosslinked Proteins

The position of a protein spot below the diagonal does not permit unambiguous identification because in many instances two or more proteins have the same or overlapping mobilities. The 50S proteins comprising each region on the diagonal are indicated in Figure 4. Individual spots are cut from the diagonal gel, the proteins (which are either already radioactive being derived from cells grown in [^{35}S]sulfate or radiolabeled with iodine) are eluted, mixed with non-radioactive total protein and analyzed by two-dimensional polyacrylamide/urea gel electrophoresis (26,27). Superimposition of the autoradiograph of the dried gel containing the stained 70S protein pattern shows directly the identity of the crosslink first characterized on the diagonal gel. Examples of the identification procedure are shown in Figure 5. The urea gel pattern for 70S protein is shown in Figure 8.

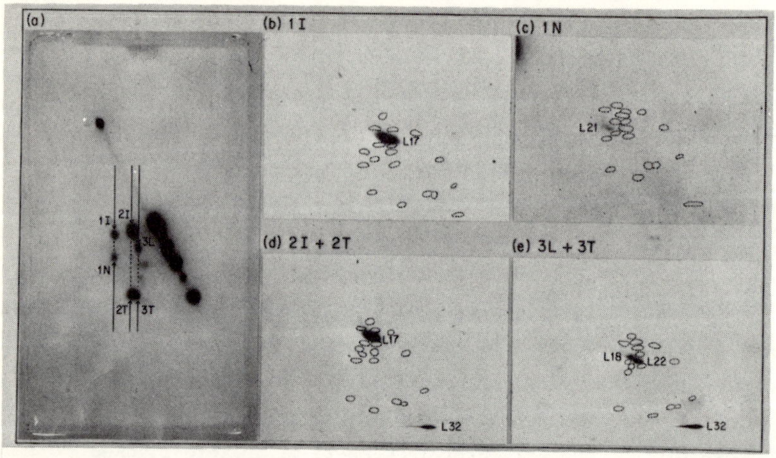

Figure 5. Analysis of a purified function of crosslinked 50S protein shown to contain the crosslinks: L17-L21, L17-L32, L18-L32 and L22-L32. (a) Diagonal gel electrophoresis of a slice from a polyacrylamide/urea gel electrophoresed at pH 5.5 which contained the crosslinks indicated by arrows in Figure 4. The spots indicated by arrows were eluted, mixed with non-radioactive protein and analyzed by gel electrophoresis in a two-dimensional polyacrylamide/urea gel.

Fractionation of Crosslinked Proteins

The complexity of the diagonal gel pattern of total protein from crosslinked subunits or the partial overlapping of certain spots from crosslinks of nearly the same molecular weight make it useful to fractionate the crosslinked protein prior to diagonal gel analysis. Two methods have been used: first, successive extraction of crosslinked subunits with increasing concentrations of LiCl; second, electrophoresis of the proteins in polyacrylamide/urea gels a pH 5.5. Each 0.5 cm slice of such a gel is then analyzed by diagonal gel electrophoresis. Figure 6 shows a gallery of diagonal gels

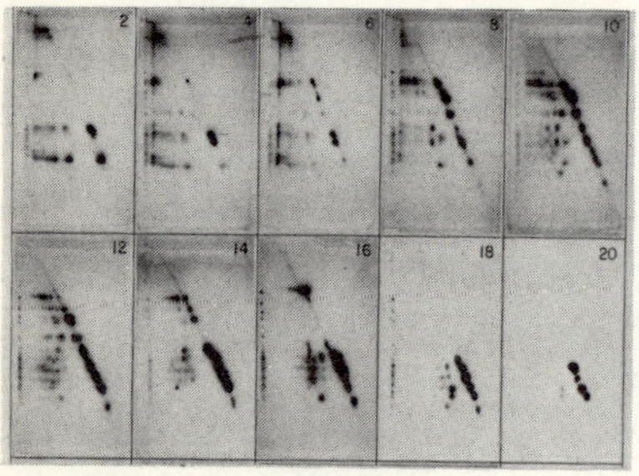

Figure 6. A gallery of diagonal gels of crosslinked 70S ribosomal proteins. Proteins extracted from crosslinked 70S ribosomes were fractionated by polyacrylamide/urea gel electrophoresis using a buffer containing bis-Tris·acetate at pH 5.5 (27).

representing the fractionation of crosslinked 70S protein by urea gel electrophoresis. The simplification achieved is apparent; the changes in intensity of members of a pair in sequential slices also aids in identification.

The Specificity and Yield of Disulfide Crosslinks

Disulfide crosslinking has been criticized as being open to possible artifacts due to disulfide interchange or random oxidation (10,28). Free sulfhydryl groups remain following oxidation and these could catalyze interchange or be substrates for random oxidation. A number of experiments have been done to show that such is not the case. Crosslinked protein samples are routinely treated with iodoacetamide to block free sulfhydryl groups. The specificity of a typical diagonal gel pattern of 30S protein crosslinked in the intact subunit is evident by comparison in Figure 7 to a "random" pattern formed by deliberate oxidation of proteins following extraction from the ribosome. When radioactive S4 or the dimer S5-S8 were mixed with non-radioactive, crosslinked subunits there was no appearance of radioactivity in any other species (21).

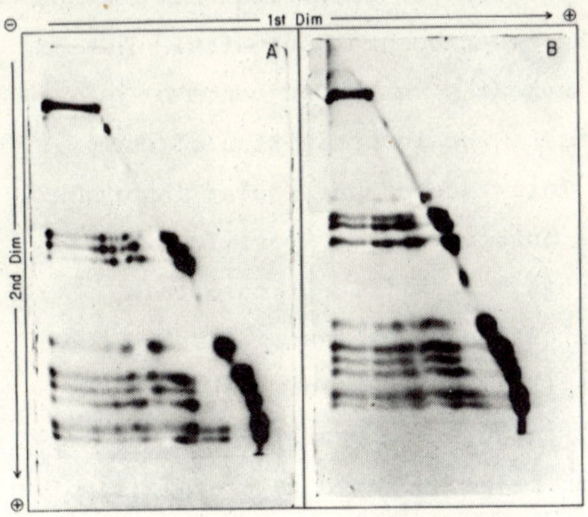

Figure 7. Specific and random diagonal gel patterns of 30S proteins modified with 2-iminothiolane. Panel A: protein extracted from crosslinked 30S subunits by LiCl/urea in the presence of iodoacetamide (21). Panel B: protein extracted from modified, reduced, non-alkylated 30S subunits prior to intentional oxidation (crosslinking) by H_2O_2.

All of the sulfhydryl groups added by modification with 2-iminothiolane are oxidized during crosslinking (21). However, no more than 20% of the total protein extracted from crosslinked subunits is found below the diagonal. Because oxidation converts sulfhydryls to disulfides, the majority of which are not intermolecular, we conclude that intramolecular disulfide bond formation is a major competing reaction. Different crosslinks are found in different yields. The yield of a given crosslink may be influenced by a number of factors including the relative reactivity of lysine residues toward amidination and of the sulfhydryl groups introduced, competition by intramolecular disulfide formation, and competition among several equally favorably situated neighboring proteins for dimer formation with the given protein. Slight differences in conformation within the population of ribosomes, conformational flexibility of the particle (29,30), and differences in composition due to the partial loss of weakly bound proteins, may also influence the yield of crosslink formation. The positive identification of a crosslink is interpreted to mean that two proteins are neighbors in the ribosome. All the crosslinks seen in a diagonal representing the population do not exist in each individual ribosome particle.

THE 30S:50S INTERFACE

The interaction between the large and small ribosomal subunits plays a central role in the mechanism of protein synthesis. The association of many ligands such as messenger RNA and various protein factors is thought to involve neighborhoods of each ribosomal subunit at or near the interface between the two ribosomal subunits. The necessary association of ribosomal subunits in the sequence of events culminating in protein synthesis, makes the investigation of the structure of the ribosomal subunit interface of particular importance.

Direct chemical evidence for the involvement of specific proteins at the ribosomal subunit interface has been difficult to obtain. Several years ago we attempted to crosslink proteins of the 30S subunit to proteins of the 50S subunit (31), by treating 70S ribosomes with relatively short crosslinking reagents: glutaraldehyde and formaldehyde. Protein S16 was found to be crosslinked to protein(s) of the 50S subunit. Indirect evidence concerning the occurrence of the same protein on both ribosomal subunits suggests that such proteins are at the subunit interface. Protein S20 is identical with protein L26 and it is considered to be a unit protein in 70S ribosomes (32). The protein remains bound, in part, to both subunits during dissociation of 70S ribosomes and is considered to be at the subunit interface (33). We have found that proteins S5 and S9, like S20, are found also, in part, bound to 50S subunits.

A typical two-dimensional polyacrylamide/urea gel electrophoresis pattern of stained proteins extracted from 70S ribosomes is shown in Figure 8, panel A. The system resolves all the proteins of 70S

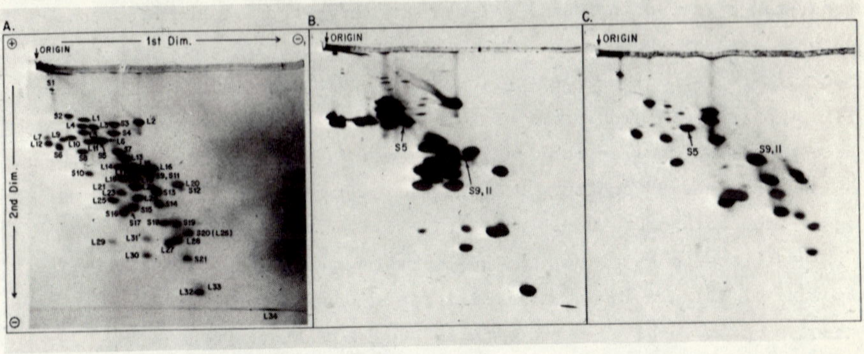

Figure 8. The presence of S5 and S9/S11 in two-dimensional polyacrylamide/urea gel (20,27) of total 50S protein. Panel A: proteins from 70S ribosomes. Panel B: proteins from purified 50S subunits. Panel C: proteins from purified 30S subunits.

ribosomes except for S9 and S11 which migrate as a single spot. Panels B and C show similar two-dimensional gels of proteins extracted from purified 50S and 30S subunits, respectively. Each of the purified subunits was free from contamination by the other subunit as judged by sucrose density gradient centrifugation. The gels represented here show that 50S subunits have a significant amount of a stained protein that migrates to the same position as the 30S protein S9 and/or S11. No other 30S proteins are readily apparent in the 50S protein pattern. Protein L26, considered to be identical to S20, was barely visible in agreement with the work of Hardy (32), and was definitely less intense than S9/S11. Comparison of the [^{35}S]-labeled tryptic peptides of the material extracted from the spot with those of individual 50S and 30S ribosomal proteins showed that the major component was S9 with a smaller amount of S11. Peptide mapping also showed the presence of S5 in 50S subunits. Small amounts of protein S5 would not be readily evident from staining since its position on most two-dimensional gels overlaps that of L6.

Experiments in which [^{35}S]-labeled 30S subunits and non-radioactive 50S subunits were mixed in 10 mM Mg^{++} to form 70S ribosomes showed that some radioactive 30S protein was transferred to the 50S subunit. The amount of radioactivity associated with the repurified 50S subunits was 1.4% of the total introduced as [^{35}S]-labeled 30S subunits, equivalent to the transfer of approximately 30% of a single protein of average sulfur content from the 30S subunit to the 50S particle. Proteins were extracted from the radioactive purified 50S subunits, and separated by two-dimensional polyacrylamide/urea gel electrophoresis. Autoradiography showed that most of the radioactivity was found in two specific regions of the gel corresponding to the proteins S5, S9 and S11. The radiolabeled proteins were analyzed in a polyacrylamide/dodecyl sulfate gel system in order to distinguish unambigously between S9 and S11, and S5 and L6. Most of the transferred radioactivity was associated with proteins S9 and S5.

These results show that the 30S subunit proteins S5, S9 and small amounts of S11 are found in significant amounts on the 50S subunit. The absence of any other 30S proteins in the 50S subunit, except for S20 (L26) which has been reported previously (34), suggests that the occurrence there of S5, S9 and S11 is not adventitious. Specificity of the S9 site on the 50S subunit is indicated strongly by the identification of a specific set of crosslinked dimers between S9 and 50S proteins: S9-L5, S9-L27, S9-L32 and/or L31 and S9 to one or more of the group L11, L14 and L17. The other proteins, S5 and S11, have not as yet been investigated in crosslinking studies of the 50S subunit. The most likely explanation of the occurrence of the four 30S proteins, S5, S9, S11 and S20, in the 50S subunit is that they are normally located in that region of the 30S subunit which interacts with the 50S subunit when the two particles associate to form 70S ribosomes: at the subunit interface. Protein S5 is close to S9 in the 30S subunit as indicated by the identification of a S5-S9 crosslink (19). The presence of S5 and S9 at the interface was suggested earlier by Marsh and Parmeggiani (35) who found that the capacity of 30S core particles to associate with 50S subunits was restored specifically by addition of S5 and S9.

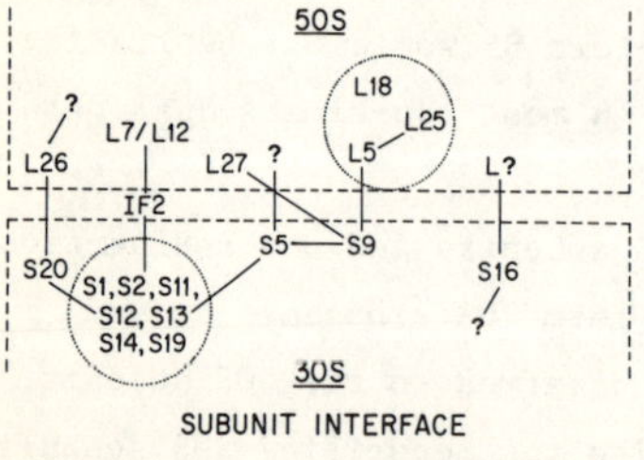

Figure 9.

The scheme shown in Figure 9 represents an arrangement of proteins at or near the interface region of each subunit and shows the crosslinks S9-L5 and S9-L27 across the interface. Lines connect S5 and S20 to the 50S subunit on the assumption that their occurrence in both subunits is due to their primary location at the 30S interface. Protein S16 has been shown to be crosslinked to one or more 50S proteins which were not identified (31). No 30S crosslinked neighbors of S16 have yet been determined.

Complexes containing initiation factor IF2 and 70S ribosomes have been crosslinked. IF2 was crosslinked to 50S proteins L7/L12 (36), and to the 30S proteins represented by the circle in Figure 9, although direct crosslinked pairs between IF2 and individual 30S proteins were not identified (11). Proteins L5, L18 and L25 are related insofar as they bind specifically to 5S RNA (37). They are enclosed in the small circle in the 50S portion of the scheme. The location of the 5S complex at the interface is consistent with the crosslink S9-L5. 50S subunits lacking 5S RNA do not associate with 30S subunits (38).

CROSSLINKING OF THE 30S SUBUNITS

S1 Crosslinks

Crosslinks containing S1 were not detected in earlier studies since the composition of the diagonal gels had been optimized to resolve monomeric proteins and crosslinks with molecular weights not exceeding 55,000. The characteristic properties of S1, high molecular weight (68,000) and relatively greater acidity can, however, be used to facilitate investigation of S1 crosslinks. Figure 10 shows an SDS diagonal gel of crosslinked 30S proteins contained in an acidic

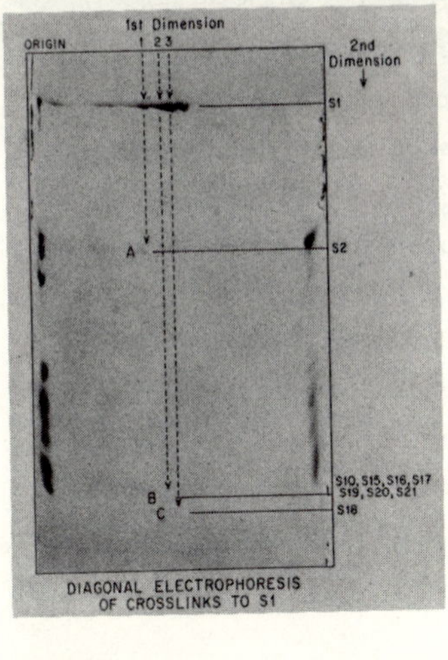

Figure 10. Diagonal gel electrophoresis of crosslinks containing S1. Crosslinked total protein from [^{35}S]-labeled 30S subunits was fractionated by electrophoresis in a polyacrylamide gel containing urea at pH 8.7 (27). A slice containing negatively charged protein was analyzed in a diagonal gel containing 7.5% and 13.5% acrylamide in the first and second dimensions, respectively. Protein S1 and three spots below the diagonal are indicated by arrows.

fraction of a polyacrylamide/urea gel. Protein S1 and three smaller proteins fall below the diagonal. There is one strong crosslink to S1 and two weaker crosslinks. The position of the strongly crosslinked protein on the SDS diagonal gel coincides with that of S2. This was confirmed by extraction of the radioactive protein from the spot and two-dimensional urea gel electrophoresis with non-radioactive protein. One of the weaker crosslinked proteins has been identified as S18 and the other as S10.

There is functional and structural evidence relating S1 to the initiation of protein synthesis and to the 3' end of the 16S RNA (39-43). It was shown in this laboratory that initiation factors IF1, IF2 and IF3 could be crosslinked to S1 along with several other proteins (11-13). It was not determined whether the crosslinks to the factors were direct or bridged by other proteins. Nevertheless, S2 was among the proteins found in the group crosslinked to IF2. The S1-S2 crosslink is consistent with crosslinks between the initiation factors and S1 and S2, and with the location of S1 in the initiation region of the ribosome. One of the other proteins found crosslinked to S1 is S18. While not found crosslinked to initiation factors, S18 is found near another protein, S21, which is both crosslinked to initiation factor IF3 and to the 3' end of the 16S RNA (43). Protein S1 itself was found crosslinked to the 3' end of the 16S RNA in the study cited.

Partial reconstitution of 3M-NH_4Cl protein-deficient core particles showed that two or more proteins among the group S2, S9, S10 and S14 are involved in the binding site for S1 (44). The S1-S2 and S1-S10 crosslinks suggest that S2 and S10 are required in the partial reconstitution experiments. Subsequent work showed a major role for S9 in restoring S1 binding capacity (45). Protein S9 has not been detected crosslinked to S1 nor have any of the other proteins implicated for S1 binding been found in crosslinks with S9.

When the N-terminal forty-three amino acid fragment of S4 is absent, proteins S1, S2, S10, S18 and S21 fail to bind to reconstituted particles (46). The group of proteins includes the three crosslinks to S1 and two of the proteins found to contribute to S1 binding. The scheme shown below (Fig. 11) represents all these results which bear on the protein composition of the S1 binding region. The solid lines indicate crosslinks, the slashed lines indicate the reconstitution of S1 binding and the dotted lines the proteins dependent upon the N-terminal fragment of S4 for incorporation into reconstituted particles. With the exception of S4, all of the proteins are added late in the assembly sequence (45,47).

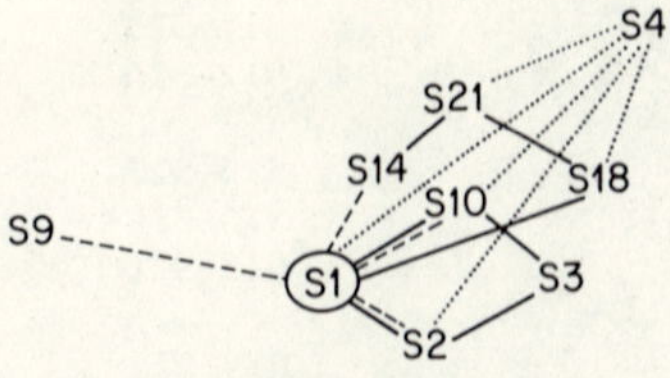

Figure 11.

A location for S1 on the 30S subunit has also been assigned from immunoelectron microscopy. Lake (48) places S1 on the platform in his model; the Berlin group, whose model does not have a conspicuous platform, finds an S1 antigenic site on the upper one-third of the subunit (33). The S1-S2 crosslink is consistent with the Lake model, although the antigenic sites are separated by the cleft, but not with the Berlin model. On the other hand, the S1-S10 crosslink is consistent with the Berlin model but not with the Lake model. The location assigned to S18 in both models would not be inconsistent with the observed S1-S18 crosslink.

Two binding sites for S1 have been reported (44). The subunits for the crosslinking studies reported here were derived from ribosomes washed in high salt concentrations and contained less than one copy of S1 per 30S particle. Similar results have been obtained with 30S subunits derived from tight couples that contained S1 in amounts equimolar with the particle. The limited number of crosslinks is consistent with a specific binding site for S1, but suggests that the bulk of the mass of S1 is not in close contact with other ribosomal proteins and that it may either be in contact with RNA or be extended away from the ribosome.

30S Crosslinks

All 30S crosslinks identified through the use of 2-iminothiolane are listed in Table 1. The systematic investigation of S1 crosslinks described above led also to the confirmation of the majority of those published previously (20). In addition three new crosslinked dimers were identified: S11-S21, S13-S21 and S14-S21. Each of the three

TABLE 1. 30S Crosslinked Dimers.

S1-S2	S4-S5*	S6-S18*	S12-S13
S1-S10	S4-S6		S12-S20
S1-S18	S4-S8	S7-S8	
	S4-S9	S7-S9*	S13-S17
S2-S3*	S4-S12	S7-S13	S13-S19*
S2-S5*	S4-S13		S13-S21
S2-S8*	S4-S17	S8-S11	
		S8-S13	S14-S19*
S3-S4	S5-S8*	S8-S15*	S14-S21
S3-S5	S5-S9		
S3-S9	S5-S13	S11-S13	S18-S21*
S3-S10*		S11-S21	
S3-S12			

proteins crosslinked to S21 is found crosslinked to initiation factors (12), as is S21. The new crosslinks add to the information on the organization of the initiation neighborhood. A trimer containing S11, S18 and S21 was reported some time ago (49). Those crosslinks identified in other laboratories using other reagents are indicated in Table 1 by single asterisks. Two crosslinks, S8-S15 (9) and S14-S19 (50), have never been observed using 2-iminothiolane.

The conditions used for the modification of 30S subunits with 2-iminothiolane prior to crosslinking led to the addition of 42 sulfhydryl groups per subunit. All proteins were modified to some extent (25). When twice the number of sulfhydryls were added the pattern of crosslinked proteins beneath the diagonal was largely unchanged with respect to the position and relative intensity of spots. For this reason we think that we are near establishing an exhaustive list of 30S proteins susceptible to crosslinking with 2-iminothiolane that two proteins do <u>not</u> form a crosslink will eventually contribute to the construction of models of ribosome protein topography.

We have not attempted to quantify the absolute or relative yields of crosslinks. Inspection of a diagonal gel of crosslinked 30S, 50S or 70S proteins shows clearly that yields of different crosslinks vary greatly. The 30S pairs S5-S8, S7-S9 and S13-S19 are formed in higher yield than most others and have been found in different laboratories with different crosslinkers and procedures. The rest of the crosslinks occur in yields lower than those cited. In no case have we observed the quantitative conversion of two proteins into a crosslinked dimer.

The scheme below (Fig. 12) shows an arrangement of the 30S proteins based on the crosslinking data of Table 1. It is a two-dimensional projection of the crosslinking data and is not intended to be interpreted as a three-dimensional model. A major principle or constraint in constructing the scheme was to accommodate those proteins with many crosslinked neighbors. Such proteins were treated in one of two ways. Either they were assumed to be elongated (to have two different crosslinked domains), or they were considered to occupy "central" locations with their crosslinked neighbors surrounding them. Protein S16 has not been identified in a crosslink, while S15 and S20 have been found only in one crosslinked pair. Proteins S2, S3 and S4 have each been divided into two domains designated "A" and "B". Published data from immunoelectron microscopy has in fact represented distant antigenic sites for each of these proteins (33,48). Protein S13 is represented as a central protein with nine crosslinked neighbors. These neighbors have been divided into two groups, three on the left and six on the right. The left and right neighborhoods thus distinguished are characterized by many crosslinks within each. Only the S8-S11 crosslink, besides S13 itself, bridges the two groups. The "A" domains of S2, S3 and S4 are placed on the left, the "B" domains on the right. The group on the right contains most of the proteins crosslinked to initiation factors, including S1 and its crosslinks. Protein S7 in the left domain has been found crosslinked to IF3 (14) and to other crosslinks to which are in the right region; protein S9, which promotes the binding of S1 but is not found crosslinked to it, is placed in the left domain, opposite to S1.

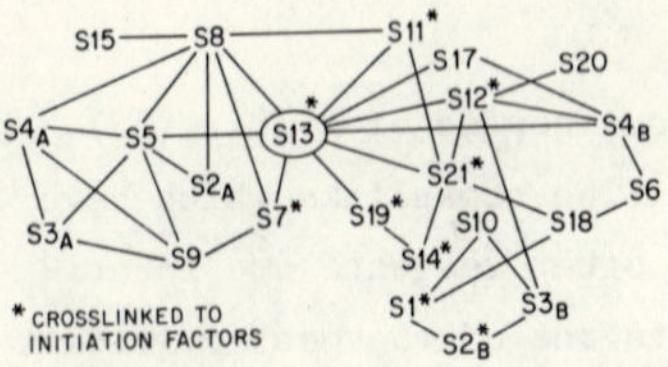

Figure 12.

CROSSLINKING OF THE 50S SUBUNIT

The 50S crosslinks identified thus far through the use of reversible disulfide crosslinking with 2-iminothiolane are listed in Table 2. There are fifty-four dimers. In addition, proteins L5, L6, L10 and L11 have all been found crosslinked to two and three copies of L7/L12,

TABLE 2. 50S Crosslinked Dimers.

L1-L31′	L4-L11	L6-L7/12	L10-L11*	L13-L14	L17-L21	L22-L31′
	L4-L14	L6-L10	L10-L13	L13-L17	L17-L27	L22-L32
L2-L5		L6-L11	L10-L14	L13-L21	L17-L31′	
L2-L6	L5-L7/12	L6-L17	L10-L21		L17-L32*	L23-L31′
L2-L7/12	L5-L11		L10-L31′	L14-L17		L23-L32
L2-L9*	L5-L17	L7/12-L10*	L10-L32		L18-L31′	
L2-L10	L5-L23	L7/12-L11*		L16-L23	L18-L32	L27-L29
L2-L11	L5-L25	L7/12-L14	L11-L14	L16-L25		
L2-L17	L5-L31′		L11-L17	L16-L27	L21-L31′	
		L9-L28	L11-L21		L21-L32	
L3-L5						

consistent with the presence of three to four copies of L7/L12 (52) and its occurrence as a stable dimer (53). The crosslinks which have been identified in other laboratories with other reagents are identified by asterisks. The crosslink L17-L32 is one of the most prominent found with 2-iminothiolane.

The conditions used in these crosslinking studies involved the addition of almost 100 sulfhydryl groups to the 50S subunit, between two and three per protein on the average. The investigation, while less advanced as that for the 30S subunit, suggests certain general conclusions. Certain proteins, particularly many of the smaller, basic proteins are seldom found in crosslinks. Other proteins are found in many different crosslinks just as was found for the 30S subunit. The number of crosslinks for L2 is seven; for L5, eight; for L10, nine; for L7/L12, six; and for L17, ten. Of these there is evidence either from immunoelectron microscopy or solution studies for elongation of L2, L11 and L7/L12 (34,54,55). For the other multiple crosslinked proteins we must postulate that they also are elongated or that they are centrally located with several neighboring proteins, each with favorably situated reactive lysines such that a small number of sulfhydryls (two to four) on the central protein can generate up to ten different crosslinked dimers with roughly equal frequency.

The extensive number of 50S crosslinks discourages their depiction in a single scheme as was done for the 30S subunit. Instead selected groups of crosslinks are represented as they illustrate specific points and relate to other types of functional and structural data (56-59). Protein L17 appears in the most different dimers, ten, one of which, L32, is one of the dimers in highest yield. It has also been identified using dimethyl suberimidate as a crosslinking reagent (60). We have represented the central role of L17 in Figure 13. Its direct crosslinks occur in three different independently defined neighborhoods. Proteins L2 and L27 are in the peptidyl transferase neighborhood; L10 and L11 are related to L7/L12 and GTPase activity (61-63) and L5 is part of the 5S RNA complex so indicated by the dotted circle in this and other figures. Two of the proteins found in the 5S RNA-protein complex, L5 and L18, are related to L17, the former by a direct crosslink, the latter by a "bridged" crosslink. The third 5S protein, L25, is crosslinked to L5.

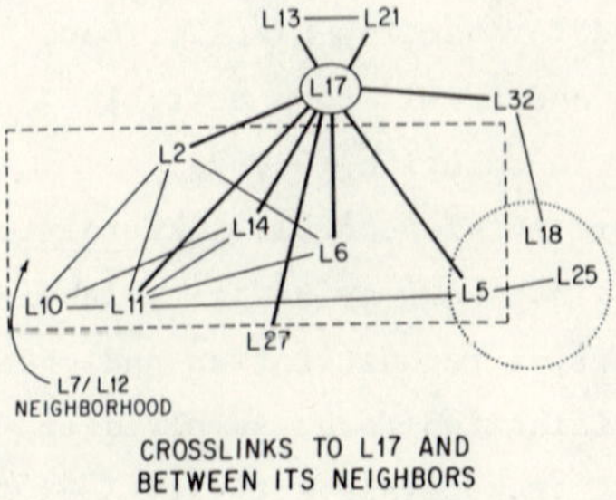

Figure 13.

Protein L2 also appears in multiple crosslinks which interrelate the peptidyl transferase, GTPase and 5S neighborhoods as shown in Figure 14. Proteins L2 and L6 have been found necessary for the linking of the 5S complex to the 50S particle (64) and the crosslinks between L2 and L6 and those shown are consistent with this observation. Fahnestock has shown that the protein equivalent to L2 from Bacillus stearothermophilus is involved not only in peptidyl transferase but also in subunit association and the binding of elongation factor G (65). The crosslink between L2 and L7/L12 also relates L2 to the interface and factor binding regions (36)(see Figure 9).

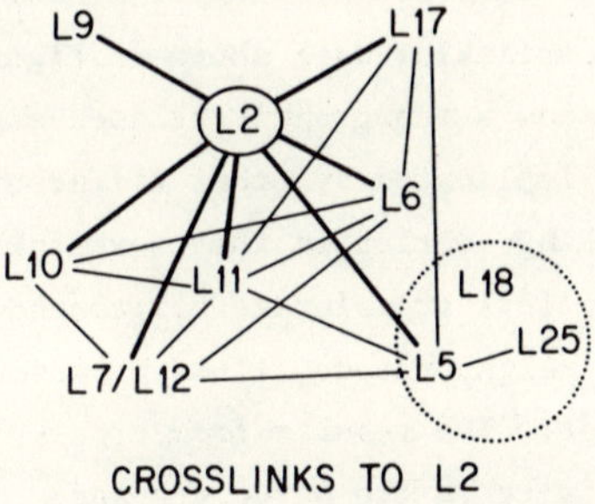

Figure 14.

Proteins L7/L12 are elongated, occur in multiple copies and are located in models based on immunoelectron microscopy, in one of the conspicuous protrusions from the major mass of the 50S particle (33,48). They are certainly not remote from interactions with other proteins despite their location on a protuberance. As shown earlier in Figure 9 they are crosslinked to initiation factor IF2 and therefore assigned a

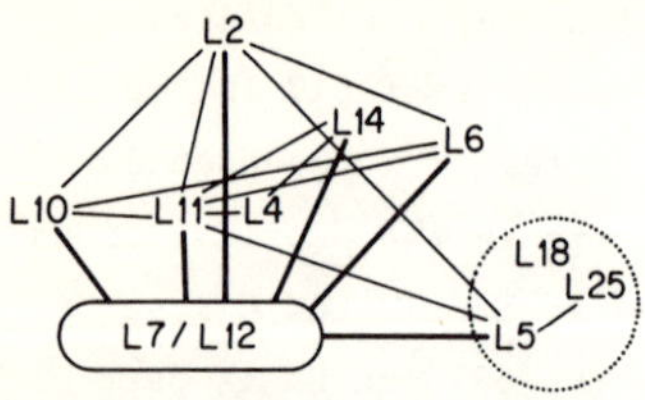

Figure 15.

location in the interface region. As shown in Figure 15 they are crosslinked to six other 50S proteins. Of these L5 is part of the 5S complex; L2 and L11 are near the peptidyl transferase center; L10 and L11 are required for binding L7/L12; (63), and L10 and L7/L12 form a stable complex in solution (66,67).

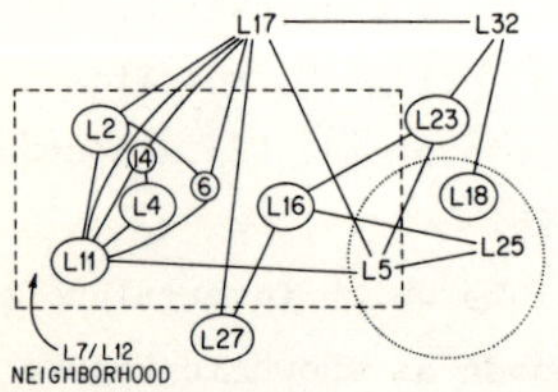

Figure 16.

Figure 16 shows crosslinks related to proteins shown by affinity labeling, partial reconstitution and chemical modification to be involved in peptidyl transferase activity (34,56, 59,68,69). The peptidyl transferase proteins are circled and the proteins from Figure 15 in the L7/L12 neighborhood are enclosed by slashed lines. The crosslinking data shown in Figures 13-16 show a topographic relatedness between the peptidyl transferase proteins implicated by other different experimental approaches. There is a striking overlap in the crosslinking map between the set of proteins in the peptidyl transferase neighborhood and the L7/L12, or GTP/factor interaction neighborhoods, links between the 5S complex and both of the former exist. The results from crosslinking reinforce the concept of cooperativity in the structure and function of the ribosome. The crosslinks identified thus far were not investigated on the basis of possible functional roles. Yet the proteins implicated as important in ribosome function seem often to be highly accessible to crosslinking. Very likely they are located on the surface of the 50S subunit towards the interface. They may be clustered in a protein rich region which is highly accessible to crosslinking.

PERSPECTIVE

What does the occurrence of a crosslink between two proteins in the ribosome signify? It does not mean necessarily that the two proteins are in close physical contact. The length of the reagent itself (see Fig. 2) even without taking into account the lengthy lysine side chains, suggests that intermolecular crosslinks could be formed between two proteins separated by RNA. The formation of an intermolecular crosslink does indicate that certain regions of the two proteins involved are within 15 Å of each other. This does not imply that their centers of mass are equally close. There is more and more evidence from immunoelectron microscopy, neutron scattering and studies on single proteins in solution that many ribosomal proteins are not globular but are elongated with axial ratios of up to 10:1. The next step in the use of crosslinking for the mapping of ribosomal proteins will require identification of the crosslinked amino acid residues in each protein dimer. Such an investigation is fraught with difficulties since it must first be established that two proteins are crosslinked at unique positions. Fortunately there is now extensive information on the amino acid sequences of many of the ribosomal proteins (33,34). One of the first crosslinked dimers identified, S18-S21, was formed with p-phenylenebismaleimide (70). The sequences of both proteins have been established (33). The unique cysteine residues of each protein are the most likely point of crosslinking. More recently a single crosslink in the abundant dimer S5-S8 has been demonstrated and the crosslinked amino acids identified (71). The identification of the sites of crosslinking for other neighboring pairs of proteins will greatly extend the resolution of crosslinking as a tool for determining the protein topography of the ribosome.

ACKNOWLEDGEMENTS

We would like to thank our colleagues Dr. Andreas Sommer, Dr. Thomas Fanning and Mr. Jan Cover who have contributed to the work described. We thank Ms. Beverley Haskins and Ms. Kathy Amann for careful and arduous typing of this manuscript. This work has been supported by a grant from the United States Public Health Service (grant no. GM 17924).

REFERENCES

1. Traut, R. R., and Monroe, R. E. (1964) J. Mol. Biol. 10, 63-72.
2. Pellegrini, M., and Cantor, C. R. (1977) Molecular Mechanisms of Protein Biosynthesis (Weissbach, H. and Pestka, S., Eds.) pp. 203-244, Academic Press, New York, N.Y.

3. Nicholson, A. W., and Cooperman, B. S. (1978) FEBS Lett. 90, 203-208.

4. Breitmeyer, J. B., and Noller, H. F. (1976) J. Mol. Biol. 101, 297-306.

5. Davies, G. E., and Stark, G. R. (1970) Proc. Natl. Acad. Sci. U.S.A. 66, 651-656.

6. Bickle, T. A., Hershey, J. W. B., and Traut, R. R. (1972) Proc. Natl. Acad. Sci. U.S.A. 69, 1327-1331.

7. Lutter, L. C., Bode, U., Kurland, C. G., and Stöffler, G. (1974) Mol. Gen. Genet. 129, 167-176.

8. Clegg, C., and Hayes, D. H. (1974) Eur. J. Biochem. 42, 21-28.

9. Expert-Bezançon, A., Barritault, D., Milet, M., Guérin, M. F., and Hayes, D. H. (1977) J. Mol. Biol. 112, 603-629.

10. Peretz, H., Towbin, H., and Elson, D. (1976) Eur. J. Biochem. 63, 83-92.

11. Bollen, A., Heimark, R. L., Cozzone, A., Traut, R. R., Hersey, J. W. B., and Kahan, L. (1975) J. Biol. Chem. 250, 4310-4314.

12. Heimark, R. L., Kahan, L., Johnston, K., Hershey, J. W. B., and Traut, R. R. (1976) J. Mol. Biol. 105, 219-230.

13. Langberg, S. R., Kahan, L., Traut, R.R., and Hershey, J. W. B. (1977) J. Mol. Biol. 117, 307-319.

14. van Duin, J., Kurland, C. G., Dondon, J., and Grunberg-Manago, M. (1975) FEBS Lett. 59, 287-290.

15. Traut, R. R., Bollen, A., Sun, T-T., Hershey, J. W. B., Sundberg, J. and Pierce, L. R. (1973) Biochemistry 12, 3266-3275.

16. Lutter. L. C., Ortanderl, F., and Fasold, H. (1974) FEBS Lett. 48, 288-292.

17. Sun, T-T., Bollen, A., Kahan, L., and Traut, R. R. (1974) Biochemistry 13, 2234-2340.

18. Sommer, A., and Traut, R. R. (1974) Proc. Natl. Acad. Sci. U.S.A. 71, 3946-3950.

19. Sommer, A. & Traut, R. R. (1975) J. Mol. Biol. 97, 471-481.

20. Sommer, A., and Traut, R. R. (1976) J. Mol. Biol. 106, 995-1015.

21. Lambert. J. M., Jue, R., and Traut, R. R. (1978) Biochemistry 17, 5406-5416.

22. Kenny, J. W., Sommer, A., and Traut, R. R. (1975) J. Biol. Chem. 250, 9434-9436.

23. Traut, R. R., and Kenny, J. W. (1977) Protein Crosslinking (Friedman, M., Ed.) Part A, pp. 215-233, Plenum Publishing Corp., New York, N.Y.

24. Kenny, J. W., and Traut, R. R. (1979) J. Mol. Biol. 127, 243-263.

25. Jue, R., Lambert, J. M., Pierce, L. R. and Traut, R. R. (1978) Biochemistry 17, 5399-5406.

26. Knopf, U. C., Sommer, A., Kenny, J., and Traut, R. R. (1975) Mol. Biol. Rep. 2, 35-40.

27. Kenny, J. W., Lambert, J. M., and Traut, R. R. (1979) Methods in Enzymology (Grossman, L. and Moldave, K., Eds.) 59, 534-550, Academic Press, New York, N.Y.

28. Kurland, C. G. (1977) Molecular Mechanisms of Protein Biosynthesis (Weissbach, H., and Pestka, S., Eds.) pp. 81-116, Academic Press, New York, N.Y.

29. Ghosh, N., and Moore, P. B. (1979), Eur. J. Biochem. 93, 147-156.

30. Lam, M. K. T., Changchien, L-M., and Craven, G. R. (1979) J. Mol. Biol. 128, 561-575.

31. Sun, T-T., Traut, R. R., and Kahan, L. (1974) J. Mol. Biol. 87, 509-522.

32. Hardy, S. J. S. (1975) Mol. Gen. Genet. 140, 253-274.

33. Stöffler, G., and Wittmann, H. G. (1977) Molecular Mechanisms in Protein Biosynthesis (Weissbach, H., and Pestka, S., Eds.) pp. 117-202, Academic Press, New York, N.Y.

34. Brimacombe, R., Stöffler, G., and Wittmann, H. G. (1978) Ann. Rev. Biochem. 47, 217-249.

35. Marsh, R. C. and Parmeggiani, A. (1973) Proc. Natl. Acad. Sci. U.S.A. 70, 151-155.

36. Heimark, R. L., Hershey, J. W. B., and Traut, R. R. (1976) J. Biol. Chem. 251, 7779-7784.

37. Horne, J. R., and Erdmann, V. A. (1972) Mol. Gen. Genet. 119, 337-344.

38. Dohme, F. and Nierhaus, K. H. (1976) Proc. Natl. Acad. Sci. U.S.A. 73, 2221-2225.

39. Steitz, J. A. (1979) Biological Regulation and Development, Vol. 1, Gene Expression (Goldberger, R. F., Ed.) pp. 349-399, Plenum Press, New York, N.Y.

40. Kenner, R. A. (1973) Biochem. Biophys. Res. Comm. 51, 932-938.

41. Dahlberg, A. E. (1974) J. Biol. Chem. 249, 7673-7678.

42. Dahlberg, A. E., and Dahlberg, J. E. (1975) Proc. Natl. Acad. Sci. U.S.A. 72, 2940-2944.

43. Czernilofsky, A. P., Kurland, C. G. and Stöffler, G. (1975) FEBS Lett. 58, 281-284.

44. Laughrea, M., and Moore, P. B. (1978) J. Mol. Biol. 122, 109-112.

45. Laughrea, M., and Moore, P. B. (1978) J. Mol. Biol. 121, 411-430.

46. Changchien, L-M., and Craven, G. (1978) J. Mol. Biol. 125, 43-56.

47. Held, W. A., Ballou, B., Mizushima, S., and Nomura, M. (1974) J. Biol. Chem. 249, 3103-3109.

48. Lake. J. A. (1978) Advanced Techniques in Biological Electron Microscopy II (Koehler, J. K., Ed.) pp. 173-211, Springer-Verlag, Berlin.

49. Shih, C. T., and Craven, G. R. (1973) J. Mol. Biol. 78, 651-663.

50. Bode, U., Lutter, L. C. and Stöffler, G. (1974). FEBS Lett. 45, 232-236.

51. Zimmerman, R. A., Muto, A., and Mackie, G. A. (1974) J. Mol. Biol. 86, 433-450.

52. Subramanian, A. R. (1975) J. Mol. Biol. 95, 1-8.

53. Wong, K-P. (1974) Biochem. Biophys. Res. Comm. 61, 78-84.

54. Giri, L., Dijk, J., Labischinski, H. and Bradaczek, H. (1978) Biochemistry 17, 745-749.

55. Osterberg, R., Sjöberg, B., Liljas, A., and Pettersson, I. (1976) FEBS Lett. 66, 48-51.

56. Cooperman, B. S. (1978) Bioorganic Chemistry Vol. IV, pp. 31-113, Academic Press, New York.

57. Traut, R. R., Heimark, R. L., Sun, T-T., Hershey, J. W. B., and Bollon, A., (1974) Ribosomes (Nomura, M., Tissières, A., Lengyel, P., Eds.) pp. 271-308, Cold Spring Harbor Laboratory, New York.

58. Pellegrini, M., and Cantor, C. R. (1977) Molecular Mechanisms of Protein Biosynthesis (Weissbach, H., and Pestka, S., Eds.) pp. 203-244, Academic Press, New York.

59. Dohme, F., and Fahnestock, S. R. (1979) J. Mol. Biol 129, 63-81.

60. Expert-Bezançon, A., Barritault, D., Clegg, J. C. S., Milet, M., Khouvine, Y., and Hayes, D. H. (1975) FEBS Lett. 59, 64-69.

61. Maassen, J. A., and Möller, W., (1974) Proc. Natl. Acad Sci. U.S.A. 71, 1277-1280.

62. Schrier, P. I., Maassen, J. A., Möller, W., (1977) Biochem. Biophys. Res. Comm. 53, 90.

63. Highland, J. H., Howard, G. A., (1975) J. Biol. Chem. 250 (3), 831-834.

64. Gray, P. N., Garrett, R. A., Stöffler, G., Monier, R. (1972) Eur. J. Biochem. 28, 412-421.

65. Auron, P. E., Erdelsky, J., and Fahnestock, S. R., (1978) J. Biol. Chem. 253, 6893-6900.

66. Osterberg, R., Sjöberg, B., Pettersson, I., Liljas, A., and Kurland, C. G., (1977) FEBS Lett. 73, 22.

67. Pettersson, I., Havoy, S. J. S., and Liljas, A. (1976) FEBS Lett. 64, 135-138.

68. Jaynes, E. N., Jr., Grant. P. G., Giangrande, G., Wieder, R., and Cooperman, B. S. (1978) Biochemistry 17, 561.

69. Moore, V. G., Atchison, R. E., Thomas, G., Moran, M. and Noller, H. F. (1975) Proc. Natl. Acad. Sci. U.S.A. 72, 844-848.

70. Chang, F. N., and Flaks, J. G. (1972) J. Mol. Biol. 68, 177-180.

71. Allen, G., Capasso, R., and Gualerzi, C. (1979) J. Biol. Chem., in press.

SCATTERING STUDIES OF THE THREE-DIMENSIONAL ORGANIZATION OF THE *E. COLI* RIBOSOME

PETER B. MOORE

Departments of Chemistry and
Molecular Biophysics & Biochemistry,
Yale University,
New Haven, CT 06520

INTRODUCTION

In biology function follows form. It follows that a knowledge of the three dimensional organization of the ribosome will be necessary in order to understand its mechanism of action in protein synthesis. As our grasp of its structural details increases, insights into its mode of action should result.

Accordingly it is not surprising that considerable effort has already been invested in ribosome structure determination. Moreover, we can be sure that work in this area will not cease until an atomic resolution description is in hand; nothing less is likely to be fully satisfying. The monumental task of determing the ribosome's structure at high resolution, however, has hardly begun for lack of crystals suitable for crystallographic study, the only approach which can lead to that kind of information. Recent success in obtaining microscopic ordered aggregates of E. coli 30S subunits (1) gives hope that this impasse may soon be broken.

The structural studies on the ribosome carried out so far have had as their goal a description of the particle at a resolution far below atomic. The motivation for these studies has been and remains the conviction that a low resolution understanding will contribute insights worth having. Most of the substrates and ligands of the ribosome are macromolecules which can be visualized by the low resolution methods currently in use. A knowledge of where they bind to the ribosome and their spatial relationship to the components of the ribosome and to each other, even at low resolution, should provide valuable hints about the nature of the events which transpire on the ribosome's surface during protein synthesis.

The purpose of this article is to review the contributions made to our understanding of the structure of the ribosome over the past five years by one of the methods used for its study, small angle, solution scattering. Attention will be confined to work done on intact ribosomes and ribosomal subunits. Only brief mention will be made of investigations of protein-deficient particles, and none at all of the growing and interesting literature on scattering from individual ribosomal components and simple complexes, discussions of which will be found elsewhere in this volume. The purpose of this review is to provide non-specialists with an evaluation of what recent scattering results have contributed to our understanding of how the intact ribosome is organized.

The reader will note the absence of any discussion of organisms other than E. coli in this article. While this could be interpreted as reflecting the author's bias, it is also a true reflection of the state of this field. Regretably little scattering work has been done on ribosomes from other organisms; data on eucaryotic ribosomes are virtually non-existent.

Background; Small Angle Scattering

The scattering given by a solution of macromolecules differs from that of pure solvent and analysis of the differences can provide information about solute structure. The kinds of information one can obtain have to do with the size and shape of the macromolecule and include its molecular weight, molecular volume, radius of gyration and shape, as an equivalent ellipsoid (or better). Of these quantities and concepts the one likely to be least familiar is "radius of gyration". The radius of gyration of a particle is the distance between its center of gravity and its mass elements measured as a root mean square average. Limited information can also be obtained about internal organization. The technique is called "small angle scattering" since when x-rays or thermal neutrons are used as the probing radiation, the difference in scattering between solution and solvent is most intense in the regions right around the direction of flight of the unscattered radiation, i.e. at low scattering angles (2,3). The data on which one works are simply the intensity of the macromolecular scattering contribution measured as a function of scattering angle. The technical details of the applications of these techniques to the study of the ribosome and its components have been reviewed recently elsewhere (4,5).

Strengths and Limitations of the Method

The primary virtue of solution scattering is that it gives information about the 3-dimensional organization of a macromolecule as it exists free in solution, in which regard it has a distinct advantage over electron microscopy, for example, which requires fixation, staining, dehydration, etc. of the specimen. Indeed, in the case of the ribosome the solution scattering data are the best indication we have of the 3-dimensional structure of the fully hydrated particle.

It is nevertheless necessary to recognize the limitations of the method. First and most obvious, it is not a high resolution method; no macromolecule will ever be described at atomic resolution as the result of solution scattering experiments. Second, the models for a macromolecule which emerge from solution scattering experiments are representations of the average structure of all the particles in a given sample. We know that ordinary ribosome preparations are heterogeneous, certainly with respect to the protein composition (6), and recently it has been suggested with respect to surface chemical reactivity (7) as well. What is not known is the significance of these kinds of heterogeneity for the 3-dimensional structure of the particle at low resolution. It follows that one cannot be sure whether the average structure will turn out to be a good representation of all ribosomes, which vary slightly about this mean, or whether it is merely a mathematical curiosity which no individual particle ever closely resembles at any time. Existing evidence (see below) suggests the latter possibility is unlikely, but the data still are not conclusive. Third, it is extremely hard to get useful information from scattering data obtained on solutions in which the structure of interest is aggregating. Ribosomes and ribosomal subunits tend to form dimers and higher oligomers in solutions containing magnesium ion at physiological concentrations. For this reason most solution scattering data has been collected on particles in buffers whose Mg^{++} concentration near 1 mM, well below the optimum for protein synthesis. We know that alterations of magnesium concentration over the range of 1-10 mM affect ribosome conformation (8,9,10), and again we do not know how big these changes are, measured in Ångstroms. The magnesium concentration problem has had significant practical effects. Efforts of the Madison group to measure 70S scattering in the late 1960's (11), which required high magnesium concentrations to ensure 70S couple stability, failed due to aggregation and it is only recently that satisfying results have been obtained on 70S particles (see below).

Early Scattering Results

The first major investigation of the ribosome by small angle x-ray scattering was carried out by Anderegg and his collaborators at Madison in the late 1960's and early 1970's. This work has been authoritatively reviewed by Van Holde and Hill (12) and will not be covered in detail here. Suffice it to say that this group determined radii of gyration for the 30S and 50S subunits (about 70 Å and 75 Å respectively) and presented equivalent ellipsoid models for the two particles. The 30S subunit is best represented as an ellipsoid of revolution having axes of 55 Å by 220 Å by 220 Å and the corresponding ellipsoid for the 50S subunit has dimensions of 115 Å by 230 Å by 230 Å. Their equivalent ellipsoid model for the 30S subunit fit the measured data so well as to lend credence to the idea that the true shape of the subunit must be quite close to that ellipsoid (13). However, it has proved impossible to find an equally satisfying ellipsoid model for the 50S subunit; it cannot be well approximated as a simple solid. Over the years these equivalent ellipsoid shapes for the ribosome have been useful to many investigators as a basis for their thinking about the organization of the particle.

THE PAST FIVE YEARS

What distinguishes the scattering experiments of the past few years from what has gone before is their heavy reliance on neutron radiation instead of x-rays, a trend whose coming was hinted at by a single brief and very preliminary presentation at the Cold Spring Harbor meeting of 1973 (14). Since the principles of data collection and interpretation are identical for x-ray and neutron scattering, and the quality of structural information which results the same for both, one might well ask what the fuss is all about.

The reason for studying an object with different kinds of radiation is that each has a qualitatively different physical interaction with matter. This difference in turn means that the way a particle "looks" will depend on the radiation used to study it. At low resolution, with a nucleoprotein like the ribosome, one might be interested in localizing the protein in nucleic acid parts of the particle. These two substances scatter differently due to differences in atomic composition and structure. Moreover, their relative prominence as scatters is different for x-rays and neutrons. One might expect, therefore, that comparative studies using both kinds of radiation would allow us to say something about the relative disposition of protein and nucleic acid in the ribosome.

There is an additional fact about neutrons which makes them specially interesting. Different isotopes of the same element often scatter neutrons to differing degrees. This means that two chemically identical molecules, differing only in isotopic composition, can sometimes be distinguished in a neutron experiment, a trick impossible with x-rays. It happens that one of the pairs of distinguishable isotopes is the ^1H-^2H pair (15).

This isotopic effect can be exploited in two ways. First, it is possible either by in vivo labelling or by reconstitution to deuterium label specific, chemically identified, parts of the ribosome. Comparisons between the scatter of the labeled particles and their unlabeled counterparts can reveal the location of the deuterated component(s). Second, the scattering observed in a solution scattering experiment is the result of the differences which exist between the macromolecule of interest and the solvent in which it is immersed. One would anticipate, therefore, that the scattering given by a macromolecule will vary if the solvent's scattering properties are altered. As an extreme example, if we place a ribosome in a solvent whose composition is such that its average scatter per unit volume, more properly its "scattering length density", is the same as the ribosome's, no small angle scatter will be seen; the particle will have "disappeared". The changes in the shape of the scattering profile (not its absolute magnitude) induced by varying the solvent reflect the internal inhomogeneities in the structure, e.g. the protein and nucleic acid distributions in the ribosome, whose extent and nature can be deduced therefrom. This method of studying a macromolecule is called "contrast variation" and is based on theoretical developments made in the 1960's (16,17,18,19,20). To get full benefit from such experiments it is useful to be able to vary the scattering length density of the solvent from a value below that of the lowest density region of the macromolecule to a value greater than that of the highest region.

Contrast variation experiments can be done using x-rays, but in order to significantly alter the scattering properties of the solvent, usually water, one must add electron-dense, small molecule solutes often in high concentration. High concentrations of small molecule solutes can perturb macromolecular structure, and in addition, it is not practical to make a water solution electron dense enough to match RNA in x-ray scattering properties, which is a severe limitation when dealing with nucleoproteins. In the neutron case, solvent scattering properties can be altered by the substitution of H_2O by D_2O its ^2H analogue, an isotopic substitution which is rather benign in its effects on macromolecular structure, and the scattering length density range covered by H_2O/D_2O mixtures includes both nucleic acids and proteins. Thus when experimental facilities suitable for neutron small angle work became available in the late 1960's and early 1970's the enthusiasm for using them to study ribosomes and other nucleoproteins was considerable, because a whole class of small angle experiments, some previously impossible to carry out, could be undertaken with ease.

The Shape of the Ribosome

In a typical contrast variation experiment scattering data are collected over a wide range of angles on samples of the molecule in question in a series of solvents which differ in scattering length density. In general a systematic change in the shape of the scattering profile will be observed as a function of the scattering length density of the solvent. It is possible to extrapolate from the measured trend and construct a scattering profile which corresponds to that which would be observed if the difference between the particle's average scattering length density and that of the solvent were infinite. This extrapolated profile is interesting because at infinite contrast, the fluctuations in scattering length density inside the particle due to its chemical inhomogeneities, which are finite, become inconsequential. Thus the extrapolated profile is that which would be given by a particle whose shape is exactly the same as the one of interest but which has no internal structure whatever.

Extrapolated profiles are extremely useful in studies of macromolecular shape. In a typical molecular shape study based on solution scattering results what one tries to do is to find the shape with no internal structure (i.e., the uniform solid) whose calculated scattering curve best fits the measured data. In the pre-contrast variation era it was always tacitly assumed that at low scattering angles the shape of the molecule dominated the observed scatter and that contributions due to its internal features could be ignored. This assumption is not rigorously correct, and to the extent that internal structure did contribute to the small angle scatter, the uniform shapes arrived at as interpretations of the data were in error. [The equivalent ellipsoid models for the ribosomal subunits (11) resulted from model building exercises of this kind.] Shape analyses carried out on extrapolated scattering profiles are not liable to errors of this kind.

Over the past several years a sophisticated technique for deriving shapes from extrapolated scattering curves has been developed by Stuhrmann and his collaborators (21). Figure 1 shows the results of the application of their technique to neutron data on the 50S ribosomal subunit. The 50S subunit based on its scattering profile, appears to be a (crudely) hemispheric object whose cut face is distinctly trilobed. Those familiar with the electron microscopic appearance of the 50S subunit in negative stain will be struck by the similarity of those images (22,23) with the model shown in Figure 1. The similarity of the two kinds of images gives us assurance that electron microscopic techniques are fairly representing the particle in solution. Given the amount of effort that has been put into interpreting electron micrographs of ribosomes in the past several years this assurance is welcome.

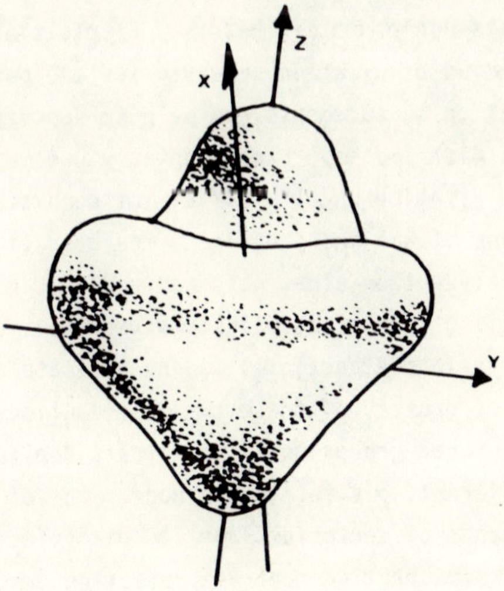

Figure 1. A view of the 50S subunit as determined by solution scattering. This view of the model of the 50S subunit deduced by Stuhrmann et al. (reproduced from (21)) from scattering data shows a view which resembles the "chair" shape which is characteristic of negative stained images of 50S subunits.

Two limitations of this analysis used should be pointed out. First, due to hydrogen exchange between the ribosome and its surrounding solvent, the extrapolated neutron scattering profile which was the basis for the model does not correspond precisely to the profile of a scatterer lacking on internal structure. Thus the separation of shape scatter from internal structûre scatter, while good by previous standards, is not perfect. Second, it is not possible to specify the precision of the model, i.e., one has only a rough idea of how large and varied the class of shapes might be which would fit the data equally well. What is clear, however, is that the shape shown is consistent with the scattering data and that other, simpler shapes such as spheres or ellipsoids of revolution do not fit the data as well. Thus, strictly speaking, what can be concluded is not that the solution scattering proves that the electron microscopic appearance is the correct representation of the subunit but rather that the electron microscopic appearance of the 50S subunit is consistent with the scattering data.

The Distribution of Protein and RNA in the 50S Ribosomal Subunit

The ribosome has marked internal fluctuations in scattering length density because of its nucleoprotein nature. Fluctuations of this kind contribute to the radius of gyration observed for the particle. Accordingly radius of gyration measurements can help to specify how the two chemical species are distributed. Experimentally one must measure the particle's radius of gyration under a variety of conditions in which the relative contributions of RNA and protein to the overall scatter vary significantly and analyze the values obtained. What can be gained are estimates of the radii of gyration of the distributions of the two macromolecular species in the particle, and an estimate of the distance between the centers of gravity of those two distributions.

In the 1974-1975, three groups published papers dealing with this problem by three different, but related methods. One of the groups had carried out measurements of deuterium labelled particles (24), the second had completed some neutron contrast variation experiments (25) and the third reported an experiment in which radius of gyration was measured using different kinds of radiation (26,27,28). All came to the conclusion that there exists a very large difference in the way protein and RNA are arranged in space in the 50S subunit; the subunit is nowhere near a homogeneous mixture of the two species. The existence of these differences in distribution imply that there must be regions, large compared to the overall volume of the particle, containing almost nothing but RNA and correspondingly large regions which are nearly pure protein. On two important points, however, there was disagreement. The second and third groups concluded that a very large difference exists in the radius of gyration of the RNA and protein distributions and almost no separation between their centers of mass, while the first suggested a polar ribosome with an RNA-rich end and a protein-rich end, but little difference in radius of gyration between the two distributions. On the other hand, the first two groups agreed on the absolute radius of gyration of the whole 50S particle under a number of conditions while the third found a significantly higher value, a most unsatisfactory state of affairs.

The conflicts in this area have been partially resolved by a collaborative experiment in which contrast variation measurements were done on deuterium labeled particles (29). This experiment revealed some problems with the assumptions used in interpreting the original deuterium labelling data and resolved the differences between the position presented by Moore et al. (24) and that of Stuhrmann et al. (25). Thus there is now general agreement that the protein and RNA distributions in the 50S subunit have roughly coincident centers of mass, the separation is less than 15 Å, but that the protein part of the particle has a much larger radius of gyration than the RNA part (90-100 Å vs 55-60 Å). The roughly 5 Å difference in the absolute magnitude of the radius of gyration of the 50S particles prepared by the Euro-American group and the Russian workers remains a mystery. Because a number of measurements were made on the Russian material using Western instruments there is no doubt the difference is a real one; preparative problems may be involved. A collaborative effort should be undertaken to sort out this issue.

Protein and RNA in the 30S Ribosomal Subunit

Radius of gyration measurements analogous to those done on the 50S subunit have been carried out on the 30S subunit. Happily these data have led to much less controversy. In the case of the 30S subunit the distributions of protein and RNA are much closer to identical in size and shape. It would be reasonable to think of the subunit as being characterized by interdigitating distributions of proteins and RNA which are nearly coextensive with the boundaries of the particle itself. There is general agreement that the distributions of the two macromolecular species are roughly concentric and that there is a tendency of the surface of the particle to be protein-rich while the center tends to be protein-poor (30,31,32). The radius of gyration of the protein distribution is about 78 Å while that of the RNA distribution is about 61 Å.

The Significance of the Radius of Gyration Values

While knowledge of radius of gyration values for protein and RNA in the two subunits by no means permits the specification of a unique model for the distribution of the two species in the ribosome, it is useful to consider some crude models to see what the measured data imply. As a model for a highly segregated nucleoprotein we might take a spherical virus with its nucleic acid core and surrounding protein shell. Such structures have concentric mass distributions, low RNA radii of gyration and high protein radii of gyration. [It should be noted that spherical virus-like models have the smallest possible difference in protein and RNA radius of gyration consistent with the requirement that the two distributions be concentric, but completely segregated.] Modeling both subunits in this way we can readily calculate radii of gyration for both subunits as a whole and for the two parts of each, the exact values obtained depend on assumptions made about hydration. What is rapidly discovered is: 1) that the difference in radius of gyration between RNA and protein in the 50S subunit is <u>larger</u> than can be generated by any reasonable virus-like model and 2) that the corresponding difference for the 30S subunit is <u>smaller</u> than is found for RNA core-protein shell models. The second finding is decisive. It means that the protein and RNA distributions in the 30S subunit must interpenetrate significantly, as already suggested. The first conclusion may seem impossible on first consideration, but in fact is not. Were the 50S subunit possessed of a compact RNA core with lumps of protein at its north and south poles, a very large radius of gyration for the protein distribution would result accompanied by a small radius of gyration for the RNA, and no separation of centers of mass. In fact bipolar structures with moderate interpenetration of the RNA and protein at the two ends can be readily accomodated by the observed data. Thus the 50S values are neither impossible nor terribly restrictive.

The 70S Couple

Stuhrmann, Koch and their colleagues have recently published the results of some neutron scattering experiments on the 70S ribosome which constitute the best data we have on that particle. Two series of experiments have been done: 1) a contrast variation experiment on fully protonated particles (32), and 2) a contrast variation experiment on 70S particles reconstituted from deuterated and protonated subunits (33). Both investigations led to the same conclusions about the organization of the 70S particle. The radius of gyration of the 70S particle at infinite contrast is around 83 Å ($\pm$ 2 Å). In the 70S couple, as a whole, the protein and RNA distributions are concentric. The radius of gyration of the protein parts is about 98 Å while that of the RNA parts is in the neighborhood of 70 Å. The difference in radii for the two parts of the 70S particle is intermediate between the differences seen for the two subunits separately and is almost exactly that one would predict if the formation of 70S couples occurred with no disturbance of the RNA and protein distributions found in the isolated subunits. Assuming the subunits are constant in configuration, as the data suggest, one calculates a value for the separation of their centers of mass in the 70S particle of 76 ± 15 Å (32). Using couples composed of both protonated and deuterated subunits it is possible to measure the radii of gyration of the individual subunits as they exist in the 70S couple directly, by contrast matching. The values found in the couple are within 1 Å of the values found for free subunits under similar contrast conditions, consistent with the results on unlabelled particles. Making no assumptions about subunit radius of gyration, an estimate for their separation in the partially deuterated couples is obtained of 88 ± 15 Å, a value within error that found for the fully protonated 70S couple (33).

Besides what they tell us about the 70S particle, these results are an important commentary on the likely magnitude of the changes in ribosome configuration which can take place under physiological ionic conditions. The free subunit data with which the 70S results are compared were all obtained in low magnesium buffers, whereas the 70S data were measured in 10 mM Mg^{++}. In addition the subunits in the 70S experiments, of course, were coupled to each other instead of free. These differences in conditions are highly significant in terms of the chemical and functional properties of the subunits. Yet no change is seen in any of the radius of gyration parameters within the accuracy of the data.

The Effect of Protein Factors on Ribosome Configuration

Among the factors which might induce conformational changes in the ribosome one would certainly have to include the various macromolecules which interact with it during protein synthesis. The effect of two of these, IF-3 and ribosomal protein S1 have been investigated by small angle techniques. (The status of S1 as a factor can be debated; the weakness of its binding to the subunit is incontestable (34).)

In the case of IF-3 the first report was a claim that the radius of gyration of the 30S subunit diminishes dramatically upon IF-3 binding (35), which would have been an extremely significant finding except for one fact; no one has been able to reproduce the result. Beaudry et al. (31) carried out a neutron radius of gyration study which demonstrated no change and x-ray experiments have been done with the same outcome

(36). It appears that the original result was spurious.

The radius of gyration of the 30S subunit has been examined in the presence and absence of bound S1 (37). Both x-ray and neutron scattering measurements have been made. It was possible in the latter type of experiment, using partially deuterated S1, to measure the scatter given by the 30S subunit in the presence of S1 under conditions where S1 made no contribution to the overall scatter. The conclusions drawn from these studies were two: 1) that the binding of S1 does not alter the radius of gyration of the rest of the 30S subunit and 2) that S1's binding site on the subunit must be far from the center of mass of the particle.

Tentative Conclusions on Ribosomal Conformational Changes

It is clear from chemical modification data that the interaction of protein factors with the ribosome and the interaction of the two ribosomal subunits in forming the 70S couple are accompanied by distinctive alterations in chemical reactivity, as well as in spectroscopic properties, (9,10,38,39,40,41,42) which show that conformational changes are occurring. On the other hand, the scattering data summarized above, which bear on these conformational changes, are all negative. The fact that the scattering results have been negative, however, should not be interpreted as contradicting the chemical evidence. What it tells us is that the changes we see as <u>large</u> alterations in chemical reactivity correspond to <u>small</u> alterations in the geometry of the particle. This trend in the results in turn suggests that the average ribosome, about whose shape and structure we are slowly gaining information by scattering techniques, in fact may be a fair representation of the ribosome in all of its normal allowed configurations.

Secondary and Tertiary Structure in the Ribosome

Our knowledge about the details of the internal architecture of the ribosome is rudimentary to say the least. Most of what one might offer for suggestions as to how the ribosome is structured in detail would be based more on analogy with other, better known nucleic acid and protein structures than on any hard data. A few hints, however, have been gleaned from scattering studies over the years. Those who have examined the x-ray scattering patterns given by ribosomes in solution and in gels have consistently called attention to shoulders and peaks in the intermediate angular range of the pattern having Bragg spacings near 45 Å, 30 Å and 25 Å (12,43). They are a conspicuous feature of the scattering from ribosomal gels; in dilute solution they are weak shoulders on the central maximum due to the shape scatter of the particle. Most have interpreted these features as being due to regions in the ribosome with regular or near regular, side by side packing of RNA helices. The possibility has always existed, however, that some aspect of the overall shape of the particle might give rise to shoulders of this kind in the scattering profile. Recent experiments done on subunits, particles partially stripped of proteins and ribosomes from eukaryotic organisms, all of which show these features, but which differ in shape and protein content make it reasonable to conclude that it is the packing of the RNA part of the ribosome which gives rise to them (26,28,44).

Internal Structure: Positions of Proteins in the 30S Subunit

One of the more interesting aspects of the scattering work done on the ribosome recently has been the progress made in determining the locations of individual protein molecules within the ribosome by neutron scattering. The possibility of making such measurements was pointed out in 1972 (45) but the first data did not appear until 1975 (46).

Experiments of this kind exploit the fact that the ribosome can be reconstituted. This capability makes it possible to construct particles with specific macromolecular components inserted in heavily deuterated form. Particles which contain two such deuterium labelled components have solution scattering profiles which include a strong interference contribution, absent from the profiles of unlabelled particles. These contributions, which are measurable, reveal the distribution of the lengths of the vectors joining deuterium atoms in the two labelled proteins. The average vector length estimated from such data is closely related to the distance between the centers of mass of the labelled proteins and triangulation can be performed on such distances to produce a three dimensional map of subunit positions in an aggregate like the ribosome. Some information about protein shape and orientation is available in the data as well. [For technical details see (47) and (48).]

With the exception of one suggestive, but inconclusive, feasibility study on a protein pair in the 50S subunit (49) all the data obtained so far relates to the 30S subunit. Table 1 gives second moments (M_{ij}) and first moments ($\bar{r}$) of the length distributions measured for 36 protein pairs. The first moment of a length distribution is the average vector length. The second moment is the average value of the vector lengths

TABLE 1. A Numerical Summary of the Scattering Data on 30S Protein Positions

Pair	M_{ij} (Å)^2	σ_{ij}	$\bar{r}$ (Å)	$\sigma_{\bar{r}}$
S2-S5	14850	4225	120.0	20.5
S3-S4	5511	344	70.8	3.1
S3-S5	4862	270	66.9	3.0
S3-S7	12620	901	107.1	6.4
S3-S8	8342	493	87.3	4.1
S3-S9	3542	241	56.5	2.7
S3-S10	5395	559	65.2	4.8
S3-S11	15000	954	118.3	1.4
S3-S12	7815	1067	86.0	8.7
S4-S5	2870	253	50.9	3.0
S4-S7	13795	709	111.3	4.7
S4-S8	6644	362	77.5	3.0
S4-S9	10149	428	97.2	3.6
S4-S10	9063	474	93.3	4.2
S4-S11	14981	1448	119.4	9.7
S4-S12	2764	200	51.6	2.5
S4-S15	9173	927	95.2	8.5
S5-S6	11099	2412	103.7	16.3
S5-S7	13233	1364	112.7	12.3
S5-S8	1534	697	37.3	9.8
S5-S9	9093	1072	94.4	10.2
S5-S10	12840	1399	115.3	9.4
S5-S11	8868	800	92.2	7.1
S5-S12	3367	178	55.1	2.0
S5-S15	4702	283	67.9	4.4
S6-S11	3633	173	58.2	1.9
S6-S15	2647	97	48.6	1.3
S7-S8	13670	645	114.0	4.8
S7-S9	1536	119	37.0	1.7
S7-S12	12091	1410	109.3	11.5
S8-S9	12998	1988	112.6	17.7
S8-S11	12362	1276	105.3	9.0
S8-S12	3230	222	53.2	3.4
S8-S15	2393	240	47.7	3.5
S9-S10	1959	176	42.7	2.4
S11-S12	13902	1118	114.2	7.5

Each pair distance measurement provides information about the distribution of the lengths of the vectors between two proteins in the 30S subunit. The measured data permit estimates to be made of M_{ij}, the second moment of the length distribution, which is simply the average value of the squares of the length of all possible interprotein vectors, and $\bar{r}$, the average vector length. σ_{ij} is the error in the estimate of M_{ij} and $\sigma_{\bar{r}}$ the error in the estimate of r. These numerical values summarize data which has been or will be reported elsewhere (46,52,53,54, 55).

squared. While the significance of first moments is clear it may seem strange that second moments get equal billing in Table I. The reason they do is that $M_{ij} = R_i^2 + R_j^2 + d_{ij}^2$ where R_i and R_j are the radii of gyration of the two labelled proteins in situ and d_{ij}, the distance between their centers of mass (50,51) and it is upon this relationship that mapping is based.

It should be emphasized that accompanying the increase in the amount of experimental data collected has been a corresponding development of the theory for interpreting those data. When work began it appeared it would be necessary to make assumptions about protein shapes and orientations in order to extract estimates of the distance between a pair of proteins from the measured data. These assumptions had the potential for generating serious errors in interpretation if the labelled proteins examined had strange shapes, e.g. intertwined or were exceptionally elongated or irregular. Further theoretical developments have shown that in fact no assumptions whatever need to be made in order to obtain an accurate three dimensional map from data of this kind (48,50). Thus the three dimensional map of the 30S subunit which is currently emerging will ultimately give protein positions whose accuracy is limited not by the quality of the assumptions used, but only by the quality of the data.

An Eleven Protein Map of the 30S Subunit

There are 29 data sets relating proteins S3, S4, S5, S7, S8, S9, S10, S11 and S12 listed in Table 1, more than sufficient to place these 9 components in three dimensions. In addition there are 3 data sets relating S15 to proteins in this set of 9 as well as 2 such data sets for S6. The distance between S6 and S15 has been measured and preliminary data for S6-S8 (~ 100 Å) and for S11-S15 (~ 95 Å) are also available. This information, while not quite enough to position S15 and S6 uniquely, is sufficiently restrictive to permit us to suggest tentative positions for these proteins also. Figure 2 is a stereo pair of a characteristic view of the spatial arrangement of proteins specified by the data. Proteins are represented for convenience as spheres whose volume corresponds to their anhydrous volume, to the scale of the drawing. [The data by no means imply that they are in fact spherical!] The shaded proteins are the nine we can position with certainty while S6 and S15 are shown unshaded to indicate their less secure status. Table 2 lists the coordinates of the centers of mass positions of the central group of 9 proteins in this map and the error estimates for those coordinates (52). Rough coordinates for S6 and S15 are, in Å, (77.2, 90.0, -58.0) and (79.6, 91.2, -6.3) respectively. S15's position is quite well determined; the errors for its coordinates are around 10 Å. S6's z coordinate is equally well determined, but its x and y coordinates have huge errors. Thus S6 is constrained with respect to its

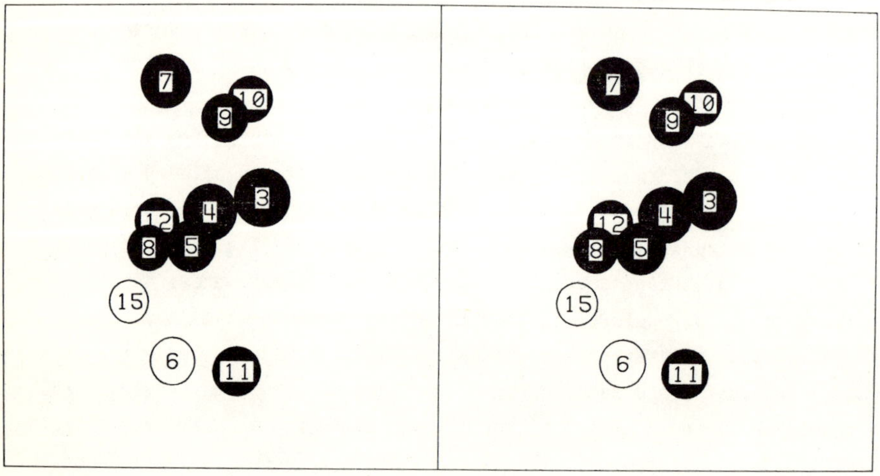

Figure 2. A stereoscopic pair showing the arrangement of proteins in the 30S subunit deduced from the neutron data.

TABLE 2. Coordinates of Components in the 30S Subunit

Component	x	σ_x	[Å] y	σ_y	z	σ_z
S3	0	—	0	—	0	—
S4	68.3	5.3	0	—	0	—
S5	40.7	6.5	45.4	5.8	0	—
S7	17.6	10.3	0.8	14.8	102.1	3.8
S8	48.1	7.4	71.4	4.4	18.9	8.9
S9	-5.3	9.2	-8.8	13.2	59.8	6.5
S10	6.8	10.0	-41.8	14.2	53.4	14.2
S11	42.2	13.0	72.3	14.1	-85.8	8.8
S12	79.3	6.1	33.9	6.9	19.3	7.1

Coordinates were obtained by triangulation and least squares refinement using the second moments listed in Table 1 which relate the 9 components in question. The technique used for computing these positions and their errors has been described elsewhere (48,52).

vertical position on the S7-S11 axis, but is quite unconstrained in the plane perpendicular to that axis. The S7 to S11 distance, the longest distance in the map, is about 200 Å. These eleven components account for about half the total protein in the 30S subunit. The neutron map is quite flat. Viewing it so that the S3, S7 and S11 plane is edge on, it about 70 Å thick, consistent with equivalent ellipsoid model deduced from x-ray data.

Comparisons of the Neutron Map with other Data

The neutron map is informative enough to make comparison with other data on ribosome structure and function interesting. A few of the more important points which have arisen in our own comparisons will be mentioned below. A more complete exposition will appear elsewhere (52).

First, of the two maps which exist showing the positions of antigenic determinants, the current neutron map appears to favor the UCLA version (see (56) for review) over the Berlin version (see (57) for review). This conclusion comes from computer graphics exercises in which we have attempted to fit the neutron map into the two electron microscopic maps bringing protein centers of gravity as close as possible to corresponding antigenic determinants. Figure 3 is a typical outcome of such a hand fitting exercise with the UCLA map. It will be noted that a good degree of correspondence is obtained in this case.

Figure 3. A stereoscopic view of the neutron map superimposed on the UCLA electron microscopic map. The superposition was carried out by hand and is an attempt to minimize the discrepancy between the positions of protein centers and antigenic determinants by eye. Antigenic sites are denoted by uncircled numbers. Neutron positions are designated by numbers enclosed by spheres. The spheres have volumes corresponding to anhydrous protein molecular weights to the scale of the figure. The contour to contour distance is 20 A. Note that there are antigenic determinants for S9 and S7 obscured by the neutron S7 and that the neutron S11 obscures determinants for S6 and S15.

An important restriction in these fitting exercises is that the neutron map fit within the outline of the electron microscopic shapes. It is our inability to find orientations of the neutron map which are consistant both with the Berlin antigenic determinant locations and also lie within their shape on which our preference for the UCLA model rests. It must be emphasized that as further neutron data accumulate and new proteins are added to our model this preference could change.

Second, three distances have been measured both by neutron scattering and by fluorescence energy transfer (58). Two of the protein pairs in question, S7-S9 and S8-S15, involve near neighbors. They were found to have separations of 30 to 40 Å by both techniques. The third pair, S4-S15, was measured at around 65 Å by fluorescence, but appears to have a separate near 90 Å by neutron scattering. [The length distribution found by neutron scattering runs from about 50 Å to 120 Å.] Thus fluorescence methods are certainly able to give a useful indication of protein separations.

Third, the conclusion Huang et al. (58) drew from their fluorescence distances with respect to protein proximity and interaction in assembly has been extended by the neutron data. Proteins which interact strongly in reconstitution are usually close together in the mature particle. The reverse is not true, however. Some proteins which are near neighbors in the mature particle do not interact strongly in assembly.

Fourth, protein pairs which crosslink in high yield are near neighbors by neutron criteria as well. Consistency with the crosslinking data is not complete, however. There are reports of crosslinked pairs in the literature where the proteins in question do not appear to come close enough together to crosslink by neutron criteria. These conflicts will have to be resolved by further experimentation.

Finally, the neutron map is beginning to make some sense functionally. S12, S5 and S4 mutants in which interact in governing the particle's response to streptomycin are near neighbors (59). It is likely that the interactions seen between these proteins genetically reflect their proximity.

The Limitations of the Map

Several points should be born in mind in thinking about this map. First, like all scattering results it represents the average particle in the population measured. Second, data were collected in low magnesium buffer. (A single experiment done to test the effect of magnesium on intrasubunit distances showed no change.) Third, placement of one of the proteins in the group of 9, S11 leaves something to be desired. Because of the near planar arrangement of the proteins whose relationships to S11 were used to place it, there are two positions which can be entertained for S11, the one shown for it in Figure 2 and a second with S11 about 35 Å from S7. S6's tentative position is strongly dependent on S11's location. Thus a shift in the location of the latter would affect our thinking about the position of S6 as well. The position of S11 in Figure 2 is significantly more probable than the one close to S7, but its superiority is not so great as to exclude the alternative altogether, which raises a fourth and final point. The map is the product of a refinement procedure designed to minimize discrepancies between the positions of the proteins and the measured data. Thus as new data are collected and improvements made in our methods of analysis, the positions of all proteins are subject to adjustment. This means that this map must be regarded as provisional; it is a statement of what we are able to do within the limits of the data and methods of analysis currently available.

PROSPECTS FOR THE FUTURE

Small angle techniques have contributed to our understanding of ribosomal structure in four areas in the past few years. First, they have given rise to a substantial increase in our knowledge of the overall size and shape of the 70S particle and its subunits in solution. These advances have depended on the application of new kinds of analysis to data collected in novel ways. Nevertheless they can be regarded as logical extensions of the work Hill et al. (11) were doing in the late 1960's. Second, solution scattering techniques have provided us with virtually everything we currently know about the global distribution of RNA and protein in the ribosome and have supplied some hints about the manner in which these substances pack in situ. Third, the data we have indicates that the conformational changes which accompany protein synthesis are likely to be subtle. No situation examined so far has given scattering data suggesting a measurable change in the radius of gyration of the ribosome or any appreciable internal reorganization. The conformational change seen in hemoglobin as it passes from the deoxy- to oxy-state may be the sort of alteration which we should expect in the ribosome, i.e., something modest. Fourth, scattering experiments have begun to provide us with explicit information about the location of components in intact ribosome.

It is easy to anticipate some of the directions likely to be explored in the near future. Clearly a number of lines pursued in recent years have not been completed. We need shapes for both the 30S subunit and the 70S couple. The study of conformational variations should be continued, fruitless as it has proven so far, simply because of our need to understand what these changes amount to. Obviously, the 30S protein map must be completed and a comparable effort undertaken on the 50S subunit. Finally, it is hardly necessary to point out the need for more scattering studies on eucaryotic ribosomes. All we have so far are a few radii of gyration (5). Many of the experiments done on ribosomes from E. coli would be easy to repeat on eucaryotic particles. They should be done if for no other reason than to establish whether the picture of the ribosome which emerging from procaryotic studies has general significance.

The mapping studies done so far only on the 30S subunit have proven the validity of scattering methods as a means for locating protein positions. Theory indicates that much more than position alone can be extracted from the data. Certainly radii of gyration for the individual proteins in the ribosome can be obtained. In addition there is information in the data about protein shape and orientation which we do not yet know how to extract. Progress in this area may be hoped for. It is also likely that this scattering method will suffice both to determine the positions of factors and tRNA on the ribosome and to measure the influence of ionic condition, couple formation and ligand binding on the organization of the particle. Thus the full exploitation of isotopic labelling techniques should lead to a picture of the ribosome which is far more than a set of positions.

Looking at the ribosome field from the standpoint of an investigator who has been involved both as a biochemist and as an advocate of scattering studies, one can find cause for unease in the past record. All too often the scattering fraternity has provided biochemists and geneticists with data irrelevant to the problems on which they were working. The result has been that scattering data have had an impact on ribosome research which was small compared to the effort expended to get them. This "information gap" is closing rapidly at present. The data obtained in the past few years have enabled us to begin a useful dialogue with these using other approaches and the prospects for productive interactions in the future are excellent.

ACKNOWLEDGEMENTS

A large number of people have contributed to the neutron distance measurements described in this review: Drs. D. M. Engelman, B. P. Schoenborn, J. A. Langer, D. G. Schindler and V. Ramakrishnan. Without their efforts there would have been nothing to report. This work has been supported by grants from the National Institutes of Health (AI-09167) and the National Science Foundation (PCM78-10361). All but two of the data sets reported in Table 1 were collected at Brookhaven National Laboratory under the auspices of the Department of Energy. The remaining two were obtained at the Institut Laue-Langevin.

REFERENCES

1. Clark, M., Hammonds, M., Langer, J., Eisenberg, D., and Lake, J., personal communication.

2. Jacrot, B. (1976) Rev. Prog. Phys. 39, 911-953.

3. Kratky, O. and Pilz, I. (1978) Quarterly Rev. Biophys. 11, 39-70.

4. Koch, M. H. J. and Stuhrmann, H. B. (1979) Methods in Enzymology 59, Moldave, K. and Grossman, L., eds., pp. 670-706, Academic Press, New York.

5. Damaschun, G., Müller, J. J., and Bielka, H. (1979) Methods in Enzymology 59, Moldave, K. and Grossman, L., eds., pp. 706-750, Academic Press, New York.

6. Voynow, P. and Kurland, C. G. (1971) Biochemistry 10, 517-524.

7. Lam, M. K. T., Changchien, L.-M., and Craven, G. R. (1979) J. Mol. Biol. 128, 561-576.

8. Zamir, A., Miskin, R., and Elson, D. (1971) J. Mol. Biol. 60, 347-364.

9. Ginzburg, I., Miskin, R., and Zamir, A. (1973) J. Mol. Biol. 79, 481-494.

10. Ginzburg, I. and Zamir, A. (1975) J. Mol. Biol. 93, 465-476.

11. Hill, W. E., Thompson, J. D., and Anderegg, J. W. (1969) J. Mol. Biol. 44, 89-102.

12. Van Holde, K. E. and Hill, W. E. (1974) Ribosomes, Nomura, M., Tissieres, A., and Lengyel, P., eds., pp. 53-92, Cold Spring Harbor Laboratory, Cold Spring Harbor, New York.

13. Hill, W. E. and Fessenden, R. J. (1974) J. Mol. Biol. 90, 719-726.

14. Moore, P. B., Engelman, D. M., and Schoenborn, B. P. (1974) Ribosomes, Nomura, M., Lengyel, P., and Tissieres, A., eds., pp. 601-613, Cold Spring Harbor Laboratory, Cold Spring Harbor, New York.

15. Bacon, G. E. (1975) Neutron Diffraction, 3rd edition, Oxford University Press.

16. Benoit, H. and Wippler, C. (1960) J. Chim. Phys. 57, 524-527.

17. Stuhrmann, H. B. and Kirste, R. G. (1965) Z. Phys. Chem., Frankfurt 46, 247-250.

18. Stuhrmann, H. B. and Kirste, R. G. (1967) Z. Phys. Chem., Frankfurt 56, 334-337.

19. Cotton, J. P. and Benoit, H. (1975) J. Phys., Paris 36, 905-910.

20. Luzzati, V., Tardieu, A., Mateu, L., and Stuhrmann, H. B. (1976) J. Mol. Biol. 101, 115-127.

21. Stuhrmann, H. B., Koch, M. H. J., Parfait, R., Haas, J., Ibel, K., and Crichton, R. R. (1977) Proc. Nat. Acad. Sci., USA 74, 2316-2320.

22. Tischendorf, G. W., Zeichhardt, H., and Stöffler, G. (1975) Proc. Nat. Acad. Sci., USA 72, 4820-4824.

23. Lake, J. A. (1976) J. Mol. Biol. 105, 131-160.

24. Moore, P. B., Engelman, D. M., and Schoenborn, B. P. (1974) Proc. Nat. Acad. Sci., USA 71, 172-176.

25. Sturhmann, H. B., Haas, J., Ibel, K., DeWolf, B., Koch, M. H. J., Parfait, R., and Crichton, R. R. (1976) Proc. Nat. Acad. Sci., USA 73, 2379-2383.

26. Serkyuk, I. N. and Grenader, A. K. (1975) FEBS Letters 59, 133-136.

27. Serdyuk, I. N. (1975) Brookhaven Symp. Biol. 27, IV-49-IV-60.

28. Serdyuk, I. N. (1979) Methods in Enzymology 59, Moldave, K. and Grossman, L., eds., pp. 750-775, Academic Press, New York.

29. Crichton, R. R., Engelman, D. M., Haas, J., Koch, M. H. J., Moore, P. B., Parfait, R., and Stuhrmann, H. B. (1977) Proc. Nat. Acad. Sci., USA 74, 5547-5550.

30. Moore, P. B., Engelman, D. M., and Schoenborn, B. P. (1975) J. Mol. Biol. 91, 101-120.

31. Beaudry, P., Petersen, H. V., Grunberg-Manago, M., and Jacrot, B. (1976) Biochem. Biophys. Res. Commun. 72, 391-397.

32. Stuhrmann, H. B., Koch, M. H. J., Parfait, R., Haas, J., Ibel, K., and Crichton, R. R. (1978) J. Mol. Biol. 119, 203-212.

33. Koch, M. H. J., Parfait, R., Haas, J., Crichton, R. R., and Sturhmann, H. B. (1978) Biophys. Struct. Mechanism 4, 251-262.

34. Laughrea, M. and Moore, P. B. (1977) J. Mol. Biol. 112, 399-421.

35. Paradies, H. H., Franz, A., Pon, C. L., and Gualerzi, C. (1974) Biochem. Biophys. Res. Comm. 59, 600-607.

36. Moore, P. B. and Wahba, A. J., unpublished results.

37. Laughrea, M., Engelman, D. M., and Moore, P. B. (1978) Eur. J. Biochem. 85, 529-534.

38. Ghosh, N. and Moore, P. B. (1979) Eur. J. Biochem. 93, 147-156.

39. Ewald, R., Pon, C., and Gualerzi, C. (1976) Biochemistry 15, 4786-4791.

40. Michalski, C. J., Sells, B. H., Chowdhury, R. M., and Wahba, A. J. (1978) Eur. J. Biochem. 89, 589-594.

41. Hogan, J. J. and Noller, H. F. (1978) Biochemistry 17, 587-601.

42. Spitnik-Elson, P., Schechter, N., Abramovitz, R., and Elson, D. (1976) Biochemistry 15, 5246-5253.

43. Dolgov, A. D., Iranov, D. A., Kapitnova, K. A., and Mokulskii, M. A. (1974) Mol. Biol. (USSR) 8, 410-418.

44. Serdyuk, I. N., Grenader, A. K., and Koteliansky, V. E. (1977) Eur. J. Biochem. 79, 505-510.

45. Engelman, D. M. and Moore, P. B. (1972) Proc. Nat. Acad. Sci., USA 69, 1997-1999.

46. Engelman, D. M., Moore, P. B. and Schoenborn, B. P. (1975) Proc. Nat. Acad. Sci., USA 72, 3888-3892; Engelman, D. M., Moore, P. B., and Schoenborn, B. P. (1975) Brookhaven Symp. Biol. 27, IV-20-IV-37.

47. Moore, P. B. and Engelman, D. M. (1979) Methods in Enzymology 59, Moldave, K. and Grossman, L., eds., pp. 629-639, Academic Press, New York.

48. Moore, P. B. and Weinstein, E., J. Appl. Cryst., in press.

49. Hoppe, W., May, P., Stöckel, P., Lorenz, S., Erdmann, V. A., Wittmann, H. G., Crespi, H. L., Katz, J. J., and Ibel, K. (1975) Brookhaven Symp. Biol. 27, IV-38-IV-48.

50. Moore, P. B., Langer, J. A., and Engelman, D. M. (1978) J. Appl. Cryst. 11, 479-482.

51. Stöckel, P., May, R., Strell, I., Cejka, Z., Hoppe, W., Heumann, H., Zillig, W., Crespi, H. L., Katz, J. J., and Ibel, K. (1979) J. Appl. Cryst. 12, 176-185.

52. Schindler, D. G., Langer, J. A., Engelman, D. M., and Moore, P. B., manuscript in preparation.

53. Moore, P. B., Langer, J. A., Schoenborn, B. P., and Engelman, D. M. (1977) J. Mol. Biol. 112, 199-234.

54. Langer, J. A., Engelman, D. M., and Moore, P. B. (1978) J. Mol. Biol. 119, 463-485.

55. Ramakrishnan, V. and Moore, P. B., unpublished results.

56. Lake, J. A. and Kahan, L. (1975) J. Mol. Biol. 99, 631-644; Lake, J. A. (1978) Advanced Techniques in Biological Electron Microsocpy II, Koehler, J. K., ed., pp. 173-211, Springer-Verlag, Berlin.

57. Brimacombe, R., Stöffler, G., and Wittmann, H. G. (1978) Ann. Rev. Biochem. 47, 271-303 (1978).

58. Huang, K. H., Fairclough, R. H., and Cantor, C. R. (1975) J. Mol. Biol. 97, 443-470.

59. Jaskunas, S. R., Nomura, M., and Davies, J. (1974) Ribosomes, Nomura, M., Tissieres, A., and Lengyel, P., eds., pp. 333-368, Cold Spring Harbor Laboratory, Cold Spring Harbor, New York.

INTERACTIONS AMONG PROTEIN AND RNA COMPONENTS OF THE RIBOSOME

ROBERT A. ZIMMERMANN

Department of Biochemistry
University of Massachusetts
Amherst, MA 01003

INTRODUCTION

The capacity of isolated ribosomal components to reassociate spontaneously into active 30S and 50S subunits under appropriate solution conditions has focused attention on the specific protein-RNA interactions that underlie both assembly reactions and the structural integrity of the ribonucleoprotein particles. The binding sites for approximately one-third of the proteins of the 30S subunit have now been identified in the 16S RNA and significant progress has been made on the localization of analogous sites in the 23S RNA to which a dozen or so of the 32 different 50S subunit proteins attach. Interactions of three of the latter proteins with 5S RNA are also under intensive investigation. There is substantial evidence that several more ribosomal proteins associate specifically with ribosomal RNA although, with a few exceptions, their binding sites have not been elucidated. In addition, many of the weak interactions that have been observed are undoubtedly shored up by cooperative association with neighboring components in partially assembled or intact ribosomal particles.

Current perspectives on protein-RNA interactions in the ribosome will be summarized in the following pages. Emphasis will be placed upon the location and properties of protein binding sites in the ribosomal RNA and an attempt will be made to envision possible relationships between nucleic acid conformation and functional activity. The former topic has been greatly illuminated by the availability of an unambiguous primary sequence for the 16S RNA of *Escherichia coli* and the latter has been strongly influenced by knowledge of the complete three-dimensional structure of tRNA. A number of new developments will also receive special attention, including the chemical crosslinking of protein and RNA in ribosomal subunits, the organization of the RNA molecules in the intact subparticles, and the recently initiated investigation of protein-RNA association in the eukaryotic ribosome. This article will nonetheless be based mainly on results obtained with ribosomal components from prokaryotes, in particular *E. coli*, and any departures will be noted explicitly. The last full review of protein-RNA interaction in the ribosome appeared five years ago (1) and, although there have been updates in the intervening years (2-4), there is much new food for thought.

PROTEIN-RNA INTERACTIONS

Methods of Investigation

Most studies of ribosomal protein-rRNA interaction have been carried out under conditions similar to those originally optimized for assembly of the *E. coli* 30S subunit (5). A buffer consisting of 10-50 mM Tris·HCl, pH 7.6, 20 mM $MgCl_2$ and 300-400 mM KCl (TMK) is frequently used and the complexes are generally heated at 30-42° for 15-60 minutes. The two most important ingredients in this procedure are probably Mg^{2+} and temperature, and the reasons for this can be understood, at least in part, through the following considerations. RNA molecules, even those as large as the 16S and 23S RNAs, are remarkable in that they can spontaneously attain a relatively unique and uniform conformation under given solution conditions. The structures most active in binding ribosomal proteins appear to be favored by high Mg^{2+} concentration (6) and, if the configuration of the polynucleotide chain becomes disorganized during isolation, incubation at moderate temperatures will speed its renaturation. Although the proteins may also undergo conformational adjustments during the incubation step (7), the primary effects appear to involve the RNA (6,8). This argument has been largely validated in the case of 5S RNA (9), and will be developed further with respect to 16S RNA in the sections on protein binding sites. Furthermore, while the structure the RNA molecule assumes upon heating in 20 mM Mg^{2+} should probably not be equated with its "native" state, there is reason to believe that the configuration of 16S RNA in these circumstances may be very similar to that in the intact ribosomal particle (10). In any case, if protein and RNA are separately incubated in Mg^{2+}, they will in general combine quite readily at 0° and the complexes themselves are usually stable to substantial Mg^{2+} depletion. The 5S RNA will even interact with its cognate proteins at very low Mg^{2+} concentrations once its active conformation has been "set" by incubation in TMK buffer (11).

A number of techniques appropriate to the analysis of protein-RNA complexes, including sucrose gradient centrifugation, polyacrylamide gel electrophoresis, gel filtration chromatography and methods of quantitation have been thoroughly described elsewhere and require no further discussion here (1,12). An important new approach consists in the use of affinity chromatography on columns containing rRNA covalently attached to agarose (13). When *E. coli* 5S RNA is coupled to agarose, for instance, the matrix quite efficiently ensnares proteins L5, L18 and L25 from unfractionated mixtures of ribosomal proteins, while the immobilization of eukaryotic 5S and 5.8S RNAs in like manner has permitted the identification of RNA-binding proteins from the 80S ribosome (14-16). In another development, variations on the traditional membrane filter assay have been exploited in the measurement of equilibrium constants for interactions of a number of ribosomal proteins with 16S and 5S RNAs (17,18).

Although criteria for evaluating the specificity of protein-RNA interactions have been stressed in the past (1), their importance justifies repetition here in slightly modified form.

1. A given protein should interact with only one kind of RNA in the presence of a mixture containing, for example, 16S, 23S and 5S RNAs.
2. The molar ratio of protein and RNA in a site-specific complex should reach a plateau value that does not surpass 1:1 in the presence of excess protein.
3. Upon limited RNase digestion of the complex, the protein should remain attached to discrete fragments of the RNA that arise from limited regions of the nucleic acid chain.
4. The protein moiety of the complex should be directly incorporated into ribosomal particles without dissociation when incubated with an unfractionated mixture of subunit proteins.

Application of these standards is particularly important in establishing the significance of previously uncharacterized interactions.

Interaction of 30S Subunit Proteins with 16S RNA

There is general agreement that proteins S4, S7, S8, S15, S17 and S20 of the *E. coli* 30S subunit associate with specific binding sites in the 16S RNA (19-24). The specificity of interaction has been confirmed in each case by two or more laboratories according to the criteria listed above and formation of these complexes does not depend strongly on the mode of preparation of the components. Moreover, the sites to which each protein attaches in the 16S RNA molecule have been defined (see below).

The interaction of other proteins with 16S RNA is of course a possibility and, since proteins and RNA are very closely packed in the intact ribosome, intimate contacts between them must be the rule. Failure to observe association between isolated protein and RNA molecules may simply reflect their low affinity for one another under the artificial conditions of the <u>in vitro</u> binding assay. In addition, the particular methods used to extract and purify the ribosomal components may in themselves alter the configuration of the macromolecules in such a way that their ability to interact is diminished. This inference can be illustrated most lucidly by two recent observations. First, 16S RNA extracted from 30S subunits with urea and acetic acid has been reported to associate with S3, S5, S9, S12, S13 and S18 in addition to the established RNA-binding proteins and to possess a lower electrophoretic mobility than 16S prepared by the more conventional phenol/SDS procedure (25-27). Second, a new method of fractionating ribosomal proteins has been developed which exploits their solubility at high salt concentrations but does not entail the prolonged exposure to urea and acetic acid that characterized most previous purification schemes (28). When the RNA-binding properties of the salt-extracted proteins were investigated, it was found that S2, S13 and S19 interacted weakly, and S5 strongly, with 16S RNA obtained by standard methods (29). More interestingly, it was shown that a stable [S13-S19] complex isolated by the new procedure associated specifically with the 16S RNA to form a ribonucleoprotein containing equimolar amounts of the three components (30). Together, these findings underscore the delicate nature of some protein-RNA interactions and indicate that the physical state of the isolated components can be an important factor in the detection of complex formation. Moreover, many of the associations that are perceived as weak when investigated by individual binding assays may in fact be considerably enhanced by cooperative interaction with other proteins in the 30S subunit (1). The newly identified complexes must nevertheless be subjected to further scrutiny and it will be important to ascertain that the proteins involved bind to defined segments of the RNA molecule. However, since S13 has long been known to exhibit an affinity for 16S RNA (19,23), and because binding of S5 has now been described by two groups (27,29), it would seem appropriate to classify them as RNA-binding proteins along with S4, S7, S8, S15, S17 and S20.

Interaction of 50S Subunit Proteins with 23S and 5S RNAs

Of the 32 different proteins of the *E. coli* 50S subunit, L1, L2, L3, L4, L6, L13, L16, L20, L23 and L24 each form specific complexes with the 23S RNA (31,32) while L5, L18 and L25 have been shown to interact specifically with 5S RNA (11,33,34). Here agin, the binding stoichiometry was measured in each case (11,32), and the positions of the interaction sites in the RNA molecules have been investigated (see below). The use of proteins isolated by the high-salt procedure revealed that L11 and L15 can also associate stoichiometrically with the 23S RNA (29), and it has been found that L14 binds to a large subfragment of the 23S RNA, although not to the parent molecule (35). A protein complex consisting of L7/L12 and L10 in a molar ratio of 4:1 also associates specifically with the 23S RNA (30). The suggestion that the interaction is mediated primarily by L10 (36) has been strengthened by evidence indicating that L10 alone is capable of binding to 23S RNA (37).

Complexes of L5, L18 and L25 with the 5S RNA have been the subject of numerous chemical and physicochemical studies in recent years. Among the main findings are the following. 1) The 5S RNA binds stoichiometric amounts of L5, L18 and L25 and therefore contains no more than one specific site of attachment for each of the proteins (11). 2) Association constants for the three interactions range over two orders of magnitude, from about 10^6 M^{-1} for the L5-5S RNA complex to about 10^8 M^{-1} for the L18-5S RNA complex (18). In addition, L18 elicits a strong cooperative stimulation of L5 binding (11). Other cooperative effects among the 5S RNA-binding proteins have been inferred from fluorescence studies (38). 3) Although the 5S RNA does not undergo extensive structural changes when complexed with proteins, L18 does appear to induce a significant alteration in the secondary, and perhaps tertiary, structure of the RNA as measured by changes in its CD spectrum and its ability to bind ethidium bromide (18,38,39). Protein L25 promotes little or no change in the CD spectrum of the 5S RNA, however, but has been found to enhance its thermal stability (18,39; T.L. Marsh and R.A.Z., unpublished). 4) The binding sites for L5, L18 and L25 have been elucidated by enzymatic digestion and chemical modification of 5S RNA and protein-5S RNA complexes (40-42). Several of these topics will be discussed more fully in subsequent sections, as will the properties of a rather similar protein-5S RNA complex from *Halobacterium cutirubrum* (43).

Interspecific Interactions

It has been known for some time that hybrid 30S subunits can be reconstituted with components isolated from different species of bacteria (44, 45). At the very least, these results implied that the sites of mutual interaction in both proteins and RNA must be highly conserved. This view was confirmed with respect to protein-RNA interaction by studies in which RNA-binding protein from *E. coli* were shown to form specific complexes with rRNAs from other members of the enteric-vibrio group of bacteria, as well as from several species of *Bacillus* (22,46).

Only recently, however, has it been appreciated how great an evolutionary gulf is spanned by such heterologous associations. Thus, it has been demonstrated that 30S subunit proteins from *E. coli* bind specifically to the 16S RNAs of prokaryotes as diverse as *Anacystis nidulans*, a cyanobacterium, *Clostridium pasteurianum*, a primitive anaerobe, and *Chromatium vinosum*, a purple sulfur bacterium (47). Even more surprisingly, *E. coli* proteins interact with RNAs from several archaebacteria, a recently classified group of prokaryotes whose ribosomal RNAs exhibit no more sequence homology to those of enteric bacteria than they do to those of eukaryotes (48). The most highly conserved interactions are those involving S8, S15 and L1, suggesting that these proteins, or their antecedents, played a key role in the evolution of ribosome assembly mechanisms. Since the RNAs to which the proteins bind are very dissimilar in primary structure, their binding capacities must be attributed to common features of secondary structure or of tertiary folding (47). Indeed, it has been found that the binding sites for S8 and L1 in the ribosomal RNAs of *Bacillus stearothermophilus* differ in sequence by 30-40% from the analogous regions in *E. coli* (49,50). Nevertheless, both S8-specific RNAs can be arranged in a similar secondary structure, as can the two L1-specific segments.

Protein-RNA Interactions in the Eukaryotic Ribosome

The investigation of interactions among isolated protein and RNA components of the eukaryotic ribosome has only recently begun. Attention has focused primarily on the small eukaryotic rRNAs, and the proteins that interact with 5S and 5.8S RNA from rat liver have now been studied in two laboratories (14-16). In both cases, RNA-binding proteins were isolated on affinity columns consisting of 5S or 5.8S RNA linked to agarose via an adipic acid spacer. When mixtures of ribosomal proteins are passed through such columns in TMK buffer, it is assumed that only those capable of specific association with the RNA are retained, although additional proteins may become bound through protein-protein interaction. Retained proteins can then be eluted and identified by two-dimensional gel electrophoresis. In this way, Metspalu and coworkers found that rat-liver 5S RNA associated mainly with proteins they identified as L6 and L18, while smaller amounts of L7, L8 and L35 were found as well (14). Rat-liver 5.8S RNA also strongly interacted with L6 and L18, but retained significant quantities of L5 and L7. Using a similar approach, Ulbrich and Wool determined that rat-liver 5S RNA bound L6 and L19 along with traces of other proteins (15), and that 5.8S RNA from the same source retained principally L6, L8 and L19 from the 60S subunit and S9 and S13 from the homologous 40S subunit (16). Since L18 and L19 are very difficult to distinguish on the two-dimensional gels used by both groups, it is likely that Metspalu and collaborators mistook L19 for L18 (16). Assuming this to be so, the results from the two laboratories indicate that rat-liver proteins L6 and L19 bind strongly to *both* 5S and 5.8S RNAs, which suggests in turn that the small RNAs are in close proximity, presumably at the peptidyl transferase center of the eukaryotic large subunit (14,16). It is of interest that no *E. coli* ribosomal proteins were retained by the agarose-bound 5.8S RNA from rat liver (16) despite earlier reports that L18 and L25 from the *E. coli* 50S subunit were capable of association with yeast 5.8S RNA (51). While the two 5.8S RNAs are assumed to be similar in structure, the binding experiments were performed under different conditions. This discrepancy must therefore be resolved by directly comparing the capacity of the small RNAs to bind the prokaryotic proteins with the same techniques.

LOCATION AND CHARACTERISTICS OF PROTEIN BINDING SITES ON PROKARYOTIC rRNAs

The distribution and properties of protein binding sites on *E. coli* 16S, 23S and 5S RNAs have been investigated extensively in several laboratories. Binding site locations have been determined mainly by the isolation and analysis of protein-specific RNA fragments from RNase digests of free rRNA, protein-rRNA complexes and partially unfolded ribosomal particles. The different approaches have been discussed previously and will not be described in detail in the present review (1,12). One general conclusion that has emerged from these studies concerns the question of "protection". Certain RNA fragments can be obtained only by digestion of protein-RNA complexes and not of free RNA, implying that the protein confers resistance to RNase hydrolysis upon, or "protects", portions of the RNA that contain its binding site. Nonetheless, a number of homogeneous rRNA fragments have been isolated from digests of free rRNA, which indicates that these molecules are generally quite tightly folded and hence rather inaccessible to nucleolytic enzymes. The existence of compact RNase-resistant domains in rRNA must be attributable to fairly dense networks of secondary and tertiary contacts within the polynucleotide chains. In any case, the presence of a protein binding site in a given rRNA fragment can only be ascertained with confidence by confirming its ability to reassociate specifically with the protein in question (52). Such assays have now become a standard feature of binding-site investigations.

Protein Binding Sites on the 16S RNA

The principal binding sites for proteins S4, S17 and S20 are located within the 5'-proximal 35% of the 16S RNA molecule (Fig. 1). Fragments specific for these proteins have been isolated by a variety of different techniques and those that associate with S4 have been subjected to particularly close scrutiny (23,53-63). All of the S4-specific fragments characterized to date lie between residues 10 and 580 in the 16S RNA sequence (64,65; see Noller, this volume) and most are discontinuous, with numerous "hidden breaks" and excisions ranging from a few, to over 100, nucleotides. The gaps are probably spanned by secondary RNA-RNA interactions that link the stems of the many hairpin loops believed to occur in this region or by tertiary RNA-RNA interactions that join more distant segments of the nucleic acid chain (56-59,61,63). Analysis of one S4-specific fragment by small-angle x-ray scattering suggests that this region can be adequately represented by a compact, oblate ellipsoid with dimensions of 13.2 x 13.2 x 3.2 nm (66). The binding site for S20 has been further localized to residues 90-350 and it was found that a 100-nucleotide segment situated between residues 245 and 350 was particularly well protected from RNase attack by this protein (58). Although S17 also associates with the 5' one-third of the 16S RNA, its binding site has not been further characterized (56,59,63).

The structural bases for protein interaction within the 5' portion of the 16S RNA are not at all clear. Some interesting inferences can nonetheless be drawn from the binding capacities of various fragments, or groups of fragments, isolated from this region. First, segments between residues 245-360 and 515-530 are often missing or recovered in only low yield from many of the S4-specific RNAs (57,58,60,61). A good example of this is provided by "component A" derived by digestion of the S4-16S RNA complex with RNase A. Deproteinized component A, which was found to contain three continuous RNA fragments corresponding to segments 21-244, 364-515 and 531-568, was able to reassociate stoichiometrically with S4, but not with S20 or S17 (57). This finding indicated that the excised regions were dispensable for S4 binding and suggested that the protein makes contact with widely separated portions of the RNA chain (56,57). Further, the main excision, extending from residue 245 to residue 363, corresponds precisely to the sequence most strongly protected by S20, explaining the failure of this protein to interact with component A and confirming the location of its primary binding site. The lack of interaction between S17 and component A may mean either that S17 also associates

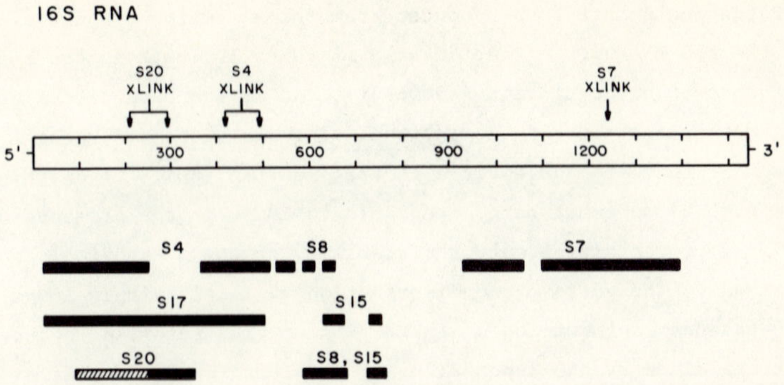

Figure 1. Location of protein binding sites on the 16S RNA. Solid bars represent protein-specific RNA fragments aligned according to their placement in the complete 16S RNA sequence. Cross-hatched segment of the S20-specific RNA may not be necessary for protein-RNA association. See text for details of binding site structure. Positions of UV-induced cross-links, discussed in a later section, are described in refs. 72 and 131. Numbers within the 16S RNA denote distances in nucleotide residues from the 5' terminus of the molecule.

with residues 245-363, or that it simply cannot bind without the cooperative stimulation of S20 even if its site of attachment is present (57,63). A second relevant observation concerns two continuous fragments comprising residues 12-296 and 305-504 that can be isolated from T_1 digests of the 16S RNA. Although neither one is capable of binding S4, S20 or S17 individually, the two RNA segments can reassociate spontaneously to form an "8S complex" that interacts with all three proteins (56,63). This experiment considerably strengthens the evidence for multiple contacts between S4 and its binding site. It is noteworthy in this regard that S4 appears to associate more weakly with the 8S RNA complex than with component A (63). The former lacks the sequence 531-568, which may therefore supply a third point of interaction between S4 and the polynucleotide chain. Finally, the inability of S20 to associate with either component of the 8S RNA complex can be attributed to the fact that neither one of them contains the presumed protein binding site intact. Association of the two RNA fragments, however, apparently restores the integrity of the region between residues 245 and 350 and reconstitutes the structural features required for recognition by this protein (63).

It is clear from the foregoing that the 5'-proximal domain of the 16S RNA is structurally complex. In addition to the intricacies suggested by the protein binding data, the same conclusion can be reached by a consideration of its extraordinary resistance to nuclease digestion. Very similar fragments have been produced from the 5' region by partial hydrolysis of the 16S RNA with either RNase A or RNase T_1, both in the absence and presence of bound protein (23,56-61), and it has been determined that S4 does not protect any of the internal cleavage sites for RNase A within this segment that are susceptible in free 16S RNA (60). If anything, the RNA seems to be somewhat more accessible to RNase in the presence of S4 (60,63). To a certain extent, the stability of the 5'-proximal RNA can be ascribed to the pattern and distribution of local hairpin loops and the spontaneous re-formation of the 8S RNA complex referred to above can even be explained by the reannealing of complementary sequences at the 3' end of fragment 12-296 and the 5' end of fragment 305-504 (63). The discovery that discrete internal segments were almost totally missing from many of the S4-specific fragments, however, led inescapably to the realization that tertiary interactions must occur between RNA segments more widely separated along the polynucleotide chain (57). Direct evidence for tertiary RNA-RNA contacts was provided by experiments in which dissociated fragments from the S4 binding site that were separated by over 100 nucleotides in the primary structure could be shown to recombine when heated in the presence of 20 mM Mg^{2+} (59). Another RNA-RNA interaction was detected by very mild hydrolysis of the S4-16S RNA complex (61). In this case, RNA bound to the proteins was found to contain sequences corresponding to residues 820-900 from the middle of the molecule in addition to the expected 5'-proximal fragments. Since the same RNA segments were found to remain associated after limited digestion of the 16S RNA alone, it seems likely that the interaction of residues 820-900 with the 5' region can be maintained solely by RNA-RNA contacts. Whether or not the 80-nucleotide segment from the central portion of the 16S RNA comprises part of the S4 binding site is still open to question, although a recent electron microscopic study of the S4-16S RNA complex suggests that the protein can at least stabilize contacts among three noncontiguous portions of the RNA which are located at the 5' terminus, between residues 250-600, and in the vicinity of residue 900 (67).

Proteins S8 and S15 bind to neighboring, or perhaps even overlapping, sequences that lie between residues 570 and 760, toward the middle of the 16S RNA (Fig. 1). Various portions of this region have been isolated by partial hydrolysis of S8-, S15-, or S8,S15-16S RNA complexes with RNases A and T_1 (55,56,58,68-70). One of the protected RNAs was found to consist of three main subfragments, spanning residues 571-619, 620-681 and 720-761, for which a stable secondary structure can be predicted with rather

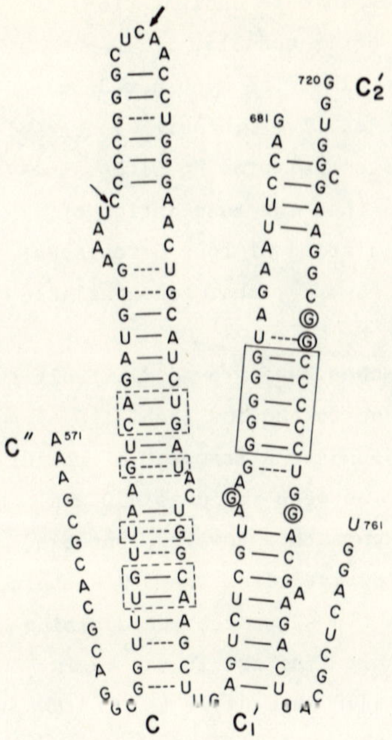

Figure 2. Structure of the S8/S15 binding region. The secondary structure of fragments protected from RNase hydrolysis by S8 and S15 exhibit a number of interesting features, including a "hyphenated" two-fold symmetry in the lefthand stem, a long run of G:C base pairs in the righthand stem, and a large excision from residue to 682 to 719 (68). Kethoxal modification of circled G residues is blocked by S15 (69). Heavy and light arrows mark sites of major and minor RNase A scissions. C", C, C_1' and C_2' denote sections of the 16S RNA primary structure (64,65).

high confidence. As shown in Figure 2, the 5' subfragment is highly complementary to the 5' half of the central subfragment and together with it gives rise to the hairpin loop at the left of the diagram. The remaining portion of the central subfragment can in principle base-pair with the 3' subfragment to form the stem of a second hairpin loop that is likely to be quite stable even in the absence of its 40 distal nucleotides. A somewhat smaller fragment from the S8/S15 region, which contains the principal base-paired segments depicted in Figure 2, has also been described (69). Both of the protected RNAs readily reassociate with protein S15, but only the first has been shown to interact with S8 as well (68,69).

It has been argued that protein S8 binds mainly to the 5'-proximal stem, since it has been recovered from digests together with the complementary sequences 586-602 and 634-652 (58,70). Indeed, electron micrographs of the S8-16S RNA complex have revealed a single hairpin-like structure near the center of the RNA molecule (67). However, limited hydrolysis of the S8-16S RNA complex from *B. stearothermophilus* resulted in the protection of segments analogous to both stem regions illustrated in Figure 2 (49). Because S15 alone also appears to confer resistance upon both double-stranded sections (58,68), it is possible that the binding sites for S8 and S15 each include portions of the two adjoining hairpin structures. The dearth of extended single-stranded sequences in the S8/S15 region strongly suggests that the two proteins associate primarily with helical segments of the RNA. Since the upper part of the 5'-proximal hairpin is accessible to nuclease (68), or else completely excised (69), in the presence of S8 and S15, it is not likely to be in close proximity to either protein. The hyphenated two-fold axis in the 5'-proximal stem and the run of five G:C pairs in the 3'-proximal stem command greater attention as potential binding sites, however. It is of great interest in this regard that the association of S15 with its cognate fragment inhibits the kethoxalation of four G residues, located to either side of the G:C stack (see Fig. 2), that are available for reaction in the free RNA (69).

The site or sites to which protein S7 attaches have proved difficult to prepare from S7-16S RNA complexes, probably due to the low stability of the interaction (23). Although S7 remained bound to a fragment of roughly 8S that includes most of the 16S RNA sequence between residues 950 and 1500, no further progress was made via this approach. However, through the analysis of a ribonucleoprotein fragment derived from the 30S subunit, the S7 binding site has now been localized to two sequences encompassing residues 930-1057 and 1096-1392 (71). From this finding, it can be inferred that S7, as S4, interacts with an extended segment of the 16S RNA. S7 probably makes more than one contact with the nucleic acid chain as well although it becomes crosslinked to only one specific residue in the 3'-proximal region, U_{1239}, after UV irradiation of the 30S subunit (72).

Protein Binding Sites on the 23S RNA

The positions of binding sites for 50S subunit proteins on the 23S RNA are depicted in Figure 3. These studies were initially aided by the availability of two large fragments covering almost the entire 23S RNA that can be isolated from the 50S subunit following limited hydrolysis with solid-phase pancreatic RNase (73). The 13S fragment contains roughly 1200 nucleotides and arises from the 5'-terminal region of the 23S RNA while the 18S fragment spans the remaining 1800 or so nucleotides (74). The 18S RNA can be further subdivided by digestion with T_1 RNase into an 800-nucleotide 8S fragment and a 1000-nucleotide 12S fragment which encompass the central and 3'-terminal segments of the 23S RNA, respectively (75). The binding capacities of these fragments were then tested by incubating each one with individual 50S subunit proteins and, when complex formation occurred, the specificity of the interaction was assessed (36, 74-76). In this way, it was shown that the attachment sites for L4, L11, L20 and L24, as well as for the [L7/12-L10] complex are located within the 5'-terminal fragment, that L2, L13 and L23 interact with the central fragment and that L1, L3 and L6 associate with the 3'-terminal fragment

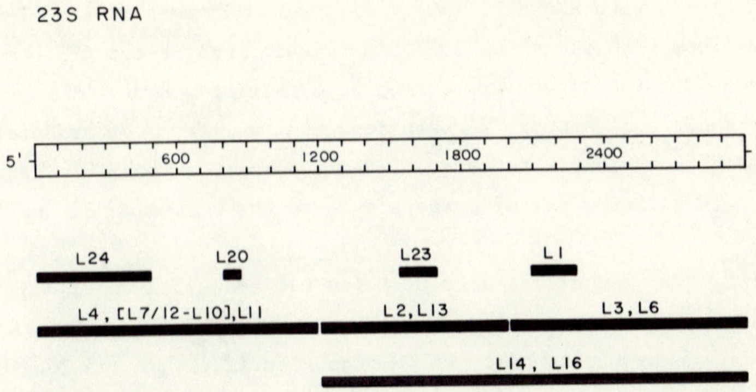

Figure 3. Location of protein binding sites on the 23S RNA. Solid bars correspond to positions of various RNA fragments according to as yet incomplete sequence data (85). See text for details of binding site placement and structure. Numbers within the 23S RNA represent approximate distances in nucleotide residues from the 5' end of the molecule.

(Fig. 3). Although L14 and L16 interact with the 18S RNA, their binding sites have not yet been assigned to either of the smaller subfragments derived from it.

The binding sites for proteins L1 and L24 have now been much more thoroughly characterized by the isolation of protected RNA fragments from digests both of individual protein-23S RNA complexes and of partially unfolded 50S subunits. L24 has been recovered in association with fragments comprising a sequence of 480 nucleotides, beginning only 9 residues from the 5'-terminus of the 23S RNA (77,78). As for the 5' end of the 16S RNA, this region appears to possess a similar conformation in the presence and absence of L24, although the protein markedly reduces its sensitivity to RNase digestion (78). Recent evidence suggests that the L24-specific fragments are also stabilized by RNA-RNA interactions between sequences that are separated by several hundred nucleotides in the 23S RNA primary structure (79,80).

Protein L1 interacts with and strongly protects a sequence of about 115 nucleotides that is located some 800-900 bases from the 3' terminus of the 23S RNA (81-85). Somewhat weaker protection is afforded another 30-35 nucleotides that adjoin the 3' end of the main binding region. Virtually identical sequences were recovered in association with L1 and L9 following limited hydrolysis of 50S subunits (86). Whether or not L9 actually binds to this region of the RNA is an open question, since its presence could be explained equally well by a strong association with L1. *E. coli* L1 also protects a 115-nucleotide segment of *B. stearothermophilus* 23S RNA from hydrolysis with RNase T_1 (50). Although alignment of the L1-specific sequences from both bacterial 23S RNAs revealed only about 60% homology, the fragments can be fitted to a common base-pairing scheme. This analysis provides additional support for the idea that at least some ribosomal proteins recognize specific features of RNA secondary or possibly tertiary structure in addition to, or perhaps even to the exclusion of, nucleotide sequence.

RNA fragments associated with proteins L20 and L23 have also been analyzed and tentatively positioned in the 23S RNA structure. The RNA fragment protected by L20 consists of 40-50 nucleotides that are located approximately 800 residues from the 5' end of the intact RNA molecule (81,85). The sequences protected by L23 span 100-150 bases near the middle of the 23S RNA (81,85). It is gratifying that the binding site locations inferred from the protection experiments are all consistent with assignments based on the association of proteins with the 13S, 8S and 12S RNA segments (Fig. 3). Smaller fragments specific for L2, L3, L4, L6, L13 and L16 have not yet been reported. Although this may in part reflect the rather weak interaction of several of these proteins with the 23S RNA (32), it is possible that some of the proteins associate with complex sites built up of sequences at some distance from each other in the primary structure. If so, we shall have to learn a great deal more about the secondary and tertiary folding of the RNA chain before the nature of such binding sites can be fully appreciated.

Protein Binding Sites on the 5S RNA

Attempts to delineate protein binding sites on the 5S RNA have only recently yielded unambiguous results. An early report implied that proteins L18 and L25 protected the segment between residues 69-120 from hydrolysis by RNase A (87). The short sequence 1-11, which is complementary to sequence 110-119 (see Fig. 4), was usually recovered together with the longer 3'-proximal fragment by virtue of secondary interactions. This view was generally supported by experiments using RNase T_1, from which it was inferred that the segment 58-116 contained binding sites for the two proteins (40). However, in neither case was it convincingly demonstrated that L18 and L25 remained bound to, or could specifically reassociate with, RNA fragments from the putative binding region.

Sequences 1-11 and 69-120 have now been shown to exhibit a high intrinsic resistance to RNase even in the absence of protein (41). Although L25 conferred no added protection upon the RNA, it did remain strongly

5S RNA

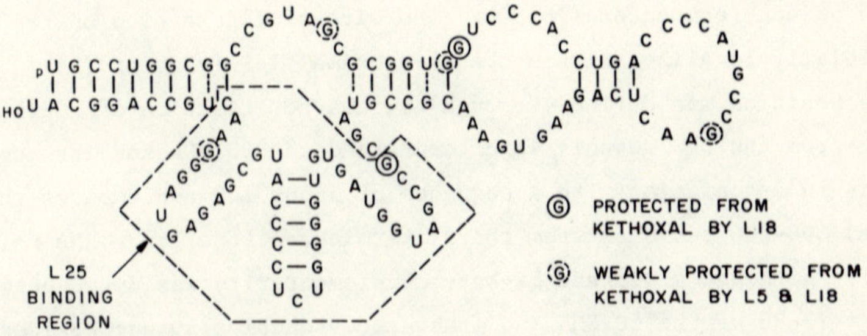

Figure 4. Model of 5S RNA secondary structure with L25 binding site. The L25-binding sequence was identified by the association of L25 with products arising from the 5S RNA after digestion with RNase A (41). G residues protected from kethoxalation by L18 and L5 are distributed throughout the 5S RNA suggesting that the binding sites for these proteins are more extended than that for L25 (42). Secondary structure of 5S RNA is that proposed by Fox and Woese (88).

attached to the resistant fragment after hydrolysis of the L25-5S RNA complex. Under somewhat harsher digestion conditions, L25 was recovered with a smaller RNA fragment spanning residues 69-110. Figure 4 shows the location of this region in the Fox-Woese model of 5S RNA secondary structure (88). A similar study of the L18-5S RNA complex revealed that while the protein increases the overall resistance of the RNA to nuclease digestion, it is released either before, or concomitant with, formation of the highly stable RNA core (41). Moreover, L18 was unable to reassociate with the isolated fragment, whereas L25 did so quite efficiently. These results indicate that the primary binding site for L25 occurs between residues 69 and 110 in the 5S RNA. They also suggest, if only indirectly, that attachment of L18 entails recognition of a more extended region of the 5S RNA and may require that the tertiary structure of the nucleic acid chain be intact. The latter inference is consistent with the observation that G_{24} and G_{69}, which are located on opposite sides of the main axis of the RNA model (Fig. 4), are strongly protected from kethoxal modification by L18 (42). Moreover, a much larger fragment of the 5S RNA, encompassing residues 41-120, also failed to bind L18, though it stably associated with L25 (R. Villems, personal communication). Models of the L18-5S RNA complex consistent with this view have been postulated on the basis of small-angle x-ray scattering measurements (89). In one interpretation, both protein and RNA are envisioned as very elongated structures whose surfaces could be in contact over a sizable area.

When L5 and L18 are bound to 5S RNA, the resulting complex can interact with the 23S RNA to form a ribonucleoprotein particle containing equimolar amounts of each component (76,90). The site of interaction on the 23S RNA was initially localized within the 3'-proximal 18S fragment (91). A more precise position was defined through the characterization of a ribonucleoprotein from the 50S subunit which contained the 5S RNA and its cognate proteins in association with a sequence of about 120 nucleotides that is situated 600-700 residues from the 3' terminus of the large RNA molecule (85,86). Although a perfect 12-base complementarity was noted between residues 72-83 of the 5S RNA and a sequence roughly 160 nucleotides from the 5' terminus of the 23S RNA, its significance in 5S RNA-23S RNA interaction is still unclear (78,92).

The first attempt to characterize a protein binding site on a eukaryotic 5S RNA has recently been reported (93). A protein-5S RNA complex released from yeast 60S subunits by treatment with EDTA was digested with pancreatic RNase and the RNA fragments that remained associated with the protein were subjected to sequence analysis. The protected RNA consisted of 42 residues from the 3' end of the 5S RNA and 12 residues from the 5' end which were presumably base-paired with the 3' terminus of the former fragment. These sequences correspond quite closely to those that associate with L25 in *E. coli* (see Fig. 4), raising the possibility that the binding site for at least one large-subunit protein has been conserved in both prokaryotes and eukaryotes. It will be interesting to learn whether the 5S RNA-binding protein from yeast corresponds to any of the rat-liver proteins found to associate with homologous 5S RNA by affinity chromatography.

STRUCTURE AND RNA-BINDING REGIONS OF RIBOSOMAL PROTEINS

Structure of RNA-Binding Proteins

Although it has been known for some time that the RNA-binding proteins of the *E. coli* ribosome are generally small and rather basic, only recently have their secondary and tertiary structures been probed in detail. A variety of physical techniques have been used in these studies, prominent among which are various hydrodynamic approaches, small-angle x-ray scattering, circular dichroism, nuclear magnetic resonance and neutron scattering. The preponderance of the evidence suggests that most ribosomal proteins contain moderate amounts of α helix and β structure (94,95) as well as nonpolar contacts among amino acid side chains that are characteristic of intramolecular tertiary interactions (96). Moreover, the majority of isolated proteins in solution have been found to possess rather asymmetric shapes including several that are extremely elongated (see ref. 4). Although the latter results were unexpected, they did not arouse much controversy, largely because immune electron microscopy suggested that antibody binding sites for a number of proteins spanned distances of 100 Å or more in both 30S and 50S subunits (see ref. 97). Nonetheless, there are important inconsistencies in some of the physical measurements which have not yet been resolved. Small-angle x-ray scattering investigations, for instance, indicate that S8 and L25 are both highly asymmetric (98,99) whereas hydrodynamic techniques, and in the case of S8, neutron scattering studies *in situ*, suggest that these proteins are more nearly globular (100-102). Furthermore, a recent evaluation of the radii of gyration of S4 and S7 provide support for the view that these two proteins, generally held to be elongated (see ref. 4), are capable of assuming a globular conformation in solution (103). Such uncertainties in protein shape make it difficult to construct satisfactory models of protein-RNA complexes at the present time.

RNA-Binding Regions of Ribosomal Proteins

There is still relatively little known about the structural features of ribosomal proteins that are involved in recognition of, and interaction with, rRNA. Earlier work indicated that the modification of lysine, tyrosine, methionine, and the C terminus of S4 impaired its ability to interact with 16S RNA (104-107, see ref. 1). Although the RNA-binding properties of S7 were also reduced by treatments that affected methionine and lysine (105), the extreme C-terminal region of this protein is apparently not required for complex formation since S7 from *E. coli* strain MRE600 binds just as well to 16S RNA as that from strain K (22), even though it contains 24 fewer residues at its C terminus (108).

Considerable progress has been made in the isolation of protein fragments capable of interacting with ribosomal RNA, however. In particular, large segments of S4, S8 and L18 have been generated either by chemical procedures or by digestion of specific protein-RNA complexes with proteolytic enzymes (109-115). In the case of trypsin digestion, it is noteworthy that a number of RNA-binding proteins are completely resistant to enzymatic attack when associated individually with their binding sites whereas others are completely hydrolyzed and released (109,110). Thus, of a number of 30S subunit proteins that bind to 16S RNA, only S4 yields a discrete fragment that retains the specific RNA-binding capacities of the intact protein (109-111).

The large trypsin-resistant fragment isolated from partial digests of the S4-16S RNA complex consists of residues 47-203 (110,111). Although the protected fragment reassociates specifically with 16S RNA, none of the smaller fragments produced by trypsin digestion of S4 were able to do so. While the 46 N-terminal residues of S4 therefore play no role in the attachment of S4 to 16S RNA, it is more striking that over 75% of the molecule *is* essential for the interaction. This observation is consistent with the large size and complexity of the S4 binding site in the 16S RNA (see above). Attempts to reassemble 30S subunits containing the trypsin-resistant fragment of S4 resulted in defective particles missing S1, S2, S10, S18 and S21 (110,113). When a slightly longer fragment of S4, encompassing residues 31-203, was used in the reconstitution studies, however, a 30S particle containing all of the subunit proteins resulted (112,113). These particles are capable of binding poly(U) but are inactive in the poly(U)-stimulated binding of tRNA. Taken together, the S4 fragment studies suggest that residues 1-30 contribute to the binding of tRNA, residues 31-46 influence late subunit assembly reactions via either protein-protein interactions or the facilitation of conformational changes in the 16S RNA, and residues 47-203 comprise the region of primary interaction with the 16S RNA (111,113).

Hydrolysis of the L18-5S RNA complex with trypsin yields a large protected fragment that lacks only 17 amino acids from the N-terminus of the 117-residue protein (115). Although this fragment efficiently reassociates with 5S RNA, a somewhat smaller product spanning residues 26-117 does not. Here, as in the case of S4, protein-RNA interaction requires that over 75-80% of the protein be intact, suggesting that the specific conformation of the polypeptide chain may play an important role in its ability to associate with RNA. Addition of L5 to the 5S complex leads to the complete protection of L18 from trypsin attack. Moreover, the L18 fragment does not stimulate the binding of L5 to 5S RNA, nor can the ternary complex of L5, L18 fragment, and 5S RNA form a stable complex with 23S RNA (R.A. Garrett, personal communication). The N-terminus of L18 may therefore directly interact with L5 in the ternary complex and in so doing, provoke structural changes permitting its interaction with the 23S RNA.

CONFORMATIONAL CHANGES IN INTERACTING RIBOSOMAL COMPONENTS

The binding of ribosomal proteins to ribosomal RNA requires close physical contact among the interacting components, particularly as subunit assembly progresses and the ribonucleoproteins assume their compact native structure. Consequently, it is possible that such associations bring about conformational changes in either protein or RNA that improve the packing. Here again, most of the available data derives from the interaction of S4, and other 30S subunit proteins, with 16S RNA, although the attachment of certain 50S subunit proteins to their binding sites in the 23S and 5S RNAs have provided important information on conformational alterations as well.

As discussed in the preceding sections, protein S4, which is considered to be a highly elongated molecule by most--but not all--criteria, interacts with an extended binding site in the 16S RNA, most probably via multiple contacts with the polynucleotide chain. In addition, S4 plays a key role in 30S subunit assembly, facilitating the association of several other proteins with the nascent ribonucleoprotein particle (113). Although S4 might therefore be expected to influence the structure of the RNA, the evidence points strongly in the opposite direction. Furthermore, there appears to be no gross alteration in the conformation of the protein when it interacts with the RNA molecule (111).

An impressive array of data, both chemical and physical, suggests that the S4-binding region of the 16S RNA is in itself highly structured and that it changes very little upon interaction with the protein, at least under the conditions of 30S subunit reconstitution *in vitro*. For instance, the susceptibility of the S4 binding site to pancreatic and T_1 RNases, as well as its accessibility to kethoxal, is very similar in the presence and absence of the protein and, indeed, these chemical probes indicate that the nucleic acid chain may be somewhat more exposed or "open" in the complex than when free in solution (60,63,116). Moreover, the S4-16S RNA complex differs very little from free 16S RNA either in a number of physical properties (117-119) or in size and shape as determined by electron microscopy (120,121). Finally, S4 does not significantly alter the circular dichroism spectrum or radius of gyration of specific S4-binding fragments of the 16S RNA (116,122, A.S. Spirin, personal communication). Once bound, however, the protein does help to maintain the cooperative, tightly-folded organization of the RNA, both during thermal denaturation and at reduced Mg^{2+} concentrations (116,118).

All of these results indicate that the 5' end of the 16S RNA, at least, attains a specific, compact conformation in TMK buffer by virtue of its own intramolecular secondary and tertiary interactions and that S4 "fits" this structure without substantial perturbation. This view is reinforced by the observation that a "denatured" form of the 16S RNA can be produced by extensive dialysis against water or by rapidly cooling from 60° in TMK buffer (8,123). The denatured species differs from native 16S RNA in that it is unable to bind S4, it migrates more slowly than 16S RNA on both sucrose gradients and polyacrylamide gels, and it is degraded much more readily when treated with RNase. More importantly, these changes were all reversed by heating the denatured RNA at 40° at 20 mM Mg^{2+}. In particular, pre-incubation restored the ability of this RNA to bind S4 even at 0°. Thus, the protein binding site was spontaneously re-formed under reconstitution conditions, presumably through rearrangements in the intramolecular structure of the 16S RNA alone. It has been reported that 16S RNA extracted from 30S subunits with urea and acetic acid differs substantially from phenol-extracted 16S RNA in structure and protein-binding capacities (25-27). In addition, it was maintained that the binding of S4, as well as other 30S subunit proteins, could convert phenol-extracted 16S RNA to a structure typical of urea/acetic acid 16S RNA. The significance of the urea/acetic acid RNA remains uncertain at present, however. Although it may represent a distinct metastable conformation of the nucleic acid molecule, it appears to be rather similar to "denatured" forms of 16S RNA investigated by other laboratories in its lower electrophoretic mobility, greater RNase sensitivity and destructured S4 binding site (8,25,26,123). In any case, after incubation at 37°-40° in high Mg^{2+}, 16S RNA prepared by the urea/acetic acid procedure and by phenol extraction appear to be indistinguishable either by differences in RNase sensitivity or in a number of physical properties (123,124).

Evidence for the induction of conformational changes in the 16S RNA by other 30S proteins is scant. Although S7, S8 and S20 have all been reported to effect small changes in the optical properties of the RNA molecule, the data are not conclusive (125). Moreover, a comparison of the diffusion constants of the 16S RNA and of a number of individual protein-16S RNA complexes indicates that S8 causes little or no change in the RNA, whereas S7 and S15 bring about only a slight increase in folding (118).

Complexes of 5S RNA with proteins L5, L18 and L25 are particularly appropriate for the study of structural changes owing to the small size of the RNA and the cooperative interactions that occur among the RNA-binding proteins (11,38). Indeed, association of the proteins has been found to evoke a 15-25% increase in the amplitude of the main circular dichroism band of the 5S RNA, most of which is attributable to L18 (18,39,126). Although this change could result from increases in base-pairing or base-stacking, it might equally well arise from an alteration in the structural regularity of existing double-stranded segments of the RNA molecule. Nevertheless, the shift is a dramatic one and could be related to the strong cooperative influence of L18 on L5 binding (11). Further effects of L18 on the organized structure of the 5S RNA were revealed by its ability to stimulate the release of ethidium bromide from the nucleic acid chain (38) and to stabilize the RNA molecule against thermal denaturation (39). Circular dichroism measurements suggest that the attachment of L5 and L25 does not lead to any major changes in the secondary structure of either RNA or protein (18,39). L25 does, however, increase the accessibility of G_{69} to RNase cleavage and kethoxal modification (41,42). Moreover, this protein profoundly affects the minimum temperature at which the native, or A, configuration of 5S RNA can be converted to the stable denatured, or B, form (T.L. Marsh and R.A.Z., unpublished). Although it is quite difficult to mold these somewhat disjointed observations into a consistent description of protein-5S RNA interaction, they do at least demonstrate that L18 and L25 are capable of altering the conformation of 5S RNA in defined and measurable ways and offer promise that the structural dynamics of ribonucleoprotein complex formation can eventually be understood. The attainment of this goal may be facilitated by comparison of the above results with similar studies on a 5S ribonucleoprotein released from the 50S subunit of *H. cutirubrum* which has been reported to contain proteins analogous to *E. coli* L5 and L18 (127). This complex was also found to exhibit a reduced fluorescence enhancement in the presence of ethidium bromide and an increased circular dichroism at 268 nm relative to the free RNA (43).

Information on conformational changes in the 23S RNA upon protein binding is limited at the moment. The interaction of L24 with fragments of its binding site have been investigated by several physicochemical techniques and it was concluded, mainly on the basis of circular dichroism spectra that L24 brings about a small amount of base unstacking but very little change in base pairing (128). In another study, association of the L5,L18-5S RNA complex with 23S RNA was determined to promote a slight opening of the structure of the larger rRNA molecule (78). This appears to rule out the possibility that the 5S ribonucleoprotein plays a role in organizing widely separated regions of the 23S RNA during 50S subunit assembly.

PROTEIN-rRNA CROSSLINKING

In the past few years, a number of experiments have been reported in which ribosomal proteins were crosslinked to ribosomal RNA by various photochemical and chemical procedures. For example, covalent crosslinks have been produced by irradiation of reconstituted protein-RNA complexes, of isolated subunits and of undissociated ribosomes with UV light (129-136). Dye-sensitized photooxidation has also been used to covalently attach ribosomal proteins to both 16S and 23S RNAs in undissociated 70S ribosomes (137). In all such cases, the crosslinks are presumed to be direct so that in principle, analysis of the attachment sites should provide information on the physical proximity of the interacting macromolecules in the parent ribonucleoproteins. In only one case, however, has a positive identification been made of the oligonucleotide and oligopeptide actually involved in the crosslink (138).

Individual complexes between proteins S4, S7 and S20 with 16S RNA of *E. coli*, have all been crosslinked with UV light (129-131). An effort was then made to identify the peptides involved by first isolating the covalent complexes free of unbound protein and RNA by CsCl centrifugation and then fingerprinting the peptides released from the complexes by trypsin. In the case of both S4 and S7, three or four characteristic spots were either reduced in quantity or absent from the respective peptide maps. It was assumed without additional evidence that such peptides, which originated from widely separated regions of the polypeptide chains, failed to appear in the fingerprint because they remained attached to 16S RNA. There are drawbacks in this kind of negative analysis, however, since a reduction in the yield of specific peptides could result from causes other than crosslinking, such as photochemical alteration of amino acid residues or decreases in the susceptibility of the polypeptide chain to enzymatic cleavage in the presence of attached RNA (138). Crosslinked oligonucleotides were also sought in the irradiated complexes. RNase T_1 digests of the S4 and S20-16S RNA complexes were fractionated by electrophoresis on polyacrylamide gels and in each instance, one oligonucleotide band was decreased and a new, more slowly migrating component containing the same segment of the RNA, was observed (131). The new band was assumed to correspond to the crosslinked protein-RNA fragment complex, although it was not possible to verify the presence of the protein in such material. The RNA segments identified in this way correspond to residues 422-500 in the case of S4 and to residues 221-296 in the case of S20. These segments both lie within the binding sites for S4 and S20 determined by partial RNase hydrolysis of protein-16S RNA complexes (see Fig. 1).

Irradiation of ribosomal subunits or intact ribosomes from *E. coli* with UV light at low doses results in the formation of specific crosslinks

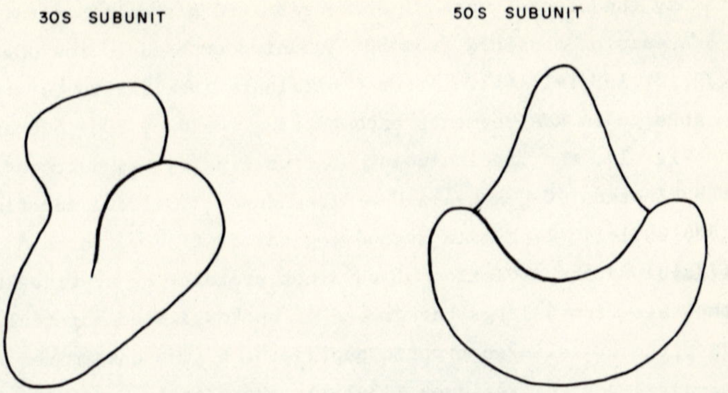

Figure 5. Regions of protein-rRNA crosslinking in 30S and 50S subunit models derived by immune electron microscopy. Stippled zones encompass the antibody binding sites of proteins most frequently crosslinked to rRNA both photochemically and with bifunctional reagents. Proteins in this category are S3, S4, S5, S7, S9, S11, S13, S18 and S21 from the 30S subunit and L2, L4, L13 and L16 from the 50S subunit. Note that the stippling coincides with the subunit interface region in both cases. The 30S subunit model is adapted from Lake (151); the 50S subunit model is that of Stöffler and collaborators (97).

between S7 and the 16S RNA, and of L2 and L4 with the 23S RNA (132,135, 138). Higher doses provoke much more extensive crosslinking and concomitant unfolding of the subunits (132,139,140). Thus, even though most ribosomal proteins can eventually be linked to rRNA at high levels of irradiation, the significance of this finding cannot be easily assessed because a certain proportion of the crosslinking reactions may occur after the ribosomal particles have lost their compact, native configuration and could therefore be spurious. The same sort of objection can be raised to protein-RNA crosslinks formed by dye-sensitized photooxidation since the particles were found to be functionally inactive following irradiation in the presence of methylene blue (137). In summary, it is quite clear that S7 becomes crosslinked to the 16S RNA in the 30S subunit at UV doses which do not greatly perturb the subunit structure; under somewhat harsher conditions, S3, S4, S9 and S11 also become crosslinked (136), but this result must be regarded with caution. A similar pattern holds for the 50S subunit, in which L2 and L4 can be photochemically crosslinked to 23S RNA at high efficiency whereas L3, L16 and others become attached only under more severe conditions. Portions of the ribosomal subunits where the highest levels of crosslinking occur are illustrated in Figure 5.

The potential for deriving precise information on the structural organization of protein and RNA in the ribosome by UV crosslinking has been confirmed by the elegant work of Brimacombe and his collaborators on the S7-16S RNA complex obtained from 30S subunits exposed to low doses of light (72,132,138,141,142). After preliminary localization of the S7 attachment site to RNA segments encompassing residues 900-1400 of the 16S RNA (see Fig. 1), the precise point of crosslinking was narrowed down to the octanucleotide -C-U-A-C-A-A-U-G- (residues 1233-1240) and finally to U_{1239} (72,138,141). The main methodological difficulty overcome in these studies involved the isolation of covalent protein- or peptide-RNA fragment complexes from a large background of uncrosslinked material. At the same time, the crosslinked tryptic peptide of S7 was determined to be -ser-met-ala-leu-arg- (residues 113-117). Here again, positive identification of the crosslinked peptide was possible. A sample of S7 attached to a small fragment of ^{32}P-labelled RNA was digested with trypsin and fingerprinted. The ^{32}P-containing peptides were eluted from the tryptic map and identified by amino acid analysis. The absence of met from the otherwise well-defined peptide suggested that the crosslink was mediated by this residue.

Chemical methods for crosslinking protein and RNA have been less extensively exploited than the UV technique largely due to the paucity of suitable reagents. Two well-known reactions, involving periodate oxidation and formaldehyde, have been used to a limited extent in the investigation of intact subunits, however, and both are expected to produce "zero-length" crosslinks. In the first approach, the 3'-terminal ribose of the ribosomal RNA is oxidized with periodate _in situ_. The resulting aldehydes are believed to react with amino groups of neighboring proteins to form Schiff's base intermediates which can then be stabilized by reduction with borohydride. The first studies of this sort showed that proteins S1 and S21 could be crosslinked to the 16S RNA in this fashion although the yields were rather low (143,144). By contrast, a recent report indicates that S3 and S4 are the major crosslinked components in similar circumstances and that a number of other 30S subunit proteins, with the notable exception of S1, become covalently bound as well (145). The results imply either that the position of the 3' end of the 16S RNA in the subunit is not securely fixed, at least under the conditions of the experiment, or that crosslinking is not limited to the 3' terminus of the nucleic acid molecule (145). The validity of the periodate/borohydride technique as a probe of ribosome topography is thus open to question. Treatment of *E. coli* ribosomal subunits with formaldehyde leads to the formation of rather unstable adducts between a number of ribosomal proteins and ribosomal RNA

(146). Owing to the instability of the adducts, however, as well as to the apparently low specificity of the crosslinking reaction, it is doubtful that formaldehyde fixation will contribute materially to the elucidation of protein-RNA contacts in the ribosome.

The value of chemically induced protein-RNA crosslinks in the ribosome may be realized in the near future through the use of specially adapted bifunctional reagents. A number of these have been described recently and several were successfully tested on intact ribosomal particles (147-152). Although the relative merits of the different agents remains to be determined, each appears to be quite selective and, together, they encompass a broad range of chemical properties, varying in length, flexibility, susceptibility to cleavage, and chemical reactivity.

It is of interest that certain proteins are crosslinked to RNA by almost all of the techniques described in this section. In the 30S subunit, S3, S4, and S7 are attached to 16S RNA by UV, dye-sensitized photooxidation, formaldehyde, and various bifunctional reagents, and S5, S9/S11, S13, S18 and S21, by a number of these. In the 50S subunit, L2 is very susceptible to crosslinking with the 23S RNA, while L4, L13, L16 are bound in several cases. Only 2,3-butanedione treatment, of all the methods used, gives a significantly different pattern of protein-RNA linkage (150). Such consistencies are difficult to explain given the diversity of the techniques employed, but they may be related to the accessibility of the proteins in the subunit, to their shape, or to the unique distribution of their reactive groups. Moreover, as shown in Figure 5, the most readily crosslinked proteins appear to be clustered in regions of the subunit models that correspond to their mutual interface. In contrast, over one-third of the 30S subunit proteins and more than one-half of the 50S subunit proteins have been crosslinked by only one or, in several cases, none of the methods employed. This is particularly striking in the case of the two RNA-binding proteins S8 and S15 of the 30S subunit, which, together with S6 and S18 comprise a ribonucleoprotein assembly domain near the middle of the 16S RNA (1). These proteins are all located in the lower portion of the 30S subunit (151) which must either be inaccessible or poor in close protein-RNA contacts. The S8 and S15 binding sites are also rather small and probably provide only limited targets for the various crosslinking agents.

Although the method of UV crosslinking has been to used to advantage in studies of the covalent S7-16S RNA complex, it is not of general applicability to the definition of protein-RNA contacts in the ribosome as a whole. The only additional complexes that can clearly be generated under conditions mild enough to avoid disruption of ribosome structure are those of L2 and L4 with the 23S RNA. It appears that short-range, bifunctional reagents may prove to be much more versatile in generating the large number of protein-RNA crosslinks that will be needed to delineate the arrangement of the rRNA chain relative to the protein components of the ribosome. It must be stressed, however, that while protein-RNA crosslinking obviously occurs in regions where the two kinds of molecules are in close proximity, the points of covalent attachment do not necessarily represent sites of recognition and interaction. This reservation is particularly relevant when isolated protein-RNA complexes are to be investigated. In such cases, large segments of the RNA will undoubtedly enjoy considerable flexibility and may therefore fall within crosslinking range of amino acid sequences that they do not normally adjoin in the intact subunit (138).

STRUCTURAL ORGANIZATION OF rRNA IN THE RIBOSOME

Small-angle neutron scattering has shown that the radius of gyration of the 16S RNA in the *E. coli* 30S subunit is significantly smaller than that of its protein complement, indicating that the RNA is tightly packed toward the interior of the particle (152,153). In addition, hydrodynamic and electron microscopic studies of free 16S RNA suggest that the nucleic acid chain is itself capable of assuming a compact conformation of roughly the same size and shape as the 30S subunit (119,121). Whether or not the RNA molecule actually contains all of the information necessary for attaining the structure typical of the subunit is not entirely clear, however, since the association of proteins S4, S7, S8 and S15 leads to additional folding and a reduction in susceptibility to RNase and a number of chemical reagents (118,154). Therefore, while the RNA is evidently quite compact in the free state, its tertiary structure may be further constrained by interaction with key assembly proteins.

On the basis of cooperative protein-RNA associations and the composition of ribonucleoprotein particles reconstituted from specific RNA fragments, it was possible to distinguish three domains in the 16S RNA which correspond to residues 1-550, 575-900, and 920-1500 (1). The first of these associates with proteins S4, S16, S17 and S20, the second, with proteins S6, S8, S15 and S18, and the third, with proteins S7, S9, S13 and S19. Figure 6 illustrates an a preliminary effort to delineate the corresponding ribonucleoprotein neighborhoods according to the 30S subunit model of Lake (151). In this representation, the 5'- and 3'-proximal domains

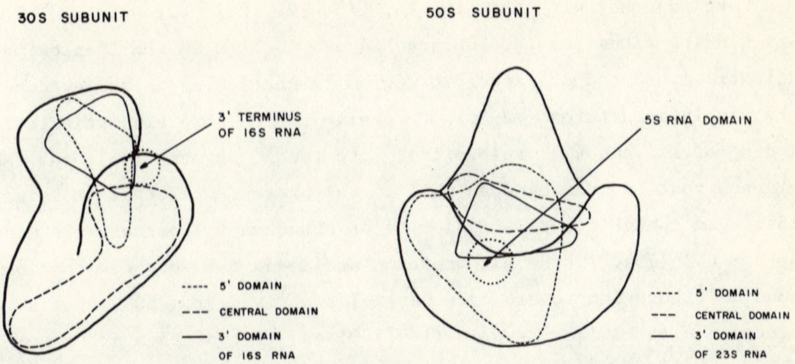

Figure 6. Possible locations of rRNA domains in 30S and 50S subunit models. The different regions are defined by the positions of binding sites for antibodies directed against proteins known to associate with various segments of the rRNA chains subject, of course, to the reservation that the antibody binding sites of a given protein need not necessarily be in the immediate vicinity of the RNA. In the 30S subunit, the 5' domain is specified by S16, S17 and S20; the central domain, by S6, S8, S15 and S18; and the 3' domain, by S7, S9, S13 and S19. The 3' terminus of the 16S RNA was located independently by immune electron microscopy (I.N. Shatsky and A.A. Bogdanov, personal communication). In the 50S subunit, the 5' domain is specified primarily by L4, L20 and L22, as well as by L11; the central domain, by L2 and L13; and the 3' domain, by L1, L3 and L6. The 5S RNA was located via its three binding proteins, L5, L18 and L25. Subunit models are as in Figure 5.

exhibit substantial overlap in the upper half of the subunit, while the central domain is situated in the lower half.

Residues 1-550 of the 16S RNA are very resistant to nuclease digestion in the 30S subunit (62,155-157) and are virtually inaccessible to kethoxal modification as well (158-160). The 5' domain is thus almost certainly buried within the subunit core where, together with S4, it seems to be responsible for maintaining the structural integrity of the particle (160, 161). The central and 3' domains, however, are considerably more susceptible to RNase cleavage and kethoxalation (157,159,160), suggesting that they lie closer to the subunit surface and, consequently, can participate in functional activities connected with the binding of mRNA, tRNA and the 50S subunit (160). The 50 or so 3'-terminal residues are particularly exposed as judged by their sensitivity to nucleases (156,157,162), availability for chemical modification (158,163), and reactivity toward antibodies (164,165).

In the 50S subunit, the RNA is even more highly concentrated at the center of the particle than in the 30S subunit (166,167). Here again the 5' end of the RNA appears to be intimately involved in the organization of a core together with proteins L4, L13, L20, L21, L22 and L24. The primary binding sites for L4, L20 and L24 are located in the 5'-proximal 1000 nucleotides and the L24-specific region in particular is believed to be very highly structured (78-80). Proteins L20 and L24 also stimulate the binding of L21 and L22 respectively, to the 5' segment (35), and there is evidence that L13 can bind as well in the presence of other proteins (147, 168), even though its principal site of attachment lies near the middle of the 23S RNA (75). These six proteins are among the very few that remain bound to ribonucleoprotein core particles derived from 50S subunits by treatment with LiCl or trypsin (169,170) and five of them are essential for the early stages of 50S subunit reconstitution in vitro (171,172). Electron microscopy of one such core particle showed it to possess a compact structure and overall dimensions very similar to those of the intact 50S subunit (173). Thus, the 5' region of the 23S RNA appears to make an important contribution both to the assembly of the 50S subunit and to the maintenance of its RNA-rich interior. The boundaries of 5'-proximal, central and 3'-proximal domains of the 23S RNA are rather difficult to define in the 50S subunit model, but all three seem to be concentrated near the meridian (Fig. 6). Although the distribution of the RNA chains in the ribosomal subunits remains a matter of speculation for the time being, this topic will undoubtedly receive considerable attention in the future owing to an increasing appreciation of the pivotal role of the RNAs in ribosome structure and function.

ACKNOWLEDGEMENT

Work in the author's laboratory is supported by grants PCM78-16027 from the N.S.F. and GM-22807 from the N.I.H. He is also the recipient of U.S.P.H.S. Research Career Development Award GM-00129.

REFERENCES

1. Zimmermann, R.A. (1974) in Ribosomes (Nomura, M., Tissières, A., and Lengyel, P., Eds.), pp. 225-269, Cold Spring Harbor Laboratory, Cold Spring Harbor, N.Y.

2. Brimacombe, R., Nierhaus, K.H., Garrett, R.A., and Wittmann, H.G. (1976) Prog. Nucleic Acid. Res. Mol. Biol. 18, 1-44.

3. Erdmann, V.A. (1976) Prog. Nucleic Acid Res. Mol. Biol. 18, 45-90.

4. Brimacombe, R., Stöffler, G., and Wittmann, H.G. (1978) Ann. Rev. Biochem. 47, 217-249.

5. Traub, P., and Nomura, M. (1969) J. Mol. Biol. 40, 391-413.

6. Schulte, C., Morrison, C.A., and Garrett, R.A. (1974) Biochemistry 13, 1032-1037.

7. Schulte, C., and Garrett, R.A. (1972) Mol. Gen. Genet. 119, 345-355.

8. Muto, A., and Zimmermann, R.A. (1978) J. Mol. Biol. 121, 1-15.

9. Richards, E.G., Lecanidou, R., and Geroch, M.E. (1973) Eur. J. Biochem. 34, 262-267.

10. Spirin, A.S. (1978) Proc. 12th FEBS Meeting, Dresden, in press.

11. Spierer, P., and Zimmermann, R.A. (1978) Biochemistry 17, 2474-2479.

12. Zimmermann, R.A. (1979) Meth. Enzymol. 59, 551-583, Academic Press, N.y.

13. Burrell, H., and Horowitz, J. (1977) Eur. J. Biochem. 75, 533-544.

14. Metspalu, A., Saarma, M., Villems, R., Ustav, M., and Lind, A. (1978) Eur. J. Biochem. 91, 73-81

15. Ulbrich, N., and Wool, I.G. (1978) J. Biol. Chem. 253, 9049-9052.

16. Ulbrich, N., and Wool, I.G. (1979) J. Biol. Chem. 254, in press.

17. Spicer, E., Schwarzbauer, J., and Craven, G.R. (1977) Nucleic Acids Res. 4, 491-499.

18. Spierer, P., Bogdanov, A.A., and Zimmermann, R.A. (1978) Biochemistry 17, 5394-5398.

19. Mizushima, S., and Nomura, M. (1970) Nature 226, 1214-1218.

20. Schaup, H.W., Green, M., and Kurland, C.G. (1970) Mol. Gen. Genet. 109, 193-205.

21. Schaup, H.W., Green, M., and Kurland, C.G. (1971) Mol. Gen. Genet. 112, 1-8.

22. Garrett, R.A., Rak, K.H., Daya, L., and Stöffler, G. (1971) Mol. Gen. Genet. 114, 112-124.

23. Muto, A., Ehresmann, C., Fellner, P., and Zimmermann, R.A. (1974) J. Mol. Biol. 86, 411-432.

24. Held, W.A., Ballou, B., Mizushima, S., and Nomura, M. (1974) J. Biol. Chem. 249, 3103-3111.

25. Hochkeppel, H.-K., Spicer, E., and Craven, G.R. (1976) J. Mol. Biol. 101, 155-170.

26. Hochkeppel, H.-K., and Craven, G.R. (1976) Nucleic Acids Res. 3, 1883-1902.

27. Hochkeppel, H.-K., and Craven, G.R. (1977) Mol. Gen. Genet. 153, 325-329.

28. Littlechild, J.A., and Malcolm, A.L. (1978) Biochemistry 17, 3363-3369.

29. Littlechild, J., Dijk, J., and Garrett, R.A. (1977) FEBS Lett. 74, 292-294.

30. Dijk, J., Littlechild, J., and Garrett, R.A. (1977) FEBS Lett. 77, 295-300.

31. Stöffler, G., Daya, L., Rak, K.H., and Garrett, R.A. (1971) Mol. Gen. Genet. 114, 125-133.

32. Garrett, R.A., Müller, S., Spierer, P., and Zimmermann, R.A. (1974) J. Mol. Biol. 88, 553-557.

33. Horne, J.R., and Erdmann, V.A. (1972) Mol. Gen. Genet. 119, 337-344.

34. Yu, R.S.T., and Wittmann, H.G. (1973) Biochim. Biophys. Acta 324, 375-385.

35. Spierer, P., and Zimmermann, R.A. (1976) J. Mol. Biol. 103, 647-653.

36. Dijk, J., Garrett, R.A. and Müller, R. (1979) Nucleic Acids Res. 6, 2717-2729.

37. Petersson, I. (1979) Nucleic Acids Res. 6, 2637-2646.

38. Feunteun, J., Monier, R., Garrett, R., Le Bret, M., Le Pecq, J.B. (1975) J. Mol. Biol. 93, 535-541.

39. Bear, D.G., Schleich, T., Noller, H.F., and Garrett, R.A. (1977) Nucleic Acids Res. 4, 2511-2526.

40. Zimmermann, J., and Erdmann, V.A. (1978) Mol. Gen. Genet. 160, 247-257.

41. Douthwaite, S., Garrett, R.A., Wagner, R., and Feunteun, J. (1979) Nucleic Acids Res. 6, 2453-2470.

42. Garrett, R.A., and Noller, H.F. (1979) J. Mol. Biol., in press.

43. Nazar, R.N., Willick, G.E., and Matheson, A.T. (1979) J. Biol. Chem. 254, 1506-1512.

44. Nomura, M., Traub, P., and Bechmann, H. (1968) Nature 219, 793-799.

45. Higo, K., Held, W., Kahan, L. and Nomura, M. (1973) Proc. Natl. Acad. Sci. USA 70, 944-948.

46. Daya-Grosjean, L., Geisser, M., Stöffler, G., and Garrett, R.A. (1973) FEBS Lett. 37, 17-20.

47. Thurlow, D.L., and Zimmermann, R.A. (1978) Proc. Natl. Acad. Sci. USA 75, 2859-2863.

48. Woese, C.R., and Fox, G.E. (1977), Proc. Natl. Acad. Sci. USA 74, 5088-5090.

49. Stanley, J., and Ebel, J.-P. (1977) Eur. J. Biochem. 77, 357-366.

50. Stanley, J., Sloof, P., and Ebel, J.-P. (1978) Eur. J. Biochem. 85, 309-316.

51. Wrede, P., and Erdmann, V.A. (1977) Proc. Natl. Acad. Sci. USA 74, 2706-2709.

52. Zimmermann, R.A., Muto, A., and Mackie, G.A. (1974) J. Mol. Biol. 86, 433-450.

53. Schaup, H.W., Sogin, M., Woese, C., and Kurland, C.G. (1971) Mol. Gen. Genet. 114, 1-8.

54. Schaup, H.W. and Kurland, C.G. (1972) Mol. Gen. Genet. 114, 350-357.

55. Zimmermann, R.A., Muto, A., Fellner, P., Ehresmann, C., and Branlant, C. (1972) Proc. Natl. Acad. Sci. USA 69, 1282-1286.

56. Zimmermann, R.A., Mackie, G.A., Muto, A., Garrett, R.A., Ungewickell, E., Ehresmann, C., Stiegler, P., Ebel, J.-P., and Fellner, P. (1975) Nucleic Acids Res. 2, 279-302.

57. Mackie, G. A., and Zimmermann, R.A. (1975) J. Biol. Chem. 250, 4100-4112.

58. Ungewickell, E., Garrett, R., Ehresmann, C., Stiegler, P., and Fellner, P. (1975) Eur. J. Biochem. 51, 165-180.

59. Ungewickell, E., Ehresmann, C., Stiegler, P., and Garrett, R. (1975) Nucleic Acids Res. 2, 1867-1888.

60. Ehresmann, C., Stiegler, P., Carbon, P., Ungewickell, E., and Garrett, R.A. (1977) FEBS Lett. 81, 188-192.

61. Ungewickell, E., Garrett, R.A., Ehresmann, C., Stiegler, P., and Carbon, P. (1977) FEBS Lett. 81, 193-198.

62. Barritault, D., and Hayes, D.H. (1977) Biochimie 59, 463-472.

63. Mackie, G.A., and Zimmermann, R.A. (1978) J. Mol. Biol. 121, 17-39.

64. Brosius, J., Palmer, M.L., Kennedy, P.J., and Noller, H.F. (1978) Proc. Nat. Acad. Sci. USA 75, 4801-4805.

65. Carbon, P., Ehresmann, C., Ehresmann, B. and Ebel, J.P. (1978) FEBS Lett. 94, 152-156.

66. Österberg, R., Sjöberg, B., Garrett, R.A., and Ungewickell, E. (1977) FEBS Lett. 80, 169-172.

67. Cole, M.D., Beer, M., Koller, T., Strycharz, W.A., and Nomura, M. (1978) Proc. Natl. Acad. Sci. USA 75, 270-274.

68. Zimmermann, R.A., and Singh-Bergmann, K. (1979) Biochim. Biophys. Acta, in press.

69. Müller, R., Garrett, R.A., and Noller, H.F. (1979) J. Biol. Chem. 254, 3873-3878.

70. Schaup, H.W., Sogin, M., Kurland, C.G., and Woese, C.R. (1973) J. Bacteriol. 115, 82-87.

71. Yuki, A., and Brimacombe, R. (1975) Eur. J. Biochem. 56, 23-34.

72. Zwieb, C., and Brimacombe, R. (1979) Nucleic Acids Res. 6, 1775-1790.

73. Allet, B., and Spahr, P.F. (1971) Eur. J. Biochem. 19, 250-255.

74. Spierer, P., Zimmermann, R.A. and Mackie, G.A. (1975) Eur. J. Biochem. 52, 459-468.

75. Spierer, P., Zimmermann, R.A., and Branlant, C. (1976) FEBS Lett. 68, 71-75.

76. Spierer, P., Wang, C.-C., Marsh, T.L., and Zimmermann, R.A. (1979) Nucleic Acids Res. 6, 1669-1682.

77. Branlant, C., Krol, A., Sriwidada, J., Fellner, P., and Crichton, R.R. (1973) FEBS Lett. 35, 265-273.

78. Branlant, C., Sri Widada, J., Krol, A., and Ebel, J.-P. (1977) Eur. J. Biochem. 74, 155-170.

79. Sloof, P., Hunter, J.B., Garrett, R.A., and Branlant, C. (1978) Nucleic Acids Res. 5, 3503-3513.

80. Krol, A., Machatt, M.A., Branlant, C., and Ebel, J.P. (1978) Nucleic Acids Res. 5, 4933-4947.

81. Branlant, C., Krol, A., Sriwidada, J., Ebel, J.-P., Sloof, P., and Garrett, R. (1975) FEBS Lett. 52, 195-201.

82. Sloof, P., Garrett, R., Krol, A., and Branlant, C. (1976) Eur. J. Biochem. 70, 447-456.

83. Branlant, C., Krol, A., Sriwidada, J., Ebel, J.-P., Sloof, P., and Garrett, R.A. (1976) Eur. J. Biochem. 70, 457-469.

84. Branlant, C., Korobko, V., and Ebel, J.-P. (1976) Eur. J. Biochem. 70, 471-482.

85. Branlant, C., Sri Widada, J., Krol, A., and Ebel, J.-P. (1977) Nucleic Acids Res. 4, 4323-4345.

86. Branlant, C., Krol, A., Sriwidada, J., and Brimacombe, R. (1976) Eur. J. Biochem. 70, 483-492.

87. Gray, P.N., Bellemare, G., Monier, R., Garrett, R.A., and Stöffler, G. (1973) J. Mol. Biol. 77, 133-152.

88. Fox, G.E., and Woese, C.R. (1975) Nature 256, 505-506.

89. Österberg, R., and Garrett, R.A. (1977) Eur. J. Biochem. 79, 67-82.

90. Gray, P.N., Garrett, R.A., Stöffler, G., and Monier, R. (1972) Eur. J. Biochem. 28, 413-421.

91. Gray, P.N., and Monier, R. (1972) Biochimie 54, 41-45.

92. Herr, W., and Noller, H.F. (1975) FEBS Lett. 53, 248-252.

93. Nazar, R.N. (1979) J. Biol. Chem., in press.

94. Lemieux, G., Lefevre, J.-F., and Daune, M. (1974) Eur. J. Biochem. 49, 185-194.

95. Allen, S. H., and Wong, K.-P. (1978) Biophys. J. 23, 359-373.

96. Morrison, C.A., Bradbury, E.M., Littlechild, J., and Dijk, J. (1977) FEBS Lett. 83, 348-352.

97. Stöffler, G., and Wittmann, H.G. (1977) in Molecular Mechanisms of Protein Biosynthesis (Weissbach, H., and Pestka, S., Eds.) pp. 117-202, Academic Press, New York.

98. Österberg, R., Sjöberg, B., and Littlechild, J. (1978) FEBS Lett. 93, 115-119.

99. Österberg, R., Sjöberg, B., and Garrett, R.A. (1976) FEBS Lett. 65, 73-76.

100. Giri, L., Littlechild, J., and Dijk, J. (1977) FEBS Lett. 79, 238-244.

101. Giri, L., Franz, A., and Dijk, J. (1979) Biochemistry 18, 2520-2525.

102. Moore, P.B., Langer, J.A., Schoenbrun, B.P., and Engelman, D.M. (1977) J. Mol. Biol. 112, 199-234.

103. Serdyuk, I.N., Zaccai, G., and Spirin, A.S. (1978) FEBS Lett. 94, 349-352.

104. Daya-Grosjean, L., Garrett, R.A., Pongs, O., Stöffler, G., and Wittmann, H.G. (1972) Mol. Gen. Genet. 119, 277-286.

105. Daya-Grosjean, L., Reinbolt, J., Pongs, O., and Garrett, R.A. (1974) Mol. Gen. Genet. 44, 253-256.

106. Amons, R., Möller, W., Schiltz, E., and Reinbolt, J. (1974) FEBS Lett. 41, 135-138.

107. Lemieux, G. (1974) Can. J. Biochem. 52, 1038-1043.

108. Reinbolt, J., Tritsch, D., and Wittmann-Liebold, B. (1978) FEBS Lett. 91, 297-301.

109. Schulte, C., Schiltz, E., and Garrett, R. (1975) Nucleic Acids Res. 2, 931-942.

110. Changchien, L.-M., and Craven, G.R. (1976) J. Mol. Biol. 108, 381-401.

111. Newberry, V., Yaguchi, M., and Garrett, R.A. (1977) Eur. J. Biochem. 76, 51-61.

112. Changchien, L.-M., Schwarzbauer, J., Cantrell, M., and Craven, G.R. (1978) Nucleic Acids Res. 5, 2789-2799.

113. Changchien, L.-M., and Craven, G.R. (1978) J. Mol. Biol. 125, 43-56.

114. Bruce, J., Firpo, E.J., and Schaup, H.W. (1977) Nucleic Acids Res. 4, 3327-3340.

115. Newberry, V., Brosius, J., and Garrett, R. (1978) Nucleic Acids Res. 5, 1753-1766.

116. Bear, D.G., Schleich, T., Noller, H.F., Douthwaite, S., and Garrett, R.A. (1979) FEBS Lett. 100, 99-102.

117. Folkhard, W., Pilz, I., Kratky, O., Garrett, R., and Stöffler, G. (1975) Eur. J. Biochem. 59, 63-71.

118. Bogdanov, A.A., Zimmermann, R.A., Wang, C.-C., and Ford, N.C., Jr. (1978) Science 202, 999-1001.

119. Allen, S.H., and Wong, K.-P. (1979) J. Biol. Chem. 254, 1775-1777.

120. Vasiliev, V.D., Koteliansky, V.E., Shatsky, I.N., and Rezapkin, G.V. (1977) FEBS Lett. 84, 43-47.

121. Vasiliev, V.D., Selivanova, O.M., and Koteliansky, V.E. (1978) FEBS Lett. 95, 273-276.

122. Österberg, R., Sjöberg, B., Garrett, R.A., and Littlechild, J. (1978) Nucleic Acids Res. 5, 3579-3587.

123. Ungewickell, E., Garrett, R.A., and Le Bret, M. (1977) FEBS Lett. 84, 37-42.

124. Allen, S.H., and Wong, K.-P. (1978) J. Biol. Chem. 253, 8759-8766.

125. Seals, A.A., and Champney, W.S. (1976) Biochem. Biophys. Res. Commun. 72, 753-760.

126. Fox, J.W., and Wong, K.-P. (1978) J. Biol. Chem. 253, 18-20.

127. Smith, N., Matheson, A.T., Yaguchi, M., Willick, G.E., and Nazar, R.N. (1978) Eur. J. Biochem. 89, 501-509.

128. Tritton, T.R., and Crothers, D.M. (1976) Biochemistry 15, 4377-4385.

129. Ehresmann, B., Reinbolt, J., and Ebel, J.P. (1975) FEBS Lett. 58, 106-111.

130. Ehresmann, B., Reinbolt, J., Backendorf, C., Tritsch, D., and Ebel, J.P. (1976) FEBS Lett. 67, 316-319.

131. Ehresmann, B., Backendorf, C., Ehresmann, C., and Ebel, J.P. (1977) FEBS Lett. 78, 261-266.

132. Möller, K., and Brimacombe, R. (1975) Mol. Gen. Genet. 141, 343-355.

133. Gorelic, L. (1975) Biochemistry 14, 4627-4633.

134. Gorelic, L. (1975) Biochim. Biophys. Acta 390, 209-225.

135. Baca, O.G., and Bodley, J.W. (1976) Biochem. Biophys. Res. Commun. 70, 1091-1096.

136. Turchinsky, M.F., Broude, N.E., Kussova, K.S., Abduraschidova, G.G., Muchamedganova, E.V., Schatsky, I.N., Bystrova, T.F., and Budowsky, E.I. (1978) Eur. J. Biochem. 90, 83-88.

137. Zook. D.E., and Fahnestock, S.R. (1978) Biochim. Biophys. Acta 517, 400-406.

138. Möller, K., Zwieb, C., and Brimacombe, R. (1978) J. Mol. Biol. 126, 489-506.

139. Gorelic, L. (1976) Biochemistry 15, 3579-3590.

140. Gorelic, L. (1976) Biochim. Biophys. Acta 454, 185-192.

141. Rinke, J., Yuki, A., and Brimacombe, R. (1976) Eur. J. Biochem. 64, 77-89.

142. Zwieb, C., and Brimacombe, R. (1978) Nucleic Acids Res. 5, 1189-1206.

143. Kenner, R.A. (1973) Biochem. Biophys. Res. Commun. 51, 932-937.

144. Czernilofsky, A.P., Kurland, C.G., and Stöffler, G. (1975) FEBS Lett. 58, 281-284.

145. Rinke, J. and Brimacombe, R. (1978) Mol. Biol. Rep. 4, 153-156.

146. Möller, K., Rinke, J., Ross, A., Buddle, A., and Brimacombe, R. (1977) Eur. J. Biochem. 76, 175-187.

147. Oste, C., Parfait, R., Bollen, A., and Crichton, R.R. (1977) Mol. Gen. Genet. 152, 253-257.

148. Ulmer, E., Meinke, M., Ross, A., Fink, G., Brimacombe, R. (1978) Mol. Gen. Genet. 160, 183-193.

149. Oste, C., and Brimacombe, R. (1979) Mol. Gen. Genet. 168, 81-86.

150. Bäumert, H.G., Sköld, S.-E., and Kurland, C.G. (1978) Eur. J. Biochem. 89, 353-359.

151. Lake, J.A. (1978) in Advanced Techniques in Biological Electron Microscopy II (Koehler, J.K., Ed.), pp. 173-211, Springer-Verlag, Berlin.

152. Beaudry, P., Petersen, H.V., Grunberg-Manago, M., and Jacrot, B. (1976) Biochem. Biophys. Res. Commun. 72, 391-397.

153. Stuhrmann, H.B., Koch, M.H.J., Parfait, R., Haas, J., Ibel, K., and Crichton, R.R. (1978) J. Mol. Biol. 119, 203-212.

154. Kopylov, A.M., Teterina, N.L., Potapov, A.P., Vasilenko, S.K., and Bogdanov, A.A. (1976) Bioorg. Khim. 2, 1642-1651.

155. Fellner, P., Ehresmann, C., Ebel, J.-P. and Blasi, O. (1970) Eur. J. Biochem. 13, 583-588.

156. Rinke, J., Ross, A., and Brimacombe, R. (1977) Eur. J. Biochem. 70, 189-196.

157. Santer, M., and Shane, S. (1977) J. Bacteriol. 130, 900-910.

158. Noller, H.F. (1974) Biochemistry 13, 4694-4703.

159. Chapman, N.M., and Noller, H.F. (1977) J. Mol. Biol. 109, 131-149.

160. Herr, W., Chapman, N.M., and Noller, H.F. (1979) J. Mol. Biol. 130, 433-449.

161. Garrett, R.A., Ungewickell, E., Newberry, V., Hunter, J., and Wagner, R. (1977) Cell Biol. Internat. Rep. 1, 487-502.

162. Bowman, C.M., Dahlberg, J.E., Ikemura, T., Konisky, J. and Nomura, M. (1971) Proc. Natl. Acad. Sci. USA 68, 964-968.

163. Schreiber, J.P., Hsiung, N., and Cantor, C.R. (1979) Nucleic Acids Res. 6, 181-193.

164. Thammana, P., and Cantor, C.R. (1978) Nucleic Acids Res. 5, 805-823.

165. Politz, S.M., and Glitz, D.D. (1977) Proc. Natl. Acad. Sci. USA 74, 1468-1472.

166. Serdyuk, I.N., and Grenader, A.K. (1977) Eur. J. Biochem. 79, 495-504.

167. Crichton, R.R., Engelman, D.M., Haas, J., Koch, M.H.J., Moore, P.B., Parfait, R., and Stuhrmann, H.B. (1977) Proc. Natl. Acad. Sci. 74, 5547-5550.

168. Chen-Schmeisser, U., and Garrett, R. (1976) Eur. J. Biochem. 69, 401-410.

169. Homann, H.E., and Nierhaus, K.H. (1971) Eur. J. Biochem. 20, 249-257.

170. Kühlbrandt, W., and Garrett, R.A. (1978) FEBS Lett. 94, 207-212.

171. Spillmann, S., Dohme, F., and Nierhaus, K.H. (1977) J. Mol. Biol. 115, 513-523.

172. Spillmann, S., and Nierhaus, K.H. (1978) J. Biol. Chem. 253, 7047-7050.

173. Boublik, M. (1978) Cytobiologie 18, 309-319.

STRUCTURAL ORGANIZATION OF THE *ESCHERICHIA COLI* RIBOSOME AND LOCALIZATION OF FUNCTIONAL DOMAINS

G. STÖFFLER, R. BALD, B. KASTNER, R. LÜHRMANN, M. STÖFFLER-MEILICKE, G. TISCHENDORF

Max-Planck-Institut für Molekulare Genetik, Abt. Wittmann
and B. Tesche Institut für Elektronenmikroskopie am Fritz-Haber-Institut
Berlin-Dahlem, Germany

INTRODUCTION

The aim of research on ribosomes is to elucidate the structure and function of this large multimolecular assembly. The Escherichia coli ribosome is made up of 53 separate protein molecules and comprises three molecules of rRNA. In order to understand how the ribosome performs its task in protein synthesis, our aim should be to define the three-dimensional structure of the ribosome and its subunits, to specify the precise coordinates, and to define the exact function of each of the molecules in the ribosome. This knowledge may then allow us to localize the various functional domains where mRNA, tRNAs and protein synthesis factors interact with the ribosome during protein synthesis.

Since ribosome crystals suitable for X-ray diffraction are not yet available, a number of elegant techniques have been applied or especially developed, each of them contributing considerable knowledge to the spatial arrangement of the proteins and the rRNA molecules within the ribosome (1-3). One of these approaches, namely the mapping of the locations of individual ribosomal proteins by immuno electron microcopy, using antibodies against specific proteins, has considerably extended our knowledge of the spatial arrangement of ribosomal proteins, and has indirectly provided some ideas as to the location of functional domains on the ribosomal surface. In this article we shall describe our recent work on the morphology of ribosomes, on the location of individual proteins, and our attempts to localize functional domains and antibiotic binding sites on the ribosomal surface in a direct manner.

THE MORPHOLOGY OF NEGATIVELY STAINED ESCHERICHIA COLI RIBOSOMES AND RIBOSOMAL SUBUNITS

Electron microscopy of ribosomes has proved to be an extremely useful technique for studying their morphology (for references see 2-6).

70S ribosomes have globular, polygonal or slightly prolate contours (Fig. 1a). The structures of the 30S and the 50S subunits are readily recognizable. The small subunit is arranged horizontally transverse with respect to the 50S subunit. The groove and the lobes of the 30S subunit are oriented towards the seat of the 50S particle. Thus, a channel is formed between the two subunits (Figs. 1a,7c).

Electron micrographs of negatively stained 30S subunits show asymmetric and quasisymmetric images (Fig. 1b) which represent different projections of a single three-dimensional structure. The 30S subunit is an elongated, polar and asymmetric structure (Fig. 1b). A groove perpendicular to the long axis divides the 30S subunit into a rounded head which is somewhat bent with a body which is approximately twice as long and narrows towards the opposite end. Two lobes of unequal size protrude from the body and break the apparent symmetry in the images (Figs. 1b, 7a).

The images of ribosomal particles as seen on electron micrographs may be affected by a number of parameters, such as the ribosome preparation procedure, the functional activity, the mode of specimen deposition or the staining and dehydration conditions. Despite the considerable differences in the preparative steps used by the various investigators, the actual electron micrographs show a remarkable degree of similarity, at least for 30S subunits and 70S ribosomes (7-14). In contrast, the three-dimensional models proposed by the different groups vary considerably, although they do share some common structural elements (2,3,6).

Electron micrographs of 50S subunits show different images such as crowns, kidneys with an asymmetric notch, and rounded forms. Three protuberances arise from the plane of a hemispherical base (Fig. 2). In contrast to electron micrographs of 30S subunits and 70S ribosomes, there are significant differences between the actual electron micrographs of large subunits published so far. The differences relate to the sizes, shapes and origins of these three protuberances. Boublik (13,15) and Spiess (16) emphasized that the two lateral protuberances are

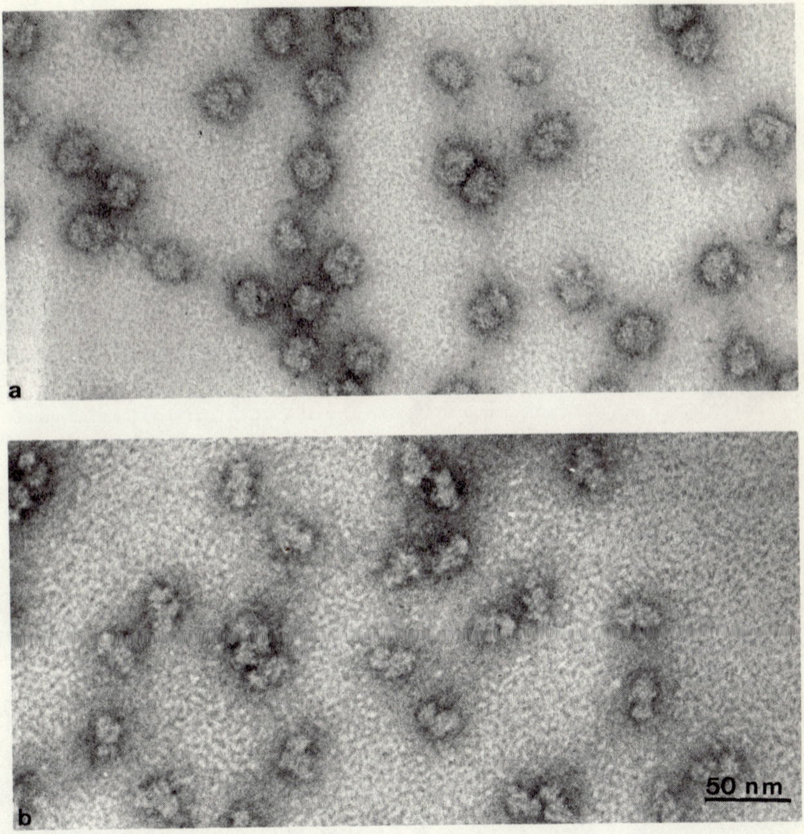

Figure 1. A field of E. coli 70S ribosomes (a) and of 30S ribosomal subunits (b), negatively contrasted with uranyl acetate by the double layer carbon technique.

unequal in size. According to Lake (11), one of the protuberances resembles a rod-like appendage (100 X 40 Å). A filamentous structural element arising from the 50S subunit has been described by Nanninga (17), which could be completely removed by mild treatment with nucleases. A rod-like structure can be seen in images of 50S subunits as published by Tischendorf et al. (18), but was neither described nor introduced into the proposed model. For all the above reasons we have now systematically investigated the morphology of three preparations of 50S subunits, and of 50S core particles (Pi-cores) lacking only proteins L7/L12 (19). 50S subunit preparations dissociated with initiation factors IF-3 and IF-1 at 6 mM Mg^{++} (20) are abundant in highly active subunits, and should thus be closest to native particles. Most of these particles showed the asymmetrical crown images and thus the appendage (Fig. 2a). 50S subunits

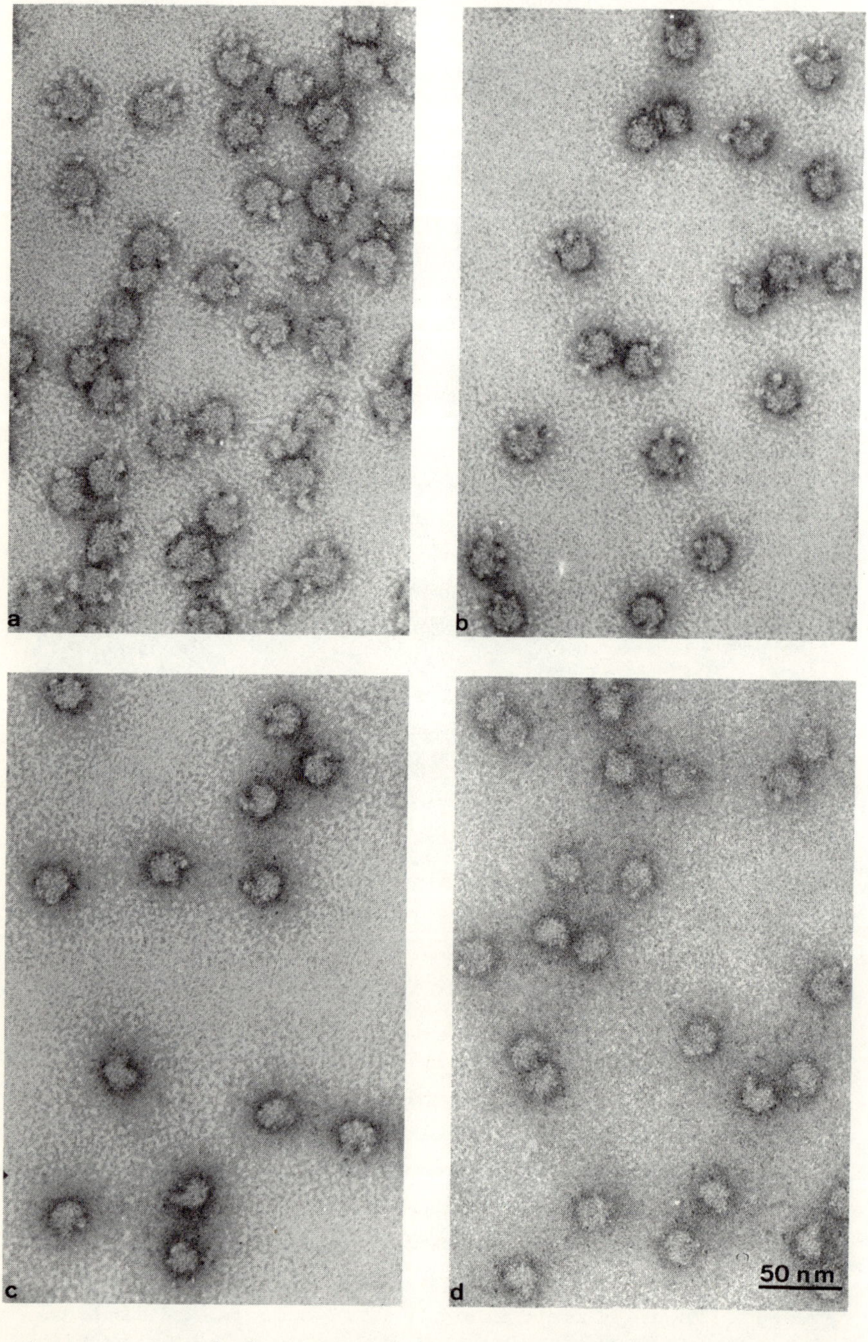

Figure 2. Negatively contrasted 50S subunits dissociated with initiation factors (a), centrifuged (b), ethanol precipitated (c). Pi-cores are shown in (d).

prepared by high speed centrifugation (Fig. 2b) exhibited structural features similar to subunits dissociated by initiation factors (compare Fig. 2a with 2b). Ethanol precipitated ribosomes showed symmetrical crowns without appendages, and kidney forms contributed almost 50% of the total images. Therefore, ethanol precipitation either removes or obscures the visualization of the appendage and also significantly alters the distribution of the various projections, a result that agrees with Spiess (16). A similar image frequency was, however, also observed in a 50S preparation which showed, at least in retrospect, the appendage (18). Pi-cores are shown in Figure 2d. They lacked exclusively proteins L7/L12 and could be reconstituted with any preparation of pure L7/L12 to full activity. Again, the appendage was missing and a random distribution of particle projections was observed, a result in agreement with others (15,21).

Although negative staining is frequently used for contrast enhancement of ribosomes, it has some disadvantages which will cause distortion of the specimen. For example, the ribosome surface is not only surrounded with electron dense stain, but the stain also reacts to different degrees with certain surface areas, depending on their charge. Moreover, drying of highly hydrated particles leads to shrinkage, and even the stain would itself undergo shrinkage, and thus a redistribution would occur on the specimen surface upon radiation.

Thus, instead of overinterpreting electron micrographs of ribosomes, obtained by similar preparation procedures, new methods have to be applied which provide a better structure preservation during specimen preparation and improve the electron optical resolution. This has been achieved by a new technique for tungsten shadowing of freeze-dried ribosomal particles with a low specimen damage electron impact evaporator (6,22,23).

HIGH RESOLUTION SHADOW CASTING OF ESCHERICHIA COLI RIBOSOMES AND RIBOSOMAL SUBUNITS

We have applied high resolution shadowing, in different ways, for investigating the morphology of E. coli ribosomes and their subunits. The most complete studies have been undertaken with unidirectional shadowing (Figs. 3-5). Furthermore, stereo-electron micrographs have been made (Fig. 4). Finally, portrait and rotatory shadowing have been applied (see Fig. 6). The shadows of the electron micrographs show

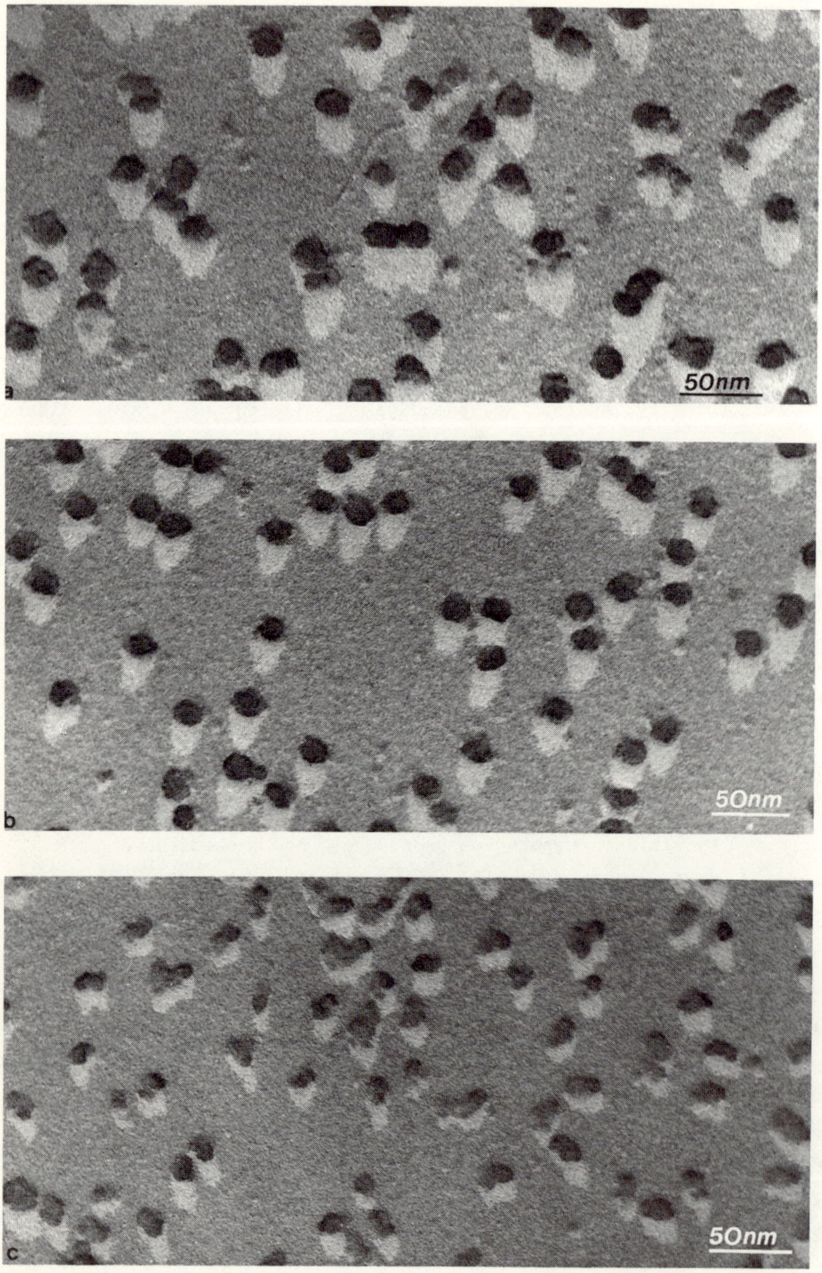

Figure 3. Freeze-dried and tungsten-shadowed E. coli 70S ribosomes (a), 50S subunits (b), and 30S subunits (c).

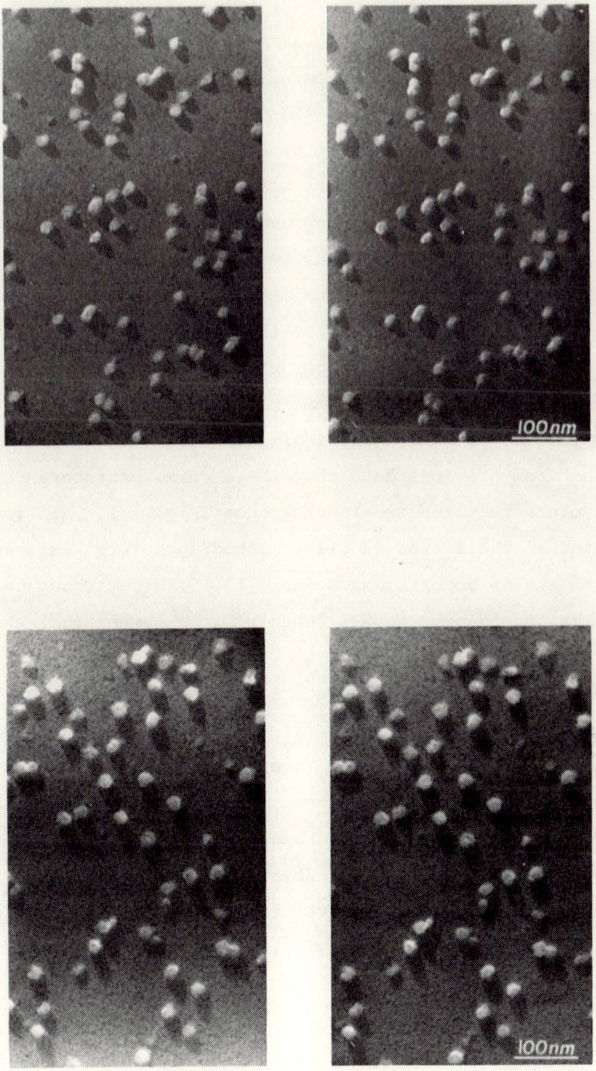

Figure 4. Stereo pair pictures of tungsten-shadowed 50S subunits (a) and 70S ribosomes (b).

sharply delineated boundaries and typical structures (Figs. 3-5). We attempted to derive three-dimensional models from these shadowgraphs. We proceeded in principle as previously illustrated (24). Shadowcasts were prepared of macroscopic models made as described by ourselves, and other authors, and they were compared with the shadowcasts found in the micrographs. Interpretational difficulties, or ambiguities as exemplified previously (24) , could then be eliminated by invoking the results of either the two-directional portrait or rotatory shadowcasting or both. The evaluation of stereo-shadowgraphs turned out to be extremely useful (Fig. 4a). The main conclusions drawn from all these experiments are summarized in the three-dimensional models depicted in Figure 7. We confirmed that two lobes arise from the body of the 30S subunit, but one of them is significantly larger than the other (Figs. 3c,6,7a). Thus, the 30S subunit has an asymmetric structure.

The evaluation of shadowgraphs of 50S subunits confirmed our supposition from previous data that the central protuberance is in a different plane from the two lateral ones (Fig. 7b). This is best demonstrated on the stereo-electron micrographs (Fig. 4a) and agrees with the models of Boublik and Spiess (13,15,16). Without considering the rod-like appendage, the 50S subunit has a symmetrical structure in its gross morphology.

The appendage is now considered. Its visualization should be straightforward, since the thin, fine-grain tungsten layer gives a resolution of 6 Å (22,23) and biomolecules with similar dimensions as the appendage, e.g. antibodies have been visualized by this technique (25). In a comparative experiment, we shadowed both 50S subunits prepared by high speed centrifugation (Figs. 3b,5a) and by ethanol-precipitation (Fig. 5b). Shadows that could have been cast by the appendage were found in each preparation (Figs. 3b,5a,5b). The frequency of particles showing this structure was approximately 10%, and was not lower in ethanol precipitated samples (Fig. 5).

The reasons for this discrepancy between results obtained with negatively stained (Fig. 2b,c) and shadowed 50S subunits (Figs. 3b and 5) are unknown. More has to be learned about the nature, the origin, and the rigidity or flexibility of the appendage. From the shadows of 50S subunits showing an appendage-like structure it is obvious that it is not one of the three main protuberances; there is evidence that this structures arises from the seat region in the proximity of the main protuberance.

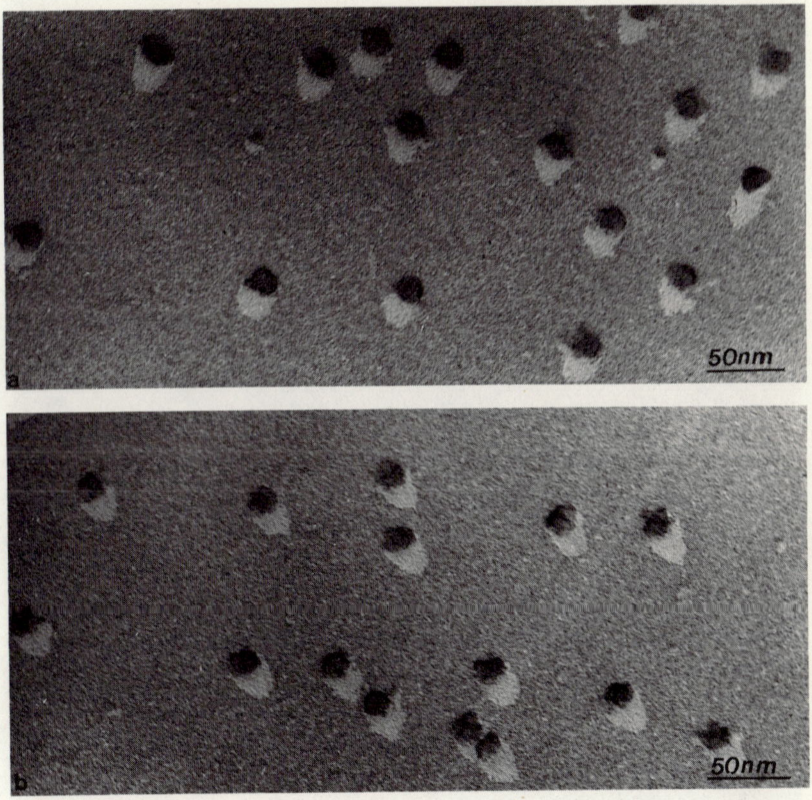

Figure 5. Shadowgraphs of 50S subunits centrifuged (a) and ethanol precipitated (b), obtained with the same 50S subunit preparations shown in Figures 2a and 2c. Electron micrographs depicted in Figures 3-6 were taken in a Zeiss EM 10A with an instrumental magnification of 100.000 at 80 kV (for details see 22 and 23). Electron micrographs shown in Figures 3-6 are from B. Tesche, G. Tischendorf and G. Stöffler (unpublished results).

Shadowgraphs obtained with 70S ribosomes clearly demonstrate that the small subunit is oriented horizontally transverse with respect to the 50S particle (Fig. 3a). The groove and the lobes of the 30S subunit are oriented towards the seat of the 50S particle. The proposed arrangement shown in Figure 7c is also experimentally supported by the stereo-electron micrographs (Fig. 4b) and is also in agreement with the subunit arrangement proposed by Boublik et al. (12).

The models shown in Figure 7 are, at the moment, the best approximation of our data to a three-dimensional ribosome structure. The portrait shadowcasting and rotatory evaporation techniques are being further applied to refine and, if necessary, to alter these models. We

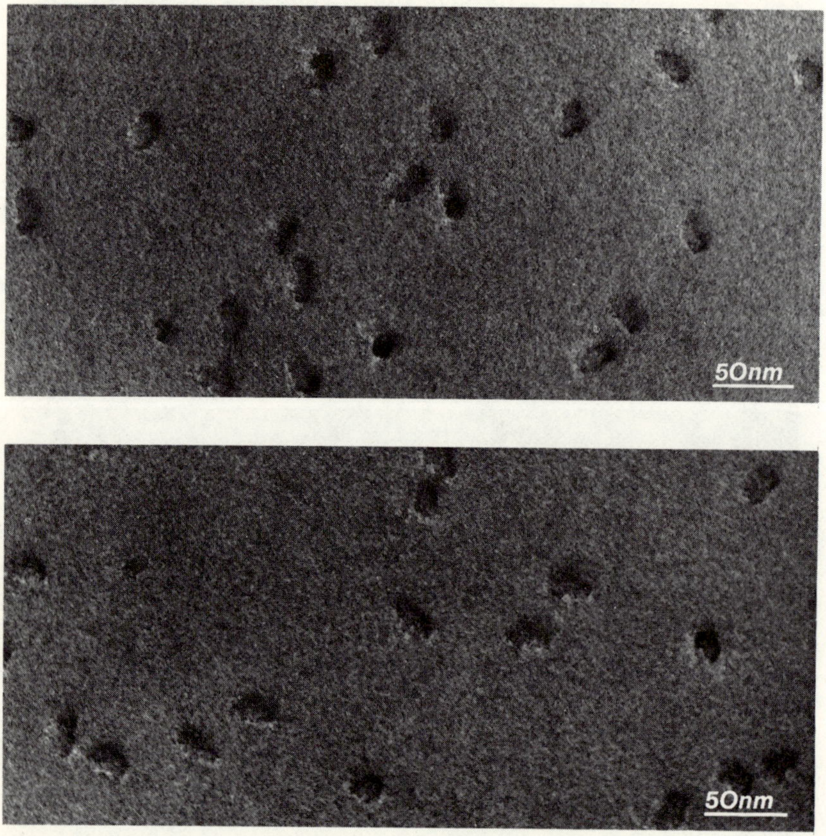

Figure 6. *Two typical fields of freeze-dried E. coli 30S subunits after multidirectional shadowing.*

give it as our view that the present state of the art can be developed further.

Optical diffraction and computer image processing procedures have already been applied to electron micrographs of negatively stained 30S subunits, and a tilt series of subunits from +60° to -52.5° has been published (26). Results of the three-dimensional reconstruction from this tilt series are in good agreement with our 30S subunit model (V. Knauer, H.J. Schramm, W. Hoppe, G.W. Tischendorf, G. Stöffler and H.G. Wittmann, unpublished results). Rotatory shadowing of ribosomal particles will enhance the resolution, reduce the irradiation damage and eliminate problems related to artifacts in negatively stained and air-dried specimens; these particles should be considerably better suited to image processing procedures.

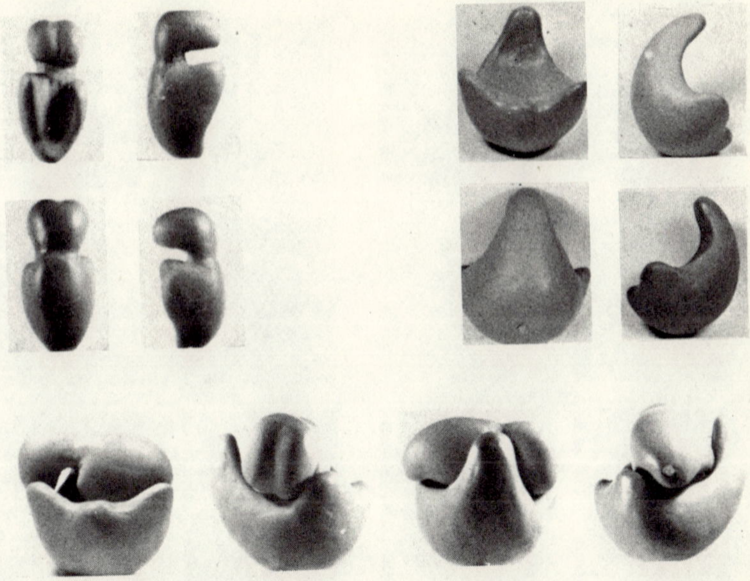

Figure 7. Three-dimensional models of 30S subunits (a), 50S subunits (b) and 70S ribosomes (c). The four views presented were derived by successive rotation through an angle of 90°. These models are modified from previous ones (2,3,10) and contain results derived from negatively contrasted and tungsten-shadowed ribosomes. In addition to features described in the text, a cleavage on the head of the 30S subunit should be noted. According to rotatory-shadowed 30S subunits (Fig. 6) the indentation between the lobes could be shallower or possibly not exist. A minor hump-like projection between the two lateral protuberances of the 50S subunit can be seen. The rod-like appendage has not been included in the model of 50S subunits (unpublished results of B. Tesche, G. Tischendorf and G. Stöffler).

PROTEIN TOPOGRAPHY OF ESCHERICHIA COLI RIBOSOMAL SUBUNITS AS INFERRED FROM IMMUNO ELECTRON MICROSCOPY

Immuno electron microscopy combines the use of antibodies against individual ribosomal proteins with electron microscopy. Two ribosomal subunits joined by a bivalent IgG-molecule allow the localization of the points of antibody attachment with reasonable accuracy in negatively stained preparations.

Immunochemical Prerequisites for Immuno Electron Microscopy

The main prerequisites are as follows: (a) Specific antisera to each of the 53 ribosomal proteins must be available (27-29). (b) The antisera should only react with their cognate ribosomal proteins. The only excep-

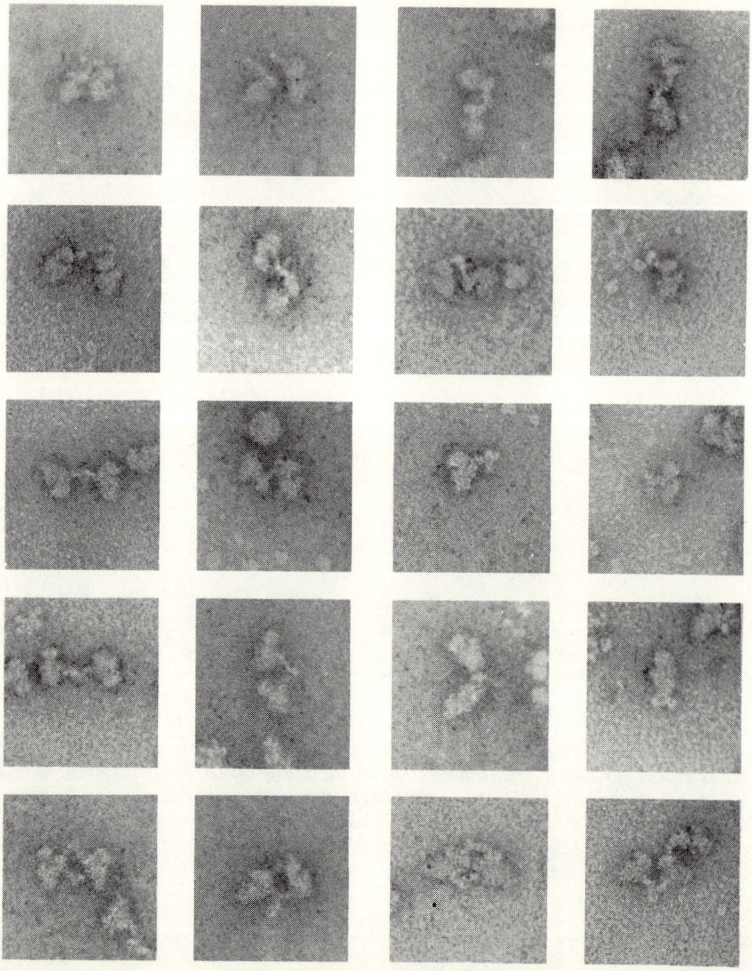

Figure 8. Electron micrographs of 30S·IgG·30S or 30S·IgG complexes showing typical examples of antibody binding sites which can be localized without knowing the exact three-dimensional structure of the 30S subunit.

tions are the protein pairs L7/L12 and S20/L26 (27-29). (c) The antisera must be highly pure. IgG antibodies specific for each of the 30S subunit proteins and some of the 50S subunit proteins have been purified by affinity chromatography with the cognate antigen bound to the agarose (9). The monospecificity of the antisera was established as described (27,28) and, in addition, by a modified immunoelectrophoresis which is

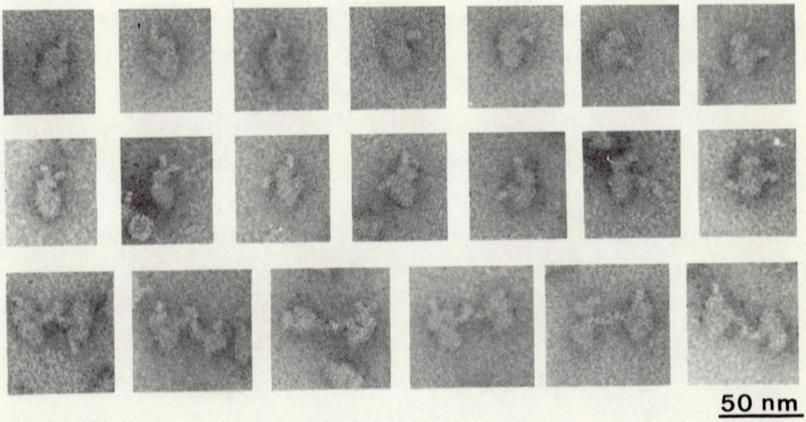

Figure 9. Selected electron micrographs decorated with monovalent Fab-fragments (rows a and b) and double-decorated simultaneously by bivalent IgG and monovalent Fab molecules (row c).

also based on antibody-binding to the immobilized antigen (30). (d) Loss of any acidic antibody directed against basic determinants has to be avoided. This was achieved by purification of the IgG by gel-filtration rather than ion-exchange chromatography (31). (e) Each ribosomal protein should have antigenic determinants on the ribosomal surface which are accessible for antibody binding. This was demonstrated by five independent techniques (29,32,33). (f) The antiserum against a given protein should react exclusively with the cognate protein in the ribosome. These matters have recently been considered in great detail (2,3). It was concluded that whereas crossreactivity does not occur, contamination of antibody is a problem and, therefore we used antibodies purified by affinity chromatography (see below).

Principles of Protein Mapping by Immuno Electron Microscopy

Firstly, we shall attempt to demonstrate the principles for the localization of antibody binding sites on 30S subunits by simply correlating the antibody attachment site to typical structural features (landmarks), visible on any two-dimensional ribosomal particle image.

The addition of bivalent IgG, specific against an individual protein, to ribosomal subunits leads to the formation of subunit pairs ("dimers") which have a higher S-value than the monomeric subunit and can thus be separated from the monomer and unreacted antibody by sucrose gradient centrifugation (7,18,32) or gel filtration (34,35). The dimers are then negatively stained by the double layer carbon technique. Images of subunits connected by one or more bivalent IgG molecules are readily recognized. Subunits connected by an antibody are arranged in a rotatory symmetry (7,18,34,35). Each antibody molecule leads to a constrained orientation of the antibody-subunit complex by different adsorption to the support, and a subunit dimer combined by the same bivalent IgG antibody provides a redundancy which is very useful for the interpretation of the data (7,18,34,35). Thus immuno electron microscopy confirmed the assumption that various 50S subunit images are different projectional forms of one subunit structure (18,34,35).

Needless to say knowledge of the exact three-dimensional structure (and not of a model) is a prerequisite for the three-dimensional localization of an antibody attachment site. Various three-dimensional models have been proposed, but the exact structure of ribosomal subunits and ribosomes is not yet known (see above). Therefore, before proceeding, we emphasize that protein mapping by immuno electron microscopy is still of value without this knowledge. Antibody binding to the head, to the neck, to the body, to the same or the opposite side of the groove, to the upper or lower pole of the 30S subunit are shown in Figure 8, and are independent of a model. Antibody binding to the platform, as described by Lake (5), may, with few exceptions, correspond to binding on the larger lobe in our model. The occurrence of proteins which are elongated in situ can be best demonstrated by pairs of subunits simultaneously connected by two antibody molecules (Fig. 8) which bind at remote sites. Subunit pairs combined by antibodies which bind in close vicinity allow the conclusion that the two exposed antigenic sites are located close to one another. Such "double sites" can be rarely recognized if subunit pairs are connected by one or the other antibody molecule alone (e.g. Fig. 8, last row, first on the left side).

The data summarized in Figure 8 demonstrate that many conclusions concerning the location of antibody attachment sites can be drawn without knowing the exact three-dimensional structure (see also ref. 2).

Since monovalent Fab fragments can be visualized after negative staining (9), we attempted to decorate ribosomal subunits with Fabs against one protein (Fig. 9a,b) and to pair these subunits with a bivalent antibody specific for another protein (Fig. 9c). This technique enables us to determine the locations of two proteins relative to one another. In special cases, e.g. labelling with IgG-Fab combinations such as S14/S20 or S4/S7 or S2/S11, it may allow us to determine the absolute hand of the 30S subunit, and application to the 50S subunit should be even more informative (see below).

The Locations of All 21 Small Subunit Proteins

Immuno electron microscopical studies of protein locations on the 30S subunit of E. coli were reported in 1974 by Lake and collaborators (8) and by the Berlin group (7), and by 1975, 19 protein locations had been published (10). Preliminary locations of all 21 proteins have now been reported by both groups (2,5,6).

In order to exclude artifacts arising from contaminated antisera, we have now reinvestigated the localization of each of the 21 small subunit proteins with specific antibodies, purified by affinity chromatography (G. Tischendorf and G. Stöffler, unpublished results). Electron micrographs of this study are shown in Figure 10, and an attempt at a three-dimensional localization is summarized in Figure 13a. There are relatively few differences from our previous data (2,3,6,10); the differences mainly concern proteins S6, S11 and S15 (for details see Legend to Figure 13a).

Proteins S1, S4, S7, S12, S15, S18 (and with some reservations S11 and S21) should have highly elongated structures within the intact 30S subunit since they revealed multiple antibody binding sites at widely separated points at the subunit surface. Proteins S2, S3, S5, S6, S10, S17 and S19 revealed clearly separated binding sites (Fig. 13a), but due to the size of the Fab-arm it cannot be conclusively decided whether this group of proteins is elongated or not. It should be stressed that even proteins which were mapped at a single site (e.g. S8, S13, S14 and S20) are not necessarily globular (vide 2,3).

The Localization of Large Subunit Proteins

The first ribosomal proteins to be localized by immuno electron microscopy were the 50S subunit proteins L1, L19 and L 27 (34). Subsequently the locations of a total of 19 proteins have been reported by the Berlin group (18,35,36,37). Other groups have so far only attempted to localize the position of proteins L7/L12 (15,21). The locations of 23 proteins on the surface of the 50S subunit have been so far determined; at least six proteins, viz. L1, L2, L6, L11, L18 and L20 have elongated structures in situ (Fig. 13b). A few electron micrographs of

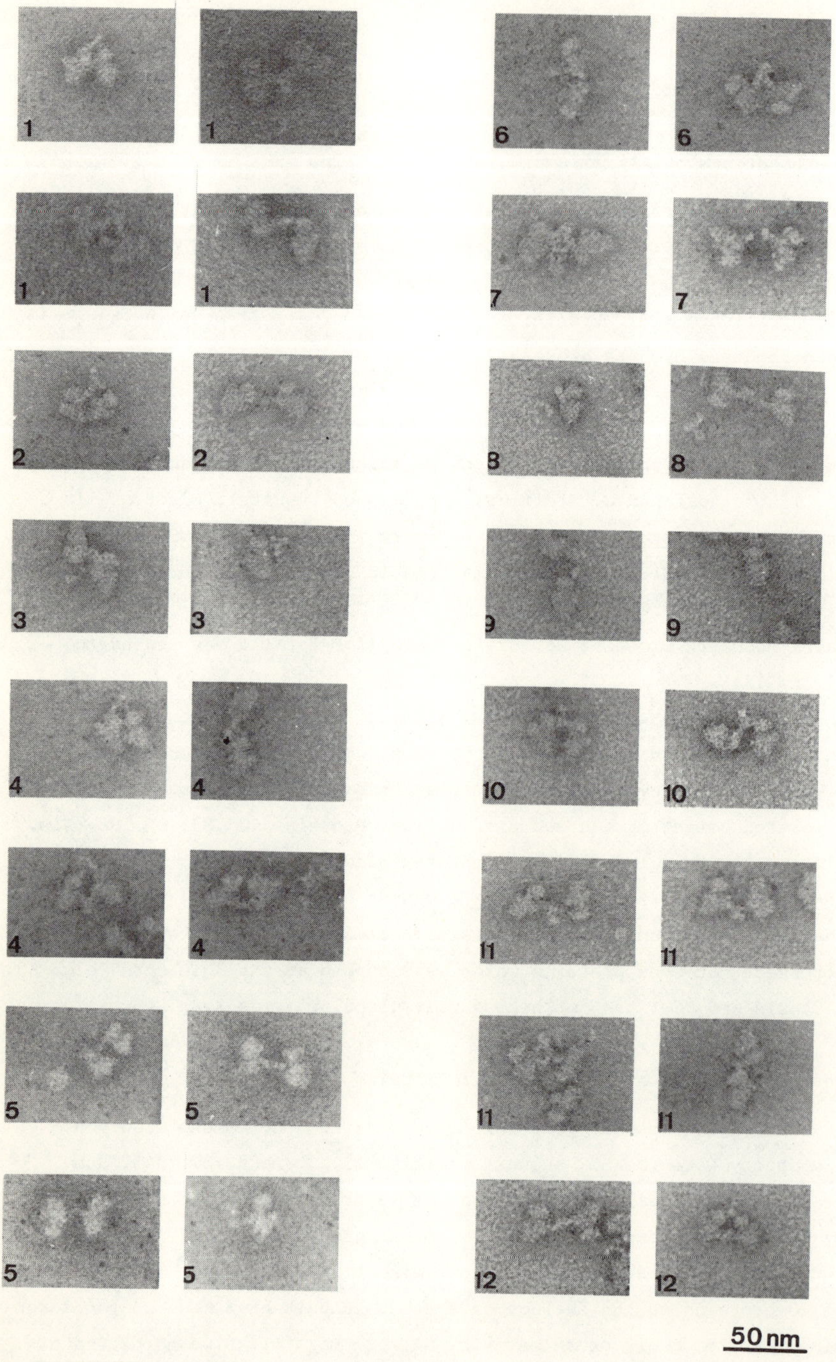

Figure 10. Selected electron micrographs of small subunits reacted with antibodies to ribosomal proteins S1-S21. The numbers indicate the antibody used in the experiment.

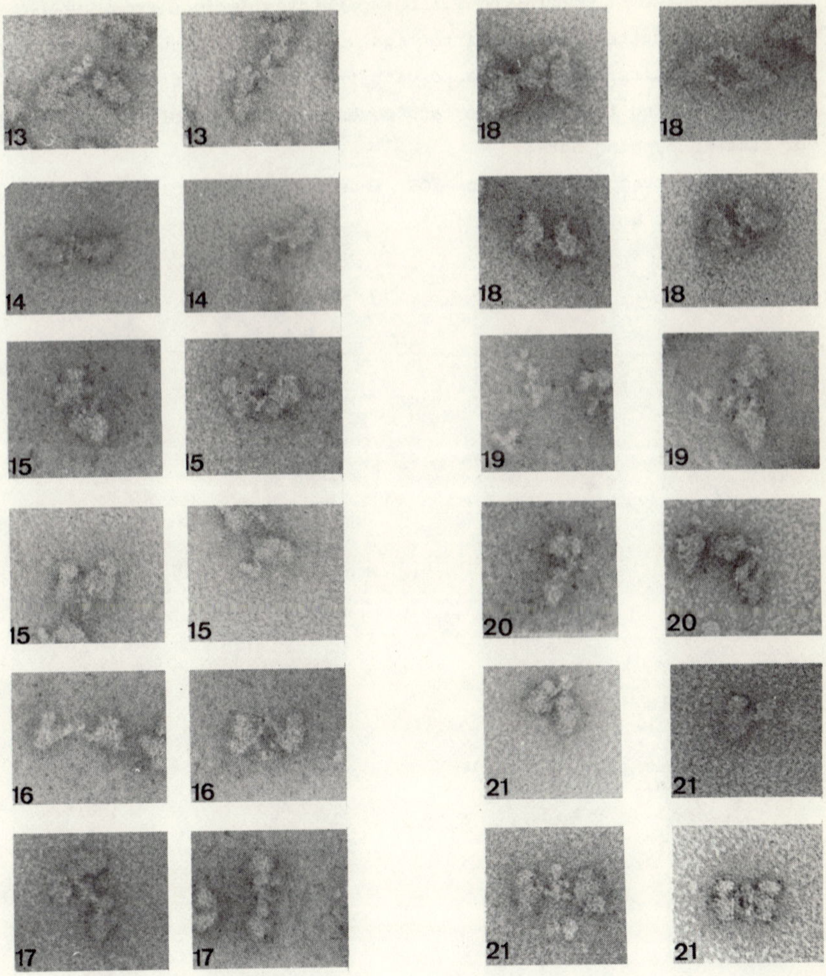

50S subunit pairs connected by antibodies to individual 50S proteins are shown (Fig. 11). The more regular structure of the 50S subunit makes the exact placement of a given antibody site on the three-dimensional model more difficult. An attempt to localize the 23 proteins is summarized in Figure 13b. These locations should be considered as preliminary, since only some of the proteins were mapped with specific antibodies, purified by affinity chromatography.

The application of the double decorating technique with IgGs and Fabs (see Fig. 9) is also being applied to further substantiate the protein localizations.

Recently the structures of various particles derived from 50S ribosomal subunits have been investigated (38). The structures of these depleted subunits were correlated with the locations of the ribosomal proteins missing from the core, and revealed a reasonable agreement with our protein mapping data.

Antibody attachment sites on 70S ribosomes have so far only been determined for a few proteins (37).

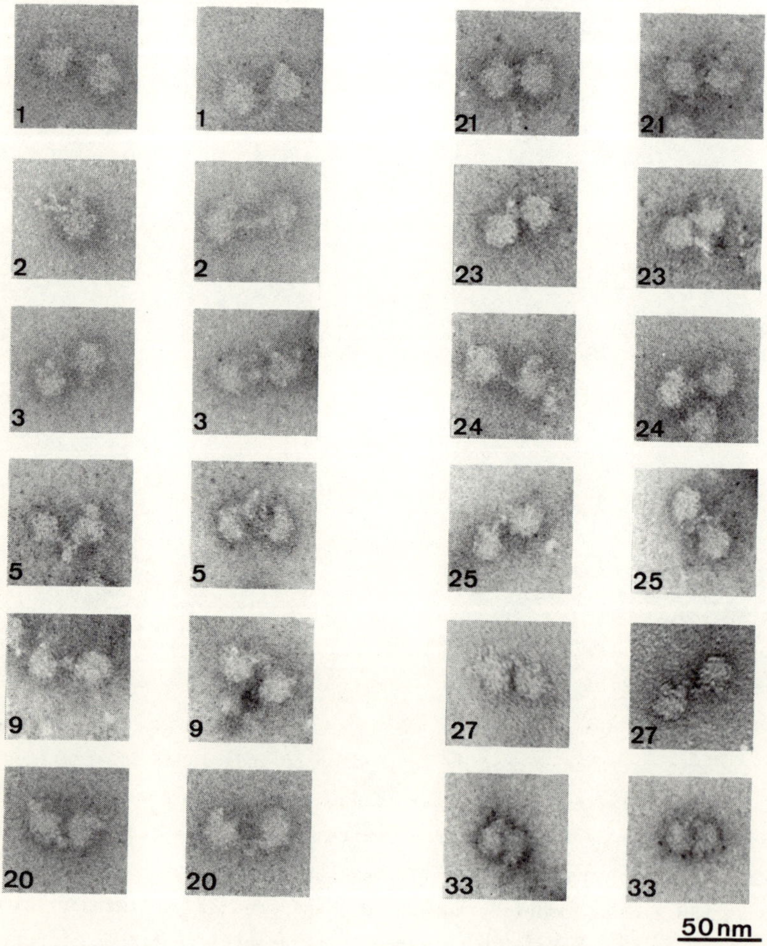

Figure 11. Selected electron micrographs showing the locations of some 50S proteins which have not been documented previously (for additional electron micrographs see refs. 6,18,29,34-37). The numbering system is the same as in Figure 10 (Electron micrographs shown in Figures 8-11 are unpublished results of G. Tischendorf and G. Stöffler).

The Locations of Proteins L7/L12

Proteins L7/L12 have several unique properties: (a) They are present in four copies per ribosome; (b) the amino acid sequences of L7 and L12 are identical (only the amino-terminal seryl-residue is acetylated); (c) proteins L7/L12 are required in all steps of factor-dependent GTP hydrolysis occurring on the ribosome viz. with IF-2, EF-Tu, EF-G, RF-1 and RF-2 (reviewed in 1-3).

Proteins L7/L12 have been mapped by immuno electron microscopy in quite different positions by three groups of investigators (10,15,21). Multiple binding sites, forming an interrupted garland around the central protuberance (10), on both lateral protuberances (15), and on the rod-like appendage (21), have been reported. The localization of an antigenic site of L7/L12 on the appendage by immuno electron microscopy is well documented (21), but, this finding is, however, in contrast to recent experiments of two other groups (16,38). One group showed that loss of the appendage is not associated with the loss of L7/L12, but results from partial depletion of proteins L16 and L28 (38). Spiess (16) indicated that the loss of the appendage is due to ethanol precipitation of 50S subunits, a finding in agreement with our results shown in Figure 2. This author indicated, however, that any protein depletion is unlikely under the conditions used.

By considering that proteins L7/L12 are represented in four copies per ribosome, and that these proteins may have elongated conformations (see H.G. Wittmann, J. Littlechild and B. Wittmann-Liebold, this volume), and by assuming that each copy may be located on a different region of the 50S subunit, the results described above are not necessarily in contradiction. Therefore, we have now systematically reinvestigated the location of proteins L7/L12 on 50S subunits with an appendage (high speed centrifuged) and those showing no appendage (ethanol precipitated). We first added L7/L12-specific IgG to 50S subunits, in amounts comparable to those used for the localization of all other ribosomal proteins: subsequently, in the electron microscope, we found lattices of 50S subunits and antibodies, and this independent of the 50S preparation used (Fig. 12a). This finding is in agreement with the first attempt to localize these proteins, as early as in 1973 by Wabl (34). A localization of the antibody attachment site on the subunit surface is, however, not possible from such lattices.

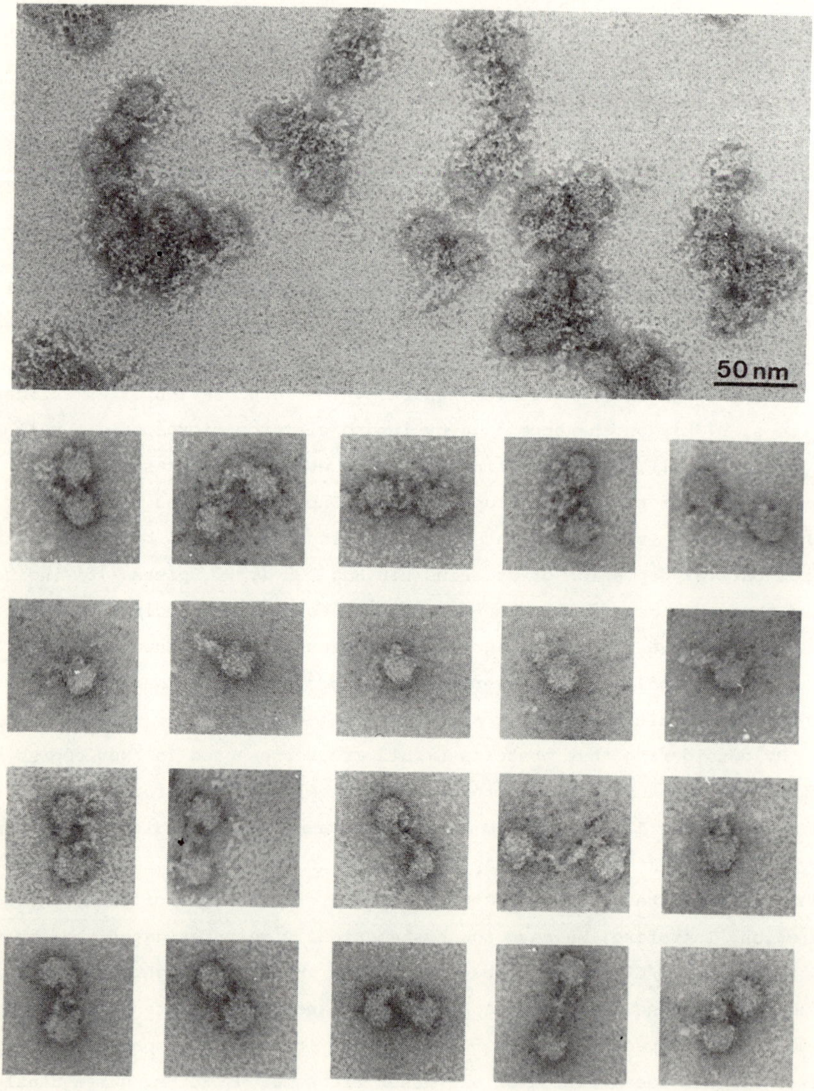

Figure 12. *50S subunits reacted with anti-L7/L12. For detailed description see Text (R. Lührmann, G. Tischendorf and G. Stöffler, unpublished results).*

To keep this aggregation to minimum, dimer formation has been achieved using very low antibody concentrations. If such experiments were performed with centrifuged 50S subunits, the antibodies were found to be preferentially attached to the rod-like appendage (Fig. 12b,c). Binding to the appendage was best seen when an IgG was bound to a monomeric 50S subunit (Fig. 12c). A small percentage of these ribosomes, however, showed antibody attachment sites at the central protuberance (Fig. 12d). When ethanol precipitated 50S subunits, incubated with anti-L7/L12, were investigated by electron microscopy, the resulting dimers showed antibody binding sites located solely at the central protuberance (Fig. 12e).

We next compared the amounts of L7/L12 present in centrifuged and ethanol-precipitated subunits: These proteins are present in reduced amounts in ethanol-precipitated particles, as judged by immunochemical and electrophoretical analyses. Sucrose gradient centrifugation indicated that the proteins L7/L12 which remain, appear to be less accessible for antibody binding on ethanol-precipitated 50S subunits. It should be noted that Pi-core particles were completely depleted of proteins L7/L12 and gave also no dimers with anti-L7/L12 during sucrose gradient centrifugation.

Finally, a comparison of the functional properties of these two subunit preparations has been initiated. Ethanol precipitated subunits showed reduced activity in the EF-Tu and EF-G dependent steps of *in vitro* protein synthesis, as measured individually. However, in the poly(U)-dependent poly(phe)-synthesis, performed with purified EF-Tu and EF-G and with precharged phe-tRNA, both preparations were equally active. The same was found for the poly(U)-dependent non-enzymic phe-tRNA binding. Thus each of these experiments excludes effects of L7/L12 present in the supernatant. Pi-cores, used as a control, were inactive in all the functions tested; they could, however, be functionally reconstituted by the addition of any preparation of proteins L7/L12 (R. Lührmann, R. Kittler, G. Tischendorf and G. Stöffler, unpublished results).

Needless to say, further experiments are required to resolve the chemical nature of the appendage and to localize proteins L7/L12 definitely.

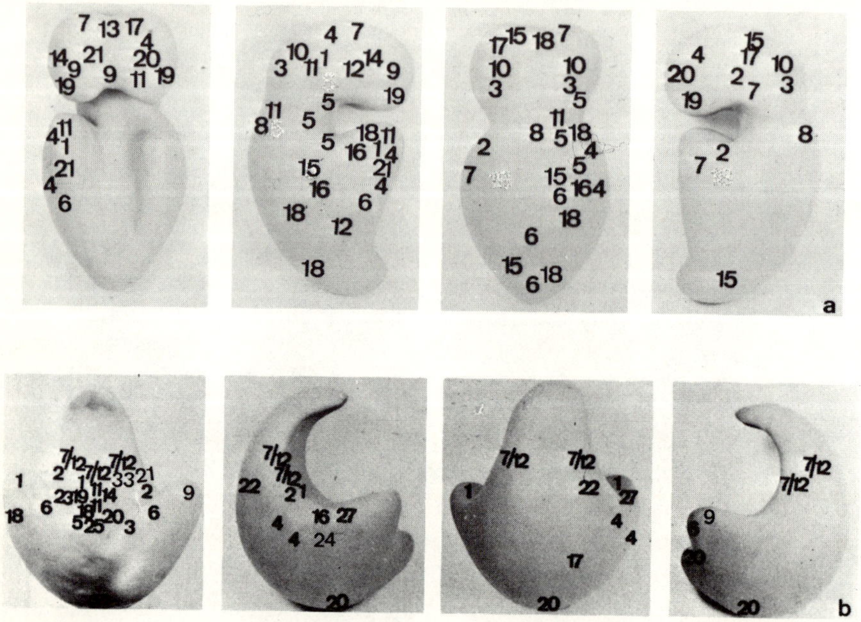

Figure 13. a) Three-dimensional model of the *E. coli* subunit with the locations of the centers of the antibody binding sites for individual proteins. New antibody binding sites have been detected for proteins S6, S15 and S18. One S18 site (on the body) had to be eliminated; specific antibody to proteins S11 and S15 revealed predominantly antibody binding sites on the body; it may be that sites S11A and S11B, and S15A (see ref. 2) were due to contaminating antibody and have to be eliminated. The location of protein S8, though still occurring at a single site at the neck region may be slightly different. The locations of proteins S1, S2, S6 and S21 are being reinvestigated since these proteins are occasionally displaced from the ribosome by their cognate antibodies (G. Tischendorf and G. Stöffler, unpublished results).

b) Three-dimensional model of 50S subunits with preliminary locations of the antibody binding sites for 23 individual ribosomal proteins. The binding site of anti-L7/L12 on the rod-like appendage (see Fig. 12) is not in this scheme.

Note: Where several antibody sites are closely clustered (e.g. in the seat region of 50S or the head of 30S), the relative locations of the individual sites within these clusters are not precisely distinguishable.

The Determination of the Polarity of Elongated Ribosomal Proteins

The polypeptide chain of ribosomal protein S18 consists of 74 amino acids. Enzymatic cleavage of this protein at its two glutamyl-residues yields three fragments, amino acids 1-19, 20-34, and 35-74 (39). Specific antibodies against these fragments were prepared by direct immunization with each of the fragments. In addition, fragment-specific antibodies have been fractionated from antisera directed against the entire protein S18 on immuno-adsorbent columns, bearing the selected polypeptide fragments of protein S18. Antibodies to the amino-terminal fragment 1-19 revealed a single site on the head of the small subunit (Fig. 14a), whereas antibody to the fragment 35-74 gave three binding sites on the body (Fig. 14b,c). Antibody to the fragment 20-34 did not react with 30S subunits, as judged by sucrose gradient centrifugation.

Similar studies have been performed with polypeptide fragments of proteins S4, S7, L1, L7, L11, L21 and L27. Preliminary results indicate that protein S8 has a rather compact structure. The carboxy-terminal end of protein S4 has been located at the large lobe (see also reference 9).

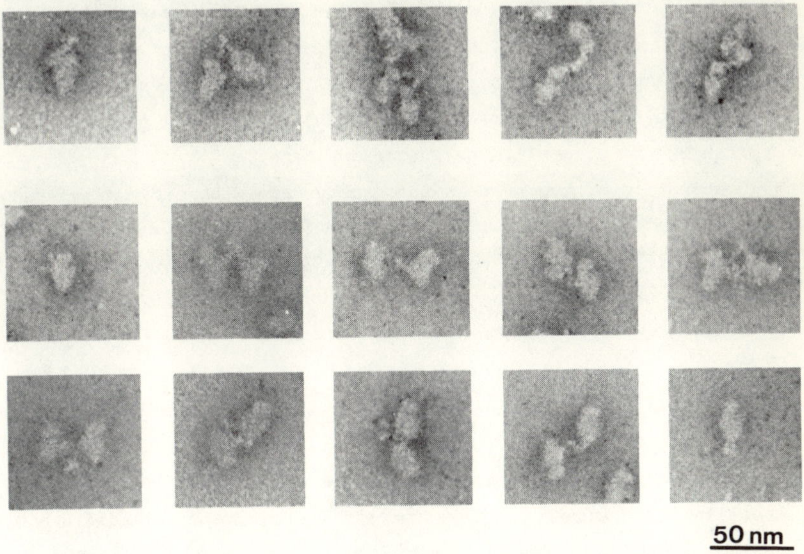

Figure 14. Selected electron micrographs of small subunits reacted with antibodies against defined polypeptide-fragments of protein S18. Antibodies specific for the amino-terminal fragment 1-19 were used in (a), and antibodies specific for a fragment spanning from amino acids 35-74 in (b,c) (data from G. Stöffler, K.H. Rak, G. Tischendorf and M. Yaguchi).

THE LOCALIZATION OF FUNCTIONAL DOMAINS

Enough information now exists about the structural organization of the E. coli ribosome to ask how the accummulated data on the functional involvement of individual proteins can be used to locate functional domains on the ribosome. This approach is limited by the existence of several three-dimensional ribosomal models and by certain discrepancies between the protein mapping results of Lake and ourselves (see above). A more aggravating obstacle is the existence of elongated proteins.

We began our chemical studies under the supposition that ribosomal proteins are globular. Antibodies to individual ribosomal proteins represented, in theory, an ideal tool for the structure-function correlation. We found that monovalent antibody specifically inhibited distinct steps of protein synthesis in vitro (for references 2,29). Since monovalent antibody fragments can be visualized by electron microscopy (9, Fig. 9), we could have then compared the specific inhibitory effect of each Fab with its antibody attachment site, determined by immuno electron microscopy, thus obtaining both topographical and functional data from a single Fab. The notion of globular proteins already had to be revised in 1974 and made this approach to structure-function correlation almost impossible. However, the use of specific antibodies to defined fragments of elongated ribosomal proteins, which bind at a single site only, re-opens this approach, although an enormous effort is required to accumulate the necessary data.

Combination of the protein topography data with data on the functional contribution of individual ribosomal proteins, as have been obtained with affinity labelling, reconstitution and crosslinking experiments, could also provide a means to deduce the localization of functional domains on the ribosome (2,5).

To give an example, various combinations of proteins S1, S3, S4, S5, S11, S12, S13, S18 and S21 have been shown by several approaches to be juxtaposed at the decoding site (1,2,3). In this selected example, most of the functional data were derived from crosslinking and affinity labelling experiments. In this context it must be emphasized, that the antigenic site could be far apart from the chemically reacted protein region. Considering the topography of these proteins, seven of them have antigenic sites on both the head and the body (Fig. 13a). These proteins must thread through the neck; there they could be either not at the surface at all and thus inaccessible for both chemical reagents (e.g.

affinity analogues) and for antibodies. Alternatively, they could be exposed in the depth of the groove, thus being accessible for reaction with a small chemical reagent, but inaccessible for binding with an antibody. By inspecting the locations of the antigenic sites of this group of proteins (Fig. 13a), one can find two clusters on the subunit surface where at least one site of each of these proteins is present ("minimal areas of common exposure", see ref. 2). One cluster of antigenic sites can be localized on the head slightly above the groove, another on and around the large lobe. Therefore, the decoding site could correspond to either of the antigenic clusters or it could even be located in the groove where no antibody binding sites have been found (Fig. 13a). For all the reasons mentioned above, such indirect localizations of functional domains should be treated with caution (2,3,5).

A better method is to localize functional regions directly by immuno electron microscopy. This may be done either by using antibodies specific for the molecule which interacts with the ribosome, e.g. antibodies to factors or antibiotics, or by using antibodies specific for haptens which have been covalently bound to the ribosomal ligand to be localized.

In general, the direct localization of ribosomal ligands requires that the molecule is covalently bound to its ribosomal site, either by affinity labelling or crosslinking, prior to reaction with the specific antibody. Otherwise the ligand may well be displaced from its specific site on the ribosome. If haptenized molecules are used, the hapten should be attached in close proximity to the functional center in question.

LOCALIZATION OF FUNCTIONAL DOMAINS ON THE 50S RIBOSOMAL SUBUNIT

Many antibiotics have been found which inhibit protein synthesis in various steps by interacting with the ribosome (40 and references therein). On a molecular level, the sites of interaction of a given drug on the ribosome have been predominantly elucidated by two techniques: reconstitution and affinity labelling (for references see 1,2,40,41). By using these techniques, those ribosomal components which should be located in the proximity of a drug binding site have been identified (2,40,41). Deduced localizations of several antibiotic binding sites have been published previously (36).

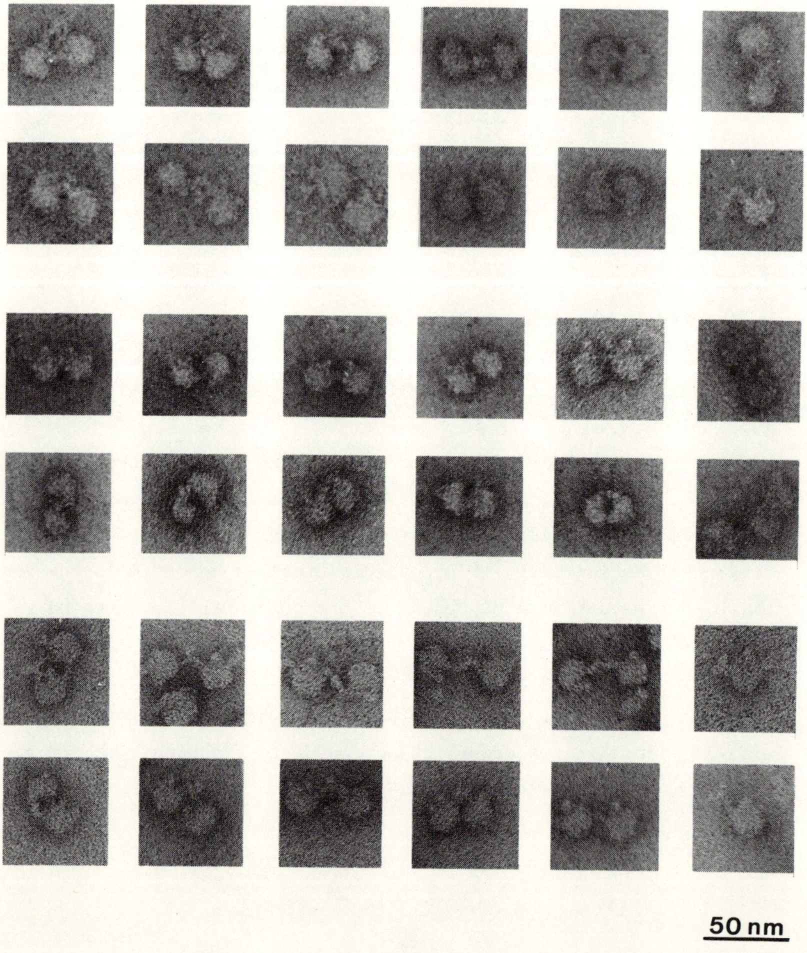

Figure 15. Localization of antibiotic binding sites on the 50S ribosomal subunit. Selected electron micrographs showing the thiostrepton binding site on 50S ribosomal subunits are shown in rows a and b (R. Bald, E. Cundliffe, M. Stöffler-Meilicke, G. Tischendorf, G. Stöffler, unpublished results). Micrographs showing the location of the chloramphenicol binding site are given in rows c and d. A second binding site located just on the opposite side has been seen in few electron micrographs when 50S·chloramphenicol complexes were incubated with the antibody (24,48, R. Bald, G. Tischendorf and G. Stöffler, unpublished results). Electron micrographs showing the location of the puromycin binding site are in rows e and f (R. Lührmann, R. Bald, M. Stöffler-Meilicke, G. Stöffler, unpublished results).

Localization of the Binding Sites of Thiostrepton, Chloramphenicol and Puromycin

We have now attempted to localize the binding sites of three antibiotics, viz. thiostrepton, chloramphenicol and puromycin on the 50S subunit by immuno electron microscopy, using specific antibodies to these drugs.

Thiostrepton has a potent inhibitory effect on several steps of protein synthesis: initiation, enzymatic AA-tRNA binding, translocation and termination (40). The binding constant of thiostrepton to ribosomes is so high that its reaction is almost irreversible. Accordingly, we raised antisera against thiostrepton and have incubated 50S·thiostrepton complexes with the antibody and investigated the resulting dimers by electron microscopy. A single antibody binding site was found and located in the center of the seat region of the 50S subunit (Fig. 15a,b). The drug binding site is in close proximity to the antigenic site of protein L11 (site A). Protein L11 is essential for binding thiostrepton to ribosomes, as has been shown in several bacterial ribosomes by several investigators (see E. Cundliffe, this volume).

The binding sites of chloramphenicol and puromycin have been localized by similar immunochemical procedures. However, the ribosomes were treated with affinity analogues of these antibiotics, because the binding constants to these antibiotics to ribosomes are relatively low and in consequence they tend to be displaced by the antibodies from their binding sites.

Extensive studies have been performed with 50S subunits labelled with iodoamphenicol, an affinity analogue of chloramphenicol (42). The affinity labelled 50S subunits, which were used for immuno electron microscopy, have also been analyzed with respect to the nature of the labelled components. Proteins L13, L2 and L27, but not L16, were found to be the main targets (R. Bald and G. Stöffler, in preparation). The drug binding site has been located (Fig. 15c,d) and was found in the proximity of proteins L2, L16, L27 (37,48). The location of protein L13 has not yet been established. Due to the known action of chloramphenicol (see 40), this site may be considered as the region where the peptidyl transferase center is located. By considering the locations of ribosomal components, labelled by tRNA-analogues which were modified at the α-amino group on the aminoacyl end of tRNA (2), these data would be in support of our direct localization of the peptidyl transferase center.

The mode of action of puromycin is one of the best understood of all antibiotics. Puromycin inhibits by substituting for the incoming coded AA-tRNA, and serves as an acceptor of the nascent peptide chain of ribosome-bound peptidyl-tRNA (40).

N-iodoacetyl- and N-bromoacetylpuromycin, as well as photoaffinity labels have been applied in affinity labelling experiments (40,41,43,44). The 50S proteins L6 and L2 (44), L23 (45,46) and L11, L2, L3, L17, L18 and L27 (47) have been identified as reacting with one of these labels. In our mapping experiments we used N-bromoacetylpuromycin, which is not an ideal affinity probe. Therefore, our experiments should mainly be considered as a feasibility-study, rather than as an attempt to localize the drug binding site. The probe was covalently bound to 50S subunits, which were then incubated with antibodies specific to N^6,N^6-dimethyl-adenosine which crossreact with puromycin; the resulting dimers were used for electron microscopy.

A single antibody binding site has been observed and could be preliminarily localized in the seat region, close to the lateral protuberance (Fig. 15e,f). Though an exact placement of this site is not yet possible, the experiments demonstrated that the puromycin site can be localized on the 50S subunit surface (Fig. 15e,f).

Our experiments clearly demonstrate that antibiotics or their affinity analogues can be localized on the ribosomal surface by immuno electron microscopy. The problems in such experiments are almost exclusively connected with the specificity of the affinity analogues (see 41). If the specificity of an affinity probe is sufficiently certain, the localization of such a probe by immuno electron microscopy should then be generally applicable.

LOCALIZATION OF FUNCTIONAL DOMAINS ON THE 30S RIBOSOMAL SUBUNIT

Localization of the IF-3 Binding Site on the 30S Subunit

Besides other functions, initiation factor IF-3 prevents 30S subunits from associating with 50S subunits to form 70S monosomes, by binding to a specific site on the small subunit (49).

Recently Emanuilov et al. (50) have localized the eukaryotic initiation factor 3 on ribosomal subunits from rabbit reticulocytes directly, by comparing native subunits and subunits stripped of eIF-3 in the electron microscope. Because of the high molecular weight of the eukaryotic factor (approximately 500.000 Daltons) the factor could be directlv visualized by electron microscopy and located on the 40S subunit. Initiation factor IF-3 of E. coli has only a molecular weight of 20.000 (51). A direct visualization of IF-3 bound to 30S subunits, by negative staining failed. No differences in the contours between 30S subunits and 30S·IF-3 complexes were observed (R. Lührmann, G. Tischendorf and G. Stöffler, unpublished results). A deduction of the location of the binding site of IF-3 on E. coli 30S subunits has been attempted using the various crosslinking data (2,5).

We tried to localize the IF-3 on the small subunit by immuno electron microscopy with antibodies specific to IF-3. 30S·IF-3 complexes were formed and IF-3 was crosslinked to the 30S subunit with dimethylsuberimidate. The 30S subunit pairs connected by IF-3 bind to the entire front-neck region. IF-3 seems to be bound in the groove-region of the 30S subunit, since antibody binding sites can be seen from the head-neck junction down to the lobes (Fig. 16).

The Localization of mRNA on the 30S Subunit

A ribosome bound mRNA cannot be directly visualized by the negative staining technique. We thus attempted to use a synthetic mRNA analogue which can be covalently bound to ribosomes by photoactivation and, in addition, can be modified with haptens.

Poly(4-thiouridylic acid), with a chain length varying between 30-40 nucleotides (52) has been substituted with N-bromoacetylpuromycin such that statistically every sixth to seventh nucleotide had reacted. The coding activity for phe-tRNA of the modified polynucleotide was assessed and found to be 80% of that of unmodified polythiouridylic acid.

A complex of 30S ribosomal subunits (from the kasugamycin-resistant strain ksg A19), phe-tRNA, and the mRNA analogue was formed, irradiated (52) and purified. The purified complexes with the covalently bound mRNA-analogue were incubated with antibodies specific for N^6,N^6-dimethyladenosine and ribosome-antibody complexes were investigated in the electron microscope.

The electron micrographs shown in Figure 17 indicate that most of the antibody attachment sites are located in the groove between the head and the body of the 30S subunit. Due to the length of the mRNA-analogue an unique antibody binding site cannot be expected.

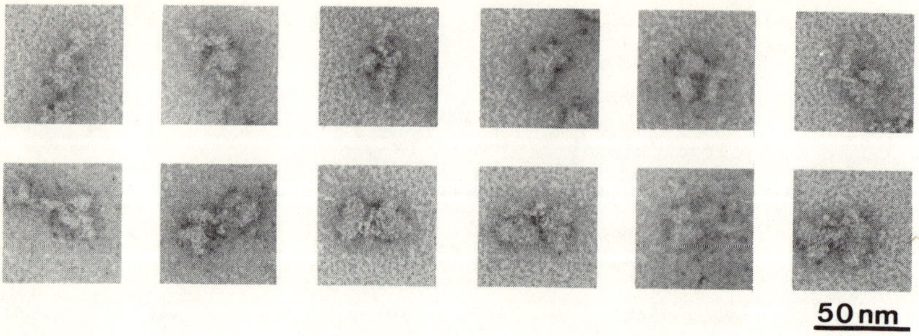

Figure 16. Selected electron micrographs showing the location of initiation factor IF-3 on crosslinked 30S·IF-3 complexes with antibodies specific to IF-3 (R. Lührmann, G. Tischendorf and G. Stöffler, unpublished results).

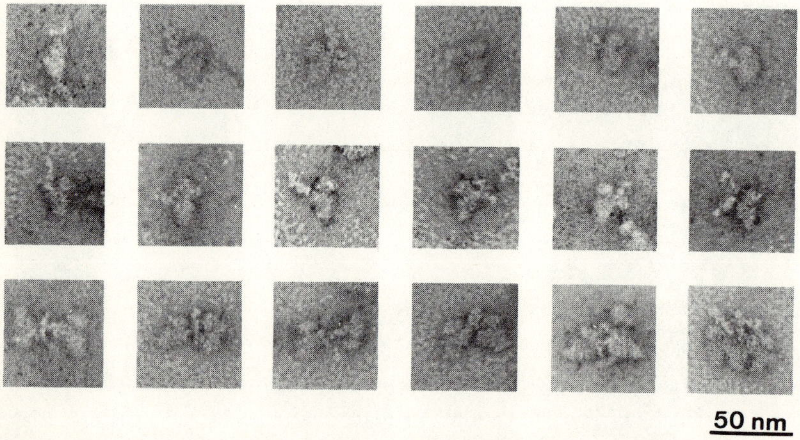

Figure 17. Selected electron micrographs showing the location of mRNA binding site on the 30S subunit; for experimental details see Text (R. Lührmann, R. Bald, G. Tischendorf and G. Stöffler, unpublished results).

Localization of the Two Adjacent N^6,N^6-Dimethyladenosines of 16S rRNA on Ribosomes

Investigation of the nucleotide sequence of a number of mRNAs revealed that they carry certain purine-rich sequences proximal to, and "upstream" from the initiation-triplet; these purine sequences were, to a greater or lesser extent, complementary to a sequence at the 3'-end of 16S rRNA (see 53). There is evidence that such complementary stretches could serve for the correct alignment of mRNA in the initiation complex. Crosslinking of both, initiation factors, and ribosomal proteins that are involved in

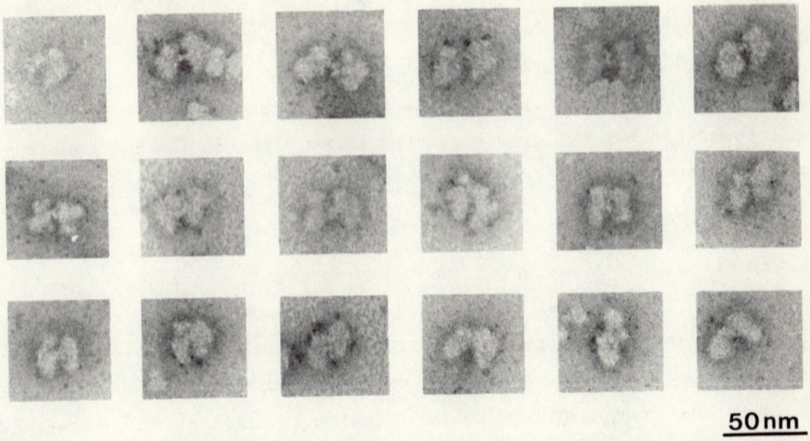

Figure 18. Selected electron micrographs showing the location of the two adjacent N^6,N^6-dimethyladenosines on 30S subunits (48,57, J. Dieckhoff, R. Bald, G. Tischendorf and G. Stöffler, unpublished results).

initiation, to the 3'-end of 16S rRNA (see 2,3), as well as inhibition of initiation complex formation by oligonucleotides complementary to this region of 16S rRNA, further substantiated this hypothesis (54,55).

The rRNA of small ribosomal subunits from E. coli and other organisms contains two adjacent N^6,N^6-dimethyladenosines at positions 24 and 25 from the 3'-end. These methylated adenosines are thus only 15 nucleotides apart from the center of the pyrimidine-rich 16S RNA region, which is involved in initiation (53). These two methylated adenosines have been localized with anti-m^6_2A on the platform, at the level of its junction with the body (14). In accordance with the results of Thammana and Cantor (56), who used antibodies against m^6_2A, we found that anti-$m^6_2Apm^6_2A$ failed to react with functionally active 30S subunits as did our antibody preparations to m^6_2A and to m^6A (48,57). Functionally inactive 30S subunits did, however, react with anti-$m^6_2Apm^6_2A$ and examples of subunit pairs connected by antibody are shown in Figure 18. In most of the dimers the two subunits are very close together, and arranged such that the hollows of the combined subunits were facing each other (Fig. 18). In some of these subunit pairs the antibody is not visible at all. The connection of the two subunits by an antibody molecule can only be deduced by the arrangement of the subunits in a rotatory symmetry (Fig. 18). Electron micrographs of other subunit pairs show the connecting antibody (Fig. 18), but the actual antibody attachment site is only seen in very few dimers (Fig. 18).

30S ribosomes in which the 3'-end of 16S RNA has been modified with 2,4-dinitrophenyl-aminobutyric acid hydrazide were reacted with antibody to DNP. Preliminary electron microscopic studies revealed that the actual 3'-end is not burried in the groove. Antibody attachment sites were found between the groove and the large lobe of the 30S subunit (R. Lührmann, M. Stöffler-Meilicke, G. Stöffler, unpublished results).

CONCLUDING REMARKS

The direct localization of the chloramphenicol binding site described above is so far the most straightforward approach to localize one of the important functional sites of the ribosome, namely the peptidyl transferase center. Supposing that (a) iodoamphenicol is a true and reliable affinity analogue of chloramphenicol (which is still controversial) and (b) that the three-dimensional localization of the antibody binding site has been correctly interpreted, a matter which is not trivial especially on 50S subunits (see above), the peptidyl transferase center is situated in the angle formed between the central and the left lateral protuberance of the 50S subunit, in the proximity of proteins L16, L2 and L27 (Fig. 13b).

In contrast, none of the approaches used for localizing functional domains on the 30S subunit was designed to determine the precise position of the ribosomal decoding site but rather the mRNA-binding region. Our data on the direct localization of the 3'-end of 16S rRNA, of the mRNA and of the initiation factor IF-3, indicate that all these sites are located in the groove between the head and the body on the 30S subunit.

Our data agree with the localization of the decoding region as found by affinity immuno electron microscopy by Keren-Zur, Boublik and Ofengand (58), and with the data of Politz and Glitz (14). Attempts to localize this region by combining data from protein mapping with affinity labelling studies using mRNA or codon analogues gave results which are also in agreement (2). The data are also consistent with the idea that mRNA threads through the channel formed between the small and the large subunit in the 70S ribosome, as first proposed by Rich (see 59).

The various applications of immuno electron microscopy described in this article enlarged the body of our knowledge about the protein topography in the E. coli ribosome. Moreover, they contributed to our knowledge about the locations of functional domains on the ribosome. The experiments presented, encouraged us to believe that eventually, we should be able to experimentally determine how the mRNA and the two tRNAs fit on a 70S ribosome.

With the advent of immuno electron microscopy as a tool of molecular topography, molecular species of almost any kind can be located in rather complex macromolecular assemblies. By further improving the various labelling techniques, several of the most exciting problems of modern biology can be tackled.

DEDICATION

This article is dedicated to Prof. Dr. Karl Fellinger, Vienna, on the occasion of his 75th birthday.

ACKNOWLEDGEMENTS

The authors are indebted to Heinz-Günther Wittmann and Elmar Zeitler for their support and encouragement, to Claudio Gualerzi for his help and advice and to Richard Brimacombe and Roger Garrett for many helpful discussions and for a careful reading of the manuscript. We wish to express our special gratitude to Renate Hasenbank and to Karl-Heinz Rak for their extraordinary help during the preparation of the manuscript. This work was supported by a grant of the Deutsche Forschungsgemeinschaft to G.S.

REFERENCES

1. Nomura, M., Tissières, A. and Lengyel, P. (1974) Ribosomes, Cold Spring Harbor Laboratory

2. Stöffler, G. and Wittmann, H.G. (1977) Molecular Mechanisms of Protein Biosynthesis, Weissbach, H. and Pestka, S., eds., pp. 117-202, Academic Press, New York

3. Brimacombe, R., Stöffler, G. and Wittmann, H.G. (1978) Ann. Rev. Biochem. 47, 217-249

4. van Holde, K.E. and Hill, W.E. (1974) Ribosomes, Nomura, M., Tissières, A. and Lengyel, P., eds., pp. 53-91, Cold Spring Harbor Laboratory

5. Lake, J.A. (1978) Advanced Technique in Biological Electron Microscopy II, Koehler, J.K., ed., Vol. 2, pp. 173-211, Springer-Verlag Berlin·Heidelberg·New York

6. Wittmann, H.G., Nierhaus, K.H., Sieber, G., Stöffler, G., Tesche, B., Tischendorf, G.W., Wittmann-Liebold, B. (1978) International Congress on Electron Microscopy, Toronto, Vol. III, pp. 459-469

7. Tischendorf, G.W., Zeichhardt, H. and Stöffler, G. (1974) Molec. Gen. Genetics 134, 209-223

8. Lake, J.A., Pendergast, M., Kahan, L. and Nomura, M. (1974) Proc. Nat. Acad. Sci. USA 71, 4688-4692

9. Tischendorf, G.W. and Stöffler, G. (1975) Molec. Gen. Genetics 142, 193-208

10. Tischendorf, G.W., Zeichhardt, H. and Stöffler, G. (1975) Proc. Nat. Acad. Sci. USA 72, 4820-4824

11. Lake, J.A. (1976) J. Mol. Biol. 105, 131-159

12. Boublik, M., Hellmann, W. and Kleinschmidt, A.K. (1977) Cytobiologie 14, 293-300

13. Boublik, M. and Hellmann, W. (1978) Proc. Nat. Acad. Sci. USA 75, 2829-2833

14. Politz, S.M. and Glitz, D.G. (1977) Proc. Nat. Acad. Sci. USA 74, 1468-1472

15. Boublik, M., Hellmann, W. and Roth, H.E. (1976) J. Mol. Biol. 107, 479-490

16. Spiess, E. (1978) FEBS Letters 91, 289-292

17. Meyer, M., Bont, W.S., De Vries, M. and Nanninga, N. (1974) Eur. J. Biochem. 42, 259-268

18. Tischendorf, G.W., Zeichhhardt, H. and Stöffler, G. (1974) Molec. Gen. Genetics 134, 187-208

19. Hamel, E., Koka, M. and Nakamoto, T. (1972) J. Biol. Chem. 247, 805-814

20. Naaktgeboren, N., Roobol, K. and Voorma, A.O. (1977) Eur. J. Biochem. 72, 149-156

21. Strycharz, W.A., Nomura, M. and Lake, J.A. (1978) J. Mol. Biol. 126, 123-140

22. Tesche, B. (1975) Vakuum Technik 24, 104-110

23. Tesche, B. (1978) Mikroskopie 34, 29-34

24. Tesche, B., Tischendorf, G.W. and Stöffler, G. (1978) Ninth International Congress on Electron Microscopy, Toronto, Vol. II, pp.240-241

25. Tesche, B., Tischendorf, G.W. and Stöffler, G. (1979) Hoppe Seyler's Z. Physiol. Chem. 360, 384-385

26. Knauer, V., Schramm, H.J. and Hoppe, W. (1978) Ninth International Congress on Electron Microscopy, Toronto, Vol. II, pp. 4-5

27. Stöffler, G. and Wittmann, H.G. (1971) Proc. Nat. Acad. Sci. USA 68, 2283-2287

28. Stöffler, G. and Wittmann, H.G. (1971) J. Mol. Biol. 62, 407-409

29. Stöffler, G. (1974) Ribosomes, Nomura, M., Tissières, A. and Lengyel, P., eds., pp. 615-667, Cold Spring Harbor Laboratory

30. Zubke, W., Stadler, H., Ehrlich, R., Stöffler, G. and Wittmann, H.G. (1977) Molec. Gen.Genetics 158, 129-139

31. Sela, M. and Mozes, E. (1966) Proc. Nat. Acad. Sci. USA 55, 445-452

32. Stöffler, G., Hasenbank, R., Lütgehaus, M., Maschler, R., Morrison, C.A., Zeichhardt, H. and Garrett, R.A. (1973) Molec. Gen. Genetics 127, 89-110

33. Morrison, C.A., Tischendorf, G., Stöffler, G. and Garrett, R.A. (1977) Molec. Gen. Genetics 151, 245-252

34. Wabl, M. (1973) Ph.D. thesis, Freie Universität Berlin

35. Wabl, M.R. (1974) J. Mol. Biol. 84, 241-247

36. Stöffler, G. and Tischendorf, G.W. (1974) Topics in Infectious Diseases, Drews, J. and Hahn, F.E., eds., Vol. 1, pp. 117-143 Springer-Verlag Wien, New York

37. Tischendorf, G.W., Tesche, B. and Stöffler, G. (1976) Sixth European Congress on Electron Microscopy, Jerusalem, Vol. 2, pp. 524-526

38. Boublik, M., Spiess, E., Roth, H.E., Hellmann, W. and Jenkins, F. (1978) Cytobiologie 18, 309-319

39. Yaguchi, M. (1975) FEBS Letters 59, 217-220

40. Vázquez, D. (1979) Molecular Biology Biochemistry and Biophysics, Vol. 30, Springer-Verlag, Berlin·Heidelberg·New York

41. Cooperman, B.S. (1978) Bioorganic Chemistry, Vol. IV, pp. 81-115, Academic Press, New York·San Francisco·London

42. Pongs, O., Bald, R. and Erdmann, V.A. (1973) Proc. Nat. Acad. Sci. USA 70, 2229-2233

43. Stahl, J., Dressler, K. and Bielka, H. (1974) FEBS Letters 47, 167-170

44. Pongs, O., Bald, R., Erdmann, V.A. and Reinwald, E. (1975) Topics in Infectious Diseases, Drews, J. and Hahn, F.E., eds., Vol. 1, pp. 179-190, Springer-Verlag Wien, New York

45. Cooperman, B.S., Jaynes, E.N., Brunswick, D.J. and Luddy, M. (1975) Proc. Nat. Acad. Sci. USA 72, 2974-2978

46. Jaynes, E.N., jr., Grant, P.G., Giangrande, G., Wieder, R. and Cooperman, B.S. (1978) Biochemistry 17, 561-569

47. Nicholson, A.W. and Cooperman, B.S. (1978) FEBS Letters 90, 203-208

48. Tischendorf, G.W., Bald, R.W., Kittler, R., Lührmann, R., Tesche, B. and Stöffler, G. (1977) Abstracts 11th FEBS Meeting Copenhagen, A2 9, 204

49. Grunberg-Manago, M. and Gros, F. (1977) Progr. Nucleic Acids Res. Mol. Biol. 20, 209-284

50. Emanuilov, I., Sabatini, D.D., Lake, J.A. and Freienstein, C. (1978) Proc. Nat. Acad. Sci. USA 75, 1389-1393

51. Brauer, D. and Wittmann-Liebold, B. (1977) FEBS Letters 79, 269-275

52. Fišer, I., Scheit, K.H. and Küchler, E. (1977) Eur. J. Biochem. 74, 447-456

53. Steitz, J.A., Sprague, K.U., Steege, D.A., Yuan, R.C., Laughrea, M. and Moore, P.B. (1977) Nucleic Acid-Protein Recognition, Vogel, H.J., ed., pp. 491-508, Academic Press, New York·San Francisco·London

54. Taniguchi, T. and Weissmann, C. (1978) Nature, 275, 770-772

55. Eckhardt, H. and Lührmann, R. (1978) Hoppe Seyler's Z. Physiol. Chem. 359, 1079

56. Thammana, P. and Cantor, C.R. (1978) Nucleic Acids Res. 5, 805-823

57. Dieckhoff, J., Bald, R., Tischendorf, G.W. and Stöffler, G. (1979) Hoppe Seyler's Z. Physiol. Chem. 360, 248

58. Keren-Zur, M., Boublik, M. and Ofengand, J. (1979) Proc. Nat. Acad. Sci. USA 76, 1054-1058

59. Rich, A. (1974) Ribosomes, Nomura, M., Tissières, A. and Lengyel, P., eds., pp. 871-884, Cold Spring Harbor Laboratory

RIBOSOME STRUCTURE AND FUNCTIONAL SITES

JAMES A. LAKE

Molecular Biology Institute and
Department of Biology
University of California,
Los Angeles, CA 90024

INTRODUCTION

During the last 20 years, significant advances have been made in understanding ribosome structure using such techniques as protein crosslinking, neutron diffraction, three-dimensional reconstruction, affinity labeling, genetic studies, RNA and protein protection and modification studies, reconstitution experiments and immune electron microscopy.

Currently these approaches are providing mutually overlapping information. When the results on protein localization by immune electron microscopy are systematically compared with results from other techniques, a generally consistent set of data emerges. This article will emphasize the results of immune electron microscopy and compare them with other physical and functional studies in order to present an overview of what is most reliably known about ribosome structure, about the locations of ribosomal proteins, and about functionally important regions of the structure.

ELECTRON MICROSCOPIC STUDIES OF RIBOSOME STRUCTURE

Electron microscopy of ribosomes has been used to study the gross structure of ribosomes for more than 20 years. In general, the resolution of micrographs has improved in parallel with improved methods of specimen preservation.

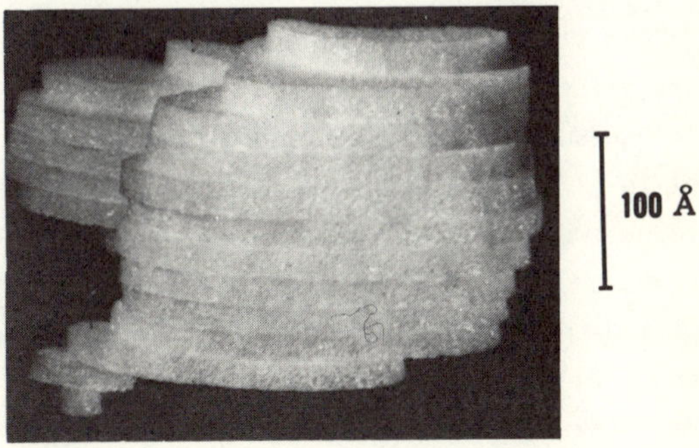

Figure 1.
Three-dimensional reconstruction of the ribosome of E. invadens at approx. 100Å resolution (1). The small subunit is behind the approximately circular profile of the large subunit in this view of the monomeric ribosome.

The earliest studies on *E. coli* utilized both shadowing with heavy metal (2) and negative contrasting (3). Quite early on it was realized that the occurrence of *in vivo* ribosome crystals (4,5) raised the possibility of performing quantitative structural studies on ribosomes and the first three-dimensional reconstruction of a ribosome was completed several years later (1,6; see Fig. 1). Subsequently higher resolution maps (7) have been completed; however, progress has been difficult because of the inability to crystallize ribosomes *in vitro*. Recently, however, ordered arrays of *E. coli* ribosomal subunits, suitable for three-dimensional reconstruction studies, have been obtained (8; also see the section on "*in vitro* arrays.") These promise to accelerate progress in this area.

Progress using immune electron microscopy has been, overall, more rapid than that obtained by three-dimensional reconstruction. In 1970, Yanagida and Ahmad-Zadeh (9) used immune labeling techniques to map the location of specific proteins in bacteriophage T4 and four years later immune electron microscopy was used to map 50S ribosomal proteins in two dimensions (10). It was thought at that time that electron microscopy was not applicable to the *E. coli* small subunit since Lubin (11) had been unable to obtain reproducible views of this particle. Nevertheless, Lake *et al*. (12) were first to apply immune electron microscopy to the small subunit, and introduced the approach of mapping proteins in three dimensions. Subsequently, immune electron microscopy has been used to map proteins on the 70S ribosome (13), to map specific nucleotides of 16S RNA (14) and to map ligands covalently bound to subunits (15). A related technique has been used to map the location of eIF-3 in three dimensions on the eukaryotic ribosome (16). That study also indicated a conservation of small subunit structural features, such as the platform and the cleft, in both eukaryotes and prokaryotes.

Small and Large Subunits and Monomeric Ribosomes

The structures of the small and large subunits of the *E. coli* ribosome are illustrated in Figure 2. The 30S subunit (Fig. 2A,B) is divided into two unequal parts by an indentation and a region of accumulated negative stain (17). The parts are referred to as the upper one third and the lower two thirds. A region of the subunit, called the platform, extends from the lower two thirds of the small subunit and forms a cleft between it and the upper one third (18). An asymmetric 30S model similar to that shown in Figure 2, but with a less prominent cleft has been derived by Vasiliev (19) from heavy metal shadowed subunits. The platform and cleft

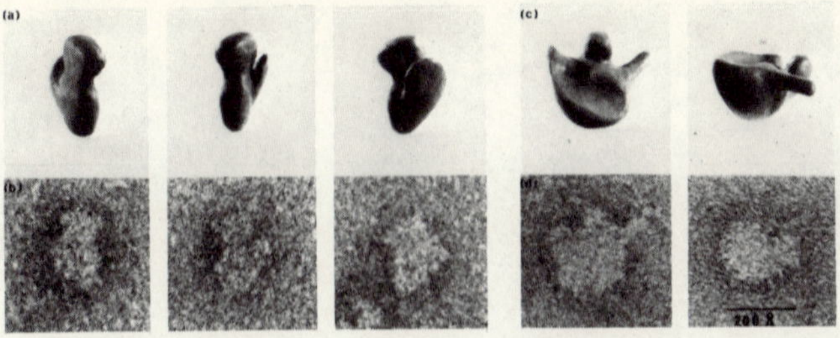

Figure 2. A comparison of the 3-dimensional models with electron micrographs for small subunits (A,B) and for large subunits (C,D). Adapted from Lake (13).

have been confirmed in E. coli ribosomes by Boublik et al. (20), and in eukaryotic ribosomes by Boublik et al. (20) and Emanuilov et al. (16). Tischendorf et al. (21) have proposed a symmetric model that lacks a platform and cleft because they have not incorporated the view shown in the electron micrograph in Figure 2B, frame 3, into their model. As a result their protein mappings are ambiguous. Antibody-labeling experiments using protein S13 as a marker (18) and tilting and heavy metal shadowing experiments (22) support our asymmetric model. In addition, the tilting and shadowing experiments established an absolute hand for the small subunit. Some of the shadowing results are reproduced in Figure 3.

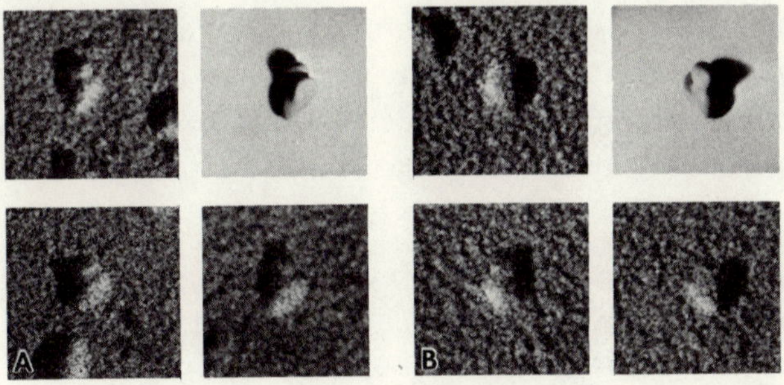

Figure 3. Tungsten-tantalum shadowing of 30S subunits. The shadow at the level of the one third-two thirds partition is present (A) if the platform is oriented toward the observer and absent (B) if the platform faces away. Adapted from Leonard and Lake (22).

The large subunit, like the small subunit, is asymmetric (Fig. 2C,D) (12,13). It consists of a central protuberance (Fig. 2C, first frame) and protrusions inclined approximately 50° from the central protuberance. One of these projections, the "L7/L12 stalk," is at the right. In a projection approximately orthogonal to this (shown in the right frames of Fig. 2C,D), the large subunit is characterized by a notch on the upper surface. The model of Boublik et al. (20) is also an asymmetric structure in general agreement with our results; however, Tischendorf et al. (21) have proposed a symmetric model and, therefore, their large subunit protein maps also differ fundamentally from ours.

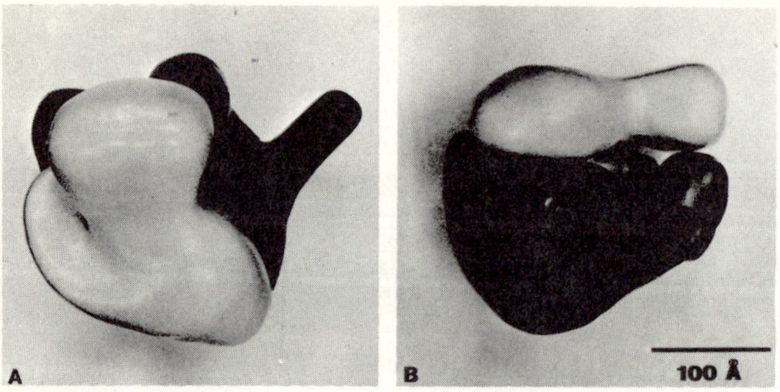

Figure 4. Model of the 70S E.coli ribosome showing relative orientation of the subunits (the 30S subunit is light, the 50S subunit is dark). Adapted from (13).

In the monomeric ribosome the small subunit is positioned asymmetrically on the large subunit as shown in Figure 4. The platform of the small subunit is in contact with the large subunit (13), so that the one third-two thirds partition of the small subunit is approximately aligned with the notch of the large subunit (12). An alternative orientation, in which the long axis of the small subunit is rotated 90° with respect to that shown in Figure 4, has been proposed by two groups (21,23). Their conclusion was based on the interpretation of micrographs of a single view of the subunit (the one shown in Figure 4B); 70S antibody-labeling studies were not performed. While the possibility of an alternative 70S model should be kept in mind, we think an alternative is unlikely, since the model in Figure 4 is derived from two different, mutually perpendicular views of the ribosome, and the detailed orientation of the 30S subunit was deduced by antibody labeling studies of 70S ribosomes using anti-S13 IgGs (13).

Using these three-dimensional structures, preliminary locations of most ribosomal proteins have been mapped in collaboration with L. Kahan, W. Strycharz and M. Nomura and their locations are providing insight into the functioning of the ribosome. In this paper, most proteins that will be described have been both implicated in important functional events and well documented by our immune mapping experiments.

ANTIBODY LABELING STUDIES OF SMALL SUBUNIT PROTEINS

The orientation of the small subunit on the large subunit makes possible an operational distinction between 30S protein locations on the "exterior" (i.e., the side away from the interface) of the subunit and those at the interface. Ultimately, the spatial distinctions between exterior and interior proteins are translated into functional differences. In the view of the subunit shown in Figure 4B, the 30S subunits are characterized by a platform and a cleft. In this projection exterior proteins map at the top, i.e., on the side opposite the platform.

Two distinctive views, corresponding to a positioning of the small subunits on the carbon substrate in approximately orthogonal orientations, are shown in the electron micrographs of Figure 5. In the view illustrated in the first three frames of each row in Figure 5, the platform is on

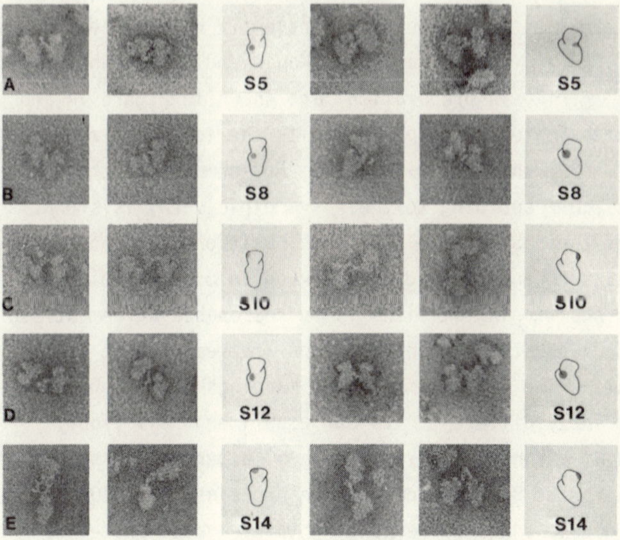

Figure 5. A gallery of electron micrographs of subunits reacted with antibodies directed against "external" small subunit proteins. Also included as external proteins are S3 and S19-I, not shown. The first three frames of each row illustrate subunits in the quasisymmetric projection and the second three frames illustrate subunits in the asymmetric projection. The antibody binding site is diagrammed in the third frame of each set of three.

the right side at the level of the one third-two thirds partition. Exterior proteins (S5, for example) map on the side opposite the platform in this projection. In the projection diagrammed in the sixth frame of each row, S5 maps on the concave profile of the subunit at the one third-two thirds partition, and S14, for example, maps on the concave profile of the subunit at the top of the upper one third. The locations of these proteins, as well as some others, are shown in Figure 6.

At present we have mapped the locations of nine 30S ribosomal proteins in comprehensive studies (12,18; Kahan and Lake, manuscript in preparation). These are proteins S3, S5, S8, S10, S11, S12, S13, S14 and S19. In addition, we have performed preliminary studies on the remaining proteins (Lake and Kahan, preliminary results; see e.g., 24). For the purposes of comparing our results with those provided by other techniques we

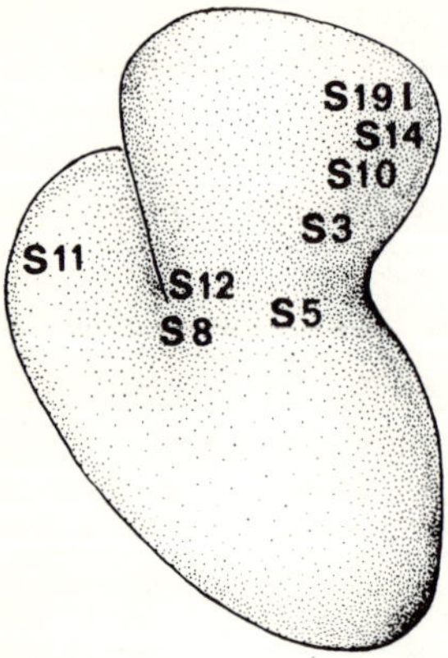

Figure 6. Locations of some of the 30S proteins discussed in this chapter. This view corresponds to the surface of the subunit exposed to the cytoplasm, i.e., the exterior surface.

have, however, restricted our discussion to this set of nine well mapped proteins.

COMPARISON WITH OTHER STRUCTURAL STUDIES

The locations recently determined for S3, S8, S10 and S12 (Kahan and Lake, manuscript in preparation) are shown in Figure 7A and serve as a convenient example for comparison with the locations reported by Tischendorf et al. (21) shown in Figure 7B. For comparison the quasi-symmetric projection (corresponding to their symmetric view) is shown. With the exception of their S12A site (which is positioned at about the same level as a non-S12 dependent binding observed in our control experiments), the vertical positioning of sites is in reasonable agreement suggesting that the IgG preparations of both groups are reacting at similar regions of the subunit. The three-dimensional mappings of both groups, however, are significantly different. In our experiments all (four) proteins were mapped at a single site, although S3 extended a distance of approximately 60 Å. In the Berlin model, three proteins, S3, S10 and S12, were mapped as double sites corresponding to elongated proteins, and the remaining protein, S8, was mapped at a single site. The sites mapped for S3, S8 and S10 are symmetrically positioned about the central mirror plane of the Berlin model. Their double sites, S3A, S3B and S10A, are located at mirror related positions and their single S8 site is positioned on the mirror plane. S12A and S12B occupy non-symme-

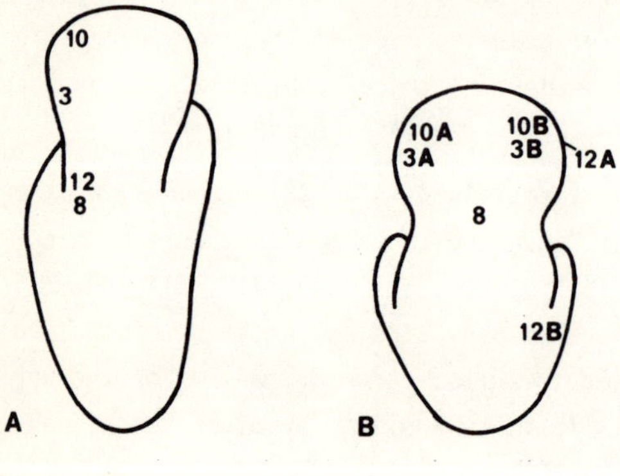

Figure 7. Comparison of the locations of small subunit proteins S3, S8, S10 and S12 in the quasisymmetric (symmetric) projection (A) in the map of Kahan and Lake (manuscript in preparation) and (B) in the map of Tischendorf et al. (21).

trical sites in their map. These results are most likely related to the use of a mirror symmetric model for interpreting their electron microscopic images. Recent immune mapping and electron microscopic studies by other groups of both E. coli (14,15,25) and of eukaryotic small subunits (16; see also 26) have exclusively used asymmetric models similar to the one used in our studies.

Neutron Diffraction

Another useful source of structural information is neutron diffraction. Neutron diffraction has provided information about both gross morphology and protein locations (27,28,29) and has been particularly useful for measuring distances between proteins. By diffracting reconstituted subunits containing pairs of deuterated proteins, one is provided with the spectrum of all distances connecting the atoms of the two proteins. The first step in the analysis of these data is to calculate the distances between the centers of mass of the proteins. Schindler et al. (30) have measured the distances between centers of mass for nine proteins, six of which overlap with our completed set. The distances from neutron diffraction and the distances measureed from the approximate centers of the protein location in our map are listed in Table 1.

TABLE 1. Comparison of distances between ribosomal proteins determined by neutron diffraction and by electron microscopy.

Protein Pairs	Neutron (Å)	Immune Electron Microscopy (Å)	Difference (Å)
S3 - S5	64.9	43	+22
S3 - S8	88.2	92	- 4
S3 - S11	65.2	37	+28
S3 - S11	118.3	145	-27
S3 - S12	86.0	90	- 4
S5 - S8	40.0	55	-15
S5 - S10	115.3	67	+48
S5 - S11	92.2	130	-38
S5 - S12	55.1	55	0
S8 - S10	——	112	——
S8 - S11	105.3	82	+23
S8 - S12	53.2	25	+28
S10 - S11	——	152	——
S10 - S12	——	125	——
S11 - S12	114.2	87	+27

A linear regression analysis between the two sets of data showed that a linear relationship, significant at 99% confidence level, exists between them.

Since antigenic determinants are exposed only at the surface of the subunit and not necessarily distributed evenly over the surface of the protein, somewhat different distances are being compared. However, for the set of proteins compared in Table 1 the agreement between the two sets of data is quite impressive. A linear regression analysis determined that the data are best related by a line of slope 0.94 (using neutron distances as the independent variable) and application of the "t test" suggested that the relationship is significant at the 99% confidence level.

A second type of comparison has been performed by Schindler et al. by directly comparing their three-dimensional map with the immune mapping results (see the chapter on neutron diffraction in this volume). In Figure 8 the common proteins of the neutron and our immune electron microscopy maps are compared in the view of the external surface of the subunit. (The immune locations are shown by shaded lines and the neutron locations are shown by small numbers on a white background.) The orientation selected is that chosen by Schindler et al. to provide the best agreement when considering all their proteins and our set of preliminary mappings. The agreement is quite good with the sole exception of S11. They suggest that an alternative model with S11 positioned near the location mapped by immune electron midroscopy would also be consistent with their data, although the location shown in Figure 8 gives a significantly better fit to the neutron diffraction data.

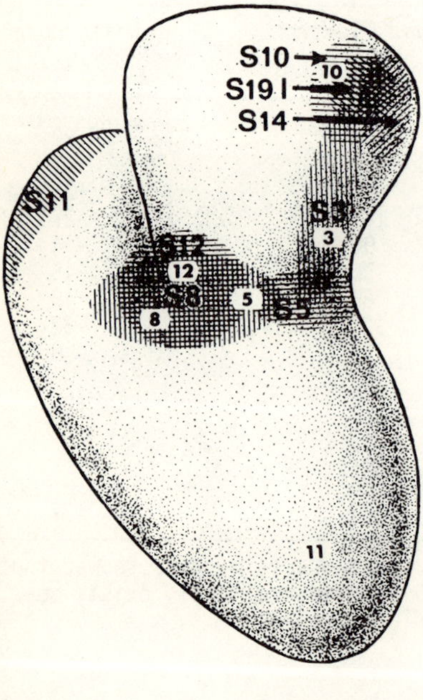

Crosslinking Results

A wealth of information has been provided by crosslinking studies. Much of the work published has used reagents that preferentially react with the epsilon amino groups of lysines. In order to compare these

Figure 8. A comparison of the results of neutron diffraction (small numbers on white background) with the immune electron microscopy protein locations (shown by shaded lines). This corresponds to the exterior view of the small subunit. Adapted from Kahan and Lake (manuscript in preparation).

results in a general way, we have considered all the possible pairings of the nine ribosomal proteins discussed in this paper and divided them into two groups according to whether their antigenic determinants approach within 30 Å (± 10 Å), or not. The 30 Å figure was chosen in order not to miss any possible crosslinks since the longest of the crosslinking reagents approach 20 Å. There exist 36 possible pairs of subunits among the group of nine proteins. Crosslinks between eleven of the 36 possible pairs have been reported and are shown in the left two columns of Table 2.

TABLE 2. Comparison of the results of crosslinking experiments with closest distances between determinants mapped by immune electron microscopy.

Crosslinks reported		Crosslinks not reported	
$\delta \lesssim 30$ Å	$\delta \gtrsim 30$ Å	$\delta \lesssim 30$ Å	$\delta \gtrsim 30$ Å
S3 - S5[8]	S5 - S13[8]	S3 - S8	S3 - S11
S3 - S10[8,6]	S8 - S11[8]	S3 - S14	S3 - S13
S3 - S12[8]	S8 - S13[8]	S3 - S19	S5 - S10
S5 - S8[1,2,3,5,6,7,9,10]	S12 - S13[8]	S5 - S12	S5 - S11
S11 - S13[8]		S8 - S12	S5 - S14
S13 - S19[1,4,5,6,8,10]		S10 - S14	S5 - S19
S14 - S19[2]		S10 - S19	S8 - S10
		S11 - S19	S8 - S14
			S8 - S19
			S10 - S11
			S10 - S12
			S10 - S13
			S11 - S12
			S11 - S14
			S12 - S14
			S12 - S19
			S13 - S14

[1]Lutter et al. (1972)
[2]Bode et al. (1974)
[3]Clegg and Hayes (1974)
[4]Lutter et al. (1974)
[5]Sun et al. (1974)
[6]Lutter et al. (1975)
[7]Sommer and Traut (1975)
[8]Sommer and Traut (1976)
[9]Peretz et al. (1976)
[10]Expert-Bezancon (1977)

Correlation of the two types of data was significant at the 90% confidence level for all crosslinks and significant at the 99% level for crosslinks reported by two or more groups.

Seven of these 11 pairs are consistent with our mappings and 4 pairs have determinants that we map as being greater than 30 Å apart. There is a general correlation between the protein distances and the results of crosslinking studies. In particular, those pairs that have been reported by two or more independent groups (S3-S10, S5-S8 and S13-S19) are all included in the group found to be close in the electron microscopic map. The relationship between the two types of data is not necessarily exact, however. For example, two proteins that have 'close' determinants may have improperly positioned or oriented lysine residues, or even may lack them entirely in their neighboring regions. The other type of exception can also happen when, for example, surface determinants are 'distant' but interior regions of the proteins are close. Nevertheless, there should be a correlation between the crosslinking and the immune electron microscopy results; i.e., 'close' proteins should, in general, be more frequently crosslinked than 'distant' ones; hence, we can use a statistical analysis to ask if there is a correlation between cross-linking and immune mapping results. Using the χ^2 test we find that we reject the hypothesis that "no relationship exists between the set of crosslinks and the crosslinks predicted from the immune map" at the 90% level of significance ($\chi^2 = 3.15$), and reject this hypothesis at the 99.5% level of significance ($\chi^2 = 10.00$) if only those crosslinks observed by two or more groups are considered. In other words, a positive correlation between crosslinking and immune electron microscopic data is very likely if only those crosslinking results obtained by two or more groups are considered and is possibly significant if all crosslinking data are used.

Assembly Map

One of the earliest suggestions of the spatial arrangements of ribosomal proteins was provided by the "assembly map" (41,42). Although it is widely recognized that the order of protein assembly into the small subunit may frequently be related to direct protein-protein interactions, it is also appreciated that other interactions are also possible (43) and that proteins that are near neighbors need not interact physically during assembly. The set of seven exterior proteins considered here, S3, S5, S8, S10, S12, S14 and S19, occupy positions in the assembly as emphasized in Figure 9A. Interactions among all members of the group are shown by heavier arrows. For comparison the assembly map interactions are illustrated on the exterior surface of the small subunit in Figure 9B. For this set of proteins, only adjacent protein pairs; e.g., S12-S5, S8-S5, S5-S3, S10-S3, S14-S10 and S19-S14, are involved in assembly interactions, although some adjacent pairs; e.g. S3-S14, S10-S19 and S3-S19, are not related by assembly interactions. Hence, it appears that for this protein set that proximity of proteins is a necessary, but not sufficient, requirement for assembly interactions. Whether this will be true for the remaining 30S proteins remains to be determined. The agreement with the assembly map found for the particular set of external proteins considered in this chapter is, however, exceptionally good.

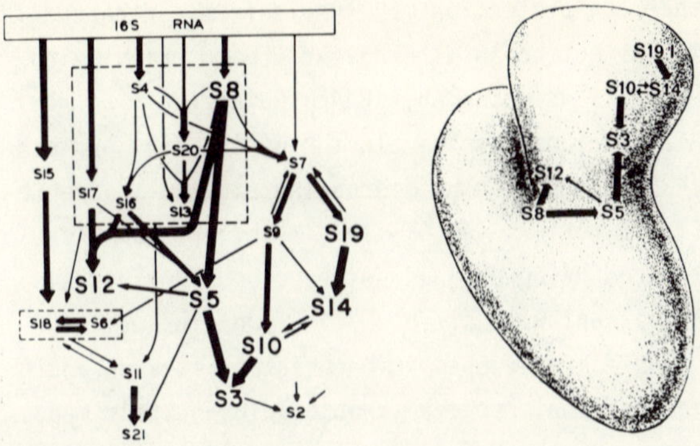

Figure 9. A comparison of the assembly map (left) with the location of exterior proteins (right). The groups of proteins being composed are indicated with larger letters in the assembly map. Adapted from Kahan and Lake (manuscript in preparation).

FUNCTIONAL INTERACTION OF SMALL SUBUNITS

Exterior Proteins

The functional properties of small subunit proteins can be correlated in a rather general way with their positions on the subunit surface.

In particular, several different types of experiments suggest that the cluster of proteins located at the concave edge of the upper one-third of the subunit containing S3, S10, S14 and S19, may be involved in aspects of tRNA binding. Single component omission experiments (44) suggested an important role for proteins S3, S10, S14 and S19 in poly-U dependent phe-tRNA binding. When these proteins were omitted, poly-U dependent phe-tRNA binding was reduced from 12 to 33 percent of that obtained with subunits reconstituted with all 30S proteins present. Particularly interesting is their finding that AUG dependent fmet-tRNA binding in the presence of IF2 and IF1 was significantly less (1.6 to 3 times less) affected by the omission of these proteins than phe-tRNA binding. Single component omission of S3, S10, S14 and S19 caused relatively little change in the sedimentation properties of the reconstituted subunits (26S, 28S, 28S and 27S, respectively) and suggests that the effect on tRNA binding could be direct. The sole remaining protein to cause a significant reduction in phe-tRNA binding when it was omitted, S8, also significantly altered the sedimentation value of the subunits.

Other experiments have also suggested an involvement of these proteins in tRNA binding. Rummel and Noller (45) found that bound phe-tRNA partially protected proteins S3, S14, S18, S19 and S21 from trypsin digestion and in related studies showed that histidine-specific photo-oxidation in the presence of Rose Bengal led to a loss of tRNA binding and mRNA binding that was attributable to modification of S3 (46).

These authors point out that the protection may result from a direct shielding effect of the tRNA's but could alternatively be attributed to indirect conformational alterations caused by tRNA binding.

The proteins, S5 and S12, included in the cluster of exterior proteins have been implicated in tRNA recognition by misreading studies. Streptomycin has been employed (for a review, see 47) and its effects on misreading have been characterized using defined heteropolynucleotides (48). Among cultures surviving treatment with streptomycin, two classes of mutants have been found. These are streptomycin-resistant strains and resistant strains that, in addition, require streptomycin in their media for growth, (streptomycin-dependent strains). Using the 30S reconstitution technique, the alteration was shown to be in S12 both in resistant strains (49) and in dependent strains (50). Revertants from streptomycin dependence, although able to grow in the presence or absence of streptomycin, have an altered S5 and S4 (51); however, the misreading properties of these revertants have not been studied. Some mutations to streptomycin resistance cause restriction of the suppression of nonsense mutations. By observing the degree of restriction of an amber codon with related amber and ocher suppressors, Gorini (52) concluded that the streptomycin-resistance mutation (S12) involved an interaction with some part of the tRNA other than the anticodon. Similarly, protein S5 is altered by the mutation to spectinomycin resistance (53). Although the spectinomycin-resistant mutants do not cause misreading (54,55) demonstrated that the spectinomycin-resistant mutation could restore the suppressor activity of restricted streptinomycin-resistant strains. Thus, mutations in both proteins, S5 and S12, alter tRNA recognition. Modifications in both affect streptomycin-induced misreading, and alterations in both can modify, (i.e., restrict [S12] or restore [S5]), the suppressor activity of mutant strains.

Proteins on the 30S Platform

The platform is a region of the small subunit with a unique shape and location that seems to be associated with a special structural role. This evolutionarily conserved feature is present in such diverse prokaryotes and eukaryotes as mammals, e.g., rat liver and rabbit reticulocytes (16), and *E. coli* (18). In the 70S ribosome, one surface of the platform faces the cytoplasm, whereas the other faces the large subunit. Because of its thinness (~ 30 Å), it is possible for a protein to be exposed on both the exterior and the interface surfaces of the platform.

Proteins S1, S11 and S18 map on the platform at sites summarized in two views of the subunit in Figure 10. The locations of S1 and S18 are very preliminary.

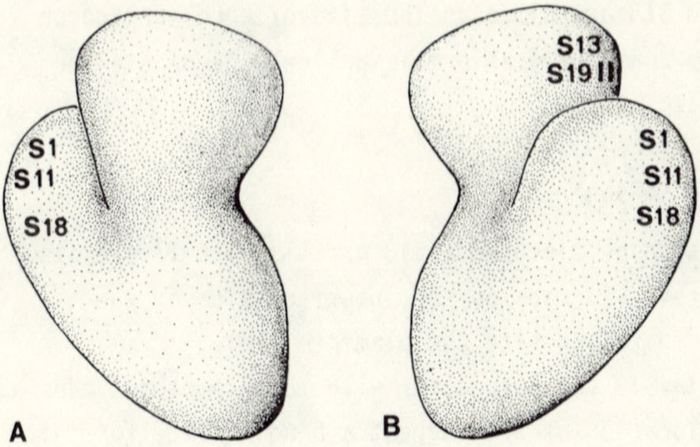

Figure 10. Diagrammatic representation of the location of protein S1, S11, S13, S18 and S19 on the exterior (A) and interface (B) surfaces of the small subunit.

Several lines of experimentation have linked these proteins to mRNA binding. Single-component omission experiments (56) demonstrated that streptomycin-induced misreading is insensitive to the omission of most proteins with two exceptions, S11 and S12. In the absence of S11, misreading (i.e., incorporation of isoleucine, tyrosine, and serine in a poly-U directed system) was increased by a factor of 6.5. Decreased misreading in the absence of S12 was consistent with an S12-tRNA interaction at a region other than the codon-anticodon site. Since the omission of no other protein significantly increased the error frequency, it was concluded that S11 may be participating directly in the selection of the correct tRNA, i.e., in the codon-anticodon interaction (56).

Proteins S1 and S18 have been cross-linked to mRNA using a variety of affinity-labeling analogs [these are comprehensively reviewed in (57)]. The reagents employed have included poly-U, UUUU, UGA, GUUU, electrophilic derivatives of AUG, AAAAAAAA, poly(s^4U), poly(Br^5U). The most frequently labeled protein is S18 and the second most frequently labeled protein is S1, (other labeled proteins include S4, S12, and S21). Recently, two additional proteins, S3 and S5, have been labeled with a photoaffinity tag bound to the 3' terminal ribose of the mRNA (58).

Additional evidence placing S1 near the codon comes from the suggestion (59) that the 3' end of the 16S RNA can make between 4 and 7 base pairs with a leader sequence located about 10 nucleotides before (i.e., on the 5' side of) the initiation codon, from the experiments supporting this suggestion (60,61), and from the isolation of protein S1 complexed to this base-pairing region (62). Consistent with this, dimethyl As located on the 16S RNA near the leader-sequence pairing region have been mapped (14) on the platform by immune electron microscopy and the platform has been localized as the decoding site by affinity immune electron microscopy (15).

Thus, a considerable body of evidence exists to suggest that the platform proteins S1, S11 and S18 are close to the site of codon-anticodon interaction. We conclude that the platform is a likely candidate for the codon-anticodon binding site.

Initiation Factor Binding Region

The final 30S proteins to be discussed, S13 and S19 site II are also shown in Figure 10. Protein S13 maps on the interface side of the subunit on the upper one third opposite the platform (18).

The location of protein S13 taken together with those proteins mapping on the platform and the location of S12 suggest a binding site for initiation factors IF-1, IF-2 and IF-3. IF-3, required for the binding of natural mRNA to the 30S subunit, has been efficiently crosslinked to S7 (63), and to S11, S12, S13 and S19 (64). IF-2 stimulates the binding of formylmethionyl-tRNA^{fMet} to the 30S subunit and catalyzes the hydrolysis of GTP on the 70S ribosome [reviewed in (65) and (66)]. In the presence of IF-1 and IF-3 it has been efficiently crosslinked to S1, S2, S11, S13, S14 and S19 with the strongest immunological reactions against S13 and S19 and next strongest against S11 and S12 (67). (IF-1 has been crosslinked to S1, S12, S13 and S19 (68) in the presence of IF-2 and IF-3.) Consideration of the results of both crosslinking and protein localization suggests that IF-3, IF-2 and IF-1 are positioned across the groove between S13, S19-II and S12 on the one-third side and S11 on the platform side in the general region indicated in Figure 11. (Our preli-

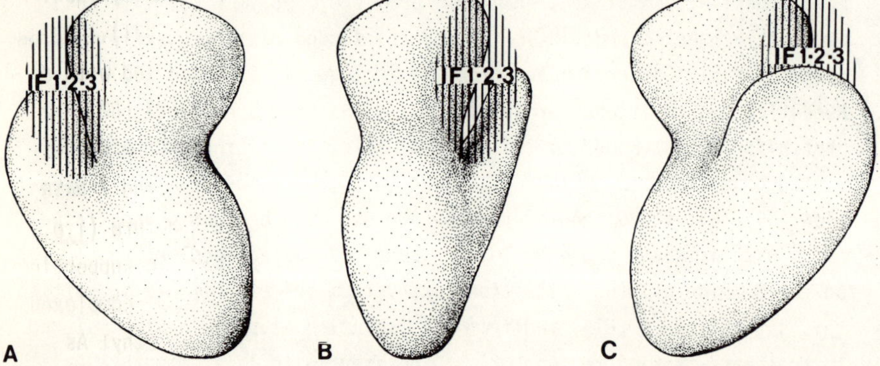

Figure 11. Locations suggested for IF-1, IF-2 and IF-3 (A) on the exterior, (B) in the cleft, and (C) at the interface of the 30S subunit.

minary mapping of S7 is also consistent with this interpretation.) The binding of IF-3 is consistent with the location of the codon-anticodon interaction; the binding of IF-2, in turn, provides an indication of the binding site for formylmethionyl-$tRNA^{fMet}$.

ANTIBODY LABELING OF LARGE SUBUNIT PROTEIN

Proteins L7/L12

Ribosomal proteins L7 and L12 have several unique properties [for reviews, see (69) and (70)]. Proteins L7 and L12 have identical amino acid sequences except that the amino terminus of L7 is acetylated, whereas that of L12 is not. L7 and L12 are the only ribosomal proteins present in multiple copies and the total number of L7 and L12 copies per ribosome is most likely four. Both proteins are intimately involved in elongation factor Tu (EF-Tu), in elongation factor G (EF-G)-dependent GTP hydrolysis, and in IF-2-dependent GTP hydrolysis. Hence, localization of the L7/L12 stalk provides information about the ribosome binding sites of factors EF-Tu, EF-G and IF-2.

Electron micrographs of 50S subunits reacted with antibodies against L7/L12 are shown in Figure 12A (71) and for comparison subunits labeled with antibodies against L27 are shown in 12B. In the quasisymmetric projection, shown in the first two frames, L7/L12 is localized on the L7/L12 stalk and the location of L7/L12 in the asymmetric projection, shown in the fourth and fifth frames of Figure 12A, is diagrammed in the sixth frame. Using several approaches in addition to direct immune mapping including doubly labeled subunits and removal and rebinding of

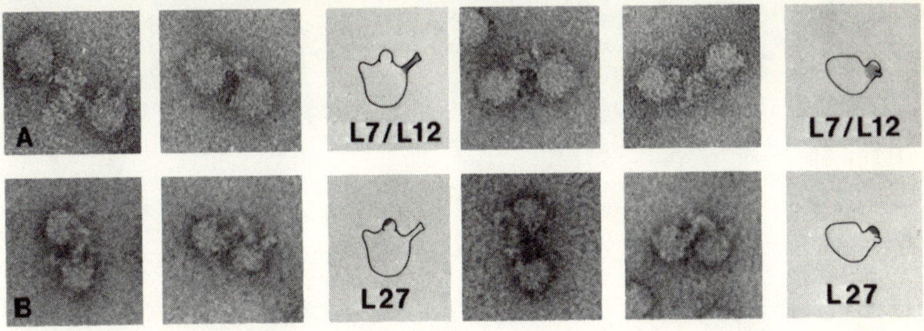

Figure 12. Electron micrographs of 50S subunits reacted with antibodies against (A) protein L7/L12 and against (B) protein L27. Subunits in the quasisymmetric projection are shown or illustrated in the first three frames of a row and subunits in the asymmetric projection are shown in the second three frames of each row.

L7 and L12, shown in Figure 13, we (71) demonstrated that L7 and L12 map exclusively on the elongated appendage called the L7/L12 stalk.

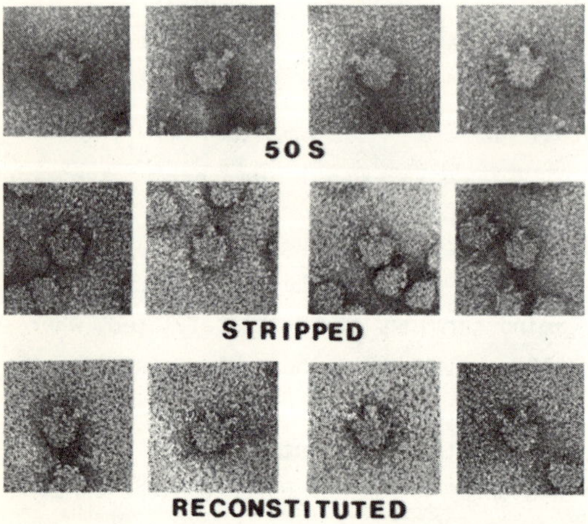

Figure 13. Electron micrographs of large ribosomal subunits containing all ribosomal proteins (first row), all ribosomal proteins except L7/L12 (second row) and subunits containing all ribosomal proteins that have been reconstituted from L7/L12 deficient cores (lower row). Adapted from (71).

The relationship of proteins L7 and L12 to the monomeric 70S ribosome is shown in Figure 14. For reference, some small-subunit protein locations are shown. The large subunit is illustrated in the conformation principally observed in fields of dissociated 50S subunits (Figure 14A) and in the conformation principally observed in fields of 70S ribosomes (Figure 14B). In either conformation, however, L7 and L12 are quite distant from the convex edge (at the left) of the small subunit. Marsh and Parmeggiani (72) showed that a subunit depleted of S3, S5, S10 and S14 (in addition to S1, S2 and S9) is unable to stimulate GTP hydrolysis dependent upon EF-G and the 50S subunit. Activity could be restored by adding S5 and S9. This is consistent with

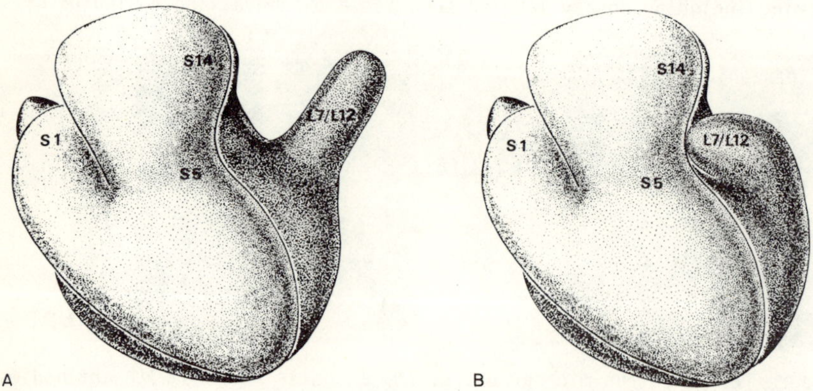

Figure 14. The location of the L7/L12 stalk relative to the small subunit. (A) The morphology of the stalk as observed in fields of large subunits. (B) The morphology principally observed in fields of 70S ribosomes. The location of a few small subunit proteins are included for reference.

the location of L7/L12 adjacent to the region of the small subunit containing S5. The L7/L12 stalk location at the 30S/50S subunit interface is consistent with the IF-2 binding site discussed in the section "Initiation-Factor Binding Region." However, since IF-2 as well as EF-Tu and EF-G are relatively large, factor binding sites could be positioned over a sizable region of the 70S ribosome and still be consistent with the reported factor L7/L12 stalk crosslinks [EF-G, (73) EF-Tu, (74)].

Other 50S Proteins

Preliminary locations of the four additional large subunit proteins, L17, L21, L23 and L27; are shown schematically in Figure 15. At present there is little information relating either L17 or L21 to functional events occurring on the 50S subunits.

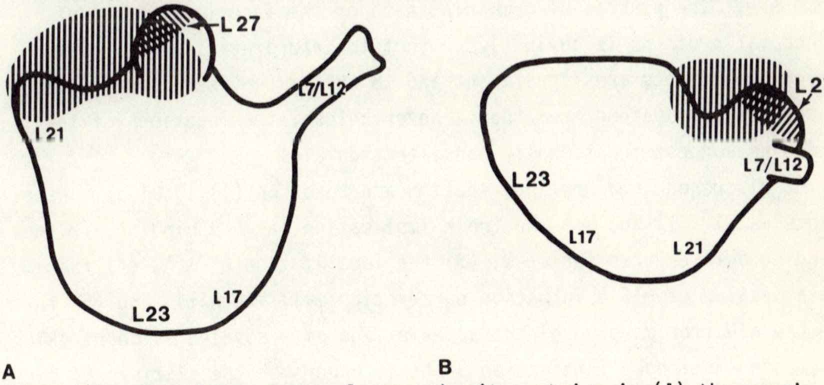

Figure 15. Locations of some large subunit proteins in (A) the quasi-symmetric and (B) the asymmetric projections of the subunit. The vertical stripes correspond to a possible region of contact of the acceptor end of the tRNA assuming the codon-anticodon interaction occurs on the platform. The diagonal stripes correspond to the localization of protein L27. From the work of Strycharz, Nomura and Lake (76).

Protein L27 is a candidate to be near the peptidyl transferase. It is among the proteins labeled by modified aminoacyl-tRNA affinity labels bound to the peptidyl site and to the aminoacyl site [for reviews, see (75) and (57)]. The most common affinity labels for peptidyl-site studies have been electrophilic derivatives of N-acylphenylalanyl-tRNA^{Phe}. Using these affinity lables, the principal labeled proteins are L2, L11, L18 and L27. Under conditions where aminoacyl-site labeling is favored, the predominantly labeled proteins are L2, L16 and L27. The third suggestion that L27 might be at the peptidyl transferase comes from acyl-methionyl-tRNA^{fMet} affinity labeling of ribosomes, where L2 and L27 have been the major product. Hence, L27 consistently seems to be labeled near the peptidyl transferase by affinity analogs of initiator tRNAs and of tRNAs in both the aminoacyl and the peptidyl sites. The electrophilic nature of the labels should, however, be kept in mind. In contrast, L23 is not significantly labeled by peptidyl-tRNA analogs, but is labeled with radioactive puromycin. Rather unexpectedly, however, small subunit proteins are also labeled, the most significant reaction being with S14. This suggests that the L23 labeling might be occurring somewhere other than at the peptidyl transferase although a fortutious labeling of L27 by electrophilic derivatives cannot be ruled out.

RIBOSOME FUNCTIONAL REGIONS

Given the locations of the ribosomal proteins discussed in this paper, one can begin to ask where the functional regions of the ribosome are. The process of combining data on the functional roles of ribosomal proteins is obviously subject to interpretation regarding which experiments are significant and to what extent structural inferences can be obtained from them. Nevertheless, the locations of the proteins suggest functionally consistent domains.

As discussed in a previous section and based on (1) localizations of proteins S1, S11 and S18 and their implication in mRNA binding, (2) the binding site deduced for IF-3, (3) the localization of m_6^2A, (4) the localization of eIF-3 in native eukaryotic small subunits, and (5) the immune electron mapping of the acceptor end of a covalently bount tRNA; the codon-anticodon interaction probably occurs on the platform of the 30S subunit.

The region of the 30S subunit to which IF1, IF2 and IF3 have been crosslinked is indicated in Figure 11. The location of IF2 suggests an approximate location for initiator tRNA binding, since its binding is dependent upon the presence of both the initiation codon and IF2. In order for the anticodon of a tRNA in the A or the P site to contact the codon on the platform, the tRNA can only contact a limited region of the 50S subunit. Using the constraints of tRNA size and shape, Lake (13) reasoned that the peptidyl transferase was probably located within the area indicated by vertical striations in Figure 15. However, a flexing of the tRNA at the TΨC corner as suggested by Richter et al. (77) would not greatly modify this region. The location of L27 is also indicated by diagonal striping. Proteins L23 and S14, both labeled by an affinity analogue of puromycin but not labeled in tRNA affinity experiments, are located in regions quite far from the area of diagonal striations.

Protein L27, on the other hand, is one of the primary molecules labeled by both peptidyl and aminoacyl-tRNA affinity labels and is presumed to be near the peptidyl transferase. Hence, the cross-hatched region in Figure 12 where both regions overlap is suggested as the peptidyl transferase site.

This location of the peptidyl transferase, which is separated from proteins L7/L12, is consistent with evidence suggesting that L7/L12 are spatially removed from the peptidyl transferase. Removal of L7/L12, for example, does not alter the peptidyl transferase activity of the cores (78). Large subunit cores with L7/L12, L10 and L11 removed are active in the fragment reaction (79,80) and are capable of binding antibiotics reacting with the peptidyl transferase center (81). The erythromycin binding site, reported to be near the peptidyl transferase (82,83), has been measured to be approximately 70 Å from L7/L12 by singlet-singlet energy transfer (84). (The labeled region of L7/L12 in these experiments was not known, however, so this does not preclude the possibility that some regions of L7/L12 could be significantly closer to the peptidyl transferase center that 70 Å.) Inhibition of both EF-G binding and EF-Tu-associated GTPase activity, but not of peptidyl transferase activity, by binding of the antibiotic siomycin (85) has also suggested a separation of L7/L12 from the peptidyl transferase center. Thus, the results of biochemical experiments suggesting separation of the peptidyl transferase from L7/L12 are in agreement with our mapping results and with the suggested location for the peptidyl transferase. In order for a tRNA to simultaneously contact the codon, the IF2 binding region, and the peptidyl transferase, it might be positioned in the A or the P site approximately as shown in Figure 16. For comparison, the approximate locations of some external small subunit proteins are also marked.

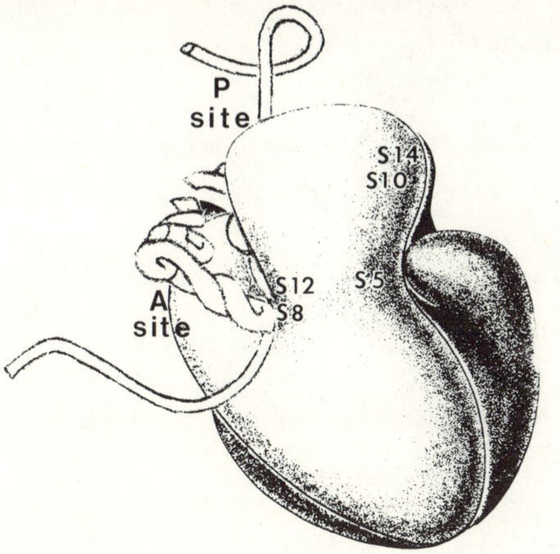

Figure 16. Model of the monomeric ribosome showing suggested aminocyl and peptidyl-tRNA binding sites (76).

The 30S proteins mapping on the exterior surface of the subunit seem to be the most significant cluster of proteins. Their three-dimensional locations are consistent with the results obtained by other techniques including neutron diffraction, crosslinking and the assembly map, suggesting that their locations are quite reliable and accurate. The region on the upper one-third consisting of S3, S5, S10, S14 and S19, as discussed, was implicated by several types of studies in tRNA binding. Those proteins at the level of the one third-two thirds partition S5 and S12, and presumably S4, are thought to participate in the process of codon recognition, although some evidence suggests they are not directly involved in the codon-anticodon interaction. Hence (with the possible exception of S8 whose role in tRNA binding is obscured by its important role in subunit assembly) all of the external proteins are implicated in the process of tRNA binding and recognition. Thus far 7 small subunit proteins have been mapped on the exterior surface of the ribosome. This represents fully one-third of the total 30S proteins and strongly implicates this region of the subunit in tRNA binding. In view of the considerable experimental evidence suggesting a role in tRNA binding for these proteins, their separation from the presumed locations of the A and P tRNA-binding sites is quite remarkable. It could, perhaps, be argued that these proteins might alter tRNA binding indirectly or that, although they might extend into the tRNA binding region, they might not be sufficiently exposed to permit antibody labeling. The simplest alternative, however, is that they directly interact with tRNAs. Certainly in the case of protein S5 at least, the information obtained from neutron diffraction makes the extension of a significant portion of S5 much beyond a diameter of 40 Å unlikely (86,87). Whereas a single tRNA binding protein mapping outside the region suspected to be the A and P tRNA binding sites could be easily explained, it was extraordinary to find a set of 7 proteins all strongly implicated in tRNA binding (fully one third of all 30S proteins) mapping outside this region. Just as perplexing was the observation that they were located on the external surface of the small subunit, since conventional wisdom dictated that tRNA binding sites occur at the interface between subunits in order for the anticodon to contact the message (on the 30S subunit) and the acceptor end of the tRNA to contact the peptidyl transferase (on the 50S subunit).

The cluster of proteins that is involved in tRNA binding but is located on the exterior of the 30S subunit suggested that another tRNA binding site might exist. As a result, a reconsideration of the elongation cycle led to the conclusion that during protein synthesis initial codon recognition occurs with the aminoacyl-tRNA bound in a structurally distinct recognition tRNA binding site (88). Although initially suggested by antibody labeling results, the concept of the recognition site is supported by the fundamental experiments defining the elongation cycle, quite independently of the results of our antibody-labeling studies.

THE RECOGNITION SITE

Initial Codon Recognition Occurs at the Recognition tRNA Binding Site

The broad outlines of the chain elongation cycle of protein synthesis are quite well understood [for reviews, see (89), (90)]. Typically, the major events that occur are summarized as (1) codon recognition, (2) peptide-bond formation, and (3) translocation. In Figure 17A, a cycle of elongation has been completed and the n + 1 codon is available to be read by the correct tRNA. In the first step of the next elongation cycle (initial codon recognition; Figure 17B) the codon is recognized and the correct aminoacyl-tRNA·EF-Tu·GTP complex is bound by the ribosome. In the second step (peptide-bond formation; Figure 17C) the GTP is cleaved to GDP and Pi, the EF-Tu·GDP complex and Pi are released from the ribosome, the peptide bond is completed, and the n + 1 tRNA occupies the aminoacyl site. The third event, translocation, refers to a switching of the peptidyl-tRNA from the aminoacyl site (Figure 17C) to the pepticyl site (Figure 17D). This step utilizes EF-G and the cleavage of GTP and results in the release of EF-G, GDP, Pi, and the uncharged tRNA.

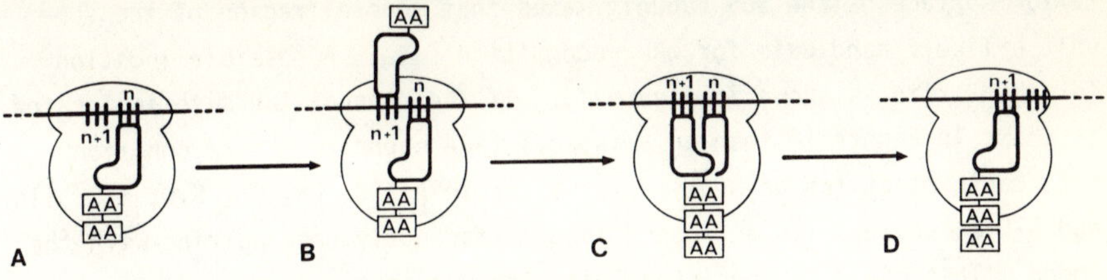

Figure 17. The elongation cycle of protein synthesis. (A) Peptidyl-tRNA in the P site; (B) aminoacyl-tRNA in the R site; (C) peptidyl-tRNA in the A site; and (D) peptidyl-tRNA in the P site. From Lake (88).

Each of the three states shown in Figure 17 is a functionally distinct state (and, therefore, a physically distinct state) of the elongation cycle. The three states are functionally distinguishable according to the following criteria. If peptidyl-tRNA is reactive with puromycin it is said to be in the peptidyl site (Figure 17) and if peptidyl-tRNA is not reactive with puromycin then it is in the aminoacyl site (Figure 17). In a similar way, with an aminoacyl-tRNA in the recognition site (i.e., in the presence of both EF-Tu and GMPP(CH_2)P the nonhydrolyzable analog of GTP), peptidyl transfer does not occur, yet the aminoacyl site of the peptidyl transferase is vacant, since the peptidyl-tRNA in the peptidyl site is puromycin reactive (91). Just as the aminoacyl and peptidyl tRNA binding sites refer to tRNAs bound to the ribosome in defined physical states, the term recognition site describes the distinct physical state of the aminoacyl-tRNA when bound to the ribosome during initial condon recognition.

The Recognition Site is Probably on the External Surface of the Small Subunit

The existence of the recognition site rests on the current biochemical understanding of the elongation cycle, quite different data from those provided by our mapping studies. Information on the elongation cycle itself provides few clues to the location of the recognition site other than to suggest that the recognition and aminoacyl sites are distinct. Distinct may refer to only a separation of a few angstroms between the sites or could refer to a separation of many angstroms. The observation that puromycin can release the peptidyl-tRNA while an aminoacyl-tRNA is bound to the recognition site implies that the CCA end of the tRNA bound to the recognition site is separated from the aminoacyl site by at least 5-10 Å.

The existence of the cluster of 7 tRNA binding proteins on the exterior surface of the 30S subunit makes that general region of the subunit a likely candidate for the recognition site. A possible location for the R site is shown in Figure 18. If the recognition site is located as shown in Figure 18 then an aminoacyl-tRNA bound to the recognition site can contact (or be near) external proteins S3, S5, S8, S10, S12, S14 and S19-I, whereas its anticodon is simultaneously base pairing with the codon. This would be consistent with the previously discussed effects of these proteins on tRNA binding and with the possible location of the codon on the platform. In addition, the D stem of a tRNA in the recognition site would be near proteins S5 and S12, consistent with the roles of those proteins in restricting the suppression of nonsense mutations and with the role of a single base substitution in the D stem in converting $tRNA^{trp}$ into a nonsense suppressor tRNA (92).

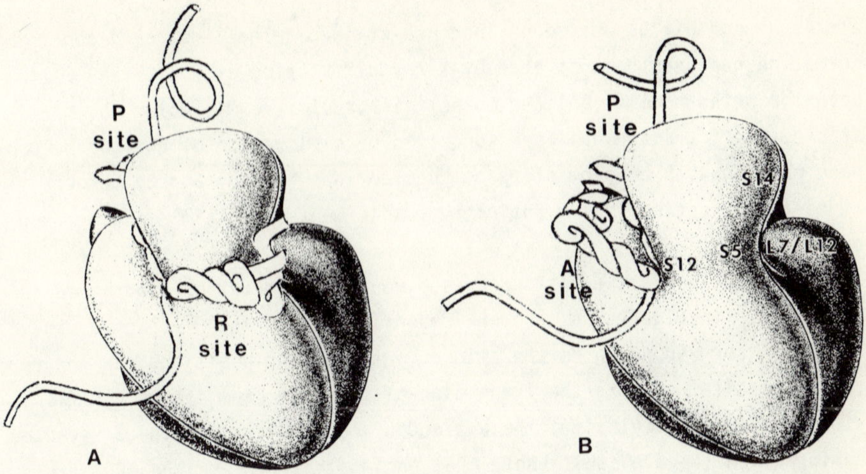

Figure 18. Suggested locations of tRNA binding sites on the ribosome. (A) Recognition site binding (Corresponding to Figure 17B); (B) aminoacyl-site binding (Corresponding to Figure 9C). The tRNAs are approximately drawn to scale.

EF-Tu has been crosslinked to the L7/L12 stalk (74) in the presence of GMPP(CH_2)P (i.e., under conditions probably corresponding to recognition site binding). In addition, other evidence, such as the lack of EF-Tu binding cores stripped of L7/L12 [for a review see (90)], suggests an interaction of the L7/L12 stalk with EF-Tu. Thus, the recognition site as illustrated in Figure 18, on the exterior side of the small subunit is consistent with its proximity to proteins L7 and L12. However, EF-Tu is relatively large so that other locations on the small subunit could also be consistent with the reported factor-L7/L12 stalk crosslink.

A Model for Switching from the Recognition Site to the Aminoacyl Site

If the recognition tRNA site is located on the exterior of the 30S subunit, then a mechanism is needed for switching the tRNA from the recognition site to the aminoacyl site. This mechanism for tRNA movement to the aminoacyl site must also be consistent with the observation (93, 94) that no measurable messenger movement, relative to the ribosome, occurs during recognition to aminoacyl site switching and that the message is shifted three nucleotides only during the aminoacyl to peptidyl site switching.

Evidence suggests that the essential element of the switching mechanism could involve the tRNA itself. The anticodon loop is thought to assume two conformations referred to as the 5' - and the 3'-stacked conformations. In the 5' stack, the five bases at the 5' end of the loop are stacked in a quasihelical manner. In the 3' stack, the five bases at the 3' end of the anticodon loop are stacked. The 3'-stacked conformation has been demonstrated by X-ray diffraction of yeast $tRNA^{Phe}$ in both the orthorhombic (95) and monoclinic crystal forms (96). The 5'-stacked configuration has been suggested by exploration of anticodon loop base pairing with complementary polynucleotides (97,98) and by temperature-jump kinetics (99). Definitive experiments are still lacking, however.

The switching in the anticodon loop would then occur as diagrammed in Figure 19. In Figure 19A, the anticodon loop is in the 5'-stacked conformation corresponding to the aminoacyl-tRNA being positioned in the recognition site. During the transition of the tRNA from the recognition site to the aminoacyl site, the anticodon loop switches to the 3'-stacked conformation (Figure 19B). Note that during this process the mRNA (in black) and the anticodon have not moved, although the acceptor end of the tRNA has moved a large distance (see Figure 18). An interesting aspect of this detailed model is that essential elements of the translation machinery are properties of the tRNA (100).

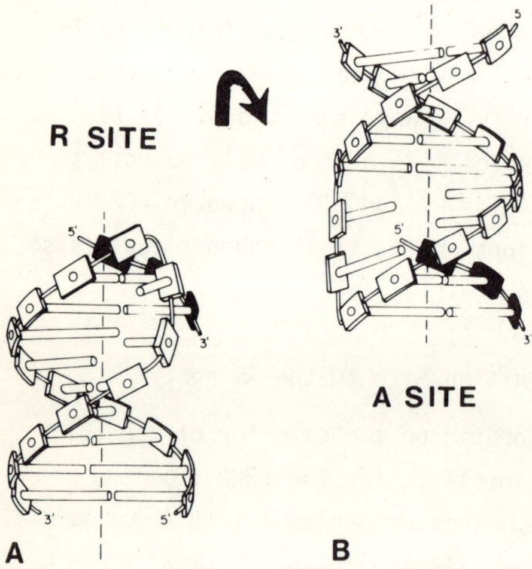

Figure 19. Diagramatic representation of the switching of the anticodon loop from the 5'-stacked conformation (A) to the 3'-stacked conformation. Note that during switching the codon (black) and the anticodon remain fixed, although the acceptor stem of the tRNA (not shown) moves a considerable distance. From (88).

The switching mechanism is basically different from the model proposed by Woese (101) and also applies to a different step in the elongation cycle. In Woese's model, the switching between the aminoacyl and peptidyl sites, i.e., translocation, is described, whereas in this model the switching between the recognition and aminoacyl sites is described. Also, the tRNA remains fixed on the ribosome in the model of Woese and the mRNA moves, whereas in this model the message remains fixed and the tRNA moves. Recent evidence [cited in 102] is inconsistant with the Woese model.

IN VITRO ARRAYS OF E. COLI RIBOSOMAL SUBUNITS

In the introductory section, we traced the progress of structural studies on ribosomes and noted that the rapid advances in detailing the E. coli ribosome structure have resulted in large part from the convergence of different experimental approaches and that progress would be accelerated if ordered arrays of E. coli ribosomes could be formed in vitro. Recently, Clark et al. (8) have produced in vitro ordered helical arrays of E. coli small ribosomal subunits which promise to be useful for three dimensional reconstruction. An electron micrograph of a field of these arrays is shown in Figure 20.

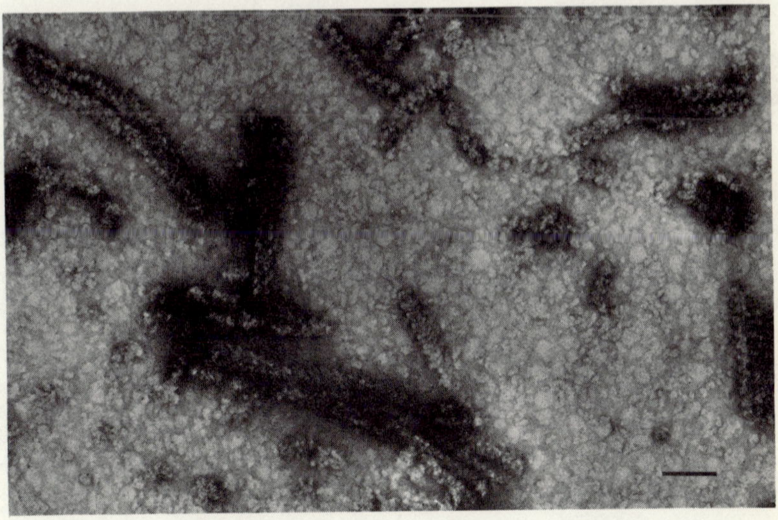

Figure 20. Electron micrograph of the contents of a crystallization well illustrating helical ribbons of 30S subunits. The contents were negatively stained six months after the wells were set up. The scale bar is 1000 Å. Adapted from Clark et al. (8).

Two distinct image types are seen in the fields. The wider of the two has a width of about 320 Å and the narrower is about 180 Å wide. Images of twisted ribbons (not shown) suggest that the two image types correspond to two different projections of the same ribbon. Optical diffraction patterns of both image types suggested that both have a central glide plane, an axial repeat of 220 Å and correspond to subunits packed along a two-fold screw axis.

Ordered arrays of large subunits have also been obtained as illustrated in Figure 21. In these arrays the pseudorepeat distance (between adjacent layers) is approximately 245 Å and the true repeat (between alternate layers) is 490 Å.

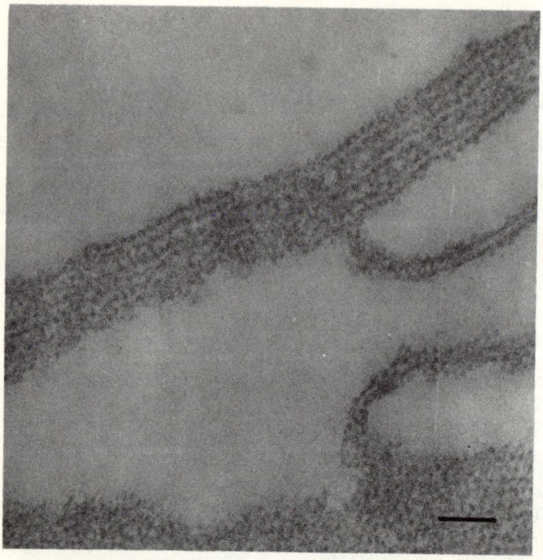

Figure 21. Electron micrograph of positively stained sections of the contents of a crystallization well of 50S subunits. The contents were fixed in glutaraldehyde, embedded in epon, and stained after sectioning with uranyl-acetate. Subunits are dark. The scale bar is 1000 Å.

At present we are optimistic about obtaining larger and better ordered arrays. We find (8) that under the storage conditions used, both large and small subunits can remain in crystallization wells for as long as six months without significant alteration in their sedimentation velocities in sucrose gradients or alteration in their composition as detected on native, agarose-acrylamide gels (103,104).

CONCLUSION

In conclusion, the localization of ribosome proteins is providing data that were unobtainable a few years ago. We now know where many ribosomal proteins, including the multiple copy proteins L7 and L12, are located; and we have an idea of the location of the codon-anticodon interaction site and of the aminoacyl- and peptidyl-tRNA binding sites. It is known that the existence of a recognition site is necessary, given the current understanding of the elongation cycle (obviously its existence is not dependent on the correctness of the molecular model for switching from the recognition site to the aminoacyl site), and we have suggested that it may be located on the exterior of the small subunit. In the future we hope that the localization of other ribosomal proteins and the localization of ribosomal ligand binding sites will lead to an increased understanding of the molecular events of protein synthesis. We also hope that we will be able to apply the techniques of three dimensional reconstruction and X-ray crystallography to the continuing study of ribosome structure.

ACKNOWLEDGEMENTS

I thank M. Hwang, A. Kost, and S. Pressman-Nelson for excellent Electron Microscopy and photography. Supported by grants from the National Science Foundation and from the National Institute of General Medical Sciences.

REFERENCES

1. Lake, J. A., and Slayter, H. S. (1972) J. Mol. Biol. 66, 271-282.

2. Hall, C. E., and Slayter, H. S. (1959) J. Mol. Biol. 1, 329-332.

3. Huxley, H. E., and Zubay, G. (1960) J. Mol. Biol. 2, 10-18.

4. Byers, B. (1967) J. Mol. Biol. 26, 155-167.

5. Morgan, R. S., and Uzman, B. G. (1966) Science 152, 214-215.

6. Lake, J. A., and Slayter, H. S. (1970) Nature 227, 1032-1037.

7. Unwin, P. T. N. (1977) Nature 269, 118-122.

8. Clark, M. W., Hammons, M. Langer, J. A., and Lake, J. A. (1979) J. Mol. Biol., in press.

9. Yanagida, M., and Ahmad-Zadeh, C. (1970) J. Mol. Biol. 51, 411-421.

10. Wabl, M. R. (1974) J. Mol. Biol. 84, 241-247.

11. Lubin, M. (1968) Proc. Natl. Acad. Sci. USA 61, 1454-1461.

12. Lake, J. A., Nonomura, Y., and Sabatini, D. D. (1974a) in Ribosomes (M. Nomura et al., ed.) p. 543, Cold Spring Harbor Laboratory, Cold Spring Harbor, New York.

13. Lake, J. A. (1976) J. Mol. Biol. 105, 131-159.

14. Politz, S. M., and Glitz, D. G. (1977) Proc. Natl. Acad. Sci. USA 74, 1468-1472.

15. Keren-Zur, M., Boublik, M., and Ofengand, J. (1979) Proc. Natl. Acad. Sci. USA 76, 1054-1058.

16. Emanuilov, I., Sabatini, D. D., Lake, J. A., and Freienstein, C. (1978) Proc. Natl. Acad. Sci. 75, 1389-1393.

17. Lake, J. A., Pendergast, M., Kahan, L., and Nomura, M. (1974b) Proc. Natl. Acad. Sci. 71, 4688-4693.

18. Lake, J. A., and Kahan, L. (1975) J. Mol. Biol. 99, 631-644.

19. Vasiliev, V. D. (1974) Acta Biol. Med. Germ. 33, 779-793.

20. Boublik, M., and Hellmann, W. (1978) Proc. Natl. Acad. Sci. USA 75, 2829-2833.

21. Tischendorf, G. W., Zeichhardt, H., and Stöffler, G. (1975) Proc. Natl. Acad. Sci. USA 72, 4820-4824.

22. Leonard, K. R., and Lake, J. A. (1979) J. Mol. Biol. 129, 155-163.

23. Boublik, M., Hellmann, W., and Roth, H. E. (1976) J. Mol. Biol. 107, 479-490.

24. Lake, J. A. (1978) in Advanced Techniques in Biological Electron Microscopy II (J. Koehler, ed.) p. 173, Springer-Verlag, Berlin.

25. Vasiliev, V. D., Selivanova, O. M., and Koteliansky, V. E. (1978) FEBS Letters 95, 273-276.

26. Bielka, H., and Stahl, J. (1978) International Reviews of Biochemistry in Amino Acid and Protein Biosynthesis (V. Arnstein, ed.) pp. 79-168, University Park Press, Baltimore.

27. Engelman, D. M., Moore, P. B., and Schoenborn, B. P. (1975) Proc. Natl. Acad. Sci. USA 75, 1389-1393.

28. Stuhrmann, H. B., Koch, M. H. J., Parfait, R., Haas, J., Ibel, K., and Crichton, R. R. (1977) Proc. Natl. Acad. Sci. USA 74, 2316-2320.

29. Langer, J. A. Engelman, D. M., and Moore, P. B. (1978) J. Mol. Biol. 119, 463-485.

30. Schindler, D. G., Langer, J. A., Engelman, D. M., and Moore, P. B., manuscript submitted

31. Lutter, L. C., Zeichhardt, H., Kurland, C. G., and Stöffler, G. (1972) Mol. Gen. Gent. 119, 357-366.

32. Bode, U., Lutter, L. C., and Stöffler, G. (1974) FEBS Letters 45, 232-236.

33. Clegg, C., and Hayes, D. (1974) Eur. J. Biochem. 42, 21-28.

34. Lutter, L. C., Bode, U., Kurland, C. G., and Stöffler, G. (1974) Mol. Gen. Genet. 129, 167-176.

35. Sun, T. T., Bollen, A., Kahan, L., and Traut, R. R. (1974) Biochemistry 13, 2334-2340.

36. Lutter, L. C., Kurland, C. G., and Stöffler, G. (1975) FEBS Letters 54, 144-150.

37. Sommer, A., and Traut, R. R. (1975) J. Mol. Biol. 97, 471-481.

38. Sommer, A., and Traut, R. R. (1976) J. Mol. Biol. 106, 995-1015.

39. Peretz, H., Towbin, H., and Elson, D. (1976) Eur. J. Biochem 63, 83-92.

40. Expert-Bezancon, A., Barritault, D., Milet, M., Guerin, M.-F., and Hayes, D. H. (1977) J. Mol. Biol. 112, 603-629.

41. Mizushima, S., and Nomura, M. (1970) Nature 226, 1214-1218

42. Held, W. A., Ballou, B., Mizushima, S., and Nomura, M. (1974) J. Biol. Chem. 249, 3101-3111.

43. Nomura, M., and Held, W. A. (1974) in Ribosomes (M. Nomura, Lengyel, P., and Tissières, A., eds.) pp. 193-223, Cold Spring Harbor Laboratory, Cold Spring Harbor, New York.

44. Nomura, M., Mizushima, S., Ozaki, M., Traub, P., and Lowry, C. V. (1969) Cold Spring Harbor Symp. Quant. Biol. 34, 49-61.

45. Rummel, D. P., and Noller, H. F. (1973) Nature New Biology 245, 72-75.

46. Noller, H. F., Chang, C., Thomas, G., and Aldridge, J. (1971) J. Mol. Biol. 61, 669-679.

47. Gorini, L. C. (1974) in Ribosomes (M. Nomura, et al., eds.) p. 791, Cold Spring Harbor Laboratory, Cold Spring Harbor, New York.

48. Davies, J., Jones, D. S., and Khorana, H. G. (1966) J. Mol. Biol. 18, 48-60.

49. Ozaki, M., Mizushima, S., and Nomura, M. (1969) Nature 222, 333-339.

50. Birge, E. A., and Kurland, C. G. (1970) Mol. Gen. Genet. 109, 356-369.

51. Hasenbank, R., Guthrie, C., Stöffler, G., Wittmann, H. G., Rosen, L., and Apirion, D. (1973) Mol. Gen. Genet. 127, 1-18.

52. Gorini, L. C. (1969) Cold Spring Harbor Symp. Quant. Biol. 34, 101-111.

53. Bollen, A., Davies, J., Ozaki, M., and Mizushima, M. (1969) Science 165, 85-86.

54. Anderson, P., Davies, J., and Davies, B. D. (1967) J. Mol. Biol. 29, 203-212.

55. Kuwano, M., Endo, H., and Ohrishi, Y. (1969) J. Bacteriol. 97, 940-943.

56. Nomura, M. (1970) Bacteriol. Rev. 34, 228-277.

57. Cooperman, B. S. (1978) Bioor. Chem., in press.

58. Towbin, H., and Elson, D. (1978) Nucleic Acids Res. 5, 3389-3408.

59. Shine, J., and Delgarno, L. (1975) Nature 254, 34-38.

60. Argetsinger-Steitz, J. A., and Jakes, K. (1975) Proc. Natl. Acad. Sci. USA 72, 4734-4738.

61. Dunn, J. J., Buzash-Pollert, E., and Studier, W. F. (1978) Proc. Natl. Acad. Sci. USA 75, 2741-2745.

62. Dahlberg, A., and Dahlberg, J. (1975) Proc. Natl. Acad. Sci. USA 72, 2940-2944.

63. Van Duin, J., Kurland, C. G., Dandon, J., and Grunberg-Mango, M. (1975) FEBS Letters 59, 287-290.

64. Heimark, R. L., Kahan, L., Johnston, K., Hershey, J. W. B., and Traut, R. R. (1976) J. Mol. Biol. 105, 219-230.

65. Lucas-Lenard, J., and Lipmann, F. (1971) Annu. Rev. Biochem. 40, 409-448.

66. Haselkorn, R., and Rothman-Denes, L. B. (1973) Annu. Rev. Biochem. 42, 379-438.

67. Bollen, A., Heimark, R. L. Cozzone, A., Traut, R. R., Hershey, J. W. B., and Kahan, L. (1975) J. Biol. Chem. 250, 4310-4314.

68. Langberg, S., Kahan, L., Traut, R. R., and Hershey, J. W. B. (1977) J. Mol. Biol. 117, 307-319.

69. Möller, W. (1974) in <u>Ribosomes</u> (M. Nomura, A. Tissières and P. Lengyel, eds.) Cold Spring Harbor Laboratory, Cold Spring Harbor, New York, pp. 711-731.

70. Weissbach, H., and Pestka, S. (1977) in <u>Molecular Mechanisms of Protein Biosynthesis</u> (H. Weissbach and S. Pestka, eds.) Academic Press, New York.

71. Strycharz, W. A., Nomura, M., and Lake, J. A. (1978) J. Mol. Biol. 126, 123-140.

72. Marsh, R. C., and Parmeggiani, A. (1973) Proc. Natl. Acad. Sci. USA 70, 151-155.

73. Acharya, A. S., Moore, P. B., and Richards, F. M. (1973) Biochem. 12, 3108-3114.

74. San José, C., Kurland, C. G., and Stöffler, G. (1977) FEBS Letters 71, 133-137.

75. Traut, R. R., Heimark, R. L., Sun, T.-T., Hershey, J. W. B., and Bollen, A. (1974) in <u>Ribosomes</u> (M. Nomura <u>et al</u>., eds.) pp. 271-238, Cold Spring Harbor Laboratory, Cold Spring Harbor, New York.

76. Lake, J. A. (1979) in <u>Transfer RNA: Structure, Properties, and Recognition</u> pp. 393-411, Cold Spring Harbor Press, Cold Spring Harbor, New York.

77. Richter, D., Erdmann, V. A., and Sprinzl, M. (1973) Nature New Biology 246, 132-135.

78. Hamel, E., Koka, M., and Nakamoto, T (1972) J. Biol. Chem. 247, 805-814.

79. Howard, G. A., and Gordon, J. (1974) FEBS Letters 48, 271-274.

80. Ballesta, J. P. G., and Vazques, D. (1972) in RNA Viruses and Ribosomes (Proc. 8th Meeting Fed. Eur. Biochem. Soc.) pp. 281-289, North Holland, Amsterdam.

81. Ballesta, J. P. G., and Vazques, D. (1974) FEBS Letters 48, 266-270.

82. Olenick, N. L., Carcoran, J. W. (1970) Proc. Int. Congr. Chemother., VIth (Tokyo, 1969), vol. 1, pp. 202-208, University Park Press, Baltimore.

83. Pestka, S. (1974) Antimicrob. Agr. Chemother. 5, 255-267.

84. Langlois, R., Lee, C. C., Cantor, C. R., Vince, R., and Pestka, S. (1976) J. Mol. Biol. 106, 297-313.

85. Modolell, J., Cabrer, B., Parmeggiani, A., and Vazquez, D. (1971) Proc. Natl. Acad. Sci. USA 68, 1796-1800.

86. Engelman, D. M., Moore, P. B., and Schoenborn, B. P. (1975) Proc. Natl. Acad. Sci. USA 72, 3888-3892.

87. Langer, J. A., Engelman, D. M., and Moore, P. D. (1978) J. Mol. Biol. 119, 463-485.

88. Lake, J. A. (1977) Proc. Natl. Acad. Sci. USA 74, 1903-1907.

89. Leder, P. (1973) Adv. Protein Chem. 27, 213-242.

90. Kaziro, Y. (1978) Biochim. Biophys. Acta 505, 95-135.

91. Lucas-Lenard, J., Tao, P., and Haenni, A. (1970) Cold Spring Harbor Symp. Quant. Biol. 34, 455-462.

92. Hirsh, D. (1970) Nature 228, 57-60.

93. Gupta, S. L., Waterson, J. Sopori, M. L., Weissman, S. M., and Lengyel, P. (1971) Biochem. 10, 4410-4421.

94. Thach, S. S., and Thach, R. E. (1971) Proc. Natl. Acad. Sci. USA 68, 1791-1795.

95. Kim, S. H., Suddath, F. L., Quigley, G. J. McPherson, A., Sussman, J. L., Wang, A. H. J., Seeman, N. C., and Rich, A. (1974) Science 185, 435-440.

96. Robertus, J. D., Ladner, J. E., Finch, J. T., Rhodes, D., Brown, R. S., Clark, B. F. C., and King, A. (1974) Nature 250, 546-551.

97. Eisinger, J., and Spahr, P.-F. (1973) J. Mol. Biol. 73, 131-137.

98. Uhlenbeck, O. C. (1972) J. Mol. Biol. 65, 25-41.

99. Yoon, K., Turner, D. H., and Tinoco, I. (1975) J. Mol. Biol. 99, 507-518.

100. Crick, F. H. C. (1968) J. Mol. Biol. 38, 367-379.

101. Woese, C. (1970) Nature 226, 817-820.

102. Nierhaus, K. (1978) Nature 274, 743-744.

103. Dahlberg, A. (1974) J. Biol. Chem. 249, 7673-7678.

104. Laughrea, M., and Moore, P. (1977) J. Mol. Biol. 112, 399-421.

A COMPUTER-AIDED COMPARATIVE ANALYSIS OF TWO INDEPENDENT IMMUNO ELECTRON MICROSCOPIC MODELS OF THE 30S RIBOSOME

PHILIP T. GAFFNEY and GARY R. CRAVEN

Laboratory of Molecular Biology
University of Wisconsin
1525 Linden Drive
Madison, WI 53706

INTRODUCTION

Electron microscopy has been applied to the Escherichia coli ribosome to reveal shapes of the large and small subunits in various configurations (1,2,3,4,5). In addition, the arrangement of ribosomal proteins on the surface of these particles has been mapped using the technique called immuno electron microscopy (1,2,5). This technique involves the binding of antibodies, specific for a single ribosomal protein, to the appropriate subunit. Dimers of the subunits, produced by reaction with the bivalent antibodies, are isolated and examined by electron microscopy after negative staining. The points of attachment of the Fab arms indicate the surface locations of the ribosomal protein being studied.

Mapping of the 50S ribosomal proteins is incomplete, but two independent laboratories have located the surface antigenic sites of all 21 proteins in the 30S subunit. Stöffler's laboratory has found 45 antigenic sites (2,6); whereas, Lake and Kahan, in a preliminary mapping, have identified 27 sites for the same proteins (7,8).

The work presented in this review analyzes the similarities and differences between the results of these two laboratories. We have used computerized multidimensional scaling as a technique to aid in these comparisons (9,10,11). It enables the construction of "naked" protein models, two-dimensional projections and graphical representations of data inconsistencies. We find that the models derived from the two independent sets of data are in considerable disagreement with each other.

We also compare the immuno electron microscopic results from both laboratories with those obtained by other techniques which yield information about the arrangement of proteins in the ribosome. These techniques include protein crosslinking (12,13,14), chemical and enzymatic protection (15,16), fluorescent energy transfer (17), and neutron diffraction (18,19). All these methods give information on distances between ribosomal proteins. The extent to which these distances correspond with those derived from each of the immuno electron microscopic models may be an indication of that model's accuracy. We find that both models are in agreement with most of the data from these studies.

CONSTRUCTION OF MODELS

Most techniques used to gather information about the arrangement of ribosomal proteins have produced data giving interprotein distance relationships. Crosslinking and protection studies identify close distances, while fluorescence energy transfer and neutron diffraction can register a range of values from close to distant. The approach we have employed to compare these data with that from immuno electron microscopy is to use the immuno-EM data to generate three-dimensional scale models of the protein species and then measure the distances between the 21 proteins. The 210 closest distances between pairs can be directly measured and compared with the close distances found from crosslinking, chemical protection, fluorescent energy transfer and neutron scattering.

The ribosomal scale models are constructed in two steps. First, three-dimensional plaster models are built from the immuno-EM micrographs (one was constructed from photographs of the model of Tischendorf (2), the other was a gift from James Lake) and the protein antigenic sites marked on them. The distances between all antigenic sites are then measured and used as input in a multidimensional scaling program. The program's output is the coordinates of all antigenic sites in three dimensions. These coordinates are then used to construct "naked" scale models of the proteins as previously described (9).

DESCRIPTION OF THE MODELS

Figure 1 is a view of the "naked" protein model derived from the multidimensional scaling coordinates of the 45 antigenic sites from Stöffler's laboratory, and Figure 2 is a similar view of the model developed from the 27 antigenic sites found by Lake and Kahan. (Note: Antigenic sites for protein S4 have been deleted (20).) Two of the most notable differences

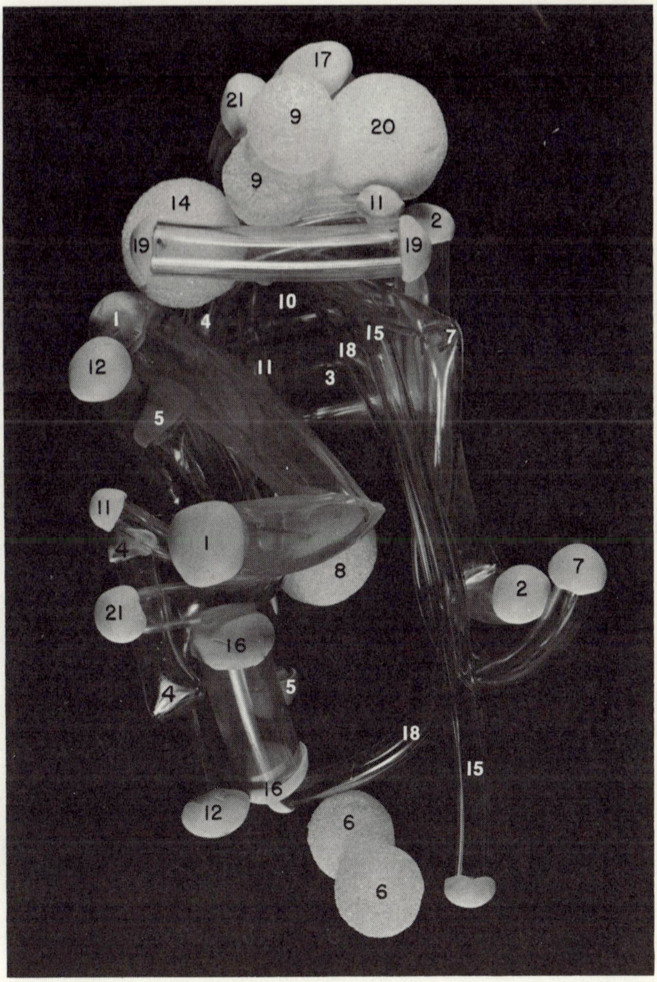

Figure 1. A view of the "naked" protein model of the 30S ribosome derived from the immuno-EM data of Stöffler's laboratory.

between the models are their overall shape and the number of elongated proteins. Lake and Kahan's model shows much more asymmetry, while that of Stöffler's group displays dramatically the extended nature of many of the component proteins and the complexity of some of their pathways. The latter model is also thicker than that of Lake and Kahan. The "neck"

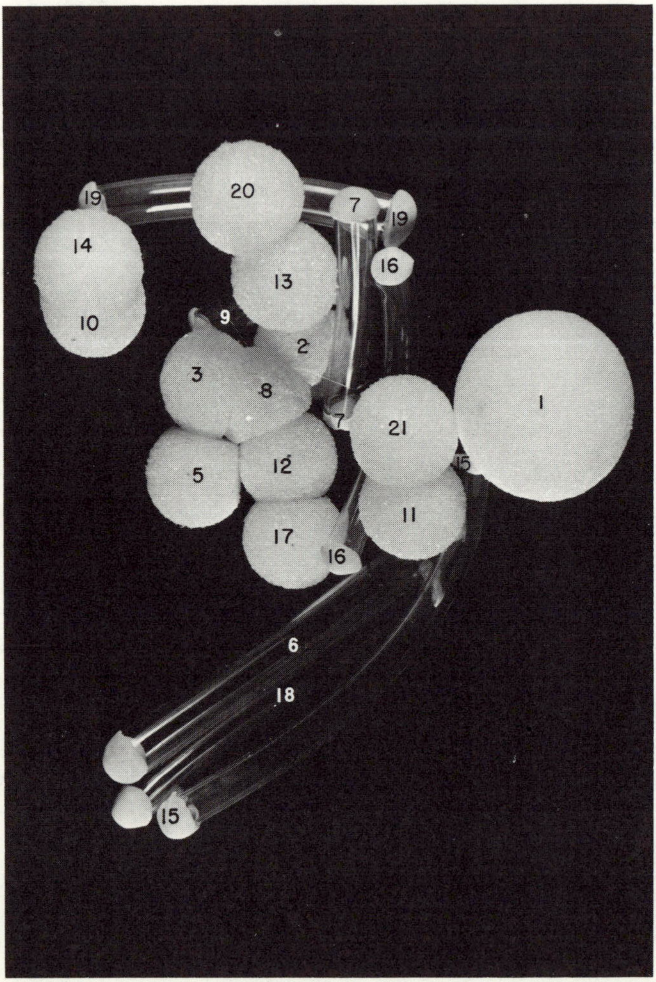

Figure 2. A view of the "naked" protein model of the 30S ribosome derived from the immuno-EM data of Lake and Kahan.

region with its cleft is preserved in both "naked" protein models and a concentration of protein in the top half and, to a lesser extent, to one side of each model can be seen. Lake and Kahan's model has seven proteins that can be considered elongated, while there are 12 of this type in the model from Stöffler's laboratory. Some proteins in the latter model have two antigenic sites so close together that they could be considered to be part of the same globular protein: S6, S9, S16 and S17.

There is some difference between the two groups in the estimation of the length of the 30S subunit. One group estimated it as 180-200 Å (2), and the other as 240-260 Å (21). Variation in the calculations is probably caused by the different techniques used for sample preparation (4,22). Our analyses here use a length of 220 Å for both models to aid in the comparison. This number is the average of the above estimates and the size found by X-ray scattering in dilute solutions (23). The dimensions of the models in Figures 1 and 2 with their lengths scaled to 220 Å are 220 Å x 115 Å x 90 Å for the Stöffler model, and 220 Å x 155 Å x 90 Å for the Lake and Kahan model. The model of Lake and Kahan shown in Figure 2 is wider than originally described (21), but this is in large part due to the unlikely assumption that S1 is globular with a diameter of 65 Å. This protein, if globular, cannot fit within the confines of the 25 Å thick "platform" of the model. A second antigenic site for S1 was recently added to the other model (6), making it elongated with a length of about 140 Å and a diameter of 36 Å. This is consistent with the observation that, in isolation, S1 assumes an extensively elongated form having a length between 220 Å and 280 Å (24,25,26,27).

Both models have regions where there seem to be too many proteins to fit into the existing volume. The phenomenon is particularly noticeable when building the "naked" protein models to scale. The protein crowding can be seen in the head region of the Stöffler model and in the central part of the Lake and Kahan model. Certainly, the larger dimensions measured by Lake, 250 Å x 175 Å x 100 Å, could more readily accommodate this concentration of protein.

The lengths, diameters and axial ratios of the 30S ribosomal proteins have been calculated for both models (Table 1). The lengths were obtained for both models using the computer printouts, making allowances for the radii of the proteins and the particular contours that a protein was assumed to follow between antigenic sites. Proteins with particularly tortuous pathways were measured directly from the models. Some of the proteins in both models are surprisingly long. Proteins S15, S18 and S21 in the model from Stöffler's group are longer than they could be if they were a fully extended alpha helix. This indicates either error in the identification of their antigenic sites or the existence of unusually extended polypeptide structures within the environment of the small ribosomal subunit.

TABLE 1

Shapes and sizes of the 30S ribosomal protein in the two immunoelectron microscopy models

Protein	Shape	Length Å	Maximum Length as α Helix Å	Diameter Å	Axial Ratio	MW x 10^{-4}
S1	Elongated	140	890	36	3.9	6.60
S1*	Globular	-		65	-	
S2	Elongated	100	380	28	3.6	2.80
S2*	Globular	-		49	-	
S3	Elongated	75	385	32	2.3	2.85
S3*	Globular	-		49	-	
S4	Elongated	225	305	17	13.2	2.26
S5	Elongated	90	235	23	3.9	1.75
S5*	Globular	-		42	-	
S6	Globular	-		40	-	1.50
S6*	Elongated	130	205	18	7.2	
S7	Elongated	205	265	16	12.8	1.97
S7*	Elongated	75		25	3.0	
S8	Globular	-		37	-	1.21
S8*	Globular	-		37	-	
S9	Globular	-		39	-	1.46
S9*	Elongated	70	195	24	2.9	
S10	Elongated	85	190	21	4.0	1.42
S10*	Globular	-		39	-	
S11	Elongated	260	220	13	20.0	1.61
S11*	Globular	-		40	-	
S12	Elongated	125	185	17	7.4	1.36
S12*	Globular	-		38	-	
S13	Globular	-		38	-	1.30
S13*	Globular	-		38	-	
S14	Globular	-		39	-	1.45
S14*	Globular	-		39	-	
S15	Elongated	210	135	11	19.1	1.00
S15*	Elongated	130	135	14	9.6	
S16	Elongated	50	125	22	2.3	0.92
S16*	Elongated	90	125	17	5.3	
S17	Globular	45		24	1.9	0.96
S17*	Globular	-		34	-	
S18	Elongated	240	111	10	24.0	0.90
S18*	Elongated	120	111	14	8.6	
S19	Elongated	75	140	19	3.9	1.03
S19*	Elongated	105	140	16	6.6	
S20	Globular	-		34	-	0.96
S20*	Globular	-		34	-	
S21	Elongated	185	105	11	16.8	0.78
S21*	Globular	-		32	-	

*Data from Lake and Kahan's model

COMPARISON OF THE TWO "NAKED" PROTEIN MODELS WITH ESTABLISHED NEAR-NEIGHBOR RELATIONSHIPS

Most studies on the arrangement of ribosomal proteins give information on near-neighbor relationships among the proteins. Chemical crosslinking, fluorescence energy transfer, chemical and enzymatic protection, and affinity labelling give results which indicate short distances between ribosomal proteins. In order to compare these results with those from immuno electron microscopy, the shortest distances between all protein pairs in both models were calculated. This was done in some instances by reading from the computer printout the distance in relative units between the antigenic sites of two proteins and converting it to Å, allowing for the fact that the antigenic sites are on the surface of the proteins rather than in the center. The radii of the proteins were then subtracted to find the distance between their closest points (see ref. 9). For other protein pairs, the path taken had to be considered. For example, in the Stöffler model, both antigenic sites of protein S15 are far from the S8 antigenic site. However, S8 is close to S15 in the neck region of the model. In such instances, reference was made to the "naked" protein models and the interprotein distances were measured directly.

Table 2 compares the results obtained using immuno electron microscopy with 65 near-neighbor pairs determined by chemical crosslinking (for review, see ref. 28), fluorescence energy transfer (17), neutron scattering (19), chemical and enzymatic protection (15,16), and affinity labelling (29,20,31,32). We have summarized these results in Table 3 and have broken them down into somewhat arbitrary categories: "strong agreement", "fair agreement", "neither agreement nor disagreement", and "disagreement". The data from the "near-neighbor" experiments appear to agree well with that from immuno electron microscopy as approximately 80% are in agreement or fair agreement with immuno-EM results from both models and only approximately 5% are in definite disagreement. Around 15% of the "near-neighbor" data are neither supported nor contradicted by the immuno-EM information. This classification is necessary owing to the degree of experimental uncertainty involved in the immuno-EM technique, the assumptions involved in the construction of our "naked" protein models, as well as the uncertainty behind the various different experimental approaches used in determining the 65 "close" protein pairs.

In our comparison of the "near-neighbor" distance information with that from immuno-EM studies in Table 2, only four definite inconsistencies are

Table 2

Summary of the Closest Distances between Ribosomal Proteins in Immuno-EM Models Determined to be Near-Neighbors by Other Techniques

Near Neighbor Pairs	Interprotein Distance Stöffler	Interprotein Distance Lake & Kahan
S1-S4	0	-
S1-S18	5	15
S1-S21	0	0
S2-S3	0	0
S2-S4	0	-
S2-S5	25	15
S2-S8	0	0
S3-S4	0	-
S3-S5	0	0
S3-S9	30	0
S3-S10	0	10
S3-S12	25	0
S3-S13	35	0
S3-S14	25	10
S4-S5	0	-
S4-S6	55	-
S4-S7	0	-
S4-S8	20	-
S4-S9	20	-
S4-S10	0	-
S4-S12	0	-
S4-S13	0	-
S4-S15	0	-
S4-S16	0	-
S4-S17	0	-
S4-S18	5	-
S4-S19	15	-
S4-S20	0	-
S4-S21	30	-
S5-S8	0	0
S5-S9	50	10
S5-S10	15	30
S5-S13	45	20
S6-S8	60	40
S6-S18	10	0
S7-S8	15	0
S7-S9	10	0
S7-S11	0	30
S7-S12	40	0
S7-S13	0	5
S7-S14	5	65
S7-S15	0	45
S7-S19	30	5
S8-S11	0	30
S8-S13	60	10
S8-S15	0	35
S9-S10	25	40
S9-S13	5	0
S10-S13	20	15
S11-S12	0	15
S11-S13	0	50
S12-S13	25	20
S12-S20	50	40
S12-S21	15	25
S13-S17	0	40
S13-S19	0	20
S13-S20	0	0
S14-S19	0	0
S15-S16	50	5
S15-S17	0	25
S15-S19	40	70
S16-S20	105	30
S17-S20	0	70
S18-S20	30	105
S19-S21	0	40

Table 3

Summary of Results Given in Table 2 for all Protein Pairs

Separation	Classification	Number of Pairs		Percent	
≤ 20 Å	Strong agreement	45	28*	69	60*
25-35 Å	Fair agreement	10	8*	15	17*
40-65 Å	Neither agreement nor disagreement	9	8*	14	17*
≥ 70 Å	Disagreement	1	3*	2	6*

*Data from Lake's model

revealed. These are S20-S16 in the Stöffler model and the pairs S20-S17, S20-S18 and S15-S19 in the Lake and Kahan model. We cannot evaluate whether these discrepancies are due to the misplacement of antigenic sites, errors in the original determination of near-neighbor relationships, or the undetected elongated shape of the component proteins.

COMPARISON OF THE STÖFFLER MODEL WITH THE LAKE AND KAHAN MODEL

We have compared the two protein models shown in Figures 1 and 2 in several different ways. In one instance, we computed or measured all the shortest distances between the 210 pairs of the Stöffler model and compared these data with the shortest distances for the 190 pairs between the 20 proteins in the Lake and Kahan model. These results are tabulated in the matrix referred to as Table 4. The similarity between the data was analyzed by calculating a standard correlation coefficient. The correlation is only 0.17, indicating a very slight degree of similarity between the shortest distance data of the two alternative models. This method of comparison is, however, most certainly too rigorous in view of the large number of assumptions behind the construction of these models. Perhaps a more reasonable approach is to determine the percentage of the data that correlates within a distance of 40 Å.

Thus, we find a total of 68 of the 190 protein pairs, or 36%, differ between the two models by 40 Å or more. This is a seriously high degree of discrepancy, especially in view of the many elongated proteins in the two models which should greatly increase the probability of apparent agreement between nearest distances. This disappointing lack of correlation can be better appreciated by the construction of a plot of the 190 shortest distances. This is shown in Figure 3 in which the shortest distances of the Lake and Kahan model are simply plotted against the nearest distances in the Stöffler model. Were there complete agreement

TABLE 4

Matrix of the shortest distances (Å) between the 210 protein pairs in the Stöffler model and the 190 protein pairs in the model of Lake and Kahan

	S1	S1*	S2	S2*	S3	S3*	S4	S4*	S5	S5*	S6	S6*	S7	S7*	S8	S8*	S9	S9*	S10	S10*
S2	35	25																		
S3	10	45	0	0																
S4	0	-	0	-	0	-														
S5	0	55	25	15	0	0	0	-												
S6	45	0	40	40	120	50	55	-	50	40										
S7	0	30	0	0	0	0	0	-	35	15	55	55								
S8	10	35	0	0	25	0	20	-	0	0	60	40	15	0						
S9	0	30	50	0	30	0	20	-	50	10	145	70	10	0	55	0				
S10	0	75	0	25	1	10	0	-	15	30	135	100	0	55	35	35	25	40		
S11	0	0	30	15	0	35	0	-	0	30	60	0	0	30	0	30	0	20	0	70
S12	0	25	30	0	25	0	0	-	0	0	20	25	40	0	15	0	15	5	20	40
S13	0	45	0	0	35	0	0	-	45	30	160	80	0	5	60	10	5	0	20	15
S14	0	80	40	50	25	10	0	-	20	35	150	100	5	65	60	40	5	35	20	0
S15	20	0	0	15	0	40	0	-	30	40	15	0	0	45	0	35	20	50	0	105
S16	0	5	45	10	65	10	0	-	0	15	45	5	60	0	30	0	75	0	70	65
S17	20	25	0	15	30	0	0	-	80	0	165	25	0	20	55	0	0	20	10	55
S18	5	15	0	55	0	55	5	-	30	45	10	0	0	50	5	45	15	70	0	90
S19	25	40	25	10	45	15	15	-	45	40	160	80	30	5	65	35	0	0	40	0
S20	30	70	0	10	40	10	0	-	80	50	155	105	0	5	65	50	0	15	35	10
S21	0	0	25	15	20	25	30	-	20	45	55	10	25	25	5	25	0	40	30	85

	S11	S11*	S12	S12*	S13	S13*	S14	S14*	S15	S15*	S16	S16*	S17	S17*	S18	S18*	S19	S19*	S20	S20*
S12	0	15																		
S13	0	50	25	20																
S14	0	85	0	40	0	35														
S15	20	0	45	30	5	75	25	105												
S16	0	0	0	0	75	5	75	55	50	5										
S17	0	20	45	0	0	40	15	55	0	25	100	0								
S18	20	5	50	25	10	95	20	115	0	0	20	30	20	25						
S19	0	55	25	50	0	20	0	0	40	70	85	10	0	70	50	90				
S20	0	75	50	40	0	0	25	10	15	100	105	30	0	70	30	105	0	15		
S21	0	0	15	25	0	45	0	80	30	0	10	5	15	35	35	25	0	40	15	70

*Data from the Lake and Kahan model

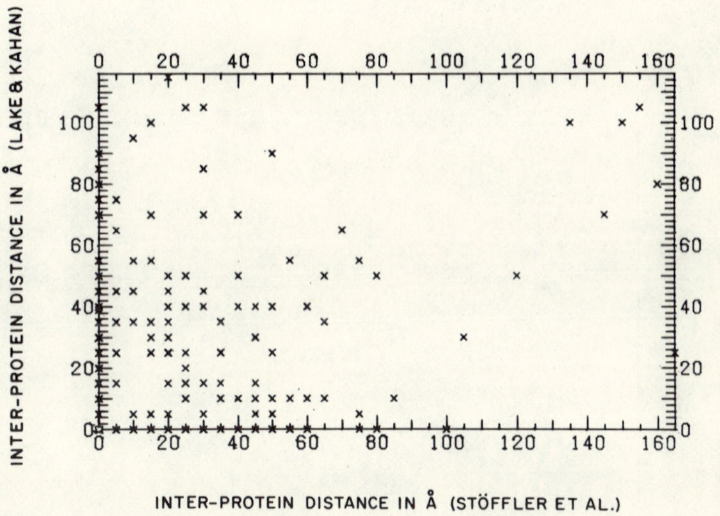

Figure 3. Plot of closest distances between proteins in the two models.

between the two models, all points would lie on a diagonal line. In contrast, as dramatically illustrated by the scatter of points in this diagram, there appears to be a general lack of correlation between the two models.

There are a very large number of pairs which are near one another in one of the models and relatively far from one another in the other model. Table 5 summarizes the 68 protein pairs which have discrepancies of 40 Å or more. Unfortunately, these discrepancies do not appear to break down into any readily detectable classes which might reveal some systematic reason for the differences. All 20 proteins are involved in the observed discrepancies and the number of discrepant pairs they are each involved in ranges from 2 to 11. Eight proteins are members of eight discrepant pairs or more. These are S6, S10, S13, S14, S16, S17, S18 and S20.

Some of the discrepancies seen in Table 5 are so great they deserve special comment. The greatest deviation, 140 Å, occurs with the S6-S17 pair. This is probably due to a combination of two prominent differences in the identification of antigenic sites by the two laboratories. Stöffler's group has located two antigenic sites for protein S17. These

Table 5

Summary of Discrepancies between Shortest Interprotein Distances of 40 Å or Greater

Protein Pair	Stöffler Distance(Å)	Lake and Kahan Distance(Å)	Discrepancy (Å)
S1-S5	0	55	55
S1-S6	45	0	45
S1-S10	0	75	75
S1-S13	0	45	45
S1-S14	0	80	80
S1-S20	30	70	40
S2-S9	50	0	50
S2-S18	0	55	55
S3-S6	120	50	70
S3-S14	0	40	40
S3-S16	65	10	55
S3-S18	0	55	55
S5-S9	50	10	40
S5-S17	80	0	80
S6-S11	60	0	60
S6-S9	145	70	75
S6-S13	160	80	80
S6-S14	150	100	50
S6-S16	45	5	40
S6-S17	165	25	140
S6-S19	160	80	80
S6-S21	155	105	50
S6-S21	55	10	45
S7-S10	0	55	55
S7-S12	40	0	40
S7-S14	5	65	60
S7-S15	0	45	45
S7-S16	60	0	60
S7-S18	0	50	50
S8-S9	55	0	55
S8-S13	60	10	50
S8-S17	55	0	55
S9-S16	75	0	75
S9-S18	15	70	55
S9-S21	0	40	40
S10-S11	0	70	70
S10-S15	0	105	105
S10-S17	10	55	45
S10-S18	0	90	90
S10-S19	40	0	40
S10-S21	30	85	55
S11-S13	0	50	50
S11-S14	0	85	85
S11-S19	0	55	55
S11-S20	0	75	75
S12-S14	0	40	40
S12-S17	45	0	45
S13-S15	5	75	70
S13-S16	75	5	70
S13-S17	0	40	40
S13-S18	10	95	85
S13-S21	0	45	45
S14-S15	25	105	80
S14-S17	15	55	40
S14-S18	20	115	95
S14-S21	0	80	80
S15-S16	50	5	45
S15-S20	105	30	75
S16-S17	100	0	100
S16-S19	85	10	75
S16-S20	105	30	75
S17-S19	0	70	70
S17-S20	0	70	70
S18-S19	50	90	40
S18-S20	30	105	75
S19-S21	0	40	40
S20-S21	15	70	55

two sites are fairly close together, residing on the upper "head" region. On the other hand, Lake and Kahan have found a single site located near the base of the "neck" region. In addition, Stöffler's studies have found two very close sites for protein S6 situated at the very lower base or bottom of the ribosome. In contrast, Lake and Kahan have two S6 antigenic sites, one of which is also at the lower base position, but the other site is identified in the upper "platform" region. Thus, in the Stöffler model, S6 and S17 reside at the extreme poles of the ribosome, approximately 165 Å apart; whereas in the Lake and Kahan model, they come within 25 Å somewhere around the "neck" region.

Protein S18 is involved in five pairs which have discrepancies of 85 Å or more. The S10-S18 pairs differ between the two models by 90 Å, the S13-S18 pair by 85 Å, the S14-S18 pair by 95 Å, the S19-S18 pair by 90 Å, and the S20-S18 pair by 105 Å. These are all very large discrepancies and can be accounted for by a discrepancy in the location of antigenic sites for protein S18. Both laboratories have found antigenic sites for proteins S10, S13, S14, S19 and S20 in the upper "head" section. In addition, both groups have located an S10 site at the bottom of the ribosome. The divergence develops from identification of a second site by Lake and Kahan in the "platform" which is considerably different from a second site found by Stöffler at the upper part of the body below the "neck", and even more in contrast to a third site identified in the upper "head" section. It is this third site which brings all five of these proteins close to S18 in the Stöffler model.

A situation completely analogous to the S18 problem exists for protein S15. There are also large differences between the calculated interprotein distances of S15 and the five proteins discussed above for S18. Once again, this is due to a major discrepancy involving the identification of a second antigenic site for S15. Both laboratories have one S15 antigenic site located at the bottom of the particle, but a second site has been positioned on the "platform" by Lake and Kahan. The placement of this site disagrees substantially with a second site found by Stöffler on the top of the "head" of the model. This discrepancy in the identification of a second S15 antigenic site is reflected in the shortest distance calculations for the pairs in Table 5.

The output from the multidimensional scaling programs used in this analysis can also be used as an aid in the comparison of the two models. The programs we have employed have the capacity to compute the coordinates of the antigenic sites in two dimensions as well as the three dimensions we used for the construction of the models in Figures 1 and 2. The analysis shows that both models have a high degree of "two dimensionality", which is reflected in very low Stress values (see ref. 10). We have constructed an interpretive drawing of the two-dimensional coordinates for both models as seen in Figure 4. These drawings help to visualize the backbones of the two protein models which greatly simplifies the task of comparison. The two models have been oriented so that the "platform" proteins in the Lake and Kahan model are on the right and the proteins found on the "high shoulder" region of Stöffler's model are also on the right.

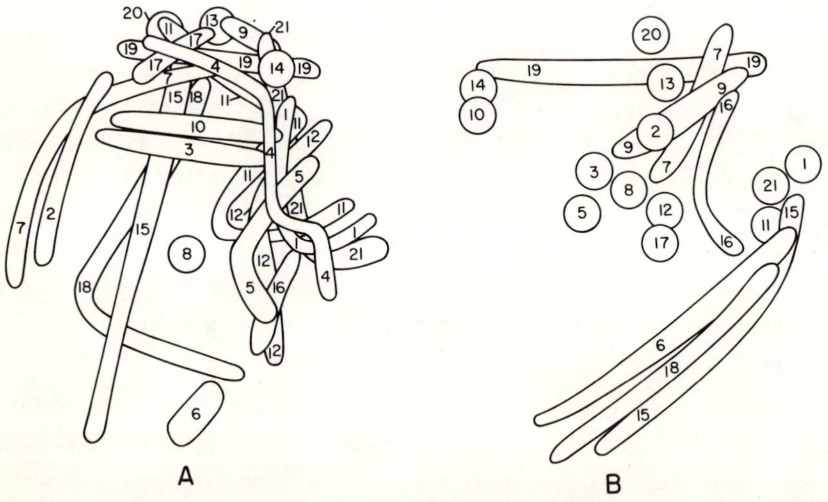

Figure 4. Two-dimensional representations of the immuno electron microscopy data. A, the data obtained from Stöffler's group, and B, the data derived from Lake and Kahan.

The problems we have discussed above regarding proteins S6, S17, S15 and S18 can be readily appreciated by reference to Figure 4. Furthermore, several less obvious discrepancies become apparent. With these orientations, several important problems emerge. For example, protein S14 is on the far left in the Lake and Kahan model and on the far right in the

Stöffler model. A similar but reverse indictment can be made about protein S20, which is on the right in the Lake and Kahan model and on the left in the Stöffler model. The same can be said for protein S2 and several of the antigenic sites for S7. Thus, there appears to be a genuine discrepancy between the two laboratories in the determination of some "left-right" assignments.

Another area of difference between the results of the two groups is in the number of antigenic sites. Lake and Kahan have found a total of 27 antigenic sites, whereas the experiments from Stöffler's laboratory implicate 45 antigenic sites. These discrepancies, as well as all others discussed above, are listed in Table 6. For this summary, it is apparent

Table 6

Summary of Observed Discrepancies between the Two Immuno Electron Microscopic Models of the 30S Ribosome

Protein	Discrepancy Description
S1	Stöffler has one additional site
S2	Stöffler has one additional site; "left-right" disagreement
S3	Stöffler has one additional site
S4	Stöffler has four sites, Lake and Kahan none
S5	Stöffler has two additional sites
S6	Lake and Kahan have site in "platform" not matched by Stöffler
S7	"Left-right" disagreement for two Stöffler Sites
S8	NO DISCREPANCY
S9	NO DISCREPANCY
S10	Stöffler has one additional site
S11	Stöffler has three additional sites
S12	Stöffler has one additional site
S13	NO DISCREPANCY
S14	"Left-right" disagreement
S15	Stöffler second site in "head", Lake and Kahan second site in "platform"
S16	Lake and Kahan second site in "head", not matched by Stöffler
S17	Stöffler has two sites in "head", Lake and Kahan have one site in "neck"
S18	Stöffler second site in "head", Lake and Kahan have one site in "platform"
S19	NO DISCREPANCY
S20	"Left-right" disagreement
S21	Stöffler has one additional site

that substantial discrepancies exist for all proteins except four (S8, S9, S13 and S19). There seems to be serious problems about the placement of antigenic sites on the surface of the 30S ribosome for 17 of the 21 proteins.

CONCLUDING STATEMENT

Immuno electron microscopy is a powerful and direct technique for revealing the protein structure of the ribosome. However, this approach to ribosome architecture depends greatly on interpretation to obtain data. The two major laboratories using this technique have based their interpretations on two considerably different models of the shape of the 30S ribosome. Up to now, this had generally been thought to be the major problem between the two laboratories. Our analysis based on "naked" protein models derived from the immuno electron microscopic data of the two groups indicates that the degree of discrepancy between the two is far more serious than previously recognized. We have summarized major discrepancies in the antigenic site assignments of 10 of the 21 proteins. These are S2, S4, S6, S7, S14, S15, S16, S17, S18 and S20. In addition, we point out that seven of the remaining proteins show discrepancies in the number of identified antigenic sites. This leaves only four proteins that have no observable discrepancies in the positioning of antigenic sites. It is hoped that with future refinements of the immuno electron microscopy methodology, these discrepancies will be resolved and a single definitive model of the 30S ribosome will emerge.

ACKNOWLEDGEMENTS

We thank Dr. Michael Sukoviak for advice on the application of multidimensional scaling programs. This work was supported by the Graduate School and the College of Agriculture and Life Sciences of the University of Wisconsin-Madison, and by research grants GM15422 and GM24109 from the National Institutes of Health.

REFERENCES

1. Lake, J.A., Pendergast, M., Kahan, L., and Nomura, M. (1974) Proc. Natl. Acad. Sci. U.S.A. 71, 4688-4692.
2. Tischendorf, G.W., Zeichhardt, H., and Stöffler, G. (1975) Proc. Natl. Acad. Sci. U.S.A. 72, 4820-4824.
3. Leonard, K.R., and Lake, J.A. (1979) J. Mol. Biol. 129, 155-163.
4. Vasiliev, V.D. (1974) Acta Biol. Med. Germ. 33, 779-793.
5. Boublik, M., Hellmann, W., and Roth, H.E. (1978) J. Mol. Biol. 107, 479-490.
6. Stöffler, G. (1978) Personal communication.

7. Lake, J.A. and Kahan, L. (1974) J. Mol. Biol. 99, 631-644.

8. Lake, J.A. (1977) FEBS Federation of European Biochemical Societies 11th Meeting, Copenhagen, 43, Symposium A2, 121-130.

9. Gaffney, P.T., and Craven, G.R. (1978) Proc. Natl. Acad. Sci. U.S.A. 75, 3128-3132.

10. Gaffney, P.T., and Craven, G.R. (1979) Methods in Enzymology, Vol. LIX, pp. 602-611, Academic Press, New York.

11. Kurskal, J.B., Young, F.W., and Seery, J.B. (1978) How to Use KYST, a Very Felxible Program to do Multidimensional Scaling and Unfolding, Bell Laboratories, Murray Hill, New Jersey.

12. Sommer, A., and Traut, R.R. (1976) J. Mol. Biol. 106, 995-1015.

13. Lutter, L.C., Kurland, C.G., and Stöffler, G. (1975) FEBS Lett. 54, 144-150.

14. Expert-Bezançon, A., Barritault, D., Milet, M., Guérin, M.F., and Hayes, D.H. (1977) J. Mol. Biol. 112, 603-629.

15. Changchien, L.M., and Craven, G.R. (1977) J. Mol. Biol. 113, 103-122.

16. Changchien, L.M., and Craven, G.R. (1979) Biochemistry 18, 1275-1281.

17. Huang, K., Fairclough, R.H., and Cantor, G.R. (1975) J. Mol. Biol. 97, 443-470.

18. Engelman, D.M., Moore, P.B., and Schoenborn, B.P. (1975) Proc. Natl. Acad. Sci. U.S.A. 72, 3888-3892.

19. Moore, P.B., Langer, J.A., Schoenborn, B.P., and Engelman, D.M. (1977) J. Mol. Biol. 112, 199-234.

20. Winkelmann, D., and Kahan, L. (1979) J. Supramol. Res., in press.

21. Lake, J.A. (1976) J. Mol. Biol. 105, 131-159.

22. Lake, J.A. (1978) Advanced Techniques in Biological Electron Microscopy II, ed. J.K. Koehler, pp. 173-211, Springer-Verlag, Berlin Heidelberg.

23. Hill, W.E., Thompson, J.D., and Anderegg, J.W. (1969) J. Mol. Biol. 44, 89-102.

24. Giri, L., and Subramanian, A.R. (1977) FEBS Lett. 81, 199-203.

25. Österberg, R. Sjöberg, B., and Littlechild, J. (1978) FEBS Lett. 93, 115-119.

26. Labischinski, H., and Subramanian, A.R. (1979) Eur. J. Biochem. 95, 395-366.

27. Laughrea, M., and Moore, P.B. (1977) J. Mol. Biol. 112, 399-421.

28. Brimacombe, R., Stöffler, G., and Wittmann, H.G. (1978) Ann. Rev. Biochem. 47, 217-249.

29. Pongs, O., Stöffler, G., and Lanka, E. (1975) J. Mol. Biol. 99, 301-315.

30. Fiser, I., Sheit, K.H., Stöffler, G., and Kuechler, E. (1975) FEBS Lett. 56, 226-229.

31. Schreiner, G., and Nierhaus, K.H. (1973) J. Mol. Biol. 81, 71-82.

32. Pongs, O., and Rossner, E. (1976) Nucleic Acids Res. 3, 1625-1634.

THE REACTION OF ANTIBODIES WITH RIBOSOMAL SUBUNITS AND ASSEMBLY INTERMEDIATES

DONALD WINKELMANN and LAWRENCE KAHAN

Department of Physiological Chemistry,
University of Wisconsin,
1215 Linden Drive,
Madison, WI 53706

INTRODUCTION

It has long been realized that the exquisite specificity of the antibody-antigen reaction could be exploited to great advantage in the study of ribosome structure and function. Antibodies which react specifically with each of the *Escherichia coli* ribosomal proteins have been prepared in a number of laboratories (1,2,3,4,5). These antibodies have been utilized for the detection, identification and correlation of the individual proteins e.g. the characterization of the *E. coli* ribosomal proteins (6), the identification of ribosomal proteins in crosslinking experiments (3,7,8), the detection of products of *in vitro* coupled transcription-translation of ribosomal protein genes (9), and the identification of homologous ribosomal proteins in other procaryotes and eucaryotes (2,27).

In addition, antibodies have been used to specifically label ribosomal proteins and RNA *in situ*. Antibodies have been used to test models of ribosome structure derived from electron micrographs (4,10,11) and to map components on the surface of the ribosomal subunits (4,5, 10,12,13,14,15,16). The inhibition of functional activities by antibodies and antibody fragments has been used to define components which are near active sites within the ribosome (17). Specific changes in the immunochemical accessibility of proteins during *in vitro* assembly have been used to demonstrate interactions between proteins within the small ribosomal subunit (18). These studies have also shown that antibodies can be used to define and quantitate subpopulations of assembly intermediates and subunits.

ANALYSIS OF ANTIBODY-RIBOSOME REACTIONS

Two experimental measurements are required to interpret the experimental observations of in situ studies in terms of ribosomal structure and components. It is necessary to measure the extent of the antibody-antigen reaction, both to determine the reactivity of the antibody population and to determine the fraction of the ribosomal particles which are reactive, in order to assess the functional significance of the observations. In addition, the specificity of the antibody-antigen reaction must be determined, since the specificity of the reactivity of the components in situ may differ greatly from their reactivity as isolated components. As summarized below, the methods used to assay these parameters are widely varied, and, in some cases, are either inadequate or inappropriate.

Aggregation Assays

Several approaches have been used to detect antibody-ribosome reactions. The most commonly used methods rely on antibody dependent aggregation of ribosomal particles (13,19,20). Aggregated particles can be separated from monomeric particles and unbound immunoglobulins by a variety of sedimentation techniques, the most frequently chosen of which is zone sedimentation in sucrose gradients. The predominant aggregate observed is an IgG linked subunit dimer, with smaller amounts of higher aggregates formed to a variable extent if multiple antigenic determinants are accessible. The amount of aggregates is either measured directly or estimated from the decreased recovery of monomeric particles. This type of assay has one important advantage. It is usually possible to characterize both products and reactants after the reaction has been completed; thus, subunit degradation (often caused by nuclease contamination of the antibody preparation) is revealed by changes in the sedimentation profile of the unreacted, monomeric particles.

The drawbacks of aggregation based assays are numerous and formidable. The separation of products from free antibody and unreacted particles permits dissociation of the aggregates, particularly if the separation requires several hours. Instead of forming discrete aggregate peaks, the products may not be completely separated from the monomeric particles. The problem of aggregate dissociation may be minimized by the use of vertical rotors for rapid separation; however, even the most rapid separations require 15-20 minutes (20). In addition, aggregation assays require a secondary reaction in addition to the primary antibody-antigen reaction, i.e. the reaction of a second subunit with an antibody-subunit complex. This reaction may be thermodynamically unfavorable, slow, or highly sensitive to inhibition by excess antibody. Thus aggregation assays may not give an accurate quantitative measure of the extent of the primary antibody-ribosome interaction.

Double Antibody Assays

Antibody binding to ribosomes may be directly measured by the inclusion of antibody-ribosome complexes in an IgG-anti-IgG precipitate (15,18,19). Radiolabeled ribosomal particles are reacted with rabbit IgG antibody, which in turn is precipitated with goat anti-rabbit IgG. This assay quantitatively measures antibody bound particles regardless of their state of aggregation. Furthermore, the second antibody reaction and separation can be completed in less than four minutes, minimizing the dissociation problem (18). Since ribosome dimer formation is not required, the amount of ribosomal particles used is far less than required in the aggregation assay. This reduction in subunit concentration decreases the amount of antibody necessary to obtain a measurable reaction.

The advantage of the antibody binding assay lies in the ready quantitation of antibody bound particles. If the antibody population were homogeneous with respect to affinity for ribosomal particles, the binding data would be expected to fit the equation:

$$\frac{[\mathit{Particles}]_{bound}}{[\mathit{Particles}]_{free}} = \frac{K\ F\ [\mathit{Antibody}]_{total}}{1 + K\ [\mathit{Particles}]_{free}} \qquad (1)$$

where K is the equilibrium constant and F is the fraction of the total antibody which can react with the particles tested. When the initial specific antibody concentration is much greater than the particle concentration the above equation simplifies to:

$$\frac{[\mathit{Particles}]_{bound}}{[\mathit{Particles}]_{free}} = K\ F\ [\mathit{Antibody}]_{total} \qquad (2)$$

Thus, the ratio of bound to free particles is expected to be a linear function of antibody concentration. K and F are , of course, determined by both the antibody and the particles tested; however, the same antibody population may be tested against a series of particles in order to detect changes in the immunochemical accessibility of individual components.

The homogeneity of the particles may also be assessed from the linearity of the antibody binding titrations. If only a fraction of the particle population reacts with antibody the ratio of bound to free particles will plateau at a level that is determined by the size of the reactive fraction.

An example of antibody binding data plotted according to equation (2) for the binding of anti-S4 with subribosomal particles and the 30S ribosomal subunit is shown in Figure 1. It is clear that the product,

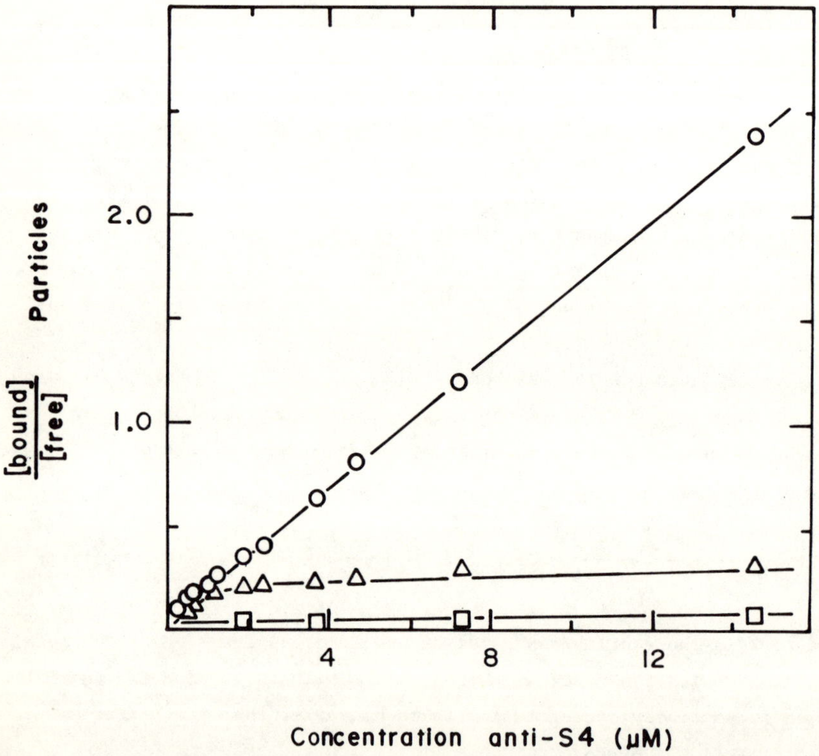

Figure 1. The reaction of anti-S4 with 30S subunits (□), RI* (△) and RI[4,7,8,15,16,17,19,20] (O) particles. Reconstituted subribosomal particles were prepared as previously described (20). 1.2 pmole of ^{3}H-30S subunits, ^{3}H-RI* or ^{3}H-RI[4,7,8,15,16,17,19,20] particles were reacted with increasing concentrations of anti-S4 IgG diluted with non-immune IgG to give a total of 3.6 x 10^{-7} mmole rabbit IgG per point. Reaction with anti-S4 was for 5 min at 40°C followed by 40 min at 0°C. An equivalent amount of goat anti-rabbit IgG IgG was added simultaneously to all tubes and incubated 2 min at 24°C to completely precipitate the rabbit IgG and anti-S4-ribosome complexes. The samples were diluted 30-fold with reaction buffer (10 mM $MgCl_2$, 100 mM KCl, 10 mM Tris, pH 7.4) mixed and centrifuged 1.5 min in a Beckman Microfuge B. The radioactivity of the supernatant and the immunoprecipitate was then measured. Data is plotted according to equation (2). The RI* particle contains 16S RNA, S4, S5, S6, S7, S8, S9, S11, S12, S13, S15, S16, S17, S18, S19 and S20.

K F, for anti-S4 differs among the ribosomal particles in this example. The interpretation of these results will be discussed below.

DETERMINATION OF ANTIBODY SPECIFICITY

It is essential when studying the reaction of antibodies with ribosomal particles to examine the specificity of that interaction. One basic premise of all studies of antibody ribosome interactions is that the antibodies react with the desired protein or hapten and no other *in situ*. However, this may not always be the case. There are two general types of interfering reactions which might be expected to violate this premise. First, contaminants of the purified antigen used for immunization may produce contaminant antibodies which react with the contaminant *in situ*. Second, antibodies may crossreact with a second ribosomal component (*e.g.* some anti-S20 antibodies might crossreact with the hexapeptide leu-thr-ala-gln-ile-asn found in both S15 (21) and S20 (22)). As summarized below, a variety of approaches have been utilized to test the specificity of antibody-ribosome reactions. In evaluating these approaches, one must keep in mind the possibility that the immunogen may not have any exposed antigenic determinants *in situ* and may be contaminated with other ribosomal proteins or their degradation products.

Absorption With Immunogen

The common experimental demonstration of specificity is the inhibition of the *in situ* reaction by pre-absorption of the antibody with purified immunogen (13,17,20). Even if the absorption is done at antibody-antigen equivalence, this control can fail both in the case of true crossreaction, since the reactive determinants are present on the immunogen, and in the case of contaminant antibody, since the minor contaminant in the immunogen may absorb a minor contaminant in the antibody preparation. Failure of this control is, of course, more likely if the immunogen is weakly reactive or unreactive *in situ*, requiring high concentrations of antibody for detection of the reaction. In this circumstance a very minor contaminant antibody may react *in situ* and be absorbed in the control experiment, yet be present in amounts which are not detectable by the usual immunodiffusion or immunoelectrophoretic tests.

Affinity Purification

A second approach, very similar to the absorption approach, is the purification of anti-protein antibodies by affinity chromatography (20,23). Unfortunately, this approach suffers the same drawbacks as the absorption approach. Neither true cross-reactions nor minor contaminants can be excluded by this control. It should be noted that either of the above approaches may detect a contaminant reaction and thus invalidate the results obtained with a particular antibody preparation; however, the successful absorption of the antibody preparation is not sufficient proof of specificity.

The affinity approach may be made more sensitive to contaminants if purified peptide fragments are utilized as the immunoabsorbent rather than the intact protein. Since contaminant proteins generally give peptides which differ in properties from the immunogen derived peptides, the traces of contaminant in the immunogen may be removed in the purification process. However, it is not always possible to prepare, purify and characterize the necessary peptides.

Heterologous Substitution

A third approach requires the reconstitution of hybrid 30S subunits containing all _E. coli_ components except the single _B. stearothermophilus_ functional homologue of the _E. coli_ protein in question (13). Such hybrid ribosomes are expected to be immunochemically unreactive in those cases where the antibodies do not strongly crossreact with the _B. stearothermophilus_ homologue. Even in the case of strongly crossreacting proteins such as the homologues of S19 and S11, the reactivity _in situ_ is so weak that the experiment may still be performed (our unpublished results). The elimination of the observed antibody-ribosome reaction when hybrid ribosomes are utilized is consistent with _in situ_ recognition of the substituted protein. The only alternative explanation is a conformational alteration in the subunit which decreases the immunochemical reactivity of another protein but does not affect the functional activity of the subunit. However, this specificity demonstration has a practical problem; the _B. stearothermophilus_ protein may be contaminated with the homologue of a contaminant of the _E. coli_ protein (_e.g._ BS4 may be contaminated with BS7 just as ES4 is often contaminated with ES7 (20,23)). The hybrid ribosomes prepared under these circumstances might show a decreased reactivity with contaminant antibodies (_e.g._ anti-S7 in an anti-S4 in the example above) giving a misleading result.

IMMUNOCHEMICAL ACCESSIBILITY OF S4 IN SITU

The importance of the above considerations in the interpretation of in situ antibody binding experiments is graphically illustrated by the various studies of S4 in situ structure and function. S4 was reported by Stöffler and co-workers to be immunochemically exposed in the 30S subunits as measured by double antibody sandwich precipitation (using a 24 hour incubation with goat anti-rabbit IgG serum rather than the purified IgG fraction) and by aggregate formation assays (19). Lake et al. subsequently reported the immunoelectron microscopic mapping of S4 on the 30S subunit surface, relying on an aggregation assay to perform immunoabsorption and hybrid reconstitution specificity tests (13). Stöffler and Tischendorf reported contradictory immunoelectron microscopic mapping results (14), but later reported more detailed studies (23,24) which mapped S4 at locations similar to those reported by Lake et al. However, our subsequent studies have shown that, in fact, S4 is immunochemically inaccessible in the intact 30S subunit (20, see also Figure 1). In order to confirm this inaccessibility, antibodies were isolated by three different methods from eight rabbit antisera in order to obtain the widest range of antibody specificities. Some of these antibody preparations did react with 30S subunits; this reaction was not decreased by the complete removal of the anti-S4 antibody population and thus represented the reaction of contaminant antibodies. Other anti-S4 antibody preparations did not react with 30S subunits but did react with 30S subunit assembly intermediates. These anti-S4 preparations may be used to monitor changes in S4 accessibility during in vitro 30S ribosomal subunit assembly.

S4 is accessible on several reconstitution intermediates including the complex of 16S RNA with S4, S7, S8, S15, S16, S17, S19 and S20 (RI[4,7,8,15,16,17,19,20])(20). The reaction of anti-S4 with reconstitution intermediates was detected both by aggregation assays and by antibody binding assays. The specificity of the reaction of anti-S4 with reconstitution intermediates was demonstrated by absorption of the antibody preparations at equivalence with pure S4 and by substitution of the B. stearothermophilus homologue for S4 in the subribosomal particles (Fig. 2).

The significance of the reaction of anti-S4 with reconstituted subparticles depends both on the relationship of the structure of the particles to that of the functional subunit and on the homogeneity of

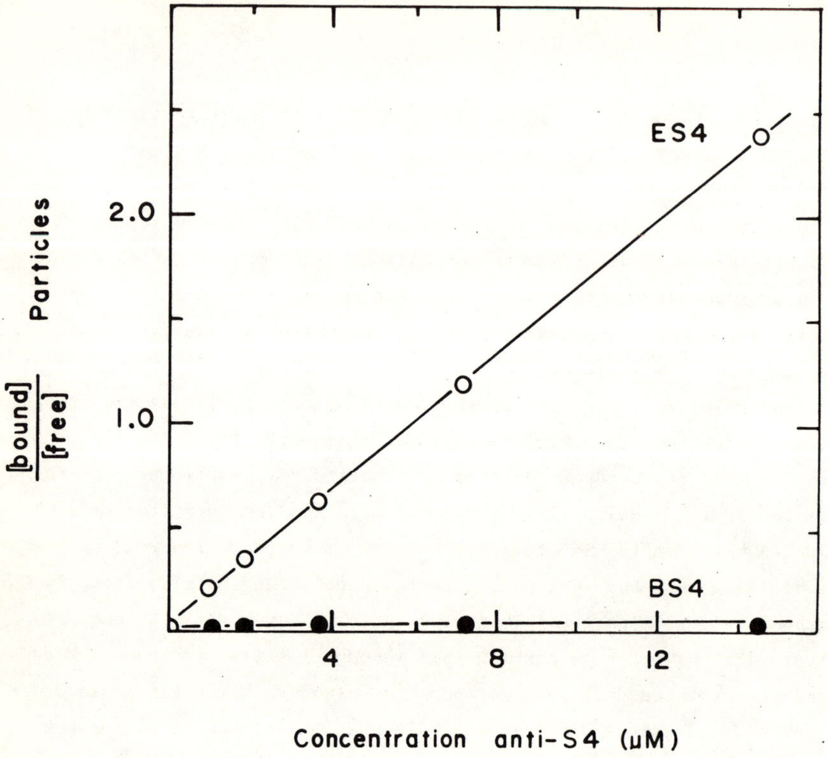

Figure 2. The reaction of anti-S4 with the RI[4,7,8,15,16,17,19,20] particle containing E. coli S4 (ES4) (O) or B. stearothermophilus S4 homologue (BS4) (●) substituted for the E. coli protein. Particles were prepared as previously described (20). Reaction conditions were identical to those described in Figure 1.

the particle preparations. The RI[4,7,8,15,16,17,19,20] and RI* particles could be completed by the addition of the missing proteins under the appropriate conditions to form functionally active 30S subunits, indicating that the majority of the particles are structurally correct assembly intermediates. However, while greater than 80% of the RI[4,7,8,15,16,17,19,20] particles were reactive in the antibody binding assay with a constant apparent association constant and thus represent a homogeneous population by this assay, only 20-25% of the RI* particles reacted with anti-S4 despite all efforts to increase this level of reaction (Fig. 1). We have concluded that the anti-S4 reactive RI* particles are a minor subpopulation. The majority (80%) of the RI* particles apparently do not react with anti-S4. These observations illustrate one of the difficulties inherent in the interpretation of antibody-ribosome reactions: ribosome structural heterogeneity gives rise to immunochemically reactive sub-populations.

EFFECTS OF S5 AND S12 ON S4 ACCESSIBILITY

We have identified the proteins responsible for the large difference in anti-S4 reactivity between the RI[4,7,8,15,16,17,19,20] and RI* particles (18, see also Figure 1). The inclusion of S5 or S12 in the RI[4,7,8,15,16,17,19,20] causes a large decrease in immunochemical reactivity of S4. Conversely, the omission of S5 and S12 from the RI* particle results in an increase in reactivity to the level of the RI[4,7,8,15,16,17,19,20]. Of the seven additional proteins in the RI* particle only S5 and S12 affect S4 accessibility in these reconstitution intermediates.

There are several possible interpretations of the effects of S5 and S12 on S4 accessibility. There may be direct protein-protein interactions between these proteins which interfere with antibody-antigen recognition. The close proximity of these proteins as suggested by the results of immunoelectron microscopy (see J. A. Lake, this volume), neutron scattering (see P. B. Moore, this volume) and the phenotypic interdependence of mutations in the respective genes (25), are consistent with this hypothesis. It is also possible that changes in the particle structure resulting from the addition of S5 and S12 might result in formation of interactions between S4 and other proteins or 16S RNA which inhibit antibody binding. Finally, the addition of these proteins or structural changes resulting from their addition might hinder the approach of the antibody to S4 without direct interaction of any component with S4 itself.

CONCLUDING REMARKS

S4 is the first example of a ribosomal protein which is inaccessible to antibody on the 30S ribosomal subunit surface. This lack of accessibility implies that the location and functional roles of S4 in the intact subunit are not amenable to immunochemical dissection. However, S4 may be mapped by immunoelectron microscopy in appropriate protein deficient subparticles, and ribosomal component interactions within the subunit can be probed with anti-S4 antibodies.

Demonstration of the specificity of in situ antibody-ribosome reactions will remain a difficult problem as long as the antibodies are produced by classical immunization procedures. One solution to these specificity problems is the production of monoclonal antibodies by the lymphocyte hybridoma technique (26). The use of monoclonal antibodies would eliminate completely the problem of contaminant antibodies, while reducing the chance of a true crossreaction to the probability that a particular antigenic determinant is present on two proteins. Such monospecific antibodies should yield information which is free of questions of reaction specificity, and at the same time, is at the level of resolution of individual antigenic determinants rather than individual proteins.

ACKNOWLEDGEMENTS

This work was supported by research grants GM 22150 and GM 07215 from the National Institutes of Health.

REFERENCES

1. Stöffler, G., and Wittmann, H. G. (1971) Proc. Natl. Acad. Sci. U.S.A. 68, 2283-2287.
2. Higo, K., Held, W., Kahan, L. and Nomura, M. (1973) Proc. Natl. Acad. Sci. U.S.A. 70, 944-948.
3. Hawley, D. A., Slobin, L. I., and Wahba, A. J. (1974) Biochem. Biophys. Res. Commun. 61, 544-550.
4. Strycharz, W. A., Nomura, M. and Lake, J. A. (1978) J. Mol. Biol. 126, 123-140.
5. Boublik, M., Hellmann, W. and Roth, H. E. (1978) J. Mol. Biol. 127, 479-490.
6. Wittmann, H. G., Stöffler, G., Hindennach, I, Kurland, C. G., Randall-Hazelbauer, L., Birge, E. A., Nomura, M., Kaltschmidt, E., Mizushima, S., Traut, R. R. and Bickle, T. A. (1971) Mol. Gen. Genet. 111, 327-333.
7. Lutter, L. C., Zeichhardt, H., Kurland, C. G. and Stöffler, G. (1972) Mol. Gen. Genet. 119, 357-366.
8. Sun, T.-T., Traut, R. R., and Kahan, L. (1974) J. Mol. Biol. 87, 509-522.
9. Kaltschmidt, E., Kahan, L., and Nomura, M. (1974) Proc. Natl. Acad. Sci. U.S.A. 71, 446-450.
10. Lake, J. A. and Kahan, L. (1975) J. Mol. Biol. 99, 631-644.
11. Lake, J. A. (1976) J. Mol. Biol. 105, 131-159.
12. Wabl, M. R. (1974) J. Mol. Biol. 84, 241-247.

13. Lake, J. A., Pendergast, M., Kahan, L., and Nomura, M. (1974) Proc. Natl. Acad. Sci. U.S.A. 71, 4688-4692.

14. Stöffler, G. and Tischendorf, G. W. (1974) In Topics in Infectious Diseases. Drug Receptor Interactions in Antimicrobial Chemotherapy (J. Drews and F. E. Hahn, eds.), Vol. 1, pp. 117-143, Springer-Verlag, Wien and New York.

15. Politz, S. M. and Glitz, D. G. (1977) Proc. Natl. Acad. Sci. U.S.A. 74, 1468-1472.

16. Keren-Zur, M., Boublik, M., and Ofengand, J. (1979) Proc. Natl. Acad. Sci. U.S.A. 76, 1054-1058.

17. Lelong, J. C., Gros, D., Gros, F., Bollen, A., Maschler, R., and Stöffler G. (1974) Proc. Natl. Acad. Sci. U.S.A. 71, 248-252.

18. Winkelmann, D., and Kahan, L. (1979), In preparation.

19. Stöffler, G., Hasenbank, R., Lutgehaus, M., Maschler, R., Morrison, C. A., Zeichhardt, H., and Garrett, R. A. (1973) Mol. Gen. Genet. 127, 89-110.

20. Winkelmann, D. and Kahan, L. (1979) J. Supramol. Struc., in press.

21. Morinaga, T., Funatsu, G., Funatsu, M. and Wittmann, H. G. (1976) FEBS Lett. 64, 307-309.

22. Wittmann-Liebold, B., Marzinzig, E. and Lehmann, A. (1976) FEBS Lett. 68, 110-114.

23. Tischendorf, G. W., and Stöffler, G. (1975) Mol. Gen. Genet. 142, 193-208.

24. Tischendorf, G. W., Zeichhardt, H., and Stöffler, G. (1975) Proc. Natl. Acad. Sci. U.S.A. 72, 4820-4824.

25. Wittmann, H. G. and Wittmann-Liebold, B. (1974) In Ribosomes (M. Nomura, A. Tissières, and P. Lengyel, eds), pp. 93-114. Cold Spring Harbor Laboratory, Cold Spring Harbor, New York.

26. Köhler, G., and Milstein, C. (1975) Nature 256, 495-497.

27. Stöffler, G., Wool, I. G., Lin, A., and Rak, K.-H. (1974) Proc. Natl. Acad. Sci. U.S.A. 71, 4723-4726.

ANALYSIS OF THE ASSEMBLY AND FUNCTION OF THE 50S SUBUNIT FROM *ESCHERICHIA COLI* RIBOSOMES BY RECONSTITUTION

KNUD H. NIERHAUS

Max-Planck-Institut für Molekulare Genetik, Abt. Wittmann,
Berlin-Dahlem, Germany

INTRODUCTION

Any reaction between three molecules is a rare event, due to the low frequency of simultaneous collision of the reactants. It follows that the self-assembly of the small and the large ribosomal subunits (22 and 34 components, respectively) must consist of a series of consecutive reactions. The product of this self-assembly process is the highly defined and compact subunit containing one copy of each component (1). The only exception is the 50S protein L7/L12 which exists in four copies per 50S subunit (2).

The series of reactions is not simply a concatenation of analogous processes. Obviously, some reactions have to surmount a drastically higher activation energy barrier than others, leading to the accumulation of stable precursor particles in the course of the *in vivo* assembly. Two precursor particles have been described for the 30S subunit (P_1 30S and P_2 30S, respectively), and three for the 50S subunit (P_1 50S, P_2 50S and P_3 50S, respectively; see ref. 3 for review). In good agreement, the analysis of the *in vitro* assembly of the 30S subunit intensively performed in Nomura's group has also revealed two intermediate particles RI_{30} and RI^*_{30}. The first of these, RI_{30}, is very similar to the corresponding particle P_1 30S found *in vivo* with respect to S-value and protein content (see ref. 4 for review). The similarities between both particles suggest strongly that the rate limiting step in both *in vivo* and *in vitro* assembly is very similar if not the same. Nevertheless the activation energy for the rate limiting step *in vitro* (160 kJ/mol = 38 kcal/mol ; ref. 5) is much higher than that *in vivo*. This is indicated by the fact that the 30S reconstitution needs temperatures above 35°C (5) whereas *E. coli* cells grow even at temperatures of about ten degrees. The many specific features of the *in vivo* assembly, such as the "assembly gradient" following the RNA synthesis from the 5'-end towards the 3'-end (see below), the processing of RNA and methylation of ribosomal proteins, and possibly the existence of specific precursor proteins not present on the mature particle (ref. 3, review) may account for these different levels of activation energy. Furthermore, regulation mechanisms may establish additional rate limiting steps *in vivo* by limiting the amount of a particular component essential for the assembly.

PARTIAL RECONSTITUTION OF THE 50S SUBUNIT

Methods for the partial reconstitution of the 50S subunit (6-8) were developed at the same time as those for the 30S subunit. Core particles were derived from the 50S subunit by salt treatment using either isopycnic CsCl gradients (8) or washing procedures with NH_4Cl/EtOH (9) or LiCl (10,11). The NH_4Cl/EtOH and LiCl procedures have the advantage that large amounts of cores and split proteins can be conveniently obtained. With these techniques it could be demonstrated that proteins L7/L12 are involved in the binding of elongation factors (see ref. 12 for review), L6, L11 and L16 in the reconstitution of the peptidyltransferase center (13-15; see below), and L16 in addition in the binding of chloramphenicol (16) and virginiamycin S (17). Furthermore, the binding of L7/L12 to cores lacking these proteins was shown to be dependent on L10 and L11 (18).

An early dilemma which limited the scope of such experiments was the large number of proteins (at least two thirds of the 32 proteins) still present on the cores. Nevertheless, at least one LiCl procedure (11) could overcome this handicap: 50S subunits incubated in 4 M LiCl under defined conditions release more than 25 proteins. The corresponding 4.0c core, containing less than 10 proteins and 23S RNA, could be reconstituted to active particles using the partial reconstitution technique (19). However, this finding did not initiate a boom of studies on assembly and function, since at the same time a method was discovered for the total reconstitution of the 50S subunit from *E. coli* (19). The latter technique was more useful for the analysis of the 50S assembly, whereas the 4.0c cores became important for specific problems in the early assembly (see below).

TOTAL RECONSTITUTION OF THE 50S SUBUNIT

Procedures for Total Reconstitution

When only the partial reconstitution procedure was available for the *E. coli* 50S subunit many groups attempted the total reconstitution of this subunit, starting from separated RNA and protein fractions, but without any success. One possible reason for this frustrating experience could be a rate limiting step with a very high activation energy during the course of the assembly which should, therefore, be performed at incubation temperatures above 50°C. Above this temperature, however, ribosomal proteins of the large subunit from *E. coli* start to denature, and are not anymore able to assemble an active particle. Consequently, Nomura's group shifted their attention to a thermophilic organism, *Bacillus stearothermophilus*, and succeeded in the first total reconstitution of a large ribosomal subunit (20). However, this technique has never attained the same importance for the analysis of ribosomal structure and function as the corresponding technique for the small subunit. The reason is that the ribosome from *E. coli* has become the "model ribosome" of the prokaryotic type, since by far the greatest amount of genetic and biochemical data have been obtained from the *E. coli* ribosome. Furthermore, a one to one correlation between the ribosomal components of *E. coli* and *B. stearothermophilus* has not yet been established for the respective large subunits in contrast to the small ones (21,22). Nevertheless, one finding of a preliminary assembly study is of interest. When the reconstitution mixture of *B. stearothermophilus* 50S subunit components was incubated at 30°C instead at the normal 60°C, an accumulation of reconstitution intermediate particles was observed (23). Thus, similar to the 30S case, the 50S assembly seems also to proceed via reconstitution intermediates.

We have suggested above that the 50S assembly needs to surmount a high activation energy barrier. If this hypothesis is valid, a one-step reconstitution procedure for the *E. coli* 50S subunit is in principle not possible, as long as the *E. coli* components cannot be maintained in an active form at temperatures above 50°C. Fortunately, an alternative procedure was discovered, consisting of a two-step incubation (19,24). (23S+5S)RNA and the total proteins of the 50S subunit (TP50) from *E. coli* are incubated for 20 minutes at 44°C, 4 mM Mg^{2+} and 400 mM NH_4^+; then the Mg^{2+} concentration is raised to 20 mM Mg^{2+}, and the incubation is continued for 90 minutes at 50°C. The conditions of the second incubation are identical to those of the partial reconstitution procedure. As will be described in detail below, the rate limiting steps of the 50S assembly are two conformational changes, both requiring different ionic optima and incubation conditions. It follows that the total reconstitution of the 50S subunit from *E. coli* necessarily involves a two-step incubation.

Recently, a modification of the two-step procedure has been described (25) which deviates from the original procedure (24) in the following manner (the parameters of the original procedure are given in parenthesis): RNA is isolated from 70S ribosomes (50S subunits); TP50 is acetic acid extracted, without a final transient urea dialysis (with urea dialysis); TP50 is present in a 2.2 to 4 fold molar excess over RNA in the reconstitution mixture (1.2 to 1.4 excess); the first step incubation is performed at 7.5 mM Mg^{2+} for 30 minutes (4 mM Mg^{2+}/20 minutes) and the second incubation for 120 minutes (90 minutes). These discrepancies could be resolved in a collaboration with R. Amils who visited our laboratory. The higher Mg^{2+} dependence was due to the modified RNA preparation, the higher protein excess to the omission of the urea dialysis step. When the TP50 fraction prepared by the modified procedure was subjected to a urea dialysis according to the original prescription, the protein excess required for optimal activity was shifted to the lower values of 1.2 to 1.4 moles. In all experiments, the activity of the particles obtained by the original procedure was significantly higher than that of the particles prepared by the modified procedure (Nowotny, Amils and Nierhaus, unpublished observation). The latter effect could also be demonstrated by F. Dohme in C. Cantor's laboratory (Dohme, personal communication). The original procedure routinely yields highly active particles, provided that intact 23S RNA is used and that the following precaution is observed: During the final dialysis for the removal of urea, the TP50 has to be checked at frequent intervals in order to avoid precipitation, and the dialysis should be stopped at once as soon as opalescence is visible (protein concentration ∿100 e.u./ml; for details see ref. 26).

Kinetic Discontinuities in the *in vitro* Assembly

In order to identify and characterize possible reconstitution intermediates, a kinetic experiment was performed during the first-step incubation. One part of the aliquots taken at various times was analysed on sucrose gradients, and another part was subjected to a standard second-step incubation and tested for poly(Phe) synthesis after complementation with native 30S subunits. Figure 1A strikingly demonstrates that discrete intermediate particles are formed. The first one (S-value:33S) forms spontaneously at 0°C after mixing (23S+5S) RNA and TP50 under the condition of the first incubation, the second one (41S) is the main component present between 5 and 10 minutes incubation, whereas after 20 minutes the third one (48S) predominates. The area below each peak was integrated and taken as a measure of the amount of the corresponding particle formed. With these data a kinetic picture of the appearance of the respective particles could be drawn (Fig. 1B), which clearly reveals the sequential order of the appearance of the three particles. The three intermediates were isolated and their protein content determined. The first and the second particle contained the same complement of RNA and protein, i.e. 23S and 5S RNA and about 20 proteins, whereas all 50S proteins were present on the third particle. Thus, the first and the second particles represent different conformations of the same complex. Accordingly, the reconstitution intermediate particles were designated as $RI_{50}(1)$, $RI^*_{50}(1)$ and $RI_{50}(2)$, respectively (24).

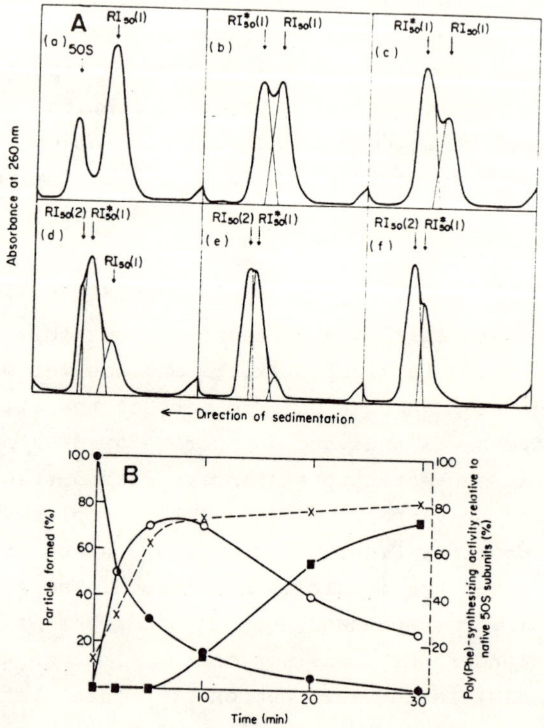

Figure 1. Kinetic analysis of the first-step incubation according to ref. 24. A, an aliquot of the first incubation mixture containing 2 A_{260nm} units of particles was withdrawn at various times and subjected to a sucrose gradient centrifugation (4 to 20% sucrose in the ionic milieu of the first incubation; 5 h at 160,000 x g and 4°C). (a), 0 min; this sample contains in addition 0.5 A_{260nm} units of 50S subunits as a control; (b) 2 min; (c), 5 min; (d), 10 min; (e), 20 min; (f), 30 min. B, the areas from the peaks of the various particles were integrated. The sum of the areas of one sucrose gradient profile was taken as 100% "particles formed" and used for the kinetics of appearance of the various particles. ●—●, $RI_{50}(1)$; ○—○ $R^*_{50}(1)$; ■—■, $RI_{50}(2)$. x—x, activity in the poly(U) system; before testing in the poly(U) system the samples were incubated according to the standard second-step condition (90 min at 50°C in the presence of 20 mM Mg^{2+}).

None of the intermediate particles showed any functional activity. However, when aliquots from the kinetic experiment were subjected to the standard second-step incubation as described above, and the activity obtained superposed on the particle curves in Figure 1B, it became strikingly apparent that the appearance of activity parallels the formation of the $RI^*_{50}(1)$ particle. Therefore, this particle is the essential product of the first incubation and an obligatory intermediate in the course of the assembly of active 50S subunits. Furthermore, the conformational change $RI_{50}(1) \rightarrow RI^*_{50}(1)$ seems to be the rate limiting step of the first incubation.

The transition $RI^*_{50}(1) \rightarrow RI_{50}(2)$ can occur in either the first or the second incubation. Since the $RI_{50}(2)$ particle contains a full complement of proteins, it differs from the 50S subunit only in conformation. It follows that the rate limiting step of the second incubation is also a conformational change (24).

In order to learn more about these rate limiting steps, further kinetic experiments were performed in both incubation steps at various temperatures. The evaluation of the data confirmed and extended our conclusions. The rate limiting reaction of the second incubation is a first-order and unimolecular reaction (i.e. a conformational change). Most likely, the same is true for the first incubation. The Arrhenius activation energies were calculated as 290 kJ/mol (= 70 kcal/mole) and 225 kJ/mol (= 54 kcal/mol) for the rate limiting steps of the first and second incubations, respectively (27).

Surprisingly, the activation energy required is higher (290 kJ/mol) under the less stringent condition of the first incubation (20 minutes/44°C) than that (225 kJ/mol) required during the second incubation (90 minutes/50°C). The most likely explanation for this is that the rate limiting step of the first incubation consists of a series of conformational changes; for example, three consecutive conformational changes each requiring 100 kJ/mol could well occur at 44°C.

The latter hypothesis was tested in an electron microscopy study using the shadow cast technique. Again a kinetic ex-

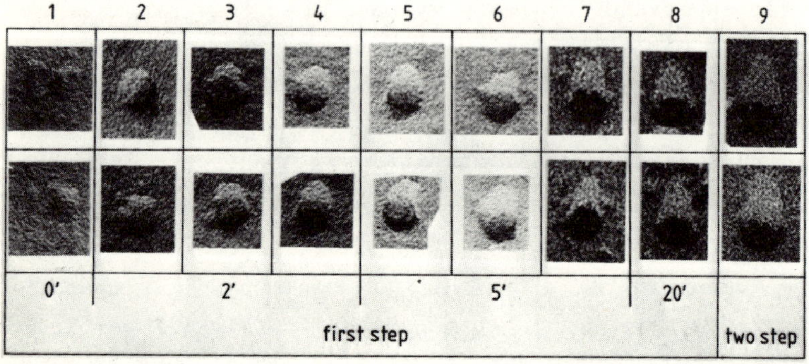

Figure 2. A gallery of consecutive assembly stages derived from a kinetic analysis of the first-step incubation (according to ref. 28). Sample no. 1, $RI_{50}(1)$; sample nos. 2-6, $RI^{*}_{50}(1)$; samples nos. 7 and 8, $RI_{50}(2)$; sample no. 2, reconstituted 50S subunits.

periment was performed during the first incubation, and one half of the aliquots taken at various times was inspected in the electron microscope, the other half being subjected to a second incubation and activity test as described above. The EM pictures demonstrated that the particles became more dense, compact and protuberant (indicated by the prolonged shadow) in the course of the assembly. Using these criteria, a series of consecutive assembly stages could be selected, which cinematographically illustrate the assembly of the 50S subunit (Fig. 2). $RI_{50}(1)$ particles are seen in image no. 1, $RI^{*}_{50}(1)$ particles are represented in nos. 2 to 6; nos. 7 and 8 demonstrate $RI_{50}(2)$ particles, and no. 9 shows reconstituted 50S subunits. The various images obtained from the $RI^{*}_{50}(1)$ particles are not different projections of one and the same structure, because the populations of the various images changed in the course of the assembly. That is to say, the amounts of the flat structures (nos. 2 and 3) decreased, whereas those of the more prominent structures (nos. 5 and 6) increased during incubation times of up to 10 minutes. Thus, a series of conformational changes occurs during the reaction $RI_{50}(1) \rightarrow RI^*_{50}(1)$ (ref. 28).

Components Important for the Early Assembly

As mentioned above, both the $RI_{50}(1)$ and $RI^*_{50}(1)$ particles contain the same complement of RNA (23S+5S) and proteins (about 20), but differ drastically in their S-values (33S and 41S, respectively). The question is whether all these components are required for this conformational change.

To analyse this question we used an operational definition for the $RI^*_{50}(1)$ conformation: *viz.* a particle which can be formed exclusively in the first-step incubation and is able to form active particles in the second one (in presence of TP50) has adopted the $RI^*_{50}(1)$ conformation. Thus, we had to identify the minimal set of ribosomal components which were able to form active particles after the following experimental sequence:

$$\text{23S RNA + minimal set} \xrightarrow{\text{1. incub.}} \text{+ TP50} \xrightarrow{\text{2. incub.}} \text{active particle.}$$

Activity after the second-step incubation indicates $RI^*_{50}(1)$ formation.

The result of this experiment was that 23S RNA and five proteins (L4, L13, L20, L22, and L24) were essential for $RI^*_{50}(1)$ formation, and protein L3 had a stimulatory effect. These seven components were sufficient to generate a particle with $RI^*_{50}(1)$ conformation (29). Three interesting conclusions could be drawn from these results:

1. The *in vivo* precursor particles P_1 50S (S-value: 32S) and P_2 50S (43S) have been isolated and their protein content determined (30). All the proteins important for $RI^*_{50}(1)$ formation are present on the P_1 50S particle, except the stimulatory protein L3. Furthermore, the *in vivo* addition of proteins in the course of assembly has been studied (31). Again, all the important $RI^*_{50}(1)$ proteins belong to the first set of ten which are assembled *in vivo*. This striking correspondence indicates that the early assembly follows very similar pathways both *in vivo* and *in vitro*.
2. The 23S RNA can be split by nuclease treatment into three large fragments (the 13S, 8S and 12S fragments, respectively, ordered from the 5'- to the 3'-end). All the essential proteins can bind independently to the 5'-fragment, whereas only L3 binds to the 3'-fragment.

Clearly, the early assembly proteins bind *in vivo* to the 5'-region of the 23S RNA, and determine the first important conformational changes before the 23S RNA transcript has been finished. Thus, the *in vivo* assembly proceeds in parallel with the RNA synthesis, forming an "assembly gradient" along the 23S RNA from the 5'-end towards the 3'-end.

3. The early assembly proteins important for $RI^*_{50}(1)$ formation are, with the exception of L24, all present on the 4.0c core, which is derived from the 50S subunit by incubation in 4 M LiCl. This core contains about 10 proteins (11,19). Thus, the proteins which assemble first are the last ones to be split off the 50S subunit. This finding indicates that the order in which the proteins are assembled corresponds, to a first approximation, to the reverse of the sequence in which the proteins can be split off the 50S subunit.

As described in the preceding section, the reaction $RI_{50}(1) \rightarrow RI^*_{50}(1)$ consists of a series of conformational changes. In fact, it could be shown, that the early assembly group of proteins (L3, L4, L13, L20, L22, and L24) is not solely responsible for the drastic S-value shift. In addition to these six proteins, L21 and L23 strongly influence the conformation, whereas L2 and L17 (which are also early proteins) do not show any effect in sucrose gradient analysis (32).

The observation that the 4.0c core does not contain L24 has initiated a further investigation. The 4.0c core can be reconstituted in the second-step incubation to an active particle (19), i.e., the 4.0c core has at least the $RI^*_{50}(1)$ conformation. It follows that the $RI^*_{50}(1)$ conformation can be maintained without L24 which is essential for generating this conformation. The suspicion is at hand that L24 is an assembly protein exclusively involved in the early assembly.

This suspicion could be corroberated showing a lack of dependence on L24 when the 4.0c core was reconstituted to an active particle, whereas a total reconstitution using the identical set of proteins and (23S+5S)RNA was fully dependent on the presence of L24. Thus, L24 is essential for the early assembly but dispensable for the late assembly and the functions of the mature 50S subunit (33). Supporting these results, a thermosensitive mutant has been described showing an impaired assembly *in vivo* of the 50S subunit at 42°C due to an altered L24 (ref. 34).

A similar analysis led to the same conclusions with respect to protein L20 (ref. 35). Thus, L20 and L24 are highly specialized proteins which play an essential role exclusively in the early assembly.

Components Important for the Late Assembly

Table 1 summarizes the assembly process of the large subunit and gives the protein content of the reconstitution intermediates. Those proteins important for RI^*_{50}(1) formation and designated as "early assembly group" are enclosed in a box. The "late assembly group" completes the protein complement of the RI_{50}(1) particle, and consists of at least 8 proteins (L6, L15, L16, L25, L27, L28, L30, and L32).

TABLE 1. Assembly *in vitro* of the 50S subunit from *E. coli* ribosomes.

23S + 5S RNA + 18 proteins	L1 L2 [L3/L4] L5 L9 L10 L11 [L13] L17 L19 [L20] L21 [L22] L23 [L24] L29 L33	0°C, 4 mM Mg^{2+} →	RI_{50}(1)	(33S)
	RI_{50}(1)	44°C, 4 mM Mg^{2+} →	RI^*_{50}(1)	(41S)
RI^*_{50}(1) + 8 proteins	L6 L15 L16 L25 L27 L28 L30 L32	44°C, 4 mM Mg^{2+} → or 50°C, 20 mM Mg^{2+}	RI_{50}(2)	(48S)
	RI_{50}(2)	50°C, 20 mM Mg^{2+} →	50S subunit	

Not assigned: L7/L12, L14, L18, L26, L31 and L34

0°C, 4 mM Mg^{2+} or 44°C, 4 mM Mg^{2+} defines the ionic conditions of the first step, and 50°C, 20 mM Mg^{2+} those of the second step of the 2-step reconstitution procedure. Proteins important for RI^*_{50}(1) formation are enclosed by a square. The S values of the reconstituted intermediates are given in parenthesis.

Protein L16 is of particular interest, since this protein has been shown to be involved in many ribosomal functions. It is essential for peptidyltransferase activity (15,36), and is involved in the binding of various antibiotics such as chloramphenicol (16,37), virginiamycin S (17), and erythromycin (36). The multi-functionality of L16 points to the possibility that this protein is not directly involved in all these functions, but rather controls the active conformation of the ribosome. Thus, L16 is possibly one of those components which are responsible for the conversion from inactive $RI_{50}(2)$ particles to active 50S subunits. Accordingly, we tested whether a 50S-derived core particle lacking L16 changes its conformation upon addition of purified L16.

Incubation of 50S subunits with 1.3 M LiCl yields a 1.3 c core which has lost about a dozen proteins (11,36) including L16. This core particle is inactive with respect to the above mentioned functions, but regains all the activities after reconstitution (under second step conditions) with a five- to ten-fold excess of L16. Kinetic reconstitution experiments performed at various temperatures demonstrated that L16 in fact induces a conformational change in the core particle with an Arrhenius activation energy of 125 kJ/mol (30 kcal/mol; ref. 38). This value is considerably less than the activation energy required for the rate limiting step of the second incubation (225 kJ/mol ≙ 54 kcal/mol; ref. 27). The difference in the activation energy values implies that L16 is not solely responsible for the rate limiting step of the second incubation, but is rather just one essential component of this reaction. However, these results do not rule out the possibility that L16 is in addition directly involved in one or the other function.

Another interesting component of the late assembly is 5S RNA. This component is already present in the first precursor P_1 50S (39,30) and, correspondingly, in the first reconstitution intermediate $RI_{50}(1)$. However, the assembly can proceed normally in the absence of 5S RNA as far as the stage of the $RI_{50}(2)$ particle, whereas the last assembly step $RI_{50}(2) \rightarrow 50S$ occurring in the second incubation requires the presence of 5S RNA. Furthermore, 5S RNA can be integrated at any assembly stage, indicating that the binding site for the 5S RNA is exposed throughout the whole assembly process.

Finally, a functional analysis was performed with a totally reconstituted particle lacking 5S RNA. This particle showed full EF-G dependent GTPase activity, but the activities with respect to peptidyltransferase and tRNA binding to the ribosomal P-site were drastically reduced, and no activity was found with regard to tRNA binding to the A-site. It follows that 5S RNA plays no role in EF-G dependent GTPase activity, whereas it is important but not essential for peptidyltransferase activity and tRNA binding to the P-site, and is possibly directly involved in tRNA binding to the A-site (39). The latter finding supports evidence suggesting an interaction of the G-T·Ψ·C sequence in tRNA loop IV with a G·A·A·C sequence in 5S RNA at the ribosomal A-site (for reviews see 40, 41).

Prerequisites for Elucidation of the Assembly Map

Assembly mapping experiments require highly purified proteins which are active in reconstitution assays. A large number of procedures for the isolation of ribosomal proteins has been described (for reviews see 42,43), but in our hands these procedures did not yield active proteins. Therefore, we developed an isolation procedure (44) which gave satisfying results with respect to both purity and activity. The method consists of three steps:

1. The proteins were prefractionated by washing 50S subunits with NH_4Cl/EtOH and LiCl.
2. The protein fractions were separated by gel filtration (Sephadex G-100) in the presence of 6 M urea.
3. Suitable fractions from step 2 were pooled and subjected to chromatography on carboxy-methyl-cellulose in the presence of 6 M urea at pH 7.5. The proteins were eluted with a KCl gradient.

In order to obtain active proteins, it is important to remove urea degradation products by washing the urea solutions with bentonite before use. Furthermore, the urea solutions must be kept at 0° to 4°C and should be used within three weeks. The chromatography at pH 7 avoids a possible "adsorptive denaturation" of the proteins, since at pH 7 the proteins are only loosely bound to the column matrix.

Interestingly, this procedure allowed in addition the isolation of multi-protein complexes capable of withstanding the 6 M urea treatment. Two observations indicated that these are genuine protein complexes and not simply artefacts caused by different conformations of one of the proteins, eluting at different positions. Firstly, the complexes elute from the

G-100 column at a volume corresponding to a higher molecular weight than that of at least the lighter component. Secondly, the complexes form a sharp, distinct peak in the chromatography step, eluting at a conductivity value different to that of either of the single components.

The complexes found with this and other techniques are compiled in Table 2. The results of other techniques, such as cross-linking and fragmentation of ribosomes, which also point to protein neighborhoods in the ribosome are also included. The large number of protein complexes accounting for

TABLE 2. Protein neighborhoods in the 50S subunit.

Protein complexes	Crosslinked proteins (for reviews see 45, 46)		Proteins present in small fragments of the 50S subunit
L1-L3 (ref. 44)	L2-L15	L5-L17	L1-L9 (ref. 49)
L2-L15 (ref. 44)	L2-L7/L12	L5-L23	L1-L23 (ref. 50)
L2-L17 (ref. 44)	L2-L9	L5-L25	L1-L3-L9-L23-L24 (ref. 50)
L3-L23 (ref. 47)	L2-L10	L7/L12-L10	L3-L4-L24 (ref. 50)
L5-L6-L11 (ref. 44)	L2-L11	L7/112-L11	L5-L18 (ref. 50)
L7/L12-L10 (ref. 48)	L2-L17	L10-L11	L5-L18-L25 (ref. 49)
L10-L11 (ref. 44)	L3-L5	L11-L14	
L13-L19 (ref. 47)	L4-L11	L17-L21	
L14-L19-L24 (ref. 44)	L4-L14	L17-L32	
L15-L27 (ref. 44)	L5-L7/L12	L18-L32	
L15-L17-L27 (ref. 44)		L22-L32	
L16-L27 (ref. 47)	L2-L13-L20		
L18-L19 (ref. 44)	L2-L20-L27-L32/33		
L19-L22 (ref. 44)	L9-L11-L14		

about 2/3 of the 50S proteins strikingly indicates that protein-protein interaction is an important element of the 50S structure. Furthermore, if these complexes are formed before they enter the 50S particle during assembly, they would represent an important feature of the strong cooperative character of the assembly process.

The Assembly Map of the 50S Subunit

Ribosomal proteins which can bind to their respective rRNA independently of other proteins are referred to as "binding proteins". Eleven proteins (L1, L2, L3, L4, L6, L13, L16, (possibly L17), L20, L23, and L24) have been reported as binding proteins (see ref. 51 for review). Recently, proteins L11 and L15, purified by a gentle isolation technique (43), were also identified as binding proteins (52).

In the series of assembly mapping experiments, we first tested the binding capacities of the 50S proteins to 23S RNA under the conditions of the first (4 mM Mg^{2+}) and second step incubations (20 mM Mg^{2+}). A protein was incubated with 23S RNA under the respective condition; the protein bound to the 23S RNA was separated from the non-bound portion on a sucrose gradient, precipitated by trichloroacetic acid, and analysed by SDS-gel electrophoresis. The results revealed notable differences from the previous group of "binding proteins": 17 proteins interacted with 23S RNA, namely L1, L2, L3, L4, L7/L12, L9, L10, L11, L15, L16, L17, L18, L20, L22, L23, L24,

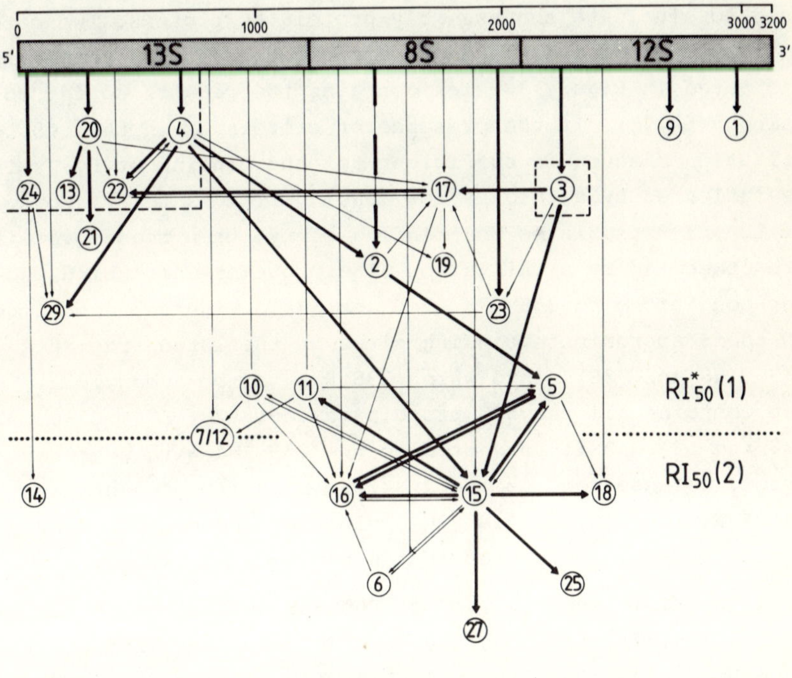

Figure 3. Assembly map of the 50S subunit determined under the conditions of the first-step incubation. The assembly dependences of the proteins present on the $RI^*_{50}(1)$ particle have been determined and confirmed with various protein preparations (54). The assembly dependences of the proteins found in addition on the $RI_{50}(2)$ particle have been determined twice with one protein preparation and will be rechecked with a second one (55).

and L29. Seven of these proteins (L1, L2, L3, L4, L9, L23 and L24) bound to 23S RNA at a level of about one copy per 23S RNA molecule, whereas L20 could bind with more than one copy. Some proteins bound less strongly to 23S RNA at 20 mM Mg^{2+} than at 4 mM Mg^{2+} (L2, L3, L4, L7/L12, L9, L10, L15, L16 and L18). Finally, it was found that the binding of L2, L7/L12, L10, L11, L15, L16 and L18 was preparation sensitive, i.e. the binding ability changed notably from one preparation to another (53).

There are two possible reasons for the striking discrepancies with the previous set of binding proteins: 1) We have used binding conditions which are the same as those optimal for the reconstitution of 50S subunits, and 2) the proteins in our experiments were active in reconstitution assays.

The same experimental procedure (first step condition) was used to elucidate the assembly, protein by protein. Two categories of binding dependence were established and these are indicated in Figure 3 which contains the results so far obtained (54,55). If the presence of protein Lx in the reconstitution mixture is sufficient for the binding of Ly, this is indicated by a thick arrow leading from Lx to Ly. However, if Lx only stimulates the binding of Ly, then these proteins are connected by a thin arrow. The proteins which bind under our conditions to 23S RNA are located in Figure 3 according to their approximate binding sites on the large fragments (13S, 8S and 12S), into which 23S RNA can be subdivided. This map contains all the 50S proteins except L28, L30, L31, L32, L33 and L34. L2 is connected to the 23S RNA with a broken arrow, because some L2 preparations bind independently to the 23S RNA with one copy per RNA molecule, whereas other L2 preparations could only bind in the presence of L4.

A "partial assembly map" of the 50S subunit has been reported (56,57), in which the dependence of the binding of seven proteins (L5, L10, L11, L21, L22, L28 and the unresolved mixture of L17/27) on other proteins was studied. A comparison of these data with our results shows agreement for only two of the protein interdependences (L20 to L21 and L16 to L5), but not for any of the others. The reasons for these discrepancies are possibly the same as those described above, regarding the RNA-binding proteins.

When two proteins are interrelated in the assembly map, the simplest interpretation is that these proteins are neighbors, and this type of topographical correlation in the case of the 30S assembly map (58) has been noted (59). However, the reverse conclusion need not always be valid; a neighborhood relationship between two proteins may not necessarily imply an assembly dependence. Accordingly, many but not all of the protein neighborhoods compiled in Table 2 correspond to a binding dependence in the assembly map. For example, L5, L18 and L25 were found on small fragments isolated from the 50S subunit, and all these three proteins depend on L15 for assembly. As a further example, L2 and L15 were found in a protein complex, and both proteins promote the binding of L5, again a physical neighborhood between L2 and L15.

The importance of the assembly map is more than just the topographical aspect. Some previous observations can be substantiated by assembly relationships. A comparison of the *in vitro* assembly (Table 1) with the sequence in which the proteins can be washed off the 50S subunit by increasing LiCl concentrations suggested that the one sequence was roughly the reverse of the other (24). This interrelationship is strikingly reflected by the assembly map (Fig. 4). The proteins are indeed split off the 50S subunit in just the opposite sequence to that in which they are assembled. Another example of the application of the assembly map is the protein complement of the two-step reconstituted particle lacking 5S RNA, this particle does not contain proteins L5, L16, L18 and L25 (39). The absence of L5, L18 and L25 is very reasonable, since these proteins are the only ones which can bind to 5S RNA (60,61), and it is clear that 5S RNA is required to bind these three proteins tightly to the assembled particle during the second step incubation. The assembly map also offers a plausible explanation for the absence of L16 from the particle lacking 5S RNA, since L5 strongly supports the binding of L16.

Inspection of the assembly map reveals assembly domains in the 50S subunit. One domain is located around protein L20 and consists of L4, L13, L20, L21, L22, L24 and possibly L29.

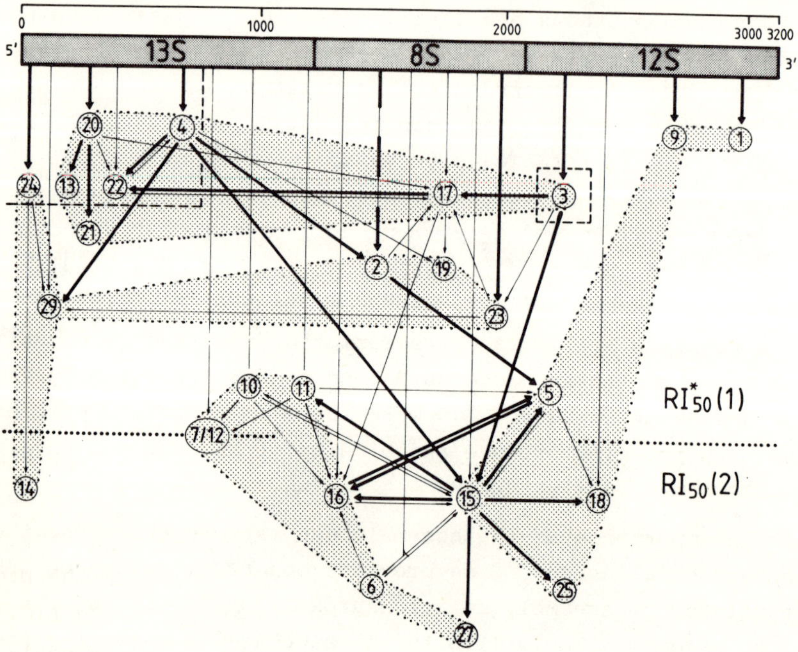

Figure 4. Comparison of assembly map and LiCl splitting. The zones (from bottom to top) indicate the protein families which can be split off consecutively with increasing LiCl concentrations: 1.3 M; 3 M; 4 M; proteins present on the 4.0c core (according to 13, 19 and Nowotny and Nierhaus, unpublished results).

Three of these proteins bind strongly to the RNA, and their binding sites are located near to the 5'-end. All the proteins essential for RI^*_{50}(1) formation (L4, L13, L20, L22 and L24) are present in this group, and, as outlined above, at least two of these proteins (L20 and L24) are exclusively involved in the early assembly, and play a role neither in the late assembly nor in the functions of the mature 50S subunit. Thus, the "L20 domain" forms a core or nucleus in the early assembly. A second domain surrounds L15 and consists of L5, L6, L7/L12, L10, L11, L15, L16, L18, L25 and L27. None of these proteins binds tightly to the 23S RNA, indicating that the "L15 domain" is a late assembly domain. In fact, most of the late assembly group (L6, L15, L16, L25 and L27, but without L28, L30 and L32 which are not yet integrated in the assembly map) is present in this domain. The "L15 domain" contains functionally important proteins, e.g., L7/L12 are involved in the binding of elongation factors (for review see ref. 62), and L15, L16 and L18 are essential for peptidyltransferase activity (see below). Thus, the "L15 domain" represents a late assembly domain and a functional cluster.

The Reconstitution of the Peptidyltransferase Center

Formation of the peptide bond takes place in the 50S subunit (63) and is designated as peptidyltransferase activity. This activity transfers the peptidyl residue from the peptidyl-tRNA present in the P-site to the aminoacyl-residue of the aminoacyl-tRNA in the A-site. The result is that the peptidyl-tRNA, now prolonged by one aminoacyl residue, is located in the A-site. An involvement of an enzymatic activity in this reaction is likely. However, it is questionable whether "peptidyltransferase" correctly describes the enzymatic activity involved. Various evidence points to the destabilization of the ester bond between peptidyl residue and tRNA as being the minimal enzymatic requirement for peptide bond formation (see ref. 64 for discussion).

About half of the 50S components, including 5S and 23S RNA, have been implicated in the peptidyltransferase activity (Table 3). The method which has been most extensively used, and which has indicated most of the components, is the affinity labelling technique. Affinity labelling is a topographical technique, i.e. it identifies components at or near the peptidyltransferase center. It is possible therefore that the large number of components identified with this technique may not even include the structural element exerting the activity. Further, it cannot be excluded that experiments using chemical modifications of ribosomal components affect the assembly and/or the function of a neighboring component. Thus, the observed effects cannot be unequivocally assigned to the modified components.

On the other hand, reconstitution experiments raise another ambiguity, in that components identified with this technique can be involved in either the assembly or the function of the activity under observation. However, the unique

TABLE 3. Literature references to components at or near the peptidyltransferase center.

Component	Identified by		
	Affinity labelling	Reconstitution	Other methods
L2	65, 66, 67, 68	76	-
L4	77	-	79
L6	69	14	-
L11	70	13	80
L13	68	-	-
L14	66	-	-
L15	68	80	81
L16	71, 72, 68	16, 15, 36	80
L18	70	-	-
L23	73	-	-
L24	74	-	-
L27	72, 65, 66, 67, 68	-	-
L32/L33	74	-	-
23S RNA	75	-	-
5S RNA	-	78, 39	-

advantage of this technique is that the components essential for the function must necessarily be included in the set of components essential for the reconstitution of that function. Therefore, we analysed by partial and total reconstitution all the ribosomal components concerned in the peptidyltransferase activity (64).

Firstly, partial reconstitution experiments were performed with 1.3c LiCl-cores and highly purified proteins, in order to settle an early controversy on the role of protein L11. Reconstitution experiments had identified this protein as being possibly involved in peptidyltransferase activity (13). This finding was supported by experiments using affinity labelled Phe-tRNA (70) or specific antibodies against L11 (80). However, L11 can be removed to a large extent from the ribosome, without affecting the peptidyltransferase activity (82,83). Furthermore, a thiostrepton-resistant mutant of *B. subtilis* lacking a protein corresponding to L11 has been described (84). The experiments performed with the 1.3c core now demonstrated that none of the proteins L6, L10, L11 and

L16, when added to the core particle one at a time in equimolar amounts was able to restore peptidyltransferase activity. However, all four proteins together gave excellent activity, and L16 plus two out of the remaining three proteins were found to be sufficient for activity. Thus, L16 is responsible for the activity, and at least two of the other proteins are essential for the assembly of L16. Clearly, the earlier reconstitution experiments had measured the L11- (and L10-) dependent binding of L16.

Surprisingly, the dependence of the L16 binding on L6, L10 and L11 can be bypassed by the addition of a five- to ten-fold excess of L16 to the core. However, L16 is loosely bound under these conditions, and can be easily washed off the core by sucrose gradient centrifugation. In contrast, when the reconstitution is performed with equimolar amounts of all four proteins, L16 becomes tightly bound to the core. These results also confirm experiments in which an excess of L16 alone was able to restore activity to an inactive LiCl core (15,36).

The next experiments were performed with 4.0c cores containing about ten proteins, in the presence of all the other proteins except one (single omission tests). L16 and L18 were found to be essential for reconstitution of the peptidyltransferase activity, and omission of L25 and L27 strongly affected the activity. The ultimate experiment was a total reconstitution using 23S and 5S RNA together with all the proteins in a highly purified state, but with a single protein omitted. The results are presented in Table 4: Seven proteins and 23S RNA are essential for the reconstitution of the peptidyltransferase center. L20 and L24 are in parenthesis, since they are early assembly proteins (see above) and therefore out of the question as candidates for the peptidyltransferase. Thus, the peptidyltransferase must be one of the remaining five proteins (L2, L3, L15, L16 and L18). Seven further proteins and also 5S RNA strongly affect the peptidyltransferase activity, whereas L6 and the remaining proteins show little or no effect.

A comparison with the components previously discussed (Table 3) reveals that four proteins of the "essential" and "highly stimulating" groups (Table 4) have not up to now been

TABLE 4. Importance of ribosomal components for reconstitution of the peptidyltransferase center, determined by single protein omission tests. The proteins are divided into groups, according to the percentage reduction of activity by omission of the respective component.

Essential (90-100%)	Highly stimulating (60-80%)	Moderately stimulating (∿10%)	No effect
L2	L4	L6	L1
L3	(L6,L10,L11)[c]	L10	L5
L15	L13	L11	L7/L12
L16	L14[a]	L22	L9
L18	L17	L30	L19
(L20)[b]	L21		L23
(L24)[b]	L25		L28
23S RNA	L27		L29
	5S RNA		L31
			L32
			L33

[a]*The L14 preparation used was not highly purified, and its assignment is questionable.* [b]*(L20) and (L24), these proteins are essential for the early assembly but are not involved in peptidyltransferase activity.* [c]*(L6, L10, L11), any two of these proteins are essential for the binding of L16, if this protein is not present in 5- to 10-fold molar excess over the other ribosomal components during reconstitution.*

identified, (namely L3, L17, L21 and L25), whereas five of the previous group (L6, L11, L23, L24 and L32/L33) are not directly involved in peptidyltransferase activity. It is striking that most of the proteins important for reconstitution of the peptidyltransferase activity are either present in the "L15 domain" (L15, L16, L18, (L6, L10, L11), L25, L27) or else directly support the binding of L15 (L3, L4).

Comparison of *in vivo* and *in vitro* Assembly

The above outlined assembly process indicated that important principles are common for both *in vivo* and *in vitro* assembly. On the other hand, the striking differences between the *in vivo* and *in vitro* process cannot be overlooked. Above all, the incubation requirements for the *in vitro* assembly (20 min 44°C → 50 min/50°C) differ drastically from the half-time of ribosomal biosynthesis which is in the range of a few minutes at 37°C. The lower activation energy required for the *in vivo* assembly must be attributable to one or more of the specific features of the *in vivo* assembly.

1. The "assembly gradient" follows the RNA synthesis towards the 3'-end (29).
2. There is an interdependence between the 30S and 50S assembly, indicated by assembly mutants with an altered component in one subunit but with assembly defects in both subunits (see ref. 3 for review).
3. The RNA molecules of the precursor particles are longer than those in the mature particle (3).
4. The 16S and 23S RNA are methylated during the assembly process (3).
5. Some proteins are modified (methylated) during the assembly (85,86). Inhibition of this methylation *in vivo* leads to a severe assembly defect, predominantly in the 50S subunit (87).
6. Specific assembly proteins may be found on the precursors but not on the mature subunits. Such proteins have been identified in the biosynthesis of eukaryotic cytoplasmic ribosomes (for review see ref. 88). Evidence for the existence of corresponding proteins in prokaryotic species has been described for *Bacillus subtilis* (89) and is consistent with the results of a genetic approach (90).

The relative importance of all those individual points for lowering the activation energy *in vivo* remains to be elucidated.

CONCLUSIONS

At least three principles govern the selection of a protein's primary structure: The folding process, the maintenance of the tertiary structure and the function of the active center. Similar principles but on a higher level of organization can be expected to be valid in the ribosome. Here we have to replace "folding process" by "assembly" and "tertiary structure" by "quaternary structure". The large number of ribosomal components allows their specialization. Thus, we can distinguish mere assembly proteins (L20, L24), while other proteins may be responsible for stabilizing the quaternary structure, and again others for the functions, although overlapping may occur. For example, L20 and L24 also stabilize the $RI^{*}_{50}(1)$ structure (see above). The diversification of the ribosomal components with respect to different tasks concerning assembly, structure and function is reflected in the assembly domains. The "L20-domain" forms a core or nucleus at the beginning of the assembly process, and many of the components of this domain fullfil assembly tasks (at least L20 and L24). In contrast, the "L15-domain" is mainly devoted to ribosomal function, at least as far as the peptidyltransferase activity is concerned.

As mentioned in the INTRODUCTION, the assembly must be a sequential process as a result of the many constituents involved. The question is how rigid must this sequential process be? For example, the proteins essential for the $RI^{*}_{50}(1)$ particle formation all (with the exception of L13) belong to the set of RNA-binding proteins. Do they in fact bind independently to the 23S RNA during the early assembly, or does an "assembly starting protein" exist which drastically enhances the binding constant for the following protein(s) in the course of the assembly, thus selecting the RNA molecule used for each individual ribosomal biosynthesis? This question can be analyzed quantitatively. Evaluation of a series of reconstitutions with a constant amount of TP50 and increasing amounts of RNA provided strong evidence that only one protein (or protein complex) starts the assembly *in vitro* (Nowotny and Nierhaus, unpublished observation). It is tempting to assume that the starting protein is one of the early assembly group essential for the formation of the $RI^{*}_{50}(1)$ particle. However, alternative possibilities can be imagined.

In spite of the many details of the 50S assembly described above, we still have only a vague idea of the mechanisms of this complicated process. One protein seems to start the assembly by binding near the 5'-end of the 23S RNA, thereby drastically increasing the affinities for the members of the early assembly group, which also bind near the 5'-end, with the exception of L3. This assembly core or nucleus adopts a conformation which facilitates the binding of further components, following the RNA synthesis towards the 3'-end (as far as the formation of $RI^{*}_{50}(1)$). Protein complexes formed before assembly may enhance the cooperativity of this process. Again, it is possible that a single protein (similar to the "starter protein") induces a conformational change, generating binding sites for the late assembly group, (and giving rise to $RI_{50}(2)$). A final conformational change activates the particle, forming the 50S subunit.

ACKNOWLEDGEMENTS

I thank H. G. Wittmann and R. Brimacombe for criticisms and discussions and J. Belart for preparation of the manuscript.

REFERENCES

1. Hardy, S.J.S. (1975) Mol. Gen. Genet. 140, 253-274.
2. Subramanian, A.R. (1975) J. Mol. Biol. 95, 1-8.
3. Schlessinger, D. (1974) Ribosomes, pp. 393-416, Cold Spring Harbor Laboratory, New York.
4. Nomura, M., and Held, W.A. (1974) Ribosomes, pp. 193-224, Cold Spring Harbor Laboratory, New York.
5. Traub, P., and Nomura, M. (1969) J. Mol. Biol. 40, 391-413.
6. Staehelin, T., and Meselson, M. (1966) J. Mol. Biol. 16, 245-249.
7. Hosokawa, K., Fujimura, R.K., and Nomura, M. (1966) Proc. Nat. Acad. Sci. USA 55, 198-204.
8. Staehelin, T., Maglott, D., and Monro, R.E. (1969) Cold Spring Harbor Symp. Quant. Biol. 34, 39-48.
9. Hamel, E., Koka, M., and Nakamoto, T. (1972) J. Biol. Chem. 247, 805-814.
10. Atsmon, A., Spitnik-Elson, P., and Elson, D. (1969) J. Mol. Biol. 45, 125-135.
11. Homann, H.E., and Nierhaus (1971) Eur. J. Biochem. 20, 249-257.
12. Möller, W. (1974) Ribosomes, pp. 711-731, Cold Spring Harbor Laboratory, New York.
13. Nierhaus, K.H., and Montejo, V. (1973) Proc. Nat. Acad. Sci. USA 70, 1931-1935.
14. Dietrich, S., Schrandt, I., and Nierhaus, K.H. (1974) FEBS Lett. 47, 136-139.
15. Moore, V.G., Atchison, R.E., Thomas, G., Moran, M., and Noller, H.F. (1975) Proc. Nat. Acad. Sci. USA 12, 844-848.
16. Nierhaus, D., and Nierhaus, K.H. (1973) Proc. Nat. Acad. Sci. USA 70, 2224-2228.
17. de Béthune, M.-P., and Nierhaus, K.H. (1978) Eur. J. Biochem. 86, 187-191.
18. Highland, J.H., and Howard, G.A. (1975) J. Biol. Chem. 250, 831-834.
19. Nierhaus, K.H., and Dohme, F. (1974) Proc. Nat. Acad. Sci. USA 71, 4713-4717.
20. Nomura, M., and Erdmann, V.A. (1970) Nature 228, 744-748.
21. Higo, K., Held, W., Kahan, L., and Nomura, M. (1973) Proc. Nat. Acad. Sci. USA 70, 944-948.
22. Isono, K., Isono, S., Stöffler, G., Visentin, L.P., Yaguchi, M., and Matheson, A.T. (1973) Mol. Gen. Genet. 127, 191-195.

23. Fahnestock, S., Held, W.A., and Nomura, M. (1972) The First John Innes Symposium on Generation of Subcellular Structures, pp. 179-217, North-Holland, Amsterdam.

24. Dohme, F., and Nierhaus, K.H. (1976) J. Mol. Biol. 107, 585-599.

25. Amils, R., Matthews, E.A., and Cantor, C.R. (1978) Nucl. Ac. Res. 5, 2455-2470.

26. Nierhaus, K.H., and Dohme, F. (1979) Meth. Enzymol. 59, 443-449.

27. Sieber, G., and Nierhaus, K.H. (1978) Biochemistry 17, 3505-3511.

28. Sieber, G., Tesche, B., and Nierhaus, K.H. (1979), manuscript submitted.

29. Spillmann, S., Dohme, F., and Nierhaus, K.H. (1977) J. Mol. Biol. 115, 513-523.

30. Nierhaus, K.H., Bordasch, K., and Homann, H.E. (1973) J. Mol. Biol. 74, 587-597.

31. Pichon, J., Marvaldi, J., and Marchis-Mouren, G. (1975) J. Mol. Biol. 96, 125-137.

32. Nierhaus, K.H., Spillmann, S., and Dohme, F. (1977) Translation of Natural and Synthetic Polynucleotides, pp. 349-355, Publishers of Poznań Agricultural University, Poznań.

33. Spillmann, S., and Nierhaus, K.H. (1978) J. Biol. Chem. 253, 7047-7050.

34. Cabezón, T., Herzog, A., Petre, J., Yaguchi, M., and Bollen, A. (1977) J. Mol. Biol. 116, 361-374.

35. Nowotny, V., and Nierhaus, K.H. (1979), manuscript submitted.

36. Teraoka, H., and Nierhaus, K.H. (1978) J. Mol. Biol. 126, 185-193.

37. Pongs, O., Bald, R., and Erdmann, V.A. (1973) Proc. Nat. Acad. Sci. USA 70, 2229-2233.

38. Teraoka, H., and Nierhaus, K.H. (1978) FEBS Lett. 88, 223-226.

39. Dohme, F., and Nierhaus, K.H. (1976) Proc. Nat. Acad. Sci. USA 73, 2221-2225.

40. Erdmann, V.A. (1976) Progress in Nucleic Acid Research and Molecular Biology, vol. 18, pp. 45-90, Academic Press, New York.

41. Ofengand, J. (1978) Molecular Mechanisms of Protein Biosynthesis, pp. 7-79, Academic Press, New York.

42. Wittmann, H.G. (1974) Ribosomes, pp. 93-114, Cold Spring Harbor Laboratory, New York.

43. Dijk, J., and Littlechild (1979) Meth. Enzymol. 59, 481-502.

44. Wystup, G., Teraoka, W., Schulze, H., Hampl, H., and Nierhaus, K.H. (1979) manuscript submitted.

45. Kenny, J.W., and Traut, R.R. (1979) J. Mol. Biol. 127, 243-263.

46. Expert-Bezançon, A., Barritault, D., Milet, M., and Hayes, D.H. (1976) J. Mol. Biol. 108, 781-787.

47. Dijk, J., and Littlechild, J. (1979) Meth. Enzymol. LIX, Part G, 481-502.

48. Pettersson, I., Hardy, S.J.S., and Liljas, A. (1976) FEBS Lett. 64, 135-138.

49. Newton, I., and Brimacombe, R. (1974) Eur. J. Biochem. 48, 513-518.

50. Roth, H.E., and Nierhaus, K.H. (1975) J. Mol. Biol. 94, 111-121.

51. Zimmermann, R.A. (1974) Ribosomes, pp. 225-269, Cold Spring Harbor Laboratory, New York.

52. Littlechild, J., Dijk, J., and Garrett, R.A. (1977) FEBS Lett. 74, 292-294.

53. Marquardt, O., Roth, H.E., Wystup, G., and Nierhaus, K.H. (1979) manuscript submitted.

54. Roth, H.E., and Nierhaus, K.H. (1979) manuscript submitted.

55. Röhl, R., and Nierhaus, K.H. (1979) unpublished observation.

56. Spierer, P., and Zimmermann, R.A. (1976) J. Mol. Biol. 103, 647-653.

57. Spierer, P., Wang, C.-C., Marsh, T.L., and Zimmermann (1979) Nucl. Ac. Res. 6, 1669-1682.

58. Mizushima, S., and Nomura, M. (1970) Nature 226, 1214-1218.

59. Kurland, C.G. (1977) Molecular Mechanism of Protein Biosynthesis, pp. 81-116, Academic Press, New York.

60. Gray, P.N., Garrett, R.A., Stöffler, G., and Monier, R. (1972) Eur. J. Biochem. 28, 412-421.

61. Horne, J., and Erdmann, V.A. (1972) Mol. Gen. Genet. 119, 337-344.

62. Möller, W. (1974) Ribosomes, pp. 711-731, Cold Spring Harbor Laboratory, New York.

63. Monro, R.E. (1967) J. Mol. Biol. 26, 147-151.

64. Hampl, H., Schulze, H., and Nierhaus, K.H., manuscript in preparation.

65. Sonnenberg, N., Wilchek, M., and Zamir, A. (1973) Proc. Nat. Acad. Sci. USA 70, 1423-1426.

66. Oen, H., Pellegrini, M., Eilat, D., and Cantor, C.R. (1973) Proc. Nat. Acad. Sci. USA 70, 2799-2803.

67. Sopori, M., Pellegrini, M., Lengyel, P., and Cantor, C.R. (1974) Biochemistry 13, 5432-5439.

68. Czernilofsky, A.P., Collatz, E.E., Stöffler, G., and Küchler, E. (1974) Proc. Nat. Acad. Sci. USA 71, 230-234.

69. Pongs, O. (1974) Acta Biol. Med. Germ. 33, 629-633.

70. Hsiung, N., Reines, S.A., and Cantor, C.R. (1974) J. Mol. Biol. 88, 841-855.

71. Pongs, O., Bald, R., and Erdmann, V.A. (1973) FEBS Lett. 35, 137-140.

72. Pellegrini, M. (1973) Ph.D. Dissertation, Columbia University, New York.

73. Cooperman, B.S., Jaynes, E.N., Brunswick, D.J., and Luddy, M.A. (1975) Proc. Nat. Acad. Sci. USA 72, 2974-2978.

74. Eilat, D., Pellegrini, M., Oen, H., Lapidot, Y., and Cantor, C.R. (1974) J. Mol. Biol. 88, 831-840.

75. Barta, A., Küchler, E., Branlant, C., Sriwidada, J., Krol, A., and Ebel, J.P. (1975) FEBS Lett. 56, 170-174.

76. Fahnestock, S.R. (1975) Biochemistry 14, 5321-5327.

77. Dohme, F., and Fahnestock, S. (1979) J. Mol. Biol. 129, 63-81.

78. Erdmann, V.A., Fahnestock, S., Higo, K., and Nomura, M. (1971) Proc. Nat. Acad. Sci. USA 68, 2932-2936.

79. Wittmann, H.G., Stöffler, G., Apirion, D., Rosen, L., Tanaba, K., Tamaki, M., Dekio, S., Otaka, E., and Osawa, S. (1973) Mol. Gen. Genet. 127, 175-189.

80. Tate, W.P., Caskey, C.T., and Stöffler, G. (1975) J. Mol. Biol. 93, 375-389.

81. Ballesta, J.P.G., Montejo, V., Hernandez, F., and Vâzquez, D. (1974) Eur. J. Biochem. 42, 167-175.

82. Ballesta, J.P.G., and Vâzquez, D. (1974) FEBS Lett. 48, 266-270.

83. Howard, G.A., and Gordon, J. (1974) FEBS Lett. 48, 271-274.

84. Wienen, B., Ehrlich, R., Geisser, M., Stöffler, G., Smith, I., Weiss, D., Vince, R., and Pestka, S. (1979) J. Biol. Chem., in press.

85. Alix, J.H., and Hayes, D. (1974) J. Mol. Biol. 86, 139-159.

86. Chang, C.N., and Chang, F.N. (1975) Biochemistry 14, 468-477.

87. Alix, J.M., Hayes, D., and Nierhaus, K.H. (1979) J. Mol. Biol. 127, 375-395.

88. Hadjielov, A.A., and Nikolaev, N. (1976) Progr. Biophys. Mol. Biol. 31, 95-144.

89. Guha, S., Roth, H.E., and Nierhaus, K.H. (1975) Mol. Gen. Genet. 138, 299-307.

90. Bryant, R.E., and Sypherd, P.S. (1973) J. Bacteriol. 117, 1082-1092.

PART 2: EVOLUTION

COMPARATIVE STUDIES ON THE STRUCTURE OF RIBOSOMAL PROTEINS, WITH EMPHASIS ON THE ALANINE-RICH, ACIDIC RIBOSOMAL, 'A' PROTEIN

ALASTAIR T. MATHESON

Department of Biochemistry and Microbiology, University of Victoria, Victoria, British Columbia, Canada, V8W 2Y2

WIM MÖLLER and R. AMONS

Laboratory of Physiological Chemistry, Sylvius Laboratories, State University of Leiden, Wassenaarseweg 72, Leiden, The Netherlands

MAKOTO YAGUCHI

Division of Biological Sciences, National Research Council of Canada, Ottawa, Ontario, Canada, KlA OR6

INTRODUCTION

Since the first sequences of ribosomal proteins (r-proteins) were published in 1972-73 (1-4), considerable data has become available on the primary structure of a large number of r-proteins from a wide range of organisms. In this chapter we would like to review the comparative biochemistry of certain r-proteins from both the large and small subunits of the ribosome. Such comparative studies may give an insight into the evolution of the translational apparatus and shed some light on possible mechanisms of the evolution of living cells. In addition, a search for conserved regions in these r-protein molecules may indicate which part of the structure is most essential for function and, therefore, less amenable to change in the face of forces which cause alterations in a protein molecule during evolution.

We will mainly confine our discussion to the structure of the r-proteins from sources other than *Escherichia coli*, since a comprehensive review on *E. coli* r-proteins is presented elsewhere in this book (H.G. Wittmann, J. Littlechild, and B. Wittmann-Liebold). In particular, we will concentrate on the acidic or 'A' protein (equivalent to L7/L12 in *E. coli*), since extensive data is now available on this protein. Additional information on the evolution of r-proteins will be found in the chapter by S. Osawa and H. Hori.

In this chapter we will use the phylogenetic divisions proposed by Woese and Fox (5), since the sequence data on the 'A' protein offers support for this classification. Woese and his colleagues (6) have proposed that there are at present three lines of descent: the eubacteria, which include blue-green bacteria as well as gram-positive and gram-negative bacteria; the urcaryotes, which include animal, plant, fungal and slime mold cells; and the archaebacteria, which include the extreme halophiles, the methanogens, *Sulfolobus* and *Thermoplasma*.

GENERAL PROPERTIES OF THE RIBOSOME

The general properties of the 70S ribosomes from eubacteria and the 80S ribosomes from urcaryotes are well characterized (see ref. 7). Much less information, however, is available on the ribosomes from archaebacteria. The most thoroughly studied ribosomes from this group are those from the extreme halophile *Halobacterium cutirubrum*. These ribosomes show many of the properties of other bacterial ribosomes. The basic ribosomal unit is a 70S structure with 50S and 30S subunits, which contain 60% RNA and 40% protein (8). Like the eubacterial r-RNA, the sedimentation values for the archaebacterial r-RNA are 5S, 16S, and 23S (no 5.8S RNA is present). However, the primary structure of the r-RNA from both the extreme halophile and the methanogens is quite distinct from the equivalent r-RNA from eubacterial sources (9-11). The r-proteins from both the extreme halophile and the methanogens appear to be much more acidic than those from eubacteria and urcaryotes (12-14, A. Böck, personal communication). The number of r-proteins in the ribosomes from the extreme halophile appears to be equivalent to the number found in eubacteria (15), although a more thorough study is required before this can be accepted as certain for archaebacteria.

COMPARATIVE STUDIES ON RIBOSOMAL PROTEINS FROM THE SMALL RIBOSOMAL SUBUNIT

Eubacteria

In addition to the extensive sequence data available on all the *E. coli* 30S r-proteins (see chapter by Wittmann, Littlechild, and Wittmann-Liebold) and on 20 of the 30S r-proteins from *Bacillus stearothermophilus* (see ref. 16), more sequence data is now available on the equivalent proteins from *Bacillus subtilis* (Higo and Osawa, personal communication). Table 1 summarizes the sequence data now available on the 30S r-proteins from *B. stearothermophilus* and *B. subtilis*. For an extensive discussion on this data the reader is referred to the review by Yaguchi *et al.* (16) and the chapter by Osawa and Hori in this volume.

TABLE 1. A comparison of the N-terminal sequence of the 30S r-proteins from *E. coli* (EC), *B. stearothermophilus* (BT), and *B. subtilis* (BS).

		10 20
S2	EC	ATVSMRDMLKAGVHFGHQTR
	BT	SVISMKQLLESGVHFGXZTR
		30
S3	EC	GQKVHPNGIRLGIVKPGNSTGFANTKEFADNL
	BT	GQKVNPIGLRIGIIRNGZSRWYSZ-KXYABLL
		40 50
S4	EC	ARYLGPKLKLSRREGTDLFLKSGVRAIDTKCKIEQAPGQHGARKPR-LSDYGVQLRE
	BS	ARYTGPSWKLSRRLGISL---SGTGKE
	BT	ARYTGPMWKISRRLGISL---SGTGKELQ--KRPYPPGQHGPGQRRKLSEYGLQLXE
S5	EC	-AHIEKQAGELQEKLIAVNRVSKTVKGGRIFSFTALTVV
	BS	MRXIDPXKLELEERLVTVNRVAKVVKGGR-FRFAAL-VV
	BT	MRRIDPNKLELEERVVAVNRVAKVVKGGR
S6	EC	MRHYEIVFMVHPDQSEQVPGMI-ERYTAAI
	BS	MRKYEVMYIIRPNIDE
	BT	MRKYEIMYIIRPDMDDQARQSLVERFDNVL
S7	EC	PRRRVIGQRKILPDPKFGSEL
	BS	PRKGPVAKRDVLPBGIYB
	BT	PRRGPVAKRDVLPDPIYNSKL
		30 40
S8	EC	SMQDPIADMLTRIRNGQAANKAAVTMPSSKLKVAIANVLKEEGFIEDF
	BS	VMTDPIADMLTRIRNADNVRHEKLEIPABK
	BT	VMTDPIADMLTRIRNANMVRHEKLEVPASKIKREIAEILKREGFIRDY
S9	EC	AENQYYGTGRRKSSAARVFIKPGNGKIVINQRSLEQYFGRETARM
	BT	AQVQYYGTGRRKSSVARVRLVPGDGRIIVNKQDIREYIPTEALIZ
S10	EC	MQNQRIRIRLKAFDHRLIDQATAEIVETAKRTGAQVRGPI
	BS	-AKQKIRIRLKAYDHRILBZSAZKIVZTA
	BT	-AKEKIRIRLKAYDHRILDQSAEKIVETAKRSGAKVXGXI
S11	EC	AKAPIRARKRVRKQVSDGVAHIHASFNNTI
	BT	ARRTNTRKRRVRKNIDTGIAHIRSTFNNTI
S12	EC	ATVNQLVRKPRARKVAKSNVPALEACPQKRGVGTRVY
	BS	PTINQLIRKGREHKVENSK
	BT	PTINQLVRKGREHKVVKSKSPALNKCYDSFKKZZTBV
S13	EC	ARIAGINIPDHKHAVIALTSIYGVGKTRSKAILAAAGIAED
	BS	ARIAGVDIPRDKRVVISLTYVFGIGRTRAQQVLKEAGVSED
	BT	ARIAGVDIPRDKRVVISLTYIYGIGK
S14	EC	AKQSMKAREVKRVALADKYFAKRAELKAIISD
	BT	AKKSMIAKQKRTPKFKVRAYTRTERRGKRHSV
S15	EC	SLSTEATAKIVSEFGRDANDTGSTEVQV
	BS	AITQERKNQLINEFKVHEBB
	BT	ALTQERKBEIIEQFHVRENDTGSPZVZV

		10 20 30 40
S16	EC	MVTIRLARHGAKKRPFYQVVVADSRNARNGRFIERVGFFNPIASEK
	BS	AVKIRLKRMGAKKSPFYRIVVADSRSPRDGRFIETVGTYNPVAKPA
	BT	AVKIRLKRMGAKKKPFYRIVVADSRSPRBGRFIE
S17	EC	TDKI-RTL-QGRVVSDKMEKSIVVAIERFVKHPIYGKFI
	BS	SERNQRKVYQGRVVSDKMDKTIGVVVETYKKHTLYGKRV
	BT	SERNQRKVYVGRVVSDKMDKTIVVLVZXY
S18	EC	A-R-YFR--RRKFCRFTAQGVQEIDYKDIA
	BT	AGRKGGRGKRRKVCYFTANNITHIDYKDVD
S19	EC	PRSLKKGPFIDLHLLKKVEKAVESGDKKPLRTWS
	BS	ARSLKKGPFVBBH
	BT	GRSLKKGPFCDEHLMKKIEKLNATGQKZVIKTXS
S20	EC	ANIKSAKKRAIQSEKARKHNASRRS
	BS	ANIKNAIKRVHNAEIKRAHNA
	BT	ANIKSAIKRAKTSEKRRAHBASRKS
S21	EC	PVIKVRGNEPFDVALRRF
	BT	PM------EVFEQALKNS

The one letter code for the amino acids, used throughout this chapter, is as follows: A (ala), C (cys), D (asp), E (glu), F (phe), G (gly), H (his), I (ile), K (lys), L (leu), M (met), N (asn), P (pro), Q (gln), R (arg), S (ser), T (thr), V (val), W (trp), Y (tyr), B (asx), Z (glx). References to the sequence data from *E. coli* are contained in the chapter by Wittmann, Littlechild, and Wittmann-Liebold. The references for *B. stearothermophilus* (3, 4, 16, 18, 19) and *B. subtilis* (16-18) are as indicated. S2, S3, S4, etc. refers to the *E. coli* protein.

Archaebacteria

Table 2 summarizes the N-terminal sequence data available on the 30S r-proteins from the extreme halophile *H. cutirubrum*. These proteins are almost all acidic (15) and in most cases this is reflected in the amino acid composition of the N-terminal regions shown in Table 2.

TABLE 2. Amino-terminal sequence of ten ribosomal proteins from the 30S subunits of *Halobacterium cutirubrum* (preliminary unpublished data from M. Yaguchi, L.P. Visentin, A.T. Matheson, C. Roy, and A.R. Strøm).

	1 10 20 30
S4	S D- E- L E- F I E- E- G L Q R+ S Q I D- E- F F A E- Z(-) L A R+ A G Y Z(-)
S5	T N H+ Q K+ R+ L S V P K+ S Z(-) P V Z(-) R+ K+ T E+ T F T S L S G S G P
S9	A L P G E- N T K+ F Y E- T P N H+ P Y Q G Z(-) R+ I A Z(-) Z(-) S B(-) L L X
S11	A R+ M H+ T R+ R+ R+ G S S G S D- R+ P T A D- E- P P Z(-) Z(-) S B(-) V B(-) Z(-) B(-)
S13	A T L Y D- A P V D- E- L I D- E- L A A B(-) L Z(-) T R+ L B(-) Z(-) P P G A B(-)

```
      1                 10                  20                  30
        -   -       -     - +     -   - -     + -         +
S14   P V Y V D Y D V P A D L Q E R A L E S E E V A R D T G S V K
              -   -         - - -       +           +
S16   A I G L N V T E P E G T C S D E D C P F H G N L S V R G Q V
  -   -         - -(-)      + +      (-)  +     +   -
S17   M E I E I L G Q E D B P L L H R T N V Z F K I V H N D A T P
      -       - -    (-)  - -   -       +
S20   T A N D P L S D E L S Z I D D A E S V G H L T H T
  -   -   -   -  (-)   (-)-      (-)
S5A   S E N E S D T E S Z L L Z B S P P Z
```

The S numbers indicate the position on 2D-gels (15), and do not indicate equivalency to r-proteins from other sources.

In a few proteins, however, the N-terminal region is extremely basic, as is the case with protein S11, where 5 arginines are located in the first 15 residues of this protein.

When the amino acid sequence of the proteins in Table 2 is compared to the amino acid sequence data from the eubacteria (Table 1), no obvious sequence homology between the two groups is apparent. Protein S11 from *H. cutirubrum* may be equivalent to protein S11 from *E. coli* and *B. stearothermophilus* since all three proteins have a cluster of basic residues in the N-terminal region (Table 3).

TABLE 3. Possible sequence homology between the archaebacterial S11 protein from *H. cutirubrum* (HC) and the equivalent eubacterial proteins from *E. coli* (EC) and *B. stearothermophilus* (BT).

```
                            10                  20
            +       +   + + +   + +       -       +   +
EC  S11   A*K A P I R A R K R V R K Q V S D G V A H I H A
          _             _   _                   _
            +   +       + + +           - +       - -
HC  S11   A R M H - T - R R R G S S G S D R P T A D E P P
          ___           _____           _       _
            + +       + + + +   + +     -         +   +
BT  S11   A R R T N T R K R R V R K N I D T G I A H I R S
          ___       _     ___           _       _
```

The underlined amino acids indicate regions of homology between the HC protein and the putative, equivalent protein from EC and/or BT. A* = modified A.

Urcaryotes

The amino acid sequences of the N-terminal region of three rat liver r-proteins S4, S6, and S8 have been determined (20) and are shown in Table 4. No obvious sequence homologies are evident between these proteins and the 30S r-proteins from eubacteria (Table 1) or archaebacteria (Table 2).

TABLE 4. The N-terminal amino acid sequence of S4, S6, and S8 from the 40S subunit of rat liver (20).

```
        1                 10                  20
          +     + + +   + +         + +         - +
S4      A R G P K K H L K(R)V A A P K(H)W M L D K L L G F V A(P)(T)
        30
          +         - +
        P(K)L G(P)(D)K L
        1                 10                  20
          +                       +     -       - - - + +   +       -
S6      M K L N I(S)F P A T G(S)Q K L L E V D D E(R)K L(R)X F Y(E)
        30
        +
        K X M A
        1                 10                  20
              + -         + + + +         + + +
S8      G I(S)R D N W H K R(R)K(T)G G K R K P Y
```

COMPARATIVE STUDIES ON RIBOSOMAL PROTEINS FROM THE LARGE RIBOSOMAL SUBUNIT

The ribosomal 'A' protein and the 5S RNA-binding proteins, which are part of the large ribosomal subunit, will be discussed later in this chapter.

Eubacteria

Sequence data on eubacterial 50S ribosomal proteins are available mainly from *E. coli* and are discussed elsewhere in this book (see Wittmann, Littlechild, and Wittmann-Liebold).

Archaebacteria

No sequence data is available on the 50S r-proteins from archaebacteria except for the ribosomal 'A' protein and the 5S RNA-binding proteins which will be discussed later. Almost all the r-proteins from the 50S subunit of *H. cutirubrum* have been purified and the amino acid composition determined (Matheson *et al.*, unpublished data). The N-terminal sequence analysis has been started on these proteins.

Urcaryotes

The sequence data available on the r-proteins from the 60S ribosomal subunit of urcaryotes are shown in Table 5. Protein 44 from yeast is the first urcaryotic r-protein to be completely sequenced. This protein may have homologous regions with *E. coli* L14 and rat liver L37a (21).

The seven rat liver r-proteins shown in Table 5 all have a unique sequence. A comparison of the rat liver r-protein sequences with those from *S. cerevisiae* indicate that RL - L37 is homologous to Y55 from yeast ribosomes and that RL - L7a and/or RL - L37a may be related to Y44 from yeast (20). There are also some similarities in the sequence of rat liver and *E. coli* r-proteins, but the significance of these similarities remains to be determined.

TABLE 5. Amino acid sequence data on the r-proteins from the 60S ribosomal subunit of urcaryotes.

S. cerevisiae

SC - Y44 (21)

```
                  10                  20
        +   + +       +   +     + + +     + +
V N V P K T R K T Y C K G K T C R K H T Q H K V T

        30                  40                  50
    +     +               + + +   - + +
Q Y K A G K A S L F A Q G R K*R Y D R K Q S G F G

                  60                  70
      +       + + +   +     + +       +   -     +
G Q T K*P V F H K K A K T T K K V V L R L E C V K

        80                  90                  100
  +   +           + + +   -       - + +   +
C K T R A Q L T L K R H F E L G G E K K Q K G Q A

L Q F
```

SC - Y55 (20)

```
                  10                  20
  +             + + +   +   +         +     + +
G K G T P S F G K R H N K S H T L C N R C G R R S

        30
  +     +
F H V Q K
```

Rat liver (20)

RL - L6

```
                  10                    20
    - +   - +   - + + -     +       + +
A G E K A E K P D K K E (Q) K P A A K K A
```

RL - L7a

```
                  10                  20
  +   + +   +   + +                 + +   -   + +
P K G K K A K G K K V A P A P A V V K K Q E A K K

            30
                     +         +
V (V) N (P) L/T F(Q) K (S) (P) K N F G
```

RL - L18

```
                    10
    -    +  +   + -     +        +  +     +
G V D I (R) H N K E (S) K V (S) (R) K L P K L
```

RL - L27

```
                  10                  20
  +     +     +           +             +
G K F M K L G K V V L V L K G X Y X G X K A V I V

+
K N
```

RL - L30

```
                      10                              20
      + +     + +         -               +
V A A K K L/T K K (S) L/T E (S) I (N) (S) R L/T Q L/T V

                          30
  +          +              +
M K (S) (G) (K) Y V L/T G Y K Q
```

RL - L37

```
                        10
  +                   + + +   +      +          +
X K (G) T (S) (S) F G K R R N K (T) (H) T L (G) R

20                  30
+ +   +   +         +
R K X K G K G A L Q K
```

```
                               10                   20
               + +   + +           +       +             + +
RL - L37a    A K R T K K V G I V G K Y G T R Y G A (S) L R K

                   + +
             M (V) K K I
                               10
                 + +     +   + +       + +   +
RL - L39     S S H K T F R I K R F L A K K Q K Q
```

K* in SC - Y44 indicates an unidentified amino acid derivative related to mono-methyl-lysine.

COMPARATIVE STUDIES ON THE 5S RNA-BINDING PROTEINS

The 5S RNA-protein complex, present in the large ribosomal subunit, has been extensively studied. The 5S RNA is thought to be involved with the binding of aminoacyl-tRNA to the ribosome (22). It has also been postulated that it may be involved in subunit interaction (23).

In this section we will discuss the structural features of the 5S RNA-binding proteins from *E. coli*, *H. cutirubrum*, and yeast. A more comprehensive review of the evolution of the 5S RNA-complex is reported elsewhere (24).

Eubacteria

The 5S RNA-complex has been isolated from *E. coli* and contains three proteins EC - L5, EC - L18, and EC - L25 (25). These proteins were shown earlier to bind to 5S RNA (see ref. 22). The complete amino acid sequences of EC - L5 (26), EC - L18 (27), and EC - L25 (28, 29) have been reported. Additional information on these proteins will be found in the chapter by Wittmann, Littlechild, and Wittmann-Liebold.

Archaebacteria

The 5S RNA-protein complex from the extreme halophile *H. cutirubrum* has been isolated (30) and its properties extensively studied (31, 32). Two proteins, HC - L13 and HC - L19, are associated with the 5S RNA in the native complex. The N-terminal regions of these proteins have been sequenced (30). Protein HC - L13 shows some sequence homology to EC - L18, while HC - L19 appears to be equivalent to EC - L5 (Table 6).

The N-terminal region of HC - L13 is extremely basic, although the overall protein is quite acidic (30). Over 50% of the total arginine content is located in the first 15% of this protein. EC - L18 is also very basic in this region with 9 out of a total of 13 arginine residues

TABLE 6. Sequence homologies between the 5S RNA-binding proteins of *E. coli* and *H. cutirubrum*.

		20 30
EC	L18	A T R A R R K L Q E L G A T R L V V H R T P R H I Y A
		40 50
		Q V I A P N G S E V L V A
		10 20
HC	L13	A T G P R Y K V P - M R R R R E V - - R T D Y H Q R L
		30
		L L K A V G G P - P L L A
		20 30
EC	L5	V K K L M T E F N Y N S V M E V P R V E K I T L N M G
		40
		V G E A I A
		10 20
HC	L19	S E T D S T D F H - E - - M R E P R I E K V V V H M G
		30
		V G Q G G V

Underlined amino acids indicate regions of sequence homology.

present in the first 33 residues. Additional evidence to suggest that HC - L13 and EC - L18 are equivalent proteins has been provided by both physical and chemical studies on these proteins (32).

Urcaryotes

The 5S RNA-protein complex can be released from the 60S ribosomal subunit of urcaryotes using EDTA (33-37). In urcaryotes, the 5S RNA-complex contains a single protein. This protein has been identified as RL - L5 in rat liver (35, 36), and in yeast as SC - L3 (37). If, however, the 5S RNA-binding proteins in rat liver are identified by affinity chromatography (using bound 5S RNA), RL - L6 and RL - L19 specifically bind to the 5S RNA, while RL - L5 does not bind (36). The reason for two proteins associating with 5S RNA during affinity chromatography, and only a single, quite different protein being present when the 5S RNA complex is released from the ribosome, remains to be resolved.

The 5S RNA-binding protein in yeast, SC - L3, has been partially sequenced (37). Amino acid sequences homologous to regions in EC - L18, EC - L5, and also HC - L13 have been located in SC - L3. Nazar *et al.* (37) have suggested that this single protein in yeast may be equivalent to two or more of the 5S RNA-binding proteins in eubacteria and archaebacteria.

COMPARATIVE STUDIES ON RIBOSOMAL 'A' PROTEINS

The ribosomal 'A' protein (equivalent to L7/L12 in *E. coli*) is an very acidic, ubiquitous protein present in the large ribosomal subunit. It occurs in several copies per ribosome and has unique structural and functional properties. In this section we will discuss mainly the comparative aspects of this protein in eubacteria, urcaryotes and archaebacteria. For reviews on the structure and function of the 'A' protein and its natural environment on the ribosome, the reader is referred to Möller (38), Matheson *et al.* (11), Wool (39), and Leader (40).

RECENT ADVANCES IN THE STRUCTURE AND FUNCTION OF 'A' PROTEINS IN EUBACTERIAL RIBOSOMES

Structure of *E. coli* L7/L12

In solution, the protein L7/L12 exists as a stable dimer (41) with an extremely elongated shape (42, 43). Crystallization of two separate protein domains in L7/L12 from *E. coli* has been reported (44), but information on the three-dimensional structure and mode of alignment of the two chains in the dimer is not yet available.

Recently, two conflicting models for the L7/L12 dimer in solution have been proposed (45, 46). Both models consist of two side-by-side monomer chains positioned in a parallel (46) or antiparallel (45) fashion. The designation "parallel" indicates the amino termini are on one end of the elongated dimer and the carboxyl termini on the other; the "antiparallel" pairing implies that the N-terminal region of one molecule is near the C-terminal end of the other. Direct assignment of the crosslinking residues in the L7/L12 dimer (47) should help to decide which model is correct.

The N-terminal fragment (residues 1-73) of L7/L12 binds strongly to L7/L12-depleted 50S cores, and inhibits EF-G-dependent hydrolysis of GTP (48). Gudkov and Behlke (49) concluded from sedimentation studies that this part of L7/L12 is primarily responsible for the protein's strong dimerization. Sedimentation equilibrium studies give a value closer to a monomer for the fragment 1-73 (Möller, W. and Reinalda, H., unpublished data).

The seryl amino-terminus of *E. coli* L12 has been found to be unreactive to chemical modification and appears, therefore, to be buried (50). On the other hand, earlier work on the equivalent 'A' protein from *Bacillus stearothermophilus* (51), based on the same competitive labelling technique, gave a methionyl amino-terminus which was highly reactive. The significance of the different behaviour of these amino terminal groups is not clear at the present time.

Neighborhood relations of *E. coli* L7/L12

Protein L7/L12 and protein L10 form a stable protein complex (52) which consists of two L7/L12 dimers and an L10 monomer (53-55). Small angle X-ray scattering studies suggest that the shape of the complex corresponds to an ellipsoid having semi-axes of 90 Å, 45 Å, and 6 Å (53). Such a body constitutes an elongated and somewhat flattened structure. In the 50S ribosome, protein L7/L12 is a neighbor of proteins L10 and L11 (56, 57). Functional studies show that elongation-factor-dependent binding requires prior assembly of L7/L12 to the ribosome, and this in turn is dependent on the presence of L10 (58-62). The first evidence that the complex L7/L12 - L10 binds in a 1:1 molar ratio to 23S RNA was given by van Dijk, Littlechild, and Garrett (63). Subsequently, it was shown that protein L11 also binds to 23S RNA in the 5'-one third of the RNA molecule, and that the binding to the RNA of proteins L7/L12 - L10 and L11 is mutually dependent (64). In addition, the chemical modification studies of Pettersson support the view that L10 possesses one binding site for tetrameric L7/L12 and another for 23S RNA (65).

Efforts to map L7/L12 on the ribosome by means of immunoelectron-microscopy have given conflicting results. Tischendorf *et al.* (66) mapped L7/L12 like a ribbon wound around the center protuberance of their symmetrical 50S model. Boublik *et al.* (67) favor a model that positions L7/L12 on the two side crests of the crown in the 50S subunit. On the other hand, the results of Strycharz *et al.* (68) strongly support a model in which the L7/L12 is located at a single region of the large subunit which they named the L7/L12 stalk, and which extends from one side of the central protuberance. The strength of the asymmetric, single L7/L12 stalk model is based on the use of statistical criteria employed to decide which of the two latter models is most likely. By counting the frequency of occurrence of antibody-attachment to either a single site or to two sites, Strycharz *et al.* have concluded that L7/L12 forms one single stalk on the large subunit.

Role of *E. coli* L7/L12 in GTPase activity and the binding of factors to ribosomes

Acharya *et al.* (69) cross-linked EF - G to ribosomes with suberimidates and recovered a protein-complex containing, among other components, EF - G and L7/L12. However, it was not established whether L7/L12 was cross-linked directly to EF - G or via another, uncharacterized protein. Maassen (70) synthesized a heterofunctional cross-linker, 4-[6-formyl-3-azido-phenoxy]butyrimidate, which makes the assignment of neighbor relations unambiguous. It was found that 50S proteins L1, L3, L11, sometimes L7/L12 and the 30S proteins, S3 and S4, can be direct neighbors of EF - G in the ternary complex of 70S ribosomes, EF - G and guanine nucleotides (71). Protein L11 reacts strongly with photoaffinity analogues of GTP and GDP if EF - G is present (72). Cross-linking of EF - Tu to ribosomes also indicates an extended neighborhood containing L1, L5, L7/L12, L15, L20, L30, and L33 (73).

From the cross-linking experiments it would appear that EF - G approaches several ribosomal proteins, but not ribosomal RNA. To what extent this cross-linking pattern reflects a multiprotein attachment site, or a heterogeneity in the quaternary structure of the ribosome related to the functional state of the elongation cycle, remains to be seen. In any case, the proteins that cross-link to EF - G as well as to EF - Tu are part of a functional cluster of proteins which is involved in GTP hydrolysis and map in a central region of the particle (74). Protein L7/L12 has been implicated in the ribosomal-binding site of initiation factor IF - 2, elongation factors EF - G and EF - Tu, and the release factors RF - 1 and RF - 2 (38, 74).

Incubation of L7/L12, together with elongation factor Tu, is reported to lead to significant hydrolysis of GTP (75). Attempts in several laboratories to reproduce these results have not been successful (J.A. Maassen, unpublished data; A. Parmeggiani, private communication).

A tantalizing problem that remains to be solved is the reason for the occurrence of four copies of L7/L12 on the ribosome (76, 78). Four copies of the 'A' protein have also been found in the ribosome from *Bacillus stearothermophilus* (79). However, no more than one dimer of L7/L12, bound to one mole of L7/L12-depleted-50S core particles, is needed to obtain particles as active in EG - G-dependent GTPase activity as intact ribosomes (Table 7).

A similar result was obtained for EF - Tu-uncoupled GTPase (78). By means of a sensitive radioimmunoassay of L7/L12, it was found that four moles of L7/L12 bind maximally to one mole of cores. Intact 50S and 70S

TABLE 7. Reconstitution of 50S P_0 and P_{0-30} cores (59) with split proteins and purified L7. 0.5 nmol of the ribosomal particles were incubated in 0.5 ml 2× concentrated assay buffer for 5 min at 37°C with 41 μg of L7 (3.4 nmol), 60 μg of 50S P split proteins (4.7 nmol of L7/L12) or 180 μg of 50S P_{37} split proteins (2.5 nmol of L7/L12) as indicated in the table (the amounts of L7/L12 in the split fractions were determined by radioimmunoassay). The amount of L7/L12 in the cores was determined by radioimmunoassay. The EF - G-dependent GTPase was measured in parallel experiments. The reaction mixtures contained 25 pmol of ribosomal particles, 50 pmol of 30S ribosomes, 5 eq. of L7, 2 μg of P_0 split protein and 7 μg of P_{37} split protein as indicated. 100% corresponds to 41 nmol of GTP hydrolyzed. Assay buffer: 50 mM Tris HCl (pH 7.4). 10 mM Mg (O Ac)$_2$; 50 mM NH_4Cl and 7 mM 2-mercaptoethanol. Data from Schrier (78).

Experimental conditions	L7/L12 bound	EF - G dependent GTPase
	eq.	%
1. 50S	3.8	100
2. P_0 cores + L7	0.1	7
3. P_0 cores + L7	2.4	93
4. P_0 cores + P_0 split	4.6	97
5. P_{0-30} cores + L7	0.1	6
6. P_{0-30} cores + L7	0.3	18
7. P_{0-30} cores + P_{37} split	4.4	96

ribosomes also contain four molecules of L7/L12 per mole of ribosomes. Furthermore, a comparison of the binding characteristics of intact L7/L12 and the residue 1-73 fragment, which binds strongly to the 50S core, indicates that there may be two binding sites, one for each of the L7/L12 dimers, present on the ribosome (78).

Outlook and speculation

The two-dimer model of L7/L12 on the ribosome would fit a cooperative mechanism of free energy transduction based on subunit-subunit interactions within a tetramer of L7/L12. Such an organization of this part of the particle may impose strong protein-protein interactions which might result from large contact areas between the helical domains of these α helical-rich proteins. Reporter groups on known positions of L7/L12 may tell whether the two dimers of L7/L12 change with respect to their relative position during the elongation step of protein synthesis. Since a dimer of L7/L12 seems sufficient (Table 7) to support hydrolysis of GTP on the ribosome, tetrameric L7/L12 may just strengthen the interaction with the elongation factors and thereby facilitate the transient passage of ingoing and outgoing t-RNA. The finding that 20% methanol releases much of the need for L7/L12 in EF - G- or EF - Tu-dependent GTP hydrolysis (80) may point to the possibility of a direct role of L7/L12 in contributing to a water-free environment for catalysis of GTP hydrolysis in the protein domains between the subunits. The high solubility of L7/L12 in 50% ethanol (59), the large content of alanine in L7/L12 (25 mol %), and the presence of long alanine-rich stretches in peptide "antifreeze" (81) is compatible with such a mechanism. In this context it would be desirable to search for a hydrophobic pocket in the 50S ribosome which is dependent on the absence or presence of L7/L12.

Primary structure of eubacterial 'A' proteins

The complete amino acid sequence of eubacterial 'A' proteins from *E. coli* (77), the moderate halophile NRCC 11227 (11, 82, P. Falkenberg *et al.*, unpublished data), and *Bacillus subtilis* (83) has been reported. In addition, the N-terminal sequence of four other bacterial 'A' proteins is known; the moderate halophile *Vibrio costicola* (82), the anaerobe *Clostridium pasteurianum* (84), the moderate thermophile *B. stearothermophilus* (84, 85), and the psychrophile *Arthrobacter glacialis* (84). As shown in Table 8, there is extensive sequence homology present in the eubacterial 'A' proteins and where changes have occurred, they are usually conservative. These bacterial 'A' proteins form a well defined group called the eubacterial (or procaryotic) group (11). The sequence homology varies between 86% for *C. pasteurianum* and *B. stearothermophilus* to 28% homology between *E. coli* and *A. glacialis* (Table 9). There appears to be distinct differences in the N-terminal amino acid sequence of gram-negative and gram-positive bacteria (84).

When the complete amino acid sequence of *E. coli*, NRCC 11227, and *B. subtilis* 'A' proteins is compared (Table 8), the extensive homology noted in the N-terminal portion of these proteins is evident throughout the entire molecule with 65% homology between *E. coli* and NRCC 11227, 57% between NRCC 11227 and *B. subtilis*, and 54% between *E. coli* and *B. subtilis*.

TABLE 8. The amino acid sequence of eubacterial 'A' proteins from *E. coli* (77), *V. costicola* (82), NRCC 11227 (82, P. Falkenberg *et al.*, unpublished data), *C. pasteurianum* (84), *B. stearothermophilus* (84, 85), *B. subtilis* (83), and *A. glacialis* (84).

(Charge marks printed above residues are shown after the residue: + positive, - negative, (-) in parentheses as printed; * as printed.)

		1									10										20					
Gram negative																										
E. coli		S	I	T	K+	D-	Q	I	I	E-	A	V	A	A	M	S	V	M	D	V	V	E-	L	I	S	A
V. costicola		S	I	T	N	E-	Q	I	L	D-	A	I	A	A	M	S	V	M	Q	V	V	E-	L	I	E-	A
NRCC 11227		A	L	T	Q	E-	D-	I	I	N	A	V	A	E-	M	S	V	M	E-	V	A	E-	L	V	S	A
Gram positive																										
C. pasteur.			M	S	K+	E-	Q	I	I	Q	A	I	K+	G	M	T	V	L	E-	L	N	E-	L	V	K+	S
B. stearo.			M	T	K+	E-	Q	I	I	Q	A	V	K+	N	M	T	V	L	E-	L	N	E-	L	V	K+	A
B. subtilis		A	L	N	I	E-	E-	I	I	A	S	V	K+	E-	A	T	V	L	E-	L	N	D-	L	V	K+	A
A. glacialis	A	K+	L	S	T	E-	D-	L	L	E-	Q	F	K+	G	L	T	L	I	E-	L	S	E-	F	V	K+	A

					30										40										
E. coli	M	E-	E-	K+	F	G	V	S	A	A	A	A	V	A	V	A	A	G	P	V	E-	A	-	-	A
V. costicola	M	E	E	K+	F	G	V	S	A	A	A	A	V												
NRCC 11227	M	E-	E-	K+	F	G	V	S	A	A	A	A	V	V	A	G	P	G	G	G	E-	A	E-	E-	A
C. pasteur.	I	E-	E-	E-	F	G	V	S	A	A	A	P	V												
B. stearo.	I	E-	E-	E-	F	G	V	T	A	A	A	P	V												
B. subtilis	I	E-	E-	E-	F	G	V	T	A	A	A	P	V	A	V	A	G	G	A	A	A	G	G	A	A
A. glacialis	F	E-	E-	T	F	E-	V	G	A	A	A	P	V												

		50										60										70			
E. coli	E-	E-	K+	T	E-	F	D-	V	I	L	K+	A	A	G	A	N	K+	V	A	V	I	K+	A	V	R+
NRCC 11227	E-	E-	Q	T	E-	F	B(-)	L	V	L	T	S	A	G	E-	K+	K+	V	B(-)	V	I	K+	V	V	R+
B. subtilis	E-	E-	-	-	E-	F	D-	L	I	L	A	G	A	G	S	Q	K+	I	K+	V	I	K+	V	V	R+

							80												90						
E. coli	G	A	T	G	L	G	L	K+*	E-	A	K+	D-	L	V	E-	-	S	-	A	P	A	A	L	K+	E-
NRCC 11227	E-	I	T	G	L	G	L	K+*	E-	A	K+	A	A	V	-	B(-)	-	G	A	P	A	T	L	K+	E-
B. subtilis	E-	I	T	G	L	G	L	K+	E-	A	K+	E-	L	V	E-	N	T	P	K+	P	L	E-	V	K+	E-

```
                      100                 110                 120
                      + - -   -     + +     - -       -   -   +
E. coli         G V S K D D A E A L K K A L E E A G A E V E V K
                      +(-)(-) (-)-  +   +   - -           -   +
NRCC 11227      G M S K Z B G B E A K T K L E E A G A S V E L K
                      + - -   - -   +   +   - -           -   +
B. subtilis     G I A K E E A E E L K A K L E E V G A S V E V K
```

K* in residue 81 is methyl lysine.

TABLE 9. Matrix of amino acid sequence homology in N-terminal region (first 38 residues) of eubacterial type 'A' proteins (Table 8).

	EC	VC	NRCC	CP	BT	BS	AG
E. coli	-	79%	74%	47%	55%	39%	28%
V. costicola	79%	-	68%	50%	50%	37%	31%
NRCC 11227	74%	68%	-	53%	58%	55%	38%
C. pasteurianum	47%	50%	53%	-	86%	66%	49%
B. stearothermophilus	55%	50%	58%	86%	-	74%	46%
B. subtilis	39%	37%	55%	66%	74%	-	46%
A. glacialis	28%	31%	38%	49%	46%	46%	-

CHARACTERIZATION OF 'A' PROTEINS IN THE 60S RIBOSOMES FROM URCARYOTES

Review articles closely related to this part of the chapter have been written recently by Wool (39) and Leader (40).

Isolation of acidic phosphoproteins from 60S ribosomes

Martini and Gould (86) first discovered a large subunit protein in reticulocytes which resembled the 'A' protein of *E. coli* with respect to its acidity and molecular weight. This protein (L28) was the only protein strongly phosphorylated in the 60S subunit when monosomes were phosphorylated *in vitro*. Major credit for the discovery of the occurrence *in vivo* of acidic phosphoproteins in the 60S ribosomes should go to Zinker and Warner (87). The presence of phosphorylated acidic r-proteins (A1 and A2) in yeast has also been reported by Osawa's group (18) and Planta's group (88).

Leader and Coia (89) reported a phosphorylated acidic r-protein in the large subunit of Krebs II ascites cells, while three proteins of identical molecular weight but phosphorylated to different extents have been detected in the 60S ribosomes of Hela cells (90). The inaccessibility of these acidic proteins to phosphorylation *in vitro* on 80S ribosomes suggested a localization at the interface between the 40S and 60S subunits. A slightly larger acidic protein was also detected in minor amounts (90).

Two very acidic 60S r-proteins, eL7 and eL12 were isolated from the 60S ribosome of *Artemia salina* and characterized with respect to molecular weight, amino acid composition and comparative peptide maps (91). The proteins are very similar in structure and differ by a single phosphate group. They were shown to be involved in elongation-factor-2-dependent hydrolysis of GTP (92). For this reason the proteins are presently designated eL12(P) and eL12. In addition to eL12/eL12(P), two other 60S acidic proteins from *A. salina* were isolated [eL12' and eL12'(P)]. These proteins, although immunologically related to eL12/eL12(P), differ from it in amino acid composition and chain length (92).

Phosphorylated acidic r-proteins have also been isolated from rat liver ribosomes (93, 94). It is very likely that the two phosphoproteins eL12'(P) and eL12(P) in the 60S ribosomal subunit of *A. salina* correspond to proteins P_1 and P_2 in the 60S subunit of rat liver.

Copy number of the acidic phosphoprotein

Using a sensitive radioimmunoassay and antibodies against eL12, 1.8 copies of eL12 and 0.9 copy of eL12' were found, on the average, in the 80S ribosome of *A. salina*. Copy numbers were estimated in ribosomes where the ratio of eL12/eL12(P) to eL12'/eL12'(P) was two to one; moreover, eL12'/eL12'(P) was one-fourth as active as eL12/eL12(P) in the radioimmunoassay, based on anti-eL12 IgG as antibody and $[^3H]$eL12 as antigen (95). *A. salina* 80S ribosomes contain about 2.7 copies of eL12-type protein. This number is close to the three copies found in 80S ribosomes of yeast (96). It is less clear what determines the relative amounts of the "major" and "minor" acidic 60S phosphoproteins in 60S ribosomes of different organisms. Although it was initially suspected that the larger acidic protein was a precursor of the smaller one, subsequent sequence studies have revealed that the two proteins are products of different but related genes (97). Analytical ultracentrifugation studies show that eL12 and eL12' from *A. salina* and eL12 from rat liver dimerize in free solution (H. Reinalda and W. Möller, unpublished data).

Primary structure of urcaryotic 'A' proteins

The complete amino acid sequence (except for residues 20 and 34) of the 'A' protein from *A. salina* (98) and most of the sequence of the equivalent protein from yeast (T. Itoh and S. Osawa, unpublished data) is now available. In addition, the N-terminal sequence of the 'A' protein from wheat germ (84) and rat liver (B. Wittmann-Liebold *et al.*, as reported in ref. 39) has been published. The amino acid sequence of these four urcaryotic 'A' proteins is shown in Table 10. They form a well defined group of 'A' proteins with properties quite distinct from eubacterial 'A' proteins (see Table 8). For example, they contain two residues of tyrosine, an amino acid absent in the 'A' proteins from eubacteria.

TABLE 10. The amino acid sequence of urcaryotic 'A' proteins from *A. salina* (98), *S. cerevisiae* (Itoh and Osawa, unpublished data), wheat germ (84), and rat liver (39). Residues 36-109 in *S. cerevisiae* are preliminary data.

```
                                    10                  20
                  +                               -       -   - +
A. salina         M R Y V A A Y L L A A L T G N A D P T X1A D I E K I L S
                  +                                           -   +       -
S. cerevisiae     M K Y L A A Y L L L V Q G G N A A P S A A D I K A V V E
                  +L                                          -   +
Wheat germ        M K /F I A A Y L L A Y L G G N S S P S A A D V K D I L N
                                                                 -
Rat liver         (M)X Y V A(G)Y L L A A L G G N(W)N P(W)A(A)D I(G)(M)I L L

                     30                  40                    50
                          -                +               -   +   +-    -
A. salina         S V G I E X2N P S Q L Q K V - - - M N E L K G K D - L E
                          -   - -          +       -         -   +       -
S. cerevisiae     S V G A E V D E A - - - R I N E L L S S L E G K G S L E
                          -   - -          +   -             -   +   +
Wheat germ        A V G A E A N E E - - - K L E F L L A E L K G K

Rat liver         X V

                                60                   70                  80
                          -       +
A. salina         A L I A E G Q T K L A S M P T G G A P A A A A G G A A A
                  -       -     + +
S. cerevisiae     E I I A E G Q K K F A T V P T G G A S S A A ( S1P1G2A3)

                                    90                  100
                          -   + -   + + - - + + - -     - - - - - -
A. salina         A P A A E A K E A K K E E K K E E S E E E D E D M G F G
                          (-)(-)                + - - -   (-) (-)        +
S. cerevisiae         A ( B1Z3A3G2)           - - K E E E ( B3S1Z1G1M1F1K1) G

                       110
                      -
A. salina         L F D
                      -
S. cerevisiae     L F D
```

Within the urcaryotic group of 'A' proteins there is considerable sequence homology (Table 11). This varies from 53% between yeast and wheat germ to 43% between yeast and *A. salina* (using the first 49 residues for these calculations). This latter value increases to about 54% homology when the complete sequence of these two proteins is compared. No comparisons were made to the rat liver 'A' protein because of the preliminary nature of the data.

Microheterogeneity is found in residue 3 of wheat germ where both leucine and phenylalanine are present.

TABLE 11. Matrix of the amino acid sequence homology in the N-terminal region of urcaryotic and archaebacterial 'A' proteins.

	A. sal.	Wheat germ	*S. cer.*	*M. thermo.*	*H. cut.*
Urcaryotes					
A. salina	-	51%(25/49)	43%(21/49)	20%(10/49)	20%(10/49)
Wheat germ	51%	-	53%(26/49)	31%(15/49)	20%(10/49)
S. cer.	43%	53%	-	39%(19/49)	31%(15/49)
Archaebacteria					
M. thermo.	20%	31%	39%	-	56%(27/48)
H. cut.	20%	20%	31%	56%	-

CHARACTERIZATION OF 'A' PROTEINS IN THE 50S RIBOSOMES FROM ARCHAEBACTERIA

Isolation and general properties of the 'A' protein

The ribosomal 'A' protein has been isolated and purified from two archaebacteria, *Halobacterium cutirubrum* (99, Oda - unpublished data, 11) and *Methanobacterium thermoautotrophicum* (14). These proteins show several unusual properties compared to the 'A' protein from the other bacteria that belong to the eubacterial group. The 'A' protein from archaebacteria has two tyrosine residues, an amino acid absent in eubacteria but present in urcaryotes (Table 12). It also contains significantly less valine than the eubacterial protein and is much more acidic than the equivalent protein from either eubacteria or urcaryotes. There is a large reduction in the lysine content and a substantial increase in aspartic acid (*H. cutirubrum*) or glutamic acid (*M. thermoautotrophicum*).

TABLE 12. Amino acid composition of ribosomal 'A' proteins.

	Eubacteria		Archaebacteria		Urcaryotes	
	E. coli (77)	*B. subtilis* (83)	*M. thermo.* (14)	*H. cut.* (99)	Wheat germ (84)	*A. salina* (100)
Asx	5.8	4.9	6.9	19.5	8.1	7.4
Thr	2.5	3.3	3.9	2.0	2.0	2.9
Ser	5.0	2.5	1.6	3.7	9.0	6.4
Glx	14.2	18.9	25.0	18.4	12.6	17.3
Pro	1.7	2.5	2.2	2.8	2.1	4.0
Gly	6.7	9.0	5.6	8.1	10.4	14.0
Ala	23.3	17.2	25.8	24.2	20.3	18.5
Val	13.3	11.5	5.8	6.1	5.0	2.4
Met	2.5	0.0	2.4	0.9	2.1	3.7
Ile	5.0	7.3	4.1	2.7	4.8	3.8
Leu	6.7	9.8	7.0	7.2	8.8	8.1
Tyr	0.0	0.0	2.0	1.8	2.0	1.7
Phe	1.7	1.6	1.4	0.8	3.5	1.4
His	0.0	0.0	1.1	0.0	0.2	0.0
Lys	10.8	10.7	3.6	1.0	8.0	8.0
Arg	0.8	0.8	1.9	0.9	1.4	0.7
A (Asx + Glx)	20.0	23.8	31.9	37.9	20.7	24.7
B (His + Lys + Arg)	11.6	11.5	6.6	1.9	9.4	8.7
B/A	0.58	0.48	0.21	0.05	0.45	0.35

Primary structure of the 'A' proteins from archaebacteria.

The 'A' protein from *H. cutirubrum* shows a substantial amount of sequence homology (56%) to the equivalent protein from *M. thermoautotrophicum* (Table 13) in the N-terminal portion of these proteins (first 49 residues). In the residues where the amino acids are not homologous, the change can be accounted for by a point mutation in all residues but one (14). The comparison of these sequences to those from eubacteria and urcaryotes will be discussed in the next section.

TABLE 13. Amino acid sequence of the 'A' protein from *H. cutirubrum* (99, 101; Oda, unpublished data) and *M. thermoautotrophicum* (14).

	10 ... 20 ... 30
M. thermo.	M E Y I Y A A M L L H T T G K E I N E E N V K S V L E A A G
H. cut.	M E Y V Y A A L I L N E A D E E L T E D N I T G V L E A A G
	40 ... 50 ... 60
M. thermo.	A E V D D A R V K A L I A A L E D V
H. cut.	V D V E E S R A K A L V A A L E D V D I E E A V E E A A A A
	70 ... 110
H. cut.	P A A A P A A S G S B(−) D E A A A S (B_{12}(−)Z_7(−)$G_4P_1A_4$) G L D L F G

COMPARATIVE STUDIES ON THE 'A' PROTEINS FROM EUBACTERIA, URCARYOTES AND ARCHAEBACTERIA

Immunochemical and functional interchangeability of 'A' proteins in eubacteria and urcaryotes

The early immunochemical interchangeability studies are in hindsight not as convincing as first thought (102). The first report concerning a cross-reaction between an acidic r-protein from the 60S ribosomes of rat liver and anti-*E. coli* L7/L12 was not verified, presumably due to difficulties in reisolating such a protein (39). The two acidic r-proteins that were isolated from *A. salina* 60S ribosomes gave no evidence of immunological cross-reaction with *E. coli* L7/L12 (92). Subsequently, a similar negative result was reported for acidic r-proteins from yeast (103), although a variable cross-reactivity was obtained by Visentin, Yaguchi, and Matheson (84).

Howard *et al.* (104) raised antibodies against *E. coli* L7/L12 in rabbits and reported a cross-reaction with 60S ribosomal proteins from chicken liver. Towbin and Gordon (personal communication) have subsequently been unable to raise similar cross-reacting antibodies in goats, instead of rabbits. Inhibition by anti-*E. coli* L7/L12 of reactions catalyzed by EF - 1 and EF - 2 in rat liver (104) and in mouse ascites cells (105) are difficult to evaluate, since no control was done to show that the inhibition could be overcome by addition of an excess of *E. coli* L7/L12.

In the *Artemia* system such controls have been reported in which anti-eL12 inhibited EF - 2-dependent GTP hydrolysis and polyphenylalanine synthesis. Preincubation of anti-eL12 with eL12 releases the inhibition of EF - 2-dependent GTP hydrolysis in the homologous *Artemia* system; on the other hand, anti-*E. coli* L7/L12 did not inhibit EF - 2-dependent GTP hydrolysis of *Artemia* ribosomes, in agreement with the lack of cross-reactivity of *E. coli* L7/L12 and *Artemia* eL12 (92). In addition, the rat liver 60S proteins P_1 and P_2 failed to cross-react with *E. coli* L7/L12 (93). A cross-reaction between an acidic phosphoprotein (L_γ) from ascites cells and *E. coli* L12 was seen by Leader and Coia (106). However, this cross-reaction was observed when antibodies were raised against pieces of polyacrylamide gels containing L_γ; in subsequent experiments anti-*E. coli* L7/L12 failed to cross-react with a split-protein fraction containing L_γ (40). In summary, immunology has not yet given a clear-cut indication that urcaryotic cells contain acidic ribosomal proteins which cross-react with *E. coli* L7/L12.

The functional interchangeability between *E. coli* L7/L12 and urcaryotic acidic ribosomal proteins via hybrid reconstitution is also not free of criticism. Hybrids of 60S yeast core-particles, supplemented with *E. coli* L7/L12, showed considerable activity in elongation-factor-dependent reactions (107). In addition, *E. coli* L7/L12 stimulated EF - 1-dependent activity of 60S *Artemia* cores depleted of acidic proteins (100). However, the extent of binding of phe-t-RNA to such 60S reconstituted particles was reported to be variable and critically dependent on the heat-activation of the particle (100). Subsequent reconstitution experiments on the *Artemia* system failed to circumvent this difficulty, although a slight but definitely positive reconstitution of several elongation-factor-dependent reactions occurred. Reconstitution was due to the assembly of eL12 into the 60S ribosomal core (108). At this stage, further work is required on the reconstitution of urcaryotic ribosomes and the involvement of urcaryotic 'A' proteins in elongation-factor-dependent reactions.

A new approach to the problem of exchangeability of a particular ribosomal protein is the successful incorporation of *Artemia* eL12 into 60S ribosomes of oocytes after microinjection of the protein eL12 into the oocyte (109). A clear exchange reaction *in vivo* of *E. coli* L7/L12 - L10 with oocyte ribosomes was not observed (109).

To summarize, there is some evidence from reconstitution studies that the 'A' protein of ribosomes is involved in the factor-dependent reactions. More extensive reconstitution experiments in homologous systems are required. Inhibition of EF - 2-dependent GTPase by a homologous antibody against the major urcaryotic 60S acidic protein, indicates that the EF - 2-dependent GTPase is blocked by specific binding of the antibody to this acidic protein on the ribosome. Finally, it would be worthwhile to consider that in urcaryotic systems the 'A' protein may have other functions.

Comparative sequence studies on ribosomal 'A' proteins

There are now complete or partial sequence data on 12 'A' proteins covering the three groups or lines of descent used as a classification scheme in this chapter. Figure 1 is a schematic outline of the regions of sequence homology found when the 'A' proteins from these three groups are compared.

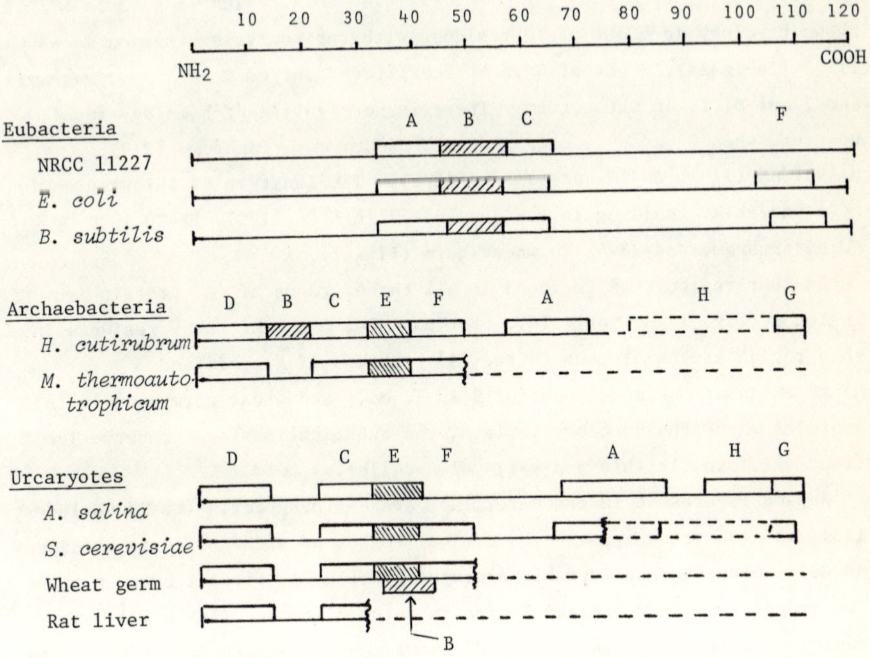

Figure 1. A schematic outline of the primary structural similarities between the three classes of ribosomal 'A' protein. The scale indicates the location of the amino acids in the protein, a solid line indicates regions of the protein where the sequence is known, dotted lines, where the sequence determination is in progress. The regions (A-H) indicate areas of sequence similarity between the various proteins. The complete sequence data for eubacteria is given in Table 8, for archaebacteria in Table 13, and for urcaryotes in Table 10.

Regions common to all three groups

All 'A' proteins, where sufficient sequence data is available, have been shown to contain an alanine-rich region (region A, Fig. 1 and Table 14). This region also contains glycine and proline, serine and/or threonine and frequently valine. Its universal occurrence would suggest a special role for this region in the functioning of the ribosome. The glycine and proline content will give this region considerable flexibility and it may serve as a flexible joint for the 'A' protein dimer. The polypeptide chain at either end of this alanine-rich region in *E. coli* L7/L12 is extremely susceptible to mild proteolysis (A. Liljas, personal communication).

Region B is present in all three groups of 'A' protein but not in all 'A' proteins studied thus far. For example, this region in *E. coli* L7/L12 shows homology in 9 out of 10 residues with the equivalent region in NRCC 11227 (Table 14), 7 out of 8 in *B. subtilis*, 7 out of 9 in *H. cutirubrum*, and 7 out of 10 in wheat germ. There is no significant homology found, however, when *E. coli* is compared to *M. thermoautotrophicum* (2/9), *A. salina* (1/9), and *S. cerevisiae* (2/9). The location of this region is also variable, residing in residues 46-57 in NRCC 11227, 12-20 in *H. cutirubrum*, and 33-42 in wheat germ (Fig. 1).

Another region that is found in all three groups of 'A' protein is region C (Fig. 1 and Table 14). For example, 6 out of the 9 residues in this region are homologous in the 'A' proteins of *E. coli* and *M. thermoautotrophicum*, 7 out of 9 in *E. coli* and wheat germ, and in 'A' proteins where the homology is less, the substitutions are conservative (leu, val, and ile; ala and val; val and ile).

Region F, present in the C-terminal area of eubacteria (residues 103-113), is also present in the N-terminal region of archaebacteria (residues 38-48). Although considerable sequence homology is present between these two groups, the charge distribution is quite different (Table 14). In addition, this region in NRCC 11227 shows the same charge distribution as in the other two eubacterial 'A' proteins, but shows no significant sequence homology to this region in archaebacteria.

There also appears to be some sequence homology in this region to the 'A' protein from wheat germ where glu (40), leu (42), ala (44), and leu (46) line up with the equivalent positions in eubacterial and archaebacterial 'A' proteins. There is also some homology between the 'A' protein of yeast and this region in the equivalent protein from the other two groups.

These last three regions (B, C, and F), like region A, appear to be common to all three groups of 'A' proteins. The significance of these common regions, however, remains to be determined.

TABLE 14. Regions of sequence similarity in ribosomal 'A' proteins.

1. Region A (Fig.)

```
Eubacteria
                               40
                                           -
NRCC 11227       S A A A A V V A G P G G G E
                                           -
E. coli          S A A A A V A V A A G P V E
                                                     -
B. subtilis      T A A A P V A V A G G A A A G G A A E

Archaebacteria
                         60                  70
                 -                             -
H. cutirubrum    E A A A A P A A A P A A S G S D

Urcaryotes
                         70                  80
                                                       -
A. salina        P T G G A P A A A A G G A A A A P A A E

Rat liver        S V P A G G A V A V S A A P G S A A P A A ----

                 P A A A G
                             70                     83
S. cerevisiae    A T V P T G G A S S A A (S1P1G2A3)A
```

2. Region B (Fig. 1)

```
Eubacteria
                         50
                 -   - -   - -     -   -
NRCC 11227       E A E E A E E Q T E F D
                 -         - - +   -   -
E. coli          E A - - A E E K T E F D
                           - -     -   -
B. subtilis      G A - - A E E - - E F D

Archaebacteria
                           15
                 -       - - -     -   -
H. cutirubrum    E A - - D E E L T E - D
                           + -     -   -
[M. thermo.      T T - - G K E I N E - E]

Urcaryotes
                         35
                 -         - - +   -
Wheat germ       E A - - N E E K L E F L
                 -                   +
[A. salina       E X2- - N P S Q L Q K V]
                 -       - -   +     -
[S. cerevisiae   E V - - D E A R I N E L]
```

3. Region C (Fig. 1)

Eubacteria

```
                             60
               -                 -
NRCC 11227     D L - - V L T S A G E
               -           +60
E. coli        D V - - I L K A A G A
               -             60
B. subtilis    D L - - I L A G A G S
```

Archaebacteria

```
                                 30
                           -
H. cutirubrum  N I T G V L E A A G V
                   +       -
M. thermo.     N V K S V L E A A G A
```

Urcaryotes

```
                               30
               -   + -
Wheat germ     D V K D I L N A V G A
               -   - +
A. salina      D I E K I L S S V G I
               -   +       -
S. cerevisiae  D I K A V V E S V G A
               -
Rat liver      D I(G)(M)I L L X V
```

4. Region D (Fig. 1)

Eubacteria — not present

Archaebacteria

```
                                 10
                 -                     -   -
H. cutirubrum  M E Y V Y A A L I L N E A D
                 -                 +
M. thermo.     M E Y I Y A A M L L H T T G
```

Urcaryotes

```
                                 10
                 +
Wheat germ     M K L/F I A A Y L L A Y L G G
                 +
A. salina      M R Y V A A Y L L A A L T G
                 +
S. cerevisiae  M K Y L A A Y L L L V Q G G
Rat liver      M X Y V A(G)Y L L A A L G G
```

5. Region E (Fig. 1)

Archaebacteria

```
               30
                   -   - -       +
H. cutirubrum  G V D V E E S - - - R
                   -   - -       +
M. thermo.     G A E V D D A - - - R
```

```
Urcaryotes
                          35
                    −   − −           +
S. cerevisiae       G A E V D E A - - - R
                        −                 +
A. salina           G I E X2N P S Q L Q K
                    −   − −       +
Wheat germ          G A E A N E E - - - K
```

6. Region F (Fig. 1)

```
Eubacteria
                                  110
                      −       + +     − −
E. coli             A E A L K K A L E E A
                      − −     +   +   − −
B. subtilis         A E E L K A K L E E V
                      − −     +   +   − −
[NRCC 11227         G E E A K T K L E E A]

Archaebacteria
                      40
                      +               − −
H. cutirubrum       A K A L V A A L E D V
                      +               −
M. thermo.          V K A L I A A L E A V

Urcaryotes
                                45
                                    +   +
Wheat germ          L E F L L A E L K G K
                                  − +   +
[A. salina          V - - - M N E L K G K]
                        −           − +
[S. cerevisiae      I N E L L S S L E G K]
```

7. Region G (Fig. 1)

```
Archaebacteria
                          110
H. cutirubrum       G L D L F G

Urcaryotes
                            110
                                  −
A. salina           G F G L F D
                      106
                                  −
S. cerevisiae           G L F D
```

8. Region H (Fig. 1)

```
Archaebacteria
                    (-)(-)
H. cutirubrum       (B12Z7G4P1A4)78-106

Urcaryotes
                                      100
                    − − + + − −     − − − − − −
A. salina           E E K K E E S E E E D E D
                    (-)(-)-      +
S. cerevisiae       (B4Z4E3G3A3K2M1F1S1)84-105
```

Regions common to archaebacteria and urcaryotes

There do not appear to be unique regions between the 'A' proteins of eubacteria and urcaryotes, and between eubacteria and archaebacteria. There are, however, regions that are unique to archaebacteria and urcaryotes. These regions are D, E, G, and H (Fig. 1 and Table 14).

Region D, the N-terminal region in the 'A' proteins from both archaebacteria and urcaryotes, contains a N-terminal methionine and shows considerable sequence homology between the two groups. In positions where the amino acids are not homologous, the changes have been conservative (ile or leu, leu or ala, val or leu). This region also contains the two tyrosine residues absent in eubacterial 'A' proteins (Table 12).

At the C-terminal end of these 'A' proteins, we again find a region of sequence homology (region G) which is absent in eubacteria. The conservation of both the N-terminal and the C-terminal regions and the similar length of these proteins (∿110 amino acids) strongly support the hypothesis that the 'A' proteins in these two groups of organisms are very similar (see below).

Region E, containing the conserved arginine residue (lysine in *A. salina*), shows a great deal of sequence similarity in the 'A' proteins of the two archaebacteria and the urcaryote, *S. cerevisiae*. Of special interest is the conservation of charge distribution in region E, equivalent to residues 32, 34, 35, and 37 in *H. cutirubrum*. The amount of sequence similarity with the 'A' protein from *A. salina* is much reduced, although certain characteristics remain.

Finally, there appears to be a very acidic region (region H) near the C-terminal end of the 'A' protein from archaebacteria and urcaryotes, although only the 'A' protein from *A. salina* has been fully sequenced in this region. Region H is not present in eubacteria.

These results clearly indicate the 'A' proteins from the archaebacteria show much more homology to the equivalent protein from urcaryotes than they do to eubacterial 'A' protein. There is only 15-20% homology between the 'A' proteins of eubacteria and the equivalent protein from archaebacteria, while the proteins from archaebacteria and urcaryotes appear to be much closer phylogenetically (Table 11). For example, the amount of homology between the 'A' protein of the archaebacteria *M. thermoautotrophicum* and the urcaryote *S. cerevisiae* (39%) is almost the same as between the two urcaryotes *S. cerevisiae* and *A. salina* (43%).

The common features of the 'A' protein from archaebacteria and urcaryotes can be summarized as follows:

1. The 'A' proteins are of similar size (109-112 amino acids) while the eubacterial protein is significantly larger (120-122 amino acids).
2. The amount of sequence homology between certain 'A' proteins in the archaebacterial and urcaryotic groups is almost as large as the homologies between certain proteins within each group (Table 11).
3. There are a number of regions of sequence similarities (regions D, E, G, and H) which are unique to the 'A' proteins from archaebacteria and urcaryotes. These regions include a common N-terminal region, which contains the two tyrosine residues absent in the 'A' proteins from eubacteria, and a common C-terminal region.
4. The sequence around the single, conserved arginine residues (equivalent to residue 37 in *H. cutirubrum*) shows a large amount of conservation in the 'A' proteins from archaebacteria and urcaryotes (region E, Fig. 1 and Table 14) and is completely different from the conserved region around the equivalent arginine residue in eubacteria (residue 73 in *E. coli* 'A' protein).

As pointed out by Woese *et al.* (6), although the archaebacteria contain a large number of eucaryotic features, they also contain many features unique to the archaebacteria. From this data they concluded (5) the archaebacteria represented a third line of descent, quite separate from the eubacteria (procaryotes) and urcaryotes (eucaryotes).

However, from the amino acid sequence data on the ribosomal 'A' proteins discussed in this chapter, it is clear that these proteins fall into two major groups, an eubacterial type and an urcaryotic-archaebacterial type. It is also evident from the sequence data that the archaebacterial 'A' proteins are as close to the urcaryotic 'A' proteins, within this latter group, as the gram-positive eubacterial 'A' proteins are to the gram-negative 'A' proteins in the former group. In addition, the sequence data indicate that the archaebacteria (*H. cutirubrum*) are closer to gram-negative bacteria (*E. coli*) than to gram-positive bacteria (*B. subtilis*). From these results, it is possible to construct a phylogenetic tree (Fig. 2) with two major lines of descent; a eubacterial line, containing the gram-negative and gram-positive bacteria and an eucaryotic archaebacterial line which contains a further subdivision into the 80S urcaryotic line and the 70S archaebacterial line.

Several important questions, however, remain unanswered. There is no doubt from the sequence data that the archaebacteria are completely different from the eubacteria, as was earlier indicated by Woese and his co-workers (5, 6). What is not clear, however, is whether the archaebacteria are ancient, unicellular forms of life, as suggested by Woese's group (6), or whether this group is more recent and should be

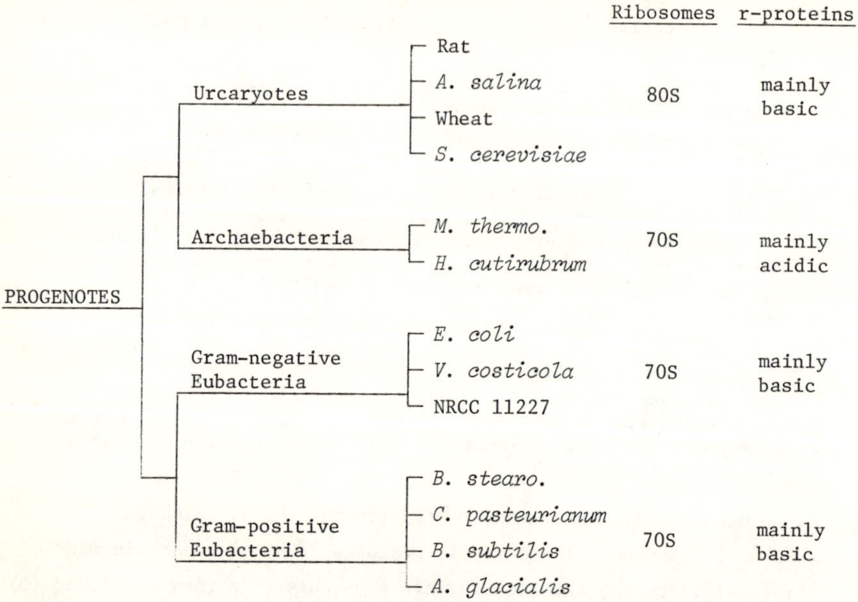

Figure 2. A phylogenetic tree based on the amino acid sequence data from the ribosomal 'A' proteins discussed in this chapter.

called by some other name such as metabacteria as suggested by Osawa and Itoh (see chapter in this book). The observation that the archaebacterial and urcaryotic 'A' proteins show considerable similarity in their primary structure might suggest that their separation into two subgroups may be a 'recent' division if one assumes urcaryotes are more recent than eubacteria (110).

If the urcaryotes have branched off from the archaebacteria, two important modifications have occurred in the ribosome. The archaebacterial, 70S ribosome, containing mainly acidic r-proteins, have evolved into a larger 80S urcaryotic ribosome containing mainly basic r-proteins. One might postulate, therefore, that the r-proteins from urcaryotes may be more acidic than those from eubacteria, reflecting their origin. This appears to be true for the ribosomal 'A' protein. From their mobility on 2-D acrylamide gels (15), the urcaryotic 'A' proteins are more acidic than those from eubacteria, but less acidic than those from archaebacteria (84). Of the four 'A' proteins that have been sequenced from urcaryotes, the one from yeast is the most acidic. This is the protein that showed the closest homology to the archaebacterial 'A' proteins.

The discovery of intervening sequences in eucaryotic (urcaryotic) DNA has lead to extensive discussion on the evolution of procaryotes (eubacteria) and eucaryotes (111, 112). Doolittle, for example, has suggested that rather than assume the procaryotic genome is the primitive form, it is possible that the eucaryotic genome, with its surplus DNA ('gene in pieces'), may in fact be the original form. If this hypothesis is accepted, it would raise further questions as to the primitive nature of archaebacteria. Recent studies on the size of the genome of the archaebacteria *M. thermoautotrophicum* indicates it is less than half the size of the *E. coli* genome (113) but still has all the information required for a free-living organism. Based on this criteria, it would again suggest the archaebacteria may be less primitive than either the eubacteria or urcaryotes.

CONCLUSION

As pointed out in the Introduction, comparative sequence studies on r-proteins may give an insight into the regions in these proteins essential for their function. The data now available on the ribosomal 'A' protein indicate certain areas of sequence homology, such as the alanine-rich region and the highly conserved regions around the 'single' arginine residue, are common to all the 'A' proteins. It should now be possible to specifically modify these areas and observe the effects of these changes on the function of this protein in the ribosome.

The comparative sequence data on the r-proteins should also prove to be a valuable indication of phylogenic relationships similar to those obtained from sequence studies on the 16S rRNA (114). Earlier comparative sequence data on various proteins gave somewhat limited results, since these proteins did not have a wide distribution in nature. The ribosomal 'A' protein does not suffer from this limitation. Further comparative sequence studies on this protein should provide valuable information for phylogenetic studies.

REFERENCES

1. Terhorst, C., Möller, W., Laursen, R., and Wittmann-Leibold, B. (1972) FEBS Letters 28, 324-328.

2. Funatsu, G., Nierhaus, K., and Wittmann, H.G. (1972) Biochim. Biophys. Acta 287, 282-291.

3. Yaguchi, M., Roy, C., Matheson, A.T., and Visentin, L.P. (1973) Can. J. Biochem. 51, 1215-1217.

4. Higo, K., and Loertscher, K. (1974) J. Bact. 118, 180-186.

5. Woese, C.R., and Fox, G.E. (1977) Proc. Natl. Acad. Sci. U.S.A. 74, 5088-5090.

6. Woese, C.R., Magrum, L.J., and Fox, G.E. (1978) J. Mol. Evol. 11, 245-252.

7. Ribosomes (1974) Nomura, M., Tissiéres, A., and Lengyel, P., eds., Cold Spring Harbor Laboratory, New York.

8. Bayley, S.T., and Kushner, D.J. (1964) J. Mol. Biol. 9, 654-669.

9. Fox, G.E., Magrum, L.J., Balch, W.E., Wolfe, R.S., and Woese, C.R. (1977) Proc. Natl. Acad. Sci. U.S.A. 74, 4537-4541.

10. Nazar, R.N., Matheson, A.T., and Bellemare, G. (1978) J. Biol. Chem. 253, 5464-5469.

11. Matheson, A.T., Yaguchi, M., Nazar, R.S., Visentin, L.P., and Willick, G.E. (1978) In: Energetics and Structure of Halophilic Microorganisms, Caplan, S.R., and Ginzburg, M., eds., pp. 481-501, Elsevier/North-Holland, Amsterdam.

12. Visentin, L.P., Chow, C., Matheson, A.T., Yaguchi, M., and Rollin, F. (1972) Biochem. J. 130, 103-110.

13. Bayley, S.T. (1966) J. Mol. Biol. 15, 420-427.

14. Matheson, A.T., Yaguchi, M., Balch, W.E., and Wolfe, R.S. (submitted to Nature).

15. Strøm, A.R., and Visentin, L.P. (1973) FEBS Letters 37, 274-280.

16. Yaguchi, M., Visentin, L.P., Nazar, R.N., and Matheson, A.T. (1978) In: Biochemistry of Thermophily, Friedman, S.M., ed., pp. 169-191, Academic Press, New York.

17. Itoh, T., Kosugi, H., Higo, K., and Osawa, S. (1975) Mol. Gen. Genet. 139, 293-301.

18. Higo, K., Itoh, T., and Osawa, S. (1978) In: Evolution of Protein Molecules, Matsubara, H., and Yamanaka, T., eds., pp. 197-207, Japan Sci. Soc. Press, Tokyo.

19. Vassos, A., Higo, K., and Marcus, L. (1976) Biochim. Biophys. Acta 427, 371-374.

20. Wittmann-Liebold, B., Geissler, A.W., Lin, A., and Wool, I.G. (submitted for publication).

21. Itoh, T., and Wittmann-Liebold, B. (1978) FEBS Letters 96, 399-402.

22. Erdmann, V.A. (1976) Prog. Nucleic Acid Res. Mol. Biol. 18, 45-90.

23. Azad, A.A., and Lane, B.G. (1973) Can. J. Biochem. 51, 1669-1672.

24. Matheson, A.T., Nazar, R.N., Willick, G.E., and Yaguchi, M. (1980) In: Genetics and Evolution of Transcriptional and Translational Apparatus, Osawa, S., *et al.*, eds., Kodansha Scientific, Tokyo (in press).

25. Chen-Schmeisser, U., and Garrett, R.A. (1977) FEBS Letters 74, 287-291.

26. Chen, R., and Ehrke, G. (1976) FEBS Letters 69, 240-245.

27. Brosius, J., Schiltz, E., and Chen, R. (1975) FEBS Letters 56, 359-361.

28. Dovgas, N.V., Markova, L.F., Mednikova, T.A., Vinokurov, L.M., Alakhov, Yu.B., and Ovchinnikov, Uu.A. (1975) FEBS Letters 53, 351-354.

29. Bitar, K.G., and Wittmann-Liebold, B. (1975) Hoppe-Seyler's Z. Physiol. Chem. 356, 1343-1352.

30. Smith, N., Matheson, A.T., Yaguchi, M., Willick, G.E., and Nazar, R.N. (1978) Eur. J. Biochem. 89, 501-509.

31. Nazar, R.N., Willick, G.E., and Matheson, A.T. (1979) J. Biol. Chem. 254, 1506-1512.

32. Nazar, R.N., Willick, G.E., and Matheson, A.T. (1978) Feb. Proc. 37, 2139.

33. Blobel, G. (1971) Proc. Natl. Acad. Sci. U.S.A. 68, 1881-1885.

34. Peterman, M.L. (1974) Methods Enzymol. 30, 346-349.

35. Terao, K., Takahashi, Y., and Ogata, K. (1975) Biochim. Biophys. Acta 402, 230-237.

36. Ulbrich, N., and Wool, I.G. (1978) J. Biol. Chem. 253, 9049-9052.

37. Nazar, R.N., Yaguchi, M., Willick, G.E., Rollin, C.F., and Roy, C. (submitted to Eur. J. Biochem.).

38. Möller, W. (1974) In: Ribosomes, Nomura, M., Tissiéres, A., and Lengyel, P., eds., pp. 711-731, Cold Spring Harbor Press, Cold Spring Harbor, New York.

39. Wool, I.G. (1979) Ann. Rev. Biochem. 48, 719-754.

40. Leader, D.P. (1979) In: Protein Phosphorylation in Regulation, Cohen, P.C., ed., Elsevier/North Holland, Amsterdam (in press).

41. Möller, W., Groene, A., Terhorst, C., and Amons, R. (1972) Eur. J. Biochem. 25, 5-12.

42. Wong, K.P., and Paradies, H.H. (1974) Biochem. Biophys. Res. Comm. 61, 178-184.

43. Österberg, R., Sjöberg, B., Liljas, A., and Petterson, I. (1976) FEBS Letters 66, 48-51.

44. Liljas, A., Eriksson, S., Donner, P., and Kurland, C.G. (1978) FEBS Letters 88, 300-304.

45. Gudkov, A.T., Behlke, J., Vtiurin, N.N., and Lim, V.I. (1977) FEBS Letters 82, 125-129.

46. Luer, C.A., and Wong, K.P. (1979) Biochemistry 18, 2019-2027.

47. Maassen, J. A.(1979) Biochemistry 18, 1288-1292.

48. Van Agthoven, A.J., Maassen, J.A., Schrier, P.J., and Möller, W. (1975) Biochem. Biophys. Res. Comm. 64, 1184-1191.

49. Gudkov, A.T., and Behlke, J. (1978) Eur. J. Biochem. 90, 309-312.

50. Hasnain, S., Visentin, L.P., and Kaplan, H. (1977) Eur. J. Biochem. 80, 35-41.

51. Visentin, L.P., and Kaplan, H. (1975) Biochemistry 14, 463-468.

52. Pettersson, I., Hardy, S.J.S., and Liljas, A. (1976) FEBS Letters 64, 135-138.

53. Österberg, R., Sjöberg, B., Pettersson, I., Liljas, A., and Kurland, C.G. (1977) FEBS Letters 73, 22-24.

54. Gudkov, A.T., Tumananova, L.G., Venyaminov, S.Yu., and Khechinashvilli, N.N. (1978) FEBS Letters 93, 215-218.

55. Pettersson, I., and Liljas, A. (1979) FEBS Letters 98, 139-143.

56. Expert-Bezancon, A., Barritault, D., Milet, M., and Hayes, D.H. (1976) J. Mol. Biol. 108, 781-787.

57. Kenny, J.W., and Traut, R.R. (1979) J. Mol. Biol. 127, 243-263.

58. Kischa, K., Möller, W., and Stöffler, G. (1971) Nature, 233, 62-63.

59. Hamel, E., Koka, M., and Nakamoto, T. (1972) J. Biol. Chem. 247, 805-814.

60. Schrier, P.I., Maassen, J.A., and Möller, W. (1973) Biochem. Biophys. Res. Comm. 53, 90-98.

61. Highland, J.H., and Howard, G.A. (1975) J. Biol. Chem. 250, 831-834.

62. Stöffler, G., Hasenbank, R., Bodley, J.W., and Highland, J.H. (1974) J. Mol. Biol. 86, 171-174.

63. Dijk, J., Littlechild, J., and Garrett, R.A. (1977) FEBS Letters 77, 295-300.

64. Dijk, J., Garrett, R.A., and Müller, R. (1979) Nucleic Acid Res. 6, 2717-2728.

65. Pettersson, I. (1979) Nucleic Acids Res. 6, 2637-2645.

66. Tischendorf, G.W., Zeichard, H., and Stöffler, G. (1975) Proc. Natl. Acad. Sci. U.S.A. 72, 4820-4824.

67. Boublik, M., Hellmann, W., and Roth, H.E. (1976) J. Mol. Biol. 107, 479-490.

68. Strycharz, W.A., Nomura, M., and Lake, J.A. (1978) J. Mol. Biol. 126, 123-140.

69. Acharya, A.S., Moore, P.B., and Richards, F.M. (1973) Biochemistry 12, 3108-3114.

70. Maassen, J.A. (1979) Biochemistry 18, 1288-1292.

71. Maassen, J.A., and Möller, W. (submitted for publication).

72. Maassen, J.A., and Möller, W. (1978) J. Biol. Chem. 253, 2777-2783.

73. San José, C., Kurland, C.G., and Stöffler, G. (1976) FEBS Letters 71, 133-137.

74. Stoffler, G., and Wittmann, H.G. (1977) In: Molecular Mechanisms of Protein Biosynthesis, Weissbach, H., and Pestka, S., eds., pp. 117-202, Academic Press, New York.

75. Donner, D., Villems, R., Liljas, A., and Kurland, C.G. (1978) Proc. Natl. Acad. Sci. U.S.A. 75, 3192-3195.

76. Subramanian, A.R. (1975) J. Mol. Biol. 95, 1-8.

77. Terhorst, C., Möller, W., Laursen, R., and Wittmann-Liebold, B. (1973) Eur. J. Biochem. 34, 138-152.

78. Schrier, P. (1977) Doctoral dissertation, Leiden University, Netherlands.

79. Marquis, D., and Fahnestock, S.R. (1978) J. Mol. Biol. 119, 557-567.

80. Ballesta, J.P.G., and Vazquez, D. (1972) FEBS Letters 28, 337-342.

81. DeVries, A.L., and Lin, Y. (1977) Biochim. Biophys. Acta 495, 388-392.

82. Falkenberg, P., Yaguchi, M., Rollin, C.R., Matheson, A.T., and Wydro, R. (1979) Biochim. Biophys. Acta 578, 207-215.

83. Itoh, T., and Wittmann-Liebold, B. (1978) FEBS Letters 96, 392-394.

84. Visentin, L.P., Yaguchi, M., and Matheson, A.T. (1979) Can. J. Biochem. 57, 719-726.

85. Visentin, L.P., Matheson, A.T., and Yaguchi, M. (1974) FEBS Letters 41, 310-314.

86. Martini, O.H.W., and Gould, H.J. (1973) Biochim. Biophys. Acta 295, 621-629.

87. Zinker, S., and Warner, J.R. (1976) J. Biol. Chem. 251, 1799-1807.

88. Kruiswijk, T., Planta, R.J., and Krop, J.M. (1978) Biochim. Biophys. Acta 517, 378-389.

89. Leader, D.P., and Coia, A.A. (1977) Biochem. J. 162, 199-200.

90. Horak, I., and Schiffmann, D. (1977) Eur. J. Biochem. 79, 375-380.

91. Möller, W., Slobin, L.I., Amons, R., and Richter, D. (1975) Proc. Natl. Acad. Sci. U.S.A. 72, 4744-4748.

92. Van Agthoven, A.J., Maassen, J.A., and Möller, W. (1977) Biochem. Biophys. Res. Comm. 77, 989-998.

93. Tsurugi, K., Collatz, E., Todokoro, K., Ulbrich, N., Lightfoot, H.H., and Wool, I.G. (1978) J. Biol. Chem. 253, 946-955.

94. Arpin, M., Madjar, J.J., and Rebout, J.P. (1978) Biochim. Biophys. Acta 519-537-541.

95. Van Agthoven, A.J., Kriek, J., Amons, R., and Möller, W. (1978) Eur. J. Biochem. 91, 553-565.

96. Kruiswijk, T., Planta, R.J., and Mager, W.H. (1978) Eur. J. Biochem. 83, 245-252.

97. Amons, R., and Möller, W. (1980) In: Genetics and Evolution of Transcriptional and Translational Apparatus, Osawa, S., *et al.*, eds., Kodansha Scientific, Tokyo.

98. Amons, R., Pluijms, W., and Möller, W. (1979) FEBS Letters (in press).

99. Oda, G., Strøm, A.R., Visentin, L.P., and Yaguchi, M. (1974) FEBS Letters 43, 127-130.

100. Möller, W., Slobin, L.I., Amons, R., and Richter, D. (1975) Proc. Natl. Acad. Sci. U.S.A. 72, 4744-4748.

101. Duggleby, R.G., Kaplan, H., and Visentin, L.P. (1975) Can. J. Biochem. 53, 827-833.

102. Stöffler, G., Wool, I.G., Lin, A., and Rak, K.H. (1974) Proc. Natl. Acad. Sci. U.S.A. 71, 4723-4726.

103. Sánchez-Madrid, F., Conde, P., Vázquez, D., and Ballesta, J.P.G. (1979) Biochem. Biophys. Res. Comm. 87, 281-291.

104. Howard, G.A., Smit, R.L., and Gordon, J. (1976) J. Mol. Biol. 106, 623-637.

105. Grasmuk, H., Nolan, R.D., and Drews, J. (1977) Eur. J. Biochem. 79, 93-102.

106. Leader, D.P., and Coia, A.A. (1978) Biochem. J. 176, 569-572.

107. Richter, D., and Möller, W. (1974) In: Lipmann Symposium, Richter, D., ed., pp. 524-533, de Gruyder, New York, Berlin.

108. Van Agthoven, A.J. (1979) Doctoral dissertation, Leiden University, The Netherlands.

109. Kalthoff, H., and Richter, D. (1979) Biochemistry (in press).

110. Schopf, J.W. (1975) Endeavour 34, 51-58.

111. Darnell, J.E. (1978) Science 202, 1257-1260.

112. Doolittle, W.F. (1978) Nature, 272, 581-582.

113. Mitchell, R.M., Loeblich, L.A., Klotz, L.C., and Loeblich, A.R. (1979) Science 204, 1082-1084.

114. Balch, W.E., Fox, G.E., Magrum, L.J., Woese, C.R., and Wolfe, R.S. (1979) Microbiol. Rev. 43, 260-296.

MOLECULAR EVOLUTION OF RIBOSOMAL COMPONENTS

SYOZO OSAWA and HIROSHI HORI

Department of Biochemistry
and Biophysics,
Research Institute for Nuclear
Medicine and Biology,
Hiroshima University,
Hiroshima, Japan

INTRODUCTION

In 1965, Zuckerkandl and Pauling (1) found that the amino acid sequences in hemoglobin from various mammals are similar but not identical, and the difference is nearly proportional to the time elapsed since they had separated from their common ancestor. This would mean that (i) the rate of amino acid substitution in these homologous molecules during evolution is constant, and (ii) the observed amino acid substitutions among these hemoglobin molecules are mostly "selectively neutral" because they are nearly the same in their function. Since then, similar studies have been conducted using other proteins, supporting the conclusions derived from the hemoglobin studies (ref. 2). Especially important among such studies are the extensive amino acid sequence comparisons among cytochrome c molecules from various sources including primitive eukaryotes such as yeasts or protists, whose evolutionary positions could not have been determined from paleontological means by the lack of their fossil records (ref. 3). Comparing the number of amino acid substitutions in cytochrome c between yeasts and mammals with the number, for example, between mammals and fishes whose separating point has been known from fossil records, one can calculate the branching point between mammals (or fishes) and yeasts to be 1.2×10^9 years ago. In this way, phylogenic trees of various organisms have been constructed. This line of works thus opened a new area in which phylogeny or, in a more broad sense, mechanism of evolution can be studied at the molecular level.

One of the limitations to use hemoglobin or typical cytochrome c in such a study lies in the fact that their occurrence is limited only to higher animals or to eukaryotes. To understand the evolutionary process of organisms including both eukaryotes and bacteria, one should find out certain informational macromolecules that occur commonly, and function in the same way, throughout all organisms. The ribosomal components, rRNA's and ribosomal proteins, could be such candidates, because the ribosome occurs as the sole protein synthesizing machinary in all organisms and has no, or nearly no, functional species-specificity in protein synthesis. In the present review, we will examine if the ribosomal components can be used in the study of the biologically meaningful phylogeny among organisms.

CONSTRUCTION OF PHYLOGENIC TREE BY 5S RNA

Comparisons of homologous site of the homologous RNA or protein molecules from different organisms are a prerequisite to recognize nucleotide or amino acid substitutions during evolution and the construction of phylogenic tree. A simple alignment of the sequences, e.g., from the 5'-end to 3'-end of RNA, or N-terminal to C-terminal of protein, is usually meaningless except for the comparison of closely related organisms, because during evolution rearrangements of sequences by deletions, additions, etc. seem to have occurred fairly extensively. Therefore, it is indispensable to have reasonable sequence alignments for all homologous molecules to be compared. Indeed, the evaluation of evolutionary stability of the molecules and the construction of phylogenic tree will depend largely upon a proper alignment.

Since RNA, unlike protein, has no codon degeneracy or masked multihits, evolutionary relationships deduced from nucleotide sequences are more reliable than those deduced from proteins (4). Recently, 5S RNA sequences from a considerable number of organisms have been reported (ref. 5). A total of 61 are known to date. These sequences provide with sufficient information for the study of phylogenic relationships among a range of organisms much wider than in the case of cytochrome c.

Fundamentally the same secondary structure can be constructed for all 5S RNA's by the method of Tinoco et al. (6, 7-9). This is to be expected if their function has been kept nearly unchanged during evolution. Examinations of these structures enable one to classify the known 5S RNA's into four types (Fig. 1; see also refs. 9, 10). The first, to which all eukaryotic 5S RNA's belong, may be called the eukaryotic type, having 120 nucleotides. The eukaryotic 5S RNA differs from all the bacterial 5S RNA's (i) by having a well-conserved loop at position 83-94, (ii) by lacking the hairpin structure that exists in the 5S RNA's from eubacteria (typical bacteria and blue green algae; ref. 13), and (iii) by lacking the sequences which, in eubacteria, probably interact with 23S rRNA and are well conserved (12; the sequences with broken underlines

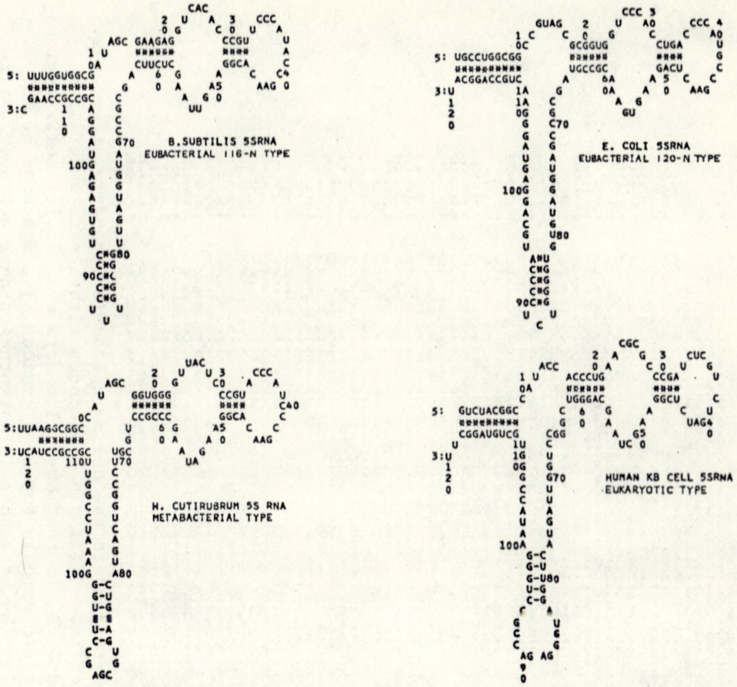

Figure 1. Models of the secondary structures of the four types of 5S RNA. From ref. 9 except for *Halobacterium* 5S RNA model.

in Fig. 2). An apparent exception is seen in the 5S RNA's from plants, which were reported to have 116-118 nucleotides. However, their secondary structure and alignment clearly show that, qualitatively, they belong to the eukaryotic type. The second type is what we may call the metabacterial type (see later), to which 5S RNA from *Halobacterium cutirubrum* belongs, possessing 121 nucleotides. This model is unique to *Halobacterium* and cannot be formed by any known 5S RNA from either eukaryotes or eubacteria [Two other models were constructed by Matheson et al. (10), which also show the uniqueness of this 5S RNA]. The properties are somewhat closer to those of eukaryotic rather than eubacterial 5S RNA in the following points: (i) The presence of fairly well-formed loop at position 87-93, which is well developed in the eukaryotic type at position 83-94 and is lacking in the eubacterial types, (ii) the absence of the sequences involving in the interaction with bacterial 23S rRNA, and (iii) others (see also ref. 10). The terminal helix of the *Halobacterium* 5S RNA is less complete than other types of 5S RNA's; the first to third nucleotides from the 5'-end do not base-pair with those from the

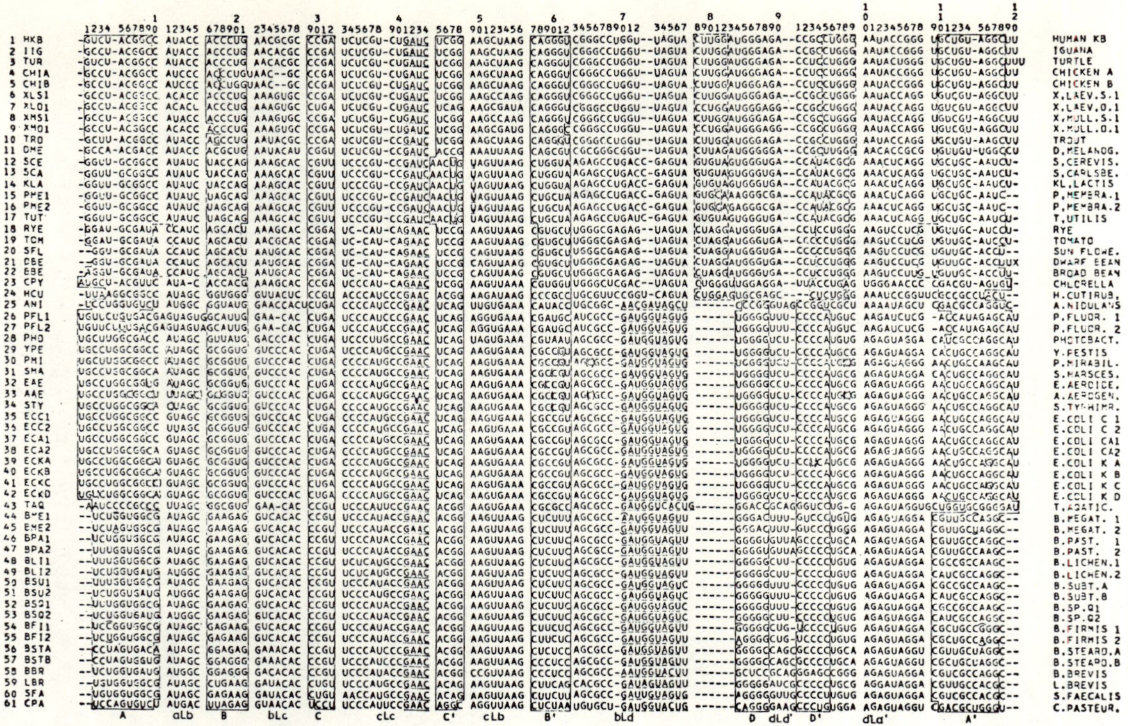

Figure 2. Alignment of 5S RNA's. The squared-off sequences correspond to the helix regions in secondary structures. The underlined sequences probably interact with the GTψC loop of tRNA (11). The sequences with broken underlines probably interact with 23S rRNA (12). A, A', B, B', C, C', and D, D' in the lowest line are symbols for helix regions. aLb, bLc, cLc', c'Lb', b'Ld, dLd', and d'La' in the lowest line are symbols for loop regions. Y, Sequence of CHI (nos. 4 and 5) that was originally reported had C in this place. The sequences in parentheses have not been determined definitively. Sequences of PMI (no. 30), SMA (no. 31), EAE (no. 32), AAE (no. 33), and STY (no. 34) were not completely determined; and, therefore, they were reconstituted from the oligonucleotide maps in the literature. For 5S RNA primary sequences, see ref. 5 (from ref. 9 with several additional sequences).

3'-end. The third type is the eubacterial 120-N type, having 120 nucleotides. The 5S RNA's from gram-negative bacteria belong to this type. The 5S RNA's from gram-positive bacteria (*Bacillus* species and *Clostridium pasteurianum*) possess 116 (sometimes 117) nucleotides and form the fourth type, the eubacterial 116-N type. Characteristic differences between the eubacterial 120-N type and 116-N type may be seen in the regions of A, A', aLb and d'La' (see Fig. 2).

Because all 5S RNA's reveal basically the same secondary structure, it is reasonable first to juxtapose all the 5S RNA secondary structure to obtain the alignments for most parts of the sequences. Alignments for several regions between different types of the models are not possible by the simple juxtaposition technique and therefore needs manual alignments (9). For example, two gaps each were inserted into 5'- and 3'-ends of the 116-N type 5S RNA sequences. This gives an excellent result for total alignment. No further gap insertions into the remaining regions are necessary both for the 120-N type and 116-N type sequences. Note the great similarity between 120-N type and 116-N type of the "5'-3'-terminal helix" when the gaps are inserted. This suggests that the terminal helix is evolutionary rather stable in eubacterial evolution. The reasonableness of the two-gap insertion into the 5'-end of the 116-N type sequence is additionally supported by the following fact: The 5'-end of the 5S RNA precursor molecule has a short stretch pAUU in *E. coli* (120-N type) and a spacer segment consisting of 21 nucleotides in *B. subtilis* (116-N type) (see Fig. 3). By comparing the 5'-terminal regions of these two precursor molecules, one finds long sequences of great similarity - i.e., *AUU*UGCCUGGCGG in *E. coli* and *AUUUG*UUUGGUGG in *B. subtilis*. This strongly suggests that these two segments are the homologous counterparts; UG at the 5'-end of the *E. coli* mature molecule corresponds to the *B. subtilis* precursor *UG* which connects with the 5'-terminus of the mature molecule. Thus it is reasonable that, in place of *UG*, two gaps are inserted into the mature 5'-end of *B. subtilis* 5S RNA.

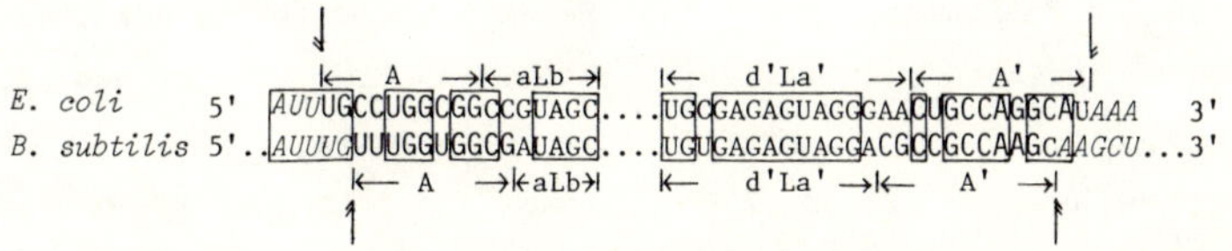

Figure 3. Partial alignment of *E. coli* and *B. subtilis* precursor 5S RNA sequences. Italic and gothic letters represent the precursor regions and the base-paired regions, respectively. From ref. 9.

The alignments of several regions between eubacteria and eukaryotes were tentatively done simply to obtain the best matching, because of a considerable discontinuity in these regions between the eubacterial and eukaryotic secondary structure models. The alignments so obtained allow the comparison of homologous sequences throughout the molecules. To obtain the rate of nucleotide substitution in the course of evolution, two 5S RNA sequences are compared and the rate of the nucleotide substitution per site, Knuc, may be calculated as follows (14):

$$\text{Knuc} = -\,(3/4)\ \ln\,[\,1 - (4/3)\,\lambda\,] \qquad [1]$$

where λ is the fraction of sites which differ from each other. The Knuc values so obtained can be used for determination of the branching order and the relative evolutionary distance in the construction of phylogenic trees. All of the pairs of organisms are rearranged in the order of increasing Knuc values, and the pair to be formed first is decided simply by choosing the pair with the smallest Knuc values. The value of 1/2 Knuc of the pair is then taken to settle the branching point between them. The branching points between two or more pairs are determined from the average number of 1/2 Knuc between the pairs. The Knuc values may be taken as the relative time scale in the phylogenic tree (Fig. 4). The lowest node represents the divergence of two populations, one of which developed into the eubacterial group, and the other into a special bacterial group (e.g., *Halobacterium*; see later) and eukaryotes. The time of divergence of eubacteria and eukaryotes is 1.5 times as great as that of human and yeast. Assuming that the human and yeast divergence occurred $(1.2 \pm 0.075) \times 10^9$ years before, the divergence time of eubacteria and eukaryotes goes back to about 1.8×10^9 years (15, 16, 9). However, due to certain discontinuities between eukaryotic and bacterial sequences or between 120-N type and 116-N type, some uncertainties cannot be avoided in constructing the tree and in determining the divergence time of the early branches. In bacteriology, enterobacteria, *Pseudomonas* and blue green algae are classified into gram-negative organisms and *Bacilli* into gram-positive. The phylogenic tree also shows the same relationship. The branching point of *Thermus aquaticus* is not definite at present. The sequence similarity of this species is greatest with *B. stearothermophilus* (70 % similarity), while it is about 64 % with *E. coli* group (17). However, *T. aquaticus* is a gram-negative bacterium and its 5S RNA has 120 nucleotides possessing AU at the 3'-end, which is a characteristic of the 120-N type. Thus this bacterium may be more

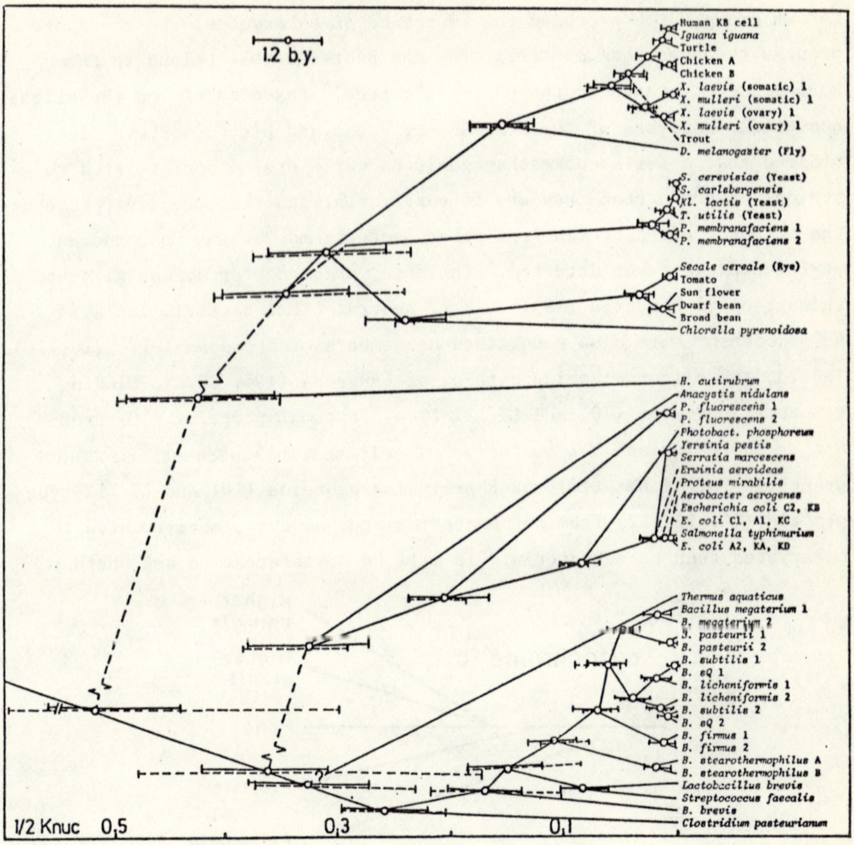

Figure 4. Phylogenic tree of 5S RNA
├──●──┤ Range of standard error; ├---o---┤ Range of 1/2 Knuc values of all pairs at the branching point. Redrawn from Fig. 4 of ref. 9.

related to the 120-N type than to the 116-N type, even the similarity percentage reveals the reverse relationship.

Until recently, it has been a general belief that all bacteria (prokaryotes) belong to one stem which originated from an ancestor common to eukaryotes and prokaryotes. However, the 5S RNA sequence of *Halobacterium cutirubrum* indicates that its secondary structure is unique and resembles the eukaryotic type rather than eubacterial types. Its Knuc value also suggests that this organism is more related to eukaryotes than to eubacteria and that its emergence occurred from the main stem after the separation of eubacteria (9). This means that the emergence of bacteria from the main stem occurred at least twice during early evolution; first the emergence of eubacteria and then that of *Halobac-*

terium group which preceded the emergence of eukaryotes. We therefore propose the name "*Metabacteria*" for the bacteria that belong to the *Halobacterium* branch in the phylogenic tree. Based mainly on the oligonucleotide analyses of the 16S rRNA's, Woese and his co-workers (18) claimed that a series of methanogenic bacteria are no more related to typical bacteria than they are to eukaryotes, and that the divergence of the methanogenic bacteria from other bacteria may be the most ancient phylogenic event yet detected. The name, *Archaebacteria*, was given to this group of bacteria (19). Later, several other bacteria including *Halobacterium* were also recognized as members of *Archaebacteria* (19, 20). Indeed, the oligonucleotide catalog of 16S rRNA (19), 5S RNA-binding ribosomal proteins (10) and L7/L12-type ribosomal proteins (10) from *Methanobacterium* and *Halobacterium* closely resemble each other. The properties of 5S RNA (10), 5S RNA-binding proteins (10) and L7/L12-type protein (10, 21, 22) from *Halobacterium* are, however, more related to eukaryotes than to eubacteria. It will be of interest to see whether

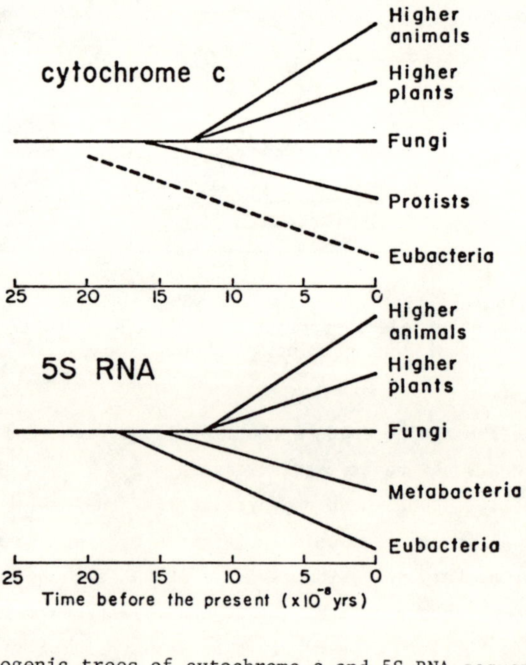

Figure 5. Phylogenic trees of cytochrome c and 5S RNA sequences. Only the main branches are shown. The cytochrome c tree was redrawn from the cladogram by Sneath (23). The branching point between higher animals and yeasts is shown to be 1.3 x 10^9 years before, according to the original cladogram. In the 5S RNA tree, the time of 1.2 x 10^9 years before is adopted for their branching point. By the lack of typical cytochrome c in eubacteria, the eubacterial branch in the cytochrome c tree is derived from the sequence data of *Rhodospirillum rubrum* cytochrome c_2.

the methanobacterial 5S RNA can form the second type of secondary structure like *Halobacterium*, when the sequence data become available.

The tree shows that fungi and plants separated from animals almost at the same time. The detailed examination indicates that first fungi diverged from the animals/plants, and slightly after that, plants and animals separated. However, plants could have first diverged from fungi/animals, because (i) Knuc of animals/plants and animals/fungi are practically the same (0.61 vs. 0.63, both with a standard error of about 17%; Knuc of plants/fungi = 0.75); (ii) all 5S RNA's of animals and fungi have a GAUC sequence at about position 41-44, whereas the corresponding sequence in plants is GAAC, which is almost common to all the eubacterial 5S RNA's; and (iii) human 5.8S RNA reveals 75 % identity with yeast 5.8S RNA, whereas it is less similar to plant 5.8S RNA. It is noteworthy that the phylogenic tree of eukaryotic kingdom deduced from the 5S RNA data agrees very well with the tree from cytochrome c data (ref. 23; see Fig. 5). The cytochrome c tree contains some protists (*Euglena, Crithidia* and *Tetrahymena*), which are still absent in the 5S RNA tree. An approximate time of their divergence from other eukaryotes can be estimated as $1.3 - 1.6 \times 10^8$ years ago (3, 23), which is close to that of metabacteria. It will be interesting to examine the phylogenic relationship between protists and metabacteria, when the 5S sequence data of protists become available.

THE RATE OF NUCLEOTIDE SUBSTITUTION OF rRNA AND AMINO ACID SUBSTITUTION OF RIBOSOMAL PROTEINS

The construction of phylogenic tree, determination of divergence time among organisms, and the rate of nucleotide substitution in RNA and that of amino acid substitution in proteins, all depend upon the assumption that the substitution rate is constant during evolution. As will be discussed below, whether or not this assumption is strictly applicable to all the ribosomal components is still not certain and the verification awaits for further evidence on the structure and function of ribosomes from a wide range of organisms. Nevertheless, at the present stage of this field, it would be of significance to analyse the available data from the molecular evolutionary point of view.

Earlier work by Nomura et al. (24) clearly demonstrated that 16S rRNA from different bacterial species interacts with *E. coli* ribosomal proteins to give functionally active 30S ribosomes. Later, it was shown that most of the *E. coli* ribosomal proteins have functional counterparts among *Bacillus stearothermophilus* proteins by substitution of individual proteins in *in vitro* reconstitution of 30S ribosomes (25). The N-terminal amino acid sequences of 30S ribosomal proteins revealed a high degree of similarity among *E. coli*, *B. stearothermophilus* and *B. subtilis* (ref. 26). Furthermore, immunological cross-reaction was found in many 30S ribosomal proteins (25, 27). Finally, it was shown that *E. coli* 5S RNA is functionally active in the reconstitution of *B. stearothermophilus* 50S ribosomes (28). All these facts suggest that the ribosomal components have been highly conserved during eubacterial evolution. However, there exist gross structural differences between the eukaryotic and eubacterial ribosomes; the eukaryotic ribosome can be characterized by (i) the larger sizes of its subunits and two main rRNA species, (ii) the occurrence of 5.8S RNA, (iii) the occurrence of more numbers of ribosomal proteins. Furthermore, recent studies have indicated that the sequences of some eukaryotic ribosomal proteins including L7/L12-type protein have only a little similarity, if any, with those of eubacterial ribosomal proteins (21, 29; unpublished data). In *Halobacterium*, the ribosomal size and the number of ribosomal proteins are of the eubacterial nature, but the sequence of L7/L12-type protein and 5S RNA-binding proteins have considerable similarities to those from eukaryotes (ref. 10). Therefore, drastic changes might have occurred in the ribosome at least twice, i.e., first, an extensive sequence alteration in ribosomal proteins somewhere between eubacterial and metabacterial emergence(or upon eubacterial emergence), and then, upon eukaryotic emergence, the development of 5.8S RNA, an addition of more ribosomal proteins, and an increase in size of two main rRNA species. In contrast to these changes, 5S RNA can take essentially the same secondary structure, and its sequence seems to have been well conserved throughout organisms during evolution. Thus, except for 5S RNA, the constancy of the rate of nucleotide substitution in rRNA's or amino acid substitution in ribosomal proteins could have been temporarily broken at the time of drastic ribosomal structural alterations. Even in the case of 5S RNA, there is a proposal that eubacterial 5S RNA and eukaryotic 5.8S RNA are of the same evolutionary origin and that their functions in protein synthesis are similar. The function of eukaryotic 5S RNA may be involved in initiator tRNA-binding to the 80S ribosomes (30). This proposal is mainly based on the observation that *E. coli* 5S RNA-binding proteins, L18 and L25, form a specific complex with yeast 5.8S RNA and not with yeast 5S RNA. From the functional point of view, it is possible that in eubacterial 5S RNA functions in the binding of both initiator tRNA and other tRNA's, and that in eukaryotes the ribosomal function of tRNA-binding is differentiated so that 5S RNA binds only to initiator tRNA, and the newly developed 5.8S RNA takes a part for the binding of other tRNA's. However, it is unlikely that eukaryotic 5.8S RNA evolved from eubacterial 5S RNA, because neither an appreciable sequence identity (31) nor a similarity in their secondary structure exists between eubacterial 5S RNA and eukaryotic 5.8S RNA. A certain discontinuity between eubacterial and eukaryotic 5S RNA secondary structures might be

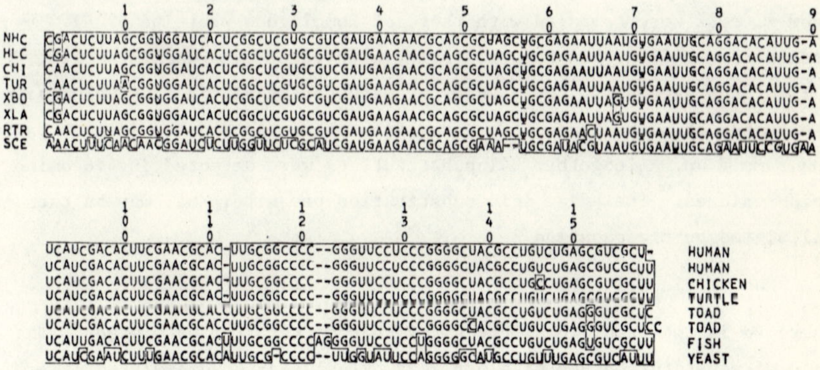

Figure 6. Alignment of 5.8S RNA. U, G: Methylated U or G. Ψ: Pseudo U. The sequences were taken from ref. 5.

correlated with the possible functional differentiation in eukaryotic ribosomes mentioned above. Therefore, some uncertainties cannot be avoided in extending the rate of nucleotide substitution of eukaryotic 5S RNA to that of bacterial 5S RNA.

The rate of nucleotide substitution per site per year, k(nuc) in RNA may be calculated as

$$k(nuc) = Knuc / 2T \qquad [2]$$

where T is the number of years that has elapsed since the evolutionary divergence of the two molecules from their common ancestor (14). The Knuc value of 5S RNA between human and yeast is about 0.69 ± 0.11 and the time of their divergence was 1.2×10^9 years ago according to the cytochrome c data. It then follows that k(nuc) of 5S RNA is $(2.9 \pm 0.5) \times 10^{-10}$. The Knuc value of 5.8S RNA between animals and yeast is about 0.31 ± 0.05, and the k(nuc) of 5.8S RNA is then $(1.3 \pm 0.2) \times 10^{-10}$ (26, 32; for alignment of the sequences, see Fig. 6).

It may be noted here that there are several gene loci for rRNA's in bacteria (except for *Mycoplasma* in which only one locus seems to exist) and this number is much larger in eukaryotes. However, there is evidence that different genes in one organism seem to produce rRNA's whose nucleotide sequences are more or less the same (ref. 26). This suggests that there is some mechanism by which the diversification of duplicate genes by mutation is prevented. We do not know the nature of the mechanism and how this affects the assumption of constant rate of nucleotide substitution in 5S RNA and 5.8S RNA.

To obtain the rate of amino acid substitution of ribosomal proteins during eubacterial evolution, N-terminal sequences of 17 30S proteins from *E. coli* were compared with those of homologous proteins of *B. stearothermophilus*, and when avilable, with those of *B. subtilis* (26, 32; for the sequences, see a paper by Matheson et al. in this volume). The number of sites that differ from each other was counted in each pair and summed up. Altogether 241 substitutions were detected in 476 amino acid residues. The amino acid substitution per site, Kaa, may be then calculated by the equation

$$Kaa = -\ln(1 - Pd) \qquad [3] \quad \text{(ref. 1)}$$

where Pd is the fraction of the sites that differ from each other. The rate of substitution per site per year, k(aa), may be calculated as

$$k(aa) = Kaa / 2T \qquad [4] \quad \text{(ref. 1)}$$

Since T between *E. coli* and *Bacillus* is 1.2×10^9 when calculated from 5S RNA data, then the average k(aa) of the N-terminal regions is about $(2.8 \pm 0.19) \times 10^{-10}$ per site per year from Eqs. [3] and [4].

The protein L7/L12 is so far the only protein available for the comparison of complete amino acid sequence between two bacteria, i.e., *E. coli* (33) and *B. subtilis* (34; for the sequences, see a paper by Matheson et al. in this volume). Using the same technique as in the case of N-terminal regions, k(aa) of L7/L12 was calculated to be $(3.3 \pm 0.8) \times 10^{-10}$, indicating that the rate of amino acid substitution in L7/L12 is about comparable with k(aa) of the N-terminal regions of 30S proteins. Recently, the major parts of the sequences of L7/L12-type protein from yeast have been determined (Itoh, unpublished), and compared with the corresponding sequences of the homologous protein from *Artemia salina*, revealing 50 % identity (for the two sequences, see a paper by Matheson et al. in this volume). The k(aa) value was then calculated to be $(3.0 \pm 0.5) \times 10^{-10}$. It is interesting that, despite a gross difference of their sequences and as yet unsettled homology between eubacterial and eukaryotic L7/L12-type proteins, the rate of amino acid substitution

TABLE 1. The rate of amino acid (or nucleotide) substitutions of ribosomal components and other proteins

Proteins (or RNA's)	k(aa) or k(aa') $\times 10^{-10}$/site/yr.
Bacterial ribosomal proteins *	3
Bacterial ribosomal protein L7/L12	3
Eukaryotic L7/L12-type protein **	3
5S RNA	4
5.8S RNA	3
Fibrinopeptides	90
Growth hormone	37
Pancreatic RNase	33
Immunoglobulins	32
Lactalbumin	25
Hemoglobin chains	14
Myoglobin	13
Animal lysozyme	10
Gastrin	8
Melanotropin β	7
Trypsinogen	5
Insulin	4
Cytochrome c	3
Glyceraldehyde 3-PO_4 dehydrogenase	2
Histone IV	0.06

* N-terminal regions.
** 75 residues from N-terminal.
Mainly from refs. 26, 32.

is close to each other. This might be expected, because the role of these proteins in protein synthesis in bacteria and in eukaryotes is approximately the same.

To directly compare the values of k(nuc) with those of k(aa), k(nuc) was converted to the rate of amino acid substitution, k(aa'), by using the equation (ref. 62):

$$Pd(aa)=1-(1-\lambda)^2[1-(1/4)\lambda] \quad [5]$$

and is shown in Table 1 with the known data for other representative proteins. Eq. [5] depends on the assumption that the probability of nucleotide substitution is the same for all sites. However, there is evidence that the rate of nucleotide substitution is much higher at the third base of a triplet codon than the first and second bases. Therefore, the k(aa') for 5S RNA and 5.8S RNA should be somewhat smaller than those actually obtained here. From the table, it is evident that the substitution rates of 5S RNA, 5.8S RNA and ribosomal proteins are close to each other and are the slowest among the listed informational macromolecules except for histone IV (ref. 26, 32; Table 1).

The question how the sequences of two main rRNA species are conserved during evolution can be answered only partly, since the complete sequence has been determined only for *E. coli* 16S rRNA (35, 36). One thing apparent is that there must have occurred appreciable structural changes in these rRNA's during evolution, because all the eukaryotic 18S rRNA's have a longer chain length than the bacterial 16S rRNA's. In this connection, it is noteworthy that the 28S rRNA has a tendency of increasing molecular size and GC-content in the direction from primitive organisms to advanced animals, while no such tendency is seen in 18S rRNA. This would suggest that at least rRNA of the 28S-type has evolved to increase its chain length even after the eukaryotic emergence.

Despite the molecular size difference between bacterial and eukaryotic rRNA's, there is evidence that the sequences of certain regions in 16S(18S) rRNA are strongly conserved. This is seen in the 3'-terminal regions of 16S(18S) rRNA's from various organisms (Fig. 7; 37-39). The features specific to the eukaryotic 18S rRNA are (i) the deletion of the CCUYY (or CCU in the case of slime mold) sequence near the 3'-end which occurs in the eubacterial 16S rRNA, and (ii) the substitution of the dinucleotide of UU in *E. coli* (the 15th and the 16th nucleotides from 3'-end) by AA (or AU in slime mold) in eukaryotes (38). It has been suggested that the initiation of protein synthesis in *E. coli* involves the formation of base pairing between a purine-rich tract preceding the initiation codon of mRNA's and a pyrimidine-rich region including the sequence CCUYY near the 3'-end of 16S rRNA. The CCUYY sequence is universally found in all the eubacterial 16S rRNA's. However the eukaryotic 18S rRNA's do not contain a CCUYY sequence near the 3'-end. On the contrary they contain a purine-rich region GCGGAAGG (or GCGGAUGG) which is complementaty to a sequence in the 5' non-coding regions of many eukaryotic mRNA's (38). These facts might represent the evolution of 18S rRNA in coordination with the evolution of mRNA upon the eukaryotic emergence.

About 300 nucleotide sequences at the 3'-terminal region of *Zea mays* chloroplast 16S rRNA have recently been sequenced and compared with that of *E. coli* (40). Over 80 % of the sequences is identical between them. Furthermore, the CCUYY sequence near the 3'-end, which is a characteristic for eubacterial 16S rRNA, is also found in the chloroplast 16S rRNA at the corresponding region. The time of divergence between eubacteria and *Zea mays* goes back to about 1.8×10^9 years which is much older than the divergence time of 1.2×10^9 years between *Zea mays* and other eukaryotes, and yet a greater similarity of the chloroplast 16S rRNA with *E. coli* 16S rRNA and the eubacterial feature of the 3'-terminal sequences are discernible. These facts suggest a high stability of these regions in rRNA of the 16S size, and also the bacterial origin of the chloroplast 16S rRNA.

```
1                                                           YUCCUUUCU  (3') B, SUBTILIS
2                                                          YUCCUUUCUA  (3') B, STEAROTH,
3                                                       GXXYCUCUCCUUA  (3') PS, AERUGINOSA
4  GUGAAGUCGUAACAAGGUAACCGUAGGGGA'A'CCUGCGGUUGGAUCACCUCCUUA  (3') E, COLI
5  GUGAAGUCGUAACAAGGUAGCCGUACUGGAAGGUGCGGCUGGAUCACCUCCUUU  (3') CHLOROPLAST
6                                 GA'A'CUCGCGGAAGGAUCA-----UUA  (3') YEAST
7                                  A'A'CCUGCGGAUGGAUCA---UUUUA  (3') SLIME MOLD
8                                  A'A'CCUGCGGAAGGAUCA-----UUG  (3') WHEAT
9                                  A'A'CCUGCGGAAGGAUCA-----UUA  (3') BOMBIX MORI
10                AAGGUUUCUGUAGGUGA'A'CCUGCGGAAGGAUCA-----UUA  (3') RAT 1
11                AAGGUUUCCGUAGGUGA'A'CCUGCGGAAGGAUCA-----UUA  (3') RAT 2
12                                 A'A'CCUGCGGAAGGAUCA-----UUA  (3') MOUSE
```

```
CHLOROPLAST  CCGGGCCTTGTACACACCGCCCGTCACACTATAGGAGCTGGCCAGGTTT-GAAGTCATTA
E.COLI       CCGGGCCTTGTACACACCGCCCGTCACACCATGGGAG-TGGGTTGCAAAAGAAGTAGGTA
                .     1400,          .          .          .          .

             CCCTTAACCGTAAGGAGGGGGA-TGCCTAAGGCTAGGCTTGCGACTGGAGTGAAGTCGTA
             G-CTTAACCTTCGGGAGGGCGCTTACCACTTTGT-GATTCATGACTGGGGTGAAGTCGTA
                  1450.          .          .          .          .
                                                         ↑
             ACAAGGTAGCCGTACTGGAAGGTGCGGCTGGATCACCTCCTTTTCAGGGAGAGCTAATGC
             ACAAGGTAACCGTAGGGGAACCTGCGGTTGGATCACCTCCTTACCTTAAAGAAGCGTACT
                .     1510.          .          .          .    1550.

             TTATGCTTATTGGGTATTTTGGTTTGACACTGCTTCACGCCCAAAAAGAAGGCAGCTACG
             TTGCAGTGCTCACACAGATTGTCTGATGAAAATGAGCAGTAAAACCTCTACAGGCTTGTA
                .          .          .          .    1600.          .

             TCTGAGCTAAACTTGGATATGGAAGTCTTCTTTCGTTTAGGGTGAAGTAAGACCAAGCTC
             GCTCAGGTGGTTAGAGCGCACCCCTGATAAGGGTGAGGTCGGTGGTTCAAGTCCACTCAG
                .          .          .    1650.          .          .

             ATGAGCTTATTATCCTAGGTCGGAACAAATTAGTTGATAGTGATAGGATCC
             GCCTACCA
                .
```

Figure 7. Alignment of the 3'-terminal sequences of 16S(18S) rRNA's. Upper half: A' = m_2^6A. (A'), Methylation not finally proved. Slime mold = *Dictyostelium discoideum*. Unboxed indicate the nucleotides different from those of *E. coli*. Chloroplast (*Zea mays*) sequences (no.5) were transcribed from the rDNA sequences (ref. 40). Sequences of no.1-3, from ref. 37. No.4, 7-9 and 12, from ref. 38. A part of the No.4 sequences, parhaps in somewhat modified form (see ref. 38). Lower half: Nucleotide sequences of the 3'-terminal regions of 16S rRNA gene from *Zea mays* chloroplast in comparison with those from *E. coli*. ————: 3'-terminal sequences of *E. coli* 16S rRNA gene. ════: tRNA^{Ile} gene of *E. coli*. Boxed indicate the nucleotides common in chloroplast and *E. coli* (from ref. 40).

As was shown before, 5S RNA, 5.8S RNA and eubacterial ribosomal proteins are evolutionally rather stable as a whole. Then the next question is, in the essential molecules or molecular complexes such as the ribosome, whether the substitutions of nucleotides or amino acids in rRNA's or ribosomal proteins are qualitatively limited so that the molecule can take a specific higher order structure to support its function, and/or it can specifically interact with other molecules. In the following, the pattern of nucleotide substitution in 5S RNA and amino acid substitution in bacterial ribosomal proteins will be examined to answer a part of the above question.

5S RNA: First, ancestral sequences of 5S RNA at all the nodes in the phylogenic tree were deduced so that the total number of substitutions in the molecule is reduced to a minimum. The complete descent for all the 5S RNA species shows 189 and 184 substitutions in the base-paired and nonbase-paired regions, respectively. The nucleotide substitutions in 5S RNA during evolution seem to have occurred predominantly as transitions. The ratio of overall transitions to transversions is about one. If no special mutational pressures or selective mechanisms are operating, the frequency of transversions should be twice as much as that of transitions (the ratio should be 0.5). Especially noteworthy is that the U → C and C → U transitions occurred with frequencies much higher than expected in both base-paired and nonbase-paired region, while C → A and U → G transversions are much less. Also, the C → U change is significantly higher in the base-paired regions than the nonbase-paired regions (unpublished data). This could be due to the occurrence in the base-paired regions, but not in the nonbase-paired regions, of GU pairing in place of GC pairing. Of course it is difficult from this kind of observations to decide the neutrality or non-neutrality of the individual substituions; in a certain site, U → C could be neutral and U → G be not, while in certain other site the situation be reversed. It depends on how the substitution affects the function of the molecule. Generally speaking, however, transition, a change between nucleotides of a chemically similar nature, could be more neutral than transversion, a change between less similar nucleotides.

Eubacterial Ribosomal Proteins: Using alignments of N-terminal regions of 30S ribosomal proteins from *E. coli*, *B. stearothermophilus* and *B. subtilis**, the nature of amino acid substitutions was examined (26, 32, 41). Table 2 shows the number of amino acid substitutions, n_r (i, j), at the corresponding positions in the alignments of homologous proteins from *E. coli* and *B. stearothermophilus*. The n_r (i, j) with $i = j$ and n_r (i, j) with $i \neq j$ were counted, where i is the amino acid of *B. stearothermophilus* and j is that of *E. coli*. It is meant by $i = j$ or $i \neq j$ that the amino acid of *B. stearothermophilus* is equal to or not

* The sequences are almost the same between *B. stearothermophilus* and *B. subtilis* (89 % identity), and fairly big differences can be seen between *E. coli* and these two *Bacilli*, the identity being in both cases 54 %. These facts, together with the quality of substitutions, indicate that most of the amino acid substitutions between *E. coli* and *B. stearothermophilus* do not seem to be the results of selection by high temperature (ref. 26).

TABLE 2. Substitution frequencies in ribosomal proteins of *Bacillus stearothermophilus* and *Escherichia coli*

	W	Y	F	L	I	M	V	A	P	G	T	S	N	D	Q	E	H	K	R	C	X	B	Z	–	O	U
W (Trp)	0	0	0	1	0	0	0	0	0	2	0	0	0	0	0	0	0	0	0	0	0	0	0	0	0	0
Y (Tyr)	0	8	6	1	0	0	1	0	1	0	0	0	0	0	1	0	0	0	0	0	0	0	0	0	1	0
F (Phe)	0	0	6	1	0	0	0	1	0	0	0	0	0	0	0	0	0	0	1	0	0	0	0	0	0	0
L (Leu)	0	0	0	15	3	1	1	3	0	0	0	0	0	1	0	1	0	0	0	0	0	0	0	0	0	0
I (Ile)	0	0	1	5	17	1	12	1	0	0	1	0	1	0	0	0	0	3	0	0	0	0	0	0	0	1
M (Met)	0	0	0	1	0	6	1	0	0	0	0	0	0	0	1	0	1	1	0	0	0	0	0	0	0	0
V (Val)	0	1	0	0	5	0	16	2	1	0	2	1	1	1	1	0	0	1	2	0	0	0	0	0	0	0
A (Ala)	0	1	0	0	0	1	0	21	0	1	1	1	0	0	1	0	0	1	0	0	1	0	0	0	0	0
P (Pro)	0	0	0	0	1	0	2	4	13	1	0	0	0	0	0	0	0	0	0	0	0	0	1	0	0	0
G (Gly)	0	0	0	0	0	0	0	0	2	26	0	0	0	0	0	0	0	1	3	1	0	0	0	0	0	0
T (Thr)	0	0	0	1	0	0	1	1	1	0	8	2	0	1	2	0	0	0	4	0	0	0	0	0	1	0
S (Ser)	0	0	0	0	0	0	0	2	0	0	1	15	1	1	0	0	0	0	0	0	0	0	0	1	1	0
N (Asn)	0	0	0	0	1	0	1	0	0	1	0	0	4	1	1	0	1	0	0	0	0	1	0	0	0	0
D (Asp)	0	0	0	0	0	0	0	0	1	0	0	1	1	10	0	1	1	1	0	0	0	1	0	0	1	0
Q (Gln)	0	0	0	0	0	0	0	0	0	0	2	0	0	1	6	3	0	2	1	0	0	0	0	1	0	0
E (Glu)	0	0	0	1	1	0	0	3	0	0	0	1	0	1	2	8	0	2	0	0	0	1	0	0	0	0
H (His)	0	0	0	0	0	0	0	0	0	1	0	0	0	0	0	0	5	0	0	0	0	0	0	0	0	0
K (Lys)	0	0	0	0	1	0	2	5	0	1	2	0	1	0	2	0	2	25	9	0	0	1	4	1	0	0
R (Arg)	0	0	1	0	2	0	0	4	1	0	1	0	0	2	1	0	2	8	32	0	1	0	1	0	0	0
C (Cys)	0	0	0	0	1	0	0	0	0	0	0	0	0	0	0	0	0	0	0	0	0	0	0	0	0	0
X ( )	0	0	0	1	1	0	2	1	0	0	1	0	0	0	0	0	1	0	3	0	2	0	0	0	1	0
B (Q/E)	0	0	0	0	0	0	1	1	0	0	0	0	2	0	0	0	0	0	0	0	0	0	2	0	0	0
Z (N/D)	0	0	0	0	1	1	0	1	0	0	1	0	0	1	1	2	0	1	0	1	0	0	1	0	0	0
– (Gap)	0	0	1	1	0	1	0	0	0	0	0	0	0	0	0	0	0	2	0	1	0	0	0	0	0	0
O (S/C)	0	0	0	0	0	0	1	2	0	0	0	3	0	0	0	0	0	1	0	0	0	0	0	0	1	0
U (L/I)	0	0	0	0	0	0	0	0	0	0	0	0	0	0	0	0	0	0	0	0	0	0	0	0	0	0

Each diagonal element is the number of homologous sites. The off-diagonal elements are $n_r(i, j)$. For example, 6 tyrosine residues (Y) in *B. stearothermophilus* are changed to 6 phenylalanine (F) in *E. coli*. Each box encloses the similar group of amino acids according to their physico-chemical properties. Group I (W, Y, F, L, I, M, V): non-polar, large a.a. Group II (V, A, P, G, T, S): non-polar or slightly polar, relatively small a.a. Group III (T, S, N, D, Q, E, H): polar, relatively small a.a. Group IV (Q, E, H, K, R): polar, relatively large a.a. Group V (C): special a.a. The a.a. pairs under dotted lines are only incompletely determined. From ref. 26.

requal to that of *E. coli* at the homologous site. Here, 20 amino acid species are classified into five groups according to their physico-chemical properties (see five boxes in Table 2). The number in Table 2 shows that most of the amino acid substitutions in ribosomal proteins occurred between amino acids of similar physico-chemical properties. The situation is also essentially the same between *E. coli* and *B. subtilis*. These facts would be in favor of the view that the amino acid substitutions in eubacterial ribosomal proteins are mostly neutral.

McLachlan (42, 43) and Grantham (44) pointed out that, in many protein species, the substitutions of amino acids during evolution occur between similar amino acids; the fixation of mutation between dissimilar amino acids is generally rare. Therefore, the tendency of amino acid substitutions in eubacterial ribosomal proteins may not be very much different from that in other proteins.

BACTERIAL TAXONOMY BY RIBOSOMAL COMPONENTS

It was shown in the first section of this paper that the 5S RNA is a molecule suited to study the phylogenic relationships among organisms. This molecule is especially useful in the phylogenic studies in bacteria, because the molecules such as hemoglobin and typical cytochrome c, which have been widely used in molecular phylogeny in eukaryotes, do not exist in bacteria, and the bacterial proteins of a general occurrence have not yet been studied extensively. The phylogenic positions of various bacterial species can be seen in the phylogenic tree of 5S RNA (see Fig. 4), although there exist some uncertainties in the early branches. Except for *Thermus aquaticus*, all eubacteria studied can phylogenically be classified into two groups, one containing enterobacteria, photobacteria, *Pseudomonas* and blue green algae, and the other containing *Bacilli*, *Lactobacillus*, *Streptococcus* and *Clostridium*. In bacteriology, all the eubacteria in the first group, which have the 120-N type 5S RNA, belong to gram-negative and those in the second group, which have the 116-N type 5S RNA, to gram-positive. Thus the classification by 5S RNA of bacteria at this level agrees very well with that by the gram-staining.

Generally speaking, the taxonomic positions of individual species within each group seem to be reasonable. In the first (gram-negative) group, blue green alga is the most separated one from other gram-negative bacteria; all the enterobacteria are very closely related, etc. In the second (gram-positive) group, *Bacillus* species are moderately related to each other, and the extent of resemblance among these *Bacilli* seems to agree with that deduced in classical bacteriology, except for *B. brevis* which is rather unrelated to other members of *Bacillus* when deduced from the 5S RNA data.

It should be stressed here that enterobacteria, which are classified into many different genera, seem to be very close to each other so that meaningful phylogenic relationships cannot be established from the 5S RNA sequences. For example, the 5S RNA sequence of *Salmonella typhimurium* differs from that of *E. coli* by two nucleotide substitutions in 120 nucleotide length. The sequence of *Aerobacter aerogenes* also contains only three nucleotide changes as compared with *E. coli* sequence. On the other hand, many species, which are taxonomically classified into the same genus, *Bacillus*, seem to be only moderately close to each other so that reliable relationships can be established. Even *B. subtilis* and

B. licheniformis, which seem to be most closely related to each other among the listed *Bacillus* species, have 5S RNA sequences with 8 nucleotide substitutions between them. The sequence of *B. megaterium* contains 17 nucleotide substitutions as compared with *B. stearothermophilus*. These facts suggest that taxonomic criteria such as Genus, Family, etc. in bacteriology are rather conventional and not at all equivalent in different bacterial groups. Such relationships in enterobacteria and in *Bacilli* are further evidenced from the analyses of ribosomal proteins by carboxymethyl-cellulose chromatography (45-48) or two-dimensional gel electrophoresis (49, 50). For example, the location of individual ribosomal proteins from *E. coli* and *Salmonella* is very similar in the chromatographic profile; only seven proteins can be distinguished among some 50 protein species (45, 47, 48). On the contrary, one can find almost no common ribosomal proteins among *Bacillus subtilis*, *B. megaterium* and *B. cereus* (45). The observations mentioned above indicate that, in molecular taxonomy of bacteria, the sequences of 5S RNA are useful for moderately separated bacteria; this molecule is, however, not suited for the phylogenic studies of closely related species, because 5S RNA is a molecule with no nucleotide substitution for million of years. On the other hand, ribosomal protein composition is a useful characteristic for studies of phylogenic taxonomy of closely related species such as those in *Enterobacteriaceae**. Thus the proximity of about 70 bacterial strains belonging to the family *Enterobacteriaceae* was quantified by co-chromatographing differentially labelled ribosomal proteins from two strains in various combinations (47, 48). They could be classified into 12 groups (conventionally called "species" here), each of which includes bacteria whose ribosomal protein composition reveals more than 90 % similarity. The fraction of proteins which differ chromatographically from each other between two species (Pd) is then estimated by dividing the number of different proteins (dpp) by the number of proteins compared (Npp), i.e., Pd = dpp/Npp. Supposing chromatographically distinguishable allelic proteins A, A' and A'' from three bacteria, it is possible that protein A' differs from A by one chance, while A'' differs from A by more than one, etc. Therefore, Pd

---

* The ribosomal proteins can also be used for the phylogenic taxonomy in eukaryotes (60, 61), where the difference of 5S RNA sequences is too small to estimate their phylogenic relationship. For example, chicken and mammalian 5S RNA sequences reveal over 96 % identity, while 24 out of some 70 ribosomal proteins can be distinguished electrophoretically between chicken and rat liver (60).

should be converted to the mean fraction, Kpp, according to Eq. [1] where Kaa is replaced by Kpp.

The phylogenic tree that has been constructed by using the Kpp values is shown in Figure 8 (48). The method of determination of the branching points in the figure is the same as in the case of 5S RNA. Generally, the rate of amino acid substitution (Kaa) (or Knuc in the case of 5S RNA) is used to estimate phylogenic relationships. However, the probability of the occurrence of a net charge difference between two homologous proteins due to a single amino acid substitution has been theoretically and empirically estimated to be 0.3 or 0.4 (51). Thus it is reasonable to assume that the chromatographic difference of ribosomal proteins, on average, due to about three amino acid exchanges. Kpp is then roughly proportional to Kaa. The phylogenic relationships seen in Figure 8 are in general agreement with those of intergeneric recombination (52), DNA hybridization (53-55), immunology of alkaline phosphatase (56, 57), and sequences of tryptophan synthetase subunits (58).

The 16S(18S) rRNA has been used by Woese and his co-workers for the molecular taxonomy of various organisms, especially of methanogenic bacteria (13, 18, 20, 59).

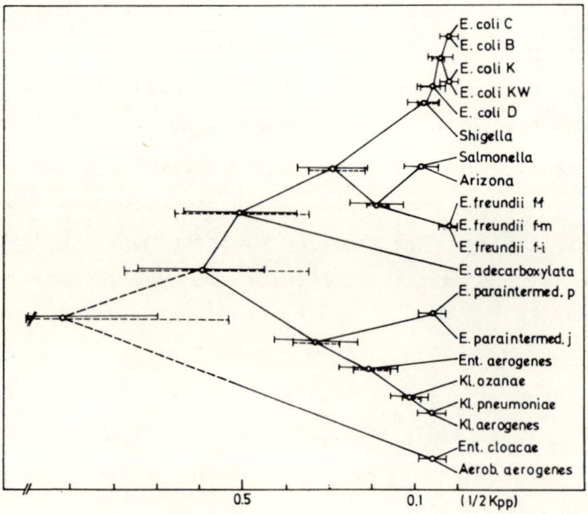

Figure 8. Phylogenic tree of enterobacteria. The Kpp values were taken as the relative time scale in the abscissa. ├---o---┤ Range of the standard error of Kpp. ├───●───┤ Range of 1/2 Kpp of all pairs at the branching point. From ref. 48.

ACKNOWLEDGEMENTS

We wish to express our appreciation to Drs. Motoo Kimura and Tomoko Ohta of the National Institute of Genetics, Mishima, who introduced us the field of molecular evolution. Thanks are also due to Drs. Eiko Otaka, Takuzi Itoh and Ken-ichi Higo of our laboratory who collaborated in the works cited in this article. The works from our laboratory were supported by grants 220623 and 348375 from the Ministry of Education.

REFERENCES

1. Zuckerkandl, E., and Pauling, L. (1965) In: Evolving Genes and Proteins, eds., Bryson, V., and Vogel, H. J., pp. 97-166, Academic Press, New York.

2. Kimura, M., and Ohta, T. (1974) Proc. Natl. Acad. Sci. USA, 71, 2848-2852.

3. Schwartz, R. M., and Dayhoff, M. O. (1978) In: Atlas of Protein Sequence and Structure, ed., Dayhoff, M. O., vol. 5, Suppl. 3, pp. 29-44, National Biomedical Research Foundation, Washington, D. C.

4. Holmquist, R., Cantor, C., and Jukes, T. H. (1973) J. Mol. Biol. 64, 145-161.

5. Erdmann, V. A. (1978) Nucleic Acids Res. 5, r1-r14.

6. Tinoco, I. Jr., Uhlenbeck, O. C., and Levine, M. D. (1971) Nature, 230, 362-367.

7. Fox, G. E., and Woese, C. R. (1975) Nature, 256, 505-507.

8. Hori, H. (1976) Mol. Gen. Genet. 145, 119-123.

9. Hori, H., and Osawa, S. (1979) Proc. Natl. Acad. Sci. USA, 76, 381-385.

10. Matheson, A. T., Yaguchi, M., Nazar, R. N., Visentin, L, P., and Willick, G. E. (1978) In: Energetics and Structure of Halophilic Microorganisms, eds., Caplan, S. R., and Ginzburg, M., pp. 481-501, Elsevier/North-Holland, Amsterdam.

11. Erdmann, V. A., Sprinzl, M., and Pongs, O. (1973) Biochem. Biophys. Res. Commun. 54, 942-948.

12. Herr, W. and Noller, H. F. (1975) FEBS Lett. 53, 248-252.

13. Woese, C. R., and Fox, G. E. (1977) Proc. Natl. Acad. Sci. USA, 74, 5088-5090.

14. Jukes, T. H., and Cantor, C. R. (1969) In: Mammalian Protein Metabolism, ed., Munro, H, N., pp. 21-132, Academic Press, New York.

15. Kimura, M., and Ohta, T. (1973) Nature New Biol. 243, 199-200.

16. Hori, H. (1975) J. Mol. Evol. 7, 75-86.

17. Nazar, R. N., Matheson, A. T., and Bellemare, G. (1978) J. Biol. Chem. 253, 5464-5469.

18. Balch, W. E., Magrum, L. J., Fox, G. E., Wolfe, R. S., and Woese, C. R., (1977) J. Mol. Evol. 9, 305-311.

19. Woese, C. R., Magrum, L. J., and Fox, G. E. (1978) J. Mol. Evol. 11, 245-252.

20. Magrum, L. J., Luehrsen, K. R., and Woese, C. R. (1978) J. Mol. Evol. 11, 1-8.

21. Amons, R., Agthoven, van A., Pluijms, W., Möller, W., Higo, K., Itoh, T., and Osawa, S. (1977) FEBS Lett. 81, 308-310.

22. Amons, R., Pluyms, W., and Möller, W. (1979) FEBS Letters, in press.

23. Sneath, P. H. A. (1974) In: Evolution in the Microbial World, eds., Carlile, M. J. and Skehel, J. J., pp. 1-39, Cambridge Univ. Press, Cambridge.

24. Nomura, M., Traub, P., and Beckmann, H. (1968) Nature, 219, 793-799.

25. Higo, K., Held, W., Kahan, L., and Nomura, M. (1973) Proc. Natl. Acad. Sci. USA, 70, 944-948.

26. Hori, H., Higo, K., and Osawa, S. (1977) J. Mol. Evol. 9, 191-201.

27. Isono, K., Isono, S., Stöffler, G., Visentin, L. P., Yaguchi, M., and Matheson, A. T. (1973) Mol. Gen. Genet. 127, 191-195.

28. Wrede, P., and Erdmann, V. A. (1973) FEBS Lett. 33, 315-319.

29. Itoh, T., and Wittmann-Liebold, B. (1978) FEBS Lett. 96, 399-402.

30. Wrede, P., and Erdmann, V. A. (1977) Proc. Natl. Acad. Sci. USA, 74, 2706-2709.

31. Cedergren, R. J., and Sankoff, D. (1976) Nature, 260, 74-75.

32. Hori, H., Higo, K., and Osawa, S. (1977) In: Molecular Evolution and Polymorphism, ed., Kimura, M., pp. 240-260, Mishima.

33. Terhorst, C., Möller, W., Laursen, R., and Wittmann-Liebold, B. (1973) Eur. J. Biochem. 34, 138-152.

34. Itoh, T., and Wittmann-Liebold, B. (1978) FEBS Lett. 96, 392-394.

35. Brosius, J., Palmer, M. L., Kennedy, P. J., and Noller, H. F. (1978) Proc. Natl. Acad. Sci. USA, 75, 4801-4805.

36. Carbon, P., Ehresmann, C., Ehresmann, B., and Ebel, J. P. (1978) FEBS Lett. 94, 152-156.

37. Shine, J., and Dalgarno, L. (1975) Nature, 254, 34-38.

38. Hagenbüchle, O., Santer, M., Steitz, J. A., and Mans, R. J. (1978) Cell, 13, 551-563.

39. Alberty, H., Raba, M., and Gross, H. J. (1978) Nucleic Acids Res. 5, 425-434.

40. Schwarz, Z. S., and Kössel, H. (1979) Nature, in press.

41. Higo, K., Itoh, T., and Osawa, S. (1978) In: Evolution of Protein Molecules, eds., Matsubara, H., and Yamanaka, T., pp. 197-207, Japan Sci. Soc. Press, Tokyo.

42. McLachlan, A. D. (1971) J. Mol. Biol. 61, 409-424.

43. McLachlan, A. D. (1972) J. Mol. Biol. 64, 417-437.

44. Grantham, R. (1974) Science, 185, 862-864.

45. Otaka, E., Itoh, T., and Osawa, S. (1968) J. Mol. Biol. 33, 93-107.

46. Osawa, S., Takata, R., and Dekio, S. (1970) Mol. Gen. Genet. 107, 32-38.

47. Osawa, S., Itoh, T., and Otaka, E. (1971) J. Bacteriol. 107, 168-178.

48. Hori, H., and Osawa, S. (1978) J. Bacteriol. 133, 1089-1095.

49. Geisser, M., Tischendorf, G. W., Stöffler, G., and Wittmann, H. G. (1973) Mol. Gen. Genet. 127, 111-128.

50. Geisser, M., Tischendorf, G. W., and Stöffler, G. (1973) Mol. Gen. Genet. 127, 129-145.

51. Nei, M., and Chakraborty, R. (1973) J. Mol. Evol. 2, 323-328.

52. Demerec, M., and Ohta, N. (1964) Proc. Natl. Acad. Sci. USA, 52, 317-323.

53. Falkow, S., and Citarella, R. V. (1965) J. Mol. Biol. 12, 138-151.

54. McCarthy, B. J., and Bolton, E. T. (1963) Proc. Natl. Acad. Sci. USA, 50, 156-164.

55. Schildkraut, C. L., Marmur, J., and Doty, P. (1961) J. Mol. Biol. 3, 595-617.

56. Cocks, G. T., and Wilson, A. C. (1972) J. Bacteriol. 110, 793-802.

57. Steffen, D. L., Cocks, G. T., and Wilson, A. C. (1972) J. Bacteriol. 110, 803-808.

58. Li, S.-L., Drapeau, G. R., and Yanofsky, C. (1973) J. Bacteriol. 113, 1507-1508.

59. Fox, G. E., Magrum, L. J., Balch, W. E., Wolfe, R. S., and Woese, C. R. (1977) Proc. Natl. Acad. Sci. USA, 74, 4537-4541.

60. Ramjoué, H.-P. R., and Gordon, J. (1977) J. Biol. Chem. 252, 9065-9070.

61. Fischer, N., Stöffler, G., and Wool, I. G. (1978) J. Biol. Chem. 253, 7355-7360.

62. Kimura, M., and Ohta, T. (1972) J. Mol. Evol. 2, 87-90.

JUST SO STORIES AND RUBE GOLDBERG MACHINES: SPECULATIONS ON THE ORIGIN OF THE PROTEIN SYNTHETIC MACHINERY

CARL R. WOESE

Department of Genetics and Development,
Urbana, IL 61801

PROLOGUE

Its organizers have given this Symposium the laudable but immodest objective of generating a volume that serves both as a source book and as a guide for future research on ribosomes. Comprehensive reviews would seem to fulfill the first requirement. The second is another matter. To project the future course in a useful way one's present understanding, however rudimentary, has to be basically correct. The basis for understanding translation is the well-known A-site-P-site model. To this other concepts have been joined -- e.g. the idea of tRNA as an "adaptor", the idea of translation being some sort of "indirect templating" process (whatever these terms mean). Then there are prejudices concerning how the translation function is divided between tRNA and ribosome, and what roles ribosomal proteins play in the process as opposed to ribosomal RNAs.

This conceptual potpourri is no paradigm for the future; it already serves little purpose. The field today seems inundated by facts. No single investigator even attempts to keep abreast of all developments any longer. Yet, the problem is not facts *per se*; it is a conceptual structure incapable of weaving the facts into a coherent, comprehensible picture. Our paradigm for translation is ill-defined, yet it severely constrains our thinking. It predicts little and is difficult to prove or disprove. Over the years it has evolved in an *ad hoc* way, without significant refinement. These are not the characteristics of a vital, correct paradigm.

Given this situation even the most complete reviews are of limited value. The facts that go into them are necessarily selected (just as there is selection of what facts are included or emphasized in the original reports). More often than not important facts that do not accord with the existing paradigm -- and so are perceived as unimportant -- are weeded out in the process. If we are to produce a treatise to guide future research, we need to do more at this point than document facts; we must reexamine our whole picture of translation.

The organizers of this Symposium have asked me to speak on the topic "Speculations on the Origin of the Protein Synthetic Machinery", which I have appropriately retitled "Just-So-Stories, and Rube Goldberg Machines". The topic is a challenging but frustrating one. It is challenging because in order to address it properly one is forced into the much-needed reexamination of our concept of translation and its relationship to Biology as a whole. It is frustrating for two reasons: For one, unavoidably I will have to present a Just-So Story. What do we really know about how translation works at the molecular level? We know nothing! How then, does one explain the evolution of an unknown mechanism? By a Just-So Story! My second reason is that this presentation will at best elicit a ho-hum response; the field is atune to a paradigm that sees little value in understanding how translation evolved. From the set of codon assignments on down all facets of the translation mechanism are taken as arising by "historical accident", as being unrepeatable evolutionary events. This can and has given rise to the prejudice that the translation apparatus is basically a Rube Goldberg Machine -- some incongruous assemblage of parts, where knowing even ninety percent of the mechanism would not permit one to predict the remaining ten percent. The Rube Goldberg view not only generates disinterest in the mechanism's evolution, but also leads to a feeling that there is no point in attempting to think, to theorize about a mechanism that is unknowable _a priori_; one's approach needs to be "strictly empirical". If this presentation serves no other purpose, I should like it to raise the issue of the design of the translation apparatus: Is it really a Rube Goldberg Machine? Is there a simple mechanism at its core? By what principles does it achieve its low noise level? Does it possess an understandable evolutionary structure?

The presentation henceforth consists of three parts: The first is a brief conceptual review of the field, designed to reveal the nature of the current paradigm. The second is an attempt to get at certain basic features of the mechanism, while the third uses these to construct a Just-So Story, the origin and evolution of the whole.

DEVELOPMENT OF THE CONCEPT

The roots of our picture of translation are in the idea that proteins are formed by direct templating (1). Although this idea has long been discredited, its residuum still appears to shape our thought. In formulating the Adaptor Hypothesis Crick did not--as is commonly believed--reject templating, merely direct templating (2). He replaced the latter with "indirect templating", through the intermediary of a molecular prosthesis for the amino acid, an "adaptor". When tRNAs were then discovered, biologists accepted these as "adaptors" -- despite their being some ten to twenty times larger than had been postulated! [With an "adaptor" this complex it is debatable whether "indirect templating" is a meaningful concept.] And so the die was cast; protein synthesis was seen as a "templating" process, albeit "indirect".

The templating, of course, was found not to be a general aligning of (adapted) amino acids followed by their random joining; it was more in the nature of a linearly propagating wave of synthesis, more like typing or a tape reading process (3). By the arrival on the scene of message RNA, the stage was well prepared for seeing the ribosome as the mechanism that mediates this wave of templating. In this way the relationship of ribosome to tRNA became that of a machine to some unit processed by that machine.

Whatever specific understanding we do have of the workings of the translation mechanism is centered around the A-site-P-site picture, which has become a dogma (4-5). This hypothesis appears to have arisen as an explanation for the fact that two tRNAs can simultaneously exist on the bacterial ribosome. In our minds, and in those of biology students and informed laity, there exist distinct images of a tRNA entering one depression (A site) on the ribosome, and passing to another (P site), from which it is ultimately ejected. In the process the growing peptide is transferred from one tRNA to the next in line.

This explanation has long ago passed the point (if there ever was one) of being a useful hypothesis. It is deceptively simple: It lulls us into thinking we understand when we do not; it is marginally predictive, at best; it is so ill-defined it is near impossible to disprove. Yet the model is quite constraining conceptually. To give a prime example -- for over a decade we have believed and taught that a tRNA is physically moved from an A site to a P site; this "translocation", is so thoroughly accepted that we have invented an enzyme, "translocase", to oversee it. Significantly, we have never felt an obligation to try to prove the idea experimentally. For all we know "translocation" could be a fiction -- an unfortunate one if it keeps us from uncovering the mechanics of translation.

In approaching the mechanism, the ribosome, we have always tended to focus on the proteins; there are so many different kinds (6-7)! It is perfectly reasonable _a priori_ (if the mechanism's evolution is not considered) to assume that its protein components define the function of the ribosome. A decade or more of research on ribosomal reconstitution was predicated on this tacit assumption (8-9). Unfortunately, no aspect of ribosome function (other than catalysis of peptidyl transfer) has so far been shown to reside in a ribosomal protein; no individual proteins have all-or-none effects on ribosome function (except for those whose absence prevents assembly of the particle). It is time to question the original prejudice.

The dogma that tRNA is simply an "adaptor" has had, I feel, a particularly stultifying effect on our understanding of translation. The adaptor conjures up strong visual imagery of something that is both passive and static. It leads us to see a tRNA being passively processed by a machine, something moving down an assembly line. Ribosome and tRNA thus exist in a sort of dichotomous relationship, and the field then easily divides into two conceptually separate camps. And then we picture tRNA the adaptor as having a single functional configuration. How many times have we seen reference made to _the_ (tertiary) structure of tRNA? This tendency was particularly pronounced among the X-ray crystalographers working on this problem (10-11). Given that tRNA comprises four arms connected by what is in effect a universal joint, one indeed must have strong prejudices to assume _a priori_ that the molecule has a single functionally significant configuration.

The historical evidence is by now reasonably clear. The ways in which we look at translation have proven unproductive of an understanding of the process. Emphases have been wrongly placed. Interpretation has lagged behind accumulation of data. Progress in the field is largely a function of technological advance coupled to serendipity. The time has come to discharge the old preconceptions and to attempt to look at translation in new, imaginative ways. The sooner we do so the faster will be our progress.

THE TRANSLATION APPARATUS TODAY

Understanding translation begins with the simple realization that the essence of the process is movement, the sequential processing of message RNA with the concomitant sequential output of a polypeptide. Any such mechanism will have moving parts; the biologist's task is to identify them and define the movement.

Although nothing can be said with certainty about the actual molecular mechanics of translation, we are finally coming to realize where the basis for the mechanism lies. As stated above, to this point in time attention has been focussed on the protein components of the mechanism, except for tRNA (which is treated as is message RNA -- i.e. as something processed _by_ the mechanism). It is now becoming apparent that the interactions that define translation (as opposed to those that facilitate it, make it more precise, and so on) probably involve its RNA components, not its protein components. This point can be argued from several perspectives. From an evolutionary viewpoint a chicken-and-egg paradox results from assuming that the defining interactions in translation reside in the protein components: A translation mechanism is already required to evolve those proteins that would be required to build that translation mechanism in the first place (12). The paradox could be avoided, of course, if nucleic acids were to define the mechanics of translation initially. From an experimental viewpoint ribosome reconstitution experiments, as mentioned, give no evidence that the protein components define translation. Moreover, the phenomenon of "non-enzymatic" translation shows that the process can occur in the absence of certain proteins originally thought to be important if not essential to it (13-14). RNA presents a different picture: rRNA sequence tends to be highly conserved phylogenetically, which seems difficult to rationalize were it to act merely as a scaffolding that positions the ribosomal proteins (15). Also minor modifications in rRNA sequence tend to produce sharp, all-or-none effects on translation -- which is not generally the case for the proteins (16-17).

Ribosomal RNA can no longer be viewed as a static structural element, as a scaffolding upon which the function defining proteins are positioned. Ribosomal RNA is capable of mechanism. The initial suggestion of this was made on theoretical grounds (18): It was proposed that tRNA was not the static adaptor it is conventionally taken to be, but rather a small machine, a ratcheting device that brings about the movement of

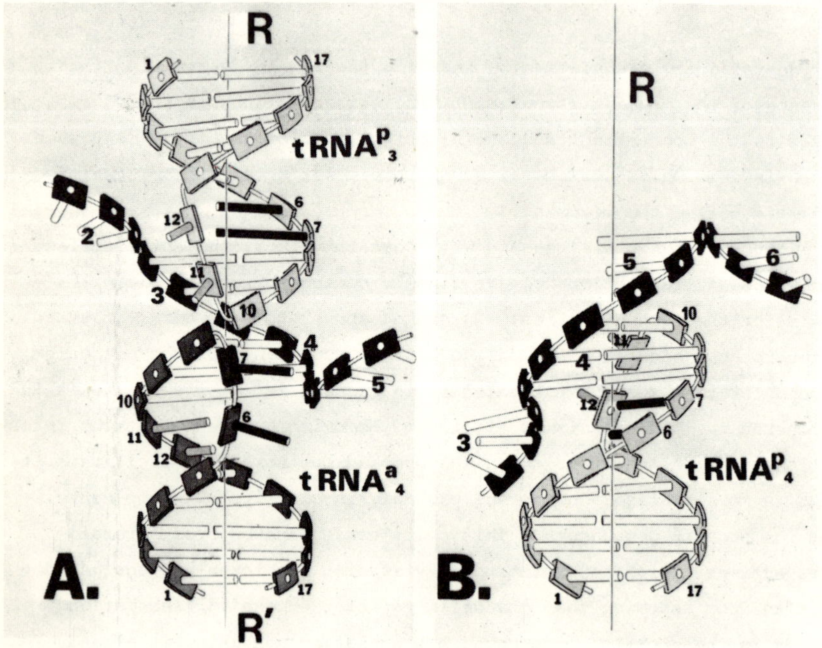

Figure 1. Molecular Mechanics of Translation. The anticodon arm of tRNA, by undergoing a transition from the conformation shown in A (hf or -) to that shown in B (FH or +), can exist in a "translation complex" with the preceeding or succeeding tRNA respectively. Each transition moves the message RNA one codon by what in effect is a ratcheting action. See refs. 18 and 32 for details.

message RNA in translation by means of allosteric transitions in the anticodon arm -- see Figure 1. [Although proposed almost a decade ago, the suggestion that tRNA is an allosteric part in the translation mechanism appears to have been too novel to be generally acceptable: the discussions of tRNAs role in translation based upon its three dimensional structure, for example, never took serious note of this possibility (19-20); also, the citation index indicates a lack of overt interest. In retrospect this episode serves to demonstrate the powerful constraining effect of intrenched ideas -- in this case the view of tRNA as static and passive. That a conceptual change is now in process is indicated by recent publications that incorporate the molecular ratcheting principle for tRNA (21-22).]

The idea of an allosteric anticodon arm suggests that in general the various "loop" structures in ribosomal RNAs and tRNAs are allosteric (23-24). The common arm of tRNA, which is isomorphic with the anticodon arm, is one reasonable candidate for allosteric transitions (18). Another seems to be the central helix in the 5S rRNA, the so-called "tuned helix" (25-26). Such a paradigm would envision ribosomal and transfer RNAs, then, as collections of allosteric elements, that through their state transitions define and provide the basis for the movement and specificity in the translation process.

Although many other detailed properties of the translation apparatus could be listed (and are elsewhere in this volume), the simple recognition of the possible functional importance of ribosomal RNA -- together with the general properties discussed in the next section -- are the crucial known characteristics of translation as far as evolution of the apparatus is concerned.

EVOLUTIONARY CONSIDERATIONS

The complexity of the translation apparatus is an important clue to its evolution: The mechanism comprises over a hundred macromolecular components. This implies two things: (1) It is impossible to evolve such a mechanism in a single step; its evolution must have occurred in stages, and therefore, (2) translation should be possible using a far simpler apparatus than now exists. Another of its general properties also carries an important evolutionary implication: Translation as we know it is a very precise process, codon misreadings occurring at the level of several parts in ten thousand or less (27). That degree of accuracy could not have been present initially; it too must have evolved.

From these inferences we can construct the following general evolutionary picture: In an unspecified way some rudimentary form of translation arises. This mechanism (the makeup of which is discussed below) was far simpler than its modern counterparts. Since the accuracy of tRNA charging, for example, relies upon sophisticated aminoacyl synthetases -- enzymes capable of error correction -- primitive translation must have been inaccurate at very best. It may well have been worse than that; codon-amino acid pairings could have been ambiguous (28-29). In other words, inability of the mechanism to distinguish among certain (related) amino acids and various (related) codons as well, could result in "group" codon assignments -- the amino acids in question being assigned <u>as a group</u> to the corresponding group of codons. In this way the number of kinds of amino acids placed into protein might be far larger than the twenty now utilized, but the number of <u>distinguishable</u>

amino acids would be far smaller than twenty (28).

Such a mechanism would synthesize what could be called "statistical proteins" (12, 28-29) -- each message RNA being translated into a family of related but non-identical polypeptide sequences. It might seem that statistical proteins would be of no use to the cell. I claim they would be. They are far from random sequences, being somewhat analogous to (the variable regions in) antibodies today. If they existed, such proteins would probably be rather small. The type of cell that could be fashioned from them would be very different from cells today; it would be remarkable for a low degree of biological specificity.

Subsequent evolution of this mechanism would then proceed through stages in which the ambiguous (group) assignments were refined to individual (unique) codon assignments, and the individual assignments further refined in terms of their accuracy. In summary then, the evolution of translation is a boot-strapping process. I can envision its having three major stages (24) -- (1) the initial establishing of the rudimentary mechanism (2) the subsequent evolution to individual, unique codon assignments, and (3) the final stage, in which the process becomes a very accurate one.

One can readily see that a translation process evolving in this way would place general constraints on the nature of the cell at each stage, through limiting the kinds of proteins that could evolve -- both in size and general properties. The nature of enzymatic pathways, of genetic repair processes, and so on, would all be affected. It follows that these constraints would limit the size of the cell's genome as well; mutation rates as low as 10^{-8} per base pair per replication are a modern luxury, dependent upon sophisticated polymerases and repair systems (30). What we are seeing at this early stage of evolution is biological specificity coming into existence. The evolution of biological specificity goes hand in hand with the evolution of an increasingly accurate linkage between genotype and phenotype (31-33). The name Progenote is appropriate for primitive entities that are in the process of completing, of refining the coupling between genotype and phenotype (33).

It is important to look for evidence for the existence of these hypothetical progenotes. Such entities are unlikely to exist on the planet today, but their existence might be inferred from the properties of their descendants: Were two lines of descent to separate at a late progenote stage, for example, they would be basically similar, but they would have encountered separately the set of problems associated with completing the genotype-phenotype link. Therefore, they would differ from one another mainly in the details, the "fine tuning", of their component systems: Many aspects of control of gene expression should be different in the two cases (each line having separately increased its genome size); biochemical pathways should differ somewhat, especially the highly specific ones; polymerases and repair systems should differ; and, of course, the detailed aspects of the translation apparatus should be unique in each case.

So long as the biologist believed that there were two basic kinds of life on this planet, eucaryotes and procaryotes, the issue could not be convincingly argued. Any profound differences between the two at the molecular level could always be attributed to evolutionary events that happened during the assumed transition from procaryote to eucaryote -- i.e. at relatively advanced stages of evolution. However, we now know that the basic division among living forms is not a bipartite one. Instead of eucaryote vs procaryote, the fundamental genealogical distinction is tripartite (34-35) -- eucaryotes (their cytoplasmic aspect) vs true bacteria (eubacteria) vs archaebacteria. No two of these three are closer to one another than either is to the remaining one. Moreover, archaebacteria and true bacteria are of the same order of complexity, which makes it impossible to explain away their differences in terms of one evolving as an advanced form of the other.

The nature of the molecular differences among representatives of the three primary lines of descent, does indeed suggest that they separated from one another at a late progenote stage. For instance, archaebacterial cell walls appear unrelated to eubacterial cell walls -- suggesting each kingdom to have separately evolved the wall (35-37). Lipids in the archaebacteria (ether linked branched chain glycerides) are quite unlike their counterparts in either true bacteria or eucaryotes (38-40) -- suggesting independent evolution of aspects of biochemistry. The RNA polymerases of the archaebacteria do not resemble eubacterial polymerases in the details of their subunit structure (41). And the details of the translation apparatus appear idiosyncratic in each of the three major lines of descent (35). This last can be seen in the pattern of post-transcriptional modification of bases in rRNA and tRNA. In 16S (18S) rRNA for example, the same locales in the molecule tend to be modified in all cases, but the nature of the modification, even its exact position, are often different in the various cases -- as Table 1 shows.

TABLE 1. Modified Sequences in 16S Ribosomal RNA (42-44).

Sequence[a]	Found in[b]		
	Eubacteria	Archaebacteria	Eucaryotes (18S)
1. a) Ċ'-CCG	++	o	o
b) Ċ-CCG	o	+	++
c) Ċ-CĊG	o	+	o
d) C(C,Ċ)G	o	±	o
2. a) ȦȦG	+	o	o
b) ȦȦCCUG	+	+	++
c) ȦȦUCUG	o	+	o
3. a) U̇AACAAG	++	o	o
b) UȦȦCAAG	o	+	o
c) UAȦCAGG	o	+	++
4. a) AUĠĊAACG	+	o	o
b) AAĠĊAACG	+	o	o
c) AUṄCAACG	o	+	o
d) ACṄCAACG	o	+	o
e) ACṄ'CAACACG	o	o	++
5. a) CCĠCG	++	±	o

[a]A superscript dot denotes a modification. To the extent they are known the modifications are these: (1a) Ċ' is 4Me 2'OMe; (1b) Ċ is 2'OMe, as is the first Ċ in (1c); the remaining Ċ (1c and 1d) are cleaved by pancreatic nuclease, but differ in electrophoretic mobility from C. (2) all Ȧ are N,N di-Me. (3a) U̇ is cleaved by pancreatic nuclease. The first Ȧ in (3b) is cleaved by U2 nuclease, the second Ȧ in 3b and c is N,N di-Me. The Ġ in (4a and 4b) is 2Me; the Ċ is 5Me; Ṅ and Ṅ' are unspecified modifications, distinguishable electrophoretically from one another and from 2Me G. Ġ in (5a) is 7Me G.

[b]++, +, and ± signify occurrence in most, many, or a few members respectively of the group in question.

Similarly, tRNA base modifications tend to differ from one kingdom to another: for instance, archaebacterial tRNAs contain neither ribothymidine nor dihydrouridine, although they do possess the corresponding structures, the "common arm" and the "D loop", which contain unique counterparts of these modified bases (42, R. Gupta and C. Woese, unpublished results). The situation strongly suggests that many of the modifications, at particular loci in tRNA and rRNA, evolved separately in the various kingdoms -- implying that the common ancestor of all had not yet evolved modifications at these loci. It remains to be determined whether post-transcriptional modifications in general influence the precision of translation. If so, the case for the progenote is considerably strengthened.

According to the present view then, the bulk of the course of evolution of the translation apparatus has been involved not with establishment of the mechanism per se, but with refinement of the mechanism, through stages of greater and greater precision. The bulk of the mechanism today, should then be concerned not with the basic mechanics of translation, but with its accuracy. It appears likely that the earliest branches in the universal phylogenetic tree arose during the later stages in the evolution of translation, before full accuracy was achieved. Since evolution has in this way recorded the final stages in translation's development, we should by comparative study be able to elucidate them.

We now turn to the Just So Story, a detailed proposal for the origin and evolution of translation. If one accepted the conventional paradigm -- which sees the ribosome as the mechanism, tRNA as a thing processed by the mechanism -- it would be difficult to account for the origin of translation; too many parts are required to begin with. However, the molecular ratcheting mechanism of Figure 1 suggest an acceptable way to evolve translation, for it permits one to dispense in principle with the ribosome initially (32, 45).

There is little point in assuming, as has sometimes been done, that primitive translation occurred via a mechanism basically different than that now employed in the process; any transition between two basically different mechanisms presents as many problems as evolving one mechanism from scratch. The rudiments of translation as we know them consist of four interactions: (1) amino acid recognition and attachment, (2) peptidyl transfer, (3) codon recognition, and (4) mRNA movement. A fanciful primitive tRNA can be designed that appears to possess inherently all these properties.

The tRNA molecule today is remarkable in that two of its four arms are isomorphic: Both the anticodon arm and the common arm consist of a stalk of five base pairs which delimits a loop of seven bases. [Also the 5' two bases of the loop are generally pyrimidines in both cases.] I interpret this to suggest that the common arm evolved from the anticodon arm (or vice versa). In other words, tRNA today evolved from a smaller primitive version. In any case it seems necessary to make such an assumption, for tRNA today seems too large to have been primitive.

I will assume then, that the primitive version of tRNA was about the size and shape of the anticodon arm of today's tRNA. Two of the translation functions, codon recognition and mRNA movement can be easily rationalized for such a structure in terms of the ratcheting mechanism discussed above (Fig. 1). The mechanism has two details worth stressing: The model derives the codon size from the size of the anticodon loop. Loops of size seven lead to triplet codons (32, 45). Furthermore, the mechanism automatically maintains correct reading frame; even if the one or two bases immediately distal to a given codon were to pair -- making in effect a tetra- or penta-codon -- the additional base pair(s) would be eliminated when the ratchet mechanism flips from one conformation to the other (32, 45). Crick *et al.*, have used this feature of the ratchet mechanism to design a rather ingenious primitive coding scheme -- which invokes a penta-nucleotide codon read in a partially overlapping fashion, which in addition yields a comma-free code (22).

The amino acid recognition function is more difficult to rationalize in any case, because it generates the chicken-and-egg paradox if synthetases have to be invoked (12, 45). To avoid the paradox the primitive tRNA must be able to discriminate among amino acids in its own right, no matter how crude the distinction. This function should reside in the anticodon loop, but not necessarily be confined to the anticodon *per se*. It is reasonable then to attach the amino acid somewhere in the loop, not at a remote site as occurs today in enzyme mediated tRNA charging.

Almost a decade ago I made a specific proposal for the primitive amino acid attachment site, using the following argument (45): The base immediately adjacent (3') to the anticodon in tRNA today -- in almost all cases an A residue -- is unusual in that it is highly modified in most instances. The nature of the modification tends to correlate with the anticodon composition (46). In some cases the modification involves an amino acid, threonine. [The threonine tRNA is one of those so modified.] The presence of these anticodon-adjacent substitutions has a profound effect on the physical properties of tRNA (29, 47). The unique nature of this substitution suggests that it evolved for quite different reasons and in a different way than the other modifications of tRNAs; this modification may not represent a mere "fine tuning" of tRNA function. I feel this particular substitution to be a vestige of an aboriginal attachment of the amino acid and growing peptide chain; specifically, that attachment would be to the amino group of an anticodon-adjacent A residue (45).

Thus amino acid attachment and peptidyl transfer would occur in the immediate vicinity of the mRNA ratcheting mechanism. In fact peptidyl transfer might somehow serve as the pall for the ratchet (45). [Model building has demonstrated the proposed interactions to be sterically feasible.]

If so simple a device can be invested with the translation functions, then what is the purpose of employing so elaborate a mechanism today? The answer, of course, has been given -- precision. A simple device can not avoid thermal buffeting; to function accurately, it must be nearly immune to thermal noise, and so must be properly large (23). There is little point in compounding the Just-So-Story by adding a chapter here on the transition from the primitive noise-ridden apparatus to its modern, noise-free counterpart. That is more usefully done when something concrete is known about the molecular mechanics of ribosome function -- particularly rRNA function. The present scheme certainly does not readily suggest a specific origin for the ribosome.

In closing, I would like to add one note concerning the hypothetical predecessor to the primitive mechanism postulated here. The latter is unsatisfying in that it appears out of nowhere. The question that should be addressed is whether there were interactions, themselves not actually translation, that were predecessors to that process. In the formulation presented here the tape reading movement, the ratcheting, is taken as basic to translation. [By way of contrast we tend to picture the origin of gene replication in a manner that introduces the tape reading mechanics as not fundamental to the process, as a later addition that makes the process more precise.] Ratcheting seems an inherent property of nucleic acids (RNA) in certain loop-stalk configurations -- configurations simple enough perhaps to have been spontaneously formed in a primitive environment (48). Among the likely interactions among these structures would be the formation of couples paired through complementary sequences in their respective loops -- as two tRNAs are known to pair together when their anticodons are complementary (49). This process might also lead to the formation of polypeptides in a non-translational way analogous to the primitive mechanism just discussed -- i.e. by a ratcheting mechanism that involved no mRNA, but instead direct contact of two "tRNAs" whose "anticodons" were complementary. The peptides produced would be interesting in that they would be of alternating sequence, repeats of a dimeric unit (45).

It has been pointed out that alternating sequence polypeptides could serve either of two functions in a primitive context -- one as membranes, β-sheets, that have a hydrophobic and hydrophilic side (50), the other as an alternating basic--nonbasic sequence that acts in nucleic acid replication, either to helicize single stranded nucleic acid or perhaps unwind double stranded nucleic acid (51).

Since we have been charged by the organizers of this symposium with pointing future directions in research, I will attempt in closing to do so.

Translation is a multifaceted problem. For one, it is a problem in molecular mechanics. We have before us what could be the most elegant mechanism ever created. [When it is understood -- which will be in the 1980's -- it is likely to revolutionize the teaching of and reorient interest in Biology.] The essence of this mechanism appears to be in its RNA components. Given the breakthroughs in nucleic acid sequencing that have recently occurred, the ribosomal RNAs now come within the purview of the biologist, and spectacular strides in understanding these molecules should occur within the next few years. Transfer RNA appears to be the center of the mechanism; it is not as previously supposed, something processed <u>by</u> the mechanism in an assembly line fashion. The last five years or so have seen significant progress in understanding the overall structure of tRNA in isolation. The real challenge now comes in trying to understand its conformations during translation -- the tRNA-tRNA interactions and tRNA-ribosome interactions at various stages in the process. Also important are the ribosomal RNA secondary structures; the dynamics of translation is written in terms of their interconversions.

Then, too, translation is a problem in biological specificity. The process is incredibly accurate -- distinguishing among fifty or so codons and twenty amino acids, with an accuracy of better than one part in a thousand. By what mechanisms is this accomplished? Can such accuracy be explained solely in terms of a classical picture, a model, of the machine? To what extent is the accuracy a static (lock and key) as opposed to a dynamic feature of the mechanism? The problem of speed vs accuracy vs size (complexity) of mechanism should come into focus. Attention should return to mutations, antibiotics, and so on, that alter speed and accuracy of the process. Specific alterations in tRNA and rRNA primary structure and mRNA composition, should be useful in defining the basis of accuracy.

In addition to the details of translation accuracy, there exists a larger question, the general nature of biological specificity -- for translation is indeed a prime example of, a paradigm for, biological specificity. Is our molecular, lock-and-key, static formulation an optimal one, or are there very different, more unifying ways to conceptualize biological specificity?

Above all translation is a problem in evolution. Translation defines the gene. It (indirectly) limits the genome size. It defines, again by limiting, the types of proteins that can evolve. Translation is central to the evolution of the cell, of biological specificity. Comparative studies of the translation mechanism will be used not only in defining the secondary structure and so on of the rRNAs, but to define stages in the evolution of the mechanism and so stages in cellular evolution. The ultimate goal of Biology is to explain how living systems arose on this planet; the evolution of translation holds the key to the problem.

REFERENCES

1. Pauling, L. and Delbruck, M. (1940) Science. 92, 77-79.
2. Crick, F. H. C. (1957) Discussion in: The Structure of Nucleic Acids and Their Role in Protein Synthesis. Biochem. Soc. Symp. No. 14, pp. 25-26. Cambridge University Press.
3. Dintzis, H. M. (1961) Proc. Natl. Acad. Sci. U.S. 47, 247-261.
4. Watson, J. D. (1964) Bull. Soc. Chem. Biol. 46, 1399-1425.
5. Erb, R. W., Nau, M. M., and Leder, P. (1969) J. Mol. Biol. 38, 441-444.
6. Waller, J-P. (1964) J. Mol. Biol. 10, 319-328.
7. Wittmann, H. G. (1974) In Ribosomes, M. Nomura, A. Tissiers, and P. Lengyel (eds). Cold Spring Harbor Laboratory. Cold Spring Harbor, New York. pp. 93-114.
8. Traub, P. and Nomura, M. (1968) Proc. Natl. Acad. Sci. U.S.A. 59, 777-782.
9. Normura, M. and Held, W. A. (1974) In Ribosomes, M. Nomura, A. Tissiers, and P. Lengyel (eds). Cold Spring Harbor Laboratory, Cold Spring Harbor, New York. pp. 193-224.
10. Roberts, J. D., Ladner, J. E. Finch, J. T. Rhodes, D., Brown, R. S., Clark, B. F. C., and Klug, A. (1974) Nature 250, 546-551.
11. Kim, S. H., Suddath, F. L., Quigley, G. J., McPherson, A. Sussman, J. L., Wang, A. H. J., Seeman, N. C., and Rich, A. (1974) Science 185, 435-440.
12. Woese, C. R. (1967) The Genetic Code: the Molecular Basis for Genetic Expression. Harper and Row, New York.

13. Pestka, S. (1968) J. Biol. Chem. 243, 2810-2820.

14. Gavrilova, L. P. and Spirin, A. S. (1974) FEBS Lett. 39, 13-16.

15. Woese, C. R., Fox, G. E., Zablen, L. B., Uchida, T., Bonen, L., Pechman, L., Lewis, B. J., and Stahl, D. A. (1975) Nature 254, 83-86.

16. Thomas, G., Sweeney, R., Chang, C., and Noller, H. F. (1975) J. Mol. Biol. 95, 91-102.

17. Noller, H. F. and Chaires, J. B. (1972) Proc. Natl. Acad. Sci. U.S.A. 69, 3115-3118.

18. Woese, C. R. (1970) Nature 226, 817-820.

19. Rich, A. (1974) In Ribosomes, M. Nomura, A. Tissiers, and P. Lengyel, eds. Cold Spring Harbor Laboratory. Cold Spring Harbor, New York. pp. 871-884.

20. Pongs, O. in Transfer RNA, S. Altman, ed. MIT Press, Cambridge, Mass. pp. 78-104.

21. Lake, J. A. (1977) Proc. Natl. Acad. Sci. USA 74, 1903-1907.

22. Crick, F. H. C., Brenner, S., Klug, A., and Piecznik, G. (1976) Origins of Life 7, 389-397.

23. Woese, C. (1972) Brookhaven Symp. 23, 326-365.

24. Woese, C. R. (1972) in Exobiology pp. 301-341. C. Ponnamperuma (ed.) North-Holland Pub. Co., Amsterdam.

25. Fox, G. E. and Woese, C. R. (1975) J. Mol. Evol. 6, 61-76.

26. Fox, G. E. and Woese, C. R. (1975) Nature 256, 505-507.

27. Loftfield, R. B. (1963) Biochem. J. 89, 83-87.

28. Woese, C. R. (1965) Proc. Natl. Acad. Sci. USA 54, 1546-1552.

29. Woese, C. R., Dugre, D. H., Dugre, S. A., Kondo, M., and Saxinger, W. C. (1966) Cold Spring Harbor Symp. Quant. Biol. 31, 723-736.

30. Drake, J. W. (1970) The Molecular Basis of Mutation, Holden-Day, San Francisco.

31. Woese, C. R. (1970) Symp. Soc. Gen. Microbiol. 20, 39-54.

32. Woese, C. R. (1973) Naturwiss. 60, 447-459.

33. Woese, C. R. and Fox, G. E. (1977) J. Mol. Evol. 10, 1-6.

34. Woese, C. R. and Fox, G. E. (1977) Proc. Natl. Acad. Sci. USA 74, 5088-5090.

35. Woese, C. R., Magrum, L. J., and Fox, G. E. (1978) J. Mol. Evol. 11, 245-252.

36. Kandler, O. (1979) Naturwiss. 66, 95-105.

37. Thauer, R. K. and Fuchs, G. (1979) Naturwiss. 66, 89-94.

38. Kates, M. (1972) In Ether Lipids, Chemistry and Biology, F. Snyder ed. Academic Press, New York pp. 351-398.

39. Tornabene, T. G., Wolfe, R. S., Balch, W. E., Holzer, G., Fox, G. E. and Oro, J. (1978) J. Mol. Evol. 11, 259-266.

40. Langworthy, T. A., Mayberry, W. R., and Smith, P. F. (1974) J. Bacteriol. 119, 106-116.

41. Zillig, W., Stetter, K. O., and Tobien, M. (1978) Eur. J. Biochem. 91, 193-199.

42. Balch, W. E., Fox, G. E., Magrum, L. J., Woese, C. R., and Wolfe, R. S. (1979) Microbiol. Rev. 43, No. 2 (in press).

43. Magrum, L. J., Luehrsen, K. R. and Woese, C. R. (1978) J. Mol. Evol. 11, 1-8.

44. Woese, C. R., Sogin, M. L., Stahl, D. A., Lewis, B. J., and Bonen, L. (1976) J. Mol. Evol. 7, 197-213.

45. Woese, C. R. (1970) Bioscience 20, 471-485.

46. Nishimura, S. (1978) In Transfer RNA, S. Altman, ed. MIT Press, Cambridge, Mass. pp. 168-195.

47. Gefter, M. L. and Russell, R. L. (1969) J. Mol. Biol. 39, 145-147.

48. Woese, C. R. (1969) Progr. Mol. Subcell. Biol. 1, 5-46.

49. Eisinger, J. (1971) Biochem. Biophys. Res. Commun. 43, 854-861.

50. Orgel, L. E. (1972) ISR J. Chem. 10, 287-292.

51. Woese, C. R. (1973) J. Mol. Evol. 2, 205-208.

PART 3: FUNCTION

SOLUBLE FACTORS IN PROTEIN SYNTHESIS

HERBERT WEISSBACH

Roche Institute of Molecular Biology,
Nutley, NJ 07110

INTRODUCTION

An integral part of the protein synthetic process involves the interaction of soluble factors with the ribosome. In retrospect the ease of obtaining sufficient quantities of these factors in a high state of purity provided the initial impetus for the large number of studies on their structure and function, especially in prokaryotes. Thus, our knowledge in this area proceeded quite rapidly and it has only been in recent years that the emphasis has shifted to the ribosome itself. An attempt will be made in this chapter to review studies on the structure and function of the soluble factors involved in the initiation, elongation and termination steps of protein synthesis, stressing the prokaryotic system. Since a recent detailed monograph has appeared (1), this article will attempt to summarize in rather broad terms the currently accepted role of the soluble factors.

PROKARYOTIC INITIATION

The initiation process involves the interaction of the ribosomal subunits, mRNA and initiator tRNA. The final product is a 70S (or 80S in eukaryotes) initiation complex containing the initiator tRNA on the P site of the ribosomes. There were several findings that led to our present understanding of this process. Waller (2) first observed that there was a non-random distribution of amino acids in the N-terminal position of total *Escherichia coli* proteins, and that methionine was present at the N-terminal position of 50% of the proteins. Marcker *et al*. (3) first noted the presence of a blocked species of acylated tRNA that was identified as formylmethionyl-tRNA (fmet-tRNA) and implicated in the initiation process. The requirement for specific initiation factors was apparent when it was observed that high salt-washed ribosomes were active in the translation of poly(U), but not with a natural mRNA such as

MS2 RNA. The ribosomal salt wash was then shown to contain three factors (IF-1, IF-2, IF-3) that were needed for the translation of a natural mRNA (4,5).

Figure 1 summarizes the overall reactions that are believed to be involved in the initiation process in prokaryotes (for recent review see 6). There are pools of 30S and 50S subunits in the cell that arise from

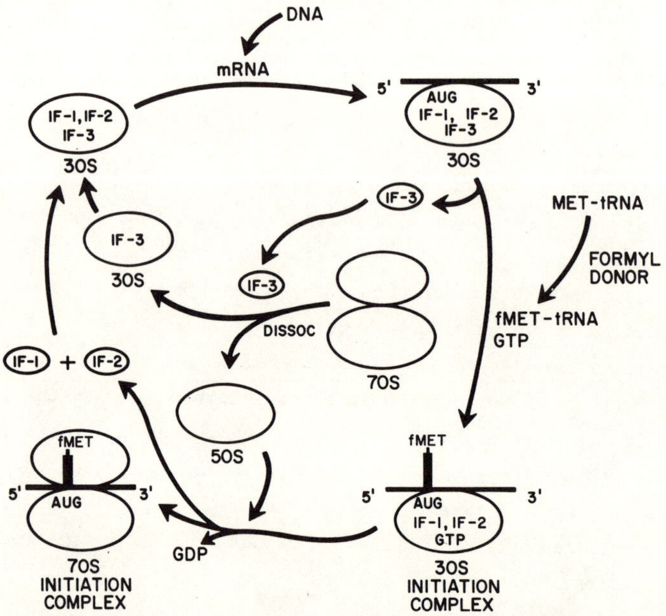

Figure 1. Steps involved in the initiation process in prokaryotes. The figure has been taken from Revel (6). The tRNA is symbolized by a solid bar.

the dissociation of 70S subunits and it is known that IF-3 facilitates this dissociation process. The initial reaction involves the interaction of IF-3, and very likely IF-1 and IF-2, with the smaller (30S) ribosomal subunit. mRNA and initiator tRNA (fmet-tRNA) then bind to the 30S subunit to form a 30S initiation complex. The sequence of these reactions is still not completely resolved but, for the sake of discussion, the figure depicts that mRNA is bound to the 30S particle prior to the initiator tRNA. As will be discussed below, it is accepted that IF-3 is primarily responsible for mRNA binding and IF-2 is involved in the binding of the fmet-tRNA to the ribosomal subunit. IF-3 is very likely released during the formation of the 30S initiation complex but IF-1 and IF-2 remain bound to the subunit. In a subsequent reaction, the 50S subunit interacts with the 30S initiation complex to form the 70S initiation complex. This reaction is accompanied by GTP hydrolysis and the release of IF-1 and IF-2 (perhaps as an IF-2·GDP complex). The fmet-tRNA is located at a puromycin sensitive site (P site) of the ribosome and thus the initiation complex is ready for the peptide chain to be elongated, i.e. the binding of the second AA-tRNA to the A site on the ribosome. Some of the properties of the factors involved in the initiation reaction and more details on the individual steps are described below.

General Properties of IF-1, IF-2 and IF-3.

IF-1 is a heat-stable, basic protein of about 9,000 daltons and is readily purified from a high salt wash of ribosomes by phosphocellulose column chromatography and Sephadex. An excellent procedure for the purification of all the initiation factors has recently been published by Hershey et al. (7). The factor is generally assayed by its ability to stimulate fmet-tRNA binding to ribosomes in the presence of IF-2 and the initiator codon AUG. Two laboratories have reported the amino acid composition of IF-1 (7,8), but significant differences were obtained which have not yet been explained. The protein is not inactivated by sulfhydryl reagents.

IF-2 is the largest of the initiation factors having a mol wt of 118,000. However, highly purified preparations generally show the presence of two major bands on gel analysis, having mol wts of 118,000 (IF-2α) and 98,000 (IF-2β), (7,9-13). There is ample evidence that the smaller mol wt material is a degradation product of the large protein (7,9). Although both are equally active in fmet-tRNA binding to ribosomes, a surprising finding was that only the high mol wt form (IF-2α) was active in the in vitro synthesis of β-galactosidase directed by transducing phage DNA carrying the lactose operon (14). Unlike IF-1 or IF-3, significant amounts of IF-2 are found in E. coli supernatant fractions (15), although the factor is routinely isolated from the ribosomal salt wash. IF-2 is heat labile, acidic and sensitive to sulfhydryl reagents.

IF-3 is a basic protein of mol wt about 22,000, although two forms of the protein with slightly different mol wts have been isolated (16,17). The two species appear to be products of the same gene with the lower mol wt form missing a hexapeptide (18-20). At present, there is no substantial evidence that the two forms differ in biological activity.

INDIVIDUAL STEPS IN THE INITIATION PROCESS

The scheme in Figure 1 depicts the generally accepted reactions leading to the formation of the 70S initiation complex. The individual steps are briefly discussed below.

Dissociation of Ribosomes and Binding of Initiation Factors to the 30S Subunit.

It is now apparent that in order for a 70S ribosome to participate in the initiation of protein synthesis it must dissociate into 30S and 50S subunits. IF-3 possesses dissociation factor activity and functions either by causing the dissociation of the 70S ribosome directly (21) or by preventing the association of the 30S and 50S subunits during the reversible dissociation of the 70S ribosome (22,23) as described below.

$$70S \rightleftharpoons 50S + 30S \xrightleftharpoons{IF\text{-}3} 30S\cdot IF\text{-}3$$

The latter possibility seems more attractive at this time. The IF-3 may function by binding at a site on the surface of the 30S subunit where the 50S subunit interfaces. As an example, IF-3 has been reported to bind to 16S RNA. The effect of IF-3 on 30S and 50S association could be explained if the segment of the 16S RNA, where IF-3 binds, normally base pairs to 23S RNA during 70S ribosome formation (24). One IF-3 molecule is bound per 30S subunit and the binding is stimulated by the presence of IF-1 (25), which enhances both the rate and extent of dissociation in the presence of IF-3. In fact it should be stressed that the interaction of the initiation factors with the 30S subunit is a highly cooperative process. This cooperative effect is also apparent in the binding of IF-2 to the 30S subunit. In the absence of IF-1 and IF-3 this binding is quite unstable (26). It is obvious from these studies that the three initiation factors do not compete for the same sites on the 30S subunit. IF-1 can also interact with the 30S subunit by itself but the binding is stabilized by the presence of IF-2 and IF-3 (27). The ability of all three factors to bind to the 30S subunit *in vitro* also agrees with the finding that all three initiation factors are found associated with the 30S subunit in cell-free extracts.

There is also good evidence that the initiation factors form complexes with each other. IF-2·IF-3 and IF-1·IF-3 complexes have been detected (28,29) and although they represent only minor fractions of the total amount of initiation factors in the ribosomal wash, their presence might be indicative of what occurs on the 30S subunit.

Binding of mRNA to the 30S Subunit.

Once the initiation factors are bound to the 30S subunit the binding of mRNA and fmet-tRNA can occur. It is not clear which occurs first since the in vitro results indicate that each can bind, under certain conditions, independent of the other. However, unless both mRNA and fmet-tRNA are present the resulting complex is not stable (30-33). Although the binding of mRNA to the 30S subunit may be, in part, due to a cooperative effect of all three factors, IF-3 is directly responsible for the formation of the mRNA·30S complex. How IF-3 accomplishes this is not clear. As Shine and Dalgarno (34) first proposed, the mRNA 30S interaction involves base pairing between a pyrimidine sequence at the 3' end of 16S RNA and a purine rich sequence on the mRNA located between the 5' end of the mRNA and the initiator AUG. IF-3 has been reported to bind to MS2 RNA (35), which may facilitate the mRNA·30S interaction.

Binding of fmet-tRNA to the 30S Subunit.

The binding of fmet-tRNA to the 30S subunit requires IF-2 and GTP and involves the intermediate formation of an fmet-tRNA·IF-2·GTP ternary complex. Despite the fact that this complex can be demonstrated independent of the ribosome (26,36-38), there are other indications that IF-2 bound to the 30S subunit functions in ternary complex formation (28). However, the ability to form a ternary complex independent of ribosomes has facilitated studies on the requirements for this reaction. As an example, the specificity of the initiator tRNA has been carefully examined. There is a rather stringent requirement for fmet-tRNA since all other naturally occurring AA-tRNA's do not react with IF-2. Most N-blocked AA-tRNA species are also not active with the exception of N-acetyl-Phe-tRNA (39). However, base sequence studies have not provided any clue to the specificity for fmet-tRNA, and the answer may only be apparent when three-dimensional models are available.

Both IF-1 and IF-3 stimulate the IF-2 dependent binding of fmet-tRNA to the 30S subunit (26,40). How these factors function is not clear although all of the results could be explained by an effect of IF-1 and IF-3 in enhancing the binding of IF-2 to the 30S subunit. The components that comprise the 30S initiation complex have also been identified. There is one mole of IF-2 present per mole of fmet-tRNA (41,42). GTP is also part of the complex (43), as is IF-1 (30,42,44-46), but IF-3 is very likely released when the complete 30S initiation complex is formed (47). It appears that the binding of the initiator tRNA is responsible for the release of the IF-3 and is suggestive evidence that the mRNA binds to the 30S subunit prior to the initiator tRNA.

Formation of the 70S Initiation Complex.

The addition of 50S subunits to the 30S initiation complex results in the formation of the 70S initiation complex. IF-1 and IF-2 are released at this stage and GTP hydrolysis occurs. The release of IF-2 from the 30S initiation complex is evident by the fact that, in the presence of 50S subunits (forming a 70S complex), IF-2 functions catalytically in fmet-tRNA binding (26,42,48). This catalytic effect of IF-2 requires GTP and IF-1 as would be expected.

Although the primary roles of IF-3 and IF-2 are established (mRNA binding and fmet-tRNA binding, respectively), the role of IF-1 in the formation of the 70S·initiation complex is still very vague. Several possibilities have been proposed. If IF-2·GDP is released from the ribosome, IF-1 could function by releasing the GDP or exchanging it with GTP (42,44), similar to the function of EF-Ts in elongation. IF-1 also stimulates the binding of IF-2 to the 30S subunit (26). At present, this finding could explain the stimulatory effect of IF-1 on the initiation process, since the proper binding of IF-2 is necessary for the subsequent reactions (50S subunit joining, GTP hydrolysis, etc.) leading to the formation of the 70S·initiation complex.

As will be discussed below, GTP hydrolysis is required for IF-2 to recycle off the ribosome (26,40,42,44), but GTP also appears to increase the affinity of IF-2 for the 30S subunit (41).

EUKARYOTIC INITIATION

Studies on initiation of protein synthesis in eukaryotes have recently received considerable attention since it is apparent that the process differs significantly from the prokaryote system. In addition, there is good evidence for translational regulation at the initiation step. To begin with it might be useful to summarize the basic properties of the eukaryotic system that are different from the prokaryotic process. These are listed below:

1. lack of formylation of the initiator met-tRNA

2. as many as 8 factors are required
3. the initiator tRNA appears to bind to the smaller subunit before the mRNA
4. ATP is required for mRNA binding
5. regulation is mediated by a protein kinase system

Eukaryotic Initiation Factors.

The eukaryote factors are designated EIF (eukaryote initiation factors), based on nomenclature proposed in 1977 (49). The early studies by Staehelin and coworkers (50,51) and Anderson and coworkers (52-54) showed that the initiation reaction in reticulocyte lysates requires a large number of factors. Subsequently, other laboratories have been concerned with the purification of these factors (55-62) and there is, in general, good agreement with regard to the number and nature of the factors and the basic steps in the process. Although eight factors have been identified (see below, Fig. 2), it should not be assumed that there are no other factors that affect this process. Some points about the individual

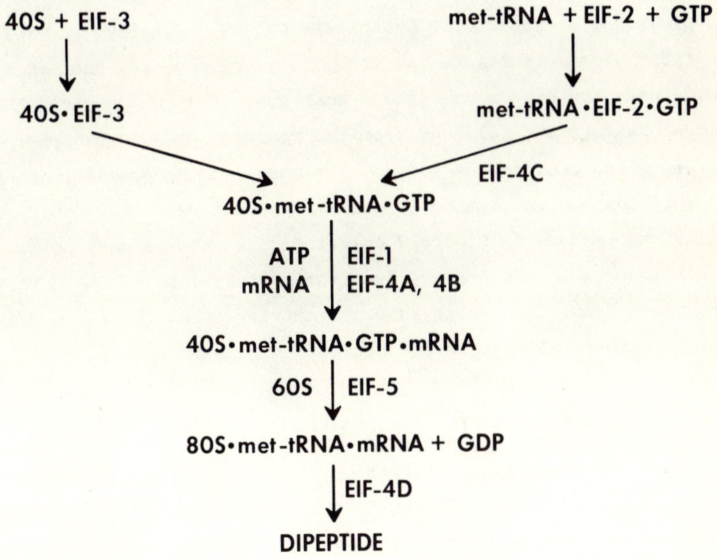

Figure 2. Steps involved in initiation in eukaryotes. Scheme taken from Benne and Hershey (63).

factors should be noted. Both EIF-2 and EIF-3 are multi-subunit factors; EIF-2 having three subunits of 55,000, 52,000 and 37,000 daltons (52,60, 64) and EIF-3 having nine subunits (55,59,64). Although these factors perform similar functions to the corresponding prokaryotic factors (IF-2, IF-3), it is remarkable that they are so different from their monomeric prokaryotic counterparts. EIF-1 also appears different from IF-1, and the other eukaryote factors have at this time no prokaryote equivalent.

Steps in Eukaryotic Initiation.

The individual steps in the initiation process are summarized in Figure 2 (63). Although the eukaryotic initiator tRNA, met-$tRNA_i$, can be formylated using the bacterial enzyme (N^{10}-formyl-H_4-folate-met-tRNA transformylase), (65) it is not formylated in eukaryotic systems since the transformylase is lacking from eukaryotic extracts. Thus, the initiation complex contains met-tRNA and not fmet-tRNA. A 40S·EIF-3 complex is pictured as reacting with an met-tRNA·EIF-2·GTP complex in the presence of EIF-4C to form a 40S·met-tRNA·GTP complex prior to the binding of mRNA. The interaction of mRNA with this 40S complex requires ATP and three additional factors, EIF-1, EIF-4A, and EIF-4B, besides the EIF-3 which is already part of the complex. Upon addition of the 60S subunit, an 80S initiation complex is formed. This reaction requires EIF-5 and as in the prokaryotic system, GTP hydrolysis occurs. Finally, the formation of the first peptide (measured as met-puromycin) requires another factor, EIF-4D. There are also some results that suggest at which steps the individual factors are released from the ribosome. As an example, EIF-2 and EIF-3, which bind to the 40S subunit early in the process, are released when the 80S initiation complex is formed (63). It should also be noted that the scheme in Figure 2 only indicates the primary point at which each initiation factor may function. It is likely, as in the prokaryotic system, that there is cooperativity among the factors and that each factor will enhance the activity of other factors.

Role of GTP and ATP in Eukaryote Initiation.

As noted above, eukaryote initiation requires both GTP and ATP. The role of GTP is very likely similar to that in prokaryotes. It is required for the formation of the 40S initiation complex and is hydrolyzed during the formation of the 80S complex. EIF-2 appears to be required for this hydrolysis, although EIF-5 may also be involved at this step since this factor possesses GTPase activity (55). ATP hydrolysis is associated with mRNA binding to 40S subunits (64,66). As yet, none of the factors involved in this step (EIF-1, EIF-3, EIF-4A, EIF-4B) has been shown to have ATPase activity, so that the function of ATP or the individual factors associated with mRNA binding is still unknown.

Regulation of Eukaryote Initiation.

Well before the details of the initiation process in eukaryotes were understood, it was known that in reticulocyte lysates protein synthesis proceeded for only a short time in the absence of hemin (67-70). The lack of hemin results in the conversion of a proinhibitor to an inhibitor (71-73) in the system. Recent studies from several laboratories have provided information about the mechanism of this inhibition and how hemin functions. The available evidence indicates that the active inhibitor is a cAMP independent protein kinase that phosphorylates the 38,000 mol wt subunit of EIF-2 and, by so doing, prevents the formation of a 40S initiation complex (74-77).

How phosphorylation inhibits the action of EIF-2 has not been fully elucidated. It now appears that EIF-2 can interact with another factor (ESP) that stimulates ternary complex formation at low levels of EIF-2 (78). The phosphorylated form of EIF-2 may not be able to associate with ESP, which could explain the observed inhibition.

There is still some doubt as to how hemin functions in preventing the action of the protein kinase. The inhibitor (EIF-2 kinase) can phosphorylate itself, which may activate the protein. This auto-phosphorylation has been reported to be prevented by hemin (79). cAMP dependent kinases are also present in reticulocyte lysates and it has been proposed that the phosphorylation of EIF-2 kinase is mediated by a cAMP dependent protein kinase (80).

Other Regulatory Systems.

It now appears that there are other mechanisms by which the initiation process can be regulated that are not under hemin control. Addition of oxidized glutathionine (81-83) or double stranded RNA (dsRNA) (84-92) also activate a protein kinase(s), even in hemin containing reticulocyte lysates, that phosphorylates EIF-2. Double stranded RNA has also been shown to be required for the synthesis of a novel nucleotide when incubated with reticulocyte lysates and ATP. This nucleotide, pppA2'p5'A2'-p5'A, (93-95) may function by activating an endonuclease (96).

The effect of the antiviral agent interferon may, in part, be on the initiation of protein synthesis. Interferon in the presence of dsRNA has been shown in animal cells to induce the synthesis of a protein kinase

(96-101) and also stimulate the synthesis of pppA2'p5'A2'p5'A. Although far from proven, one could imagine that interferon, produced in cells after viral infection, could interact with dsRNA (produced in the virus-infected cell) and activate both a kinase that inhibits EIF-2 and an endonuclease that degrades mRNA (92). The reader is referred to a recent review on the regulation of eukaryotic initiation (102).

ELONGATION

Once the 70S (or 80S) initiation complex is formed, amino acids are added to the growing polypeptide chain one at a time in a sequence determined by the base sequence of the mRNA. The steps involved in the elongation process are schematically presented in Figure 3. More specifically, this figure shows how the second amino acid is added to the 70S

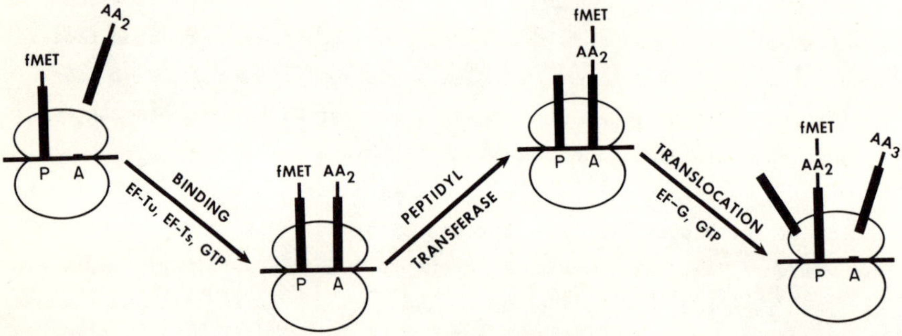

Figure 3. Steps involved in peptide chain elongation. The tRNA is symbolized by a solid bar. AA = amino acid.

initiation complex containing fmet-tRNA on the P site of the ribosome, but these reactions can be used as a model for any amino acid added during elongation. In the first step in elongation, an AA-tRNA specified by the code word on the mRNA binds to the empty A site on the ribosome. Two soluble factors, EF-Tu and EF-Ts, as well as GTP, are required for this reaction. A new peptide bond is then formed by transfer of the amino acid from the tRNA on the P site (formylmethionine in this case) to the free amino group of the AA-tRNA on the A site. This reaction is catalyzed by a complex (peptidyl transferase) that is part of the ribosome. In vitro, it does not require either a soluble factor or an exogenous energy source. In the final reaction, the peptidyl-tRNA on the A site and the mRNA move to the P site, the deacylated tRNA on the P site is released, and a new code word appears in the A site. This translocation reaction requires another soluble factor, EF-G, and GTP.

PROKARYOTIC FACTORS INVOLVED IN AA-tRNA BINDING TO RIBOSOMES

Early studies by Lipmann and coworkers revealed the presence of two factors (called T and G), which were required for the elongation process (103,104). Subsequently, the T factor was separated into two components, one unstable (EF-Tu) and the other stable (EF-Ts) (105), which have been purified to homogeneity.

General Properties of EF-Tu.

As will be discussed below, EF-Tu represents as much as 5% of the soluble proteins of E. coli (106). Although there are two EF-Tu genes on the E. coli chromosome (tuf a and tuf b) (107), their gene products appear very similar (108,109). Recently, mutants of both EF-Tu genes have been found (110). Relatively simple purification procedures have been available for several years and large amounts of this protein can be isolated with ease and crystallyzed (111,112). It was originally noted that the purified factor had a $\frac{280\ nm}{260\ nm}$ absorbance ratio of only 1.1 indicating the presence of a 260 nm absorbing material associated with the protein (111). It was then demonstrated that EF-Tu as isolated contained one equivalent of non-covalently bound GDP (111). As will be discussed below, the interaction of EF-Tu with guanosine nucleotides is an essential feature of the mechanism by which this factor functions in bringing the AA-tRNA to the ribosome. The stability of EF-Tu is significantly greater when it is complexed with a guanosine nucleotide.

EF-Tu is a protein with a mol wt of $\sim$44,000 (111,112). Amino acid analysis shows no unusual features. There are three cysteines present, one of which is essential for guanosine nucleotide binding and a second for AA-tRNA interaction with the protein (113). More recently, part of the amino acid sequence of EF-Tu has been obtained (114). The N-terminal amino acid is blocked, but at this time the blocking group has not been identified.

General Properties of EF-Ts.

EF-Ts from E. coli is a protein of about 30,000 mol wt (112,115). It is found in crude extracts associated with EF-Tu and purification procedures take advantage of the fact that GDP can dissociate this complex (see below) to yield EF-Tu·GDP and free EF-Ts (112,115). The amino acid composition of EF-Ts has shown that this protein lacks tryptophan and has 1 cysteine residue (115).

Interactions of EF-Tu with EF-Ts and Guanosine Nucleotides.

The amount of EF-Tu in the *E. coli* cell appears to be in the same range as tRNA (106). EF-Ts is in lesser amounts and, very likely, the bulk of the EF-Ts in the cell is found in the form of an EF-Tu·EF-Ts complex. This corresponds to the original T factor described by Allende *et al.* (103). The complex is very stable and can be chromatographed on Sephadex with very little dissociation. This complex was determined to have a mol wt of about 70,000 or close to the sum of the individual factors (116). The complex is, however, dissociated in the presence of excess GDP as shown below.

$$\text{EF-Tu}\cdot\text{EF-Ts} + \text{GDP} \rightleftharpoons \text{EF-Tu}\cdot\text{GDP} + \text{EF-Ts}$$

Since this is a reversible reaction, EF-Ts will cause the displacement of GDP from EF-Tu·GDP, and catalyze an exchange of added GDP with GDP bound to EF-Tu. This exchange reaction provides a convenient assay for both EF-Tu as well as EF-Ts (117,118). In a like manner, EF-Tu will also react with GTP to form an EF-Tu·GTP complex. However, GDP binds more tightly to EF-Tu than GTP [K_{diss} being 8×10^{-9} for GDP and 3×10^{-7} for GTP (111,119,120)]. Despite the much higher affinity of EF-Tu for GDP than GTP, in the presence of a large excess of GTP (as may occur *in vivo*), EF-Ts will catalyze the exchange of GTP with GDP bound to EF-Tu by the following reactions:

$$1.\quad \text{EF-Tu}\cdot\text{GDP} + \text{EF-Ts} \rightleftharpoons \text{EF-Tu}\cdot\text{EF-Ts} + \text{GDP}$$

$$2.\quad \text{EF-Tu}\cdot\text{EF-Ts} + \text{GTP} \rightleftharpoons \text{EF-Tu}\cdot\text{GTP} + \text{EF-Ts}$$

$$\text{Net}\quad \text{EF-Tu}\cdot\text{GDP} + \text{GTP} \xrightleftharpoons{\text{EF-Ts}} \text{EF-Tu}\cdot\text{GTP} + \text{GDP}$$

It appears that EF-Ts, GDP and GTP react at a common site on EF-Tu since all three components protect one of the three sulfhydryl groups on EF-Tu (113). Although an EF-Tu·EF-Ts·guanosine nucleotide ternary complex has not been detected with the *E. coli* factors, such a complex has been reported using elongation factors purified from a thermophilic organism (121).

Interaction of EF-Tu·GTP with AA-tRNA.

The EF-Tu·guanosine nucleotide complexes are involved in the overall function of EF-Tu in AA-tRNA binding to ribosomes. EF-Tu·GTP is the form of EF-Tu that interacts with AA-tRNA to form an AA-tRNA·EF-Tu·GTP ternary complex (122). An unusual characteristic of the ternary complex is that it is not retained on a nitrocellulose filter, whereas the binary complexes (EF-Tu·GTP and EF-Tu·GDP) are (123). This provides a rapid assay for ternary complex formation. It is important to note that EF-Tu·GDP does not interact to a significant degree with AA-tRNA. However, all naturally occurring AA-tRNA's that are involved in protein synthesis react with EF-Tu·GTP with the important exception of the initiator tRNA, fmet-$tRNA_f^{met}$ (formylated or not). Other structural requirements are the presence of an aminoacyl moiety on the tRNA, although the hydroxy acid is also active (124). N-blocked aminoacyl-tRNA's are not good substrates.

The observation that EF-Tu·GTP, but not EF-Tu·GDP, can react with AA-tRNA (123) led to the postulation that the nucleotide ligand influenced the conformation of EF-Tu. Thus, GTP associated with EF-Tu exposes a binding site for AA-tRNA not accessible in EF-Tu·GDP. There is now experimental evidence supporting the view that the guanosine nucleotide ligand can affect the conformation of EF-Tu. Tritium exchange experiments showed that the tritium on EF-Tu exchanged more slowly when EF-Tu·GDP was used than EF-Tu·GTP (125). In addition, the binding of a hydrophobic fluorescent dye to EF-Tu has shown that EF-Tu·GTP enhances the fluorescence to a greater extent than EF-Tu·GDP (126). Other experiments have used ORD analysis and intrinsic fluorescence to study conformational changes in EF-Tu (127). It appears from these studies that the binding of either GDP or GTP to EF-Tu causes large conformational changes and that the conformation of EF-Tu·GDP and EF-Tu·GTP also differ.

Interaction of the Ternary Complex with the Ribosome.

The ternary complex (AA-tRNA·EF-Tu·GTP) is the substrate that interacts with the ribosome resulting in the transfer of the AA-tRNA to the acceptor site (A site) on the ribosome. GTP hydrolysis accompanies this reaction and the EF-Tu·GDP formed is released from the ribosome and converted to EF-Tu·GTP in the presence of EF-Ts and GTP as described above (equations 1 and 2). The correct codon on the mRNA is essential for GTP hydrolysis to take place. As an example, if a Phe-tRNA·EF-Tu·GTP complex is incubated with ribosomes containing poly(U), hydrolysis occurs, but not if poly A is used. In the presence of a non-hydrolyzable GTP analogue, such as GDP-PCP, AA-tRNA binding to the ribosome will take place but the EF-Tu is not released from the ribosome (128). Although many ribosomal proteins may be involved in this interaction, it is quite clear that proteins L7L12 play a special role in the recognition of EF-Tu and other soluble factors that interact with the ribosome (129-132).

Figure 4 summarizes the above results and outlines the steps that appear to be involved in the binding of an AA-tRNA to the A site on the

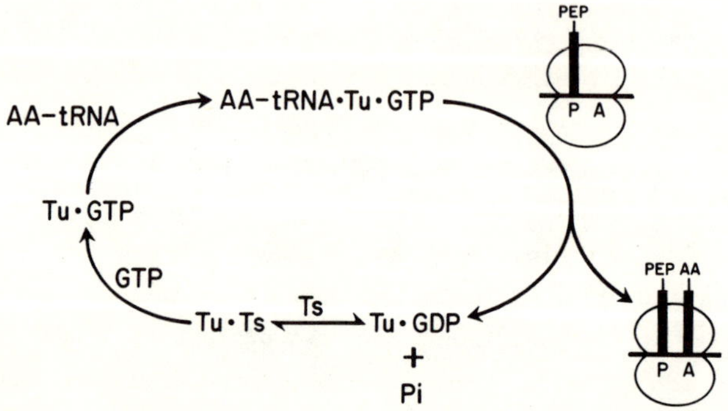

Figure 4. Role of EF-Tu and EF-Ts in AA-tRNA binding to ribosomes. The P site on the ribosome is pictured as having a peptidyl-tRNA bound to it. Tu, Ts, and PEP are abbreviations for EF-Tu, EF-Ts, and peptidyl, respectively.

ribosome during peptide chain elongation. As noted above, EF-Tu·GTP is the species of EF-Tu that interacts with AA-tRNA to form a ternary complex, AA-tRNA·EF-Tu·GTP. This can react with a 70S ribosome having a peptidyl-tRNA on the P site and a free A site with transfer of the AA-tRNA to the A site and release of EF-Tu·GDP from the ribosome. Since EF-Tu·GDP cannot interact with AA-tRNA, the conversion of EF-Tu·GDP to EF-Tu·GTP is essential for EF-Tu to recycle during peptide chain elongation. This is accomplished as shown in the Figure. The EF-Tu·GDP interacts with EF-Ts to form an EF-Tu·EF-Ts complex that is then converted to EF-Tu·GTP in the presence of excess GTP.

Inhibitors of EF-Tu Action.

Kirromycin, an antibiotic isolated from cultures of a strain of *Streptomyces*, appears to inhibit protein synthesis by interacting directly with EF-Tu (133). This antibiotic stimulates the formation of EF-Tu·GTP and the binding of the ternary complex to the ribosome (134). However, in the presence of kirromycin, EF-Tu is not released from the ribosome (135). In addition, the GTP hydrolysis that is seen when the AA-tRNA·EF-Tu·GTP complex reacts with the ribosome, can be obtained with EF-Tu in the absence of ribosomes when kirromycin is present (134). These findings suggest that the catalytic site for GTP hydrolysis that occurs upon

AA-tRNA binding to ribosomes is on EF-Tu and not on the ribosome. These results, if extended to other reactions in protein synthesis that involve soluble factors and GTP hydrolysis, would suggest that in all cases the catalytic site for GTP hydrolysis is on the soluble factor (e.g. IF-2, EF-Tu, EF-G and possibly release factor), and not on the ribosome. A major question posed by these findings is how the ribosome activates the GTPase site on the soluble factor.

Pulvomycin is another antibiotic that interacts with EF-Tu. This compound appears to prevent the interaction of EF-Tu·GTP with AA-tRNA to form a ternary complex (136). It is apparent that inhibitors such as kirromycin and pulvomycin will be of great use in understanding the mechanism by which EF-Tu functions in binding AA-tRNA to the ribosome.

EUKARYOTIC FACTORS INVOLVED IN AA-tRNA BINDING TO RIBOSOMES

During the past few years the situation in eukaryotes regarding the binding of AA-tRNA to ribosomes has been greatly clarified. The eukaryotic factor first studied by Schweet and coworkers (137-139), now called EF-1, has been known to exist in tissues in the form of high mol wt aggregates ranging in weight from 2×10^4 to over 1×10^6 (140-151). The size of the aggregates that are obtained varies depending on the purification procedures used, especially the initial steps in extracting the factor from the tissue. It was also clear from the early studies that in some tissues the high mol wt aggregates had lipid associated with them (144), which very likely helped to keep the aggregates intact. Elastase and carboxypeptidase A can cause the disaggregation of large aggregates of EF-1, although the exact mechanism or physiological significance of these observations is not known (152,153). The large aggregates of EF-1 appear to be more stable than the lower mol wt species.

A protein of about 50,000 mol wt has been purified from a large number of tissues (144,146,148,150,151,154,155-159) that corresponds in function to the *E. coli* factor EF-Tu, in that it reacts with GTP and AA-tRNA to form a ternary complex. Although some large aggregates have initially been reported (144,151) to contain only this 50,000 mol wt species (referred to as $EF\text{-}1_L$ or $EF\text{-}1_\alpha$), several laboratories have described the isolation of two additional factors, called $EF\text{-}1_\beta$, and $EF\text{-}1_\gamma$, that are present in the high mol wt aggregates of EF-1 (159-164). Very likely all eukaryotic tissues contain $EF\text{-}1_\beta$ and $EF\text{-}1_\gamma$ and the difficulty in detecting them has been due to the low amounts present in some of the EF-1 preparations or the isolation procedures used. EF-1_β has a mol wt of about 27,000, under denaturing conditions, but may be a dimer (162). It corresponds to the E. coli factor EF-Ts in that it can catalyze a guanosine nucleotide exchange with EF-1_α·GDP, (159,161-163) and stimulate AA-tRNA binding and polymerization (161,162,164). However, the role of EF-1_γ (mol wt also about 50,000) is still not clear. There are indications that EF-1_β and EF-1_γ form a complex that has essentially the same activity as EF-1_β (159,160,165). Since EF-1_α and EF-1_β appear to correspond to EF-Tu and EF-Ts, it is most probable that the sequence of reactions leading to AA-tRNA binding to ribosomes with these factors is similar to that for the prokaryotic factors shown above in Figure 4. One basic difference that has been observed in some laboratories (not all) is that EF-1_α appears to be bound to the ribosome after the AA-tRNA is placed on the A site (157, 166,167). As noted before in the prokaryotic studies, EF-Tu is rapidly released from the ribosome as EF-Tu·GDP at this stage. It is also surprising that in the studies with ribosome bound EF-1_α the factor acted stoichiometrically in AA-tRNA binding in the absence of EF-2, but catalytically when EF-2 was present. The EF-2 may unblock the A site on the ribosome by catalyzing translocation, permitting the ribosomal bound EF-1_α to interact with another AA-tRNA (157,166). The significance of this difference between the prokaryotic and eukaryotic systems remains uncertain.

Other Observations Concerning EF-Tu and EF-1.

Of all the soluble factors required for prokaryotic protein synthesis, EF-Tu has recently received the most attention. Genetic studies have shown that there are two genes (tuf a and tuf b) on the E. coli chromosome containing the genetic information for EF-Tu (107). In E. coli, EF-Tu can account for as much as 5% of the soluble protein (106), with the bulk of the synthesis from the tuf a gene (168). The synthesis of EF-Tu is also under stringent control (169-171).

EF-Tu and EF-Ts are also two of the host-derived subunits of Q_β replicase (172) and appear to function in the initiation of RNA synthesis catalyzed by this enzyme (173). For this function EF-Tu and EF-Ts do not have to dissociate (174). There has also been a report that EF-Tu stimulates rRNA synthesis (175), and although such a regulatory role would be very attractive, the evidence is not yet conclusive. Other observations that are provocative are the findings that EF-Tu is associated with the cell membrane (176) and that the protein has some properties that are similar to actin (e.g. ability to bind to DNase and form fibers) (177).

There have been reports that the size of the aggregates of EF-1 change during the development of the brine shrimp, Artemia salina (178). This might be related to a role of this factor in the developmental process, although the fact that certain proteases (e.g. elastase, carboxypeptidase) can cause dissociation of EF-1 aggregates makes one cautious in interpreting these findings (152,153). EF-1 is also found in the nucleus (179,180) and is known to bind tightly to nucleic acids (179). However, as of now there are no findings to indicate a role of EF-1 in eukaryotic RNA synthesis or transport from the nucleus. More work will undoubtedly be done in this area.

PROKARYOTIC TRANSLOCATION, ELONGATION FACTOR G

EF-G was the first prokaryotic protein synthesis factor shown to be involved in the hydrolysis of GTP in the presence of ribosomes (103,104, 182). It later became clear that this factor was required for translocation, perhaps the most complex of all the reactions of protein synthesis. As shown in Figure 5, translocation involves the movement of the peptidyl tRNA from the A to the P site on the ribosome. In addition, the mRNA

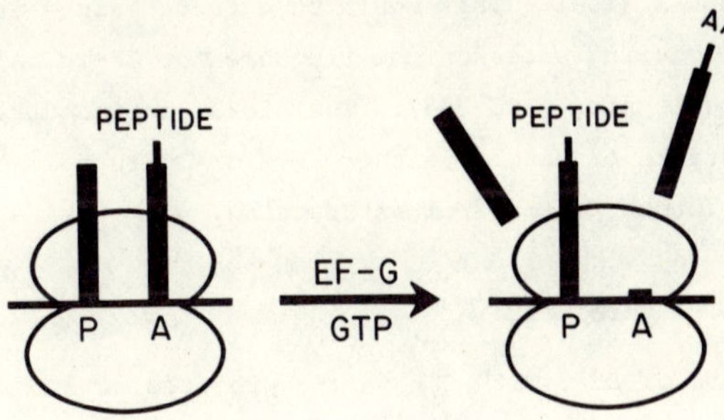

Figure 5. Translocation during peptide chain elongation. tRNA is symbolized by the solid bar.

moves three bases allowing another code word to enter the A site, the deacylated tRNA originally on the P site is released from the ribosome, and GTP is hydrolyzed. As yet, the sequence of these events is not known.

Properties of EF-G.

EF-G consists of a single polypeptide chain of molecular weight about 80,000 (182-186). The protein is quite acidic, has six half cysteines per mole and a 280 nm/260 nm ratio of over 1.75 (183,185,186). Thus, unlike EF-Tu, the protein as isolated does not contain a bound nucleotide. The amount of EF-G in the cell is stoichiometric to the amount of ribosomes (188) and very likely represents about 0.5% of the soluble protein. It crystallyzes easily in the form of needles or crystalline plates.

Interaction of EF-G with Guanosine Nucleotides and 50S Ribosomal Subunits.

The interaction of EF-G with guanosine nucleotides yields stable complexes only in the presence of ribosomes or 50S subunits. Incubation of EF-G with either GTP or GDP plus ribosomes yields an EF-G·GDP·50S complex. The complex is rather unstable but can be detected by gel filtration or filter binding (189-191). However, the complex can be markedly stabilized by the addition of fusidic acid. This antibiotic fixes the complex on the ribosome by forming an EF-G·GDP·50S·fusidic acid complex (191-197). It has been shown that in the presence of fusidic acid one round of translocation can occur (197,198), and therefore this antibiotic appears to inhibit protein synthesis by preventing the release of EF-G from the ribosome after it has functioned in translocation.

In vivo studies also demonstrated that fusidic acid inhibits AA-tRNA binding to ribosomes (199). This would be expected since it is known in *E. coli* that the binding sites on the ribosome for EF-Tu and EF-G are very close or overlapping (200-203). Therefore, any condition that results in the stable binding of either EF-G or EF-Tu to the ribosome will prevent the other factor from interacting.

EF-G Dependent GTP Hydrolysis.

The interaction of EF-G with GTP in the presence of ribosomes results in a rapid uncoupled hydrolysis of GTP. This was first demonstrated by Lipmann and coworkers (181,204). Although GTP hydrolysis can be obtained with 50S subunits alone (195,196), 30S subunits markedly stimulate this uncoupled hydrolysis. Rohrbach *et al.*, (186) have used $H_2{}^{18}O$ to study the mechanisms of the GTP hydrolysis. ^{18}O was only incorporated into the orthophosphate derived from the γ phosphate of GTP. Thus, the hydrolysis of GTP occurs between the γ phosphorous atom and the oxygen bound to the β and γ phosphorous atoms.

Translocation.

The process of translocation can be assayed in different ways. It is most easily assayed by binding an AA-tRNA to the A site on the ribosome and measuring puromycin releasable product. In the presence of EF-G and GTP, translocation to the P site will occur and the AA-tRNA will become puromycin reactive. The release of the deacylated tRNA on the P site in the pre-translocation stage has also been studied. It is known that this reaction requires the same components as those required for the overall reaction, i.e. EF-G and GTP, and the A site occupied with AA-tRNA or peptidyl-tRNA. Accurate rate studies have not been possible so it is unclear whether translocation preceeds release of the deacylated tRNA or whether the release of the tRNA is the initial step. Each translocation step results in the movement of three bases on the mRNA (205). The presence of a new code word appearing in the A site after translocation has been seen in studies using synthetic as well as natural mRNAs.

The above results on the interaction of EF-G with guanosine nucleotides and the ribosome do not explain how EF-G functions in the translocation reaction, and there is no satisfactory model as to how translocation occurs. The fact that a slow rate of translocation can occur in the absence of EF-G and GTP suggests that the basic process is intrinsic to the ribosome (206,207). In some way, EF-G and GTP markedly stimulate the ribosome-dependent process. Virtually all of the theories that have been proposed to explain the translocation process involve either changes in the configuration of the mRNA [inch worm theory (208)] or ribosome [locking-unlocking model (209)]. It is obvious that new approaches are needed to gain more insight into this process. It may turn out that the translocation event will be the most difficult and last of all the steps in protein synthesis to be fully understood.

EUKARYOTIC TRANSLOCATION, ELONGATION FACTOR 2

The eukaryote factor, corresponding to EF-G, is EF-2 (137-139). Since the overall function of EF-2 is similar to EF-G, only the significant differences between the two factors will be summarized. EF-2 is a protein monomer of 110,000 mol wt determined by amino acid composition (210). It has been purified by several laboratories (210-213) and catalyzes translocation judged by the same criteria used in the studies of EF-G. However, the characteristics of some of the partial reactions are different. One notable example is that EF-2 will interact with GTP and

GDP to form a stable binary complex, even in the absence of ribosomes (213-216). The interaction of the isolated EF-2·GTP complex with the ribosome results in rapid hydrolysis of GTP. As in the prokaryote system, the ternary complex formed is a ribosome·EF-2·GDP complex. The eukaryote complex is more stable than the prokaryotic ribosome EF-G·GDP complex. Thus fusidic acid has less of an effect on the stability of the eukaryotic complex than on the prokaryote complex.

Another interesting difference between EF-G and EF-2 is that the latter is inhibited by diphtheria toxin. Strauss and Hendee (217) and Collier and Pappenheimer (218) first showed that the primary action of diphtheria toxin was on protein synthesis, and subsequently it was shown that the site of inhibition was EF-2 (219-221). It is now clearly established that the toxin catalyzes an ADP ribosylation from NAD to EF-2 (222) and that the modified EF-2 is inactive. However, not all of the partial reactions of EF-2 are affected by the ADP ribosylation with NAD. ADP ribosylated EF-2 can still interact with guanosine nucleotides in the presence or absence of ribosomes (216), but cannot function in polymerization. A 15 amino acid peptide has been isolated from EF-2 that contains the ADP ribose (223,224) although the exact nature of the amino acid involved in this interaction is still unknown.

PROKARYOTIC PEPTIDE CHAIN TERMINATION

The overall reaction is schematically presented in Figure 6. The completed peptide, still attached to tRNA, is pictured on the P site of

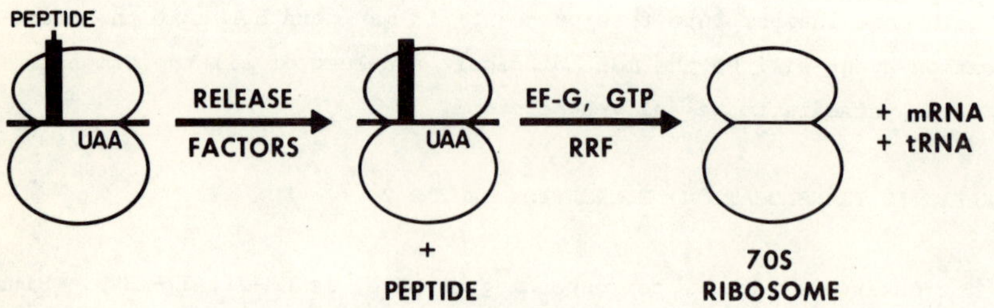

Figure 6. Steps involved in termination.

the ribosome with a terminator code word on the A site. In the presence of a release factor, the completed protein is released, leaving a deacylated tRNA still attached to the ribosome-mRNA complex. In subsequent reactions the tRNA is released, and the ribosome dissociates from the mRNA. EF-G and another factor, called ribosome release factor (RRF), are required for these final steps.

Ganoza (225) first obtained evidence that there was a factor(s) in the supernatant that was required for peptide chain termination but distinct from the elongation factors. Capecchi (226) later purified a protein factor that was required for termination using a mRNA from bacteriophage R17 carrying an amber mutation. Subsequently, Caskey et al. (227) showed that release factor activity could be assayed by measuring the release of fmet-tRNA from an fmet-tRNA·AUG·ribosome complex, dependent on a terminator code word such as UAA, UAG or UGA. Two protein factors were isolated, RF-1 and RF-2, that catalyzed peptide chain termination, but with different codon specificities (228-230). RF-1 is active with UAA or UGA, whereas RF-2 uses UAA or UGA (231,232).

Properties of RF-1, RF-2; Interaction with Ribosomes.

RF-1 and RF-2 have been purified to homogeneity (229,230). Their molecular weights have been reported to be 44,000 and 47,000 respectively (229). It has been calculated that each E. coli cell has between 500-700 molecules of each factor (229). Recently, it has been reported that RF-1 and RF-2 have common antigenic determinants (233). Both release factors can bind to the ribosome to form a stable complex in the presence of the appropriate terminator code word. Ribosomes depleted of L7L12 do not interact with RF-1 indicating that proteins L7L12 are involved in proper recognition on the ribosome of the release factors (132). Even in the absence of ribosomes, equilibrium dialysis studies have provided evidence for release factor-terminator codon interaction (234).

Role of Peptidyl Transferase in Termination.

Several lines of evidence indicate that the ribosome peptidyl transferase center is intimately involved in the termination reaction. Changes in ionic conditions (removal of K and NH_4) (235) or the presence of antibiotics (chloramphenicol, puromycin, etc.) that affect the ability of ribosomes to catalyze peptide bond formation or codon recognition (tetracycline, streptomycin, etc.) also inhibit termination (228,235, 236). The peptidyl transferase inhibitors prevent the release of fmet in the standard release assay suggesting that the peptidyl transferase is involved in the actual release of the completed peptide. Since the peptidyl transferase has been shown to form ester bonds in the presence of an alcohol (236), it appears that the peptidyl transferase center can transfer peptides to a variety of acceptors, including amino acids during peptide bond formation, to alcohols to form esters or to water when the peptide is hydrolyzed during termination. In some way the release factor interaction with the terminator codon on the A site activates the peptidyl transferase, resulting in the release (or hydrolysis) of the peptide attached to the tRNA on the ribosome.

Role of GTP Hydrolysis and RF-3.

A clear role of GTP hydrolysis in the termination reaction in prokaryotes, catalyzed by the release factors, has not been established. However, there are several observations that suggest GTP hydrolysis may also be involved in the release of the completed polypeptide from the ribosome. As noted below, there is a requirement for GTP in the eukaryotic process. Also, all of the soluble factors (IF-2, EF-Tu and EF-G) that interact with the ribosome and are involved in GTP hydrolysis, interact with 50S subunit proteins L7L12. Release factors also require proteins L7L12 for activity in a standard fmet-tRNA release assay (132). Finally, a factor that stimulates the termination reaction (referred to as RF-3) has been observed (230). Although this protein has no release factor activity itself, it appears to lower the Km of the terminator codons in the assay, stimulate the binding of RF-1 and RF-2 to the ribosome and facilitate the dissociation of the RF·UAA·ribosome complex (237). The activity of RF-3 is either inhibited or stimulated by guanosine nucleotides, depending on the concentration of the trinucleotide codon. Capecchi and Klein have reported that RF-3 and EF-Tu are the same (234), although Milman _et al_. (230) have not observed EF-Tu activity in their purified RF-3 preparations.

EUKARYOTIC PEPTIDE CHAIN TERMINATION

A eukaryotic release factor was first demonstrated in rabbit reticulocyte extracts (238). Using the fmet-tRNA release assay, the reaction required a tetranucleotide such as UAAA, UAGA, UGAA or UAGG and was stimulated by GTP (239). Purified factors have been obtained also from guinea pig liver, hamster liver and insects (240). Reticulocyte RF, based on gel filtration, has a mol wt of 115,000, but gel analysis indicates two subunits suggesting that the active factor may be a dimer (241). The requirement for a tetranucleotide in the _in vitro_ assay differentiates the prokaryotic and eukaryotic systems, and in addition, it appears that eukaryotic cells contain only one release factor capable of recognizing all the terminator code words, as compared to the two factors seen in _E._

coli. A ribosome-dependent GTP hydrolysis is also associated with purified preparations of reticulocyte RF, which is stimulated by the terminator oligonucleotide (e.g. UAAA) (241). No peptidyl-tRNA is required on the ribosome for GTP hydrolysis so that the GTP hydrolysis does not appear to be related to the release of the peptide by peptidyl transferase.

Summary of Steps Involved in Peptide Chain Release.

A model of the steps involved in the termination reaction has been proposed by Caskey (241) and is shown in Figure 7. Although the intermediate steps in this model have not been established, it is in accord

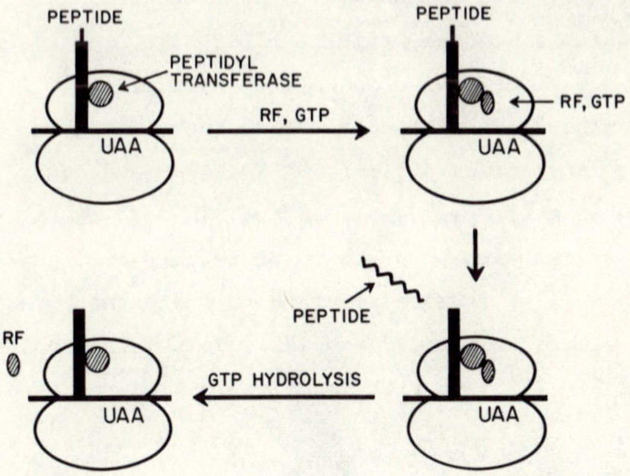

Figure 7. Partial reactions in the termination process. Figure taken from Caskey (241).

with most of the experimental data from both pro- and eukaryote studies. The basic points in this scheme include the binding of the appropriate RF to the ribosome in the presence of a terminator code word in the A site, activation of peptidyl transferase resulting in the release of the peptide, and GTP hydrolysis resulting in the dissociation of the RF from the ribosome. The latter reaction has not been observed in prokaryotes and is assumed to occur based on the GTP requirement in the eukaryotic system.

Reactions Occurring After the Peptide is Released.

The release of the completed peptide still leaves a ribosome attached to the mRNA with a deacylated tRNA on the P site (see Fig. 6). Hirashima and Kaji (242) first described a factor, called ribosome release factor,

(RRF), that appeared to be involved in termination and could be assayed by its ability to release labelled mRNA from ribosomes (243). The physiological significance of RRF has been confirmed by studies showing that it is required for the release from ribosomes of R17 RNA at an amber codon (243) and for the DNA-directed *in vitro* synthesis of β-galactosidase (244).

INTERACTION OF THE SOLUBLE FACTORS WITH THE RIBOSOME, ROLE OF GTP HYDROLYSIS

I refer the reader to other sections in this volume to obtain more detailed information on the ribosomal components involved in the interaction of soluble factors and tRNA with the ribosome.

Studies in which ribosomal proteins L7L12 have been specifically removed from the 50S subunit have shown that none of the GTP requiring reactions take place. It seems clear that these ribosomal proteins are involved in the interaction of IF-2, EF-Tu, EF-G and, very likely, RF with the 50S subunit (129-132,245). Since, in most cases, the factor is complexed with GTP as part of a binary or ternary complex (the RF complex has not been shown), it is more accurate to state that proteins L7L12 recognize the soluble factor·GTP complex. This is summarized schematically in Figure 8. The factor·GTP complex is pictured as having a high

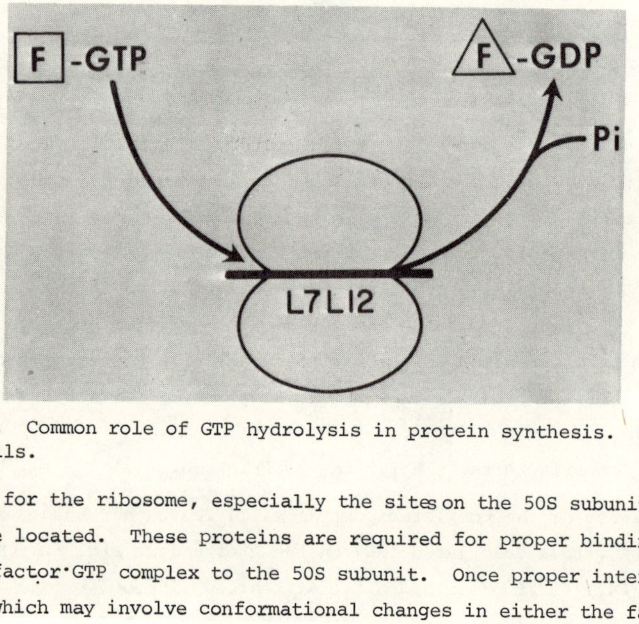

Figure 8. Common role of GTP hydrolysis in protein synthesis. See text for details.

affinity for the ribosome, especially the sites on the 50S subunit where L7L12 are located. These proteins are required for proper binding of the soluble factor·GTP complex to the 50S subunit. Once proper interaction occurs, which may involve conformational changes in either the factor

(125-127) or ribosome or both, GTP hydrolysis takes place. After hydrolysis, the factor·GDP complex dissociates from the ribosome. Conformational changes could explain the different behaviors of the factor·GTP and GDP complexes. Whether the GTPase active center is on the soluble factor or the ribosome is not clear. Studies with kirromycin (see above) indicate that it is on the EF-Tu (134), and this may be indicative that other soluble factors have the capacity to hydrolyze GTP when activated by ribosomes. However, a 5S RNA·protein complex from ribosomes has been reported to have both GTPase and ATPase activity, but as yet the physiological role of this complex has not been determined (247,248).

SEARCH FOR OTHER FACTORS

Although considerable progress has been made in identifying the soluble factors required for translation in both prokaryotic and eukaryotic cells, the reader should not be left with the impression that there is little new information to be gained or that we have identified all of the required factors. Most of the studies on the present factors have utilized synthetic templates, and usually only partial reactions have been examined. More should be done in systems using natural templates. In addition, there are already examples of factors whose roles in translation have not been fully elucidated (249-252) and, quite likely, there are a large number of soluble factors required for translation that are still unknown. They may only be identified when more defined DNA-directed or mRNA-directed systems are studied. Experiments along these lines in E. coli have been initiated on the DNA-directed in vitro synthesis of β-galactosidase (253-256) with the specific goal of identifying new factors required for gene expression and to learn more about the mechanism of action of existing factors. The ultimate aim of these studies would be to obtain the expression of a bacterial gene in a completely defined system.

REFERENCES

1. Molecular Mechanisms of Protein Biosynthesis. (H. Weissbach and S. Pestka, eds.), Academic Press, Inc., New York (1977).

2. Waller, J.P. (1963) J. Mol. Biol. 7, 483-496.

3. Marcker, K. and Sanger, F. (1964) J. Mol. Biol. 8, 835-840.

4. Salas, M., Hille, M.B., Last, J.A., Wahba, A.S. and Ochoa, S. (1967) Proc. Natl. Acad. Sci. U.S.A. 57, 387-394.

5. Revel, M., Lelong, J.C., Brawerman, G. and Gros, F. (1968) Nature (London) 219, 1016-1021.

6. Revel, M. (1977) in: Molecular Mechanisms of Protein Biosynthesis. (H. Weissbach and S. Pestka, eds.) pp. 246-321, Academic Press, Inc., New York.

7. Hershey, J.W.B., Yanov, J., Johnston, K. and Fakunding, J.L. (1977) Arch. Biochem. Biophys. 182, 626-638.

8. Lee-Huang, S., Sillero, M.A.G. and Ochoa, S. (1971) Eur. J. Biochem. 18, 536-543.

9. Kolakofsky, D., Dewey, K. and Thach, R.E. (1969) Nature (London) 223, 694-697.

10. Lelong, J.C., Grunberg-Manago, M., Dondon, J., Gros, D. and Gros, F. (1970) Nature (London) 226, 505-510.

11. Mazumder, R. (1971) FEBS Lett. 18, 64-66.

12. Fakunding, J.L., Traugh, J.A., Traut, R.R. and Hershey, J.W.B. (1972) J. Biol. Chem. 247, 6365-6367.

13. Miller, M.J. and Wahba, A.J. (1973) J. Biol. Chem. 248, 1084-1090.

14. Eskin, B., Treadwell, B., Redfield, B., Spears, C., Kung, H.F. and Weissbach, H. (1978) Arch. Biochem. Biophys. 189, 531-534.

15. Kung, H.F., Spears, C., Schulz, T. and Weissbach, H. (1974) Arch. Biochem. Biophys. 162, 578-584.

16. Lee-Huang, S. and Ochoa, S. (1971) Nature New Biol. 234, 236-239.

17. Lee-Huang, S. and Ochoa, S. (1973) Arch. Biochem. Biophys. 156, 84-96.

18. Suryanarayana, T. and Subramanian, A.R. (1977) FEBS Lett. 79, 264-268.

19. Brauer, D. and Wittmann-Liebold, B. (1977) FEBS Lett. 79, 269-275.

20. Springer, M., Graffe, M. and Hennecke, H. (1977) Proc. Natl. Acad. Sci. U.S.A. 74, 3970-3974.

21. Davis, B.D. (1974) in: Ribosomes (Nomura, M., Tissieres, A. and Lengyel, P., eds.) pp. 705-710, Cold Spring Harbor Lab., New York.

22. Kaempfer, R. (1970) Nature (London) 228, 534-537.

23. Kaempfer, R. (1972) J. Mol. Biol. 71, 583-598.

24. Van Duin, J., Kurland, C.G., Dondon, J., Grunberg-Manago, M., Branlant, C. and Ebel, J.P. (1976) FEBS Lett. 62, 111-114.

25. Miall, S.H. and Tamaoki, T. (1972) Biochemistry 11, 4826-4830.

26. Lockwood, A.H., Sarkar, P. and Maitra, U. (1972) Proc. Natl. Acad. Sci. U.S.A. 69, 3602-3605.

27. Grunberg-Manago, M., Godefroy-Colburn, T., Wolfe, A.D., Dessen, P., Pantaloni, D., Springer, M., Graffe, M., Dondon, J. and Kay, A. (1973) in: Regulation of Transcription and Translation in Eukaryotes (E. Bautz, et al., eds.), 24th Mosbach Colloq., pp. 213-249, Springer-Verlag, Berlin and New York.

28. Groner, Y. and Revel, M. (1971) Eur. J. Biochem. 22, 144-152.

29. Hershey, J.W.B., Remold-O'Donnell, E., Kolakofsky, D., Dewey, K.F. and Thach, R.E. (1971) in: Methods in Enzymology (K. Moldave and L. Grossman, eds.) Vol. 20, pp. 235-247, Academic Press, Inc., New York.

30. Leder, P. and Nau, M.M. (1967) Proc. Natl. Acad. Sci. U.S.A. 58, 774-781.

31. Noll, M., Noll, H. and Lingrel, J.B. (1972) Proc. Natl. Acad. Sci. U.S.A. 69, 1843-1847.

32. Noll, M., Hapke, B. and Noll, H. (1973) J. Mol. Biol. 80, 519-530.

33. Greenshpan, H. and Revel, M. (1969) Nature (London) 224, 331-335.

34. Shine, J. and Dalgarno, L. (1974) Proc. Natl. Acad. Sci. U.S.A. 71, 1342-1346.

35. Johnson, B. and Szekely, M. (1977) Nature (London) 267, 250-252.

36. Rudland, P.S., Whybrow, W.A. and Clark, B.F.C. (1971) Nature New Biol. 231, 76-78.

37. Lockwood, A.H., Chakraborty, P.R. and Maitra, U. (1971) Proc. Natl. Acad. Sci. 68, 3122-3126.

38. Van der Hofstad, G.A.J.M., Foekens, J.A., Van den Elsen, P.J. and Voorma, H.O. (1976) Eur. J. Biochem. 66, 181-192.

39. Lucas-Lenard, J. and Lipmann, F. (1967) Proc. Natl. Acad. Sci. U.S.A. 57, 1050-1057.

40. Fakunding, J.L. and Hershey, J.W.B. (1973) J. Biol. Chem. 248, 4206-4212.

41. Dubnoff, J.S., Lockwood, A.H. and Maitra, U. (1972) J. Biol. Chem. 247, 2884-2894.

42. Benne, R., Naaktgeboren, N., Gubbens, J. and Voorma, H.O. (1973) Eur. J. Biochem. 32, 372-380.

43. Thach, S.S. and Thach, R.E. (1971) Nature New Biol. 229, 219-221.

44. Benne, R., Ebes, F. and Voorma, H.O. (1973) Eur. J. Biochem. 38, 265-273.

45. Hershey, J.W.B., Dewey, K.F. and Thach, R.E. (1969) Nature (London) 222, 944-947.

46. Vermeer, C., Boon, J., Talens, A. and Bosch, L. (1973) Eur. J. Biochem. 40, 283-293.

47. Vermeer, C., van Alphen, W., van Knippenberg, P. and Bosch, L. (1973) Eur. J. Biochem. 40, 295-308.

48. Chae, Y.B., Mazumder, R. and Ochoa, S. (1969) Proc. Natl. Acad. Sci. U.S.A. 63, 828-833.

49. Anderson, W.F., Bosch, L., Cohn, W.E., Lodish, H., Merrick, W.C., Weissbach, H., Wittmann, H.G. and Wool, I.G. (1977) FEBS Lett. 76, 1-10.

50. Schreier, M.H., Erni, B. and Staehelin, T. (1977) J. Mol. Biol. 116, 727-753.

51. Staehelin, T., Trachsel, H., Erni, B., Boschetti, A. and Schreier, M.H. (1975) FEBS Proc. Meet. 39, 309-323.

52. Safer, B., Anderson, W.F. and Merrick, W.C. (1975) J. Biol. Chem. 250, 9067-9075.

53. Merrick, W.C., Kemper, W.M. and Anderson, W.F. (1975) J. Biol. Chem. 250, 5556-5562.

54. Prichard, P.M. and Anderson, W.F. (1974) in: Methods in Enzymology, (K. Moldave and L. Grossman, eds.), Vol. 30, pp. 136-141, Academic Press, Inc., New York.

55. Safer, B., Adams, S.L., Kemper, W.M., Berry, K.W., Lloyd, M. and Merrick, W.C. (1976) Proc. Natl. Acad. Sci. U.S.A. 73, 2584-2588.

56. Kemper, W.M., Berry, K.W. and Merrick, W.C. (1976) J. Biol. Chem. 251, 5551-5557.

57. Sundararajan, T.A. and Thach, R.E. (1966) J. Mol. Biol. 19, 74-90.

58. Bose, K.K., Chatterjee, N.K. and Gupta, N.K. (1974) Methods Enzymol. 29, 522-529.

59. Benne, R. and Hershey, J.W.B. (1976) Proc. Natl. Acad. Sci. U.S.A. 73, 3005-3009.

60. Benne, R., Wong, C., Luedi, M. and Hershey, J.W.B. (1976) J. Biol. Chem. 251, 7675-7681.

61. Benne, R., Luedi, M. and Hershey, J.W.B. (1977) J. Biol. Chem. 252, 5798-5803.

62. Benne, R., Brown-Luedi, M.L. and Hershey, J.W.B. (1978) J. Biol. Chem. 253, 3070-3077.

63. Benne, R. and Hershey, J.W.B. (1978) J. Biol. Chem. 253, 3078-3087.

64. Schreier, M.H. and Staehelin, T. (1973) in: Regulation of Transcription and Translation in Eukaryotes (E. Bautz, et al., eds.), 24th. Mosbach Colloq. pp. 335-349, Springer-Verlag, Berlin and New York.

65. Caskey, C.T., Redfield, B. and Weissbach, H. (1967) Arch. Biochem. Biophys. 120, 119-123.

66. Marcus, A. (1970) J. Biol. Chem. 245, 955-961.

67. London, I.M., Clemens, M.J., Ranu, R.S., Levin, D.H., Cherbas, L.F. and Ernst, V. (1976) Fed. Proc. 35, 2218-2222.

68. Zucker, W.V. and Schulman, H.M. (1968) Proc. Natl. Acad. Sci. U.S.A. 59, 582-589.

69. Hunt, T., Vanderhoff, G. and London, I.M. (1972) J. Mol. Biol. 66, 471-481.

70. Mathews, M.B., Hunt, T. and Brayley, A. (1973) Nature New Biol. 243, 230-233.

71. Maxwell, C.R., Kamper, C.S. and Rabinovitz, M. (1971) J. Mol. Biol. 58, 317-327.

72. Gross, M. and Rabinovitz, M. (1972) Proc. Natl. Acad. Sci. U.S.A. 69, 1565-1568.

73. Gross, M., Rabinovitz, M. (1972) Biochim. Biophys. Acta 287, 340-352.

74. Farrell, P.J., Balkow, K., Hunt, T., Jackson, R.J. and Trachsel, H. (1977) Cell 11, 187-200.

75. Levin, D.H., Ranu, R.S., Ernst, V. and London, I.M. (1976) Proc. Natl. Acad. Sci. U.S.A. 73, 3112-3116.

76. Kramer, G., Cimadevilla, J.M. and Hardesty, B. (1976) Proc. Natl. Acad. Sci. U.S.A. 73, 3078-3082.

77. Gross, M. and Mendelewski, J. (1977) Biochem. Biophys. Res. Commun. 74, 559-569.

78. de Haro, C.D. and Ochoa, S. (1978) Proc. Natl. Acad. Sci. U.S.A. 75, 2713-2716.

79. Trachsel, H., Ranu, R.S. and London, I.M. (1978) Proc. Natl. Acad. Sci. U.S.A. 75, 3654-3658.

80. Datta, A., deHaro, C.D., Sierra, J.M. and Ochoa, S. (1977) Proc. Natl. Acad. Sci. U.S.A. 74, 1463-1467.

81. Kosower, N.S., Vanderhoff, G.A., Benerofe, B., Hunt, T. and Kosower, E.M. (1971) Biochem. Biophys. Res. Commun. 45, 816-821.

82. Clemens, M.J., Safer, B., Merrick, W.C., Anderson, W.F. and London, I.M. (1975) Proc. Natl. Acad. Sci. U.S.A. 72, 1286-1290.

83. Ernst, V., Levin, D.H. and London, I.M. (1978) Proc. Natl. Acad. Sci. U.S.A. 75, 4110-4114.

84. Levin, D. and London, I.M. (1978) Proc. Natl. Acad. Sci. U.S.A. 75, 1121-1125.

85. Ehrenfeld, E. and Hunt, T. (1971) Proc. Natl. Acad. Sci. U.S.A. 68, 1075-1078.

86. Graziadei, W.D. III and Lengyel, P. (1972) Biochem. Biophys. Res. Commun. 46, 1816-1823.

87. Darnbrough, C., Hunt, T. and Jackson, R.J. (1972) Biochem. Biophys. Res. Commun. 48, 1556-1564.

88. Hunter, T., Hunt, T., Jackson, R.J. and Robertson, H.D. (1975) J. Biol. Chem. 250, 409-417.

89. Lenz, J.R. and Baglioni, C. (1978) J. Biol. Chem. 253, 4219-4223.

90. Beuzard, Y. and London, I.M. (1974) Proc. Natl. Acad. Sci. U.S.A. 71, 2863-2866.

91. Hovanessian, A.G., Brown, R.E., Kerr, I.M. (1977) Nature 268, 537-540.

92. Farrell, P.J., Sen, G.C., Dubois, M.F., Ratner, L., Slattery, E. and Lengyel, P. (1978) Proc. Natl. Acad. Sci. U.S.A. 75, 5893-5897.

93. Kerr, I.M., Brown, R.E. and Hovanessian, A.G. (1977) Nature 268, 540-542.

94. Kerr, I.M., Brown, R.E. (1978) Proc. Natl. Acad. Sci. U.S.A. 75, 256-260.

95. Clemens, M.J. and Williams, B.R.G. (1978) Cell 13, 565-572.

96. Roberts, W.K., Hovanessian, A., Brown, R.E., Clemens, M.J. and Kerr, I.M. (1976) Nature 264, 477-480.

97. Sen, G.C., Taira, H. and Lengyel, P. (1978) J. Biol. Chem. 253, 5915-5921.

98. Zilberstein, A., Kimchi, A., Schmidt, A. and Revel, M. (1978) Proc. Natl. Acad. Sci. U.S.A. 75, 4734-4738.

99. Cooper, J.A. and Farrell, P.J. (1977) Biochem. Biophys. Res. Commun. 77, 124-131.

100. Ohtsuki, K., Dianzani, F. and Baron, S. (1977) Nature 269, 536-538.

101. Lewis, J., Falcoff, E. and Falcoff, R. (1978) Eur. J. Biochem. 86, 497-509.

102. Ochoa, S. and deHaro, C. Ann. Rev. Biochem., in press.

103. Allende, J.E., Monro, R. and Lipmann, F. (1964) Proc. Natl. Acad. Sci. U.S.A. 51, 1211-1216.

104. Nishizuka, Y. and Lipmann, F. (1966) Proc. Natl. Acad. Sci. U.S.A. 55, 212-219.

105. Lucas-Lenard, J. and Lipmann, F. (1966) Proc. Natl. Acad. Sci. U.S.A. 55, 1562-1566.

106. Furano, A.V. (1975) Proc. Natl. Acad. Sci. U.S.A. 72. 4780-4784.

107. Jaskunas, S.R., Lindahll, L., Nomura, M. and Burgess, R.R. (1975) Nature 257, 458-462.

108. Furano, A.V. (1977) J. Biol. Chem. 252, 2154-2157.

109. Miller, D.L., Nagarkatti, S., Laursen, R.A., Parker, J. and Friesen, J.D. (1978) Molec. Gen. Genet. 159, 57-62.

110. Vande Klundert, J.A.M., Van der Meide, P.H., Vande Putte, P. and Bosch, L. (1978) Proc. Natl. Acad. Sci. U.S.A. 75, 4470-4473.

111. Miller, D.L. and Weissbach, H. (1970) Arch. Biochem. Biophys. 141, 26-37.

112. Arai, K.I., Kawakita, M. and Kaziro, Y. (1972) J. Biol. Chem. 247, 7029-7037.

113. Miller, D.L., Hachmann, J. and Weissbach, H. (1971) Arch. Biochem. Biophys. 144, 115-121.

114. Laursen, R.A., Nagarkatti, S. and Miller, D.L. (1977) FEBS Lett. 80, 103-106.

115. Hachmann, J., Miller, D.L. and Weissbach, H. (1971) Arch. Biochem. Biophys. 147, 457-466.

116. Miller, D.L. and Weissbach, H. (1970) Biochem. Biophys. Res. Commun. 38, 1016-1022.

117. Ertel, R., Redfield, B., Brot, N. and Weissbach, H. (1968) Arch. Biochem. Biophys. 128, 331-338.

118. Miller, D.L. and Weissbach, H. (1974) in: Methods in Enzymology (K. Moldave and L. Grossman, eds.), Vol. 30, pp. 219-232, Academic Press, Inc., New York.

119. Miller, D.L., Cashel, M. and Weissbach, H. (1973) Arch. Biochem. Biophys. 154, 675-682.

120. Arai, K., Kawakita, M. and Kaziro, Y. (1974) J. Biochem. (Tokyo) 76, 293-306.

121. Arai, K.I., Arai, N., Nakamura, S., Oshima, T. and Kaziro, Y. Eur. J. Biochem. 92, 521-531 (1978).

122. Shorey, R.L., Ravel, J.M., Garner, C.W. and Shive, W. (1969) J. Biol. Chem. 244, 4555-4564.

123. Gordon, J. (1968) Proc. Natl. Acad. Sci. U.S.A. 59, 179-183.

124. Fahnestock, S., Weissbach, H. and Rich, A. (1972) Biochim. Biophys. Acta 269, 62-66.

125. Printz, M.P. and Miller, D.L. (1973) Biochem. Biophys. Res. Commun. 53, 149-156.

126. Crane, L.J. and Miller, D.L. (1974) Biochemistry 13, 933-939.

127. Arai, K.I., Kawakita, M., Kaziro, Y., Maeda, T. and Ohnishi, S.I. (1974) J. Biol. Chem. 249, 3311-3313.

128. Shorey, R.L., Ravel, J.M. and Shive, W. (1971) Arch. Biochem. Biophys. 146, 110-117.

129. Brot, N., Yamasaki, E., Redfield, B. and Weissbach, H. (1972) Arch. Biochem. Biophys. 148, 148-155.

130. Brot, N., Marcel, R., Yamasaki, E. and Weissbach, H. (1973) J. Biol. Chem. 248, 6952-6956.

131. Lockwood, A.H., Maitra, U., Brot, N. and Weissbach, H. (1974) J. Biol. Chem. 249, 1213-1218.

132. Brot, N., Tate, W.P., Caskey, C.T. and Weissbach, H. (1974) Proc. Natl. Acad. Sci. U.S.A. 71, 89-92.

133. Wolf, H., Chinali, G. and Parmeggiani, A. (1974) Proc. Natl. Acad. Sci. U.S.A. 71, 4910-4914.

134. Chinali, G., Wolf, H. and Parmeggiani, A. (1977) Eur. J. Biochem. 75, 55-65.

135. Wolf, H., Chinali, G. and Parmeggiani, A. (1977) Eur. J. Biochem. 75, 67-75.

136. Wolf, H., Assmann, D. and Fischer, E. (1978) Proc. Natl. Acad. Sci. U.S.A. 75, 5324-5328.

137. Bishop, J.O. and Schweet, R.S. (1961) Biochem. Biophys. Acta 49, 235-236.

138. Arlinghaus, R., Favelukes, G. and Schweet, R. (1963) Biochem. Biophys. Res. Commun. 11, 92-96.

139. Hardesty, B., Arlinghaus, R., Shaeffer, J. and Schweet, R. (1963) Cold Spring Harbor Symp. Quant. Biol. 28, 215-222.

140. Felicetti, L. and Lipmann, F. (1968) Arch. Biochem. Biophys. 125, 548-557.

141. Gasior, E. and Moldave, K. (1965) J. Biol. Chem. 240, 3346-3352.

142. Schneir, M. and Moldave, K. (1968) Biochim. Biophys. Acta 166, 58-67.

143. Collins, J.F., Moon, H.M. and Maxwell, E.S. (1972) Biochemistry 11, 4187-4194.

144. Moon, H.M., Redfield, B. and Weissbach, H. (1972) Proc. Natl. Acad. Sci. U.S.A. 69, 1249-1252.

145. McKeehan, W.L. and Hardesty, B. (1969) J. Biol. Chem. 244, 4330-4339.

146. Iwasaki, K., Mizumoto, K., Tanaka, M. and Kaziro, Y. (1973) J. Biochem. (Tokyo) 74, 849-852.

147. Jerez, C., Sandoval, A., Allende, J., Henes, C. and Ofengand, J. (1969) Biochemistry 8, 3006-3014.

148. Golinska, B. and Legocki, A.B. (1973) Biochim. Biophys. Acta 324, 156-170.

149. Tarrago, A., Allende, J.E., Redfield, B. and Weissbach, H. (1973) Arch. Biochem. Biophys. 159, 353-361.

150. Bollini, R., Soffientini, A.N., Bertani, A., Lanzani, G.A. (1974) Biochemistry 13, 5421-5425.

151. Drews, J., Bednarik, K. and Grasmuk, H. (1974) Eur. J. Biochem. 41. 217-227.

152. Twardowski, T., Redfield, B., Kemper, W.M., Merrick, W.C. and Weissbach, H. (1976) Biochem. Biophys. Res. Commun. 71, 272-279.

153. Twardowski, T., Hill, J.M. and Weissbach, H. (1977) Arch. Biochem. Biophys. 180, 444-451.

154. Iwasaki, K., Nagata, S., Mizumoto, K. and Kaziro, Y. (1974) J. Biol. Chem. 249, 5008-5010.

155. Nagata, S., Iwasaki, K. and Kaziro, Y. (1977) J. Biochem. (Tokyo) 82, 1633-1646.

156. Slobin, L.I. and Möller, W. (1976) Eur. J. Biochem. 69, 351-366.

157. Kemper, W.M., Merrick, W.C., Redfield, B., Liu, C.K. and Weissbach, H. (1976) Arch. Biochem. Biophys. 174, 603-612.

158. Ejiri, S., Taira, H. and Shimura, K. (1973) J. Biochem. (Tokyo) 74, 195-197.

159. Ejiri, S., Murakami, K. and Katsumata, T. (1977) FEBS Lett. 82, 111-114.

160. Motoyoshi, K., Iwasaki, K. and Kaziro, Y. (1977) J. Biochem. (Tokyo) 82, 145-155.

161. Slobin, L.I. and Möller, W. (1978) Eur. J. Biochem. 84, 69-77.

162. Grasmuk, H., Nolan, R.D. and Drews. J. Eur. J. Biochem., in press.

163. Nagata, S., Motoyoshi, K. and Iwasaki, K. (1976) Biochem. Biophys. Res. Commun. 71, 933-938.

164. Iwasaki, K., Motoyoshi, K., Nagata, S. and Kaziro, Y. (1976) J. Biol. Chem. 251, 1843-1845.

165. Motoyoshi, K. and Iwasaki, K. (1977) J. Biochem. (Tokyo) 82, 703-708.

166. Grasmuk, H., Nolan, R.D. and Drews, J. (1977) Eur. J. Biochem. 79, 93-102.

167. Grasmuk, H., Nolan, R.D. and Drews, J. (1976) Eur. J. Biochem. 71, 271-279.

168. Pedersen, S., Reeh, S.V., Parker, J., Watson, R.J., Friesen, J.D. and Fiil, N.P. (1976) Molec. Gen. Genet. 144, 339-343.

169. Furano, A.V. and Wittel, F.P. (1976) J. Biol. Chem. 251, 898-901.

170. Lindahl, L., Post, L. and Nomura, M. (1976) Cell 9, 439-448.

171. Chu, F., Miller, D.L., Schulz, T., Weissbach, H. and Brot, N. (1976) Biochem. Biophys. Res. Commun. 73, 917-927.

172. Blumenthal, T., Landers, T.A. and Weber, K. (1972) Proc. Natl. Acad. Sci. U.S.A. 69, 1313-1317.

173. Landers, T.A., Blumenthal, T. and Weber, K. (1974) J. Biol. Chem. 249, 5801-5808.

174. Brown, S. and Blumenthal, T. (1976) Proc. Natl. Acad. Sci. U.S.A. 73, 1131-1135.

175. Travers, A. (1973) Nature (London) 244, 15-18.

176. Jacobson, G.R. and Rosenbusch, J.P. (1976) Nature 261, 23-26.

177. Beck, B.D., Arscott, P.G. and Jacobson, A. (1978) Proc. Natl. Acad. Sci. U.S.A. 75, 1250-1254.

178. Slobin, L.I. and Möller, W. (1975) Nature 258, 452-454.

179. Kolb, A.J., Redfield, B., Twardowski, T. and Weissbach, H. (1978) Biochim. Biophys. Acta 519, 398-405.

180. Rao, M.S., Rothblum, L.I. and Busch, H. (1978) Cell Biol. Intnl. Reports 2, 25-32.

181. Conway, T.W and Lipmann, F. (1964) Proc. Natl. Acad. Sci. U.S.A. 52, 1462-1469.

182. Kaziro, Y. and Inoue, N. (1968) J. Biochem. (Tokyo) 64, 423-425.

183. Parmeggiani, A. (1968) Biochem. Biophys. Res. Commun. 30, 613-619.

184. Leder, P., Skogerson, L.E. and Nau, M.M. (1969) Proc. Natl. Acad. Sci. U.S.A. 62, 454-460.

185. Kaziro, Y., Inoue-Yokosawa, N. and Kawakita, M. (1972) J. Biochem. (Tokyo) 72, 853-863.

186. Rohrbach, M.S., Dempsey, M.E. and Bodley, J.W. (1974) J. Biol. Chem. 249, 5094-5101.

187. Kaziro, Y., Inoue, N., Kuriki, Y., Mizumoto, K., Tanaka, M. and Kawakita, M. (1969) Cold Spring Harbor Symp. Quant. Biol. 34, 385-393.

188. Gordon, J. (1970) Biochemistry 9, 912-917.

189. Brot, N., Spears, C. and Weissbach, H. (1969) Biochem. Biophys. Res. Commun. 34, 843-848.

190. Parmeggiani, A. and Gottschalk, E.M. (1969) Cold Spring Harbor Symp. Quant. Biol. 34, 377-384.

191. Kuriki, Y., Inoue, N. and Kaziro, Y. (1970) Biochim. Biophys. Acta 224, 487-497.

192. Bodley, J.W., Zieve, F.J., Lin, L. and Zieve, S.T. (1969) Biochem. Biophys. Res. Commun. 37, 437-443.

193. Bodley, J.W., Zieve, F.J., Lin, L. and Zieve, S.T. (1970) J. Biol. Chem. 245, 5656-5661.

194. Bodley, J.W., Zieve, F.J. and Lin, L. (1970) J. Biol. Chem. 245, 5662-5667.

195. Bodley, J.W. and Lin, L. (1970) Nature (London) 227, 60-62.

196. Brot, N., Spears, C. and Weissbach, H. (1971) Arch. Biochem. Biophys. 143, 286-296.

197. Okura, A., Kinoshita, T. and Tanaka, N. (1971) J. Antibiot. 24, 655-661.

198. Modolell, J., Cabrer, B. and Vazquez, D. (1973) J. Biol. Chem. 248, 8356-8360.

199. Cundliffe, E. (1972) Biochem. Biophys. Res. Commun. 46, 1794-1801.

200. Cabrer, B., Vazquez, D. and Modolell, J. (1972) Proc. Natl. Acad. Sci. U.S.A. 69, 733-736.

201. Miller, D.L. (1972) Proc. Natl. Acad. Sci. U.S.A. 69, 752-755.

202. Richman, N. and Bodley, J.W. (1972) Proc. Natl. Acad. Sci. U.S.A. 69, 686-689.

203. Richter, D. (1972) Biochem. Biophys. Res. Commun. 46, 1850-1856.

204. Nishizuka, Y. and Lipmann, F. (1966a) Arch. Biochem. Biophys. 116, 344-351.

205. Erbe, R.W. and Leder, P. (1968) Biochem. Biophys. Res. Commun. 31, 798-803.

206. Pestka, S. (1968) Proc. Natl. Acad. Sci. U.S.A. 61, 726-733.

207. Gavrilova, L.P. and Spirin, A.S. (1971) FEBS Lett. 17, 324-326.

208. Hardesty, B., Culp, W. and McKeehan, W. (1969) Cold Spring Harbor Symp. Quant. Biol. 34, 331-344.

209. Spirin, A.S. (1969) Cold Spring Harbor Symp. Quant. Biol. 39, 197-207.

210. Robinson, E.A. and Maxwell, E.S. (1972) J. Biol. Chem. 247, 7023-7028.

211. Galasinski, W. and Moldave, K. (1969) J. Biol. Chem. 244, 6527-6532.

212. Raeburn, S., Collins, J.F., Moon, H.M. and Maxwell, E.S. (1971) J. Biol. Chem. 246, 1041-1048.

213. Henriksen, O., Robinson, E.A. and Maxwell, E.S. (1975) J. Biol. Chem. 250, 720-724.

214. Baliga, B.S. and Munro, H.N. (1972) Biochem. Biophys. Acta 277, 368-383.

215. Bermek, E. and Matthaei, H. (1971) Biochemistry 10, 4906-4912.

216. Chuang, D.M. and Weissbach, H. (1972) Arch. Biochem. Biophys. 152, 114-124.

217. Strauss, N. and Hendee, E.D. (1959) J. Exp. Med. 109, 145-163.

218. Collier, R.J. and Pappenheimer, A.M., Jr. (1964) J. Exp. Med. 120, 1019-1039.

219. Collier, R.J. (1967) J. Mol. Biol. 25, 83-98.

220. Goor, R.S. and Pappenheimer, A.M., Jr. (1967) J. Exp. Med. 126, 899-912.

221. Goor, R.S. and Pappenheimer, A.M., Jr. (1967) J. Exp. Med. 126, 913-921.

222. Honjo, T., Nishizuka, Y., Hayaishi, O. and Kato, I. (1968) J. Biol. Chem. 243, 3553-3555.

223. Robinson, E.A., Henriksen, O. and Maxwell, E.S. (1974) J. Biol. Chem. 249, 5088-5093.

224. Van Ness, B.G., Howard, J.B. and Bodley, J.W. (1978) J. Biol. Chem. 253, 8687-8690.

225. Ganoza, M.C. (1966) Cold Spring Harbor Symp. Quant. Biol. 31, 273-278.

226. Capecchi, M.R. (1967) Proc. Natl. Acad. Sci. U.S.A. 58, 1144-1151.

227. Caskey, C.T., Tompkins, R., Scolnick, E., Caryk, T. and Nirenberg, M. (1968) Science 162, 135-138.

228. Scolnick, E., Tompkins, R., Caskey, T. and Nirenberg, M. (1968) Proc. Natl. Acad. Sci. U.S.A. 61, 768-774.

229. Klein, H.A. and Capecchi, M.R. (1971) J. Biol. Chem. 246, 1055-1061.

230. Milman, G., Goldstein, J., Scolnick, E. and Caskey, T. (1969) Proc. Natl. Acad. Sci. U.S.A. 63, 183-190.

231. Benzer, S. and Champe, S.P. (1962) Proc. Natl. Acad. Sci. U.S.A. 48, 1114-1121.

232. Capecchi, M.R. and Klein, H.A. (1970) Nature (London) 226, 1029-1033.

233. Ratliffe, J.C. and Caskey, C.T. (1977) Arch. Biochem. Biophys. 181, 671-677.

234. Capecchi, M.R. and Klein, H.A. (1969) Cold Spring Harbor Symp. Quant. Biol. 34, 469-477.

235. Vogel, Z., Zamir, A. and Elson, D. (1969) Biochemistry 8, 5161-5168.

236. Caskey, C.T., Beaudet, A.L., Scolnick, E.M. and Rosman, M. (1971) Proc. Natl. Acad. Sci. U.S.A. 68, 3163-3167.

237. Goldstein, J.L. and Caskey, C.T. (1970) Proc. Natl. Acad. Sci. U.S.A. 67, 537-543.

238. Goldstein, J.L., Beaudet, A.L. and Caskey, C.T. (1970) Proc. Natl. Acad. Sci. U.S.A. 67, 99-106.

239. Beaudet, A.L. and Caskey, C.T. (1971) Proc. Natl. Acad. Sci. U.S.A. 68, 619-624.

240. Ilan, J. (1973) J. Mol. Biol. 77, 437-448.

241. Caskey, C.T. (1977) in: Molecular Mechanisms of Protein Biosynthesis, (H. Weissbach and S. Pestka, eds.) pp. 443-465, Academic Press, Inc., New York.

242. Hirashima, A. and Kaji, A. (1972) Biochemistry 11, 4037-4044.

243. Ogawa, K. and Kaji, A. (1975) Eur. J. Biochem. 58, 411-419.

244. Kung, H.F., Treadwell, B.V., Spears, C., Tai, P.C. and Weissbach, H. (1977) Proc. Natl. Acad. Sci. U.S.A. 74, 3217-3221.

245. Hamel, E. and Nakamoto, T. (1972) J. Biol. Chem. 247, 6810-6817.

246. Brot, N., Marcel, R., Yamasaki, E. and Weissbach, H. (1973) J. Biol. Chem. 248, 6952-6956.

247. Horne, J.R. and Erdmann, V.A. (1973) Proc. Natl. Acad. Sci. U.S.A. 70, 2870-2873.

248. Gaunt-Klöpfer, M. and Erdmann, V.A. (1975) Biochim. Biophys. Acta 390, 226-230.

249. Glick, B.R. and Ganoza, M.C. (1976) Eur. J. Biochem. 71, 483-491.

250. Van Der Meer, J.P. and Ganoza, M.C. (1975) Eur. J. Biochem. 54, 229-237.

251. Menninger, J.R., Walker, C., Tan, P.F. and Artherly, A.G. (1973) Molec. Gen. Genet. 121, 307-324.

252. Menninger, J.R. (1979) J. Bact. 137, 694-696.

253. Kung, H.F., Spears, C. and Weissbach, H. (1976) Arch. Biochem. Biophys. 174, 100-104.

254. Kung, H.F., Spears, C. and Weissbach, H. (1975) J. Biol. Chem. 250, 1556-1562.

255. Kung, H.F., Redfield, B., Treadwell, B.V., Eskin, B., Spears, C. and Weissbach, H. (1977) J. Biol. Chem. 252, 6889-6894.

256. Kung, H.F., Chu, F., Caldwell, P., Spears, C., Treadwell, B.V., Eskin, B., Brot, N. and Weissbach. (1978) Arch. Biochem. Biophys. 187, 457-463.

STRUCTURE OF tRNA DURING PROTEIN BIOSYNTHESIS

B. F. C. CLARK

Chemistry Department,
University of Aarhus,
Denmark

INTRODUCTION

Transfer RNA functions in protein biosynthesis by carrying esterified activated amino acids to the ribosomal site in the correct order for polymerization into a polypeptide. During this central biological function in the transmission of genetic information the tRNA plays a role as an adaptor by decoding the genetic information carried by mRNA. Indeed, the tRNA can be considered as a molecular interface between the protein and nucleic acid languages.

The total tRNA in a cell makes up about 1% of dry weight in the case of a bacterial cell so that there are about 4 x 10^5 tRNA molecules of perhaps 55 different types in such a cell. Thus, the tRNA content corresponds to a concentration of about 0.5 mM. Estimates of how much of the cellular tRNA is charged with an amino acid are rather difficult to obtain because of the lability of the aminoacyl bond to the tRNA. Probably, estimates such as about 80% of the tRNA being charged[1] are on the low side since it is likely that the catalytic activity of the adequate amount of aminoacyl-tRNA synthetase molecules (activating enzymes) in the cell[2] keeps the tRNA fully charged provided that sufficient free amino acids are available.

Now that the three dimensional structure has been determined to high resolution (2.5Å) for one tRNA species[3,4,5] and that more than 100 primary structures[6,7], all with predictable secondary structures in clover leaf forms[8], are known there is considerable motivation for attempting to explain tRNA function in terms of structure. Although the availability of the crystal structure of a tRNA provides a new meaningful basis for discussion of function it must be remembered that this structure gives only one conformational model of the tRNA in its most stable or resting state[9]. How this conformation changes during biological function especially during protein biosynthesis is an intriguing problem currently engaging many different types of research approach using X-ray crystallography, chemical modification studies, enzyme kinetic measurements and physical chemical techniques such as proton magnetic resonance, electron spin resonance, laser light scattering and fluorescence.

In this chapter I shall consider the role of tRNA in protein biosynthesis with emphasis on binding of tRNA to the ribosome. It is impossible for this chapter to be all embracing of the extensive tRNA literature in the space available so I apologize for being arbitrarily selective in reference citations. More extensive literature surveys will be found in recent reviews by Rich and RajBhandary[10], by Sprinzl and Cramer[11], and by Clark[6,12].

BIOLOGICAL ROLE OF tRNA IN PROTEIN BIOSYNTHESIS.

The current knowledge about the biological function or, more precisely biological activities of tRNA in protein biosynthesis is summarised in Table 1. Even in this area of molecular biology some of the activities listed have not been characterized adequately enough to describe them as clearly defined functions.

TABLE 1. Biological Activities of tRNA in Protein Biosynthesis.

1. Activation of amino acids	
Elongation	Initiation
2. Recognition by EF-Tu	6. Recognition of initiator tRNA by IF
3. Location in A-site of the ribosome	7. Location in I-site (part of P-site) of the ribosome
4. Decoding mRNA	
5. Signal for "magic spot"	8. Recognition by transformylase
9. Regulation a) Feedback Inhibitor b) Suppression	

In normal protein biosynthesis each tRNA species is charged with an amino acid (activity 1 of Table 1) by an aminoacyl-tRNA synthetase ("activating enzyme"). The tRNA is linked to its specific amino acid at the 2' or 3' hydroxyl group on its 3'-terminal adenosine by its specific (cognate) aminoacyl-tRNA synthetase (activity 1 of Table 1)[2,11]. The precise position of attachment depends on the particular enzyme species. The charged species is then carried to the ribosome in the form of a ternary complex made with the elongation factor Tu (EF-Tu) and GTP (activity 2). In similar fashion the unique tRNA species, the initiator tRNA, a special class of methionine tRNA, is thought to be carried to the ribosome by an initiation factor (activity 6). It is as yet uncertain whether GTP is required to make the initiator tRNA complex off the ribosome. The aminoacyl-tRNA is located by an uncharacterized mechanism in the A-site of the ribosome (activity 3) where it decodes mRNA via its anticodon triplet (activity 4). In contrast, the initiator tRNA, formylmethionyl-$tRNA_f^{Met}$ in prokaryotes, and methionyl-$tRNA_f^{Met}$ (sometimes called Met-$tRNA_i$) in eukaryotes, is located in the initiation (I)-site on the small ribosomal subunit (activity 7) for decoding the initiator triplet codon. This site becomes part of the ribosomal P-site. Since the prokaryotic Met-$tRNA_f$ has to be formylated for its initiator activity, it is also recognized by a special enzyme for this, the transformylase (activity 8).

When prokaryotic cells are starved for amino acids an unusual role has been detected for uncharged tRNA[6]. The uncharged tRNA is bound to the ribosomal A-site as if in mRNA decoding, but sets off a signal for the formation, by the so-called stringent factor, of unusual guanosine nucleotide derivatives, ppGpp and pppGpp, called "magic spots" (activity 5).

Assorted roles for tRNA[6] in the regulation of protein biosynthesis have been collected as activity 9. These include the well characterized suppressor of nonsense mutations [13,14] and a possible role as a feedback inhibitor in yeast. Obviously, restricting the amount of one specific decoding aminoacyl-tRNA will control protein biosynthesis at this point in translation. However, despite the large literature on the subject, especially in connection with control of protein synthesis in carcinogenic states, the regulatory role of tRNA has not been clearly characterized[6,10].

Because of our present structural knowledge of tRNA it is now appropriate to study the molecular mechanism of protein biosynthesis in structural terms. We wish to know what happens to the tRNA structure during charging of the tRNA with an amino acid, what happens to the tRNA on the ribosome and how the tRNA gets on and off the ribosome. Possible steps where conformational changes in the tRNA structure could occur during the elongation step of protein biosynthesis are summarised in Table 2.

TABLE 2. Steps in protein biosynthesis where changes in tRNA conformation could occur.

1) Charging	4) Peptide bond formation
2) Binding to EF-Tu:GTP	5) Translocation of peptidyl-tRNA
3) Decoding the mRNA	6) Ejection of discharged tRNA
a) Anticodon loop conformation	
b) Binding to ribosome at another part of tRNA	

PRIMARY STRUCTURE OF tRNA

The technique of X-ray crystallographic analysis is still not developed to a stage where crystallographers can build a three dimensional macromolecular model from a high resolution (~2.5Å) electron density map without a knowledge of the primary structure (the linear order of subunits) of the molecule. The technology concerned with RNA sequencing in recent years has benefitted from the tremendous development in rapid DNA sequencing so that even samples of unradioactive tRNAs in amounts as small as 10 μg can permit a total primary structural determination. Previously, we saw a great burst in tRNA sequencing activity by using the Sanger fingerprinting technique on ^{32}P-labelled tRNAs. Today, the difficult step is the purification of the tRNA rather than the actual sequencing. In the past most sequenced tRNAs have been isolated from bacteria or yeast, but with the modern technology, animal cells also provide readily available source material. Especially is it predictable that most new sequences in the future will be for tRNAs from animal sources. Although it is generally accepted that there are of the order of 55 different tRNA gene types in a cell, for example in the case of bacteria making up about 0.05% of the genome, only about 30 tRNA sequences are known for the most popular source material, *E. coli*. Even in the case of *E. coli* we do not yet know the sequences of tRNAs chargeable with all of the amino acids since no *E. coli* $tRNA^{Pro}$ has been determined[6].

The information from 105 different tRNA sequences known in June 1979 and listed in Table 3, (Clark[6], and 5 new structures, Gauss *et al.*[7]) has been conveniently incorporated into a standard "clover leaf" form as shown in Figure 1. This remarkable feature of all the primary structures was first proposed by Holley and his colleagues[8] and is based on Watson-Crick base pairing which forms double stranded helical stretches (the secondary structure). Thus, we have small (class 1) and large (class 2) tRNAs dependent upon the size of the extra arm (Table 3). The 5 new structures not listed in my previous review[6] are Dm Met_f[15], ym Phe[16], yp Glu_1^{1}[7], Tu Leu[18] and Ncm Tyr[19]. Actually, these recent sequence determinations[15-19] give good illustrations of the most recent developments of the rapid RNA sequencing methods referred to previously in this article and in an earlier review[6]. To correct a discrepancy of one sequence difference in my previous review, I should point out that Sf Met_f should have been included as a separate sequence under class 1B but in fact was omitted in a typographical error. However, except for some differences in minor base modification Sf Met_f is the same sequence as Ec Met_f, (A. Delk and J.C. Rabinowitz, personal communication).

TABLE 3. Classes of tRNA according to arm sizes and structure correlations.

Class 1 (88)	
A (51)	4 base pairs in b-stem (D stem) 5 bases in extra loop and containing m^7G
(and with A9)	Ec Ala$_1$, Sw Ala$_1$, Sw Ala$_2$, Ec Arg$_1$, Ec Asn, Ec Asp$_1$, Ec Gly$_1$, Ec Gly$_3$, Ec + Sal His$_1$, Ec Ile$_1$, Ec Lys, Bsu Lys$_1$, y Lys, Rbl + Dm Lys$_{2A(2B)}$, Rbl Lys$_3$, Svt Lys$_4$, An Met$_f$, Ec Met$_m$, y Met$_3$, Mye + Rbl Met$_4$, Ec Phe, Mp Phe, Bs Phe, Bsu Phe, y Phe, Wg + Ps Phe, Egc Phe, Bc Phe, Rbl + Hup Phe, T4 Pro, Ec Thr, Bsu Thr, T4 Thr, Ec + su$^+$ Trp, Ec Val$_1$, Ec Val$_{2A}$, Ec Val$_{2B}$, Bs Val$_{2A}$, Mye + Rbl + Hup Val$_1$.
(and with m^1G9 or G9)	Rl Asn, y Cys, Ec + Sf Met$_{f1}$, Ec Met$_{f2}$, Bsu Met$_f$, Tt Met$_f$, Nc Met$_f$, Dm Met$_f$, Mye + Rbl + St + Xl + Smg + Hup Met$_f$, yp Phe, y Trp, Chi + Bl + Rbl Trp$_1$.
B (6)	4 base pairs in b-stem (D stem) 5 bases in extra loop III without m^7G y Ala$_1$, Tu Ala$_1$, y Arg$_3$, Hay Lys, Mp Met$_f$, Bl + Rbl Trp$_2$.
C (12)	4 base pairs in b-stem (D stem) 4 bases in extra loop T4 Arg, y Asp$_1$, Ec Glu$_1$, Ec Glu$_2$, yp Glu$_1$, Ec + Sal Gly$_1$, Sta Gly, Wg Gly, y Gly, Sw Gly$_1$, Sw Gly$_2$, ym Phe.
D (19)	3 base pairs in b-stem (D stem) small extra arm with no. of bases (3-5) y Arg$_2$ (5), Ec Cys (4), Ec Gln$_1$ (5), Ec Gln$_2$ (5), T4 Gln (5), y Glu$_3$ (4), Ec Gly$_2$ (4), T4 Gly (4), Tu Ile (5), Ncm Met$_f$ (4), Mul Pro (5), y Thr$_{1A}$ (5), y Thr$_{1B}$ (5), y + su$^+$ Tyr (5), Tu Tyr (5), y Val$_1$ (5), y Val$_{2A}$ (5), y Val$_{2B}$ (5) Tu Val (3).
Class 2 (17)	3 base pairs in b-stem (D stem) large extra arm with no. of bases (13-21) Ec + Sal Leu$_1$ (15), Ec Leu$_2$ (15), T4 Leu (14), y Leu$_3$ (13), y Leu$_4$ (13), Tu Leu (13), Ec Ser$_1$ (16), Ec Ser$_3$ (21), T4 Ser (18), y Ser$_1$ (14), y Ser$_2$ (14), Rl Ser$_2$ (14), Rl Ser$_3$ (14), Ec + su$_3^+$ Tyr$_1$ (13), Ec Tyr$_2$ (13), Bs Tyr (13), Ncm Tyr (16).

The different tRNA species are referred to by their chargeable amino acids.

List of abbreviations used in Table 3.

su = Suppressor
An = *Anacystis nidulans*
Bc = Bean chloroplast
Bl = Beef liver
Bs = *Bacillus stearothermophilus*
Bsu = *Bacillus subtilis*
Chi = Chicken
Dm = *Drosophila melanogaster*
Ec = *Escherichia coli*
Eg = *Euglena gracilis*
Egc = *Euglena gracilis* chloroplasts
Hay = Haploid yeast
Hup = Human placenta
Mp = Mycoplasma
Mul = Murine leukemia virus
Mye = Myeloma
Nc = *Neurospora crassa*
Ncm = *Neurospora crassa* mitochondria
Ps = *Pisum sativum*
Rbl = Rabbit liver
Rl = Rat liver
Sal = *Salmonella typhimurium*
Sf = *Streptococcus faecalis*
Smg = Sheep mammary gland
St = Salmon testis
Sta = Staphylococcus
Stf = Starfish
Svt = Svt 2 cells
Sw = Silk worm
T4 = Bacteriophage T4
Tt = *Thermus thermophilus*
Tu = *Torulopsis utilis*
Wg = Wheat germ
Xl = *Xenopus laevis*
y = Yeast, undefined and *Saccharomyces cerevisiae*
ym = Yeast mitochondria
yp = Yeast, *Saccharomyces pombe*

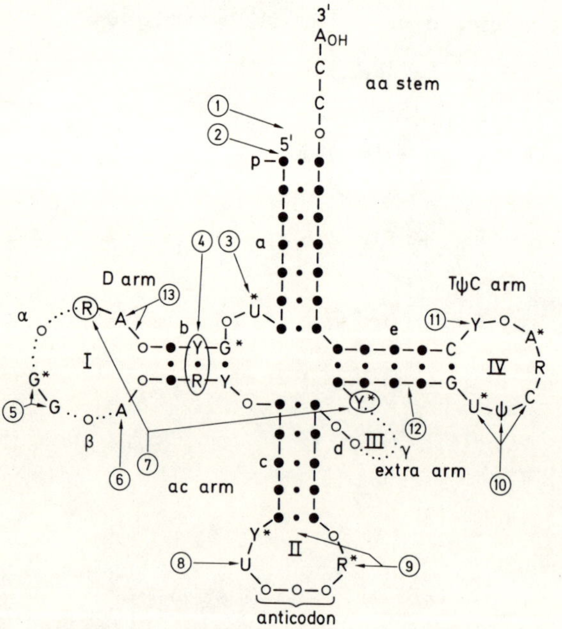

Figure 1. Class 1 generalized cloverleaf. Solid circles ● · ● indicate bases involved in helical stems containing Watson-Crick base pairs. Open circles o signify non-Watson-Crick base paired bases in cloverleaf arrangement. Starred nucleosides are positions where modifications of that nucleoside can possibly occur. The locations of the exceptions are given by the arrowheads. The locations with circled numbers are described accordingly in the text and refer to known exceptions to the generalized cloverleaf.

In Figure 1 the information from the small class 1 sequences has been incorporated into a standard generalized cloverleaf. To obtain this information 76 of the 88 class 1 species indicated in Table 3 have been used: clear sequence exceptions based on functional properties, such as the initiator Met_f species and the cell wall Sta Gly, to the generalized form were omitted.

As shown in Figure 1 the Watson-Crick base pairs give rise to four double helical stem regions a, b, c, and e, three of which are closed by non-base paired loop regions I, II and IV. These four stems give rise to the clover leaf arrangement of the secondary structure. Whether there is a functional role for the secondary structure is still unclear. Another point of nomenclature illustrated in Figure 1 is that a stem plus a loop is also called an arm. Most of the tRNA cloverleaf forms have remarkedly constant regions. There is a phosphate at the 5'-end, and at the 3'-end where the amino acid is attached there is a common sequence CpCpA. In addition, stems a, c, and e contain 7,5 and 5 base pairs, respectively, and loops II and IV each contain 7 non-base paired nucleotides. Further points of nomenclature as summarised in Figure 1 have been defined previously[6]. The variable regions are confined to loops I and III and stem d.

There are also many invariant and semi-invariant nucleotide positions in the generalized structure. The invariant positions are shown by nucleoside letters whereas the semi-invariant positions are shown by R (signifying a purine nucleoside A or G) or Y (signifying a pyrimidine nucleoside C or U). The dotted stretches in the diagram are regions of variable length. For example, and including class 2 tRNAs, the extra arm can vary from 3 to 21 nucleotides in length. In contrast the D loop I is much less variable in length - from only 7 to 11 nucleotides long when the base pair b4, even though not of the Watson-Crick type, is considered as part of the D stem. For convenience the variable dotted regions can be named α, β and γ.

Until recently what the invariant and semi-invariant nucleosides represented was an interesting but open question. It was not possible to decide whether these positions in the primary structure were conserved for metabolic or for structural reasons. Now that a three-dimensional structure has been determined[3,4] the primary and secondary structural information can be viewed in a new light and we can see a structural role for most of these special nucleosides. One invariant position which has not been properly explained in terms of structure is the constant Ψ in the TΨC loop. As shown in Table 3 the class of small tRNAs (class 1) can be conveniently subdivided according to structural characteristics.

The tertiary structure has been determined for yeast $tRNA^{Phe}$, which belongs to the subclass 1A but the structure appears relevant to subclass 1B and will probably also accommodate subclass 1C and subclass 1D structures. For example a short extra arm, only 4 nucleotides in length, but making all the tertiary interactions, can be built with the excision of the residue U47, which is not involved in the tertiary bonding found for yeast $tRNA^{Phe}$.

Exceptions to the Generalized Structure

Several exceptions to the generalized structure have previously been enumerated[6,12] and their locations in the generalized cloverleaf including some new exceptions are shown in Figure 1. Apart from these, non-Watson-Crick base pairs sometimes occur in helical stems. The most frequent non-Watson-Crick base pairs occurring in stems are G·U, G·Ψ and A·Ψ. In stems G·Ψ is considered equivalent to G·U and A·Ψ to A·U. When the simplified classification of tRNA structures into two classes is used then the other non-Watson-Crick base pairs which occur in stems are U·U, C·C, G·A, C·A, Ψ·U and Ψ·Ψ. No more than one of these unusual base pairs is normally found in a stem.

The most important exceptions to the standard cloverleaf concern tRNAs which have special functional roles such as initiator tRNAs and glycine tRNAs involved in cell wall metabolism. These special exceptions can be considered to point out positions in the cloverleaf which could be concerned with these special functions. However, this kind of interpretation is likely to be too simplistic without a proper knowledge of the tertiary structures involved.

New additions to the previous exceptions' list[6] are as follows: in location (5) Ncm Met_f has A-G instead of the usual G*-G; in location (6) Ncm Tyr has a G instead of A in position 21. Positions are numbered according to the commonly adopted numbering system for y $tRNA^{Phe}$ structure as used in the compilation by Gauss et al.[7]; in location (7) Ncm Met_f has A15-C48 instead of the coordinate base pair ; in location (10) Ncm Met_f has U*-Ψ-C replaced by U-G-C. Recently, further exceptions at locations (12) and (13) have been reported[16,19]. In the sequence of ym Phe there is an extra U residue inserted between positions 51 and 52 of the TΨC stem giving a looped out nucleotide. At location (13) in Figure 1 Ncm Tyr contains U-U instead of A14 in the D loop. This is equivalent to replacement of A by U and an insertion of a U residue between positions 13 and 14. Finally, there is some preliminary sequence information on human mitochondrial tRNA genes

(B. Barrell, personal communication) indicating further exceptions to Figure 1 even changing the hitherto sacrosanct length of loop IV.

Possible Significance of the Exceptions to the Generalized Structure for Function

The significance and role of modified bases and non Watson-Crick base pairs are not clearly defined. However, the use of G·U base pairs at ends of helices to relieve backbone strain and as recognition points within helices has been reviewed previously[6]. What was earlier considered to be a well defined role for Ψ in conferring repressor activity[1,6] upon bacterial $tRNA_1^{His}$ is now debatable since the tRNA gene product is now thought not be the direct cause of repression (O. Uhlenbeck, personal communication). There does appear to be a connection between the usual modified base after the anticodon in position 37 and the type of base in the first position of the codon, but the actual structural functional relationship needs further clarification. This feature and the occurrence of other modified bases have been reviewed recently by McCloskey and Nishimura[20].

The special role of prokaryotic initiator tRNAs may indeed be related to their exceptional structures at locations (2) and (4) in Figure 1. Location (2) concerns the lack of a Watson-Crick base pair and at location (4) there is an A·U instead of Y·R. How these changes affect the overall structure is not yet known. It is worth noting that location (4) of Figure 1 may be concerned with a ribosomal interaction since substitution of G24 by A24 converts Ec $tRNA^{Trp}$ into a UGA suppressor tRNA with no change in the anticodon[21]. A reasonable explanation is that this is evidence of a ribosome-mediated effect on the codon-anticodon interaction rather than a transmitted effect through the tRNA backbone. The latter would be difficult to understand in terms of our current knowledge of the tertiary structure. In contrast to the situation with prokaryotic tRNAs there appears at present to be no special sequence peculiarities just for eukaryotic initiator tRNAs.

Since the Sta $tRNA^{Gly}$ species do not play a role in protein biosynthesis their lack of constant features should point out parts of the generalized tRNA involved in protein biosynthesis. It is possible that the common sequence of the TΨC loop missing in Sta $tRNA^{Gly}$ is involved in ribosomal site binding of the tRNA and this will be discussed later. The lack of the constant G*-G of the D loop in the cell wall tRNAs also implicates this constant feature in protein biosynthesis. However, rather than being a recognition point for some enzyme it may have a structural role because we know that it is concerned in holding the tertiary structure of yeast $tRNA^{Phe}$ together.

From the lack of a purine after the anticodon in the cell wall tRNAs we may infer that the purine is a structural requirement concerning proper fitting during the anticodon-codon decoding interaction. However, this feature may also be influenced by the possible sixth Watson-Crick base pair in the ac stem of cell wall tRNAs.

Analysis of Primary Structure and Significance for Function

Because of the extensive data on tRNA nucleotide sequences the question naturally arises whether a comparison of sequences showing common sequences (sequence homologies) can be interpreted in terms of tRNA functional sites. Different tRNAs recognized by the same enzyme have been compared to determine the common parts of their primary structures. It is now known where these common parts lie in the three-dimensional structure so we can evaluate the results in a more sensible way than before the tertiary structure was known. Especially sequence homologies have been considered in terms of ribosomal binding sites on tRNA and recognition sites of isoaccepting species of tRNA by the same activating enzyme.

Hints at the possible ribosomal recognition come from the lack of invariant regions in cell wall biosynthesising tRNAs as discussed in the previous section. A study of sequence homologies of isoaccepting tRNA species has been less fruitful. No obvious common similarities have been observed so far[2,6]. Clearly, such a proposition for recognition based on a consideration of similarities among sequences arranged in cloverleaf forms is not correct. We need to know how sequence information specifies spatial structure before much progress will be achieved in our understanding of the recognitory interaction between nucleic acids (tRNAs) and proteins or between nucleic acids (tRNAs) and nucleic acids (perhaps rRNAs).

TERTIARY STRUCTURE OF tRNA AND FUNCTIONAL PROPERTIES

At present the only available method for determing the total three-dimensional structure (tertiary structure) of a macromolecule such as tRNA is X-ray crystallographic analysis. The crystallization of tRNA[22] heralded intensely competitive research efforts in several laboratories resulting in the independent tertiary structural elucidation for only one tRNA species[6,9,10,23].

Normally, the X-ray crystallographic determination of a structure goes through several stages, each needing to be successful for progress to the final solution, in the following order: 1) the production of reproducible crystals of suitable size (about 0.2-0.5 mm in each dimension), 2) the introduction of heavy atoms into the crystals to provide isomorphous heavy atom derivatives, 3) the collection of X-ray data from the native and at least two derivatives in order to solve the crystallographers' phase problem, 4) the production of an electron density map and the fitting of an atomic model to the map; the model nowadays can also be built by computer graphics, saving time over building a physical model using an optical comparator and Kendrew skeletal atomic models, 5) the refinement of the model using computer techniques, and 6) further collection of higher resolution native data followed by further refinement of the atomic model. The determination of the y $tRNA^{Phe}$ structure is currently at stage 6.

The three-dimensional structure of the yeast $tRNA^{Phe}$ in the monoclinic form was solved to 2.5Å resolution in Cambridge by the method of isomorphous replacement[3], using 5 heavy atom derivatives. Independent structural determinations using the orthorhombic form[4] and also the monoclinic form[5] have yielded the same structure in broad outline. A photograph of the Kendrew skeletal model built to fit the electron density is shown in Figure 2.

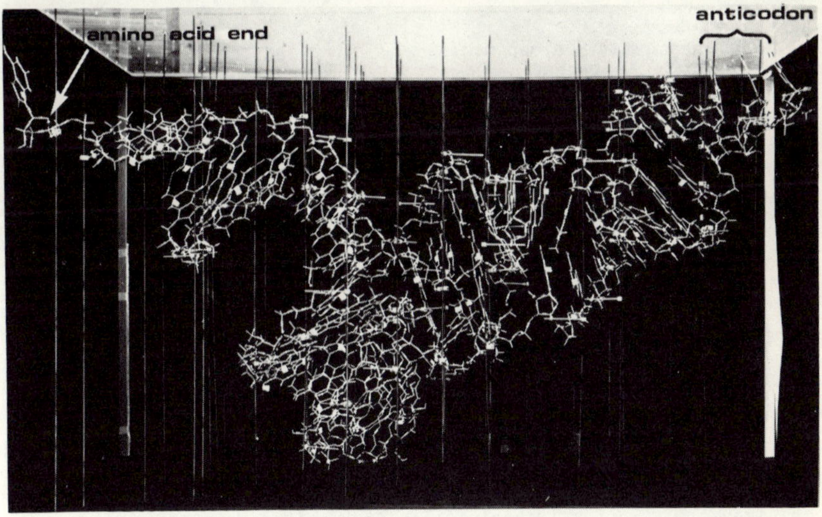

Figure 2. Photograph of a 2.5Å model of y $tRNA^{Phe}$ built at Aarhus University with the help of Dr. J.E. Ladner. The two most significant functional parts are labelled. Actually, on the model the distance from the anticodon to the terminal 3'-end A is about 80Å. The ribosomal site is so far unknown. Reproduced from ref. 6 with permission.

Many features of the molecular stereochemistry have been revealed. In addition to the base pairs of the cloverleaf arrangement of y $tRNA^{Phe}$, many other interactions (see Fig. 3) which fix the tertiary structure of yeast $tRNA^{Phe}$ have been deduced.

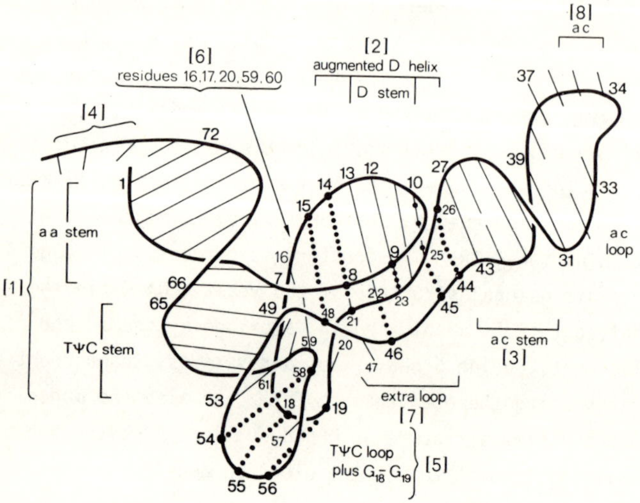

Figure 3. A schematic diagram of the folding and the tertiary interactions between bases of y $tRNA^{Phe}$ designed by A. Klug (see ref. 3) and reproduced from ref. 6 with permission. The ribose-phosphate backbone is represented by a continuous line. Base pairs in the double helical stems are represented by long lightly weighted lines, and non-paired bases by shorter lines. Base interactions additional to those in the cloverleaf form are indicated by dotted lines. Substructures referred to in the text are given by numbers in square brackets.

From the X-ray studies we now have a picture of the ordered complexity of a folded RNA molecule, a complexity as great as that of a protein. An interesting by-product is a detailed picture of the stereochemistry of a G·U base pair in a double helical stem (positions 4·69 in Fig. 3), where the pairing appears to be that predicted by the "Wobble" hypothesis[24].

The structural determination has shown that the hydrogen bonding potential is more varied than simple Watson-Crick type pairing, that the planar bases can form structural sheets which involve H-bonding of 3 nucleotides, and that the ribose residues can participate in forming structurally important H-bonds between themselves, with bases and with phosphate oxygens. In addition the presence of single stranded loops, a core region where four backbone chains run along together, and regions of reasonably standard RNA-type double helix make it of interest to determine in detail the torsional angles used and to identify what restrictions can be placed on particular angles when predicting an unknown structure. Our knowledge of this tRNA structure should be invaluable for our understanding of the folding of larger RNAs such as rRNA and mRNA.

A prominent set of H-bonds stabilizes the T-joint made in the structure by the two long double helices which meet approximately at right angles. Another set reinforces the links between bases at the bends of the D and TΨC loops. The structural integrity of the molecule as a whole depends upon the central part where elements of four chains come close together and a number of base triples or pairs are found. These additional base pairs of triples are stacked or intercalated so as to make their H-bonds inaccessible to water molecules. The tertiary structure is further stabilized by the stacking of non-paired bases on each other or on base pairs as for example in the anticodon loop. Indeed about 90% of the bases can be said to be internal as hydrophobic groups are internal in proteins.

Detailed model building during the fitting of the 2.5Å map dictated that the pucker of ten of the sugars was changed to the C2'-endo conformation[3]. Although the resolution of the electron density map is still not sufficient to show the shape of a sugar unequivocally, the conformation can be deduced from the restrictions imposed on it by the relative dispositions of the two phosphate groups on either side and of the base that emanates from it. A good example is that of nucleotide m^1A58 whose sugar is C2'-endo in the 2.5Å[3,9] model.

The structural features which appear to have functional significance in the molecule depicted in the model (Fig. 2 and schematically in Fig. 3) can be briefly described in terms of 3 major skeletal substructures: [1] the long double helix formed by the aa stem and TΨC stem stacking on top of each other; [2] the augmented D helix forming the central part, or "core", or "thorax" of the molecule, consisting of the D stem, augmented laterally by interactions with the short "stretcher" region 8-9, with the extra loop III and with a part of the D loop; [3] the anticodon stem tilted off by about 20° to the D stem and apparently hinged to it by hydrogen bonds between nucleotides A44 and m^2_2G26 and between G45 and m^2G10. The molecule has been described as L-shaped[23], but the two major long double helical stretches, i.e. [1] and [2] + [3] above, are arranged as the letter T. Thus the T-form of tRNA is a tertiary structural arrangement of the secondary structure represented by the cloverleaf form. The L-shape just reflects one side view of the molecule shown in Fig. 2.

To these skeletal substructures described above are joined five regions with established, or potential, functional roles. In Fig. 3 these are numbered as follows: [4] at one end of the long double helix [1] there is the 3'-end CpCpA to which the amino acid is attached; [5] at the other end of [1] the invariant TΨC loop is tightly folded and interacts with a part (constant G*-G) of the D loop in a complex set of interactions[3,4]. In short U59 stacks on to the pair G15·C48, T54 is base paired with m^1A58 in a reverse Hoogsten type, G57 intercalates between G18 and G19 giving rise to a stack of 4 purines 58, 18, 57 and 19, and interacts with the backbone of the D loop, C56 forms a Watson-Crick base pair with G19, and G18 appears to interact with Ψ55 and with the TΨC backbone on either side of it. The interaction of the TΨC loop with the invariant G18 and G19 clearly masks the potential of the T-Ψ-C-G sequence for its proposed role of binding to the ribosomal A-site in a Watson-Crick base pairing interaction during peptide bond formation[25]. The question of a possible conformational change exposing the T-Ψ-C-G will be discussed later. [6], the variable regions in the D loop perhaps form a discrimination site (or part of one) for enzymes to distinguish between tRNAs (not necessarily the amino acyl-tRNA synthetase recognition). The two sections α and β of the D loop (see Fig. 1) are near each other and form a surface patch on the molecule; they each contain a variable number (1 to 3) of nucleotides and, either on their own or together with the nearby non-base paired residues 59 and 60, could well provide a general enzyme discrimination site. The variable extra loop (or, at least, a part of it, e.g. U47 in $tRNA^{Phe}$), which is numbered region [7] in Fig. 3, may also play such a role in some cases. Finally, the decoding function of tRNA is provided for by region [8], the anticodon loop, containing the anticodon. This loop contains a stack of five bases on the 3'-end side continuing from the anticodon stem. In addition the two other bases Cm32 and U33 are also stacked separately on the end of the anticodon stem but with a sharp bend between U33 and Gm34[26,27].

Recent developments in refining the structures determined independently by the MRC group in Cambridge[28,29] and by the MIT group[30,31] have given evidence locating important metal ion binding sites and of particular biological importance, strong magnesium binding sites and spermine sites in the y $tRNA^{Phe}$ structure.

Both groups agree[28,30] that the strong magnesium binding sites are found near 1) the phosphate oxygens of U8 and A9; 2) near phosphate oxygens of G20 and A21; and 3) at a site near the phosphate of G19, atoms N7 and O6 of G20 and O4 of U59. In addition a good case has been made for the location of a magnesium ion in the anticodon loop between Cm32 and A38. Two spermine molecule positions have been proposed by Quigley *et al.*[30]; in the major groove at one end of the anticodon stem and near phosphate 10 where the polynucleotide chain makes a sharp turn. The positions of these cations may have important biological implications. Certainly, spermine and magnesium ions often appear to be important in maintaining the overall folding of tRNA. There is also experimental evidence[32] that varying combinations of polyamine and magnesium in the tRNA structure can induce structural changes.

Finally, new data collection and refinement have resulted in a slight modification of the structure in the D loop by the MRC group[29] to their 2,5Å model. Now the base D16 is turned inwards rather than pointing out into solution so that it points towards C60, i.e. to what Klug has called the variable pocket[3].

Generality of y $tRNA^{Phe}$ Tertiary Structure and Other Structural Determiations

The elucidation of the tertiary structure of y $tRNA^{Phe}$ naturally led one to ask whether other tRNA primary structures could be folded in the same way. In particular, could other bases be substituted into the base triple positions of y $tRNA^{Phe}$ so as to give equivalent interactions? Many changes of sequence in class 1 can indeed be compensated by concomitant changes elsewhere, so that local geometry in the 3-dimensional structure remains essentially unchanged. Such changes have been described before[6].

A good structural case can be made after substituting the class 1A y $tRNA^{Phe}$ with base combinations giving equivalent tertiary interactions and assuming that a non-Watson-Crick base pair can be accommodated in position b4 of Figure 1, for the structure shown in Figure 2 being directly relevant for class 1B, 1C and 1D sequences of Table 3[6]. There is no firm structural information on any class 2 tRNAs although some speculative model building studies have been reported for some bacterial $tRNA^{Tyr}$ species (ref. 33, R.S. Brown and J.R. Rubin, personal communication and S. Altman, personal communication). So far there has been a lack of reasonable crystals of class 2 tRNA species, but very recently (R. Stroud, personal communication) promising crystals of Ec $tRNA^{Leu}$ diffracting to better than 4Å have been obtained. I need not stress the obvious importance of determining a class 2 tRNA structure for obtaining information about ribosomal binding since the 2 classes of tRNA, while differing in chain length, both bind to the same ribosomal site during protein biosynthesis.

The uniqueness of the y $tRNA^{Phe}$ crystals, being the only ones giving very high resolution diffraction patterns from what has been shown by other methods to be a native structure, is still rather frustrating. Although there has been no suspicion that the y $tRNA^{Phe}$ structure is not a general one for other tRNAs, a supposition supported by solution studies, clearly it would be important (and reassuring) to be able to determine other tRNA structures. Currently, a number of other tRNA species are being investigated, but none of them except perhaps y $tRNA^{Gly}$ give high resolution diffraction patterns capable of an independent structural analysis. In the case of y $tRNA^{Gly}$ the Princeton group agrees from their model building studies that their crystallization conditions have denatured the molecule. Nevertheless, their structure will be interesting perhaps pointing the way to a type of conformational change. Other standard elongator tRNA structures under investigation are y $tRNA^{Asp}$ (in Strasbourg) and Ec $tRNA^{Arg}$ (in Aarhus, Dk, and Madison, Wisc.) again using a structural comparison with y $tRNA^{Phe}$. Finally, the specially functioning initiator tRNAs are being worked on in an attempt to explain how they bind to different ribosomal sites. So far reasonable progress has been made, but not in a manner completely independent of the y $tRNA^{Phe}$, on y $tRNA_f^{Met}$ (in Chicago) and on Ec $tRNA_f^{Met}$ (at M.I.T.). At a resolution of 4.5Å the structure of y $tRNA_f^{Met}$ shows only minor conformational differences (in the ac loop) from y $tRNA^{Phe}$ so a structural explanation of its initiator function needs further investigation[34].

It must be emphasized that the crystal structure of y $tRNA^{Phe}$ is the most stable "resting form" of tRNA. This crystal form also appears to be the most favourable conformation for naked tRNA in solution.

SOLUTION STRUCTURE OF tRNA

The topography of the model deduced from the crystallographic study agrees with a companion study of the chemical reactivity of y $tRNA^{Phe}$ in solution[35]. There is thus no reason to presume that the structure in the crystal departs significantly from that in solution. This presumption is also confirmed by physical chemical studies, such as NMR of tRNA in solution. Information from the chemical studies on y $tRNA^{Phe}$ have been combined with other results from the literature to produce a composite picture[6] of base accessibility in $tRNA^{Phe}$. The bases which react chemically can be correlated with exposed positions in the three-dimensional structure of tRNA. Those which do not react are either in the double helical regions or are involved in maintaining the tertiary structure. These results confirm the usual structural assumptions made about reactivity in chemical studies and hence give confidence for extending such work to other tRNA structures. Furthermore, since the chemical studies are carried out on tRNAs in solution rather than in the crystalline state, they also support the contention that there is no significant change in the conformation of y $tRNA^{Phe}$ upon crystallization.

In addition, physical chemical studies especially using the NMR technique have identified the H-bonds involved in the tertiary interactions between bases[36,37]. These studies are now also interpreted in terms of the same structure existing in solution as in the crystal form. Some of the difficulties encountered by workers attempting to interpret spectra in terms of three-dimensional structure have recently been reviewed by Ladner[9]. At any rate the $tRNA^{Phe}$ conformation identified with the crystal form appears also to be the most stable form of lowest potential energy in solution so that if the conformation does become changed during function once it is released by its reacting macromolecule then it reverts to the crystal form's conformation.

POSSIBLE tRNA STRUCTURAL CHANGES DURING FUNCTION IN PROTEIN BIOSYNTHESIS

Although we have strong evidence confirming that the tRNA conformation is the same in solution as in the crystal form, it is unclear how the structure changes its conformation for its role in protein biosynthesis. Nevertheless, it is likely that the conformation does change and is affected by the many different interactions which the tRNA undergoes with other protein and nucleic acid components. To gain more insight into putative conformational changes which can occur at steps in protein biosynthesis as listed in Table 2, we need information on the different tRNA-protein complexes formed during the steps of protein biosynthesis from chemical, X-ray diffraction and other physical chemical studies. With the current lack of definitive direct evidence on conformational changes, especially lacking from tRNA: protein complexes relevant to protein biosynthesis researchers have taken recourse to a study of model systems. Unfortunately, the most easily studied and therefore the most commonly used model systems tend to use naked tRNA with the accompanying problem of perturbing the very stable crystal form which pertains to naked tRNA in solution. Such model systems may have little connection with biological reality. Nevertheless, the general and reasonable assumption is made that if a pertubation of the crystal form of tRNA is observed it can indicate the potential change which another macromolecule could induce during biological function. This philosophy justifies the discussion of the meaning of model systems in this section.

Charging

There is a surprising lack of hard facts concerning the possible difference in conformations between uncharged and charged tRNA (referred to as step 1 of Table 2). The literature contains evidence both in favor and against a change in conformation. Those in favor suggest that the methods used where negative results have been obtained are not capable of measureing the possibly subtle changes which occur. As Sprinzl and Cramer[11] have pointed out, the thermodynamic barrier between the two conformations is perhaps so small that the native form of aminoacyl-tRNA is not preserved during the preparation and isolation procedure. It does seem that large structural rearrangements which would be readily detectable do not occur.

Some of the changes detected by physical chemical measurements[11] comparing charged and uncharged tRNA may possibly be due to perturbations in the shell of counterions surrounding the tRNA. Aminoacylation of the tRNA may cause a small rearrangement of the structure giving rise to an alteration in the divalent cation binding sites and the distribution of counter ions. What is needed to solve this problem is an X-ray crystallographic analysis of aminoacyl-tRNA, but so far no suitable crystals have been obtained for a charged tRNA.

Very recently careful proton magnetic resonance studies[38] of minor nucleosides in a y Phe-tRNA analogue have indicated that residue T54 can adopt a less stable configuration compared with the uncharged tRNA suggesting that aminoacylation produces a small rearrangement of the tertiary structure especially affecting local stacking interactions. The analogue y Phe-tRNA used in the studies was a $tRNA^{Phe}$ containing a 3'-terminal 3'-deoxy-3'-amino A instead of A so that the charged form was stable enough to carry out the experiments without significant hydrolysis of the amino acid linkage to the tRNA. Although the criticism can be levelled at these analogue experiments that their very nature using non native species produces artifacts, their results seem to support a recent proposal by Potts *et al.*[39] from laser light scattering that a subtle conformational change occurs in the tRNA structure upon charging it.

Binding to EF-Tu:GTP

EF-Tu provides a good contrast in the problem of specific recognition of aminoacyl-tRNA to an aminoacyl-tRNA synthetase because EF-Tu recognizes all aminoacyl-tRNAs whereas a synthetase recognizes at most a set of isoaccepting species. The molecular details of how EF-Tu forms a ternary complex with an aa-tRNA and GTP, how the ternary complex binds to the ribosome and how the EF-Tu:GDP is released after hydrolysis of the GTP are not known.

The generally accepted cyclic scheme[40] for the role of EF-Tu in carrying an aa-tRNA to the ribosome is shown in Figure 4.

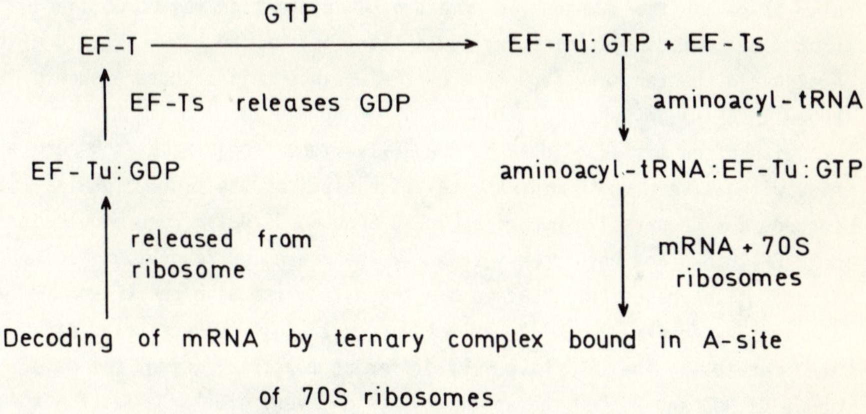

Fig. 4. Cyclic scheme for role of EF-Tu in protein biosynthesis. Adapted from ref. 6 with permission.

The regeneration of EF-Tu:GTP is thought to be brought about catalytically by another protein factor, EF-Ts, which binds to EF-Tu to make the bimolecular EF-T. Interestingly, there is a large amount of EF-Tu in bacterial cells - about 5% of total protein giving a concentration of about 0.3 mM which is of the same order as the aminoacyl-tRNA concentration. Perhaps the high amount of EF-Tu is needed to sequester the aminoacyl-tRNA during protein biosynthesis[6] to protect it against hydrolysis.

So far it has not been possible to determine the chemical reactivity of an aminoacyl-tRNA complexed with EF-Tu and GTP because the chemical reagents inactivate the protein too rapidly. However, there is evidence that the nuclease sensitive regions of aminoacyl-tRNA are not different in the ternary complex with EF-Tu and GTP[40]. Again it is reasonable to conclude that no large changes of conformation occur on binding to EF-Tu. In addition attempts at crosslinking aa-tRNA to EF-Tu in the ternary complex from positions 8 and 47 of the tRNA have been unsuccessful[40]. Actually, at present the only part of the aa-tRNA which is clearly interacting with the EF-Tu is the amino acid and the 3'-CCA end. Since the naked structure of tRNA is so stable, perhaps all the EF-Tu needs to do to make a stable ternary complex is to interact with and make rigid the flexible single-stranded 3'-end of the aa-tRNA without significantly altering the rest of the tRNA. However, to most of us this interpretation of limited factual data is rather unlikely and there is also suggestive evidence[40] for other recognition points on the aa-tRNA. Much effort is currently being expended into trying to crystallize a ternary complex in order to solve this problem of recognition. Meanwhile some useful information may be obtained from model building studies trying to fit together the known structure of a tRNA and the tertiary structure of EF-Tu since this structure should be known in the near future.

Quite often mistaken ideas of how EF-Tu might recognize aa-tRNA can arise without keeping in mind the relative sizes of the macromolecules concerned. For comparative purposes Figure 5 shows 5Å Balsa wood models of both EF-Tu:GDP and uncharged y $tRNA^{Phe}$. These are low resolution models to emphasize shapes but the data are taken from the high resolution X-ray analyses by Morikawa *et al.*[41] and Ladner *et al.*[3], respectively. Although the protein and the tRNA have very different molecular weight ratios of about 45,000 and 25,000 the spatial sizes are very similar, viz. 75Å x 50Å x 45Å for EF-Tu and 80Å x 50Å x 30Å for $tRNA^{Phe}$.

In connection with ribosomal binding of aa-tRNA, an interesting speculation on the role of EF-Tu in providing a checking mechanism to increase fidelity of decoding has been proposed by Sprinzl and Cramer[11]. Based on some biochemical studies with analogues of normal aminoacyl-tRNAs to prevent aminoacyl migration, they have suggested that EF-Tu holds the aminoacyl-tRNA in the ternary complex with GTP so that the amino acid is linked to the 2'-OH group of the 3'-terminal A. Only after codon checking has taken place would aminoacyl migration occur prior to peptide bond formation.

Binding of aa-tRNA to the Ribosome

Much work is in progress using physical chemical techniques, affinity

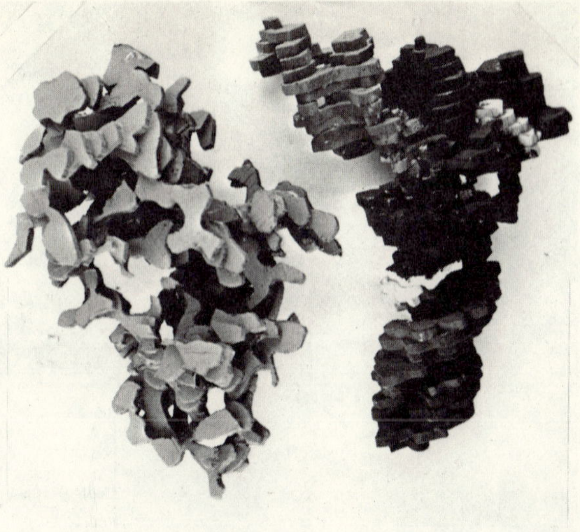

Figure 5. A photograph of 5Å Balsa wood models of EF-Tu:GDP containing essentially all of the EF-Tu molecule on the left and y $tRNA^{Phe}$ (with the ac arm pointing downwards and the aa stem pointing upwards to the left) on the right. Reprinted from ref. 6 with permission.

labelling and chemical crosslinking to identify the protein and nucleic acid environments for the tRNAs bound in the different ribosomal sites. What is known about the ribosomal environment of the bound aa-tRNA is described in detail elsewhere in this volume by J. Ofengand. In this chapter I shall consider the binding in terms of the tRNA structure.

A simplified schematic version summarizing the general view of the elongation step of protein biosynthesis is shown in Figure 6.

This cyclic scheme[6] is really self-explanatory and is only a current version of a scheme usually attributed to J.D. Watson. For more details on elongation the reader is referred to another chapter by H. Weissbach in this volume. Another elongation factor (of molecular weight about 80,000d in *E. coli* and called EF-G) is involved in the translocation step. Whether or how EF-G interacts with aa-tRNA is not known.

Although much evidence is being collected about the ribosomal environment for the bound aa-tRNA, we have little idea so far about the conformation of the aa-tRNA on the ribosome. Indeed, recent work has done little to clarify the situation.

For convenience the binding of tRNA to the ribosome referred to as steps 3 and 4 of Table 2 can be considered in terms of binding of different parts of tRNA with as yet unknown temporal sequence. Clearly, there is binding of the 3'-CCA end, as in step 4 for peptide bond formation, and there are

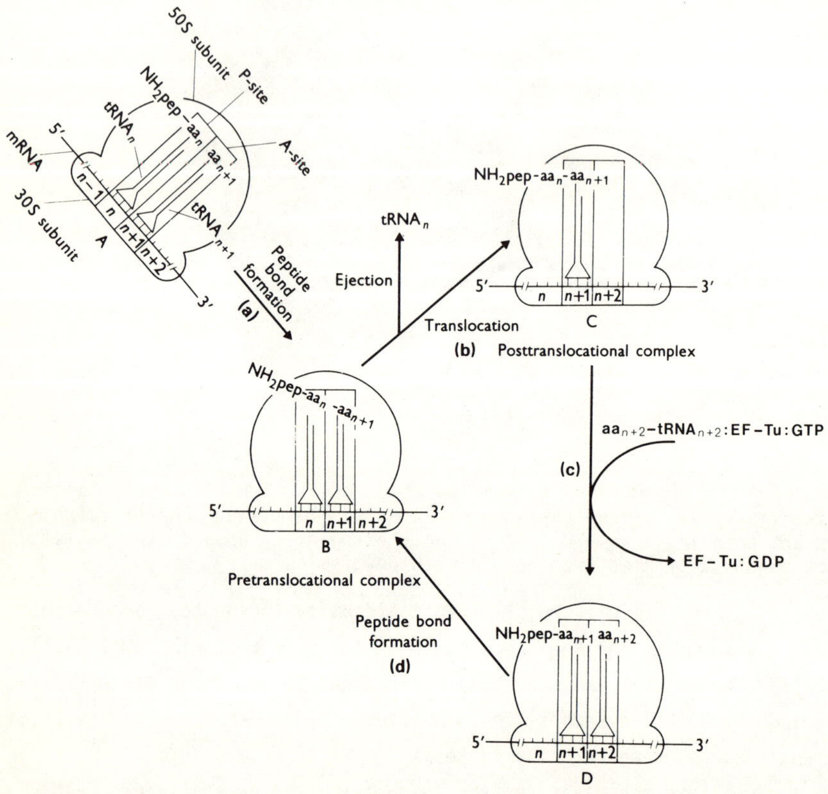

Figure 6. Cyclic scheme for peptide bond formation on the ribosome in terms of two tRNA binding sites.

probably other binding sites in the body of the tRNA such as the TΨC region postulated from sequence homology studies (step 3(b)). At the opposite end of the tertiary structure from the aa end is the anticodon site for decoding the mRNA on the ribosome (step 3(a)).

Binding of the aa-end of tRNA to the Ribosome

Since the 3'-CCA end of the tRNA where the aa is attached is single-stranded it is rather unstructured and flexible. In the crystal structure except for the terminal A the single-stranded region is stacked in a reasonable helix. It is unknown how this structure is changed on binding aa-tRNA to the ribosome. Indeed, different structures are possible in the A- and P-sites, and since each of the single stranded regions could be stretched out by the ribosomal interactions to about 20Å, there is much leeway at present for constructing plausible models for peptide bond formation on the ribosome. What we know about the peptidyl transferase enzyme substrate binding sites has come from the use of analogues of and fragments of aa-tRNA and can be described in terms almost devoid of tertiary structural information. In brief at the A-site on the ribosome only aa-A is needed for peptide bond formation.

The aa cannot be N-blocked, but the amino group can be replaced by a hydroxyl group for peptidyl transferase activity. There is also strong specificity for the nucleoside although analogues, such as puromycin containing N_6-dimethyl A and with the amino acid attached in an amide rather than an ester bond, are acceptable. It appears that 3'-aas are necessary so aa-tRNA bound with the aa attached to the 2'-position undergoes migration of the aa group prior to peptide bond formation (for a detailed discussion see ref. 11). At the P-site on the ribosome an acylated aa-tRNA is needed and a piece of tRNA probably at least 3 units long; thus C-C-A-fMet is a substrate, but C-A-fMet has only a very low affinity for the P-site (see refs. 11 and 42).

Binding of aa-tRNA to the Ribosome not Necessarily Dependent on the Peptidyl Transferase Site or Decoding

Based on oligonucleotide inhibition studies there is evidence that the binding of the aminoacyl-tRNA to the ribosome requires the exposure of sequence T-Ψ-C-G for complementary binding to the ribosomal 5S RNA[25], and this appeared to reinforce the sequence homology comparisons described earlier. Should this be really the case a change in conformation of the tRNA is necessary from the structure represented in Figures 2 and 3 where the common sequence is unavailable for further base pairing because of its involvement in the tRNA tertiary structural design. Further evidence for the tightness of the tertiary structure in the regions including the T-Ψ-C-G which interacts with the D loop comes from monitoring various intermediate conformations of tRNA seen on heating the tRNA until it melts completely, i.e. with all of its tertiary and secondary structure destroyed. Results from both proton magnetic resonance studies[43] and from chemical reactivity studies[44,45] at different temperatures confirm that the interaction between the TΨC and D loops is the stablest part of the tertiary structure. Of course if T-Ψ-C-G must become exposed for binding to the ribosome the change in conformation could be effected by ribosomal proteins.

Although the ribosomal 5S RNA[46] contains C43-G-A-A46 which is complementary with T-Ψ-C-G, it has recently been shown that G44 is inaccessible in the ribosome structure to chemical modification by kethoxal[47] and that A45 and A46 are also inaccessible to chemical modification in the ribosome (V. Erdmann, personal communication). Thus if the aa-tRNA binds to 5S rRNA in a complementary way then there must also be a rearrangement of ribosome structure as well as the tRNA's. This does not exclude a recognition of this part of the tRNA by other than Watson-Crick base pairing.

Similarly, one could explain the experimental results of Schwarz and Gassen[48] by suggesting that C-G-A-A binding to the presumed T-Ψ-C-G region of the tRNA dependent on an oligonucleotide complementary to the anticodon need not be interpreted in terms of Watson-Crick base pairing but by a combination of stacking and stereochemical fitting with the minimum of base pairing interactions. This explanation could reconcile the clear lack of evidence from physical chemical measurements for a large structural rearrangement expected on exposing the T-Ψ-C-G stimulated by a codon-anticodon interaction (see refs. 6, 38, 52). Although Wagner and Garrett have also detected a small change in the TΨC loop from a tRNA's accessibility to chemical modification stimulated by a complementary codon, they feel that it is a secondary effect with unclear biological significance, perhaps even due to a biologically inactive form. Indeed, Wagner and Garrett[49] find a larger conformational change around the extra loop region in that dependent upon the presence of a complementary codon the region around m^7G46 becomes more accessible to chemical modification. What form this rearrangement takes and what is its biological significance for ribosomal binding is still unclear. At any rate now we do seem to have evidence for long range effects being transmitted through the tRNA, presumably by conformational change due to effector interactions. Perhaps this approach will tell us more of the ribosomal facilitation of such changes.

An additional point of interaction of the tRNA for ribosomal binding may well be the D stem region at base pair b2 (see Fig. 1 for nomenclature) as suggested by the work on UGA suppressor tRNA by Buckingham and Kurland[50] and will be discussed in the next subsection on decoding.

Decoding of mRNA

Clearly, when a peptide bond is formed on the ribosome two tRNAs lie side by side on the ribosome giving rise to our conception of ribosomal sites. Currently, it is difficult to think in terms of different dynamic states rather than static states. Much more information is needed about the ribosomal stereochemical environment of the peptidyl and aminoacyl-tRNAs and this subject will receive intensive research efforts in future.

At present it is not completely certain that the tRNA decodes or interacts with the mRNA in both P- and A-sites. However, evidence such as that described for the two-site model of protein biosynthesis by Roufa *et al.*[51] or for P-site decoding by Wurmbach and Nierhaus[66] suggests that this generally accepted assumption is correct. In this respect the primary decoding event is normally thought of as taking place in the A-site.

There are technical difficulties in measuring anticodon-mRNA interactions under natural conditions. As a next best thing Grosjean *et al.*[52], have measured the binding constants for the interactions of a large number of tRNA pairs (more than 60) which have complementary anticodons. The binding constants have been measured by a temperature-jump relaxation technique. From their results Grosjean *et al.*[52] have made the striking conclusion that the stability of a tRNA pair in complex does not increase regularly according to the number of G·C base pairs involved (i.e. not in order of increasing H-bonds as might have been the case). Presumably there must be some effect that stabilises A·U pairs. Furthermore, Grosjean (personal communication) has concluded that stacking effects conditional on the sequence of bases are probably important especially as there is an effect due to the nature of the hypermodified base next to the 3'-end of the anticodon (see also ref. 20). Thus there is a context effect in the decoding interaction. Although these studies[52] have given valuable information also on stability of non-Watson-Crick interactions especially in support of Wobble-base interactions, one still is uncertain how far the system of interacting tRNAs can reflect the native tRNA decoding of mRNA.

In a different approach using oligonucleotide stimulated binding of tRNA to the ribosome Möller *et al.*[53] have shown that in some cases anticodon-complementary tetranucleotides terminating in 3'-A show efficient presumptive binding to the complementary tRNA but are inactive as codons for binding tRNA to the ribosome. They have concluded that there is no straightforward relationship between effective binding of a complementary oligonucleotide to the tRNA and the coding activity of the respective oligonucleotide.

Another view of the decoding interaction has recently been proposed by Lagerkvist[54]. He bases his proposal on a reconsideration of tRNA decoding patterns and contradicts the findings of Grosjean *et al.*[52] as too artefactual by suggesting that more G·C base pairs in codon-anticodon interactions lead to an enhanced stability as in double helical DNA. Lagerkvist's proposal arose out of some of his group's earlier work where it was found that under the conditions of *in vitro* protein biosynthesis only one valine tRNA could decode all four valine codons, GUN (N=any nucleotide). The third position of the codon did not appear to have any discriminatory function. Thus the idea of codon families was introduced where a family of codons is a set of 4 codons differing only in the third base and all coding for the same amino acid. This decoding, called "two out of three" by Lagerkvist, has been proposed as an alternative method of reading of the genetic code to the Wobble Hypothesis[24] and to occur under certain conditions *in vivo*. Clearly, in two out of three reading only one tRNA is needed to decode a family of codons whereas the Wobble Hypothesis requires at least two tRNAs to decode such a family. Because the genetic code is not made up only in such families Lagerkvist has introduced mixed families and the idea that the need for more discrimination arises only with more A·U base pairs being involved. Something to distinguish non-family codons is needed because, if two out of three reading was general, translational errors would occur on a large scale. It is difficult to decide whether Lagerkvist is on the right lines because his evidence is also taken from an *in vitro*, i.e. unnatural, system. One drawback of Lagerkvist's proposal is that the role of the hypermodified base is not considered. *I also feel that one role of the ribosome is to equalise all the codon-anticodon interactions so that a very complicated differential rate of decoding does not occur with each different tRNA binding to the ribosome.*

Very recently *in vitro* studies by Goldman *et al.*[55] using $tRNA^{Leu}$ species for decoding have purported to show more wobble than allowed in the Wobble Hypothesis[24] but still with discrimination in the third base of the codon possibly contradicting Lagerkvist's proposal[54].

Under *in vitro* conditions of protein synthesis and the triplet binding assay certainly one tRNA can decode all four codons differing only in the third base. This has been demonstrated, for example, with pure *E. coli* tRNAs specific for tryptophan (decoding UGN, Buckingham and Kurland[48]) and for arginine (decoding CGN, Kruse *et al.*[56]). Actually, the non-specific decoding effect by $tRNA^{Trp}$ is inhibited by the presence of other tRNAs such as $tRNA^{Cys}$ when $tRNA^{Trp}$ is used to decode the Cys codon. Clearly, *in vitro* events are not necessarily reflecting normal *in vivo* events.

In the case of the non-specific decoding by $tRNA^{Trp}$ very interesting information arose about the requirement for tRNA fitting on the ribosome for proper decoding[48,57]. Interestingly, the UGA suppressor $tRNA^{Trp}$ gives more non-specific two out of three decoding of the Cys codons than does the wild type (su^-) $tRNA^{Trp}$ *in vitro*[50]. Both the UGA suppressor $tRNA^{Trp}$ and the su^- $tRNA^{Trp}$ have the same anticodon (CCA)[21] but the suppressor tRNA contains U11·A24 instead of U11·G24 in the su¯tRNA at position b2 of the D stem (Fig. 1). It is difficult to speculate on how this base change could affect the decoding activity of the anticodon through the tertiary structure to the anticodon which is contained in an arm that does not interact with the rest of the tRNA (Figs. 2 and 3). Most likely the structural environment of base pair b2 is in some way recognized by the ribosomal interaction which then influences the decoding. This implicates ribosomal components in the anticodon's decoding interaction, perhaps another way of increasing the specificity of translation.

There is also physical chemical evidence for a structural change in the D loop (and thus implicating the interacting TΨC loop) from fluorescence measurements of a probe in the D loop of y $tRNA^{Phe}$ bound to the ribosome with and without oligonucleotides complementary to the anticodon[58]. The interpretation of effects involving the interaction with an oligonucleotide is difficult because the oligonucleotide may change the tRNA structure (although as discussed later, it is unlikely in the case of the anticodon loop) and the location of the oligonucleotide may not be predictable because binding need not depend on Watson-Crick base pairing (as discussed earlier). These effects in the presence of ribosomes need further investigation for clarification. Likewise, although codon dependent binding of Phe-tRNA to the ribosomal A-site has been interpreted by Kirillov *et al.*[59] in terms of binding properties of two conformers of the tRNA there is no indication of how these two conformers differ in structure.

The anticodon loop conformation during decoding

The anticodon loop is another place where conformational changes possibly occur during protein synthesis. The hinge-like region of the structure where the anticodon arm is attached to the rest of the structure (see Fig. 3 around residues 26 and 44) could provide a point of directing conformational change in the anticodon loop or for conformational change concomitant with movement of the tRNA from the A-site to the P-site (Fig. 6) during protein synthesis. As can be seen in the diagrammatic Figure 7, the two bases following the anticodon to the 3'-side are stacked on the anticodon. This conformation found in the crystal form (Fig. 3) can be the called 3'-S conformation. There is a break in the regular stacking to the 5'-side of the anticodon but the two preceding pyrimidines are stacked on each other and on to the anticodon stem.

This structure is similar to that originally proposed by Fuller and Hodg-

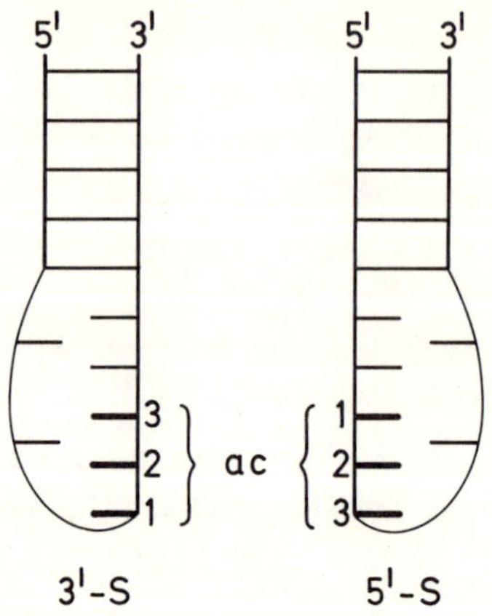

Figure 7. Possible conformations of the anticodon loop dependent on stacking bases on either end of the anticodon.

son[60] from model building studies. In the crystal form stacking of the anticodon, with somewhat more overlapping of bases than in a regular RNA double helix, appropriately exposes the anticodon bases for interaction with a mRNA codon. An alternative conformation for the anticodon loop where the 2 bases to the 5'-side of the anticodon are stacked on it (the 5'-S form) is shown in Figure 7. This has also been called the anti-Fuller Hodgson conformation and used by Woese[61] to define the P-site of the ribosome in a model for protein biosynthesis. Actually, there is recent evidence (Kuechler, personal communication) from fluorescent studies of tRNA crosslinked to mRNA that the anticodon conformation does not change when the tRNA is translocated from A-site to P-site. However, this is still rather controversial and other workers find structural changes in tRNAs bound to different ribosomal sites. Further experimentation is needed to decide this point unambiguously. The present state of knowledge surrounding this uncertainty has been detailed recently in an excellent comprehensive review by Cantor[62]. He also gives evidence on distances between tRNAs bound to the ribosome in different sites from fluorescence measurements.

In summary physical chemical studies (discussed in refs. 6 and 62) suggest that the 3'-S form of the anticodon loop is not changed on binding to a complementary oligonucleotide and that the increased binding constant of the tetranucleotide ending in 3'-A over the triplet is due to stacking forces rather than additional Watson-Crick base pairing increasing stability

Another scheme has been proposed by Lake[63] involving tRNA conformational changes which could incorporate these two conformations of the anticodon loop during protein synthesis. In his proposal the 5'-S conformation exists in a third ribosomal binding site, the recognition or R site where codon recognition (checking) is supposed to occur. The anticodon loop conformation flips to the normal 3'-S conformation for the A-site binding of the aa-tRNA and this also brings the amino acid into the correct environment for peptide bond formation. Although there is no what we might call conclusive evidence on these speculations, recently Urbanke and Maas[64] by investigating the fluorescence of the yW base in y $tRNA^{Phe}$ by a temperature jump method have concluded that they have evidence for a transition between two conformations such as illustrated in Figure 7. Reassuringly, their low temperature state of the tRNA is the 3'-S or crystal form conformation. Of course, at present they cannot decide whether their different conformations are related to what could occur in Lake's scheme.

Role of Metal Ions and Spermine

As described earlier, magnesium and spermine binding sites[20,30] have been located in the crystal structure of y $tRNA^{Phe}$. These cations appear to play a role in holding the folded tRNA structure together. Indeed, many physical chemical studies involving melting the tRNA structure under various ionic conditions confirm this suggestion. Thus the sensitivity of the tRNA structure to Mg^{2+} conditions[32,65] could well be an indication that selective ribosome-mediated removal of structurally crucial cations such as Mg^{2+} or spermine will trigger unfolding of the aa-tRNA which could return to the crystal form upon restoration of these special cations[6,9].

So far very little is known about the metal ion role in the tRNA's ribosomal environment. Much more effort using sophisticated physical chemical techniques is necessary.

From what I have described it is obvious that, in this topic of tRNA conformational changes during peptide bond formation on the ribosome, at present *we see through a glass darkly*. Much inspiration and investigative effort will be required in the near future to make definitive progress in this biologically important field so that clear molecular details of the *tRNA and ribosome face to face* will be elucidated.

ACKNOWLEDGEMENT

I am grateful for research and travel support from the Danish Natural Science Research Council (grant nos. 511-7073 and 511-10622) and for additional research support from the Novo Fund.

My thanks are also due to Lisbeth Heilesen, Arne Lindahl and Orla Jensen for helping me with the manuscript preparation.

REFERENCES

1. Lewis, J.A., and Ames, B. (1972) J. Mol. Biol. 66, 131-142.
2. Bruton, C.J. (1979) in Nonsense Mutations and tRNA Suppressors, eds., J.E. Celis and J.D. Smith, pp. 47-68, Academic Press, London.
3. Ladner, J.E., Jack, A., Robertus, J.D., Brown, R.S., Rhodes, D., Clark, B.F.C., and Klug, A. (1975) Proc. Nat. Acad. Sci. USA 72, 4414-4418.
4. Quigley, G.J., Wang, A., Seeman, N.C., Suddath, F.L., Rich, A., Sussman, J.L., and Kim, S.H. (1975) Proc. Nat. Acad. Sci. USA 72, 4866-4870.
5. Stout, C.D., Mizuno, H., Rubin, J., Brenner, T., Rao, S.T., and Sundaralingam, M. (1976) Nucleic Acids Res. 3, 1111-1123.
6. Clark, B.F.C. (1979) in Nonsense Mutations and tRNA Suppressors, eds. J.E. Celis and J.D. Smith, Academic Press, London.
7. Gauss, D.H., Grüter, F., and Sprinzl, M. (1979) Nucleic Acids Res. 6, r1-r19.
8. Holley, R.W., Apgar, J., Everett, G.A., Madison, J.T., Marquisee, M., Merrill, S.H., Penswick, J.R., and Zamir, A. (1965) Science 147, 1462-1465.
9. Ladner, J.E. (1978) in Biochemistry of Nucleic Acids II, International Review of Biochemistry, ed., B.F.C. Clark, Vol. 17, pp. 1-18, University Park Press, Baltimore, Md.
10. Rich, A., and BajBhandary, U.L. (1976) Ann. Rev. Biochem. 45, 805-860.
11. Sprinzl, M., and Cramer, F. (1979) in Progress in Nucleic Acid Res. and Molecular Biol., ed., W.E. Cohn, Vol. 22, pp. 1069, Academic Press, New York.
12. Clark, B.F.C. (1978) in Transfer RNA, ed., S. Altman, pp. 14-47, MIT Press, Cambridge, Mass.
13. Smith, J.D. (1979) in Nonsense Mutations and tRNA Suppressors, eds., J.E. Celis and J.D. Smith, pp. 109-125, Academic Press, London.
14. Piper, P.W. (1979) in Nonsense Mutations and tRNA Suppressors, eds., J.E. Celis and J.D. Smith, pp. 155-171, Academic Press, London.
15. Silverman, S., Heckman, J., Cowling, G.J., Delaney, A.D., Dunn, R.J., Gillam, I.C., Tener, G.M., Söll, D., and RajBhandary, U.L. (1979) Nucleic Acids Res. 6, 421-433.

16. Martin, R.P., Sibler, A.P., Schneller, J.M., Keith, G., Stahl, A.J.C., and Dirheimer, G. (1978) Nucleic Acids Res. 5, 4579-4592.

17. Wong, T.W., McCutchan, T., Kohli, J., and Söll, D. (1979) Nucleic Acids Res. 6, 2057-2068.

18. Murasugi, A., and Takemura, S. (1978) J. Biochem. (Japan) 83, 1029-1038.

19. Heckman, J.E., Alzner-Deweerd, B., and RajBhandary, U.L. (1979) Proc. Nat. Acad. Sci. USA 76, 717-721.

20. McCloskey, J.A., and Nishimura, S. (1977) Accounts Chemical Res. 10, 433-410.

21. Hirsh, D. (1971) J. Mol. Biol. 58, 439-459.

22. Clark, B.F.C., Doctor, B.P., Holmes, K.C., Klug, A., Marcker, K.A., Morris, S.J., and Paradies, H.H. (1968) Nature, 1222-1224.

23. Rich, A., and Kim, S.H. (1978) Scientific American 238, 52-62.

24. Crick, F.H.C. (1966) J. Mol. Biol. 19, 548-555.

25. Erdmann, V.A., Sprinzl, M., and Pongs, O. (1973) Biochem. Biophys. Res. Commun. 54, 942-948.

26. Jack, A., Ladner, J.E., and Klug, A. (1976) J. Mol. Biol. 108, 619-649.

27. Quigley, G.J., and Rich, A. (1976) Science 194, 796-806.

28. Jack, A., Ladner, J.E., Rhodes, D., Brown, R.S., and Klug, A. (1977) J. Mol. Biol. 111, 315-328.

29. Hingerty, B., Brown, R.S., and Jack, A. (1978) J. Mol. Biol. 124, 523-534.

30. Quigley, G.J., Teeter, M.M., and Rich, A. (1978) Proc. Nat. Acad. USA 75, 64-68.

31. Sussman, J.L., Holbrook, S.R., Warrant, R.W., Church, G.M., and Kim, S.H. (1978) J. Mol. Biol. 123, 607-630.

32. Bolton, P.H., and Kearns, D.R. (1977) Biochim. Biophys. Acta 477, 10-19.

33. Brennan, T., and Sundaralingam, M. (1976) Nucleic Acids Res. 3, 3235-3251.

34. Schevitz, R.W., Podjarny, A.D., Krishnamachari, N., Hughes, J.J., and Sigler, P. (1979) Nature 278, 188-190.

35. Robertus, J.D., Ladner, J.E., Finch, J.T., Rhodes, D., Brown, R.S., Clark, B.F.C., and Klug, A. (1974) Nucleic Acids Res. 1, 927-932.

36. Robillard, G.T., Tarr, C.E., Vosman, F., and Berendsen, H.J.C. (1976) Nature 262, 363-369.

37. Reid, B.R., and Hurd, R.E. (1977) Accounts Chemical Res. 10, 396-402.

38. Davanlou, P., Sprinzl, M., and Cramer, F. (1979) Biochem., in press.

39. Potts, R., Fournier, M.J., and Ford, N.C. (1977) Nature 268, 563-564.

40. Miller, D.L. (1978) in Gene Expression, Symposium 12 of the 11th FEBS Meeting, Copenhagen, 1977, eds., B.F.C. Clark, H. Klenow and J. Zeuthen, Vol. 43, pp. 59-68, Pergamon Press, Oxford.

41. Morikawa, K., la Cour, T.F.M., Nyborg, J., Rasmussen, K.M., Miller, D.L., and Clark, B.F.C. (1978) J. Mol. Biol. 125, 325-338.

42. Vazquez, D. (1978) in Amino Acid and Protein Biosynthesis II, International Review of Biochemistry, ed., H.R.V. Arnstein, Vol. 18, pp. 169-232, University Park Press, Baltimore, Md.

43. Kastrup, R.V., and Schmidt, P.G. (1978) Nucleic Acids Res. 5, 257-269.

44. Rhodes, D. (1977) Eur. J. Biochem. 81, 91-101.

45. Goddard, J.P., and Lowdon, M. (1978) Eur. J. Biochem. 89, 531-541.

46. Brownlee, G.G., Sanger, F., and Barrell, B.G. (1968) J. Mol. Biol. 34, 379-412.

47. Noller, H.F., and Garrett, R.A. (1979) J. Mol. Biol., in press.

48. Schwarz, U., and Gassen, H.G. (1977) FEBS Letters 78, 267-270.

49. Wagner, R., and Garrett, R.A. (1979) Eur. J. Biochem., in press.

50. Buckingham, R.H., and Kurland, C.G. (1977) Proc. Nat. Acad. Sci. USA 74, 5496-5498.

51. Roufa, D.J., Skogerson, L.E., and Leder, P. (1970) Nature 227, 567-570.

52. Grosjean, H.J., de Henau, S., and Crothers, D.M. (1978) Proc. Nat. Acad. Sci. USA 75, 610-614.

53. Möller, A., Schwarz, U., Lipecky, R., and Gassen, H.G. (1978) FEBS Letters 89, 263-266.

54. Lagerkvist, U. (1978) Proc. Nat. Acad. Sci. USA 75, 1759-1762.

55. Goldman, E., Holmes, W.M., and Hatfield, G.W. (1979) J. Mol. Biol. 129, 567-585.

56. Kruse, T.A., Clark, B.F.C., and Sprinzl, M. (1978) Nucleic Acids Res. 5, 879-892.

57. Vacher, J., and Buckingham, R.H. (1979) J. Mol. Biol. 129, 287-294.

58. Robertson, J.M., Kahan, M , Wintermeyer, W., and Zachau, H.G. (1977) Eur. J. Biochem. 72, 117-125.

59. Kirillov, S.V., Makhno, V.I., Odinzov, V.B., and Semenkov, Y.P. (1978) Eur. J. Biochem. 89, 305-313.

60. Fuller, W., and Hodgson, A. (1967) Nature 215, 817-821.

61. Woese, C.R. (1970) Nature 226, 817-820.

62. Cantor, C.R. (1979) in tRNA Workshop 1978, eds., J. Abelson, P. Schimmel, and D. Söll, Cold Spring Harbor Laboratory, Long Island, New York.

63. Lake, J.A. (1977) Proc. Nat. Acad. Sci. USA 74, 1903-1907.

64. Urbanke, C., and Maas, G. (1978) Nucleic Acids Res. 5, 1551-1560.

65. Stein, A., and Crothers, D.M. (1976) Biochem. 15, 160-168.

66. Wurmbach, P., and Nierhaus, K.H. (1979) Proc. Nat. Acad. Sci. USA 76, 2143-2147.

INITIATION OF PROTEIN SYNTHESIS AS SEEN IN 1979

MARIANNE GRUNBERG-MANAGO

Institut de Biologie Physico-Chimique,
13, rue Pierre et Marie Curie,
75005 Paris, France

INTRODUCTION

The initiation step ensuring the recognition of a correct starting triplet in the mRNA is critical in phasing the readout of this mRNA. At this step the initiator tRNA specific for methionine interacts with the mRNA at the specific initiator triplet and thereby ensures that translation begins at the proper site.

Initiation is believed to be the rate-limiting step in protein synthesis, during which a discrimination occurs between different mRNAs and different cistrons in mRNA (1). It is therefore expected to be an essential control point in translation. Important progress has been made by using in vitro reconstructed systems to determine the pathways and the structure of the components involved in initiation. Progress has been somewhat more rapid with bacterial systems but in recent years our understanding of eukaryotic systems has advanced considerably and the components and pathways are now well characterized. The similarities between the overall mechanisms of protein synthesis in the two systems are striking and the prokaryotic system has been a valuable model for our understanding of the eukaryotic system. However there are important differences at the initiation step, particularly in the mechanism of selection by the ribosome of the starting triplet. While this review will mainly deal with the prokaryotic system, this particular difference from the eukaryotic system will be discussed ; it may be fundamental for our understanding of the control of differentiation in eukaryotes. In recent years many reviews have been written on the initiation of protein synthesis (2-7). I will try here to focus on recent developments.

PROKARYOTES

Components involved

Initiation complex formation on 70S ribosomes in Escherichia coli involves, in addition to initiator tRNA and mRNA, the cooperative interaction of three protein factors and is stimulated by GTP. The sequence of events can be summarized as follows :

step 1 : 70S $\rightleftharpoons$ 30S + 50S
step 2 : 30S + fMet-tRNA^{fMet} + mRNA + GTP   30S fMet-tRNA^{fMet} mRNA
step 3 : 50S + 30S fMet-tRNA^{fMet} mRNA GTP   70S fMet-tRNA^{fMet} mRNA + GDP + Pi

These reactions are greatly influenced by divalent and monovalent cations ; increasing Mg^{2+} concentration shifts the equilibrium of step 1 toward ribosome association while monovalent cations have the opposite effect. Under physiological ionic conditions and with wild type ribosomes, the equilibrium is towards the 70S, and two proteins, called initiation factors IF1 and IF3 shift the equilibrium towards the subunits (step 1). IF1 serves to increase the rate constant for the dissociation of the 70S ribosome, while IF3 shifts the equilibrium strongly towards dissociation by binding to the 30S particle (8).

The second step requires the third initiation factor IF2 to direct fMet-tRNA^{fMet} binding, while mRNA binding is stimulated by IF3. The third step is accompanied by the release of all three factors. IF3 is released before the junction of the 50S subunit, IF1 during this process, whereas IF2 is released afterwards and is the only factor requiring GTP hydrolysis for its release.

The three steps result in the formation of the initiation complex which contains a 70S ribosome, mRNA and fMet-tRNA^{fMet}. Our view on the components and pathways involved has not essentially changed during the past two years.

Initiator tRNA

Recently a three dimensional structure of E.coli initiator tRNA has been reported (9). These results indicate that basically the same kind of tertiary folding occurs in the E.coli tRNA^{fMet} as in yeast tRNA^{Phe} ; there may, however, be a small difference in the structure of the anticodon loop. It has also been realized that some but not all aatRNAs have common features with fMet-tRNA^{fMet} when their amino acid is formylated or acetylated (10). Different tRNAs have more or less affinity for the P or A site of the ribosome. Formylation of the aatRNA decreases its affinity for the A site while increasing it for the P site. Initiator factors either as a complex with initiator tRNA or on the ribosome (see discussion below) increase the affinity for the P site.

A satisfactory structural explanation of the functional differences between the initiator tRNA or its analogues and other tRNAs remains, however, to be elucidated.

Ribosomes

The equilibrium between the 70S ribosome and its subunits depends on the structure of the ribosome and the ratio of divalent to monovalent metal ions in the medium (11). Divalent metals such as alkaline earths are interchangeable. Even the transition metals which do not damage RNA bases promote association to 70S particles ; the rate constants are comparable and somewhat larger than those for the alkaline earths. These results are in accordance with a model in which the principal effect of divalent ions is that of neutralizing non-specifically the charge of the sugar-phosphate chain. This does not exclude the possibilities of further sites, involving the ribosomal proteins, in which other divalent metal ions can replace Mg^{2+}. Monovalent cations favour dissociation with the order of efficiency : Li $>$ Na $>$ K $>$ NH_4 (11-14). Cosolvents can partially replace divalent cations, presumably by interacting with hydrophobic groups of the proteins and stabilizing an active conformation of the subunits (14). The fact that cosolvent can partially replace divalent cations led to the postulate that Mg^{2+}, in addition to a non-specific role of neutralizing the negative charge of the sugar-phosphate backbone and thus decreasing electrostatic repulsion, binds to the ribosomal proteins thus stabilizing an active conformation of the subunits. Under physiological conditions some E.coli mutants, particularly those which are streptomycin resistant, show a shift of equilibrium toward dissociation (15).

It has also been realized that for optimum initiation complex formation there is an interdependence between optimum ionic strength and pH. This has been explained by the polyanionic character of the ribosomal particles (16). At acidic pH and low ionic strength the apparent pK of the amino groups of the protein is abnormally high because of the existence of an acid microenvironment due to the net negative charge of the ribosomal particles. At high ionic strength the negative charges are neutralized and the pK is shifted back to normal values. At alkaline pH, even at low ionic strength, the amino groups of proteins are less protonized due to the increase in pH of the microenvironment ; thus explaining the decrease in requirement for Mg^{2+}.

Messenger RNA

In prokaryotic cells, RNA transcripts are frequently polycistronic, possessing a number of translational start and stop signals in vivo. While most studies in vitro with so-called natural mRNA are limited to the RNA genome of the bacteriophages Qβ, MS2 and R17, the translation of cellular mRNA differs in important respects. In phage mRNA non-translated intercistronic regions exist between the different cistrons. Besides playing a part in ribosome recognition, such untranslated nucleotide sequences presumably eliminate possible interference between ribosomes involved in termination and initiation. In cellular mRNA the situation might differ somewhat from that of the phage mRNA. In mRNA coded by the gene of the tryptophan operon the intercistronic region between trpB and trpA (the last two genes of the operon) consists of only two untranslated nucleotides ; the UGA terminator codon of tryp B overlaps the tryp A AUG initiation codon (17).

Moreover, in vivo translation of prokaryotic mRNA is coupled to transcription which might regulate protein expression. The latter may also be regulated by formylation of $Met\text{-}tRNA^{fMet}$. It has been reported that in the presence of trimethoprim (an antibiotic inhibiting formylation) the first cistron in a lac operon is expressed preferentially with respect to the last one (15). This phenomenon can fail to occur in vitro even in a transcription-translation system where the rate of protein synthesis is much slower than in vivo (18).

Initiation factors

In crude cell extracts the initiation factors are localized primarily on native 30S ribosomal subunits and cosediment with ribosomes on centrifugation. They are dissociated by high salt and differ from the structural ribosomal proteins by the fact that they are only transiently associated with ribosomes at specific periods during protein synthesis and are subsequently released. Three general types IF1, IF2 and IF3 are known and have been purified to homogeneity from the ribosomal wash (19). In the presence of phage mRNA they are required for the binding of initiator tRNA to the ribosome or the maximum rate of protein synthesis *in vitro*. Specific antibodies towards the three initiation factors have been prepared in rabbits. The initiation factors are antigenically distinct from each other and from ribosomal proteins and other molecules in crude cell lysates (20). There is no reason to suspect the existence of additional factors, since a reconstructed system with the three purified factors is more active than the crude lysate (M. Grunberg-Manago and J. Dondon, unpublished experiments). However, as already mentionned, the reaction rates of protein synthesis obtained *in vitro* (18,21) and reflecting initiation are much slower than the rates of protein synthesis in intact cells. It is therefore impossible to conclude definitely that all the components for protein synthesis have been identified.

The properties of the factors have been previously extensively discussed (2,3) and are summarized in Table 1.

TABLE 1. Initiation factors from *Escherichia coli*

Factors	Molecular weight	Function
IF1	9,000	Increase dissociation rate Promotes IF2 and IF3 functions
IF2a IF2b	115,000 90,000	Binds fMet-tRNA ; GTPase association activity
IF3a IF3b	21,000	mRNA binding ; antiassociation

IF2 is indispensable for initiator tRNA binding stimulated by GTP. Moreover, IF2 helps the association of the two subunits and has a GTPase activity in the presence of ribosomes (8). IF3, as was discussed, prevents the association of the two subunits and is required for the binding or translation of phage mRNA or synthetic poly A,U,G. IF1 helps the action of the two other factors. IF1 and IF3 are small basic proteins. IF3 exists in two molecular forms (22) which have been sequenced : one is 181 amino acid long and the other a shorter form of the same protein lacking 6 amino acid residues at the N terminus (23). It is therefore likely that they are both products of the same gene, a finding consistent with the results of Springer *et al*. (24). In addition to these two forms, a minor species has been detected in which the N terminal amino-group of methionine is methylated. IF2 also exists in two forms, IF2a and IF2b, with molecular weights of 115.000 and 90.000. Sometimes a third minor form with a molecular weight of 70.000 has been observed (Petersen, H.U., Dondon, J., and Grunberg-Manago, M., unpublished experiments) ; all three forms react with antibodies against IF2a and are probably products of the same gene ; they are active in promoting complex formation. However it has been reported that only IF2a in a highly purified DNA coupled system stimulates the synthesis of active β-galactosidase (25).

In B. stearothermophilus and Caulobacter crescentus only two initiation factors have been found (3,26,27). They correspond to IF2 and IF3. This may point to a somewhat dispensable role of IF1 in the other prokaryotic systems.

Sequence of initiation complex formation

A major problem in defining the pathways of initiation is to determine the order of binding of initiator tRNA and mRNA to the 30S ribosomal subunit. It is also essential to know whether the factors first bind to the ribosome or whether IF2 and IF3 respectively recognize initiator tRNA and messenger RNA and promote the binding of the complex to the ribosome. Experimental data are not easy to interpret.

The three factors IF1 IF2 and IF3, bind cooperatively to 30S subunits in the absence of mRNA and fMet-tRNA and are found on native 30S subunits in cell extracts ; GTP stabilizes the binding. It would then be tempting to define the first step as a cooperative binding of factors to ribosomes (Fig. 1).

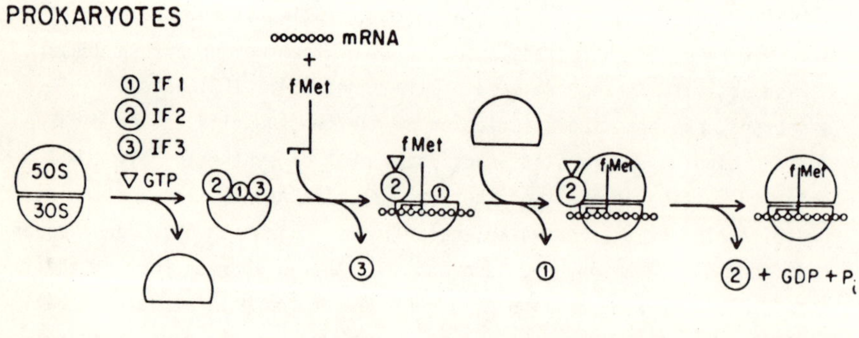

Figure 1. Pathways for initiation of protein synthesis in prokaryotes (from 5).

On the other hand, a number of workers regard IF2 as the fMet-tRNA recognition or carrier protein, and establish a parallel with EF-Tu which is responsible for the recognition and positioning of internal amino acid tRNA in the presence of GTP. The analogy, however, is questionable : whereas EF-Tu forms a stable ternary complex with GTP and aminoacyl-tRNA in the absence of the ribosome, a similar complex between fMet-tRNA GTP and IF2, if it exists, is not stable enough to allow its isolation. There is evidence, however, for binary interactions between IF2 and fMet-tRNA^{fMet} and IF2 and GTP. IF2, specifically inhibits the spontaneous chemical hydrolysis of the amino ester bond of fMet-tRNA (28) ; GTP is not needed for this protection. It was also reported that GTP protects IF2 against thermal and sulphydryl reagents inactivation (2,3). There is, moreover, some parallel between IF2 and EF-G, which also interacts rather weakly with GTP. They both show an active uncoupled GTPase activity, i.e. GTP hydrolysis in the presence only of ribosomes and EF-G or IF2, whereas EF-Tu has very little uncoupled GTPase activity. It is possibly artificial to try to order the different steps and that the whole process is highly cooperative. The same holds for the order of addition to the 30S subunit, initiator tRNA preceding that of mRNA or vice versa. It is possible that either order of binding depends on mRNA structure (29). More sophisticated kinetic analyses are required to elucidate the precise order of addition of initiation complex to ribosomes (30). Furthermore it will have to be proven that the in vitro situation reflects the in vivo one and to determine whether a unique pathway for assembly of the 30S initiation complex exists, or whether several pathways operate simultaneously.

Recycling of initiation factors

A number of experimental approaches indicate that initiation factors are only transiently associated with ribosomes and leave them after initiation complex is formed : the polysome fraction which contains the bulk of the ribosomes active in protein synthesis does not contain the factors. The levels of initiation factors are substantially below the levels found for elongation factors but appear to be of the order of initiator tRNA found in the cell (A. Danchin, unpublished experiments). Initiation factor levels have been determined by a radioimmune assay in lysates of rapidly growing cells. The levels of IF1, IF2 and IF3 are equimolar each being present as about one copy per 7 ribosomes (20).

The ejection of the initiation factors from the ribosome has been shown directly by analysis of initiation complexes formed with radioactive factors (3). IF3 appears to leave the 30S subunit when the preinitiation complex mRNA fMet-tRNA is formed. As soon as the factor is ejected from the 30S, the 50S can join and the junction is helped by IF2.

IF1 is displaced from the 30S initiation complex during this junction. The recycling of IF2 requires both the junction of both ribosomal particles and GTP hydrolysis. The role of GTP hydrolysis is not clear but may be that of providing the rapid release of IF2 so that the elongation cycle of protein synthesis can proceed.

There are two distinct GTPase activities promoted by IF2 when both ribosomal subunits are present : IF2 can hydrolyse GTP in the presence of ribosomes but in the absence of the other initiation components (uncoupled GTPase) ; the product GDP is a potent inhibitor of the hydrolysis (2,3). There is an interdependence between the ionic strength of the medium and optimum pH. The lower the ionic strength, the higher the optimum pH. This again can be explained by the polyanionic environment of the GTPase catalytic site (31).

When initiator tRNA, fMet-$tRNA^{fMet}$, and polymer containing the initiator codon AUG, are added to the ribosomes and IF2, the hydrolysis of GTP is increased (coupled GTPase) and the behaviour of the reaction as a function of ionic strength is significantly changed (31).

Selection of mRNA initiation sites

The step at which mRNA binds to ribosomes is likely to be the rate limiting reaction in protein synthesis. A number of features influencing mRNA binding have been identified and summarized in recent reviews (3,4). Examination of the known initiator regions of prokaryotic cistrons indicates that nearly all cistrons begin with the initiator codon AUG, a few utilize GUG (Table 2) and intracistronic reinitiation following nonsense mutations is known to involve UUG and CUG as well. The initiation triplet (AUG or GUG) is not located at the 5' end of the mRNA with one exception : λpR repressor cistron, and ribosomes select internal initiation codons in polycistronic mRNA. The initiation codons of fMet-$tRNA^{fMet}$ (AUG or GUG) do not code only for this tRNA, but also for Met-$tRNA^{Met}$ or Val-$tRNA^{Val}$ which position their aminoacids internally in the polypeptide chain. It is evident that some other feature in mRNA should distinguish the correct initiation codon from the many other potential initiation triplets either in or out of phase in the mRNA. Base sequences of initiator regions in mRNA were determined by analyzing mRNA fragments protected from nuclease digestion by ribosomal initiation complexes (32). Based on these sequences and the sequence of the 3' terminus of the 16S RNA which they determined, Shine and Dalgarno presented, in 1974, a hypothesis attributing the selection specificity to the interaction between a polypurine region located in the intercistronic region in the vicinity of the 5' end of the initiation triplet, and the 3' terminal sequence of the RNA of the small ribosomal subunit (33). Indeed, all known initiation regions contain a common purine rich sequence GGAG complementary to the sequence CUCC of the 16S RNA.

TABLE 2. Initiation sequences recognized by E.coli ribosomes (from 4).[a]

mRNA	Ribosome binding site
R17 A	GAU UCC UAG GAG GUU UGA CCU **AUG** CGA GCU UUU AGU G
MS2 A	GAU UCC UAG GAG GUU UGA CCU **GUG** CGA GCU UUU AGU G
Qβ A	UCA CUG AGU AUA AGA GGA CAU **AUG** CCU AAA UUA CCG CGU
R17 coat	CC UCA ACC GGG GUU UGA AGC **AUG** GCU UCU AAC UUU
Qβ coat	AAA CUU UGG GUC AAU UUG AUC **AUG** GCA AAA UUA GAG ACU
f2, MS2 coat	CC UCA ACC GGA GUU UGA AGC **AUG** GCU UCC AAC UUU ACU
R17, MS2 replicase	AA ACA UGA GGA UUA CCC **AUG** UCG AAG ACA ACA AAG
Qβ replicase	AG UAA CUA AGG AUG AAA UGC **AUG** UCU AAG ACA G
f1 coat	UUU AAU GGA AAC UUC CUC **AUG** AAA AAG UCU UU
f1 gene 5	A AGG UAA UUC ACA **AUG** AUU AAA GUU GAA AU
f1 gene (?)	A AAA AAG GUA AUU CAA **AUG** AAA UU
T3' *in vitro*	AAC AUG AGG UAA CAC CAA **AUG** AUU UUC ACU AAA GAG
T7 gene 0.3 site a	AAC UGC ACG AGG UAA CAC AAG **AUG** GCU AUG UCU AAC AUG
T7 gene 0.3 site b	GUA CGA GGA GGA UGA AGA GUA **AUG** U
λP_R	pppAUG UAC UAA GGA GGU UGU **AUG** GAA CAA CGC
λC_I	TTG CGG TGA TAG ATT TAA CGT **ATG** AGC ACA AAA AAG
φX174 G	TTT CTG CTT AGG AGT TTA ATC **ATG** TTT CAG ACT TTT ATT
φX174 F	CCT ACT TGA GGA UAA AUU **AUG** UCU AAU AUU CAA ACU
φX174 D	ACC ACT AAT AGG TAA GAA ATC **ATG** AGT CAA CTT ACT
φX174 H	ACT TAA GTG AGG TGA TTT **ATG** TTT GGT GCT ATT
φX174 B	UAA AGG UCC AGG AGC UAA AGA **AUG** GAA CAA CUC ACU
φX174 J	ACG TGC GGA AGG AGT GAT GTA **ATG** TCT AAA GGT AAA
araB	UUU UUU GGA UGG AGU GAA ACG **AUG** GCG AUU GCA AUU
trp leader	CAC GUA AAA AGG GUA UCG ACA **AUG** AAA GCA AUU UUC GUG
trpE	GAA CAA AAU UAG AGA AUA ACA **AUG** CAA ACA CAA AAA CCG
trpA	GAA AGC ACG AGG GGA AAU CUG **AUG** GAA CGC UAC GAA UCU
lacZ	AAU UUC ACA CAG GAA ACA GCU **AUG** ACC AUG AUU ACG GAU
lacI	AGU CAA UUC AGG GUG GUG AAU **GUG** AAA CCA GUA ACG
galE	AUA AGC CUA AUG GAG CGA AUU **AUG** AGA GUU CUG GUU ACC
galT	TAT CCC GAT TAA GGA ACG ACC **ATG** ACG CAA TTT AAT CCC
	16 S RNA 3' end ...A U U C C U C C A C U A G

[a]Underlining indicates contiguous bases complementary to the 3'-oligonucleotide of E.coli 16S rRNA. Dots indicate G.U base pairs. Initiator triplets are indicated by bold type. The sequence of the MS2 coat protein RNA is that given by 34.

The trpA ribosome binding site not preceded by an intercistronic sequence but nevertheless contains the common Shine and Dalgarno sequence in the segment of the trpB gene. The only known exception, phage λ repressor mRNA which begins with the initiation codon AUG and therefore lacks the Shine and Dalgarno sequence, is inefficiently translated. It is now believed that binding of mRNA to the 30S ribosomal subunit is primarily stabilized by two independent RNA-RNA interactions : 1) fMet-tRNA anticodon loop and initiator codon ; 2) the polypurine-rich portion of mRNA located in the vicinity of the 5' end of the initiation codon, pairing with the pyrimidine-rich 3' terminus of 16S RNA. The number of base pairs possibly involved in this last interaction varies from 3 to 9, the average being 4-5. The number of nucleotides separating the complementary region from the initiator triplet also varies with an average of 10 from middle to middle (Table 2). In vitro studies supporting the Shine and Dalgarno hypothesis have been extensively summarized previously (3,4). Among these is the isolation from the initiation complex of a hybrid between the ribosome-protected mRNA segment and the 3'OH terminal fragment of 16S RNA (35). I shall now discuss recent evidence related to the Shine and Dalgarno hypothesis.

Studies with different mutants

It is obvious that information as to how essential are the two RNA-RNA interactions and the different proteins involved in initiation in vivo can only be obtained with the help of specific mutants. Mutants in the two regions of mRNA involved in initiation have now been isolated and studied. The first series of mutants are those affecting the gene 0.3 of bacteriophage T7 (36).

This gene codes for an early mRNA about 600 nucleotides-long and specifying two proteins ; one of these, the 0.3 protein, about 100 amino acids long is made in large amounts during T7 infection. The initiation site on the 0.3 mRNA has been identified by sequencing both the 0.3 protein and the fragment of mRNA which is protected from nuclease digestion by ribosomes during initiation complex formation.

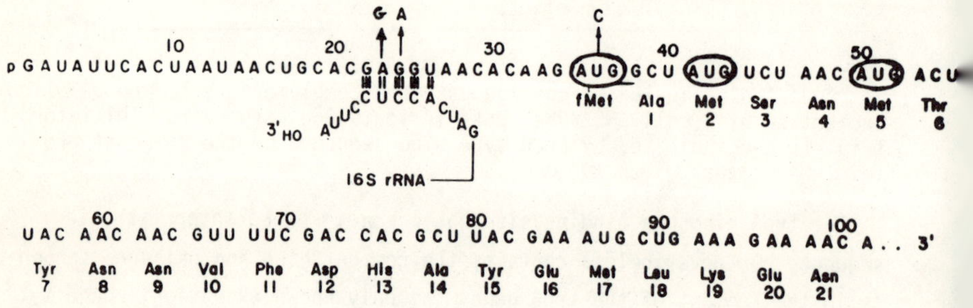

Figure 2. Nucleotide and amino acid sequences near the beginning of the 0.3 protein (from 36).

The Shine and Dalgarno sequence has been identified as GAGGU corresponding to a 5 base complement between the 0.3 RNA of T7 and the 16S RNA of E.coli (fig. 2).

Two mutants in the 0.3 gene having a greatly reduced rate of synthesis for 0.3 protein have been isolated. One of the mutations reduces complementarity in the Shine and Dalgarno sequence by a G-A transition (nucleotide 24) in the center of this sequence. The other mutant has a U-C transition at nucleotide 36 in the initiation codon. It is evident that a single mutation either in the Shine and Dalgarno sequence or in the initiation codon drastically affects the rate of protein synthesis. Furthermore the mutation in the Shine and Dalgarno sequence shifts in vitro the position of the ribosome binding site. Initiation of protein synthesis is greatly reduced and recognition of the correct initiator triplet is less specific since protein synthesis begins not only at the correct AUG but also at two other AUGs (position 2 and 5) specifying internal methionine in the wild type (Studier, personal communication).

In contrast, the mRNA mutated in the initiation codon binds to ribosomes in vitro at the same site and with the same efficiency as wild type mRNA, but the rate of synthesis of 0.3 protein is drastically reduced both in vitro and in vivo. The Shine and Dalgarno interaction that results in ribosome binding is strong enough in the absence of an AUG but the initiator codon is necessary for starting protein synthesis. The small amount of protein synthesized by this mutant begins at the in phase AUG that normally specifies the methionine at position 2 of the wild type protein (3 bases apart from the normal initiator triplet AUG).

Spontaneous phenotypic revertants were also isolated that retained the original mutation but had acquired a second one. The suppressor one mutant of the Shine and Dalgarno sequence has acquired an additionnal mutation A $\longrightarrow$ G at nucleotide 23, which restores a new four-base complementarity to the 16S rRNA. This suppressor mutation shifts the ribosome protection site back to the wild type position and restores the initiator triplet specificity. The suppressor mutation thus restores both the wild type rate of synthesis and the wild type position for ribosome binding. This suggests that pairing between the mRNA and 16S RNA may be important both in selecting the site in mRNA at which ribosomes bind and in determining the efficiency of initiation of protein chains. However, an important influence of mRNA context has been established for the recognition of the nonsense codons and for the codon:anticodon interaction (Grosjean, H. and Buckingham, R.H., review in preparation).One may therefore speculate that the situation is perhaps similar for initiation. To confirm that it is specifically the lack of Shine and Dalgarno complementarity which abolishes protein initiation,more mutants need to be studied (37,38).

Sequence analysis of the initiator region of the lactose repressor gene (lac I) indicates that the protein is initiated at a GUG codon preceded by a sequence substantially complementary to the 16S RNA. Of the three in phase GUG triplets clustered in this region, the real initiator is that codon which is most appropriately situated relative to the polypurine stretch. Studies of intracistronic reinitiation following nonsense mutations in the lac I gene compared to the true lac I initiator region indicate that mRNA complementarity to the rRNA is weak at the restart sites (3,4).

Other structural feature in prokaryotic mRNA appear to enhance or prevent initiation

Conformational masking of potential initiation sites is clearly a major negative determinant (4). Structural features that appear to enhance initiation include : occurence of a pyrimidine residue immediately preceding the AUG codon (39) and occurrence of an A residue immediately following the AUG codon (thus permitting a 4 base pair interaction with E.coli fMet-tRNA). This result was obtained by Taniguchi et al. (40) who compared the efficiency of ribosome binding, using initiator tRNA and Qβ mRNA isolated from wild type phage and mutants of the Qβ coat protein cistron, where the third position of the initiator AUG and the next residue (the first position of the succeeding alanine codon) were altered. Even triplet AUA followed by an A allowed some binding of initiator tRNA. Unfortunately Taniguchi et al. (40) were able to study only binding and not protein synthesis and it is difficult to predict the possible repercussion of a 4 base-pair interaction in initiator tRNA on the in phase triplet reading of mRNA.

Inhibition by nucleotide analogues of the Shine and Dalgarno sequence

Further evidence in favor of the importance of the Shine and Dalgarno interaction for the recognition of the initiation site on mRNA comes from studying in vitro the effect of oligonucleotide analogues of Shine and Dalgarno sequences. The hypothesis of Shine and Dalgarno predicts that if the 3' terminal region of the 16S tRNA is blocked, binding of the ribosome to the initiator region of the mRNA should be diminished or prevented while binding of the nucleotide AUG should not be affected. Taniguchi and Weissmann (41) have shown that the oligonucleotide A-G-A-G-G-A-G-G-U_{OH}, 8 bases of which are complementary to the sequence A-C-C-U-C-C-U-U-A_{OH} at the 3' terminal region of the 16S RNA, binds tightly to the 30S ribosomal subunit and is released as a complex with the 3' terminal 49-nucleotides of 16S rRNA after digestion with cloacin DF13.

Preincubation of ribosomes with all the components of the initiation complex and the above mentioned oligonucleotide leads to a 50 to 87 % inhibition of the initiation complex when the molar ratio of oligonucleotide to QβRNA varies from 2.5 to 10.

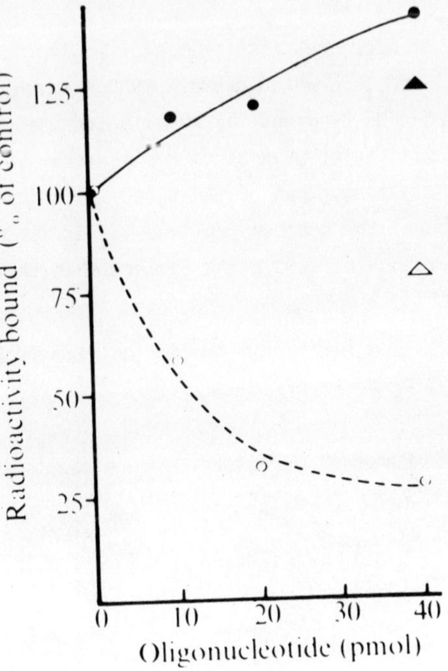

Figure 3. Effect of oligonucleotides on the binding of 70S ribosomes to Qβ(^{3}H)RNA and (^{32}P)pAUG, a mixture of (A-G-A-G-G-A-G-G-U_{OH} + G-G-G-A-G-A-A-G-U_{OH})+ Qβ(^{3}H)RNA (o) or (^{32}P)pAUG (●) and G-G-A-G-A-A-G-G_{OH} + Qβ(^{3}H)RNA (Δ) or (^{32}P)pAUG (▲) (from 41).

We have also shown that the synthesis of the coat protein is inhibited by this oligonucleotide (42) (Table 3). Also interesting is the observation that the binding of the triplet AUG is stimulated rather than inhibited in the presence of the oligonucleotide (Fig. 3), suggesting an allosteric effect of the oligonucleotide on the conformation of the ribosome, which is thus ready to bind the AUG initiator triplet (41, 43).

TABLE 3. Inhibition of QβRNA binding (from 42)

Additions	Exp	H3 QβRNA bound	% inhibition
Control	1	12	-
Control	2	10	-
- fMet-tRNA	1	1	-
+ 20 pmoles (A-G-A-G-G-A-G-G-U_{OH})	1	6	50
	2	3.7	63

Role of protein S1 and initiation factors in the recognition of initiator regions in mRNA

The interaction of ribosomes with initiator region is modulated by proteins such as S1 and IF3, which promote mRNA binding, and IF2 which is required for fMet-tRNA binding. The relative contribution of each protein to the specificity of mRNA binding may differ with different mRNA and different cistrons. Thus of the three R17 initiator regions, ribosome recognition of the coat or replicase sites is several fold more dependent on the presence of initiation factors than that of the maturation site (44).

TABLE 4. Effect of S1 and initiation factors on recognition of R17 initiator regions (from 44).

Expt. No.	Ribosomes	Factors	Ratio A:coat:replicase
I	30S	IF2,IF3	1:2.0:0.5
	30S	-	1:0.2:0.05
	30S (-S1)	IF2,IF3	1:0.3:0.04
	30S (-S1)	-	1:0.4:0.03
II	30S	Crude	1:7.1:2.7
	30S	IF2,IF3	1:3.6:0.8
	30S	-	1:0.2:0.2
	30S (-S1)	IF2,IF3	1:0.8:0.2
	30S (-S1)	-	1:0.1:0.06

What do S1 and IF3 recognize ? Could they operate independently from ribosomes and recognize free mRNA ? S1 and IF3 are known to bind to various mRNAs in the absence of ribosomes (45 ,47) but it is difficult to determine whether these interactions are specific and not purely electrostatic. In the case of S1, which is known to be in molar excess of ribosomes and remains bound to the ribosome throughout protein synthesis, the complex between free S1 and mRNA is likely to be an artefact (48).

A variety of hypotheses has been brought forward to account for the effect of S1 : viz. to facilitate the binding of mRNA to ribosomes (49,50). The first and most simple hypothesis is that an RNA binding site on S1 forms part of the messenger binding site on the ribosome (43,50). From data on the binding of partially denatured mRNA to ribosomes it was postulated that S1 aids mRNA binding by influencing the tertiary structure of the mRNA. Dahlberg,A., and Dahlberg,J. (51) reported evidence that S1 binds to the ribosome in the region of the 3' terminus of 16S RNA (the colicine fragment) and this suggests that the role of S1 is to melt out a hairpin with the messenger. In accordance Szer et al. (52) have shown that S1 has a melting activity. Steitz et al. (42) suggested that S1 may stabilize this base paired region once it is formed.

Draper and Von Hippel (52) found that S1 has two binding sites for nucleic acids. Site I was shown to bind either DNA or RNA and polymers while site II was found to be highly specific for RNA binding. In addition, Laughrea and Moore (1977) have shown that 30S ribosomal subunits in inactive conformation (54) can bind two copies of S1 protein while only one binds per active conformation. We have also separated two classes of S1 on DNA columns : one has a high affinity for the small subunit and the other much less affinity (Bonnet, D., Petersen, H.U., Danchin, A., and Grunberg-Manago, M., unpublished results). However, it is possible to explain our results in the light of Von Hippel's data since these workers observed that one binding site was more easily denatured that the other. The presence of two binding sites opens the possibility that S1 could bind ribosomal RNA and mRNA simultaneously in vivo. The definition of the exact location of S1 binding sites on the 16S RNA and the elucidation of their functional significance will have to await further study.

Since IF3 does not stay on the ribosome throughout protein synthesis it is more difficult than for S1 to discard the possibility of free IF3 recognizing mRNA. Does IF3, either free or bound to the ribosome, recognize the initiator triplet or some other feature of mRNA ? In the presence of the AUG triplet, initiator complex formation on the 30S subunit does not require IF3, IF1 and IF2 being sufficient, whereas with other mRNAs, including synthetic poly AUG, IF3 is highly stimulatory (3).

TABLE 5. Stimulation of fMet-tRNA binding to 30S subunits by initiation factors (from 3).

Additions	Poly (A,G,U)	AUG
IF2	511	553
IF2 + IF3	3167	956
IF2 + IF1	2597	2923
IF1 + IF2 + IF3	5517	2102

The results are reported as counts per min, following the subtraction of blanks (minus all factors).

In the presence of 70S ribosomes, IF3 is required with triplet AUG but this may be explained by its dissociation activity. This suggests that the role of IF3 in mRNA recognition is not that of interacting specifically with the triplet AUG, but consists of affecting some other feature in mRNA : perhaps promoting recognition of the Shine and Dalgarno sequence. A hypothesis has been proposed which explains the two IF3 activities, dissociation activity and mRNA recognition, as the result of a sole function. This postulates that the interaction between ribosomal subunits involves complementary base pairing between 16S and 23S rRNAs, possibly at their 3' termini (55). The hypothesis is supported by a reasonable complementarity of the two sequences and by the higher susceptibility of 16S RNA to nuclease in the 30S compared to the 70S ribosomal particle (3). Upon dissociation the base pairs between 16S and 23S are replaced by self complementary base pairs, existing in both 16S and 23S RNA. IF3 has two effects due to its localization on the ribosome near the 3' termini of 16S RNA. It prevents the interaction of 16S RNA with the 23S RNA therefore shifting the equilibrium towards the 30S mRNA interaction ; also by stabilizing the 16S termini conformation it helps its interaction with the Shine and Dalgarno region in mRNA.

One may speculate that initiation complex formation is more or less dependent on either one of the two RNA-RNA interactions depending on the nature of the mRNA : when the mRNA-rRNA interaction is weak, mRNA binding to ribosomes has to be mainly stabilized by the IF2-dependent codon-anticodon interaction, IF3 and S1. When the interaction between mRNA and rRNA is strong, the requirement for initiator factors appears to be less strict, though nevertheless important. The apparent affinity of initiation factors for the mRNA-rRNA complex varies with the half life of the latter. This may, however, be explained in terms of the instability of the mRNA-RNA complex itself without necessarily supposing that the intrinsic affinity of initiation factors for the complex varies widely.

Localization of protein factors on the ribosome

From the preceding discussion, it is evident that precise structural information on the initiation complex is required in order to elucidate the mechanism of mRNA binding.

Efforts have been made to identify the 30S ribosomal proteins and the region of 16S rRNA where initiation occurs. The following experimental approaches have been used :

Initiation factors bound to the 30S subunit have been crosslinked to ribosomal proteins with bifunctional reagents and the crosslinked proteins identified (3). Another crosslinking technique which has a relatively high yield and has the further advantage of using no bridging molecule is irradiation at 365 nm in the presence of a photo-sensitizer (56). For IF3 the results refer to experiments on binary 30S-IF3 complexes (3). Results for IF1 and IF2 were obtained in solutions containing 30S and all three factors, either IF1 or IF2 being radioactive (Table 6). The results can be summarized as follows : all three initiation factors were crosslinked to a number of common proteins suggesting that the factors bind contiguously on the ribosomal surface (57). An IF2-IF1 crosslink in a trifactor complex was also observed (57). The major crosslinked proteins were S1, S11, S12, S13 and S19 (3).

The 3' terminus of 16S rRNA may be chemically crosslinked to some of the same proteins and to all three bound initiation factors as well. By irradiation at 365 nm IF3 could also be crosslinked to 16S RNA (56). It was found that the major part is crosslinked to the 3' terminus of 16S RNA but another site exists in the 12S part of the 16S RNA (42). Thus the factors bind near the sequence of rRNA involved in base pair interactions with mRNA.

The other approach to the understanding of the role of IF3 and S1 and their relation to the 3'OH end of 16S RNA is to use 30S ribosomal subunits where the 3' end of 16S RNA has been cleaved by colicin E3 treatment and either removed (30S(-E3)subunits) or left associated. Colicin treatment results in a nucleotide cleavage at about the 50th nucleotide from the 3' terminus of the 16S RNA.

TABLE 6. Crosslinking of initiation factors (from 3).[a]

S protein	Factor IF1[a]	Factor IF2[b]	Factor IF3[c]
1	+	+	+
2		+	?
3			?
4			?
7			++[d]
9			++[f]
11		++	++
12	++	++	++[c,e]
13	+	++	?
14	+	+	
18			?
19	++	++	++
20			?
21			+

[a]++ major croslinked protein ; + minor crosslinked protein ; ? not always found.

If the fragment is left associated, IF3 binds as well as to non-treated 30S subunits, but when the fragment is removed following treatment by heating, the apparent affinity of IF3 is reduced and becomes much more pressure-sensitive (58). Under similar conditions S1 binding is unaffected (54).

The demonstrable pressure dependence of the IF3-30S(-E3) complex indicates that the molar volume of such an interaction is positive and much larger than when IF3 binds to either control or cleaved, 30S subunits. One explanation is that IF3 binding to 30S(-E3) results in the trapping of excess solvent in a groove normally occupied by the missing RNA. This would confirm the localization of IF3 near the 3'OH end of 16S RNA, at the Shine and Dalgarno interaction site. In order to define more precisely the position of IF3 in relation to the initiation triplet AUG, initiation complexes were studied using electrophilic AUG analogues reactive either on their 5' or 3' extremities. Comparison of the labelling results obtained with either AUG or AUGU in initiation complexes suggests that proteins IF3, S1 and S21 are mainly located in the neighborhood of the 5' extremity, distant from the 3' end of the triplet (59).

Localization of the gene for IF3 and gene organization around it on the *E.coli* chromosome

Although each of the initiation factors shows stimulatory activity in one or more of the assays devised, it does not alone constitute a proof that the factors are actually necessary *in vivo*. It is difficult to reproduce *in vivo* conditions *in vitro*, and assay systems with highly purified components may lead to artefactual stimulation by some components. The best way to approach this problem is to obtain conditional mutant genes for the factors and show that protein synthesis is affected under the non-permissive condition.

A thermosensitive mutant with a thermolabile IF3 was isolated (60). Unfortunately this mutant was also modified in the activity of the small subunit of phenylalanyl-tRNA synthetase and it proved difficult to separate genetically the two defects. Nevertheless, this mutant has been used to isolate λ transducing phages complementing its thermosensitivity. One of these phages (λp2) was shown to carry the structural genes for initiation factor IF3 (*infC*), the α and β subunits of phenylalanyl-tRNA synthetase (*pheS* and *pheT*), the threonyl-tRNA synthetase (*thrS*) and for a 12 000 MW protein of unidentified function (*P12*) (24,61,62). This grouping of genes whose products are involved in translation raised the question as to whether they are transcribed independently or not. Cotranscription of genes involved in translation has already been observed ; this possibility was investigated for *infC* and aminoacyl-tRNA synthetase genes. Starting from the original phage (λp2) carrying the whole cluster of genes a set of deleted phages was isolated, carrying only some of the cluster. The genetic content of these deleted phages was studied by complementation of suitable mutants and by polyacrylamide gel analysis of proteins synthesized by UV irradiated and subsequently infected cells (62). The comparison of the genetic content of these phages and the structure of their DNA, studied with restriction endonucleases, gave the order of the *E.coli* genes on λp2 as *thrS* - *infC* - *P12* - *pheS* - *pheT* (62,63). Thus *infC* is located in the middle of the cluster. This implies that *infC* is apparently not immediately adjacent to the structural genes of the other initiation factors IF1 or IF2 whose genetic loci are still unknown.

A segment of *E.coli* DNA between an EcoRI site and a HindIII site (B1 fragment) of λp2 was cloned in pBR322 and in the λ vector NM540. Both plasmid and phage carrying fragment B1 expressed all the genes originally identified on λp2, showing that this fragment probably contains all the genes with their promotors (63). The data obtained with the deleted phages showed that the EcoRI site of B1 together with thrS on one side and the HindIII site of B1 together with P12, pheS and pheT can be deleted without affecting infC expression. This result indicates that infC is probably expressed independently of its neighboring genes. This enables us to obtain expression of infC from a plasmid derived from pBR322 carrying a 2.2 x 10^6 MW fragment of E.coli DNA not coding for thrS, pheS or pheT. Further analysis of the deletion phages has localized the gene on a 10^6 MW DNA fragment within the former fragment (63).

EUKARYOTES

Differences between prokaryote and eukaryote initiation

Although many parallels are evident between prokaryotic and eukaryotic translational systems there are significant differences (5,6) which are summarized in Table 7. The machinery of protein initiation is much more complex in eukaryotes, not only with respect to ribosomes but also because of the multiplicity of initiation factors. At the present time as many as 10 initiation factors have been identified, 2 of which are complex fractions. Moreover, ATP hydrolysis is involved in mRNA binding (ATP also stimulates initiation complex formation in a S.thermophilus cell-free system by an unknown mechanism) (26).

The most striking dissimilarities in initiation are due to the considerable structural differences between prokaryotic and eukaryotic messenger RNAs. In contrast to prokaryotic organisms, translation in eukaryotes cannot be initiated as soon as a portion of a cistron has been transcribed. The mRNA must first be transported from its site of synthesis in the nuclear chromatin to the ribosomes in the cytoplasm. During this process a large amount of eukaryotic mRNA is discarded. The mRNA molecules which appear in the cytoplasm differ from prokaryote mRNA in a number of structural features. As discussed above, many prokaryotic messages are polycistronic with initiation occuring efficiently at internal sites. Thus in prokaryotes there is no barrier to internal initiation although initiation at 5' proximal and internal cistrons may proceed via somewhat different mechanisms (15) at least with some cellular mRNA.

In striking contrast, eukaryotic ribosomes appear unable to bind to internal regions of messenger RNA. Eukaryotic messages are functionally monocistronic with initiation limited in most cases to a single 5' proximal site (6,64). Further differences include the presence of a blocked and methylated cap structure at the 5' terminus (7 methylguanosine attached at its 5'-hydroxyl by a triphosphage bridge to the 5'-hydroxyl of the RNA) (65), internal methyl residues (66), and a rather long stretch of adenylic acid residues at the 3' terminus (67) of nearly all eukaryotic messages. In addition, they are found in the cytoplasm as ribonucleo-protein complexes.

Nearly all mRNAs, either cellular or viral, are capped at the 5' end and polyadenylated at the 3' end which confers metabolic stability to this RNA. Among cellular mRNA, capping is more universal than polyadenylation ; thus, for example, the mRNAs for the histones lack this modification (68). However, some viral mRNAs such as picornavirus lack the cap structure (64).

TABLE 7. Differences between prokaryotic and eukaryotic protein synthesis initiation

Prokaryotes	Eukaryotes
. Ribosomes smaller, less complex	. Greater complexity and larger sizes
. Initiation factors : 3	. Initiation factors : 10
. Ternary complex GTP initiator tRNA IF2 ?	. Ternary complex GTP initiator tRNA eIF2
. Initiator tRNA methionine formylated	. NH2 free of methionine initiator tRNA
. AUG, GUG codons	. AUG codon
. $Gram^+$ ATP ? $Gram^-$ no ATP	. ATP for mRNA binding
. Polycistronic mRNA (with initiation on internal sites)	. A single mRNA site located near 5' PO4 but not at a fixed distance (11 to 200 nucleotides) ; heterogeneity of non-coding sequences, cap 5', polyA 3', m^6A internal residues
. Selective mRNA discrimination $gram^+$ and $gram^-$	. Less specificity in translation
. Shine and Dalgarno sequence	
. Fragments of mRNA initiation rebind to ribosomes	. 80S mRNA initiation fragments do not rebind to ribosomes
. Same mRNA region protected by 30S and 70S	. The 40S protect longer regions than 80S for mRNA sequences

A further notable difference is the high degree of messenger discrimination shown by prokaryotic but not eukaryotic ribosomes. Extracts from gram positive bacteria generally failed to translate messenger RNA from phage infecting gram negative bacteria (69,70). It has been shown that the specificity lies in the 30S subunit and not in the other factors involved in initiation (71). Eukaryotic cellular or viral mRNAs seldom possess the features needed for efficient binding to E.coli or gram$^+$ ribosomes. In contrast to the discrimination shown by prokaryotic ribosomes against heterologous messages, including some from prokaryotes, eukaryotic extracts are able to translate a wide variety of prokaryotic messages but their efficient translation requires addition of a cap structure (72,73). The following mRNAs prepared in vitro and capped with the vaccinia enzyme were translated in the wheat germ system and synthesized correct proteins : the monocistronic 8S cro mRNA, the λ polycistronic transcript containing at its 5' the cro mRNA, the polycistronic mRNA derived from the gal operon. For a given amount of mRNA the cap-modified cro mRNA was shown to synthesize 5 fold more protein than did hemoglobin mRNA in the wheat germ system. The prokaryotic mRNA contains therefore all the information necessary for efficient recognition and initiation by eukaryotic translational components except for the cap structure. However in polycistronic prokaryotic mRNA the cap-dependent translation appears restricted to the 5' proximal coding sequence (73). This confirms that protein initiation sites located "too far" from the cap structure may not be efficiently recognized by the eukaryotic translational components (this point will be discussed further below). The cap structure is not, however, required for an efficient translation of all mRNA (picorna virus mRNA which are not capped are nevertheless very efficiently translated) (64). The translation of heterologous T7 mRNA is completely independent of cap modification (74) ; however, T7 mRNAs are relatively poorly translated compared to the other cap-modified prokaryotic mRNA.

Finally, an important difference is that the segments of initiator regions protected by 40S ribosomes against nuclease digestion are significantly longer at their 5' ends than those protected by 80S ribosomes (75-77). This is in contrast to the nearly equal sized initiator regions protected by E.coli 30S and 70S ribosomes (77,78). In addition, in a heterologous system using f1 mRNA and reticulocyte ribosomes it has been found that the regions of f1 mRNA protected by 40S subunits are not only larger but more heterologous than those protected by 80S ribosomes (77). Furthermore, protected initiation fragments from 80S ribosomes which do not have the cap structure bind very inefficiently to the ribosomes (79) in contrast to the 70S protected segments which rebind efficiently to the 70S ribosomes (9,77).

Eukaryotic factors and pathway

Several recent reviews described in details the properties of the different mammalian initiation factors and the initiation pathways (2, 5,6,7,80). Only a summary will be reported here.

Most work on initiation has been carried out with components from rabbit reticulocytes. Three main groups have been involved : T. Staehelin, H. Trachsel and M. Schreir, F. Anderson and B. Merrick,initially and J. Hershey with R. Benne later.

Table 8. Soluble protein factors from rabbit reticulocytes (from 5)

Initiation	Molecule weight	Number of polypeptides	Functions
eIF1	15,000	1	promotes mRNA binding
eIF2	150,000	3	forms ternary complex with Met-tRNA and GTP
eIF3	ca. 700,000	ca. 9	promotes Met-tRNA and mRNA binding ; dissociation
eIF4A	49,000	1	promotes mRNA binding
eIF4B	80,000	1	promotes mRNA binding
eIF4C	17,500	1	promotes Met-tRNA binding
eIF4D	16,500	1	stimulates Met-puromycin
eIF5	150,000	1	required for 80S complex formation
CO-EIF2	22,000	1	stimulates ternary complex formation
cap binding	24,000	1	binds to cap of mRNA

Eight initiation factors, named eIF1, 2, 3, 4A, 4B, 4C, and 5 were purified to near homogeneity using globin synthesis as a test (80-83). Two of the factors eIF2 and eIF3 are multiprotein complexes. eIF2 contains 3 non-identical subunits (in 1:1:1 stoichiometry), eIF3 contains about 9 to 11 subunits, of total molecular weight 724 000. All the other factors are single polypeptides, (ranging from 28 000 to 140 000 MW) two of which are present in two copies. Two additional proteins are involved in the initiation process : a 24 000 protein which appears to bind the cap portion of the mRNA (84) and a 22 000 MW protein referred to as co-eIF2 (old nomenclature : co-EIF1) (85) which interacts with eIF2 during ternary complex formation.

As with prokaryotes initiation begins with the dissociation of the 80S ribosomes into subunits (5,86), for which the multicomponent eIF3 may be responsible (Fig. 4).

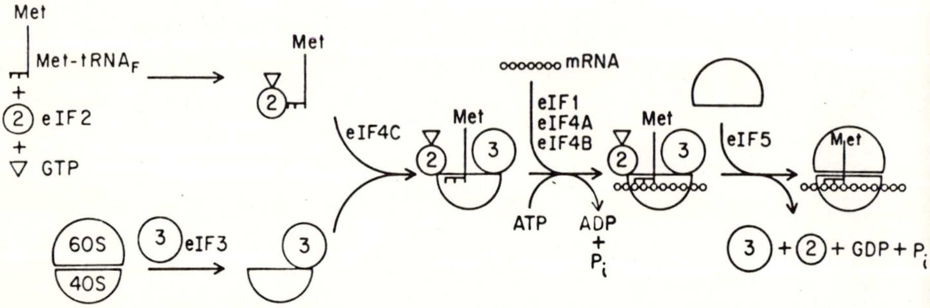

Figure 4. Pathways for initiation of eukaryotes protein synthesis (from 5).

In the absence of initiation factors even at low Mg very few free 40S ribosomes are observed, since 40S ribosomes have a tendency to form dimers (M. Grunberg-Manago and T. Staehelin, unpublished results). All of the protein components of eIF3 bind to the 40S ribosome in the absence of other initiation components. None of the other initiation factors bind to 40S subunits in the absence of other initiation components (86).

In a parallel set of reactions the initiator tRNA (Met-tRNA^{fMet}) forms a stable ternary complex with GTP and eIF2 (GDP is a potent inhibitor of ternary complex formation) (88). Additional factors might stimulate ternary complex formation but it is not clear whether they influence the extent of GDP inhibition or not. It is believed that the Met-tRNA-eIF2-GTP complex binds to 40S subunits in the absence of other components of initiation. eIF3 and eIF4C enhance the binding.

Stable binding of mRNA is promoted by eIF4A and eIF4B and needs the other components previously described. In addition mRNA binding requires ATP which is hydrolyzed. If any of the factors is omitted or if ATP is replaced by a non-hydrolyzable analogue, mRNA is released from the ribosome. This may indicate a high cooperativity where the two factors and ATP would be required for a single reaction. Junction of the 40S preinitiation complex with the 60S subunit requires the presence of eIF5 and the hydrolysis of GTP to GDP + P : eIF2 and eIF3 are released at this stage. It is not clear whether the hydrolysis of GTP and the release of eIF2 and eIF3 occur before or after the junction. Perhaps the function of eIF5 is to facilitate the removal of initiation factors from the surface of 40S subunit to allow junction with the 60S subunit.

Molecular aspects of message discrimination

How do eukaryotic ribosomes select initiation regions in messanger RNA ? Is the Shine and Dalgarno hypothesis applicable to eukaryotic ribosomes ? The sequences directly preceding each eukaryotic initiator AUG have been closely examined for complementarity to the 3' end of 18S RNA (89,90). Eight terminal nucleotides have been conserved throughout a wide range of eukaryotic mRNAs. They also exhibit remarkable homology with the same region from E.coli 16S rRNA. The striking difference is that the eukaryotic 18S RNAs do not contain the Shine and Dalgarno CCUCC sequence found in 16S RNA.

TABLE 9. 3' terminal sequence of 16S and 18S RNAs

$m^6_2Am^6_2A$ CCUGCGGUUGGAUCACCUCCUU A_{OH}	E. coli 16S
[$m^6_2Am^6_2A$]CCUGCGGΔUGGAUCA — UUUU[A]$_{OH}$	D. discoideum 18S
[$m^6_2Am^6_2A$]CCUGCGGAAGGAUCA ——— UU A_{OH}	B. mori 18S
[$m^6_2Am^6_2A$]CCUGCGGAAGGAUCA ——— UU A_{OH}	Mouse 18S
[$m^6_2Am^6_2A$]CCUGCGGAAGGAUCA ——— UU G_{OH}	Wheat 18S

[a]Ths four eukaryotic sequences are continuous covalent structures, the sequences are written with a gap, so that homologous regions can be aligned with the terminal sequence of E.coli 16S RNA. The underlined nucleotides indicate the changes from E.coli 3' terminal sequences of 16S and 18S RNAs. From 89.

Thus the eukaryotic ribosome lacks the very sequence involved in binding prokaryotic mRNA during initiation of translation. This could explain the difference in specificity between prokaryotic and eukaryotic ribosomes. No regularly positioned mRNA sequence which could pair with the terminus of the 18S RNA is evident. It has been suggested that eukaryotic ribosomes contain a purine-rich region complementary to a pyrimidine sequence occuring in the 5' non-coding region of many eukaryotic messenger molecules (89). However, there is no constancy in the position of the pyrimidine-rich sequence within mRNA molecules. In the 5' non-coding region of other eukaryotic mRNA (satellite tobacco mosaïc virus vesicular stomatites virus) this sequence is lacking. Furthermore, as was discussed above, prokaryotic mRNAs when capped are well translated in eukaryotes. The structural features of eukaryotic mRNA are also found in prokaryotes. Therefore an interaction comparable to that postulated by Shine and Dalgarno for prokaryotes appears unlikely in eukaryotes. The AUG sequence by itself is obviously insufficient to specify the unique site in each message for ribosome attachment. The suggestion that ribosomes and factors simultaneously recognize the capped end of mRNA and the AUG codon is difficult to reconcile with the variation in nucleotide sequence and length of 5' terminal regions.

A scanning mechanism was proposed by Kozak to explain initiation in eukaryotes. This author (6) proposed that a 40S ribosomal subunit with the ternary complex Met-tRNA-IF2-GTP and IF3 binds first near the 5' end of a message irrespective of sequence and subsequently advances until it encounters the first AUG triplet. At this point base pairing between codon and anticodon prevent further movement and, as was discussed above, the 60S subunit joins the complex. Such a scanning mechanism explains the inability to initiate within the interior of a message and the non-rebinding of fragments protected against digestion by 80S ribosomes. However, the difference of protection by 40S and 80S ribosomes could be merely explained by the fact that only 40S ribosomes carry initiation factors, particularly eIF3, which, as a particle, could protect mRNA regions against nuclease digestion. Furthermore, there is increasing evidence that initiation does not obligatorily start with the most proximal AUG of the 5' end of mRNA such as Simian virus 40 VP1 mRNA, Avian sarcoma virus, Sindbis virus RNA, Turnip Yellow mosaïc virus (6,90,91) and more direct evidence to strenghten the hypothesis is needed. Interestingly enough it has been shown that ribosomes can bind to the internal initiation site for coat protein cistron in RNA 3 of Alfalfa mosaïc virus (Pinck, L., Franck, A., and Fritsche, C., personal communication). This cistron near the 3' end of the mRNA is silent in this RNA and is not translated. The failure of a second cistron in an eukaryotic mRNA to be translated does not necessarily arise from an inability of ribosomes to bind at an internal initiation site. However, it appears that ribosomes do bind but that the AUG initiation triplet might not be positioned at the proper ribosomal site.

As noted above for prokaryotes, eukaryotic messages also appear to differ in their requirement for certain initiation factors providing a potential mechanism for regulating translation. Most of the _in vitro_ translation studies on message discrimination have been done in two systems : one, derived from rabbit reticulocytes, in which β-globin mRNA is found to be a much better message than α-globin (92), and the other, obtained from mouse ascites cells, in which encephalo myocardites viral RNA is much more efficient than either mouse cellular or rabbit globin mRNA (93,94). The major findings of these studies are the following :

Message discrimination occurs at initiation,

Discrimination is observed with excess message, when several mRNA's are simultaneously translated. Under these conditions the message which has the highest initiation efficiency outcompetes the other messages for a limiting component of the initiation machinery.

The efficiencies of translation of each message, taken individually, are very similar (92,95).

Message competition can be relieved by adding crude initiation factors eIF4B and, to a lesser extent eIF4A (92,95).

The most straightforward explanation of these phenomena, discussed extensively by Kabat and Chappell (91), is that there exists a limiting amount of messenger discriminatory factor which preferentially binds the efficient messages when presented with an excess of mRNA, thereby shifting the polypeptide output of the system toward those coded by the most efficient mRNAs. Increasing the concentration of discriminatory factor then allows the weaker mRNA's to be translated. The messenger discriminatory factor was identified as eIF4B, with eIF4A acting in a synergistic way but the active protein might be the cap binding protein 24.000 (84). Furthermore, the insensitivity of discrimination to the ribosome concentration has been interpreted to mean that the discriminatory binding occurs between free factor and mRNA molecules. For translation of natural messages, none of the recognized initiation factors are completely dispensable, but differential requirements are observed.

The most obvious consequence of this assumption is that one should be able to isolate the mRNA factor complex, and to demonstrate the different affinities of the factor for low efficiency and high efficiency messages. This has led several research groups to undertake binding studies with eIF4B and various messages. The results from different laboratories vary somewhat : this may be explained by _in vitro_ conditions and the stage of purification of the initiation factors. It has been shown that eIF4B or a contaminant of this factor can bind mRNA with a preference for EMC RNA over globin mRNA (96). In cases where the mRNA initiation factor complex has been observed it yet remains to be shown that the interaction is specific.

In conclusion, our interpretation of the present binding data is that, in physiological buffer, the interaction of mRNA with the competition relief factor may not be strong enough to allow quantitative isolation of the complex. On the other hand, eIF3, which does not relieve competition between α- and β-globin mRNAs, seems to bind mRNA in a stable and specific way. An ability of eIF3 to discriminate among messages has been suggested (97). Binding of mRNA by eIF2 and by the eIF2.GTP.Met-$tRNA_f$ complex has also been reported (98-101). None of these interactions is discriminatory, but they might facilitate the subsequent binding of the discriminatory factor to mRNA.

Translation of various capped cellular and viral messages requires more eIF4B than uncapped picorna virus mRNA (102). Inhibition of cellular protein synthesis when picorna virus infect the cells may be due to the inactivation of eIF4B which is limiting for reading cap mRNA but not for picorna viral mRNA (102).

However, screening of purified initiation factors for cap binding activity suggested that the only component able to interact specifically with the cap was a minor polypeptide that copurified with eIF4B (84). Whatever the factor, the question remains open whether that factor is limiting *in vivo* thus providing a mechanism for differentially controlling expression of mRNA.

CONCLUSION AND PERSPECTIVES

The challenging task of identifying the molecular bases underlying differential expression of both prokaryotic and eukaryotic messenger RNA lies ahead. If one compares the extent of our present knowledge on transcription and translation processes we find that we have much less information on the molecular mechanism possibly regulating the latter mechanism.

In prokaryotes, initiation could be modulated by the stability of the Shine and Dalgarno interaction, and the levels of initiation factors. But all our knowledge of initiation factors and S1 comes from *in vitro* experiments. In the future more physiological experiments will be possible with conditionnal mutants for initiation factors. The cloning of the gene for IF3 in a plasmid and the use of hybrid λ phage carrying *E.coli* structural genes will hopefully yield further information concerning both the role of the factor and the regulation of its synthesis. Purified coupled transcription translation systems using cellular mRNA will certainly also yield essential information, particularly if the rate of synthesis can be increased.

A second feature of translation which needs further development and experimentation is the measurement of reaction kinetics. Rate constants for the various reactions in proposed pathways are required. Finally, the structural studies on ribosomes have not yet reached the degree of precision which is needed to shed new light on precise molecular mechanisms. Ultimately we hope to explain all of the steps of protein synthesis in terms of well-understood chemical forces. For eukaryotes we still need to have precise structural information concerning the localization on ribosomes of the different components of initiation. Now that most of these are defined, the application of techniques developed for bacterial ribosomes are suitable for the study of eukaryotic components and much new data on the structure of mammalian initiation complex can soon be expected.

Experiments are in progress to insert a structural gene such as the β globin coding sequence into a viral capsid gene such as VP1, VP2 or VP3 in SV40 DNA (103). The recombinant genome multiplies efficiently in CV1 monkey Kidney cell culture and is transcribed to yield cytoplasmic polyadenylated hybrid mRNAs containing the β globin sequence. It is thus possible to assess the effect of varying the distance of the β globin coding sequence from the 5' end of the mRNA and studying whether the initiation will start with the initiation triplet of the virus capsid or β globin or other proteins. Another possible promising approach is the setting up of an active yeast cell free system for protein synthesis which might lead to the construction of conditional mutants (104).

It is only by coupling results from genetic and physiological studies with data on molecular mechanisms that we will arrive at a comprehensive understanding of the translational process and its control.

ACKNOWLEDGEMENTS

I am very much indebted to Dr. R. H. Buckingham, Dr. M. Springer, and Dr. Th. Godefroy-Colburn for helpful discussion during the writing of the manuscript. I am grateful to Dr. J.W.B. Hershey for providing his review before publication. I thank H. Costinesco and M. Jouannaud for preparing the manuscript.

This work was supported by the following grants: Centre National de la Recherche Scientifique (G.R. 18) and Délégation Générale à la Recherche Scientifique et Technique (79.7.0158), Ligue Nationale contre le Cancer (Comite de la Seine) and Commissariat à l'Energie Atomique (C.E.A.).

REFERENCES

1. Lodish, H. F. (1974) Nature 251, 385-388.
(1976) Annual Rev. of Biochem. 45, 39-72.

2. Grunberg-Manago, M., and Gros, F. (1977) Initiation mechanisms of protein synthesis, Progress in Nucleic Acid Research and Molecular Biology 20, pp. 209-284, Academic Press, New-York.

3. Grunberg-Manago, M., Buckingham, R. H., Cooperman, B. S., and Hershey, J. W. B. (1978) Structure and function of the translation machinery, in Relation between structure and function in the prokaryotic cell, Symp. Soc. Gen. Microbiol. 28, pp. 27-110.

4. Steitz, J. A. (1979) Genetic signals and nucleotide sequences in messenger RNA, in Biological Regulations and Development 1, pp. 349-369, Plenum, New-York, Goldberger, R. F., ed.

5. Hershey, J. W. B. (1979) Gene expression : translation and the behaviour of proteins, in The translational machinery components and mechanism, vol. 4, in press.

6. Kozak, M. (1978) Cell 15, 1109-1123.

7. Merrick, W. C., Peterson, D. T., Safer, B., Lloyd, M., and Kemper, W. M. (1977) Eukaryotic initiation of protein synthesis, in Gene Expression, Proc. 11th FEBS Meeting 43, pp. 17-26.

8. Godefroy-Colburn, Th., Wolfe, A.D., Dondon, J., Grunberg-Manago, M. Dessen, P., and Pantaloni, D. (1975) J. Mol. Biol. 94, 461-478.

9. Schevitz, R. W., Podjarny, A. D., Krishnamachari, N., Hughes, J. J., Sigler, P. B., and Sussman, J. L. (1979) Nature 278, 188-190.

10. Leon, M., Dondon, J., Labouesse, J., Grunberg-Manago, M., and Buckingham, R. H. (1979) Eur. J. Biochem., in press.

11. Wishnia, A., Boussert, A., Graffe, M., Dessen, P., and Grunberg-Manago, M. (1975) J. Mol. Biol. 93, 499-515.

12. Wishnia, A,, and Boussert, A. S. (1977) J. Mol. Biol. 116, 577-591.

13. Zamir, A., Miskin, R., Vogel, Z., and Elson, D. (1974) in : Methods in Enzymology, Academic Press, vol. 30 Nucl. Acids and Protein Synthesis, pp. 406-426.

14. Hui Bon Hoa, G., Begard, E., Maurel, P., Grunberg-Manago, M., and Douzou, P. (1980) submitted to Biochemistry.

15. Petersen, H.U., Joseph, E., Ullmann, A., and Danchin, A. (1978) J. of Bacteriol. 135, 453-459.

16. Beaudry, P., Begard, E., Douzou, P., and Grunberg-Manago, M. (1978) Eur. J. Biochem. 92, 613-619.

17. Platt, T., and Yanofsky, C. (1975) Proc. Natl. Acad. Sci. USA 72, 2399-2403.

18. Kung, H.F., Eskin, B., Redfield, B., and Weissbach, H., submitted to Arch. of Biochem. Biophys.

19. Hershey, J.W.B., Yanov, J., Johnston, K., and Fakunding, J.L. (1977) Arch. of Biochem. Biophys. 182, 626-638.

20. Howe, J.G., Yanov, J., Meyer, L., Johnston, K., and Hershey, J.W.B. (1978) Arch. of Biochem. Biophys. 191, 813-820.

21. Kung, H., Redfield, B., Treadwell, B.V., Eskin, B., Spears, C. and Weissbach, H. (1977) J. Biol. Chem. 252, 6889-6894.

22. Suryanarayana, T., and Subramanian, A.R. (1977) FEBS Letters 79, 264-268.

23. Brauer, D., and Wittmann-Liebold, B. (1977) FEBS Letters 79, 269-275.

24. Springer, M., Graffe, M., and Hennecke, H. (1977) Proc. Natl. Acad. Sci. USA 74, 3970-3974.

25. Eskin, B., Treadwell, B., Redfield, B., Spears, C., Kung, H., and Weissbach, H. (1978) Arch. Biochem. Biophys. 189, 531-534

26. Kay, A.C., Graffe, M., and Grunberg-Manago, M. (1975) FEBS Letters 58, 112-118

27. Kay, A.C., Graffe, M., and Grunberg-Manago, M. (1976) Biochimie 58, 183-199

28. Petersen, H.U., Grunberg-Manago, M., Roll, Th., and Clark, B.F.C., unpublished results

29. Gualerzi, C., Risuleo, G., and Pon, C.L. (1977) Biochemistry 16, 1684-1689

30. Weiel, J., Hershey, J.W.B., and Levison, S.A. (1978) FEBS Letters 87, 103-106

31. Beaudry, P., Sander, G., Grunberg-Manago, M., and Douzou, P. (1979) Biochemistry 18, 202-207

32. Steitz, J.A. (1969) Nature 224, 957-964

33. Shine, J., and Dalgarno, L. (1974) Proc. Natl. Acad. Sci. USA 71, 1342-1346

34. Mi Jou, W., Haegeman, G., Ysebaert, M., and Fiers, W. (1972) Nature 237, 82-88

35. Steitz, J.A., and Jakes, K. (1975) Proc. Natl. Acad. Sci. USA 72, 4734-4738

36. Dunn, J.J., Buzash-Pollert, E., and Studier, F.W. (1978) Proc. Natl. Acad. Sci. USA 75, 2741-2745

37. Feinstein, S.I., and Altman, S. (1978) Genetics 88, 201-219

38. Fluck, M.M., Salser, W., and Epstein, R.H. (1977) Molec. Gen. Genet. 151, 137-149

39. Ganoza, M.C., Frazer, A.R., and Neilson, T. (1978) Biochemistry 17, 2769-2775

40. Taniguchi, T., and Weissmann, C. (1978a) J. Mol. Biol. 118, 533-565

41. Taniguchi, T., and Weissmann, C. (1978b) Nature 275, 770-772

42. Grunberg-Manago, M. (1979) Frontiers in Bioorganic Chemistry and Molecular Biology, I.U.P.A.C., Pergammon Press, in press.

43. Eckardt, H., and Lührmann, R. (1979) Hoppe Seyler's Z. Physiol. Chem. 360, 253

44. Steitz, J.A., Wahba, A.J., Laughrea, M., and Moore, P.B. (1977) Nucl. Acids Res. 4, 1-15

45. Tal, M., Aviram, M., Kanarek, A., and Weiss, A. (1972) Biochim. Biophys. Acta 281, 381-392

46. Sabol, S., Sillero, M.A.G., Iwasaki, K., and Ochoa,S. (1970) Nature Lond. 228, 1269-1273

47. Vermeer, C., Boon, J., Talens, A., and Bosch, L. (1973) Eur. J. Biochem. 40, 283-293

48. Van Duin, J., and van Knippenberg, P.H. (1974) J. Mol. Biol. 84, 185-195

49. Van Dieijen, G., Vanderlaken, C., van Knippenberg, P., and van Duin, J. (1975) J. Mol. Biol. 93, 351-366

50. Van Dieijen, G., van Knippenberg, P., and van Duin, J. (1976) Eur. J. Biochem. 64, 511-518.

51. Dahlberg, A., and Dahlberg, J. (1975) Proc. Natl. Acad. Sci. USA 72, 2940-2944.

52. Draper, D.E., and Von Hippel, P.H. (1978) J. Mol. Biol. 122, 321-338.

53. Szer, W., Hermoso, J.M., and Boublik, M. (1976) Biochem. Biophys. Res. Commun. 70, 957-964.

54. Laughrea, M., and Moore, P.B. (1977) J. Mol. Biol. 112, 399-421.

55. Van Duin,J., Kurland, C.G., Dondon, J., Grunberg-Manago, M., Branlant, C., and Ebel, J.P. (1976) FEBS Letters 62, 111-114.

56. Cooperman, B.S., Dondon, J., Finelli, J., Grunberg-Manago, M., and Michelson, A.M. (1977) FEBS Letters 76, 59-63.

57. Langberg, S., Kahan, L., Traut, R.R., and Hershey, J.W.B. (1977) J. Mol. Biol. 117, 307-319.

58. Laughrea, M., Dondon, J., and Grunberg-Manago, M. (1978) FEBS Letters 91, 265-268.

59. Pongs, O., Petersen, H.U., Grunberg-Manago, M., Lauka, E., Bald, R., and Stöffler, G., submitted to J. Mol. Biol.

60. Springer, M., Graffe, M., and Grunberg-Manago, M. (1977) Mol. Gen. Genet. 151, 17-26.

61. Hennecke, H., Springer, M., and Böck, A. (1977) Mol. Gen. Genet. 152, 205-210.

62. Springer, M., Graffe, M., and Grunberg-Manago, M. (1979) Molec. Gen. Genet. 169, 337-343.

63. Plumbridge, J., Springer, M., Graffe, M., Gourseau, R., and Grunberg-Manago, M., manuscript in preparation.

64. Smith, A.E. (1977) Gene Expression, Proc. 11th FEBS Meeting 43, pp. 37-46.

65. Shatkin, A.J. (1976) Cell 9, 645-653.

66. Schibler, U., Kelley, D.E., and Perry, R.P. (1977) J. Mol. Biol. 115, 695-714.

67. Brawerman, G. (1976) Prog. Nucl. Acids Res. Mol. Biol. 17, 117-148.

68. Grady, L.J., North, A.B., and Campbell, W.P. (1978) Nucl. Acids Res. 5, 697-712.

69. Steitz, J.A. (1973a) J. Mol. Biol. 73, 1-16.

70. Stallcup, M.R., Sharrock, W.J., and Rabinowitz, J.C. (1976) J. Biol. Chem. 251, 2499-2510.

71. Lodish, H.F. (1970) Nature 226, 705-707.

72. Paterson, B.M., and Rosenberg, M. (1979) Nature 279, 692-696.

73. Rosenberg, M., and Paterson, B.M. (1979) Nature 279, 696-701.

74. Anderson, C.W., Atkins, J.F., and Dunn, J.J. (1976) Proc. Natl. Acad. Sci. USA 73, 2752-2756.

75. Kozak, M., and Shatkin, A.J. (1976) J. Biol. Chem. 251, 4259-4266

76. Legon, S. (1976) J. Mol. Biol. 106, 37-53.

77. Legon, S., Model, P., and Robertson, H.D. (1977) Proc. Natl. Acad. Sci. USA 74, 2692-2696.

78. Steitz, J.A. (1973b) Proc. Natl. Acad. Sci. USA 70, 2605-2609.

79. Kozak, M., and Shatkin, A.J. (1978) Cell 13, 201-212.

80. Schreier, M.H., Erni, B., and Staehelin, T. (1977) J. Mol. Biol. 116, 727-753.

81. Trachsel, H., Erni, B., Schreier, M.H., and Staehelin, T. (1977) J. Mol. Biol. 116, 755-767.

82. Safer, B., Adams, S.L., Kemper, W.M., Berry, K.W., Lloyd, M., and Merrick, W.C. (1976) Proc. Natl. Acad. Sci. USA 73, 2584-2588.

83. Benne, R., Brown-Luedi, M.L., and Hershey, J.W.B. (1978a) J. Biol. Chem. 253, 3070-3077.

84. Sonenberg, N., Morgan, M.A., Merrick, W.C., and Shatkin, A.J. (1978) Proc. Natl. Acad. Sci. USA 75, 4843-4847.

85. Dasgupta, A., Majumdar, A., George, A.D., and Gupta, N.K. (1976) Biochem. Biophys. Res. Commun. 71, 1234-1241.

86. Benne, R., and Hershey, J.W.B. (1978) J. Biol. Chem. 253, 3078-3087.

87. Benne, R., and Hershey, J.W.B. (1976) Proc. Natl. Acad. Sci. USA 73, 3005-3009.

88. Benne, R., Amesz, H., Hershey, J.W.B., and Voorma, H.O. (1979) J. Biol. Chem. 254, in press.

89. Hagenbüchle, O., Santer, M., Steitz, J.A., and Maus, R.J. (1978) Cell 13, 551-563.

90. Shine, J., Czernilofsky, A.P., Friedrich, R., Bishop, J.M., and Goodman, H.M. (1977) Proc. Natl. Acad. Sci. USA 74, 1473-1477.

91. Briand, J.P., Keith, G., and Guilley, H. (1978) Proc. Natl. Acad. Sci. USA 75, 3168-3172.

92. Kabat, D., and Chappell, M.R. (1977) J. Biol. Chem. 252, 2684-2690.

93. Wigle, D.T., and Smith, A.E. (1973) Nature New Biol. 242, 136-140.

94. Blair, G.E., Dahl, H.H.M., Truelsen, E., and Lelong, J.C. (1977) Nature 265, 651-653.

95. Golini, F., Thach, S.S., Birge, C.H., Safer, B., Merrick, W.C., and Thach, R.E. (1976) Proc. Natl. Acad. Sci. USA 73, 3040-3044.

96. Baglioni, C., Simili, M., and Shafritz, D.A. (1978) Nature 275, 240-243.

97. Kennedy, D.S., and Heywood, S.M. (1976) FEBS Letters 72, 314-318.

98. Shafritz, D.A., Weinstein, J.A., Safer, B., Merrick, W.C., Weber, L.A., Hickey, E.D., and Baglioni, C. (1976) Nature 261, 291-294.

99. Kaempfer, R., Hollender, R., Abrams, W.R., and Israeli, R. (1978a) Proc. Natl. Acad. Sci. USA 75, 209-213.

100. Kaempfer, R., Rosen, H., and Israeli, R. (1978b) Proc. Natl. Acad. Sci. USA 75, 650-654.

101. Barrieux, A., and Rosenfeld, M.G. (1978) J. Biol. Chem. 253, 6311-6314.

102. Rose, J.K., Trachsel, H., Leong, K., and Baltimore, D. (1978) Proc. Natl. Acad. Sci. USA 75, 2732-2736.

103. Mulligan, R.C., Howard, B.H., and Berg, P. (1979) Nature 277, 108-114.

104. Gasior, E., Herrera, F., Sadnik, I., Mc Laughlin, C.S., and Moldave, K. (1979) J. Biol. Chem. 254, 3965-3969

RNA·RNA INTERACTIONS DURING POLYPEPTIDE CHAIN INITIATION

JOAN ARGETSINGER STEITZ

Department of Molecular Biophysics and Biochemistry,
Yale University,
New Haven, CT 06510

INTRODUCTION

The history of how we have acquired our current knowledge of the structure and function of ribosomes can be divided into different eras characterized by distinctly different approaches to the problem. In the early 1960's - when the discovery of mRNA was new, the first primary structures of tRNAs were being elucidated, and the genetic code was yet a mystery - the role of RNA molecules in the translational apparatus seemed foremost. Then, with the discovery of many ribosomal proteins and numerous initiation and elongation factors (each with intriguingly different functions), attention in the late 1960's shifted to proteins and protein-nucleic acid interactions. Interest refocused on the RNA components of the ribosome only sometime after 1970, prompted by technological advances allowing determination of extensive RNA (and later DNA) sequences. With the resulting realization that mRNA and rRNA molecules probably assume preferred secondary and tertiary conformations has come an appreciation of the direct contribution of RNA sequences and structures to the synthesis and functioning of ribosomes.

My goal in this article is to update (with as little repetition as possible) a review written two years ago (1) in which I emphasized the central role of RNA·RNA interactions during the initiation of protein synthesis by E. coli ribosomes. At that time evidence for the importance of a direct mRNA·rRNA interaction was primarily biochemical. Today, these data are augmented by elegant studies of a number of initiator region mutants. Likewise, the catalogue of initiation sites bound by E. coli ribosomes has multiplied 2.5-fold, and its analysis reveals some previously unrecognized features. Finally, we have much more information concerning the generalizability to other prokaryotes of mechanisms elucidated for initiation in E. coli; the many studies that have appeared on ribosome-mRNA interactions in eukaryotes convince me that the fundamental recognition process may be quite different in higher cells.

OVERVIEW OF EARLY WORK

Ever since the genetic code revealed that AUG and GUG specify fMet-tRNA (the prokaryotic initiator) as well as internal methionine and valine, respectively, there has been the question of how ribosomes distinguish real initiator triplets from the many such internal codons in every messenger RNA. Early experiments isolated and compared the sequences of ribosome binding sites - the 30 nucleotide-long mRNA regions protected from ribonuclease digestion in a 70S ribosome initiation complex. Disappointingly, few homologies could be identified even when the list of sequenced initiator regions grew to near a dozen. The elusive initiation signal was assumed to be recognized by one or more of the initiation factors or by 30S ribosomal proteins (see ref. 2).

Only in 1974 was it first seriously suggested and a second RNA·RNA interaction (in addition to the obvious $tRNA_f^{Met}$·initiator codon match) was central to ribosome recognition of initiation sites. Shine and Dalgarno (3) sequenced the 3' end of E. coli 16S rRNA and found it to be pyrimidine-rich and complementary to polypurine tracts which preceded mRNA initiator codons known at that time. They proposed that the formation of Watson-Crick base pairs between the mRNA and 16S rRNA allows ribosomes to detect true initiator regions.

Shortly thereafter, biochemical evidence supporting Shine and Dalgarno's hypothesis emerged (4). Namely, a ribosome binding site when recovered from a detergent-dissociated 70S initiation complex was found to be associated with the 50 3'-terminal residues of 16S rRNA, generated by colicin E3 treatment. Characterization of two such mRNA·rRNA complexes (5) confirmed that the predicted regions engaged in intermolecular base pairing, but revealed that the isolated complexes melt at surprisingly low temperatures. The latter experiments emphasized that additional molecular contacts (codon-anticodon and mRNA-initiation factor or -ribosomal protein) must augment and stabilize the initiator region-ribosomal RNA interaction during initiation.

CURRENT RIBOSOME BINDING SITE CATALOGUE

The proposal of Shine and Dalgarno is strongly supported by today's catalogue of ribosome binding sites (Table 1). It shows the following: 1) Each of 74 initiator regions from bacteriophage or E. coli mRNAs contains a sequence complementary to the 3' end of 16S rRNA. This is true

TABLE 1

mRNA INITIATOR REGIONS

RECOGNIZED BY *E. COLI* RIBOSOMES		Reference
R17 A protein	UCC UAG GAG GUU UGA CCU AUG CGA GCU UUU	6
MS2 A protein	UCC UAG GAG GUU UGA CCU GUG CGA GCU UUU	7
MS2, f2 coat	UCA ACC GGA GUU UGA AGC AUG GCU UCC AAC	8,9
R17 coat	UCA ACC GGG GUU UGA AGC AUG GCU UCU AAC	6
MS2 lysis	UGC AAG GUC UCC UAA AAG AUG GAA ACC CGA	8,10
MS2, R17 replicase	CAA ACA UGA GGA UUA CCC AUG UCG AAG ACA	8,6
Qβ A protein	CUG AGU AUA AGA GGA CAU AUG CCU AAA UUA	11
Qβ coat	CUU UGG GUC AAU UUG AUC AUG GCA AAA UUA	12,14
Qβ replicase	UAA CUA AGG AUG AAA UGC AUG UCU AAG ACA	13,14
λ Cro	AUG UAC UAA GGA GGU UGU AUG GAA CAA CGC	15
λ o	CUA UCA ACA GGA GUC AUU AUG ACA AAU ACA	16
λ cI	CGG UGA UAG AUU UAA CGU AUG AGC ACA AAA	17,18
λ cII	GCG UUG AUA AGC UUA CAU AUG GUU CGU GCA	19
λ cII (nonfunctional)	←GUC UAG AAA UAC UUA CAU AUG GUU CGU GCA	20
	IS2	
φX174 A	AAU CUU GGA GGC UUU UUU AUG GUU CGU UCU	21
φX174 A*	UUG CUG GAG GCC UCC ACU AUG AAA UCG CGU	21
φX174 B	AGG UCU AGG AGC UAA AGA AUG GAA CAA CUC	22
φX174 C	GUG GAC UGC UGG CGG AAA AUG AGA AAA UUC	21
φX174 D	ACU AAU AGG UAA GAA AUC AUG AGU CAA GUU	23
φX174 E	GCG UUG AGG CUU GCG UUU AUG GUA CGC UGG	21
φX174 F	CUU ACU UGA GGA UAA AUU AUG UCU AAU AUU	21,22
φX174 G	CUG CUU AGG AGU UUA AUC AUG UUU CAG ACU	24
φX174 H	ACU UAA GUG AGG UGA UUU AUG UUU GGU GCU	21
φX174 J	UGC GGA AGG AGU GAU GUA AUG UCU AAA GGU	23
φX174 K	CAC UUU CGG AUA UUU CUG AUG AGU CGA AAA	21
G4 A	AUC AAA CGG AGG CUU UUC AUG UUU AAA GUA	25
G4 A*	UCU UGG AGG AGU UCA ACU AUG AAG UCU CGA	25
G4 B	AGG CCU AGG GAA UAA AGA AUG GAA CAA UUC	25
G4 K	CAA GUA CGG AUA UUU CUG AUG AAA CCA AAA	25
G4 C	GUG GAC UGC UGG UGG AAA AUG AGG AAA UUC	25
G4 D	ACC ACA AAG GAA ACU GAA AUG UCU AAA UCA	25
G4 E	CUA UCG AGG CUU GCG UAU AUG GAA CAC UGG	25
G4 J	CUU UUA AGG AGU UAU GUA AUG AAA AAA UCA	25
G4 F	CUA UUU AAG GAU ACA AAA AUG UCU AAC GUU	25
G4 G	AGC CAA AAG GAC UAA CAU AUG UUC CAG AAA	25
G4 H	UGA AAU AAG GAU UAU CCU AUG UUU GGC UCU	25
f1 V	A AGG UAA UUC ACA AUG AUU AAA GUU	27
fd I	GGA UUG GGA UAA AUA AAU AUG GCU GUU UAU	26
fd II	UUA UCA ACC GGG GUA CAU AUG AUU GAC AUG	26
fd III	UUG GAG AUU UUC AAC GUG AAA AAA UUA	26
fd IV, f1 unidentified	UUA AAA AAG GUA AUU CAA AUG AAA UUG UUA	26,27
fd V	CGC AUA AGG UAA UUC AAA AUG AUU AAA GUU	26
fd VI	UAA UAA GGA GUC UUA AUC AUG CCA GUU CUU	26
fd VII	UCG UUC CGG CUA AGU AAC AUG GAG CAG GUC	26
fd VIII, f1 coat	UUU AAU GGA AAC UUC CUC AUG AAA AAG UCU	26,27
fd IX	AAU CGC UGG GGG UCA AAG AUG AGU GUU UUA	26
fd X	GCA UUU GAG GGG GAU UCA AUG AAU AAU UAU	26
T7 0.3 site a	UGC ACG AGG UAA CAC AAG AUG GCU AUG UCU	28
T7 0.3 site b	CGA GGA GGA UGA AGA GUA AUG U	28
T4 rII B	C CUA AUA AGG AAA AUU AUG UAC AAU	29
thrA	GGU AAC GAG GUA ACA ACC AUG CGA GUG UUG	30

thr leader	ACA GAG UAC ACA ACA UCC *AUG* AAA CGC AUU	30
phe leader	UCA CUU AAG GAA ACA AAC *AUG* AAA CAC AUA	33
pheA	UAA CAA AAA GGC AAC ACU *AUG* ACA UCG GAA	33
lipoprotein	AUC UAG AGG GUA UUA AUA *AUG* AAA GCU ACU	34
recA	GGC AUG ACA GGA GUA AAA *AUG* GCU AUC G	35
trpC	GCG GCA CGA GGG UAA AUG *AUG* CAA ACC GUU	36
trpL-C hybrid	GUA AAA AGG GUA UCG ACG *AUG* CAA ACC GUU	36
trpB	AUA UUA AGG AAA GGA ACA *AUG* ACA ACA UUA	37
trp leader	GUA AAA AGG GUA UCG ACA *AUG* AAA GCA AUU	38
trpE	CAA AAU UAG AGA AUA ACA *AUG* CAA ACA CAA	38
trpA	AGC ACG AGG GGA AAU CUG *AUG* GAA CGC UAC	39
araB	UUU GGA UGG AGU GAA ACG *AUG* GCG AUU GCA	40
galE	AGC CUA AUG GAG CGA AUU *AUG* AGA GUU CUG	41
galT	CCC GAU UAA GGA ACG ACC *AUG* ACG CAA UUU	42
lacI	CAA UUC AGG GUG GUG AAU *GUG* AAA CCA GUA	43
lacZ	UUC ACA CAG GAA ACA GCU *AUG* ACC AUG AUU	44
β-lactamase	UAU UGA AAA AGG AAG AGU *AUG* AGU AUU CAA	45
ribosomal L14	ACA UUA GCG GAG CCU AAA *AUG* AUC CAA GAA	46
ribosomal S12	AAA ACC AGG AGC UAU UUA *AUG* GCA ACA GUU	46
ribosomal L11	CAA CUU GAG GAA UUU AUA *AUG* GCU AAG AAA	47
ribosomal L1	GUA GUG GAG GAC UAA GAA *AUG* GCU AAA CUG	47
ribosomal L10	CAU CCA GGA GCA AAG CUA *AUG* GCU UUA AAU	47
ribosomal L7/L12	UAU UCA GGA ACA AUU UAA *AUG* UCU AUC ACU	47
RNA polymerase β	AGC GAG CUG AGG AAC CCU *AUG* GUU UAC UCC	47
	16S 3' end $_{HO}$A U U C C U C C A U A G$_{5'}$	3
RECOGNIZED BY OTHER RIBOSOMES		
S. typhimurium hisL	GCA UUC AUC GGA AUU UUU *AUG* ACA CGC GUU	31
S. typhimurium hisG	GAU AAA GAG GAA CGC AGA *AUG* UUA GAC AAC	31
S. typhimurium hisC	CGC CCU CAA GGA GCA AGC *AUG* AGC ACU GAA	32
S. typhimurium trpB	GCA UAU UAA GGA AAA AAC *AUG* ACA ACA CUU	48
P. aeruginosa phage PP7 coat	UUU UCA UGG GAG GAC AUA *AUG* UCC AAA ACC	49
S. marescens trpG	UGC CAA GGA GGU GUU CUG *AUG* GCC GAU	48

In the cases where ribosome-binding site sequences from two closely-related phages are identical, they are not duplicated. Initiator triplets are italicized. mRNA regions of contiguous complementarity (including G·U base pairs) to the 3' end of 16S rRNA are underlined. In the 5 non-coli regions where the 3'-end sequence of the corresponding 16S rRNA has not been determined (S. typhimurium and S. marescens) complementarity only to the conserved CUCC is indicated by dashed underlining.

even of the many initiators which directly abut or overlap the previous cistron; the complementary nucleotides simply reside within the coding region of the upstream gene. 2) The complementarity to 16S can be as short as 3 residues or as long as 9 residues; the average is 4.8. 3) The complementarity region is centered an average of 9.8 nucleotides 5' to the initiator AUG or GUG; but the intervening distance varies considerably in individual cases. (Note that the numbers for the average length and position of the complementarity have not changed by lengthening the list, and that we still have only two GUG initiators.) 4) When more than one AUG or

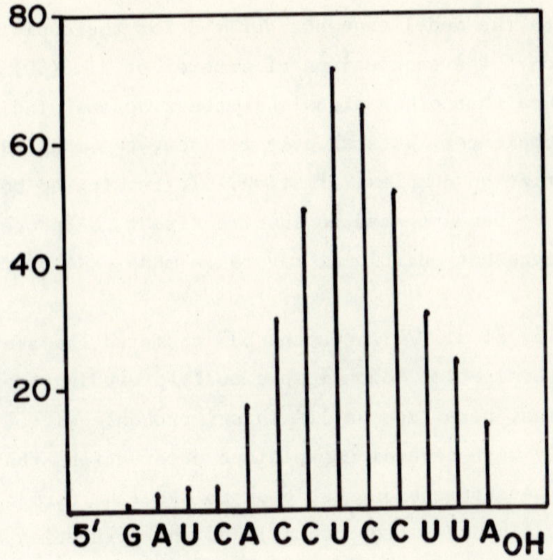

Figure 1. Utilization of residues at the 3' end of 16S rRNA in pairing with mRNAs. The data are compiled from the 74 sequences in Table 1.

GUG is present in the 30-nucleotide ribosome-protected region, that preceded by the longest, most appropriately positioned complementarity is invariably used for initiation. 5) If RNA secondary structures can be drawn for the region around the initiator codon, they are usually predicted to be unstable under physiological conditions; an open conformation is clearly essential if the mRNA is to engage in intermolecular base pairing when binding to the ribosome. 6) Finally, note from the plot in Fig. 1 that the rRNA region most highly utilized for the predicted mRNA·rRNA interaction is the CUCC sequence that appears just in from the extreme 3' end of the 16S molecule. That these residues make an essential contribution to initiator region recognition by ribosomes has been elegantly confirmed by Taniguchi and Weissmann (83); as predicted, an oligonucleotide complementary to ACCUCCUU (see Table 2) inhibits fMet-tRNA binding to 70S ribosomes when directed by Qβ RNA but not by AUG trinucleotide!

A recent computer analysis of 32 of the sequences listed in Table 1 reveals some intriguing additional properties (50). When aligned by the AUG or GUG triplet, homology between these sites is detected not only in the region complementary to the 3' end of 16S rRNA (where there is a huge peak, as expected) but also along the entire sequence recovered as a ribosome-protected piece! The low but significant degree of homology suggests that proper intermolecular contacts involving any portion of the 30-nucleotide segment can contribute favorably to the mRNA-ribosome interaction. Most satisfyingly, the one message in the entire catalogue which is known to function without a Shine-Dalgarno sequence, the λ cI mRNA, exhibits good agreement with the model sequence derived for the coding region 3' to the initiator codon. The conclusions of Scherer et al. (50) therefore reemphasize the idea that other signals in the ribosome binding site region not only supplement, but can even replace, the mRNA·rRNA interaction during initiation complex formation. It remains to be seen whether any of these will be demonstrated to involve direct RNA-protein contacts or whether they represent additional favorable mRNA·rRNA or mRNA·tRNA interactions.

Finally, scrutiny of all 74 sequences has prompted the realization that UAG is never the termination codon most closely preceding an initiator triplet (J.F. Atkins, pers. commun.). This improbable situation correlates tantalizingly with several unexplained observations that link together UGA and UAA suppression, antibody to RF2 (the termination factor which recognizes UGA and UAA stop codons) and the initiation of certain proteins (see ref. 10). Perhaps the role of termination signals in the initiation process (whether or not the cistron is the first on the mRNA) should be reexamined.

MUTANTS IN RIBOSOME BINDING SITES

A growing roster of initiation mutants underscores the importance of RNA·RNA interactions in the initiation process on E. coli ribosomes. In fact, all mutants that have emerged so far exert their effects directly on: 1) the $tRNA_f^{Met}$·mRNA interaction, 2) the 16S rRNA·mRNA interaction, or 3) appear to act indirectly via the mRNA secondary structure itself — likewise a perturbation of RNA·RNA interactions.

The Shine-Dalgarno hypothesis is elegantly supported by the recent analysis of several mutants affecting translation of the bacteriophage T7 gene 0.3 protein (51). Most interesting is a transition mutation localized by Dunn et al. (51) in the middle of the region complementary to 16S rRNA; it is predicted to disrupt the 5-base pair mRNA·rRNA interaction as follows:

$$\begin{matrix} GAGGU \\ CUCCA \end{matrix} \rightarrow \begin{matrix} GA^{A}GU \\ CU_{C}CA \end{matrix}$$

In vivo expression of the 0.3 gene is reduced to about 10% and ribosome binding, likewise inefficient, is shifted to a region centered on an AUG triplet appearing five codons downstream in the translated portion of the message. (This observation dovetails nicely with a rule enunciated above: if more than one AUG occurs in a ribosome protected region, that utilized is preceded by the longest most appropriately positioned complementarity to 16S rRNA.) One "revertant" of this gene 0.3 mutant changes not the mutated but an adjacent nucleotide, thereby cleverly allowing a shifted but contiguous 4 base pair mRNA•rRNA interaction: $\begin{matrix} GGAG^{U} \\ CCUC_{C} \end{matrix}$; in vivo function and the original ribosome protection pattern are restored.

The most plentiful ribosome binding site mutations are those that alter the codon-anticodon interaction at the initiator triplet itself. These are interesting in that their phenotypes vary - so far, in a way that can be reasonably correlated with the strength of the predicted mRNA•rRNA interaction for each affected initiator region. 1) In the T7 0.3 gene, an AUG → ACG change reduces translation over 10-fold, but ribosome binding to the appropriate region remains high (51); recall that this initiation site exhibits a strong complementarity to the 16S 3' end (28). 2) An AUG → AUA transition in the initiator triplet of the T4 rIIB cistron results in temperature-sensitive gene expression - no function at 37° C but low levels (10-15%) at 25° C (29). Presumably a two base pair codon•anticodon interaction suffices at low temperature when there is a strong (6 base pair) complementarity to 16S rRNA. 3) In the bacteriophage Qβ coat protein cistron an in vitro-contrived AUG → AUA switch has a more profound effect (52); ribosome binding in vitro is reduced to 10% that of the wild-type level, but here the complement to 16S rRNA is only 4 base pairs (including one G•U pair). Taniguchi and Weissmann (52) also noted that in vitro initiation complex formation could be increased by altering the first base of the second codon of the Qβ coat cistron; this they interpret in terms of a potential 4 base codon•anticodon interaction with the initiator tRNA.

In addition to the previous list (1) of situations where mRNA secondary structures appeared to act as negative effectors of initiation under various physiological conditions, we now have two new examples of this phenomenon. 1) The first concerns the op3 mutant of bacteriophage f2, which was isolated as a nonsense (UGA) mutant in a fourth overlapping RNA phage cistron encoding a lysis function (10). The altered nucleotide in op3 happens to be located in the second codon of the f2 replicase gene and results in inefficient translation of the replicase polypeptide both in vivo and in vitro. The best (but not the only) explanation for this effect is that the nucleotide change lengthens and strengthens a marginally stable hairpin loop structure whose formation would render both the replicase initiator codon and the preceding Shine-Dalgarno sequence

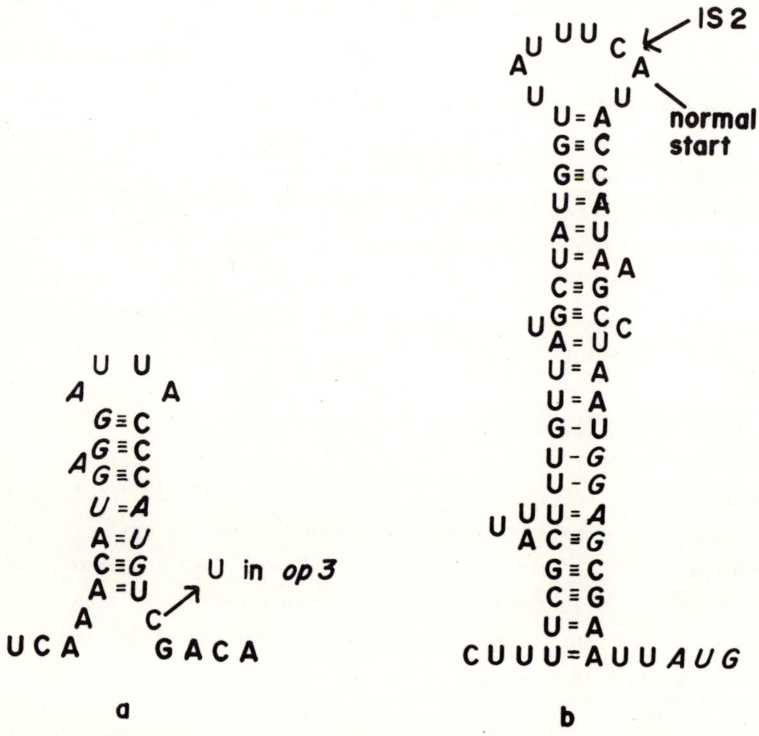

Figure 2. RNA secondary structures surrounding initiator regions for the f2 replicase cistron (a) and the galE gene (b). The sequences are from ref. 6 for (a) and refs. 54 and 55 for (b). Stabilities calculated according to ref. 75 for the two structures are -6.1 kcal/mol for (a) and -20.7 kcal/mol for (b). The initiator AUGs and regions complementary to the 3' end of 16S rRNA are indicated by italics.

inaccessible to ribosomes (see Fig. 2a). 2) In the second new situation, which is not really a mutant but was genetically contrived, inefficient translation can likewise be explained by our ability to draw an extensive RNA secondary structure that would interfere with ribosome attachment. Normally the three genes of the gal operon are expressed in about equal amounts. However, when gal is transcribed from a λ prophage promoter located thousands of nucleotides upstream, galE protein appears in much lower amounts than the galT and galK products (53). Inspection of the DNA sequence preceding the RNA start site (54) reveals that the gal promoter region is largely complementary to the sequence within the beginning of the message (see Fig. 2b). Formation of such a structure, which cannot be assumed by either the normal gal mRNA or when an IS2 insertion has occurred at the site indicated by the arrow (55), could explain why galE is abnormally translated from the monster message and regains activity upon IS2 insertion (M. Gottesman, pers. commun.).

mRNA·rRNA PAIRING IN OTHER PROKARYOTES

Very early on, Shine and Dalgarno (56) examined 16S rRNAs from bacteria other than E. coli to ask whether different 3'-terminal sequences might explain previously observed species-specific translational effects (57). This (56) and subsequent studies (58) revealed that Ps. aeruginosa and B. stearothermophilus 16S RNAs possess the highly utilized CUCC sequence, and that C. crecentus 16S has pyUCC; but interestingly this stretch does not always appear at the same exact distance from the extreme 3' terminus of the rRNA molecule. The portion of the B. stearothermophilus 3' end which is nonidentical to E. coli 16S can explain thermophilic ribosome binding to several non-initiator regions in the RNA phage RNAs (58). However, the few true non-coli mRNA initiator regions that have now been sequenced (see bottom of Table 1) all possess the complement to the common CUCC sequence. This strongly suggests that these other bacteria utilize the same mRNA·rRNA base pairing mechanism as E. coli.

Another set of organisms whose 16S rRNAs have recently been scrutinized are the methanogens. Here Woese and his collaboratores (59) have argued that these unusual bacteria may constitute a third form of life, distinct from both prokaryotes and eukaryotes. Certainly, methanogenic metabolism differs from that of the two standard groups in many persuasive ways. However, the fact that the 3' termini of the 16S rRNAs from ten different species of methanogens is $GAUCACCUCCU_{OH}$ (59) - identical to E. coli except at the very end (see Table 1) - suggests to me that the methanogenic bacteria fit nicely into the prokaryotic roster (at least as regards their translational apparatus).

Finally, one might ask about the 3' termini of the small subunit ribosomal RNAs in cellular organelles such as chloroplasts or mitochondria, which have long been attributed prokaryotic origin. The first sequence corresponding to the end of such an RNA - from Zea mays chloroplasts - has just appeared (60). It is $GAUCACCUCCUUU_{OH}$ - again identical to E. coli 16S except for a terminal A → U switch! This sequence can now be added to the already rather convincing list of features shared by the translational systems of eukaryotic organelles and bacterial cells.

INITIATION IN EUKARYOTES

The most striking feature of the 3' termini of 18S rRNAs from a wide range of eukaryotic species (61,62) is that none possesses the CCUCC sequence most highly utilized by bacterial ribosomes for binding mRNAs (see Fig. 1). This finding seems even more significant when one learns that the bacterial 16S and the eukaryotic 18S rRNAs are otherwise highly homologous in the sequences of their 50 3'-terminal nucleotides; as shown in Fig. 3, all fold into a common secondary structure with two dimethyl-A residues (conserved in both pro- and eukaryotes) appearing in the bend of the hairpin loop! The fact that the very sequence which interacts with mRNA during bacterial initiation is deleted from the 18S rRNAs strongly suggests that eukaryotic ribosomes (perhaps deliberately) have developed some alternative scheme for recognizing their own mRNAs.

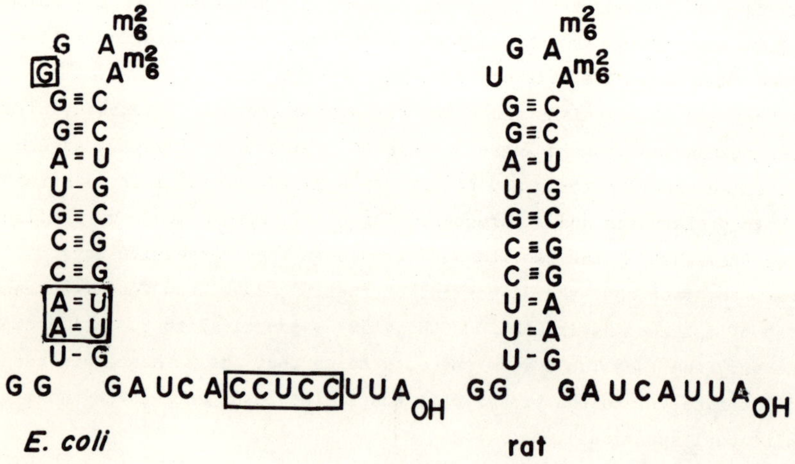

Figure 3. Homology at the 3' ends of prokaryotic 16S (3) and eukaryotic 18S rRNAs (61,62). Differences in the E. coli sequence are boxed in. Other eukaryotes (B. mori, D. discoideum and wheat germ) show a few base substitutions compared to rat, but the secondary structure and the lack of the CCUCC sequence near the 3' end are conserved (62; P. Gage, pers. commun.). The polypurine tract on the 3' side of the helix is the 18S region that has most often been suggested to be involved in base pairing with eukaryotic mRNAs; note that such interaction would require unfolding the stable helical region pictured.

Both compilation of a growing list of eukaryotic mRNA initiator regions and studies on the mechanism of initiation complex formation by eukaryotic ribosomes support the above contention. [See Kozak (63) for a superb up-to-date review of eukaryotic initiation.] Of course, each new eukaryotic mRNA sequence has been eagerly scrutinized for the presence of a region complementary to the distinctive 3' terminus of 18S rRNA; the results can at best be described as disappointing. Although complementarities do appear [and are often seductively long (64)] their position relative to the initiator codon is exceedingly variable (see ref. 61 and 65). Thus, on a purely theoretical basis, it is difficult to argue for a eukaryotic mRNA•rRNA interaction comparable to that in prokaryotes.

Moreover, an experiment which should give a positive result if the 40S ribosomal subunit contains an rRNA region poised to interact via base pairing with mRNA turns out negative. Specifically, if one prepares a total pancreatic RNase digest of any large RNA molecule and mixes it with bacterial ribosomes, the 30S subunit will selectively and *quantitatively* bind all oligonucleotides in the digest which possess a 6-nucleotide or longer complementarity to the 3' end of 16S rRNA (see Table 2). This works for Qβ (66) or MS2 (J. Weliky and J. Steitz, unpublished observations) RNA with E. coli ribosomes. (See Table 2.) However, when the analogous experiment is performed using washed Ehrlich ascites cell ribosomes (J. Weliky, J. Slonczewski, and J. Steitz, unpublished observations), no selective binding of oligonucleotides from either a T_1 or pancreatic RNase digest of MS2 RNA to the small ribosomal subunit occurs. [Note that the portion of the 18S sequence most often implicated in an mRNA•rRNA interaction should bind T_1 oligonucleotides (see Fig. 3); there are several with extensively complementary sequences in the MS2 catalogue (67).] Although the result may be negative for other reasons, these observations support the growing notion that mRNA•rRNA interactions (at least involving the 3' end of 18S rRNA) do not contribute importantly to eukaryotic initiation.

How then, do eukaryotic ribosomes recognize initiator regions? One unique characteristic of eukaryotic messengers is that it is always the AUG closest to the 5' end that is utilized for initiation (see ref. 63); the only exceptions to this rule I know of are the SV40 late mRNAs (68; W. Fiers and S. Weissman, pers. commun.). Usually, even when several cistrons are present on an mRNA, only the AUG proximal to the 5' end is translationally active. Elegant genetic manipulations of the beginning of the yeast cytochrome c cistron confirm that the first AUG is obligatorily

TABLE 2

OLIGONUCLEOTIDES BOUND BY E. COLI RIBOSOMES

from Qβ RNA (66)	from MS2 RNA
A G A G G A G G U	A G G A G G U
A G G G G G U	C G G A G G C
A A G G A G C	

$_{HO}$ A U U C C U C C A C U A G 16S rRNA 3' end

used (69). Moreover, eukaryotic mRNAs often possess a 5' cap moiety, whose presence (on the messages that have them) markedly enhances ribosome attachment. Perhaps not surprisingly, then, even prokaryotic mRNAs can be converted into good templates for eukaryotic ribosomes by the addition of a 5' cap (70).

Based on the above pecularities of eukaryotic mRNAs and on analyses of the effects of drugs on initiation in the wheat germ system, Marilyn Kozak (63) has proposed that 40S subunits attach to the 5' end of a message and move along scanning for the first AUG. Recall here that another difference between eukaryotic and prokaryotic initiation is that the charged $tRNA_f^{Met}$ *obligatorily* binds to 40S ribosomes before the mRNA, whereas either the initiator tRNA or the mRNA can attach first to the E. coli 30S subunit (see refs. 1 and 63); the codon-anticodon interaction could thereby serve to fix the mobile eukaryotic ribosome at the first AUG. Observations that additional ribosomes can pile up behind an 80S complex formed at the first AUG are also readily interpreted in terms of this idea (71,72,73). Very recently, an elegant series of experiments using circular mRNA molecules has provided overwhelming evidence for the attach-and-slide model (74); whereas E. coli ribosomes bind avidly to such templates, circular RNAs are ignored by eukaryotic ribosomes until a single nick is introduced. Use of a sliding mechanism and the apparent absence of an mRNA·rRNA interaction in eukaryotic initiation may seem revolutionary, but distinct molecular mechanisms should perhaps be anticipated in cells that carry out transcription and translation in different compartments (eukaryotes) as opposed to cells where these two processes are intimately coupled (prokaryotes).

The realization that there are significant differences in the way eukaryotes and prokaryotes recognize their mRNAs prompts some skepticism about the ability of eukaryotic genes to function when cloned in prokaryotic organisms. Presuming that a cloned DNA is transcribed in a bacterial cell, how likely is it that the eukaryotic initiator region will compete favorably with host messages and be translated by prokaryotic ribosomes? In general, unless a eukaryotic gene has been deliberately fused onto an E. coli ribosome binding site sequence, protein production is appallingly low (see refs. 76 and 77). Several notable exceptions have emerged in studies where the desired chaemeric plasmid was selected on the basis of its ability to express a particular function in E. coli; these are all experiments where either a cDNA or a genomic region from a lower eukaryotic (both lacking intervening sequences) were cloned (78,79,80,81). To my knowledge in only one instance has the sequence of the functional eukaryotic initiator region been determined (R. Davis, pers. commun.); in the yeast his3 gene, the initiator AUG happens to be preceded (with 5 residues in between) by a four-base complementarity to the 3' end of 16S rRNA (well within the acceptable limits set by Table 1). It will be amusing to learn how many other dually functional genes exhibit similar properties.

UNRESOLVED PROBLEMS

It is distressing to find that my list of questions concerning initiation on E. coli ribosomes has not changed much in two years time. We still do not understand the relative contributions to translational efficiency of the two central RNA•RNA interactions versus other ribosome-mRNA contacts. More specifically, we would like to know exactly how the activity of a particular initiator region correlates both with the length (or strength) of its complementarity to 16S rRNA and with the distance between this region and the AUG. Presumably, intensive mutant analysis now underway will soon provide direct answers to these particular questions. Meanwhile, not even the relatively simple determination of when during the course of elongation the mRNA•rRNA bonds are broken has been accomplished.

Most perplexing is the growing realization that some degree of complementarity to 16S rRNA may be *necessary*, but certainly is not *sufficient* to define an initiation locus in an E. coli mRNA. Why (now that extensive sequences are becoming available) can we identify many initiator codons appropriately juxtaposed to complementary regions and yet detect no translational initiation? Equally puzzling are the recent observations of

Roberts et al. (82), who find that changing the distance between the 5' end of a hybrid λ Cro message and its *intact* ribosome binding site by only a few nucleotides can drastically alter the level of in vivo Cro expression; here we may be stepping into the mire of mRNA stability, but other explanations are possible. I suspect the answers to these more profound questions lurk in the secondary and tertiary structures of mRNA molecules. Until we begin to understand how the 3D shape of an RNA is determined by its sequence, the details of the specific contacts between messengers and other cellular macromolecules may well remain mysterious.

ACKNOWLEDGEMENTS

I am most grateful to Joan Weliky for her assistance in many spheres, and to Margaret Rosa, Nancy Batter and Jim Spilsbury for their help with the manuscript. Unpublished work from my laboratory was supported by grants AI 10243 from the NIH and NSF grant PCM 74-01136.

REFERENCES

1. Steitz, J. A. (1979) in Biological Regulation and Development (Goldberger, R. F., ed.), vol. 1., pp. 349-399, Plenum Press, New York.
2. Steitz, J. A. (1975) in RNA Phages (Zinder, N. D., ed.), pp. 319-352, Cold Spring Harbor Laboratory, New York.
3. Shine, J. & Dalgarno, L. (1974) Proc. Nat. Acad. Sci. USA 71, 1342-1346.
4. Steitz, J. A. & Jakes, K. (1975) Proc. Nat. Acad. Sci. USA 72, 4734-4738.
5. Steitz, J. A. & Steege, D. A. (1977) J. Mol. Biol. 114, 545-558.
6. Steitz, J. A. (1969) Nature 224, 957-964.
7. Fiers, W., Contreras, R., Duerinck, F., Haegeman, G., Merregaert, J., Min Jou, W., Raeymaekers, A., Volckaert, G., Ysebaert, M., Van de Kerckhove, J., Nolf, F. & Van Montagu, M. (1975) Nature 256, 273-278.
8. Min Jou, W., Haegeman, G., Ysebaert, M. & Fiers, W. (1972) Nature 237, 82-88.
9. Gupta, S. L., Chen, J., Schaefer, L., Lengyel, P. & Weissman, S. M. (1970) Biochem. Biophys. Res. Commun. 39, 883-888.
10. Atkins, J. F., Steitz, J. A., Anderson, C. W. & Model, P. (1979) Cell, in press.
11. Staples, D. H., Hindley, J., Billeter, M. A. & Weissmann, C. (1971) Nature New Biol. 234, 202-204.
12. Hindley, J. & Staples, D. H. (1969) Nature 224, 964-967.
13. Staples, D. H. & Hindley, J. (1971) Nature New Biol. 234, 211-212.

14. Steitz, J. A. (1972) Nature New Biol. 236, 71-75.

15. Steege, D. A. (1977) J. Mol. Biol. 114, 559-567.

16. Scherer, G. (1978) Nucleic Acids Res. 5, 3141-3155.

17. Ptashne, M., Backman, K., Humayun, M. Z., Jeffrey, A., Maurer, R. Meyer, B. & Sauer, T. (1976) Science 194, 156-161.

18. Walz, A., Pirrotta, V. & Ineichen, K. (1976) Nature 262, 665-669.

19. Schwarz, E., Scherer, G., Hobom, G. & Kössel, H. (1978) Nature 272, 410-414.

20. Rosenberg, M., Court, D., Shimatake, H., Brady, C. & Wulff, D. L. (1978) Nature 272, 414-423.

21. Sanger, F., Coulson, A. R., Friedmann, T., Air, G. M., Barrell, B. G., Brown, N. L., Fiddes, J. C., Hutchinson III, C. A., Slocombe, P. M. & Smith, M. (1978) J. Mol. Biol. 125, 225-246.

22. Ravetch, J. V., Model, P. & Robertson, H. D. (1977) Nature 265, 698-702.

23. Barrell, B. G., Air, G. M. & Hutchinson III, C. A. (1976) Nature 264, 34-41.

24. Robertson, H. D., Barrell, B. G., Weith, H. L. & Donelson, J. E. (1973) Nature New Biol. 241, 38-40.

25. Godson, G. N., Barrell, B. G., Staden, R. & Fiddes, J. C. (1978) Nature 276, 236-247.

26. Schaller, H., Beck, E. & Takanami, M. (1978) in The Single-Stranded DNA Phages (Denhardt, D. T., Dressler, D. & Ray, D. S., eds.), pp. 139-163, Cold Spring Harbor Laboratory, New York.

27. Pieczenik, G., Model, P. & Robertson, H. D. (1974) J. Mol. Biol. 90, 191-214.

28. Steitz, J. A. & Bryan, R. A. (1977) J. Mol. Biol. 114, 527-543.

29. Belin, D., Hedgpeth, J., Selzer, G. B. & Epstein, P. H. (1979) Proc. Nat. Acad. Sci. USA 76, 700-704.

30. Gardner, J. F. (1979) Proc. Nat. Acad. Sci. USA 76, 1706-1710.

31. Barnes, W. M. (1978) Proc. Nat. Acad. Sci. USA 75, 4281-4285.

32. Barnes, W. M. & Kohno, T., personal communication.

33. Zurawski, G., Brown, K., Killingly, D. & Yanofsky, C. (1978) Proc. Nat. Acad. Sci. USA 75, 4271-4275.

34. Pirtle, R. M., Pirtle, I. L., Inouye, M. (1978) Proc. Nat. Acad. Sci. USA 75, 2190-2194.

35. Sancar, A. and Rupp, W. D. (1979) Nucleic Acids Res., submitted.

36. Christie, G. & Platt, T. (1979) J. Mol. Biol., submitted.

37. Platt, T. & Pinkham, J. (1979) in preparation.

38. Bertrand, K., Korn, L., Lee, F., Platt, T., Squires, C. L., Squires, C. and Yanofsky, C. (1975) Science 189. 22-26.

39. Platt, T. & Yanofsky, C. (1975) Proc. Nat. Acad. Sci. USA 72, 2399-2403.

40. Lee, N. & Carbon, J. (1977) Proc. Nat. Acad. Sci. USA 74, 49-53.

41. Musso, R., de Crombrugghe, B., Pastan, I., Sklar, J., Yot, P. & Weissman, S. (1974) Proc. Nat. Acad. Sci. USA 71, 4940-4944.

42. Grindley, N. D. F. (1978) Cell 13, 419-426.

43. Steege, D. A. (1977) Proc. Nat. Acad. Sci. USA 74, 4163-4167.

44. Maizels, N. (1974) Nature 249, 647-649.

45. Sutcliffe, J. G. (1978) Proc. Nat. Acad. Sci. USA 75, 3737-3741.

46. Post, L. E., Arfsten, A. E., Reusser, F. & Nomura, M. (1978) Cell 15, 215-229.

47. Post, L. E., Strycharz, G. D. Nomura, M., Lewis, H. & Dennis, P. L. (1979) Proc. Nat. Acad. Sci. USA 76, 1697-1701.

48. Selker, E. & Yanofsky, C. (1979) J. Mol. Biol. 130, 135-143.

49. Bassel, B. A. & Mills, D. R. (1979) Nucleic Acids Res. 6, 2003-2016.

50. Scherer, G. F. E., Walkinshaw, M. D., Arnott, S. & Morré, D. J. (1979) in preparation.

51. Dunn, J. J., Buzash-Pollert, E. & Studier, F. W. (1978) Proc. Nat. Acad. Sci. USA 75, 2741-2745.

52. Taniguchi, T. & Weissmann, C. (1978) J. Mol. Biol. 118, 533-565.

53. Merril, C., Gottesman, M., Court, D. & Adhya, S. (1978) J. Mol. Biol. 118, 241-245.

54. Musso, R. E., Di Lauro, R., Rosenberg, M. & de Crombrugghe, B. (1977) Proc. Nat. Acad. Sci. USA 74, 106-110.

55. Ghosal, D., Sommer, H. & Saedler, H. (1979) Nucleic Acids Res. 6, 1111-1122.

56. Shine, J. & Dalgarno, L. (1975) Nature 254, 34-38.

57. Stallcup, M. R., Sharrock, W. J. & Rabinowitz, J. C. (1975) J. Biol. Chem. 251, 2499-2510.

58. Sprague, K. U., Steitz, J. A., Grenley, R. M. & Stocking, C. E. (1977) Nature 267, 462-465.

59. Fox, G. E., Magrum, L. J., Balch, W. E., Wolfe, R. S. & Woese, C. R. (1977) Proc. Nat. Acad. Sci. USA 74, 4537-4541.

60. Schwarz, Z. & Kössel, H. (1979) Nature 279, 520-522.

61. Hagenbüchle, O., Santer, M., Steitz, J. A. & Mans, R. (1978) Cell 13, 551-563.

62. Alberty, H., Raba, M. & Gross, H. J. (1978) Nucleic Acids Res. 5, 425-434.

63. Kozak, M. (1978) Cell 15, 1109-1123.

64. Ziff, E. B. & Evans, R. M. (1978) Cell 15, 1463-1475.

65. Both, G. W. (1979) FEBS Letts. 101, 220-224.

66. Taniguchi, T. & Weissmann, C. (1979) J. Mol. Biol. 128, 481-500.

67. Fiers, W., Contreras, R., Duerinck, F., Haegeman, G., Iserentant, D., Merregaert, J., Min Jou, W., Molemans, F., Raeymaekers, A., Van den Berghe, A., Volckaert, G. & Ysebaert, M. (1976) Nature 260, 500-507.

68. Ghosh, P. K., Reddy, V. B., Swinscoe, J., Choudary, P. V., Lebowitz, P. & Weissman, S. M. (1978) J. Biol. Chem. 253, 3643-3647.

69. Sherman, F. & Stewart, J. (1975) in Organization and Expression of the Eukaryotic Genome (Bernardi, G. & Gros, F., eds.), pp. 175-191, American Elsevier, New York.

70. Paterson, B. M. & Rosenberg, M. (1979) Nature 279, 692-701.

71. Filipowicz, W. & Haenni, A.-L. (1979) Proc. Nat. Acad. Sci. USA, in press.

72. Ahlquist, P., Dasgupta, R., Shih, D. S., Zimman, D. & Kaesberg, P. (1979) submitted.

73. Legon, S. (1979) J. Mol. Biol., in press.

74. Kozak, M. (1979) Nature, in press.

75. Gralla, J. & Crothers, D. M. (1973) J. Mol. Biol. 73, 497-510.

76. Fraser, T. H. & Bruce, B. J. (1978) Proc. Nat. Acad. Sci. USA 75, 5936-5940.

77. Mercereau-Puijalon, Royal, A., Cami, B., Garapin, A., Krust, A., Gannon, F. & Kourilsky, P. (1978) Nature 275, 505-510.

78. Struhl, K. & Davis, R. (1977) Proc. Nat. Acad. Sci. USA 74, 5255-5259.

79. Ratzkin, B. & Carbon, J. (1977) Proc. Nat. Acad. Sci. USA 74, 487-491.

80. Vapnek, D., Hautala, J. A., Jacobson, J. W., Giles, N. H. & Kushner, S. (1977) Proc. Nat. Acad. Sci. USA 74, 3508-3512.

81. Chang, A. C. Y., Nunberg, J. A., Kaufman, R. J., Erlich, H. A., Schimke, R. T. & Cohen, S. (1978) Nature 275, 617-624.

82. Roberts, T. M., Kacich, R. & Ptashne, M. (1979) Proc. Nat. Acad. Sci. 76, 760-764.

83. Taniguchi, T. & Weissmann, C. (1978) Nature 275, 770-772.

THE TOPOGRAPHY OF tRNA BINDING SITES ON THE RIBOSOME

JAMES OFENGAND

Roche Institute of Molecular Biology,
Nutley, NJ 07110

INTRODUCTION

Although the two site model (1) for the interaction of tRNA with the ribosome during protein synthesis has ably served to rationalize the accumulated data for the past 15 years, surprisingly little is known about its molecular workings. For example, the way in which codon-anticodon base pairing triggers binding of aminoacyl-tRNA (AA-tRNA) to the ribosome, the molecular mechanism of the translocation process, and the existence of other possibly more transient sites, are still only objects for speculation and hypothesis. Even the location of the amino-acyl (A) and peptidyl (P) sites on the ribosome is largely unknown. This review will be concerned mainly with the latter question, namely, with those experiments that try to define in topographic terms the location of the A and P sites on the ribosome and their relationships to each other. This information will be needed before the problem of molecular mechanism can be tackled. The question of other sites, such as the postulated initial AA-tRNA binding site of Lake (2), the A-like site involved in stringent factor-tRNA recognition (3), and the possible existence of an initiator tRNA binding site distinct from the P site will also be considered.

Topographic questions of this kind are well suited to analysis by affinity labelling, a method which can provide an unambiguous determination of close contact points between two macromolecules. This method was first used in the ribosome field by Bochkareva _et al._ (4), Pellegrini _et al._ (5), and Czernilofsky and Kuechler (6) in their pioneering studies of the peptidyl transferase center. It has subsequently been applied to ribosomes in many ways and has been reviewed before from varying aspects (7-11). Here we will emphasize the studies dealing specifically with the location of the various parts of the tRNA molecule on the ribosome. The results will be correlated with the existing data on structural organization as obtained by electron and immunoelectron microscopy and used to outline the functional domains for tRNA.

GENERAL CONSIDERATIONS

Since the length of each arm of the tRNA molecule is approximately one-third that of the diameter of the ribosome, (Fig. 1), localization of

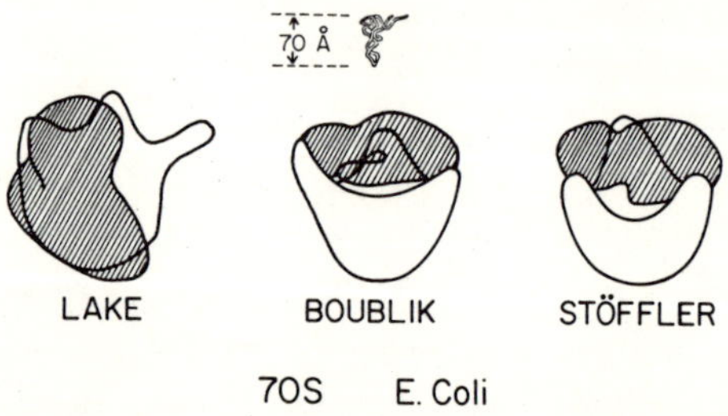

Figure 1. (11) Structure of the *E. coli* ribosome and tRNA to the same scale. The tRNA structure is the crystal structure of yeast $tRNA^{Phe}$ (12). The ribosome structures are from Stoffler and Wittmann (13), Lake (14), and Boublik *et al*. (15).

the tRNA considered as a whole cannot be done with any degree of precision. For a higher resolution analysis, it is useful to consider the tRNA as consisting of three parts, namely an aminoacyl end, an anticodon, and a central region loosely defined as that part of the tRNA where the two arms of the L-shaped structure come together. It can be approximated quite well as everything except the anticodon loop and stem and the amino acid loop and stem. There is a functional rationale to this division as well, since the amino acid end and anticodon together provide the translational capacity, while the central region is probably occupied with ribosomal fixation and movement from the A to the P site.

The anticodons of tRNAs in both A and P sites are close together. Energy transfer measurements between the Y base of yeast $tRNA^{Phe}$ and a proflavine substituted at the Y base position, in two tRNAs bound to the same ribosome, show these two residues to be 14-24 Å apart (16). In other experiments, irradiation at 320 nm crosslinked the Y base of $tRNA^{Phe}$ in both A and P sites to poly U. Moreover, tRNA crosslinked to poly U in the A site could still be translocated to the P site (11). This shows that the tRNA anticodon is closely associated with mRNA codons at both A and P sites, and that these are likely to be adjacent.

Clearly the two amino acids must be close enough for peptide bond formation. The potential flexibility of the CCA end of each tRNA would allow a maximum distance of ca. 50Å (12), but the available evidence suggests that the ends are not that far apart (see Table 1 and Fig. 2 below). Therefore, although we will attempt to separate the ribosomal components near to the anticodon and aminoacyl end into those which are part of the A and P sites, respectively, as well as the existing data will allow, the assignments are not critically important, given the level of resolution available. On the other hand, the central region of the tRNA is the part expected to show the maximum movement on translocation from A site to P site so that a distinction in the ribosomal location of this part of the A and P sites should not only be possible, but should be the most revealing in terms of the relative topography of the two sites.

We shall assume throughout that the crystal structure of yeast $tRNA^{Phe}$ corresponds to the solution structure for all tRNAs (12), and furthermore that there is no change on binding to the ribosome. This latter statement is in detail certainly not true (see article by B.F.C. Clark, this volume) but is likely to approximate the truth if only physical proximity to the ribosome is of concern rather than the detailed conformation and tertiary structural interactions of the tRNA. The only structural evidence available to date that bears on this point is that the $^{4}S_{8}$-C_{13} UV-induced crosslink which requires a precise orientation of at least these two residues does not interfere with A site binding, translocation, or peptide bond formation (reviewed in 17), has no effect on induction of (p)ppGpp synthesis (18) and can be formed <u>in situ</u> at normal rates by irradiation of both A and P site tRNA-ribosome complexes (J. Ofengand, unpublished results).

A second assumption which we make, knowing that it also is not completely true, is that affinity labelling and immunoelectron microscopy (IEM) of ribosomal proteins, the technique used for localization of the proteins on the ribosome, detect the same region of a protein. If the protein is elongated, as many are (multiple antigenic sites will be noted in Figs. 2 and 3), the affinity labelled site and the antigenic site may be far apart. Given this situation, a localization becomes more convincing when at least one antigenic site for each of a set of affinity labelled proteins are observed to cluster in a single region, and this principle has been used below in assigning functional domains.

Unfortunately a similar localization of ribosomal RNA on the ribosome does not exist, except as noted below, and can so far only be deduced indirectly by the IEM localization of proteins known to associate with specific regions of the RNA (see 13 for an attempt along these lines). Thus affinity labels which react primarily with ribosomal RNA so far provide limited topographic information. A way to circumvent this problem by the AIM (affinity immunoelectron microscopy) technique was recently described (19). This method allows direct visualization of the crosslinking site by IEM, limited only by the distance between the antigenic site and the crosslink, and by the inherent limitations of the IEM technique.

Although this review will be mainly concerned with affinity labelling results since they give direct information about contact regions, it should be noted that there is a large amount of additional information, based on indirect evidence of various kinds which also deals with the same subject as the topic of this review. These mostly earlier studies used such techniques as specific chemical, enzymatic, or immunological blocking of the binding sites, deletion or addition of specific ribosomal proteins, or protection by bound tRNA from an inactivating process (reviewed in 13,20-22; see also 23-26). However, these methods were not usually able to distinguish between a direct action of the reagent on the binding site from an indirect effect due to a modification elsewhere, and therefore were subject to alternative interpretations. Correlation of the results of these earlier studies with the newer affinity labelling data can be found in numerous reviews (7,9,13,20-22) and will not be repeated here.

THE AMINOACYL END

Historically, the aminoacyl end of tRNA was the first to be studied because of the ease with which it could be activated. The α-amino group of the aminoacyl moiety provides a single, readily acylatable group, the probe can be made radioactive by the use of amino acid, and the functionality of the tRNA at least in aminoacylation is guaranteed since only those molecules are detected. The ease of cleavage of the aminoacyl ester bond to the tRNA after fixation of the activated amino acid to the ribosome makes characterization of the crosslinked ribosomal components more straightforward. Cleavage of the aminoacyl ester bond also eliminates a potential problem with some tRNAs which is the concomitant attachment of the affinity label to the aliphatic amino group of 3-(3-amino-3-carboxypropyl)uridine (acp^3U) which is found at position 47 in several E. coli tRNAs. Although crosslinking to the ribosome can occur in this situation (see below), it will not be detected since the radioactive amino acid tag is removed from the tRNA.

Specificity Of Crosslinking

The N-acylated AA-tRNAs, being similar to peptidyl-tRNA, bind preferentially to the P site in the absence of factors. They cannot interact with EFTu because of the absence of a free α-amino group (see article by H. Weissbach, this volume) and therefore are not bound enzymatically to the A site. Many workers have assumed therefore that non-enzymatic binding, especially of N-acylaminoacyl-tRNAs, is P site specific, but this need not be so. AA-tRNA and even acyl AA-tRNA can be bound to the A site at sufficiently high Mg^{++} concentrations, especially if deacylated tRNA is present to block the P site and the ribosomes are present in large excess (27-29). Deacylated tRNA can be present inadvertently due either to partial deacylation or failure to achieve complete aminoacylation initially. Also the use of unfractionated tRNA instead of a purified tRNA species provides many other unacylated tRNA species which can compete for the same P site codons, individually poorly but collectively rather well.

In view of this, P site specific binding must be shown directly. The usual way has been to demonstrate that the crosslinked tRNA was still capable of peptide bond formation, either with another AA-tRNA in the A site (30-32), or with puromycin (33-35). Most earlier studies with chemical affinity labels which attempted the former control were unfortunately complicated by an inability to know whether crosslinking preceded peptide bond formation or vice versa (5,36-38), and use of puromycin as a control was equally complicated (39). Blockage of the crosslinking reaction by prior reaction with puromycin has also been used (34,40). It is not adequate to show only that non-covalently bound tRNA can react with puromycin as ribosome-tRNA complexes cannot be assumed to be homogeneous, and usually the fraction of complexes crosslinked is less than the fraction of functionally unreactive complexes.

It is more difficult to demonstrate that crosslinking has occurred from the A site since the usual test for A site binding, EFG-dependent reaction with puromycin, can only be used before the crosslinking event has occurred. Another test for A site affinity labelling from the aminoacyl end that has been applied is tetracycline inhibition of crosslinking (37,41), but this is not reliable as tetracycline may perturb the ribosome at the P site as well. The dependence of crosslinking on EFTu and

GTP (32,42) is only applicable if the α-amino group of AA-tRNA is not blocked by the affinity probe. So far, the only example in aminoacyl end studies is the use of Lys-tRNA substituted at the ε-amino group. This derivative is recognized by EFTu because of its free α-amino group, and is bound specifically to the A site (32). Alternatively, N-acylated AA-tRNAs originally in the P site have been placed in the A site by peptide bond formation with an AA-tRNA. If only the AA-tRNA is radioactive, subsequent crosslinking of radioactivity to the ribosome must have been due to peptide bond formation before crosslinking, and this has been taken as proof of A site occupancy (30,31,35). Both of the last two tests may fail if some EFG is present as a contaminant, since translocation to the P site may occur (32) before crosslinking does. A true A site reaction is best verified, therefore, as EFTu and GTP-dependent crosslinking which is altered by translocation before crosslinking (32,43).

Initiation factor IF2-dependent crosslinking of derivatives of fMet-tRNA can be considered as evidence for crosslinking to a functional I site. The I site is defined here as the site occupied by initiator tRNA or its analog AcPhe-tRNA (44) when bound to 30S or 70S ribosomes in an IF2-dependent manner. Two examples of such crosslinking are known (45,46), one of which (45) was puromycin reactive as well. They are usually considered as P site labels, but we shall consider them to be at the I site, in order to examine the degree of congruence between the aminoacyl parts of the I and P sites.

The 50S Subunit Domain

Affinity probes from the aminoacyl end of tRNA reacted almost exclusively with the 50S subunit (Table 1). A wide variety of both chemically and photochemically reactive reagents, which also vary in length from one C atom to ca. 11 C atom equivalents were used. Despite this, there is a strong uniformity of results. Six proteins, L2, L11, L15, L16, L18, L27 and the 3'-2/3 of 23S RNA are the major sites labelled in all cases. The 23S RNA was crosslinked from the P or I site with reagents 2d,7,8a and 8b. In the only case carefully studied, reagent 2d, crosslinking to RNA did not occur from the A site (32). The major product with many of the other reagents was also 23S RNA. However, in these cases, it was not clear from which tRNA site crosslinking took place. Regarding the proteins, L2, L11, and L18 are definitely in the P site, with the probable inclusion of L15 and L27. L16 is likely to be only in the A site, although this conclusion rests on only one set of experiments (41). The A site definitely contains L27, and possibly L11, L18 and L2 also. The I site contains L15, L27 and possibly L2.

TABLE 1. Ribosomal Components Labelled From The Aminoacyl End Of tRNA With Affinity And Photoaffinity Probes[a]

No.	Reagent[b,c]	AA-tRNA	Site[d]	Labelled Components Major	Labelled Components Minor	Reference
1a	-COOØNO_2	Phe	P/A	L27	L2,14-16	6,39,47
1b		Met[f]	I	L15,27	-	45,47,48
2a	$\text{-COCH}_2\text{Br}$	Phe	P/A	L2,27	L14-17	5,36,37,41
			A?	L16	-	41
			P/A	23S RNA	-	49
2b		Met[f]	P,I?	L2	L16,14, 20,27	33
2c		(Gly)_n-Phe n=1-16	P/A	L2,26/27,32/33 24 (with increasing n)		50
2d		Lys	P	23S RNA	L2,13-15	32
			A	L27	-	32
3	$\text{-COCH}_2\text{I}$	Phe	P/A	23S RNA	-	38,51
4	$\text{-COC(N}_2\text{)CO}_2\text{Et}$	Phe	P/A	23S RNA	-	52
5	$\text{-COCH}_2\text{CH}_2\text{COØ}$	Phe	P/A	23S RNA[e]	-	53
6	$\text{-COØ(NO}_2\text{)NN}_2$	Phe	P/A	23S RNA	50S Protein	54
				S3,7,14 (with 30S only)	-	54
7	$\text{-COCH}_2\text{CH}_2\text{ØCOØ}$	Phe	P	23S RNA[e]	-	34
8a	$\text{-COCH(NHtBoc)-CH}_2\text{ØNN}_2$	Phe	P	23S RNA[e]	-	35,55
8b		Met[f]	I	23S RNA[e]	-	46
8c		(Gly)_n-Phe n=2,4	P/A	23S RNA[e]	-	35
8d		Phe-Phe	P/A	23S RNA[e]	-	35
9	$\text{-COCH}_2\text{NHØ(NO}_2\text{)-NN}_2$	Phe	P,A?	L11,18	L2,5,27, 33	31
10	$\text{-COCH}_2\text{ØOØ(NO}_2\text{)-NN}_2$	Phe	P,A?	L11,18	L2,5,27, 33	30

TABLE 1. (continued)

No.	Reagent[b,c]	AA-tRNA	Site[d]	Labelled Components Major	Labelled Components Minor	Reference
11	$-COCH_2CH_2CH_2ØN-(CH_2CH_2Cl)_2$	Phe	P/A	23S RNA	-	4,56

(a) approximately in order of increasing length except as noted for 2c and 8c. All reactions are with α-NH_2 except 2d which is with ε-NH_2.

(b) Ø; phenyl. All substituents are p- except NO_2 in 6,9,10 which is m- to N_3.

(c) __; atom displaced by NH_2 except in 5 and 7 where simple addition occurs.

(d) P/A, P and/or A. I, initiation factor dependent site. ?, likely but not certain.

(e) 3'-2/3 (18S) fragment.

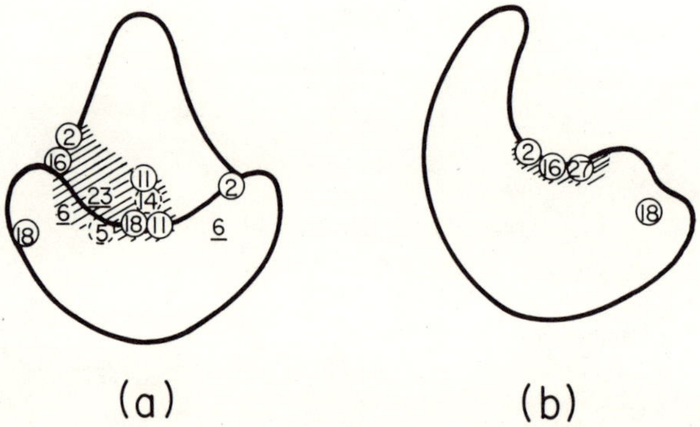

Figure 2. (11) Proteins on the surface of the E. coli 50S subunit that are in the vicinity of the aminoacyl end of tRNA. (a) front view. (b) left side view. Location of the proteins and overall shape of the particle is according to Brimacombe et al. (57). Circled numbers represent proteins near to the peptidyl transferase center according to the data of Table 1. Solid circles are major proteins and dashed circles are minor proteins. The criteria for selection are given in the text. Underlined numbers denote proteins affinity-labelled by puromycin or an analogue (58-60). The shaded area denotes the putative peptidyl transferase center.

Five of the six major proteins have been localized on the 50S subunit by immune electron microscopy (Fig. 2). The Stöffler model has been used in this figure since most of the IEM identification of proteins has been done by these workers. It agrees at least in a general sense with the Boublik version. The location of only L27 is known for the Lake model. The figure shows in addition the positions of L5 and L14. They are the only two minor proteins occurring more than once in the Table that have been mapped. Proteins L6 and L23 have also been identified as part of the peptidyl transferase center by affinity labeling with puromycin or analogs (see article by B. Cooperman, this volume). The location of the 18S fragment of the 23S RNA is not known although in keeping with its being a frequent target of affinity probes, many of the proteins with which it is associated have antigenic sites within the shaded region of Fig. 2 (57).

The major proteins labelled by affinity analogs of AA-tRNA cluster in two groups; L16, L27 and L11, L18. L2 apparently spans both groups. While it is tempting to consider these two groups as the A and P sites, respectively, this cannot be the case since L27 is a definite I site protein as well as an A site protein. Furthermore, the two regions are separated by some 70Å, too long to be spanned even by a pair of extended CCA ends. In fact, just because this distance is so large, at least some of the antigenic sites localized here cannot correspond in position to their affinity labelling site. They must be examples of the non-correspondence of antigenic sites with crosslinking sites that was alluded to earlier.

Comparison of the I-site and P-site labelling shows only that the 18S piece of 23S RNA is shared by both sites. L15 has not yet been mapped and L27, a definite I site protein, has not been definitively placed at the P site. Nevertheless, there is no reason at this time to believe that the aminoacyl domain of the I site, if it exists, is in a different locale than the same domain of the P site. In any case, the aminoacyl end in both sites can react with puromycin.

There is no direct information available on the location of the stringent factor-tRNA site on the ribosome other than its A-like properties (3). L11 is required in order to synthesize (p)ppGpp (61) but not to bind stringent factor to the ribosome (62), while L10 is required for stringent factor binding (63). There is also the report (62) that a particular subset of ribosomal proteins which weakly stimulate (p)ppGpp synthesis in place of the intact ribosome and tRNA map in the head region of the 30S subunit. It will be interesting to apply the direct affinity labelling approach to this site, and steps in that direction are being taken (18).

The 30S Subunit Domain

Since the peptidyl transferase activity of the ribosome is localized to the 50S subunit (64), it is at first surprising to find any part of the 30S subunit within the aminoacyl end domain. However, since the tRNA probably lies between the two subunits in view of its protection from nuclease attack upon binding to the ribosome (65), it follows that the 30S subunit must also be close to the aminoacyl ends. Proteins S3, S7, and S14 were found to be in this domain by affinity labeling (54, Table 1) but the significance of non-enzymatic labeling of isolated 30S subunits is lessened by the observation that only 50S subunits were labelled when the tRNA was bound to 70S ribosomes. However, more recently, Cooperman has observed labelling of the 30S proteins S18 (66) and S14 (60) with puromycin or an analog, and has even localized the S14 crosslinking site by the AIM technique using antibody to puromycin (67). The site corresponds to that previously identified with anti-S14 (68). In this case, the IEM technique and affinity labelling method identified the same topographic site.

In the Stöffler model, (see Fig. 3), determinants of S3,S7,S14, and S18 all cluster on the left side of the head region. This is one of the contact areas of the 30S with the shaded region of the 50S that corresponds to the aminoacyl domain (Figs. 1 and 2). Labeling of these proteins with aminoacyl analogs is therefore reasonable. The situation is less clear in the case of the Lake model since only S7 appears within range from either the A or P sites (68). However, close contact of the aminoacyl end to the 30S subunit is predicted in another situation. In Lake's detailed model for the initial AA-tRNA binding site, the R site (2), S14 but not S3, S7, or S18 should be within range of the aminoacyl end.

The most striking distinction between the R and A site in Lake's model is the location of the aminoacyl end. Since the R site is simply EFTu-dependent binding in the presence of GDPCP, the non-hydrolyzable analog of GTP, it should be a simple matter to test for its existence with any one of the analogs of AA-tRNA described in Table 1. Thus far, only one negative report has appeared (32). Although no differences in A and R site labelling were found, the total labelling was low, so that a final conclusion could not be drawn in this case. With probes at other sites in the tRNA, labelling of both A and putative R sites has been obtained, but the identifications have not yet been made (43; also see below).

Stereospecificity Of The Aminoacyl End

This topic has been reviewed recently (69-72) so that only the subject of ribosomal specificity for the 2'- or 3'- aminoacyl end of tRNA will be considered here. Aminoacyl tRNA can occur in one of two isomeric forms, that is with its amino acid attached either to the 2' or 3' hydroxyl of the 3'-terminal ribose. Normally, the rate of isomerization is very rapid, so that a single isomeric form cannot be isolated. By replacing one of the OH groups with H or NH_2, or by opening the furanose ring between the 2' and 3' C atom, isomerization can be blocked, and the resulting isomeric AA-tRNAs can be used to test for stereospecificity at either the A or P site.

The first experiments along this line used small fragments from the aminoacyl end, and showed that the A site strongly preferred the 3'-isomer (73,74). Many subsequent experiments with intact modified AA-tRNAs have confirmed this. The companion question, namely, the specificity for the donor site, has not yet been answered. All of the non-isomerizable acetylaminoacyl-tRNAs tested inhibited the binding of control acetylaminoacyl-tRNA to the P site, but none were active as peptidyl donors. Presumably, the modifications to the 3'-ribose needed to block isomerization also interfere with function at the donor site.

Does the specificity of the A site for the 3'-isomer of AA-tRNA have any meaning? EFTu can recognize both isomers, and place both in the A site, although the 2'-isomer is preferred in both reactions (72). If the preference for one of a pair of fixed isomers is an indication of specificity, it is possible that with an unmodified AA-tRNA able to freely isomerize, the process of EFTu complexation fixes the tRNA in the 2'-position in the A site. Transacylation on the ribosome must then be postulated in order to provide an active peptide acceptor. It has been proposed that the hydrolysis of GTP controls this process by releasing EFTu from the ribosome and from its interaction with the aminoacyl end of tRNA (69), but a logical reason from this mechanism was obscure. Recently, it has been suggested that this model would allow time for proofreading, since with such a mechanism, the amino acid would not be correctly positioned for peptide bond formation until cognate codon-anticodon interaction had triggered GTP hydrolysis and EFTu release (72). In the absence of such a system, the α-amino group of incorrect AA-tRNAs transiently bound to the A site is thought by these workers to be sufficiently reactive to rather frequently form peptide bonds with peptidyl-tRNA in the P site.

III THE ANTICODON

Except for one recent crosslinking study in which the tRNA anticodon was directly attached to the ribosome (75), all of our information about the site of codon-anticodon interaction comes from affinity labelling studies with analogs of messenger RNA. Only 30S subunits were labelled, in keeping with the fact that the 30S subunit, but not the 50S, is known to associate with mRNA (76).

Specificity Of Crosslinking

It is more difficult to ascertain that an mRNA analog was crosslinked at the codon binding site (decoding site) than it was for the aminoacyl end above, as similarly well developed criteria are not available. Three tests for specificity have been applied in addition to the obvious one of dependence on the presence of the reactive group and/or light for activation. The first most general one is the inhibition of crosslinking by competition with another mRNA (77-80). This eliminates random reaction from solution but does not distinguish between binding to the decoding site and some other polynucleotide-binding site. A more specific test for location in the decoding site that has been used is the specific binding of the cognate tRNA to ribosome-mRNA covalent complexes freed of non-covalently attached mRNA (78-86). For this test to be meaningful, however, two criteria need to be met. First, the extent of binding to control ribosomes crosslinked in a mock reaction or preferably with a different heterologous mRNA codon must be taken into account. Second, the moles of tRNA bound must be comparable to the moles of codon crosslinked, if any real test of specificity is to be obtained from such an experiment. This has only been approached once so far (78). The best way to test for specificity is to determine the dependence of crosslinking on the presence of the cognate, versus non-cognate, tRNAs in the reaction mixture. A strong dependence of crosslinking on tRNA has only been achieved in two cases (see Table 2).

Localization of the codon binding site to the P or A site has also been attempted by experiments in which tRNA was bound to the mRNA-ribosome covalent complex in what was then deduced to be either the P or A site (78,81-84,86). Despite claims to the contrary, none of these experiments were successful because the ratio of tRNA bound to codons crosslinked was usually 0.1-0.2 (81-84,86), and not greater than 0.5 (78). Consequently, no information was available about the greater part of the covalently-bound codon, since homogeneity of these complexes could not be assumed and in most cases more than one site was crosslinked.

The 30S Domain

The derivatives of oligo- and polynucleotides used to label the decoding site are listed in Table 2, together with the ribosomal components reacted. After the polynucleotides, oligonucleotides are grouped according to their reactive site. The 5'-end activated (on the 5'-phosphate) oligonucleotides are first, followed by the 3'-terminal probes (on the base), and then the ribose modified compounds. Despite the wide variety of reactive groups, the use of both chemical and photochemical affinity labels, and the different stereochemistry of the probes, the list of ribosomal components labelled is quite self-consistent. It seems clear from this comparison that the decoding site on the 30S subunit

TABLE 2. Ribosomal Components Labelled Near The Decoding Site By mRNA Analogs

No.	Reagent[a]	Cognate tRNA Dependence[b]	Site	Labelled Components[c] Major	Minor	Reference
1	poly(U)	+	P/A	S1	16S RNA	87-89
2	poly(^{4}S)		P/A	S1,18,21; 16S RNA		77,90,91
3	$BrCH_2CONHØ$-pApUpG		P/A	S4,18[d] (S4,12,18)	S21 (S11,13, 21)	81,82
4	$BrCH_2CONHØ$-pUpGpA		P/A	S18 (S4,18)		84
5	GpUpUpU-5-($NHCOCH_2I$)		A?	S1,18		78
6	ApUpGpU-5-($NHCOCH_2Br$)		P/A	S1	S18	86
7	UpUpUpU-2'-(NHCOØCOCHO)	±	P/A	(16S RNA; segment I)		85,92
8	$(Ap)_7$A-2',3'-O-(>CHØN(CH_3)CH_2CH_2Cl)		P/A	16S RNA	30S protein	79
9	$(Ap)_n$A-2',3'-C-(>$NNHCOCH_2NHCOØN_3$) n=5-7	+	P/A	S3,5; 16S RNA		80

(a) Ø, phenyl. All substituents para.
(b) Crosslinking required the presence of the cognate tRNA.
(c) All reactions with 70S ribosomes except those in ( ) which were done only with 30S.
(d) Crosslinked amino acid is Cys-10 (83).

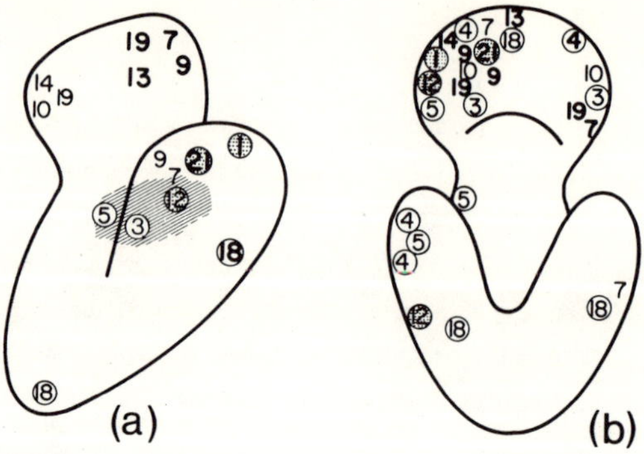

Figure 3. (11) Proteins on the surface of the E. coli 30S subunit that are in the vicinity of mRNA or the 3' segment of 16S RNA. Location of the proteins and overall shape of the particles were derived by IEM according to (a) Lake (68) and Winkelmann and Kahan (93) and (b) Brimacombe et al. (57). Circled numbers locate proteins near to mRNA according to the data of Table 2. Only major proteins have been considered. Plain numbers represent proteins associated with the 3' third of the 16S RNA on the basis of protein-RNA binding studies (94), analysis of ribonucleoprotein fragments (95,96) and both chemical and photochemical crosslinking (97-99). See also (13,57) for reviews. Stippled circles represent proteins with both properties. Boldface type signifies proteins on the surface toward the viewer and light type, those on the surface away from the viewer. The approximate location of the m_2^6A residues near the 3' end of the 16S RNA is shown by shading (100).

contains S1, S3, S4, S5, S12, S18, S21, and segment I of 16S RNA. Two proteins, S11 and S13, are not included in this list as they were minor proteins found only once, and only with 30S subunits.

These proteins, which mark the decoding site of the 30S subunit, have been localized by IEM (Fig. 3). They are shown inside circles on the figure, and can be clearly seen to cluster in both models. The Boublik model (101) of the 30S is similar to the Lake model, but is not shown since no protein localizations have been done. The Stöffler model (Fig. 3b) which has more identified proteins also illustrates well the problem of multiple antigenic sites. Initially, it appeared as if the circled proteins were scattered throughout the particle. However, upon closer inspection, there is at least one antigenic site for each of the circled proteins in a cluster at the upper left crest. In the Lake model, the cluster appears to be localized on and around the platform. This cluster region should mark the decoding site.

The location of segment I, approximately residues 430-475, can so far only be approximated by the location of those proteins thought to be associated with it, such as S4, and possibly S13 and S17 (57). These proteins are found in the same general region as the other decoding site proteins, in agreement with the labelling results using reagent 7 (see Table 2).

In all of the above cases, the decoding site was identified by virtue of its codon binding properties. The tRNA anticodon binding property of the decoding site has also been exploited. By a fortuitious set of circumstances, direct crosslinking of the anticodon of tRNA to the ribosome can be induced by irradiation at suitable wavelengths (40,75), and study of this system has led to some important findings about the tRNA anticodon domain (43). The pertinent facts are briefly set out below, and in detail elsewhere (75).

1. Irradiation at 310-325 nm crosslinked unmodified AcVal-$tRNA_1^{Val}$ of *E. coli* to the ribosome with single-hit kinetics and in up to 70% yield.

2. The crosslinking reaction was dependent on the presence of the cognate mRNA codons, and took place in the P site since the crosslinked tRNA was still puromycin-reactive.

3. The same Val-tRNA, placed in the A site by EFTu, did not crosslink. Therefore, the reaction was P site specific.

4. The site of crosslinking in the tRNA was the 5'-anticodon base, 5-carboxymethoxyuridine in the case of *E. coli* $tRNA_1^{Val}$ or $tRNA_1^{Ser}$, or 5-methoxyuridine in the case of *B. subtilis* $tRNA^{Val}$ or $tRNA^{Thr}$.

5. Crosslinking was exclusively to the 16S RNA.

6. The crosslinked base of the 16S RNA was in the 3'-third of the molecule (102).

7. The crosslink was directly between the two RNA species. The presence of either protein (102) or mRNA (J. Ofengand, unpublished results) spacers has been ruled out.

8. The structure of the crosslink is probably that of a cyclobutane dimer between the 5'-anticodon base and a pyrimidine in the 16S RNA because the crosslink is rapidly split by low doses of short wavelength UV light to regenerate the original reactants (43). This property is one of the characteristic features of Pyd-Pyd cyclobutane dimers that distinguishes it from other photochemically-induced nucleotide adducts that are also photochemically-cleavable (reviewed in 75).

Attachment of the tRNA anticodon to the 3'-third of 16S RNA is in agreement with earlier IEM studies which showed that proteins known to be associated with that part of the 16S RNA were also observed to cluster in the region of the mRNA marker proteins. These are shown as plain numbers in fig. 3. In particular, S1, S12, and S21, which possess both properties, are grouped together.

The availability of a tRNA covalently linked via its anticodon to the 30S subunit made it possible to locate the vicinity of the decoding site by a direct approach (19). The procedure used was acylation of the α-amino position of valine on Val-tRNA with the dinitrophenyl (DNP) group, linkage to the ribosome by irradiation, and after isolation of the 30S subunits, formation of 30S dimers by reaction with anti-DNP antibody. The dimers, which were 1:1 covalent complexes of tRNA and 30S, could then be separated from unreacted 30S by centrifugation, and examined in the electron microscope. From the antibody locations, it was concluded that the area in the cleft region of the 30S (see Figs. 1 and 7) should be the decoding site. This result is in general agreement with the location of the mRNA marker proteins shown in Fig. 3, although the fit to Fig. 3(a) is better than to 3(b).

The close contact between the 5'-base at the tip of the tRNA anticodon and a residue of 16S RNA which is implicit in the finding of a direct crosslink upon UV irradiation suggests that some special structural arrangement exists between these two residues on the 30S subunit. The marked preference for crosslinking from the P site indicates in addition that this should be a site-specific structure. A hypothesis for this interaction which would still allow codon-anticodon recognition to occur is shown in Fig. 4. In this figure we assume that base-pairing takes place at both P and A sites simultaneously, although there is no evidence for this (see below). A loop of 16S RNA must also be part of the decoding site since crosslinking, probably by cyclobutane dimer formation, can occur (dashed lines). We postulate that it may be necessary to stabilize, by base stacking, the anticodon of P site bound tRNA either in addition to normal base-pairing, or because of the lack of it, and a loop of 16S RNA may be used for that purpose.

This model fits in well with the fact that it is necessary for mRNA to kink between the two adjacent codons in the P and A sites in order for two tRNA anticodons in the crystal structure conformation to form antiparallel base pairs (20,104,105). Furthermore, the consequence of this kinking is that the 5'-anticodon base in the P site has a different environment than the other two bases of the codon (104). This fact appears to be expressed as the ability to be crosslinked upon irradiation.

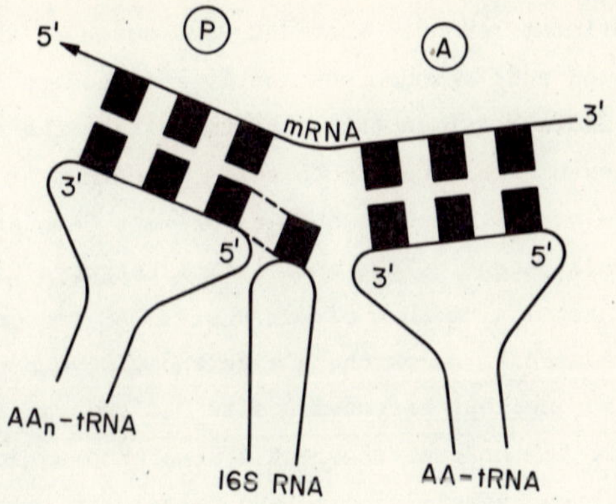

Figure 4. (43) Schematic diagram of tRNAs in both P and A sites engaged in simultaneous translation of adjacent codons. AA_n-tRNA, peptidyl tRNA in the P site; AA-tRNA, aminoacyl-tRNA in the A site. The arrangement of tRNA and mRNA shown on the figure is dictated by the fact that mRNA is translated from the 5' to the 3' end (103), and codon-anticodon base pairing is anti-parallel (69). The arrow shows the direction of mRNA movement past a stationary ribosome. A loop of 16S rRNA is inserted such that one of its pyrimidine bases is brought sufficiently close to the 5'-anticodon base of the P site bound tRNA to allow cyclobutane dimer formation (dashed lines).

Codon-Anticodon Interaction At The P Site

As pointed out above, simultaneous codon-anticodon (CAC) base pairing at both A and P sites requires mRNA to kink between the two adjacent codons (20,104,105). Obviously if only one codon is paired at a time, this steric problem would vanish. This could be accomplished if after base-pairing in the A site, and translocation together with its associated mRNA codon to the P site, the anticodon and codon were to dissociate. Dissociation need not move the pairs apart very far, only enough to straighten out the kink. If this hypothesis were correct, one of two further situations should exist. (A) the process of entering the P site triggers dissociation so that in the P site proper, CAC recognition does not exist. (B) CAC interaction persists in the P site until some event, possibly the binding of the EFTu-AA-tRNA complex itself, induces its disruption. In the former case, the well-known direct codon-dependent binding to the P site is viewed as requiring a transient binding to the A site where codon selection occurs, followed by a rapid factor-independent translocation to the P site, whence puromycin reaction can occur. In the latter case, direct codon-dependent P site binding is possible, since disruption only occurs in the process of filling the A site.

Several experiments suggest that CAC does occur in the P site. Pongs et al. (82) showed that a codon covalently attached to the ribosome could direct P site binding of fMet-tRNA. Assuming that the codon was not sufficiently flexible to occupy both sites, and that it is obligatory for the tRNA to move with its codon, this experiment demonstrates P site CAC interaction. More recently, the antibiotics tetracycline and viomycin were used in an attempt to show direct P site CAC recognition (29). Tetracycline was used to block the A site and viomycin was used to inhibit translocation, but neither affected P site CAC interaction. If the effect of antibiotics on the transient A site binding and movement is the same as for the known stable binding and translocation, then this experiment shows that transient A site binding does not occur during P site CAC recognition.

A different kind of experiment was recently described by Kuechler (11). He was able to crosslink the wY base of yeast $tRNA^{Phe}$ to poly U by irradiation when the tRNA was positioned at either the A site or the P site, and both crosslinking reactions were equally efficient. If crosslinking is a measure of CAC interaction, then this experiment shows clearly that P site recognition occurs. However, since the chemistry of the crosslink is unknown, this is not certain. Nevertheless, another experiment is suggestive. After A site crosslinking, the tRNA-mRNA covalent complex could be translocated to the P site. Thus since the covalent complex formed during CAC base-pairing in the A site can be accommodated at the P site, it is at least likely that they are still base-paired.

None of the experiments described above bear on the question of simultaneous base-pairing. The crosslinking approach seems most likely to supply the answer, since if two tRNAs on the same ribosome could be crosslinked to poly U with similar efficiencies, it would be strong evidence in favor of simultaneous CAC interaction at both sites.

The Kuechler experiment also bears on the question of the conformation of the anticodon loop at the A and P sites. If the only two conformations considered are the 5' stack and the 3' stack (106), only the 3' stack should exist in both A and P sites because it is only in this configuration that the wY base is close enough to mRNA to crosslink (2,107). This result should rule out models of translocation which involve switching of anticodon conformation from 5' to 3' or vice versa (107). Of course, other different conformations are not ruled out. A more precise analysis of anticodon loop conformation will have to wait until the stereochemistry of the crosslink is known.

THE CENTRAL REGION

Affinity Labelling Studies

As pointed out in the Introduction, crosslinking from this part of the molecule should provide the most useful topographic information about the location of the A and P sites. Despite this, only our own laboratory has so far described the use of affinity labels attached to bases in this potentially important region. The particular minor bases chosen for study, 4-thiouridine (^4S) and 3-(3-amino-3-carboxypropyl)uridine (acp^3U), are readily derivatized (108) and occur only once in the appropriate tRNA. The probes used were aryl azides, which can be photoactivated above 300 nm, thus minimizing the risk of damage to the ribosomes. Probes at these sites do not prevent non-enzymatic P site binding, EFTu-dependent A site binding, or initiation factor dependent binding of fMet-tRNA (108), nor do they affect the ability of such modified tRNAs to be recognized by stringent factor (18). Consequently, tRNAs modified at these positions can be used to study the topography of tRNA at both P and A sites.

The position and stereochemistry of the probes in the overall tRNA structure is shown in Fig. 5. There are several points to note. First, the p-azidophenacyl (APA) group, 9Å long, fits readily into the tRNA structure, with the reactive nitrene pointed outward. Second, its reactive end does not extend beyond the overall perimeter of the tRNA so that any crosslinking obtained must reflect a very close approach to the ribosomal surface. Third, although the APA probe is bound close to the body of the tRNA, it nevertheless defines one of its two topographical surfaces, that on the right side in view (b). Fourth, the reactive end of the p-azidonitrophenylglycine (NAG) probe is freely available, and defines the other topographical surface of the tRNA, namely that to the left in view (b). Fifth, since both nucleotides occur in the same tRNA, $tRNA^{Phe}$ of E. coli, labelling of each nucleotide separately should yield complementary information.

Crosslinking With APA-^4S Modified tRNA (43,109-111)

Specific crosslinking to the ribosome was obtained with this probe. When the APA-modified Phe-tRNA, Phe-$tRNA^{APA}$, was enzymatically bound to

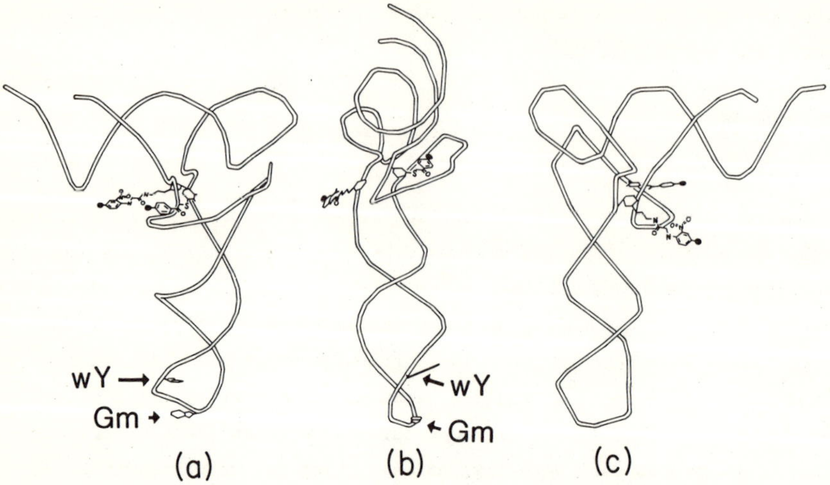

Figure 5. (11) Stereochemistry of affinity probes attached to the central region of tRNA. Three views of the structure of yeast $tRNA^{Phe}$ (12) with N-(2-nitro-4-azidophenyl)glycyl-acp^3U_{47} (NAG-acp^3U) in place of U_{47} and S-(p-azidophenacyl)-^4S (APA-^4S) in place of U_8. In views (a) and (b) APA-^4S is on the right, and in view (c) it is the upper structure. The continuous double line shows the ribose phosphate backbone. APA and NAG, drawn to scale, are shown in their maximally extended form. Other less extended conformations are also possible as a result of bond rotations. The azide group (solid ball) is not to scale. The 5'-anticodon base, Gm_{34}, and wY_{37} are shown in views (a) and (b).

the A site, only labeling of 30S protein(s) was found. This was shown both by SDS treatment and by 8 M urea dissociation. The 50S subunit and 16S RNA were not crosslinked. The specificity of crosslinking was shown by the absolute requirement for mRNA (poly U), EFTu, and GTP. Other controls proved that true photoaffinity labelling had occurred since both the aryl azide function and irradiation were required. True A site binding was shown in two ways. First, following peptide bond formation with unmodified AcPhe-tRNA in the P site, the crosslinked product attached to the 30S was found to consist of AcPhe and Phe in a ratio of 1:2. Thus at least 50% of the crosslinkable Phe-$tRNA^{APA}$ was functionally in the A site. Recall that no crosslinking was found in the absence of EFTu or GTP. Second, translocation to the P site with EFG before irradiation blocked all crosslinking. This second control showed that crosslinking did not occur inadvertently from the P site due to contaminating amounts of EFG. It also showed that the crosslinking reaction was A site specific. Apparently, close contact between 30S proteins and the central fold of the tRNA can only occur when the tRNA is bound at the A site. The 30S proteins which were crosslinked have been identified by immunological methods (L. Kahan, unpublished results). Virtually all of the crosslinked tRNA was associated with S19. A trace amount was also associated with S2. No other proteins were labelled (S15 was not tested). In the Lake model (Fig. 3a), antigenic sites for S19 are found on both sides of the head near to the platform. In the Stöffler model Fig. 3(b), the two sites span the frontal lobe of the head. This result will be discussed further below.

The existence of a distinct R site was looked for by means of this probe. The first experiments showed that replacement of GTP by the non-hydrolyzable analog, GDPCP, completely blocked crosslinking although it only decreased non-covalent binding in half. This result could be interpreted as support for the R site. However, it could also mean (a) the ribosome or tRNA has a different conformation when EFTu is bound in the complex; or (b) ribosome-bound EFTu interferes directly with the crosslinking reaction. In subsequent experiments, increasing the Mg^{++} concentration from 5 to 15 mM was found to allow some crosslinking, GDPCP, although it was still 30S specific. This indicates that (a) may be the correct interpretation. In any case, the availability of proteins crosslinked from a Phe tRNA-ribosome-EFTu-GDPCP complex should allow a direct test of the R site hypothesis. If again protein S19 were to be found, the specific version proposed by Lake would require reconsideration.

Similar specific binding to the 30S subunit was observed when $tRNA_1^{Val}$ was modified in the same way, or when a longer 12 Å probe with the same reactive end was placed on either $tRNA^{Phe}$ or $tRNA^{Val}$. The labeling of the 50S subunit which was observed previously with APA-$tRNA^{Val}$ (42) using salt-washed ribosomes was not detected in these experiments which used tight couple ribosomes. This observation emphasizes that ribosomes in subtly different conformational states may give different crosslinking results.

When fMet-$tRNA^{APA}$ modified with APA in the same way was used to study interaction at the initiator tRNA binding site on 70S ribosomes (112) the 30S subunit was the primary target in a reaction strongly dependent on initiation factors. This result should be compared to the failure of Phe-$tRNA^{APA}$ to crosslink when translocated to the P site. The result might be explained by appropriate differences in the structures of Phe-tRNA and fMet-tRNA when on the ribosome but the possibility should also be considered that the I site on 70S ribosomes is not identical to the P site, at least in so far as the central region of the tRNA is concerned.

Crosslinking With NAK- or NAG-Modified tRNA (43,110,111)

$tRNA^{Phe}$ photoactivated by labelling with 6-(2-nitro-4-azidophenylamino)caproate (NAK) on the opposite face of the tRNA (refer to Fig. 5) could also be crosslinked to the ribosome. This probe has the same reactive end as NAG, but the nitrene is 20 Å from the uridine instead of the 14 Å found with NAG. It gave a several-fold greater crosslinking yield than NAG but otherwise was similar. With this probe, labelling occurred from the I and P sites and both 50S and 30S subunits were labelled. In the previous case, the complementary APA-^{4}S probe was unreactive at the P site. Crosslinking was specific since it was dependent on the presence of the probe, on irradiation, and pre-photolysis of the probe blocked crosslinking. It was all in the P site since in all cases, the crosslinked tRNA was still capable of reaction with puromycin. All of the crosslinking to 30S was to proteins. They have not been identified. Some of the 50S crosslinking appeared to be to 23S RNA but the quantitative aspects have not been determined.

The most interesting aspect was the variation in labelling pattern in the presence and absence of IF2 and GDPCP. When no initiation factor was present, that is P site binding, both 50S and 30S subunits were equally labelled when either GTP or GDPCP was present. When IF2 was added with GTP, crosslinking to both subunits decreased but the 30S was most affected. On the other hand when IF2 was added with GDPCP in place of GTP, the 50S labelling increased 3-fold, with little or no effect on the 30S labelling. Non-covalent binding did not change in any of these three situations. Apparently, retention of IF2 on the ribosome somehow caused the probe region of the P site bound tRNA to more closely approach the 50S surface. It has been suggested that IF2 occupies the A site when it is bound to the ribosome (113). If so, it could have this effect.

This alteration of labelling pattern in the presence of IF2 suggests that the I and P sites for the central part of the tRNA may be different. The results described above for APA-modified Phe- and fMet- tRNAs further support this notion. Although the findings are still preliminary, The possibility that there are at least partly different binding domains for the I and P sites should be kept in mind.

It is not clear whether or not crosslinking with NAK-modified tRNA can occur from the A site. EFTu-dependent binding to the A site proceeded readily, as was shown for NAG several years ago (112). However, there was no significant EFTu- or poly(U)-dependent crosslinking, although there was some apparently non-specific reaction with both subunits which required the arylazide as well as photo-activation. At this stage, it seems more likely that crosslinking from NAK- or NAG-acp^3U is largely a P and I site phenomenon.

Ribosomal Interaction of the GTΨCG Sequence

The GTΨCG sequence of the central region of tRNA, common to almost all non-initiator tRNAs (114) is thought to be involved in ribosomal binding. The idea, first suggested in 1965 (115), was supported by the inhibitory effect of added TΨCG on the binding of AA-tRNA to the ribosome (116,117). Examination of the site specificity for this inhibition showed it to be A site and 50S specific (118-120). It was at first thought that the ribosome binding site was specific for the TΨCG sequence but later studies showed, first, that the TΨCG sequence in tRNA could be replaced by f^5Uf^5UCG with no loss of ribosome binding ability (121) and, second, that UUCG, and even some unrelated tRNA fragments, were also effective inhibitors of ribosome binding (122).

The ribosomal component involved has been thought to be 5S RNA for three reasons. First, the association constant for the binding of TΨCG to a 5S RNA-protein complex was three times greater than the equivalent K_a for UUCG, and oxidation of two unspecified adenine residues of the 5S RNA blocked the preferential binding of TΨCG (123). Second, prokaryotic 5S RNAs have conserved the complementary sequence, CGAAC, at or near the same place in their linear sequence (69,124). Third, reassembled 50S particles lacking 5S RNA lost their A site binding capacity, but retained some P site binding ability (125).

However, neither the CGAAC sequence of the 5S RNA in the ribosome nor the GTΨCG sequence of the tRNA is exposed to solvent. The G residue of the CGAAC sequence is chemically unreactive (126,127), and a number of studies ranging from crystal structure analysis (12), chemical modification (128,129), and oligonucleotide binding (130,131), to NMR analysis (132), have shown that the GTΨCG sequence is not exposed in solution. Clearly, if the interaction with 5S RNA and tRNA occurs, both sequences must become exposed only as a result of EFTu-dependent A site binding. That this can occur, at least for the tRNA, was indicated by experiments in which the presence of the cognate codon stimulated the binding of the complementary oligonucleotide, CGAA (133,134). The change in chemical reactivity of the G residue of the TΨCG sequence previously ascribed to the presence of the cognate codon (135) has now been found to be a secondary effect. The primary effect of CAC interaction appears to be exposure of m^7G (136).

To sum up, the evidence is reasonably good that tRNA fragments, and particularly UUCG and its modified forms, can inhibit A site binding probably by acting on the 50S subunit. There is suggestive evidence that CAC interaction can expose the TΨCG sequence. There is only circumstantial evidence that the binding site on the 50S subunit is the 5S RNA. Indeed, recent studies have shown that the oligonucleotide CCG, which contains the CG of the CGAAC sequence, can be excised from 5S RNA, and the resulting two large fragments reconstituted into a 50S particle which is as active as intact 5S RNA-containing 50S for poly U-dependent polyphenylalanine synthesis (C. Cantor, pers. communic.)

In any case, for topographical purposes, this interaction is not very helpful. Since the 5S RNA location can so far only be approximated by that of its closely associated proteins, L5, L18, and L25 (124), there is considerable room for error.

CONCLUSIONS AND SPECULATIONS

Relative Orientation Of The Two RNAs On The Ribosome

An important experiment has been performed by A. E. Johnson and C. Cantor (pers. communic.). By derivatizing the ^4S of two different tRNAs with appropriate fluorescent groups, and placing both tRNAs on the same ribosome, they were able to determine that the dye molecules were <30 Å apart, that is, in the range of the thickness of a tRNA duplex. The simplest way to accomplish this while at the same time keeping the amino acid ends close together and the anticodons adjacent is for the two tRNAs in the A and P sites to be side by side, and in the same orientation. This is schematically shown in Fig. 6b. A possible alternative configuration that may also satisfy the data (Fig. 6c) has the plane of the two tRNAs more or less perpendicular to each other (104,105). Models in which the central regions of the two tRNAs are widely separated (2,68, 104) as in Fig. 6a appear inconsistent with the experimental results. This is an important conclusion since it severely limits the possible arrangements of two tRNAs on the ribosome.

Which tRNA Is In The P Site And Which In The A Site?

The results of affinity labelling from the central region of the tRNA

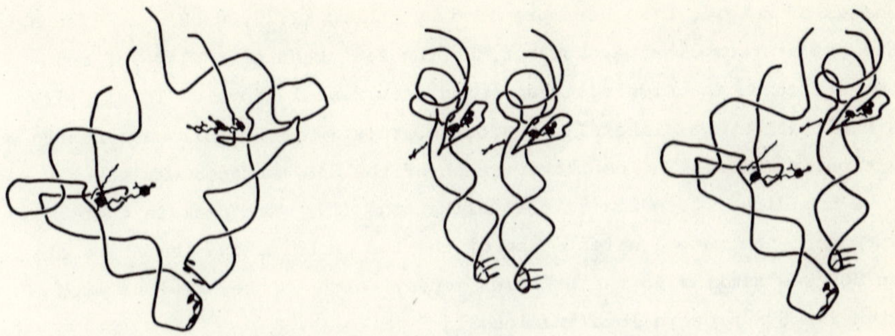

Figure 6. Possible arrangements of two tRNAs that keep both aminoacyl ends and anticodons close together. The yeast $tRNA^{Phe}$ crystal structure (12) was used with APA and NAG added as described in the legend to Figure 5.

(see above) provide strong evidence that the tRNA on the left hand side in Fig. 6b is in the P site and the one on the right hand side is in the A site. The reasoning is that only the APA probe from the right-hand tRNA could make contact with the ribosomal surface. If the A site tRNA were on the left, the other tRNA would be in the way. In fact, the APA probe in the P site did not crosslink. This argument should hold for Fig. 6c as well, since even though there is space between the tRNAs, the area should be free of a ribosomal surface so that translocation can occur. The NAK probe results are in agreement with this notion. Since this probe is now on the surface of the tRNA opposite to APA, the tRNA on the left, i.e. in the P site, should be the one able to make contact with the ribosome. Indeed, those were our findings. P or I site bound NAK-modified tRNA crosslinked readily, but A site bound tRNA did not. It is particularly important to note that these conclusions are independent of the choice of any particular ribosome model, and thus are equally applicable as stereochemical constraints to all ribosome models.

Location Of The Two tRNAs On The Ribosome

The various facts accumulated about tRNA topography in the preceding parts of this review have been put together to generate a speculative model of the tRNA-mRNA-ribosome complex (Fig. 7). This model should not be considered as a detailed proposal, but only as a means of graphically summarizing the topographic facts discussed above. Other models, in varying degrees of detail, have been proposed by others (2,13,20,68). All such attempts are complicated, however, by the fact that no single 70S monosome structure is universally accepted (see Fig. 1 above). To simplify this situation, we, like all previous workers (2,13,20,68), have chosen a particular model. The Boublik version of the 70S monosome was taken since our direct decoding site localizations (19) were done in that system and the mutual orientation of the two subunits and the shape of the 50S was similar to the Stöffler model, which is the only one with extensive 50S protein localizations.

The peptidyl transferase center was localized as described above (see Fig. 2). The putative peptidyl transferase center marked by the protein antigenic sites is too large, and therefore a decision had to be made regarding the true site of the aminoacyl end. The left side region was chosen in view of the localization to this area of antibody to chloramphenicol which had been crosslinked to the ribosome at the peptidyl transferase center (13).

The decoding site for both tRNAs was placed in the cleft of the 30S according to the approximate localization by AIM (19). This result agrees well with the location of the mRNA marker proteins according to Lake (Fig. 3a), although not as well with the localizations of Stöffler (Fig. 3b). On the other hand, the 30S structure seen in our EM experiments resembled the Lake model. Both tRNA anticodons were kept close together for the reasons given earlier. At this level of structural resolution, it is not necessary to know if there is simultaneous codon-anticodon recognition in both A and P sites.

The side-by-side (Fig. 6b) or angular orientation (Fig. 6c) of the two tRNAs was maintained and the A site was considered to be on the right for the reasons discussed above. Finally, the model was required to explain the exclusive 30S labelling from the short APA probe in the A site, and the labelling of both 50S and 30S subunits with the 20 Å probe from the P site. These latter requirements eliminated from consideration a model in which the two tRNAs lay down on the 50S essentially in the positions shown with the A site on top because there appeared to be no way for the NAK probe to reach the 30S in this situation.

The arrangement chosen, with the plane of the two tRNAs at an approximate 45° angle, was the best fit to all of the above constraints. However, the model is in only fair agreement with the exclusive crosslinking of the APA probe at the A site to S19. As noted above, the two antigenic sites for S19 are located further to the left on both sides of the head of the 30S. This apparent discrepancy might be resolved by shifting both tRNAs further to the left, or by assuming that the antigenic sites do not indicate the whereabouts of the entire S19 polypeptide chain.

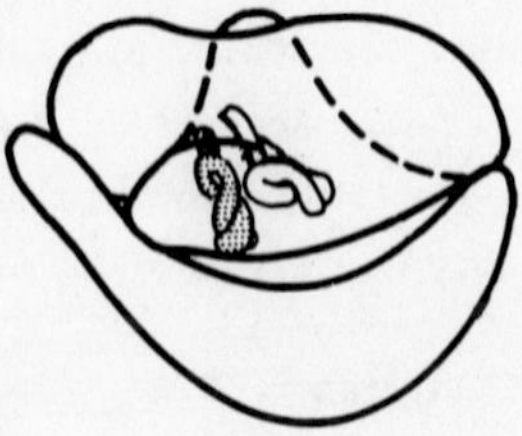

Figure 7. Hypothetical model of the 70S ribosome-mRNA-tRNA complex. The Boublik (15,101) 70S ribosome structure was assumed, and the tRNA dimensions are from Kim (12). The P site tRNA (shaded) is on the left. The facts considered in placement of the tRNA are discussed in the text.

The location of the mRNA is not indicated on the figure. Clearly it must be at the decoding site adjacent to the tRNA anticodons. Location of the region 5' to the decoding site can be deduced to be near to the top of the large projection, or platform, of the 30S since the 3'-end of 16S RNA, which is only 6 bases from the center of the Shine-Dalgarno sequence (see article by J. A. Steitz, this volume) has been recently localized to that vicinity by IEM techniques (137). Thus it is possible that the course of the mRNA runs under or through the cleft of the platform, following the linear order of Shine-Dalgarno sequence, P site codon, and A site codon, to exit either through the central channel between the subunits or through the right rear channel. Such an arrangement is approximately in keeping with the size of the ribosome-protected fragments of mRNA on either side of the initiation codon (138; J.A. Steitz, this volume).

Although the model has been discussed in some detail, it should again be emphasized that it is merely a device to correlate the existing facts. Its primary function should be to incite new speculation and experimentation rather than to serve as a very specific prediction of the true nature of the functioning ribosome.

ACKNOWLEDGEMENTS

I wish to thank my colleagues, M. Boublik and F.-L. Lin for stimulating discussions about the tRNA-mRNA model, and in particular M. Boublik for preparing the 3-dimensional model of the 70S couple used in Fig. 7.

REFERENCES

1. Watson, J.D. (1964) Bull. Soc. Chim. Biol. 46, 1399-1425.

2. Lake, J.A. (1977) Proc. Natl. Acad. Sci. U.S.A. 74, 1903-1907.

3. Cashel, M. (1975) Ann. Rev. Microbiol. 29, 301-318.

4. Bochkareva, E.S., Budker, V.G., Girshovich, A.S., Knorre, D.G. and Teplova, N.M. (1971) FEBS Lett. 19, 121-124.

5. Pellegrini, M., Oen, H. and Cantor, C.R. (1972) Proc. Natl. Acad. Sci. U.S.A. 69, 837-841.

6. Czernilofsky, A.P. and Kuechler, E. (1972) Biochim. Biophys. Acta 272, 667-671.

7. Pellegrini, M. and Cantor, C.R. (1977) in: Molecular Mechanisms of Protein Biosynthesis (H. Weissbach and S. Pestka, eds.) pp. 203-244, Academic Press, Inc., New York.

8. Zamir, A. (1977) in: Methods Enzymol. (W.B. Jakoby and M. Wilchek, eds.) Vol. 46, pp. 621-637, Academic Press, Inc., New York.

9. Cooperman, B.S. (1978) in: Bioorganic Chemistry (E. Van Tamelen, ed.) Vol. IV, pp. 81-115, Academic Press, Inc., New York.

10. Kuechler, E. (1978) in: Theory and Practice in Affinity Techniques (P.V. Sundaram and F. Eckstein, eds.) (in press), Academic Press, Inc., New York.

11. Kuechler, E. and Ofengand, J. (1979) in: Transfer RNA, pt. 1: Structure, Properties, and Recognition (P. Schimmel, D. Soll and J. Abelson, eds.) (in press), Cold Spring Harbor Laboratory, Cold Spring Harbor, New York.

12. Kim, S.-H. (1978) in: Transfer RNA (S. Altman, ed.) pp. 248-293, M.I.T. Press, Cambridge, Massachusetts.

13. Stöffler, G. and Wittmann, H.G. (1977) in: Molecular Mechanisms of Protein Biosynthesis (H. Weissbach and S. Pestka, eds.) pp. 117-202, Academic Press, Inc., New York.

14. Lake, J.A. (1976) J. Mol. Biol. 105, 131-159.

15. Boublik, M., Hellmann, W. and Kleinschmidt, A.K. (1977) Cytobiologie 14, 293-300.

16. Fairclough, R.H. and Cantor, C. (1979) J. Mol. Biol. (in press).

17. Ofengand, J., Delaney, P. and Bierbaum, J. (1974) in: Methods Enzymol. (L. Grossman and K. Moldave, eds.) Vol. 29, 673-684, Academic Press, Inc., New York.

18. Ofengand, J. and Liou, R. (1978) Nucleic Acids Res. 5, 1325-1334.

19. Keren-Zur, M., Boublik, M. and Ofengand, J. (1979) Proc. Natl. Acad. Sci. U.S.A. 76, 1054-1058.

20. Pongs, O. (1978) in: Transfer RNA (S. Altman, ed.) pp. 78-104, M.I.T. Press, Cambridge, Massachusetts.

21. Brimacombe, R., Nierhaus, K.H., Garrett, R.A. and Wittmann, H.G. (1976) Prog. Nucleic Acid Research Mol. Biol. 18, 1-44.

22. Pongs, O., Nierhaus, K.H., Erdmann, V.A. and Wittmann, H.G. (1974) FEBS Lett. 40, S28-S37.

23. Fanning, T.G., Cantrell, M., Shih, C.-Y.T. and Craven, G.R. (1978) Nucleic Acids Res. 5, 933-950.

24. Shimizu, M. and Craven, G.R. (1976) Eur. J. Biochem. 61, 307-315.

25. Thomas, G., Sweeney, R., Chang, C. and Noller, H.F. (1975) J. Mol. Biol. 95, 91-102.

26. Ginzburg, I., Miskin, R. and Zamir, A. (1973) J. Mol. Biol. 79, 481-494.

27. De Groot, N., Panet, A. and Lapidot, Y. (1971) Eur. J. Biochem. 23, 523-527.

28. Watanabe, S. (1972) J. Mol. Biol. 67, 443-457.

29. Wurmbach, P. and Nierhaus, K.H. (1979) Proc. Natl. Acad. Sci. U.S.A. 76, 2143-2147.

30. Hsiung, N., Reines, S.A. and Cantor, C.R. (1974) J. Mol. Biol. 88, 841-855.

31. Hsiung, N. and Cantor, C.R. (1974) Nucleic Acids Res. 1, 1753-1762.

32. Johnson, A. (1979) in: Transfer RNA. Structure, Properties, and Recognition (P. Schimmel, D. Soll and J. Abelson, eds.) (in press), Cold Spring Harbor Laboratory, Cold Spring Harbor, New York.

33. Sopori, M., Pellegrini, M., Lengyel, P. and Cantor, C.R. (1974) Biochemistry 13, 5432-5439.

34. Barta, A., Kuechler, E., Branlant, C., Sri Widada, J., Krol, A. and Ebel, J.P. (1975) FEBS Lett. 56, 170-174.

35. Sonenberg, N., Wilchek, M. and Zamir, A. (1977) Eur. J. Biochem. 77, 217-222.

36. Oen, H., Pellegrini, M., Eilat, D. and Cantor, C.R. (1973) Proc. Natl. Acad. Sci. U.S.A. 70, 2799-2803.

37. Pellegrini, M., Oen, H., Eilat, D. and Cantor, C.R. (1974) J. Mol. Biol. 88, 809-829.

38. Yukioka, M., Hatayama, T. and Morisawa, S. (1975) Biochim. Biophys. Acta 390, 192-208.

39. Czernilofsky, A.P., Collatz, E.E., Stöffler, G. and Kuechler, E. (1974) Proc. Natl. Acad. Sci. U.S.A. 71, 230-234.

40. Schwartz, I. and Ofengand, J. (1978) Biochemistry 17, 2524-2530.

41. Eilat, D., Pellegrini, M., Oen, H., De Groot, N. Lapidot, Y. and Cantor, C.R. (1974) Nature 250, 514-516.

42. Schwartz, I., Gordon, E. and Ofengand, J. (1975) Biochemistry 14, 2907-2914.

43. Ofengand, J., Lin, F.-L., Hsu, L., Keren-Zur, M. and Boublik, M. (1979) in: Annals N.Y. Acad. Sci. (in press), New York Academy of Sciences, New York.

44. Lucas-Lenard, J. and Lipmann, F. (1967) Proc. Natl. Acad. Sci. U.S.A. 57, 1050-1057.

45. Hauptmann, R., Czernilofsky, A.P., Voorma, H.O., Stöffler, G. and Kuechler, E. (1974) Biochem. Biophys. Res. Commun. 56, 331-337.

46. Sonenberg, N., Wilchek, M. and Zamir, A. (1976) Biochem. Biophys. Res. Commun. 72, 1534-1541.

47. Bauer, K., Czernilofsky, A.P. and Kuechler, E. (1975) Biochim. Biophys. Acta 395, 146-151.

48. Collatz, E., Kuechler, E., Stöffler, G. and Czernilofsky, A.P. (1976) FEBS Lett. 63, 283-286.

49. Breitmeyer, J.B. and Noller, H.F. (1976) J. Mol. Biol. 101, 297-306.

50. Eilat, D., Pellegrini, M., Oen, H., Lapidot, Y. and Cantor, C.R. (1974) J. Mol. Biol. 88, 831-840.

51. Yukioka, M., Hatayama, T. and Omori, K. (1977) Eur. J. Biochem. 73, 449-459.

52. Bispink, L. and Matthaei, H. (1973) FEBS Lett. 37, 291-294.

53. Kuechler, E. and Barta, A. (1977) in: Methods Enzymol. (W.B. Jakoby and M. Wilchek, eds.) Vol. 46, pp. 676-683, Academic Press, Inc., New York.

54. Girshovich, A.S., Bochkareva, E.S., Kramarov, V.M. and Ovchinnikov, Yu. A. (1974) FEBS Lett. 45, 213-217.

55. Sonenberg, N., Wilchek, M. and Zamir, A. (1975) Proc. Natl. Acad. Sci. U.S.A. 72, 4332-4336.

56. Bochkareva, E.S., Budker, V.G., Girshovich, A.S., Knorre, D.G. and Teplova, N.M. (1973) Mol. Biol. (Moscow) 7, 278-288.

57. Brimacombe, R., Stöffler, G. and Wittmann, H.G. (1978) Ann. Rev. Biochem. 47, 217-249.

58. Pongs, O., Bald, R., Erdmann, V.A. and Reinwald, E. (1975) in: Topics in Infectious Diseases (J. Drews and F.E. Hahn, eds.) pp. 179-190, Springer-Verlag Wien, New York.

59. Jaynes, E.N. Jr., Grant, P.G., Giangrande, G., Wieder, R. and Cooperman, B.S. (1978) Biochemistry 17, 561-569.

60. Grant, P.G., Strycharz, W.A., Jaynes, E.N. Jr. and Cooperman, B.S. (1979) Biochemistry 18, 2149-2154.

61. Parker, J., Watson, R.J., Friesen, J.D. and Fiil, H.P. (1976) Mol. Gen. Genet. 144, 111-114.

62. Christiansen, L. and Nierhaus, K.H. (1976) Proc. Natl. Acad. Sci. U.S.A. 73, 1839-1843.

63. Howard, G.A., Gordon, J., Farnung, K. and Richter, D. (1976) FEBS Lett 68, 211-214.

64. Harris, R.J. and Pestka, S. (1977) in: Molecular Mechanisms of Protein Biosynthesis (H. Weissbach and S. Pestka, eds.) pp. 413-442, Academic Press, Inc., New York.

65. Kuechler, E., Bauer, K. and Rich, A. (1972) Biochim. Biophys. Acta 277, 615-627.

66. Nicholson, A.W. and Cooperman, B.S. (1978) FEBS Lett. 90, 203-208.

67. Cooperman, B.S. (1979) in: Annals. N.Y. Acad. Sci. (in press), New York Academy of Sciences, New York.

68. Lake, J.A. (1978) in: Advanced Techniques in Biological Electron Microscopy II. (J.K. Koehler, ed.) pp. 173-211, Springer-Verlag, Berlin and Heidelberg.

69. Ofengand, J. (1977) in: Molecular Mechanisms of Protein Biosynthesis (H. Weissbach and S. Pestka, eds.) Chapter 1, pp. 7-79, Academic Press, Inc., New York.

70. Hecht, S.M. (1977) Tetrahedron 33, 1671-1696.

71. Sprinzl, M. and Cramer, F. (1979) Prog. Nucleic Acid Research Mol. Biol. 22, 1-69.

72. Sprinzl, M. and Wagner, T. (1979) in: Transfer RNA: Structure, Properties, and Recognition (P. Schimmel, D. Soll and J. Abelson, eds.) (in press), Cold Spring Harbor Laboratory, Cold Spring Harbor, New York.

73. Hussain, Z. and Ofengand, J. (1973) Biochem. Biophys. Res. Commun. 50, 1143-1151.

74. Chladek, S., Ringer, D. and Zemlicka, J. (1973) Biochemistry 12, 5135-5138.

75. Ofengand, J., Liou, R., Kohut, J., Schwartz, I. and Zimmermann, R.A. (1979) Biochemistry 18, (in press).

76. Takanami, M. and Okamoto, T. (1963) J. Mol. Biol. 7, 323-333.

77. Fiser, I., Scheit, K.H. and Kuechler, E. (1977) Eur. J. Biochem. 74, 447-456.

78. Lührmann, R., Gassen, H.G. and Stöffler, G. (1976) Eur. J. Biochem. 66, 1-9.

79. Budker, V.G., Girshovich, A.S., Grineva, N.I., Karpova, G.G., Knorre, D.G. and Kobets, N.D. (1973) Dokl. Akad. Nauk SSSR 211, 725-728.

80. Towbin, H. and Elson, D. (1978) Nucleic Acids Res. 5, 3389-3407.

81. Pongs, O. and Lanka, E. (1975) Proc. Natl. Acad. Sci. U.S.A. 72, 1505-1509.

82. Pongs, O., Stöffler, G. and Lanka, E. (1975) J. Mol. Biol. 99, 301-315.

83. Yaguchi, M., Lanka, E., Dworniczak, B., Kiltz, H.-H. and Pongs, O. (1978) Eur. J. Biochem. 92, 243-251.

84. Pongs, O. and Rossner, E. (1976) Nucleic Acids Res. 3, 1625-1633.

85. Wagner, R. and Gassen, H.G. (1975) Biochem. Biophys. Res. Commun. 65, 519-529.

86. Pongs, O., Stöffler, G. and Bald, R.W. (1976) Nucleic Acids Res. 3, 1635-1646.

87. Schenkman, M.L., Ward, D.C. and Moore, P.B. (1974) Biochim. Biophys. Acta 353, 503-508.

88. Fiser, I., Margaritella, P. and Kuechler, E. (1975) FEBS Lett. 52, 281-283.

89. Margaritella, P. and Kuechler, E. (1978) FEBS Lett. 88, 131-134.

90. Fiser, I., Scheit, K.H., Stöffler, G. and Kuechler, E. (1974) Biochem. Biophys. Res. Commun. 60, 1112-1118.

91. Fiser, I., Scheit, K.H., Stöffler, G. and Kuechler, E. (1975) FEBS Lett. 56, 226-229.

92. Wagner, R., Gassen, H.G., Ehresmann, Ch., Stiegler, P. and Ebel, J.P. (1976) FEBS Lett. 67, 312-315.

93. Winkelmann, D. and Kahan, L. (1978) Fed. Proc. 37, 1739.

94. Zimmermann, R.A. (1974) in: Ribosomes (M. Nomura, A. Tissieres and P. Lengyel, eds.) pp. 225-269, Cold Spring Harbor Laboratory, Cold Spring Harbor, New York.

95. Morgan, J. and Brimacombe, R. (1972) Eur. J. Biochem. 29, 542-552.

96. Rinke, J., Ross, A. and Brimacombe, R. (1977) Eur. J. Biochem. 76, 189-196.

97. Kenner, R.A. (1973) Biochem. Biophys. Res. Commun. 51, 932-938.

98. Czernilofsky, A.P., Kurland, C.G. and Stöffler, G. (1975) FEBS Lett. 58, 281-284.

99. Rinke, J., Yuki, A. and Brimacombe, R. (1976) Eur. J. Biochem. 64, 77-89.

100. Politz, S.M. and Glitz, D.G. (1977) Proc. Natl. Acad. Sci. U.S.A. 74, 1468-1472.

101. Boublik, M. and Hellmann, W. (1978) Proc. Natl. Acad. Sci. U.S.A. 75, 2829-2833.

102. Zimmermann, R.A., Gates, S.M., Schwartz, I. and Ofengand, J. (1979) Biochemistry 18, (in press).

103. Schweet, R. and Heintz, R. (1966) Ann. Rev. Biochem. 35, 723-758.

104. Rich, A. (1974) in: Ribosomes (M. Nomura, A. Tissieres and P. Lengyel, eds.) pp. 871-884, Cold Spring Harbor Laboratory, Cold Spring Harbor, New York.

105. Sundaralingam, M., Brennan, T., Yathindra, N. and Ichikawa, T. (1975) in: Structure and Conformation of Nucleic Acids and Protein-Nucleic Acid Interactions (M. Sundaralingam and S.T. Rao, eds.) pp. 101-115, University Park Press, Baltimore, Maryland.

106. Fuller, W. and Hodgson, A. (1967) Nature 215, 817-821.

107. Woese, C. (1970) Nature 226, 817-820.

108. Ofengand, J., Schwartz, I., Chinali, G., Hixson, S.S. and Hixson, S.H. (1977) in: Methods Enzymol., Vol. 46 (W. Jakoby and M. Wilchek, eds.) pp. 683-702, Academic Press, Inc., New York.

109. Hsu, L., Keren-zur, M. and Liou, R. (1978) Fed. Proc. 37, 1405.

110. Lin, F.-L., Hsu, L. and Ofengand, J. (1979) Fed. Proc. 38, 228.

111. Ofengand, J., Lin, F.-L. and Liou, R. (1979) XIth Int. Congr. Biochem. Toronto, p. 89.

112. Schwartz, I., Tejwani, R. and Ofengand, J. (1976) Xth Int. Congr. Biochem., Hamburg, p. 121.

113. Grunberg-Manago, M. and Gros, F. (1977) Prog. Nucleic Acid Research Mol. Biol. 20, 209-284.

114. Siddiqui, M.A.Q., Ofengand, J., Garel, J.-P. and Drabkin, H. (1979) in: CRC Handbook of Microbiology (A.I. Laskin and H.A. Lechevalier, eds.) (in press) CRC Press, West Palm Beach, Florida.

115. Zamir, A., Holley, R.A. and Marquisee, M. (1965) J. Biol. Chem. 240, 1267-1273.

116. Ofengand, J. and Henes, C. (1969) J. Biol. Chem. 244, 6241-6253.

117. Shimizu, N., Hayashi, H. and Miura, K.-I. (1970) J. Biochem. (Tokyo) 67, 373-387.

118. Richter, D., Erdmann, V.A. and Sprinzl, M. (1973) Nature (London) New Biol. 246. 132-135.

119. Erdmann, V.A., Sprinzl, M., Richter, D. and Lorenz, S. (1974) Acta Biol. Med. Ger. 33, 605-608.

120. Grummt, F., Grummt, I., Gross, H.J., Sprinzl, M., Richter, D. and Erdmann, V.A. (1974) FEBS Lett. 42, 15-17.

121. Ofengand, J., Bierbaum, J., Horowitz, J., Ou, C.-N. and Ishaq, M. (1974) J. Mol. Biol. 88, 313-325.

122. Sprinzl, M., Wagner, T., Lorenz, S. and Erdmann, V.A. (1976) Biochemistry 15, 3031-3039.

123. Erdmann, V.A., Sprinzl, M. and Pongs, O. (1973) Biochem. Biophys. Res. Commun. 54, 942-948.

124. Erdmann, V.A. (1976) Prog. Nucleic Acid Research Mol. Biol. 18, 45-90.

125. Dohme, F. and Nierhaus, K.H. (1976) Proc. Natl. Acad. Sci. U.S.A. 73, 2221-2225.

126. Noller, H.F. and Herr, W. (1974) J. Mol. Biol. 90, 181-184.

127. Delihas, N., Dunn, J.J. and Erdmann, V.A. (1975) FEBS Lett. 58, 76-80.

128. Ofengand, J. (1971) in: Methods Enzymol. (K. Moldave and L. Grossman, eds.) Vol. XX C, pp. 150-168, Academic Press, Inc., New York.

129. Rhodes, D. (1977) Eur. J. Biochem. 81, 91-101.

130. Uhlenbeck, O.C. (1972) J. Mol. Biol. 65, 25-41.

131. Pongs, O., Bald, R. and Reinwald, E. (1973) Eur. J. Biochem. 32, 117-125.

132. Kastrup, R.V. and Schmidt, P.G. (1978) Nucleic Acids Res. 5, 257-269.

133. Schwarz, U., Menzel, H.M. and Gassen, H.G. (1976) Biochemistry 15, 2484-2490.

134. Möller, A., Manderschied, U., Lipecky, R., Bertram, S., Schmitt, M. and Gassen, H.G. (1979) in: Transfer RNA: Structure, Properties, and Recognition (P. Schimmel, D. Soll and J. Abelson, eds.) (in press), Cold Spring Harbor Laboratory, Cold Spring Harbor, New York.

135. Wagner, R. and Garrett, R.A. (1978) FEBS Lett 85, 291-295.

136. Wagner, R. and Garrett, R.A. (1979) Eur. J. Biochem. (in press).

137. Glitz, D.G., Olson, H.M. and Politz, S.M. (1979) XIth Int. Congr. Biochem. Toronto, p. 85.

138. Steitz, J.A. (1979) in: Biological Regulation and Development (R.F. Goldberger, ed.) Vol. 1, pp. 349-399, Plenum Press, New York.

FUNCTIONAL SITES ON THE *E. COLI* RIBOSOME AS DEFINED BY AFFINITY LABELING

B. S. COOPERMAN

Department of Chemistry,
University of Pennsylvania,
Philadelphia, PA 19104

INTRODUCTION

Several reviews have recently appeared summarizing the results of affinity labeling studies of the ribosome (1-4). In this review we wish to critically examine what reliable information these studies have yielded as to the location of functional sites on the ribosome. Although many affinity labels have been shown to incorporate into ribosomal RNA, in few cases have the sites of affinity labeling been precisely placed within the primary sequence of an RNA chain. Moreover, at present little is known regarding the placement of portions of ribosomal RNA within the ribosome. By contrast, many affinity labeling incorporation sites have been localized to specific ribosomal proteins, and a fair amount is known about the location of these proteins in the ribosome. For these reasons, our discussion will be confined to an analysis of studies in which ribosomal proteins have been labeled.

Affinity Labeling and Photoaffinity Labeling

The general subjects of affinity labeling and photoaffinity labeling have also been recently reviewed (5-9), and the reader is directed to these references for more detailed treatment of matters we will here only touch upon. The logic of an affinity labeling experiment is described in Equations (1)-(3). A reactive group, X (electrophilic or photolabile), is attached to the natural ligand, L. If the resulting modified ligand L'X retains high affinity for the native ligand receptor site and forms a non-covalent complex, R·L'X, (1) then, at low L'X concentration, covalent attachment via the reaction (2), a first order process, will proceed at a much faster rate than attachment via the reaction (3), a second order process, and specific attachment to the ligand site will be achieved.

$$R + L'X \rightleftharpoons R\cdot L'X \qquad (1)$$

$$R\cdot L'X \longrightarrow R\text{-}L' + X \qquad \text{affinity labeling} \qquad (2)$$

$$R + L'X \longrightarrow R\text{-}L' + X \qquad \text{non-specific labeling} \qquad (3)$$

In the majority of affinity labeling studies we will be discussing, X has been an electrophilic group (e.g., bromoacetyl), so that reaction (2) takes place spontaneously with nucleophiles at, or accessible from, the binding site, or a photolabile group (e.g., α-carbonyl-diazo, aryl azide) in which case reaction (2) takes place on illumination with an appropriate light source. In a limited number of studies, it has been shown that some naturally occurring ligands [e.g., puromycin, poly(U)] with no special photolability, can be induced to covalently incorporate into ribosomes on direct U.V. irradiation by as yet unknown mechanisms.

The great advantage of electrophilic labels is that they often yield high levels of covalent incorporation, which is obviously of practical importance for detection. However, they suffer the severe difficulty of absolutely requiring nucleophiles for reaction. Thus a weak or nonspecific binding site of high nucleophilicity may be labeled in preference to a specific binding site of low nucleophilicity. In addition, their reactions with ribosomes are often slow and difficult to control, which can pose problems for some types of study. By contrast, photolyses of aryl azides and α-carbonyl diazo compounds give rise to highly reactive intermediates, aryl nitrenes and α-carbonyl carbenes, respectively. These intermediates, though having some electrophilic character, nevertheless display wide chemical reactivity, being capable, for example, of inserting into C-H bonds. One therefore, has a better chance of obtaining affinity labeling results which accurately reflect non-covalent ligand binding when using these reagents. In addition, the covalent insertion reactions are rapid and easy to control, being totally light-dependent. The main disadvantage of photoaffinity labeling reactions is that they often proceed in low yield. With aryl azide and α-diazo carbonyl derivatives, the quantum yields for the reactions to give nitrenes and carbenes are high, and photoaffinity labeling can be carried out with a light fluence low enough that there is little photo-induced denaturation of the ribosome (10). In studies using photo-induced incorporation of the native ligand, this is not always the case and is a point that must be carefully checked (11,12). However, this latter approach, when feasible, presents the obvious advantage that chemical derivatization is not needed, so that at least non-covalent binding occurs, by definition, at the correct ribosome site, a point that has to be demonstrated when a derivatized ribosomal ligand is used.

Localizing Incorporation Sites

In general, affinity labeling experiments are performed with radioactive affinity labeling reagents. The principal approaches for determining the extent of radioactive labeling of individual proteins are one- and two-dimensional polyacrylamide gel electrophoresis (PAGE) (cellulose acetate has also been used) and immunological analysis. The one-dimensional gels do not fully resolve all the ribosomal proteins, but they do allow measurement of a labeling pattern and the response of such a pattern to changes in the experimental protocol in a fairly rapid fashion. The two-dimensional gels resolve all the ribosomal proteins and can be employed to identify labeled proteins, but are considerably more time-consuming. Since covalent incorporation of an affinity label can sometimes lead to altered electrophoretic mobility, even two-dimensional PAGE analysis is sometimes insufficient for unambiguous identification. In these cases immunological techniques employing purified antibody preparations to individual ribosomal proteins can provide the necessary confirmations. The principal techniques which have been applied in studies on labeled ribosomal proteins include specific immunoprecipitation ("sandwich") using goat antibody to rabbit γ-globulin (13), immunodiffusion using loose agarose solutions (14) and cosedimentation in a sucrose gradient (15). The immunological techniques are excellent for definitive identification of labeled proteins but generally have not been used for rigorously quantifying results, perhaps because of high procedural backgrounds. They thus complement well two-dimensional PAGE analysis.

Measurement of Covalent Incorporation

Measurement of true covalent incorporation in an affinity labeling experiment is a non-trivial problem which has often not been well-resolved in affinity labeling studies. Determination of overall labeling of the ribosome requires removal of all non-covalently bound material, and techniques such as dialysis, co-sedimentation in a sucrose gradient or through a sucrose cushion, millipore filtration or acid-precipitation have been used for this purpose. Since these procedures are not necessarily vigorous enough to remove tightly bound non-covalent material, appropriate controls must be performed. For photoaffinity labels, it has been common to use "labeling" measured in the absence of irradiation as a background value. Although this approach seems reasonable, it makes the tacit assumption that procedures which are adequate to remove unphotolyzed non-covalently bound material will necessarily be adequate to remove photolyzed non-covalently bound material. This is, unfortunately, not always true, as has recently been demonstrated (16-18). A better control is to use pre-photolyzed material, but this may present practical problems which make it hard to do correctly. A fuller discussion of this issue may be found in reference 18. For electrophilic labels, which are generally slow to react, estimates of background are commonly determined by measuring "labeling" values at zero-time, before covalent reaction has had time to proceed, or with radioactive native ligand lacking an electrophilic group. These procedures are also not without objection, since some ribosomal ligands bind non-covalently in a time-dependent manner [e.g., mRNA (19), tRNA (20)], and the non-covalent binding properties of a derivative need not exactly parallel those of the native ligand.

This is not to say that the situation is hopeless, but only that a potential problem exists which must be carefully dealt with. In several cases in our own laboratory we have found that high-salt washing affords a good measure of overall covalent labeling. Strongly bound non-covalent material can even comigrate with a protein during PAGE analysis and in such cases the use of denaturing SDS gels might be appropriate (16). What is clear is that published estimates of covalent labeling for experiments lacking such controls must be viewed with suspicion.

Overall Labeling vs. Incorporation Into Proteins

Once it is clear that covalent labeling has been achieved, control experiments are necessary to investigate whether such labeling is first, site-specific and second, related to the function of the native ribosomal ligand. Since the principal purpose of the affinity labeling experiments described here is to localize ribosomal ligand binding sites to specific ribosomal proteins, such control experiments should be evaluated by measuring incorporation into individual proteins rather than by measuring overall ribosome labeling, for two reasons. First, since measures of overall labeling often overestimate true covalent incorporation, as discussed above, effects which are observed by such measures need not be indicative of effects on covalent incorporation. Second, because of the complexity of the ribosome structure, it is frequently the case that the fraction of radioactivity incorporated into the most highly labeled protein is considerably less than half of even the total true covalent incorporation into the ribosome. An example is provided by our data on puromycin photoincorporation (Figure 1) into protein extracted from a 70S ribosome. Although L23 is clearly the most highly labeled protein, incorporation into it represents only 15% of total incorporation, 25% of incorporation into TP70 and 40% of incorporation into TP50 (11). In this case, it clearly would be unjustified to interpret changes in overall labeling as necessarily reflecting changes in the labeling of L23.

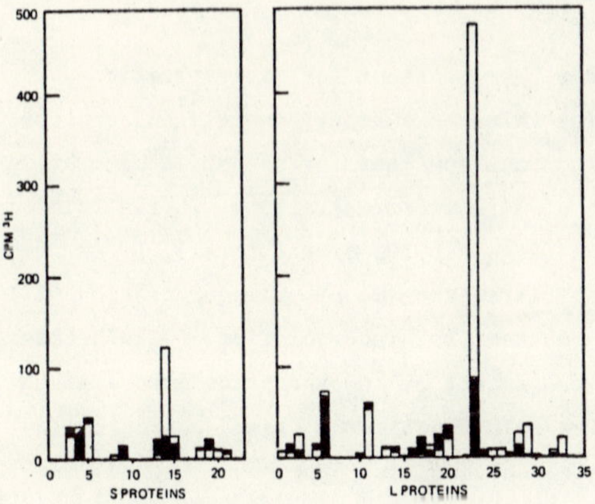

Figure 1: Two-dimensional gel electrophoresis pattern of ribosomal protein labeling at high and low puromycin concentration. (□)0.1 mM puromycin, sp act. 1.68 Ci/mmol; (■) 2.0 mM puromycin, sp act. 0.08 Ci/mmol. (See ref. 11.)

The points made above are obvious, and I discuss them in such detail only because of the large number of publications in which control experiments are carried out only at the level of overall labeling. This is understandable, since overall labeling is experimentally much easier to determine than incorporation into individual proteins, but regretable nonetheless. If it can be shown that incorporation into a single protein accounts quantitatively for all or most of overall labeling, then overall labeling could be used to assess control experiments. However, this would be highly unusual for ribosome affinity labeling studies, and even here could give misleading results. For example, the lack of change in overall labeling on changing the experimental protocol could result from an increase of incorporation into one protein and a decrease of incorporation into another, rather than from no change at all.

SITE-SPECIFIC AND FUNCTIONAL SITE VS. RANDOM LABELING

It is to be expected that in affinity labeling experiments, some of the observed covalent incorporation will arise from random, second order reactions of portions of the ribosome with affinity-labeling reagents coming from solution, rather than from a site on the ribosome. In addition site-specific labeling may occur at sites distinct from the functional site of interest. Below we discuss several approaches which may be used to distinguish site-specific from random labeling, and to obtain evidence that the site specifically labeled is at or near a functional site.

Concentration Dependence of Affinity Labeling

The observed pseudo-first order rate constant for site-specific incorporation reaches a saturating value as a function of ligand concentration, whereas the corresponding rate constant for random incorporation increases linearly as a function of ligand concentration. In the initial zero-order phase of incorporation (e.g. <10-20% of the total), by an argument familiar to enzyme kineticists, the amount of incorporation will be a direct measure of the rate constant for incorporation. Within this initial phase, the finding that the amount of incorporation into a given protein reaches a saturating value as a function of ligand concentration is evidence for site-specific labeling, and also provides a method for estimating the dissociation constant (K_D) of the ligand for that site. On the other hand, if incorporation increases linearly as a function of ligand concentration, one may conclude either that incorporation is random, or is occurring from a site having a K_D value greater than the highest concentration of ligand studied (i.e., a weak site). These experiments are often carried out with a constant amount of radioactive affinity label and increasing amounts of non-radioactive affinity label. In this case, the amount of radioactivity incorporated into the protein decreases as a function of added ligand if labeling is site-specific, and remains constant if it is random. An example of such a measurement is shown in Figure 2, and the data for incorporation into the principal peak (corresponding to protein L23) are plotted as a saturation function in Figure 3.

The great advantage of this approach is that it is totally general and can be applied to study any affinity labeling reaction. On the other hand, it suffers from the disadvantage that site specific labeling determined in this manner need not be the same as functional-site labeling. Evidence for functional-site labeling is provided by a demonstration that the K_D for incorporation is similar to a biologically relevant K_D. For example, we have used the similarity of the K_D value for puromycin labeling of L23 (0.6mM) to the K_m value for puromycin in a peptidyl transferase assay (0.2mM) to argue that L23 is at or near the peptidyl transferase center (11).

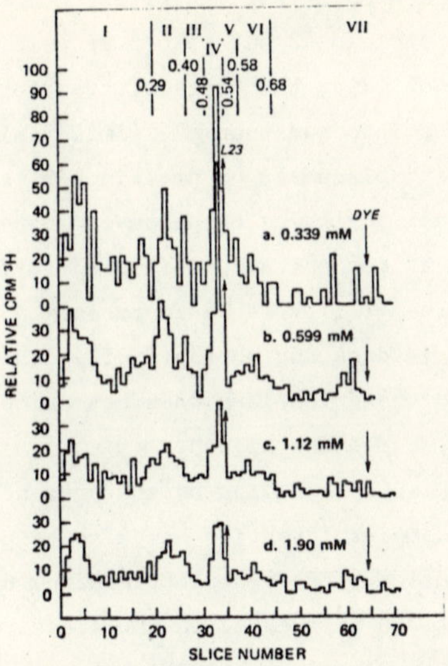

Figure 2. Polyacrylamide gel pattern of labeled proteins from 50S particles as a function of puromycin concentration. The specific activities of puromycin were: (a)890 Ci/mol; (b)504 Ci/mol; (c)269 Ci/mol; 159 Ci/mol. (See ref. 11.)

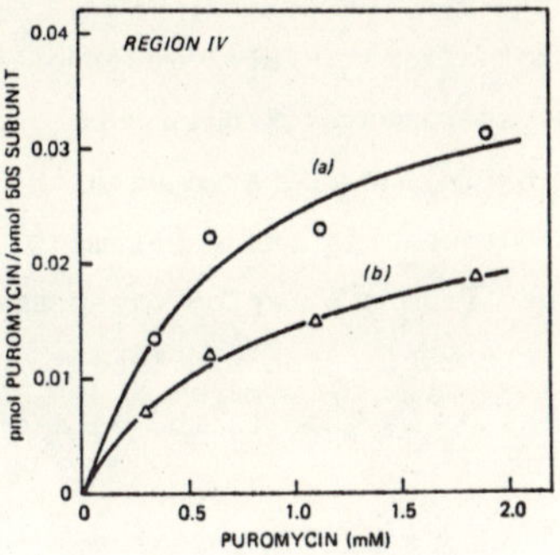

Figure 3. Saturation of region IV labeling with increasing puromycin concentration. (See ref. 11.)

Competitive and Cooperative Effects of Added Ribosomal Ligands

The effects of added ribosomal ligands of known function on labeling can be used to obtain evidence that such labeling occurs at a functional site. Thus, the demonstration by Fiser et al. (21) that R17 mRNA competes for protein S1 photoaffinity labeling by poly s^4U supports placement of S1 at the mRNA binding site. In a like manner, Oen et al. (22) have interpreted the result that peptidyl transferase inhibitors compete for bromoacetyl PhetRNA^{Phe} labeling of protein L2 but not of protein L27 as strengthening the case for placement of L2 at the peptidyl transferase center while at the same time calling into question the similar placement of L27. We have argued for a similar placement of protein L23 from the result that structural and functional analogues of puromycin block puromycin photo-incorporation into L23 at concentrations related to their K_m values in a peptidyl transferase assay (11). It is important to note that when a competition experiment is done for photoaffinity labeling, it is essential to demonstrate that decreased labeling does not arise from internal light filtering by the added, competing ligand.

The dependence of affinity labeling on the presence of a third component necessary for ternary complex formation has also been used as an indication of functional labeling. For example, labeling by tRNA affinity labels has been shown to be stimulated in the presence of cognate mRNA (23-25), and reciprocally, labeling by mRNA has been shown to be stimulated by cognate tRNA (19,26,27).

Mutants and Stereoisomers

Ribosomes isolated from mutant strains can be used to test site-specific and functional labeling. In our laboratory, Arlene Minnella has recently shown that a lincomycin photoaffinity label incorporates less extensively into ribosomal protein in experiments with ribosomes derived from a lincomycin-resistant strain (28) as compared with those derived from wild-type. In a negative but unambiguous experiment, Michael Luddy has shown that ribosomes derived from a streptomycin-resistant strain which lack the high affinity streptomycin site are labeled by a streptomycin photoaffinity label in the same way and to the same extent as wild-type ribosomes.

Many ribosomal ligands contain one or more asymmetric centers, and frequently only one or a limited number of the possible stereoisomers are biologically active. For a group of stereoisomeric affinity labels, correlation of the extent of affinity labeling with biological activity would be a good test of functional labeling. This approach has been used by us in an indirect way to show that only the biologically active stereoisomer of chloramphenicol stimulates puromycin labeling of protein S14 (12).

Gain or Loss of Ribosomal Function

A gain in ribosomal function as a result of affinity labeling can be used to provide evidence for functional site labeling. A loss or inhibition of function may also be significant, although it must be recognized that inhibition of function can reflect an allosteric effect. A particularly cogent example of a gain in function is provided by the work of Hsiung and Cantor (24,29). In their experiments two different nonradioactive aryl azide derivatives of N-acylPhetRNA^{Phe} which function as peptidyl donors were separately photoincorporated into E. coli ribosomes and then treated with either radioactive PhetRNA^{Phe} or puromycin. In each case, radioactivity was found incorporated into both proteins L11 and L18. Since incorporation of radioactivity can only come about as a result of peptide bond formation, these results demonstrate the functionality of tRNA molecules covalently linked to the ribosome, and provide direct evidence for placement of L11 and L18 at the peptidyl transferase center. A similar experiment with N-bromoacetylPhetRNA^{Phe} resulted in incorporation of radioactivity into proteins L2, L15/16 and L27 (23). However, the interpretation of the results of these experiments is more ambiguous, since in this case it was not certain that covalent incorporation into the ribosome preceeded peptide bond formation.

Other examples of this type of study are provided by experiments correlating losses of peptidyl transferase activity with affinity labeling by peptidyl transferase antibiotics (11,30,31) and gains in tRNA binding capability with labeling by mRNA affinity labels (19,26,32,33). Such correlations are commonly used as evidence that those proteins labeled to the highest extents in these experiments are located at the relevant functional sites. Although such interpretations are plausible, they can only be made with confidence if the data obtained satisfy certain important conditions. First, it is typical of these experiments that losses of function are quantitatively correlated with covalent modification. As any non-covalently bound material which is not removed could seriously distort the results obtained, the importance of removing all such material is obvious, particularly as affinity labeling often proceeds in low yield. This is not always straightforward. For example, complete removal of a ligand which is rather tightly bound non-covalently might require extensive washing and result in unacceptable losses of the functional activity under investigation. Second, unambiguous placement of a particular protein at the relevant functional site is only possible if the labeling of that protein is clearly shown to account for all or most of the total covalent labeling, for reasons already discussed. It is our view that many of the studies of this kind reported in the literature fail to satisfy one or both of these criteria, and thus draw what may be unwarranted conclusions.

Before leaving this section we would like to point out, because of the availability of partial and total reconstitution procedures for both the 30S (34) and 50S subunits (35,36), it is possible to construct and functionally test subunits in which only a single protein is affinity labeled. Although this approach has not as yet been exploited in affinity labeling studies, it has been used by several groups employing chemical modification (37-39). An elegant demonstration is provided by the work of Baxter and Zahid (40) who have shown that 50S subunits reconstituted with a modified protein L16 in which only a single histidine has been ethoxyformylated show essentially complete loss of peptidyl transferase activity.

AFFINITY LABELING RESULTS

Tables 1 and 2 present an up to date list of all of the proteins which have been reported to be labeled in affinity labeling experiments. It is obvious by inspection that these results taken as a whole will tell us little about the identity of proteins located at particular functional sites unless the effort is made to distinguish more reliable from less reliable results. For example, eighteen 50S proteins, or more than half the total, have been reported labeled by affinity labels which may be assumed to be directed toward the peptidyl transferase center of the ribosome, i.e., tRNA molecules derivatized at the 3'end and various peptidyl transferase antibiotics.

In what follows we shall attempt to make such distinctions, using three general criteria. First, how convincingly have the identities of labeled proteins been established? Second, what is the evidence that labeling is site-specific and functional site-specific? Third, are the proteins labeled highly nucleophilic?

Proteins of High Nucleophilicity

Because of the total or partial electrophilicity of affinity labeling reagents, the significance of affinity labeling experiments in which highly nucleophilic proteins are labeled should be viewed with considerable skepticism. These proteins may be identified from studies on the reaction of small electrophilic reagents with the ribosome. N-ethyl maleimide (NEM) has been most extensively studied in this regard (31,41-46). Ghosh and Moore (46) have recently shown that the detailed labeling pattern of ribosomal proteins with NEM is quite sensitive both to the mode of ribosome preparation and to whether an intact 70S ribosome or an isolated subunit is the target. Thus it is not surprising that the agreement between the results of different groups working with the reagent is far from complete. Nevertheless it is a fair generalization that in both 30S subunits and 70S ribosomes, S1, S18, and S21 are the most reactive proteins toward NEM, whereas in 30S subunits, but not in 70S ribosomes, S9/11, S12, S13, S19 and perhaps S4 are also highly reactive. Similar results have been obtained for ribosome modification by both iodoacetamide (30,44) and iodoacetic acid (47). The most reactive L proteins toward NEM modification are L2, L8/10, L11, L17 and L27, but there is no general agreement in the literature as to differences in reactivity of L proteins in 50 subunits as compared with 70S ribosomes. Studies with iodoacetamide (44) and N-bromoacetylphenylalanine (31) are largely in accord, except that the latter showed high reactivity in L1 and little in L27.

TABLE 1. Summary of Affinity Labeling Results for 50S Proteins

ligand / L protein	3'-end tRNA	puromycin	chloramphenicol	lincomycin	GDP,GTP	streptomycin
1	+ +					
2	+ +		+ +			
5	+	+			+	
6	+(?)	+				
11	+ +				+ +	
14	+			+ +	+(?)	
15	+ +				+(?)	
16	+ +		+ +			
17	+					
18	+ +	+ +			+	+ +(?)
20	+					
22	+ +					
23		+ +				
24	+ +		+ +			
27	+ +		+ +			
30					+	
34,32/33	+ +					
34	+					
References	23-25,29 48,49	10-13,17, 18,85,87	30,31,88	86	16,55-57	60

+ + major labeled protein
\+ minor labeled protein
(?) identification uncertain

TABLE 2. Summary of Affinity Labeling Results for 30S Proteins

ligand / S protein	3'-end tRNA	mRNA	puromycin	chloramphenicol	lincomycin	streptomycin	tetracycline
1	+	+ +				+ +	
3	+ +	+ +					
4		+ +	+ +				+ +
5					+ +		
6				+ +			
7	+					+ +	+
12		+ +					
13	+ +					+ +	+
14	+ +		+ +			+	+
18	+ +	+ +	+ +			+	+ +
21		+					
References	23,91,92	14,15, 19,21,26, 27,32,33, 50-54,93	10-13 17,18 85	88	86	59,60	89,90

+ + major labeled protein
+ minor labeled protein

FUNCTIONAL SITES

Peptidyl Transferase Center

In Table 4 the data pertaining to the peptidyl transferase center are evaluated more critically. For brevity a letter code is used to describe the data obtained. I and G refer to methods of protein identification, immunological or gel electrophoretic, respectively. SSP refers to a demonstration of site specificity, either by saturation of labeling as a function of affinity label concentration or by competition by native ligand for affinity labeling by a derivative, with incorporation measured at the individual protein level. SSO would refer to such a demonstration measured by overall labeling. Triplets beginning with F refer to evidence for functional site labeling as described by the middle letter. K refers to binding constant measurement, C to competition, T to ternary complex formation, M to evidence from mutants, and G or L to gain or loss of ribosomal function on affinity labeling. The third letter, O or P, again describes whether evaluation is at the overall or individual protein level.

TABLE 3. Peptidyl Transferase Affinity Label Reagents

A. Reactive group attached to the α-amino group of Phe-tRNA^{Phe} and/or Met-tRNA^{fMet}

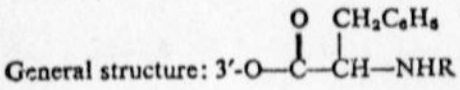

R	References
	22,23,48
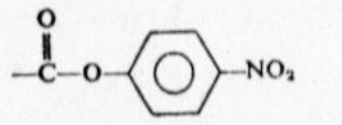	25,49
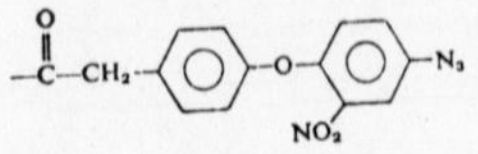	24
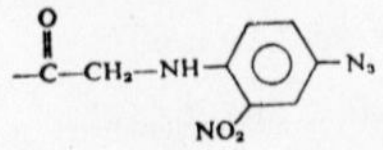	29

B. Puromycin and Derivatives

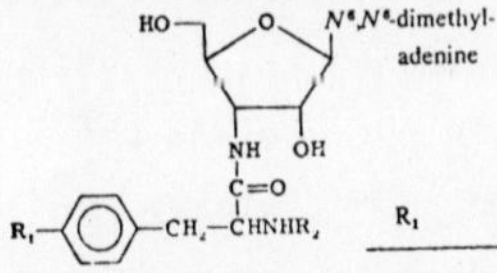

	R_1	R_2	References
Puromycin	CH_3O	H	10-13,18
p-Azidopuromycin	N_3	H	17,18,85

C. Chloramphenicol and Derivatives

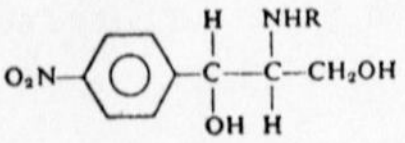

	R	References
chloramphenicol	$\overset{O}{\|}CCHCl_2$	
iodoamphenicol	$\overset{O}{\|}CCH_2I$	30
bromamphenicol	$\overset{O}{\|}CCH_2Br$	31

D. Lincomycin and Derivative

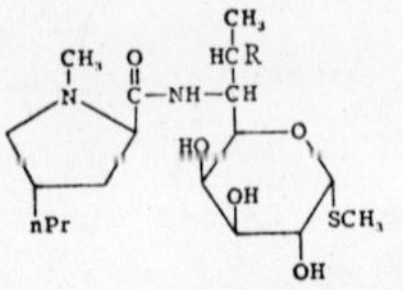

	R	References
lincomycin	OH	18,86
S-(3-Carbethoxy-3-diazo) acetonyl-7-thiolincomycin	$SCH_2\overset{O}{\|}CCN_2CO_2Et$	

The tRNA studies were performed by Cantor and his co-workers and by Kuechler and his co-workers, using the tRNA derivatives shown in Table 3. The identifications of L15 and L27 are secure. That of L2 is also secure, because this protein is so well resolved by two-dimensional gel electrophoresis. On the other hand, proteins L11, L16, and L18 fall in crowded regions of the two-dimensional gel so their identification by gel analysis alone must be considered tentative. In fact, since L15 and L16 migrate close to one another on the two-dimensional gel, and both are reported labeled by electrophilic reagents, it is possible that what has been reported as L16 labeling actually represents L15 labeling, as has been openly acknowledged (23). Three of the proteins labeled, L2, L11, and L27 are among the five most nucleophilic proteins in the 50S particle. However, the extensive controls performed in these experiments make it highly probable that these proteins are in fact involved at the peptidyl transferase center.

Our own work with puromycin and p-azidopuromycin, as summarized in Table 4, has provided strong evidence for the presence of proteins L18 and L23 at the puromycin binding site on the 50S particle. The photolabeling of the ribosome with both of these compounds is clearly site-specific, and for both there is evidence for functional site-specificity as well. Puromycin itself photoincorporates into L23 and p-azidopuromycin, by a purely nitrene-dependent mechanism, photoincorporates into both L18 and L23. That L23 labeling occurs by two different photochemical mechanisms increases our confidence that it is part of the puromycin binding site. Two electrophilic derivatives of chloramphenicol have been found to label L2, L27 and L16 (30,31). Again, the identification of L16, made only by gel analysis, is insecure. Although controls for these experiments are not extensive, so that taken in isolation the results would be questionable, they do parallel the results obtained in the tRNA experiments. The diazo-containing lincomycin derivative shown in Table 3 has provided the first information on the binding site of this antibiotic. The derivative incorporates via a carbene-dependent mechanism and labels a single 50S protein to a high extent. By two-dimensional gel analysis this protein appears to be L14, but our results would not exclude the possibility of L15 as the true site of incorporation. Preliminary experiments have shown this labeling to be decreased both in the presence of lincomycin and in ribosomes isolated from lincomycin-resistant mutant strains.

Putting together all these results, the proteins implicated by affinity labeling studies on peptidyl transferase are L2, L11, L15, L18, L23, L27, and, less securely, L14 and L16, and the definition of this group represents a significant success for the affinity labeling technique. The question which now poses is whether one can go further and make distinctions within this group between proteins which are more and less directly involved. Present evidence permits informed speculation at least for the tRNA experiments, where we would argue that L2, L11, and L18 are more directly implicated. The selection of proteins L11 and L18 is based on the less ambiguous nature of the important experiment demonstrating the peptidyl donor capability of the covalently bound tRNA obtained using the aryl azide derivatives as compared to that obtained using the bromoacetyl derivative, as discussed above. Also, photolabile reagents give inherently more reliable results than electrophilic ones. Of the proteins labeled by electrophilic reagents, L2 is more strongly implicated by the experiment showing that peptidyl transferase antibiotics block N-bromoacetylPhetRNA^{Phe} labeling of L2, but not L27.

mRNA Binding Site

A detailed analysis of mRNA affinity labeling studies is presented in Table 5 where the letter code is the same as that used in Table 4. The affinity labels listed include five oligonucleotides of which four are electrophilic and one photolabile. Pongs and his co-workers have employed three bromoacetyl derivatives of trinucleotides. In two of them, denoted BrAUG and BrUGA, the bromoacetyl group is attached to the 5'end (14,32,33, 50,51) and in one, denoted AUGBr, the bromoacetyl group is attached to the 3'end (51). Similarly, Lührmann et al (26) have employed an affinity label denoted GUUUI, in which an iodoacetyl group is attached to the 3'end of GUUU, and Towbin and Elson (19) have studied affinity labels denoted $A_{6,7}$ azide, in which a nitrophenyl azide is attached to the 3'end of A_6 or A_7. In addition, Kuechler and his co-workers have studied the photo-incorporation of Poly U (27,53) and poly s^4U (15,21,54). Six proteins have been labeled to major extents in these studies and the identification of each is reasonably secure. Four of the six, S1, S4, S12, and S18 are highly nucleophilic. Good control experiments have been done in some of the experiments in which S1 is labeled. Specifically, S1 labeling of poly s^4U has been shown to be inhibited by R17 mRNA, S1 labeling of poly U has been shown to be stimulated in the presence of $tRNA^{Phe}$, and S1 but not S18 labeling by GUUUI has been shown to be stimulated in the presence of $tRNA^{Val}$. On the other hand, in the experiments with BrAUG and BrUGA, no controls evaluated at the protein level have been presented for the labeling of S4, S12, and S18. In the experiments with BrAUG, which are the most extensive, only S18 is labeled in freshly prepared 70S ribosome. Protein S4 is labeled in cold-denatured 70S ribosomes, and protein S12 is only labeled in 30S subunits. The results with S12 and S18 in particular parallel the reactivity of these proteins toward NEM, raising the suspicion that the observed labeling is rather non-specific. Experiments have been presented showing the functionality of ribosomes "labeled" with BrAUG but are difficult to evaluate because it is not clear what fraction of the measured overall "labeling" corresponds to true covalent labeling of the individual proteins. Photolysis of ribosomes in the presence of $A_{6,7}$ azide results in the labeling of two proteins, S3 and S5, neither of which is highly nucleophilic. In addition the labeling of S3 and S5 is stimulated by the addition of the cognate $tRNA^{Lys}$. In summary, the available affinity labeling evidence implicates S1, S3, and S5 at the mRNA binding site, but is ambiguous with respect to the involvement of S4, S12, and S18.

TABLE 5. Major Labeled Proteins in mRNA Affinity Labeling Studies

mRNA Affinity Label	Br AUG Br UGA	AUG Br	$A_{6,7}$ azide	poly s^4U	poly U	GUUUI
Reagent Type	electrophilic	electrophilic	photolabile	photolabile	photolabile	electrophilic
References	14,32,33,50,51	52	19	15,21,54	27,53	26
S protein						
1*		G,I,FGO		G,I,FCP,FLO	G,FTP	G,I,FTP,FCO,FGO
3			G,FTP,FCO,FGO			
4*	G,I,FGO,FLO,FGO					
5			G,FTP,FCO,FGO			
12*	G,I,FGO					
18*	G.I,FCO,FLO,FGO					

*highly nucleophilic protein

Symbols see Table 4

Other Sites

The affinity labeling data on other sites is much more limited and will only be discussed briefly. We have shown that both puromycin and chloramphenicol have binding sites on the 30S subunit, and that, in the presence of chloramphenicol, puromycin photoincorporates in a site-specific manner into protein S14 (12). Girshovich et al. have shown that an aryl azide derivative of PhetRNA^{Phe} photoincorporates into 30S subunits in a poly U dependent manner, with S14 as the protein labeled to the highest extent, and S3 and S7 labeled to lesser extents (91). Since puromycin is a structural analogue of the 3'end of tRNA, these studies together support the notion that tRNA, when bound to the 30S subunit, has a specific binding site for its 3'end of which S14 forms a part. In other studies S18 has been shown to be labeled by bromoacetylPhetRNA^{Phe}. This labeling is competed for by tRNA but is not stimulated by poly U (23). In addition, both p-azidopuromycin (17,18,85) and tetracycline (89,90), an inhibitor of aminoacyltRNA binding, photoincorporate into S4 and S18 and N-ethyl-2-diazomalonylpuromycin photoincorporates into S14 and S18. Thus there is some evidence for implicating S4 and S18 as part of the putative site, but, as in our discussion of the mRNA site, the high nucleophilicity of these two proteins makes such an assignment problematic in the absence of rigorous controls.

Two groups have published photoaffinity labeling studies on ribosome·EF-G·guanine nucleotide ternary complexes employing arylazide derivatives of either GDP or GTP, with apparently contradictory results. Maasen and Möller have used arylazide derivatives of both GDP (55) and GTP (16) in which, in each case, the arylazide was attached to the terminal phosphate. For both derivatives overall ribosomal protein labeling was rather nonspecific. However, with the GDP derivative, addition of fusidic acid led to selective enhancement in the labeling of proteins L11, L5>L18, L30.

With the GTP derivative, addition of a non-hydrolyzable GTP analogue led to specific blocking of the labeling of proteins L11>L5, L14/L15 and L18. In both cases labeled proteins were identified by two-dimensional PAGE. Girshovich et al. (56,57) have studied two aryl azide derivatives of GTP, one of which is very similar to that described by Maasen and Möller, while in the other the aryl azide is attached to the ribose moeity. With both derivatives, Girshovich et al. find predominant labeling of EF-G and only minor, non-specific labeling of ribosomal proteins, in accord with the demonstration by Rohrbach and Bodley (58) of a binary EF-G·GTP complex. In commenting on these differences, Maasen and Möller (16) point out that the methods used by Girshovich et al. to show covalent binding [trichloroacetic acid (TCA) precipitation, cosedimentation] might not have been stringent enough to eliminate sticky, non-covalent binding, and show, using a TCA precipitation procedure, that they can get apparent labeling of EF-G, which disappears when an SDS gel procedure is used instead. Clearly, more work is needed to resolve this dispute. Finally, two derivatives of streptomycin have been used in photoaffinity labeling studies. The nitroazidophenyl derivative of Girshovich et al. (59) labels S7>S14, S16/17 and our dinitrophenyl derivative (60) labels S1 and probably S13, but in neither case is there clear evidence that labeling takes place at the unique, tight streptomycin binding site (83,84).

CORRELATION OF AFFINITY LABELING RESULTS WITH OTHER STUDIES

It is of interest to correlate affinity labeling results with those obtained from other studies. In a topographical sense, the two most relevant types of studies are the immunoelectron microscopic and neutron diffraction (30S particle) localizations of ribosomal proteins and protein cross-linking studies. In a functional sense, the most relevant studies are those measuring functional properties of reconstituted ribosomes, and less directly, of ribosomes isolated from mutant strains.

Peptidyl Transferase Center

Only Stöffler and his co-workers have published localization studies for the antigenic sites of a large number of 50S proteins (61,62), including seven of the eight proteins (not L15) implicated in affinity labeling studies of peptidyl transferase. It is striking that determinants for all seven of these proteins, L2, L11, L14, L16, L18, L23 and L27 fall within a limited area of the 50S subunit, as shown in Figure 4a. In addition, the following cross-links have been found; L2-L11, L11-14,

L16-23, and L16-L27 (63,64). Recent studies of Lake and his co-workers using the immunoelectron microscopic techniques (65-67) (Fig. 4b) place L23 quite far away from L27, in clear disagreement with the results of Stöffler and possibly inconsistent with the L16 cross-linking results as well. We return to this point below. From a functional standpoint, the recent single omission reconstitution results of Nierhaus and his co-workers (68) provide an excellent set of data with which to compare affinity labeling results.

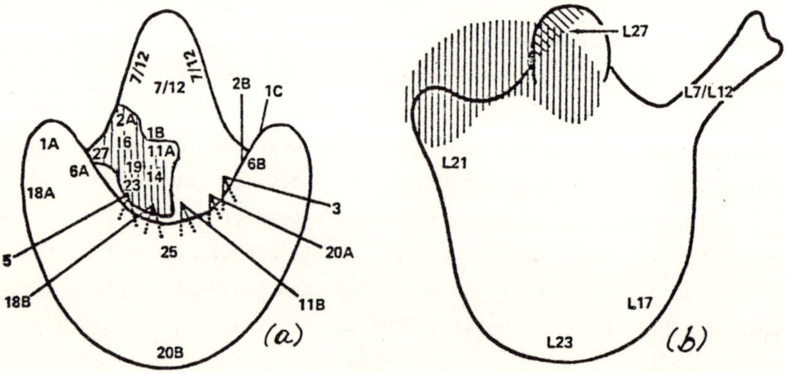

Figure 4. Structural models for 50S subunit and location of protein antigenic sites. (a) Model of Stöffler and co-workers. (b) Model of Lake and co-workers.

They find five proteins to be essential for peptidyl transferase activity and not involved in early assembly, L2, L3, L15, L16 and L18. Four of these proteins (all but L3) are implicated by the affinity labeling technique. In addition, they find an additional seven proteins which are important for activity, of which two, L14 and L27, have also been implicated by affinity labeling. By contrast, L11 and L23 do not seem to be important for peptidyl transferase. These results are consistent with several other partial reconstitution studies, some with chemically modified proteins, showing L2 and L16 to be important or essential for peptidyl transferase activity (38-40,82) and L11, L15, and L16 to be essential for the binding of the antibiotics thiostrepton (69,70), erythromycin (71) and chloramphenicol (72), respectively. Finally, several lincomycin-resistant mutants have recently been isolated having altered proteins L14 and L15 (28). In summary, both topographical and functional studies are in excellent accord with the results of the affinity labeling studies. Of the eight proteins implicated, direct functional roles appear to be possible for six, and two others, L11 and L23, though clearly present in a topographical sense, appear not to be involved functionally.

mRNA and Other Sites

In both the Stöffler et al. (62) and Lake et al. (67) models for the 30S subunit, antigenic determinants for proteins S3 and S5 are close together, and this agrees with the S3-S5 distance measurement from neutron diffraction analysis (73) and the finding of an S3-S5 cross-link (74). On the other hand, the two models disagree with respect to S1, Stöffler et al. placing it fairly close to one of the S5 determinants, whereas Lake et al. place it on the platform, quite far away from S3 and S5.

The protein most clearly implicated as being important for mRNA binding in functional studies is S1 (75-77,79). Protein S3 has been directly implicated in mRNA binding by its ability to restore full poly U binding to 30S particles photo-oxidized in the presence of Rose Bengal (37), but the magnitude of the effect is small. Reconstitution experiments have also shown S3 (34,78) and S5 (37) to be important for poly U -stimulated $tRNA^{Phe}$ binding, although such experiments obviously do not show a direct involvement in mRNA binding. With respect to the other three proteins which might be involved, S4, S12, and S18, the immunoelectron microscopic and neutron diffraction studies are consistent with an S3-S4-S5-S12 neighborhood. The Lake et al. model puts S1 and S18 on the platform close to one another but far from this neighborhood, whereas in the Stöffler et al. model S1 and an S18 determinant are adjacent to it. The relevant cross-links which have been found are S1-S18, S3-S4, S3-S5, S3-S12, S4-S5, S4-S12 (64). Several mutants affecting response to aminoglycoside antibiotics have altered S4, S5, and S12 proteins, and S4 is the site for the ram mutation, but these results need not have a direct bearing on the mRNA binding site. In addition, reconstitution experiments measuring the effect of added native proteins on the restoration of mRNA binding activity of chemically modified 30S subunits have led to the definition of a group of 7 proteins as possibly important (95,96), including most convincingly S1 and S12 and, less clearly, S4 and S18. In summary, the clearest overlaps between the affinity labeling and other results for the mRNA binding site are for proteins S1 and possibly S3. The other four proteins appear to be reasonably close to one or the other of these two proteins.

Two affinity labeling studies support S14 as being part of the binding site for the 3'end of tRNA on an isolated 30S subunit. Experiments utilizing partial reconstitution, single omission, and protection against enzyme digestion (34,78,79,94) are in accord, each of them showing S14 to be important for tRNA binding to the 30S subunit.

Finally, L11 is the protein labeled to the largest extent by a GTP affinity analogue, and is necessary for thiostrepton binding (69), an antibiotic which inhibits GTPase activity. Although L11 cannot be a catalytic center for GTPase activity on the ribosome, since mutant strains lacking L11 on the ribosome have been found (70), it may be close to the GTPase center on the ribosome.

IMMUNOELECTRON MICROSCOPY STUDIES ON AFFINITY-LABELED RIBOSOMES

Recent studies have shown that complexes between an affinity-labeled ribosomal subunit and antibody to the affinity label can be visualized by electron microscopy in much the same way that such complexes have been visualized with antibodies to ribosomal protein. Thus, Keren-Zur *et al.* (80) have used anti-DNP antibody to locate the position of a dinitrophenyl group attached at the 3'end of an aminoacylated $tRNA^{Val}$ which has been photoincorporated into the 30S subunit through a base near the anticodon loop, Stöffler *et al.* have used antichloramphenicol to localize positions of chloramphenicol affinity labeling on the 50S subunit (62), and, in collaboration with Glitz's laboratory, we have localized the site of puromycin incorporation into the 30S subunit (81). As described above, the principal protein labeled in this latter study is S14, and it is gratifying that the dominant location found corresponds very closely to the position of S14 determined by immunoelectron microscopy. This study is now being extended to an examination of the puromycin-labeled 50S subunit. Since the dominant protein labeled is L23, we should soon be in a position to choose between the two models for the location of this protein (Figure 4). Further extensions should permit many other interesting questions to be explored. For example, are the incorporation sites of p-azidopuromycin on the 30S (S4, S18) and 50S (L18, L23) subunits close or identical to those seen with puromycin? How close is the site of lincomycin incorporation (L14) on the 50S subunit to that seen for puromycin? We confidently expect that these and other related applications of this approach will lead to better definition of functional sites on the ribosomes in the coming years.

ACKNOWLEDGEMENT

This work was supported by grants from the National Science Foundation PCM-84278 and the National Institutes of Health AI-14717.

REFERENCES

1. Pellegrini, M. and Cantor, C.R. (1977) in Molecular Mechanisms of Protein Synthesis, (H. Weissbach and S. Pestka, eds.) pp. 203-244, Academic Press, New York.
2. Zamir, A. (1977) Methods Enzymol. 46, 621-637.
3. Cooperman, B.S. (1978) in Bioorganic Chemistry. A Treatise to Supplement Bioorganic Chemistry, An International Journal, (E.van Tamelen, ed.) Vol. IV. pp. 81-115, Academic Press, New York.
4. Kuechler, E. and Ofengand, J. (in press) in Transfer RNA: Structure, Properties and Recognition, (P. Schimmel, D. Soll, and J. Abelson, eds.) Cold Spring Harbor, New York.
5. Singer, S.J. (1967) Advan. Protein Chem. 22, 1-54.
6. Johnson, A.E. and Cantor, C .R. (1977) Methods Enzymol. 46, 180-194.
7. Cooperman, B.S. (1976) in Aging, Carcinogenesis and Radiation Biology. The Role of Nucleic Acid Addition Reactions, (K.C. Smith, ed.)pp. 315-340, Plenum Press, New York.
8. Bayley, H. and Knowles, J.R. (1977) Methods Enzymol. 46, 69-114.
9. Chowdhry, V. and Westheimer, F.H. (1979) Annu. Rev. Biochem. 48, 293-325.
10. Cooperman, B.S., Jaynes, E.N., Brunswick, D.J., and Luddy, M.A. (1975) Proc. Natl. Acad. Sci. U.S.A. 72, 2974-2978.
11. Jaynes, E.N., Jr., Grant, P.G., Giangrande, G., Wieder, R., and Cooperman, B.S. (1978) Biochemistry 17, 561-569.
12. Grant, P.G., Cooperman, B.S., and Strycharz, W.A. (1979) Biochemistry 18, 2154-2160.
13. Grant, P.G., Jaynes, E.N., Jr., Strycharz, W.A., and Cooperman, B.S. (1979) Biochemistry 18, 2149-2154.
14. Pongs, O., Stöffler, G., and Lanka, E. (1975) J. Mol. Biol. 99, 301-315.
15. Fiser, I., Scheit, K.H., Stöffler, G., and Kuechler, E. (1975) FEBS Lett. 56, 226-229.
16. Maasen, J.A. and Möller, W. (1978) J. Biol. Chem. 253, 2777-2783.
17. Nicholson, A.W. and Cooperman, B.S. (1978) FEBS. Lett. 90, 203-208.
18. Cooperman, B.S., Grant, P.G., Goldman, R.A., Luddy, M.A., Minnella, A., Nicholson, A.W., and Strycharz, W.A. (1979) Methods Enzymol. 59, 796-815.
19. Towbin, H. and Elson, D. (1978) Nucleic Acids Res. 5, 3389-3407.
20. Pestka, S. (1966) J. Biol. Chem. 241, 367-372.
21. Fiser, I., Scheit, K.H., and Kuechler, E. (1977) Eur. J. Biochem. 74, 447-456.
22. Oen, H., Pellegrini, M., and Cantor, C.R. (1974) FEBS Lett.45, 218-222.
23. Pellegrini, M., Oen, H., Eilat, D., and Cantor, C.R. (1974) J. Mol. Biol. 88, 809-829.
24. Hsiung, N., Reines, S.A., and Cantor, C.R. (1974) J. Mol. Biol. 88, 841-855.

25. Czernilofsky, A.P., Collatz, E.E., Stöffler, G., and Kuechler, E. (1974) Proc. Natl. Acad. Sci. U.S.A. 71, 230-234.

26. Lührmann, R., Gassen, H.G., and Stöffler, G. (1976) Eur. J. Biochem. 66, 1-9.

27. Margaritella, P. and Kuechler, E. (1978) FEBS Lett. 88, 131-134.

28. Hummel, H., Piepersberg, W., and Bock, A. (1979) Mol. Gen. Genet. 169, 345-347.

29. Hsiung, N. and Cantor, C.R. (1974) Nucleic Acids Res. 1, 1753-1761.

30. Pongs, O., Bald, R., and Erdmann, V.A. (1973) Proc. Natl. Acad. Sci. U.S.A. 70, 2229-2233.

31. Sonenberg, N., Wilcheck, M., and Zamir, A. (1973) Proc. Natl. Acad. Sci. U.S.A. 70, 1423-1426.

32. Pongs, O. and Lanka, E. (1975) Proc. Natl. Acad. Sci. U.S.A. 72, 1505-1509.

33. Yaguchi, M., Lanke, E., Dworniczak, B., Kiltz, H.-H., and Pongs, O. (1978) Eur. J. Biochem. 92, 243-251.

34. Nomura, M., Mizushima, S., Ozaki, M., Traub, P., and Lowry, C.V. (1969) Cold Spring Harbor Symp. Quant. Biol. 34, 49-61.

35. Nierhaus, K.H. and Dohme, F. (1979) Methods Enzymol. 59, 443-449.

36. Amils, R., Matthews, E.A., and Cantor, C.R. (1979) Methods Enzymol. 59, 449-461.

37. Thomas, G., Sweeney, R., Chang, C., and Noller, H.F. (1975) J. Mol. Biol. 95, 91-102.

38. Fahnestock, S.R. (1975) Biochemistry 14, 5321-5327.

39. Dohme, F. and Fahnestock, S.R. (1979) J. Mol. Biol. 129, 63-81.

40. Baxter, R.M. and Zahid, N.D. (1978) Eur. J. Biochem. 91, 49-56.

41. Ginzburg, I. and Zamir, A. (1973) J. Mol. Biol. 79, 481-494.

42. Ginzburg, I. and Zamir, A. (1975) J. Mol. Biol. 93, 465-476.

43. Ginzburg, I. and Zamir, A. (1976) J. Mol. Biol. 100, 387-398.

44. Bakardjieva, A. and Crichton, R.R. (1974) J. Mol. Biol. 143, 599-606.

45. Modolell, J. and Vazquez, D. (1976) in Ribosomes and RNA Metabolism, (J. Zelinka and J. Balan, eds.), Vol. 2, pp. 259-268.

46. Ghosh, N. and Moore, P.B. (1979) Eur. J. Biochem. 93, 147-156.

47. Changchien, L.-M., and Craven, G.R. (1977) J. Mol. Biol. 113, 103-122.

48. Sopori, M., Pellegrini, M., Lengyel, P., and Cantor, C.R. (1974) Biochemistry 13, 5432-5439.

49. Collatz, E., Kuechler, E., Stöffler, G., and Czernilofsky, A.P. (1976) FEBS Lett. 63, 283-286.

50. Pongs, O. and Lanka, E. (1975) Hoppe Syeler's Z. Physiol. Chem. 356, 449-458.

51. Pongs, O. and Rossner, E. (1976) Nucleic Acids Res. 3, 1625-1634.

52. Pongs, O., Stöffler, G., and Bald, R.W. (1976) Nucleic Acids Res. 3, 1635-1646.

53. Fiser, I., Margaritella, P., and Kuechler, E. (1975) FEBS Lett. 52, 281-283.

54. Hauptmann, R., Czernilofsky, A.P., Voorma, H.O., Stoffler, T., and Keuchler, E. (1974) Biochem. Biophys. Res. Commun. 56, 331-337.

55. Maasen, J.A. and Möller, W. (1974) Proc. Natl. Acad. Sci. U.S.A. 71, 1277-1280.

56. Girshovich, A.S., Pozdnyakov, V.A., and Ovchinnikov, Yu.A. (1976) Eur. J. Biochem. 69, 321-328.

57. Girshovich, A.S., Kurtskhalia, T.V., Pozdnyakov, V.A., and Ovchinnikov, Yu. A. (1977) FEBS Lett. 80, 161-163.

58. Rohrbach, M.S. and Bodley, J.W. (1976) Biochemistry 15, 4565-4569.

59. Girshovich, A.S., Bochkareva, E.S., and Ovchinnikov, Yu.A. (1976) Mol. Gen. Genet. 144, 205-212.

60. Luddy, M.A. and Cooperman, B.S. (1979) Manuscript in preparation.

61. Stöffler, G. and Wittmann, H.G. (1977) in Molecular Mechanisms of Protein Synthesis, (H. Weissbach and S. Pestka, eds.) pp.117-202, Academic Press, New York.

62. Stöffler, G., et al., this Volume.

63. Kenny, J.W., and Traut, R.R. (1979) J. Mol. Biol. 127, 243-263.

64. Traut, R.R., et al. this Volume.

65. Lake, J.A. (1978) in Advanced Techniques in Biological Electron Microscopy, (Koehler, J.K., ed.), Vol. II, pp. 173-211, Springer Verlag, New York.

66. Lake, J.A. (in press) in Transfer RNA: Structure, Properties, and Recognition, (P. Schimmel, D. Soll, and J. Abelson, eds.), Cold Spring Harbor, New York.

67. Lake, J.A., this Volume.

68. Nierhaus, K.H., this Volume.

69. Highland, J.H., Howard, G.A., Ochsner, E., Stöffler, G., Hasenbank, R., and Gordon, J. (1975), J. Biol. Chem. 250, 1141-1145.

70. Cundliffe, E., this Volume.

71. Teraoka, H. and Nierhaus, K.H. (1978) J. Mol. Biol. 126, 185-193.

72. Nierhaus, D. and Nierhaus, K.H. (1973) Proc. Natl. Acad. Sci. U.S.A. 70, 2224-2228.

73. Moore, P.B., this Volume.

74. Traut, R.R. and Kenny, J.W. (1977) in "Protein Crosslinking", (M. Friedman, ed.), Part A, pp. 215-233, Plenum Press, New York.

75. Linde, R., Khanh, N.Q., and Gassen, H.G. (1979) Methods Enzymol. 60, 417-436.

76. Van Duin, J., Von Diejen, G., Zipori, P., and Van Prooijen, W. (1979) Methods Enzymol. 60, 436-446.

77. Hermoso, J.M. and Szer, W. (1979) Methods Enzymol. 60, 446-454.

78. Randall-Hazelbauer, L.L. and Kurland, C.G. (1972) Mol. Gen. Genet. 115, 234-242.

79. Rummel, D.P. and Noller, H.F. (1973) Nature New Biol. 245, 72-75.

80. Keren-Zur, M., Boublik, M., and Ofengand, J. (1979) Proc. Natl. Acad. Sci. U.S.A. 76, 1054-1058.

81. Olson, H.M., Grant, P.G., Glitz, D., and Cooperman, B.S., submitted for publication.

82. Moore, V.G., Atchinson, R.E., Thomas, G., Morgan, M., and Noller, H.F., (1975) Proc. Natl. Acad. Sci. U.S.A. 72, 844-848.

83. Chang, F.N. and Flaks, J.G. (1972) Antimicrob. Ag. Chemother. 2, 294-307.

84. Schreiner, G. and Neirhaus, K.H. (1973) J. Mol. Biol. 81, 71-82.

85. Nicholson, A.W. and Cooperman, B.S., manuscript in preparation.

86. Minnella, A. and Cooperman, B.S., manuscript in preparation.

87. Pongs, O., Bald, R., Erdmann, V.A., and Reinwald, E. (1975) in Topics in Infectious Diseases , Vol. I (J. Drews and F.E. Hahn, eds.) pp. 179-190, Springer-Verlag, New York.

88. Pongs, O. and Messer, W. (1976) J. Mol. Biol. 101, 171-184.

89. Tritton, T.R., private communication.

90. Goldman, R.A. and Cooperman, B.S., manuscript in preparation.

91. Girshovich, A.S., Bochkareva, E.S., Kramarov, V.A., and Ovchinnikov, Yu. A. (1974) FEBS Lett. 42, 213-217.

92. Girshovich, A.S., Bochkareva, E.S., and Pozdnyakov, V.A. (1974) Acta Biol. Med. Germ. 33, 639-648.

93. Schenkman, M.L., Ward, D.C., and Moore, P.B. (1974) Biochim. Biophys. Acta 353, 503-508.

94. Shimizu, M., and Craven, G. R. (1976) Mol. Biol. Rep. 3, 105-111.

95. Chang, C., and Craven, G. R. (1977) J. Mol. Biol. 117, 401-418.

96. Cantrell, M., and Craven, G. R. (1977) J. Mol. Biol. 115, 389-402.

ANTIBIOTICS AND PROKARYOTIC RIBOSOMES: ACTION, INTERACTION, AND RESISTANCE

ERIC CUNDLIFFE

Department of Biochemistry,
University of Leicester,
Leicester, England

INTRODUCTION

Studies of the modes of action of antibiotics upon protein synthesis have always carried with them an implied article of faith: namely, that such antibiotics would ultimately be found to possess single "primary" modes of action from which other "secondary" effects would be shown to follow. According to this viewpoint, antibiotics would be highly-specific biochemical tools which could be used to elucidate the mechanism of protein synthesis and to unravel the complexities of ribosomal structure and function. The aim of this article is to outline the extent to which such hopes have been justified. The three sections are designed to indicate progress in the determination of the modes of action of antibiotics against prokaryotic ribosomes, to discuss what is known about antibiotic binding sites in ribosomes and to summarise our knowledge of the manner in which ribosomes can be rendered resistant to the action of antibiotics.

Since it is obviously impossible in the space available to cite all those who deserve credit for the results being discussed, review articles (1 - 3) will be used to direct the reader to relevant literature, particularly in the first section. No attempt will be made to catalogue antibiotics and their structures or to produce an encyclopaedic account of their actions. This has been done elsewhere (1 - 3).

In order to discuss the biochemical modes of action of antibiotics upon protein synthesis it is necessary to employ a mechanistic model for ribosome function. The one which has dominated the field, and is still preferred, is the so-called "translocation-type" model first suggested by Watson (4). According to this scheme, ribosomes possess two tRNA-binding sites and peptide bond formation involves transfer of the growing peptide chain from tRNA bound in one such site (the P site or donor site) onto mRNA-cognate aminoacyl-tRNA bound in an adjacent site (the A site or acceptor site). A fundamental tenet of this model is that the ribosome is functionally asymmetric and that the peptidyl transferase reaction always involves donor substrates in the same site (the P site). The foundation stone for the study of antibiotic action upon ribosomes was laid by Traut and Monro (5) who showed that puromycin could be used as a ubiquitous acceptor substrate in the peptidyl transferase reaction and that this enzymic activity is a property of the 50S ribosomal subunit. Puromycin therefore reacts with, and releases, nascent peptides or other suitable donor substrates from the ribosomal P site but not from the A site. Details of the mechanism of protein synthesis are given in references (1 - 3).

MODES OF ACTION OF ANTIBIOTICS AGAINST PROKARYOTIC RIBOSOMES

Effects upon Initiation and Termination of Protein Synthesis

Surprisingly few antibiotics inhibit specifically polypeptide chain initiation in prokaryotic systems although *a priori* this process might have been expected to afford a prime target for selectively toxic agents. However, as shown by N. Tanaka and colleagues, kasugamycin inhibits the binding of fmet-tRNA into the 30S initiation complex and thus appears to be a specific inhibitor of initiation. These workers also showed similar effects with viomycin; however this drug is also a potent inhibitor of peptide chain elongation and its primary action appears to be upon translocation (reviews 1 -3). Triphenylmethane dyes, such as aurintricarboxylic acid, inhibit *in vitro* the binding of mRNA to the 30S ribosomal subunit, a process which is not known to be affected by any antibiotics. Although aurintricarboxylic acid is not absolutely specific in this action, and some preparations are less specific than others, it is possible to choose concentrations of this compound which produce a specific inhibition of polypeptide chain initiation *in vitro* (reviews 1 - 3). Another drug which inhibits initiation of protein synthesis *in vitro* is edeine (a mixture of basic peptides) although *in vivo* the primary effect of this drug may be upon DNA metabolism. Largely due to the efforts of Kurylo-Borowska and Szer, this drug is believed to act in both prokaryotic and eukaryotic systems by preventing the binding either of mRNA and/or of initiator tRNA to the 30S or 40S ribosomal subunit, prior to involvement of the larger subunit (reviews 1 - 3). Pactamycin was shown by Goldberg and colleagues to be a specific inhibitor of polypeptide chain initiation in eukaryotic extracts when employed at carefully selected concentrations but to inhibit chain elongation at higher drug input. In bacterial systems, however, effects on initiation and elongation "overlap" and the drug cannot be used as a specific inhibitor of initiation (reviewed in 1 - 3).

No specific inhibitors of polypeptide chain termination are known. The peptide-release reaction involves the peptidyl transferase and is inhibited by antibiotics such as chloramphenicol and sparsomycin which act against that enzyme. Also, terminator codons are recognised in the ribosomal A site by release factors which can be prevented from doing so by drugs such as thiostrepton or chlortetracycline which block the A site. However, every drug which has been shown to inhibit termination assays in vitro has also been shown to inhibit other aspects of protein synthesis. Thus it remains to be seen whether negamycin, shown by Umezawa and coworkers to cause misreading of mRNA codons and to affect the initiation, elongation and termination phases of polypeptide synthesis, shows a selective action upon the latter under any circumstances. No other drug is known to achieve this (reviews 2,3).

Effects of Antibiotics upon the Puromycin Reaction in vivo

Various antibiotics halt protein synthesis in intact bacterial protoplasts and "freeze" polyribosomes without significant loss of their nascent peptide chains. Subsequent addition of puromycin can be used to elucidate their modes of action. Not all antibiotics are suitable for use here, either because they are not sufficiently active against polyribosomes or because they cause degradation of these structures (see below). Also, it is usually necessary to employ drugs at quite high concentrations in this system, usually around 1-5 x 10^{-4}M. It may be worth noting, however, that for E.coli the intracellular concentration of ribosomes exceeds 10^{-5}M.

Some antibiotics halted protein synthesis in vivo without affecting the subsequent release of nascent peptides from ribosomes by puromycin. Clearly, such drugs were not affecting the peptidyl transferase reaction nor were they inhibiting translocation, since that would leave nascent chains in the ribosomal A sites from which puromycin could not release them. Accordingly, it was concluded by Cundliffe and McQuillen (6) that chlortetracycline, pactamycin and bottromycin, prevented the binding of amino acyl-tRNA into the ribosomal A site. Similar observations were subsequently made using thiostrepton (7) and fusidic acid (8) and prompted similar conclusions. In contrast, chloramphenicol, sparsomycin and erythromycin inhibited the puromycin reaction _in vivo_ (6). To determine whether they achieved this directly (by inhibiting peptidyl transferase) or indirectly (by inhibiting translocation) the effects of these antibiotics were examined further in protoplasts pre-treated with chlortetracycline. Continued inhibition of the puromycin reaction by chloramphenicol and sparsomycin indicated that these drugs inhibit the peptidyl transferase directly. Failure of erythromycin to inhibit the puromycin reaction under these conditions was taken to imply that this drug normally inhibits a reaction (namely translocation) which cannot proceed following pre-treatment with chlortetracycline (6). Demonstration of this effect does not depend upon any peculiarity in the mode of action of chlortetracycline and similar results have since been obtained, with erythromycin and sparsomycin, in protoplasts pre-treated with bottromycin, pactamycin or thiostrepton (7,9,10). Other antibiotics which behave similarly to sparsomycin _in vivo_ and are evidently inhibitors of the peptidyl transferase reaction include althiomycin, amicetin and, less potently, gougerotin (9,10).

Fusidic acid sequesters elongation factor EFG and GDP on the ribosome and this leads to inhibition of the binding of aminoacyl-tRNA into the A site. The significance of this observation is discussed later. Complexes stabilised by fusidic acid do not persist indefinitely, however, and in protoplasts treated with the drug a slow ribosomal cycle is eventually established. Under these conditions, release of nascent peptides by puromycin was inhibited by spectinomycin, lincomycin and celesticetin. Again, the implication is that each of these three drugs must inhibit either peptidyl transferase activity or translocation and from the kinetics of inhibition it was concluded (10) that they affect translocation.

TABLE 1. Modes of action of antibiotics deduced from studies of the puromycin reaction _in vivo_

Drug	Reaction inhibited
Chlortetracycline, Pactamycin, Bottromycin, Thiostrepton, Fusidic acid	Binding of aminoacyl-tRNA into A site
Chloramphenicol, Sparsomycin, Amicetin, Althiomycin, Gougerotin	Peptidyl transferase
Erythromycin, Spectinomycin*, Lincomycin*, Celesticetin*	Translocation

*Tentative conclusions: in the case of lincomycin _in vitro_ evidence suggests peptidyl transferase is inhibited (see below)

Inhibition of the Peptidyl Transferase Reaction *in vitro*

A variety of systems have been used to assay peptidyl transferase activity *in vitro*. Usually, puromycin is used as acceptor substrate but the donor substrate can vary widely. The following (incomplete) list gives some idea of the range of donor substrates and the manner in which they are introduced to the ribosome:

Polypeptidyl-tRNA present on native polyribosomes.

Polyphenylalanyl-tRNA synthesized *in vitro* in response to polyU.

Initiator-tRNA bound to ribosomes in response to phage mRNA or the AUG triplet codon in the presence of initiation factors and GTP.

N-acetylphe-tRNA bound to ribosomes in the presence of polyU with or without initiation factors and GTP.

Oligonucleotide "fragments" carrying N-acyl aminoacids at their 3'terminus. These are used with 70S ribosomes, or with 50S subunits in the presence of organic solvents such as methanol, in the absence of mRNA, GTP or factors. This technique has been used extensively by Vazquez and coworkers.

It is, perhaps, not surprising that these systems are differently affected by certain drugs. However, the antibiotics sparsomycin, chloramphenicol, althiomycin, amicetin, blasticidin S and gougerotin usually inhibit peptidyl transferase assay systems *in vitro* regardless of their characteristics (reviews 1 - 3). In particular, as shown by Pestka and coworkers, these drugs are potent inhibitors of the puromycin reaction upon native polyribosomes *in vitro* (11,12) which correlates well with their action *in vivo* (Table 1). With the exception of althiomycin, which was not tested, these antibiotics also inhibited the binding of aminoacyl-oligonucleotides (acceptor "fragment" substrates for peptidyl transferase) ostensibly into the ribosomal A site while stimulating binding of the corresponding N-acylated substrate (donor "fragment") supposedly into the P site (13,14). Thus some or all of these drugs may inhibit the peptidyl transferase reaction by preventing recognition of acceptor substrates by the enzyme.

Other antibiotics which inhibit the peptidyl transferase reaction *in vitro* include lincomycin and the streptogramin A drugs (synonyms for the latter include vernamycin A, virginiamycin M and others) but here the picture is somewhat complicated because they are inactive against polyribosomes. These drugs cause degradation of polyribosomes *in vivo* (15) with the accumulation of monosomes blocked "early" in translation. The peptidyl transferase reaction is inhibited *in vitro* with N-acyl aminoacids as donated substrates (16) but there is no effect upon peptidyl-puromycin formation with native polyribosomes (11). Lincomycin and streptogramin A also differ from sparsomycin and the other drugs mentioned above in their ability to inhibit binding to ribosomes of both donor and acceptor "fragment" substrates (13,14). If the binding of such substrates is truly site-specific, these data may imply that lincomycin and streptogramin A can prevent recognition by peptidyl transferase of both donor and acceptor substrates. These two antibiotics apparently inhibit peptidyl transferase activity only on ribosomes bearing nascent peptides shorter than some unspecified, but critical, chain length. Thus, as shown by Contreras and Vazquez (17), removal of nascent chains with puromycin allows binding of these drugs to polyribosomes. As indicated above, the classification of lincomycin as a translocation inhibitor as a result of its action _in vivo_ is not consistent with this body of data.

In contrast to the drugs mentioned above, whose mode of action is reasonably clearly defined, there are various others which have been shown to inhibit peptidyl transferase assays on occasion, but which have also exhibited other apparently contradictory effects. Space does not permit discussion of all these drugs but the macrolides deserve further comment. The macrolide antibiotics possess a lactone ring of 12 - 16 atoms substituted with mono- or disaccharides including amino sugars. It is convenient here to refer to erythromycin (and oleandomycin), the carbomycin-group (including niddamycin and leucomycin) and the spiramycin group (tylosin, relomycin, angolamycin and others). This classification is based not only upon the structures of these drugs but also upon their actions. Other macrolides fall outside these groups and could be considered separately (e.g. lankamycin, narbomycin, methymycin and others). The literature concerning the actions of these drugs is quite confusing and it seems clear that no single, simple statement can describe them. Thus, _in vivo_, erythromycin and oleandomycin (author's unpublished data) stabilise polyribosomes and apparently inhibit translocation (Table 1) whereas macrolides of the carbomycin and spiramycin groups cause rapid degradation of polyribosomes _in vivo_ (15,18) and are inactive against native polyribosomes _in vitro_ (11). Unexpectedly, however, erythromycin was also found to be inactive against polyribosomes _in vitro_ (19). This serves to emphasise the point (discussed below) that with antibiotics which bind preferentially to free ribosomes some of their effects (e.g. inhibition of polyribosomes) may be concentration-dependent.

In considering the effects of the macrolides upon the peptidyl transferase reaction in vitro, two generalizations can be made. Firstly, in any given assay system it is rare for the macrolides to behave similarly and secondly, individual macrolides may behave quite unpredictably in different systems. This latter point may reflect the fact that not all "peptidyl transferase" assays measure exclusively this activity. In particular, those which function in the absence of the 30S ribosomal subunit or which require organic solvent (usually added last) may require donor substrates to bind into the ribosomal P site in the presence of added drug. In contrast, in other assay systems the donor substrate may be firmly bound to the ribosome prior to addition of drugs. Usually, the carbomycin-group of macrolides are potent inhibitors of the puromycin reaction in systems employing N-acylated aminoacids or dipeptides as donated substrates. These drugs also inhibit the binding to ribosomes of both donor and acceptor "fragment" substrates (13) and generally resemble in their actions streptogramin A and lincomycin. Thus, the carbomycin-group of macrolides appear to inhibit the peptidyl transferase reaction. Although the spiramycin group of macrolides also inhibit the binding to ribosomes of both donor and acceptor substrates (13,14), they do not always inhibit model peptidyl transferase reactions (reviews 1 - 3). Further, when added to a system synthesizing polylysine, these drugs caused a marked accumulation of dilysyl-tRNA as if they were inhibiting translocation (20). This latter effect was particularly prominent with erythromycin which, in other systems, sometimes stimulated and sometimes inhibited peptide bond formation (21). Erythromycin also stimulated weakly the binding of both donor and acceptor "fragments" to ribosomes (13). It is concluded from these data that erythromycin differs markedly from other macrolides in its action and may not be a peptidyl transferase inhibitor at all. As discussed below, there is some evidence from in vitro studies to support the contention, originally derived from studies of its action in vivo, that erythromycin inhibits translocation.

Inhibition of Translocation in vitro

The classic studies of the translocation reaction were those of Haenni and Lucas-Lenard (22) who synthesized N-acetyl-diphe-tRNA in the absence of factor EFG in order to obtain a translocatable substrate in the ribosomal A site and then followed its conversion to a puromycin-sensitive state after addition of EFG. Since then, various other systems have been developed to assay translocation but they usually focus attention upon that phase or extent of the puromycin reaction which requires the presence of factor EFG.

The first drug postulated to inhibit translocation was erythromycin as a result of its effects in vivo (6). Support for this conclusion was provided by Kaji and colleagues who showed that the drug inhibited in vitro the puromycin reaction dependent upon the presence of EFG and GTP but not the reaction independent of these additives (23). Also, the drug inhibited release of deacylated tRNA from ribosomes incubated with EFG and GTP. Subsequently, it was found that the ability of erythromycin to inhibit translocation assays varied with the nature of the peptidyl moiety being translocated (24) and that under certain circumstances peptidyl-tRNA could be displaced from ribosomes in the presence of the drug (25; also 19). Others, however, did not observe loss of peptidyl-tRNA from ribosomes in vitro (26) or in vivo (6). It was further suggested (25) that erythromycin allows peptidyl-tRNA to be translocated out of the A site so that, with deacylated tRNA retained in the P site, it ends up loosely bound elsewhere. However, acceptance of this hypothesis would require proof that a number of processes assayed separately in vitro occurred all upon the same small fraction of the ribosomes present. Accordingly, it is concluded here that erythromycin is, in all probability, an inhibitor of translocation as suggested by its effects both in vivo (6) and in vitro in a system actively synthesizing protein (26). Possibly, the drug achieves this by sequestering deacylated tRNA in the ribosomal P site. However, in the absence of unequivocal data, the mode of action of erythromycin is not established beyond question.

There is less doubt, however, concerning the inhibition of translocation by viomycin (closely-related to the capreomycins and tuberactinomycins) and various aminoglycosides including neomycin, kanamycin, gentamicin and hygromycin B but excluding streptomycin. These antibiotics acted similarly in a number of assay systems employed by Modolell and Vazquez and colleagues and by N. Tanaka and coworkers. Protein-synthesising systems were inhibited by these drugs so that peptidyl-tRNA was left in a puromycin-insensitive state, not by direct inhibition of the peptidyl transferase reaction but, rather, by interference with translocation (27-30). There was no effect upon the uncoupled hydrolysis of GTP by ribosomes and factor EFG but, with the possible exception of hygromycin B, which was not tested, they inhibited turnover of [ribosome-EFG-GDP] complexes formed in the presence of fusidic acid (27,31). Evidently, viomycin and the aminoglycosides interact with both ribosomal subunits (see below) where they compete with each other for binding sites (30). When ribosomal subunits were treated separately with aminoglycosides (in particular, neomycin) and then re-associated with untreated partners, it was concluded that effects upon either subunit could lead to inhibition of translocation (31). Viomycin and hygromycin B also inhibited release of peptidyl-tRNA from the ribosomal A site following partial removal of NH_4^+ ions (27,29). Conceivably, this effect might be related to inhibition of translocation due to fixation of peptidyl-tRNA in the A site and, with hygromycin B, might result in misreading due to stabilization of non-cognate aminoacyl-tRNA in the A site. In any event, viomycin is a potent inhibitor of translocation and does not cause misreading of mRNA codons whereas translation continues in the presence of the aminoglycosides and is accompanied by misreading (see below).

Other drugs claimed to inhibit translocation include the novel aminoglycoside, apramycin, a potent inhibitor which also causes low levels of misreading (32). The action of the bottromycins is less clear. Bottromycin A_2 was reported to inhibit translocation *in vitro* by uncoupling site-shifting of peptidyl-tRNA, which was inhibited, from ejection of deacylated tRNA, which was not (review 1). However, *in vivo* the Merck product "bottromycin", which is a mixture of A_1 plus B_1, was a most effective inhibitor of protein synthesis but failed to inhibit the puromycin reaction (see above). Accordingly, this drug appears to inhibit the aminoacyl-tRNA binding reaction. It remains to be seen whether bottromycin and bottromycin A_2 behave differently in similar assay systems. Also lincomycin, celesticetin and spectinomycin were postulated to inhibit translocation *in vivo* (Table 1), although, as noted above, studies with lincomycin *in vitro* do not support this conclusion.

Inhibition of Binding of Aminoacyl-tRNA into the A site

Several antibiotics which inhibit the peptidyl transferase reaction prevent the interaction of acceptor "fragment" substrates with the transferase enzyme (see above). However, such antibiotics may not need to bind into the A site to achieve this since, for example, streptogramin A and lincomycin also prevent the binding of donor "fragments" to ribosomes apparently *via* action at a single site. Nor does inhibition of acceptor "fragment" binding necessarily imply complete blockade of the A site;

thus, of the peptidyl transferase inhibitors discussed above, only streptogramin A (33,34) and sparsomycin (33,35) have been shown to prevent aminoacyl-tRNA from binding into the A site. The classical inhibitors of this process are the tetracyclines. When [aminoacyl-tRNA : EF Tu : GTP] complexes interact with the ribosome-mRNA complex in the presence of chlortetracycline, GTP is hydrolyzed and EF Tu-GDP is released but aminoacyl-tRNA does not bind to the ribosome (1 - 3). This observation can be rationalized according to a model whereby GTP hydrolysis occurs on the 50S ribosomal subunit while chlortetracycline prevents codon-anticodon recognition on the 30S subunit. Other drugs which prevent binding of aminoacyl-tRNA without affecting the associated GTPase event include the peptidyl transferase inhibitors streptogramin A (33) and sparsomycin (35); in contrast thiostrepton is a potent inhibitor of both processes (33). The action of the latter drug and of fusidic acid is discussed in detail below.

Chlortetracycline and thiostrepton also inhibit the synthesis of guanosine tetra- and pentaphosphates upon ribosomes _in vitro_ in response to stringent factor and mRNA-cognate deacylated tRNA (36). Synthesis of these regulatory nucleotides occurs _in vivo_ in response to starvation of stringent bacteria for aminoacids and results in the cessation of synthesis of stable RNA. Evidently, deacylated tRNA is required to bind into the ribosomal A site in order that stringent factor can be activated or allowed to function.

As a result of their effects _in vivo_, both bottromycin and pactamycin were claimed to be inhibitors of the binding of aminoacyl-tRNA to ribosomes (Table 1). Neither antibiotic has been reported to exert this effect _in vitro_ although, as discussed above, the appropriate preparation of bottromycin may not have been tested. While it is known that pactamycin can inhibit polypeptide chain elongation in bacterial extracts, in addition to effects upon initiation complexes, the mechanism by which elongation is affected is not known.

Inhibition of Ribosomal GTPase Reactions

Elongation factor EFG together with washed 70S ribosomes or 50S subunits catalyzes _in vitro_ an "uncoupled" hydrolysis of GTP: this reaction is inhibited by thiostrepton (37) and, under certain conditions, by fusidic acid (38). Since EFG is known to function during translocation, this led to the widespread belief that these two antibiotics inhibit this process, an idea which was strengthened when EFG isolated from fusidic acid-resistant strains of E.coli was found not to respond to the drug (38-40). Eventually, it was established by Bodley et al. (41) that fusidic acid allows a single round of GTP hydrolysis to occur per ribosome before stabilizing a complex of [ribosome - EF G - GDP]. Thus, the drug only inhibits the uncoupled GTPase reaction when GTP is present in molar excess. Thiostrepton was then shown by Weisblum and Demohn (42) and by Bodley et al. (43) to inhibit the assembly of such complexes, which can be formed from either GTP or GDP in vitro. Conversely, ribosomes bearing EF G and GDP in the presence of fusidic acid are protected from irreversible inactivation by thiostrepton (44).

These observations assumed particular significance when thiostrepton was found to inhibit the binding of aminoacyl-tRNA into the ribosomal A site in vivo (7) without affecting translocation. Concurrently, thiostrepton and the related compounds siomycin and thiopeptin were shown by Modollel et al. (33) and by Kinoshita et al. (45) to inhibit with equal potency in vitro EF Tu-dependent binding of aminoacyl-tRNA to ribosomes, the GTPase event associated with binding and the uncoupled hydrolysis of GTP catalyzed by ribosomes and EF G. Results similar to those obtained in vivo were reproduced in vitro when thiostrepton, finally, was tested in a fully synthetic system (26,46). Given that ribosomes possess a single binding site for thiostrepton, located on the 50S subunit (47), these observations suggested that factors EF Tu and EF G utilise alternately a single "GTPase site" on the 50S subunit which is inactivated by thiostrepton (7,33,45). Since the drug does not inhibit translocation, it was additionally postulated that thiostrepton has access to this site only on post-translocation ribosomes, i.e. when the A site is empty (7). At first it was assumed, at least by this author, that the elongation factors activated a latent GTPase activity of the ribosome. Now, however, since EF Tu hydrolyzes GTP off the ribosome in the presence of kirromycin (see below) it is clear that this factor at least carries GTPase capability. In contrast, EF G has not been shown to possess hydrolytic activity off the ribosome. The notion that ribosomes might possess a single "GTPase center" was strengthened when thiostrepton was found to inhibit GTP hydrolysis by IF2 and ribosomes (48).

The single GTPase model carries with it the explicit notion that factors EF G and EF Tu cannot interact with the GTPase simultaneously and might therefore be incapable of binding to the ribosome concurrently. Accordingly, the model predicted that ribosomes bearing EF G and GDP in the presence of fusidic acid should not bind aminoacyl-tRNA from its complex with EF Tu and GTP. This was established directly *in vitro* (49-52) concurrently with the demonstration that *in vivo* fusidic acid resembled thiostrepton in inhibiting this process exclusively, with no effect upon translocation (8). A similar effect was later shown when the drug was added to a cell-free protein-synthesising system (46). Accordingly, it is concluded that in those cases where fusidic acid appeared to inhibit translocation *in vitro*, as for example in the important studies of Haenni and Lucas-Lenard (22), it did so indirectly by sequestering EF G and GDP upon ribosomes which were not participating in translocation. In support of this interpretation is the finding (53) that the effects of fusidic acid *in vitro* can be overcome by addition of EF G and GTP to excess. To date, thiostrepton and the related compounds siomycin, sporangiomycin and thiopeptin are the only agents known to inhibit either the uncoupled hydrolysis of GTP by ribosomes and EF G or the assembly of [ribosome-EF G - GDP] complexes in the presence of fusidic acid.

The action of Kirromycin

This antibiotic affects the conformational state of elongation factor EF Tu which is normally modulated by ribosomes, aminoacyl-tRNA, GTP and GDP. Thus, when [aminoacyl-tRNA : EF Tu : GTP] complexes bind to ribosomes in the presence of kirromycin, GTP hydrolysis occurs but EF Tu is unable to undergo the change of conformation required for its release. The factor therefore remains on the ribosome and aminoacyl-tRNA so bound cannot take part in the peptidyl transferase reaction. Since added puromycin can react, however, it is evident that the peptidyl transferase itself is not directly inhibited. Other effects of kirromycin can also be related to its action upon EF Tu; thus, in the presence of the drug, EF Tu can promote binding of aminoacyl-tRNA to ribosomes in the absence of GTP and can itself hydrolyse GTP off the ribosome. These effects have been documented by Parmeggiani and coworkers (54,55).

Misreading of the Genetic Message

Various aminoglycoside antibiotics including streptomycin, neomycin, kanamycin, gentamicin, hygromycin B and apramycin, cause misreading of mRNA codons *in vitro* and *in vivo*. Detailed examination of patterns of misreading, produced *in vitro*, revealed important differences between the effects of streptomycin on the one hand and those of neomycin, kanamycin and gentamicin on the other. Thus, misreading induced by streptomycin was restricted to predictable aminoacid substitutions, whereas the other aminoglycosides produced "random" miscoding (56). The latter drugs also showed multiphasic effects when inhibition of protein synthesis and misreading were examined over a broad range of drug concentrations (57). For example (58), low concentrations of gentamicin inhibited synthesis with very little misreading; raising the drug concentration caused a marked stimulation of misreading and substantial relief of inhibition; finally with high concentrations of gentamicin, inhibition of synthesis increased. To rationalize these effects it was first suggested by Davies and Davis (57) that neomycin, kanamycin and gentamicin would be found to possess multiple ribosomal binding sites. This prediction has now been confirmed (see below) and offers a ready explanation for failure to isolate single-step mutants showing high level resistance to these drugs. In contrast, streptomycin binds to a single ribosomal site and resistant mutants are readily obtained. The misreading effect with streptomycin _in vitro_ is maximal at about a 1:1 drug to ribosome ratio and generally parallels the level of inhibition of synthesis. Thus, in the presence of streptomycin, post-initiation ribosomes continue to track along mRNA albeit at a reduced rate and with demonstrable misreading. Following detachment from mRNA, "streptomycin-ribosomes" (which are irreversibly affected by the drug) are doomed to a futile existence in which they repeatedly form unstable initiation complexes with mRNA and fmet-tRNA. The instability of such complexes means that streptomycin-affected ribosomes do not get sequestered on old mRNA but recycle continuously. Thus, even after removal of unbound drug, they are able to block the functioning of newly-synthesised ribosomes on new mRNA. Hence the killing action of streptomycin. This mechanism was largely expounded by Davis and colleagues and is reviewed in (59).

Misreading of mRNA codons induced by hygromycin B and by apramycin is not nearly so extensive as that observed with the other aminoglycosides and has not been studied in detail.

ANTIBIOTIC BINDING SITES IN RIBOSOMES

Binding of Native Antibiotics to Ribosomes

The binding of radiolabelled antibiotics to ribosomes or subunits can be examined using a variety of techniques. Some of these involve physical separation of ribosomes from unbound drug, for example by centrifugation, filtration, precipitation or exclusion, and are therefore applicable only in those cases where binding is tight enough to prevent dissociation. Similar considerations apply to assays involving removal of unbound drug by adsorption, e.g. onto activated charcoal or ion exchange resin. Binding equilibria can be examined more rigorously by equilibrium dialysis or in some cases by spectrofluorimetry. The latter technique has the advantage that radiolabelled drug is not required. Dissociation constants (K_d) for binding of antibiotics either to intact ribosomes or to subunits usually fall within the range 5×10^{-7} M to 5×10^{-6} M although there are exceptions (Table 2).

TABLE 2. Binding of antibiotics to E.coli ribosomes

Drug	Sites per Ribosome and Location		K_d	Reference
Thiostrepton	1	50S	-	47
Erythromycin	1	50S	1×10^{-8} M	60
Vernamycin A	1	50S	2×10^{-8} M	61
Streptogramin B alone	1	50S	6×10^{-8} M	62
plus Streptogramin A	1	50S	1×10^{-8} M	
Virginiamycin S	1	50S	$4\text{-}7 \times 10^{-7}$ M	63,64
Dihydrostreptomycin	1	70S	8×10^{-7} M	65
	1	30S	1×10^{-6} M	
Chloramphenicol	1	50S	$1\text{-}3 \times 10^{-6}$ M	66,84
Gougerotin	1	50S	1.5×10^{-6} M	67
Tuberactinomycin O	1	50S	1×10^{-6} M	68
	1	30S	4×10^{-6} M	
Kanamycin	1	50S	2.5×10^{-6} M	30
	1	30S	2×10^{-6} M	
Lincomycin	1	50S	$3\text{-}6 \times 10^{-6}$ M	69
Virginiamycin M	1	50S	3×10^{-6} M	70
Kasugamycin	1	30S	2×10^{-5} M	71
Tetracycline	1	30S	2×10^{-5} M	72

Thiostrepton is quite remarkable in binding so tightly to 50S subunits or 70S ribosomes that a value for K_d cannot be determined by equilibrium dialysis; although binding of this drug is non-covalent, it is effectively irreversible short of disintegrating the 50S subunit. Many antibiotics bind to 70S ribosomes with 1:1 stoichiometry and this primary target site can usually be located unambiguously on one or other subunit, which is not to say that weaker "secondary" interactions do not also occur notably at high drug concentrations. However, some antibiotics bind to ribosomes with 2:1 stoichiometry with a strong binding site on each subunit; conceivably, such drugs may exert dual effects upon ribosomal function.

Many pairs of antibiotics compete for ribosomal binding sites and, at least superficially, there is a correlation with their modes of action. Thus, peptidyl transferase inhibitors such as chloramphenicol, lincomycin and gougerotin compete with each other but not with thiostrepton nor with the translocation inhibitors neomycin, kanamycin, gentamicin, viomycin and tuberactinomycin O. These latter drugs also compete with each other for binding sites on both ribosomal subunits. The well-known synergism with streptogramins A and B can be explained by the observation that the affinity of $[^3H]$-streptogramin B for the ribosome is increased over five-fold in the presence of streptogramin A (Table 2).

By hindsight, it might have been expected that the presence or absence of elongation factors, the site-location of peptidyl-tRNA and the conformational states of individual ribosomes would affect the binding of antibiotics to polyribosomes. Thus, as shown by Contreras & Vazquez (69), polyribosomes contain heterogeneous binding sites for certain drugs, although chloramphenicol, sparsomycin, althiomycin, thiostrepton and, doubtless, many others bind perfectly well to polyribosomes and inhibit their function. Other antibiotics bind preferentially or exclusively to free ribosomes; examples are lincomycin, the streptogramins and macrolides of the carbomycin and spiramycin groups. Effects of erythromycin are concentration-dependent and this drug only inhibits polyribosome function, which it manifestly does *in vivo*, at high drug-input.

Binding to Ribosomes of Antibiotic Affinity Analogues

The use of affinity-labelled probes for the identification of components involved in specific ribosomal sites has been reviewed critically by Cooperman (73) and is also discussed elsewhere in this volume. Comments here will be restricted to the use of radioactive affinity-analogues of antibiotics. Given the use of proper controls to establish the relevance of labelling patterns obtained, this technique affords a powerful and direct tool with which to characterise antibiotic-binding sites in ribosomes. Such controls ought to include a demonstration that each analogue binds to the same site(s) as the parent molecule and shares with it a common mode of action. In general, photoactivated affinity analogues are preferable to others since the specificity of their non-covalent interactions with ribosomes can be established prior to irradiation. In addition to producing alternative analogues, it is also desirable to employ individual probes under a range of reaction conditions since these can influence the labelling patterns obtained. Relatively few rigorous studies have been carried out to date and the results are discussed here only briefly.

Some results obtained with puromycin and its analogues are given in Table 3.

TABLE 3. Ribosomal components labelled by puromycin and its analogues

Reagent	Ribosomal components labelled	Reference
Puromycin*	L23 S14 RNA50	Jaynes et al. (74)
N-ethyl-2-diazomalonyl-puromycin*	S18 S14	Jaynes et al. (74)
p-azido puromycin*	L11 S18 major L2 L3 L17 L18 L27 minor S3 S4 S5	Nicholson & Cooperman (75)
p-azido puromycin*	L6 L13 L18 L22 L25 major L11 L23 L27 L30 minor	Krassnigg et al.(76)
N-iodoacetyl puromycin	L6 major; L2 minor	quoted in Pongs et al. (77)
Bap-Pan-Phe	23S RNA	Greenwell et al.(78)

*photoirradiated: RNA50 = RNA from 50S subunits
p-azidopuromycin and Bap-Pan-Phe are covenient trivial names (see 75, 76, 78 for details)

The parent molecule binds upon photoirradiation to both ribosomal subunits, labelling proteins and RNA, whereas photoactivated analogues gave different patterns of labelling of proteins. One analogue behaved differently in different hands. Also, two electrophilic analogues of puromycin, which did not require photoirradiation, labelled separately protein L6 and 23S RNA. Although these results are not in total agreement, there are some similarities in the labelling patterns obtained and they are rendered more intelligible by comparison with data obtained from cross-linking studies and from immune electron microscopy. Thus, several of the 50S and 30S proteins labelled are neighbors within their respective subunits and various proteins from either group have been located at the subunit interface. For further discussion see Stöffler (79) and contributions elsewhere in this volume.

Other antibiotics and analogues have not been examined in similar depth although Girshovich et al. (80) synthesised an aryl azido-derivative of streptomycin which, upon photoirradiation, labelled proteins S7 and S14 primarily and S16/S17 to a lesser extent. Reagents which did not require photoactivation were used by Pongs and colleagues; these included an iodoacetyl hydrazone of streptomycin which labelled protein S4 (81) and iodoacetyl chloramphenicol which bound to protein L16 (82). In contrast, in the hands of Sonenberg et al. (83) the corresponding bromoacetyl-derivative of chloramphenicol reacted with proteins L2 and L27.

The use of this technique to characterize drug binding sites is still in its infancy although, now that the ground rules have been established, a rapid accumulation of meaningful results can be expected

Antibiotic Binding Sites examined by Partial Reconstitution of Ribosomes

The use of the splitting-reconstitution technique in the identification of functional sites within ribosomes is discussed elsewhere in this volume. This method involves sequential addition of split proteins to core particles until a given function is restored. In a similar way, ribosomal proteins essential for the creation or completion of antibiotic binding sites can be identified. Although this method does not indicate to which ribosomal component(s) a drug actually binds, it provides information complementary to that obtainable by the use of drug affinity analogues.

The reconstitution technique has been exploited extensively by Nierhaus and colleagues. They have identified protein L16 as being essential for the binding to _E.coli_ 50S core particles of chloramphenicol (84) virginiamycin S (64) and, together with protein L15, erythromycin (85). To date, protein L15 stands alone in binding a drug off the ribosome (K_d for erythromycin binding to L15 is 2×10^{-5} M) and protein L16 probably enhances this binding within the ribosome (K_d 70S = 10^{-8} M). Since chloramphenicol is believed to inhibit the peptidyl transferase reaction (as discussed above) it is reassuring that, in a number of laboratories, protein L16 has been shown to be essential for this activity. Other proteins implicated in this way in drug binding are protein L11 for thiostrepton (86) and proteins S3 or

S5 for dihydrostreptomycin (65).

Binding of Thiostrepton to Ribosomes

Thiostrepton binds to 50S ribosomal subunits or intact ribosomes essentially irreversibly although the drug is released if ribosomes are extracted with chloroform. Protein L11 was first implicated in the binding of thiostrepton to the 50S subunit when Highland et al. (86) showed that 4M LiCl cores, which contained 23S RNA plus seven proteins, would bind the drug avidly if supplemented with the purified protein. We have extended these studies and find that [^{35}S]-thiostrepton binds with high affinity to a 23S RNA-L11 complex (87). Following digestion of such complexes with T_1 ribonuclease, about 50 nucleotides of RNA remain bound to protein L11 and the ability to bind thiostrepton is maintained. It is not clear whether the protected RNA is intact or is itself fragmented and it remains to be seen whether fragments of protein L11 can support binding of thiostrepton to 23S RNA. The protected RNA fragment(s) contains the site into which the "thiostrepton-resistance" methylase from *Streptomyces azureus* (the organism which produces thiostrepton - see below) introduces a single methyl group. Together with the evident stabilization of specific RNA-protein complexes by thiostrepton, this should allow precise localisation of the binding site for protein L11 within 23S RNA.

When 23S RNA and protein L11 were examined separately by equilibrium dialysis for their ability to bind [^{35}S]-thiostrepton, weak interaction was observed with 23S RNA (K_d approx. 4 x 10^{-7} M) but not with the protein (M.J.R. Stark, unpublished data). The dominant role of 23S RNA in determining the response to thiostrepton can be seen in ribosomes of *Streptomyces azureus*. These contain a protein, serologically-related to ribosomal protein L11 of *E.coli* (G. Stöffler, personal communication), which can restore to *E.coli* core particles the ability to bind thiostrepton with high affinity. However, *S.azureus* ribosomes do not interact detectably with [^{35}S]-thiostrepton (M.J.R. Stark and E.Cundliffe, unpublished data). The drug does, however, bind to the ribosomes of certain thiostrepton-resistant mutants of *Bacillus megaterium* (K_d approx. 2 x 10^{-7} M; M.J.R. Stark, unpublished data). These mutants possess ribosomes which have lost a protein "BM-L11", the serological homologue of protein L11 of *E.coli*, and although drug binding is much weaker than that observed with wild-type ribosomes, it is nonetheless tighter than that with 23S RNA. It is also as tight as the binding of many other antibiotics to wild-type ribosomes (Table 2). Accordingly it should prove possible to identify other ribosomal proteins which contribute to the "low-affinity" site for thiostrepton.

RIBOSOMAL RESISTANCE TO ANTIBIOTICS

Resistances determined by Ribosomal Proteins

Various bacterial mutants, resistant to antibiotics, possess ribosomes in which mutational alteration of a single protein has been shown to be the resistance determinant. This can be established _in vitro_ by reconstitution of ribosomal subunits containing components derived from both mutant and wild-type or by genetic analysis. The paradigm case, examined by Nomura and coworkers, involves resistance to streptomycin determined by ribosomal protein S12 in _strA_ mutants of _E.coli_ (89). Other ribosomal proteins shown to be resistance determinants in _E.coli_ include S5 in _spcA_ mutants resistant to spectinomycin as shown by Bollen et al. (90) and L4 in _eryA_ or L22 in _eryB_ mutants resistant to erythromycin as shown by Wittmann et al. (91). Mutations in protein L4 may also result in resistance to macrolides other than erythromycin (92) and protein L6, together with another unidentified cellular component, is altered in certain gentamicin-resistant strains (93). These latter are interesting since the ribosomes, while not induced to misread mRNA by gentamicin, are not resistant to inhibition by the drug _in vitro._ Also _ksgC_ mutants which are resistant to kasugamycin contain an altered form of protein S2 (94).

There may be many ways of attaining a given phenotype as a result of multiple mutations affecting the ribosome. Thus _eryA_ strains (altered in protein L4) can be rendered sensitive to erythromycin by the introduction of _strA_ (protein S12) or _spcA_ (protein S5) mutations (95). In the extreme case, as shown by Dabbs and Wittmann (96), a streptomycin-dependent strain altered in protein S8 can revert to drug-independence _via_ mutations affecting almost any ribosomal protein. Finally, in contrast to mutants which possess altered ribosomal proteins, there are the thiostrepton-resistant mutants of _Bacillus megaterium_ (88) and _B.subtilis_ (97) whose ribosomes have lost a protein serologically-homologous with _E.coli_ protein L11 (see below).

Resistances determined by Ribosomal RNA

As first shown by Helser et al. (98), resistance to kasugamycin in ksgA mutants of E.coli results from failure to methylate geminal adenine residues close to the 3' terminus of 16S rRNA. A methylase enzyme present in wild-type can act upon 30S cores from ksgA mutants so that, following reconstitution with split proteins, they are sensitive to the drug. Evidently this enzyme introduces both methyl groups into each of 2 adjacent residues of N^6,N^6-dimethyladenine (99).

In contrast, strains of Staphylococcus or Streptococcus which exhibit the MLS phenotype (resistance to macrolides, lincosamides and streptogramin B-type antibiotics) have been shown by Lai and Weisblum (100) to possess 23S rRNA which is over-methylated in comparison with wild-type. In such strains sub-inhibitory concentrations of erythromycin, a macrolide, induce the synthesis of a plasmid-encoded enzyme which produces approximately two residues of N^6,N^6-dimethyladenine within 23S rRNA. Ribosomes containing such methylated 23S RNA are resistant to macrolides in vitro. The presence of N^6,N^6-dimethyladenine in 23S RNA from Streptomyces erythreus, which produces erythromycin, indicates that ribosomes of this organism may also owe their resistance to the drug to a similar mechanism (101).

The only other example known where drug-resistance in bacterial mutants is determined by ribosomal RNA relates to viomycin-resistant strains of Mycobacterium smegmatis. These mutants, which were produced by mutagenesis, were examined by Yamada et al. (102) and found to carry lesions either in 23S RNA or in 16S RNA although the nature of the modifications are not known.

Resistance to Thiostrepton

Cells of E.coli do not take up thiostrepton and therefore resistance to this drug has been studied in spontaneously-arising mutants of Bacillus megaterium (88) and Bacillus subtilis (97). Several such strains from either organism have been subjected to a rigorous immunological examination and in each case a protein, "BM-L11" or "BS-L11" respectively, serologically-related to E.coli protein L11, appeared to have been lost from ribosomes of the mutants. The B.subtilis mutants regain a protein formally analogous to BS-L11 upon reversion to thiostrepton-sensitivity. The B.megaterium mutants do not revert, but reconstitution of their ribosomes with protein BM-L11 restores wild-type behavior. This involves a two-fold stimulation of their protein-synthetic activity in vitro and has allowed the physiological role of protein BM-L11 to be investigated. Only one of the partial reactions associated with polypeptide chain elongation, namely EF G-dependent hydrolysis of GTP, was stimulated by addition of protein BM-L11 to ribosomes of these mutants (M.J.R. Stark and E. Cundliffe, unpublished data). Protein L11 of E.coli can replace protein BM-L11 in producing similar effects. Accordingly, we conclude that proteins L11 and BM-L11 are related functionally as well as immunologically. These data support the earlier suggestion from Möller and colleagues (103) that protein L11 is involved in the EF G-dependent GTPase reaction on ribosomes, although not indispensably. Protein BM-L11 does appear indispensable, however, for another ribosomal function not directly involved in protein synthesis. The thiostrepton-resistant mutants of B.megaterium and of B.subtilis exhibit relaxed control of RNA synthesis and thiostrepton-sensitive revertants of the latter show stringent control (104). The B.megaterium mutants do not produce guanosine tetra- or pentaphosphates in vivo when starved for isoleucine nor do ribosomes from these strains support synthesis of guanosine polyphosphates in vitro unless supplemented with protein BM-L11 (M.J.R. Stark and E. Cundliffe, unpublished data). To date, this ancillary function of the ribosome is the only process for which an absolute requirement for protein BM-L11 has been demonstrated.

Certain organisms which produce antibiotics are faced with the problem of defending themselves against their own toxic products. This applies to Streptomyces azureus which produces thiostrepton. Ribosomes from this organism fail to bind thiostrepton (as discussed above) and function perfectly well in vitro in the presence of saturated aqueous solutions of the drug (author's unpublished data). Ribosomes from other Streptomyces species are very sensitive to thiostrepton. Reconstitution of hybrid ribosomes containing RNA and proteins from different sources revealed that 23S rRNA from S.azureus determines resistance to thiostrepton (105). Ribosomal proteins from this organism contain an immunological homologue of E.coli protein L11 (G. Stöffler, personal communication) and can be used to reconstitute a high affinity binding site for thiostrepton with rRNA from other organisms (105). Extracts of S.azureus contain an RNA-pentose methylase (106) which introduces a single methyl group into 23S rRNA from other organisms thereby forming 2'-O-methyladenosine. Intact ribosomes or 50S subunits are not substrates for the "thiostrepton-resistance" methylase which acts instead upon 50S core particles or upon 23S RNA extracted from ribosomes with LiCl plus urea and subsequently deproteinised at pH 2 (J. Thompson and E. Cundliffe, unpublished data). The enzyme is only poorly-active upon RNA extracted by the phenol-SDS method. Following methylation of 23S RNA, hybrid 50S subunits can be reconstituted using ribosomal proteins from Bacillus stearothermophilus. Such reconstituted particles are resistant to thiostrepton (105; see also Table 4).

TABLE 4. Action of thiostrepton-resistance methylase from S.azureus on 23S ribosomal RNA

Source of 23S RNA (treatment)	Binding of ^{35}S-THS (pmol./pmol.particles) after reconstitution	Complex* formation (pmol./pmol.particles) after reconstitution	
		minus THS	plus THS
S.coel. (unmethylated)	0.98	0.78	0.02
S.coel. (methylated)	0.07	0.89	0.92
E.coli (unmethylated)	1.04	0.82	0.06
E.coli (methylated)	0.08	0.77	0.76
S.azureus	0.06	0.75	0.76

RNA was incubated with S.azureus methylase with and without S-adenosyl-methionine. Hybrid 50S particles were then reconstituted with B.stearothermophilus proteins.

*Assay for function reconstitution : formation of complexes between reconstituted particles, EFG and [ring-^{3}H]-GTP in the presence of fusidic acid. Thiostrepton (THS), a specific inhibitor of complex formation, was used in three-fold molar excess over particles.

The methylated residue introduced into E.coli 23S RNA by the thiostrepton-resistance methylase is located within that fragment of RNA protected by protein L11 from digestion with T_1 ribonuclease (J. Thompson and E. Cundliffe, unpublished data). Determination of the nucleotide sequence around the methylated residue will allow precise localization within 23S RNA of the binding site for protein L11. Enzymes of this type afford highly specific probes and should prove invaluable in locating binding sites within ribosomal RNA. It is noteworthy that RNA-pentose methylation is rare in bacteria while being commonplace in eukaryotes. Although the Uniformity of Nature may reside only in the eye of the Beholder, it is conceivable that patterns of pentose methylation in ribosomal RNA may underly the discriminatory action of certain antibiotics against either 70S or 80S ribosomes.

CONCLUDING REMARKS

Studies of the biochemical basis of antibiotic action upon ribosomes have contributed fundamentally to our knowledge of the mechanism of protein biosynthesis. If this article does justice to the achievements of the many investigators working in the field, it will be evident that the unravelling of the complexities of structure-function relationships within the ribosome is also being greatly aided by studies of the molecular basis of antibiotic action.

ACKNOWLEDGEMENT

The author is indebted to the Medical Research Council of Great Britain for research support.

REFERENCES

1. Cundliffe, E., (1972) in The Molecular Basis of Antibiotic Action, pp. 278-379, John Wiley & Sons, London.
2. Vazquez, D., (1979) Antibiotic Inhibitors of Protein Biosynthesis, Springer-Verlag, Berlin Heidelberg New York.
3. Pestka, S., (1977) in Molecular Mechanisms of Protein Biosynthesis, pp. 467-553, Academic Press, New York.
4. Watson, J.D. (1964) Bull. Soc. Chim. Biol. 46, 1399-1425.
5. Traut, R.R., and Monro, R.E. (1964) J. Mol. Biol. 10, 63-72.
6. Cundliffe, E., and McQuillen, K. (1967) J. Mol. Biol. 30, 137-146.
7. Cundliffe, E. (1971) Biochem. Biophys. Res. Commun. 44, 912-917.
8. Cundliffe, E. (1972) Biochem. Biophys. Res. Commun. 46, 1794-1801.
9. Cundliffe, E. (1972) in Molecular Mechanisms of Antibiotic Action on Protein Biosynthesis and Membranes, pp. 242-261, Elsevier Publishing Company, Amsterdam.
10. Burns, D. J. W., and Cundliffe, E. (1973) Eur. J. Biochem. 37, 570-574.
11. Pestka, S. (1972) J. Biol. Chem. 247, 4669-4678.

12. Pestka, S., and Hintikka, H. (1971) J. Biol. Chem. 246, 7723-7730.

13. Celma, M.L., Monro, R. E., and Vazquez, D. (1970) FEBS Lett. 6, 273-277.

14. Harris, R., and Pestka, S. (1973) J. Biol. Chem. 248, 1168-1174.

15. Cundliffe, E. (1969) Biochemistry 8, 2063-2066.

16. Monro, R.E., and Vazquez, D. (1967) J. Mol. Biol. 28, 161-165.

17. Contreras, A., and Vazquez, D. (1977) Eur. J. Biochem. 74, 539-547.

18. Ennis, H. L. (1972) Antimicrob. Ag. Chemother. 1, 197-203.

19. Tai, P. C., Wallace, B. J., and Davis, B. D. (1974) Biochemistry 13, 4653-4659.

20. Mao, J. C-H., and Robishaw, E. E. (1971) Biochemistry 10, 2054-2061.

21. Mao, J. C-H., and Robishaw, E. E. (1972) Biochemistry 11, 4864-4872.

22. Haenni, A. L., and Lucas-Lenard, J. (1968) Proc. Natl. Acad. Sci. 61, 1363-1369.

23. Igarashi, K., Ishitsuka, H., and Kaji, A. (1969) Biochem. Biophys. Res. Commun. 37, 499-504.

24. Tanaka, S., Otaka, T., and Kaji, A. (1973) Biochim. Biophys. Acta 331, 128-140.

25. Otaka, T., and Kaji, A. (1975) Proc. Natl. Acad. Sci. 72, 2649-2652.

26. Cannon, M., and Burns, K. (1971) FEBS Lett. 18, 1-5.

27. Modolell, J., and Vazquez, D. (1977) Eur. J. Biochem. 81, 491-497.

28. Cabañas, M. J., Vazquez, D., and Modolell, J. (1978) Biochem. Biophys. Res. Commun. 83, 991-997.

29. Cabañas, M. J., Vazquez, D., and Modolell, J. (1978) Eur. J. Biochem. 87, 21-27.

30. Misumi, M., Nishimura, T., Komai, T., and Tanaka, N. (1978) Biochem. Biophys. Res. Commun. 84, 358-365.

31. Campuzano, S., Vazquez, D., and Modolell, J. (1979) Biochem. Biophys. Res. Commun. 87, 960-966.

32. Perzynski, S., Cannon, M., Cundliffe, E., Chahwala, S. B., and Davies, J. (1979) Eur. J. Biochem. in press.

33. Modolell, J., Cabrer, B., Parmeggiani, A., and Vazquez, D. (1971) Proc. Natl. Acad. Sci. 68, 1796-1800.

34. Hill, R. N. (1969) J. gen. Mic. 58, viii.

35. Ravel, J. M., Shorey, R. L., and Shive, W. (1970) Biochemistry 9, 5028-5033.

36. Pedersen, F. S., Lund, E., and Kjeldgaard, N. O. (1973) Nature New Biol. 243, 13-15.

37. Pestka, S. (1970) Biochem. Biophys. Res. Commun. 40, 667-674.

38. Tanaka, N., Kinoshita, T., and Masukawa, H. (1968) Biochem. Biophys. Res. Commun. 30, 278-283.

39. Tocchini-Valentini, G. P., and Mattoccia, E. (1968) Proc. Natl. Acad. Sci. 61, 146-151.

40. Bernardi, A., and Leder, P. (1970) J. Biol. Chem. 245, 4263-4268.

41. Bodley, J. W., Zieve, F. J., Lin, L., and Zieve, S. T. (1970) J. Biol. Chem. 245, 5656-5661.

42. Weisblum, B., and Demohn, V. (1970) FEBS Lett. 11, 149-152.

43. Bodley, J. W., Lin, L., and Highland, J. H. (1970) Biochem. Biophys. Res. Commun. 41, 1406-1411.

44. Highland, J. H., Lin, L., and Bodley, J. W. (1971) Biochemistry 10, 4404-4409.

45. Kinoshita, T., Liou, Y-F., and Tanaka, N. (1971) Biochem. Biophys. Res. Commun. 44, 859-863.

46. Celma, M. L., Vazquez, D., and Modolell, J. (1972) Biochem. Biophys. Res. Commun. 48, 1240-1246.

47. Sopori, M. L., and Lengyel, P. (1972) Biochem. Biophys. Res. Commun. 46, 238-244.

48. Springer, M., Dondon, J., Graffe, M., Grunberg-Manago, M., Lelong, J. C., and Gros, F. (1971) Biochimie 53, 1047-1057.

49. Richman, N., and Bodley, J. W. (1972) Proc. Natl. Acad. Sci. 69, 686-689.

50. Cabrer, B., Vazquez, D., and Modolell, J. (1972) Proc. Natl. Acad. Sci. 69, 733-736.

51. Miller, D. L. (1972) Proc. Natl. Acad. Sci. 69, 752-755.

52. Richter, D. (1972) Biochem. Biophys. Res. Commun. 46, 1850-1856.

53. Burns, K., Cannon, M., and Cundliffe, E. (1974) FEBS Lett. 40, 219-223.

54. Wolf, H., Chinali, G. and Parmeggiani, A. (1977) Eur. J. Biochem. 75, 67-75.

55. Fasano, O., Bruns, W., Crechet, J-B., Sander, G., and Parmeggiani, A. (1978) Eur. J. Biochem. 89, 557-565.

56. Davies, J., Jones, D. S., and Khorana, H. G. (1966) J. Mol. Biol. 18, 48-57.

57. Davies, J., and Davis, B. D. (1968) J. Biol. Chem. 243, 3312-3316.

58. Tai, P-C., and Davis, B. D. (1979) Biochemistry 18, 193-198.

59. Davis, B. D., Tai, P-C., and Wallace, B. J. (1974) in Ribosomes, pp. 771-789, Cold Spring Harbor Laboratory.

60. Pestka, S. (1974) Antimicrob.Ag. Chemother. 6, 474-478.

61. Ennis, H. L. (1971) Biochemistry, 10, 1265-1270.

62. Contreras, A., and Vazquez, D. (1977) Eur. J. Biochem. 74, 549-551.

63. Parfait, R., de Béthune, M. P., and Cocito, C. (1978) Molec.Gen. Genet. 166, 45-51.

64. de Béthune, M. P., and Nierhaus, K. H. (1978) Eur. J. Biochem. 86, 187-191.

65. Schreiner, G., and Nierhaus, K. H. (1973) J. Mol. Biol. 81, 71-82.

66. Contreras, A., Barbacid, M., and Vazquez, D. (1974) Biochim. Biophys. Acta 349, 376-388.

67. Barbacid, M., and Vazquez, D. (1974) Eur. J. Biochem. 44, 445-453.

68. Misumi, M., and Tanaka, N. (1978) Biochem. Biophys. Res. Commun. 82, 971-976.

69. Contreras, A., and Vazquez, D. (1977) Eur. J. Biochem. 74, 539-547.

70. Cocito, C., and Di Giambattista, M., (1978) Molec. Gen. Genet. 166, 53-59.

71. Okuyama, A., Tanaka, N., and Komai, T. (1975) J. Antibiotics 28, 903-905.

72. Tritton, T. R. (1977) Biochemistry 16, 4133-4138.

73. Cooperman, B. S. (1978) in Bio-organic Chemistry, vol. 4, pp. 81-115, Academic Press, New York.

74. Jaynes, E. N., Grant, P. G., Giangrande, G., Wieder, R., and Cooperman, B. S. (1978) Biochemistry 17, 561-569.

75. Nicholson, A. W., and Cooperman, B. S. (1978) FEBS Lett. 90, 203-208.

76. Krassnigg, F., Erdmann, V. A., and Fasold, H. (1978) Eur. J. Biochem. 87, 439-443.

77. Pongs, O., Nierhaus, K. H., Erdmann, V. A., and Wittmann, H. G. (1974) FEBS Lett. 40, suppl. S28-S37.

78. Greenwell, P., Harris, R. J., and Symons, R. H. (1974) Eur. J. Biochem. 49, 539-554.

79. Stöffler, G. (1974) in Ribosomes, pp. 615-667, Academic Press, New York.

80. Girshovich, A. S., Bochkareva, E. S., and Ovchinnikov, Y. A. (1976) Molec. Gen. Genet. 144, 205-212.

81. Pongs, O., and Erdmann, V. A. (1973) FEBS Lett. 37, 47-50.

82. Pongs, O., Bald, R., and Erdmann, V. A. (1973) Proc. Natl. Acad. Sci. 70, 2229-2233.

83. Sonenberg, N., Wilckek, M., and Zamir, A. (1973) Proc. Natl. Acad. Sci. 70, 1423-1426.

84. Nierhaus, D., and Nierhaus, K. H. (1973) Proc. Natl. Acad. Sci. 70, 2224-2228.

85. Teraoka, H., and Nierhaus, K. H. (1978) J. Mol. Biol. 126, 185-193.

86. Highland, J. H., Howard, G. A., Ochsner, E., Stöffler, G., Hasenbank, R., and Gordon, J. (1975) J. Biol. Chem. 250, 1141-1145.

87. Thompson, J., Cundliffe, E., and Stark, M. (1979) Eur. J. Biochem. in press.

88. Cundliffe, E., Dixon, P., Stark, M., Stöffler, G., Ehrlich, R., Stöffler-Meilicke, M., and Cannon, M. (1979) J. Mol. Biol. in press.

89. Ozaki, M., Mizushima, S., and Nomura, M. (1969) Nature 222, 333-339.

90. Bollen, A., Davies, J., Ozaki, M., and Mizushima, S. (1969) Science 165, 85-86.

91. Wittmann, H. G., Stöffler, G., Apirion, D., Rosen, L., Tanaka, K., Tamaki, M., Takata, R., Dekio, S., Otaka, E., and Osawa, S. (1973) Molec. Gen. Genet. 127, 175-189.

92. Tanaka, K., Tamaki, M., Itoh, T., Otaka, E., and Osawa, S. (1971) Molec. Gen. Genet. 114, 23-30.

93. Kühberger, R., Piepersberg, W., Petzet, A., Buckel, P., and Böck, A. (1979) Biochemistry 18, 187-193.

94. Okuyama, A., Yoshikawa, M., and Tanaka, N. (1974) Biochem. Biophys. Res. Commun. 60, 1163-1169.

95. Saltzman, L., and Apirion, D. (1976) Molec. Gen. Genet. 143, 301-306.

96. Dabbs, E. R., and Wittmann, H. G. (1976) Molec. Gen. Genet. 149, 303-309.

97. Wienen, B., Ehrlich, R., Stöffler-Meilicke, M., Stöffler, G., Smith, I., Weiss, D., Vince, R., and Pestka, S. (1979) J. Biol. Chem. in press.

98. Helser, T. L., Davies, J. E., and Dahlberg, J. E. (1971) Nature New Biol. 233, 12-14.

99. Helser, T. L., Davies, J. E., and Dahlberg, J. E. (1972) Nature New Biol. 235, 6-9.

100. Lai, C. J., and Weisblum, B. (1971) Proc. Natl. Acad. Sci. 68, 856-860.

101. Graham, M. Y., and Weisblum, B. (1979) J. Bact. 137, 1464-1467.

102. Yamada, T., Mizugichi, Y., Nierhaus, K. H., and Wittmann, H.G. (1978) Nature 275, 460-461.

103. Schrier, P. I., and Möller, W. (1975) FEBS Lett. 54, 130-134.

104. Smith, I., Paress, P., and Pestka, S. (1978) Proc. Natl. Acad. Sci. 75, 5993-5997.

105. Cundliffe, E. (1978) Nature 272, 792-795.

106. Cundliffe, E. and Thompson, J. (1979) Nature 278, 859-861.

PART 4: TRANSLATIONAL FIDELITY

THE FIDELITY OF PROTEIN SYNTHESIS

J. J. HOPFIELD†

Joseph Henry Laboratory, Department of Physics,
Princeton University, Princeton, NJ 08544
and Bell Laboratories, Murray Hill, NJ 07971

T. YAMANE

Bell Laboratories, Murray Hill, NJ 07971

The error rate in protein synthesis is a product of four major factors. Errors in making tRNA and mRNA, in charging a tRNA with its correct amino acid, and in codon recognition on the ribosome all lead to errors in protein synthesis. In addition there are errors in ribosomal processes such as mis-translocation, premature release, etc. The actual rate of making errors in vivo is somewhat uncertain due to rapid cellular degradation of ill-formed proteins and the wide range of lifetimes of functional proteins (1). The experiments of Loftfield and coworkers (2),(3) on erroneous incorporation of Val for Ile in rabbit hemoglobin give a net erroneous substitution (^{Val}/Ile) fraction for this particular error (presuming no effect on protein lifetime) of about 1/3000, but do not indicate how much contribution comes from each of the four processes. The residual OTC enzymatic activity of amber mutants is presumably due to misreading of a stop-codon (4), for it is suppressed in streptomycin resistant (error-reducing) double mutants StrA1 argF40. Thus the erroneous reading of stop as a functional amino acid occurs at a frequency of about 1/2000. From Ile/Val kinetics of the synthetase for $tRNA^{Ile}$, the probability of erroneously putting Val on $tRNA^{Ile}$ is about 1/10,000. The error rates of RNA polymerases are believed also to be about $1/10^4$. Thus all of the four major sources of simple errors in protein synthesis appear to make contributions on comparable levels, though any may dominate in generating some particular error. Unfortunately, it is not always clear whether "the errors" in a process such as tRNA charging are due to intrinsic inaccuracy in a perfect enzyme or due to occasional incorrectly made enzyme molecules. In our discussions of errors in protein elongation and tRNA charging, we will use these numbers as illustrative of the errors of a system of perfect ribosomes or synthetases, keeping in mind the limitations of this interpretation.

A simple (i.e. non-proofreading) Michaelis-Menten enzyme which recognizes a substrate A and nominally rejects a similar substrate B can be described by the kinetic scheme

$$E + A \underset{k_A}{\overset{k'_A}{\rightleftarrows}} EA \overset{W_A}{\rightarrow} E + P_A \quad \text{correct}$$

$$E + B \underset{k_B}{\overset{k'_B}{\rightleftarrows}} EB \overset{W_B}{\rightarrow} E + P_B \quad \text{error}$$

The ribosome-message complex with an empty A-site comprises a compound enzyme whose specificity is determined by the message triplet occupying the A-site. The fact that the reaction scheme has many kinetic intermediates, and several substrates in the case of the aminoacyl-tRNA synthetases does not greatly alter the general considerations as long as the pathway is unbranched. "Simple" enzymes have the property

$$\text{substrates} \leftrightarrows \text{intermediate} \rightarrow \text{products}$$

with no branches possible, while all systems with the possibility of proofreading (5) contain a branch

$$\text{substrates} \leftrightarrows \text{intermediate} \begin{array}{l} \nearrow \text{products} \\ \searrow \text{other products} \end{array}$$

We deal first with simple enzymes. If the reaction is carried out too close to equilibrium, the reactions inverse to W_A and W_B will take place, lowering the specificity. Under normal cellular conditions, however, the back-reactions are unimportant both in elongation and in charging tRNA. In this case, if the concentrations of free substrate are [A] and [B], and the error fraction f of the enzyme is defined as

$$f = \frac{\text{rate of producing erroneous } P_B}{\text{rate of producing correct } P_A} \ldots \qquad (1)$$

then

$$f = \left(\frac{[B]}{[A]}\right) \left[\left(\frac{k'_B}{k_A}\right) \left(\frac{k_A+W_A}{k_B+W_B}\right)\left(\frac{W_B}{W_A}\right)\right] \ldots \qquad (2)$$

The bracketed factor describes the intrinsic accuracy of the enzyme, and the initial factor reflects the effect of the pool size.

Eq.(2) can be given a crude stereochemical meaning if an Eyring viewpoint is taken on the reaction. The reaction is then described by the

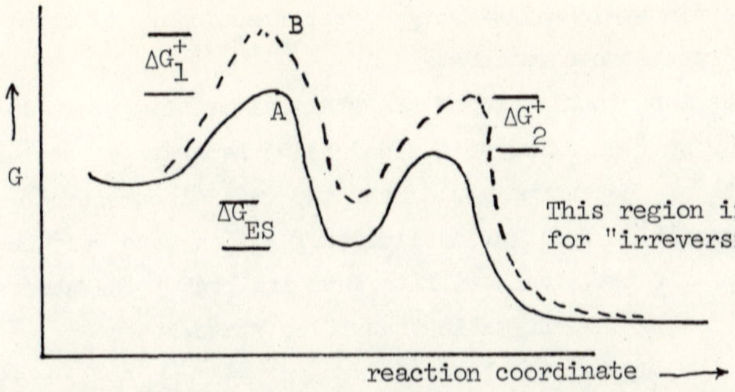

free energy as a function of reaction coordinate, and reaction rates are described by the barrier heights to the various reactions. The smallest error fraction (for [B]/[A] = 1) which can possibly be obtained from Eq.(2) is given by f_o

$$f_o = e^{-\Delta G/RT} \quad (3)$$

where ΔG is the larger of ΔG_1^{+} and ΔG_2^{+}. Thus the stereochemistry of transition states determines the accuracy (5). In this sense, accuracy in protein synthesis is always a kinetic problem. The importance of kinetics is emphasized by the fact that the peptide bonds formed or the amino acid - tRNA links are of essentially identical energies whether they are correct or incorrect. Kinetic limitations are essential to get any accuracy at all, and have already been maximized (subject to enzyme constraints) by keeping the reaction W essentially irreversible.

The competition of valine and isoleucine for an Ile-specific enzyme provides a simple illustration of the stereochemical discrimination problem. The sidechains of Val and Ile are isosteric,

Val Ile

with Val having a hydrogen in place of the Ile methyl. An enzymatic binding pocket ideally arranged to favor Ile will pack 6-9 methyl-equivalents tightly around the critical methyl, which will then contribute (6-9) x 0.27 Kcal of van der Waals energy to the binding of Ile. Val will fit into this pocket also, but will lack this ~ 2.0 Kcal of van der Waals energy. In addition, the extra hydrophobic methyl group is sequestered from water in such an environment, and the partition coefficients between aqueous and non-aqueous solvents assign about -0.5 Kcal of binding of each additional methyl to water (6). Thus the total discrimination energy available for the Val-Ile choice is optimally about 2.5

Kcal, leading to an optimum Michaelis-Menten error fraction f_o of about 1/65. for equal substrate concentrations.

This system has the functional site of the chemical reaction (COO^-) and the recognition group (the amino acid side chain) separately located. The most effective use of the enzyme would keep the initial barrier insignificant (for example in a diffusion limited reaction) and rely on $\Delta G_2^{+} = \Delta G$ for accuracy. If the sidechain fits into its pocket the same way in the ES complex as in the enzymatic transition state, then $\Delta G_{ES} = \Delta G_2^{+}$. This simple supposition leads to $W_A = W_B$ and $k'_B k_A / k_B k'_A = \frac{1}{65}$. Experiments on the Ile recognizing synthetase show, for the formation of the first irreversible intermediate (AA·AMP) as studied by hydroxamate assay or pyrophosphate exchange (7)

$$W_A = 90, \quad W_B = 70,$$

$$k'_B k_A / k_B k'_A \approx 1/30 \text{ to } 1/160.$$

Thus this enzyme appears in its earliest stages to have simple and expected stereochemical selectivity. Note that the accuracy could have been the same even if $\Delta G_{ES} < \Delta G_2^{+}$, but the detailed stereochemical implication would be different.

There is an intrinsic asymmetry in our ability to estimate ΔG. A substrate which is too small necessarily will fit into a space designed for a larger substrate with a relatively calculable loss of binding energy. However, a substrate which is too large must distort a site which is too small for it, and the distortion energy depends in detail on enzyme rigidity. Thus in the Val-Ile competition for the valine-recognizing enzyme it is both easier to achieve good accuracy (by making a rigid pocket) and difficult to estimate the likely level of accuracy.

tRNA recognition without T_u on the mRNA-ribosome complex most probably resembles the case of the Ile-enzyme. The binding to the A-site itself is very strong, and the release rate from the A-site configuration does not compete with peptide bond formation (8,9,10). Thus the general scheme is

$$\begin{array}{l} \text{AA·tRNA·mRNA·rib} \underset{W'}{\overset{W}{\rightleftharpoons}} \text{A} \xrightarrow{\Gamma} \text{peptide bond} \\ \text{A}' \\ \quad k' \upharpoonleft\downharpoonright k \\ \quad \text{AA·tRNA} \\ \quad\quad + \\ \quad \text{mRNA·rib} \end{array} \qquad (4)$$

with W' insignificant except under unusual circumstances where Γ is small. A' represents a looser state of binding at the same site as A. Kurland

and others have emphasized the probable two-step (at least) nature of binding to the A-site. The process W probably corresponds to the binding of the aminoacyl stem of tRNA to the ribosome after codon-anticodon pairing (or mispairing) has taken place. The aminoacyl terminal is rather flexibly coupled to the anticodon, and the precise anticodon geometry is not likely to be critical to process W. Further, for rapid binding of tRNA a sizable initial barrier is deleterious. Thus the intrinsic ribosomal accuracy free energy diagram most probably resembles that of the figure with ΔG_2^{+} the dominant discriminant in accuracy, a negligible activation barrier for k', and $\Delta G \approx \Delta G_{ES} \approx RT\ln {}^{k_B}/_{k_A}$. For equal tRNA concentrations of two competitive tRNA, one would then have

$$f \approx \frac{k_A+W}{k_B+W} \cdots \tag{5}$$

The form (5) should be compared with the rate of product formation,

$$\text{rate} = \frac{W[A]\,k'}{k_A+W+k'[A]} \cdots \tag{6}$$

For best accuracy, $W \lesssim k_A$ is required, while for best speed, $W \gtrsim k_A$ is required. In this and in most systems, there is a speed-accuracy tradeoff, and best speed and best accuracy are antithetical. Ninio (10) has analyzed in detail the experiments of Gorini (4) and others on accuracy in the expression of amber mutations. He noted that much data on this accuracy having to do with the effect of streptomycin, of streptomycin resistant mutants, and of error prone mutants would be understood as an unbalancing of this speed-accuracy tradeoff without an alteration of ΔG. Thus, for example, a ribosomal mutant which increases the binding energy of all tRNA to the message-ribosome complex in state A' by the same amount will leave ΔG unchanged, but increase f because k_A and k_B both decrease by the same factor.

The total ΔG is of course read. The ribosome does not distinguish the source of ΔG. If a particular tRNA has a quirk in its structure which lowers its transition state free energy arising from a region other than its anticodon, it will be more likely to have its corresponding amino acid inserted for any codon present.

This importance of the absolute binding energy to the ability to read a certain available ΔG may have been important to the development of the genetic code, as Hopfield and F. Cramer have independently speculated (unpublished). Other things being equal, GC pairs have somewhat greater binding energies than AU pairs in double helices. Before the evolutionary introduction of extensive base modifications in tRNA, this was plausibly true for codon-anticodon recognition. When the reading of the third base in the codon was introduced, it would have been relatively inaccurate to read a ΔG from the third position if GC pairs were present in the first two positions because k_A was too small. The larger k_A (and k_B) when AU pairs were present in the first two positions allows a given ΔG to produce a smaller f. With this fact in mind, it is interesting to note that the degeneracy of the genetic code is such that when GC pairs occur in the first two positions there is no need to discriminate on the basis of the third position - the correct amino acid will be chosen regardless. Conversely, when AU pairs are in the first two positions, it is always necessary to read the third position to choose the correct amino acid.

While this is a plausible speculation of relevance to the origin of the genetic code, it is probably not germane to the operation of protein synthesis today. Grosjean et al. (11) have shown that in tRNA-tRNA complementary binding the stability has rather little correlation with GC content. They have attributed this to the particular sequence and base modifications near and in the anticodons. They speculate that these details are the evolutionary consequence of the advantage (through better accuracy) of having similar k_A for all correct readings on the ribosome.

The ribosome has a major role in the accuracy of base-pairing in reading codon-anticodon interactions. If the tRNA-tRNA anticodon binding system is taken as a model of codon-anticodon recognition, then error fractions of about 1/10 would be expected (11). If the intrinsic error frequencies in the least accurate DNA and RNA polymerases (12) are taken as an estimate of the energetic cost of base mispairing, then the ribosome-free error fraction possible would be estimated as 1/400. These rather large error fractions are to be thought of in stereochemical terms as being possible only through the flexibility of the environment and/or water accessibility permitting some alternate hydrogen bonding accommodations (11,13) other than Watson-Crick oppositions. In the case of good DNA and RNA polymerases (14), the enzyme constraints (not recognitions) reduce the error fraction below 10^{-4}, and the ribosomal constraints on RNA configurations doubtless play a similar role. The exact size of ΔG and f for T_u-free protein synthesis is not clearly known. Binding studies (15), inferences from enzymatic protein elongation (16), and enzyme-free accuracy studies (17,18,19) give diverse estimates of f_o, ranging from $^1/10$ to $^1/1000$ (for equal tRNA concentrations).

The error levels which are expected in amino acid recognition for a simple enzyme are rather larger than those found in vivo, and that also appears to be the case in protein synthesis. Kinetic proofreading (5) is a scheme which permits greater net accuracy from a recognition which is stereochemically intrinsically inaccurate. Iterated use of inaccurate recognition in a "read-proofread....." mode can lead to high fidelity in recognition. From a fundamental viewpoint, an energy coupling can be used to generate a time delay which under appropriate circumstances enhances accuracy. The simplest kinetic scheme which accomplishes this end is

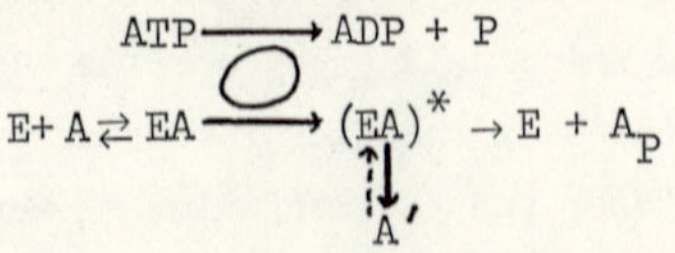

(7)

This reaction is non-Michaelis in that there is a branch in the pathway, leading either to product (A_p + ADP + P) or to abortive product (A′+ADP+P). A′ may be the same as A. ATP (or GTP....) is used as an energy source to drive the EA → $(EA)^*$ step "irreversibly". The downward arrow to A′ is the exit path for errors in proof. The branching ratio between A_p and A′ formation from state $(EA)^*$ represents the second reading, and this branching ratio must favor making A_p for correct A, while favor making B′ if B is used as substrate in order to proofread effectively. If A′ is the same as A, $(EA)^*$ must be a high energy intermediate in order that the pathway ↑ is unimportant (otherwise the scheme is really only Michaelis). The ATP or other energy source coupling is necessary to get from EA to $(EA)^*$. An energy coupling of <u>some</u> form is essential to proofreading. To obtain a second reading of a discrimination ΔG to optimal accuracy, the energy pool driving the intermediate step must itself be out of equilibrium by at least ΔG, but this is no serious limitation. Cellular ATP and GTP pools are commonly out of equilibrium by 7-10 Kcal. The non-equilibrium suppression of the concentration of A′ is an alternate form of energy coupling. When the energy driving is sufficiently strong, the error fraction of the net process is

$$f = \left(\begin{matrix}\text{error fraction in}\\ \text{making } (EA)^* \text{ and } (EB)^*\end{matrix}\right) \left(\frac{\text{ATP use}/A_P \text{ formation}}{\text{ATP use}/B_P \text{ formation}}\right).$$

The first parenthesis represents the simple enzyme fidelity; the ratio of stoichiometry ratios in the second term is the proofreading improvement of the accuracy. It is expected that the second parenthesis will be less than unity, and it can be as small as the first term without making an enzyme more accurate from a stereochemical viewpoint. For a given ΔG, the error fraction with proofreading can be as small as $(e^{-\Delta G/RT})^2$. To determine the proofreading improvement in accuracy it is necessary to determine the two stoichiometry ratios for product formation when the enzyme is acting purely on correct substrate A or incorrect substrate B.

The reaction which charges tRNA can, for the Ile synthetase, be written approximately as (20,21)

$$\mathrm{ATP{+}Ile{+}E} \rightleftarrows \mathrm{ATP\cdot Ile\cdot E} \xrightarrow{\;\nearrow P\sim P\;} \mathrm{AMP\cdot Ile\cdot E} \overset{tRNA}{\rightleftarrows} \mathrm{AMP\cdot Ile\cdot E\cdot tRNA} \xrightarrow{\;\nearrow AMP\;} \mathrm{Ile\cdot tRNA\cdot E} \rightarrow \mathrm{Ile\cdot tRNA{+}E}$$
$$\downarrow \quad \mathrm{E + Ile + tRNA} \qquad (8)$$

In principle, this branched reaction scheme is capable of proofreading both the enzymatic identification of the amino acid and the tRNA. In vivo (and in our in vitro experiments) the back reaction involving pyrophosphate is prevented by a pyrophosphatase, and the rebinding of acylated Ile·tRNA is prevented by the presence of T_u·GTP which binds tightly to charged tRNA. The following ATP/Product stoichiometries were found for this enzyme (22):

substrates		stoichiometry
Ile	$tRNA^{Ile}$	1.5
Val	$tRNA^{Ile}$	270
Ile	$tRNA^{Phe}$	40
Ile	$tRNA^{fMet}$	25
Val	$tRNA_3^{fMet}$	1000

Thus the proofreading improvement of f is by a factor of $^1/180$ in the rejection of valine and by a factor of $^1/20$ in the rejection of $tRNA^{Phe}$ and $tRNA^{fMet}$. The poorer binding of all three incorrect substrates promotes the exit path A′ for errors in proof. Similar results were found for the valine enzyme in competition for threonine. Interestingly, in 20% methanol, a circumstance known to promote mischarging, the stoichiometry for all substrates is near unity and no proofreading takes place.

There has been much discussion of this proofreading step (also known as hydrolytic editing) in the literature, trying to resolve the question of whether this enzyme is explicitly "anti-valine" by a double-seive size mechanism (23,24,25). While part of the large rejection of valine might be due to an explicit anti-valine activity in the exit path, the fact that the system proofreads both tRNA and amino acids indicates that a rather more general proofreading mechanism is involved.

The net in vivo error fraction expected is a product of three factors, the [val]/[Ile] pool concentration ratio (26) (about 6 in E. coli), the basic Michaelis accuracy in forming AMP·Ile·E (in theory and experiment both, about $^1/100$) and the proofreading improvement of $^1/180$. Thus the overall error fraction in protein synthesis for Ile → Val errors should be about $^1/3000$. While E. coli are not precisely rabbits, the comparison between this number and that found in rabbit hemoglobin suggests that this in vivo error rate is understood on the basis of quantitatively reasonable stereochemistry and kinetic proofreading of the tRNA acylation reaction.

In the presence of the $T_u\cdot GTP$ complex, normal protein elongation in *E. coli* has the kinetic scheme (27,28).

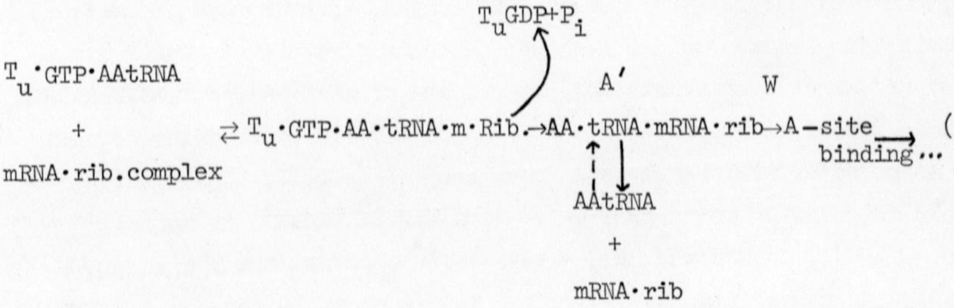

(9)

The right hand side of this diagram is the same as for non-enzymatic elongation. Hopfield pointed out that this reaction scheme has the form essential to kinetic proofreading (5). GTP provides the energy source essential to proofreading, and the path $\uparrow$ is slow compared to enzymatic arrival at the A-site both because the intrinsic rate constant is slow and because *in vivo* most acylated tRNA is in the ternary complex $T_u\cdot GTP\cdot AAtRNA$.

Thompson and coworkers (29,30) have examined the stoichiometry of peptide-bond formation and $T_u\cdot GTP$ hydrolysis for correct and incorrect codon-anticodon pairings on the ribosome. Having prepared the system with N-acetyl Phe on the P site with poly U as a message, they measured the stoichiometry for the correct match $Phe\cdot tRNA^{Phe}$ and for the incorrect match $LeutRNA_2^{Leu}$. (Anticodon GAG: misreads in the first position, and wobble pairs in the third position.) They find an average (GTP/elongation) stoichiometry of about 2 for the correct $tRNA^{Phe}$, and of about 50 for the incorrect case. The proofreading improvement in the error fraction for this particular error should then be about $^1/25$.

In addition, the experiments on $LeutRNA_2^{Leu}$ and $PhetRNA^{Phe}$ determined the relative rates of GTP hydrolysis (a measure of the intrinsic enzyme fidelity in the initial reaction) as about $^1/30$. From the known ratio of $tRNA_2^{Leu}/tRNA^{Phe} \sim {}^1/6$, they conclude that the net *in vivo* error fraction for this mistake should be about $^1/4000$.

The level of proofreading error reduction and reading accuracy are essentially the same as would be expected if a system were tuned to try to use a given (but small) ΔG as effectively as possible. This error seems a particularly difficult one to reject; the different error Leu_1^{tRNA} (anticodon CAG) making one wobble pair and one mispairing appears to have a discrimination in the initial reaction alone better than 1/1000.

Since proofreading depends on the balance of rates, we have performed a series of experiments designed to look at the effect of altering rate constants while keeping the codon-anticodon recognition invariant (31). D-Tyr can be enzymatically linked to $tRNA^{Tyr}$ and operates as a substrate in protein synthesis (32). On the other hand, the ribosome is designed to operate with L-amino acids. We therefore anticipate that when D-Tyr is used, off-kinetic constants for the release of AA·tRNA (or T_u·GTP·AAtRNA) will be faster, while rates W and Γ, which involve positioning or using the amino acid, will be slowed. Thus the (GTP/peptide bond) stoichiometry should increase on going from L-Tyr to D-Tyr if $tRNA^{Tyr}$ is normally proofread. This is indeed the case - this stoichiometry, for the enzymatic addition of L-Tyr is about 1.4, while it is about 10 for the enzymatic addition of D-Tyr. The case of D-Tyr is slightly complicated by the fact that T_u binds about 100 times less strongly to D-Tyr tRNA than to the left-handed isomer (31,36), promoting non-enzymatic ribosomal binding under low T_u conditions. The non-enzymatic rate of addition of D-Tyr is about 1/4 as fast as the non-enzymatic addition of L-Tyr, (for equal concentrations of tRNA or ternary complex) in general agreement with the scale of kinetic changes expected from the stoichiometry assay.

The delivery of AA·tRNA to the ribosome by T_u·GTP appears to have three functions. First, the binding to AA·tRNA suppresses enzymatic deacylation of tRNA after the charged tRNA leaves its synthetase. Second, the hydrolysis of this GTP and release of T_u to free the aminoacyl stem of tRNA generate the time delay and energy linkage essential to kinetic proofreading. Third, T_u seems to be an "add-on" to more primitive protein synthesizing system which operated adequately without it. It well may have played a role in the discrimination against D-amino acids at the time when it was introduced into protein synthesis, as well as promoting a quantum leap in accuracy.

Recent experiments of Jelenc and Kurland (16) show the importance of a high level of "energy pumping" in the nucleoside triphosphate pool to keep in vitro protein synthesis accurate. From the experiments it is not possible to conclude directly why this is so, but there are several ways that this effect could occur if the energetics of this pool is important to kinetic proofreading.

Codon-anticodon recognition may be proofread a second time during translocation. While peptidyl tRNA on the P-site is normally very stable, the release of an incorrect peptidyl tRNA from the P-site during or after translocation seems implied by the work of Caplan and Menninger (33,34) on mutants which accumulate peptidyl tRNA (from their failure to hydrolyze it). A second proofreading would of course further suppress errors in codon-anticodon recognition. This proofreading pathway would be distinguished by virtue of the fact that the dominant product of errors caught in proof, the "A′" of Eq.(7), should be peptidyl tRNA with the last amino acid erroneous. The formation of the peptide bond itself may be the energetic driving step for this proofreading.

The sequence of initiation of protein synthesis is such that the initiation factors may proofread the initiation codon or conformation, but there is no information as to whether in fact this happens. When a "stop" signal is present, other tRNA which erroneously bind may proofread their errors, but whether the termination factors themselves proofread seems doubtful.

The net accuracy of protein synthesis seems to come in part from the direct stereochemical discriminations of simple enzymatic processes, and in part from the way the overall kinetic scheme allows proofreading of small discriminations to occur at the expense of extra energy consumption. The biological solution to the accuracy problem seems to be not the manufacture of a super-discriminating enzyme, but rather the putting together of a system which together is more accurate than its individual steps or component parts.

While Eq.(7) describes a single proofreading, multiple configurations following the energy coupling can produce multiple proofreadings following a single energy coupling by a scheme (22)

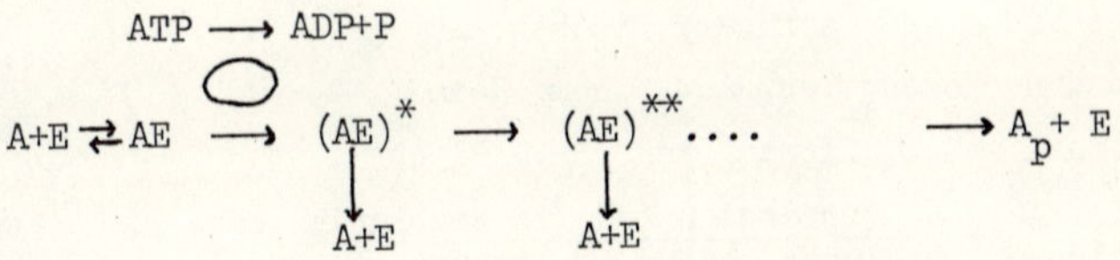

Although detailed kinetic studies would be required to demonstrate such an occurrence, the net proofreading enhancement of accuracy is still given by the stoichiometry ratios. A simple description of a proofreading involving an effectively infinite number of intermediates can occur when relevant molecular motions are diffusive (35). An enzyme or ribosome built on such a basis could obtain extreme accuracy from a limited ΔG.

†Work at Princeton supported in part by NSF grant DMR-78-05916.

1. A. L. Goldberg and A. C. St.John, Ann. Rev. Biochem. 45, 747 (1976).
2. R. B. Loftfield, Biochem. J. 89, 82 (1963).
3. R. B. Loftfield and D. Vanderjagt, Biochem. J. 128, 1353 (1972).
4. L. Gorini, Cold Spring Har. Symp. Quant. Biol. 34, 101 (1969).

5. J. J. Hopfield, Proc. Nat. Acad. Sci. (USA) 71, 4135 (1974).

6. Y. Nozaki and C. Lanford, J. Biol. Chem. 246, 2211 (1971).

7. R. Loftfield and E. A. Eigner, BBA 130, 426 (1966); J. Biol. Chem. 244, 1746 (1969).

8. C. G. Kurland, Science 109, 1171 (1966).

9. C. G. Kurland, in Molecular Mechanisms of Protein Synthesis, eds. H. Weissbach and S. Petska (Academic Press, New York) 81 (1977).

10. J. Ninio, J. Mol. Biol. 84, 297 (1974).

11. H. J. Grosjean, S. de Henau, and D. M. Crothers, Proc. Nat. Acad. Sci. 75, 610 (1978).

12. M. A. Sirover and L. A. Loeb, Cancer Res. 36, 516 (1976).

13. M. D. Topal and J. R. Fresco, Nature 263, 285 (1976).

14. L. M. S. Cheng, J. Biol. Chem. 249, 7441 (1974).

15. J. R. Meninger, private communication.

16. P. C. Jelenc and C. G. Kurland, Proc. Nat. Acad. Sci. (USA)(1979 - in press).

17. L. P. Gavrilova, O. E. Kostiashkina, V. E. Koteliansky, N. M. Rutkevitch
18. and A. S. Spirin, J. Mol. Biol. 101, 537 (1976).

19. S. Pestka, J. Biol. Chem. 244, 1533 (1969).

20. D. Söll and P. R. Schimmel, in The Enzymes, ed. P. D. Boyer (Academic Press, New York), Vol. 10, 489-538 (1974).

21. M. Yaniv and F. Gros, J. Mol. Biol. 44, 1 (1969).

22. T. Yamane and J. J. Hopfield, Proc. Nat. Acad. Sci. (USA) 74, 2246 (1977).

23. G. L. Igloi, F. von der Haar and F. Cramer, Biochem. 17, 3459 (1978).

24. J. Bonnet and J. P. Ebel, FEBS Let. 39, 259 (1974).

25. A. R. Fersht and M. M. Kaethner, Biochem. 15, 3342 (1976).

26. R. Raunio and H. Rosenqvist, Acta Chem. Scand. 24, 2737 (1970).

27. J. Lucas-Lenard and F. Lipman, Ann. Rev. Biochem. 40, 409 (1971)

28. A. S. Spirin, in Gene Expression and Its Regulation eds. F. T. Kenney, B. A. Hamkalo, G. Faveluhes, and J. T. August (Plenum Press, New York) 311-325 (1973).

29. R. C. Thompson and P. J. Stone, Proc. Nat. Acad. Sci. (USA) 74, 198 (1977).

30. R. C. Thompson, D. Dix, R. Gerson, and A. Karim (1979 - in press).

31. T. Yamane, D. L. Miller, O. C. Uhlenbeck, and J. J. Hopfield (1979 - in press).

32. R. Calendar and P. Berg, J. Mol. Biol. 26, 39 (1967).

33. J. R. Menninger, J. Biol. Chem. 253, 6808 (1978).

34. A. Caplan and J. R. Menninger, J. Mol. Biol (in press, 1979).

35. J. J. Hopfield, Proc. Nat. Acad. Sci. (USA) 75, 4334 (1978).

36. A. Pingoud, C. Urbanke and G. Maass (Tu-D-Tyr) personal communication.

ON THE ACCURACY OF ELONGATION

C. G. KURLAND

Molecular Biology Institute,
The Wallenberg Laboratory,
University of Uppsala,
Uppsala, Sweden

INTRODUCTION

The dogma that codon-anticodon interactions are solely responsible for the selection of aminoacyl-tRNA is supported by rather impressive credentials. These include the experimental verification of the messenger and adaptor hypotheses; the universal triplet assignments of the genetic code; the known sequences of many tRNA's; and a wealth of observations with suppressor tRNA species. Why then should there be continuing, and at times intense interest in the so called accuracy problem?

One reason is that there is a discrepancy amounting to orders of magnitude between the apparent accuracy of codon-anticodon interactions in model systems (1) and the accuracy of protein synthesis in vivo (2). Another problem is the common occurrence of mutants with altered ribosomes which translate at error frequencies significantly different from those of wild-type ribosomes (3). To this list may be added a remarkable suppressor mutation that alters a tRNA at a site distant from its anticodon loop and thereby changes the codon-specificity of the mutant tRNA (4). In other words, there is ample reason to suppose that the selection of an aminoacyl-tRNA may be a considerably more interesting process than the base pairing of two triplet sequences. And, this is only part of the problem.

The other part of the problem, which is most often discretely omitted from discussion, concerns the movements of mRNA. That these movements are relevant to translational accuracy follows from an important characteristic of the genetic code. Thus, the code is comma-free and the mRNA is punctuated only at its ends; this means that translation must proceed in a strictly fixed reading frame (5). Accordingly, we would expect a reasonably efficient translation apparatus to maintain the reading frame with a precision which roughly matches that of tRNA selection. Unfortunately, the very nature of reading frame errors makes them difficult to detect and we have only very rough estimates of their frequencies (6). Nevertheless, one of the objectives of this paper is to stress the physical similarities between tRNA selection and mRNA movement (7) as well as to demonstrate that the error frequencies of both processes are necessarily coupled to each other (6,7).

Finally, a complete discussion of translational accuracy would have to include considerations of the initiation and termination functions; however, these topics will not be covered. Instead, we will concentrate on the elongation steps of protein synthesis, and we will try to do so within a conceptional framework that is adequate to the scope of the problem. That is to say, we will treat the relevant processes as physical events, and not merely as epiphenomena of genetics.

GORINI'S PUZZLE

We lead into a discussion of energetics by describing the conceptional predicament into which Gorini (3) was led. There are two reasons for this oblique start. One is that the escape from Gorini's predicament is related to our central theme. The other is the unique role still played by Gorini's observations in defining the problems as well as setting the boundary conditions for their solution.

At the center of the puzzle are the many observations showing that mutant ribosomes can be classified according to their effects on the frequencies of suppression <u>in vivo</u> (3). Thus, the suppression of both nonsense as well as missense mutations by normal as well as by mutant tRNA is modulated by the mutant phenotype of the ribosome. These modulations are systematic in the sense that an altered ribosome (e.g. a streptomycin resistant ribosome) which restricts suppression in one particular case does so in all other cases, even if the degree of restriction varies. Similarly, an altered ribosome (e.g. a ribosome ambiguity mutant) which enhances suppression does so consistently but to varying degrees depending on the nature of the suppressable mutation and on the presence of different suppressor tRNA's. All of this points quite clearly to the active participation of the ribosome in the selection of tRNA species.

There is another set of observations which seems to have struck Gorini with equal force. It had been shown previously (4) that a mutant form of $tRNA^{trp}$ which normally translates the codon UGG could also translate the nonsense codon UGA; this suppressor has a normal anticodon and an alteration in its D stem. Gorini (3) could show in addition that different suppressors of a nonsense codon which bear the same anticodons were nonetheless modulated to varying degrees by different ribosome phenotypes. In other words, the modulating effects of the ribosome could not be explained simply as perturbations of the codon-anticodon interaction. Instead, tRNA discrimination depends to some extent on interactions between the ribosome and sites on tRNA separate from the anticodon.

Gorini's puzzle can now be stated simply as: how is it possible to modulate the selection of tRNA's by ribosome interactions with sites on the tRNA separate from the anticodon and at the same time preserve a primary role for the codon-anticodon interaction in the selection process? His solution was as radical as it was ineffable. It was suggested that there is a ribosomal screen which "allows binding of tRNA to ribosomes but antagonizes tRNA from reaching the stage of codon-anticodon interaction. Participation of different tRNA's in translation would depend not only on anticodon specificity but also on the type and size of the screen's mesh, thus providing an additional source of specificity for tRNA discrimination" (3).

Unfortunately, a reasonable and testable physical mechanism through which a given codon can instruct the ribosome about which tRNA's should be screened away from the codon was not presented. This omission may have diminished the impact that this otherwise remarkable paper should have had. Indeed, the ribosome screen model represents in retrospect an exceptionally clear illustration of a profound limitation on the way the problems of translation had been formulated. Thus, the dynamic aspects of selection processes were simply not recognized as relevant.

An abortive attempt to introduce dynamics to the accuracy problem had been made earlier (8), but it was Ninio (9) who convincingly demonstrated the relevance of kinetics to the analysis of Gorini's data.

Ninio's model to account for the suppression data (9) begins by dividing the selection process for aminoacyl-tRNA into two stages: a first loose association between codon and anticodon, followed by a more stable association that leads to peptide bond formation. The transition time between the two states is assumed to be a variable characteristic of each ribosome phenotype: ribosome alterations which reduce the transition time will lead to higher error frequencies and alterations which increase the transition time will lower the error frequencies.

The modulation effects of the different ribosome phenotypes on the degree of suppression obtained with a given tRNA species can now be viewed as a reflection of the degree of kinetic coupling between the sticking time of the suppressor tRNA at the codon and the transition time characteristic of the ribosomes. Accordingly, the enigmatic ribosome screen is replaced by a simple rank ordering of the ribosomes according to their characteristic transition times.

SPEED, ENERGY AND ACCURACY

In effect Ninio's model replaces a structural construct, the screen, with a temporal construct, the transition time. Most significantly it does so without recourse to any instructive interaction between codon and ribosome. Nevertheless, the selectivity of the system can be modulated by altering the kinetic characteristics of the ribosome. There is, however, a still deeper relationship between the kinetic behaviour of the system and its selectivity. Thus, it is possible to relate the speed, accuracy and energy dissipation of a selection system to a single parameter, the displacement from equilibrium. Since the displacement parameter turns out to be quite important to the execution and interpretation of experiments dealing with accuracy, we will take some space to introduce its properties. More detailed discussions of these matters can be found elsewhere (7,10,11).

We consider first a very simple, optimized enzymatic reaction between a substrate (S) and an enzyme, (E) that yields a product (P):

$$S + E \underset{k_-}{\overset{k_+}{\rightleftharpoons}} P + E \xrightarrow{k_e} \qquad (1)$$

Here, we will hold (S) constant and drive a steady state reaction at varying rates depending on the rate constant k_e, which determines the flux of product removal.

For a steady state system, the ratio of product to substrate will be:

$$\frac{(P)}{(S)} = \frac{K}{D} = \left(\frac{k_+}{k_-}\right)\frac{1}{D} \qquad (2)$$

Here, D is a measure of the departure of the system from the equilibrium concentrations of substrate and product determined by the equilibrium constant K. We refer to D as the displacement and for a scheme such as in (1) it is given by:

$$D = \left(1 + \frac{k_e}{(E)k_-}\right) \qquad (3)$$

Clearly, at equilibrium $D = 1$, and as the rate of product removal is increased so will D, the displacement.

Now, it is easily shown that the rate of product formation is proportional to (1 - 1/D). So we can conclude that the rate of product formation at a fixed substrate concentration approaches asymptotically its maximum value with increasing displacement from equilibrium. In addition, the free energy dissipation per mole of product formed is given by RT ln D. This means that the displacement uniquely describes the energetic cost of driving the reaction at a particular rate.

Next, we consider the competition between the substrate (S) and an analog (S'). We assume that the free energy change for the formation of product (P') is the same as for (P). We do so because we are not interested in the small variations in the free energy of formation of different peptide bonds. Although at equilibrium the ratio of (P')/(S') will be the same as that for (P)/(S), we assume that the interaction with enzyme for S is a factor (a) more favorable than for S'. The factor (a) can reflect any combination of binding constant or rate constant advantages of S over S' in their interactions with (E) i.e. $k_+ = a \cdot k_+'$ and $k_- = a \cdot k_-'$.

It is immediately evident that when there is no displacement from equilibrium (D = 1), there can be no selection of S over S', no matter how large (a) is. This is so because we have assumed that the free energy change in (1) for the formation of (P) is the same as that for (P'). However, as we drive the system progressively faster by increasing k_e, we can approach the selection limit: (P)/(P') = a. The approach to this limit for values of (a) much greater than one is described by the term: a(1 - 1/D) + 1/D.

In summary, D describes the effects of a perturbation of the system, such as product removal, on the steady state concentrations of substrates and products. It is, therefore, a convenient parameter to describe the dynamic state of the system. Thus, as illustrated above the displacement of a given system determines the speed, accuracy and energy dissipation with which the system responds to an external driving force. That a single parameter can be used to relate in so simple a manner these three performance characteristics of a selection system indicates how intimately they are related to each other. Now, we can apply these simple notions to protein synthesis.

We begin by representing a cycle of peptide bond formation as follows:

$$\underset{\leftharpoondown}{\overset{2}{\rightharpoonup}} \text{AT} + \text{R(n)} \underset{\leftharpoondown}{\overset{1}{\rightharpoonup}} \text{P(n)} \underset{\leftharpoondown}{\overset{2}{\rightharpoonup}} \text{R(n+1)} + \text{T} \underset{\leftharpoondown}{\overset{1}{\rightharpoonup}} \qquad (4)$$

Here, we have divided the cycle into two compartments: The first consists of the selection of an aminoacyl-tRNA (AT) by the ribosome programmed by the nth codon (R(n)) and the formation of the elongated peptidyl-tRNA held at the A-site (P(n)). The second compartment consists of the translocation of the peptidyl-tRNA and mRNA to the P-site as well as the release of the n-1th tRNA (T). The cycle is now complete with the n+1th codon at the A-site.

There are two properties of this cycle that should be obvious. One is that as they are presently written, the two selection steps corresponding to tRNA selection (1) and mRNA movement (2) overlap in the sense that they share their terminal states namely, the peptidyl-tRNA bound at the A-site and the open codon at the A-site. This means that there is a possibility of significant coupling between the errors of the selections taking place in both compartments. We return to this point below.

Second, only one external driving force can be applied to the cycle as it appears in (4) to maintain it in the steady state. This is the displacement from equilibrium of the aminoacyl-tRNA and deacylated tRNA. Thus, one free energy source to drive protein synthesis will be provided by the aminoacyl-tRNA synthetases and their consumption of ATP.

We know, of course, that there are normally two other free energy sources to drive protein synthesis: EF-Tu-dependent GTP hydrolysis and EF-G-dependent GTP hydrolysis. Indeed, it has been estimated that the displacement from equilibrium in vivo of the guanine nucleotides is of the order of 10^8 to 10^9 (7). Therefore, if these displacements, which attend tRNA selection and mRNA movement respectively, are relevant to the accuracy of translation, it should be possible to detect this with simple in vitro experiments.

The curious fact is that experiments performed with the so-called "spontaneous" system indicate that neither elongation factors nor GTP improve the accuracy of translation in vivo (12,13). However, the relevance of these particular results has been questioned on the ground that the error frequency of this experimental system is at best orders of magnitude higher than that of translation in vivo (7). Indeed, more recent experiments with a highly optimized system yielded the opposite conclusion.

It is possible to approximate in vitro the missense error frequency characteristic of in vivo translation by carefully balancing the amounts of supernatant enzymes, tRNA and ribosomes, as well as by adding a complex mixture of inorganic ions, polyamines, and a nucleoside triphosphate regenerating system (14). Under these conditions poly U can be translated with a missense error frequency in the neighborhood of 10^{-4}. Here, leucine is the major error while isoleucine and serine are incorporated at somewhat higher than the 10^{-5} level. Given such a system it becomes possible to test the effects of varying the displacement from equilibrium of the nucleoside triphosphates on the performance of the system.

The supernatant fraction used in these experiments has a limited capacity to regenerate the nucleoside triphosphates. Consequently, it is possible to compare the activities of the system with and without an exogenous regeneration system. The results show that the amount of nucleoside triphosphate (roughly 90% of total nucleotide) is increased to virtually 100% after addition of the exogenous regeneration system. In contrast to this small change, the rate of phenylalanine incorporation increases several-fold and the missense error rate decreases by approximately ten-fold. This effect cannot be explained by the change of the absolute concentration of nucleoside triphosphate (circa 10%). It was concluded, therefore, that the response of the system reflects the increase (at least ten-fold) of the ratios of nucleoside triphosphates to diphosphates and monophosphates (14). In other words, the displacement from equilibrium of the nucleotides seems to regulate the performances of the system.

Although these are the results expected from the foregoing discussion, more direct evidence was required to show that the nucleotide displacements affect ribosomes. This could be obtained by taking advantage of the well known requirement for the presence of L7/L12 in ribosomes to mediate EF-Tu and EF-G dependent GTPase (15). Thus, ribosomes lacking L7/L12 should be unable to respond to the nucleotide displacement and should, therefore, synthesize protein more slowly and less accurately than do intact ribosomes. Indeed, it had already been shown that ribosomes lacking L7/L12 are quite sluggish (16). However, in contrast to the previous study, it was now observed that the incomplete ribosomes operating in an otherwise optimized system function with an error frequency above 10^{-3}, which can be reduced to 10^{-4} by the readdition of L7/L12 (17). Thus, the coupling of the GTP displacement to the ribosome with the aid of L7/L12 reduces the error frequency of translation.

The displacement driving a single peptide elongation step is made up of that for the aminoacyl-tRNA multiplied by the square of that for the guanine nucleotides. This corresponds to an enormous amount of energy. Since the speed and accuracy of the selection steps is proportional to $(1 - 1/D)$, it would appear that the maximum performance of the system could be approached to within 99% with a displacement of only 10^2. To what possible use could multiple displacements of 10^8 or more be put? We consider one possible answer next.

PROOFREADING

The displacement parameter was introduced in the previous section as a convenient way to summarize the performance of a selection system. Now we will begin to address a central issue: the apparent discrepancy between the specificity of codon-anticodon interactions and the accuracy of translation. In so doing, the fundamental character of the displacement parameter will become more evident.

Both Hopfield (18) and Ninio (19) accept the conclusion from model systems (1) that the minimum error for codon-anticodon interactions is of the order of 10^{-2}. They then describe systems that can repeat an elementary selection step one or more times so that the error is reduced in progressive stages. Thus, with an elementary step yielding an error amounting to $1/a$, an n-step process could conceivably yield a final error of selection equal to $(1/a)^n$. Such editing or proofreading systems are fundamentally different from the class of selections summarized in scheme (1).

We can summarize a two step proofreading selection as follows:

$$\begin{array}{ccc} & & S + E + GDP + PO_4 \\ & & k_{+p} \uparrow\downarrow k_{-p} \\ S + E + GTP & \underset{k_-}{\overset{k_+}{\rightleftharpoons}} & C \underset{k_{-0}}{\overset{k_{+0}}{\rightleftharpoons}} P + E + GDP + PO_4 \overset{k_e}{\rightleftharpoons} \end{array} \qquad (5)$$

Scheme (5) is a representation of a proofreading mechanism which uses the displacement from equilibrium of the guanine nucleotides to drive the second editing step represented by the reactions governed by k_{+p} and k_{-p}. The intermediate (C) provides the branch point at which the second selection can occur, namely, the choice of rejection over the reaction governed by k_{+p} or the formation of the product by the reaction governed by k_{+0}.

If as in scheme (1) the relative advantage of (S) over (S') is (a) for both k_+/k'_+ and k'_{+p}/k_{+p}, then with sufficiently large displacements of all the substrates and products and with $k_{+0} = k'_{+0}$ this scheme will have a minimum error of $(1/a)^2$. Proofreading schemes as summarized in scheme (5) can be distinguished from conventional selections as summarized in scheme (1) by two criteria.

The first, and least useful characteristic property of proofreading systems is the excess dissipation of the nonselected substrate (GTP in the above scheme) associated with the rejection of (S'). Thus it is only when the editing step (k_{+p}) is being driven near its maximum selective level that this dissipation will be a useful experimental parameter. If on the other hand, there are multiple (m) editing steps, each of which improves the error function a fraction of the maximum (b/a), then the final error will be (b^m/a^{m+1}). Here, with appropriate values of b, m and a, the excess hydrolysis of the nonselected substrate would not be detectable within the background of experimental error.

The second, and more fundamental characteristic of proofreading schemes is their dependence on multiple displacements from equilibrium. Scheme (5) is arranged to illustrate this point. Clearly, the flux along the horizontal path from (S) through (C) to (P) will be driven by two displacements, that for the selected substrate (D_S) and that for the nonselected substrate (D_G). In contrast, the editing branch governed by k_{+p} and k_{-p} is dependent critically on D_G alone (7). Therefore, if the guanine nucleotides were maintained at their equilibrium concentrations there can be no net editing flux and scheme (5) would perform similarly to scheme (1).

Clearly, it is only with significant displacements from equilibrium for the second substrate that the editing function can be expressed. Furthermore, it has been shown (10,11) that the second substrate displacement, in our case D_G, sets the upper limit for the accuracy of the proofreading system. Thus, even if scheme (5) were expanded into an n-step process, each step with a minimum error of 1/a, the minimum error of the whole process for any value of n up to infinity, would be $1/aD_G$.

Accordingly, the large displacements for the nucleoside triphosphates estimated for E.coli (7) may reflect the thermodynamic requirements for the execution of the most accurate process in the cell, namely, the replication of DNA. If this were so, the coupling of the protein synthetic apparatus into these seemingly excessive displacements would be incidental and not a problem to concern us any longer. What should concern us instead is the experimental evidence that would permit us to decide whether or not proofreading contributes to the accuracy of translation. Frankly speaking, the evidence is by no means clear.

The only experiment directed specifically at this question is that of Thompson and Stone (20). They studied the fate of GTP when EFTu-aminoacyl-tRNA-GTP complexes are added to ribosomes programmed with poly U. The expectation is that when the ternary complex contains Phe-tRNA, the number of GTP's hydrolyzed should be close to the number of aminoacyl-tRNA's accepted by the ribosome. If on the other hand the complex contains an aminoacyl-tRNA such as that for leucine, which is prone to misread UUU codons, the GTP hydrolysis should be much greater than one per accepted amino acid. Indeed, results in the expected direction were obtained.

Nevertheless, this experiment does not show that proofreading is taking place (7). Thus, the system does not contain EF-G, and, therefore, Thompson and Stone (20) must assume that peptide bond formation is irreversible in order to draw their conclusions. This assumption is, to say the least, questionable. Second, the error rate of the system is higher than that obtained for codon-anticodon interactions in the absence of ribosomes, and the observed level of proofreading corresponds to a modest 0.5 reduction of the error function.

In summary, the Thompson and Stone experiment (20), along with the demonstration (14,17) that the displacement from equilibrium for the guanine nucleotides influences the error frequency of translation are consistent with the existence of proofreading mechanisms on the ribosome. These same data are also consistent with a role for nucleotide-dependent conventional selection mechanisms on the ribosome. The decision between these two interpretations must await further work, most likely with steady state systems rather than the binding system favored by Thompson and Stone (20). Thus, one objective of the present discussion is to illustrate the roles played by the displacement parameters in these systems, and we note that the displacements are best characterized in steady state systems.

CONFORMATIONAL SELECTION

Interest in mechanisms through which the limited specificity of a single substrate selection can be amplified by repetition is predicated on the seemingly inadequate specificities of codon-anticodon interactions. There is, however, an alternative way to interpret the data obtained in model systems for codon-anticodon interactions. Furthermore, this alternative has the attractive property that it accounts for the role of ribosome interactions with tRNA at sites distant from the anticodon.

The basic idea of this alternative model is that the previous measurements made on the relative stabilities of codon-anticodon interactions in the absence of ribosomes yield information only about the average stabilities and not the maxima. Suppose that when tRNA is interacting with a codon which properly matches its anticodon there is more than one state for the complex. Further suppose that of these states one has a much greater codon specificity than the others. If the codon-programmed ribosome were to preferentially bind tRNA in this maximized state, it would select tRNA species with a codon-specificity that exceeds the average observed in the absence of ribosomes (21).

A particularly clear example of such state dependent behavior has been described in model system studies. The interactions between the anticodon of a tRNA species and an appropriate series of triplet oligonucleotides are quite weak and relatively nonspecific (22-24). Here, the association constant for a matched triplet is approximately $10^3 M^{-1}$ and the minimum advantage of correctly matched versus mismatched olignucleotides is roughly 10:1. On the other hand, if suitable pairs of tRNA species are studied instead, the interactions between their anticodons are enhanced roughly a thousand-fold and the specificity of these interactions is increased by about a factor of 10 (22,1).

These sorts of results were predicted by Eisinger (22) who reasoned that the anticodon loop should fix the interacting triplets in a configuration that optimizes their interactions. In other words, these data indicate that by fixing the conformation of both interacting triplets in a particularly favorable conformation, the stability and specificity of their interactions can be improved. Similarly, we can consider the possibility that the ribosome can trap tRNA molecules in conformational states that maximize their codon-specificities.

In its simplest form, the conformational selection model has two critical elements. First, it is assumed that prior to peptide bond formation, the tRNA is fixed in a particular optimized conformation stabilized both by the interaction with the codon as well as by interactions directly with the ribosome. Second, the occupancy of this particular conformational state is assumed to depend on the character of the codon-anticodon interaction: if the codon and anticodon are correctly matched, the simultaneous interaction with the other ribosomal sites are favored. If, however, the codon and anticodon are mismatched the occupancy of the optimized conformation will normally be disfavored. In other words, it is possible that during the course of the association with the ribosome, the selected tRNA undergoes codon-dependent conformational rearrangements. These could facilitate multiple interactions with the codon-programmed ribosome that maximize the energetic difference between interactions of cognate and noncognate tRNA species.

There is evidence to support the idea that codon-dependent rearrangements of tRNA occur. The first such evidence was obtained by Gassen and collaborators who observed codon-dependent binding of the tetranucleotide CGAA to tRNA (25-27). Since this tetranucleotide is complementary to a sequence in the T loop of the tRNA, they suggested that the D and T loop interactions are destabilized when a properly matched codon interacts with the tRNA. However, NMR studies have so far failed to detect the postulated codon-dependent changes in these loops (28,29).

More recently, kethoxal modifications of guanine nucleotides in tRNA have been used to monitor the response to codon-anticodon interactions (30,31). These data show that there are two responses of $tRNA^{lys}$ to the interaction with its codon: A large increase in the reactivity of residue 46 (m^7G) and much smaller increases in the reactivity of G-15 as well as G-57. The latter weak response could correspond to a destabilization of the D and T loop interactions.

Such observations have led Wagner and Garrett (31) to postulate three states for the tRNA in the presence of the codon: an unperturbed state, a state in which the tertiary interaction of m^7G-46 are destabilized, and a low occupancy state in which the T and D loop interactions are destabilized. Thus, the contradiction with the NMR studies can be explained by the low occupancy of the third tRNA state that would render it undetectable at the concentrations used in the NMR studies. On the other hand, by including the CGAA probe in their experiments, Gassen et al presumably were able to raise the occupancy of the tRNA state with destabilized T and D loop interactions.

The latter inference has been verified indirectly (32). Here, the simultaneous binding of a codon analog (A_3), and CGA oligonucleotide were measured with $tRNA^{lys}$. The important new result obtained was that the presence of the CGA probe increases the amount of codon-anticodon complex. It is not yet clear what the mechanism for this effect is. Nevertheless, the data suggest that by interacting with the D loop, the CGA probe increases the occupancy of the codon-anticodon interacting state for the tRNA.

We can next ask whether these demonstrable codon-dependent conformational rearrangements influence the specificity of the tRNA selection. Here, the relevant data are more limited and less direct. Most relevant are the characteristics of the UGA suppressor described by Hirsh (4).

Wild type $tRNA^{trp}$ recognizes the codon UGG, while $tRNA^{trp}_{UGA}$ recognizes both UGG and UGA. It was suggested, therefore, that the D-stem alteration in the suppressor changes the coupling between the anticodon loop and the codon sites in the tRNA which interact with the ribosome. Therefore, this tRNA will adopt the "optimized" conformation when only two out of three of its anticodon nucleotides are properly matched by the codon (21). In other words, we might expect this tRNA to translate the cysteine-codons UGU and UGC. Indeed, this prediction was verified by the demonstration of the incorporation of tryptophan into polypeptide directed by a poly(U,G) messenger when the UGA suppressor was present and cysteine was absent from the system (33).

A reciprocal experiment also works. Thus, Vacher and Buckingham (34) have argued that if the D-stem alteration of the UGA suppressor has "relaxed" the conformational coupling between the anticodon and distant sites on the tRNA, the suppressor might be "repaired" by introducing an artificial conformational constraint on the tRNA. Accordingly, they UV-cross-linked the thio-U-8 to the D-stem of both wild type and suppressor $tRNA^{trp}$. The codon-specificities of both modified tRNA species turn out to be the same and similar to that of unmodified wild type $tRNA^{trp}$, i.e. they are restricted to UGG.

In summary, the available data seem to be consistent with two aspects of the conformational selection model: First, tRNA seems to respond to proper codon-anticodon matching by conformational changes at sites distant from the anticodon. Second, there is evidence, for at least one tRNA species, that structural changes distant from the anticodon can influence the codon specificity of the tRNA on the ribosome. If future results with other tRNA species follow this trend, then Gorini's puzzle would seem to be solved. Finally, it is worth stressing that this model is not necessarily a competitor to the kinetic models developed above. Instead, it may be that the conformational rearrangements of the tRNA observed in model systems near equilibrium are exploited on the ribosome far from equilibrium in kinetic selections such as proofreading.

READING FRAME ERRORS

The relatively small fraction of space left for a discussion of reading frame errors is a fair index of how little we know about this aspect of the accuracy problem. One reason for this situation is found in the very nature of reading frame errors: they usually cause extensive sequence changes in a protein. Another is that, with three different termination codons out of a possible total of 64, an almost certain outcome of a reading frame error will be the abortive termination of the polypeptide somewhere within the next 21 out-of-phase codons. For these reasons the reading frame error is not likely to be recovered along with the population of normally completed proteins as easily as are missense errors. Accordingly, we have limited access to proteins affected by reading frame errors.

There are, however, three very general remarks that should be made before addressing the available experimental data. First, it seems very likely, given the character of the code, that Lipmann's (35) conjecture concerning the role of tRNA in maintaining the reading frame of the mRNA will turn out to be correct. He has suggested that the codon-anticodon interaction provides the essential coupling between the translocation of tRNA and the orderly movements of mRNA. This eminently sensible suggestion has been extended in provocative detail by Woese (36).

Second, the significance of the displacement from equilibrium developed in earlier sections concerning tRNA selection applies equally well to the movements of mRNA. Thus, at equilibrium there can be no net movement of mRNA. Similarly, the precision of such movements will be dependent on adequate displacements from equilibrium of the substrates, aminoacyl-tRNA and GTP.

Third, there are no markers between codons and, in principle, any sequence of nucleotides should be translatable. Therefore, the movement of a codon into the A or P sites can have no other description than that of a conformational rearrangement. This idea will be developed in greater detail below.

Lipmann's conjecture is supported by work with what is probably still the only well characterized frameshift suppressor. This particular mutant, the suf D suppressor, was identified as an external suppressor of a frameshift mutation in the his D locus and it is caused by an insertion (37). The suggestive finding of Riddle and Carbon (38) was that the suf D mutation is in the structural gene for $tRNA^{gly}_{GGG}$ and that it is itself an insertion mutation. In particular, an extra C is inserted in the anticodon of the suf D product so that instead of the normal CCC anticodon, the suppressor anticodon loop contains the sequence CCCC. These data certainly suggest that the codon-anticodon interaction is involved in mRNA movement. The question is, how?

Prior to this discovery Woese (36) had suggested that the movement of the mRNA might be effected by a conformational rearrangement of the anticodon loop of the tRNA from one stacking configuration to another. Here, the advance of the mRNA would correspond to the coupled movement of the associated codon. Under the influence of this idea, it was possible to imagine that a tRNA with the capability of making four base pairs with the mRNA might cause the mRNA movement to correspond to four bases. In this way, the reading frame error caused by an insertion mutation might be suppressed (38).

There are, however, good reasons to be cautious about this interpretation (6). First, it seems highly unlikely a priori that mRNA can be positioned normally at the A site in a conformation that permits four base pairs to be made with a tRNA. If such extra base pairs could be formed even at relatively low frequencies with normal tRNA it would create an error catastrophe for the system. In addition, the UGA suppressor discussed above seems to behave as though it interacts only with the first two bases of codons of the form UGN. If the number of base pairs formed with the mRNA determines the distance of the mRNA movement, the UGA suppressor might be expected to cause reading frame errors with high probability. However, it is an extremely efficient suppressor of UGA codons, which suggests that it does not cause extensive reading frame errors and therefore, that the number of base pairs does not by itself determine the length of the movement. For these reasons an alternative interpretation of the suf D suppressor has been offered (6).

We note that an anticodon loop containing four C's can in principle present two different triplets to match the GGG codon: the triplet containing the first three C's and the triplet containing the second, third and fourth C's. The binding of one of these triplets (or both) to the codon might lead to significant distortion of the conformation of the tRNA-mRNA complex. Therefore, when the complex is translocated, it may be unable to move the standard distance of three nucleotides. Alternatively, the putative conformational strain may facilitate a "slippage" of the tRNA-mRNA interaction during the movement. In either case, the result would be a reading frame error, which could in turn suppress the frameshift error.

The notion that conformational distortion of the tRNA-mRNA complex at the A site may lead to reading frame errors with mutant tRNA during translocation can be extended to normal tRNA (6). For example, when a missense error occurs at the A site, the codon-anticodon interaction may be distorted by the mismatched triplets. Therefore, we might expect some missense errors to raise the incidence of reading frame errors. Indeed, precisely just such a correlation has been observed, but it was not interpreted in this way.

Atkins et al (39) have shown that there is an average of 0.03 percent wild type β-galactosidase produced in bacteria carrying any one of 16 different frameshift mutations distributed along the Z cistron. That this low level of enzyme activity is due to spontaneous frameshift suppression on the ribosome could be verified by studying the effects of altered ribosomes and streptomycin on this phenomenon (39).

For example, bacteria with ribosomes which restrict missense and nonsense errors (streptomycin resistant ribosomes) also restrict the frequency of suppression of these frameshift mutations. However, such restriction can be relieved by the presence of streptomycin in the growth medium. Furthermore, bacteria with ribosomes which enhance missense and nonsense errors (ribosome ambiguity mutants) also support higher levels of frameshift suppression. These data indicate that the ribosome phenotype influences the frequency of spontaneous reading frame errors. They show also that such errors are correlated with missense errors.

Additional support for the present interpretation is found in studies of the effects of amino acid limitation on the error frequencies of translation by rel A^- (relaxed) mutants. Not only does the missense error rate increase under these conditions, but so too does the accumulation of abnormally short polypeptides (40,41). The short polypeptides could represent products abortively terminated by out-of-phase termination codons brought into play by reading frame errors following missense.

Such an interpretation (6) leads to a more direct and striking prediction: amino acid limitation in a rel A^- background should generate frameshift suppression. This prediction has been verified recently by Foley and Gallant (42). They have shown that a frameshift mutation in β-galactosidase is suppressed by phenylalanine or isoleucine limitation. Comparison of isogenic strains shows that this type of suppression is dependent on the presence of the rel A^- mutation.

In summary, the data suggest that the occurrence of reading frame errors is coupled to that of missense errors. It seems therefore, reasonable to conclude that such error-coupling reflects the dual role of the codon-anticodon interaction in the two selection steps of elongation: aminoacyl-tRNA selection and mRNA movement. Consequently, it is not surprising that so many apparently unrelated suppression events can be at least partially understood in terms of a unifying concept, namely the notion of conformational selection.

ACKNOWLEDGEMENTS

I am grateful to Måns Ehrenberg, Pierre Jelenc, Ingvar Pettersson and Gerhart Wagner for stimulating discussions and valuable criticism.

REFERENCES

1. Grosjean, D. H., De Henau, S., and Crothers, D. M. (1978) Proc. Nat. Acad. Sci. USA 75, 610-614.
2. Edelman, P., and Gallant, J. (1977) Cell 10, 131-137.
3. Gorini, L. (1971) Nature New Biology 234, 261-264.
4. Hirsh, D. (1971) J. Mol. Biol. 58, 459-468.
5. Crick, F. H. C., Barnett, L., Brenner, S., and Watts-Tobin, R. J. (1961) Nature 192, 1227-1232.
6. Kurland, C. G. (1979) Nonsense Mutations and tRNA Suppressors (Cellis, J. and Smith, J. D., eds.) pp. 95-108, Academic Press, NY.
7. Kurland, C. G. (1978) Biophys. J. 22, 373-392.
8. Kurland, C. G. (1970) Science 169, 1171-1177.
9. Ninio, J. (1974) J. Mol. Biol. 84, 297-313.
10. Ehrenberg, M., and Blomberg, C. (1979) Quart. Rev. Biophys., submitted.
11. Blomberg, C., and Ehrenberg, M. (1979) J. Theor. Biol., submitted.
12. Gavrilova, L. P., Kostiaskina, O. E., Koteliansky, V. E., Rutkevitch, N. M., and Spirin, A. S. (1976) J. Mol. Biol. 101, 537-552.
13. Spirin, A. S., Kostiaskina, O. E., and Jonak, J. (1976) J. Mol. Biol. 101, 553-562.
14. Jelenc, P. C. and Kurland, C. G. (1979) Proc. Nat. Acad. Sci. USA 76, in press.
15. Möller,W. (1974) Ribosomes (Nomura, M., Tissieres, A. T., and Lengyel, P., eds.) pp. 711-731.
16. Kotelansky, V. E., Domogatsky, S. P., Gudkov, A. T., and Spirin, A. S. (1977) FEBS Letters 73, 6-11.
17. Pettersson, I., and Kurland, C. G. (1979), in preparation.
18. Hopfield, J. J. (1974) Proc. Nat. Acad. Sci. USA 71, 4135-4139.
19. Ninio, J. (1975) Biochemie 57, 587-595.
20. Thompson, R. C. and Stone, P. J. (1977) Proc. Nat. Acad. Sci. USA 74, 198-202.
21. Kurland, C. G., Rigler, R., Ehrenberg, M. and Blomberg, C. (1975) Proc. Nat. Acad. Sci. USA 72, 4248-4251.
22. Eisinger, J., Feuer, B., Yamane, T. (1971) Nature New Biology, 231, 126-128.
23. Högenauer, G. (1970) Eur. J. Biochem. 12, 527-532.
24. Uhlenbeck, O. C., Baller, J., and Doty, P. (1970) Nature 225, 508-510.
25. Schwarz, U., Lührmann, R. and Gassen, H. G. (1974) Biochem. Biophys. Res. Commun. 56, 807-814.
26. Schwarz, U. and Gassen, H. G. (1977) FEBS Letters 78, 267-270.
27. Schwarz, U., Menzel, H. M., and Gassen, H. G. (1976) Biochemistry 15, 2484-2490.
28. Geerdes, H. A. M., van Boom, J. H., and Hilbers, C. W. (1978) FEBS Letters 88, 27-32.

29. Davenloo, P., Sprinzl, M., and Cramer, F. (1979) Biochemistry, in press.

30. Wagner, R., and Garrett, R. A. (1978) FEBS Letters 85, 291-295.

31. Wagner, R., and Garrett, R. A. (1979) Eur. J. Biochem., in press.

32. Möller, A., Wild, U., Riesner, D., and Gassen, H.G. (1979) Proc. Nat. Acad. Sci. USA, in press.

33. Buckingham, R. H. and Kurland, C. G. (1977) Proc. Nat. Acad. Sci. USA 74, 5496-5498.

34. Vacher, J., and Buckingham, R. H. (1979) J. Mol. Biol., in press.

35. Lipmann, R. (1969) Science 164, 1024-1031.

36. Woese, C. (1970) Nature 226, 817-820.

37. Yourno, J. (1972) Nature New Biology 239, 219-220.

38. Riddle, D. L. and Carbon, J. (1973) Nature New Biology 242, 230-235.

39. Atkins, J. F., Elseviers, D., and Gorini, L. (1972) Proc. Nat. Acad. Sci. USA 69, 1192-1195.

40. Hall, B. and Gallant, J. (1972) Nature New Biology 237, 131-136.

41. O'Farrel, P. H. (1978) Cell 14, 545-557.

42. Foley, D., and Gallant, J. (1979), in preparation.

ON THE CAUSES AND PREVENTION OF MISTRANSLATION

JONATHAN GALLANT and DIANE FOLEY

Department of Genetics,
University of Washington

The specificity of the translation apparatus is by no means absolute. In cell free systems, errors of codon:anticodon recognition are easily demonstrated, and their frequency can be raised to positively alarming levels by various conditions that "detune" the ribosome (abnormal pH, cation concentrations, temperature, etc.) or by the administration of error-promoting drugs such as aminoglycoside antibiotics or alcohols (reviewed in reference 1). In whole cells, mistranslation in response to aminoglycoside antibiotics and alcohols has been abundantly documented (2-8).

The basal level of mistranslation is much lower in whole cells than in the typical cell free system, but it is high enough to estimate. Loftfield's classical measurements of isoleucine-valine mistakes in pigeon ovalbumin and rabbit α-globin yield values of 2-6 x 10^{-4} per codon (9, 10), and measurements of illegitimate cysteine incorporation into *E. coli* flagellin indicate a similar value (6).

These estimates of amino acid substitution errors reflect a composite of errors at the level of tRNA charging and at the level of aminoacyl-tRNA selection on the ribosome. An estimate of the latter, uncomplicated by tRNA charging, can be obtained by considering the frequency with which mutant nonsense codons in mRNA are mistranslated to yield active enzyme. A survey of such measurements (TABLE 1) reveals a moderately broad distribution about a mean value near 4 x 10^{-4} in *E. coli* and a few times higher in yeast. It seems safe to conclude that ribosomal ambiguity, rather than mischarging of tRNA, is the major source of translational errors *in vivo*, and that a frequency of a few times in 10,000 translation events is a reasonably hard estimate.

The fact that the normal rate of mistranslation is detectably greater than zero implies a rather interesting puzzle. Consider a common, garden-variety codon, let us say UUU. It calls for PHE-$tRNA^{phe}$ but can also be misread by incorrect species (such as LEU-$tRNA^{leu}$) at appreciable rates. The error frequency must be a monotonically increasing function of the ratio of incorrect to correct substrates available for selection. For example, if there were no PHE-$tRNA^{phe}$ in the system but the noncognate species were available, then the error frequency would be 100% at UUU codons.

TABLE 1.

GENE	ENZYME	RELATIVE ACTIVITY x 10^4 Mean, standard deviation; range; (number)	REFERENCE
phoA (E. coli)	alkaline phosphatase	3.5, 1.5; 2.0-7.0; (15)	40
phoA (E. coli)	alkaline phosphatase	4.5, 4.0; 1.65-10.5; (4)	JG, unpublished
lacZ (E. coli)	β-galactosidase	3.5, 3.1; 0-7.6; (4)	JG, unpublished
tr_5 S. cerevisiae)	tryptophan synthetase	15.7, 15.7; 4.2-46; (7)	41

Relative enzyme activities in nonsense mutants. The numbers tabulated are differential rates or specific activities in the mutants divided by the corresponding values in isogenic wild type strains. In the case of Manney's data on yeast tryptophan synthetase, four mutants have been excluded from tabulation because of strong evidence that the activity resided in the fragment, rather than in full-length enzyme produced by read-through.

It follows that limitation of the synthesis or activation of any one amino acid must necessarily increase the error frequency at codons calling for that amino acid, other things being equal. But wild type cells do not show an increased error frequency under such conditions. Evidently other things are not equal, which is to say that the translation system does not respond passively to imbalances in tRNA aminoacylation, but rather adjusts its discrimination specificity. In rel mutants, however, the increased frequency characteristic of a passive system does occur. This phenomenon demonstrates that the rel^+ (or "stringent") control system somehow modulates the accuracy of translation. We will discuss evidence for this assertion below, considering three different categories of translational error; we will then present a simple, unitary hypothesis as to the mechanism, the essence of which is E. F. Schumacher's dictum that slower can be better; and we will conclude with a brief consideration of positive feedback in mistranslation.

MISSENSE

The first clue to rel mistranslation came from the discovery that rel mutants suffered a remarkably general defect in the production of active enzymes during amino acid limitation. The list began with β-galactosidase (11), alkaline phosphatase (12) and tryptophan synthetase (13), then grew to include ornithine transcarbamylase, acid phosphatase, and D-serine deaminase (14), and has since been extended to 13 out of 14 enzymes surveyed (TABLE 2). The near uniformity of these results is revealing in light of the fact that overall protein synthesis proceeds at about the same residual rate in rel^+ and rel^-. In the cases of the first four enzymes, we have comparable measurements of enzyme synthesis and messenger RNA level, based either on enzyme-forming capacity or RNA:DNA hybridization (or both, in the case of β-galactosidase). These results, summarized in TABLE 2, make plain that amino acid limited rel mutants suffer a defect in active enzyme synthesis per unit messenger RNA. The generality of the phenomenon strongly suggests generalized mistranslation.

Recent work by O'Farrell (15) permits this statistical argument to be put in an altogether compelling form. O'Farrell found that 25% of all E. coli proteins, an adequate sample of which was resolved on 2D gels, showed an increased net synthesis during the stringent response. Therefore, the chance is 25% that a given protein will show a synthetic deficit in rel mutants relative to wild type. The chance that 13 out of 14 enzymes, chosen at random, will fall into this class is about 2×10^{-7}. We can therefore conclude, at a high level of confidence, that the enzymatic deficits observed in rel mutants reflect in part a qualitative abnormality of the monomers leading to reduced enzymatic activity.

This has been demonstrated in the case of the lacZ monomer. Hall and Gallant showed that rel mutant cells exhibit a three-fold deficit in lacZ monomer synthesis -- this protein is apparently among the 25% whose net synthesis depends on rel^+ function, probably at the level of a transcriptional effect of ppGpp -- but that the enzymatic activity per unit monomer was reduced by a further factor of three (16). At the same time, Atherly, Smith and Novick obtained preliminary evidence that β-galactosidase made in rel– cells during amino acid limitation was abnormally thermolabile, and this observation was confirmed by us (16) and Fiil et al. (17). These abnormalities, both reduced intrinsic specific activity and increased thermolability, again suggest the occurrence of translational errors.

TABLE 2. Enzyme synthesis during amino acid limitation in rel$^+$/- pairs.

ENZYME	activity rank order	ref	mRNA measurement	activity/mRNA rank order	ref
β-galactosidase	+ > -	11,16,17	c, h	+ > -	43-45
tryptophan synthetase	+ > -	13,46	h	+ >>-	13
alkaline phosphatase	+ > -	12,51	c	+ > -	12
Ornithine transcarbamylase	+ > -	14	c	+ >>-	14
acid phosphatase	+ > -	14			
D-serine deaminase	+ > -	14			
β-cystathionase	+ > -	47			
ATP: methionine-S-adenosyl-transferase	+ > -	47			
methylenetetrahydro-folate reductase	+ > -	47			
methionyl-tRNA syntetase	+ > -	48			
acetohydroxy acid synthetase	+ > -	49			
threonine deaminase	+ > -	49			
histidinol phosphate phosphatase	+ > -	50			
6-phosphogluconate dehydrogenase	+ = -	50			

Enzyme synthesis during amino acid limitation in isogenic rel$^+$/- pairs. Rank order refers to the relationship between the value measured in the two members of the isogenic pair examined. Measurements of mRNA level were done in either of two ways: capacity for enzyme synthesis after blocking new transcription initiation and restoring the missing amino acid (c); or RNA:DNA hybridization to an appropriate transducing λ phage DNA (h).

Specific evidence of missense errors has emerged from two different kinds of experiment. We examined the incorporation of trace levels of ^{35}S-cysteine into flagellin, a protein which normally contains no cysteine. The results showed that illegitimate cysteine incorporation into the protein was enhanced by partial arginine starvation, but not by partial isoleucine starvation, in a rel specific manner (6). The amino acid specificity makes sense in terms of the genetic code, for the two frequently used ARG codons CGU and CGC are related to the two CYS codons UGU and UGC via a single transition-type decoding error in the first position, a type of misreading which has been documented *in vitro* (18, 19), whereas the ILE codons cannot be confounded with the CYS codons by single base mistakes.

The second type of experiment concerns the isoionic points of a large number of proteins resolved by two-dimensional electrophoresis. O'Farrell (15) discovered that nearly all proteins made by *rel* mutant cells during histidine starvation showed a striking pattern of electrophoretic heterogeneity: the normal spots were each accompanied by a family of satellite spots of the same molecular weight, but differing from the normal species by integral reductions in positive charge. This effect was also amino acid specific: a somewhat similar pattern was observed during arginine starvation, but no charge heterogeneity was occasioned by starvation for leucine, isoleucine, or proline.

This extraordinary phenomenon was confirmed by Parker et. al. (20). Parker has since made an important further observation: starvation of a *rel* mutant for asparagine leads to a symmetrically related pattern of electrophoretic heterogeneity, wherein normal proteins are accompanied by satellite spots differing by integral *increases* in positive charge (J. Parker, manuscript in preparation).

The amino acid specificity of these effects leave little doubt that missense errors are involved. They make no sense whatever in terms of less interesting hypotheses, such as post-translational modification, but make perfect sense in terms of single base decoding errors. Both histidine and arginine are positively charged, and single base decoding errors would replace them by neutral amino acids, producing just the kind of electrophoretic heterogeneity observed. Asparagine is neutral, but decoding errors in the third base would bring in positively charged lysine. Finally, single base errors in the codons for the neutral amino acids leucine, isoleucine and proline would, in most cases, bring in other neutral amino acids and lead to no charge heterogeneity.

The variety of approaches we have reviewed, and the impressive consistency of reports from independent laboratories, constitute very persuasive evidence that *rel* mutants are prone to errors of aminoacyl-tRNA recognition when confronted by an imbalance in aminoacylation. Once again, one should remember that this is not aberrant behavior, but rather just the way a passive system would be expected to respond.

FRAMESHIFTS

About 14 months ago, my lab enjoyed a visitation by the ghost of Luigi Gorini, accompanied in more corporeal form by Charles Kurland. The two visitors suggested that rel mistranslation ought to include spontaneous frameshifting at an appreciable frequency. The prediction follows from the view that codon:anticodon recognition and translocation are intimately coupled processes, a notion which Kurland has shown is implicit in conformational selection models of tRNA recognition and movement (21 and Kurland's chapter in this volume). Experimentally, the idea is an almost unavoidable consequence of the work of Atkins, Elseviers, and Gorini on spontaneous frameshifting (22) and of the properties of the sufD frameshift suppressor tRNA (23).

We thereupon set out to look for spontaneous frameshifting in rel mutants, detected through phenotypic suppression of lacZ frameshift alleles. We expected there to be severe constraints over the alleles which might exhibit suppression, the amino acids whose limitation might produce suppression, and the degree of amino acid limitation required to detect the effect. Therefore, a broad survey was required to give the hypothesis a fair test. To accomplish the task, we devised a simple and unequivocal visual assay of lacZ frameshift suppression. We plate the tester strains on agar containing an inducer of the lac operon, cAMP, and XG (see Figure 1), a remarkably sensitive histochemical stain which turns blue at the mere mention of β-galactosidase. To produce aminoacyl-tRNA limitation, we emplace filter discs containing inhibitors of the biosynthesis or activation of specific amino acids. The inhibitor diffuses out, producing a circular zone of complete growth inhibition, surrounded by a zone of partial inhibition. At whatever radius is right for the degree of amino acid limitation necessary to produce phenotypic suppression, a striking ring of blue color, reflecting enhanced β-galactosidase activity, develops after a few days of incubation.

We have used this test to fill in a 96 box matrix of 8 different lacZ frameshift mutants subjected to limitation for each of 12 amino acids. The mutants employed, all fortunately available in a relA$^-$ genetic background, are those isolated and characterized by Austin Newton (24). FIGURE 1 illustrates the method, and demonstrates the specificity for relA: the relA$^-$ mutant shows markedly greater phenotypic suppression during limitation for PHE and for PRO than does its isogenic relA$^+$ partner.

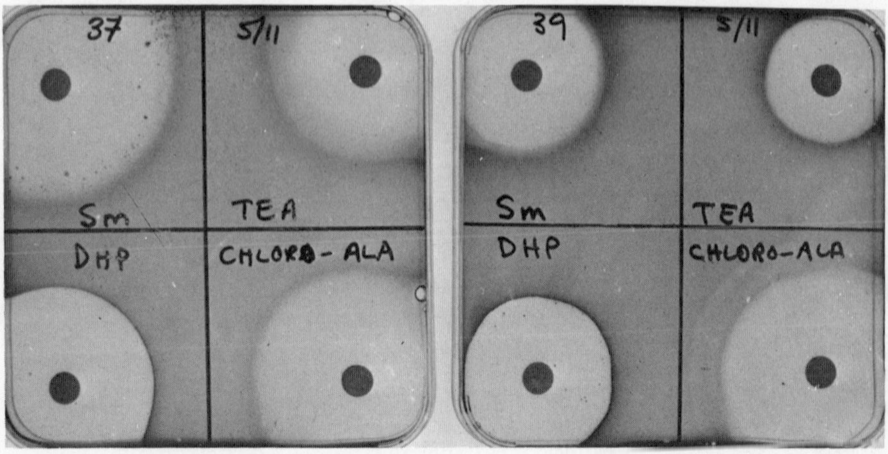

Figure 1. Phenotypic suppression assay of strains 37 (relA$^+$) and 39 (relA$^-$). The two isogenic strains were constructed as described in the legend of Table 3. The assay plates are composed of medium 63 with 1.5% Difco Noble agar, 0.2% glucose as carbon source, 2 mM IPTG, 5 mM cAMP, and 40 ug/ml XG (5-bromo-4-chloro-3-indolyl-β-D-galactosidase, from Sigma). This galactoside derivative of indigo dye was developed by Mel Cohn during the late pleistocene era, and is so sensitive an indicator of β-galactosidase activity that the background activity of most frameshift and nonsense mutants in lacZ (down to below 10^{-4} of lacZ$^+$) can be detected and rank-ordered by the intensity of blue color produced (see reference 22). Exponential cultures of the tester strains were plated (about 10^8 cells) in soft agar (0.6%). After the top agar hardened, filter discs were placed at appropriate positions, and the plates were incubated at 37° for 3-5 days. In the plates shown here, the filter discs contained, respectively, clockwise from upper left, streptomycin, thienylalanine, chloroalanine, and dehydroproline (see text and TABLE 3 for additional details). The dark rings in the regions of partial growth inhibition indicate phenotypic suppression. The black-and-white photographs do not do justice to the technique, wherein the ring of phenotypic suppression is deep blue against the faint blue surrounding color of background activity.

Quantitative data for PHE limitation are shown in TABLE 3. First, note the essential control experiment with a lacZ$^+$ revertant of the relA$^-$ strain: here PHE limitation reduces the differential rate of enzyme synthesis by a factor of 12. This is the typical result with rel$^-$ mutants, as noted earlier, comprising both the transcriptional effect (a partial requirement for ppGpp) and the general consequences of rel mistranslation. The frameshift mutant, in contrast, exhibits no such decrease but rather a two-fold increase in active enzyme synthesis. This allele specificity proves that frameshift suppression is at issue, and implies at least a 24-fold increase in frameshifting near enough to the mutant allele to restore enzymatic activity. In the $relA^{+}$ derivative of the frameshift mutant, PHE limitation elicits no change in the differential rate of enzyme synthesis. (The $relA^{+}$/- pair exhibit slightly different sensitivities to the PHE antagonist, as can be seen from the zones of inhibition in FIGURE 1. We therefore examined a range of antagonist concentrations in both strains; the appropriate comparison, in regard to enzyme synthesis, is made at inhibitor concentrations which reduce the growth rate by about the same amount.)

TABLE 4 presents the pattern of phenotypic suppression we have observed during limitation for 12 different aminoacyl-tRNA species. Four general conclusions can be drawn from the pattern. First, the enhancement of phenotypic frameshifting occasioned by aminoacyl-tRNA limitation is obviously not confined to the codon which contains the frameshift mutation itself. This follows from the fact that limitation for three or more different amino acids works for each of the suppressible alleles. Second, phenotypic frameshifting occurs preferentially in response to limitation for a special group of amino acids (or aminoacyl-tRNA's): ILE, TYR, PHE, PRO, and TRP. We suggest that these represent a special group of shifty codons at which rel errors of tRNA recognition are particularly likely to be associated with aberrant translocation. Third, slippage must occur in both directions, for frameshift alleles of both sign are suppressed. This might mean that recognition errors include both two-base and four-base misreading, with associated abnormal translocation; however, a more subtle interpretation, based on the conformational selection model of tRNA recognition, follows from Kurland's analysis of the relationship between tRNA recognition and translocation (21).

Fourth, the suppressible alleles show somewhat different patterns of response to limitation for each of the special amino acids. For example two respond to PHE and four don't; two respond better to TYR than to ILE, while one shows the converse rank order. These specificities imply that suppression is not a generalized feature of the abnormal physiology of rel mutant cells during amino acid limitation; rather, it is constrained by singularities pertaining both to frameshift allele and

TABLE 3.

TEA concentration		STRAIN: 39 lacZ$^+$ (rel$^-$)	39 (rel$^-$)	37 (rel$^+$)
None	a) % Growth inhibition	0	0	0
	b) $\Delta Z/\Delta B \times 10^4$: mean, SEM, (number determinations)	117,000, 4650 (4)	37, 1.7 (5)	104, 8.1 (3)
	c) $\Delta Z/\Delta B$ relative to control [ ]	[≡ 1.0]	[≡ 1.0]	[≡ 1.0]
0.2 γ/ml	a)	-	10%	40%
	b)	-	47.5, 0.7 (2)	113, 14 (3)
	c)		[1.28]	[1.09]
0.3 γ/ml	a)	-	-	50%
	b)	-	-	100 (1)
	c)	-	-	[0.96]
0.5 γ/ml	a)	49%	36%	-
	b)	9800, 485 (4)	73, 7.5 (3)	-
	c)	[0.084]	[1.97]	-
0.65 γ/ml	a)	-	42%	-
	b)	-	64 (1)	-
	c)	-	[1.73]	-
0.8 γ/ml	a)	-	42%	-
	b)	-	66 (1)	-
	c)	-	[1.78]	-

β-galactosidase synthesis during PHE limitation. Strains 37 (relA$^+$) and 39 (relA$^-$) were derived from Newton's frameshift mutant ICR38, using cotransduction of relA with fuc. Strain 39 lacZ$^+$ is a lacZ$^+$ revertant of strain 39. The strains were cultivated in Tris-glucose minimal medium, and the lac operon was induced (2 mM IPTG and 5 mM cAMP) for about one doubling in control and PHE limited cultures. PHE limitation was achieved by the addition of thienylalanine at the indicated concentrations; the extent of growth inhibition at each concentration is shown. Extracts, suitably concentrated, were prepared by sonication, and the enzyme activity and protein were determined by standard techniques. The entries give differential rates of enzyme synthesis in arbitrary units ± standard error of the mean, followed by the number of separate experiments in parentheses. The arbitrary units are defined as follows: one unit of β-galactosidase produces a ΔA_{420} of 1.0 per minute in a 300 ul reaction mixture (ONPG at 0.67 mg/ml) at room temperature; one unit of bacterial growth is a ΔA_{720} of turbidity of 1.0, measured in a Beckmann DB spectrophotometer, corresponding to an increase in total protein (by Lowry) of 183 μg/ml.

TABLE 4.

Antagonist (Amino Acid)	38	24	36 (+)	36-14 (+)	36-14 (-)	36-16 (-)	17 and 21
				Frameshift Mutant			
Sm	+	+	0	0	0	0	0
oMT (ILE)	++	++	++	++	+	+	0
fl-TYR (TYR)	++	++	+	++	++	++	0
TEA (PHE)	++	+	0	0	0	0	0
DHP	+	+	++	+	+	++	0
fl-TRP (TRP)	(+)	(+)	(+)	(+)	0	0	0
Antagonists of							
LEU, MET, SER	0	0	0	0	0	0	0
HIS, LYS, CYS	0	0	0	0	0	0	0
ALA	0	0	0	0	0	0	0

Patterns of phenotypic suppression of lacZ frameshift mutants. The mutants are those of Newton (24); "+" and "-" in the designation of mutants refers to reading frame, under the assumption that mutant 36 is "+" (see reference 24). Intensity of phenotypic suppression by the blue ring test is reported as follows: ++ is very strong suppression; + is strong suppression; (+) is weak but reproducible (and therefore probably significant) suppression; 0 is no suppression. The amino acid atagonists which gave positive results are: 0-methylthreonine (oMT); 3-fluorotyrosine (fl-TYR); β-2-thienylalanine (TEA); 3,4-dehydroproline (DHP); 2-fluorotryptophan (fl-TRP).

limiting amino acid. The specificities presumably reflect context effects in the message: differences in the distance between a given frameshift allele and a given shifty codon, and differences in the enzymatic activity of the suppressed lacZ protomers containing different stretches of gibberish between the positions of the frameshift allele and the point of rephasing. In retrospect, it is surprising that the constraints are as limited as they appear to be, and that as many as six out of eight frameshift alleles exhibit phenotypic suppression.

NONSENSE

Given our pleasing assay of phenotypic suppression, and with Luigi Gorini haunting the laboratory, we of course had to look for nonsense suppression as well. We tested four lacZ nonsense mutants (again present in a relA$^-$ background) and found that one of them, an amber, exhibited phenotypic suppression when limited for ILE, TYR, PHE, or LYS. TABLES 5 and 6 show quantitative results for inhibition of lysine activation. Once again, the lacZ$^+$ revertant shows a reduction in β-galactosidase synthesis, while a significant two-fold stimulation occurs in the amber mutant. TABLE 6 shows that this stimulation is specific to the relA$^-$ genotype.

TABLE 5.

	$\Delta Z/\Delta B \times 10^4$	
	Control	+ Lysine hydroxamate
YA625 (lacZ$^+$)	120,000	50,000
	160,000	90,000
YA625 (lacZ^{UAG})	57	144
	72	185

β-galactosidase synthesis in YA625 (UAG) and its lacZ$^+$ revertant, both relA$^-$. Growth, induction, assays, and units are as in TABLE 3. Lysine activation was produced by addition of lysine-hydroxamate (300 μg/ml, 30-40% growth inhibition). Each pair of entries (control and + lysine hydroxamate) represents a separate experiment, with enzyme assay in duplicate.

Among a group of four phoA nonsense mutants, we found one, an ochre, which showed an even larger stimulation of enzyme syntheiss in response to limited lysine activation; the enhanced specific activity of alkaline phosphase seen in the middle panel of FIGURE 2 corresponds to a five to six fold increase in the differential rate of synthesis. The left and right hand panels of FIGURE 2 show the appropriate controls with congenic phoA$^+$ relA$^-$ and phoA^{UAA} relA$^+$ strains, neither of which shows any increase.

Plate assays of this ochre mutant and of the other three were performed using XP, an indoxyl phosphatase substrate comparable to XG. The ochre responds to limitation for LYS, ILE, TYR, TRP, and PRO. The other three (all ambers) fail to respond to LYS but do exhibit suppression

TABLE 6.

	$\Delta Z/\Delta B \times 10^4$	
	Control	+ Lysine hydroxamate
YA624 (lacZ^{UAG}, rel$^+$)	51	58
	43	42
	38	54
	56	52
YA626 (lacZ^{UAG}, rel$^-$)	58	106
	68	110
	60	113
	62	112

β-galactosidase synthesis in YA624 (relA$^+$) and YA626 (relA$^-$). The two isogenic strains were constructed from YA625 by cotransduction of relA with fuc. Methods as in TABLE 5; each pair of entries represents a separate experiment with enzyme assays in duplicate. Enzyme synthesis in YA626 is increased significantly by growth in lysine-hydroxamate ($t = 18.15$, $p < 0.0005$).

during limitation for ILE, TYR, TRP, and PRO; two of them respond to PHE as well.

These results demonstrate conclusively that rel mistranslation is not confined to codons calling for the aminoacyl-tRNA species in short supply, since amber and ochre codons do not call for any aminoacyl-tRNA. The amino acid specificities of nonsense read-through are strikingly similar to those of frameshift suppression. For this reason, we are inclined to ascribe both phenomena to the same basic mechanism, namely aberrant translocation. Since we have already shown that phenotypic frameshifting can occur in either direction, a possible explanation is phenotypic frameshifting on either wide of the nonsense codon. However, this model suffers a serious statistical difficulty: it seems to demand an unlikely conjunction of two rare events which ought (in theory at least) to be independent. A more satisfactory interpretation would require but a single event at a shifty codon during aminoacyl-tRNA limitation.

This might be the case if the suppressible nonsense alleles belonged to a class in which spontaneous frameshifting often occurred at the nonsense codon itself; in that case, subsequent rephasing at a shifty codon located sufficiently nearby downstream could produce phenotypic

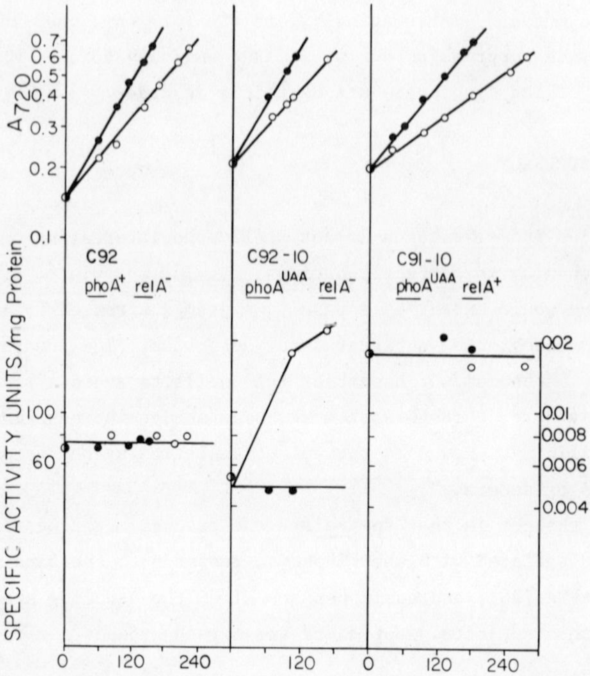

Figure 2. Alkaline phosphatase synthesis during partial inhibition of lysine activation. The S10 ochre allele of phoA (25) was introduced into the isogenic strains C91 (relA$^+$) and C92 (relA$^-$) by cotransduction of phoA with lac. The strains are constitutive for alkaline phosphatase synthesis by virtue of a phoS$^-$ allele. Exponential cultures of the three strains growing in Tris-glucose minimal medium were subcultured with (open circles) or without (filled circles) lysine hydroxamate at 300 ug/ml. The upper panels show growth of the inhibited and control subcultures, in terms of trubidity at 720 nm. The lower panels show specific activity of alkaline phosphatase; the left-hand ordinate pertains to the phoA$^+$ strain (left-hand panel); the right-hand ordinate pertains to the two phoA^{UAA} strains (middle and right-hand panels). One enzyme unit corresponds to a ΔA_{410} of 1.0 per minute in a 350 uL reaction mixture (NPP at 0.86 mg/ml) at room temperature; extracts, suitably concentrated, were prepared by sonication, and protein measured by Lowry.

suppression in a single event. The hypothesis predicts that these nonsense alleles should normally produce more than one fragment product: one due to termination at the nonsense codon itself, and a second, longer one due to frameshifting and later termination at the first out-of-phase terminator. This hypothesis would seem *ad hoc*, except that Suzuki and Garen (25) have reported precisely such a phenomenon. They found that certain nonsense mutants in phoA produced more than one fragment product. These mutants, interestingly, are the very same ones in which we find phenotypic suppression during limited aminoacylation. More detailed tests of the hypothesis are certainly in order.

MODULATION OF ACCURACY

Mistranslation in rel mutants is not really the interesting question, because its mechanism is perfectly obvious. It is, as pointed out earlier, exactly what one would expect of a passive system confronted by an imbalance of aminoacyl-tRNA substrates for selection. The interesting question is the mechanism which enables rel^+ cells to avoid a passive response of this sort, through what must be an active increase in the selection specificity. How does the rel^+ translation system accomplish this modulation of accuracy?

One's first thought is that the $relA^+$ gene product might be responsible. It is associated with the ribosome, responds to the binding of deacylated tRNA (26), and could perhaps alter the decoding specificity of the ribosome A site. But there are strong grounds for rejecting this hypothesis.

First, Pederson and Kjeldgaard find that only a tiny minority of ribosomes even have an associated relA gene product (27). This observation is controversial (see 28), but, if it is correct, it means that direct effects of the relA product on ribosomal specificity cannot be the explanation.

Second, rel mutants do not exhibit mistranslation in response to limited aminoacylation if the level of ppGpp is increased by one trick or another. One trick is carbon/energy source downshift, which leads to accumulation of ppGpp by a relA-independent pathway (29, 30). When this is done prior to the the imposition of aminoacyl-tRNA limitation, then we find that the rel translational defect vanishes: the differential rate of β-galactosidase synthesis returns to a value nearly that of its isogenic $relA^+$ partner (FIGURE 3), and the specific activity per β-galactosidase monomer is virtually identical in $relA^+$ and $relA^-$ cells (TABLE 7). Fiil et al. (17) have used a genetic trick to raise the ppGpp level in certain rel mutants, and have come to the same conclusion.

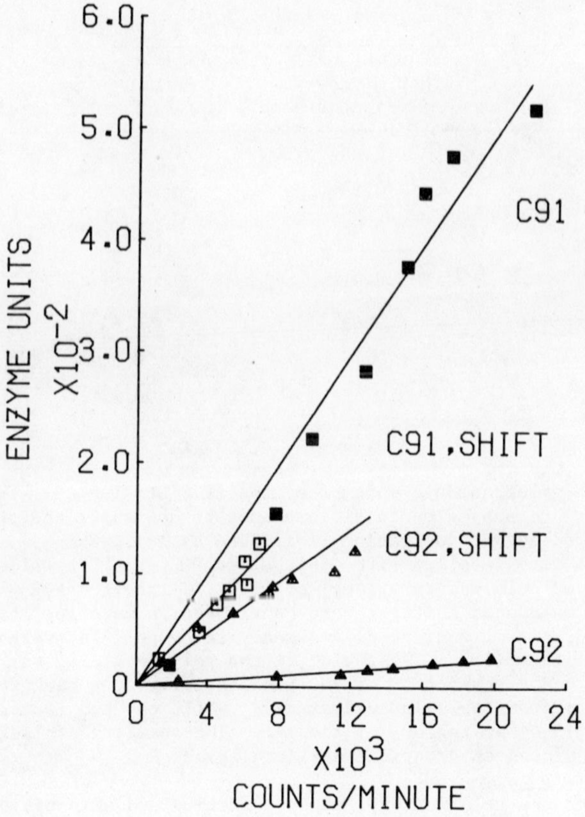

Figure 3. Differential rate of beta-galactosidase synthesis. Cultures of C91 (relA+) and C92 (relA−), growing either in excess glucose or 50 minutes after a glucose to acetate downshift, were subjected to isoleucine limitation by the addition of valine (400 ug/ml). Beta-galactosidase was induced (IPTG, 2mM) in the presence of excess cAMP (5 mM), and aliquots were labelled with ^{3}H-leucine. Samples were removed at intervals for assay of beta-galactosidase activity and TCA-precipitable leucine. The data are presented as a differential plot.

These results argue that relA+ cells are not guarded against mistranslation by any direct effect of the relA+ gene product on the ribosome, but rather by some consequence of a high ppGpp level. How could a humble small molecule effector such as this affect a process as seemingly sophisticated as aminoacyl-tRNA selection? We suggest that Jacques Ninio's analysis of "kinetic amplification" (31) provides a simple and elegant explanation.

TABLE 7.

Strain	Enzyme units per microgram β-galactosidase	Amount of β-galactosidase as % of total protein
C91, relA^{+} spoT^{-}	0.68	0.8%
	0.58	0.9%
	0.68	0.8%
	$\bar{x}$ = 0.65	$\bar{x}$ = 0.85%
C92, relA^{-} spoT^{-}	0.63	0.6%
	0.77	0.5%
	0.45	0.9%
	$\bar{x}$ = 0.62	$\bar{x}$ = 0.67%

Analysis of β-galactosidase monomer synthesis. Strains C91 (relA^{+}) and C92 (relA^{-}) were subjected to glucose-acetate downshift and the lac operon induced during isoleucine limitation as in FIGURE 3. Induced cultures were labelled with ^{3}H-labelled PHE and LEU, while parallel uninduced cultures were labelled with ^{14}C-labelled PHE and LEU. The induced and uninduced cultures were cultivated in parallel for two hours, harvested, mixed, sonicated, and analyzed by gel electrophoresis as described earlier. (16). The region of the gel containing marker β-galactosidase had a sharp peak of ^{3}H- radioactivity over background ^{14}C radioactivity from the uninduced extract, while the two isotopes tracked closely in other regions of the gel. The quantity of lacZ monomer was calculated as described earlier (16).

Ninio analyzed the relationship between kinetics and selectivity in the following, simple Michaelis-Menten system: a correct and an incorrect substrate bind reversibly to an enzymatic site, and the incorrect substrate dissociates more rapidly than the correct one; next, an irreversible catalytic step alters the state of the complex (with release of product, although this need not be so) so as to remove it from equilibrium with the free substrates. The selectivity of such a system depends on the relationship between the rates of the catalytic step and the rates of dissociation of the two substrates. In essence, if the catalytic step is fast in comparison to the dissociation rate of the incorrect substrate, then the latter will often be trapped in product due to the irreversibility of the catalytic step; if the catalytic step is delayed, then there is more time for dissociation of the incorrect substrate, and the system's selection specificity is improved.

In a little-noticed monograph, Ninio advanced the prescient suggestion that this kind of trade-off between speed and accuracy could be adjusted to one level or another through the action of a secondary element (32). We suggest that ppGpp is just such a secondary element. This modified guanine nucleotide might well inhibit the irreversible partial reactions of protein synthesis which involve GTP hydrolysis (those associated with $EF\text{-}T_u$ and EF-G) and thus improve selectivity a la Ninio.

We have performed a naive test of this general notion, using crude S_{30} extracts programmed by polyU and a mixture of ^{14}C-LEU and ^{3}H-PHE to detect the frequency of misreading. TABLE 8 shows that a partially inhibitory concentration of ppGpp does reduce the incorporation of LEU more than that of PHE, thus improving selectivity. The effect is rather

TABLE 8.

Experiment	Source of S30	Incubation condition	Incorporation (μμmoles) PHE	LEU	Both	Misreading Index
A	*relA*$^+$	control	6.2	8.4	14.6	1.35
		+ppGpp	3.8	2.6	6.4	0.68
B	*relA*$^-$	control	16.8	9.6	26.4	0.57
		+ppGpp	12.2	4.1	16.3	0.34
C	*relA*$^-$	control	4.8	5.1	9.9	1.06
		+ppGpp	4.2	2.65	6.85	0.63
		+ppG	5.8	6.3	12.1	1.09
		+ppppG	6.2	6.6	12.8	1.06
		+FUS (4 γ/ml)	3.7	2.05	5.75	0.55
		+FUS (8 γ/ml)	2.45	1.6	4.05	0.65

Misreading of polyU. Exponential cultures were cultivated for 30 minutes in the presence of rifampicin (200 ug/ml) in order to eliminate endogenous messenger RNA, concentrated by centrifugation, and crude S_{30} extracts prepared by alumina grinding essentially according to Haseltine *et al*. (42). Protein synthesis *in vitro* was measured by conventional methods during 16 minutes incubation at 37° in the presence of polyU (200 ug/ml) and a mixture of ^{3}H-PHE and ^{14}C-LEU (with a four-fold excess of the latter) at 19 mM magnesium (in order to increase mistranslation to readily detectable levels); the values shown are corrected for blank incubations without polyU, which were very low. Nucleotide effectors were at 0.9 or 1.0 mM. The misreading index is the ratio of LEU to PHE incorporation.

small, about a 40% reduction in the misreading index, but it has been observed repeatedly, and with both rel$^+$ and rel$^-$ extracts; a summary of a larger number of experiments is reported elsewhere (33) and Just Justesen, working with N.-O. Kjeldgaard, has independently made the same observation (Justesen, unpublished experiments). TABLE 8 also shows that ppG and ppppG, which do not inhibit protein synthesis, do not reduce the misreading index, while fusidic acid, an inhibitor of EF-G, has an effect similar to that of ppGpp.

The effects on misreading of ppGpp and other protein synthesis inhibitors are thus of considerable interest, and warrant more sophisticated experiments focusing on the particular partial reactions whose rates affect selection specificity.

We advance the following tentative interpretation to unify all the experimental results. A passive translation system will necessarily suffer an increase in errors of aminoacyl-tRNA recognition when the supply of different aminoacyl-tRNA species is imbalanced, and this is what happens in rel$^-$ mutants. Correct translocation is associated with tRNA recognition (21), so that misrecognition often shifts the ribosome out of the correct reading frame, especially at shifty codons, and thus gives rise to phenotypic suppression of frameshift mutations. These frameshift events usually produce premature termination at out-of-phase terminator codons, thus explaining the overproduction of small polypeptides in amino acid limited rel$^-$ cells (15, 16). Phenotypic suppression of nonsense alleles also occurs through a rephasing event at shifty codons, although this interpretation requires certain *ad hoc* assumptions which need further test. Finally, rel$^+$ cells are spared these misreading events, at least to a degree, because ppGpp delays key irreversible steps so as to modulate the trade-off between speed and accuracy. It should be emphasized that this interpretation is based on the simple Ninio scheme for Michaelis-Menten accuracy, and does not invoke kinetic proof-reading (see Hopfield's discussion elsewhere in this volume). A mechanism involving kinetic proof-reading could doubtless be devised. In fact, all of the theoretical treatments of the relationship between kinetics and accuracy are united, at a fundamental level, by the consequences of large displacements from equilibrium in a network of sequential reactions (see Kurland's chapter in this volume) and differ principally in the locations within the network of reactions which release substrate.

ERROR PROPAGATION

Some years ago, Garvin, Rosset, and Gorini (34) reported on a curious aspect of streptomycin induced mistranslation. They found that a preliminary period of growth in sub-inhibitory levels of streptomycin markedly increased the ability of bacterial cells to propagate phage nonsense mutants, and also increased the propensity of ribosomes purified from the cells to mistranslate polyU. They suggested that streptomycin produced some abnormality in ribosome assembly such that ribosomes formed in the presence of the drug were prone to mistranslation. But there is no independent evidence for this postulate. So far as we know, streptomycin induces misreading by a direct effect on preformed ribosomes.

An alternative and perhaps more satisfactory interpretation flows from Leslie Orgel's idea of error feedback (35). Orgel pointed out that there ought to be positive feedback in mistranslation events, because the translation apparatus itself depends on the specificity of protein components. When mistakes occurred in the synthesis of these components, they would sometimes suffer reduced specificity, thus generating a fresh crop of errors.

Orgel formalized the model as follows (36). Suppose, for the sake of simplicity, that "generations" of protein are discrete and distinguishable: each generation of proteins produces (through its members which are part of the translation apparatus) the next generation of proteins. The simplest possible recursion relationship to describe the situation is: $e_i = E + \alpha e_{i-1}$ where e is the agregate error frequency in a given generation, E is the residual error frequency characteristic of a "perfect" synthetic apparatus with the correct amino acid at every position, and α is a proportionality constant.

For $\alpha < 1$ the recursion relationship converges to a stable value of $e = \frac{E}{1-\alpha}$, and this is surely the case in vegetatively growing microorganisms which exhibit a stable error frequency. The addition of an error-promoting drug such as streptomycin has the effect of suddenly increasing E, thus raising e above the normal stable value during the first generation after drug addition; this in turn raises e further in the second generation, and so on. As growth in the drug proceeds, the error frequency converges more or less rapidly to a new stable value, following a trajectory determined by the value of α (FIGURE 4).

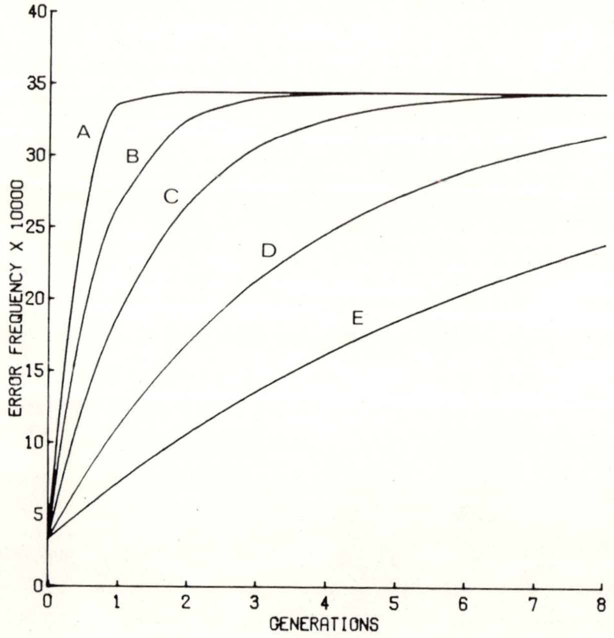

Figure 4. Predictions of the Orgel equation. The curves show the predicted kinetics of the change in error frequency following an increase in E such that the stable value of e_i goes from 3.3 x 10^{-4} to 34.5 x 10^{-4}. The two boundary conditions correspond to our measurements of mistranslation of a <u>phoA</u>^{UAA} codon during normal growth and after prolonged growth in 5 ug/ml streptomycin (see reference 8). The smooth curves between the discrete points of the model were generated by the SIMPLOT computer program. The curves are for different values of α, as follows: A, α = 0.03; B, α = 0.3; C, α = 0.5; D, α = 0.75; D, α = 0.875.

We have examined the applicability of error feedback models to a variety of data on misreading during prolonged growth in error-promoting drugs. The curves of FIGURE 5 show the Orgel model's predictions, at various values of α corresponding to FIGURE 4, for the kinetics with which alkaline phosphatase activity should increase during growth of a streptomycin-suppressible <u>phoA</u>^{UAA} mutant in streptomycin; the symbols show our experimental results (8). It can be seen that the data are in satisfactory agreement with the model for α ≃ 0.5, and values much higher or lower can be ruled out of consideration. It should be emphasized that the error frequency does reach a stable value about 13 times higher than the normal value and stays there (see reference 8).

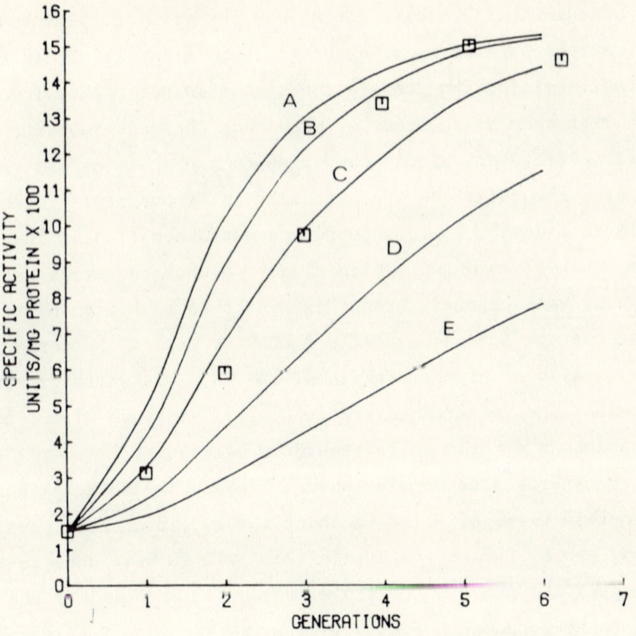

Figure 5. Predicted and observed mistranslation kinetics. The symbols show the observed increase in alkaline phosphatase specific activity during growth in 5 µg/ml streptomycin, averaged from three different experiments reported in reference 8. The curves show the increase predicted by each of the corresponding curves of FIGURE 4, with discontinuous points connected by SIMPLOT.

We find a similarly good fit, at $\alpha \simeq 0.5$, when the same ochre mutant is cultivated in different concentrations of streptomycin, or in the presence of ethanol, a drug which promotes mistranslation by a different biochemical mechanism. The aforementioned data of Garvin, Rosset, and Gorini (34) also fit the Orgel equation quite well at $\alpha \simeq 0.5$. All these data also fit a more complex error feedback expression -- the Hoffman model (37) as modified by Kirkwood and Holliday (38) -- for a particular domain of parameter values. Finally, we have shown that the Hoffman-Kirkwood and Holliday expression reduces to Orgel's simple equation for error frequencies as low as those which obtain in real life. The experimental and mathematical details of these somewhat arcane studies will be presented elsewhere (39). The conclusion we draw is that the theory of error feedback can account for a variety of otherwise puzzling data on the kinetics of misreading during growth in error-promoting drugs.

This should not be taken as supporting the error catastrophe theory of cellular senescence (35, 38). It is just the reverse. The bacterial data imply that the parameters of error feedback are well within the region of stability, for the translation system is not pushed into unstable behavior even by an order of magnitude increase in mistranslation: the error frequency simply converges to a higher, stable value and stays there; the cells continue placidly in exponential growth with no detectable cell death, i.e., no error catastrophes (7,8). These observations admit of only two explanations: either the error catastrophe theory of senescence is wrong, or, if it is right, then error feedback must operate in fundamentally different ways in bacteria and in the somatic cells of higher organisms. We have elsewhere presented arguments against the latter interpretation (8), although direct tests of error feedback in somatic cells remain to be done.

This rather far-reaching conclusion was already implicit, although not spelled out in terms of error feedback theory, in a short paper published long ago by Gorini and Kataja (2). All we have done is to fill in some gaps and add some controls, so as to make plainer the deep implications to be found in Gorini's work. The section of this volume on translational accuracy reveals that it is not only my laboratory, but the entire field, which is haunted by Gorini.

ACKNOWLEDGEMENT

This work was supported by grants from the National Institutes of Health (Grant # GM13626) and the American Cancer Society (Grant #NP279A). We are indebted to Linda Palmer for expert technical assistance in many of the experiments, to Barry Harada for technical assistance in the experiments reported in TABLE 8, and to Robert Weiss, who performed some of the many Petri plate assays of phenotypic suppression. We are grateful to Charles Kurland, John Menninger, and John Prothero for stimulating ideas.

REFERENCES

1. Woese, C. (1967) The Genetic Code, Harper and Row, New York.
2. Gorini, L., and Kataja, E. (1964) Proc. Natl. Acad. Sci. U.S. 51, 995-1001.
3. Gorini, L. (1974) Ribosomes, pp. 791-803, New York: Cold Spring Harbor Laboratory.
4. Bjare, U., and Gorini, L. (1971) J. Mol. Biol. 57, 423-435.
5. Gado, I., and Horvath, I. (1963) Life Sci. 2, 741-745.
6. Edelmann, P., and Gallant, J. (1977) Cell 10, 131-137.
7. Edelmann, P., and Gallant, J. (1977) Proc. Natl. Acad. Sci. U.S. 74, 3396-3398.
8. Gallant, J., and Palmer, L. (1979) Mech. Ageing and Dev. 10, 27-38.
9. Loftfield, R. B. (1963) Biochem. J. 89, 82-92.
10. Loftfield, R. B., and Vanderjagt, D. (1972) Biochem. J. 128, 1353-1356.
11. Fiil, N. (1969) J. Mol. Biol. 45, 195-203.
12. Foster, G. G. (1968) Studies on the Amino Acid Regulation of Messenger RNA Synthesis in Escherichia coli. M.S. thesis, University of Washington.
13. Lavalle, R., and De Hauwer, G. (1968) J. Mol. Biol. 37, 269-288.
14. Gallant, J., Erlich, H., Hall, B., and Laffler, T. (1970) Cold Spring Harbor Symp. Quant. Biol. 35, 397-405.
15. O'Farrell, P. H. (1978) Cell 14, 545-557.
16. Hall, B., and Gallant, J. (1972) Nature New Biol. 237, 131-135.
17. Fiil, N. P., Willumsen, B. M., Friesen, J. D., and von Meyenberg, K. (1977) Molec. Gen. Genet. 150, 87-101.
18. Davies, J., Gorini, L., and Davis, B. D. (1965) Mol. Pharmacol. 1, 93-106.
19. Davies, J., Jones, D. S., and Khorana, H. G. (1966) J. Mol. Biol. 18, 48-57.
20. Parker, J., Pollard, J. W., Friesen, J. D., and Stanners, C. P. (1978) Proc. Natl. Acad. Sci. U.S. 75, 1091-1094.
21. Kurland, C. G. (1979) Nonsense Mutations and tRNA Suppressors, (eds. Celis, J. E. and Smith, J. D.), pp. 95-108, Academic Press.
22. Atkins, J. F., Elseviers, D., and Gorini, L. (1972) Proc. Natl. Acad. Sci. U.S. 69, 1192-1195.
23. Riddle, D. L., and Carbon, J. (1973) Nature New Biol. 242, 230-233.
24. Newton, A. (1970) J. Mol. Biol. 49, 589-601.
25. Suzuki, T., and Garen, A. (1969) J. Mol. Biol. 45, 549-566.
26. Block, R., and Haseltine, W. A. (1974) Ribosomes, pp. 747-761, New York: Cold Spring Harbor Laboratory.

27. Pedersen, F. S., and Kjeldgaard, N.-O. (1977) Eur. J. Biochem. 76, 91-97.

28. Schmale, H., Fehr, S., and Richter, D. (1978) Hoppe-Seyler's Z. Physiol. Chem. 359, 125-127.

29. Gallant, J., and Lazzarini, R. (1976) Protein Synthesis, A Series of Advances, (ed. McConkey, E. H.), pp. 309-359, Marcell Dekker, New York.

30. Friesen, J. D., An, G., and Fiil, N. P. (1978) Cell 15, 1187-1197.

31. Ninio, J. (1975) Biochimie 57, 587-595.

32. Ninio, J. (1975) L'Evolution des Macromolecules Biologiques, (ed. Sadron, C.), pp. 51-66, Editions du C.N.R.S., Paris.

33. Gallant, J., and Foley, D. (1979) Low Molecular Weight Mediators of Macromolecular Synthesis, (eds. Koch, G. and Richter, D.), in press.

34. Garvin, R. T., Rosset, R., and Gorini, L. (1973) Proc. Natl. Acad. Sci. U.S. 70, 2762-2766.

35. Orgel, L. E. (1963) Proc. Natl. Acad. Sci. U.S. 49, 517-521.

36. Orgel, L. E. (1970) Proc. Natl. Acad. Sci. U.S. 67, 1476.

37. Hoffman, G. H. (1974) J. Mol. Biol. 86, 349-362.

38. Kirkwood, T. B. L. Q., and Holliday, R. (1975) 97, 257-265.

39. Gallant, J., and Prothero, J. Submitted to J. Theor. Biol.

40. Garen, A., and Siddiqi, O. (1962). Proc. Natl. Acad. Sci. U.S. 48, 1121-1124.

41. Manney, T. R. (1968) Genetics 60, 719-733.

42. Haseltine, W. A., Block, R., Gilbert, W., and Weber, K. (1972) Nature 238, 381-384.

43. Morris, D. W., and Kjeldgaard, N.-O. (1968) J. Mol. Biol. 31, 145-148.

44. Foley, D., Dennis, P., and Gallant, J. Manuscript in preparation.

45. Foley, D. (1979) Ph.D. Thesis, University of Washington.

46. Sokawa, Y., Sokawa, J., and Kaziro, Y. (1974) Nature 249, 59-62.

47. Sommerville, C. R., and Ahmedo, A. (1977) J. Mol. Biol. 111, 77-81.

48. Sommerville, C. R. (1976) Master's Thesis, Department of Genetics, University of Alberta.

49. Freundlich, M. (1977) Cell 12, 1121-1126.

50. Stephens, J. C., Artz, S. W., and Ames, B. N. (1975) Proc. Natl. Acad. Sci. U.S. 72, 4389-4393.

51. Hall, B. G., and Gallant, J. (1971) J. Mol. Biol. 61, 271-273.

PART 5: GENETICS AND BIOSYNTHESIS

GENETICS OF RIBOSOMAL PROTEINS AND THEIR MODIFYING AND PROCESSING ENZYMES IN *ESCHERICHIA COLI*

KATSUMI ISONO

Max-Planck-Institut für Molekulare Genetik, Abt. Wittmann, Berlin, Germany

INTRODUCTION

In contrast to the numerous biochemical studies on the structure and function of the ribosome, until recently genetic investigation in this field has been rather poor, even in the case of *Escherichia coli*. This is mainly because the number and variety of mutants making such an approach feasible has been limited: most of the mutants isolated and characterized that contain alterations in ribosomal components are those isolated as being resistant to or dependent on antibiotics such as streptomycin, spectinomycin, erythromycin, etc., and only a few ribosomal protein alterations have been observed in these mutants (for previous review, see refs. 1-3).

An alternative way to establish the locations of the genes coding for ribosomal components so as to carry out genetic studies on them has been sought by several groups. One way is to perform interspecific (e.g. *E. coli* K-12 strain vs. B strain) or intergeneric (e.g. *E. coli* vs. *Serratia marcescens*) matings and to analyze the components in question of the hybrid cells by suitable methods (see ref. 4). This method was successfully used to map several ribosomal protein genes (5-9), 5S rRNA genes (10-12) and RNA polymerase σ subunit genes (13,14). However, since this method relies on the difference between a particular component of the two mating pairs, it is not a generally applicable method. Are there, then, any other feasible methods to obtain mutants for detailed genetic analyses?

It has been well established that the ribosome of *Escherichia coli* consists of three RNA and about fifty protein molecules. The relative rate of synthesis of the individual components of the ribosome remains almost unaltered under a variety of growth conditions and the size of the pools of free (unassembled) ribosomal proteins inside the cell is very small (1-3%) (15-17). The synthesis of ribosomal RNA and proteins has been reported to show stringent response upon amino acid starvation (18,19). These facts may imply that there is a regulatory mechanism controlling the synthesis, modification and subsequent assembly into the ribosome of all the individual ribosomal components, proteins as well as RNA. How, then, can bacteria achieve such a highly coordinated synthesis of the ribosomal components?

Recent advance in the *in vitro* techniques of processing and sequencing DNA molecules (20) have opened the way for the determination of mutational alterations at the nucleotide sequence level. These techniques, when combined with the techniques for detecting RNA and protein products of the DNA, would make it possible to investigate the sites playing regulatory roles in, for example, the synthesis of ribosomal proteins. In order to do so, it is quite evident that the locations of the genes coding for individual ribosomal components should be known. To this end, the isolation of mutants with altered ribosomal proteins would be vital. Such mutants would be useful not only for the determination of the chromosomal locations of various ribosomal protein genes, but also for the cloning of these genes into phage or plasmid vectors, because one can easily detect the protein products of the genes that have been incorporated into the strains harboring suitable ribosomal protein alterations by simply analyzing them on electropherograms. Furthermore, once the locations of ribosomal genes are established, it will become much easier to isolate mutants that harbor impaired ribosomal structure causing a conditional (e.g. temperature-sensitive) malfunction of the ribosome. Studies of mutants with partial or total deletions in the ribosomal protein genes may provide further insight into the role of each ribosomal protein *in vivo*. Similarly, mutants with defects in the regulatory mechanism could be isolated as evidenced by Post et al. (21).

VARIOUS PROCEDURES OF MUTANT ISOLATION

Since the ribosome is an indispensable subcellular organelle, the only conceivable way to obtain mutants with altered ribosomal structure is to screen for conditional mutants such as temperature-sensitive mutants, cold-sensitive mutants, or suppressible (amber, ochre, etc.) mutants. Mutants of the last class would be interesting and useful only when suppressors that can be artificially regulated are used, because the insertion into and the elimination from the same bacterial strain of suppressor mutations cannot easily be done and therefore we can not readily see the effects of the mutations introduced. Temperature-sensitive nonsense suppressors may be suitable for this purpose.

A large number of independent isolates of temperature-sensitive mutants have been screened for ribosomal protein alterations (22-24). They were isolated after rather heavy mutagenesis with nitrosoguanidine and therefore it was expected that many of them would harbor multiple mutations. Surprisingly, the overall percentage of the mutants with alterations in the ribosomal proteins detectable by two-dimensional gel electrophoresis was very high (about 10%) and about 170 mutants have thus been isolated. It was found that except for a few cases the ribosomal protein alterations are not responsible for temperature-sensitivity. Altogether, mutants have been detected in 46 ribosomal proteins (Table 1). Theoretically, heavy mutagenesis followed by screening for, for example, temperature-sensitivity can be applied to any other organisms to obtain ribosomal mutants.

A unique way of isolating ribosomal mutants has been reported. A mutant dependent on streptomycin for growth was isolated which, when grown in the absence of the drug, yielded many ribosomal protein mutants (25,26). It is not clear why this mutant produces "revertants" not requiring streptomycin for growth that harbor mutational alterations in so many ribosomal proteins (almost all the ribosomal proteins amenable to detection by two-dimensional gel electrophoresis, see Table 1). It is interesting to note that many mutations detected in the temperature-sensitive mutants described above appear to be very similar to or most likely identical with those detected in the "revertants" of the above mentioned streptomycin-dependent mutant. Perhaps the two selection procedures, albeit superficially different from each other, may involve a common step(s), for example, selection towards impaired ribosome assembly.

Böck and collaborators developed a special method to isolate mutants using strains harboring temperature-sensitive defects in alanyl or valyl tRNA synthetase (27). The rationale for this method is that the mutational defects in tRNA synthetases can be compensated for by slowing down the speed of protein synthesis, which may be caused by ribosomal protein alterations (28). A large number of temperature-insensitive "revertants" were screened for ribosomal mutations. Thus, mutants with altered S5 and S20 proteins were isolated from an alanyl tRNA synthetase mutant strain (27,29) and mutants with altered S8 and S20 proteins were isolated from a valyl tRNA synthetase mutant strain (30). It was noticed that one of the "revertants" from the valyl tRNA synthetase mutant apparently lacked protein S20. As described later, several mutants each lacking one of the ribosomal proteins have recently been isolated by other methods.

Isaksson et al. (31) have reported a method to obtain temperature-sensitive mutants and characterized one of the mutants thus isolated (32,33). The mutation shows pleiotropic effects on cellular functions: temperature-sensitive growth, fusidic acid resistance and accumulation of guanosine tetra- and pentaphosphates at non-permissive temperatures. These alterations seem to be caused by a single mutation termed fusB located near lip (14 min). As will be described later this mutant harbors an alteration in the ribosomal protein S6.

The "localized mutagenesis" technique (34) has been used to isolate mutants of the strA region, where 27 ribosomal protein genes along with the genes for elongation factors and RNA polymerase subunit are located (see ref. 3). P1kc phages propagated on a wild-type strain are first treated with mutagens such as nitrous acid or nitrosoamine and then used in transductions in which $AroE^+$ transductants are selected. Transductants are subsequently screened for temperature-sensitivity and/or cold-sensitivity. Mutants with altered proteins S17, L18 and L24 have thus been isolated (35-37). It has been claimed that the mutations in ribosomal proteins S17 and L24 are responsible for the observed temperature-sensitivity (36,37).

Mutants with altered behavior toward antibiotics have been isolated which possess alterations in ribosomal proteins that have not been previously described. Until recently, it was generally assumed that mutants dependent on antibiotics for growth are confined to streptomycin. Recent reports by Dabbs (38-40) indicate, however, that it is possible to isolate mutants dependent on other antibiotics for growth. He isolated mutants dependent on spectinomycin (38), on kasugamycin (39) and on eryhtromycin (40) for growth. It seems, therefore, that mutants dependent on antibiotics for growth can be isolated if appropriate conditions are employed. Earlier Sparling et al. (41) isolated a mutant dependent on erythromycin for growth. Some possible implications of these antibiotic-dependent mutants will be described in detail later.

Mutants resistant to lincomycin have been found to possess alterations in proteins S7, L14 or L15 (42). Mutants resistant to gentamicin have alterations in protein L6 (43). Upon electrophoresis on two-dimensional slabs of the ribosomal proteins of some of these mutants, protein L6 was found to be completely missing.

A method to isolate mutants with amber mutations in the ribosomal protein genes has been described (44) and mutants thus isolated have been analyzed. The method involves use of conditions under which mutants with amber mutations in essential genes such as ribosomal protein genes can grow only at low temperatures (for details, see below).

Many of the ribosomal proteins of *E. coli* are known to undergo post-translational modification. Protein S5 (45), S18 (46) and L7 (47) are known to have acetylated amino acids at their N-termini. Proteins S11 (48), L3 (49-51), L11 (50-52), L7/L12 (47), L16 (48,53), L33 (48,53,54) and perhaps L1 and L5 (50,51) contain methylated amino acids. Protein S6 has recently been reported to have up to six glutamic acid residues added to its C-terminus after translation (55,56). Therefore, it is of considerable importance and interest to isolate mutants defective in these modification systems. The temperature-sensitive mutant collection reported by Isono et al. (22,23) contains mutants lacking the N-terminal acetyl groups of proteins S5, S18 and L7. It was also found to contain mutants with an impairment in the activity of adding glutamic acid residues to protein S6 (57) and mutants having deficient modification (methylation?) of protein L11 (58). Mutants deficient in methylation of proteins L3 and L11 have been isolated and characterized (59). Characterization of these mutants will be described below.

MAPPING OF RIBOSOMAL PROTEIN GENES

Structural Genes for Ribosomal Proteins

By isolating λ-transducing phages, it has been established that the so-called '_strA_ region' (72 min) contains the genes for 27 ribosomal proteins (see ref. 3 and Fig. 1). Similarly it has been established that the '_rif_ region' (88 min) contains 4 ribosomal protein genes.

Mapping of other ribosomal protein genes have been extensively performed using temperature-sensitive mutants harboring ribosomal protein alterations. The genes coding for proteins S6 (_rpsF_), S18 (_rpsR_) and L9 (_rplI_) are clustered at 94 min near _purA_ (60,61). Subsequent analysis of this region by isolating λ-transducing phages and infecting UV light-irradiated cells with them has revealed that this cluster contains only these three ribosomal protein genes and that no other ribosomal protein genes are present in their neighborhood (62).

The genes coding for proteins S16 (_rpsP_) and L19 (_rplS_) are clustered at 56.4 min near _tyrA_ (63,64). λ-transducing phages carrying this region have also been isolated and the ribosomal protein genes present in them similarly analyzed. It was

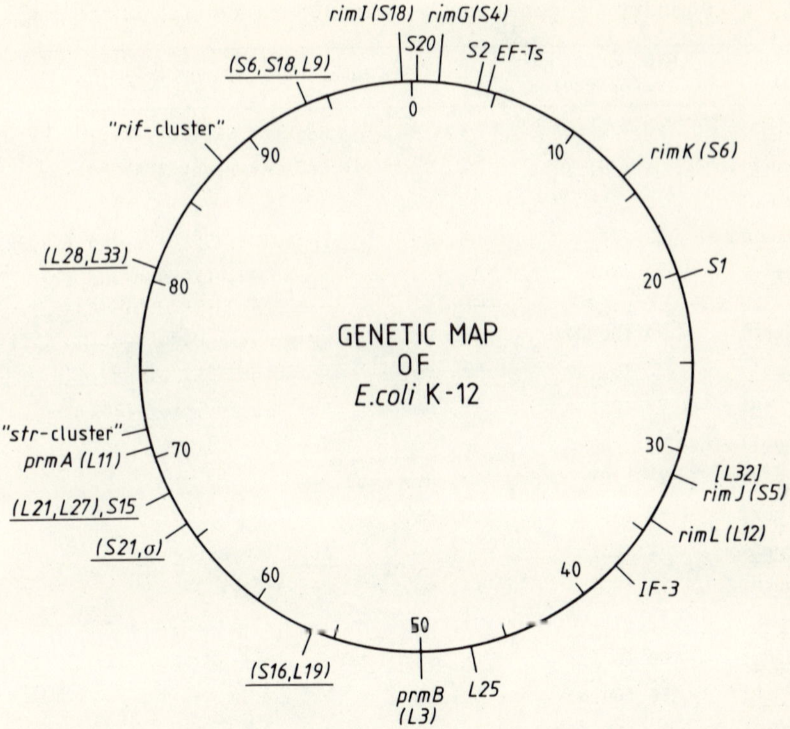

Figure 1. The genetic map of E. coli modified from Bachmann et al. (66). The str-cluster contains the genes for proteins S3, S4, S5, S7, S8, S10, S11, S12, S13, S14, S17, S19, L2, L3, L4, L5, L6, L14, L15, L16, L17, L18, L22, L23, L24, L29, L30, EF-G, EF-Tu (tufA) and the α subunit of the RNA polymerase. The rif-cluster contains the genes for proteins L1, L10, L11, L12, EF-Tu (tufB) and the ß and ß' subunits of RNA polymerase (for details of these clusters, see ref. 3 and the review by Nomura of this volume). The gene for protein L32 is only roughly placed.

found that this region contained only the above two ribosomal protein genes (62). During the course of isolating λ-transducing phages carrying a mutation mapping near tyrA that led to a gro phenotype toward λ phage propagation, Saito et al. (65) fortuitously found that the genes coding for ribosomal proteins S16 and L19 were carried by the same transducing phages. Thus, the location of these two genes has been independently confirmed.

More recently it was found that the genes coding for ribosomal proteins L21 (rplU) and L27 (rpmA) were clustered at 68.1 min near argG and that the gene for protein S15 (rpsR)

TABLE 1. Summary of genetics of the 30S ribosomal proteins[a]

Protein (gene)	Map position[b] (reference)	Mutation (reference)[c]	
S1 (rpsA)	20 (72)	-, -	Shorter S1 (72), Ksg-d (39)
S2 (rpsB)	4 (74)	Ts, Sm,	Cs^+ revertant (74)
S3 (rpsC)	72 (75)	Ts, Sm,	
S4 (rpsD)	72 (76)	Ts, Sm,	ramA (76,77), Str-ind (78,82)
S5 (rpsE)	72 (79)	Ts, Sm,	spcA (80,81), Str-ind (78) Ala^{ts}-suppressor (27)
S6 (rpsF)	94 (60,61)	Ts, Sm,	fusB (32), Modification (57)
S7 (rpsG)	72 (5)	Ts, Sm,	K-factor (5), Lin-r (42)
S8 (rpsH)	72 (75)	Ts, Sm,	Val^{ts}-suppressor (30)
S9 (rpsI)	Not known	-, Sm,	Ksg-d (39), Ery-ind (40)
S10 (rpsJ)	72 (75)	Ts, Sm,	
S11 (rpsK)	72 (75)	Ts, Sm,	
S12 (rpsL)	72 (83)	-, -,[d]	strA (84,85)
S13 (rpsM)	72 (75)	Ts, Sm,	Ksg-d (39)
S14 (rpsN)	72 (75)	Ts, Sm,	
S15 (rpsO)	68 (67)	Ts, Sm,	
S16 (rpsP)	56 (64,65)	Ts, Sm,	
S17 (rpsQ)	72 (86,87)	-, Sm,	Strain-specific (86), neaA (87), ts^- (88)
S18 (rpsR)	94 (61,89)	Ts, Sm,	Structural alteration (90) Ksg-d (39)
S19 (rpsS)	72 (75)	Ts, Sm,	
S20 (rpsT)	0 (29,91)	Ts, Sm,	Ala^{ts}-suppressor (27-29), Val^{ts}-suppressor (30)
S21 (rpsU)	66 (9,68)	-, Sm,	

[a]*For a previous summary, see ref. 2. The mutations listed in Table 1 of ref. 2 whose ribosomal protein alterations are not clear have not been incorporated in this table.*

[b]*Expressed in minute coordinates as described by Bachmann et al. (66). For the fine structure of the genes located at 72 min and 88 min, see refs. 1-3 and the review by Nomura of this volume.*

[c]*Symbols Ts and Sm indicate, respectively, temperature-sensitive mutants (refs. 22-24) and "revertants" from a special streptomycin dependent mutant (refs. 25,26). For other abbreviations see footnotes to Table 2.*

[d]*The parental strains of both temperature-sensitive mutants (refs. 22-24) and streptomycin-independent revertants (refs. 25,26) harbor S12 mutations.*

TABLE 2. Summary of genetics of the 50S ribosomal proteins

Protein (gene)	Map position (reference)	Mutation (reference)[a]	
L1 (rplA)	88 (95)	Ts, Sm,	
L2 (rplB)	72 (75)	–, Sm,	
L3 (rplC)	72 (75)	Ts, Sm,	
L4 (rplD)	72 (93,94)	–, Sm,	eryA (93,94)
L5 (rplE)	72 (75)	Ts, Sm,	
L6 (rplF)	72 (75)	Ts, Sm,	Gen-r (43)
L9 (rplI)	94 (61)	Ts, Sm,	
L10 (rplJ)	88 (95)	Ts, Sm,	
L11 (rplK)	88 (95)	Ts, Sm,	relC (96,97), Ery-ind (40)
L12 (rplL)	88 (95)	Ts, Sm,	Structural alteration (98)
L13 (rplM)	Not known	–, Sm,	
L14 (rplN)	72 (75)	Ts, Sm,	Lin-r (42), Ksg-d (39)
L15 (rplO)	72 (75)	Ts, Sm,	Lin-r (42), Ery-ind (40)
L16 (rplP)	72 (75)	Ts, Sm,	
L17 (rplQ)	72 (75)	Ts, Sm,	
L18 (rplR)	72 (75)	Ts, Sm,	Structural alteration (35)
L19 (rplS)	56 (63-65)	Ts, Sm,	
L20 (rplT)	Not known	–, –,	
L21 (rplU)	68 (67)	–, Sm,	
L22 (rplV)	72 (75)	Ts, Sm,	eryB (99)
L23 (rplW)	72 (75)	Ts, Sm,	
L24 (rplX)	72 (75)	Ts, Sm,	ts^- (36)
L25 (rplY)	47 (70)	Ts, Sm,	
L27 (rpmA)	68 (67)	–, Sm,	
L28 (rpmB)	83 (69)	Ts, Sm,	Ery-ind (40)
L29 (rpmC)	72 (75)	Ts, Sm,	
L30 (rpmD)	72 (75)	Ts, Sm,	
L31 (rpmE)	Not known	–, Sm,[b]	
L32 (rpmF)	30 (71)	Ts, Sm,	
L33 (rpmG)	83 (69)	Ts, Sm,	
L34 (rpmH)	Not known	–, –,	

[a] *Abbreviations (for details of these mutations, see text): Cs^+, cold-sensitive; Str-ind, streptomycin-independent; Ala^{ts} (Val^{ts}), temperature-sensitive alanyl (valyl) tRNA synthetase; Ksg-d, kasugamycin-dependent; Lin-r, lincomycin-resistant; Ery-ind, erythromycin-independent; ts^-, temperature-sensitive; Gen-r, gentamicin-resistant.*

[b] *Dabbs, E.R., personal communication.*

mapped at 67.8 min, on the other side of argG from the former two genes (67). Earlier, Takata and Kobata (8) and Takata (9) placed the three genes in this region by performing intergeneric matings between *E. coli* and *Serratia marcescens*. In addition to these three genes, the gene for protein S21 (rpsU) was also placed in this region, although it was slightly away from the former three (9). It was found that the gene for protein S21 was located very close (more than 95% cotransducible) to dnaG (66 min) (68). It is interesting to note that the gene rpoD coding for the σ subunit of RNA polymerase also maps in this region (13,14). It remains to be seen whether or not the two genes, rpoD and rpsU, are clustered as are the genes for other RNA polymerase subunits (α, ß and ß') with ribosomal protein genes in the strA and rif regions.

The genes coding for proteins L28 and L33 have been found to reside at 80.6 min near pyrE (69). The two genes are located on the other side of pyrE from spoT. The two genes are most likely located next to each other. It is not known whether any other ribosomal protein genes are present in their neighborhood. The gene coding for protein L25 is located at 47.3 min between nalA and fpk (70) and the gene for protein L32 near trp (71). A mutation causing a drastic change in the size of protein S1 (about 75% of the size of wild type protein) has recently been mapped at 20 min near aroA (72). Since the strain heterozygous for this region shows both wild type and mutant forms of protein S1, the authors claim that the mutation may be a large deletion in the structural gene coding for this protein (rpsA). The ribosome as well as protein S1 prepared from this mutant strain are almost as active as those from wild type strain in various *in vitro* assays (73). Protein chemical analysis indicates that the mutant protein has the same N-terminal sequence as wild type protein.

The number of the structural genes coding for ribosomal proteins in *E. coli* is 52. Of these, 47 have so far been mapped as summarized in Table 1, Table 2 and Fig. 1. This leaves the genes for proteins S9, L13, L20, L31 and L34 to be mapped. Some of the genes for initiation and elongation factors have also been mapped as shown in Figure 1 (for details, see ref. 3). The genes for initiation factors IF-1 and IF-2 and for termination factors RF-1, RF2 and RF-3 are yet to be mapped. Complete map positions of all the structural genes for the ribosomal proteins and for the factors closely related to the ribosomal functions of *E. coli* will be established before long.

Mapping and Characterization of the Mutations Affecting Ribosomal Protein Modification

In addition to the structural genes coding for ribosomal proteins in *E. coli*, there is a considerable number of genes involved in the modification and processing of ribosomal proteins. As briefly described in an earlier section, the temperature-sensitive mutant collection screened for ribosomal protein alterations (22-24) has been found to contain mutants with impaired modifications of ribosomal proteins.

Acetylation: A mutant lacking protein L7 was isolated (23). By mapping the mutation, it was found (100) that the mutation causing this alteration was located not in the rif region where the structural gene for this protein had been located (95) but it was located in the 'silent' region between trp and his (Fig. 1) which had not been covered by P1 transduction (see ref. 66). Since the only difference between proteins L7 and L12 is that the former has acetylserine at its N-terminus, while the latter has unacetylated serine (47), the mutation in this strain has most likely occurred in the gene termed rimL coding for an acetylating enzyme specific for protein L12. Indeed the post-ribosomal supernatant fraction of the mutant is unable to acetylate L12 protein *in vitro* using acetyl-CoA as an acetyl donor, whereas the wild-type counterpart efficiently acetylates the N-terminal serine residue of this protein (100). It should be emphasized that this mutant possesses normal (fully acetylated) S5 and S18 proteins both of which are known to have acetylated N-termini, indicating that the enzyme acetylating protein L12 (to convert it into L7) is highly specific toward this protein. It can not be excluded, however, that there are some other proteins in the cell which are acetylated by the same enzyme.

The mutation leading to the loss of the N-terminal acetyl

TABLE 3. Genes for the modification of ribosomal proteins[a]

Gene	Protein affected	Map position[b]	Function	Reference
prmA	L11	71	Methylation of alanine and lysine residues	59
prmB	L3	50	Methylation of lysine residues	59
rimG	S4	(2)	ramB. Modification (?)	92
rimL	L12	(34)	Acetylation of N-terminal serine	100
rimI	S18	99.3	Acetylation of N-terminal alanine	103
rimJ	S5	(31)	Acetylation of N-terminal alanine	101
rimK	S6	(14)	Addition of glutamic acid residues to the C-terminus	56,57

[a] *There have been several other loci reported which, when mutated, cause an impairment in ribosome assembly (see ref. 66). However, the nature of these mutations is not clear, and, therefore, they are not incorporated in this table.*

[b] *Map positions are indicated in minute coordinates as described by Bachmann et al. (66). Those which are bracketed are not precisely mapped yet.*

group of protein S5 has been discovered and mapped (101). The gene affected by the mutation, termed rimJ, was found to exist also in the 'silent' region between trp and his (Fig. 1). This mutation also specifically affects the acetylation of only one ribosomal protein. Furthermore, the mutation leads to a temperature-sensitive phenotype when combined with a mutation (or mutations) at or near strA either alone or in conjunction with an strA mutation. The mutation which is required for the temperature-sensitive phenotype is most likely present in all the StrA$^-$ (streptomycin-resistant) strains, because all four StrA$^-$ mutants which have been independently isolated and described contain this mutation. The mutation is, however, most likely different from strA itself, because in some cases only some of the StrA$^-$ RimJ$^-$ transductants showed temperature-sensitivity. The nature of this mutation is unclear.

In this connection it is interesting to note that Chakrabarti and Gorini (102) have reported that a certain combination of strA and rif mutations results in a temperature-sensitive phenotype. Neither mutation alone causes such a temperature-sensitivity. They interpret this to mean that a specific interaction between the ribosomal (strA) mutation and the RNA polymerase (rif) mutation renders the bacteria unable to survive at high temperatures. A difference between the former case (rimJ and strA) and the latter (rif and strA) is that in the former all the strA mutations tested possess the mutation leading to temperature-sensitivity when combined with the rimJ mutation, whereas in the latter case only a particular combination does so. Furthermore, in the latter case no segregation was observed between the strA mutation and the mutation causing temperature-sensitivity when combined with the rif mutation, whereas in the former case frequent segregation was observed between the strA mutation and the mutation causing temperature-sensitivity when combined with the rimJ mutation.

When temperature-insensitive revertants were isolated from the mutant strain harboring rimJ and strA mutations and exhibiting temperature-sensitivity, it was found that some of them had acquired an additional mutation in the structural gene coding for either protein S4 (rpsD) or S5 (rpsE) and that the remainder harbored a suppressor mutation mapping near aroE. The S4 and S5 mutations most probably exert their suppressor activity at the phenotypic level, because the rimJ phenotype itself (no acetylation of the N-terminal serine of protein S5) was still expressed. On the other hand, the third class of mutation suppresses the rimJ phenotype itself, restoring the acetylation of protein S5.

A third mutation, impairing the acetylation of the N-terminal alanine of protein S18, has been mapped at 99.3 min between dnaC and deo (103). The gene affected by this mutation was termed rimI (Fig. 1). The mutation does not affect the acetylation of other two proteins, S5 and L12, as in the case of rimL and rimJ.

It will be of interest to identify the product of these ribosomal protein modification genes and their intracellular localization as well as function in ribosome biosynthesis *in vivo*.

Methylation: Mutations impairing the methylation of protein L11 (termed prmA) and of protein L3 (termed prmB) have been identified and mapped (59). prmA is located near aroE (approximately 71 min), while prmB is located at 50 min near aroC. Earlier a mutation termed relC affecting protein L11 was reported by Friesen et al. (96,97). This mutation has most likely occurred in the structural gene for this protein (rplK) situated in the "rif-cluster" (Fig. 1). Because the prmA mutation also affects protein L11, the ribosomes of a prmA mutant were tested *in vitro* for the synthesis of guanosine tetra- and pentaphosphates (104). However, there was no detectable difference between the ribosomes from the mutant and those from the parental wild type strain, indicating that the methyl groups of protein L11 are not essential for this reaction.

Protein L11 has altogether nine methyl groups in its molecule (52) and is the most heavily methylated protein of all *E. coli* ribosomal proteins (50-53). Its N-terminal amino acid is trimethylalanine (51,52). The prmA mutant isolated by Colson et al. (59) lacks the N-terminal trimethyalanine and two residues of trimethyllysine, which implies that the enzyme coded by prmA can carry out the methylation of different amino acids of protein L11.

The L11 mutant described by Isono et al. (22) was found to be, in fact, not a structural gene mutant but instead a mutant with impaired modification (methylation?). The mutation causing altered electrophoretic mobility of this protein upon two-dimensional gel electrophoresis maps between malA and thyA, and hence it is conceivable that the mutation has occurred in prmA, because the only known modification of protein L11 is methylation (52). However, the protein of this mutant is different from the L11 protein of the prmA mutant described by Colson et al. (59) judging from their electrophoretic mobilities on two-dimensional slabs (58).

Modification of protein S6: Protein S6 of *E. coli* is unique in that it contains a stretch of up to six glutamic acid residues at the C-terminus (55). This protein is quite heterogeneous with respect to the number of the C-terminal glutamic acid residues, ranging from two to six residues, which can also be seen as a quite elongated spot on two-dimensional electropherograms. Recently, Reeh and Pedersen (56) investigated the kinetics of glutamic acid incorporation into protein S6 and found that all of the C-terminal glutamic acid residues were added stepwise after translation. His work raised the possibility of isolating mutants defective in this modification (glutamic acid adding) process.

The mutant analyzed by Hitz et al. (55), having an S6 protein which contained only two glutamic acid residues at the C-terminus, was found to harbor its mutation not in the structural gene for this protein (rpsF) situated at 94 min (see Fig. 1) but near lip (14 min) (57). This readily aroused the suspicion that the mutation might have occurred in the gene involved in the system for adding glutamic acid residues to protein S6. Then, the reason that the mutant protein S6 has only two glutamic acid residues as opposed to up to six residues is most likely due to an impairment in the glutamic acid adding system caused by the mutation near lip. As has been briefly mentioned already, the mutant isolated and described by Isaksson and Takata (32) has a slight alteration in protein S6. The mutation, termed fusB, maps near lip (14 min). The mutation may have occurred in the same gene that has been affected in the above-mentioned strain analyzed by Hitz et al. (55) as well as the several temperature-sensitive mutants with altered S6 proteins (see below), because their mutations also map near lip. The question remains whether or not the gene affected by these mutations is the structural gene for an enzyme which catalyzes the glutamic acid adding reaction.

The mutant used by Isono and Kitakawa (60,61) for mapping the structural gene for S6 has S6 protein being much less acidic. The amino acid sequence of the S6 protein of a similar (most probably identical) mutant has been analyzed. It was found that the C-terminal sequence of the mutant protein is

-Asp-Ser-Lys-Glu in contrast to the wild type sequence -Asp-Ser-Glu-Glu-Glu-Glu-Glu-Glu (105). If all the C-terminal glutamic acid residues are added to the stem of wild type protein S6 as evidenced by Pedersen (56), then it is conceivable that the mutation has occurred at the very C-terminus of the structural gene, changing one of the terminating codons (UAA or UAG) into the codon for lysine (AAA or AAG), thereby creating an unfavorable sequence (-Asp-Ser-Lys instead of -Asp-Ser) for the glutamic acid adding reaction.

The reaction of the addition of glutamic acid residues to protein S6 seems enzymologically also quite interesting, because up to date no such reaction (i.e. addition of amino acid residues to completed polypeptide chains) has been discovered. At present, it is not known whether the gene near lip, termed rimK (Fig. 1), is the gene that codes for an enzyme catalyzing the glutamic acid adding reaction or not. This is because all four mutants so far isolated seem to have the same mutation as does the mutant whose S6 protein has been analyzed (55), i.e. containing only two glutamic acid residues at the C-terminus of protein S6 as opposed to up to six residues in wild type. In this respect their mutations are leaky to the same extent. This leakiness may reflect either that mutations in rimK that totally abolish the glutamic acid adding activity are lethal to the cell, or that the gene rimK does not code for the enzyme catalyzing the glutamic acid adding reaction, but instead it specifies a co-factor needed for the reaction.

It remains to be established what roles the modifications of various ribosomal proteins described above (see Table 3) play in the biosynthesis and function of the ribosome *in vivo*.

SOME SPECIFIC TOPICS IN RIBOSOME GENETICS IN *E. COLI*

Mutations Leading to "Relaxed" Phenotype

When wild type (or stringent) bacteria are starved for the required amino acids, guanosine-tetra (ppGpp) and -pentaphosphate (pppGpp) accumulate and the synthesis of RNA ceases.

This well known phenomenon called the stringent (relaxed) response has been extensively studied (for review, see refs. 106-108), and several chromosomal sites have been shown to be involved in this cellular response. Thus, relA (109,110), relB (111,112), relC (96,97), relX (113) and spoT (114) loci have been mapped and some of the gene products identified.

Although ribosome-free synthesis of guanosine polyphosphates can occur *in vitro* (115), the involvement of ribosomes in this reaction seems essential *in vivo* at least as far as the major synthetic pathway of these compounds under amino acid starvation conditions is concerned. An evidence for that is the finding that strains with a mutational alteration in ribosomal protein L11 are relaxed, i.e. they continue to synthesize RNA upon amino acid starvation (96,97). The mutation, termed relC, has been mapped in the rif region and therefore it has most likely occurred in the structural gene coding for protein L11 (rplK) which has been located in this region (95). This indicates that protein L11 plays an important part in the stringent control of RNA synthesis *in vivo*. As described before, mutations impairing the methylation of protein L11 do not confer a Rel$^-$ phenotype (104). This finding suggests that structural alterations other than methylation in protein L11 lead to a relC phenotype.

One of the temperature-sensitive mutants which harbors an alteration in protein L11 (22) has been found to have a mutation mapping between strA and thyA (58), indicating that the mutation may have occurred in prmA which has been mapped roughly at 71 min (59). Upon two-dimensional gel electrophoresis the protein L11 of this mutant migrates to a similar position as does that of a relC mutant (58). In this respect it differs from the prmA mutant described by Colson et al. (59), since the protein L11 of the latter strain is electrophoretically indistinguishable from that of wild type. However, this mutant also shows stringent (wild type) response upon amino acid starvation as in the case of the prmA mutant described above. As will be described later, some of the revertants from erythromycin dependence apparently lack protein L11 (40). Such an apparent absence from the ribosome of the protein equivalent to protein L11 of *E. coli* has been observed with mutants of *Bacillus subtilis* (116) and *Bacillus megaterium* (117) isolated as being resistant to thiostrepton (see review by E. Cundliffe of this volume). It will be of interest to investigate if they are stringent or relaxed upon amino acid starvation.

A temperature-sensitive mutant was found that showed a relaxed phenotype (118). Genetic as well as biochemical investigations of this mutant have revealed that it contains three or four mutations that are responsible for the Rel⁻ phenotype either alone or in combination with each other. The mutations, however, are different from relA, relB and relX, all of which have been genetically well characterized. One of the mutation maps near rif (88 min) and therefore may be a relC mutation. Another mutation maps between his (44 min) and nalA (48 min) (for genetic map, see Fig. 1). A third mutation maps near proA (6 min). In addition the mutant may have a fourth and a fifth mutation involved in the Rel⁻ phenotype. They map between strA (72 min) and argH (88 min) and near trp (27 min), respectively. It is conceivable that some of these mutations function as suppressors of others, but it is not clear why so many mutations concerning Rel⁻ phenotype have been induced in this particular mutant.

The recent report by Friesen et al. (119) strongly suggests the existence of two or more synthetic pathways of guanosine polyphosphates. They inactivated the relA gene by introducing nonsense or insertion mutations and found that such mutants still produced guanosine polyphosphates at 70 to 100% of that of classical relA mutants under normal (not starved for amino acids) conditions. Moreover, their behavior upon carbon source shift-down was normal and the basal level of pppGpp + ppGpp remained unchanged. Whether the second pathway of guanosine polyphosphate synthesis requires ribosomes or not as in the case observed with *Bacillus brevis* (120) remains to be established.

Are Any of the Ribosomal Proteins Dispensable?

Since the establishment of the method to reconstitute the 30S ribosomal subunit *in vitro* from individual purified components by Nomura and coworkers (see ref. 121), experiments have been carried out to elucidate the role of each ribosomal protein by omitting the protein in question upon reconstitution or by pretreating the protein with suitable reagents prior to reconstitution. The omission of, for example, protein S12 results in a drastic decrease in the efficiency of initiation, indicating that this protein plays an important part in the initiation of protein synthesis (122). Furthermore, the omission of some of the ribosomal proteins results in the production of an intermediate particle, suggesting their importance in ribosome assembly (123). Is it feasible, then, to isolate mutants which have deletions, insertions, nonsense mutations or frameshift mutations, thereby eliminating one or more ribosomal proteins from the cell? Most probably, many of the ribosomal proteins are indispensable, but perhaps some can be partially or totally eliminated without a detectable effect.

One of the mutants isolated as revertants from a mutant harboring temperature-sensitive valyl tRNA synthetase was found to lack protein S20 (30). Another mutant isolated as being temperature-sensitive was also found to lack protein S20 (23). A similar situation was observed with revertants from an erythromycin-dependent mutant (40). Protein S20 has been shown to be one of the proteins that directly bind to 16S RNA (123). Therefore, it is conceivable that the absence of this protein from the cell causes a rather serious impact on the structure and function of the ribosome. Indeed the temperature-sensitive phenotype of the mutant lacking protein S20 (23) is most likely due to the absence of this protein (124). Various circumstantial lines of evidence suggest that the mutation in this strain is a large deletion in the gene for protein S20. The reason why this mutant shows temperature-sensitivity may be either because the ribosome lacking protein S20 can not carry out its function in pace with other cellular activities at higher temperatures, or because the mutant suffers from an assembly defect due to the mutation. The nature of the mutations in other two cases described above is not clear.

Some attempts to isolate mutants harboring nonsense mutations in the ribosomal protein genes have been reported (44, 125). The method employs a P1 phage-mediated localized mutagenesis (34) of a strain which contains a temperature-sensitive amber suppressor (125) or of a strain which is doubly lysogenized with an amber suppressor-carrying phage and a heat-inducible λ phage carrying amber mutations (44). In both methods, amber mutants with mutations in essential genes such as ribosomal protein genes can be isolated as those that can not survive at high temperatures but can grow at low (permissive) temperatures where the suppressors function. Several mutants have thus been isolated and partially characterized. However, no direct evidence has so far been obtained which rigorously demonstrates that the nonsense mutations have occurred in the ribosomal protein genes.

A mutant has been isolated which requires erythromycin for growth (40). Many revertants from this mutant strain are found to lack ribosomal proteins such as S20 (L26), L11, L15, L28, L29 and L30. The one which lacks protein L15 shows severe cold-sensitivity. Upon isolation of cold-insensitive revertants from this strain, it has been found that many pseudo-revertants which harbor L15 proteins altered not only in net charge but also in molecular weight can be isolated. Apparently the situation seems more drastic than the case reported for protein S4 of revertants from streptomycin-dependent mutant (78) or of temperature-sensitive mutants (22-24), where changes in molecular weight of S4 proteins up to 7,000 daltons have been estimated. Some of the proteins that are missing from the ribosome of the revertants derived from the erythromycin-dependent mutant (40) were not detectable in the post-ribosomal supernatant when tested with purified antibodies against the proteins in question.

As described before, some of the kasugamycin-dependent mutants apparently lack proteins L1, S1 or S9 (39). Some of the gentamicin-resistant mutants apparently lack protein L6 (43). Thiostreptone-resistant mutants of *Bacillus subtilis* (116) and of *B. megaterium* (117) have been reported to lack a protein equivalent to the protein L11 of *E. coli*. In none of these cases, however, the nature of the mutations has been investigated so as to establish what causes the loss of a particular protein (or proteins) from the ribosome (or in some cases from the cell).

Since some of the genes coding for the ribosomal proteins in *E. coli* are well separated from others (see Fig. 1) and exist as very small groups of genes (transcriptional units?), it may be of considerable interest to try to delete them by means of, for example, mu-phage excision. Bacteria may still grow, though perhaps slowly, without one or two or probably more ribosomal proteins, if those proteins are needed only to increase the functional and structural stability of the ribosome.

Antibiotic Resistance —— Some New Aspects

In 1961 Gorini and coworkers (126) isolated a streptomycin-resistant auxotrophic (arginine-requiring) strain that was observed to grow in the absence of arginine provided that streptomycin was present. This led to the discovery of conditional streptomycin dependent (CSD) mutants and subsequently to the studies on the "phenotypic suppression" caused by this now well-studied antibiotic (127). It was interpreted that the ribosomes of CSD mutants were altered in such a way that streptomycin interfered with them to cause "misreading", thereby phenotypically suppressing the auxotrophic mutations (128). From these findings it soon became generally accepted that the ribosomes of streptomycin-dependent (Str-D) mutants isolated earlier by several groups (see ref. 1) required the drug for proper translation by the mutant ribosomes. Indeed, Birge and Kurland (129) later reported that the ribosomes from such an Str-D mutant synthesized polypeptides *in vitro* much more efficiently when streptomycin was added.

Earlier mapping data of streptomycin mutations (130) indicated that streptomycin-resistance (Str-R) and streptomycin-dependence (Str-D) mutations were allelic. This and other reports have supported the notion that both mutations are single mutations occurring in the same gene. When the altered protein in Str-R and Str-D mutants was identified as protein S12 by *in vitro* reconstitution experiments (84,129), therefore, the gene coding for this protein, rpsL, became regarded as the target of both Str-R and Str-D mutations. It is, however, rather surprising that nobody, to my knowledge, has ever tried to perform more suitable genetic experiments to investigate if the above supposition is correct (see below).

S12 proteins from some streptomycin mutants have Asp at position 88 where the wild type counterpart has an unidentified amino acid (131,132) which has recently been shown to be a derivative of Asp by mass spectroscopy (132). The modification renders the bond with the preceding Lys trypsin resistant. The DNA sequence of the wild type gene rpsL clearly indicates that the code for the amino acid at position 88 is Asp (GAC) (133). A conceivable way to explain this rather unusual and perhaps potentially important situation is to postulate that the wild type protein S12 harbors a modified form of Asp at position 88 which is inhibitory to trypsin digestion, and that the modification reaction is inactivated in streptomycin mutants.

With respect to Str-D mutations, a recent experiment clearly indicates that Str-D mutations result from at least two mutations, one mapping between aroE and strA (rpsL) and the other an strA mutation (134). The former mutation is responsible for an Str-D phenotype, while the latter, when present alone, makes the cell Str-R. Since spectinomycin-dependent mutants of both *E. coli* (38) and *B. subtilis* (135) and erythromycin-dependent mutants of *E. coli* (40,41) possess at least two mutations, it is conceivable that perhaps all antibiotic-dependent mutants harbor two mutations, one conferring resistance to the drug and the other responsible for the dependence.

Two questions would emerge: what is, then, the role of the mutation that is responsible for the antibiotic-dependence, and why revertants from these antibiotic-dependent mutants possess alterations in various ribosomal proteins? In an extreme case, as described already, alterations occur in virtually all the ribosomal proteins (25,26).

It is tempting to speculate that the mutation causing the antibiotic-dependence may have some correlation with ribosome assembly. It has been observed by Garvin et al. (136) that the ribosomes synthesized in the presence of streptomycin show a high degree of misreading both *in vivo* and *in vitro*, even though the amount of streptomycin in the assay was negligible. This property is lost when the ribosomes are exposed to high salt concentrations. The authors explain that streptomycin interferes with the assembly of ribosomes, thereby imposing a conformational change in them, which causes misreading during translation. If this is the case, then it is likely that Str-D mutants require the drug for the proper assembly of ribosomes, and that this may be also the case with other antibiotics such as spectinomycin and erythromycin. Indeed, in the latter two cases, some of the mutants resistant to them are known to be cold-sensitive (137,138), which also most likely results from impaired ribosome assembly.

There have been several reports concerning the effect of antibiotic mutations which seem quite interesting and important. For example, Chakrabarti and Gorini (102,139) reported the interaction between rif and strA mutations and an unusual feature of strA mutations toward phage T7, Sanchez-Anzaldo and Bastarrachea (140) reported the existence of a gene, strM, which is frequently co-mutated with strA and causes permissiveness with regard to the streptomycin-mediated phenotypic suppression, and Yamagata and Uchida (141) described a possible interaction between spectinomycin-resistance mutation and a mutation affecting the stability of sex-factors. Furthermore, there is an indication of possible interaction between membranes and ribosomes, which was observed with sucrose-dependent spectinomycin-resistant mutants (142,143). These facts seem to suggest that the mode of action of antibiotics is not so simple as has been widely regarded, and mutants resistant to or dependent on various antibiotics can still be exploited in pursuing the function and structure of the ribosome.

ACKNOWLEDGEMENTS

I thank H. G. Wittmann, C. Pon and S. Isono for valuable suggestions and critically reading the manuscript, B. Wittmann-Liebold, E. R. Dabbs and M. Nomura for providing their unpublished results, and J.C. Belart for patiently typing the manuscript.

REFERENCES

1. Davies, J., and Nomura, M. (1972) Ann. Rev. Genet. 6, 203-234.
2. Jaskunas, S.R., Nomura, M., and Davies, J. (1974) in Ribosomes (Nomura, M., Tissières, A., and Lengyel, P., eds.) pp. 333-368, Cold Spring Harbor Lab., New York.
3. Nomura, M., Morgan, E.A., and Jaskunas, S.R. (1977) Ann. Rev. Genet. 11, 297-347.
4. Sypherd, P.S., and Osawa, S. (1974) in Ribosomes (Nomura, M., Tissières, A., and Lengyel, P., eds.) pp. 669-678, Cold Spring Harbor Lab., New York.
5. Leboy, P.E., Cox, E., and Flaks, J. (1964) Proc. Natl. Acad. Sci. USA 52, 1367-1374.
6. Takata, R., Dekio, S., Otaka, E., and Osawa, S. (1969) Mol. Gen. Genet. 105, 113-121.
7. Takata, R. (1972) Mol. Gen. Genet. 118, 363-371.
8. Takata, R., and Kobata, K. (1976) Mol. Gen. Genet. 149, 159-165.
9. Takata, R. (1978) Mol. Gen. Genet. 160, 151-155.
10. Sypherd, P., O'Neil, D.M., and Taylor, M.M. (1969) Cold Spr. Harb. Symp. Quant. Biol. 34, 77-84.
11. Matsubara, M., Takata, R., and Osawa, S. (1972) Mol. Gen. Genet. 117, 311-317.
12. Jarry, B., and Rosset, R. (1971) Mol. Gen. Genet. 121, 151-162.
13. Nakamura, Y., Owawa, T., and Yura, T. (1977) Proc. Nat. Acad. Sci. USA 74, 1831-1835.
14. Harris, J.D., Martinez, I.I., and Calendar, R. (1977) Proc. Nat. Acad. Sci. USA 74, 1836-1840.
15. Deusser, E. (1972) Mol. Gen. Genet. 119, 249-258.
16. Dennis, P.P. (1974) J. Mol. Biol. 88, 25-41.
17. Gausing, K. (1974) Mol. Gen. Genet. 129, 61-75.
18. Dennis, P.P., and Nomura, M. (1974) Proc. Nat. Acad. Sci. USA 71, 3819-3823.

19. Dennis, P.P., and Nomura, M. (1975) Nature 255, 460-465.

20. Maxam, A.M., and Gilbert, W. (1977) Proc. Nat. Acad. Sci. USA 74, 560-564.

21. Post, L.E., Arfsten, A.E., and Nomura, M. (1978) Cell 15, 231-236.

22. Isono, K., Krauss, J., and Hirota, Y. (1976) Mol. Gen. Genet. 149, 297-302.

23. Isono, K., Cumberlidge, A.G., Isono, S., and Hirota, Y. (1977) Mol. Gen. Genet. 152, 239-243.

24. Isono, S., Isono, K., and Hirota, Y. (1978) Mol. Gen. Genet. 165, 15-20.

25. Dabbs, E.R., and Wittmann, H.G. (1976) Mol. Gen. Genet. 149, 303-309.

26. Dabbs, E.R. (1978) Mol. Gen. Genet. 165, 73-78.

27. Wittmann, H.G., Stöffler, G., Piepersberg, W., Buckel, P., Ruffler, D., and Böck, A. (1974) Mol. Gen. Genet. 134, 225-236.

28. Buckel, P., Piepersberg, W., and Böck, A. (1976) Mol. Gen. Genet. 149, 51-61.

29. Böck, A., Ruffler, D., Piepersberg, W., and Wittmann, H.G. (1974) Mol. Gen. Genet. 134, 325-332.

30. Wittmann, H.G., Stöffler, G., Geyl, D., and Böck, A. (1975) Mol. Gen. Genet. 141, 317-329.

31. Isaksson, L.A., Sköld, S.E., Skjöldbrand, J., and Takata, R. (1977) Mol. Gen. Genet. 156, 233-237.

32. Isaksson, L.A., and Takata, R. (1978) Mol. Gen. Genet. 161, 9-14.

33. Takata, R., and Isaksson, L.A. (1978) Mol. Gen. Genet. 161, 15-21.

34. Hong, J.S., and Ames, B.N. (1971) Proc. Nat. Acad. Sci. USA 68, 3158-3162.

35. Berger, I., Geyl, D., Böck, A., Stöffler, G., and Wittmann, H.G. (1975) Mol. Gen. Genet. 141, 207-211.

36. Cabezón, T., Herzog, A., Petre, J., Yaguchi, M., and Bollen, A. (1977) J. Mol. Biol. 116, 361-374.

37. Herzog, A., Yaguchi, M., Cabezón, T., Corchuelo, M.-C., Petre, J., and Bollen, A. (1979) Mol. Gen. Genet. 171, 15-22.

38. Dabbs, E.R. (1977) Mol. Gen. Genet. 158, 55-61.

39. Dabbs, E.R. (1978) J. Bact. 136, 994-1001.

40. Dabbs, E.R., submitted for publication.

41. Sparling, P.F., and Blackman, E. (1973) J. Bact. 116, 74-83.

42. Hummel, H., Piepersberg, W., and Böck, A. (1979) Mol. Gen. Genet. 169, 345-347.

43. Buckel, P., Buchberger, A., Böck, A., and Wittmann, H.G.

(1977) Mol. Gen. Genet. 158, 47-54.

44. Delcuve, G., Cabezón, T., Ghysen, A., Herzog, A., and Bollen, A. (1977) Mol. Gen. Genet. 157, 149-153.

45. Wittmann-Liebold, B., and Greuer, B. (1978) FEBS Lett. 95, 91-98.

46. Yaguchi, M. (1975) FEBS Lett. 59, 217-220.

47. Terhost, C., Möller, W., Laursen, R., and Wittmann-Liebold, B. (1973) Eur. J. Biochem. 34, 138-152.

48. Chen, R., Brosius, J., Wittmann-Liebold, B., and Schäfer, W. (1977) J. Mol. Biol. 111, 173-181.

49. Muranova, T.A., Muranov, A.V., Markova, L.F., and Ovchinnikov, Yu.A. (1978) FEBS Lett. 96, 301-305.

50. Alix, J.H., and Hayes, D. (1974) J. Mol. Biol. 86, 139-159.

51. Lederer, F., Alix, J.H., and Hayes, D. (1977) Biochem. Biophys. Res. Commun. 77, 470-480.

52. Dognin, M.J., and Wittmann-Liebold, B. (1977) FEBS Lett. 84, 342-346.

53. Chang, F.N., and Chang, C.N. (1975) Biochemistry 14, 468-477.

54. Wittmann-Liebold, B., and Pannenbecker, R. (1976) FEBS Lett. 68, 115-118.

55. Hitz, H., Schäfer, D., and Wittmann-Liebold, B. (1975) FEBS Lett. 56, 259-262.

56. Reeh, S., and Pedersen, S. (1979) Mol. Gen. Genet. 173, 183-187.

57. Isono, K., and Isono , S., in preparation.

58. Isono, K., unpublished results.

59. Colson, C., Lhoest, J., and Urlings, C. (1979) Mol. Gen. Genet. 169, 245-250.

60. Isono, K., and Kitakawa, M. (1977) Mol. Gen. Genet. 153, 115-120.

61. Isono, K., and Kitakawa, M. (1978) Proc. Nat. Acad. Sci. USA 75, 6163-6167.

62. Kitakawa, M., Blumenthal, L., and Isono, K., unpublished results.

63. Kitakawa, M., and Isono, K. (1977) Mol. Gen. Genet. 158, 149-155.

64. Isono, K. (1978) Mol. Gen. Genet. 165, 265-268.

65. Saito, H., Nakamura, Y., and Uchida, H. (1978) Mol. Gen. Genet. 165, 247-256.

66. Bachmann, B.J., Low, K.B., and Taylor, A.L. (1976) Bact. Rev. 40, 116-167.

67. Kitakawa, M., Dabbs, E.R., and Isono, K. (1979) J. Bact. (in press).

68. Dabbs, E.R., unpublished results.

69. Isono, K., and Schnier, J., in preparation.

70. Schnier, J., and Isono, K. (1979) submitted for publication.

71. Isono, K., unpublished results.

72. Ono, M., Kuwano, M., and Mizushima, S. (1979) Mol. Gen. Genet. (in press).

73. Subramanian, A.R., and Mizushima, S. (1979) J. Biol. Chem. 254, 4309-4312.

74. Nashimoto, H., and Uchida, H. (1975) J. Mol. Biol. 96, 443-453.

75. Jaskunas, S.R., Fallon, A.M., and Nomura, M. (1977) J. Biol. Chem. 252, 7323-7336.

76. Rosset, R., and Gorini, L. (1969) J. Mol. Biol. 39, 95-112.

77. Zimmermann,R.A., Garvin, R.T., and Gorini, L. (1971) Proc. Nat. Acad. Sci. USA 68, 2263-2267.

78. Hasenbank, R., Guthrie, C., Stöffler, G., Wittmann, H.G., Rosen, L., and Apirion, D. (1973) Mol. Gen. Genet. 127, 1-18.

79. Apirion, D., and Schlessinger, D. (1968) J. Bacteriol. 96, 1431-1432.

80. Bollen, A., Davies, J., Ozaki, M., Mizushima, S. (1969) Science 165, 85-86.

81. Dekio, S., and Takata, R. (1969) Mol. Gen. Genet. 105, 219-224.

82. Birge, E.A., and Kurland, C.G. (1970) Mol. Gen. Genet. 109, 356-369.

83. Breckenridge, L., and Gorini, L. (1970) Genetics 65, 9-25.

84. Ozaki, M., Mizushima, S., and Nomura, M. (1969) Nature 222, 333-339.

85. Gorini, L. (1969) Cold Spring Harbor Symp. Quant. Biol. 34, 101-109.

86. Dekio, S. (1971) Mol. Gen. Genet. 113, 20-30.

87. Cannon, M., Cabezón, T., and Bollen, A. (1974) Mol. Gen. Genet. 130, 321-326.

88. Herzog, A., Yaguchi, M., Cabezón, T., Corchuelo, M.-C., Petre, J., and Bollen, A. (1979) Mol. Gen. Genet. 171, 15-22.

89. de Wilde, M., Michel, F., and Broman, K. (1974) Mol. Gen. Genet. 133, 329-333.

90. Bollen, A., Faelen, M., Lecocq, J.P., Herzog, A., Zengel, J., Kahan, L., and Nomura, M. (1973) J. Mol. Biol. 76, 463-472.

91. Friesen, J.D., Parker, J., Watson, R.J., Fiil, N.P., and Pedersen, S. (1976) Mol. Gen. Genet. 144, 115-118.

92. Zimmermann, R.A., Garvin, R.T., and Gorini, L. (1971)

Proc. Nat. Acad. Sci. USA 68, 2263-2267.

93. Takata, R., Osawa, S., Tanaka, K., Teraoka, H., and Tamaki, M. (1970) Mol. Gen. Genet. 109, 123-130.

94. Pardo, D., and Rosset, R. (1974) Mol. Gen. Genet. 135, 257-268.

95. Lindahl, L., Yamamoto, M., Nomura, M., Kirschbaum, J.B., Allet, B., and Rochaix, J.-D. (1977) J. Mol. Biol. 109, 23-47.

96. Friesen, J.D., Fiil, N.P., Parker, J.M., Haseltine, W.A. (1974) Proc. Nat. Acad. Sci. USA 71, 3465-3469.

97. Parker, J., Watson, R.J., Friesen, J.D., and Fiil, N.P. (1976) Mol. Gen. Genet. 144, 111-114.

98. Watson, R.J., Parker, J., Fiil, N.P., Falks, J.G., and Friesen, J.D. (1975) Proc. Nat. Acad. Sci. 72, 2765-2769.

99. Wittmann, H.G., Stöffler, G., Apirion, D., Rosen, L., Tanaka, K., Tamaki, M., Takata, R., Dekio, S., Otaka, E., and Osawa, S. (1973) Mol. Gen. Genet. 127, 175-189.

100. Isono, K., and Isono, S., in preparation.

101. Cumberlidge, A.G., and Isono, K. (1979) J. Mol. Biol. (in press).

102. Chakrabarti, S.L., and Gorini, L. (1977) Proc. Nat. Acad. Sci. USA 74, 1157-1161.

103. Isono, K., and Isono, S., in preparation.

104. Röhl, R., and Nierhaus, K.H. (1979) Mol. Gen. Genet. 170, 187-189.

105. Kade, B., Dabbs, E.R., and Wittmann-Liebold, B., unpublished results.

106. Cashel, M., and Gallant, J. (1974) in Ribosomes (Nomura, M., Tissières, A., and Lengyel, P., eds.) pp. 733-745, Cold Spring Harbor Lab., New York.

107. Cashel, M. (1975) Ann. Rev. Microbiol. 29, 301-318.

108. Richter, D., and Isono, K. (1977) Current Topics in Microbiology and Immunology 76, 83-125.

109. Alföldi, L., Stent, G.S., and Clowes, R.C. (1962) J. Mol. Biol. 5, 348-355.

110. Fiil, N., and Friesen, J.D. (1968) J. Bact. 95, 729-731.

111. Lavallé, R. (1965) Bull. Soc. Chim. Biol. 47, 1567-1570.

112. Diderichsen, B., Fiil, N.P., and Lavallé, R. (1977) J. Bact. 131, 30-33.

113. Pao, C.C., and Gallant, J. (1978) Mol. Gen. Genet. 158, 271-277.

114. Laffler, T., and Gallant, J. (1974) Cell 1, 27-30.

115. Christiansen, L., and Nierhaus, K.H. (1976) Proc. Nat. Acad. Sci. USA 73, 1839-1843.

116. Pestka, S., Weiss, D., Vince, R., Wienen, B., Stöffler, G., and Smith, I. (1976) Mol. Gen. Genet. 144, 235-241.

117. Cundliffe, E., Dixon, P., Stark, M., Stöffler, G.,

Ehrlich, R., Stöffler-Meilicke, M., and Cannon, M. (1979) J. Mol. Biol. (in press).

118. Schnier, J., and Isono, K., unpublished results.

119. Friesen, J., An, G., and Fiil, N.P. (1978) Cell 15, 1187-1197.

120. Sy, J., and Akers, H. (1976) Biochemistry 15, 4399-4403.

121. Held, W.A., Mizushima, S., and Nomura, M. (1973) J. Biol. Chem. 248, 5720-5730.

122. Held, W.A., Gette, W.R., and Nomura, M. (1974) Biochemistry 13, 2115-2122.

123. Held, W.A., and Nomura, M. (1973) Biochemistry 12, 3273-3281.

124. Isono, S., and Isono, K., unpublished results.

125. Nashimoto, H., Tsugawa, A., and Uchida, H. (1975) Proc. Mol. Biol. Meet. Japan, p. 12.

126. Gorini, L., Gundersen, W., and Burger, M. (1961) Cold Spr. Harb. Symp. Quant. Biol. 26, 173-182.

127. Gorini, L., and Katja, E. (1964) Proc. Nat. Acad. Sci. USA 51, 487-493.

128. Gorini, L., Jacoby, G.A., and Breckenridge, L. (1966) Cold Spr. Harb. Symp. Quant. Biol. 31, 657-664.

129. Birge, E.A., and Kurland, C.G. (1969) Science 166, 1282-1284.

130. Hashimoto, K. (1960) Genetics 45, 49-62.

131. Funatsu, G., Yaguchi, M., and Wittmann-Liebold, B. (1977) FEBS Lett. 73, 12-17.

132. Wittmann-Liebold, B., personal communication.

133. Nomura, M., personal communication.

134. Isono, S., and Isono, K., unpublished results.

135. Henkin, T.M., Campbell, K.M., and Chambliss, G.H. (1979) J. Bacteriol. 137, 1452-1455.

136. Garvin, R.T., Rosset, R., and Gorini, L. (1973) Proc. Nat. Acad. Sci. USA 70, 2762-2766.

137. Nashimoto, H., and Nomura, M. (1970) Proc. Nat. Acad. Sci. USA 67, 1440-1447.

138. Pardo, D., and Rosset, R. (1974) Mol. Gen. Genet. 135, 257-268.

139. Chakrabarti, S.L., and Gorini, L. (1975) Proc. Nat. Acad. Sci. USA 72, 2084-2087.

140. Sánchez-Anzaldo, F.J., and Bastarrachea, F. (1974) Mol. Gen. Genet. 130, 47-64.

141. Yamagata, H., and Uchida, H. (1972) J. Mol. Biol. 67, 533-535.

142. Miyoshi, Y., and Yamagata, H. (1976) J. Bact. 125, 142-148.

143. Mizuno, T., Yamada, H., Yamagata, H., and Mizushima, S. (1976) J. Bact. 125, 524-530.

ORGANIZATION OF RIBOSOMAL GENES AND REGULATION OF THEIR EXPRESSION IN *ESCHERICHIA COLI*

MASAYASU NOMURA and LEONARD E. POST

Institute for Enzyme Research,
Departments of Biochemistry and Genetics,
University of Wisconsin-Madison,
Madison, WI 53706

INTRODUCTION

The ribosome of E. coli consists of three RNA species and over 50 proteins, each present in one copy per ribosome (except L7/L12, which is present in four copies). In exponentially growing cells, there is no significant pool of free ribosomal components. Also, there is no significant degradation of ribosomal components except in slowly growing cells where rRNA is synthesized in excess, and a substantial portion is rapidly degraded. Therefore, it appears that bacterial cells have some regulatory mechanisms to insure that the amount of each ribosomal component synthesized is balanced in the same stoichiometric relation as found in the ribosome. In addition, the synthesis of ribosomal components is coordinately controlled to remain stoichiometric when the number of ribosomes synthesized varies with the growth conditions or in response to amino acid starvation (i.e., the stringent response). For details on these control phenomena, see reviews in References 1 and 2, and K. Gausing's article in this volume.

Several models can be conceived to account for the coordinate control and stoichiometric nature of the synthesis of ribosomal components. For example, one simple model would involve the promoters controlling all ribosomal protein (r-protein) genes being of equal strength and responding to the same regulatory signals. Furthermore, each cistron in a polycistronic transcription unit would be translated with equal efficiency. The promoters of rRNA operons would be expected to respond to the same regulatory signals but be of different strength than the r-protein promoters, since the rRNA gene products are not amplified by translation.

Testing such a model and understanding the molecular mechanisms of control require a detailed knowledge of the organization of genes for the ribosomal components. Since classical genetic approaches were difficult to apply in this system, our initial effort was directed to isolation of ribosomal genes on transducing phages or plasmids. Then, various modern techniques were applied to identify ribosomal genes and regulatory elements such as promoters on these isolated DNA molecules (see reviews 1 and 3). In the past several years, rapid progress has been made in this line of research. However, these extensive studies have also revealed some unexpected aspects of the ribosome genetics and added some complications to the problems of regulation of ribosome synthesis. It was discovered that rRNA genes are co-transcribed with tRNA genes (4, 5), and r-protein genes are co-transcribed with genes for RNA polymerase subunits (6, 7, 8) and elongation factors (9). DNA sequencing in ribosomal operons has begun to look for some features that might account for the coordinate control of synthesis of ribosomal components. Knowledge of the organization of r-protein genes has made it possible to design experiments which can now test possible molecular mechanisms involved in the regulation. Some recent experiments have already suggested the presence of an important post-transcriptional mechanism in the regulation of ribosome biosynthesis (10). In this article, we review mainly these recent developments. For details of earlier work, readers should consult previous reviews (1, 11).

ORGANIZATION OF rRNA GENES

Production of equimolar amounts of all three rRNA species is insured by co-transcription of genes for 16S, 23S, and 5S rRNA (12, 13, 14). There are at least seven (15) and probably only seven (16) transcription units of rRNAs. The significance of the presence of multiple (seven) rRNA operons is not known, although it is generally assumed that the multiple operons have evolved in order to supply a large amount of rRNA for fast-growing cells. The question could be examined by decreasing the number of rRNA operons by deletion. Recent experiments in this direction have shown that at least one rRNA operon (rrnE) can be deleted without any obvious deleterious effects (M. Kenerley and M. Nomura, unpublished experiments).

Table 1 shows the known mapping positions on the *E. coli* chromosome of the rRNA operons and genes carried by these operons. The experimental information used to map these rRNA operons has been reviewed (Ref. 1; see also Ref. 17).

Transcription units of rRNA genes also contain tRNA genes. All rRNA operons have either a gene for $tRNA_2^{Glu}$ or genes for $tRNA^{Ile}$ and $tRNA_{1B}^{Ala}$ in the spacer region between 16S and 23S rRNA coding sequences (4, 18, 19, 20). Several of these 16S-23S spacers have been sequenced (21, 22, 23; H. Noller, personal communication). In addition to spacer tRNAs, some,

TABLE 1. Chromosomal locations and associated tRNA genes of known rRNA operons in E. coli K12.

rRNA operon	Chromosomal* location	Spacer tRNA genes**	Distal tRNA genes**
rrnA	85 min	ile, ala	
rrnB	88 min	glu	
rrnC	83 min	glu	asp, trp
rrnD	71 min	ile, ala	(thr)
rrnE	89 min	glu	
rrnF***	74 min	--***	--***
rrnG***	56 min	--***	--***

*For the information used to map operons A through F, see Ref. 1. The rrnG was discovered on λpheA transducing phages by Zurawski and Brown (17).

**ile, ala, glu, asp, and trp are genes for $tRNA^{Ile}$, $tRNA^{Ala}_{1B}$, $tRNA^{Glu}_{2}$, $tRNA^{Asp}_{1}$, and $tRNA^{Trp}$, respectively. [For assignment of tRNA genes to the operons, see Refs. 5, 18, and 19.] (Thr) indicates a gene for $tRNA^{Thr}$, but rigorous proof for this conclusion is lacking (see Refs. 5, 19, and note in Ref. 20).

***tRNA genes associated with rrnF and rrnG have not been analyzed. Studies on rRNA operons isolated as hybrid plasmids identified two rRNA operons, called Group I and Group V, which have chromosomal locations different from rrnA through rrnE (15). These two rRNA operons may correspond to rrnF and rrnG, but the actual correspondence has not been determined. The Group I operon has been shown to carry the spacer genes for $tRNA^{Ile}$ and $tRNA^{Ala}_{1b}$ and the gene for $tRNA^{Asp}$ as a distal tRNA gene. The Group V operon has been shown to carry the spacer gene for $tRNA^{Glu}_{2}$. No distal tRNA gene was found in this operon.

but apparently not all, rRNA operons have "distal tRNA genes," located promoter-distal from the 5S rRNA gene (5). For example, genetic and physiological experiments showed that rrnC has two tRNA genes at the distal end, genes for $tRNA^{Asp}_{1}$ and $tRNA^{Trp}$ (5), and this was confirmed by DNA sequencing (R. A. Young, personal communication). Also, the $tRNA^{Trp}$ gene has been converted to an amber suppressor, and the suppressor activity was shown to be dependent on the presence of the promoter region of the rRNA operon (24). This suppressor mutation makes possible genetic manipulations on the rRNA operon, as does the fusion of rRNA operon promoters to the gal or lac operons. [The latter has recently been achieved (Y. Ota, A. Kikuchi, and M. Cashel, personal communication; J.A.H. Krueger and M. Nomura, unpublished experiments).] The association of tRNA genes with the individual rRNA operons is summarized in Table 1.

The promoter regions of four rRNA operons have been sequenced (25, 26) and studied by in vitro transcription experiments (26, 27). For each of the four sequenced operons (26,27), and one other studied operon (28), two tandem promoters were found to initiate transcription approximately 200 and 300 bases upstream from the 16S rRNA gene. The first promoters ("P_1"; i.e., the promoter farthest from the 16S rRNA gene) from all four operons show a common 15 base sequence including the Pribnow boxes but considerable sequence diversity on either side of this 15 base sequence (see Fig. 1). The second promoters (P_2) for three of the operons are identical, but the fourth is different (Fig. 1). The sequence heterogeneity covers 150 nucleotides from the transcription initiation site of the first promoter (23, 26). The significance of this heterogeneity is not known.

Although most transcripts in *E. coli* are initiated with ATP or GTP, the transcript from the second promoter is initiated with CTP in each of the four sequenced promoters (26, 27). Because of the unusual use of CTP for initiation, and lack of any precedent for tandem promoters in the *E. coli* chromosome, a serious possibility was that P_2 was an *in vitro* artifact. Two separate approaches have been used to test this possibility: hybridization of ^{32}P-labeled *in vivo* RNA to restriction enzyme fragments carrying various segments of the promoter region, followed by 5'-end analysis (29); and *in vitro* processing of 30S rRNA precursor molecules accumulated in a RNase III-deficient mutant, followed by analysis of small RNAs produced (30). These experiments have demonstrated that there are two transcription start sites for each rRNA operon *in vivo*, one initiating with ATP (or GTP), and the second initiating with CTP, and these two sites correspond to those identified in the *in vitro* studies. Preliminary estimates of relative use of the two promoters showed that the frequency of the initiation at P_2 is lower than the initiation at P_1 (29, 30), but more systematic measurements under several different physiological conditions are desirable to assess the significance of the use of the tandem promoters with different initiating nucleotides.

As mentioned above, nucleotide sequences at P_1 promoters are very similar among four rRNA operons sequenced, and a common 15 base sequence exists. This unique sequence feature is not found in P_2 promoter regions. If the stringent control system acts on the transcription initiation from both promoters, possible common features of these promoters recognized by the stringent control system are not obvious from the DNA sequence results. However, it is possible that the direct effects might be only one of them (e.g., P_1), and the other promoter function might be affected indirectly.

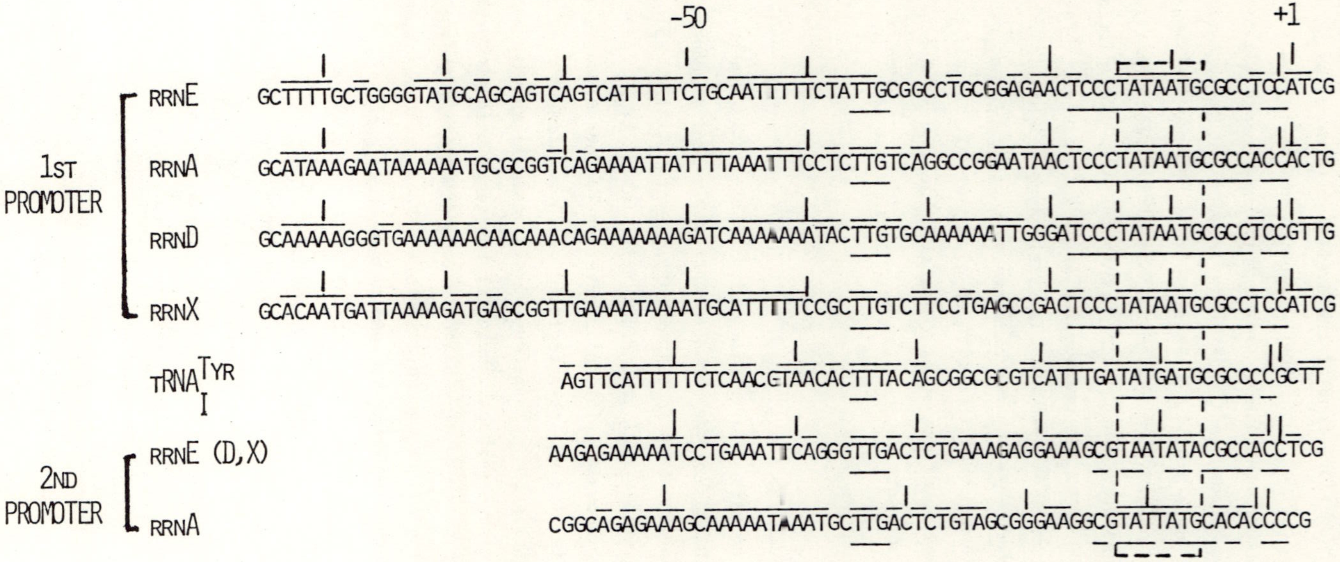

Figure 1. Promoter sequences from rRNA operons. The first promoter initiates transcription about 300 bases from the sequence coding for 16S rRNA, and the second promoter initiates transcription about 200 bases from the 16S rRNA sequence. The promoters for rrnE and rrnA are from Ref. 25, rrnD and rrnX are from Ref. 26 ("rrnX" is the hybrid rRNA operon carried on λilv5. The chromosomal origin of the promoter end of the operon is not known. See Ref. 15), and the $tRNA_1^{Tyr}$ promoter is from Ref. 84. The sequences are aligned at the Pribnow boxes, which are enclosed in the broken lines. Homologies common to the first promoter regions of the four rRNA operons are underlined. Nucleotides identical to these common sequences are underlined in the second promoter and $tRNA_1^{Tyr}$ promoter sequences. To indicate the A-T richness of some regions of these promoters A-T base pairs are overlined. The start sites for RNA synthesis are marked with +1.

ORGANIZATION OF RIBOSOMAL PROTEIN GENES

As noted in the Introduction, most of the available information on the organization of r-protein genes (and other essential genes) has been obtained by isolation of specialized transducing phages carrying r-protein genes and applying various modern approaches (reviewed in Refs. 1 and 3). Genes have been mapped by DNA-dependent protein synthesis with defined DNA fragments as template, and by analyzing proteins synthesized in ultraviolet-irradiated cells infected with transducing phages and mutants thereof. In addition, cloning of DNA fragments (from the transducing phages) into vector phages and plasmids was extensively used in the analysis. Here we shall review the results of such experiments for two clusters of r-protein genes, the *str-spc* region at 72 min and the *rif* region at 88 min of the *E. coli* chromosome. Chromosomal locations of the other r-protein genes, which are scattered around the chromosome are reviewed by K. Isono in this volume.

The *str-spc* Region

The specialized transducing phage λ*fus*3 (31) and several other transducing phages (32, 33) contain a large piece of bacterial DNA from the *str-spc* region (Fig. 2). This phage was shown to carry 27 r-protein genes, which have all been mapped by both *in vitro* protein synthesis (34, 35) and experiments using ultraviolet-irradiated cells (31). Also, λ*fus*3 carries genes for RNA polymerase subunit α (6) and elongation factors G and Tu (9). Synthesis of a protein in ultraviolet-irradiated cells requires a functional promoter directing transcription. Therefore, it is possible in this system to determine the transcriptional organization of the genes using insertion or deletion mutants of the transducing phage, and restriction fragments of the transducing phage cloned into lambdoid vectors. From such experiments, it was determined that λ*fus*3 has four operons of r-protein genes (6, 9, 31, 36, 37, 38; see legend to Fig. 1 for qualification). Unexpectedly, it was found that genes for elongation factors G and Tu were co-transcribed with genes for r-proteins S12 and S7 (9); and the gene for RNA polymerase subunit α is co-transcribed with genes for r-proteins L17, S4, S11, and probably S13 (6). The structures of these four operons are shown in Figure 2. By DNA sequencing analyses, locations of many of the genes determined in earlier work have been confirmed and some unresolved gene orders definitely determined (39; see also the legend to Fig. 2).

The gene organization determined by transducing phage analysis is also consistent with some other kinds of experiments. The order of the induction of individual r-protein genes after a nutritional shift-up (40) could

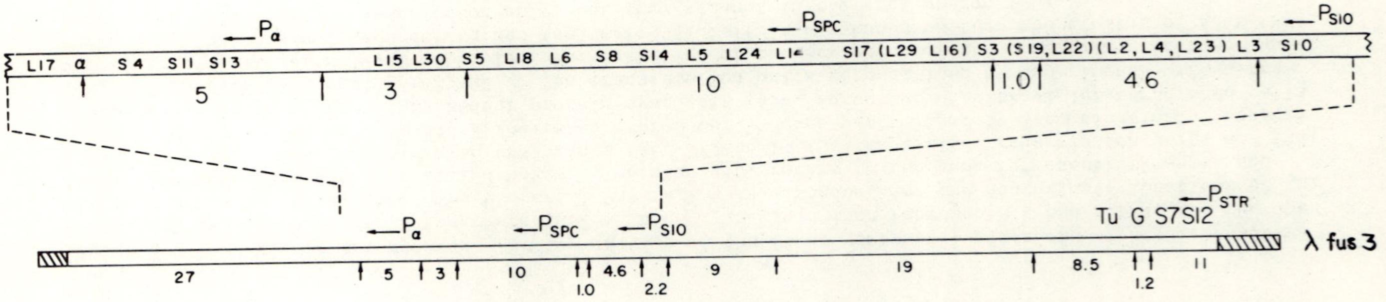

Figure 2. The λfus3 chromosome. The position of genes on the chromosome is indicated by the name of the gene product. The open area represents bacterial DNA, and the hatched area represents λ sequences. Vertical arrows indicate EcoRI sensitive sites, and numbers beneath the arrows indicate the sizes of the EcoRI fragments in percentage of the λpapa genome (1% λ = 480 base pairs). The four transcription units of r-protein genes are indicated by the positions of the four promoters (P), with transcription from right to left in all cases. All genes were detected and partially mapped by both *in vitro* DNA-dependent protein synthesis (34, 35) and infection of ultraviolet-irradiated cells (31). The mapping has been refined by DNA sequencing experiments. The positions of the genes for the following proteins have been determined by DNA sequencing: S12, S7, EF-G, L3, S3, L16, S17, L14, L24, S5, L30, S13, S11, S4, and α (39, and our unpublished experiments). The order of the promoter distal genes of the S10 operon is probably (right to left) L16, L29, S17 from coding capacity calculations (39), although no direct evidence for the exact position of L29 is available. From studies of A. Bollen (personal communication; see text), the order of other genes in the S10 operon is probably (right to left) L4, L2, L23, L22, S19, although there is not any direct physical evidence yet. The organization into transcription units was determined by experiments with ultraviolet-irradiated cells (31, 37, 38). There is still no rigorous proof that the S10 gene is co-transcribed with the other genes assigned to the S10 operon (31, 38).

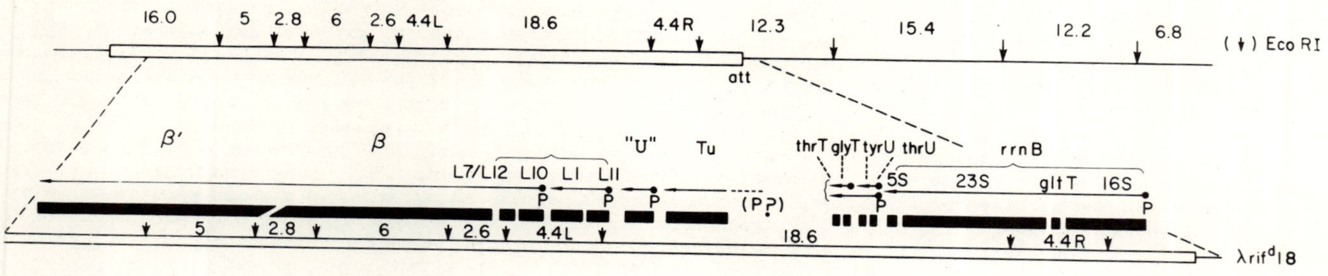

Figure 3. The λ*rif*^{d}18 chromosome. The top section of the figure shows the positions of *Eco*RI sensitive sites, and the sizes of *Eco*RI fragments in %-λ units. The thin line represents λ phage DNA, and the open box represents bacterial DNA. The bacterial DNA region is expanded, and the approximate locations of identified genes are shown as filled boxes. The name *glt*T refers to the gene for spacer $tRNA_2^{Glu}$, and *thr*T, *gly*T, *tyr*U, and *thr*U are also tRNA genes (46, 47, 48). The identified transcription units are indicated by 2 horizontal arrows, and P indicates a promoter. Genes were mapped by a combination of experiments involving *in vitro* DNA-dependent protein synthesis (49), infection of ultraviolet-irradiated cells (46, 50), and DNA sequencing (48, 52). The transcription units were defined by experiments with ultraviolet-irradiated cells (7, 8, 46, 51). The direction of transcription of the U gene (a protein of unknown function) has not been determined (see Ref. 46). It is not known whether the cluster of four tRNA genes make up a single transcription unit, but they are not in the *rrnB* operon (46).

be correlated with the order of the genes in each of the operons in this region (35), suggesting that each of the promoters for these operons is used during a nutritional shift-up to increase the rate of ribosome biosynthesis. Recently, Bollen and his collaborators isolated several amber mutants in a background with a temperature-sensitive suppressor (41; A. Bollen, personal communication). Some of these amber mutations affected the expression of several r-protein genes. In one case, the mutation strongly affected the expression of the L3 gene and less strongly that of L4, L2, L23, and L22 genes. The interpretation is that the mutation is probably in the L3 gene, and polarity affected the gene for L4, L2, L23, and L22. Similarly, another mutation probably in the S14 gene exerted polar effects on the genes for S8, L6, and L18. Although the reason for a lack of polar effect on other distal genes of the operon is not understood, the results are at least consistent with the gene orders shown in Figure 2.

The rif Region

The transducing phage λrif^{d} 18 (42) carries genes for four r-proteins (43, 44), RNA polymerase subunits β and β' (45), elongation factor Tu (a second copy) (9), several tRNAs (46, 47, 48), and also an rRNA operon (43) (Fig. 3). These genes have been mapped by in vitro protein synthesis (49), and by experiments using ultraviolet-irradiated cells infected by the transducing phage or mutant phages (4, 46, 50). The analysis in ultraviolet irradiated cells showed that this EF-Tu gene (tufB) is not co-transcribed with r-protein genes (46, 51); genes for r-protein genes L11 and L1 form one operon (7, 8, 46); and genes for L10, L7/L12, β, and β' form another operon (7, 8). DNA sequencing analysis has established the gene order and the structure of the region covering L11, L1, L10, L7/L12 genes and the 5'-end of the β gene (52).

CO-TRANSCRIPTION OF R-PROTEIN AND RNA POLYMERASE SUBUNIT GENES

Although RNA polymerase subunits α, β, and β' are co-transcribed with r-protein genes as described above, RNA polymerase subunits and r-protein synthesis are not identically controlled. Synthesis of α, β, and β' does not appear to vary with growth rate like that of r-proteins in several different nutritional environments (53, 54; but see Refs. 55 and 56). In addition, synthesis of α, β, and β' does not appear to be under stringent control (57, 58, 59). Rifampicin causes increased synthesis of β and β' but not L7/L12 (60; see also 61); and it causes increased synthesis of α but not S4 or L17 (62). Partial inactivation of a temperature sensitive RNA polymerase stimulates transcription of the β and β' genes but stimulates transcription of the r-protein genes to a considerably lesser extent (63, 64). It is clear that the regulation of transcription initiation alone is not sufficient to explain the regulation of RNA polymerase synthesis.

Since the molar amounts of RNA polymerase subunits are at least several-fold smaller than those of r-proteins, an attenuation-like mechanism has been suggested to operate at a site between the L7/L12 and β genes (65). Recent experiments by Barry _et al_. have shown that the presence of DNA carrying that region between a promoter and the _lac_Z gene in a fused operon does cause reduced expression of the _lac_Z gene (66). DNA sequencing experimentments have shown that there are 321 bases between the L7/L12 and β genes (52), and one can find structures similar to those of known attenuators in the sequence. Whether such an "attenuator" accounts for the differential regulation of β-β' and L10-L7/L12 synthesis remains to be tested. In contrast to the L7/L12-β distance, there are only 25 nucleotides between the S4 and α genes (our unpublished results). In addition, the α gene is _between_ two co-transcribed r-protein genes. Therefore, it seems unlikely that the mechanism of differential regulation of α and r-proteins is at the level of transcription. It appears more likely that some post-transcriptional process is involved in the regulation.

PROMOTERS OF R-PROTEIN OPERONS

A simple model for the coordinate control and stoichiometric nature of r-protein synthesis described in the Introduction included promoters of equal strength responding to the same regulatory signals for each r-protein operon. In fact, this has been tacitly assumed (e.g., see the discussion in Ref. 65 on the equal transcriptional activity in the r-protein genes in the _rif_ region and those in the _spc_ region). If such a hypothesis is true, one might expect some sequence similarities among the promoters. At present, promoters of five r-protein operons have been sequenced (39, 52, and our unpublished results), and the sequences are shown in Figure 4. Each of these promoters was identified by _in vitro_ transcription experiments such as those described for the _str_ and _spc_ promoters (39; A. E. Arfsten and M. Nomura, unpublished experiments). There is good evidence that this _in vitro_ system identifies authentic promoters used _in vivo_. In the case of the _str_ promoter, a mutant which abolishes expression of the _str_ operon

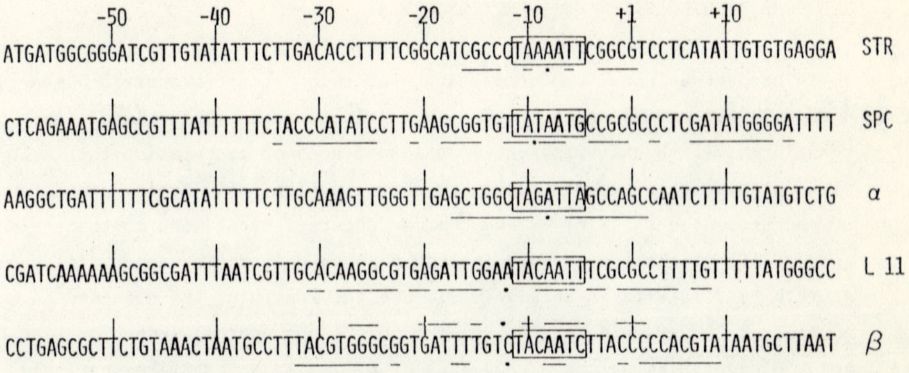

Figure 4. Comparison of promoter sequences of five r-protein operons. P_{str} and P_{spc} are from Ref. 39, and P_{L11} and P_{β} are from Ref. 52. P_{α} is the promoter preceding the S13 gene (our unpublished results) and probably directs transcription of the S11, S4, α, and L17 genes, although this has not been rigorously proven. The start sites of the five promoters are aligned at position +1. The Pribnow box sequences are boxed. Regions of dyad symmetry are indicated by overlining or underlining with a dot indicating the center of a dyad symmetry. The start site of the β promoter is inferred from its distance from the Pribnow box, to be analogous to the other four promoters. The 5'-end of *in vitro* β mRNA is pppCpC, but which of the C's is the start site has not been determined (A. E. Arfsten and M. Nomura, unpublished experiments cited in Ref. 52).

in vivo and *in vitro* was shown to be a single A-T to G-C base change in the Pribnow box of the promoter identified by the *in vitro* experiments (67). Also, when the *str* and *spc* promoters were cloned onto a multicopy plasmid, RNA products could be isolated which were initiated at the sites identified by the *in vitro* experiments (68). There is similar *in vivo* evidence that the β promoter identified *in vitro* is functional as a promoter *in vivo* (T. Ikemura and M. Nomura, unpublished experiments, cited in Ref. 52).

The comparison of the promoter sequences in Figure 4 does not show any obvious common features that might account for regulation of r-protein synthesis. They all include the components that have been noted in other *E. coli* promoters; a Pribnow box, a -35 sequence, and A-T richness in the -35 to -45 region (69, 70). Thus these examples (and the sequenced rRNA promoters discussed above) show that the promoters of ribosomal genes seem to be fundamentally similar to other promoters. However, there is no obvious unique sequence features which can be hypothesized to account for the regulation of r-protein synthesis. If the strength of the five r-protein promoters is identical and all the same regulatory mechanisms act on these promoters, as the simple model predicts, the common structural features responsible for this identical behavior are elusive.

RIBOSOME BINDING SITES OF R-PROTEIN GENES

To produce an equal amount of each r-protein from co-transcribed genes, one would expect that each cistron on a polycistronic mRNA molecule must be translated with equal efficiency. DNA sequences of translation initiation sites of several r-protein genes are shown in Figure 5. Again, comparison of the sequences shows no unique common features, even among co-transcribed genes. All of these sequences include a Shine-Dalgarno sequence (71; see article by J. Steitz in this volume), but the length of the sequence is quite variable. For example, of the genes from the spc operon, genes for L14, L24, and L30 have Shine-Dalgarno sequences of 4, 7, and 3 bases, respectively. Another sequence feature implicated in the initiation efficiency is the nucleotide following the initiation triplet AUG. Using the Qβ coat cistron initiation site, Taniguchi and Weissman showed that mutational change from AUGG to AUGA in the initiation site increases the ribosome binding efficiency by a factor of 3.2 presumably because the A residue following the AUG in the mutant mRNA has a better interaction with a complementary U residue in the anticodon loop sequence of fMet-tRNA (72). In many of the r-protein gene initiation sites, we find G following the AUG. However, we also find different nucleotides at this position among co-transcribed genes; for example, AUGA for L14 and AUGG for L24 and L30.

S 12	ACCAGGAGCTATTTA[ATG]G
S 7	CAACGGAGTATTTCC[ATG]C
L 14	TTAGCGGAGCCTAAA[ATG]A
L 24	TCTAAGGAGCGAATC[ATG]G
L 30	CTGGGGAAATAAACC[ATG]G
S 13	TAGTAGGAGTGCATA[GTG]G
S 4	TGTTGGAGAAAGAAA[ATG]G
L 11	CTTGAGGAATTTATA[ATG]G
L 1	GTGGAGGACTAAGAA[ATG]G
L 10	CCAGGAGCAAAGCTA[ATG]G
L 7/L 12	TCAGGAACAATTTAA[ATG]T
EF-G	TAAACGAGGAAACAA[ATG]G
α	CAAAGAGAGGACACA[ATG]C
β	GAGCTGAGGAACCCT[ATG]G

Figure 5. Translation initiation sites of r-protein genes and genes co-transcribed with r-proteins. The gene is indicated by the name of its protein product. The initiation codons are boxed. The brackets by the gene names indicate genes from the same transcription unit. Solid overlining or underlining indicates a sequence complementary to the 3'-end of 16S rRNA (a Shine-Dalgarno sequence). Broken underlining indicates the termination codon of the upstream gene. These sequences are from Refs. 39 and 52 and our unpublished results.

Although the sequences which determine the efficiency of a ribosome binding site are not well understood, there is nothing obvious in the sequenced translation initiation sites of r-protein genes which would account for stoichiometric synthesis from co-transcribed r-protein genes.

POST-TRANSCRIPTIONAL CONTROL IN R-PROTEIN SYNTHESIS

The simple model for regulation discussed in the Introduction assumes that the accumulation of ribosomal components is regulated at the level of mRNA synthesis, as has been observed in other bacterial regulatory systems. However, as discussed above, the sequence comparison of r-protein promoters and that of r-protein cistron initiation sites, indicate difficulty in the formulation of any simple molecular models to explain the stoichiometric relationship in the synthesis of r-proteins.

A post-transcriptional regulatory mechanism has recently been suggested from experiments examining the effect of altered dosage of r-protein genes (10). In strains diploid for the spc operon of r-proteins, spc mRNA is synthesized at a two-fold higher rate than in haploid strains, but proteins from the spc operon are synthesized at the same rate as in the haploid strains. These facts were reconciled by the observation that the excess spc mRNA is rapidly degraded. Thus E. coli apparently has a mechanism to selectively inactivate and degrade mRNA coding for an unnecessary excess of r-proteins. A model was proposed in which a r-protein, not immediately assembled into a ribosomal particle, would act as a feedback signal to inactivate and cause the degradation of its mRNA. It should be noted that the observed selective mRNA degradation could be a consequence of selective inhibition of mRNA translation by the feedback regulatory system. Alternatively, it is also possible that the feedback regulation involves a nucleolytic cleavage of mRNA as the primary reaction to inactivate the message.

Experiments similar to the above diploid experiments were also done using strains carrying a hybrid plasmid (constructed from a ColE1-amp vector plasmid, RSF2124) with cloned r-protein genes from the spc operon (73). In this case, a 10 to 20-fold increase was observed in the synthesis rate of mRNA corresponding to the cloned segments of the spc operon. Yet little or no increase was observed in the synthesis rate of corresponding r-proteins (except for the two promoter-proximal r-proteins which inexplicably showed a definite increase). Although the situation in the strains carrying the multiple-copy hybrid plasmid is somewhat complicated, the observed results are consistent with the presence of the post-transcriptional feedback regulation discussed above. The presence of a post-transcriptional mechanism was also suggested in the synthesis of L10 and L7/L12 from the experiments using hybrid plasmids with cloned r-protein genes (P. Dennis and N. Fiil, personal communication). In addition, gene dosage effects on the synthesis of r-proteins were also analyzed by several other workers (without analyzing effects on mRNA synthesis). With the genes in the str-spc region or those in the rif region, little or no gene dosage effects were observed (74; see also Ref. 75). However, with genes for S20, S21, L21, and L27 which map outside the above two regions, the presence of gene dosage effects was reported (74, 76). The significance of the latter observations is not clear but might be related to the chromosomal locations of these genes, as discussed below.

The post-transcriptional model described above can be used to explain the balanced synthesis of r-proteins. According to this model, the strength of r-protein promoters does not need to be exactly the same, nor does the translational efficiency of individual co-transcribed r-protein cistrons. The regulation can be achieved by adjusting the number of functional mRNA molecules by a feedback inactivation. Coordination using such feedback mechanisms is perhaps inherently more accurate and safer than the "simple" model described in the Introduction. First, although more than half of the r-protein genes are clustered in the str-spc region, others are located in many different places on the chromosome (see the review by K. Isono, this volume) and, therefore, their relative gene dosages change depending on the time in the cell division cycle; the genes closer to the origin of replication are duplicated before the genes more distant from the origin. Thus, mechanisms which counteract gene dosage effects by feedback regulation would be superior to the mechanism formulated in the simple model in insuring balanced synthesis of all the proteins throughout the cellular division cycle. [Since most of the r-protein genes replicate before the genes for S20, S21, and L21, it is conceivable that the feedback regulation of former genes is more important than that of the latter; this could explain the reported absence of the feedback regulation for these genes (74, 76)]. Second, if the balanced synthesis is achieved only by the structures of r-protein promoters and r-protein gene initiation sites, it is probably easily disrupted by many possible mutational alterations of these structures. On the other hand, the feedback regulation mechanisms will tolerate small alterations in the promoter efficiency or r-protein cistron initiation efficiency caused by such mutations. These considerations suggest that the feedback regulatory mechanism may have been easier to evolve than the mechanism entertained in the simple model.

FEEDBACK REGULATION IN OTHER SYSTEMS

Although details of the post-transcriptional control model involving a feedback mechanism discussed above require further studies, it is perhaps informative to compare the ribosome system with other systems where similar regulatory features have been discovered. Two systems will be mentioned here.

The best established post-transcriptional control involving a feedback mechanism is that involved in the synthesis of phage T4 gene 32 protein (77, 78). The gene 32 protein is a DNA binding protein and functions in DNA replication, repair, and genetic recombination. The protein has a strong affinity to single-stranded DNA and, to a lesser extent, to single-stranded RNA. Based on considerable experimental evidence, Gold and his co-workers have proposed that the gene 32 protein can reversibly repress the translation of its own mRNA, and that the repression takes place only after the titration of gene 32 protein onto available intracellular single-stranded DNA (77, 78). Since the protein has a general single-stranded RNA binding ability, they suggested that specific repression of the gene 32 mRNA (and not other mRNA) at the translation initiation site which is probably single-stranded in this mRNA.

Most of the r-proteins are basic and have a general affinity to negatively-charged RNA, although their affinity to rRNA (and/or other r-proteins) during ribosome assembly is probably much stronger. Thus the feedback regulation model discussed in the previous section is analogous to the gene 32 system.

Another interesting system is the regulation of the synthesis of phage P22 gene 8 protein (79). This protein functions as a scaffolding protein in the P22 phage head assembly. The first step of the assembly is the co-polymerization of gene 5 coat protein with gene 8 scaffolding protein to form the prohead. During subsequent DNA encapsidation, the scaffolding protein dissociates and then recycles, catalyzing the assembly of newly synthesized coat protein to form new proheads. King and his coworkers demonstrated that the synthesis of gene 8 scaffolding protein is regulated, and the regulation is coupled with the phage head assembly (72). They suggested that either continuous synthesis of gene 8 scaffolding protein depends on co-polymerization with coat subunits, or free scaffolding protein subunits (but not assembled subunits) inhibit their own further synthesis. Although these two possibilities were considered as distinct alternatives by the authors, they might be related if the release of the free nascent scaffolding protein from the mRNA-ribosome complex is coupled with the assembly reaction, and the regulation is achieved by the inhibition of translation of its own mRNA by the nascent protein. Such a possibility was considered in the case of the regulation of r-protein synthesis in order to explain the dominance of streptomycin sensitivity in heterozygous merodiploid E. coli cells (10). Since most r-proteins are basic and have affinity to any RNA, interaction (even "non-specific" interaction) between a newly synthesized r-protein and its own mRNA at the site of synthesis would be conceivable.

The feedback mechanism coupled with assembly is an attractive way to control production of proteins that constitute multisubunit enzymes or supramolecular structures. Although control of the synthesis of P22 scaffolding protein is a good example of such a control, this is perhaps a rare case in phage assembly systems. In many phage infected systems studied, phage structural proteins are usually overproduced, and free proteins are present in excess in infected cells. This is in clear contrast to the ribosome assembly in growing E. coli cells where very little free r-proteins are present. Phage assembly usually takes place in cells which are doomed to cellular lysis so that elaboration of control mechanisms to prevent accumulation of some harmful phage proteins may have been unnecessary. Since the presence of excess (basic) r-proteins is probably harmful to E. coli cells (see Ref. 73), efficient feedback regulatory mechanism(s) must have evolved to prevent accumulation of free r-proteins.

REGULATION OF RIBOSOME BIOSYNTHESIS

Although we have discussed in detail the feedback regulation model involved in the balanced synthesis of r-proteins, more conventional control mechanisms must also be operating at the level of transcription. The synthesis rate as well as the relative amounts of r-protein mRNA increases with the increase in growth rate in various different nutritional environments (65, 80, 81). Thus the transcriptional activity, presumably the activity of r-protein promoters, must be regulated in response to the nutritional environment, as is the activity of rRNA promoters (81; see also the discussion on the relationship between the induction times in nutritional shift-up experiments and the gene order in r-protein operons described earlier and in Ref. 35). Similarly, the stringent control system also appears to act directly on the transcription of r-protein genes (59, 82; G. D. Strycharz and M. Nomura, unpublished experiments), rather than as a secondary consequence of its primary effects on rRNA gene transcription. The post-transcriptional feedback regulation must then be superimposed on the transcriptional control, as a fine tuning mechanism. Past studies on various biosynthetic operons show that there are many ways to regulate the gene expression of simple operons (e.g., control of transcription by repressor, by positive factors, or by attenuator). Therefore, in view of many operons involved in the coordinated synthesis of ribosomal proteins, it will not be surprising if we find several more different mechanisms involved in this system, and the regulatory circuit turns out to be a more complex one.

Synthesis of rRNA is regulated but appears to exceed the amount necessary for ribosome assembly, especially in slowly growing cells (81), and the excess amounts seem to get degraded. The ability of cells to degrade excess rRNA was also demonstrated in strains carrying an rRNA operon on a multi-copy plasmid (19). Thus the coordination of accumulation of rRNA and r-proteins is usually achieved by making more rRNA and degrading the excess. Significant degradation of r-protein can be seen only in special experimental conditions (73, 74, 83); and, in these instances, the degradation appears to be too slow to be used for balanced accumulation of r-proteins.

As reviewed in this article, the organization of most genes clustered in the str-spc region and the rif region has been elucidated. By DNA sequencing many of the genes initially mapped by other techniques have now been confirmed and ambiguous gene orders elucidated. In addition, in the course of sequencing experiments, it was discovered that r-protein genes use codons in a highly non-random way (52). In the five r-protein genes sequenced completely, and the twelve partially sequenced, the codons preferentially used are recognized by the most abundant species of isoaccepting tRNAs (52; L. Post and M. Nomura, unpublished experiments). It appears that this non-random codon usage reflects a design of r-protein mRNA structure which insures a high translational efficiency and minimum translational errors in the synthesis of r-proteins (for further discussion, see Ref. 52).

The DNA sequencing experiments, combined with genetic and *in vitro* transcription experiments, have also given detailed information on the promoters for five r-protein operons and four rRNA operons. Although the

DNA sequence information did not give the answer to the question of regulation, now we have a solid basis to design more sophisticated experiments to probe into the molecular mechanisms involved in the regulation. We have already seen some unexpected results related to the regulation, such as the presence of a post-transcriptional feedback mechanism. Although the regulatory circuits connecting the many ribosomal operons are probably complex, they are now amenable to detailed experimental analysis. Thus ribosome genetics offers a great challenge to molecular biologists who are interested in the regulation of biosynthesis of supramolecular structures and in its relation to overall regulation of cellular growth.

ACKNOWLEDGEMENTS

The work done in this laboratory was supported in part by the College of Agricultural and Life Sciences, University of Wisconsin-Madison, and by grants from the National Institute of General Medical Sciences (GM-20427) and from the National Science Foundation (PCM 79-10616). This is paper no. 2377 from the Laboratory of Genetics, University of Wisconsin-Madison.

REFERENCES

1. Nomura, M., Morgan, E. A., and Jaskunas, S. R. (1977) Ann. Rev. Genet. 11, 297-347.

2. Nierlich, D. P. (1978) Ann. Rev. Microbiol. 32, 393-432.

3. Nomura, M. (1976) Cell 9, 633-644.

4. Lund, E., Dahlberg, J. E., Lindahl, L., Jaskunas, S. R., Dennis, P. P., and Nomura, M. (1976) Cell 7, 165-177.

5. Morgan, E. A., Ikemura, T., Lindahl, L., Fallon, A. M., and Nomura, M. (1978) Cell 13, 335-344.

6. Jaskunas, S. R., Burgess, R. R., and Nomura, M. (1975) Proc. Natl. Acad. Sci. USA 72, 5036-5040.

7. Yamamoto, M., and Nomura, M. (1978) Proc. Natl. Acad. Sci. USA 75, 3891-3895.

8. Linn, T., and Scaife, J. (1978) Nature 276, 33-37.

9. Jaskunas, S. R., Lindahl, L., Nomura, M., and Burgess, R. R. (1975) Nature 257, 458-462.

10. Fallon, A. M., Jinks, C. S., Strycharz, G. D., and Nomura, M. (1979) Proc. Natl. Acad. Sci. USA 76, 3411-3415.

11. Jaskunas, S. R., Nomura, M., and Davies, J. (1974) in Ribosomes (Nomura, M., Tissières, A., and Lengyel, P., eds.) Cold Spring Harbor Laboratory, Cold Spring Harbor, N.Y., pp. 333-368.

12. Dunn, J. J., and Studier, F. W. (1973) Proc. Natl. Acad. Sci. USA 70, 3296-3300.

13. Nikolaev, N., Silengo, L., and Schlessinger, D. (1973) Proc. Natl. Acad. Sci. USA 70, 3361-3365.

14. Ginsburg, D., and Steitz, J. A. (1975) J. Biol. Chem. 250, 5647-5654.

15. Kenerley, M. E., Morgan, E. A., Post, L., Lindahl, L., and Nomura, M. (1977) J. Bacteriol. 132, 931-949.

16. Kiss, A., Sain, B., and Venetianer, P. (1977) FEBS Lett. 79, 77-79.

17. Zurawski, G., and Brown, K. D. (1979) J. Mol. Biol. 130, 83-96.

18. Morgan, E. A., Ikemura, T., and Nomura, M. (1977) Proc. Natl. Acad. Sci. USA 74, 2710-2714.

19. Ikemura, T., and Nomura, M. (1977) Cell 11, 779-793.

20. Lund, E., and Dahlberg, J. E. (1977) Cell 11, 247-262.

21. Young, R. A., Macklis, R., and Steitz, J. A. (1979) J. Biol. Chem. 254, 3264-3271.

22. Morgan, E. A., Ikemura, T., Post, L. E., and Nomura, M. (1979) in Transfer RNA (Abelson, J., Schimmel, P., and Söll, D., eds.) Cold Spring Harbor Laboratory, Cold Spring Harbor, N.Y., in press.

23. Sekiya, T., and Nishimura, S. (1979) Nucleic Acids Res. 6, 575-592.

24. Morgan, E. A., and Nomura, M. (1979) J. Bacteriol. 137, 507-516.

25. deBoer, H. A., Gilbert, S. F., and Nomura, M. (1979) Cell 17, 201-209.

26. Young, R. A., and Steitz, J. A. (1979) Cell 17, 225-234.

27. Gilbert, S. F., deBoer, H. A., and Nomura, M. (1979) Cell 17, 211-224.

28. Glaser, G., and Cashel, M. (1979) Cell 16, 111-121.

29. deBoer, H., and Nomura, M. (1979) J. Biol. Chem. 254, 5609-5612.

30. Lund, E., and Dahlberg, J. E. (1979) Proc. Natl. Acad. Sci. USA, in press.

31. Jaskunas, S. R., Fallon, A. M., and Nomura, M. (1977) J. Biol. Chem. 252, 7323-7336.

32. Jaskunas, S. R., Lindahl, L., and Nomura, M. (1975) Proc. Natl. Acad. Sci. USA 72, 6-10.

33. Fiandt, M., Szybalski, W., Blattner, F. R., Jaskunas, S. R., Lindahl, L., and Nomura, M. (1976) J. Mol. Biol. 106, 817-835.

34 Lindahl, L., Zengel, J., and Nomura, M. (1976) J. Mol. Biol. 106, 837-855.

35. Lindahl, L., Post, L., Zengel, J., Gilbert, S. F., Strycharz, W. A., and Nomura, M. (1977) J. Biol. Chem. 252, 7365-7383.

36. Jaskunas, S. R., Lindahl, L., and Nomura, M. (1975) Nature 256, 183-187.

37. Jaskunas, S. R., and Nomura, M. (1977) J. Biol. Chem. 252, 7337-7343.

38. Jaskunas, S. R., Fallon, A. M., Nomura, M., Williams, B. G., and Blattner, F. R. (1977) J. Biol. Chem. 252, 7355-7364.

39. Post, L. E., Arfsten, A. E., Reusser, F. R., and Nomura, M. (1978) Cell 15, 215-229.

40. Dennis, P. P. (1974) J. Mol. Biol. 89, 223-232.

41. Delcuve, G., Cabezón, T., Ghysen, A., Herzog, A., and Bollen, A. (1977) Mol. Gen. Genet. 157, 149-153.

42. Kirschbaum, J. B., and Konrad, E. B. (1973) J. Bacteriol. 116, 517-526.

43. Lindahl, L., Jaskunas, S. R., Dennis, P. P., and Nomura, M. (1975) Proc. Natl. Acad. Sci. USA 72, 2743-2747.

44. Watson, R. J., Parker, J., Fiil, N. P., Flaks, J. G., and Friesen, J. D. (1975) Proc. Natl. Acad. Sci. USA 72, 2765-2769.

45. Kirschbaum, J. B., and Scaife, J. (1974) Mol. Gen. Genet. 132, 193-201.

46. Yamamoto, M., and Nomura, M. (1979) J. Bacteriol. 137, 584-594.

47. Rossi, J. J., Ross, W., Egan, J., Lipman, D. J., and Landy, A. (1979) J. Mol. Biol. 128, 21-47.

48. Rossi, J. J., and Landy, A. (1979) Cell 16, 523-534.

49. Lindahl, L., Yamamoto, M., Nomura, M., Kirschbaum, J. B., Allet, B., and Rochaix, J. D. (1977) J. Mol. Biol. 109, 23-47.

50. Yamamoto, M., Lindahl, L., and Nomura, M. (1976) Cell 7, 179-190.

51. Linn, T., Goman, M., and Scaife, J. G. (1979) J. Mol. Biol. 130, 405-420.

52. Post, L. E., Strycharz, G. D., Nomura, M., Lewis, H., and Dennis, P. P. (1979) Proc. Natl. Acad. Sci. USA 76, 1679-1701.

53. Engbaek, F., Gross, C., and Burgess, R. R. (1976) Mol. Gen. Genet. 143, 291-295.

54. Blumenthal, R. M., Reeh, S., and Pedersen, S. (1976) Proc. Natl. Acad. Sci. USA 73, 2285-2288.

55. Iwakura, Y., Ito, K., and Ishihama, A. (1974) Mol. Gen. Genet. 133, 1-23.

56. Iwakura, Y., and Ishihama, A. (1975) Mol. Gen. Genet. 142, 67-84.

57. Reeh, S., Pedersen, S., and Friesen, J. D. (1976) Mol. Gen. Genet. 149, 279-289.

58. Blumenthal, R. M., Lemaux, P. G., Neidhardt, F. C., and Dennis, P. P. (1976) Mol. Gen. Genet. 149, 291-296.

59. Maher, D. L., and Dennis, P. P. (1977) Mol. Gen. Genet. 155, 203-211.

60. Hayward, R. S., and Fyfe, S. K. (1978) Mol. Gen. Genet. 160, 77-85.

61. Blumenthal, R. M., and Dennis, P. P. (1978) Mol. Gen. Genet. 165, 79-86.

62. Pedersen, S., Reeh, S., and Friesen, J. D. (1978) Mol. Gen. Genet. 166, 329-336.

63. Dennis, P. P. (1977) Proc. Natl. Acad. Sci. USA 74, 5416-5420.

64. Little, R., and Dennis, P. P. (1979) J. Bacteriol. 137, 115-123.

65. Dennis, P. P. (1977) J. Mol. Biol. 115, 603-625.

66. Barry, G., Squires, C. L., and Squires, C. (1979) Proc. Natl. Acad. Sci. USA, in press.

67. Post, L. E., Arfsten, A. E., Nomura, M., and Jaskunas, S. R. (1978) Cell 15, 231-236.

68. Ikemura, T., Itoh, S., Post, L. E., and Nomura, M. (1979) Cell, in press.

69. Reznikoff, W. S., and Abelson, J. N. (1978) in The Operon (Miller, J. H., and Reznikoff, W. S., eds.) Cold Spring Harbor Laboratory, Cold Spring Harbor, N.Y., pp. 221-243.

70. Rosenberg, M., and Court, D. (1979) Ann. Rev. Genet., in press.

71. Shine, J., and Dalgarno, L. (1974) Proc. Natl. Acad. Sci. USA 71, 1342-1346.

72. Taniguchi, T., and Weissmann, C. (1978) J. Mol. Biol. 118, 533-565.

73. Fallon, A. M., Jinks, C. S., Yamamoto, M., and Nomura, M. (1979) J. Bacteriol. 138, 383-396.

74. Geyl, D., and Böck, A. (1977) Mol. Gen. Genet. 154, 327-334.

75. Olsson, M. O., and Isaksson, L. A. (1979) Mol. Gen. Genet. 169, 271-278.

76. Takata, K. (1978) Mol. Gen. Genet. 160, 151-155.

77. Gold, L., O'Farrell, P. Z., and Russel, M. (1976) J. Biol. Chem. 251, 7251-7262.

78. Lemaire, G., Gold, L., and Yarus, M. (1978) J. Mol. Biol. 126, 73-90.

79. King, J., Hall, C., and Casjens, S. (1978) Cell 15, 551-560.

80. Dennis, P. P., and Nomura, M. (1975) J. Mol. Biol. 97, 61-76.

81. Gausing, K. (1977) J. Mol. Biol. 115, 335-354.

82. Dennis, P. P. (1977) J. Bacteriol. 129, 580-588.

83. Dennis, P. P. (1974) Mol. Gen. Genet. 134, 39-47.

84. Sekiya, T., and Khorana, H. G. (1974) Proc. Natl. Acad. Sci. USA 74, 2978-2982.

REGULATION OF RIBOSOME BIOSYNTHESIS IN *E. COLI*

KIRSTEN GAUSING

Department of Molecular Biology,
University of Aarhus,
DK-8000 Arhus C. Denmark

INTRODUCTION

All studies of the regulation of ribosome synthesis aim at explaining how the cell adjusts its rate of ribosome formation to the conditions of growth. We observe that the size to which the ribosomal machinery develops in steady states of growth is such that the ribosomes are utilized at maximal or near maximal capacity and adjustments to changes in the growth condition are prompt. A great number of studies on the rate of ribosome formation have been carried out and the chief lesson of these results is that biosynthesis of ribosomes indeed is precisely controlled as regards the assembled particles as well as with respect to the individual components.

There are two main reasons why it has been and is quite difficult to reach a clear understanding of how the rate of ribosome synthesis is controlled. One obvious reason is that the number of different components involved is large. The other reason is that the fraction of energy and matter consumed in making ribosomes is high and changes in the rate of ribosome formation consequently has a significant impact on the synthesis of the other constituents of the cell and *vice versa*. In other words, models for the regulation of synthesis of r-protein must not ignore the fact that the cell has a finite capacity for protein synthesis.

This point in essence is the basis of the passive control model developed by Maaløe (1), recently re-examined and refined (2). According to the passive control model the synthesis of r-proteins is constitutive and the change in their rate of synthesis with the growth conditions is the result of different degrees of competition from the synthesis of non-r-proteins. The regulation of rRNA synthesis is not incorporated into the model but decisive evidence against it has not appeared.

The best documented control concerning ribosome synthesis is the stringent control which clearly is important when the cells are subjected to certain types of stress (3,4). The stringent control, however, is not adequate to explain the regulation of ribosome synthesis in normally growing cells (5,6,7). The alternative to the passive control model of regulation of ribosome synthesis in growing cells involves as yet unknown mechanisms that specifically control the expression of ribosomal genes. Some indication that such mechanisms exist comes from reports on protein factors that stimulate rRNA transcription (8,9,10). These factors may interact with ribosomal regulatory DNA sequences but may also, as suggested for translation factors TuTs and IF-2, interact with the RNA polymerase changing its affinity for stable RNA promotors (10).

We have not progressed much since 1974 (11) in understanding how the information about the composition of a given medium is converted into a decisive rate of ribosome synthesis. On the other hand our knowledge about steady states of growth and shifts has increased considerably in detail. Significant advance has been made in the genetic analysis of r-genes (12) and, recently, several DNA sequences of r-gene regulatory regions have been determined (13,14,15,16). In the following the emphasis is on expression of r-genes *in vivo* and results are discussed with three main aspects of the control of ribosome synthesis in mind: the overall control, the correlation between rRNA and r-protein, and the coordination of individual r-proteins.

STEADY STATES OF GROWTH

The starting point for the discussion of overall regulation of ribosome synthesis is the fact that *E.coli* can grow at steady state growth rates ranging from less than 0.1 to almost 3 generations/hr at 37° depending on the medium supplied. Table 1 recapitulates the composition of *E.coli* at different growth rates. Note that in steady state the relative rate of

TABLE 1. Cell composition and relative rates of synthesis of r-protein and rRNA in *E. coli*

Medium	$\mu(37^o)$	Amounts w/w protein/DNA	Amounts w/w RNA/protein	Amounts w/w rRNA/protein	Relative rates α_r	Relative rates F(rRNA)	Ribosome efficiency
Broth	2.7	25.7	0.50	0.44	0.28	-	1.0
Casamino acids	2.1	25.4	0.40	0.35	0.22	0.68	1.0
Glucose	1.5	25.0	0.31	0.27	0.158	0.57	1.0
Succinate	0.9	24.6	0.21	0.18	0.105	0.44	0.9
Acetate	0.5	24.4	0.17	0.15	0.083	0.35	0.65

μ is the growth rate in generations/hr. The protein/DNA values were determined in *E.coli* $15T^-$ (19,20). The RNA/protein values are virtually the same in *E.coli* ML303, $15T^-$, and B (21,19,22). The rRNA/protein values were derived from RNA/protein by subtraction of the amount of tRNA which is 13% of RNA at all growth rates (20,23). The relative rate of r-protein synthesis, α_r, has been determined in *E.coli* B and B/r (7,22.24) and F(rRNA), the fraction of the total transcript that is rRNA (including the precursor sequences), in *E.coli* B (7). Note that the composition of ribosomes arrived at by combining rRNA/protein and α_r yields a value of 37% protein in agreement with measurements on purified particles (25). The average efficiency of the ribosomes in protein synthesis was calculated from α_r, taking the limited turnover of protein at lower growth rates into account (26).

r-protein synthesis, α_r, is equal to r-protein/protein (w/w) and that practically all the rRNA and r-protein reside in mature ribosomes (17,18). The last column in Table 1 illustrates the key correlation between ribosome synthesis and the growth rate: The average rate of protein synthesis per ribosome is constant and probably maximal at medium and high growth rates. At low growth rates the average ribosome efficiency decreases significantly; this is a consequence of the finding that the ribosome content of the cells does not approach zero at low growth rates.

To get a better impression of the regulation of ribosome synthesis as a function of the growth rate, the rate of synthesis of ribosomes per min was calculated from α_r and the growth rate. Figure 1 shows that the difference in rate of synthesis of ribosomes between high and low growth rates is about 75 fold. Most of this variation in the rate of ribosome synthesis is executed at initiation of transcription of the r-genes but at low growth rates other mechanisms contribute significantly to the overall regulation.

The results of a large number of different investigations, some of which are referred to in Table 2, show that chain elongation rates are the same at different growth rates. Measurements of rates of transcription and of translation therefore directly yield estimates of frequencies of initiations of chains. Recently the rates of transcription of different classes of RNA were measured over a broad range of growth rates (7). Relative rates of transcription of rRNA and r-pro-

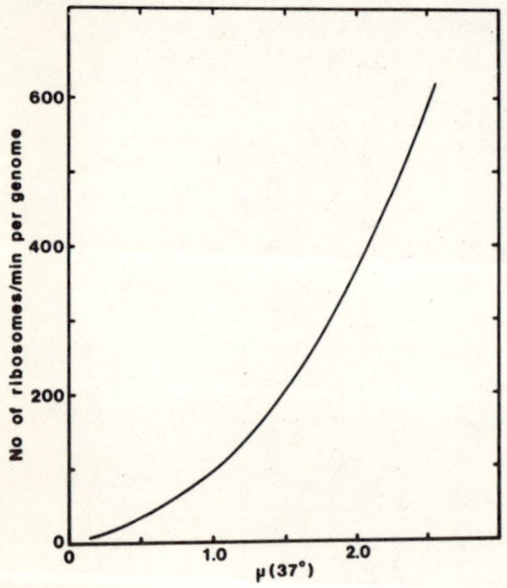

Figure 1. Rate of ribosome synthesis per genome equivalent per min. The rate of r-protein synthesis per min per weight unit of protein is calculated from α_r (Table 1) *via* the growth equation

$$\frac{d(\text{r-protein})/dt}{\text{protein}} = \alpha_r \frac{\mu \ln 2}{60} \text{(w/w min)}$$

and these values are further converted to r-protein synthesis per min per weight unit of DNA using the protein/DNA values in Table 1. The combined molecular weight of the r-proteins per ribosome is 10^6 and one genome equivalent is 3×10^9 dalton.

TABLE 2. Growth rate independent parameters

Parameters	Value at 37° x)	Refs
Peptide elongation rate	17 amino acids/sec	27,28,29
mRNA elongation rate	55 nucleotides/sec	30,31,32
rRNA elongation rate	75 nucleotides/sec	33
mRNA halflife	1.5-1.8 min(average)	27,32
Amount of mRNA/ribosome	150-200 nucleotides	34,35
Rate of mRNA synthesis/ribosome	80 nucleotides/min	7+)

x) Note that rates change with the temperature. The rates at 30° are 2/3 of the rates measured at 37°.

+) This conclusion was not stated explicitly but can be derived from $\frac{dmRNA/dt}{protein}$ (Fig. 10,ref. 7) and $\alpha_r = \frac{r\text{-}protein}{protein}$ as $\frac{dmRNA/dt}{r\text{-}protein}$ and using a molecular weight of 10^6 for the r-protein per ribosome and 325 for a nucleotide

tein mRNA were determined by specific hybridization to pulse-labelled RNA. The fraction of the total transcript which is rRNA was found to increase as a function of the growth rate (Table 1). The fraction of total transcription which is r-protein mRNA is almost independent of the growth rate showing that the regulation of transcription of rRNA and of r-protein mRNA is not strictly parallel as expected if the two types of RNA were controlled by precisely the same mechanism. The relative rate of r-protein mRNA transcription cannot be correlated directly with the relative rate of r-protein synthesis. For this comparison the rate of r-protein mRNA transcription relative to bulk mRNA transcription is needed, and a calculation of this parameter showed that the transcription of r-protein mRNA increases in proportion to α_r as a function of the growth rate (7). The same result was obtained in experiments where the amount of r-protein mRNA was measured (36,7) and both types of estimates therefore indicate that r-protein synthesis is controlled at initiation of transcription. The decrease in the efficiency of protein synthesis at low growth rates however, also plays a role. The low efficiency of the ribosomes in protein synthesis at low growth rates (Table 1) is not accompained by a decrease in the amount of mRNA per ribosome (Table 2). Consequently the molar yield of peptide per messenger decreases at low growth rates. The r-protein mRNA shares this property with bulk mRNA and it contributes markedly to the overall regulation of r-protein synthesis.

The rate of transcription of rRNA was found to balance or slightly exceed the rate of r-protein synthesis at medium and high growth rates (7). Since the rRNA chain growth rate is constant it can be seen from Figure 1 that at maximal growth rate, one round of initiation takes place at any one of the 7-8 different rRNA genes at about 1 sec intervals. This appears to be near the maximal rate physically possible considering that the distance moved by the RNA polymerase during 1 sec is 75 base pairs which is not much more than the space occupied by the RNA polymerase molecule.

At low growth rates a substantial excess of rRNA is synthesized and broken down again (35,37,7). It is not known if this excess (up to 70%) is made because the frequency of initiation at the rRNA genes cannot decrease sufficiently or whether it is a requirement for correct ribosome assembly under conditions where so few ribosomes are made per minute (Fig. 1). However the observation shows that breakdown of excess rRNA is one of the mechanisms for adjusting the amounts of rRNA and r-protein.

It has been described how low efficiency of mRNA translation and synthesis of excess rRNA contribute significantly to the overall regulation of ribosome synthesis at low growth rates. For rRNA the actual difference in rate of initiation at low and high growth rates is no more than about 25 fold. For r-protein mRNA the difference in the rate of transcription is only about 15 fold. The rate of synthesis of β-galactosidase in comparison can be induced more than a thousand fold, but in terms of maximal initiation frequencies, the r-protein genes and the *lacZ* gene are comparable. It is about 20 initiations per min in both cases. The relatively modest changes in initiation frequencies taking place at the r-genes add to the difficulties in unravelling the control mechanisms, in particular since we do not know if the regulatory mechanisms that operate in growing cells are the same as the mechanisms involved in stringent control (see later section).

At medium and high growth rates, changes in the rate of frequency of initiation play the dominant role in the regulation of ribosome synthesis and a few words should be said about the RNA polymerase. Work on RNA polymerase was reviewed recently (38) and in the present context let it suffice to note that RNA polymerase is physically quite abundant at all growth rates. The estimates vary but not more than 80% of the RNA polymerases are actively engaged in transcription at high growth rates and this fraction decreases to 25% at lower growth rates.

UPSHIFT

Adding a richer medium to an exponentially growing culture is one of the classical types of experiments carried out to study the regulation of ribosome synthesis (see 11 for refs to early work). With the recent technical advances that permit precise measurements of synthesis of r-proteins and transcription of rRNA and r-protein mRNA, the original observation that ribosome synthesis is stimulated preferentially and rapidly upon upshift has been examined in more detail.

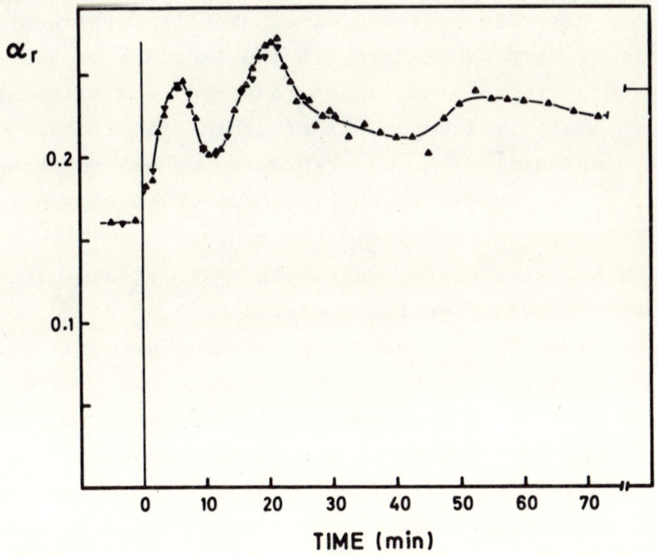

Figure 2. α_r during upshift. A mixture of 17 amino acids (minus leucine, valine, and isoleucine) and hypoxanthine was added at time zero to a culture of AS19 growing exponentially in glucose minimal medium at 30°. Before and after the shift portions of the culture were labelled for 30 sec with ^{3}H-leucine followed by a 30 min chase with excess of non-radioactive leucine. Double determinations of α_r were carried out on each sample as described (18). ▲ and ▼ refer to independent experiments.

Measurements of α_r, the relative rate of r-protein synthesis, using short labelling periods and taking frequent samples revealed an unexpected feature of ribosome synthesis (39). α_r goes through a series of rapid changes, in the following called oscillations (Fig. 2). The rate of synthesis of individual r-proteins has been followed during the first 6-10 min after an upshift and, although the results were not in complete agreement, the first increase and decrease was evident (40,41). It was also clear that the pattern varies from protein to protein, some showing more pronounced peaks in the rate of synthesis than others. The rate of synthesis of some of the aminoacyl-tRNA synthetases increases with the growth rate and interestingly the rate of synthesis of five different aminoacyl-tRNA synthetases measured during one hour after an upshift were found to oscillate in a manner quite similar to α_r (42).

An important question for the arguments about the mechanisms of regulation of ribosome synthesis is how long it takes before the rate of r-protein synthesis increases after the upshift. A very short delay in the increase of the rate of synthesis of r-proteins argues that the regulation of r-protein synthesis is independent of synthesis of other macromolecules, including rRNA. The r-protein genes are arranged in several polycistronic transcriptional units and an increased frequency of initiation of r-protein mRNA transcription (or translation of the r-protein mRNA) is expressed fully for most r-proteins during a period of 2 to 3 min at 30° (40,41,43). Using this information and the value for the halflife of mRNA it is possible to estimate the time at which the rate of initiation increases from the lag in induction of the r-proteins. The results of the three r-protein studies referred to here yield different results. The experiment shown in Figure 2 indicates that the rate of initiation is increased within 10-15 sec at all r-protein transcriptional units whereas the analysis of individual proteins were compatible with an increase within 10 sec at some of the r-protein transcriptional units but for bulk r-protein there was a delay of 20-45 sec (40,41). The experiments in Figure 2 and (40) were carried out with the permeable mutant AS19 (*E. coli*

B) whereas the experiment in (41) was done with *E.coli* B/r and it appears that the response to the upshift is swifter in AS19 than in B/r which may reflect a more rapid influx of nutrients in the permeable strain. Taken together the results suggest that the rate of synthesis of at least some of the r-proteins can increase without the preceding synthesis of other macromolecules.

The very rapid increase in the rate of RNA accumulation seen in the early studies of upshift lead to the conclusion that this increase does not require *de novo* protein synthesis, more specifically the cell content of RNA polymerases in the preshift cells is sufficient to accommodate the increased rate of RNA accumulation (44). The total rate of RNA transcription, including the substantial rate of mRNA transcription was estimated during upshift using the antibiotic rifampicin and this study confirmed the early notion that immediately upon the shift there is a sudden increase in the number of active RNA polymerase molecules (32). RNA labelling studies are hampered by the slow and variable labelling of the pools of nucleotide triphosphates and since Pato and v. Meyenburg (32) did not correct their data for the changes in the triphosphate pools during upshift, their conclusion was contested. Both Nierlich (45) and Dennis and Bremer (46) maintained that the number of active RNA polymerases increases only gradually during the upshift. In one case the rapid increase in rRNA transcription was asserted to be counteracted by a corresponding decrease in the rate mRNA transcription (45) but this notion is difficult to reconcile with the observation that the rate of protein synthesis is constant or increases during the upshift depending on the preshift growth rate (40,47). In the other case an increase in the rRNA chain growth rate was invoked (46); however, later experiments have shown that the rRNA chain growth rate is constant (33).

In an attempt to resolve the question and to see if the oscillations in α_r are accompanied by oscillations in the rates of rRNA and r-protein mRNA transcription, the total rate of RNA transcription and the relative rates of rRNA and r-protein mRNA transcription were determined during the first

15 min of an upshift similar to the one shown in Figure 2 (in preparation). The combined results of these studies are summarized in Figure 3. The size of the ATP pool was found to decrease to about 70% during the first 3-4 min after the shift and then more slowly to about 60% in fair agreement with determinations in *E.coli* K12 (48). The labelling of ATP with 2-^{3}H-adenine during 15 sec pulses increased during the shift in inverse proportion to the decrease in the size of the pool as expected. The uptake of 2-^{3}H-adenine in RNA was corrected for the increased labelling of the precursor pool and since the pulse is very short compared to the rate of degradation of unstable RNA, the corrected incorporation of 2-^{3}H-adenine in RNA represents the total rate of RNA transcription. The rate increases very rapidly after the shift and at 5 min it is 65% above the preshift rate and close to the final rate in the enriched glucose medium. The relative rate of rRNA transcription increases from 0.53 to almost 0.70 during the first 5 min after the shift (compare Table 1). The increase in the rate of total RNA transcription is in fact fully accounted for by the increase in transcription of rRNA (and tRNA) such that the rate of mRNA transcription remains

Figure 3. Rates of transcription of total RNA, rRNA, and r-protein mRNA during upshift. The upper curve in frame (a) was obtained as follows: A culture of AS19 growing exponentially in glucose minimal medium at 30° received $^{32}PO_4$ and 1.5 generations later the culture was shifted up by adding a mixture of 20 amino acids. Before and after the shift small portions of the culture were labelled for 15 sec with 2-^{3}H-adenine. Pulse labelling samples were taken every 15 sec and alternating samples were stopped with TCA and formic acid. The TCA samples were processed to determine incorporation into RNA and the formic acid samples were used to determine radioactivity in ATP (7). Radioactivity incorporated into RNA was corrected for the changes in the ATP pool by normalization to the size of the pool before the shift. The lower curve in frame (a) and the curve in frame (b) combine hybridization measurements with the curve for the total rate of RNA transcription. Upshifts were performed as described above and portions of the culture were labelled for 30 sec with 2-^{3}H-adenine. The fraction of radioactivity incorporated into RNA which is rRNA and r-protein mRNA were measured by hybridization to *λdilv*5 DNA (rRNA gene) and to *λdspc*1 DNA (14 r-protein genes) (7). The hybridization measurements yield $\frac{drRNA/dt}{dRNAtot/dt}$ and $\frac{dr\text{-}mRNA/dt}{dRNAtot/dt}$ and the results shown here were obtained by multiplication with dRNAtot/dt from the upper curve in frame (a). The results for rRNA and r-protein mRNA represent three separate experiments.

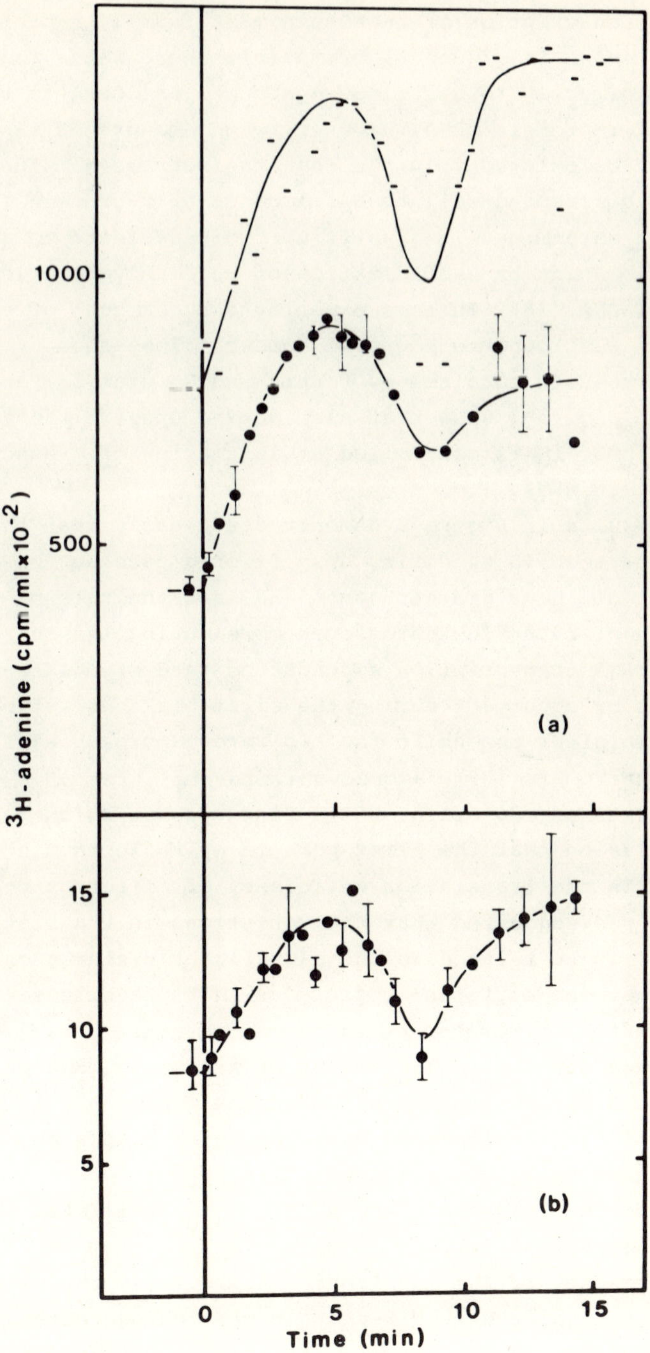

1000
500
15
10
5
(a)
(b)
0
5
10
15
Time (min)
3H-adenine (cpm/ml x 10-2)

fairly constant during the first period of the shift. The rate of transcription of r-protein mRNA from 14 r-protein genes is shown in Figure 3, frame (b). These rates cannot be compared directly to α_r, since α_r is proportional to the amount of r-protein mRNA. However, when amounts of r-protein mRNA were calculated from the results in frame (b) they were found to correspond well to α_r. Amounts of r-protein mRNA have been determined in a shift up from succinate to glucose plus amino acids by hybridization of uniformly labelled RNA to *λdspc*1 DNA (36). In this experiment the amount of r-protein mRNA was found to increase somewhat less than α_r, but both the results obtained by Dennis and Nomura (36) and the result in Figure 3 show that also during upshift the synthesis of r-protein is controlled at initiation of transcription of r-protein mRNA.

The results in Figure 3 demonstrate clearly that the oscillations seen in α_r during upshift originate at the transcription of the r-protein genes and that the rate of transcription of rRNA fluctuates even more during the shift. The rate of rRNA transcription exceeds the rate of bulk r-protein synthesis by about 40% during the first peak but between 10 and 15 min after the shift the two rates approach each other. Dennis and Nomura (36) favored an interpretation of their measurements of r-protein mRNA during upshift which leads to the conclusion that the first peak in α_r reflects minor differences in the translation efficiency of different r-protein mRNA. It was suggested that the variations in translation efficiency may be related to the significant disturbance in the balance between different r-proteins that prevails early during upshift (41). However, their measurements are also compatible with a maximum in the amount of r-protein mRNA at the time of the maximum in the α_r determinations (36). It is possible that the imbalance of the r-proteins brings about the oscillations in α_r but it is more likely that other mechanisms are involved since the $\alpha_{synthetase}$ oscillates in the same manner as α_r (42).

With respect to the question of how quickly the rates of transcription of rRNA and r-protein mRNA change after the shift, the results in Figure 3 show that a higher rate of rRNA transcription is induced without delay. The experimental accuracy is less for the determinations of rates of transcription of r-protein mRNA and the result is compatible with zero up to about a 20 sec delay in induction.

In summary the upshift involves a rapid increase in the number of active RNA polymerase molecules and this increase is accounted for by the new higher rate of initiation of rRNA. Simultaneously the rate of initiation of r-protein mRNA increases and the "induction" in both cases is very rapid. It is therefore unlikely that preceeding synthesis of macromolecules is a prerequisite for the induction. Instead the results suggest that changes in the intracellular concentrations of low molecular weight metabolites directly mediates the signal to increase the rate of rRNA and r-protein mRNA transcription. This may seem a modest conclusion but it is the nature of physiological experiments performed on growing cells that they will tell what the cell does but not much about how it does it. By the same token several intelligent models can be proposed to explain the oscillations but it will be difficult to show if any of them are correct.

STRINGENT CONTROL

When *rel*$^+$ cells are deprived of a required amino acid the rate of RNA synthesis is inhibited and the intracellular concentration of the tetraphosphate, ppGpp, goes up. This is the classical stringent response. Introducing a mutant, temperature sensitive for valine tRNA synthetase, Fiil *et al.* (3) were able to establish different degrees of amino-acid starvation by growing the *valS*^{ts} mutant at different temperatures and this system has proven very useful for detailed studies of stringency. Between 10 and 30 min after transfer of the *valS*^{ts} mutant to higher temperature the ppGpp level and the rate of RNA accumulation are fairly constant and a linear correlation between ppGpp levels and the reciprocal rate of RNA accumulation was established. Under these conditions the rate of RNA accumulation is a good measure of the rate of rRNA transcription since rRNA is not unstable during amino acid starvation (44,50,37). Amino acid starvation of stringent cells is characterized by high levels of ppGpp, a two-fold inhibition of RNA synthesis in a glucose minimal culture is accompanied by more than a four-fold increase in ppGpp level (Table 3).

The level of ppGpp decreases with increasing growth rate in exponentially growing cultures (51) but a two-fold decrease in rate of rRNA synthesis relative to a glucose grown culture is correlated with a relative small increase in the ppGpp level (Table 3). A third correlation between ppGpp level and the rate of rRNA transcription was found in cells subjected to downshift. Downshift was imposed on cultures

TABLE 3. rRNA synthesis and ppGpp levels

Condition	Strain	ΔppGpp x) (pmoles/A_{450})	Reference
Exponential growth	NF161,NF162	24 → 27	51
Amino acid starvation of *valS*^{ts}	NF342	30 → 140	3
α-Methylglucoside downshift	NF541,NF542	30 → 80	52

Condition	Strain	ppGpp (pmoles/A_{450})	Reference
Exponential growth glucose *relA*$^+$ *spoT*$^+$	NF929	30	53
Exponential growth, glucose *relA*$^+$ *spoT*$^-$	NF161	60	53
Reversal of αMG downshift	NF542		6
- by excess glucose		100	
- by amino acids		50	
Upshift	NF541		5
- by glucose		35	
- by 5 amino acids		10	
Temperature jump	NF859	600	54

x) Change in ppGpp level accompaining a two fold inhibition in the rate of rRNA synthesis relative to the rate in cells growing exponentially in glucose minimal medium.

In the lower part of the table the level of ppGpp was in all cases measured at a time when the rate of rRNA synthesis was nearly equal to the rate in cells growing exponentially in glucose minimal medium.

The results collected in this table were obtained in several different strains, all from the collection of Niels Fiil. NF342, 541, and 542 are derivatives of *E.coli* B strain AS19 and the pedigrees of the other strains are given in (53).

growing exponentially in glucose minimal medium by addition of α-methylglucoside (αMG), a competitive inhibitor of glucose uptake, and ppGpp levels and rates of transcription of rRNA were measured at different times (52). During downshift, rRNA is turning over in contrast to the situation during amino acid starvation but actual rates were measured by Molin *et al.* (52). It was found that there is a linear correlation between ppGpp levels and the reciprocal rate of rRNA transcription also during downshift but the increase in ppGpp level is more modest than during amino acid starvation (Table 3).

In all three cases referred to here there is a correlation between ppGpp level and the rate of rRNA synthesis. However, the magnitude in the change of ppGpp concentration is significantly different and it was suggested that ppGpp is not the only element controlling transcription of rRNA (3).

The results of other studies further emphasize the need for additional controls and one result even makes it doubtful that ppGpp is the "real" control nucleotide. The results listed in the lower part of Table 3 illustrate how different the level of ppGpp can be under conditions where the rate of rRNA synthesis is nearly the same. *SpoT*$^-$ mutants with a defect in the degradation of ppGpp (55,56) have a higher basal level of ppGpp during exponential growth than *spoT*$^+$ cells. The *spoT*$^-$ mutants do grow more slowly than wt cells but the growth rate (and rate of rRNA synthesis) is not reduced as much as expected from the high basal levels (53). Identical growth rates can be obtained in glucose minimal medium and with a poor carbon source supplemented with selected mixtures of amino acids. Addition of amino acids leads to a rapid decrease of the ppGpp level whereas addition of glucose to a slowly growing culture leads to a slow decline of the ppGpp level (5). During reversal of a downshift effected by α-methylglucoside the level of ppGpp was actually found to double transiently at a time when the rate of rRNA synthesis was accelerating (6). An unexpected and vigorous accumulation of ppGpp was seen when cells were transferred from low temperature (23°) to optimal growth temperatures and the larger the temperature jump, the higher were the levels of ppGpp accumulated. The rate of rRNA synthesis however, appeared to be completely unaffected by the temperature jump (54). Gallant *et al.* (54) offered several hypotheses to explain this result, in particular that the "real" inhibitor of rRNA synthesis is not ppGpp at all but perhaps another unusual nucleotide which sometimes but not always is formed in proportion to ppGpp. It is clear that the synthesis of rRNA is subject to stringent control but the mechanism of this control apparently is not as simple as originally envisioned.

Stringent control affects several other cellular processes besides the synthesis of rRNA. Using the *valS*^{ts} mutation described earlier, the synthesis of r-proteins was found to be under stringent control during the amino acid starvation (4) and the inhibition of r-protein synthesis during α-methylglucoside downshift is parallel to the inhibition of rRNA transcription (52). The stringent control of r-protein synthesis acts at transcription of r-protein mRNA (57) and it could, therefore,be expected that the non-r-proteins whose genes have been located in r-protein transcriptional units are stringently controlled. For EF-G and EF-TuA (co-transcription with S7 and S12 (12)) this, indeed, was found to be the case (58,59,60) but the synthesis of RNA polymerase subunit α (co-transcription with S4, S11, S13, and S17 (12)) and the subunits β and β' (co-transcription with L10 and L7/L12 (61,62,63)) is not inhibited very strongly during starvation of *relA*$^+$ cells and the synthesis is not stimulated in *relA*$^-$ cells as the synthesis of r-proteins (58,59,60). During exponential growth the synthesis of the r-proteins is very closely coordinated (see next section) and the synthesis of the RNA polymerase subunits are regulated in a manner similar but not identical to the r-proteins, β and β' albeit at one fifth the (molar) rate (*e.g.* 64). During amino acid starvation, the non-coordinate regulation of β + β' and L10 + L7/L12 was seen to be expressed at transcription of the corresponding mRNA segments (65) and most likely hitherto unknown regulatory principles are involved since there does not seem to be a separate promotor for the β and β' genes (61,62,63). It should also be noted that even though bulk r-protein synthesis is well correlated with the synthesis of rRNA during starvation of *relA*$^+$

and *relA*$^-$ cells, the individual r-proteins were expressed to different degrees (4).

During amino acid starvation and downshift of *relA*$^+$ cells bulk mRNA is reduced to about half of the level in exponentially growing cells whereas *relA*$^-$ cells maintain or even increase their levels of mRNA (66,67,50,52). The adjustment of mRNA to the limited rate of protein synthesis makes the transition to slow growth less traumatic for the *relA*$^+$ strains. The advantage of having stringent control can also be inferred from two other observations. It was noted a long time ago that *relA*$^+$ cells recover more rapidly from prolonged starvation than *relA*$^-$ cells (68,69). The reason for the slow recovery of *relA*$^-$ cells has not been determined but it may be caused by the accumulated relaxed particles. The accuracy of translation appears to be imparied during starvation of relaxed cells but not of stringent cells since β-galactosidase synthesized in relaxed cells consists of relatively unstable enzyme (70,53).

Numerous studies of the effect of ppGpp on transcription of rRNA and r-protein genes *in vitro* have been carried out both in purified systems and in more or less crude cell extracts. Space does not permit a fair discussion of this often controversal work. A personal view of the *in vitro* work can be summarized as follows. ppGpp inhibits rRNA and r-protein mRNA transcription but the concentrations of ppGpp used in many cases were higher than the concentration measured *in vivo*. Crude systems appear to mimic *in vivo* rRNA transcription better than purified systems and partial purification of presumed specific transcriptional factors has been started (8,9) but unfortunately this type of work is not persued very actively.

KEEPING THE BALANCE

Analysis of purified ribosomes has repeatedly shown that purified particles are heterogenous with respect to the stoichiometry of the r-proteins. Using a very mild procedure for preparation of ribosomes Hardy (71) on the other hand arrived at the conclusion that all r-proteins except L7/L12 are unit proteins *in vivo*. L7/L12 are found in at least three copies per ribosome (72) and the ratio between L7 and L12 changes with the growth conditions (73,74) but this change has no direct influence on the regulation of synthesis of L7/L12 since there is only one gene for L7/L12 (75). In a careful study Dennis (74) determined the rate of synthesis of the individual r-proteins by pulse labelling followed by a short-chase period (less than one minute) and found that the majority of the r-proteins are synthesized in a one to one ratio. Some proteins (twelve) were apparently synthesized in a 20-25% excess but as discussed by Dennis (74) this finding could be the results of a more efficient extraction of proteins not yet incorporated into mature ribosomes. Even with this reservation the study demonstrates that the synthesis of the different r-proteins is remarkably well-balanced.

Both the rRNA genes and the r-protein genes identified so far are confined to one half of the *E.coli* chromosome and are distributed on both sides of the origin of the bidirectional replication (Fig. 4). In the early genetic analyses none of the known r-protein genes mapped outside the str-spc region and it was felt that the unknown genes would map close to the str-spc region so that the balance between the different genes would be maintained at all times during the cell cycle. Figure 4 shows that this assumption cannot be sustained and

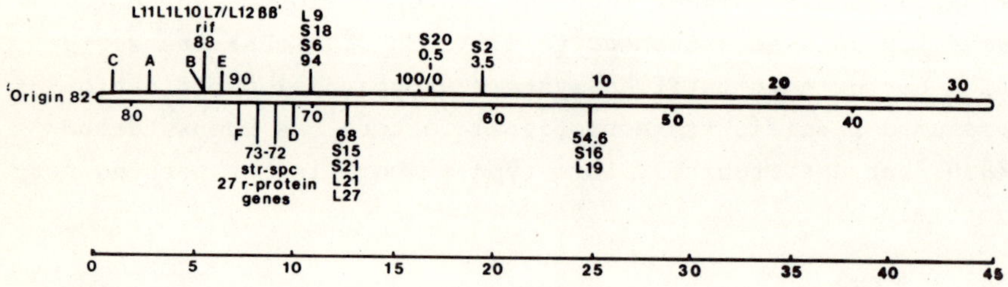

Figure 4. Positions of ribosomal genes on the *E.coli* chromosome. Numbers on the chromosome refer to map positions in minutes according to Backmann *et al.* (76). A-F indicate the positions of rRNA genes. The scale below the map shows the times of replication with the start of replication at zero min and a total replication time of 45 min at 37°. Most r-gene map positions can be found in (76,12) and the position of genes for the S16 and L19 in (77), for S6, S18, L19 in (78), and for L21, L27, S15, S21 in (79).

Table 4. Relative gene dose of r-protein genes at different growth rates

μ(37°)	rif-cluster	spc-cluster	S16 and L19	rif / spc	S16L19 / spc
0.65	1.14	1.10	0.98	1.04	0.89
1.30	1.27	1.19	0.95	1.07	0.80
2.0	1.42	1.31	0.91	1.08	0.69

The relative gene dose is the number of copies of gene N per genome equivalent and was calculated using the formula derived by Collins and Pritchard (80). The time of replication from origin to terminus is 45 minutes at all growth rates.

one may ask what is the consequence of having groups of r-protein genes placed at some distance from each other. In Table 4 the relative gene dosage for three different clusters of r-protein genes has been calculated at three different growth rates. The relative gene doses are significantly different and their relative proportions change with the growth rate. Expressed differently then it takes 20 min between the replication of the r-protein genes in the rif region and the last r-protein genes (S16 and L19 at 54.6 min). For the individual cell growing in glucose minimal medium there is an equal number of all r-protein genes during half of the cell cycle and a two-fold difference between the rif region genes and the genes at map position 54.6 min during the other half of the cell cycle. Considering the well-balanced rates of synthesis of all the r-proteins at different growth rates (74) then the calculations in Table 4 suggest that the cells possess a control function that compensates for the successive doubling of the different r-protein genes.

In principle such a "balance" control could be demonstrated by analysis of synchronous cultures. The synchronization procedure, however, is technically difficult and to date only measurements on bulk r-protein synthesis (α_r) have been carried out and α_r need not be constant during the cell cycle even if the individual r-proteins are always synthesized in balanced amounts. With 65% of the r-protein genes replicating within 5 min (Fig. 4) it can be calculated that if the duplication of the genes is reflected fully in α_r, a difference of 30% between the highest and lowest value will be seen. The measurements of α_r during the cell cycle gave conflicting results. In two cases α_r was constant during the cell cycle (81,82) whereas fluctuations in α_r up to 15% was seen in one case (83).

In view of the arguments presented above it would be interesting to follow the synthesis of a number of individual r-proteins and the rate of synthesis of r-protein mRNA corresponding to a short segment of r-protein genes during the cell cycle. Such experiments would show if a balance control exists and if it does, whether it is a transcriptional or post-transcriptional control.

A permanent imbalance between different r-genes exists in cells containing F' factors covering r-gene regions and in lysogens of specialized transducing phages with r-genes. In these two cases the increase in gene dose is 1-2. Higher increase in gene dosage prevails with plasmids constructed *in vitro*, typically ColE1 plasmid derived r-gene recombinants exist in 7-15 copies per cell.

The effect of doubling the gene dosage for r-protein genes has been investigated in different merodiploid strains with F' factors covering the str-spc region, the rif region, or the region at 68 min of the chromosome (Fig. 4). The doubling of the gene dose had no effect on the synthesis of the 30S proteins that map in the str-spc region (84,85) and the same result was found for the 50S proteins that maps in this region (Olsson, personal communication). Likewise no gene dosage effect could be detected for the proteins mapping in the rif region (84) but three of the four proteins whose genes map at 68 min were synthesized in nearly twice the normal amount in the merodiploid strain (79). These experiments were all carried out in a *recA*⁻ (recombination deficient) mutant in order to preserve the F' factor. Cultures of *recA*⁻ cells do not have well-defined growth rates since a varying fraction of the cells are non-dividing (86) and the merodiploid strains apparently grow more slowly than the haploid control strains (also *recA*⁻). Nevertheless the gene dosage experiments indicate that, with a few exceptions, the cells exert a gene dosage control on the synthesis of r-proteins, compensating for the increased gene dosis.

The nature of this gene dosage control was investigated in

the following experiment (Olsson and Gausing, in preparation). Haploid cells and cells diploid for the str-spc region were pulse labelled with ^{3}H-uridine and the RNA was hybridized to DNA from λ*dfus*2 and λ*dspc*1 carrying 27 and 14 genes from the str-spc region, respectively. DNA from a λ carrying a DNA fragment with the four rif region r-protein genes was used as control. The relative rate of str-spc mRNA transcription in the merodiploid strain was twice as high as the rate in the haploid strain, whereas the rif region r-protein mRNA was transcribed at the same rate in the two strains. This result was obtained with two different pairs of haploid and diploid strains. In one case the diploid contained an F' factor and in the other case the diploid was lysogenic for λ*dspc*1. Thus we find that r-protein mRNA transcription is proportional to the gene dose in contrast to the synthesis of the r-proteins. This result shows that the gene dosage control is a post-transcriptional control.

The effect of higher gene dose on transcription and translation has been examined for L7/L12 and the RNA polymerase subunits β and β' (Dennis and Fiil, personal communication). These three genes are in the same transcriptional unit (61,62,63). ColE1 derived plasmids carrying genes from the rif region result in a 7-12 fold increase in the gene dose for L7/L12 and ββ'. The rate of L7/L12 and ββ' mRNA transcription was increased about 6 fold in the plasmid containing strains whereas the rate of synthesis of the proteins increased at most two fold. Again the post-transcriptional control element is evident.

A new regulatiory feature has emerged: a post-transcriptional control that adjust the coordination of synthesis of r-proteins. Much more needs to be learned about this new control, but it seems probable that the same mechanism is responsible for balancing the synthesis of the r-proteins during the cell cycle.

Six of the rRNA genes are near the origin of replication and are replicated within 10 min (Fig. 4); one or two rRNA genes have not been mapped. The rRNA genes are thus relatively closely grouped and an effect on rRNA transcription of the gene doubling during the cell cycle should be easy to detect.

The fraction of rRNA in RNA, pulse labelled at different times during the cell cycle, was assayed by hybridization. This experiment showed that the relative rate of rRNA transcription is independent of cell age (37). This result indicates that the cell compensates for the doubling of the rRNA genes, in other words, that there is a gene dose control on rRNA genes.

A contradictory result was obtained in an experiment where the rRNA gene doubling was permanently established with a ColE1 plasmid carrying an rRNA gene (87). In the rDNA plasmid containing cells the rate of rRNA transcription was approximately doubled relative to the rate in control cells growing at the same growth rate. The discrepancy between the results of the cell cycle and plasmid experiment is not easy to explain, but perhaps it is wrong to compare transient changes in the rRNA-gene dose with a permanent artificial doubling of the rRNA genes. In the plasmid containing strain the total rate of RNA transcription is about 50% higher than in a normal cell at the same growth rate. It cannot be excluded that this high transcriptional activity has the same effect as partial inactivation of RNA polymerase by low concentrations of rifampicin. Rifampicin addition to wt cells stimulated the synthesis of RNA polymerase (*e.g.* 88) and it would be interesting to see if the plasmid strain has an increased level of RNA polymerase.

In addition to the gene dose effect on rRNA transcription, the experiments of Ikemura and Nomura (87) showed that there is not a corresponding increase in the rate of r-protein synthesis in the plasmid strain. Instead the excess rRNA was found to be broken down, showing that this mechanism of adjusting the amounts of rRNA and r-protein can operate at medium high growth rates as suggested previously (7). The results of the upshift experiments discussed earlier, indicated that the rate of r-protein synthesis can increase independently of increased synthesis of rRNA. The rDNA plasmid experiment suggests that r-protein synthesis is not stimulated in the presence of excessive rRNA synthesis (87). Together the two observations substantiate the conclusion that expression of rRNA and r-protein genes are under separate but parallel control.

ACKNOWLEDGEMENTS

The author would like to thank Inger Bjørndal for expert technical assistance and Birthe Hother Nielsen for typing the manuscript. The support from the Danish Science Research Council and the Carlsberg Foundation is gratefully acknowledged.

REFERENCES

1. Maaløe,O. (1969) Develop.Biol. (suppl.) 3, 33-58.
2. Maaløe,O. (1979) Biological Regulation and Development 1, 487-542.
3. Fiil,N.P., von Meyenburg,K., and Friesen,J.D. (1972) J.Mol.Biol. 71, 769-783.
4. Dennis,P.P. and Nomura,M. (1974) Proc.Natl.Acad.Sci.USA 71, 3819-3823.
5. Friesen,J.D., Fiil,N.P., and von Meyenburg,K. (1975) J.Biol.Chem. 250, 304-309.
6. Hansen,M.T., Pato,M.L., Molin,S., Fiil,N.P., and von Meyenburg,K. (1975) J.Bacteriol. 122, 585-591.
7. Gausing,K. (1977) J.Mol.Biol. 115, 335-354.
8. Block,R. (1976) In Control of Ribosome Synthesis, ed. N.O.Kjeldgaard and O.Maaløe, pp 226-238. Munksgaard, Copenhagen.
9. Muto,A. (1977) Molec.gen.Genet. 152, 161-166.
10. Travers,A. (1976) Nature 263, 641-646.
11. Kjeldgaard,N.O. and Gausing,K. (1974) In Ribosomes pp 369-392 ed. M. Nomura, A.Tissières and P.Lengyel, Cold Spring Harbor Laboratory.
12. Nomura,M., Morgan,E.A., and Jaskunas,S.R. (1977) Ann.Rev. Genet. 11, 297-347.
13. Post,L.E., Arfsten,A.E., Reusser,F., and Nomura,M. (1978) Cell 15, 215-229.
14. Post,L.E., Strycharz,G.D., Nomura,M., Levis,H., and Dennis,P.D. (1979) Proc.Natl.Acad.Sci.USA 76, 1697-1701.
15. de Boer,H.A., Gilbert,S.F., and Nomura,M. (1979) Cell 17, 201-209.
16. Young,R.A. and Steitz,J.A. (1979) Cell 17, 225-234.
17. Lindahl,L. (1975) J.Mol.Biol. 92, 15-37.
18. Gausing,K. (1974) Molec.gen.Genet. 129, 61-75.
19. Forchhammer,J. and Lindahl,L. (1971) J.Mol.Biol. 55, 563-568.
20. Skjold,A.C., Juarez,H., and Hedgcoth,C. (1973) J.Bacteriol. 115, 177-187.

21. Rosset,R., Monier,R., and Julian,J. (1964) Biochem.Biophys. Res.Commun. 15, 329-333.

22. Bennett,P.M. and Maaløe,O. (1974) J.Mol.Biol. 90, 541-561.

23. Dennis,P.P. (1972) J.Biol.Chem. 247, 2842-2845.

24. Dennis,P.P. and Bremer,H. (1974) J.Mol.Biol. 84, 407-422.

25. Tissières,A., Watson,J.D., Schlessinger,D., and Hollingworth,B.R. (1959) J.Mol.Biol. 1, 221-233.

26. Nath,K. and Koch,A.L. (1970) J.Biol.Chem. 245, 2889-2900.

27. Coffman,R.L., Norris,T.E., and Koch.A.L. (1971) J.Mol.Biol. 60, 1-19.

28. Gausing,K. (1972) J.Mol.Biol. 71, 529-545.

29. ,F., Kjeldgaard,N.O., and Maaløe.O. (1973) J.Mol. Biol. 75, 109-118.

30. Rose,J.K., Mosteller,R.D., and Yanofsky,C. (1970) J.Mol. Biol. 51, 541-550.

31. Bremer,H. and Yuan,D. (1968) J.Mol.Biol. 38, 163-180.

32. Pato,M.L. and von Meyenburg,K. (1970) Cold Spring Harbor Symp. Quant.Biol. 35, 497-504.

33. Molin,S. (1976) In Control of Ribosome Synthesis, ed. N.O.Kjeldgaard and O.Maaløe, pp 333-339. Munksgaard, Copenhagen.

34. Forchhammer,J. and Kjeldgaard,N.O. (1968) J.Mol.Biol. 37, 245-255.

35. Norris,T.E. and Koch,A.L. (1972) J.Mol.Biol. 64, 633-649.

36. Dennis,P.P. and Nomura,M. (1975) J.Mol.Biol. 97, 61-76.

37. Pedersen,S. (1976) In Control of Ribosome Synthesis, ed. N.O. Kjeldgaard and Maaløe,O. pp 345-352. Munksgaard, Copenhagen.

38. RNA polymerase (1976) ed. R.Losick and M.Chamberlin, Cold Spring Harbor Laboratory.

39. Gausing,K. (1974) Thesis for degree of Licentiatus Scientarum, University of Copenhagen.

40. Gulløv,K., von Meyenburg,K., and Molin,S. (1974) Molec. gen.Genet. 130, 271-274.

41. Dennis,P.P. (1974) J.Mol.Biol. 89, 223-232.

42. Reeh,S., Pedersen,S., and Neidhardt,N.C. (1977) J. Bacteriol. 129, 702-706.

43. Molin,S. von Meyenburg,K., Gulløv,K., and Maaløe,O. (1974) Molec.gen.Genet. 129, 11-26.

44. Maaløe,O. and Kjeldgaard,N.O. (1966) Control of Macromolecular Synthesis. W.A.Benjamin,Inc., New York.

45. Nierlich,D.P. (1972) J.Mol.Biol. 72, 751-764.

46. Dennis,P.P. and Bremer,H. (1974) J.Mol.Biol. 89, 233-239.

47. Alton,T.H. and Koch,A.L. (1974) J.Mol.Biol. 86, 1-9.

48. Beck,C., Ingraham,K., Maaløe,O., and Neuhard,J. (1973) J.Mol.Biol. 78, 117-121.

49. Stamato,T.D. and Pettijohn,D.E. (1971) Nature 234, 99-104.

50. Lazzarini,R.A. and Dahlberg,A.E. (1971) J.Biol.Chem. 246, 420-429.

51. Lazzarini,R.A., Cashel,M., and Gallant,J. (1971) J.Biol. Chem. 246, 4381-4385.

52. Molin,S., von Meyenburg,K., Maaløe,O., Hansen,M.T., and Pato,M.L. (1977) J.Bacteriol. 131, 7-17.

53. Fiil,N.P., Willumsen,B.M., Friesen,J.D., and von Meyenburg, K. (1977) Molec,gen.Genet. 150, 87-101.

54. Gallant,J., Palmer,L., and Pao,C.C. (1977) Cell 11, 181-185.

55. Laffler,T. and Gallant,J. (1974) Cell 1, 27-30.

56. Stamminger,G. and Lazzarini,R.A. (1974) Cell 1, 85-90.

57. Dennis,P.P. and Nomura,M. (1975) Nature 255, 460-465.

58. Reeh,S., Pedersen,S., and Friesen,J.D. (1976) Molec.gen. Genet. 149, 279-289.

59. Blumenthal,R.M., Lemaux,P.G., Neidhardt,F.C. and Dennis, P.P. (1976) Molec.gen.Genet. 149, 291-296.

60. Furano,A.V. and Wittel,F.P. (1976) Cell 8, 115-122.

61. Linn,T. and Scaife,J. (1978) Nature 276, 33-37.

62. Yamamoto,M. and Nomura,M. (1978) Proc.Natl.Acad.Sci.USA 75, 3891-3895.

63. Fiil,N.P., Bendiak,D., Collins,J., and Friesen,J.D. (1979) Molec.gen.Genet. 173, 39-50.

64. Pedersen,S., Bloch,P.L., Reeh,S., and Neidhardt,F.C. (1978) Cell 14, 179-190.

65. Maher,D.L. and Dennis,P.P. (1977) Molec.gen.Genet. 155, 203-211.

66. Lavallé,R. and DeHauwer,G. (1968) J.Mol.Biol. 37, 269-288.

67. Winslow,R.M. (1971) J.Biol.Chem. 246, 4872-4877.

68. Alföldi,L., Stent,G.S., Hoogs,M., and Hill,R. (1963) Z.Vererbungslehre 94, 285.

69. Fiil,N. and Friesen,J.D. (1968) J.Bacteriol. 95, 729-731.

70. Hall,B. and Gallant,J. (1972) Nature 237, 131-135.

71. Hardy,S.J.S. (1975) Molec.gen.Genet. 140, 253-274.

72. Thammana,P., Weber,H.J., Deusser,E., Maschler,R., Kurland,C.G., Stöffler,G., and Wittmann,H.G. (1973) Nature 242, 47-49.

73. Ramagopal,S. and Subramanian,A.R. (1974) Proc.Natl.Acad. Sci.USA 71, 2136-2140.

74. Dennis,P. . (1974) J.Mol.Biol. 88, 25-41.

75. Watson,R.J., Parker,J., Fiil,N.P., Flaks,J.G., and Friesen,J.D. (1975) Proc.Natl.Acad.Sci.USA 72, 2765-2769.

76. Bachmann,B.J., Low,K.B., and Taylor,A.L. (1976) Bact. Reviews 40, 116-167.

77. Isono,K. (1978) Molec.gen.Genet. 165, 265-268.

78. Isono,K. and Kitakawa,M. (1978) Proc.Natl.Acad.Sci.USA 75, 6163-

79 Takata,R. (1978) Molec.gen.Genet. 160, 151-155.

80. Collins,J. and Pritchard,R.H. (1973) J.Mol.Biol. 78, 143-155.

81. Schleif,R. (1967) J.Mol.Biol. 27, 41-55.

82. Dennis,P.D., and Young,R.F. (1975) J.Bacteriol. 121, 994-999.

83. Zaritsky,A. and von Meyenburg,K. (1974) Molec.gen.Genet. 129, 217-227.

84. Geyl,D. and Böck,A. (1977) Molec.gen.Genet. 154, 327-334.

85. Olsson,M.O. (1979) Molec.gen.Genet. 169, 259-269.

86. Capaldo,F.N., Ramsey,G., and Barbour,S.D. (1974) J. Bacteriol. 188, 242-249.

87. Ikemura,T. and Nomura,M. (1977) Cell 11, 779-793.

88. Hayward,R.S. and Fyfe,S.K. (1978) Molec.gen.Genet. 160, 77-80.

BIOSYNTHETIC REGULATION OF *rplJ, rplL, rpoB* AND *rpoC* IN *ESCHERICHIA COLI*

J. D. FRIESEN,* N. P. FIIL,** P. P. DENNIS,***
WILLA LEE DOWNING,*** G. AN*, and E. HOLOWACHUK,*

*Department of Biology,
York University,
Downsview, Ontario, Canada, M3J IP3
**Department of Microbiology & Molecular Genetics,
Harvard Medical School,
Boston, MA 02115
***Department of Biochemistry,
University of British Columbia,
Vancouver, B.C., Canada V6T IW5

I. INTRODUCTION

Until recently, approaches to studying the biosynthetic regulation of ribosome components have been primarily physiological and biochemical, perturbing the normal pattern of bacterial cellular metabolism by a variety of means and then measuring the response of selected components of cellular metabolism (1,2). These studies and others of cells in balanced growth have provided a framework of ideas upon which to build an understanding of ribosome biosynthesis . A deeper understanding depends upon a knowledge of the genetics of the system and this aspect of ribosome studies has accelerated rapidly in recent years. From the initial identification of structural genes for ribosomal and associated components (3), along with putative regulatory genes (4,5), through the isolation of specialized transducing phages (3,6-11) to their cloning on lambda phage or plasmids (12-18), these genes have been studied in ever greater detail. Moreover, genetic and biochemical studies have begun to sketch the shape of the genetic architecture with respect to clustering of genes and their division into transcriptional units (5,19-21). Despite these advances, the biosynthetic regulatory mechanisms remain obscure. Certain general regulatory paradigms have been well documented, such as growth rate control (1), and the stringent response (22-24), but details are lacking. With the advent of modern cloning and DNA manipulation techniques one can now begin to dissect the chromosomal regions which exert control of ribosomal genes. Some of these approaches are described here.

Although at least a dozen-odd regions on the Escherichia coli chromosome are known to carry structural genes for components of the transcription-translation apparatus, the work reported below will focus primarily on one of these: a cluster of genes in the so-called rif region, 88 min on the E. coli linkage map.

The importance of this region was first established when it was discovered that the mutation bestowing resistance to the drug rifampicin lay in the structural gene for the beta subunit of RNA polymerase, rpoB, (25). Subsequently it was found that the structural gene for ribosomal protein (r-protein) L12 (rplL) mapped close by (26) and that relC, which was shown to be the structural gene for L11 (rplK), was also in the immediate vicinity (27, 28). Detailed genetic studies became possible with the isolation of lambda $drif^{d}18$ (6), which has become the source of DNA for almost all subsequent detailed genetic studies. Analysis of the gene products encoded in this bacteriophage revealed that in addition to those mentioned above, the following structural genes were also present: beta-prime subunit of RNA polymerase (rpoC), L10 (rplJ), L1 (rplA), EF-Tu (tufB), rRNA (rrnB) and four transfer RNAs (20,26,29).

Using gene cloning and DNA sequencing, it has become possible to determine the likely transcriptional units in the rif region and their position relative to one another (15,20,21,30). With this ground-work one can now devise methods of focusing more directly on the control regions of this area of the chromosome; some of these methods are described here. Although we are primarily concerned with the genes of the rif region, it will be obvious that many of the methods described below are generally applicable.

II. GENETIC MAP OF THE rif REGION

A map of six genes in the rif region (rplK, rplA, rplJ, rplL, rpoB and rpoC) is shown in Fig. 1. This is a composite map derived from the work of several laboratories (15,20,21,30). The most likely position of promoters is indicated. At the end of this paper we shall elaborate on this picture . In the lower portion of Fig. 1 are drawn the structure of several plasmids that carry portions of the region, derived from restriction endonuclease digestion of lambda $drif^{d}18$ and cloned on pBR322 or ColE1. These plasmids form part of a set that has been used in studies to establish the operon structure of this region (21) and will be referred to below.

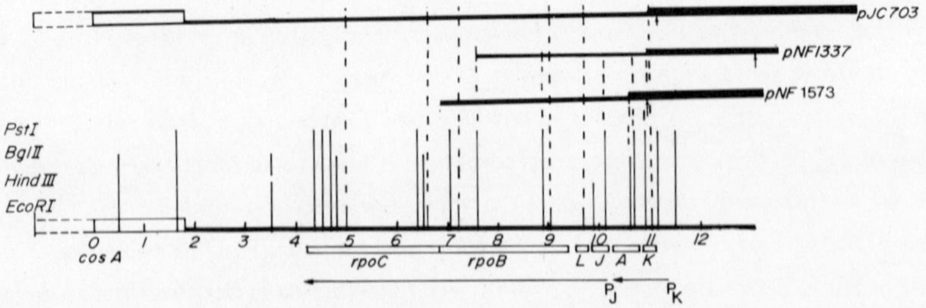

Figure 1. Map of a portion of lambda drif^{d}18. Lambda DNA is shown as thin-lined rectangles, E. coli DNA as lines of intermediate thickness and plasmid DNA as thick lines. The vertical solid lines indicate restriction endonuclease recognition sites. The vertical dashed lines represent BglII, EcoRI or PstI sites extended vertically so as to intersect with the three plasmids drawn on the upper portion of the diagram. pJC703 was constructed by cloning a partial EcoRI digest of lambda drif^{d}18 on ColEI; pNF1337 and pNF1573 were constructed by cloning partial PstI or BglII digests, respectively, on pBR322. In pJC703 the small vertical arrow represents the SmaI site; in pNF1337 and pNF1573 the small vertical arrows represent the EcoRI site. DNA sizes are given in mega-daltons (Md). P_J and P_K are promoters for rplJ and rplK, respectively. The diagram is adapted from reference 21.

III. POST-TRANSCRIPTIONAL REGULATION OF rplJ, rplL, rpoD AND rpoC

In our experience, as well as that of other researchers (31), it has generally proven difficult to clone r-protein structural genes on high copy-number plasmids such as pBR322. This is because a high gene dose, and thus presumably a high production of r-protein, is lethal to the cell for unknown reasons. One might then ask why it is possible to clone the four r-protein structural genes of the rif region. A plausible answer lies in post-transcriptional regulation of the expression of at least some of these genes, as is demonstrated in the following experiments.

We have measured the frequency of transcription of rplJ plus rplL, on the one hand, and rpoB plus rpoC on the other, in a bacterial strain carrying the composite plasmid pJC703 (18) and also in the parental, non-plasmid strain. As a control, transcription of chromosomal r-protein genes that are located in the spc region at 72 min was also examined. The results are presented in Table I and can be summarized as follows: (i) The transcription of r-protein genes in the spc region (i.e., those not present on the plasmid) is not significantly altered by the presence of pJC703; (ii) in the strain carrying pJC703 transcription frequency of rplJ and rplL is enhanced about five-fold compared either to the r-protein genes of the spc region in the strain carrying pJC703 or to rplJ and rplL in the non-plasmid control strain; (iii) transcription frequency of rpoB and rpoC in the strain carrying pJC703 is also enhanced about five-fold; (iv) transcription of rpoB and rpoC genes in the non-plasmid control strain is only about one-fifth as frequent as transcription of the proximal rplJ and rplL (compare values 0.2 and 1.1). This last observation implies the existence of a transcription attenuator located near, or within, the intergenic region between rplL and rpoB. Even in the strain carrying pJC703,this attenuator terminates about 80% of the transcripts that read through the rplJ and rplL. Hybridizations carried out with long-term labelled RNA showed that even in this case there was about a three-and-one-half-fold increase in the amount of transcription from rplJ, rplL, rpoB and rpoC (P. P. Dennis and N. Fiil, Journal of Biological Chemistry, in press). We take this to indicate that the decay rate of these mRNA species is not greatly altered.

Table 1. Transcription and translation of rplJ, rplL, rpoB and rpoC in a strain carrying pJC703

	Relative transcription frequencies[1]			Relative synthesis rates[2]	
	lambda spc1 mRNA	rplJ-rplL mRNA	rpoB-rplC mRNA	L7/L12	beta, beta-prime
Non-plasmid control	1.0	1.1	0.2	1.0	1.0
Strain with pJC703	0.8	5.5	1.1	1.2	2.3

[1]Relative transcription frequencies were determined by hybridizing pulse-labelled total cellular RNA to specific DNA probes that allow the detection of mRNA from either the entire operon (rplJ-rplL-rpoB-rpoC; pJC703, ref. 18) or of rpoB-rpoC alone (pJC720, ref. 18) (P. P. Dennis and N. P. Fiil, Journal of Biological Chemistry, in press). To obtain the frequency of transcription of the specific sequences represented by the probes, the fraction of radioactivity hybridized was corrected for the size of the respective probes. The values are normalized to 1.0 for the r-protein genes located in the spc region.

[2]Relative synthesis rates of the proteins were determined by measuring the fraction of a pulse label (2 min pulse; 2 min chase) that is associated with the protein following separtion by gel electrphoresis (44). The synthesis rates are normalized to 1.0 in the non-plasmid control strain.

We have also made estimates of the synthesis rates of the proteins translated from these mRNAs by pulse-labelling with radioactive amino acids and then separating the respective proteins by gel electrophoresis (44). In the case of L7/L12 protein, the presence of pJC703 stimulated the synthesis rate by only about 20% compared to the parental control (Table 1), whereas beta and beta-prime synthesis was stimulated about 2.3-fold in the same strain. These values are considerably less than the 5-6 fold elevation in the transcription of the same genes (see above). This indicates a mechanism that restricts translation of the mRNA derived from the rplJ-rplL-rpoB-rpoC transcription unit. This mechanism might be of importance in maintaining the stoichiometric balance of ribosome components despite the severe perturbation caused by the high gene dosage.

IV. FEEDBACK REGULATION OF rplJ

In a general way, the data presented above indicate the presence of post-transcriptional regulation of <u>rplJ</u> and <u>rplL</u>. This might be through degradation of excess mRNA or excess r-protein (32), or through control of translation efficiency. A clue to one possible mechanism for this comes from experiments with one of the plasmids, pNF1337 (21), whose structure is shown in Fig. 2(a). Attempts were made to derive deletion subsets from pNF1337 by partial digestion of the plasmid DNA with <u>Eco</u>RI, followed by ligation and transformation. Of some 45 independently- covalently closed circular (CCC) DNA isolated from cells carrying both constructed plasmids, none was observed to have the structure we call 1337 "cut-back", which is diagrammed in Fig. 2(b). This appeared to indicate that a cell harboring such a plasmid was incapable of survival. We speculated that this might be due to the disruption of a natural regulatory circuit in which L7/L12 feeds back on the transcription and/or translation of <u>rplJ</u>. The consequence of a high copy-number plasmid, such as pNF1337, possessing only a partial <u>rplL</u> gene

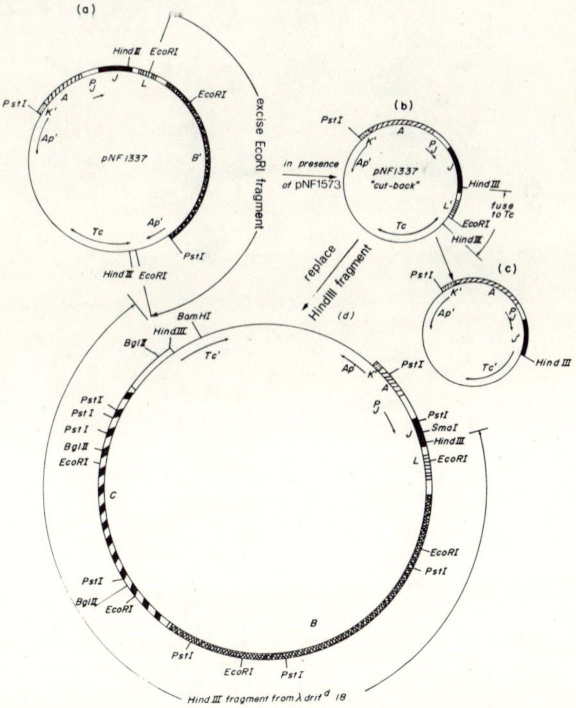

Figure 2. The structure of pNF1337, pNF1337 "cut-back" and its derivatives.The construction of pNF1337 is described in reference 21 and in the legend to Fig.1. The construction of pNF1337 "cut-back" and its derivatives are described in the text.

might be to over-burden the cell with L10 protein and thus, perhaps by the feed-back of excess L10 on the chromosomal operon, inhibit cellular growth. We reasoned that cells carrying pNF1337 "cut-back" ought to be stable if there were also present a plasmid capable of donating intact L7/L12 in order to restore the normal regulation of rplJ. Such a donor plasmid is pNF1573 (see Fig. 1). This expectation was borne out; in the presence of pNF1573 it was possible to obtain plasmids of the 1337 "cut-back" type constructed in vitro.

Covalently closed circular (CCC) DNA isolated from cell carrying both pNF1573 and pNF1337 "cut-back" was used to transform E. coli, selecting for either tetracycline resistance or ampicillin resistance. Many very small colonies appeared on the agar, with approximately 5% large ones. The very small colonies proved to be tetracycline-resistant, ampicillin-sensitive and thus carried only 1337 "cut-back", while most of the large colonies were either tetracycline-sensitive, ampicillin-resistant, indicating the presence of pNF1573 alone, or tetracycline-resistant, ampicillin-sensitive, indicating the presence of both plasmids. However, among approximately 10,000 total colonies there were six large, tetracylcine-resistant, ampicillin-sensitive ones. Plasmid DNA derived

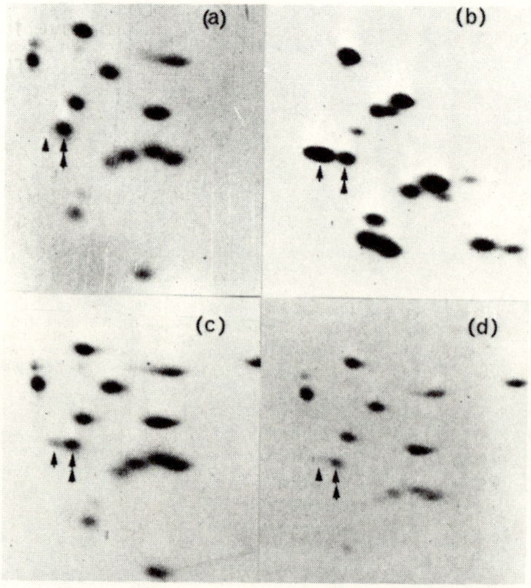

Figure 3. Failure of pNF1337 "cut-back" mutants to express rplJ. Strain NF1643 carries a chromosomal mutation in rplJ (electrophoretically altered L10). The plasmids carry wild-type rplJ (normal L10). Ribosomes from strain NF1643 carrying pNF1337 or pNF1337 "cut-back" mutants (pNF1664 and pNF1665) were extracted with acetic acid (32) and analyzed by two-dimensional polyacrylamide gel electrophoresis (45). The single arrow points to the position of wild-type L10; the double arrow points to the position of mutant L10. (a), NF1643, rplJ; (b), NF1643/pNF1337; (c), NF1643/pNF1664; (d), NF1643/pNF1665. Only the upper, right-hand portion of the conventional gel pattern (44) is shown.

from these proved to have the identical structure to pNF1337 "cut-back" and furthermore bestowed tetracycline resistance on host cells transformed with them. That is to say, these six derivatives of pNF1337 "cut-back" had the characteristics of point mutants capable of overcoming the detrimental effects of the parent plasmid. These six mutant plasmids are called pNF1661, pNF1662, pNF1663,pNF1664, pNF1665 and pNF1666 .

In order to ascertain whether rplJ is expressed from the mutant plasmids they were transformed in strain NF1643 , which carries a chromosomal mutation in rplJ, whose gene product is an electrophoretically altered L10 (21; K. Isono, personal communication). Expression of rplJ from the plasmid thus can be detected by the presence of a double L10 spot on two-dimensional gel electrophorograms (44), the normal protein expressed from the plasmid-borne gene and the altered protein from the chromosomal gene (31). These experiments showed that all six mutants were greatly reduced in the expression of L10 from the plasmid. Fig. 3 shows results for two of the mutants , pNF1664 and pNF1665. By comparing the intensity of altered and normal L10 spots in Fig. 3(b) (pNF1337) with those in Fig. 3(c) (pNF1664) or Fig. 3(d) (pNF1665) it can be seen that the latter two plasmids are much less active in expressing rplJ than is the parent plasmid, pNF1337.

We next attempted to determine whether the six mutant plasmids have retained the capacity for transcription from P_J. For this we made use of the ability of foreign promoters to activate the tetracycline (Tc) gene of pBR322 and other plasmids (see below, Section V). In plasmids of the pNF1337 "cut-back" type P_J, the promoter in question, can be spliced to the Tc gene (whose natural promoter is thereby removed) by excising the HindIII fragment, as is shown in Fig. 2(c). Only if transcription from P_J be possible will these plasmids bestow tetracycline-resistance. When this experiment was carried out all six mutant plasmids bestowed drug resistance at a level of 20ug/ml (data not shown), indicating a functional P_J. In preliminary experiments, hybridization of uniformly-labelled RNA, using lambda 21-10 (see below, Section VI) as a probe, indicated similar mRNA production from the rplJ-rplL region in all mutant plasmids (data not shown).

Measurements of L7/L12 synthesis were made in cells carrying in vitro-modified plasmids of the type shown in Fig. 2(d). These were formed by replacing the HindIII fragment of the mutant plasmids with the indicated HindIII fragment from lambda $drif^{d}18$, carrying the entire rplL, rpoB and rpoC genes. Plasmids thus constructed were transformed in strain JF1039 carrying a mutant rplL gene (21,28). In a manner similar to that described above for expression of rplJ one can

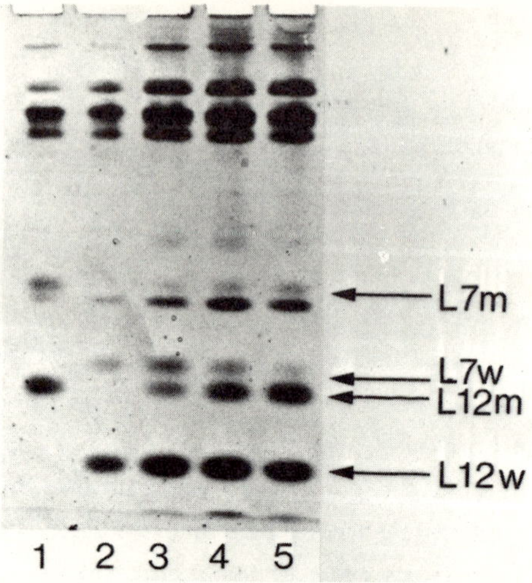

Figure 4. Expression of rplL from three pNF1337 "cut-back" mutants, pNF1662, pNF1664 and pNF1665 that had been modfied by replacing the HindIII fragment , as indicated in Fig. 2(d) .The plasmids were transformed in strain JF1039, which carries a chromosomal rplL mutation resulting in an electrophoretically altered L7/L12. The plasmids carry wild-type rplL. Ribosomes from these transformed strains were purified, the proteins extracted with acetic acid and then analyzed by the method of Li and Subramanian (33). The position of wild-type L7 and L12 is marked with a "w" and of mutant L7 and L12 with an "m". Lane 1, JF1039 rplL; lane 2, JF1039/pNF1337; lane 3, JF1039/pNF1665; lane 4,JF1039/pNF1664; lane 5, JF1039/pNF1662.

detect normal L7/L12 (plasmid gene) and altered L7/L12 (chromosomal gene) expression in electrophorograms of radio-actively labelled cell extracts. These data are shown in Fig. 4 and indicate that wild-type L7/L12,and thus the plasmid-borne rplL gene is expressed from three of the mutant plasmids tested thus far, pNF1662, pNF1664 and pNF1665.

Thus in summary, the mutant plasmids studied here have the ability to be tolerated in the absence of excess L7/L12 contributed by a donor plasmid , show reduced expression of rplJ and are probably unaffected in transcription of rplJ-rplL. In addition at least three of them show substantial expression of rplL.

For an exact localization of the mutations DNA sequencing was carried out by the method of Maxam and Gilbert (45) using appropriate fragments of plasmids pNF1661 through pNF1666. These results, shown in Fig. 5, indicate three distinct sites for single-base alterations leading to the phenotype observed in these plasmids. Two of the changes are at bases 1515 and 1516 , three are at bases 1590, 1594 and 1599 and one is at base 1640. All three regions are well- removed and down-stream from the proposed site of P_J (30). Positions 1590,1594 and 1640 lie in opposite arms of the stem-loop structure defined by the region of divergent symmetry (30), postion 1599 is at the base of the loop itself, while positions 1515 and 1516 lie centrally in the region between rplA and rplJ. Without a clearer picture of the molecular mechanism involved in the regulation of L10 production one can only speculate on the nature of these mutations. In view of the data presented earlier, it seems reasonable to guess that these are sites that govern translation frequency of rplJ.

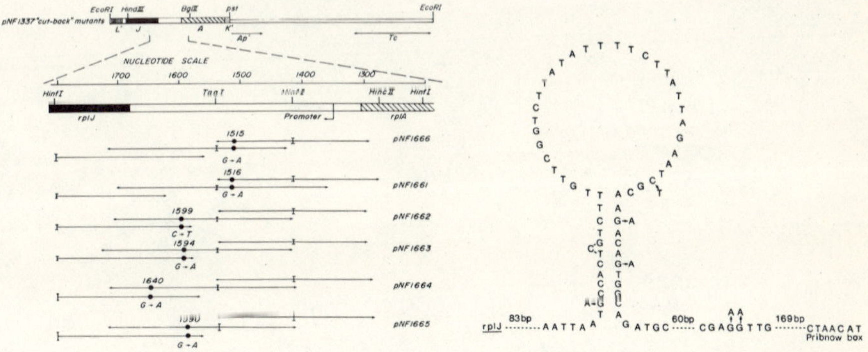

Figure 5. Base changes in the pNF1337 "cut-back" mutants. The structure of pNF1337 "cut-back" is shown in linear form with an expansion of the inter-genic region between rplA and rplJ. The position of the HinfI (1206, 1416, 1805), TaqI (1540), and HincII (1273) restriction sites used in the DNA sequence analysis are indicated. The lower left-hand portion of the figure illustrates the positions of the base changes in the mutants. The positions of 5' end-labelling with polynucleotide kinase (⌶), the extent and direction of sequence (→), and the positions of nucleotide substitutions (•) in the respective mutant plasmids are noted The positions of the base subsitution mutations were confirmed by sequencing both strands of the DNA. The right-hand portion of the figure shows the base change positions in the stem-loop structure and adjacent region.

V. PLASMID VEHICLES FOR THE DIRECT CLONING OF PROMOTERS

The availability of vehicles for the direct cloning of promoter regions that govern r-protein genes would greatly facilitate genetic and physiological studies of their biosynthetic regulation. To that end we have developed two cloning vehicles, one a plasmid and the other a lambda bacteriophage, which fulfill this function.

The plasmid promoter-cloning vehicles, whose construction is described in detail elsewhere (G. An and J.D. Friesen, in preparation), are shown in Fig. 6.

They are closely related to one another and share the characteristic of having a unique restriction endonuclease recognition site(s) near the Tc gene, whose natural promoter has been deleted. Thus, Tc is quiescent unless a DNA fragment carrying a promoter that reads toward this gene is inserted in one of the restriction sites adjacent to it. A number of promoter-carrying fragments have been cloned on members of this plasmid class, including P_{alpha} (the promoter for S4, S11, S13, L17 and alpha subunit of RNA polymerase) and P_{tufB}. However, here we shall concentrate only on those promoters of interest in the rif region, namely P_J and P_K and their associated regulatory regions.

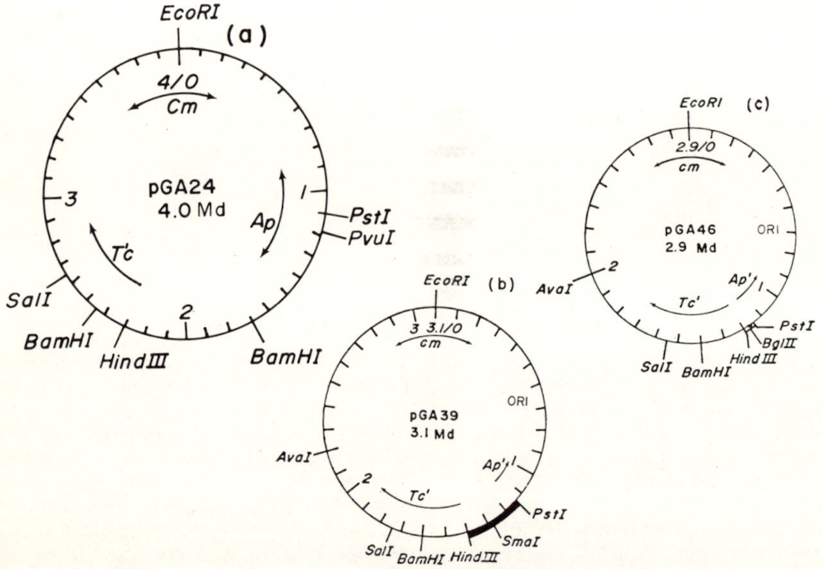

Figure 6. Plasmid promoter-cloning vehicles. This shows the structure of three plasmids that have the structural Tc gene, but lack its promoter. Plasmids are derived from pACYC177 and pACYC184 (34). The method of construction will be described elsewhere (G. An and J.D. Friesen, submitted). (a), pGA24 can accept a HindIII fragment; (b), pGA39 can accept HindIII, PstI or blunt-ended fragments; (c), pGA46 can accept HindIII, BglII, BamHI or PstI fragments. Plasmids are drawn in proportion to size.

Each of the cloning vehicles, pGA24 (Fig.6a), pGA39 (Fig. 6b) and pGA46 (Fig.6c), bestows resistance to only very low concentrations of tetracycline (less than 1 ug/ml). A promoter cloned on any of these plasmids can be recognized because it bestows resistance to a higher concentration of the drug. Thus, a HindIII fragment carrying P_J, derived from lambda drif^{d}18 via lambda 21-10 (see below, Section VI), was cloned in pGA24. This plasmid, pGA26 (Fig. 7a), bestowed resistance to 20 ug/ml tetracycline. A PstI fragment, also carrying P_J, when cloned (pGA63; Fig.7b) bestowed resistance to 20 ug/ml tetracycline; P_K, when cloned (pGA66; Fig.7e) bestowed resistance to 32 ug/ml tetracycline. A fourth plasmid, pGA43 (Fig. 7c), was derived from pGA26 by cloning in that plasmid the large HindIII fragment encompassing rplL, rpoB and rpoC. In pGA43 the entire transcriptional unit that is under the control of P_J is cloned intact. This plasmid bestows no resistance to tetracycline because read-through from P_J is prevented by the natural transcription termination signal at the end of rpoC.

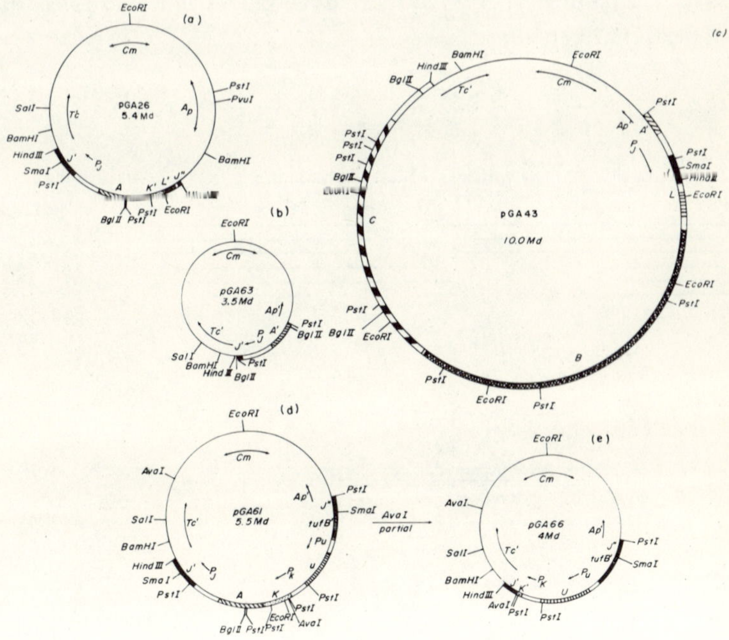

Figure 7. Plasmids in which Tc is activated by P_J or P_K. Details of construction will be provided elsewhere (G. An and J.D. Friesen, submitted). (a), pGA26 is based on the cloning vehicle pGA24; (b), pGA63 is based on pGA46; (c), pGA44 is based on pGA24 and several subsequent modifications; (e), pGA66 is based on the cloning vehicle pGA39, via pGA61 (d). Plasmids are drawn in proportion to size. A prime added to a gene name (eg, J') indicates a partial deletion of the gene.

The four plasmids described above were transformed in the "maxi-cell" strain (35). Radioactive cell extracts from these were analyzed by one-dimensional polyacrylamide gel electrophoresis (36). As expected from the known structure of the plasmids (see Fig. 7), pGA26 expressed only the amino-terminal portion of L10, pGA66 and pGA63 expressed no detectable fragment of L10, while pGA43 expressed L10 and L7/L12, as well as beta and beta-prime subunits of RNA polymerase (data not shown).

In order to assess the fidelity of expression from these cloned promoters two of the plasmids described above, pGA26 and pGA43, were subjected to in vitro micro-surgery by partial digestion with several restriction endonucleases. These deletion plasmids were transformed in a suitable host strain and their ability to bestow resistance to tetracycline was determined. In Fig.8(a) it can be seen that excision from pGA26 of the PstI fragment carrying P_J almost completely abolishes expression of Tc. Excision of any other fragment does not affect expression. The positioning of P_J in this manner agrees with that established by other methods (15,20,21,30).

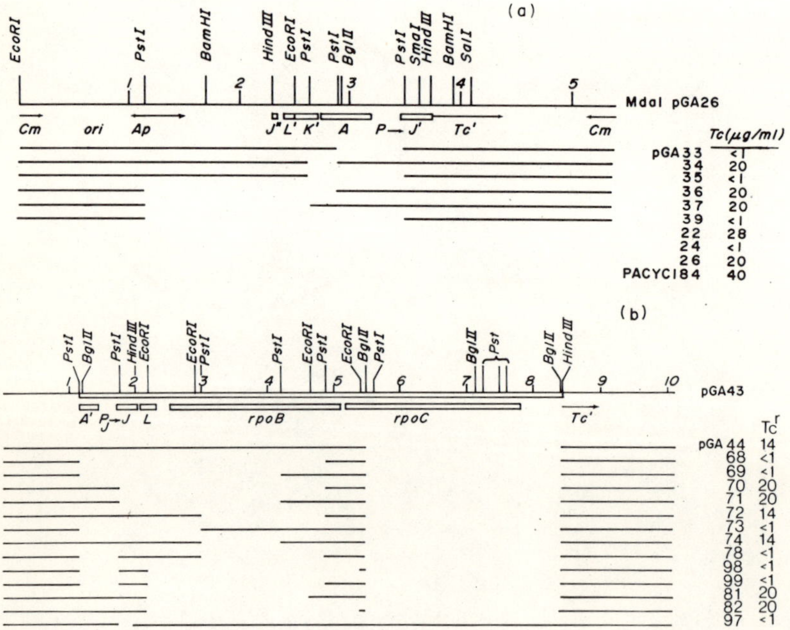

Figure 8. Deletion mapping of promoter activity. pGA26 (a) or pGA43 (b) were subjected to partial digestion with either PstI or BglII, then ligation, transformation and screening for plasmid structure. The appropriate deletion plasmids were transformed in E. coli and the transformed colonies were streaked on L-broth agar with tetracycline concentrations ranging from 1 ug/ml to 40 ug/ml. The maximum drug concentration allowing growth of bacteria carrying the various plasmids is indicated in the tables to the right of the structural diagrams. Bars indicate the portion of the parent plasmid remaining after the partial digestion. Open spaces represent deleted portions of the plasmids.

Fig.8(b) shows the characteristics of pGA43, in which the entire cluster of four genes, rplJ, rplI, rpoB, rpoC is cloned adjacent to Tc. pGA43 bestows no resistance to tetracycline. However, as is shown in the Fig. 8 (b), deletion of the left-most two BglII fragments activates Tc by removing the natural transcription stop signal at the terminus of rpoC. As was the case for pGA26, removing the PstI fragment carrying P_J, but none other, destroys expression of Tc. This appears to indicate that the only promoter that governs this group of four genes is P_J. However, further evidence to be presented below (Section VI) suggests the presence of a second promoter, too weak, apparently, to activate Tc, lying between rplL and rpoB. A comparison of the relative abilities of pGA71 and pGA72 to bestow tetracycline resistance is interesting. These two plasmids are very similar and both carry P_J. However, they differ in that the latter plasmid lacks two inter-genic regions, one between rplJ and rplL and the other between rplL and rpoB; either of these regions might carry an attenuator of transcription (see Section III) . This, apparently, is reflected in the observation that pGA72 bestows only one-half the tetracycline resistance as does pGA71.

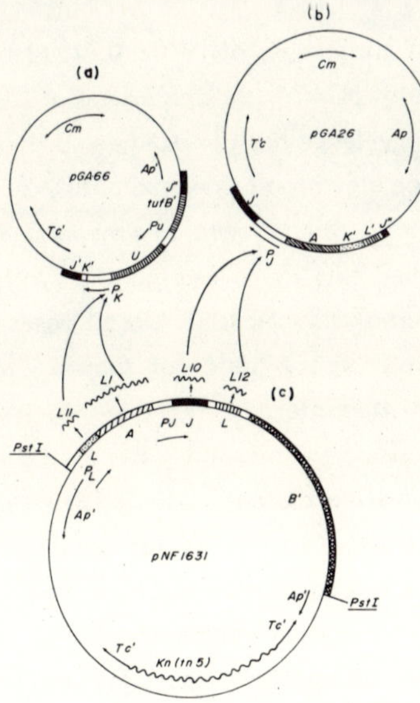

Resident plasmid	Donor plasmid	Tetra-cycline resist-ance ug/ml
pGA26	none	20
pGA26	pACYA177	20
pGA26	pGA45	14
pGA26	pNF1631	12
pGA63	none	20
pGA63	pACYC177	20
pGA63	pGA45	12
pGA63	pNF1631	12
pGA66	none	32
pGA66	pACYC177	32
pGA66	pNF1631	14

Figure 9. Diagrammatic representation of transcriptional feedback. rec^- host cells were transformed with two sets of plasmids: (1) pGA66 (Fig. 9a) plus pNF1631 (Fig. 9c); (2) pGA26 (Fig. 9b) plus pNF1631 (Fig. 9c). These transformed cells were streaked on L-broth agar containing various concentrations of tetracycline to determine the maximum drug concentration that permitted growth (recorded in the table on the right-hand side of Fig. 9). L11, L1, L10, and L12 proteins are shown; the arrows pointing from these to P_K or P_J indicate possible feedback effects.

We have used pGA26 and pGA66 to test the notion that there may exist feedback loop(s) that auto-regulate transcription in either of the operons we have been considering. For these experiments an E. coli host strain was doubly-transformed with two plasmids. In one case pGA66 (Fig. 9a), in which Tc is activated by P_K, was present together with pNF1631, a high copy-number plasmid capable of donating L11, L1, L10 and L7/L12 (Fig. 9c). In the second case the same donor plasmid was present with pGA26 (Fig. 9b), in which Tc was activated by P_J .The maximum tetracycline concentration that allowed growth of colonies on L-broth agar was determined for host cells carrying pGA26 with or without pNF1631, and for host cells carrying pGA66 with or without pNF1631 (Fig. 9, table, right-hand side). In both cases the activity of Tc was reduced approximately two-fold in the presence of the donor plasmid , pNF1631. Experiments with pGA63 (see Fig. 7b) as resident plasmid and also with pGA45 (see Fig. 8b) yielded similar results (Fig.9, table, right-hand side). Control experiments with pACYC 177 as "donor" showed no effect. Furthermore, plasmid copy-number measurements (data not shown) indicated no decrease in copy number of any of the resident plasmids as a result of the presence of the donor plasmids.

Our interpretation of these results is that the excess of L10 or L7/L12 feeds back on the transcription from P_J in the case of pGA26, and the excess of L11 or L1 feeds back on P_K in the case of pGA66. Evidence has already been presented earlier (Section III) to indicate that there is an elevation of approximately 20% in the rate of protein expression from rplJ and rplL when cloned on a medium copy-number plasmid (pJC703). We have confirmed this by measuring the synthesis rate of L1, L10, L11 and L7/L12 in cells harboring pNF1631. In all cases this was 20 to 30% higher than in the non-plasmid parent (data not shown). More evidence for this feedback effect will be presented later.

Two classes of mutants have been isolated using the plasmid system. The first consists of two independent tetracycline-sensitive mutants selected from cells carrying pGA44. One of these, pGA44-8 is sensitive to 1 ug/ml tetracycline; the second, pGA44-1, is sensitive to 4 ug/ml of the drug. Both mutations lie on the PstI DNA fragment that carries P_J (see Fig. 7). Six revertants of pGA44-1 have been isolated that restore resistance to 16 ug/ml tetracycline and that apparently reside on the host chromosome.

We have isolated from pGA78 two mutant plasmids that bestow tetracycline resistance. (Recall that pGA78 lacks P_J and thus Tc is not activated; see Fig. 8.) Both mutations lie in the EcoRI-PstI fragment encompassing the region between rplL and rpoB (see Fig. 7) and thus might be mutants that raise the activity of a possible weak promoter lying in that region (see also Section VI).

VI. A LAMBDA PHAGE VEHICLE FOR CLONING PROMOTERS

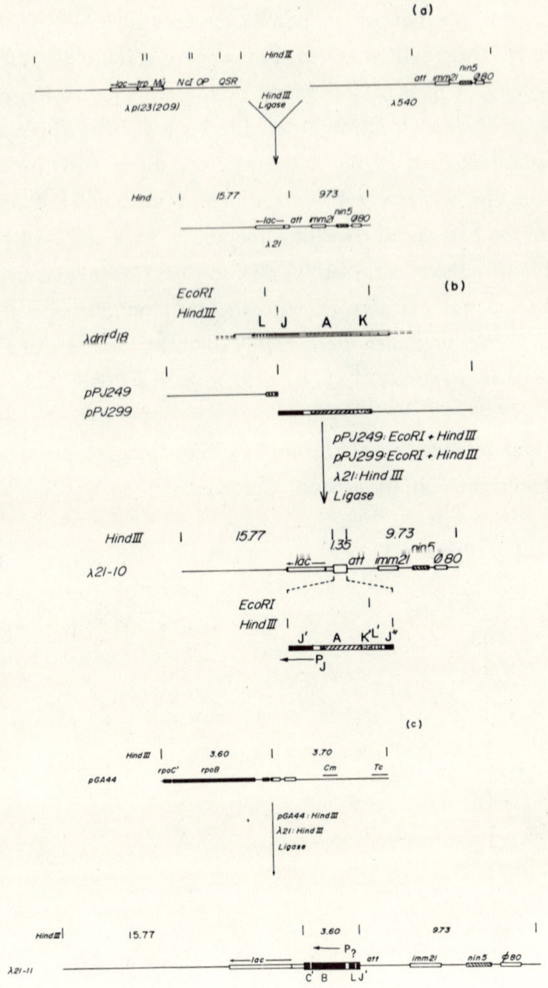

Figure 10. Construction of lambda 21 and two derivatives with cloned promoters.
(a), Lambda 21 contains the left arm of lambda p123(209) (37) and the right arm of lambda 540 (39), generated by cleavage with HindIII.
(b), Construction of lambda 21-10, carrying P_J, the rplJ promoter. The EcoRI and HindIII cleavage sites of the ribosomal protein region of lambda drif18 are shown, together with two plasmids, pPJ249 and pPJ299, which carry the EcoRI-HindIII cleavage products of this region (P. Jorgensen, personal communication). Lambda 21-10 was constructed by ligating the EcoRI-HindIII double-digestion products from the plasmids in the HindIII site of lambda 21. The order of the two cloned fragments is the reverse of their natural one.
(c), Construction of lambda 21-11 carrying $P_{beta'}$, a possible low-level promoter. The HindIII digestion products of pGA44 (see Fig.7b) were ligated in the HindIII site of lambda 21.

Lambda 21, a promoter-cloning vehicle, based on bacteriophage lambda, has been constructed (E. Holowacahuk and J.D Friesen, in preparation). This phage carries the lac operon devoid of its promoter-operator, but with an intact ribosome-binding site. A HindIII site lies approximately 700 base pairs from lacZ near the 5' end of trpB. (trpB is a remnant of the trp - lac fusion of one of the parent phages used in the construction of lambda 21; ref. 37). This bacteriophage can accept a HindIII fragment of up to 6.5 M-dal , forms light blue plaques on Xgal (38) and is lac$^-$ in lysogens. The structure of lambda 21 is shown in Fig. 10(a). Also shown are two derivatives of lambda 21; lambda 21-10 (Fig.10b) has cloned on it a DNA fragment carrying P_J plus the first 53 codons of rplJ , while lambda 21-11 (Fig.10c) has a fragment with the terminal 20 codons of rplJ plus rplL, rpoB and rpoC. We have also cloned the promoter for the alpha operon, for rrnB and for rrnC on lambda 21 (E. Holowachuk and J.D. Friesen, in preparation).

Lambda 21-10 and lambda 21-11 form dark blue plaques on a lac deletion indicator strain when plated on Xgal agar. Colonies of their lysogens are also

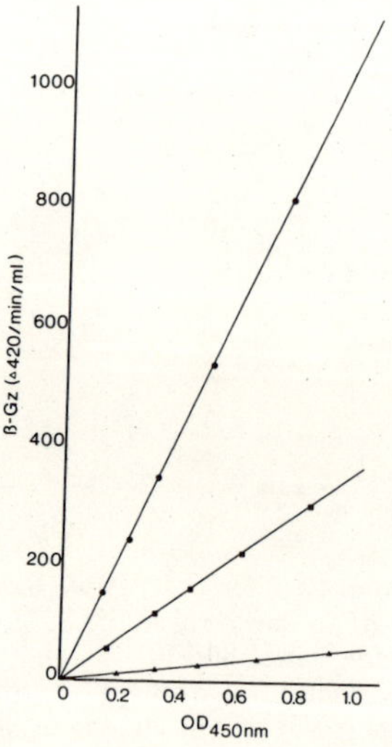

Figure 11. Relative rates of beta-galactosidase synthesis in strain JF1708 lysogenized with lambda plac5 (top curve), lambda 21-10 (middle curve), and lambda 21-11 (bottom curve) . Lysogens were incubated in M9-glucose medium. For determination of beta-galactosidase activity, samples were taken at OD_{450} of 0.15 to 0.9, and assayed as described by Miller (38). Beta-galactosidase synthesis is expressed as (delta 420/min/ml) and is plotted as a function of the optical density of the cultures.

blue on X-gal agar, indicating that a phage promoter, such as P_L, is not the site of transcription initiation. The amount of beta-galactosidase synthesized in lysogens of these phages is shown in Fig. 11. Lambda 21-10 directs the synthesis of about 20% of the amount of beta-galactosidase compared to lambda p<u>lac</u>5 in glucose-minimal growth medium. In a rich growth medium the ratio is reversed (data not shown). Lambda 21-11 directs the synthesis of only 10% of the amount of beta-galactosidase compared to lambda 21-10, which appears to reflect the relative activities of P_J and the promoter that lies between <u>rplL</u> and <u>rpoB</u>.

The proteins encoded in lambda 21-10 and lambda 21-11 were determined by infecting mini-cells (40) carrying a plasmid that expresses <u>imm</u> 21 region. (Lambda 21 and its derivatives carry <u>imm</u> 21.) Thus all phage-specific transcription is repressed. As is to be expected from its structure, only two major bands are expressed from lambda 21-10 (Fig. 12, lane 3);one of these is beta-galactosidase and the other, 51K Md is probably a fusion protein formed

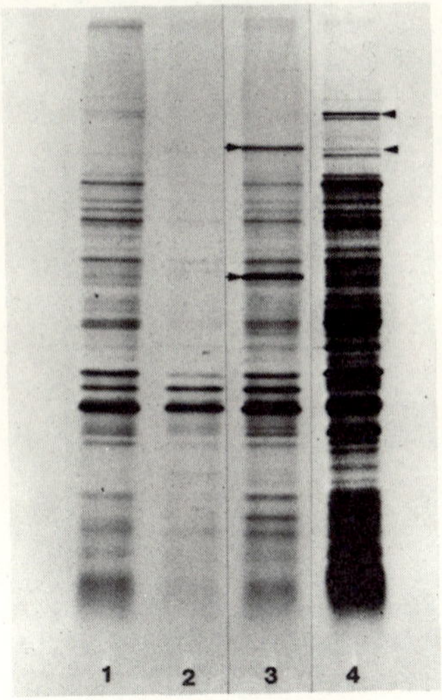

Figure 12. Electrophoretic analysis of proteins synthesized in phage-infected mini-cells. Mini-cells were prepared as described (13), infected with phage and radioactively labelled with ^{35}S-methionine for 20 min. Mini-cell extracts were prepared by boiling in SDS and were analyzed on one-dimensional, 10-20% polyacrylamide gradient gels (36). Lane 1, uninfected mini-cells. Lane 2 ,lambda 21 infection. Lane 3, lambda 21-10 infection; from top to bottom the arrows point to beta-galactsosidase and the L10-trpB fusion protein discussed in the text. Lane 4, infection with lambda 21-11; from top to bottom the arrows point to RNA polymerase beta sub-unit and beta-galactosidase. Lane 4 was exposed longer than the others in order to bring out the weak beta-galactosidase band.

between the amino-terminal portion of L10 and the carboxy-terminal portion of tryptophan synthetase beta protein. As can be predicted from its characteristics, lambda 21-11 directs the synthesis of a small amount of beta-galactosidase, and the beta subunit of RNA polymerase (Fig. 12, lane 4).

Earlier we have described experiments using a plasmid, pNF1631, capable of donating excess L10 and L7/L12. While expression of rplJ and rplL from such a donor plasmid is closely regulated, nevertheless there is a 20 to 30%-fold excess in the amount of L10 and L7/L12 in cells carrying pNF1631 (see section V). Presumably most of this excess is unbound to ribosomes and thus contributes to the free r-protein pool. In order to assess the effect of excess L10 and/or L7/L12 on the expression of beta-galactosidase when under the control of P_J, lysogens of lambda 21-10 were transformed using several different donor plasmids and then the amount of beta-galactosidase synthesized was measured (Fig. 13). These results are summarized in Table 2.

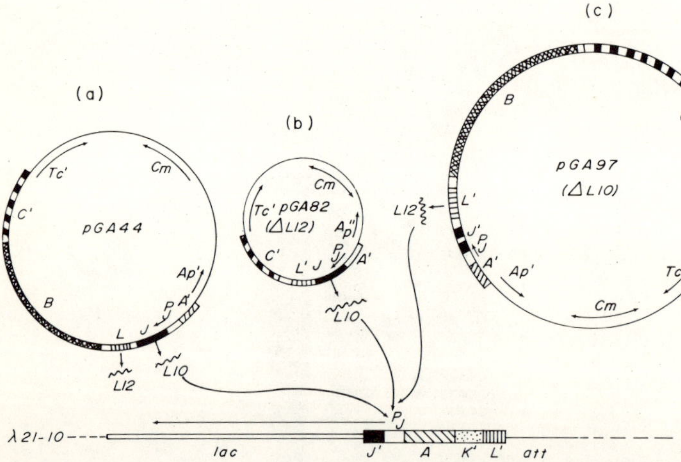

Figure 13. Schematic diagram of feedback regulation of P_J,cloned on lambda 21-10. A lysogen of lambda 21-10 was separately transformed with (a) pGA44, (b) pGA82 or (c) pGA97. The synthesis rate of beta-galactosidase in these cultures is reported in Table 2.

The presence of pNF1337 (high copy-number; see Fig.1) or pGA44 (low copy-number; Fig.7c), both capable of directing the synthesis of L10 and L7/L12, reduces the yield of beta-galactosidase by two- to three-fold (Table 2). This is in general agreement with the results reported above (Section V) when Tc was activated by P_J. Further experiments indicated that L10 alone (pGA82, rplL deletion; Fig.13b) can produce a similar degree of feedback on P_J as do pGA44

TABLE 2. Repression of lacZ under promoter P_J control by ribosomal proteins L10 and L7/L12

Prophage	Plasmid	Plasmid donor property	Doubling time(min)[a]	beta-Gz[b]	Ratio
21-10	none	none	65	475	1.00
21-10	pBR322	none	68	493	1.04
21-10	pACYC177	none	63	476	1.00
21-10	pNF1337	L10,L7/L12	65	176	0.37
21-10	pGA44	L10,L7/L12	60	203	0.44
21-10	pGA82	L10	90	160	0.34
21-10	pGA97	L7/L12	60	390	0.82
21-10	pGA61	L1,L11	63	480	1.00

[a] Cultures were grown at 37° on M9 medium containing 0.4% glucose and the appropriate drug for maintaining the donor plasmid.

[b] Beta-galasctosidase activity was determined essentially as described by Miller(38).

of pNF1337. L7/L12 alone (pGA97, rplJ deletion; Fig. 13c) has a significant, but smaller feedback effect. Plasmids with the same origin of replication as the donor plasmids, but lacking rplJ and rplL showed no effect . These results suggest a feedback regulation site near P_J that has an affinity for L10. pGA61 (Fig.7d). A plasmid capable of donating only L1 and L11, had no effect on expression from P_J (Table 2, last line).

Lambda 21-11 is about 10% as active in expressing beta-galactosidase as is lambda 21-10 (see Fig.11) This suggests the possibility of a low-level promoter to the left of rplJ, most likely between rplL and rpoB. This site is also subject to at least two control mechanisms. The first can be demonstrated by addition of the drug rifampicin (200 ug/ml) to a lysogen of lambda 21-11 in exponential growth. The rate of beta-galactosidase synthesis transiently rises two-fold (data not shown). This is reminiscent of an earlier observation (41) and might be explained as a stimulation of rpoB in response to inactivation of RNA polymerase by rifampicin. The second controlling element on lambda 21-11 is a rho-dependent one. It can be demonstrated by the observation that a lysogen of lambda 21-11 in a rho mutant (46) synthesizes ten-fold more beta-galactosidase than a rho^+ lysogen (data not shown). The position of both of the above sites is unknown at this time.

Two classes of mutants have been selected from the bacteriophages described in this section. A lysogen of lambda 21-10 transformed with pNF1337 in the cell synthesizes some three-fold less beta-galactosidase than in its absence (see

Table 2); this lysogen does not synthesize enough lactose permease (lacY gene product) to grow at 42°on melibiose as sole carbon source (38). Mutants that can grow under these conditions occur on the phage with a spontaneous frequency of 10^{-5}. The amount of beta-galactosidase synthesized by these mutants is two-fold higher than in the parental lambda 21-10; these mutations might be in the site at which P_J is feedback regulated.

The second class of spontaneous mutants is derived from lysogens of lambda 21-11, which, because of the low expression of the lac operon (Fig. 11), cannot grow at 42°on melibiose . These fall into three classes with four-, seven- and thirty-fold increase in beta-galactosidase activity. These mutants might represent up-promoter or down-attenuator mutants at a site preceding rplL and rpoB.

VII. DISCUSSION

The experimental results presented here are focused on regulation of the expression of six genes that code for elements of the transcription-translation apparatus in E. coli. These genes are rplL-rplA in one operon and rplJ-rplK-rpoB-rpoC in a second. Studies with cloned portions of these operons have yielded results that begin to hint at the shape of some of the regulatory circuits in these operons. We shall try to bring together the results of these studies, although this must be regarded as only a first approximation, since there are many gaps in the data.

(i) Perhaps the most striking observation to come from our experiments is the evidence for feedback regulatory loops, both translational and transcriptional. The first clue to this came from experiments (Section III) with r-protein and RNA polymerase genes cloned on the medium copy-number plasmid pJC703. The ten-fold increase in gene copy number was reflected in a five-fold increase in mRNA transcribed from these genes, but a barely observable increase in the amount of L7/L12 and L10. The increase in RNA polymerase subunits beta and beta-prime was greater (approximately two-fold), but still significantly less than the increase in transcription.

In attempting to explain these observations the possibility of proteolytic degradation of excess r-protein cannot be ruled out (31,32), but the protein synthesis measurements reported in Table I were made over short time periods and ought not to reflect degredation. We favour the conclusion that a significant portion of the post-transcriptional regulation is at the level of translation.

(ii) A possible mechanism for this translational control is suggested by experiments with rplJ and rplL cloned on a high copy-number plasmid, pNF1337. These indicated that L12 (or L7/L12) might regulate the expresssion of rplJ; a high rplJ gene copy number in the absence of a correspondingly high rplL copy number might be detrimental to the cell because the feedback of L7/L12 on rplJ expression is impaired. At least three mutants that overcome this disadvantage do not express rplJ, but do express rplL.These data suggest translational regulation of rplJ, although it should be noted that if one assumes the existence of a promoter immediately preceding rplL, one could explain the characteristics of the mutants on the basis of transcriptional alteration. For a variety of reasons (15,20,21,30) a promoter at this position seems unlikely.

Base sequence analysis of the mutants revealed clustering in two regions , both well removed from P_J. Perhaps the more interesting of these is the set of four single-base alterations that lie in a possible stem-loop structure some 120 base-pairs upstream from the 5' end of rplJ. One can speculate that this structure is the L7/L12 recognition site on the mRNA. Perhaps translation of the rplJ portion of the mRNA is blocked by binding free L7/L12 to this region. When the cytoplasmic concentration of L7/L12 is reduced the mRNA block is removed and rplJ translation can proceed. The mutations might block rplJ translation, even in the absence of L7/L12. Differential translation has been proposed as a regulatory mechanism in certain single-stranded RNA bacteriophages (42). The stem-loop structure is also reminiscent of transcription termination regions in phage lambda and in the trp operon (43,46). Although our data indicate that the mutants have altered translational regulation, the possibility of alteration to transcriptional regulation by a changed rho-specific termination site cannot be rigrously ruled out.

(iii) The vehicles that we have constructed for the direct cloning of promoter-bearing fragments will probably prove valuable in genetic studies of promoter function. Indeed, we have already isolated several classes of interesting mutants using these vehicles. For the present discussion the importance of these vehicles is in the demonstration that P_J or P_K cloned on them, and thereby activating either Tc or lacZ, are subject to auto-repression (Sections V and VI). The repression can be as much as three-fold. This effect is probably at the transcriptional level, since there is no reason to suspect that either Tc or lacZ mRNA possesses a binding site for r-proteins. L. Lindahl (personal communication) has also observed a transcriptional feedback, not unlike that proposed here, in another r-protein operon. Perhaps the effect is an important general one in setting the expression level of r-protein genes.

(iv) The hybridization data (Section III) , the characteristics of pGA71 and pGA72 (Section V) and the observation that expression of beta-galactosidase from lambda 21-11 is rho-influenced (Section VI) all suggest a transcription attenuator between rplL and rpoB. At present we know nothing of the details of this site, but hope that analysis of strongly-expressing mutants of pGA78 and lambda 21-11 will shed light on this.

(v) Expression of beta-galactosidase from lambda 21-11 is low, but detectable (Section VI), suggesting the existence of a weak promoter, probably between rplL and rpoB. This might account for the weak expression of rif[d] from certain fragments cloned on plasmids (21). It should be noted, however, that we have no way of knowing if this weak promoter has a real physiological significance in vivo.

In Fig. 14 we present a summary diagram of our findings. This scheme must be regarded as only an approximation to what is obviously a complex genetic control circuit. One wonders how ubiquitous are these kinds of regulatory loops. It is intriguing to speculate on a general regulatory model in which each r-protein operon is feedback-regulated by one of the gene products encoded in it and the cytoplasmic concentration of certain key non-ribosome-bound r-proteins (at least one from each operon) sets the over-all level of r-protein gene expression. The cytoplasmic r-protein concentration, in turn, might be determined by the binding of r-proteins to a key molecule, for example rRNA.

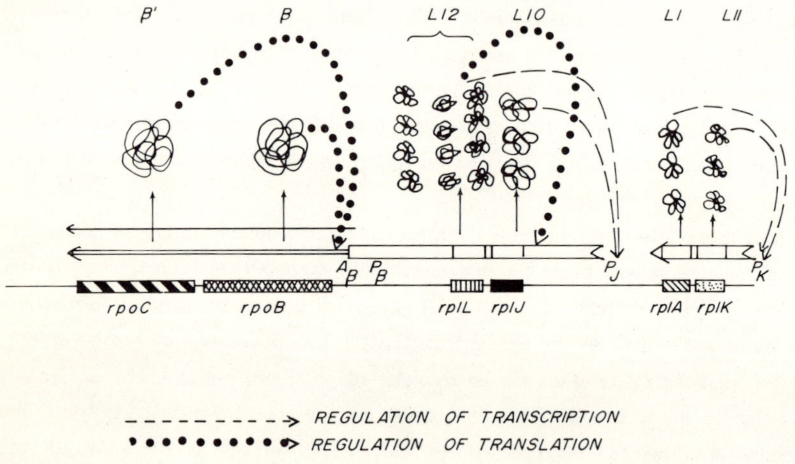

Figure 14. A highly speculative diagram of some of the possible regulatory circuits affecting P_J and P_K. Refer to text for details.

VIII. ACKNOWLEDGEMENTS

We thank Joan Sheehy and Krys Tarkowski for editorial and typographical assistance, Saleha Lakadawala for drawing and Gordon Temple for photography. This work was supported by grants from the Medical Research Council (PPD) and the National Institutes of Health (PPD), the National Cancer Institute (JDF) and the Natural Science and Engineering Reasearch Council (JDF), The Danish Research Council(NPF) and the National Insititutes of Health (Grant No. GM13017 to Jon Beckwith).

IX. REFERENCES

1. Kjeldgaard, N. O., and Gausing, K. (1974) In Ribosomes (M. Nomura, A. Tissieres, P. Lengyel, eds.) Cold Spring Harbor Laboratory, Cold Spring Harbor, N.Y., pp. 369-392.

2. Maaløe, O., and Kjeldgaard, N. O. (1966) Control of Macromolecular Synthesis. W. A. Benjamin, N.Y.

3. Nomura, M. (1976) Cell 9, 633-644.

4. Stent, G. S., and Brenner, S. (1961) Proc. Natl. Acad. Sci. USA 47, 2005-2009.

5. Laffler, T., and Gallant, J. (1974) Cell 1, 27-30.

6. Kirschbaum, J. B., and Konrad, E. B. (1973) J. Bacteriol. 116, 517-526.

7. Friesen, J. D., Parker, J., and Watson, R. J. (1976) Mol. Gen. Genet. 144, 115-118.

8. Friesen, J. D., Parker, J., Watson, R. J., Fiil, N. P., Pedersen, S., and Pedersen, F. S. (1976) J. Bacteriol. 127, 917-922.

9. Friesen, J. D., Parker, J., Watson, R. J., Bendiak, D., Reeh, S. V., and Pedersen, S. (1976) Mol. Gen. Genet. 148, 93-98.

10. Saito, H., Nakamura, Y., and Uchida, H. (1978) Mol. Gen. Genet. 165, 247-256.

11. Jørgensen, P., Fiil, N. P. (1976) In Alfred Benzon Symposium IX: Control of Ribosome Synthesis (N. O. Kjeldgaard and O. Maaløe, eds.) Copenhagen, Munksgaard, pp. 370-382.

12. Friesen, J. D., An, G. and Fiil, N. P. (1978) Cell 15, 1187-1197.

13. An, G. Justesen, J., Watson, R. J., and Friesen, J. D. (1979) J. Bacteriol. 137, 1100-1110.

14. Kenerley, M. E., Morgan, E. A., Post, L., Lindahl, L., and Nomura, M. (1977) J. Bacteriol. 132, 931-949.

15. Linn, T., and Scaife, J. (1978) Nature (London) 276, 33-37.

16. Kiss, A., Sain, B., Kiss, I., Boros, I., Udvardy, A., and Venetianer, P. (1978) Gene 4, 137-152.

17. Mackie, G. A. (1979) Gene 5, 45-58.

18. Collins, J., Fiil, N., Jørgensen, P., and Friesen, J. D. (1976) In Alfred Benzon Symposium IX: Control of Ribosome Synthesis (N. O. Kjeldgaard and O. Maaløe, eds.) Copenhagen, Munksgaard, pp. 357-369.

19. Jaskunas, S. R., Fallon, A. M., and Nomura, M. (1977) J. Biol. Chem. 252, 7323-7336.

20. Yamamoto, M., and Nomura, M. (1979) J. Bacteriol. 137, 584-594.

21. Fiil, N. P., Bendiak, D. S., Collins, J., and Friesen, J. D. (1979) Mol. Gen. Genet., in press.

22. Cashel, M., and Gallant, J. (1969) Nature (London) 221, 838-841.

23. Fiil, N. P., Meyenburg, K., and Friesen, J. D. (1977) Mol. Gen. Genet. 150, 87-101.

24. Dennis, P. P., and Nomura, M. (1975) Nature (London) 225, 460-465.

25. Boyd, D. H., Zillig, W., and Scaife, J. (1974) Mol. Gen. Genet. 130, 315.

26. Friesen, J. D., Fiil, N. P., Parker, J., and Haseltine, W. A. (1974) Proc. Natl. Acad. Sci. USA 71, 3465-3469.

27. Parker, J., Watson, R. J., Friesen, J. D., and Fiil, N. P. (1976) Mol. Gen. Genet. 144, 111-114.

28. Watson, R. J., Parker, J., Fiil, N. P., Flaks, J. G., and Friesen, J. D. (1975) Proc. Natl. Acad. Sci. USA 72, 2765-2769.

29. Jaskunas, S. R., Lindahl, L., Nomura, M., and Burgess, R. R. (1975) Nature (London) 257, 458-462.

30. Post, L. E., Strycharz, G. D., Nomura, M., Lewis, H., and Dennis P. P. (1979) Proc. Natl. Acad. Sci. USA 76, 1697-1701.

31. Fallon, A. M., Jinks, C. S., Yamamoto, M., and Nomura, M. (1979) J. Bacteriol. 138, 383-396.

32. Dennis, P. (1974) J. Mol. Biol. 88, 25-41.

33. Li, K., and Subramanian, A. (1975) Anal. Biochem. 64, 121-129.

34. Chang, A. C. Y., and Cohen, S. N. (1978) J. Bacteriol. 134, 1141-1156.

35. Sancar, A., Hack, A. M., and Rupp, W. D. (1979) J. Bacteriol. 137, 692-693.

36. Laemmli, U. K. (1970) Nature (London) 227, 680-685.

37. Casadaban, M. J. (1976) J. Mol. Biol. 104, 541-555.

38. Miller, J. H. (1972) Cold Spring Harbor Laboratory, Cold Spring Harbor, N.Y.

39. Borck, K., Beggs, J. D., Brammar, W. J., Hopkins, A. S., and Murray, N. F. (1976) Mol. Gen. Genet. 146, 199-207.

40. Roozen, K. J., Fenwick, R. G., and Curtiss III, R. (1971) J. Bacteriol. 107, 21-33.

41. Hayward, R., Tittawella, I., and Scaife, J. (1973) Nature New Biol. 243, 6-11.

42. Lodish, H. F. (1975) In RNA Phages (N. Zinder, ed.) pp. 301-318.

43. Stauffer, G. V., Zurawaski, G., and Yanofsky, C. (1978) Proc. Natl. Acad. Sci. USA 75, 4833-4837.

44. Kaltschmidt, E., and Wittmann, H. G. (1970) Anal. Biochem. 36, 401-412.

45. Maxam, A. M., and Gilbert, W. (1977) Proc. Natl. Acad. Sci. USA 74, 560-564.

46. Adhya, S., Gottesman, M. (1978) Ann Rev. Biochem. 47, 967-996.

IN VITRO SYNTHESIS AND DECAY OF GUANOSINE 3′,5′-BIS (DIPHOSPHATE) [ppGpp]

DIETMAR RICHTER

Institut für Physiologische Chemie,
Abteilung Zellbiochemie,
Universität Hamburg,
Hamburg, Germany

In auxotrophic strains of *Escherichia coli*, deprivation of an essential amino acid and consequently lack of the corresponding amino acylated tRNA initiates a regulatory process termed stringent response that is governed by the *relA* gene and defined by the curtailment of synthesis of rRNA, tRNA and some mRNA species [for review see Gallant and Lazzarini (1) and Nomura et al. (2)]. Under stringent conditions two unusual nucleotides, guanosine 3',5'-bis(diphosphate) [ppGpp] and guanosine 3'-diphosphate, 5'-triphosphate [pppGpp] are accumulated. It is generally accepted that ppGpp rather than pppGpp serves as physiological effector; it does not only restrict rRNA synthesis but also other metabolic processes such as biosynthesis of purine nucleotides, glycolytic esters, phospholipids or polyamines [for review see Cashel (3)]. In case of the inhibition of rRNA synthesis a direct effect upon RNA polymerase by ppGpp has been found (Nomura, this volume). The precise molecular mechanism of the (p)ppGpp-mediated inhibition of the other biosynthetic reactions mentioned above is totally unknown.

The observation that under conditions of amino acid starvation chloramphenicol inhibits (p)ppGpp production while it favors reinitiation of stable RNA synthesis suggested that the unusual nucleotides might be synthesized on ribosomes (4, 5). Indeed, studies with *in vitro* systems have shown that synthesis of (p)ppGpp is catalyzed by an enzyme which requires the ribosome·mRNA complex and uncharged tRNA as activator. Activation occurs only when uncharged tRNA is bound to the acceptor site of the ribosome [reviewed by Haseltine and Block (6); Lipmann and Sy (7); Richter and Isono (8)]. The enzyme, the *relA* gene product, was originally discovered by

Haseltine et al. and termed stringent factor (9). It is a ATP:GTP(GDP) pyrophosphotransferase which transfers ß, γ-phosphate as a unit from ATP into the 3'-OH group of GDP (10) or GTP (11). It is now well established that some bacterial species such as *Bacillus brevis* (12, 13) and *Bacillus stearothermophilus* (14) contain a (p)ppGpp synthesizing enzyme apparently different from the *E. coli* stringent factor. The striking difference to the *E. coli* enzyme is that the one from Bacillus functions independently of the ribosome·mRNA·tRNA complex. The *B. brevis* enzyme has been named ppGpp synthetase by Sy and Akers (13), while Fehr et al. (14) used the term stringent factor. In an effort to establish a uniform nomenclature for the ATP:GTP(GDP) pyrophosphotransferases, we propose to use the term (p)ppGpp synthetase throughout; the enzyme that depends on the ribosome·mRNA·tRNA complex for activation, i.e. *E. coli* stringent factor, will be designated as (p)ppGpp synthetase I; the other [found in Bacillus] which does not require the complex as (p)ppGpp synthetase II.

The objective in this article is to summarize our present knowledge of the *in vitro* (p)ppGpp cycle, i.e. synthesis and decay of (p)ppGpp. We will limit the discussion to a few published works rather than attempting to be comprehensive in surveying all the published papers. In particular, the report is focussed on the degradation pathway which has been elucidated only recently. For additional details, the reader is referred to earlier reviews on this topic (6-8) as well as to the published proceedings of a recent symposium (15).

SYNTHESIS OF (p)ppGpp

While the enzymatic mechanism of (p)ppGpp synthesis, donation of pyrophosphate from ATP to GTP or GDP, proceeds identically in all bacterial systems assayed so far, the (p)ppGpp synthetases involved differ in their dependence on ribosome ·mRNA·tRNA complex as well as in some of their physicochemical properties. Table 1 shows that so far only in *B. brevis* both, (p)ppGpp synthetase I and II has been found (16); it may well be that the presence of both enzymes in one and the

TABLE 1. (p)ppGpp synthesis in various bacterial strains.

	(p)ppGpp synthetase I	(p)ppGpp synthetase II
E. coli, relA^{+} (6, 9)	+	
B. subtilis (17)	+	
B. stearothermophilus (14, 17)		+
B. brevis (12, 13)	+	+
*Thermus thermophilus**	+	

*Lienert and Richter, unpublished

and the same strain is a more general phenomenon. A ribosome-independent mechanism of (p)ppGpp synthesis may exist in *relA* mutant strains known to accumulate (p)ppGpp and restrict rRNA synthesis according to nitrogen and to carbon downshift conditions but not to amino acid starvation (1, 2). Attempts to isolate either type of enzyme from *relA* mutant strains, however, have been unsuccessful so far. Since *relA*$^{+}$ strains show similar response during carbon source shift-down one should expect that *relA*$^{+}$ strains contain both enzymes, (p)ppGpp synthetase I and II. The (p)ppGpp synthetase isolated from *Streptomyces* (50-53) is not included in this table since this enzyme has a much broader substrate specificity than the other two enzymes.

(p)ppGpp Synthetase I from Escherichia coli

Subcellular distribution. ppGpp synthetase I has been found to be associated with native 70S ribosomes and with native 50S subunits (18). The enzyme is rapidly released into the supernatant fraction when 70S ribosomes are washed with NH_4Cl (9) or when converted into 50S and 30S ribosomal subunits (19). Studies of the subcellular distribution of ppGpp synthetase I show that (p)ppGpp synthesizing activity is associated with the ribosomal and particulate fraction as well (11). Distribution of (p)ppGpp synthesizing activity between the two fractions is dependent on the forces applied to disintegrate the cells (20). Association with the particular fraction may increase from 5% to more than 20% of the total (p)ppGpp synthetase I activity when cells are disrupted by osmotic lysis of spheroplasts. Although the particulate- and ribosome associated enzyme(s) show a similar catalytic mechanism in the biosynthetic reaction, the enzyme isolated from the particulate fraction occurs in molecular weights of 150,000, 300,000 and 450,000; the enzyme obtained from the ribosomal fraction appears in a single form with a molecular weight of 70,000 to 80,000. Although the heavy molecular weight forms are not convertable into one another under the conditions applied, it is likely that the enzyme present in the particulate fraction is primarily associated with membrane bound ribosomes (20) and identical with the enzyme associated with ribosomes of the cytosol.

It has been calculated that only one molecule of (p)ppGpp synthetase I is bound per every 500 ribosomes or that 10 to 20 molecules of enzyme are present per cell (6, 21). This calculation is based on numbers calculated from the ribosomal fraction and not from a crude cell lysate. One might expect a device known to sense any shortage of charged tRNA immediately to be present in a larger quantity. It would certainly be more effective if all ribosomes carried the synthetase. Alternatively, limited amounts of (p)ppGpp synthetase I could exist available to any ribosome, an assumption which seems to be unlikely since, at least *in vitro*, the enzyme is tightly bound to the ribosome. In their studies Schmale et al. (22) estimated the (p)ppGpp synthetase I concentration in crude cell extracts and found that up to 10 to 11 nmol of synthetase I and 12 to 18 nmol of ribosomes exist per gram of cells, which corresponds to 10,000 to 12,000 molecules of enzyme per cell. This number is based on the assumption that, under the experimental conditions used, 1 mg of (p)ppGpp synthetase I synthesizes 4,000 nmol of (p)ppGpp per minute, an estimate also reported by others (21, 23). This finding is in apparent conflict with data reported by Pedersen and Kjeldgaard (24) who use antibodies against purified (p)ppGpp synthetase I. This group estimates the amount of (p)ppGpp synthetase I by a combined method of immune precipitation and functional assay and finds a (p)ppGpp synthetase I : ribosome ratio of 1:200. Also this group reports a 25 to 100 times higher specific activity of the (p)ppGpp synthetase I than found by others (19, 21, 22). No doubt a direct radioimmunoassay using antibodies against (p)ppGpp synthetase I will help to determine the exact (p)ppGpp synthetase I concentration in the cell.

Interaction with ribosomes. Ribosomal subunits (50S and 30S) incubated with (p)ppGpp synthetase I and isolated by centrifugation show optimal binding of the enzyme in the presence of poly(U) as messenger RNA and 5 to 10 mM magnesium acetate (19). 50S and 30S ribosomal subunits alone do not bind the enzyme. The dependence of the complex formation on mRNA may be due to a stabilizing effect on the 70S ribosomes. Co-factors such as uncharged tRNA or nucleotides (ATP, GTP, AMPPNP, GMPPCP) do not enhance binding of the enzyme to ribosomes. NH_4^+ ions inhibit the reaction.

The existence of a ribosomal domain which functions as a common site recognized by the elongation factors has been demonstrated by competition experiments [reviewed by Möller (25)]. Similar experiments carried out with ribosomes pre-charged with unlabeled (p)ppGpp synthetase I show that the complexed ribosomes are still able to bind tritiated elongation factors, EF-Tu or EF-G. Also the complexed ribosomes are fully active in poly(U) directed polyphenylalanine synthesis (Table 2). The data coincide with experiments where ribosomes carrying EF-Tu·GMPPCP·Phe-tRNA bind tritium-labeled (p)ppGpp synthetase I (27). Apparently (p)ppGpp syn-

TABLE 2. Ribosomal bound (p)ppGpp synthetase I does not inhibit the functions of EF-G or EF-Tu.

	70S·poly(U) complex	70S·poly(U)·(p)ppGpp synthetase I complex
1. EF-T + EF-G dependent synthesis of polyphenylalanine in pmol	4.5	4.6
2. pmol of [^{3}H]EF-Tu bound per pmol of ribosomes	0.42	0.8
3. pmol of [^{3}H]EF-G bound per pmol of ribosomes	0.38	0.38

(p)ppGpp synthetase I was bound to ribosomes according to ref. 19. The complex was isolated by centrifugation and analyzed as indicated in the left part of the table [table reproduced, with permission, from Richter et al. (19)].

thetase I and elongation factors interact with ribosomal domain(s) that are not identical. This interpretation is supported by earlier reports that ribosomes depleted of the two proteins L7/L12 still bind and function with (p)ppGpp synthetase I (28, 29). Similar results are reported by Beres and Lucas-Lenard (30). The argument that the enzyme preparations used in these experiments were not pure and may contain sufficient amounts of L7/L12 could be excluded by appropriate control experiments (30). The finding by Block and Haseltine (21) is at variance with the other reports that ribosomes free of L7/L12 can be stimulated up to five-fold by addition of L7/L12 in the (p)ppGpp synthesizing reaction. The conflicting data may be explained by the finding that L10 is mandatory for ribosomal binding of (p)ppGpp synthetase I and subsequently for carrying out the (p)ppGpp synthetase I mediated synthesis of (p)ppGpp (31). Highland and Howard (32) reported earlier that treatment of 50S subunits with ammonium chloride and ethanol at 37° removes proteins L7/L12 and L10; also L10 is essential for *in vitro* incorporation of L7/L12 into ribosomal core particles (32, 33). Thus it may well be that the ribosomal core particles used in the studies by Block and Haseltine (21) also lacked L10.

(p)ppGpp synthetase I from *E. coli* is not species specific and responds to bacterial, mitochondrial, and chloroplast ribosomes (28, 30, 34). No (p)ppGpp synthesis has been found with the *E. coli* enzyme and 80S ribosomes from eukaryotes (28, 30, 34-36). These findings are at variance with a report by Pollard and Packard (37) who claim that *E. coli* (p)ppGpp synthetase I is activated by 80S ribosomes from various eukaryotic organisms. In those experiments (p)ppGpp synthesis does not depend on uncharged tRNA which suggested that the observed effect was due to a non-specific activation of the enzyme probably by basic ribosomal proteins. Indeed, it has been shown that although mouse embryo or rabbit reticulocyte ribosomes fail to activate *E. coli* (p)ppGpp synthetase I total proteins prepared from 80S ribosomes or 60S or 40S ribosomal subunits support (p)ppGpp synthesis in a non-specific reaction (36).

It has been reported earlier that in the absence of the ribosome complex (p)ppGpp synthetase I from *E. coli* may be activated by the addition of methanol (38) or other organic solvents (6). These studies have been extended to ribosomal proteins from *E. coli* (39). TP30 and TP50 as well as some but not all ribosomal proteins are capable of activating (p)ppGpp synthetase I. Interestingly equal amounts of TP60 or TP40 as well as of TP50 or TP30 are required for minimal stimulation when compared on a molar basis; maximal rate of (p)ppGpp synthesis that can be induced by addition of ribosomal proteins is similar with total ribosomal proteins from either class (36). Like their prokaryotic counterpart (39) some but not all ribosomal proteins from rat liver (out of 30 proteins tested; ref. 40) activate (p)ppGpp synthetase I. In general it is found that all the acidic proteins from both classes do not stimulate the (p)ppGpp synthesizing reaction. The most active proteins tend to be more basic. Certain proteins unrelated to ribosomal proteins, such as casein (21) or histones (40) can also stimulate (p)ppGpp synthetase I; however, they are not nearly as effective as TP40 or TP60. Christiansen and Nierhaus (39) speculate that the most stimulatory *E. coli* ribosomal proteins are involved in the binding of (p)ppGpp synthetase I. A similar enzyme has not been found in liver or other eukaryotic ribosomes (35, 37, 41) which raises the question why they should contain a "binding site". It is possible that activation of (p)ppGpp synthetase I by ribosomal proteins per se is non-specific although in general stimulation by ribosomal proteins is significantly greater than by non-ribosomal basic proteins.

Interaction with tRNA. (p)ppGpp synthetase I requires ribosomes and an uncharged tRNA capable of recognizing a codon carried by the message (42, 43). The details of this reaction including the various effects by modified tRNAs or tRNA fragments have been reviewed elsewhere (6-8). Here reports are discussed concerning structural modification of the CCA-end of uncharged tRNA (44). Only tRNA containing the complete CpCpA-end and an intact ribose residue in the terminal adenosine is active in the (p)ppGpp synthesizing process (29, 42, 44, 45). Neither tRNA-CpCpA$_{oxi}$ nor tRNA-

$CpCpA_{oxi-red}$ are active in (p)ppGpp synthesis.
While the absence of the 2'hydroxyl group of the terminal adenosine does not influence the reaction, the 3'-hydroxyl group is essential and cannot be replaced by an amino group or a hydrogen atom. tRNA in which the 3'-terminal adenosine is substituted by formycin is active in induction of (p)ppGpp synthesis but the reaction proceeds relatively slowly. Apparantly the interaction of uncharged tRNA with (p)ppGpp synthetase I and ribosomes is a highly specific process in which a free 3'-hydroxyl group on the terminal nucleotide plays a significant role.

The ribosome (p)ppGpp synthetase I complex not only recognizes the CCA-end and the intact ribose ring but also discriminates between purine and pyrimidine base at the 3'-terminal end of the tRNA molecule (46). With the modified tRNAs, tRNA-CpCpC or tRNA-CpCpCpA, the stimulation of the (p)ppGpp synthesizing reaction is low. A reduced activity with tRNA-CpCpC is also reported by Chinali et al. (45). The authors claim, however, that the lengthened tRNA-CpCpCpA is fully active in the (p)ppGpp synthesizing reaction, although they noticed a slightly reduced rate of the tRNA-CpCpCpA induced reaction at 83 mM ammonium chloride. NH_4^+ ions are supposed to inhibit the effector function of the tRNA rather than its binding to the ribosome·(p)ppGpp synthetase I complex. It has been shown earlier that binding of (p)ppGpp synthetase I is only slightly reduced at higher ammonium chloride concentration provided sufficient magnesium ions are present (19). Chinali et al. (45) further speculate that NH_4Cl serves to potentiate the discriminatory ability of the (p)ppGpp synthetase I.

(p)ppGpp synthetase I is involved in directing preferentially uncharged tRNA to the acceptor site of the ribosome (27). This binding reaction also depends on magnesium ions. Incubation of the 70S·mRNA·tRNA·(p)ppGpp synthetase I complex at suboptimal magnesium concentrations results in the release of tRNA but not of the enzyme from the ribosome. The release of tRNA is stimulated by ATP but not by β,γ-imido adenosine 5'-triphosphate suggesting that the tRNA releasing reaction is coupled to the hydrolysis of ATP. Based on these data, it has been proposed that after each round of ATP hydrolysis the tRNA is rejected by a change in the ribosomal conformation. The cycle restarts when uncharged tRNA is bound to the ribosomal acceptor site.

(p)ppGpp Synthetase I in Other Bacterial Species

E. coli (p)ppGpp synthetase I has been isolated from *Bacillus subtilis* (17, 47), *Bacillus brevis* (12, 13, 16, 47) and *Thermus thermophilus* (Lienert and Richter, unpublished). Like the *E. coli* enzyme the (p)ppGpp synthetase I from these strains shows the same specificities for ATP as pyrophosphate donor and GDP or GTP as acceptor. The enzyme is activated when uncharged tRNA is bound to the ribosome·mRNA complex (Table 3). ppGpp synthetase I from *Thermus thermophilus*, however, preferentially synthesizes pppGpp and little ppGpp, regardless of whether GDP or GTP serves as pyrophosphate acceptor. The crude enzyme preparation apparently contains a highly active nucleoside diphosphate kinase immediately phosphorylating GDP to GTP. The (p)ppGpp synthetase I from *B. brevis* (16) and *B. subtilis* (Richter, unpublished) has an apparent molecular weight of 70,000 to 80,000 which is very close to that reported for the *E. coli* enzyme. Like the *E. coli* enzyme, (p)ppGpp synthetase I from *B. brevis* is inhibited by thiostrepton and tetracycline but not by micococcin (16). The basal rate of the *B. brevis* enzyme is not stimulated by methanol whereas the *E. coli* enzyme is greatly enhanced (38). Also (p)ppGpp synthetase I from *B. brevis* prefers the homologous ribosomes over the *E. coli* ones (16).

(p)ppGpp Synthetase II

So far ppGpp synthetase II has been found in the S-100 supernatant fraction prepared from *B. brevis* (12, 13, 16, 47) and *B. stearothermophilus* (14, 17). Although this enzyme is also a ATP:GTP(GDP) pyrophosphoryl transferase, it is clearly different from (p)ppGpp synthetase I: (p)ppGpp synthetase II functions independently of the ribosome·mRNA·tRNA complex. (p)ppGpp synthetase II from Bacillus strains strictly depends

TABLE 3. Requirements of (p)ppGpp synthesis

(p)ppGpp synthetase I from	components added to the assay mixture		% of (p)ppGpp synthesized
	70S·mRNA complexes	tRNA	
B. subtilis (47)	−	−	0.7
	+	+	49.3
	+	−	2.0
	−	+	0.9
B. brevis (16)	−	−	4.7
	+	−	13.0
	+	+	49.9
*Th. thermophilus**	−	−	1.2
	+	−	2.0
	+	+	41.5

* Lienert and Richter, unpublished. The *B. brevis* data are taken, with permission, from ref. 16.

on ATP as phosphate donor (12-14) and cannot be replaced by AMPPNP; only GDP or GTP serve as good pyrophosphate acceptors (13, 17). Low activity is found with ppppG (13, 17) or ITP (13). For the brevis (p)ppGpp synthetase II magnesium ions are required as counter ions for the nucleotide substrate; excess of magnesium ions is inhibitory. The *B. stearothermophilus* counterpart, however, has a broad magnesium optimum between 3 and 30 mM. The *B. brevis* enzyme is not inhibited by thiostrepton, tetracycline or RNase clearly demonstrating that ribosomes are not essential in this reaction (12). The *B. stearothermophilus* (p)ppGpp synthetase II is significantly inhibited by ppApp or pppApp (17). More than 50% reduction is observed with 0.1 mM ppApp but not with ppGpp, pGpp, pApp, or A(5')P4(5')A (Fig. 1). Maximal inhibition occurs at 0.4 mM. The specificity of this inhibition seems to be rather high in particular because it is observed in the presence of 4 mM ATP. In the ribosome-dependent system from *E. coli* the inhibitory effect of ppApp is significantly less pronounced; at 2 mM ppApp a 50% reduction of (p)ppGpp synthesis in the *E. coli* system is observed. A recent report by Travers (48) suggests that ppApp may counteract ppGpp at the transcriptional level in *E. coli*. In *B. subtilis* ppApp is thought to play a regulatory role in the sporulation process (49). Also, a (p)ppApp synthetase has been reported from a number of *Streptomyces* strains (50-53). All these data suggest that (p)ppApp may be another candidate in regulatory processes.

At present it is not known whether the (p)ppGpp synthetase II is under metabolic control. In *B. stearothermophilus* the *in vivo* mechanism for triggering accumulation of (p)ppGpp can be achieved during carbohydrate starvation or temperature downshift; under these conditions stable RNA synthesis is restricted (Fehr and Richter, unpublished). One possible control mechanism could be the function of a heat-stable low molecular weight component recently isolated from *B. stearothermophilus* (78). This compound inhibits (p)ppGpp synthesis provided ribosomes are also present, suggesting a regulatory mechanism through the interaction with ribosomes.

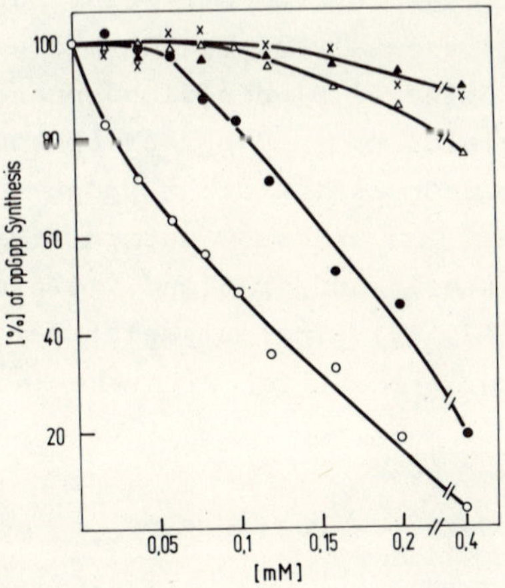

FIGURE 1. Inhibition of ppGpp synthesis by ppApp in the *B. stearothermophilus* system. 100% activity of the *B. stearothermophilus* (p)ppGpp synthetase II corresponds to 4.5 nmol of ppGpp synthesized; the assay reaction was carried out in the non-ribosomal system. Inhibitors: ppApp (o—o); pApp (● ●); pGpp (△—△); ppGpp (▲—▲); A(5')p_4(5')A (x—x) (taken from ref. 17).

DEGRADATION OF (p)ppGpp

In vivo the stringent response is reversed upon readdition of the omitted amino acid causing the rapid degradation of (p)ppGpp (4). Genetic studies reveal that decay of ppGpp is governed by the *spoT* gene. *spoT* mutant strains do not only degrade ppGpp at a significant lower rate than *spoT*⁺ strains, they also lack pppGpp (54-58). Based on these data the hypothesis has been proposed that degradation of ppGpp is preceded by its phosphorylation to ppGpp (54, 55). This hypothesis has been challenged by others postulating that phosphorylation is not a prerequisite of the degrading reaction (58, 59). No doubt these conflicting models which suggested different decay pathways, could only be elucidated by studying the decay reaction in cell-free systems. *In vivo* ppGpp is stable under conditions in which it is rapidly degraded *in vitro* (57, 60-64), which suggests that a cell-free system for studying the decay mechanism requires special conditions. Indeed as now known the ppGpp degrading enzyme is a strictly manganese-requiring enzyme. The mechanism of degradation of ppGpp proceeds at least by two different pathways. A specific ppGpp degradation is catalyzed by the product of the *spoT* gene (65), a guanosine 3',5'-bis(diphosphate) 3'-pyrophosphohydrolase which yields ppG and pyrophosphate (66). This enzyme is referred to as ppGppase. Another ppGpp degrading enzyme(s) has been found in *spoT*⁺ and *spoT* mutant strains which hydrolyzes ppGpp from the 3' and/or 5'end yielding ppGp, pGpp, and pGp as products (47, 67). The enzyme(s) may be regarded as a phosphatase(s) and is identical with the known acid phosphatase of *E. coli* (68).

ppGppase from Escherichia coli

Subcellular distribution and purification steps. ppGppase has been prepared by fractionation of crude cell extracts (69, 70). Most of the ppGppase activity is present in the ribosomal fraction. Contamination of this fraction by membrane vesicles cannot totally be excluded although little or no ppGpp degrading activity is found in the membrane fraction (65).

When crude ribosomes are dissociated at low ionic conditions and analyzed by sucrose density gradient centrifugation ppGppase activity is found in the top fraction of the gradient; significant activity is also associated with the 50S and 30S ribosomal subunits (17; Fig. 2). One explanation of this heterogeneous sedimentation profile of ppGppase may

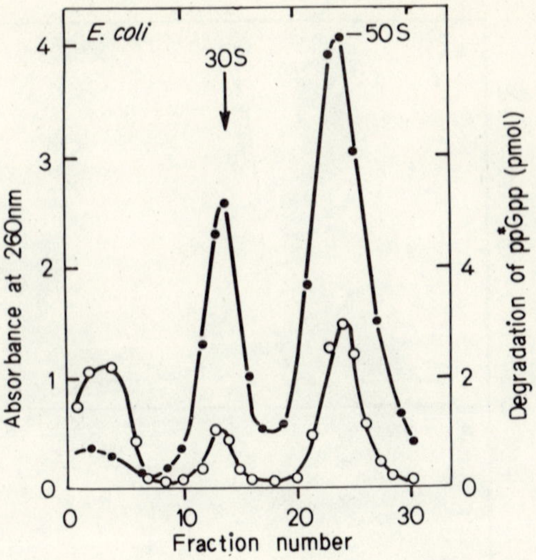

FIGURE 2. Sucrose density gradient centrifugation of the ribosomal fraction from *E. coli*. To release the ppGppase from ribosomes, 50 A_{260} units/ml of ribosomes were dialyzed against 2 l of 40 mM Tris-HCl buffer pH 7.7 containing 25 mM NH_4Cl, 2 mM dithiothreitol, 0.1 mM magnesium acetate, 0.1 mM $MnCl_2$. 1 ml of the dialyzed fraction was applied to a linear 10-30% sucrose (w/v) gradient in the same buffer as used for dialysis. After centrifugation at 20,000 rev./min for 17 h in a Beckman SW27 rotor, fractions of 1.2 ml were collected from the top. 25-µl aliquots of the fractions were assayed (17) using 100 pmol of ppGpp (0.3 Ci/mmol) in a total volume of 30 µl. (●—●) absorbance at 260, (o—o) degradation of ppGpp (taken from ref. 17).

be its properties to aggregate in buffer of low ionic strength; the aggregated material may reach a size large enough to co-migrate with either of the two ribosomal subunits. Another explanation would be that the ppGppase is bound to the interface of the two ribosomal subunits and hence detectable in both the 50S and 30S subunits fractions. Since the molecular weight of ppGppase has been determined as 65,000 (Fig. 3; ref. 17) this could imply that the enzyme is identical with one of the high-molecular weight ribosomal proteins such as S1 (73) or protein A (73, 74). Experiments with antibodies against S1 or protein A, however, show that ppGppase is not identical with either of the two ribosomal proteins (17, 75). Also antibodies against TP50 or TP30 are not inhibitory in the ppGpp degrading reaction (Richter, un-

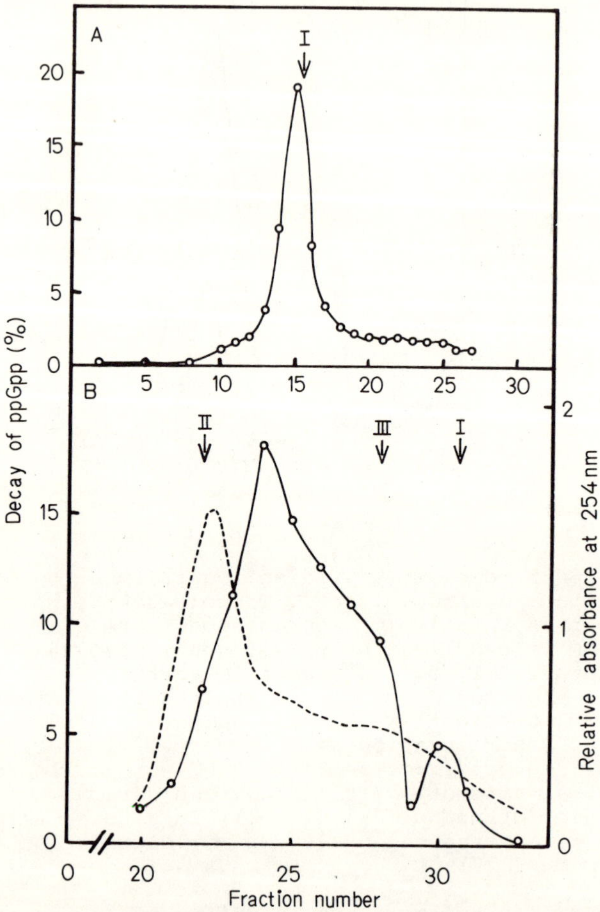

FIGURE 3. Characterization of *E. coli* ppGppase by linear sucrose density gradient centrifugation (A) and Sephadex G-150 column chromatography (B). In (A) 250 µl of ppGppase (4.4 mg/ml) extracted from ribosomes with 1.0 M potassium acetate and fractionated by $(NH_4)_2SO_4$ precipitation were applied to a sucrose density gradient (10-30% sucrose, w/v, in 20 mM Tris-HCl buffer, pH 7.8, 6 mM 2-mercaptoethanol, 750 mM potassium acetate, 2 mM $MnCl_2$) and centrifuged at 49,000 rev./min for 38 h in a Beckman SW65 Ti rotor.
Fractions of 150 µl were collected starting from the top of the gradient. Each fraction was diluted with 1.2 ml of buffer (17). The precipitated ppGppase was analyzed as indicated (17). In (B) a Sephadex G-150 column (1.0x75 cm) was equilibrated with 20 mM Tris-HCl buffer, pH 7.8, 8 mM 2-mercaptoethanol, 2 mM $MnCl_2$, 1 M potassium acetate, and 20% sucrose (w/v). 0.5-1.0 ml of ribosome-free ppGppase (3-6 mg/ml) were subjected to gel filtration and eluted with equilibration buffer. 2-ml fractions were collected: 7.5 µl aliquots were assayed in the ppGpp-degrading system. The following markers were used for calibration: (I) bovine serum albumin; (II) dextran blue; (III) lactate dehydrogenase. (o—o) Decay of ppGpp; (----) relative absorbance at 245 nm (taken from ref.17).

published).

Extraction of ribosomes with 1 M potassium releases the ppGppase (17, 65). Similar to the *E. coli* (p)ppGpp synthetase I, ppGppase forms aggregates under low salt conditions and sediments when centrifuged at low speed (17). In contrast to (p)ppGpp synthetase I, only little aggregated ppGppase can be reactivated by dialysis in buffer at high ionic strength. The aggregates can be resolubilized and most of ppGppase activity is regained only after dialysis against 2.5 M urea followed by a high-salt dialysis (17). The molecular weight of the released ppGppase has been estimated by sucrose density gradient centrifugation and Sephadex G150 gel filtration. With both methods the smallest molecular weight found is a protein of molecular weight 65,000 (17; Fig. 3). The enzyme forms dimers and tetramers with molecular weights of 130,000 and 260,000. Because of the strong tendency to form aggregates, the molecular weight of the monomeric form of the enzyme may be even lower. Attempts to analyze ppGppase in sucrose gradients containing urea have not been successful so far (Harder and Richter, unpublished).

Mechanism and specificity of the ppGpp decay reaction. Reports by Heinemeyer and Richter (69) and Sy (70) demonstrate that α-^{32}P labeled $p\overset{*}{p}Gpp$ is specifically degraded by a ribosomal fraction. In a more detailed study using double-labeled ppGpp as substrate, the mechanism of this decay reaction has been elucidated (Table 4; 66). The molar ratio of ^{32}P- and ^{3}H-labeled ppG is one with $p\overset{*}{p}[^3H]Gpp$ as substrate but not with $pp[^3H]G\overset{*}{p}p$ or

TABLE 4. Decay of double-labeled ppGpp

	products formed in nmol/50-µl assay					
	pppG		ppG		pG	
substrate	^3H	^{32}P	^3H	^{32}P	^3H	^{32}P
$pp[^3H]Gp\overset{*}{p}$	0.65	0.31	3.76	0.07	0.09	0
$pp[^3H]G\overset{*}{p}p$	0.36	0	2.93	0.02	0.06	0
$p\overset{*}{p}[^3H]Gpp$	0.30	0.41	5.56	5.86	0.52	0.55

The asterisk over 'p' indicates the position of the ^{32}P label (taken from ref. 66).

pp[^{3}H]Gp$\overset{*}{p}$ (66). The data strongly suggest that two phosphates have been released from the 3'-position of ppGpp. Furthermore the released phosphates have been identified as pyrophosphate (66). Based on these data the overall decay reaction has been proposed:

$$\text{ppGpp} \xrightarrow{\text{ppGppase}} \text{ppG} + \text{P}{\sim}\text{Pi}$$

The enzyme specifically catalyzing this reaction has been termed as guanosine 3',5'-bis(diphosphate)3'pyrophosphohydrolase [ppGppase].

The significance of the association of the ppGppase with the ribosomal fraction is not yet clearly understood. No doubt ppGppase does not require ribosomes for its activation (69, 70). Ribosome-free ppGppase is neither stimulated nor inhibited by ribosomes, however, its activity is significantly reduced in the presence of uncharged tRNA (17, 47). Ribosomes do not seem to be mandatory in this inhibitory reaction. 18S or 23S rRNA or $\text{tRNA}_{\text{oxi-red}}$ do not interfer with ppGpp decay (Richter, unpublished). Aminoacyl-tRNA inhibits ppGppase to a significantly lesser extent suggesting that uncharged tRNA plays a central role in the ppGpp cycle: On the one hand it activates (p)ppGpp synthetase I while on the other hand it suppresses ppGppase activity.

Although the ppGppase is present in the ribosomal fraction, ribosomes are not essential for ppGpp breakdown (65-71). Neither antibiotics such as chloramphenicol, puromycin or streptomycin nor RNase inhibit the ppGppase (69). On the other hand chlorotetracycline (Richter, unpublished) strongly inhibits ppGpp decay while levallorphan gave a moderate inhibition (69, 70). Only the inhibitory effect caused by tetracycline can be reversed by addition of manganese ions, which indicates that inhibition is due to chelating properties of tetracycline (72) rather than to a ribosome-specific effect. Chlorotetracycline-mediated inhibition cannot be reversed by manganese ions (Richter, unpublished). Other drugs such as oxytetracycline, rifampicin, fusidic acid, kirromycin or morphine do not interfere with the *in vitro* decay reaction (Richter, unpublished).

ppGppase depends on manganese but not on magnesium ions (67). While in the crude cell extracts Mn^{2+} are not strictly required in the decay reaction (65-70). The more purified ppGppase is irreversibly inactivated when purified in the absence of Mn^{2+} (Heinemeyer and Richter, unpublished). Obviously the crude enzyme preparation is still saturated with endogenous manganese ions, which may be lost upon prolonged purification steps. Co^{2+} or Ca^{2+} can partially replace Mn^{2+} but are considerably less effective (17). Potassium or sodium acetate (200-300 mM) stimulates the decay reaction fourfold compared to assay conditions where monovalent cations have been omitted. At these concentrations ammonium acetate stimulates the reaction twofold, while lithium chloride is inhibitory (65). In general acetate salts stimulate better than the respective chloride salts (Harder, unpublished). The pH optimum of the decay reaction is between 7.5 and 8. Kinetic studies revealed a K_m value of 0.55 mM (71) to 0.8 mM (65) for ppGpp.

Nucleotides such as pApp, ppApp or pppGpp do not inhibit the ppGpp degrading reaction at concentrations fivefold in excess over ppGpp. However, pGpp is inhibitory (69).

An et al. (71) use a fragment of a transducing lambda bateriophage carrying the *spoT* gene of *E. coli* for cloning. The clones produce nine time more *spoT* gene activity than a normal *spoT+* strain. The *spoT* gene product, the ppGppase, is isolated by ammonium sulfate extraction (71), a method that yields fractions with relatively little ppGppase activity compared to the method described above (Harder and Richter, unpublished).

The ribosome-associated ppGppase is stimulated in the presence of ATP, UTP and CTP. Analysis of the degradation products revealed that, in the presence of the nucleotides assayed, predominantly pppG is formed. Apparently a nucleoside diphosphate kinase contaminating the ribosomal fraction accelerates the decay reaction by shifting the equilibrium toward the formation of pppG (65, 67), suggesting that the nucleotide(s)-mediated effect is only indirectly correlated with the decay reaction. The preliminary observation that ATP can also stimulate ppGppase free of nucleoside diphosphate kinase activity may also be due to indirect effects (Richter, unpublished). So far there is no good evidence that nucleoside triphosphates directly enhance ppGppase activity which one should expect from *in vivo* data where several groups have shown that breakdown of ppGpp is coupled to an energy-requiring process (54, 57, 56, 58). The observation that pyrophosphate released from ppGpp by ppGppase is exchanged in the presence of ATP may also not be directly coupled to the ppGpp decay reaction (65, 66). This ATP-P~Pi exchange reaction is almost negligible when more purified ppGppase preparations are used (Richter, unpublished).

When isogenic *spoT* strains are compared, only *spoT*+ strains contain ppGppase (65). The enzyme also seems to degrade pppGpp. $\overset{*}{p}$pp[^{3}H]Gpp substrate is immediately converted into pp[^{3}H]Gpp and inorganic phosphate. Apparently the ppGppase containing fraction is contaminated by enzymes which can hydrolyse the ß-bond of nucleoside triphosphates. It has been shown before that pppGpp can be utilized by factors involved in protein synthesis (76). Unspecific hydrolysis of pppGpp can be partially reduced in the presence of excess GTP (65); under these conditions $\overset{*}{p}$ppGpp is degraded into $\overset{*}{p}$ppG. Although ppGppase converts ppGpp to pppG and pyrophosphate at conditions where unspecific GTPases are saturated with GTP, only purified ppGppase will yield precise information whether ppGpp and pppGpp are hydrolyzed with the same K_m value. A more specific enzyme has been reported recently; it only converts pppGpp to ppGpp. ppGpp is not further degraded (77).

ppGppase from Bacillus strains

In vitro degradation of ppGpp is also found in cell-free extracts prepared from *B. brevis* and *B. subtilis* (47, 75). In general the mechanism of degradation of ppGpp is found to be identical with the one reported for *E. coli*. Again ppGppase is associated with the ribosomal fraction from which it can be partially released by dissociation of ribosomes into subunits. The *B. subtilis* enzyme tends to comigrate with the 30S ribosomal subunit. The molecular weight of the ppGppase is determined to be 65,000 to 70,000.

Decay of ppGpp by Phosphatases

Another ppGpp degrading enzyme has been found in *Escherichia coli* which apparently is not identical with the *spoT* gene product and which also has been found in *spoT* mutant strains (47, 67). The enzyme(s) hydrolyzes ppGpp from the 3' and/or 5' yielding ppGp, pGpp and pGp as products; it may be regarded as phosphatase and identical with the known acid phosphatase of *E. coli* (68).

CONCLUSIONS

The metabolic pathway of the ppGpp cycle in relA^{+} strains of *E. coli* may be summarized as follows:

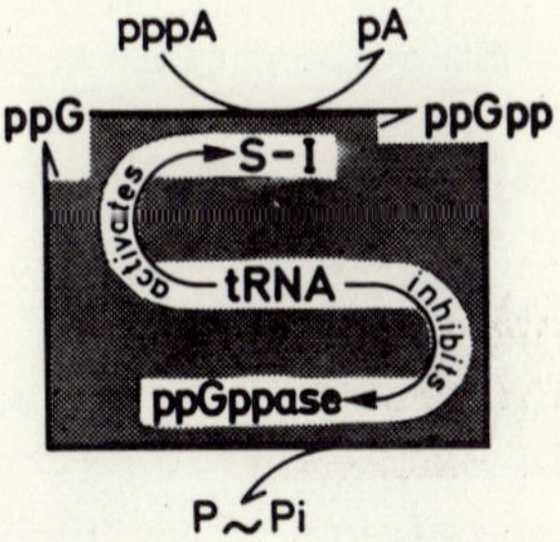

The net reaction requires two energy-rich bonds. (p)ppGpp synthetase I (S-I) is activated by uncharged tRNA bound to ribosomes, whereas ppGppase is inhibited by this process. It permits an effective control of the ppGpp level in the cell and explains the metabolic lability of the effector.

The second enzyme that synthesizes guanosine polyphosphates, (p)ppGpp synthetase II is not activated by the ribosome·mRNA·tRNA complex. The accumulation of (p)ppGpp in *relA* mutant strains during nutrition shift-down experiments may be due to (p)ppGpp synthetase II activity although this enzyme has not been found there yet. Also the metabolic control mechanism of (p)ppGpp synthetase II remains to be elucidated.

ACKNOWLEDGMENT

The author thanks S. Fehr and E. A. Heinemeyer who were involved in some of the projects reported here, and H. Kalthoff for help in editing the manuscript.

REFERENCES

1. Gallant, J., and Lazzarini, R. (1976) Protein Synthesis, A Series of Advances (ed. McConkey, E.H.), pp 309-359, Marcell Dekker, New York.

2. Nomura, M., Morgan, E.A., and Jaskunase, S.R. (1977) Ann. Rev. Genet. 11, 297-347.

3. Cashel, M. (1975) Ann. Rev. Microbiol. 29, 301-318 (ed. Starr, M.P., Ingraham, J.L., Raffel, S.), Ann. Rev. Inc., Palo Alto, Cal., USA.

4. Cashel, M. (1969) J. Biol. Chem. 244, 3133-3141.

5. Gallant, J., Margason, J., and Finch, B. (1972) J. Biol. Chem. 247, 6055-6058.

6. Haseltine, W.A., and Block, R. (1974) Ribosomes (ed. Nomura, M., Tissières, A., Lengyel, P.), pp. 747-761, Cold Spring Harbor Laboratory.

7. Lipmann, F., and Sy, J. (1976) Prog. Nucl. Acid Res. Mol. Biol. 17, 1-14.

8. Richter, D., and Isono, K. (1977) Current Topics in Microbiology and Immunology (ed. Arber, W. et al.) 76, pp 83-125.

9. Haseltine, W.A., Block, R., Gilbert, W., and Weber, K. (1972), Nature 238, 381-384.

10. Sy, J., and Lipmann, F. (1973). Proc. Natl. Acad. Sci. USA 70, 306-309.

11. Schmale, H., Farnung, K., Fehr, S., and Richter, D. (1977) Hoppe-Seyler's Z. Physiol. Chem. 358, 1613-1622.

12. Sy, J. (1976) Biochemistry 15, 606-609.

13. Sy, J., and Akers, H. (1976) Biochemistry 15, 4399-4403.

14. Fehr, S., Godt, F., Isono, K., and Richter, D. (1979) FEBS Lett. 97, 91-95.

15. Koch, G., and Richter, D. (1979) Regulation of Macromolecular Synthesis by Low Molecular Weight Mediators (ed. Koch, G., Richter, D.) Acad. Press, New York, in press.

16. Sy, J. (1979) Regulation of Macromolecular Synthesis by Low Molecular Weight Mediators (ed. Koch, G., Richter, D.) Acad. Press, New York, in press.

17. Richter, D., Fehr, S., and Harder, R. (1979) Eur. J. Biochem. 99, 57-64.

18. Ramagopal, S., and Davis, B.D. (1974) Proc. Natl. Acad. Sci. USA 71, 820-824.

19. Richter, D., Nowak, P., and Kleinert, U. (1975) Biochemistry 14, 4414-4420.

20. Randall, L.L., and Hardy, S.J.S. (1977) Eur. J. Biochem. 75, 43-53.

21. Block, R., and Haseltine, W.A. (1975) J. Biol. Chem. 250, 1212-1217.

22. Schmale, H., Fehr, S., and Richter, D. (1978) Hoppe Seyler's Z. Physiol. Chem. 359, 125-127.

23. Cochran, J.W., and Byrne, R.W. (1974) J. Biol. Chem. 249, 353-360.

24. Pedersen, F.S., and Kjeldgaard, N.O. (1977) Eur. J. Biochem. 76, 91-97.

25. Möller, W. (1974) Ribosomes (ed. Nomura, M., Tissières, A., Lengyel, P.), pp 711-731, Cold Spring Harbor Lab.

26. Kleinert, U., and Richter, D. (1975) FEBS Lett. 55, 188-193.

27. Richter, D. (1976) Proc. Natl. Acad. Sci. USA 73, 707-711.

28. Richter, D. (1973) FEBS Lett. 34, 291-294.

29. Lund, E., Pedersen, S., and Kjeldgaard, N.O. (1973) 1st Symposium on Ribosomes and RNA Metabolism, pp 307-319, Slovak Acad. Sci., Bratislava, Czechoslovakai.

30. Beres, L., and Lucas-Lenard, J. (1975) Biochim. Biophys. Acta 395, 80-90.

31. Howard, G.A., Gordon, J., Farnung, K., and Richter, D. (1976) FEBS Lett. 68, 211-214.

32. Highland, J.H., and Howard, G.A. (1975) J. Biol. Chem. 250, 831-834.

33. Schrier, P.J., Maassen, J.A., and Möller, W. (1973) Biochem. Biophys. Res. Commun. 53, 90-98.

34. Sy, J., Chua, N.H., Ogawa, Y., and Lipmann, F. (1974) Biochem. Biophys. Res. Commun. 56, 611-616.

35. Martini, O., Irr, J., and Richter, D. (1977) Cell 12, 1127-1131.

36. Martini, O., and Richter, D. (1978) Mol. Gen. Genet. 166, 291-297.

37. Pollard, J.W., and Parker, J. (1977) Nature 267, 371-373.

38. Sy, J., Ogawa, Y., and Lipmann, F. (1973) Proc. Natl. Acad. Sci. USA 73, 1839-1843.

39. Christiansen, L., and Nierhaus, K.H. (1976) Proc. Natl. Acad. Sci. USA 73, 1839-1843.

40. Fehr, S., Lin, A., Wool, I., and Richter, D. (1979) Mol. Gen. Genet., in press.

41. Silverman, R., and Atherly, A.G. (1977) Dev. Biol. 56, 200-205.

42. Haseltine, W.A., and Block, R. (1973) Proc. Natl. Acad. Sci. USA 70, 1564-1568.

43. Pedersen, F.S., Lund, E., and Kjeldgaard, N.O. (1973) Nature (New Biol. 243, 13-15.

44. Sprinzl, M., and Richter, D. (1976) Eur. J. Biochem. 71, 171-176.

45. Chinali, G., Liou, R., and Ofengand, J. (1978) Biochemistey 17, 2761-2768.

46. Richter, D. (1978) Biochem. Biophys. Res. Commun. 81, 359-365.

47. Richter, D. (1979) Regulation of Macromolecular Synthesis by Low Molecular Weight Mediators (ed. Koch, G., Richter, D.), Acad. Press, New York, in press.

48. Travers, A. (1978) FEBS Lett. 94, 345-348.

49. Rhaese, H.J., Groscurth, R., Vetter, R., and Gilbert, H. (1979) Regulation of Macromolecular Synthesis by Low Molecular Weight Mediators (ed. Koch, G., Richter, D.), Acad. Press, New York, in press.

50. Oki, T., Yoshimoto, A., Sato, S., and Takamatsu, A. (1975) Biochim. Biophys. Acta 410, 262-272.

51. Yoshinito, A., Oki, T., and Inui, T. (1978) Arch. Microbiol. 119, 81-86.

52. Murao, S., and Nishino, T. (1973) Agric. Biol. Chem. 37, 2929-2930.

53. Murao, S., and Nishino, T. (1974) Agric. Biol. Chem. 38, 2483-2491.

54. Laffler, T., and Gallant, J. (1974) Cell 1, 27-30.

55. Stamminger, G., and Lazzarini, R.A. (1974) Cell 1, 85-90.

56. Raué, H.A., and Cashel, M. (1975) Biochim. Biophys. Acta 383, 290-304.

57. Chaloner-Larsson, G., and Yamazaki, H. (1976) Can. J. Biochem. 54, 935-940.

58. de Boer, H.A., Bakker, A.J., Weger, W.J., and Gruber, M. (1976) Biochim. Biophys. Acta 432, 361-368.

59. Kari, C., Török, J., and Travers, A. (1977) Mol. Gen. Genet. 150, 249-255.

60. Hochstadt-Ozer, J., and Cashel, M. (1972) J. Biol. Chem. 247, 7067-7072.

61. Lazzarini, R.A., and Johnson, L.D. (1973) Nature New Biol. 243, 17-20.

62. Reiness, G., Yang, H.L., Zubay, G., and Cashel, M. (1975) Proc. Natl. Acad. Sci. USA 72, 2881-2885.

63. van Ooyen, A.J.J., de Boer, H.A., Ab, G., and Gruber, M. (1975) Nature 254, 530-531.

64. de Boer, H.A., Bakker, A.J., and Gruber, M. (1977) FEBS Lett. 79, 19-24.

65. Heinemeyer, E.A., and Richter, D. (1978) Biochemistry 17, 5368-5372.

66. Heinemeyer, E.A., and Richter, D. (1978) Proc. Natl. Acad. Sci. USA 75, 4180-4183.

67. Heinemeyer, E.A., Geis, M., and Richter, D. (1978) Eur. J. Biochem. 89, 125- 131.

68. Tetu, C., Dassa, E., and Boquet, P.-L. (1979) Biochem. Biophys. Res. Commun. 87, 314-322.

69. Heinemeyer, E.A., and Richter, D. (1977) FEBS Lett. 84, 357-361.

70. Sy, J. (1977) Proc. Natl. Acad. Sci. USA 74, 5529-5533.

71. An, G., Justesen, J., Watson, R.J., and Friesen, J.D. (1979) J. Bacteriol. 137, 1100-1110.

72. Albert, A., and Reese, C.W. (1956) Nature 77, 433-434.

73. Visentin, L.P., Hasnain, S., Gallin, W., Johnson, K.G., Griffith, D.W., and Wahba, A.J. (1977) FEBS Lett. 79, 258-263.

74. Subramanian, A.R., Haase, C., and Giesen, M. (1976) Eur. J. Biochem. 67, 591-601.

75. Richter, D., and Geis, M. (1978) FEBS Lett. 96, 247-251.

76. Hamel, E., and Cashel, M. (1973) Proc. Natl. Acad. Sci. USA 70, 3250-3254.

77. Somarville, and Ahmed (1979) Mol. Gen. Genet. 169, 315-323.

78. Fehr, S., Godt, F., Isono, K., and Richter, D. (1979) FEBS Lett. 97, 91-95.

PROCESSING OF RIBOSOMAL RNA TRANSCRIPTS IN BACTERIA

DAVID SCHLESSINGER

Department of Microbiology and Immunology,
Washington University School of Medicine,
St. Louis, MO 63110

INTRODUCTION

Like most RNA species, the primary transcript of bacterial rRNA is longer than the mature form and is lacking a number of base modifications. The transcript is cleaved, trimmed, and modified to produce 16S rRNA, 23S rRNA, 5S rRNA, and several tRNA species. These "processing" reactions are necessary, for neither the primary rRNA transcripts nor the intermediate, partially processed chains have been found capable of forming active ribosomes. But why the cells produce the extra sequences that necessitate processing reactions is still unknown. Therefore, the reactions themselves have remained the major interest of those interested in processing: how, when, and at what sequences do the processing reactions occur? Gradually, a set of site-specific endonucleases has been discovered which catalyze particular cleavages *en route* to the mature RNA; and modern sequencing technology has helped to define the two-dimensional structure at some processing sites. A recent review presents a comprehensive discussion of processing reactions (1). This brief survey is designed to update an earlier discussion of processing reactions in relation to ribosome formation (2).

Effective studies of pre-rRNA and its cleavage started with the discovery that large transcripts, including the 5'-triphosphate, 16S, 23S, 5S and at least three tRNA species, could be isolated from a strain with a lesion in RNase III (3-7). Little if any RNase III could be recovered from the strain (8). The full 30S pre-rRNA transcript, which is never seen in normal cells, lasts for the order of minutes in the mutant strain (4), and accumulates in large amounts in the mutant strain in the presence of chloramphenicol (9-10).

Precursors (p16S and p23S) also accumulate in normal *E. coli* treated with chloramphenicol (11), but they are smaller than those in the RNase III-deficient strain: separation of p16S and p23S RNA sequences, and the attendent release of other sequences by RNase III, still occurs easily in the wild-type cell. Very similar p16S and p23S species are found in pulse-labeled cells, and in several E. coli mutants defective in ribosome formation (see review in ref. 12). In accord with expectation, purified RNase III cleaves the 30S pre-rRNA transcripts to yield species the size and probably the composition of p16S and p23S; for the 16S-containing RNA, the oligonucleotide sequences and terminal sequences have been verified to correspond to p16S RNA from wild-type cells (5, 13-15).

[It is worth noting that the physiology of the mutant strain, AB301/105, remains incompletely understood. The strain, and the RNase III$^-$ transductant BL214, which lacks other background mutations (16), are completely temperature sensitive at 43° when grown in minimal media fortified with casamino acids; but even in those conditions, when growth is linear for some time, 16S and 23S RNA continue to form and to appear in ribosomes (T. King and D. Schlessinger, unpublished results). Since many revertants selected for growth at high temperature have regained wild-type levels of RNase III (17), the requirement for RNase III is not related solely to rRNA formation. The studies of Kennell et al that indicate some effects on mRNA metabolism (18) may be relevant. It would be useful to have studies of additional RNase III lesions, or at least a further definition of this RNase III lesion.]

The action of RNase III was early recognized to be an example of what Studier called "site-specific" cleavage: signals for cleavage were in the purified RNA (3,4) and were also found in ribonucleoprotein particles containing all the r-proteins (19).

The greatest advances in our understanding of processing have come from further structural analyses. Complementary analyses have been carried out on the 30S-pre-rRNA and cleavage fragments after treatment with RNase III or RNase P or cell extracts, and on the sequences of corresponding restriction fragments from various rrn DNA's.

Concerning RNase III, for example, two principles have emerged from the analyses: 1) site-specificity for the RNase III cleavages involves recognition sites with very considerable double-stranded character, consistent with the action of RNase III as a d.s. RNase (20). 2) The major cleavages occur, as shown in Fig. 1, at sites that include the extraordinary base-pairing of sequences far apart in the pre-rRNA, so that the 16S sequence and 23S RNA sequence each form an enormous loop closed by a stem at least 20 base pairs long (14,21).

```
          -285                                -250
rrn D       GUUGAGACGACAACGUGAAACACUUCACAGGAUGGUCGGAACAACGAAGAGAAAAAAAU

rrn E         AUCGACACGGCGGAUGUGAAUCACUUCACACAAACAGCCGGUUCGGUUGAAGAGAAAA

rrn A  ACUGACACGGAACAACGGCAAACACGCCGCCGGGUCAGCGGGGUUCUCCUGAGAACUCCGGCAGAG

rrn X       AUCGACACGGCGGAUGUGAAUCACUUCACACAAACAGCCGGUUCGGUUGAAGAG

                                  -200
rrn D       CCUGAAAUUCAGGGUUGACUCUGAAAGAGGAAAGCGUAAUAUACGCCACCUCGCGACAGU

rrn E       AUCCUGAAAUUCAGGGUUGACUCUGAAAGAGGAAAGCGUAAUAUACGCCACCUCGCGACA

rrn A       AAAGCAAAAAUAAAUGCUUGACUCUGUAGCGGGAAGGCGUAUUAUGCACACCCCGCGCCG
                        -150                                       III
                                                                    ↓
rrn D       GAGCUGAAACCCGCGUCGCAACUGCUCUUUAACAAUUUAUCAGACAAUCUGUGUGGGCAC

rrn E       GUGCGCUAAAGCGCGUCGCA

rrn A       CUGAGAAAAAGCGAAGCGGC
                -100                                               -50
rrn D       UCGAAGAUACGGAUUCUUAACGUCGCAAGACGAAAAUGAAUACCAAGUCUCAAGAGUGAA

                                                         (1)
rrn D       CACGUAAUUCAUUACGAAGUUUAAUUCUUUGAGCAUCAAACUUUUA
                                                          ↑
```

Fig. 1 (start). Sequence of 5'-terminal sequences of pre-rRNA, starting from the p_1 promoter of four rrn operons. rrn X is the same as rrn D from nucleotide -231 onward, except for an A at position -30. Sequences inferred from DNA sequencing (refs. 21, 27 and personal communications). The arrow at (1) indicates the 5'-terminus of 16S rRNA. The nucleotides from -130 to -100 are included in the stem at the base of 16S, with an RNase III cleavage site, as indicated in Fig. 1 (con.).

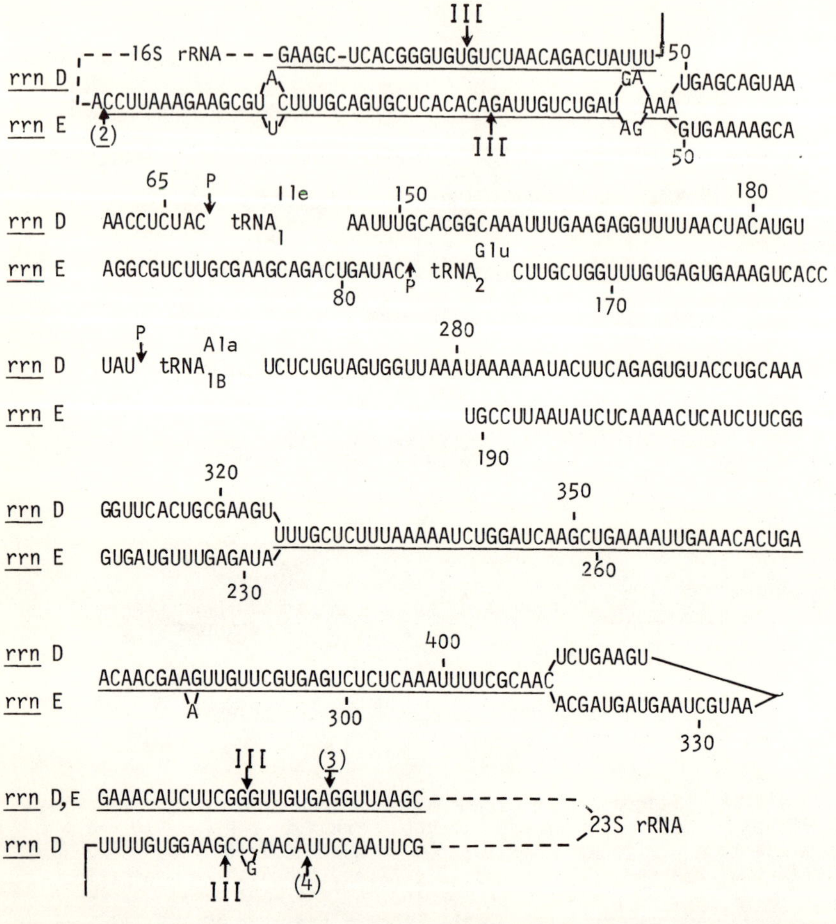

Fig. 1. (con.) Sequence of 16S-23S rRNA spacers of rrn D (29) and rrn E (28) inferred from DNA sequencing. For rrn D, a part of the paired sequences that bracket the 16S and 23S rRNA sequences in stems are also shown (at top and bottom). Three cleavage sites for RNase P (33,45) and the four major cleavage sites for RNase III (21 and R.J. Bram and J.A. Steitz, work in progress) are indicated with arrows. (2), (3), and (4) indicate the cleavage sites that generate the 3'-end of 16S rRNA, and the 5' and 3'-termini of 23S rRNA. The sequences in common between rrn D and rrn E are shown once, underlined. Note that the residue numbers (above sequences for rrn D and below rrn E) deviate because the two spacers have 438 and 354 nucleotides, respectively. Homologous sequences have been brought into register.

STRUCTURE OF rrn OPERONS AND PRE-rRNA

Sequence data are very nearly complete for representative rrn DNA's. Just as 16S and 23S rRNA from all the operons have relatively homogeneous sequences, so parts of the processed sequences are tightly conserved in all the operons examined. Table 1 and Fig. 1 indicate the common sequences, and also indicate the major known sites of enzymatic cleavage.

Among the seven detected rrn operons (22-24), the 5'-sequences can begin at p_1 or p_2 (15,25-27). At least 5 of the operons show a run of 145 nucleotides in common adjacent to the 5'-terminus of the 16S rRNA sequence (Fig. 1), and additional homologies can be found within the rest of the 5'-terminal sequence. The spacer regions between 16S and 23S have been reported for one of the four rrn operons containing a $tRNA_2^{Glu}$ (28) and for one of the three containing $tRNA_1^{Ile}$ and $tRNA_{1B}^{Ala}$ (29). In Fig. 1 it can be seen that the two show three regions of very extensive homology: 49 nucleotides just distal to the 3'-end of the 16S rRNA sequence; 20 nucleotides just before the 5'-end of the 23S rRNA sequence; and another 85 nucleotides proximal to the final sequence, separated by 8 nucleotides in rrn D and by 17 nucleotides in rrn E.

Complete sequences are not yet published for the portion of the DNA distal to the 23S rRNA, but 92 nucleotides separate 23S from 5S in rrn D; and 57 nucleotides distal to 5S is the sequence of a $tRNA_1^{Asp}$ gene (21).

rrn D is one of four operons in which tRNA species have been mapped distal to the 5S sequence as well as in the 16S-23S spacer. For rrn C, it has been demonstrated that genes for $tRNA^{Trp}$ and $tRNA^{Asp}$ are cotranscribed from the rrn promoter (24).

PROCESSING CLEAVAGES WITH SIGNALS IN NAKED RNA

With the realization that a single rrn transcript contains one to four tRNA molecules as well as 16S, 23S and 5S RNA, processing becomes more complex. One can wonder why the tRNA species are present; and correspondingly, whether there is any connection between the processing of tRNA and that of the rRNA. A number of enzymes recognize cleavage signals in RNA for the processing of one or the other.

Five principal cleavage sites are now known for RNase III. Four of these occur in two pairs, in the stems that match conserved sequences of Fig. 1 to bracket the 16S and 23S rRNA (21). There are a number of persuasive arguments that the stems exist *in vivo*: 1) Secondary structure loops around 16S and 23S rRNA sequences have been observed in single-stranded rDNA (30). 2) Double-stranded RNA fragments containing the expected oligonucleotides have been isolated from cells (31; and quoted in ref. 2). 3) None of the nucleotides at cleavage sites can be arranged in a double-stranded stem made from proximal nucleotides.

The results of Young and Steitz imply that major cleavages by RNase III cannot occur until the stems at the base of the 16S and 23S sequences (Fig. 1) have formed - i.e., complete transcription of the 16S sequence precedes cleavage by RNase III. A corresponding inference can be made for the 23S sequence. Consistent with this model is the observation that during the cleavage of 30S pre-rRNA by RNase III, p16S, p23S, and fragments from the termini of 16S and 23S rRNA tend to be formed at comparable rates (32 and unpublished data).

The precise nature of the signals for RNase III action remains unclear (see ref. 14 for some intriguing comparisons with RNase III cleavage sites in T7 mRNA); but there is at least one other RNase III-specific cleavage site in the region between the p16S and p23S sequences. That site can be detected among the products of RNase III action on 30S pre-rRNA (33), or by further cleavage of the tRNA-containing spacer fragment released by RNase III action (32). A cleavage that is probably the same has been located more precisely at about position 290 of the spacer (31). The relative extent of RNase III cleavage at various sites likely responds to other conformational features of the molecule, for it can be modified by changes in the salt conditions of the reaction (32).

Several other site-specific enzymes have been identified as active near tRNA or 5S rRNA sequences in the pre-rRNA. These include RNase P (33) and RNase "E" (34). RNase E was inferred from its apparent thermolability in a mutant that fails to cleave 5S rRNA out of a precursor form at high temperature.

RNase P has been directly shown to cleave 30S pre-rRNA, as indicated in Fig. 1, to generate mature 5'-termini of $tRNA_{1B}^{Ala}$ and $tRNA_{2}^{Glu}$ (33); but the 5'-terminus of $tRNA_{1}^{Ile}$, which would also be expected to be formed, is much less susceptible to attack. The separation of the Ala and Ile tRNA's seems to occur easily _in vivo_ (35), where either the conformation of the RNA or other conditions of the reaction are presumably more favorable.

Lund and Dahlberg have shown that a number of other activities in crude extracts cleave 30S pre-rRNA at specific sites (7). These could include RNases P2 and P4 (36).

No specific explanation has yet been proposed for the additional long homologous sequences in the spacer regions of *rrn* D and *rrn* E (Fig. 1); but it has seemed likely that they may also be involved in processing signals, perhaps by association with additional complementary sequences distal to the 23S or 5S rRNA; evidence for additional interactions has been obtained (Bram, R. and Steitz, J.A., work in progress).

PROCESSING CLEAVAGES DEPENDENT ON BOUND PROTEINS

An earlier formulation suggested a distinction between "primary" and "secondary" processing reactions (2). The former have their signals in pure RNA; they ordinarily occur during or just after transcription. The latter occur late during ribosome formation, do not occur in chloramphenicol-treated cells, and therefore probably require the binding of protein to the RNA. (The criterion is a useful operational distinction, but may not be rigorous; for example some late changes in tRNA formation, including certain methylations (12), do not occur in chloramphenicol-treated cells; but there is no evidence that they require the binding of protein factors to the pre-tRNA.)

Three studies have now supported the notion that secondary (or final) steps in the processing of pre-rRNA often and perhaps always require the binding of proteins to the RNA -- either to modify the conformation of the RNA or to provide part of the binding site for the processing enzyme. These processing steps would be blocked when the synthesis of the required proteins is blocked in chloramphenicol-treated cells.

The most extensively studied example is one in *B. subtilis*, where the generation of 5S RNA shows a step inhibited by chloramphenicol. It has been studied in detail by Pace's group (37). The precursor p5S can be matured *in vitro* by an RNase M5, yielding 5S (116 nucleotides), a 3'-fragment (42 nucleotides), and a 5'-fragment (21 nucleotides). Sogin *et al* (37) have pointed out certain symmetrical structural features in the p5S that suggest a recognition mechanism for RNase M5 to make two cuts in mirror orientation. In addition, however, at least two additional protein "factors" (possibly r-proteins) must be bound to the RNA in order for the RNase M5 reaction to take place.

Meyhack *et al.* (38) have also extended the proposal of a possible course for the reaction by cutting the mature and p5S species at a central G residue (G60). The complementary halves of the molecules were then recovered and reannealed to form substrates with one or the other precursor bond cleaved. When treated with RNase M5, the artificial substrate lacking the 5'-terminal precursor segment was cleaved much more slowly than that lacking the 3'-segment. The 5'-segment therefore contains one or more components recognized by RNase M5 during its interaction with the p5S; and the 3'-segment may well be cleaved off first (if there _is_ any sequence to the reaction _in vivo_!).

No analogous enzyme to RNase M5 has been found thus far in _E. coli_. An activity that forms this RNA from a larger molecule has been presumed because RNase III releases 5S RNA from 30S pre-rRNA in a larger fragment (5-6). An alternate source of substrate for the enzyme is a temperature-sensitive mutant that accumulates a 5S-containing RNA fragment about 300 nucleotides long at non-permissive temperature (34). Since crude protein preparations can cleave 5S RNA out of 30S pre-rRNA (7), the presumptive activity in _E. coli_, in contrast to _B. subtilis_, may _not_ require the binding of protein factors to promote its action. Consistent with the notion that the generation of 5S RNA in _E. coli_ may not require such factors, 5S RNA accumulates in wild-type cells treated with chloramphenicol, with no more than 2 extra nucleotides at its 5'-end (12).

E. coli does afford two other examples of protein-dependent cleavage of pre-rRNA. Both the 5' and 3'-termini of 16S RNA can be demonstrated to arise by cleavage reactions (13,39). In both cases, isolated pre-rRNA does not serve as a substrate, but the rRNA bound to r-proteins in a particle does. In both cases, the appearance of the true termini has been demonstrated by appropriate oligonucleotide analysis.

The 3'-terminus is formed in a 27S precursor particle (39), which likely imitates the stage of biosynthesis at which the reaction occurs in the cell. The 5'-terminus has been generated _in vitro_ from a substrate that may not be the most frequent in normal cells (13): it occurs in a particular mutant strain deficient in "RNase M16", which contains a mixture of 30S ribosomes, some with mature 16S RNA and some with a "16.3S RNA" that has already lost some of the 5'-nucleotides of the p16S RNA species and has a mature 3'-terminus. Preparations of RNase M16 from a wild-type strain cleave the 16.3S RNA in 30S particles endonucleolytically to produce a 3S RNA fragment and the mature 5'-16S terminus. A 16.3 size "intermediate" has been detected in some studies of RNA processing (see ref. 12); but the usual route to 16S RNA may employ direct cleavages of the 17S (p16S) precursor.

In the generation of the ends of 16S rRNA, as in the corresponding formation of 23S rRNA 5'- end 3'-termini by as yet unknown activities, we have no information about the nature of signals recognized for processing.

A related question of interest is the possible activity of ribosomes containing incompletely-processed rRNA species. Although neither 30S pre-rRNA nor 17S rRNA seem to form functional ribosomes (ref. 12 and unpublished results), 30S particles containing 16.3S rRNA (or likely other p16 species; reviewed in ref. 2) and all the r-proteins can be found in polyribosomes. In at least one report, ribosomes with an incompletely-matured 5'-terminus are thought to be active in protein synthesis in vivo (40); but in the well characterized BUMMER strain (13), the results are ambiguous: cells with 30S ribosomes half of which contain 16.3S rRNA grow about 30% slower than the parental in which all the rRNA is 16S.

PROCESSING MODIFICATIONS

In addition to cleavages, pre-rRNA also undergoes substantial modification of bases during its maturation. A number of methylations and other substitutions can occur even on the 30S pre-rRNA in cells treated with chloramphenicol, including changes in the 23S sequence (41) and in the spacer-tRNA's (32, 41). Others, like the N^6-dimethylations associated with sensitivity to kasugamycin, occur only on the nearly-assembled ribosomal particle (42). Thus, there is again a progression of modifying events, but the rationale for these changes is again unclear.

THE ORDER OF PROCESSING REACTIONS

The interesting question remains open whether there is an obligate order to processing reactions. At present, there are some cases in which an obligate order of reaction is strongly indicated.

In determining whether alternate processing pathways exist, it becomes important to know how much of each enzyme of interest is present. In the case of the important RNase M16, for example, which catalyzes the formation of the mature 5'-terminus of 16S RNA, it cannot be detected in extracts of the BUMMER mutant strain; yet some activity "must be present in low levels in these cells because they contain about equal amounts of (precursor) 16.3S and 16S rRNA" (13). Also, the precursor RNA is matured to 16S when the cell cultures are treated with rifampicin (13).

It is unlikely that the cell would have two different enzymes to form the 5'-terminus of 16S rRNA, one of which could substitute in a mutant lacking the other. Assuming that RNase M16 is the only enzyme that can generate the true 5'-terminus, it seems important that it cannot use as a substrate particles containing p16 rRNA (which has squences yet to be processed at both its termini). In contrast, it can use as a substrate particles containing 16.3S rRNA, which has a mature 3'-end. Dahlberg et al. infer that formation of the 3'-end (presumably by the activity reported in ref. 39) must precede the cleavage that generates the 5'-terminus.

A second possible case of a "prescribed" order for two "late", protein-dependent cleavages is that of B. subtilis p5S to 5S by RNase M5 (see above).

One might have thought that primary processing reactions, unlike the "late" steps in RNA maturation, would show little if any necessary order. RNase III, RNase P, and other enzymes are all indeed able to cleave 30S pre-rRNA in vitro (see above). Also, RNase P cleavage has been suggested to give rise to some of the unusual precursor RNA's found in the RNase III-deficient strain (33, 43); and when temperature-sensitive RNase P is introduced into the strain defective in RNase III, RNase P-catalyzed cleavage fails at the non-permissive temperature (44,45).

However, the accumulation of considerable 30S pre-rRNA as such in the RNase III-deficient mutant treated with chloramphenicol (9) suggests that RNase P sites in vivo may be hidden unless RNase III has acted. [The failure of RNase P to cleave some sites in that case is probably not related to its relatively poor action at the $tRNA_1^{Ile}$ 5'-terminus in vitro (see above); for the $tRNA_{1B}^{Ala}$ terminus would still be expected to be susceptible, leading to cleavage of the 30S pre-rRNA. Also, the 30S pre-rRNA accumulated in chloramphenicol-treated RNase III-deficient strains contains both types of spacer sequences (33).]

In a similar way, the 30S pre-rRNA accumulated in the RNase III-deficient strain contains 5S rRNA. Thus, the putative RNase(s) that cleaves out 5S in wild-type E. coli and in extracts may also fail to act when RNase III has not functioned.

Nevertheless, in spite of these puzzling discrepancies, the suggestion that tRNA processing enzymes may somehow initiate an RNase III-independent processing route (33,43) can help to explain the maturation to 16S and 23S rRNA of the unusual "25S" and "17.5S" (3-4) RNA species in RNase III-deficient strains. [30S pre-rRNA itself can give rise to rRNA in some conditions (10); but in low-phosphate medium (10) or other conditions, it seems to be largely wasted (see ref. 43).] As originally pointed out (4), maturation of these species could be catalyzed by residual RNase III or by enzymes that substitute for it. A pathway independent of RNase III seems obligate at high temperature, where the mutant cells are seemingly starved for RNase III but continue to make 16S and 23S rRNA (see above).

A critical feature of any alternate pathway can be seen in reference to Fig. 1: any precursor RNA cleaved farther away from the mature rRNA termini than the RNase III cleavage sites must be a substrate (or must somehow be further processed to a substrate) for the enzymes that produce the mature 5'- and 3'- termini. Considering the fastidious requirements of RNase M16, for example (see above), the additional sequences may inhibit the normal reactions considerably. Definitive experiments in vitro should be possible to determine the extent to which such an enzyme can function on complexes of r-proteins with RNA species different from p16 and p23S.

The challenge to further studies of processing is thus clear. In all cases, both in vitro and in vivo, there are hints that processing is sensitive to small changes in the conformation of the RNA. In no case do we understand fully the basis for the recognition of its substrate by a processing enzyme. This lack of understanding is most acute for the interesting enzymes that require RNA-protein complexes as a substrate.

WHY PROCESSING REACTIONS?

Especially since many base modifications are evolutionarily conserved (46), they may represent refinements in ribosome structure to accelerate ribosome movement, restrict miscoding, improve the fit of interacting components, etc.

Such a rationale is much more difficult to propose for cleavage reactions. Much of the "extra" precursor sequences show little cross-hybridization between bacterial species that have very similar 16S and 23S rRNA's (47). Furthermore, the precursor-specific sequences are by definition not found in mature rRNA. Thus, if they function, they must act during the transcription of rRNA and/or the formation of ribosomes. The last possibility has been an attractive one; and in particular, it has been suggested that structural features in pre-rRNA species might stabilize conformations in the rRNA that might otherwise form with some difficulty (2). Several of the recent results are consistent with this notion. The structural studies of Young and Steitz in particular show relatively long and rigid double-helical stems as a likely feature adjacent to nascent 16S and 23S rRNA sequences. Such stems would be expected to be stable even at 100° (21), and would obviously exert considerable restraint on the conformation of neighboring regions in the RNA. It is probably significant that a large part of the "extra" sequences at the 5' and 3'-ends of the 16S and 23S RNA are involved in these structures (see Fig. 1); that these sequences are conserved in at least five of the *E. coli rrn* sequences (Fig. 1); and that 30S pre-rRNA functions well in binding stoichiometric amounts of r-proteins *in vitro* (48).

The speculative possibility is thus increasingly attractive that conformational restraints in pre-rRNA help to lower the activation energy for ribosome formation from the high levels seen with mature rRNA. The cleavage reactions may represent the steps necessary to clear away construction struts when stages in ribosome formation are complete. If such a role for pre-rRNA sequences is substantiated, then the structure and not the base-sequence of these segments may be more important to compare among species. Perhaps studies of the primitive bacteria will show how processing reactions, and the involvement of tRNA's in rRNA transcripts, have also evolved by refinements of a simpler mechanism.

ACKNOWLEDGEMENTS

Joan Steitz and other friends helped to bolster the rapid writing of this review of their work and I anticipate their indulgence in its errors of haste. Results from my laboratory have been supported by NSF grant PCM 77-08387.

REFERENCES

1. Abelson, J. (1979) in Ann Rev. Biochem. 48, 1035-1069.

2. Schlessinger, D. (1974) in *Ribosomes* (Nomura, M., Tissieres, A., and Lengyel, P. eds.) pp.393-316, Cold Spring Harbor Lab., New York.

3. Dunn, J.J. & Studier, F.W. (1973) Proc. Nat. Acad. Sci. USA 70, 3296-3300.

4. Nikolaev, N., Silengo, L. & Schlessinger, D. (1973) Proc. Nat. Acad. Sci. USA 70, 3361-3365.

5. Ginsburg, D. & Steitz, J.A. (1975) J. Biol. Chem. 250, 5647-5654.

6. Hayes, F., Vasseur, M., Nikolaev, N., Schlessinger, D., Sri Widada, J., Krol, A. & Branlant, C. (1975) FEBS Letters 56, 85-91.

7. Lund, E. & Dahlberg, J.E. (1977) Cell 11, 247-262.

8. Keil, T.U. & Hofschneider, P.H. (1973) Biochim. Biophys. Acta 312, 297-310.

9. Schlessinger, D., Ono, M., Nikolaev, N. & Silengo, L. (1974) Biochemistry 13, 4268-4271.

10. Nikolaev, N., Birenbaum, M. & Schlessinger, D. (1975) Biochim. Biophys. Acta 398, 478-489.

11. Lowry, C.V & Dahlberg, J.E. (1971) Nature New Biol. 232, 52-54.

12. Pace, N.R. (1973) Bacteriol. Rev. 37, 562-603.

13. Dahlberg, A.E., Dahlberg, J.E., Lund, E., Tokimatsu, H., Rabson, A.B., Calvert, P.C., Reynolds, F. & Zahalak, M. (1978) Proc. Nat. Acad. Sci. USA 75, 3598-3602.

14. Young, R.A. & Steitz, J.A. (1978) Proc. Nat. Acad. Sci. USA 75, 3593-3598.

15. Lund, E. & Dahlberg, J.E. (1979) Proc. Nat. Acad. Sci. USA 76, in press.

16. Studier, F.W. (1975) J. Bact. 124, 307-316.

17. Apirion, D. & Watson, N. (1975) J. Bact. 124, 317-324.

18. Talkad, V., Achord, D. & Kennell, D. (1978) J. Bact. 135, 528-541.

19. Nikolaev, N., Glazier, K. & Schlessinger, D. (1975) J. Mol. Biol. 94, 301-304.

20. Robertson, H.D., Webster, R. & Zinder, N. (1968). J. Biol. Chem. 243, 82-91.

21. Young, R.A., Bram, R.J. & Steitz, J.A. (1979) in tRNA (Abelson, J., Schimell, P., & Soll, D. eds.) Cold Spring Harbor Laboratory, in press.

22. Kiss, A., Sain, B. & Venetianer, P. (1977) FEBS Letters 79, 77-79.

23. Kennerley, M.E., Morgan, E.A., Post, L., Lindahl, L. & Nomura, M. (1977). J. Bacteriol. 132, 931-949.

24. Morgan, E.A., Ikemura, T., Lindahl, L., Fallon, A.M. & Nomura, M. (1978) Cell 13, 335-344.

25. Young, R.A. & Steitz, J.A. (1979) Cell 17, 225-234.

26. Glaser, D. & Cashel, M. (1979) Cell 16, 111-121.

27. Gilbert, S.F., deBoer, N.A. & Nomura, M. (1979) Cell 17, 211-224.

28. Morgan, E.A., Ikemura, T., Post, L.E. & Nomura, M. (1979) in tRNA (Abelson, J., Schimmel, P. & Soll, D., eds.) Cold Spring Harbor Laboratory, in press.

29. Young, R.A., Macklis, R. & Steitz, J.A. (1979) J. Biol. Chem. 254,3264-3271.

30. Wu, M. & Davidson, N. (1975) Proc. Nat. Acad. Sci. USA 72, 4506-4510.

31. Robertson, H.D. & Barany, F. (1978) in Proc. of the 12th FEBS Congress, Pergamon Press, New York, in press.

32. Birenbaum, M., Schlessinger, D. & Hashimoto, S. (1978) Biochemistry 17, 298-307.

33. Lund, E., Dahlberg, J.E. & Guthrie, C,. (1979) in tRNA (Abelson, J., Schimmel, P. & Soll, D., eds.), Cold Spring Harbor Lab., in press.

34. Ghora, B.K. & Apirion, D. (1978) Cell 15, 1055-1066.

35. Ikemura, T. & Nomura, M. (1977) Cell 11, 779-793.

36. Smith, J.D. (1976) Prog. Nucl. Acid Res. Mol. Biol. 16, 25-73.

37. Sogin, M.L., Pace, B. & Pace, N.R. (1977) J. Biol. Chem. 252, 1350-1357.

38. Meyhack, B., Pace, B. & Pace, N.R. (1977) Biochemistry 16, 5010-5015.

39. Hayes, F. & Vasseur, M. (1976) Eur. J. Biochem. 61, 433-442.

40. Rosset, R., Vola, C., Feunten, J. & Monier, R. (1971) FEBS Letters 18, 127-129.

41. Dahlberg, J.E., Nikolaev, N. & Schlessinger, D. (1975) in Brookhaven Symp. Biol. No. 26 (Dunn, J.J., ed.) pp.194-200.

42. Helser, T.L., Davies, J.E. & Dahlberg, J.E. (1972) Nature New Biol. 235, 6-9.

43. Gegenheimer, P., Watson, N. & Apirion, D. (1977) J. Biol. Chem. 252, 3064-3073.

44. Gegenheimer, P. & Apirion, D. (1978) Cell 15, 527-539.

45. Robertson, H.D., Pelle, E.G. & McClain, W.H. in tRNA (Abelson, J., Schimmel, P. & Soll, D., eds.), Cold Spring Harbor Lab., in press.

46. Woese, C. this volume.

47. Kano, Y., Silengo, L. & Imamoto, F. (1976) Mol. Gen. Genet. 146, 275-283.

48. Nikolaev, N. & Schlessinger, D. (1974) Biochemistry 13, 4272-4278.

RIBOSOMES AND SPORULATION IN *BACILLUS SUBTILIS*

GLENN H. CHAMBLISS

Department of Bacteriology,
University of Wisconsin,
Madison, WI 53706

INTRODUCTION

In prokaryotes a portion of the mutations in genes encoding ribosomal components are pleiotropic. This situation has been most extensively studied in Escherichia coli but it is also observed in other organisms. In E. coli pleiotropic effects associated with ribosomal mutations include alterations in the regulation of enzyme synthesis (38, 39), reduced polypeptide chain elongation rate (51), altered translational termination properties (50), altered specificity in polypeptide chain initiation (10), and restriction of the action of transfer (t)RNA suppressors (13, 46).

Ribosomal genes of E. coli in which mutations are often pleiotropic include strA (26), spcA (26) and ramA (26). The basis for the pleiotropy in certain instances is believed to be mistranslation. The reduced ability of ribosomes of certain StrR mutants to misread appears to be involved in the restriction of transfer RNA suppressors (14). ramA mutations relieve the restriction of tRNA suppressors without reverting the drug resistance phenotype of these StrR mutants (14). Ribosomes of the revertants containing the ramA mutation are more prone to misread than the ribosomes of the original StrR mutant, corroborating the suggestion that mutational alterations in translational misreading can cause pleiotropic effects.

ORGANIZATION OF RIBOSOMAL GENES IN B. SUBTILIS

Most of the mutants of B. subtilis with altered ribosomes have been isolated as mutants resistant to antibiotics that affect ribosome function (41). Mapping studies reveal that most of the genes identified by these mutants form a cluster linked to the cysA locus (17, 41). The order of the resistance markers is cysA, thiostrepton (tsp), streptomycin (str), fusidic acid (fus), kirromycin (kir), neamine (nea) (neomycin [neo]), kanamycin (kan), erythromycin (ery) (oleandomycin [ole]), spectinomycin (spc) and lincomycin (lin) (17, 33, 41, 43). Using the technique of interspecies transformation Osawa, Tokui, and Saito (32) found that the genes for 16 ribosomal proteins, S3, S5, S8, S12, S17, S19, BL4, BL5, BL6, BL8, BL14, BL16, BL17, BL22, BL23 and BL25, are located in the cysA-linked cluster of ribosomal genes. The genes for 10 proteins, S4, S6, S13, S16, S20, BL15, BL18, BL20, BL24, and BL28 were not found in that region. In addition to the above mentioned markers the cysA-linked region of the B. subtilis chromosome contains genes for ribosomal (r)RNA and tRNA (42) as well as temperature sensitive mutations affecting elongation factors EF-Tu (12) and EF-G (27). A second region of the chromosome between purA on the left and cysA on the right contains the pactamycin resistance locus (17). This second region is now thought to contain other ribosomal genes (19, 49).

PLEIOTROPIC ANTIBIOTIC RESISTANT MUTANTS

A portion of the mutants of B. subtilis resistant to any of several of the antibiotics that inhibit protein synthesis are pleiotropically modified in their ability to sporulate. The three sporulation phenotypes that have been observed are temperature sensitive sporulation (Spo^{ts}), conditional asporogeny (Spo^{c}) and asporogeny (Spo^{-}). All three phenotypes may be different manifestations of defects to a particular sporulation process however for this review each will be treated separately.

Temperature Sensitive Sporulation

In antibiotic resistance mutants of B. subtilis exhibiting temperature sensitive sporulation (Spo^{ts}), growth and sporulation are normal and antibiotic resistant at the permissive temperature. At the restrictive temperature (46°) growth is normal and antibiotic resistant but sporulation is aborted regardless of the presence of the drug. Mutants of this type were first described by Leighton (29) who isolated streptomycin resistant, Spo^{ts} mutants. Subsequently mutants resistant to many antibiotics including erythromycin (47) and fusidic acid (28) have been found to exhibit the Spo^{ts} phenotype. The EryR, Spo^{ts} and the FusR, Spo^{ts} phenotypes demonstrate that this pleiotropic effect is not limited to mutants with altered 30S ribosomal subunits.

The proportion of antibiotic resistant mutants that are Spo^{ts} is variable and seems to depend on the antibiotic. Leighton (29) found that about 7% of the StrR mutants he isolated were Spo^{ts} whereas all of several hundred EryR mutants isolated by Tipper et al. (47) were Spo^{ts}. About two-thirds of the fusidic acid resistant mutants isolated by Kobayashi et al. (28) were defective in sporulation but not all of these were of the Spo^{ts} phenotype.

The sporulation process is characterized by a series of morphological stages culminating in the production of a mature spore. Characteristically in asporogenous mutants sporulation is normal up to a particular morphological stage and then progresses no further. In the antibiotic resistant, Spo^{ts} mutants sporulation at the restrictive temperature arrests at uniquely different stages. Leighton (29) isolated about 50 StrR, Spo^{ts} mutants. In this collection there were mutants blocked at most of the stages of sporulation although the mutant he studied most was blocked very early in the sporulation process. The temperature sensitive period began after approximately 10% of the sporulation process was complete and extended to the time when 90% of the process was complete. At the restrictive temperature sporulation of the EryR, Spo^{ts} mutants isolated by Tipper et al. (47) was also blocked in the early stages, but the block appears to be subsequent to that of the early blocked StrR, Spo^{ts} mutant studied by Leighton. The period of temperature sensitivity was also different. It extended from approximately 40% completion to 90% completion. All the EryR mutants were blocked at the same stage of sporulation. The FusR, Spo^{ts} mutant studied by Kobayashi et al. (28) was also blocked very early in the sporulation process at the restrictive temperature, possibly very close to or at the same point as that of Leighton's StrR, Spo^{ts} mutant.

Both the antibiotic resistance and Spo^{ts} phenotypes result from single mutants which map to the cysA-linked cluster of ribosomal genes. Both phenotypes cotransformed and cotransduced 100% in each case that has been reported. Revertants which sporulate normally at 46° have been isolated for the StrR, Spo^{ts} mutant and the EryR, Spo^{ts} mutants. Between 80 and 90% of the sporogenous revertants of the StrR, Spo^{ts} mutant were streptomycin sensitive (StrS) with the remainder being StrR (29). It was suggested that the latter group arose as the result of second site mutations which suppressed the Spo^{ts} phenotype. All the sporogenous revertants of the EryR, Spo^{ts} mutants were erythromycin sensitive; no second site revertants were found (47).

The map locations of the antibiotic resistance mutations suggest that they affect ribosomal components. Ribosomes of the StrR, Spo^{ts} mutant bound less than 5% as much radioactive streptomycin as did wild type ribosomes. In the Ery^R, Spo^{ts} mutants a single 50S subunit protein (L17) was found to exhibit altered electrophoretic mobility. The ribosomes of the EryR mutants also bound less radioactive erythromycin than did wild type ribosomes. The supernatant fraction of the FusR, Spo^{ts} mutant of Kobayashi et al. (28) was resistant to fusidic acid in agreement with the mutant possessing a modified EF-G although purified EF-G was not isolated and tested.

Why alteration of the translational apparatus by mutation should cause temperature sensitive sporulation is not altogether evident. One possibility is that the mutation alters the ability of the translational apparatus to carry out sporulation-specific protein synthesis. Leighton reported that protein synthesis, as measured by the incorporation of radioactive phenylalanine during a short pulse, by his StrR, Spo^{ts} mutant at the restrictive temperature decreased to levels significantly less than those of the wild type under the same conditions. Goldman and Tipper (11) also found that protein synthesis was affected in their EryR, Spo^{ts} mutants. At the restrictive temperature what appeared in autoradiograms of one dimensional acrylamide gels to be normal quantities of most proteins continued to be synthesized in stationary growth phase in the EryR mutants. Although the resolution of one-dimensional acrylamide gels is limited, the appearance of several new bands occurred during sporulation of the wild type organism. Many of these bands also appeared during stationary growth phase of the EryR mutant at the restrictive temperature, however some of these bands were missing from the mutant suggesting that certain proteins possibly required for morphogenesis were not synthesized by the mutant at the restrictive temperature. Since for the most part protein synthesis by the EryR mutant at restrictive temperature appeared to be normal and only a very limited number of proteins were not synthesized, the defect appears to be quite specific and not simply a generalized defect in protein synthesis.

The failure of the EryR, Spo^{ts} mutants to synthesize presumably sporulation-specific proteins could result from a defect in either gene transcription or mRNA translation. There is evidence that mutations to the translational apparatus of B. subtilis can affect the regulation of

RNA synthesis. Kimura et al. (27) isolated a mutant of B. subtilis in which EF-G is temperature sensitive. The control of stable RNA synthesis in this mutant is relaxed at the restrictive temperature while at permissive temperature stable RNA synthesis is under stringent control. The sporulation phenotype of this mutant was not reported. Additionally, the regulation of the synthesis of stable RNA is relaxed in thiostrepton resistant mutants of B. subtilis (44). These mutants lack a protein of the 50S subunit (44 and Cundliffe in this volume). More recently Hirochika and Kobayashi (20) reported that certain rifampicin resistance (rif^r) mutations suppress the Spo^{ts} phenotype of the FusR, Spo^{ts} mutant they are studying. In vitro RNA synthesis by extracts of RifR mutants was resistant to rifampicin suggesting that the rif^r mutation affected the polymerase molecule. This suggestion was corroborated by the map location of the rif^r mutation. This suppression of the Spo^{ts} phenotype of the FusR mutation by a rif^r mutation brings to mind the observation that rif^r mutations in E. coli interact with strA mutations to produce new phenotypes (6, 7).

The inability of the antibiotic resistant, Spo^{ts} mutants to sporulate at restrictive temperature may involve lipid metabolism. Leighton and coworkers (personal communication) find that sporulation is restored in all antibiotic resistant, Spo^{ts} mutants tested by supplementation of the cultures with ribose, glycerol, native fatty acid esters or synthetic lipids (Tweens). For a defect in lipid metabolism to be the basis for the Spo^{ts} phenotype the defect would have to be specific to the synthesis of spore membranes, since cell growth and presumably cell membrane synthesis are not noticeably affected by the antibiotic resistance mutations (29). How a ribosomal mutation could specifically affect spore membrane synthesis has yet to be demonstrated.

Conditional Asporogeny

In addition to the temperature sensitive sporulation phenotype, resistance to antibiotics that inhibit protein synthesis is sometimes accompanied by conditional asporogeny. In these mutants growth is drug resistant but sporulation remains drug sensitive as in the wild type. In the absence of the drug both growth and sporulation are normal. Resistant mutants of this type have been isolated for spectinomycin (SpcR), streptomycin, kanamycin (KanR), fusidic acid, and erythromycin. The fraction of resistant mutants exhibiting conditional asporogeny varies. Graham and Bott (15) reported these frequencies of conditional asporogeny among drug resistant mutants: spectinomycin, 16%; streptomycin, 4.9%; kanamycin, 55%; and erythromycin, 6%. Tipper et al. (47) found that each of several hundred erythromycin resistant mutants was conditionally asporogenous. All the spectinomycin resistant mutants that we have isolated were conditionally asporogenous after transfer of the spc^r marker to the parental strain by transformation.

In the presence of drug, sporulation in the different conditional mutants is blocked early. Graham and Bott (15) reported that spectinomycin blocked sporulation early possibly before stage II (i.e. before forespore formation). Sporulation in some EryR mutants proceeds normally in the presence of erythromycin until forespores are formed and then ceases (11). In other EryR mutants (8) sporulation has been reported to be blocked at stage I (i.e. before forespore formation) in the presence of erythromycin. Why these EryR mutants are blocked at different sporulation stages is not known. Sporulation of some of the EryR mutants is not only drug sensitive but also temperature sensitive (11). At the restrictive temperate sporulation is blocked before forespore formation while forespores are formed in the presence of erythromycin at permissive temperature. This observation suggests that the sporulation process(es) affected by the two conditions may be different.

Most of the drug resistance mutations involved with conditional asporogeny are located in the cysA-linked cluster of ribosomal genes, however there are exceptions. Staal and Hoch (45) isolated a StrR, conditionally asporogenous mutant in which the resistance mutation mapped to a site labeled strB unlinked to the cysA locus. It is presumed that resistance and the conditional sporulation phenotype of this type result from membrane modifications not ribosomal modifications. The existence of this type of mutant points up the importance of demonstrating that a mutation suspected of causing a ribosomal modification actually does affect the ribosomes.

Second site mutations which suppress the conditional sporulation phenotype of SpcR mutants have been isolated. Cannon and Bott (2) have isolated a large colony variant of their SpcR, Spo^c mutant that sporulates in the presence of spectinomycin. This variant contains, in addition to the original spc^r mutation, a second mutation which permits spectinomycin resistance sporulation. This second mutation cotransforms with the original spc^r mutation at a frequency of 46% suggesting that it may be in a ribosomal gene, although no modified ribosomal protein has been observed by electrophoresis. Perhaps the reason Graham and Bott (15) found that only 16% of the SpcR mutants they isolated were conditional is because of second site mutations of this type. Winek and Chambliss (unpublished results) have isolated from a SpcR, Spo^c mutant a variant capable of sporulating in spectinomycin-containing medium. This variant contained a second site mutation which suppressed the conditionally asporogenous phenotype. The suppressor mutation cotransformed 99% with the original mutation which mapped to the spcA locus (19). This high frequency of cotransformation suggests that both mutations may be in the same gene. The two-dimensional gel electrophoresis pattern of the ribosomal proteins of the original SpcR mutant was indistinguishable from that of the wild type. Protein S5 [following the nomenclature of Geisser et al. (9) for B. subtilis ribosomal proteins] of the variant had an altered electrophoretic mobility. In E. coli protein S5 is the product of the spcA locus. Thus far it has not been shown that S5 of B. subtilis is strictly analogous to E. coli S5, however the indication is that both mutations affect the gene for protein S5.

In the better studied antibiotic resistant, conditionally asporogenous mutants there is good evidence that the resistance mutation affects a ribosomal component. The ribosomes of the EryR mutants isolated by Tipper et al. (47) bind less than 1% as much erythromycin than do wild type ribosomes and protein L17 of the large subunit is altered in polarity. One of the 50S subunit proteins of the EryR, Spo^c mutant studied by Domoto et al. (8) is altered in its affinity for carboxymethyl-cellulose. The 30S subunit of the spectinomycin resistant mutant isolated in Bott's laboratory (15) contains a protein with modified electrophoretic mobility (1). The two dimensional gel electrophoretic profiles of the ribosomal proteins of the spectinomycin resistant mutant isolated by Winek and Chambliss (unpublished data) were indistinguishable from those of the wild type, but the mutant ribosomes were resistant to spectinomycin in the in vitro translation of natural messenger RNA. The supernatant fraction of the FusR, Spo^c mutant studied by Kobayashi et al. (28) was resistant to fusidic acid in the in vitro translation of poly(U).

Since the basis for the inability of these mutants to sporulate in the presence of drug has not been elucidated there are several alternative possibilities. Probably the simplest is that during sporulation total protein synthesis by the drug resistant mutants becomes drug sensitive. This may be what happens in the EryR mutants of Goldman and Tipper (11).

Although there is no direct proof that the ribosomes of these mutants are resistant to erythromycin, the indirect evidence that they are resistant is good. The extent of protein synthesis by the mutant in the presence of erythromycin is apparently normal during exponential growth and very early stationary phase. Shortly after formation of the forespore protein synthesis abruptly ceases when erythromycin is present. Unfortunately we do not know yet why protein synthesis ceases. It could be that a ribosomal modification occurs during sporulation which renders the mutant ribosomes sensitive to erythromycin.

We have found that the *in vitro* protein synthetic activity of ribosomes from sporulating SpcR, Spo^c mutant cells is just as resistant to spectinomycin as is the activity of ribosomes prepared from exponentially growing mutant cells (unpublished results). Examination by one dimensional electrophoresis of the *in vitro* synthesized products when messenger RNA from exponentially growing cells and sporulating cells served as template revealed differences in the products synthesized in response to the different mRNA preparations, but the presence of spectinomycin did not alter the pattern of products synthesized in vitro by ribosomes from either vegetative or sporulating mutant cells. It could be that the resolution of the one dimensional gels utilized was insufficient to resolve differences that might exist. So in the SpcR mutant we have studied the inhibition of sporulation by spectinomycin apparently is not the result of greatly increased sensitivity of the mutant ribosomes to the drug during sporulation.

Asporogenous Mutants

Several mutants of *B. subtilis* unable to sporulate (Spo^-) as the result of acquisition of resistance to or dependence on antibiotics that inhibit protein synthesis have been isolated. The best studied examples of mutants of this type are a StrR mutant (4), streptomycin dependent (StrD) mutants (34), and a spectinomycin dependent (SpcD) mutant (19).

The frequency of the asporogenous phenotype among StrR mutants appears to be fairly low since only one such pleiotropic mutant was found in a collection of several hundred streptomycin resistant isolates (4). Among StrD mutants asporogeny appears to occur more frequently. In the one study reported each of six streptomycin dependent isolates proved, after purification, to be streptomycin dependent and asporogenous (34). The one spectinomycin-dependent mutant of *B. subtilis* that has been isolated is asporogenous (19).

Transductional analysis of the Str^r, Spo^- mutant revealed that the str^r mutation is linked to the cysA locus and probably lies within the strA gene (4). The str^d mutations also mapped to this region of the chromosome (34). The SpcD character required two mutations (19): one conferring resistance and mapping near or within the spcA locus and the second causing drug dependence and mapping between the purA and the cysA loci.

Genetic analyses indicate that both the drug resistance and the asporogenous phenotypes result from a single mutation in the Str^r, Spo^- mutant (4). The two characters cotransformed 100%. Revertants isolated for their ability to sporulate were of two types (3). Most (36 out of 40 independent isolates) were streptomycin resistant and spore plus while the remainder appeared to be true wild type revertants. Further examination revealed that the StrR, Spo^+ revertants still retained the original str^r, spo^- mutation in their genome. So apparently they have regained the ability to sporulate as the result of a second mutation. The second site mutations of the partial revertants have yet to be mapped so it is not known if they all affect the same gene. Of the four sporogenous partial revertants examined,one contained a ribosomal protein with altered electrophoretic mobility (unpublished data). The modified protein has been tentatively identified as S4, however it is not known if it is actually analogous to S4 of the *E. coli* 30S subunit which is the ramA protein (52) and also which is modified in some streptomycin independent revertants of streptomycin dependent mutants (18). The mutation responsible for the partial reversion of this mutant and presumably for the modification of protein S4 does not cotransform with the original str^r mutation. This is consistent with the observation by Osawa, Tokui, and Saito (32) that the gene for *B. subtilis* S4 is not located in the cysA-linked cluster of ribosomal genes.

As originally isolated some of the streptomycin dependent mutants of *B. subtilis* were asporogenous while others could sporulate apparently normally (34). Transformational analyses of the dependent mutants revealed that each of the six original isolates contained a streptomycin dependence mutation which caused asporogeny. Here again it was found that second site mutations could suppress the asporogenous phenotype caused by ribosomal mutation. They cotransform from 10 to 20% with the str^d mutation suggesting that they also may affect ribosomal components.

The spectinomycin resistance mutation of the spectinomycin-dependent mutant causes conditional asporogeny (19). For that reason it was not too surprising that the spectinomycin-dependent mutant was asporogenous. However, under certain conditions the dependent mutant will grow in the absence of drug. Under these conditions without drug the mutant still retains the asporogenous character. Drug independent revertants which retain the SpcR character are of various sporulation phenotypes ranging from totally asporogenous through temperature sensitive sporulation to wild type sporulation.

The StrR, Spo$^-$ mutant apparently is blocked rather early in the sporulation process probably before stage II (5). Of the StrD, Spo$^-$ mutants examined one is blocked in stage III and another is blocked very early, apparently stage I.

The pattern of proteins synthesized by stationary phase cells of the StrR, Spo$^-$ mutant is very similar to that of proteins synthesized by sporulating wild type cells except the pattern from the mutant lacked several protein spots (<10 spots) readily discernible in the wild type pattern (5). This observation demonstrates that global protein synthesis by the asporogenous mutant is similar to that of the wild type strain but that the synthesis of certain presumably sporulation specific proteins is defective.

The reduced level of misreading by ribosomes of certain streptomycin resistant mutants of *E. coli* has been associated with the inability to synthesize specific proteins (14). The level of misreading by ribosomes of the StrR, Spo$^-$ mutant of *B. subtilis* is much less than the level observed with wild type ribosomes (4). Also ribosomes of the StrD, Spo$^-$ mutants misread much less than do wild type ribosomes (34). The same low levels of misreading of these mutants is retained in streptomycin resistant or dependent, sporogenous revertants suggesting that the alterations in misreading are not the basis for asporogeny in the original mutants (3).

PHENOTYPIC REVERTANTS OF STAGE 0 MUTANTS

Asporogenous mutants that grow normally but carry out no known sporulation functions are known as stage 0 or *spo0* mutants. These mutants are quite pleiotropic with the *spo0A* mutants being the most pleiotropic. Some of the phenotypes observed in *spo0A* mutants include: reduced transformability, loss of antibiotic and protease production (21), sensitivity to bacteriophages ϕ2 and ϕ15 (21, 24), and hypersensitivity to certain antibiotics such as polymyxin B (30) and *B. subtilis* sporulation antibiotic and, of course, asporogeny. Partial revertants of *spoOA* mutations have been isolated (16, 22, 23, 25) in which many of the pleiotropic characters are reverted to wild type but the asporogenous character is retained. The partial phenotypic revertants were isolated as variants of *spoOA* mutants which are resistant to polymyxin B (16, 25) or *B. subtilis* antibiotic (48) or bacteriophage ϕ2 or ϕ15 (23).

Several of these partial revertants have been analyzed genetically (48). Four loci were found to be involved in the partial revertants. They were labeled *abrA*, *abrB*, *abrC1* and *abrD17*. Partial revertants with mutations at the *abrB* locus were found to have alterations of ribosomal proteins. Thus far only 50S subunit proteins have been found to be affected (40, 49). The observed modifications include: lack of a ribosomal protein or altered electrophoretic mobility of one and in some cases two proteins. The *abrB* locus which is located between *purA* and *cysA* on the *B. subtilis* chromosomal map is quite removed from the *cysA*-linked cluster of ribosomal genes. It may represent a new cluster of ribosomal genes (49) or possibly a gene involved in modifying ribosomal proteins.

POLYPHOSPHORYLATED NUCLEOTIDES

Several polyphosphorylated nucleotides have been found in *B. subtilis*. Rhaese and coworkers (35) have identified both guanine and adenine containing polyphosphorylated nucleotides. Synthesis of three adenine nucleotides (adenosine tetraphosphate, adenosine pentaphosphate and adenosine hexaphosphate) was correlated with the onset of sporulation (35). Two of these nucleotides (adenosine tetraphosphate and adenosine pentaphosphate) were reported to require ribosomes from sporulating cells for in vitro synthesis (37). Polyphosphorylated nucleotides and especially adenosine hexaphosphate were proposed to be important in regulating the sporulation process (36). This proposal has become somewhat controversial. The existence of adenosine hexaphosphate could not be confirmed by other workers (31). Also the adenosine polyphosphate nucleotides as well as guanosine polyphosphate nucleotides are synthesized in response to amino acid starvation in stringent strains and are not synthesized by relaxed strains even though the relaxed strains sporulate normally (31). This latter observation contradicts the suggestion that the polyphosphorylated nucleotides are required for sporulation.

CONCLUDING REMARKS

In summary, single site ribosomal mutations cause abnormalities in the ability of *B. subtilis* to sporulate. The observation that sporulation is not blocked at the same stage in all the different mutants suggests that ribosomal mutations can affect the sporulation process unequally. In any particular mutant the stage at which sporulation is blocked is unique and constant suggesting a direct effect on sporulation or on a sporulation process possibly of regulatory importance. In all the mutants examined sporulation related protein synthesis is disturbed. In a couple of instances this is manifested as the inability to synthesize a limited number of presumably sporulation-specific proteins. This defect may cause the pleiotropic sporulation phenotypes in these cases. It could be that the configuration of the mutated ribosomes is sufficiently altered to prevent either the recognition or the translation of the mRNAs for certain sporulation-specific proteins. There is also the suggestion that RNA synthesis may be altered in certain mutants. Thirdly lipid metabolism may be disrupted in some cases. To understand how ribosomal mutations could specifically affect any or all of these will require more study.

ACKNOWLEDGEMENTS

I would like to thank those investigators who made unpublished data available. The research in my laboratory was supported by the College of Agriculture and Life Sciences of the University of Wisconsin-Madison and by grant number PCM77-09326 from the National Science Foundation.

REFERENCES

1. Cannon, J.G., and Bott, K.F. (1979) Molec. gen. Genet. 174, 149-162.
2. Cannon, J.G., and Bott, K.F. (1979) Molec. gen. Genet. (in press).
3. Campbell, K.M. (1979) Ph.D. Dissertation, University of Wisconsin, Madison.
4. Campbell, K., and Chambliss, G. (1977) Molec. gen. Genet. 158, 193-200.

5. Campbell, K., Leventhal, J., and Chambliss, G.H. (1978) in Spores VII, G.H. Chambliss and J.C. Vary, eds. p.p. 250-253. American Society of Microbiology, Washington, D.C.

6. Chakrabarti, S.L., and Gorini, L. (1975) Proc. Natl. Acad. Sci. USA 72, 2084-2087.

7. Chakrabarti, S.L., and Gorini, L. (1977) Proc. Natl. Acad. Sci. USA 74, 1157-1161.

8. Domoto, T., Kobayashi, K., and Kobayashi, Y. (1975) in Spores VI, P. Gerhardt, R.N. Costilow, and H.L. Sadoff, eds., pp. 307-313. American Society for Microbiology, Washington, D.C.

9. Geisser, M., Tischendorf, G.W., and Stöffler, G. (1973) Molec. gen. Genet. 127, 129-145.

10. Gette, W.R. (1976) Ph.D. Dissertation, University of Wisconsin, Madison.

11. Goldman, R.C., and Tipper, D.J. (1979) J. Bacteriol. 138, 625-637.

12. Goldthwaite, C., Dubnau, D., and Smith, I. (1970) Proc. Natl. Acad. Sci. USA 65, 96-103.

13. Gorini, L. (1971) Nature New Biol. 234, 261-264.

14. Gorini, L. (1974) in Ribosomes, M. Nomura, A. Tissieres, and P. Lengyel, eds., pp. 791-803, Cold Spring Harbor Laboratory, Cold Spring Harbor, New York.

15. Graham, R.S., and Bott, K.F. (1975) Molec. gen Genet. 137, 227-237.

16. Guespin-Michel, J.F. (1971) J. Bacteriol. 108, 241-247.

17. Harford, N., and Sueoka, N. (1970) J. Mol. Biol. 51, 267-286.

18. Hasenbank, R., Guthrie, C., Stöffler, G., Wittmann, H.G., Rosen, L., and Apirion, D. (1973) Molec. gen. Genet. 127, 1-18

19. Henkin, T.M., Campbell, K.M., and Chambliss, G.H. (1979) J. Bacteriol. 137, 1452-1455.

20. Hirochika, H., and Kobayashi, Y. (1978) J. Bacteriol. 136, 983-993.

21. Hoch, J.A., and Spizizen, J. (1969) in Spores IV, L.L. Campbell, ed. Anerican Society for Microbiology, Washington, D.C.

22. Ito, J. (1971) Molec. gen. Genet. 112, 104-109.

23. Ito, J. (1973) Molec. gen. Genet. 124, 97-106.

24. Ito, J., and Spizizen, J. (1971) J. Virol. 7, 515-523.

25. Ito, J., and Spizizen, J. (1972) in Spores V, H.O. Halvorson, R. Hanson, and L.L. Campbell, eds., pp. 107-112. American Society for Microbiology, Washington, D.C.

26. Jaskunas, S.R., Nomura, M., and Davies, J. (1974) in Ribosomes, M. Nomura, A. Tissieres, and P. Lengyel, eds., pp. 333-368, Cold Spring Harbor Laboratory, Cold Spring Harbor, New York.

27. Kimura, A., Muto, A., and Osawa, S. (1974) Molec. gen. Genet. 130, 203-214.

28. Kobayashi, H., Kobayashi, K., and Kobayashi, Y. (1977) J. Bacteriol. 132, 262-269.

29. Leighton, T.J. (1974) J. Mol. Biol. 86, 855-863.

30. Michel, J.F., and Millet, J. (1970) J. Appl. Bacteriol. 33, 220-227.

31. Nishino, T., Gallant, J., Shalit, P., Palmer, L., and Wehr, T. (1979) J. Bacteriol. (in press).

32. Osawa, S., Tokui, H., and Saito, H. (1978) Molec. gen. Genet. 164, 113-129.

33. Pestka, S., Weiss, D., Vince, R., Wienen, B., Stöffler, G., and Smith, I. (1976) Molec. gen. Genet. 144, 235-241.

34. Quan, G.R., Campbell, K.M., and Chambliss, G.H. (1979) Molec. gen. Genet. (in press).

35. Rhaese, H.J., Grade, R., and Dichtelmüller, H. (1976) Eur. J. Biochem. 64, 205-213.

36. Rhaese, H.J., and Groscurth, R. (1976) Proc. Natl. Acad. Sci. USA 73, 331-335.

37. Rhaese, H.J., and Groscurth, R. (1978) Eur. J. Biochem. 85, 517-528.

38. Rosenkranz, H.S. (1963) Biochem. 2, 122-125.

39. Rosenkranz, H.S., Bendich, A.J., and Carr, H.S. (1964) Biochim. Biophys. Acta 82, 110-117.

40. Shiflett, M.A., and Hoch, J.A. (1978) in Spores VII, G.H. Chambliss and J.C. Vary, eds., pp. 136-138. American Society for Microbiology, Washington, D.C.

41. Smith, I. (1977) In Molecular Mechanisms of Protein Synthesis, H. Weissbach, and S. Pestka, eds., pp. 627-700, Academic Press, New York.

42. Smith, I., Dubnau, D., Morell, P., and Marmur, J. (1968) J. Mol. Biol. 33, 123-140.

43. Smith, I., and Paress, P. (1978) J. Bacteriol. 135, 1107-1117.

44. Smith, I., Paress, P., and Pestka, S. (1978) Proc. Natl. Acad. Sci. USA 75, 5993-5997.

45. Staal, S.P., and Hoch, J.A. (1972) J. Bacteriol. 110, 202-207.

46. Strigini, P., and Gorini, L. (1974) J. Mol. Biol. 47, 517-530.

47. Tipper, D.J., Johnson, C.W., Ginther, C.L., Leighton, T., and Wittmann, H.G. (1977) Molec. gen. Genet. 150, 147-159.

48. Trowsdale, J., Chen, S.M.H., and Hoch, J.A. (1978) in Spores VII, G.H. Chambliss and J.C. Vary, eds., pp. 131-135. American Society for Microbiology, Washington, D.C.

49. Trowsdale, J., Shiflett, M., and Hoch, J.A. (1978) Nature 272, 179-181.

50. Yates, J.L., Gette, W.R., Furth, M.E. and Nomura, M. (1977) Proc. Natl. Acad. Sci. USA 74, 689-693.

51. Zengel, J.M., Young, R., Dennis, P.P., and Nomura, M. (1977) J. Bacteriol. 129, 1320-1329.

52. Zimmermann, R.A., Garvin R.T., and Gorini, L. (1971) Proc. Natl. Acad. Sci. USA 68, 2263-2267.

PART 6: EUKARYOTIC RIBOSOMES

THE STRUCTURE AND FUNCTION OF EUKARYOTIC RIBOSOMES

IRA G. WOOL

Department of Biochemistry,
University of Chicago,
Chicago, IL 60637

INTRODUCTION

The mission of research on eukaryotic ribosomes is to know the structure and function of the organelle: to be able to specify the coordinates, and to define the activity, of each of the molecules in the particles. A subsidiary aim is to compare the structure of eukaryotic and prokaryotic ribosomes. There are significant differences between the two classes of particles. Eukaryotic ribosomes are appreciably larger: they contain a greater number of proteins, 70-80 rather than 53, and they have an extra molecule of RNA as well. Moreover, the proteins and nucleic acids are, on the average, larger. The difference in size is a paradox since eukaryotic ribosomes perform the same general function, namely to catalyze the synthesis of protein; more to the point, they appear to employ appreciably the same partial reactions, although there may be real differences in the means by which the initiation of peptide synthesis is accomplished. One would like to know the nature of the evolutionary pressure for the accretion of the extra proteins in eukaryotic ribosomes - whether there are functions of the particle still to be uncovered. It has not escaped attention that knowledge of the conserved features may be of help in understanding the function of both eukaryotic and prokaryotic ribosomes.

This review supplements one on the same subject written only a year ago (1). Several of the topics dealt with in greater detail before are epitomized here; on the other hand there is more information now on the identity of the proteins that bind to ribosomal RNA and that topic is given more attention. It should be noted that the emphasis is on mammalian (particularly rat liver) ribosomes.

GENERAL CHARACTERISTICS OF EUKARYOTIC RIBOSOMES

Eukaryotic ribosomes are composed of two subunits, which together contain four molecules of RNA and perhaps as many as eighty proteins. The sedimentation coefficient of the monomer and subparticles varies between species (as do other of the characteristics) but are close to 80 for the ribosome and 60 and 40 for the subunits (2); those numbers are used to designate the particles.

The small subunit of rat liver ribosomes has a sedimentation coefficient of 36.9 (3). The particle contains one molecule of 18S RNA of molecular weight 0.7 x 10^6 (4, 5) and about 30 proteins whose combined mass is about 0.74 x 10^6 daltons (see later). Thus the molecular weight of the particle should be 1.44 x 10^6 which is almost exactly what has been measured (3). The particle might be expected then to have nearly equal amounts of protein and RNA, whereas the chemical determination (6) reveals it to contain only 45% RNA. The buoyant density of the small subparticle is 1.515, from which the percent RNA can also be calculated to be 45 (6). The values obtained by others (7, 8) in a different way are similar. The extinction coefficient for the 40S subparticle is 94.5, and the diffusion coefficient 2.0 x 10^{-7} (3).

The large subunit of rat liver ribosomes has a sedimentation coefficent of 56.3 and is composed of one molecule each of 5S, 5.8S, and 28S RNA, and 45 to 50 proteins. The molecular weight of 28S RNA is given as 1.7 x 10^6 (4, 5) and the molecular weights of 5S and 5.8S RNA, calculated from the sequence (9, 10) is approximately 39,000, and 51,000, respectively. The total mass of RNA in the particle is about 1.79 x 10^6 daltons. The mass of the protein is approximately 1.05 x 10^6. The expected molecular weight of the particle then is 2.84 x 10^6, and the actual measurement is 2.9 x 10^6 (3) which, given the uncertainties, is in reasonable agreement; a similar value had been obtained before (7). The 60S subunit has a buoyant density of 1.600 (6); the RNA content from the buoyant density is 59.4%, and the chemical determination gave the same value (6). Thus the large subparticle has an appreciably greater proportion of RNA than the 40S subunit.

Eukaryotic ribosomes are by no means a uniform group; "80S ribosomes" actually range in aggregate mass from about 3.9 for plants to 4.55 x 10^6 daltons for mammals (11-13). The change in ribosome mass is the result of an increase in the size of the large subunit from 2.4 to 3.05 x 10^6 daltons; the increase is in the mass of both RNA and protein. The 40S particle, on the other hand has not changed appreciably in size during eukaryotic evolution.

There are detailed electron microscopic studies of the morphology of eukaryotic ribosomes (14-17). Observers (16, 17) have been struck by the general similarity in the morphology of eukaryotic and prokaryotic ribosomal subunits, despite the obvious larger size of the former. The most prominent features of rat liver ribosomes are the division of the small subunit into two unequal parts; the asymmetric apposition of the small subunit to the large in ribosome couples which confers a handedness to the structure; and a notch in the profile of the large subparticle (15). The 40S subunit approximates a curved and flattened prolate elipsoid with dimensions of 230 x 140 x 115 Å (15); it is divided into "head" (one-third) and "body" (two-thirds) regions by a transverse partition seen in electron micrographs as a dense line 80 Å from one end of the subunit. The 60S subunit is seen either as a rounded object, or as asymmetric and skiff-shaped with a pointed end opposite to a blunt end (15).

A particularly promising approach to the solution of the structure of eukaryotic ribosomes is through three-dimensional reconstruction from electron micrographs of naturally occurring crystalline arrays which develop in the oocytes of lizards during hibernation (18, 19), or which can be induced in chicken embryos by cooling (15, 20-22).

ISOLATION AND CHARACTERIZATION OF EUKARYOTIC RIBOSOMAL PROTEINS

There is no way to arrive at a solution of the structure of an organelle without knowing the chemistry of the constituents. So several years ago we undertook to isolate and characterize the proteins in rat liver ribosomes. While the methods employed were in general similar to those that had served for the purification of prokaryotic (E. coli) ribosomal proteins (23-26), the effort was facilitated by the development of two ancillary procedures: a micro-method for the rapid analysis of small samples of ribosomal proteins by two-dimensional polyacrylamide gel electrophoresis (27); and an efficient means of fractionating ribosomal subunit proteins into smaller groups (28).

Rat liver ribosomal subunit proteins were separated into groups by stepwise elution from carboxymethylcellulose with LiCl at pH 6.5 (28). The proteins in the group fractions were resolved by ion-exchange chromatography on carboxymethylcellulose, or on phosphocellulose, or for the small subset of acidic proteins on DEAE-cellulose (29-33). Some of the proteins were purified entirely by ion-exchange chromatography,

TABLE 1. RAT LIVER RIBOSOMAL PROTEINS

40S Subunit		60S Subunit	
Proteins	M_r ($\times 10^{-3}$)	Proteins	M_r ($\times 10^{-3}$)
(Sa)[a]	41.5	(La)	37.9
(Sc)	33.0	(Lb)	29.8
S1	39.0	(Lf)	14.6
S2	33.1	P1	16.1
S3	30.4	P2	15.2
S3a	32.0	P3	13.0
(S3b)	30.4	L3	37.8
S4	29.5	L4	41.8
S5	22.8	L5	32.5
(S5a)	21.5	L6	33.0
S6	31.0	L7	29.2
S7	22.2	L7a	28.7
S8	26.8	L8	28.4
S9	24.3	L9	24.7
S10	20.1	L10	24.2
S11	20.7	L11	21.3
S12	14.9	L12	18.7
S13	18.6	L13	26.3
S14	17.3	L13a	24.6
S15	19.6	L14	25.8
S15a	15.7	L15	24.5
(S16)	17.1	L16	18.7
S17	18.0	L17	22.1
S18	18.5	L18	24.5
S19	17.1	L18a	21.3
S20	16.5	L19	25.3
S21	12.3	L20	16.2
S23/S24	18.8	L21	20.3
S25	17.0	L22	16.1
S26	16.5	L23	15.6
S27	14.5	(L23a)	18.0
S27a	12.8	(L25)	17.5
S28	11.3	L26	18.6
S29	11.2	L27	17.8
		L27a	18.0
		L28	17.8
		L29	20.5
		L30	14.5
		L31	15.6
		L32	17.2
		L33	15.6
		L34	15.8
		L35	17.5
		L35a	13.7
		L36	14.3
		L36a	16.2
		L37	15.4
		L37a	12.8
		L38	11.5
		L39	11.6

[a]In 40S subunit the ΣM_r = 746,000; number of proteins = 34; average M_r = 22,000. In 60S subunit the ΣM_r = 1,049,000; number of proteins = 50; average M_r = 21,000.

[b]The proteins in parentheses may not be components of the ribosome or may not be unique peptides (see text for details).

others required gel filtration as well. In that manner 34 proteins were purified from extracts of 40S subunits (29, 32, 33), 50 from the 60S subparticle (30, 31, 33), a total of 84 (Table 1). Earlier Terao and Ogata (34) had isolated 12 proteins from the 40S subparticle of rat liver ribosomes. Westermann and Bielka (35) had reported the isolation and characterization of 24 of the proteins of the small subunit of rat liver ribosomes. Some rat liver ribosomal proteins have been prepared by elution from plugs cut out of two-dimensional gels (36). A start has been made on the isolation of the proteins from the ribosomes of other eukaryotic species (37, 38).

The molecular weight of the purified proteins was determined by electrophoresis in gels containing sodium dodecyl sulfate (Table 1). The number average molecular weight for the 34 40S ribosomal subunit proteins is about 22,000, (the range is 11,200 to 39,000). The number average molecular weight for the 50 proteins isolated from 60S subparticles is 21,000 (the range is 11,500 to 41,800). The average molecular weight of an eukaryotic (rat liver) ribosomal protein is about 21,300; the average for E. coli ribosomal proteins is approximately 18,000. Although, there are some discrepancies, the results, in general, do not differ greatly from values reported earlier (34, 35, 39-44), especially when one gives consideration to the variety of methods that have been used.

The amino acid composition of the isolated rat liver ribosomal proteins tends to be similar, but in most cases they are definitely unique (29-33). There are, however, important exceptions (see next).

There are two useful procedures - genetic analysis and reconstitution of subunits - for testing whether a protein is a constituent of a prokaryotic ribosome. Neither can be used in a determination of the status of putative eukaryotic ribosomal proteins. It is vexing that one still can not say how many proteins there are in eukaryotic (mammalian) ribosomes. The original estimate was from inspection of two-dimensional polyacrylamide gel plates (45-50). It was calcualted that there were about 72; 31 in the 40S, and 41 in the 60S subunit (45). Eighty-four proteins have been isolated from rat liver ribosomes (29-33). That is 12 more than was estimated at first. Several proteins designated initially have not been isolated, indeed, were not even encountered during the purification; whereas, other proteins which were not resolved by electrophoresis were isolated. The proteins in the first subset were, in general, only rarely seen on two-dimensional gels or stained lightly; the proteins in the second group were, as a rule, difficult to resolve by electrophoresis. Proteins enumerated but not isolated include: S22, S23 or S24, S30, and S31 from the small subunit; and L1, L2, L24, L40, and L41 from the large subparticle.

Proteins S22, S30, and S31 are only rarely seen on two-dimensional electrophoretograms and were not identified during the purification of the other proteins. It is possible thay are not structural proteins of the small subunit. It is not certain whether S23 and S24 are a single protein or two distinct polypeptides. They generally are not resolved by two-dimensional gel electrophoresis; however, the configuration of the spot suggests two proteins (45). The protein that was isolated forms a single spot on two-dimensional gel electrophoresis, and a single diffuse band after electrophoresis in one dimension in a number of different concentrations of polyacrylamide either with sodium dodecyl sulfate, or in urea at pH 4.5 (29). For that reason, it is designated for the time S23/S24.

The proteins which were isolated include several (Sa, Sb, S3a, S3b, S5a, S15a, and S27a) that were not in the original classification and whose exact status is still uncertain. Sa and Sb are relatively acidic proteins found in small amounts in extracts of the 40S subunit; there is no assurance they are ribosomal proteins. Comparison of the amino acid composition, molecular weight, and tryptic peptide fingerprints suggests S3b may be a chemically modified form of S3a; and S5a, may be a derivative of S5. S15a occupies the same position on two-dimensional gels as L25; but they differ in molecular weight and amino acid composition; S15a also differs from S15. S27a is clearly different than S27: they form distinct spots on two-dimensional gel electrophoresis; they have different molecular weights (S27, 14,500; S27a, 12,800); and their amino acid compositions are dissimilar (29, 32).

The large subunit proteins L1, L2, L24, L40, and L41 designated in the beginning have not been isolated. L1, L12, and L24 were not found in any of the groups fractions (28), and hence may not be constituents of the ribosome.

On the other hand the 50 proteins isolated from the 60S subunit (30, 31, 33) include 14 (La, Lb, Lf, P1, P2, P3, L7a, L13a, L18a, L23a, L27a, L35a, L36a and L37a) that were not in the original classification. La, Lb, and Lf are acidic proteins found in small amounts and there is doubt they are structural proteins of the ribosome. Inspection of the amino acid composition, molecular weights, and tryptic peptide fingerprints raises the question of whether L23a may not be the same protein as S18. The other proteins in the group appear unique.

We were from the first sensitive to the possibility that some of the isolated proteins might be chemically modified forms of another polypeptide. The chemical modification could either be physiological, as, for example, phosphorylation, acetylation, or methylation; or it could be adventitious, having occurred during purification. Since the proteins are exposed to urea during isolation, carbamylation of amino groups can occur even though the purification is done at acidic pH: carbamylation would change the net charge and later the migration of the protein during electrophoresis. A second change we were concerned about was proteolytic cleavage at a sensitive peptide bond. Of course, the enzyme would have to be specific since we assume most of the proteins are not affected.

The decision whether an isolated eukaryotic ribosomal proteins is unique (which must be tentative until the sequence is available) has rested on the evaluation of the molecular weight, the amino acid composition, and the behavior on electrophoresis. Most of the proteins are distinct by all three criteria, however, there are exceptions.

S3a and S3b form distinct spots on two-dimensional gel plates; more-over, they differ in mass (the apparent molecular weights are 32,000 and 30,400). Nonetheless, the amino acid composition of the proteins is very close (32). The relationship of S5 and S5a is very much the same: the proteins form distinct spots on two-dimensional electrophorograms; they differ in molecular weight by 1,300 (S5, 22,800; S5a, 21,500); again the amino acid compositions are similar (29, 32). In the circumstances; it is possible that S3b is a chemically altered form of S3a, and S5a is a derivative of S5. The alteration that suggests itself is proteolysis. The possibility was tested by doing peptide fingerprints. Tryptic peptide maps of S3a and S3b are so similar as to make it likely one is derived from the other (51). The comparison for S5 and S5a has not yet been made for lack of sufficient amounts of the latter.

S16 and S19 form distinct spots on two-dimensional gel plates, however, they have the same molecular weight (17,100), and like amino acid compositions (29, 32). The tryptic peptide maps for the two proteins are similar, although there are a few peptides unique to each (51). Once again they may differ only by some chemical modification.

The small subunit protein S15 and the large subparticle protein L25 occupy distinct positions on two-dimensional gels, and differ in molecular weight (S15, 19,600; L25, 17,500), yet their amino acid composition is similar (31, 32). Tryptic peptide fingerprints of the proteins have not yet been done.

S18 and L23a are found in the same zone on two-dimensional electropherograms, their molecular weights are similar (18,500 and 18,000), and their amino acid compositions hardly differ (31, 32). Since they have the same tryptic peptide maps (51) it is likely they are one protein.

The sum of the molecular weights of 29 of the proteins isolated from the 40S ribosomal subunit is approximately 0.603×10^6 (Table 1). For the time we do not include Sa and Sc whose status is not certain; S22, S30, and S31 which were not isolated; S3b, S5a, and S16, which are likely to be related to S3a, S5, and S19.[1] The molecular weight of 18S rRNA is reported (4) to be 0.7×10^6. Thus the total mass of the molecular constituents of the small ribosomal subunit is about 1.3×10^6 daltons, which is close to the molecular weight (1.4×10^6) determined for the particle (3).

Fifty proteins have been isolated from the 60S subunit of rat liver ribosomes: the sum of the molecular weights is approximately 1.05×10^6. If one sets aside La, Lb, and Lf, whose status is not certain; and L23a and L25 which are probably the same as S18 and S15, the total mass of protein is 0.93×10^6 daltons.[1] If we take the molecular weight of 28S rRNA to be 1.75×10^6 (4) and 5S and 5.8S to be about 0.9×10^6 (9, 10), then the mass of the rRNA in the 60S subparticle of mammalian ribosomes is approximately 1.84×10^6 daltons. The mass of the components of the large subunit then is about 2.77×10^6 daltons, near the estimate of 2.9×10^6 for the molecular weight of the particle. However, the

---

[1] If the proteins whose status is uncertain (given in parenthesis in Table 1) are excluded from the count then the 40S subunit has 29 and the 60S subunit 45 for a total of 74, very close to the original estimate of 72 (45).

values for the molecular weight of eukaryotic ribosomes and subunits are not reliable, nor for that matter are the determinations of the molecular weight of ribosomal proteins and ribosomal RNA precise, and hence one must be cautious in making judgments about the stoichiometry from the available data. It is, nonetheless, unlikely there is more than one mole of most proteins, although, it is still possible that there is more than one copy of a few.

Eukaryotic ribosomes have something of the order of 16,000 amino acids. Determining the sequence of all would be a monumental task, at least by conventional methods, and its achievement clearly is not imminent. It is possible that more rapid automated methods for sequencing small amounts of protein will be developed and make the task feasible. It is also conceivable that it will become a routine matter to determine the structure of proteins by sequencing the genes that code for them. One would still need the means to prepare large amounts of eukaryotic ribosomal protein DNA; it would be necessary to know at least the amino-terminal and the carboxy-terminal sequences of the protein; and it would be necessary to be able to recognize potential intervening DNA sequences. Nonetheless, one has the impression that if the sequence of all the proteins in eukaryotic ribosomes is ever to be had it will come from a combination of the analysis of DNA and of protein. What is being done for the time is to determine the amino-terminal sequence of proteins since that gives the maximum information for the expenditure of practical amounts of time and effort (52).

One purpose in collecting the data is to determine the relatedness of individual eukaryotic ribosomal proteins since it is possible that the number was increased from the 53 in prokaryotes to the approximately 80 in eukaryotes by gene duplication. A computer comparison of the amino acid composition of rat liver ribosomal proteins suggests a number are significantly similar statistically and hence supports the possibility (53). A second reason for the analyses is to search for homologous ribosomal proteins by correlating amino acid sequences from rat liver proteins with those from other species. The comparison has already indicated that at least some proteins have been conserved.

The sequence of the amino-terminal region of 12 rat liver ribosomal proteins - S4, S6, S8, P2, L6, L7a, L18, L27, L30, L37, L37a, and L39 - has been determined (52). The sequence studies confirm the homogeneity of the proteins, and a comparison of the sequences establishes that the individual proteins are unique - they contain no significant homologous sequences. The N-terminal regions of the rat liver proteins was compared with amino acid sequences in S. cerevisiae, in A. salina, and in E. coli ribosomal proteins. It seems likely that the protein P2 (1), an acidic phosphoprotein in the large subunit that may be related in function to E. coli L7/L12 (54, 55), is homologous with A. salina eL12 and with an acidic protein from S. cerevisiae (56, 57). Thus at least that one protein has been conserved during evolution from a fungus (yeast), to an invertibrate (brine shrimp), to a mammal (rat). Moreover, the protein L37 from rat liver (52) and Y55 from yeast ribosomes (58) are probably homologous; it is possible that rat liver L7a or L37a or both are related to S. cerevisiae Y44 (59), although the similar sequences are at the amino-terminus of the rat liver proteins and in an internal region of Y44. A number of similarities in the sequences of rat liver and E. coli ribosomal proteins have been found. However, it is not yet possible to say whether they connote a common ancestry. The most interesting possibility is the relation of rat liver L37 and E. coli L16. There are similarities in the N-terminal region of rat liver L37 and two separate sequences (position 21-34 and 77-91) of E. coli L16 (52). The latter is likely to be involved in ribosomal peptidyl transferase activity. Thus it is possible that rat liver L37 serves a similar function.

It is difficult to judge the significance of the similarities in sequence between rat liver and E. coli ribosomal proteins. There are insufficient data for the eukaryotic ribosomal proteins to establish homologies. Moreover, there is a serious lack of criteria to assess the importance of limited similarities in sequence. Thus the analyses to date serve mainly as a guide to further studies of the sequence of rat liver proteins and in one instance (the similarity of EC-L16 and RL-L37) of the function of a protein. It seems certain that ascertaining homologies between eukaryotic and prokaryotic ribosomal proteins by comparison of primary sequences will be difficult since it is likely that during evolution conservative changes in amino acids, deletions, and recombination will have occurred. Some of the similarities we have noted, if they are the vestiges of real homologies rather than merely chance phenomena, would indicate those processes have happened. Immunological studies (1, 60, 61) have shown that while prokaryotic and eukaryotic ribosomal proteins share determinants, cross-reaction is not common. Indeed, it has been demonstrated that antisera specific for chicken or rat liver ribosomal proteins recognize only about 20% of common determinants (62). It is by no means surprising then that sequence homologies between rat liver and E. coli ribosomal proteins are not prominent. Homologies might be more apparent if one knew and could compare the tertiary structure of these proteins. Nonetheless, intuition leads one to surmise that the ribosome arose on a single occasion and hence eukaryotic ribosomes are likely to be related to their prokaryotic ancestors. There is also circumstantial evidence that the fundamental structure and organization of ribosomes has been conserved (1), even though the primary sequence of the components varies. To what extent those surmises on the evolution of ribosomes are correct is most likely to come from further studies of the sequence of the molecular constituents.

While it is still not possible to give the exact number of proteins in eukaryotic ribosomes, there seems little doubt that it is considerably greater than the number in prokaryotic ribosomes. What purpose do the supernumerary proteins serve? Some of the extra proteins may be for interaction with receptors in the endoplasmic reticulum (63-67), others may be involved in regulation, especially the regulation of translation of messenger RNA. In that regard it appears that the ribosome-mRNA recognition process, and the partial reactions in the initiation of protein synthesis, may be fundamentally different in eukaryotes (68-71) and could require additional ribosomal proteins. The additional proteins may participate as well in the extraordinarily complicated process of ribosome biogenesis with its need for two-way traffic beween the cytoplasm and nucleus: they could take part in assembly or in the transport of assembled subunits. Finally, it is conceivable that ribosomes have luxury proteins that serve no function. There are thiostrepton resistant mutants of Bacillus megatarium (72) and Bacillus subtilis (73) that lack ribosomal protein L11. While the former is sick the latter is not (74). Thus, it follows that B. subtilis L11 has no function in protein synthesis or at least it subserves a role capable of being assumed by one or more of the remaining proteins.

One can look at the problem the other way around and ask why are prokaryotic ribosomes so small? Perhaps, because they had to be streamlined. During rapid growth in rich medium as much as 30% of all cellular protein is ribosomal (75, 76) and the bacterium might have difficulty supporting a particle with as much protein as eukaryotic ribosomes have. The assumptions implicit in the suggestion are that ribosomes evolved only once - which seems reasonable: that prokaryotic ribosomes initially had a greater number of proteins, a number comparable to that in eukaryotic particles; and that prokaryotic ribosomes having had more opportunity to evolve responded to selective pressure by discarding proteins and reducing the size of those that were retained - the advantage being a greater number of smaller ribosomes performing the same functions as fewer larger particles. All that would have happened after divergence of primitive eukaryotes. The argument, however, does not change the fundamental nature of the problem, it merely alters the way it is put: what prokaryotic ribosomal proteins could be dispensed with without apparent loss of function; if the extra proteins could be discarded with impunity why have they been retained by eukaryotes? It is possible that in the putative process, the function of several prokaryotic ribosomal proteins were combined, as seems to have happened with the initiation factors. Eukaryotes use a greater number of initiation factors than bacteria to catalyze a similar process (71).

THE ASSOCIATION OF RIBOSOMAL PROTEINS AND RNA

The proteins and nucleic acids in ribosomes are arranged in a specific three-dimensional configuration to form structural and functional domains. Solution of the structure of those particles, and an understanding of the relation of structure to function, requires information on the interaction of the ribosomal proteins with ribosomal RNA and with nucleic acid ligands like tRNA and mRNA that associate with ribosomes. An attempt is being made to identify, by affinity chromatography, the proteins that interact with the several species of rRNA.

5S rRNA

In the initial experiments the proteins that bind to rat liver 5S rRNA were characterized (77). The nucleic acid was oxidized with periodate and coupled by its 3' terminus to Sepharose through an adipic acid dihydrazide spacer (78, 79). Ribosomal subunit proteins were passed through the affinity column and those that were bound specifically were eluted and identified by polyacrylamide gel electrophoresis. The eukaryotic 5S rRNA binding proteins revealed by this procedure were L6 and L19 (Table 2); small amounts of L7, L23a, L27 or L27a, L35a, and L39 also were bound to the affinity column (77). No proteins of the 40S ribosomal subunit associated with immobilized 5S rRNA; nor did the proteins of E. coli ribosomes bind to rat liver 5S rRNA.

Metspalu et al. (80) identified L6 and L18 as the major proteins that associate with rat liver 5S rRNA attached to Sepharose. There is reason to believe that the protein identified as L18 was in fact L19 (see later).

TABLE 2. ASSOCIATION OF RIBOSOMAL PROTEINS WITH RIBOSOMAL NUCLEIC ACIDS

5S rRNA	5.8S rRNA	Elongator-tRNAs	Initiator-tRNA
L6	L6	L6	L6
L19	L19		
(L35a)[a]	(L35a)[a]	L35a	L35a
	L7		
	S9		
	S13		
		S15	

[a]L35a binds to 5S and 5.8S rRNA in small amounts.

Experiments of this kind must be closely controlled, if for no other reason, because basic ribosomal proteins can associate nonspecifically with nucleic acids. No 60S ribosomal proteins (nor any other ribosomal proteins for that matter) bound to the activated Sepharose 4B-adipic acid dihydrazide conjugate lacking a nucleic acid, indicating the proteins did not merely associate by ion exchange chromatography on unsubstituted hydrazide groups or on reactive groups on the CNBr-activated Sepharose (77). It is more important that no 40S ribosomal protein associated with immobilized 5S rRNA, nor did the proteins of either the small (30S) or large (50S) subparticle of E. coli ribosomes, although they contain many basic proteins. The results are reassuring in that they indicate that the binding of L6 and L19 to 5S rRNA is likely to be specific, not merely the chromatography of a basic protein on a nucleic acid.

The failure of any E. coli ribosomal proteins to bind to rat liver

5S rRNA suggests that rat liver L6 and L19 are not structurally homologous with *E. coli* L5, L18 and L25 - the proteins that bind to *E. coli* 5S rRNA (79, 81-83). The results conform to the finding of Wrede and Erdmann (84) that *E. coli* L18 and L25 do not bind to yeast 5S rRNA.

A 5S rRNA-protein complex (5S rRNP) can be extracted from the large subunit of eukaryotic ribosomes with EDTA (85-88). The single protein in the complex was identified as L5 (88) and it was assumed that it was one of, or the only, protein associated with 5S rRNA in the ribosome. No L5 was bound to immobilized 5S rRNA (77). The results of the two sets of observations can be reconciled in several ways. One possibility is that EDTA unfolded the ribosomal subparticle and caused the proteins to lose their specific locations on ribosomal RNA. EDTA will randomize the distribution of proteins in *E. coli* ribosomal subparticles (89). A second possibility is that the 5S rRNP particle prepared with EDTA preserves the physiological binding whereas affinity chromatography on 5S rRNA-Sepharose gives spurious results. The circumstances in which the chromatography was done - the ionic conditions were selected to favor detection of even weak binding of protein to 5S rRNA - and the control experiments mitigate against the possibility. There is another explanation and that is that the binding site for L5 is at or near the 3' end of 5S rRNA and that site is perturbed by periodate oxidation or coupling to the adipic acid dihydrazide or both.

Ionic strength is critical in the association of proteins with nucleic acids (90). At relatively low ionic strength, nonspecific binding can occur and at relatively high ionic strength authentic binding can be inhibited. The ionic strength of the binding buffer used in the affinity chromatography experiments described here was 0.38, close to that used by Traub and Nomura (i.e. 0.37) in the reconstitution of the small subunit of *E. coli* ribosomes (90). At an ionic strength of 0.38, two proteins, L6 and L19, bind strongly to 5S rRNA; several proteins (L7, L23a, L27 or L27a, L35a, and L39) bind weakly. Since the spots were faint, i.e. small amounts of protein were bound, their significance is not certain. They might be contaminants. For example, the proteins that bind specifically to *E. coli* 5S rRNA are L5, L18, and L25 (79, 81-83). However, small amounts of a number of other *E. coli* large ribosomal subunit proteins (L1, L7, L10, L12, L27, and L30) associate with 5S rRNA and can be identified on two-dimensional gel plates (84). The latter are generally considered non-specific contaminants. The rat liver ribosomal proteins that bind weakly to 5S rRNA may result from protein-protein interaction, i.e. they may bind to L6 and L19 associated directly with 5S rRNA. The proteins may or may not associate with L6 and L19 in the ribosome. It is also possible that the binding of L6 and L19 or both alters the conformation of 5S rRNA and creates binding sites for the other proteins: that process does occur with E. coli 16S rRNA (98). Since the binding of the minor proteins would require prior binding of L6 and L19, small amounts would be expected. Finally, one must recognize that proteins that are associated physiologically with 5S rRNA might not bind during affinity chromatography because the structure of the nucleic acid has been altered, just as some proteins might bind factitiously because binding sites have been created by alterations of structure.

5.8S rRNA

The proteins that bind to 5.8S rRNA, a nucleic acid that is associated by hydrogen bonds with 28S rRNA in the 60S subunit of eukaryotic ribosomes (91), was determined next (92). The ribosomal proteins that associate with immobilized 5.8S rRNA were L19, L8, and L6 from the 60S subunit; and S13, and S9 from the small subparticle (Table 2). Small amounts of L14, L19a, L18, L27 or L27a, and L35a, and of S11, S15, S23/S24, and S26 also were bound to the affinity column. E. coli ribosomal proteins did not bind to rat liver 5.8S rRNA. The finding was somewhat surprising since E. coli ribosomal proteins L18, and L25, associate specifically with yeast 5.8S rRNA (84). There is no satisfactory explanation for the apparently discordant results, although it is true the experiments were done differently; with rat liver 5.8S rRNA by affinity chromatography; with yeast 5.8S rRNA by isolation on sucrose gradients of the complex formed from mixtures of proteins and the nucleic acid. It is possible that binding of heterologous E. coli ribosomal proteins to eukaryotic 5.8S rRNA is sensitive to small changes in the conformation of the nucleic acid and that secondary structure is better preserved in the conditions used for formation of ribonucleoprotein complex. It should be noted, however, that it has been shown that E. coli ribosomal proteins do not bind to yeast 5.8S rRNA immobilized on affinity columns (93).

Metspalu et al. (80) used affinity chromatography to identify L5, L6, L7, and L18 as the ribosomal proteins that bind to rat liver 5.8S rRNA.

While there is agreement than that L6 binds to the nucleic acid, in our experiments L8 and L19 associated with 5.8S rRNA rather than L5, L7, or L18. The discrepancies may be in part the result of the conditions employed in the binding reaction, and in part a reflection of the difficulty in making a precise identification of the ribosomal proteins. In the Metspalu experiments (80) affinity chromatography was done at 4°C, whereas the temperature in the Ulbrich experiments (92) was 18°C. We have found that the binding of ribosomal proteins to rRNA can be affected by the temperature at which the affinity chromatography is done. There were, in addition, differences in the concentration of ions in the binding buffers which is critical for the association of proteins and nucleic acids (90, 94). Moreover, conductivity of the buffers is sensitive to temperature. A partial reconciliation of the two sets of findings may be implicit in a consideration of the use of polyacrylamide gel electrophoresis for the identification of the ribosomal proteins. There are pairs of ribosomal proteins that can not be distinguished easily by two-dimensional gel electrophoresis alone (31, 32); the identification is facilitated by a combination of two-dimensional electrophoresis in urea and one-dimensional electrophoresis in gels containing sodium dodecyl sulfate. Metspalu et al. (80) used only the former; moreover the sample of ribosomal proteins was not analyzed simultaneously with a small amount of all the proteins of the relevant ribosomal subunit to provide a background to assist in the identification (27). The limitation is best exemplified by the difficulty in resolving L18, which Metspalu et al. (80) found to bind to 5.8S rRNA, and L19 which Ulbrich et al. (92) identified. L18 and L19 are poorly separated by two-dimensional gel electrophoresis and identification is especially difficult if the analysis is done without a background of all the 60S proteins. However, L18 and L19 can easily be resolved on gels containing sodium dodecyl sulfate. Ulbrich et al. (92) combined one-dimensional electrophoresis in gels containing sodium dodecyl sulfate, and two-dimensional gel electrophoresis using a background of 60S ribosomal proteins in making the identification. Thus we are confident that L19, and not L18, binds to 5.8S rRNA. Nonetheless, there seems to be real differences in their finding (80) that L5 and L7 bind to 5.8S rRNA and that of Ulbrich et al. (92) that L8 is associated with the nucleic acid. Finally, Ulbrich et al. (92) found that the small ribosomal subunit proteins S9 and S13 bind to 5.8S rRNA; there is no indication that Metspalu et al. (80) tested the possibility.

HeLa cell polysomes have been oxidized with periodate to cross-link nearby proteins to the 3'-terminus of rRNA (95). The proteins covalently bound to 5.8S rRNA were tentatively identified as L7a and L27a; neither protein associated with immobilized 5.8S rRNA. It is possible that the binding sites for L7a and L27a are at or near the 3'-end of 5.8S rRNA, and that that region of the nucleic acid is perturbed by periodate oxidation or coupling to the adipic acid dihydrazide, or both.

It would be valuable to have an independent means of verifying the results obtained by affinity chromatography. Unfortunately, there are no separate but comparable experiments. The methods used to assess the binding of prokaryotic ribosomal proteins to rRNA *in vitro*, and those employed to authenticate the *in vitro* results by analysis of the association in ribosomes (96), either have not yet been tried with eukaryotic ribosomal components or have been tried but have not been successful. Thus there are no strict criteria for evaluating divergent results, but even more important there is no way to determine how closely the results of affinity chromatography approximate the interaction of ribosomal proteins and nucleic acids *in situ*. The binding of ribosomal proteins to ribosomal nucleic acids is conditioned by the conformation of the molecules (97-99); there is no certainly that the nucleic acid retains its native configuration after it is oxidized with periodate and coupled to the affinity adsorbant. Some assurance on that score comes from experiments with prokaryotic 5S rRNA and tRNA. The proteins that associate with *E. coli* 5S rRNA have been shown by a variety of procedures to be L5, L18, and L25 (81-83); the same proteins (L5, L18, and L25) associate with *E. coli* 5S rRNA-Sepharose affinity columns (79; see also later). Moreover, the proteins that interact with *E. coli* tRNA immobilized on Sepharose (100) are exactly those one would expect from entirely different experiments. Those results provide some comfort.

The binding of 40S ribosomal proteins S13 and S9 to 5.8S rRNA, which is a component of the large subparticle, suggests that the nucleic acid may participate in the formation of subunit couples (i.e. 80S ribosomes); it follows, if the supposition is correct, that 5.8S rRNA and the proteins that bind to it, are near the ribosomal subunit interface. In that regard it has been shown (101) that 5.8S rRNA can be released from ribosomes by brief treatment at 60°C; the nucleic acid will reassociate with ribosomes in buffer containing a high concentration of salts. The results are consistant with the location of 5.8S rRNA on the surface of the 60S ribosomal subunit and hence available for association with the 40S subparticle.

L6 and L19 bind to 5S, as well as to 5.8S rRNA (Table 2). In the case of 5S rRNA the most prominent protein was L6; the ratio of the binding of L6 and L19 was 2:1. With 5.8S rRNA the most prominent protein was L19; the ratio of the binding of L6, L8, and L19 was 1:1:3·5. If the results obtained from affinity chromatography reflect protein-nucleic acid as-asociation in the ribosome, then L6 and L19 must have at least two domains for binding rRNA, one for 5S and a second for 5.8S. Since L6 also associates with tRNA (see later) it may well have a binding site for that ligand too. The results further suggest that the two nucleic acids, 5S and 5.8S rRNA, and their associated proteins L6, L8, and L19 are in close proximity in the 60S subunit. The domain may comprise a portion of the A and P sites (see later).

The large subunit of eukaryotic ribosomes contain two small ribosomal nucleic acids, 5S and 5.8S, whereas prokaryotic ribosomes have only 5S rRNA. There are reasons to suspect that the functionally homologous nucleic acids are prokaryotic 5S and eukaryotic 5.8S (84, 102). The latter is part of the transcript of rDNA with 18 and 28S rRNA; just as prokaryotic 5S rRNA is part of a transcription unit with 16 and 23S rRNA; on the other hand the genes for eukaryotic 5S rRNA are located at a separate site in the nucleus. Moreover, there are differences in the primary structure of prokaryotic and eukaryotic 5S rRNA (102-104). What is most significant is that the latter does not have the sequence C-G-A-A-C that is in all prokaryotic 5S rRNAs, and that is complementary to T-ψ-C-G in loop IV of all elongator-tRNAs and in prokaryotic initiator-tRNA; eukaryotic 5S rRNA does have the conserved sequence Y-G-C-G that is complementary to A-U-C-G of loop IV found only in eukaryotic initiator-tRNAs (see reference 102 for a summary of the evidence and earlier references). It is assumed that the T-ψ-C-G of tRNA associates with a complementary sequence in prokaryotic 5S (105, 106) and perhaps with eukaryotic 5.8S rRNA (84, 102), during binding of aminoacyl-tRNA to the ribosomal A site. The fragment T-ψ-C-G inhibits factor-dependent binding of aminoacyl-tRNA to prokaryotic (107, 108 and eukaryotic ribosomes (109); the P site binding of eukaryotic initiator-tRNA is not affected. A consistant construct is that in eukaryotes 5.8S rRNA forms part of the ribosomal A site and participates in the binding of elongator-tRNAs; that 5S rRNA forms part of the P site and takes part in the binding of initiator-tRNA.

E. coli 5S rRNA

The E. coli large ribosomal subunit proteins L5, L18, and L25 have been shown by a variety of procedures (81-83), including affinity chromatography (79), to bind independently and specifically to homologous 5S rRNA. We have confirmed the finding: when E. coli ribosomal proteins were chromatographed on 5S rRNA affinity columns L5, L18, and L25 were bound in a ratio of approximately 1:2:1 (110). The ratio was estimated by scanning the sodium dodecyl sulfate gels after the proteins were separated by electrophoresis, it may be important that the amount of L5 that bound to 5S rRNA would be underestimated if the assessment was from inspection of two-dimensional gels as is usually the case.

A number of rat liver ribosomal proteins (L6, L7, L19, L35a, and S9) bind to E. coli 5S rRNA (Table 3). Lind et al. (111) also found that rat liver ribosomal proteins bind to E. coli 5S rRNA but the identity of the proteins was not fully documented. The finding is surprising since no E. coli ribosomal proteins bind to rat liver 5S rRNA (77, 80, 111) nor to rabbit reticulocyte, nor yeast 5S rRNA (79). There is no immediate obvious explanation of the paradox. Still the idea that the function of prokaryotic 5S rRNA, which may be to participate in the binding of aminoacyl-tRNA, is split amongst two eukaryotic ribosomal nucleic acids, 5S and 5.8S, is attractive. The thesis is that bacterial

TABLE 3. RAT LIVER RIBOSOMAL PROTEINS THAT BIND TO RIBOSOMAL NUCLEIC ACIDS

Rat Liver 5S rRNA	Rat Liver 5.8S rRNA	E. coli 5S rRNA
L6	L6	L6
L19	L19	L19
	L8	
(L7)[a]		L7
(L35a)[a]	(L35a)[a]	L35a
	S9	S9
	S13	

[a]The proteins in parentheses bind to the nucleic acid in small amounts.

5S rRNA binds elongator-tRNAs, whereas eukaryotic 5S rRNA binds the initiator-tRNA and 5.8S rRNA the elongator-tRNAs. If that is the case then E. coli 5S rRNA might bind proteins that associate with either 5S or 5.8S rRNA, or both. Of the rat liver proteins that bind to E. coli

5S rRNA (Table 3): L6 and L19 bind to both rat liver 5S and 5.8S rRNA; L35a binds to both also but only in trace amounts; L7 binds to 5S in trace amounts but not to 5.8S at all; and S9 binds in significant amounts only to 5.8S rRNA. Thus the pattern of binding of rat liver proteins to E. coli 5S rRNA overlaps the binding to rat liver 5S and 5.8S rRNA. The binding of S9 to E. coli 5S rRNA is to some extent in favor of a closer similarity to eukaryotic 5.8S rRNA since that protein does not bind to eukaryotic 5S rRNA.

The similarities in the primary sequence of prokaryotic 5S rRNA and eukaryotic 5.8S rRNA are no greater than to be expected by chance (94) and hence is unlikely they share a common ancestor. However, it may be secondary structure rather than sequence that is relevant in the binding of proteins to rRNA; moreover, it may be possible to derive structurally similar binding sites from different primary sequences. It is relevant that there are similarities in the models for the secondary structure of prokaryotic and eukaryotic 5S rRNA (112). If the models reflect the native structure of 5S rRNA, the similarity could account for the binding of rat liver L6 and L19 to E. coli and rat liver 5S rRNAs. In situ 5.8S rRNA is hydrogen-bonded to 28S rRNA (91); in some models 5.8S rRNA is given a secondary structure not entirely unlike that of 5S rRNA (113); that could account for the association with E. coli 5S rRNA of rat liver proteins (L6, L19, and S9) that bind to 5.8S rRNA. Nothing of the structure accounts for the failure of E. coli ribosomal proteins to bind to rat liver 5S rRNA in the reverse experiment.

The possibility that the binding of rat liver ribosomal proteins to E. coli 5S rRNA is a non-specific artifact needs to be considered. The possibility can not be excluded with certainty but does not seem likely. First, in exactly the circumstances in which the rat liver proteins bind to E. coli 5S rRNA on affinity columns, only three of 53 E. coli proteins (L5, L18, and L25) associate with the nucleic acid and they are the very proteins demonstrated to form a complex with the nucleic acid by a variety of other procedures (81-83). If the binding of E. coli ribosomal proteins to E. coli 5S rRNA reflects an authentic interaction, then the heterologous association may also derive from a fit between binding sites on the rRNA and the protein. It is important that the rat liver ribosomal proteins that bind to E. coli 5S rRNA (L6, L7, L19, L35a, and S9) are not a random set but rather a small subgroup of the total that includes most of the proteins that bind to rat liver 5S or

5.8S rRNA or both. Of the proteins that bind to rat liver 5S or 5.8S rRNA only L8 and S13 (5.8S rRNA binding proteins) do not bind to E. coli 5S rRNA. We have inspected the properties of the binding protein - pI, number of basic and acidic residues, hydrophobicity, aromatic amino acids - and do not find anything to distinguish them from many ribosomal proteins that do not bind to 5S or 5.8S rRNA. Thus the association is likely to be the result of an affinity of the proteins for the nucleic acid rather than of ion-exchange.

It is obvious that what would be helpful is the primary sequence of the rat liver ribosomal proteins that bind to E. coli 5S rRNA and some knowledge of the actual secondary structure of the small nucleic acids in the large subunit of ribosomes. A comparison of the binding of homologous and heterologous proteins to 5S rRNA may provide important information on the nature of the rules that govern the interaction.

tRNA

The binding of aminoacyl-tRNAs to the ribosome during protein synthesis is not solely the result of codon-anticodon interaction; they associate with ribosomal proteins as well. We attempted to identify the proteins in the elongator-tRNA and initiator-tRNA domains, once again by affinity chromatography (114). In the first set of experiments a mixture of all elongator-tRNAs (except for $tRNA_m^{met}$) was immobilized on Sepharose and the rat liver ribosomal proteins that were bound was determined. Two large subunit proteins, L6 and L35a, and one small subunit protein, S15, associated with the elongator-tRNAs. In a separate experiment the proteins that were bound to initiator-tRNA ($tRNA_f^{met}$) were determined to be L6 and L35a (Table 2). Thus two proteins, L6 and L35a, associate with both elongator- and initiator-tRNAs; S15 binds only to elongator-tRNA.

Coda

A great deal has been learned of eukaryotic ribosomes since the Cold Spring Harbor Symposium in 1973. The proteins have been isolated and characterized and limited sequence information is now available, moreover, it is likely that the entire structure of some of the more important proteins will soon be available. The structure of 5S and 5.8S rRNA from a number of eukaryotic species has been determined and the prospects are good that the sequence of 18 and 28S rRNA will be known in the not too distant future. Work has begun on the interaction of the proteins and nucleic acids in the particle; that information will surely be valuable in arriving at a solution of the structure of eukaryotic ribosomes. Information is being accumulated on ribosomes from a variety of eukaryotic species, from plants to humans, and from mitochondria and chloroplast as well. The charting of the evolution of the particle has predictably benefited from the data. Not too long ago one would have guessed that the structure and function of eukaryotic ribosomes would have to be resolved without assistance from a genetic analysis, but there are now increasing numbers of authentic mutants in ribosomal proteins. They are surely going to be useful. Perhaps the most exciting advance is in the identification of the genes for ribosomal constituents. The analysis of the transcription units for rRNA is well underway and one suspects that the experiments will lead soon to an understanding of the regulation of their transcription and that the DNA will be sequenced, leading in turn to the sequence of 18 and 28S rRNA. A number of eukaryotic ribosomal protein genes have been identified: the extraordinary importance of that achievement can hardly be exaggerated. Not only is it likely to provide information on the organization and the regulation of the genes for the molecular constituents of the ribosome, but it should provide material for determining the sequence of the ribosomal proteins (from the DNA sequence), and for studying the regulation of transcription (by using the DNA as a template for coupled transcription and translation _in vitro_). In the circumstances it is not surprising that we know much more of ribosome biogenesis. Advances proceed on the supramolecular structure: there are now more refined three-dimensional reconstructions of eukaryotic ribosomes from electron micrographs of crystalline arrays; and immune-electron microscopy is being used to map the location of proteins on the surface of the particle. All that is lacking to overflow ones cup is a method for the reconstitution of ribosomal subunits from the components. One is entitled to be sanguine about prospects: clearly for eukaryotic ribosomes the future lies ahead. Some might go so far as to say the end is here even if it is not yet in sight.

ACKNOWLEDGEMENTS

I am indebted to the people with whom I have had the pleasure of working: Ekkehard Collatz, Alan Lin, Georg Stöffler, Tatsuo Tanaka, Kazuo Todokoro, Kunio Tsurugi, Norbert Ulbrich, Heinz-Günther Wittmann, and Brigitte Wittmann-Liebold. I am especially grateful to those who have allowed me to describe the results of unpublished experiments. The expenses of the research were met by grants from the National Institutes of Health (GM-21769; CA-19265).

REFERENCES

1. Wool, I. G. (1979) Ann. Rev. Biochem. 48, 719-754.

2. Petermann, M. L. (1964) in The Physical and Chemical Properties of Ribosomes, Elsevier, New York.

3. Blair, D., Hill, W. E. & Wool, I. G., unpublished data.

4. Loening, U. E. (1968) J. Mol. Biol. 38, 355-365.

5. Weinberg, R. A. & Penman, S. (1970) J. Mol. Biol. 47, 169-178.

6. Sherton, C. C. & Wool, I. G. (1974) Mol. Gen. Genet. 135, 97-112.

7. Hamilton, M. G., Pavlovec, A. & Petermann, M. L. (1971) Biochemistry 10, 3424-3427.

8. Bielka, H., Welfle, H., Böttger, M. & Föster, W. (1968) Eur. J. Biochem. 5, 183-190.

9. Brownlee, G. G., Sanger, F. & Barrell, B. G. (1967) Nature 215, 735-736.

10. Nazar, R. N., Sitz, T. O. & Busch, H. (1975) J. Biol. Chem. 250, 8591-8597.

11. Cammarano, P., Pons, S., Romeo, A., Galdiere, M. & Gualerzi, C. (1972) Biochim. Biophys. Acta 281, 571-596.

12. Cammarano, P., Romeo, A., Gentile, M., Felsani, A. & Gualerzi, C. (1972) Biochim. Biophys. Acta 281, 597-624.

13. Cammarano, P., Felsani, A., Gentile, M., Gualerzi, C., Romeo, A. & Wolfe, G. (1972) Biochim. Biophys. Acta 281, 625-642.

14. Nonomura, Y., Blobel, G. & Sabatini, D. D. (1971) J. Mol. Biol. 60, 303-323.

15. Lake, J. A., Sabatini, D. D. & Nonomura, Y. (1974) in Ribosomes (Nomura, M., Tissieres, A. & Lengyel, P., eds.) pp. 543-557, Cold Spring Harbor Lab, Cold Spring Harbor, New York.

16. Emanuilov, I., Sabatini, D. D., Lake, J. A. & Freienstein, C. (1978) Proc. Nat. Acad. Sci. USA 75, 1389-1393.

17. Boublik, M. & Hellman, W. (1978) Proc. Nat. Acad. Sci. USA 75, 2829-2833.

18. Unwin, P.N.T. & Taddei, C. (1977) J. Mol. Biol. 114, 491-506.

19. Unwin, P.N.T. (1977) Nature 269, 118-122.

20. Morimoto, T., Blobel, G. & Sabatini, D. D. (1972) J. Cell. Biol. 52, 338-354.

21. Morimoto, T., Blobel, G. & Sabatini, D. D. (1972) J. Cell. Biol. 52, 355-366.

22. Byers, B. (1967) J. Mol. Biol. 26. 155-167.

23. Hardy, S.J.S., Kurland, C. G., Voynow, P. & Mora, G. (1969) Biochemistry 8, 2897-2905.

24. Mora, G., Donner, D., Thammana, P., Lutter, L. & Kurland, C. G. (1971) Mol. Gen. Genet. 112, 229-242.

25. Hindennach, I., Stöffler, G. & Wittmann, H. G. (1971) Eur. J. Biochem. 23, 7-11.

26. Hindennach, I., Kaltschmidt, E. & Wittmann, H. G. (1971) Eur. J. Biochem. 23, 12-16.

27. Lin, A., Collatz, E. & Wool, I. G. (1976) Mol. Gen. Genet. 144, 1-9.

28. Collatz, E., Lin, A., Stöffler, G., Tsurugi, K. & Wool, I. G. (1976) J. Biol. Chem. 251, 1808-1816.

29. Collatz, E., Wool, I. G., Lin, A. & Stöffler, G. (1976) J. Biol. Chem. 251, 4666-4672.

30. Tsurugi, K., Collatz, E., Wool, I. G. & Lin, A. (1976) J. Biol. Chem. 251, 7940-7946.

31. Tsurugi, K., Collatz, E., Todokoro, K. & Wool, I. G. (1977) J. Biol. Chem. 252, 3961-3969.

32. Collatz, E., Ulbrich, N., Tsurugi, K., Lightfoot, H. N., MacKinlay, W., Lin, A. & Wool, I. G. (1977) J. Biol. Chem. 252, 9071-9080.

33. Lin, A., Tanaka, T. & Wool, I. G. (1979) Biochemistry 18, 1634-1637.

34. Terao, K. & Ogata, K. (1972) Biochim. Biophys. Acta 285, 473-482.

35. Westermann, P. & Bielka, H. (1973) Mol. Gen. Genet. 126, 349-356.

36. Goerl, M., Welfle, H. & Bielka, H. (1978) Biochim. Biophys. Acta 519, 418-427.

37. Howard, G. A., Ramjoúe, H.P.R. & Gordon, J. (1977) in Translation of Natural and Synthetic Polypeptides (Legocki, A. B., ed.) pp. 305-311, Univ. Argic., Poznan, Poland.

38. Itoh, T., Higo, K. & Otaka, E. (1979) Biochemsitry, in press.

39. Lin, A. & Wool, I. G. (1974) Mol. Gen. Genet. 134, 1-6.

40. Howard, G. A., Traugh, J. A., Croser, E. A. & Traut, R. R. (1975) J. Mol. Biol. 93, 391-404.

41. Terao, K. & Ogata, K. (1975) Biochim. Biophys. Acta 402, 214-229.

42. Martini, O.H.W. & Gould, H. J. (1975) Mol. Gen. Genet. 142, 317-331.

43. Issinger, O. C. & Beier, H. (1978) Mol. Gen. Genet. 160, 297-309.

44. Welfle, H., Goerl, M. & Bielka, H. (1978) Mol. Gen. Genet. 163, 101-112.

45. Sherton, C. C. & Wool, I. G. (1972) J. Biol. Chem. 247, 4460-4467.

46. Welfle, H., Stahl, J. & Bielka, H. (1971) Biochim. Biophys. Acta 243, 416-419.

47. Welfle, H., Stahl, J. & Bielka, H. (1972) FEBS Lett. 26, 228-232.

48. Delaunay, J., Creusot, F. & Schapira, G. (1973) Eur. J. Biochem. 39, 305-312.

49. Chatterjee, S. K., Kazamie, M. & Matthaei, H. (1973) Hoppe-Seylers Z. Physiol. Chem. 354, 481-486.

50. Lastick, S. M. & McConkey, E. H. (1976) J. Biol. Chem. 251, 2867-2875.

51. Todokoro, K. & Wool, I. G., unpublished data.

52. Wittmann-Liebold, B., Geissler, A. W., Lin, A. & Wool, I. G. (1979), submitted.

53. Reisner, A. H., personal communication.

54. Möller, W., Slobin, L. I., Amons, R. & Richter, D. (1975) Proc. Nat. Acad. Sci. USA 72, 4744-4748.

55. van Agthoven, A. J., Maassen, J. A. & Möller, W. (1977) Biochem. Biophys. Res. Comm. 77, 989-998.

56. Amons, R., van Agthoven, A. J., Pluijms, W., Möller, W., Higo, K., Itoh, T. & Osawa, S. (1977) FEBS Lett. 81, 308-310.

57. Amons, R., van Agthoven, A. J., Pluijms, W. & Möller, W. (1977) FEBS Lett. 86, 282-284.

58. Itoh, T. & Wittmann-Liebold, B., in preparation.

59. Itoh, T. & Wittmann-Liebold, B. (1978) FEBS Lett. 96, 399-402.

60. Wool, I. G. & Stöffler, G. (1974) in Ribosomes (Nomura, M., Tissiéres, A. & Lengyel, P., eds.) pp. 417-460, Cold Spring Harbor Lab, Cold Spring Harbor, New York.

61. Brimacombe, R., Stöffler, G. & Wittmann, H. G. (1978) Ann. Rev. Biochem. 47, 217-249.

62. Fischer, N., Stöffler, G. & Wool, I. G. (1978) J. Biol. Chem. 253, 7355-7360.

63. Blobel, G. & Sabatini, D. D. (1971) in Biomembranes (Manson, L. A., ed.), vol. 2, pp. 193-196, Plenum Press, New York.

64. Sabatini, D. D. & Kreibich, G. (1976) in The Enzymes of Biological Membranes (Martinosi, A., ed.), vol. 2, pp. 531-579, Plenum Press, New York.

65. Kreibich, G., Ulrich, B. L. & Sabatini, D. D. (1978) J. Cell. Biol. 77, 464-487.

66. Kreibich, G., Freienstein, C. M., Pereyra, B. N., Ulrich, B. L. & Sabatini, D. D. (1978) J. Cell. Biol. 77, 488-506.

67. Kreibich, G., Czako-Graham, M., Grebenau, R., Mok, W., Rodriquez-Boulan, E. & Sabatini, D. D. (1978) J. Supramol. Struct. 8, 279-302.

68. Kozak, M. (1978) Cell. 15, 1109-1123.

69. Kozak, M. (1979) J. Biol. Chem. 254, 4731-4738.

70. Kozak, M. (1979) Nature 280, 82-85.

71. Trachsel, H., Erni, B., Schreier, M. H. & Staehelin, T. (1977) J. Mol. Biol. 116, 755-767.

72. Cundliffe, E., Dixon, B., Stark, M., Stöffler, G., Ehrlich, R. & Geisser, M. (1979) J. Biol. Chem., in press.

73. Wienen, B., Erhlich, R., Geisser, M., Stöffler, G., Smith, I., Weiss, D., Vince, R. & Pestka, S., submitted.

74. Pestka, S., Weiss, D., Vince, R., Wienen, B., Stöffler, G. & Smith, I. (1976) Mol. Gen. Genet. 144, 235-241.

75. Maaløe, O. (1969) Dev. Biol. 3, 33-58 (Suppl.).

76. Kjeldgaard, N. O. & Gausing, K. (1974) in Ribosomes (Nomura, M., Tissiéres, A. & Lengyel, P., eds.) pp. 369-392, Cold Spring Harbor Lab, Cold Spring Harbor, New York.

77. Ulbrich, N. & Wool, I. G. (1978) J. Biol. Chem. 253, 9049-9052.

78. Burrell, H. R. & Horowitz, J. (1975) FEBS Lett. 49, 306-309.

79. Burrell, H. R. & Horowitz, J. (1977) Eur. J. Biochem. 75, 533-544.

80. Metspalu, A., Saarma, M., Villems, R., Ustav, M. & Lind, A. (1978) Eur. J. Biochem. 91, 73-81.

81. Horne, J. R. & Erdmann, V. A. (1972) Mol. Gen. Genet. 119, 337-344.

82. Gray, P. N., Bellemare, G., Monier, R., Garrett, R. A. & Stöffler, G. (1973) J. Mol. Biol. 77, 133-152.

83. Yu, R.S.T. & Wittmann, H. G. (1973) Biochim. Biophys. Acta 324, 375-385.

84. Wrede, P. & Erdmann, V. A. (1977) Proc. Nat. Acad. Sci. USA 74, 2706-2709.

85. Blobel, G. (1971) Proc. Natl. Acad. Sci. USA 68, 1881-1885.

86. Lebleu, B., Marbaix, G., Huez, G., Temmerman, J., Burney, A. & Chantrenne, H. (1971) Eur. J. Biochem. 19, 264-269.

87. Petermann, M. L. & Pavlovec, A. (1971) Biochemistry 10, 2770-2775.

88. Terao, K., Takahashi, Y. & Ogata, K. (1975) Biochim. Biophys. Acta 402, 230-237.

89. Newton, I., Rinke, J. & Brimacombe, R. (1975) FEBS Lett. 51, 215-218.

90. Traub, P. & Nomura, M. (1969) J. Mol. Biol. 40, 391-413.

91. Pace, N. R., Walker, T. A. & Schroeder, E. (1977) Biochemistry 16, 5321-5328.

92. Ulbrich, N., Lin, A. & Wool, I. G. (1979) J. Biol. Chem., in press.

93. Erdmann, V. A., personal communication.

94. Cedergren, R. J. & Sankoff, D. (1976) Nature 260, 74-75.

95. Svoboda, A. J. & McConkey, E. H. (1978) Biochem. Biophys. Res. Comm. 81, 1145-1152.

96. Brimacombe, R., Nierhaus, K. H., Garrett, R. A. & Wittmann, H. G. (1976) Progr. Nuc. Acid Res. 18, 1-44.

97. Hochkeppel, H. K., Spicer, E. & Craven, G. R. (1976) J. Mol. Biol. 101, 155-170.

98. Hochkeppel, H. K. & Craven, G. R. (1977) J. Mol. Biol. 113, 623-634.

99. Littlechild, J., Dijk, J. & Garrett, R. A. (1977) FEBS Lett. 74, 292-294.

100. Yukioka, M. & Omori, K. (1977) FEBS Lett. 75, 217-220.

101. Nazar, R. N. (1978) J. Biol. Chem. 253, 4505-4507.

102. Erdmann, V. A. (1976) in Progress in Nucleic Acid Research and Molecular Biology (Cohn, E. W., ed.) vol. 18, pp. 45-90, Academic Press, New York

103. Erdmann, V. A. (1978) Nucleic Acids Res. 5, r1-r14.

104. Erdmann, V. A. (1979) Nucleic Acids Res. 6, r29-r44.

105. Ofengand, J. & Henes, C. (1969) J. Biol. Chem. 244, 6241-6253.

106. Schwarz, U., Menzel, H. M. & Gassen, H. G. (1976) Biochemistry 15, 2484-2490.

107. Richter, D., Erdmann, V. A. & Sprinzl, M. (1973) Nature New Biol. 246, 132-135.

108. Erdmann, V. A., Sprinzl, M., Richter, D. & Lorenz, S. (1974) Acta Biol. Med. Germ. 33, 605-608.

109. Grummt, F., Grummt, I., Gross, H. J., Sprinzl, M., Richter, D. & Erdmann, V. A. (1974) FEBS Lett. 42, 15-17.

110. Ulbrich, N., Lin, A. & Wool, I. G., in preparation.

111. Lind, A., Metspalu, A., Saarma, M., Toots, I., Ustav, M., & Villems, R. (1977) Bioorg. Chem. (USSR) 3, 1138-1140.

112. Hori, H. (1976) Mol. Gen. Genet. 145, 119-123.

113. Luoma, G. A. & Marshall, A. G. (1978) Proc. Nat. Acad. Sci. USA 75, 4901-4905.

114. Ulbrich, N., Ackerman, E., Lin, A., Sigler, P. & Wool, I. G., in preparation.

TRANSLATIONAL CONTROL IN EURARYOTES

GISELA KRAMER, A. BURL HENDERSON,
NIKODEM GRANKOWSKI, and BOYD HARDESTY

Clayton Foundation Biochemical Institute,
The University of Texas at Austin,
Department of Chemistry,
Austin, TX 78712

INTRODUCTION

Posttranscriptional regulation in eukaryotic organisms may occur at a number of sites that do not involve ribosomes, such as transport, storage, modification and stability of mRNA or processing of newly synthesized peptides or proenzymes into an active form. In this review on translational control, we will consider only those regulatory events that lead directly to an effect on ribosomal reactions of protein synthesis; i.e. the steps of peptide initiation, elongation and termination. A most important and as yet not clearly answered question for the significance of translational regulation in the physiology and differentiation of eukaryotic cells involves specificity. Can regulation of reactions common to the synthesis of all proteins selectively affect the synthesis of specific proteins? The answer appears to be yes, at least under some circumstances. Peptide initiation is the rate limiting step of protein synthesis in most active systems and the point at which most translational control mechanisms operate. Factors that affect the rate of peptide initiation may have a profound influence on the relative efficiency with which different mRNA species are translated.

It appears that most, if not all of the primary regulatory mechanisms that operate at the translational level are negative in character. That is, they function by reducing the overall rate of an ongoing reaction. The search for positive regulators, such as factors that promote the translation of specific mRNA species has been the quest for the Holy Grail of translational regulation. It appears that if such factors exist they function to discriminate between classes and types of mRNA rather than individual mRNA species.

DIFFERENTIAL TRANSLATION OF mRNA SPECIES

Evidence from a number of sources demonstrate convincingly that the amount of protein synthesized in eukaryotic systems is not necessarily directly related to the molar amount of mRNA present. The various mRNAs of reovirus are synthesized in equimolar quantities in infected cells, however proteins made from these mRNA species may differ in quantity by a factor of 10 (1).

Perhaps the most thoroughly studied example of differential mRNA translation is the synthesis of the α- and β-chains of hemoglobin. These two peptides are formed on different mRNAs in nearly a 1:1 ratio in intact cells under normal conditions. The rates of peptide elongation and release are the same for the two mRNAs (2), however, Lodish (3) demonstrated that polysomes carrying β-globin mRNA contained about 1.5 times as many ribosomes as the polysomes containing α-globin mRNA. He concluded that the ratio of α-globin mRNA to β-globin mRNA must be 1.5 and that a rate-limiting step of peptide initiation must be 1.5 times more efficient with β-globin mRNA than with α-globin mRNA. These conclusions were tested by using the antibiotics cycloheximide or sparsomycin to slow down peptide elongation so that this step, rather than peptide initiation, became the rate-limiting step of the reaction sequence. Under these conditions the ratio of α-peptide to β-peptide synthesis increased to about 1.4. These data support the hypothesis that equimolar synthesis of the α- and β-globin peptides results from differences of the respective mRNA in the rate-limiting reaction of peptide initiation and led Lodish to formulate a kinetic model (4) to explain these phenomena.

It has not been possible by direct measurement to confirm the conclusion that the α-chain mRNA to β-chain mRNA ratio is 1.5:1 in intact erythroid cells. Lingrel and co-workers (5) separated mRNAs for the α- and β-chains by polyacrylamide gel electrophoresis and found that globin mRNA from both intact reticulocytes and isolated polysomes contained more β-chain than α-chain mRNA. Kramer *et al.* (6) found the α:β-globin ratio synthesized from mRNA in the ribosomal salt wash fraction of rabbit reticulocytes to be 4:1, however the corresponding value for 9S RNA isolated from salt-washed ribosomes was 1:4. These ratios remained constant under all of the conditions tested and appeared to reflect the mRNA composition. Thus it appears that the kinetic model proposed by Lodish (4) is not adequate to explain all of the phenomena observed. Nevertheless, the basic results provide a well documented example of differential translation of eukaryotic mRNA and demonstrate that the relative amounts of peptides formed may be influenced by factors that affect the general characteristics of the translational system.

Other examples of differential mRNA translation are provided by virus-mediated inhibition of host cell mRNA and by the effects of salt on the relative translational efficiency of specific mRNAs. In HeLa cells infected by polio virus, the synthesis of host cell proteins is reduced by

80% to 90% two to four hours following the initial infection, at which time the synthesis of viral protein is maximal. Pre-existing host cell mRNA is not degraded in infected cells (7). Three different mechanisms have been proposed to explain the effect of virus infection on virus versus host cell protein synthesis:

1. Virus-induced modification of initiation factors or an altered requirement for the factors.
2. Virus-induced change to an intracellular ionic composition or strength that favors preferential synthesis of viral proteins.
3. Ribosome modifications either by virus-induced phosphorylation or binding of a virus-induced protein to inhibit synthesis of host cell proteins.

Experimental evidence has been presented for each of these mechanisms. Golini *et al.* (8) reported that encephalomyocarditis (EMC) RNA is translated more efficiently than either cellular mRNA or globin mRNA in a cell-free system derived from MOPC ascites cells. However, addition of the initiation factor eIF-4B from rabbit reticulocytes restored the high efficiency of translation of globin mRNA or MOPC 10S mRNA in the presence of the EMC RNA. Rose *et al.* (9) found that vesicular stomatitis virus (VSV) mRNA is translated in lysates of uninfected HeLa cells and in lysates of polio-infected cells only when the lysates were prepared within 2 to 3 hours of infection, but not thereafter. Addition of the initiation factor eIF-4B, to lysates derived from HeLa cells 3 hours or later after infection allows translation of VSV mRNA. The authors suggest that initiation factor eIF-4B may be inactivated in infected HeLa cells and that it may bind to the cap structure of cellular mRNA. Polio mRNA does not contain a 5'-cap, and thus might be translatable without eIF-4B late after infection. It has not been possible to substantiate this hypothesis. Shatkin and co-workers did not detect cross-linking of any of the known initiation factors to the cap structure but observed cross-linking to a 24,000 dalton peptide (10). This peptide may be part of a previously undescribed initiation factor that is present in preparations of eIF-3 or eIF-4B (11).

An interesting model for shutoff of host protein synthesis has been proposed by Carrasco and Smith (12) based on their observation that viral mRNAs are translated best *in vitro* at relatively high monovalent cation concentrations, apparently through an effect on peptide initiation. In contrast to viral mRNA, host cell mRNA is translated best at lower salt concentrations. Thus, these authors propose that viruses may cause changes in the outer membrane of the host cell that result in increased intracellular salt concentrations.

Observations made by Koch and co-workers (reviewed in 13) appear to support this hypothesis. The synthesis of viral and host cell proteins was examined in intact cells that had been transferred to hypertonic medium shortly after viral infection. The pattern of mRNA translation was shifted under these conditions so that proteins normally synthesized only late after virus infection were synthesized immediately. These authors propose that different eukaryotic and viral mRNA species have a characteristic efficiency for translation and that this may be altered by different concentrations of salt.

An interesting observation involving changes in the phosphorylation pattern of ribosomal proteins during the onset of virus infection was reported by Kaerlin and Horak (14). Phosphorylation of the ribosomal proteins S2 and S16 occurred only after HeLa cells were infected with vaccinia virus. S6 was phosphorylated in infected as well as uninfected cells. Ben-Hamida and Beaud (15) reported that vaccinia virus cores inhibit translation of natural mRNA in reticulocyte lysates. The virus cores had no effect on poly(U)-directed synthesis of polyphenylalanine.

STRUCTURAL BASIS FOR DIFFERENTIAL mRNA TRANSLATION

The structure of the mRNA, the ribosome, or factors that discriminate between mRNAs might be responsible for apparent selection of specific mRNA species. Perhaps the clearest example of a system in which the structure of the mRNA appears to influence the efficiency of translation involves the RNA bacteriophages of *E. coli*, R17, Qβ and MS2. These phages contain three genes in the order of A, coat, and synthetase from the 5'- to the 3'-end of their RNA. Unlike eukaryotic mRNA, which appears to be monocistronic, each coding section of the phage mRNA has a separate initiation codon and ribosome attachment site. When intact phage RNA is used as mRNA in the cell-free translation systems, ribosomes bind primarily to the initiation site for the coat protein, apparently because the initiation sites for the synthetase and A protein are masked in double-stranded hairpin loops. The initiation sites for the synthetase and A protein become available for ribosome attachment and peptide initiation by disrupting the double-stranded regions of the RNA, either by heat or reaction with formaldehyde (16, 17). Steitz (18) isolated fragments of the phage RNA that contained 12-15 nucleotides on each side of the initiation codon for each of the three phage proteins. When these fragments were rebound to ribosomes in a cell-free system, it was found that the segment from the A protein cistron, which had the lowest efficiency of binding to ribosomes or translation in the native RNA, now bound most efficiently.

These results indicate that the efficiency of ribosome attachment to mRNA may be determined by both the secondary and tertiary structure of the RNA or its primary sequence of nucleotides.

In prokaryotes, the basis for differential initiation site selection in mRNA that contains little or no secondary structure may involve the extent of specific base pairing interaction between the mRNA and a sequence of nucleotides near the 3'-end of the 16S ribosomal RNA (19). However, there appears to be no comparable common nucleotide sequence at the 5'-side of the initiation codon in eukaryotic mRNA (20). Thus, in contrast to prokaryotes, the initiation codon of eukaryotic mRNA is unlikely to be selected by a specific base pairing between nucleotides of this region of the mRNA and bases near the 3'-end of 18S rRNA. The mechanism by which the proper, in-phase AUG codon of eukaryotic mRNA is selected is one of the more perplexing problems of eukaryotic peptide initiation. This mechanism is a likely site for translational regulation in eukaryotic cells. (See chapters by Steitz and Grunberg-Manago, this volume.)

There are two unique structural features, cap and poly(A) tails found in most, but not all eukaryotic mRNA. A 7-methyl guanosine is attached by an unusual 5'-to 5'-triphosphate linkage to the nucleoside at the 5'-terminus to form a "cap" (21). Most eukaryotic cellular and viral mRNA also have a segment of 50 to 200 adenylic acid residues at their 3'-end (22). Either of these structures somehow might function in mRNA selection and/or translational regulation. Both poly(A) tails and cap have been extensively investigated, but a firm conclusion regarding their physiological role has not been reached. Poly(A) is absent from some functionally competent mRNA species, such as histone mRNA and some viral mRNAs. Poly(A) may contribute to the stability of the mRNA molecule by protecting it from degradation by 3'-exonuclease (23).

The widespread presence of cap structures on naturally occurring eukaryotic mRNAs has led to the speculation that they may play an important role in translation, probably at the level of mRNA binding to 40S ribosomal subunits. However, some RNA species are known that are functionally competent in translation but lack cap. Poliovirus RNA isolated from polyribosomes of productively infected HeLa cells does not end in a cap structure, but rather has a pU at its 5'-terminus (24, 25). This mRNA in addition to a number of plant viral mRNAs lacking a cap structure, has been found to be faithfully translated in cell-free systems derived from a number of eukaryotic organisms. Evidence has been presented (11) that a previously unrecognized peptide of 24,000 daltons may bind to the cap structure during peptide initiation, however, it also appeared to stimulate translation of uncapped viral RNA *in vitro*. Thus its role in translational control, if any, is unclear.

CONTROL AT THE STEP OF MET-tRNA$_f$ BINDING TO 40S RIBOSOMAL SUBUNITS BY CHANGES IN THE ACTIVITY OF THE INITIATION FACTOR eIF-2

Eukaryotic peptide initiation (reviewed recently in ref. 26) involves at least seven factors, probably more, and requires the hydrolysis of GTP and ATP. The initiation factor eIF-2 forms a ternary complex with Met-tRNA$_f$ and GTP. This ternary complex can bind to 40S ribosomal subunits in the presence of other initiation factors. Evidence has been presented (27) and recently confirmed (28) that GTP hydrolysis is associated with the proper binding of the ternary complex to 40S ribosomal subunits. It is generally accepted that binding of Met-tRNA$_f$ precedes binding of mRNA in eukaryotic systems.

The initiation of protein synthesis in reticulocyte lysates is inhibited under at least seven different conditions: heme deficiency which is the most studied example as considered in detail below, when double-stranded RNA (dsRNA) is added (29), when oxidized glutathione is added (30), when the lysate has been prepared from cells incubated anaerobically or with inhibitors of oxidative phosphorylation (31), when the lysate is depleted of low molecular weight compounds by gel filtration (32), and when the lysate is subjected to high pressure (33) or heat (34). Each of these conditions causes inhibition of peptide initiation. It appears likely, but has not been definitively proven, that each condition leads to phosphorylation of the smallest subunit of eIF-2, eIF-2α (M_r = 38,000) and thereby inhibition of protein synthesis. Phosphorylation of eIF-2α can be accomplished by at least two distinct protein kinases that are activated by separate mechanisms. Heme deficiency or pressure lead to activation of the same eIF-2α kinase, which appears to involve a phosphopeptide of 95,000 to 100,000 daltons as considered below. Another eIF-2 kinase is activated by dsRNA (see below). Much less is known about the mechanism by which the other conditions cause inhibition.

Hemin Controlled Inhibitor in Reticulocytes

Protein synthesis in reticulocytes (35, 36) and their cell-free lysates (37, 38) is regulated by heme. When a reticulocyte lysate is incubated at 37°C in the absence of added hemin, the initial rate of [^{14}C]leucine incorporation is normal, but decreases abruptly after about 5 min of incubation to a low rate, typically 2% to 10% of the initial rate. This striking inhibition in the absence of added hemin has been shown in many studies to involve a block in the initiation of new peptides with no detectable effect on peptide elongation. Inhibition is associated with the disappearance of Met-tRNA$_f$ on 40S ribosomal subunits and appears to involve inhibition of the reaction by which the ternary complex formed between Met-tRNA$_f$, eIF-2, and GTP is bound to small ribosomal subunits (39, 40).

The inhibition of protein synthesis observed in lysates in the absence of heme is potentiated by millimolar concentrations of ATP, but is reversed by 2 mM GTP, or high concentrations of cAMP in the range of 2-10 mM (41). It should be noted that the concentration of cAMP necessary to activate preparations of cAMP-dependent protein kinases half-maximally is in the order of 2×10^{-7} M (42). cAMP is inactive in reversing inhibition of protein synthesis in heme-deficient lysates in this concentration range. It is assumed that the effect noted above does not involve a cAMP-dependent protein kinase as considered below.

Gross and Rabinovitz (43) partially purified what appears to be a high molecular weight complex from reticulocyte postribosomal supernatant incubated in the absence of hemin. When added exogenously to a hemin-supplemented reticulocyte lysate, this material blocked peptide initiation with the same kinetics and characteristics as observed under heme deficiency. The agent responsible for the inhibition was named the "hemin-controlled repressor" or HCR.

Phosphorylation of the Smallest Subunit of eIF-2

The mechanism by which HCR blocks peptide initiation appears to involve phosphorylation of the smallest subunit of eIF-2, eIF-2α. Highly purified preparations of HCR contain a cAMP-independent protein kinase for this peptide (44-46). It has been shown that HCR inhibition of protein synthesis in lysates may be overcome by the addition of exogenous eIF-2 (47). Farrell *et al.* (48) analyzed eIF-2 taken directly from reticulocyte lysates incubated under conditions in which HCR was activated. They found a direct correlation between phosphorylation of the smallest subunit and inhibition of protein synthesis under heme deficiency or by HCR. The eIF-2 and GTP-dependent binding of Met-tRNA$_f$ to 40S ribosomal subunits may be measured independently as a partial reaction of peptide initiation. This system was used to demonstrate directly that phosphorylated eIF-2 was inactive for the binding reaction (49-51). Typical inhibition curves are shown below in Figure 1.

No Effect of cAMP-Dependent Protein Kinases on Inhibition of Peptide Initiation

Ochoa and co-workers (52) reported that a cAMP-dependent protein kinase is involved in the activation of the cAMP-independent protein kinase for the 38,000 dalton subunit of eIF-2. They envision a regulatory system for activation of the eIF-2α kinase in which it is phosphorylated and thereby activated by the catalytic subunit of the cAMP-dependent kinase. They have reported evidence interpreted to indicate that heme regulation involves binding of heme to the regulatory subunit of cAMP-dependent protein kinase, thereby rendering it inactive (53). This is an attractive hypothesis in that it provides a basis for understanding, how protein synthesis might be directly coordinated with the myriad of cellular processes that are directly or indirectly controlled by cAMP. However, several lines of evidence weigh heavily against this model. Purified preparations of the catalytic subunit of the cAMP-dependent kinase from rabbit reticulocytes have no detectable effect on protein synthesis in reticulocyte lysates (54). Also, the activity of cAMP-dependent protein kinases may be blocked by a naturally occurring peptide inhibitor that binds tightly to the catalytic subunit of protein kinases (55). This inhibitor is active with the reticulocyte cAMP-dependent kinase in both crude and purified preparations. It would be anticipated from the cAMP-dependent system suggested by Ochoa and co-workers that addition of the inhibitor to the lysate system in the absence of heme would prevent HCR-dependent inhibition of protein synthesis. Such an effect could not be detected (54). Ernst and Levin have come to similar conclusions (56).

Inhibitor Activation by Heme Deficiency

The details of the mechanism by which HCR is activated under physiological conditions are unknown. Gross and Rabinovitz (57) demonstrated that an "inactive prorepressor" was converted to what they called the "reversible inhibitor" in the absence of heme. This activated inhibitor could be converted back to an inactive form by heme. However, in the prolonged absence of heme, the reversible inhibitor was converted into a form, called the "irreversible inhibitor" that was no longer sensitive to heme. The inactive proinhibitor could be converted directly to an irreversible inhibitor by treatment with the sulfhydryl reactive reagent N-ethylmaleimide. Trachsel *et al.* (58) and Gross and Mendelewski (59) have isolated the reversible inhibitor that can be inactivated in the presence of heme and that is converted to the irreversible form by treatment with N-ethylmaleimide. The existence of a proinhibitor, distinct from the inactive form of the reversible inhibitor, has not been established.

The active form of the reversible inhibitor appears always to be associated with a phosphorylated peptide of about 95,000 to 100,000 daltons as detected by polyacrylamide gel electrophoresis with sodium dodecylsulfate. This band corresponds in molecular weight to the active inhibitor and is concluded to be the eIF-2 kinase (58). Phosphorylation of this peptide has been assumed to occur by autophosphorylation of the active reversible inhibitor (58, 59). However, recent evidence indicates that at least one additional component is involved in the inhibitory activity of HCR (60). Enzyme fractions enriched in the 100,000 peptide that is phosphorylated during activation of HCR has appreciable endogenous activity for phosphorylation of eIF-2α and inhibition of eIF-2-dependent binding of Met-tRNA$_f$ to 40S ribosomal subunits. However, as shown in Figure 1, this latter activity could be increased by a second

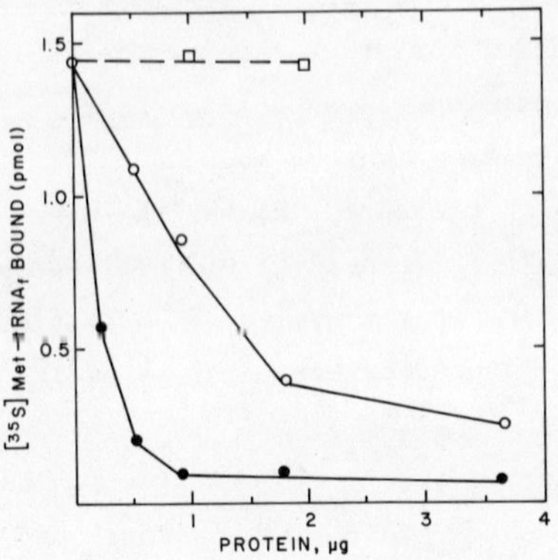

Figure 1. Synergism Between a 90K and a 100K Fraction for Inhibition of Met-tRNA$_f$ Binding to Reticulocyte 40S Ribosomal Subunits. About 2 μg eIF-2 were incubated with ATP and the 90K fraction (□- -□) or the 100K fraction (○—○) or 1 μg of the 90K fraction plus the 100K fraction (●—●) in the amounts indicated. Then binding of [^{35}S]Met-tRNA$_f$ to 40S subunits was determined with the pretreated eIF-2 as described and referenced in the text.

enzyme fraction that lacked the 100,000 dalton peptide, but was enriched in a component containing a peptide of 90,000 daltons (60). This second fraction had little or no endogenous activity for inhibition of the Met-tRNA$_f$ binding reaction to 40S subunits. The relatively simple assay system (50) used for these experiments is of special interest. Initiation factor eIF-2 is preincubated with ATP and HCR fractions, then residual ATP is eliminated in the reaction mixture by addition of glucose and hexokinase. Finally, eIF-2-dependent binding of Met-tRNA$_f$ to 40S ribosomal subunits is measured after the addition of the components necessary for this reaction. A non-hydrolyzable analog of GTP, GMP(CH$_2$)P, is used to prevent further phosphorylation of eIF-2 during the incubation with 40S ribosomal subunits.

The conclusions reached from these results were strengthened by experiments in which antibodies were used that had been raised against a homogeneous 90,000 dalton peptide found in the 90K fraction. The antibodies caused a decrease in the amount of [^{32}P]phosphate incorporated into eIF-2α as well as the 100,000 dalton peptide mentioned above (60), when they were incubated with an HCR fraction that was then used to phosphorylate eIF-2α. These experiments demonstrate that the 90,000 dalton peptide plays a role in the activation of the eIF-2α protein kinase of the HCR system. The 90K peptide may be part of a protein kinase that phosphorylates the 100K peptide and thereby activates the eIF-2α kinase, however the exact relationship between these components has not been established.

Inhibitor Activation in the Presence of Heme

The eIF-2 kinase of the HCR system can be activated in the presence of heme by a sequence of reactions involving at least two additional discrete components that are distinct from any of the factors described above (61). Both of the factors cause inhibition of protein synthesis, when they are added to reticulocyte lysates. One is an acidic, heat-stable protein, HS, with an apparent molecular weight of about 30,000 that can be activated by either heat or high pressure. HS has the unusual property of reverting to an inactive form if it is held at 0° for 24-48 h. Samples of HS have been taken through as many as six cycles of activation by heating followed by inactivation at 0° without detectable loss of its activity.

HS appears to function catalytically in the irreversible activation of the second of the two inhibitor proteins, HL, a heat-labile factor. HS has little or no detectable inhibitory activity, when it is added to a partially fractionated translational system that is deficient in HL. HS has very high specific activity for inhibition of protein synthesis in reticulocyte lysates. The protein concentration in our most highly purified preparations is below detectable levels by conventional techniques. We estimate that HS is active in the concentration range of not more than ng of protein per ml in the lysate system. Our biggest problem in the purification of HS has been that it is present apparently in very small amounts in reticulocyte lysates. We estimate that there must be less than 30 μg of recoverable HS per 100 g of protein in the postribosomal supernatant of rabbit reticulocytes. This would correspond to less than a few thousand molecules of HS per reticulocyte.

HL has been purified to apparent homogeneity in both its active and inactive forms; HL and HL_i respectively. HL_i has a molecular weight of about 43,000 as determined by the distribution of its inhibitory activity from chromatography on Sephadex G-100 and gives a single peptide band of about 21,000 daltons during SDS gel electrophoresis. The mechanism by which HL_i is converted to HL in the presence of HS is not clear. HS appears to function in catalytic amounts in an ATP-dependent reaction. Conversion of HL_i to HL does not progress efficiently, if the non-hydrolyzable analog of ATP, AMP-P(NH)P, is substituted for ATP. Active HS and active HL can be recovered in high yield from the reaction mixture following activation of HL_i. Neither component is detectably phosphorylated during the activation reaction and adenylation has not been detected in preliminary experiments. Thus, the role of ATP in this reaction is quite obscure.

The mechanism by which HS and HL inhibit protein synthesis in reticulocyte lysates or a fractionated translation system appear to involve activation of the same eIF-2α kinase that is activated in the absence of heme. The pattern of the peptides phosphorylated in the presence of activated HL is shown in Figure 2. For these experiments various combinations of HS, HL and eIF-2 were incubated in the presence or absence of a crude enzyme fraction precipitated from reticulocyte postribosomal supernatant with ammonium sulfate between the levels of 40% and 70% saturation. The phosphorylated peptides observed in the presence of this fraction have molecular weights of about 210K, 100K, 58K and 38K. Two of these bands correspond to eIF-2α and the 100K peptide that may be part of the eIF-2α kinase of the HCR system.

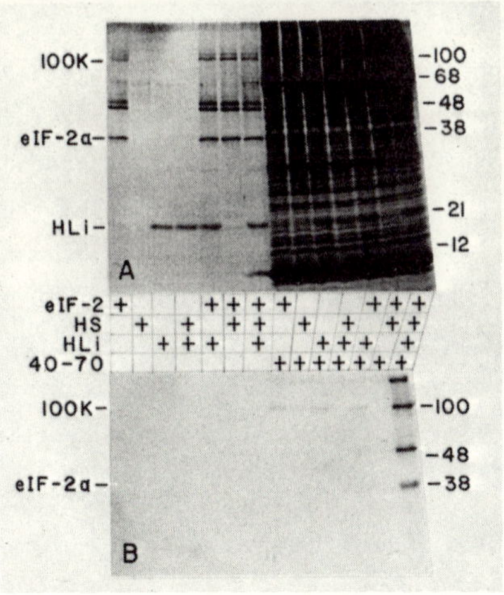

Figure 2. HL Activation of Kinase Activity in the Presence of Heme. Each reaction mixture contained the following components in a total volume of 30 μl: 20 mM Tris-HCl (pH 7.5), 5 mM $MgCl_2$, 1.5 mM DTE, 2 μg/ml hemin, 0.2 mM [γ-^{32}P] ATP with a specific activity of 353 Ci/mol, and protein fractions as indicated for each track. Protein fractions were present in the following amounts where indicated: eIF-2, 1.4 μg; 40-70% ammonium sulfate fraction of reticulocyte postribosomal supernatant, 87 μg; HL_i, 6 μg; HS, 0.4 μg. Incubation was at 37° for 5 min. Aliquots were analyzed by polyacrylamide gel electrophoresis in the presence of sodium dodecylsulfate followed by autoradiography. A. Stained Gel. B. Radioautogram.

Perhaps the strongest evidence that HS and HL promote activation of the eIF-2α kinase of the HCR system is that the inhibitory activity of either HS or HL can be blocked by antibodies raised against authentic HCR that had been purified more than 1000 fold. The results of such experiments are depicted in Figure 3. For these experiments reticulocyte lysates were incubated with HL, chromatographed on DEAE-cellulose to remove HL, then incubated with either immune or non-immune IgG. The ability of the resulting fraction to inhibit protein synthesis in the lysate system was determined.

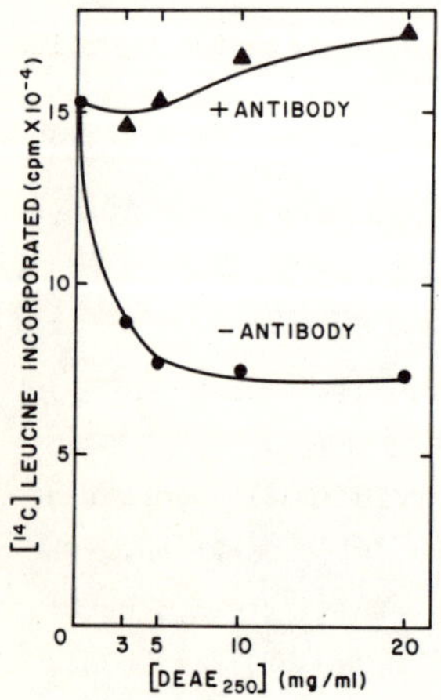

Figure 3. Effect of Anti-HCR IgG on the Inhibitor Activated by HL. A reticulocyte lysate to which exogenous HL and heme had been added was incubated 15 min at 37° under conditions in which HL inhibits protein synthesis. HL was separated from the activated inhibitor by chromatography on DEAE-cellulose. Aliquots of the resulting inhibitor fraction were incubated at 37° for 20 min with 6 μg of anti-HCR IgG (▲—▲) or non-immune IgG (●—●), then assayed in the reticulocyte lysate system for inhibition of protein synthesis.

The mechanism by which HL activates the eIF-2α kinase or its relation to the 90,000 dalton peptide described above is not known. Neither HS nor HL alone have detectable kinase activity for eIF-2α, as shown in Figure 2. They appear to provide an alternative mechanism for activation of th eIF-2α kinase of the HCR system in the presence of heme. HS and HL may constitute distal components that function with the 90,000 dalton peptide in a cascade type sequence of reactions. Proteolysis appears to be involved in that the inhibitor activity of either HS or HL in lysates can be blocked by soybean trypsin inhibitor or tosyl-lysine chloromethyl ketone, TLCK. However, HL does not appear to be labeled by [^{14}C]TLCK. It is unlikely that either HS or HL is the sensitive protease. Thus it appears that there is at least one and perhaps several additional components that function in this reaction sequence for the activation of the HCR eIF-2α kinase.

A Phosphatase That Functions with Components of the HCR System

Cognate phosphatases that carry out dephosphorylation in a phosphorylation-dephosphorylation cycle are known to function with many protein kinases. Together, these kinases and phosphatases with their respective activating and inhibiting factors form a regulatory system. Indirect evidence for an eIF-2α phosphatase in reticulocyte lysates has been presented

(62), however, this enzyme has only recently been purified and characterized (63). The enzyme was purified by precipitation with ethanol, but in contrast to phosphatases that function with the phosphoproteins involved in glycogen metabolism, it seems to be composed of a discrete molecular species with a molecular weight of about 70,000. The former appears to exist as a series of oligomers formed from 30,000 to 35,000 dalton monomeric subunits (64-66).

The reticulocyte phosphatase catalyzes the efficient release of phosphate from eIF-2α that was phosphorylated by the HCR-associated protein kinase (Fig. 4). It also will reverse the inhibition of the eIF-2 dependent binding of Met-tRNA$_f$ to 40S ribosomal subunits that is caused by HCR as shown in Table 1. These results provide strong support for our earlier conclusion that phosphorylation of eIF-2α by HCR is the causal reaction in inhibition of protein synthesis (49, 50) and demonstrate that the phosphatase used for these *in vitro* studies potentially may function in the phosphorylation-dephosphorylation cycles that regulate protein synthesis in reticulocytes.

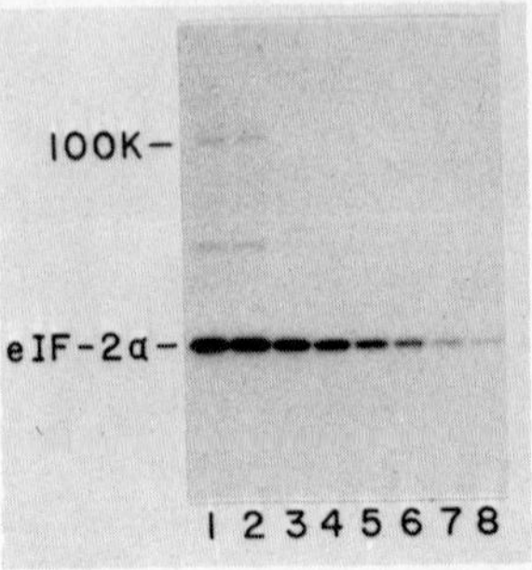

Figure 4. Dephosphorylation of eIF-2α. About 1.5 μg [^{32}P]eIF-2 (5.2 pmol [^{32}P]phosphate, 1816 cpm/pmol) were incubated with 5.6 μg protein of the phosphatase fraction for various lengths of time, then analyzed by polyacrylamide gel electrophoresis in the presence of sodium dodecylsulfate followed by autoradiography. The autoradiogram is shown. Track 1: incubation minus enzyme for 30 min; tracks 2-8: incubation with enzyme for 0, 5, 10, 20, 30, 40 and 60 min, respectively.

Table 1. The reticulocyte phosphatase fraction reverses inhibition of Met-tRNA$_f$ binding caused by HCR.[a]

Additions	[^{35}S]Met-tRNA$_f$ Bound (pmol)
None	1.08
Phosphatase	1.04
HCR	0.71
HCR, then Phosphatase	1.05

[a]Binding of [^{35}S]Met-tRNA$_f$ to reticulocyte 40S ribosomal subunits was measured under conditions and with amounts of eIF-2 with which the amount of binding was linearly dependent on the amount of active eIF-2 present. This factor was preincubated with or without HCR, then phosphatase (3.7 μg) was added for a second incubation step (10 min at 37°), before components necessary for binding of Met-tRNA$_f$ to reticulocyte 40S were added. For details of this type of assay, see (63).

The enzymatic activity of the reticulocyte phosphatase is stimulated by Mn^{2+}, but is not stimulated by Ca^{2+} or Mg^{2+}. It appears to have a rather broad substrate specificity in that it will also dephosphorylate both eIF-2 phosphorylated in its large subunit and the phosphorylated 100,000 dalton peptide observed with the eIF-2α kinase. Many of the characteristics of the reticulocyte phosphatase including its apparently rather broad substrate specificity are similar to those of phosphatases isolated from liver (65) or skeletal muscle (67). These enzymes function with the phosphoproteins involved in glycogen metabolism. A number of low molecular weight, heat-stable proteins that inhibit (68) or activate (69) these protein phosphatases have been reported. These components may be involved in the regulation of phosphatase activity. It is conceivable that these or other peptides somehow impart substrate specificity to a phosphatase catalytic subunit. Khandelwal and Zinman (69) have presented evidence for a factor that activates a protein phosphatase from liver for hydrolysis of phosphohistones but not for phosphorylase *a*. However, at this time we favor the hypothesis that regulatory specificity resides primarily on the kinase side of the phosphorylation-dephosphorylation cycle.

Double Stranded RNA in Reticulocytes

Addition of very small amounts (1-100 ng/ml) of dsRNA to the reticulocyte lysate system causes inhibition with kinetics that are strikingly similar to those observed under heme deficiency (70). Farrell *et al.*(44) presented evidence that dsRNA activates a specific protein kinase for eIF-2α and that this kinase is distinct from HCR. Antibodies prepared against reversible HCR will not prevent phosphorylation of eIF-2α by the dsRNA activated eIF-2α kinase (71). Phosphorylation of an additional protein with a molecular weight of about 67,000 was observed, when crude ribosomes were separated from a reticulocyte lysate by gel filtration and then incubated with dsRNA and [γ-^{32}P]ATP. This protein is thought to be associated with the activation of the eIF-2α kinase (44). These observations are particularly interesting in light of the effects of interferon on susceptible cells, as considered below. Interferon is induced by dsRNA. Treatment of susceptible animal cells with interferon causes their lysates to exhibit sensitivity to dsRNA similar to that observed in reticulocyte lysates. In the absence of interferon treatment, the lysates of most cells are relatively insensitive to dsRNA.

The eIF-2 kinase of the HCR system and the eIF-2α kinase that is activated by dsRNA appear to phosphorylate the same sites on eIF-2α. Samuel (72) found that three peptides formed by tryptic digestion of the small subunit of reticulocyte eIF-2 were phosphorylated after it had been incubated with the ribosomal salt-wash fraction from interferon-treated human amnion U cells, dsRNA and radioactive ATP. The same three peptides were obtained from eIF-2 that had been phosphorylated by HCR.

INTERFERONS

Interferons are species-specific glycoproteins that vary in size and structure depending on their origin. They are formed in and released from most animal cells after virus infection. Interferons will induce an "antiviral state" only in cells from the same species. Like peptide hormones and growth factors, their specificity depends upon specific receptors on the surface of target cells. Productive cell types may be triggered to produce interferon by viral double-stranded RNA (dsRNA) and some synthetic double-stranded polymers such as Poly(I)·Poly(C). Interferon-treated cells that are in the "antiviral state" are insensitive to replication of a broad range of viruses. Two different aspects of the molecular events triggered by interferon have received particular attention: 1) synthesis of 2'5'-oligoadenylates that inhibit protein synthesis, and activation of an endoribonuclease by the 2'5'-oligoadenylates, and 2) enhanced sensitivity of an eIF-2α kinase for activation by dsRNA.

An Oligonucleotide Inhibitor of Unusual Structure

Kerr and co-workers (73) detected and characterized a series of low molecular weight inhibitors of protein synthesis that are formed when extracts from interferon-treated mouse L-cells are incubated with dsRNA. The inhibitors are synthesized from ATP by an enzyme that binds dsRNA. The structure of the most prevalent compound was pppA2'p5'A2'p5'A, however 2'5'-adenylates of other sizes were detected. The 2'5'-oligoadenylate synthetase was also purified from reticulocyte lysates by affinity chromatography on dsRNA Sepharose (74). Ball (75), using interferon-treated chick embryo cells, demonstrated a 56,000 dalton peptide, the synthesis of which appears to be controlled by interferon. The peptide co-eluted from Sephadex G-100 with 2'5'-oligoadenylate synthetase activity. The oligoadenylates at subnanomolar concentrations cause inhibition of protein synthesis in lysates of sensitive cells by activation of an endonuclease (76). Very little is known about the mechanism by which the endonuclease is activated.

Lengyel's group (77) reported an endonuclease activity in extracts of interferon-treated cells that was enhanced by dsRNA and ATP. Viral mRNAs, such as those of reovirus or vesicular stomatitis virus, were more rapidly degraded than ribosomal RNA and globin mRNA, when added to extracts of interferon-treated Ehrlich ascites tumor cells, but reovirus RNA was not degraded. Nuclease activity induced by interferon also may degrade tRNA, perhaps differentially with respect to different molecular species. Revel and co-workers (78) found nuclease activity in extracts from interferon-treated L-cells. Inhibition of mRNA translation in extracts from these cells could be overcome by Leu-tRNA. One minor species of Leu-tRNA could restore translation, when Mengo virus RNA was employed as mRNA; a different Leu-tRNA species was active in promoting translation of globin mRNA in the same system. A minor species of Lys-tRNA was reported to restore EMC RNA translation in extracts from interferon-treated Friend leukemia cells (79).

It appears that 2'5'-oligoadenylates and their cognate endonuclease may play important regulatory functions in cells that apparently have not been exposed to interferon. Kerr and his co-workers (80) detected the 2'5'-oligoadenylate synthetase in a wide variety of tissues and cell types and found over a thousand fold variation in the amount of enzyme present. They observed a sharp increase in synthetase activity following withdrawal of estrogen from chick oviduct and speculate that it may function in the rapid degradation of ovalbumin mRNA and regression of this tissue following removal of the hormone. The relatively high endogenous activity of the synthetase in reticulocytes was noted above.

eIF-2α Kinase Activated by dsRNA

The phosphorylation of a protein with a molecular weight of 67,000 and of eIF-2α are two of the more striking differences seen in extracts from untreated and interferon-treated cells (81-83). The 67,000 dalton peptide may be an eIF-2α kinase that can be activated by dsRNA. The situation appears to parallel that seen in extracts of reticulocytes treated with dsRNA as considered above, except the response in reticulocytes is not dependent upon pretreatment of the intact cells with interferon. The eIF-2α kinase of reticulocytes is 68,000 daltons and is very similar or identical to the kinase detected in interferon-treated cells. These eIF-2α kinases that are activated by dsRNA appear to phosphorylate the same sites on eIF-2α as the eIF-2α kinase of the HCR system and presumably block protein synthesis by an identical mechanism (see above). However, the HCR kinase is related to a peptide of 100,000 daltons. It is not known whether interferon induces *de novo* synthesis of the 67,000 dalton peptide or somehow facilitates its activation. Ball (75) detected the *de novo* synthesis of a 56,000 dalton peptide that may be related to the 2'5'-oligoadenylate synthetase, however, the synthesis of a new peptide that might be related to the eIF-2α kinase apparently was not detected. Very little is known about the mechanism by which dsRNA activates the eIF-2α kinase. Like the 100,000 dalton peptide of the HCR system, the 67,000 dalton peptide appears to be phosphorylated when the dsRNA-sensitive eIF-2α kinase is in its active form. However, it is not known whether or not other factors are involved in this phosphorylation and activation.

SUMMARY

In eukaryotes, individual mRNA species are not necessarily translated in proportion to their molar abundance. It is not known what structural features of the mRNA itself or what protein factors contribute to this apparent selection of mRNA. However, it is clear that the relative proportion of peptides formed from a pool of mRNAs can be changed by altering translational conditions or changes in the rate of protein synthesis. Peptide initiation factor eIF-2 is a primary site for translational regulation. Phosphorylation of its smallest subunit inhibits peptide initiation by reducing its ability to facilitate binding of $Met-tRNA_f$ to 40S ribosomal subunits. Phosphorylation of eIF-2α is carried out by at least two distinct protein kinases that are activated apparently by independent mechanisms involving multiple factors acting in a cascade-type reaction sequence. These pathways are depicted in Figure 5.

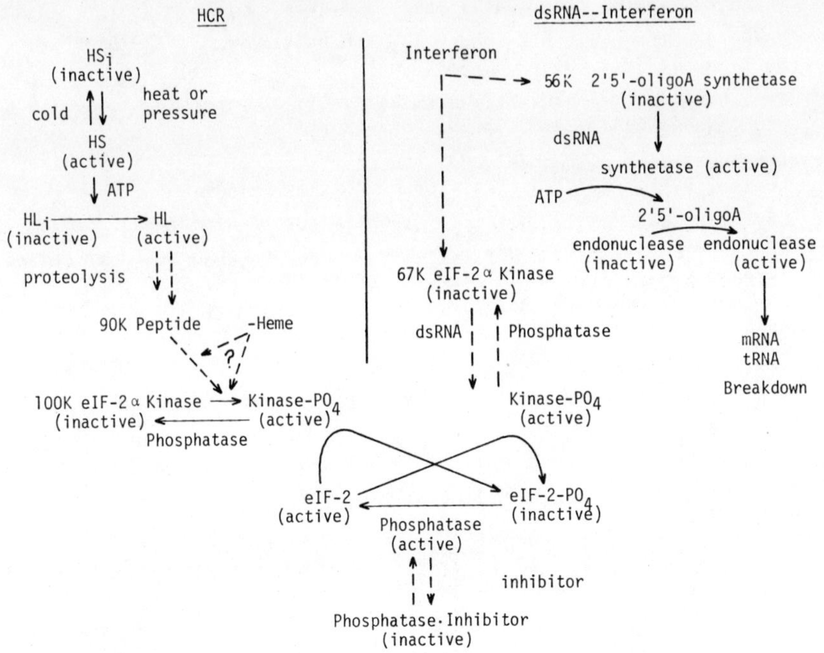

Figure 5. Phosphorylation-Dephosphorylation of eIF-2α and Activation of an Endonuclease. Solid arrows represent demonstrated reactions. Dashed arrows indicate likely reactions for which the factors or mechanisms are not known.

ACKNOWLEDGEMENTS

The authors wish to express their appreciation to M. A. Henderson for preparation of the typescript and to F. Hoffman for photographic and art work for the figures. This work was supported in part by Grant Number CA-16608, awarded to Dr. B. Hardesty and Fellowship Number 5 F32 CA-05765 awarded to Dr. B. Henderson, by the National Cancer Institute, DHEW.

REFERENCES

1. Both, G. W., Furuichi, Y., Mutukrishnan, S., and Shatkin, A. J. (1975) Cell 6, 185-195.

2. Lodish, H. F., and Jacobsen, M. (1972) J. Biol. Chem. 247, 3622-3629.

3. Lodish, H. F. (1971) J. Biol. Chem. 246, 7131-7138.

4. Lodish, H. F. (1976) Ann. Rev. Biochem. 45, 39-72.

5. Morrison, M. R., Brinkley, S. A., Gorski, J., and Lingrel, J. B. (1974) J. Biol. Chem. 249, 5290-5295.

6. Kramer, G., Pinphanichakarn, P., Konecki, D., and Hardesty, B. (1975) Eur. J. Biochem. 53, 471-480.

7. Baltimore, D. (1969) in The Biochemistry of Viruses (H. Levy, ed.), pp. 101-176, Marcel Dekker, New York.

8. Golini, F., Thach, S., Birge, C., Safer, B., Merrick, W. C., and Thach, R. (1976) Proc. Natl. Acad. Sci. USA 73, 3040-3044.

9. Rose, J., Trachsel, H., Leong, K., and Baltimore, D. (1978) Proc. Natl. Acad. Sci. USA 75, 2732-2736.

10. Sonenberg, N., Morgan, M., Merrick, W. C., and Shatkin, A. J. (1978) Proc. Natl. Acad. Sci. USA 75, 4843-4847.

11. Bergmann, J. E., Trachsel, H., Sonenberg, N., Shatkin, A. J., and Lodish, H. F. (1979) J. Biol. Chem. 254, 1440-1443.

12. Carrasco, L., and Smith, A. E. (1976) Nature 264, 807-809.

13. Koch, G., Oppermann, H., Bilello, P., Koch, F., and Nuss, D. (1976) in Modern Trends in Human Leukemia II (R. Neth, R. Gallo, K. Mannweiler, and W. C. Moloney, eds.), pp. 541-555, Lehmanns Verlag, Munich.

14. Kaerlein, M., and Horak, I. (1978) Eur. J. Biochem. 90, 463-469.

15. Ben-Hamida, F., and Beaud, G. (1978) Proc. Natl. Acad. Sci. USA 75, 175-179.

16. Lodish, H. F. (1970) J. Mol. Biol. 50, 689-702.

17. Lodish, H. F. (1971) J. Mol. Biol. 56, 627-632.

18. Steitz, J. A. (1973) Proc. Natl. Acad. Sci. USA 70, 2605-2609.

19. Shine, J., and Dalgarno, L. (1974) Proc. Natl. Acad. Sci. USA 71, 1341-1346.

20. Baralle, F. E., and Brownlee, G. G. (1978) Nature 274, 84-87.

21. Shatkin, A. (1976) Cell 9, 645-653.

22. Brawerman, G. (1974) Ann. Rev. Biochem. 43, 621-642.

23. Huez, G., Marbaix, G., Burny, A., Hubert, E., Leclercq, M., Cleuter, Y., Chantrenne, H., Soreq, H., and Littauer, U. (1977) Nature 266, 473-474.

24. Hewlett, M. J., Rose, J. K., and Baltimore, D. (1976) Proc. Natl. Acad. Sci. USA 73, 327-330.

25. Nomoto, A., Lee, Y., and Wimmer, E. (1976) Proc. Natl. Acad. Sci. USA 73, 375-380.

26. Hershey, J. (1979) in Cell Biology: A Comprehensive Treatise, Vol. V (D. Prescott and L. Goldstein, eds.), in press, Academic Press, New York.

27. Odom, O. W., Kramer, G., Henderson, A. B., Pinphanichakarn, P., and Hardesty, B. (1978) J. Biol. Chem. 253, 1807-1816.

28. Merrick, W. C. (1979) J. Biol. Chem. 254, 3708-3711.

29. Ehrenfeld, E., and Hunt, T. (1971) Proc. Natl. Acad. Sci. USA 68, 1075-1078.

30. Kosower, N. S., Vanderhoff, G. A., Benerofe, B., Hunt, T., and Kosower, E. M. (1971) Biochem. Biophys. Res. Commun. 45, 816-821.

31. Giloh, H., and Mager, J. (1975) Biochim. Biophys. Acta 414, 293-308.

32. Hunt, T. (1976) Br. Med. Bull. 32, 257-261.

33. Henderson, A. B., and Hardesty, B. (1978) Biochem. Biophys. Res. Commun. 83, 715-723.

34. Bonanou-Tzedaki, S. A., Smith, K. E., Sheeran, B. A., and Arnstein, H. R. V. (1978) Eur. J. Biochem. 84, 591-600.

35. Kruh, J., and Borsook, H. (1956) J. Biol. Chem. 220, 905-915.

36. Bruns, G. P., and London, I. M. (1965) Biochem. Biophys. Res. Commun. 18, 236-242.

37. Adamson, S. D., Herbert, E., and Godchaux, W., III (1968) Arch. Biochem. Biophys. 125, 671-683.

38. Zucker, W. V., and Shulman, H. M. (1968) Proc. Natl. Acad. Sci. USA 59, 582-589.

39. Balkow, K., Mizuno, S., and Rabinovitz, M. (1973) Biochem. Biophys. Res. Commun. 54, 315-323.

40. Legon, S., Jackson, R., and Hunt, T. (1973) Nature New Biol. 241, 150-152.

41. Balkow, K., Hunt, T., and Jackson, R. J. (1975) Biochem. Biophys. Res. Commun. 67, 366-375.

42. Beavo, J. A., Bechtel, P. J., and Krebs, E. G. (1974) Proc. Natl. Acad. Sci. USA 71, 3580-3583.

43. Gross, M., and Rabinovitz, M. (1973) Biochem. Biophys. Res. Commun. 50, 832-838.

44. Farrell, P. J., Balkow, K., Hunt, T., and Jackson, R. (1977) Cell 11, 187-200.

45. Kramer, G., Cimadevilla, J. M., and Hardesty, B. (1976) Proc. Natl. Acad. Sci. USA 73, 3078-3082.

46. Levin, D. H., Ranu, R. S., Ernst, V., and London, I. M. (1976) Proc. Natl. Acad. Sci. USA 73, 3112-3116.

47. Ranu, R. S., Levin, D. H., and Delaunay, J., Ernst, V., and London, I. M. (1976) Proc. Natl. Acad. Sci. USA 73, 2720-2726.

48. Farrell, P. J., Hunt, T., and Jackson, R. (1978) Eur. J. Biochem. 89, 517-521.

49. Pinphanichakarn, P., Kramer, G., and Hardesty, B. (1976) Biochem. Biophys. Res. Commun. 73, 625-631.

50. Kramer, G., Henderson, A. B., Pinphanichakarn, P., Wallis, M. H., and Hardesty, B. (1977) Proc. Natl. Acad. Sci. USA 74, 1445-1449.

51. Ranu, R. S., London, I. M., Das, A., Dasgupta, A., Majumdar, A., Ralston, R., Roy, R., and Gupta, N. K. (1978) Proc. Natl. Acad. Sci. USA 75, 745-749.

52. Datta, A., DeHaro, C., Sierra, J. M., and Ochoa, S. (1977) Proc. Natl. Acad. Sci. USA 74, 1463-1467.

53. Datta, A., DeHaro, C., and Ochoa, S. (1978) Proc. Natl. Acad. Sci. USA 75, 1148-1152.

54. Grankowski, N., Kramer, G., and Hardesty, B. (1979) J. Biol. Chem. 254, 3145-3147.

55. Walsh, D. A., Ashby, C. D., Gonzales, C., Calkins, D., Fischer, E. H., and Krebs, E. G. (1971) J. Biol. Chem. 246, 1977-1985.

56. Ernst, V., and Levin, D. H. (1979) Fed. Proc. 38, 781.

57. Gross, M., and Rabinovitz, M. (1972) Proc. Natl. Acad. Sci. USA 69, 1565-1568.

58. Trachsel, H., Ranu, R. S., and London, I. M. (1978) Proc. Natl. Acad. Sci. USA 75, 3654-3658.

59. Gross, M., and Mendelewski, J. (1978) Biochim. Biophys. Acta 520, 650-653.

60. Wallis, M. H., Kramer, G., and Hardesty, B. (1979) submitted to Biochem.

61. Henderson, A. B., Miller, A. H., and Hardesty, B. (1979) Proc. Natl. Acad. Sci. 76, 2605-2609.

62. Safer, B., and Jagus, R. (1979) Proc. Natl. Acad. Sci. USA 76, 1094-1098.

63. Grankowski, N., Lehmusvirta, D., Kramer, G., and Hardesty, B. (1979) submitted to J. Biol. Chem.

64. Brandt, H., Capulong, Z. L.,and Lee, E. Y. C. (1975) J. Biol. Chem. 250, 8038-8044.

65. Khandelwal, R. L., Vandenheede, J. R., and Krebs, E. G. (1976) J. Biol. Chem. 251, 4850-4858.

66. Gratecos, D., Detwiler, T. C., Hurd, S., and Fischer, E. H. (1977) Biochem. 16, 423-433.

67. Antoniw, J. F., Nimmo, H. G., Yeaman, S. J., and Cohen, P. (1977) Biochem. J. 162, 423-433.

68. Huang, F. L., Tao, S., and Glinsmann, W. H. (1977) Biochem. Biophys. Res. Commun. 78, 615-623.

69. Khandelwal, R. L., and Zinman, S. M. (1978) Biochem. Biophys. Res. Commun. 82, 1340-1345.

70. Hunter, T., Hunt, T., Jackson, R., and Robertson, H. D. (1975) J. Biol. Chem. 250, 409-417.

71. Petryshyn, R., Trachsel, H., and London, I. M. (1979) Proc. Natl. Acad. Sci. USA 76, 1575-1579.

72. Samuel, C. E. (1979) Proc. Natl. Acad. Sci. USA 76, 600-604.

73. Kerr, I. M., and Brown, R. E. (1978) Proc. Natl. Acad. Sci. USA 75, 256-260.

74. Hovanessian, A. G., and Kerr, I. M. (1978) Eur. J. Biochem. 84, 149-159.

75. Ball, L. A. (1979) Virol. 94, 282-296.

76. Kerr, I. (1979) in Modern Trends in Human Leukemia III (R. Neth, R. Gallo, K. Mannweiler, eds.), Springer Verlag, Berlin, New York; in press.

77. Sen, G. C., Lebleu, B., Brown, G. E., Kawakita, M., Slattery, E., and Lengyel, P. (1976) Nature 264, 370-373.

78. Zilberstein, A., Dudock, B., Berissi, H., and Revel, M. (1976) J. Mol. Biol. 108, 43-54.

79. Mayr, U., Bermayer, H. -P., Weidinger, G., Jungwirth, C., Gross, H., and Bodo, G. (1977) Eur. J. Biochem. 76, 541-551.

80. Stark, G. R., Dower, W. S., Schimke, R. T., Brown, R. E., and Kerr, I. M. (1979) Nature 278, 471-473.

81. Zilberstein, A., Federman, P., Shulman, L., and Revel, M. (1976) FEBS Lett. 68, 119-124.

82. Roberts, W. K., Hovanessian, A., Brown, R. E., Clemens, M., and Kerr, I. (1976) Nature 264, 477-480.

83. Lebleu, B., Sen, G. C., Shaila, S., Carbrer, B., and Lengyel, P. (1976) Proc. Natl. Acad. Sci. USA 73, 3107-3111.

ANTIBIOTIC INHIBITORS OF TRANSLATION IN EUKARYOTES

DAVID VÁZQUEZ and ANTONIO JIMÉNEZ

Instituto de Bioquímica de Macromoléculas,
Centro de Biología Molecular,
C.S.I.C. and U.A.M. Canto Blanco, Madrid
34, Spain

INTRODUCTION

Translation inhibitors have been very useful in the last thirty years to elucidate the overall mechanism of translation and to study different individual steps in the process. There are at least two types of protein synthesis systems (prokaryotic and eukaryotic). However there are a number of features which are common in both systems. Therefore translation inhibitors acting on eukaryotic systems can be broadly classified according to their specificity into those affecting selectively (a) both systems of the prokaryotic and the eukaryotic types (Table 1) and (b) only eukaryotic systems (Table 2) In general, antibiotics

TABLE 1. Inhibitors of translation acting on prokaryotic and eukaryotic systems.

Actinobolin	Guanylyl-imido-diphosphate
Adrenochrome	Nucleocidin
AHR-1911	Pactamycin
Amicetin group:	Polydextran sulphate
Amicetin	Polyvinyl sulphate
Bamicetin	Puromycin
Plicacetin	Pyrochatechol violet
Anthelmycin	Showdomycin
Aurintricarboxylic acid	Sparsomycin
Blasticidin S	Tetracycline group:
Chartreusin	Chlortetracycline
Deoxystreptamine-containing antibiotics	Doxycycline
Edeine A_1	Oxytetracycline
Fusidic acid	Tetracycline
Gougerotin	Tosylphenylalanylchloromethane
Guanylyl-methylene-diphosphate	

affecting prokaryotic-type systems are active in bacteria, blue-green algae, mitochondria and chloroplasts whereas those acting selectively on eukaryotic-type systems are active in higher cells which are known to have 80S type cytoplasmic ribosomes.

This study will be concerned with the present state of the problem of translation inhibitors active on eukaryotic systems and our own recent unpublished observations in this field. An extensive survey of the literature will not be possible in this contribution and other reviews (1-6) can be useful to readers interested in some specific aspects of the problem in a more comprehensive manner.

TRANSLATION OF mRNA

The overall reactions taking place in the process of translation by eukaryotic ribosomes can be represented schematically according to the translocation model with one or two entry sites on the small subunit. Following this model, the process of translation can be arbitrarily

TABLE 2. Inhibitors of translation acting on eukaryotic systems.

Abrin	Narciclasine group:
Alpha sarcin	Haemanthamine
Anisomycin	Narciclasine
5-azacytidine	Pretazettine
Bruceantin	PAP
Crotins	PA toxin
Curcins	Pederine
Diphtheria toxin	Phenomycin
Emetine group:	PR toxin
Emetine	Ricin
Tubulosine	Sodium fluoride
Enomycin	Tenuazonic acid
Glutarimide group:	Trichotecene antibiotics:
Actiphenol	Trichodermin group:
Cycloheximide	Fusarenon-X
Streptimidone	Trichodermin
Streptovitacin A	Trichodermol
Harringtonine group:	Trichothecin
Harringtonine	Verrucarin A group:
Homoharringtonine	Deacetoxyscirpenol
Isoharringtonine	Nivalenol
Holocanthone	Toxin T-2
Lycorine group:	Verrucarin A
Lycorine	Tylophora alkaloids:
Pseudolycorine	Cryptopleurine
MDMP	Tylocrebrine
Modeccin	Tylophorine

divided into the phases of initiation, elongation and termination (Figure 1). The most recent developments concerning the mechanism of translation in eukaryotes have been extensively reviewed recently (refs. 7 and 8 and different contributions in this volume)

INHIBITORS OF INITIATION

The process of initiation of the polypeptide chain involves four sequential steps required for the attachment of the initiator substrate Met-$tRNA_F$ to the correct position on the 80S ribosome. These steps that we will consider are (a) the recognition of eIF-3 by the 40S subunit and binding of the eIF-2·Met-$tRNA_F$·GTP complex, (b) recognition of eIF-1, eIF-4A, eIF-4B, eIF-4C and the mRNA initiation triplet, (c_1) attachment of the 60S ribosomal subunit to the initiation complex formed in step (b) and (c_2) interaction of the 3' end of Met-$tRNA_F$ with the donor site of the peptidyl transferase center (Figure 1).

Inhibitors of recognition of the initiator substrate (step (a))

Step (a) in eukaryotes involves the following reactions:interaction of the 40S subunit with eIF-3, formation of the eIF-2·Met-$tRNA_F$·GTP complex and interaction of this complex with the 40S subunit with eIF-3 bound.

There is ample experimental evidence showing that kasugamycin blocks the interaction of the initiator substrate fMet-$tRNA_F$ with the 30S ribosomal subunit in bacterial systems. Similar studies have not been performed in eukaryotic systems, although the antibiotic might be expected to act similarly in bacterial and eukaryotic systems (refs. 5, 6 and references therein).

Showdomycin (an analogue of N-ethyl-maleimide), adrenochrome and triphenylmethane dyes (such as aurintricarboxylic acid, pyrochatechol violet, azure blue B, fuchsin acidic and fuchsin basic) inhibit the formation of the eIF2-Met·$tRNA_F$·GTP complex (refs. 5,6 and references therein).

Evidence from different laboratories has shown that the translational inhibitor (heme controlled repressor) is a eIF-2 kinase that phosphorylates the small subunit of eIF-2 in cell-free systems and by doing so prevents the formation of the eIF-2·Met-$tRNA_F$·GTP complex (ref. 6 and references therein). More recent work has shown that there is a protein in rabbit reticulocyte lysates known as eIF-2 stimulating

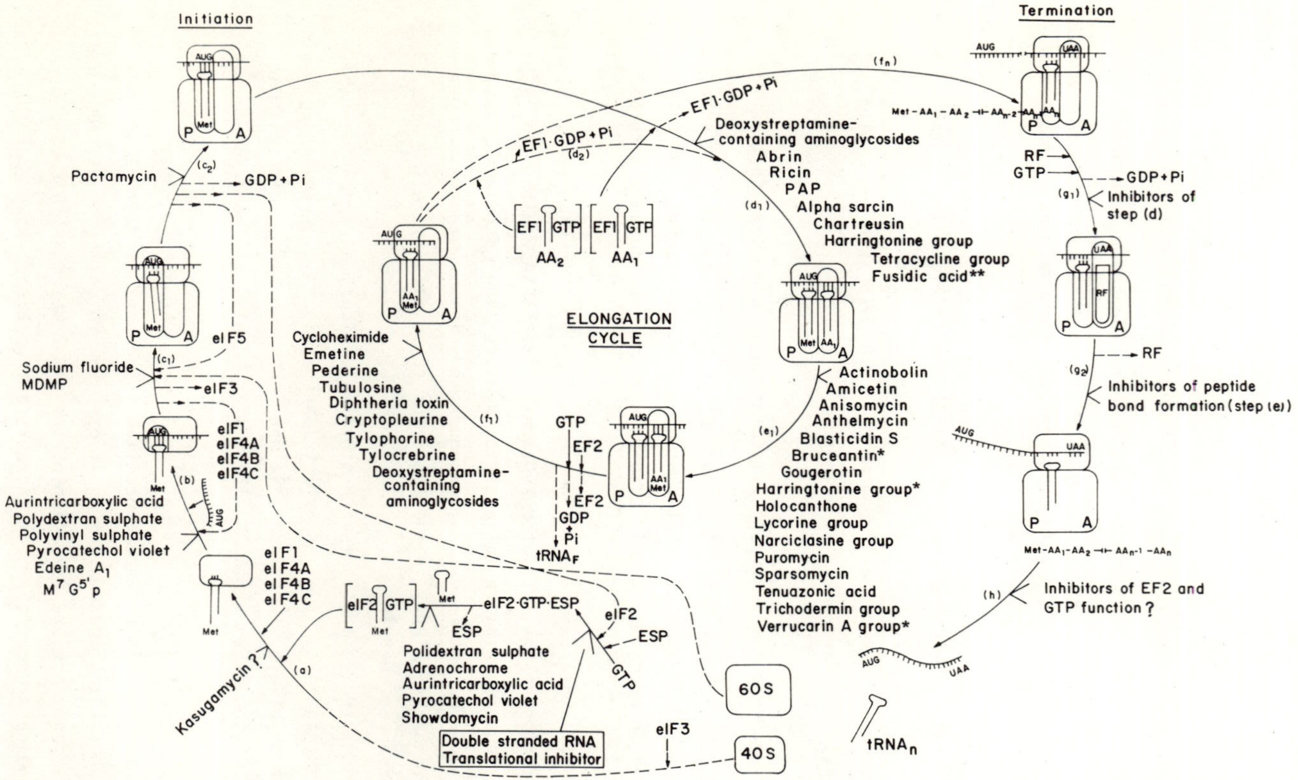

Figure 1. Translation process. Site of action of inhibitors. *Do not interact with polysomes and only block the first few rounds of peptide bod formation. **It does no inhibit aminoacyl-tRNA binding in resolved cell-free systems in the absence of EF-2; it can inhibit translocation in resolved cell-free systems containing EF-2 but not EF-1.

protein (ESP) that is required for the initial formation of the eIF2·GTP complex prior to the synthesis of the eIF-2·Met-$tRNA_F$·GTP complex. It was indeed proposed that formation of the eIF2·GTP·ESP complex is the initial reaction in the process (Figure 1, step a) although this complex has not been detected (9, 10). It appears,therefore, that phosphorylation of the smaller subunit of 2IF-2 by the translational inhibitor prevents its interaction with ESP to form the complex (10). However , phosphorylation of eIF-2 has been observed in cell-free systems from a number of eukaryotic cells but has never been described yet in intact cells.

Double stranded RNA prevents cellular and viral mRNA translation by inhibiting eIF-2 action and therefore blocking step (a) (Figure 1); this effect can be overcome in cell-free systems by addition of an excess of eIF-2 (6 and references therein). Inhibition of translation by cell-free extracts from interferon-treated cells has also been reported and it was observed that this inhibition can be reversed by the addition of $tRNA^{Leu}$; furthermore, the methylation of some viral mRNAs is impaired in interferon-treated cells (6,11 and references therein). These effects by interferon appear to be independent of the presence of double stranded RNA in the system. However in the presence of double stranded RNA and ATP the inhibitory effect of interferon is considerably increased. Also it is observed in the cell extract an enhancement in the systhesis of the oligonucleotide ppp A (2'p5'A)n (12-15) and an activation of a protein kinase (translational inhibitor) that induces phosphorylation of several proteins including the small subunit of eIF-2 (16-19). This results in an inhibition in the formation of the eIF-2·GTP complex required for step (a) as indicated above. Furthermore, addition of double stranded RNA to interferon-treated cell extracts also causes an activation of an endonuclease activity, loosely bound to ribosomes, that degrades different mRNAs at different rates (6 and references therein, 20, 21). Whereas some workers postulate that the endonuclease is activated by pppA (2'p5'A)n (21,22) it has been suggested by others that the oligonucleotide does not appear to be a mediator in the activation of the activities of the endonuclease and the translational inhibitor (16). Hence, in cell-free extracts from interferon treated cells, besides an inhibition of step (a), methylation and a degradation of mRNA, the formation of the oligonucleotide ppp A(2' p5'A)n and a further inhibitory effect on translation that can be overcome by the addition of $tRNA^{Leu}$ can be observed. The relationships between these effects is not yet clear. Furthermore it is not yet clear what the relationship is between the effects of double stranded RNA on protein synthesis inhibition in interferon treated cells and the interferon phenomenon (23).

Inhibitors of mRNA recognition (step b)

Within this group of inhibitors we can include a number of triphenylmethane derivatives (the best known of which is aurintricarboxylic acid), some aminochromes (the best known of which is adrenochrome), poly dextran sulphate and polyvinyl sulphate. All these compounds have a rather complex mode of action involving more than one interaction and inhibiting more than one step. Indeed all of them appear to interact with the 40S ribosomal subunit and therefore prevent mRNA recognition in cell-free systems, but they are hardly used in intact cells due to the permeability barrier. As indicated above these compounds also inhibit recognition of the initiator (step (a)) by preventing the formation of the ternary complex eIF-2·Met-$tRNA_F$·GTP in a reaction in which the ribosome is not involved (5,6, reviews).

Edeine A_1 is the best known member of the edeine complex of related antibiotics. In intact bacteria edeine A_1 at low concentrations reversibly inhibits DNA but in cell-free systems the antibiotic is also an effective inhibitor of initiation by interacting with the smaller ribosomal subunit (6 and references therein). The initial finding that edeine A_1 specifically inhibits the formation of the functional eIF-2·Met-$tRNA_F$·GTP·40S complex (24) is supported by experimental evidence with other systems. Indeed the antibiotic interferes with the shift of labeled mRNA from the 40S to the 80S in a wheat germ system (25) and in a rabbit reticulocyte system (26) and interferes with the AUG recognition of reovirus mRNA (27). Furthermore a direct inhibition of step (b) (Figure 1) by edeine A_1 has recently been confirmed also in other systems (28,29).

The compound $m^7G^{5'}$p(7-methylguanosine-5'-monophosphate) has been found to block translation of mRNA from different systems where mRNA is provided with the sequence $m^7G^{5'}$ppp... ("cap") at the 5' end. This inhibition is due to a blockage at the initiation phase, preventing the eIF-4B-dependent interaction of mRNA with the 40S ribosomal subunit. Studies on the structure-function of the compound have shown that the 7-methyl and 5'-phosphate groups are essential for their inhibitory activity (6 and references therein). On the other hand, eIF-2 also binds to mRNA and this binding reaction is also inhibited by cap analogues (30). Moreover the inhibition by cap analogues is strongly dependent on the potassium concentration (31).Furthermore, recent reports have shown that the translation of some uncapped mRNAs is strongly inhibited _in vitro_ by cap analogues (32,33). Indeed it appears that, although the translation of EMC RNA (an uncapped mRNA) is inhibited by m^7Gp and m^7Gppp, the inhibitory effects of these cap analogues in the translation of a capped messenger such as globin mRNA are stronger (34). Furthermore the same preferential inhibitory effects were observed with inhibitors of mRNA binding to ribosomes such as aurintricarboxylic acid but not with inhibitors of the later steps in initiation of protein synthesis such as pactamycin (34). Therefore, it was suggested that the differences in the inhibition by cap analogues between capped or uncapped mRNAs are due only to the different affinities of mRNA for the initiation machinery independent of the presence of cap in the 5' terminus; it was further observed that 2'-O-mGppp cannot be considered only as a cap analogue because it affects some step prior to mRNA binding to the ribosome (34).

Inhibitors of subunit joining (Step (c_1))

Fluoride Na^+ and K^+ salts and the D-stereoisomer of 2-(4-methyl-2, 6-dinitroanilino)-N-methyl-propionamide (D-MDMP) have been proposed as inhibitors of the joining of the 60S ribosomal subunit to the initiation complex formed on the 40S ribosomal subunit (6 and references therein).

Inhibitors of the positioning of the initiator in the donor site (Step (c_2))

The antibiotic pactamycin at low concentrations acts on the small ribosomal subunit of both prokaryotic and eukaryotic ribosomes and therefore prevents initiation of the polypeptide chain in both systems by blocking the step (c_2) (Figure 1). At higher concentrations the antibiotic also interacts with the larger ribosomal subunit and therefore has a secondary effect on elongation (5,6 and references therein). The major inhibitory effect of pactamycin on step (c_2) has been confirmed recently in rabbit reticulocyte cell-free systems using native 40S ribosomal subunits (34).

Unclassified inhibitors of initiation

The base analog 5-azacytidine primarily blocks translation in several eukaryotic systems at the level of initiation of the polypeptide chain. Since the drug is incorporated into RNA and does not inhibit initiation of translation in cell-free systems, it has been postulated that 5-azacytidine might be incorporated in a new RNA species, not yet described, some how involved in the process of initiation (6 and references therein).

Viral infection normally results in an inhibition of synthesis by the host cell (shut-off), whereas viral proteins are synthesized. The inhibition of protein synthesis after viral infection is exerted at some step not yet elucidated on the initiation level, since none of the components required for translation appears to be affected in the infected cells as tested in cell-free systems. It has recently been suggested that viral infection could interfere with some membrane function, thus altering the ionic concentration inside the cell, and that this ionic concentration might differentially affect viral and cellular protein synthesis (6 and references therein).

Some ionophore antibiotics such as valinomycin (which is known to transport monovalent cations) and A23187 (a divalent cation ionophore) are potent inhibitors of protein synthesis in intact cells. The drug phalloidin has also a similar effect and dissociates liver polysomes when administered to rats. In all these cases it was proposed that the effects of the drugs might be due to their action on the membrane, causing a leakage of ions (6 and references therein).

Hypertonicity resulting from an increased NaCl concentration in the growth medium causes an inhibition of protein synthesis in cell cultures. It was postulated that hypertonicity lead to an inhibition of peptide chain initiation since it causes polysome breakdown. It was further observed in different virus-cell systems that total cellular protein synthesis is more sensitive to hypertonic media than viral protein synthesis. This preferential inhibitory effect of hypertonicity might be due simply to a change in ionic concentration, since it has been shown that sodium ions inhibit cellular protein synthesis and enhance viral protein synthesis (6 and references therein).

When exponentially growing cultures of bacteria and eukaryotic cells are cooled to below 8°C, (temperature shift-down) initiation of protein synthesis is blocked. The compounds 3-methyleneoxindole and 2'(3'), 5'-ADP have also been proposed as inhibitors of polypeptide chain initiation (6 and references therein).

INHIBITORS OF ELONGATION

Polypeptide elongation takes place in eukaryotic systems by repeated cycles composed of three sequential steps: EF-1-dependent aminoacyl-tRNA binding (step (d)), ribosome catalyzed peptide bond formation (step (e)) and EF-2-dependent translocation (step (f)). Besides the initiation complex formed in step (c_2), two molecules of GTP, the specific aminoacyl-tRNA and the elongation factor EF-2 are required in the first elongation cycle.

There is ample experimental evidence supporting a single entry site in intact cells for GTP- and EF-1-dependent binding of aminoacyl-tRNA on the A-site of the ribosome. Non-enzymic binding of aminoacyl-tRNA to the A-site can also be studied in cell-free systems in the absence of EF-1.

Peptide bond formation (step (2)) is catalyzed by the peptidyl transferase center, which is an integral part of the 60S ribosomal subunit.

Translocation (step (f)) requires elongation factor EF-2 and involves movement of the peptidyl-tRNA from the A- to the P-site, coupled to the movement of one triplet of the mRNA, on the ribosome. Hydrolysis of one molecule of GTP is required for the release of EF-2. Spontaneous non-enzymic translocation takes place slowly in cell-free systems in the absence of elongation factors and GTP. This non-enzymic translocation is considerably accelerated in the presence of high concentrations of K^+ or NH_4^+.

The two elongation factors (EF-1 and EF-2) complexed with GTP appear to interact with a common or overlapping area of the larger ribosomal subunit and therefore do not interact simultaneously but only alternately with the ribosome. Thus the interaction of one of the elongation factors takes place after the release of the other.

Compounds interfering with aminoacyl-tRNA recognition (step (d))

Within this important category of inhibitors we can distinguish at least two groups of compounds. One of these groups includes the inhibitors of EF-1-dependent binding of aminoacyl-tRNA and the other one clusters the misreading compounds that may prevent protein synthesis by favoring the erroneous interaction of aminoacyl-tRNA in the acceptor site of the ribosome.

The antibiotics of the tetracyline group are strong inhibitors in cell-free systems of EF-1-dependent binding of aminoacyl-tRNA. These antibiotics appear to have a number of binding sites with both ribosomal subunits, at least in bacteria, but their interaction with the smaller ribosomal subunit appear to be more relevant for their mode of action (6 and references therein). Activities of tetracylines in eukaryotic cells is very small compared with their activities in bacteria. This is so because only bacteria has an active transport for these antibiotics and they are concentrated in their cytoplasm. A number of bacterial mutants resistant to tetracyclines have been found to have a modification in one of the proteins of the outer membrane (35, 37).

Fusidic acid can inhibit the step of translocation in cell-free systems when added in a large excess or alternatively a very small amount of EF-2 is added to the system. This is because the GTP·ribosome·fusidic acid·EF-2 is formed and therefore the antibiotic can sequester all EF-2 available, but the effect of fusidic acid in such systems is abolished by saturating concentrations of EF-2. However in intact cells and integrated systems for translation the antibiotic is really an inhibitor of EF-1-dependent aminoacyl-tRNA binding since this reaction cannot take place with the ribosome·fusidic acid·EF-2·GTP complex. Indeed this complex is rather stable and the GTP·EF-1·aminoacyl-tRNA complex cannot interact with its ribosomal site which is overlapping with the site occupied by EF-2·GTP (6 and references therein).

There are an increasing number of protein toxins that are known to block the elongation cycle in translation by catalytic inactivation of the 60S subunit of the eukaryotic ribosome. At least two of these compounds ricin and abrin, and probably some others are of glycoprotein nature. Three of the compounds (alpha sarcin, mitogillin and restrictocin) have been shown to cleave 320 nucleotides from the 3' terminal end of the larger ribosomal RNA from the 60S subunit. In all cases the catalytic inactivation of the 60S ribosomal subunit results in a decrease in the affinity of elongation factors EF-1 and EF-2 for the ribosome. Therefore an inhibition of aminoacyl-tRNA binding and (or) translocation by these toxins has been reported (Table 3). Inhibition of translocation is not observed in all cases since in the case of some toxins or under certain experimental conditions the pretranslocated peptidyl-tRNA bound to the A-site prevents the interaction of the EF-2·GTP complex (6 and references therein). Alpha sarcin cross-reacts with mitogillin and restrictocin and the three toxins cause the cleavage of the larger RNA in polysomes (38). Permeability mutants of abrin or ricin show cross resistance to both drugs (39). Identity of ricin and Ricinus

TABLE 3. The effects of glycoprotein and protein toxins acting or the eukaryotic ribosome on model reactions (data are taken from references 6, 38, 41, 42 and 43).

	EF-1-dependent		EF-2-dependent			
	Binding of aminoacyl-tRNA	GTP hydrolysis	Formation of the EF-2-GTP-ribosome complex	Uncoupled		Non-enzymic
				GTP hydrolysis	Translocation	translocation
Ricin A-chain	Inhibition	Inhibition	Inhibition	No effect or inhibition	No effect or inhibition or stimulation	No effect
Abrin A-chain	Inhibition	Inhibition	Inhibition	No effect or inhibition	No effect or stimulation	No effect
Alpha sarcin Mitogillin Restrictocin	Inhibition	Inhibition	Inhibition	Inhibition	No effect	---
PAP	Inhibition	Inhibition	Inhibition	No effect or stimulation	No effect or stimulation	---
Crotin II	Inhibition	Inhibition	Inhibition	No effect	No effect or inhibition	---
Modeccin	Inhibition or no effect	---	Inhibition	---	No effect or inhibition	---
Enomycin	Inhibition	---	No effect	Inhibition	No effect	---
PR toxin	?	---	?	---	---	---

communis agglutinin has been reported (40).

The antibiotic chartreusin, the mucopolysaccharide heparin and a number of catechols and aminochromes, including adrenochome, have been also reported as inhibitors of EF-2-dependent binding of aminoacyl-tRNA to 80S ribosomes (5, 6 and references therein).

It is very interesting that a high fidelity in translation of the eukaryotic ribosomes has been observed, since no significant misreading by mammalian systems was detected in the presence of polyamines, streptamine-containing aminoglycosides, some organic solvents, increasing concentrations of Mg^{++} ions and suboptimal temperatures of incubation that are known to induce misreading in bacterial cell-free systems (6 and references therein). However an increasing number of aminoglycoside antibiotics that contain the deosystreptamine or some deosystreptamine-derived moiety cause misreading in bacterial and eukaryotic cells and cell-free systems. Since these antibiotics also cause an inhibition of translocation we will discuss their effects when dealing with inhibitors of this step.

Inhibitors of peptide bond formation (step (e))

Puromycin is the most universally used and best known inhibitor of peptide bond formation. The antibiotic is well known to act as an analogue of the 3'-end of aminoacyl-tRNA (aminoacyl-adenosine) at the acceptor site of the peptidyl transferase center on the larger ribosomal subunit of both prokaryotic and eukaryotic ribosomes. Hence, the use of puromycin provides a simplified method for the study of peptide bond formation in a reaction in which the $-NH_2$ group of puromycin becomes linked to the C-terminal end of the relevant amino acid when either f-Met-$tRNA_F$ or peptidyl-tRNA ("puromycin raction") or f-Met- or Ac-Phe- or Ac-Leu-oligonucleotide ("fragment reaction") are used as donor substrates. Studies concerning structure-activity of puromycin as an acceptor substrate are well documented but the site of interaction of the antibiotic on the 60S ribosomal subunit is not yet well resolved. Thus the treatment of rat liver ribosomes with iodoacetyl-puromycin leads to labeling of proteins L28 and L29 whereas other studies suggest that proteins L21, L26, L31 and L36 (but neither L28 and L29) are involved in the peptidyl transferase center that from the functional point of view is supposed to be the site of action of puromycin (6 and references therein).

All the other inhibitors of peptide bond formation act like puromycin on the 60S ribosomal subunit. They can be widely classified in two groups including in one of them those compounds that are active in both prokaryotic and eukaryotic ribosomes and in the second group those inhibitors that are active only in eukaryotic ribosomes. Inhibitors of the first group (amicetin, gougerotin, blasticidin S, actinobolin, anthelmycin and sparsomycin) act on the acceptor site of the peptidyl transferase center. They mutually exclude in their binding site(s) to the ribosome and also prevent puromycin interaction but, unlike this antibiotic, do not have acceptor activity in peptide bond formation (6, 44). Concerning the very wide group of peptide bond formation inhibitors that selectively act on eukaryotic ribosomes they differ in their binding site, mode of action, inhibitory spectra and cross-resistance. Thus anisomycin, tenuazonic acid, bruceantin and inhibitors of the narciclasine, harringtonine, trichodermin, verrucarin A and lycorine groups (Table 2) mutually exclude in their binding sites. However bruceantin, harringtonine and verrucarin A compounds only block the first few rounds of peptide bond formation prior to polysome formation. Compounds of the lycorine group do not show cross-resistance with trichodermin and tenuazonic acid is not active on yeast ribosomes (6 and references therein). Holocanthone is chemically related to bruceantin (45) and possesses high antitumor activity. Similarly to bruceantin holocanthone inhibits protein synthesis by specifically blocking peptide bond formation. Polyphenylalanine synthesis (7×10^{-5}), polypeptide synthesis (10^{-4}) and the fragment reaction (3×10^{-6}) are also inhibited by the compound (figures in brackets indicate final M concentrations of holocanthone giving 50% inhibition) (E. Zaera, unpublished results). 80S ribosomes from a yeast mutant isolated for resistance to narciclasine are also resistant to anthelmycin and inhibitors of the trichodermin and narciclasine groups but are sensitive to gougerotin and blasticidin S (members of the 4-aminohexoxyl-cytosine group of antibiotics like anthelmicin). This pleiotropic phenotype is expressed by a single nuclear mutation. Therefore anthelmycin, which differs chemically from gougerotin and blasticidin S in lacking an aminoacyl moiety, must have a common or overlapping site of action to the one for narciclasine and trichodermin antibiotics, but not closely related to the site(s) of action of gougerotin and blasticidin S. Neither gougerotin nor blasticidin S nor anthelmycin affect the binding of either (^{14}C)trichodermin, (^{3}H) narciclasine or (^{3}H)anisomycin to eukaryotic ribosome (6 and A. González, unpublished results).

Inhibitors of translocation (step (f))

We distinguish three classes of inhibitors within this group including inhibitors acting on elongation factor EF-2, those acting on the 40S ribosomal subunit and finally the inhibitors blocking translocation by interacting on the 60S subunit.

Diphteria toxin and PA toxin, produced by <u>Corynebacterium diphtheriae</u> $(\beta)^{tox +}$ and <u>Pseudomonas aeruginosa</u>, respectively, inhibit translocation due to the production of a toxin that enzymically inactivates EF-2 in the presence of NAD^+ producing ADP-ribosyl-EF-2 and nicotinamide. ADP-ribosyl-EF-2 forms a complex with GTP that interacts with the ribosome but its affinity and the rate of hydrolysis of GTP is very low and therefore the efficiency of ADP-ribosyl-EF-2 in catalyzing translocation is very low indeed. Diphtheria toxin (MW 60,000) is active in intact cells and has two subunits, one of them fragment B (MW 38,000-39,000) being responsible for the interaction with the cell membrane and the entrance of the A fragment (MW 21,000-24,000) that is required in a free form to catalyze ADP-ribosylation of EF-2(5,6 and references therein). A subunit of MW 26,000 has been prepared from PA toxin (MW 60,000) and cannot enter in intact cells but also causes ADP-ribosylation of EF-2; however this subunit is not identical to fragment A of diphtheria toxin since it is not neutralized by fragment A-specific antiserum (46).

Deoxystreptamine-containing aminoglycoside antibiotics and cryptopleurine, tylocrebrine, tylophorine, emetine and tubulosine are included between the inhibitors of translocation acting at the level of the 40S ribosomal subunit. However deoxystreptamine-containing antibiotics not only inhibit translocation but also induce misreading and therefore we are to review both effects here. Streptomycin-induced phenotypic suppression has been reported in a <u>his 1</u> missense mutant of <u>Saccharomyces cerevisiae</u> (47) and should be confirmed by other workers. Aminoglycoside antibiotics other than streptomycin can induce high levels of misreading in <u>in vitro</u> systems from wheat germ (48), cultured human cells (49) and <u>Tetrahymena</u> (50). Moreover a number of aminoglycoside antibiotics induce phenotypic suppression of nonsense mutations in yeast cells (51, 52). The effects of these antibiotics on both phenotypic suppression and misreading in eukaryotic organisms and cell-free systems is summaryzed in Table 4. We have tested the effects of a number of aminoglycoside antibiotics on polyphenylalanine synthesis and misreading of poly(U) by yeast cell-free systems. From our data and previous reports (48,51,52) it is concluded that the antibiotics are poor inhibitors of polyphenylalanine synthesis in a number of eukaryotic cell-free systems. However hygromycin B drastically blocks polypeptide synthesis promoted by endogenous m-RNA on yeast polysomes and BMV RNA in a wheat germ system (53). Also gentamicin and paromomycin cause a strong inhibition of polypeptide synthesis in cell-free extracts from Tetrahymena (50). The strongest promoter of misreading on 80S ribosomes from yeasts is hygromycin B. In fact on the basis of total pmol of amino acids incorporated no inhibition of total polypeptide synthesis occurs, even at the highest concentrations of the antibiotic as compared to controls in its absence. Omission of non cognate amino acids from the incubations did not modify the incorporation of phenylalanine suggesting

TABLE 4. Effects of aminoglycoside antibiotics on misreading *in vitro* and phenotypic suppression *in vivo*

Antibiotic group	Misreading		Phenotypic suppresion	
	Active	Inactive	Active	Inactive
Monosubstituted deoxystreptamine antibiotics	Hygromycin B Destomycin A		Hygromycin B Destomycin A	
4,6-disubstituted deoxystreptamine antibiotics	Kanamycins A,B and C, Tobramycin, Gentamicins C_1, C^*_{1a} and C_2 Sisomicin	Gentamicin B	Kanamycins A and B, Tobramycin, Gentamicins A_1, C_{1a}, C_2, X_2. Sisomicin	Kanamycin C Gentamicins A,B,C_1, A_2
4,5-disubstituted deoxystreptamine antibiotics	Neomycin, Paromomycin, Lividomycins A, B Butirosin A, B		Neomycins B and C Paromomycin Lividomycins A and B	Butirosins A and B
Others		Streptomycin Bluensomycin		Streptomycin** Bluensomycin Kanamycin Fortimycin A

This table is essentially taken from ref. 51 collecting data presented in references 48-53 and unpublished observations by E. Zaera.

*It did not show any misreading effect on wheat germ, although it induced misreading in yeast cell-free systems.

**Only active in missense his 1 mutants (47).

that the incorporation of non cognate amino acids did not replace that of phenylalanine but took place in addition to it. Similar results have been obtained in Escherichia coli cell-free systems (54).

Although there have been suggested (45) that only those aminoglycoside antibiotics that bear a 6'-hydroxyl group on the aminosugar linked to the 4-position of 2-deoxystreptamine are highly active in promoting misreading in 80S ribosomes, there is not clear evidence to sustain this statement . Thus, sisomicin and kanamycin B bear a $6'$-NH_2 group and induce a strong misreading while kanamycin A that has a 6'-hydroxyl group (45 and E. Zaera, personal communication) only induces low levels of misreading. Gentamicin C1 has a $6'$-CH_3-NH- group and also promotes a high level of misreading (48). Moreover hygromycin B lacks the structural requirement initially proposed as essential for misreading and is the most portent inducer of mistranslation in vitro (E. Zaera, personal communication). The structural requirements proposed neither holds for the reported effects of the aminoglycoside antibiotics on phenotypic suppression, if one considers that this phenomenon is promoted by mistranslation caused by the antibiotics in yeasts. Thus kanamycin B is clearly more active than kanamycin C and sisomicin and hygromycin B are also very active (51).

Similarly to the case of bacterial ribosomes it should be expected that aminoglycoside antibiotics acting on eukaryotic systems interact with 80S ribosomes but there is direct evidence only in the case of paromomycin (48). Considering that misreading is increased with drug concentrations (51) as it also occurs with E. coli ribosomes (56) it is likely that there are multiple binding sites for all these drugs on 80S ribosomes. Indeed (^{14}C) tuberactionomycin (56) (^{3}H)kanamycin A (57) neomycin (58) and viomycin (59) have been shown to bind to both 30S and 50S subunits from E. coli.

Recent studies have shown that hygromycin B blocks the EF-2 and GTP-dependent translocation of peptidyl-tRNA from the acceptor site to the donor ribosomal site (53 and Table 5). In addition other aminoglycoside antibiotics, namely paromomycin, neomycin, gentamicins C_1 and C_{1A} and sisomicin also appear to inhibit translocation (Table 5). Although these inhibitory effects are much lower than that promoted by hygromycin B, they parallel the observed patterns of inhibition of polyphenylalanine synthesis (E. Zaera, personal communication). From the results of the effects of the kanamycin group of antibiotics on translocation it would appear that they do not block the reaction, but rather stimulate it.However these misleading results are most probably due to the stimulation of the puromycin reaction that is induced by these antibiotics (Table 5). Otherwise peptide bond formation does not seem to be drastically affected by other aminoglycoside antibiotics, and only neomycin and, to a lower extent paromomycin and gentamicin C_1 show a certain inhibition of the puromycin reaction (Table 5). However when the synthesis of peptidyl-(^3H) puromycin depends upon translocation of peptidyl-tRNA, the inhibition is considerably higher (Table 5) which indicates that it is the translocation step the main target of the antibiotics.

Hygromycin B differs from the other aminoglycoside antibiotics not only in chemical structure and a higher activity in mistranslation but also in its effects on protein synthesis in eukaryotic cell-free systems (6). Thus the inhibitory effects of hygromycin B on either polypeptide synthesis or translocation can be reversed by increasing concentrations of elongation factor EF-2 (53) while this effect is not observed for the other antibiotics (E. Zaera, personal communication). It is therefore suggested that hygromycin B might impair either the functioning or the binding of elongation factor EF-2 to the 80S ribosome.

It is difficult to correlate the effects of the aminoglycoside antibiotics on misreading, translocation and protein synthesis. Thus Paromomycin, gentamicin C_{1A} and sisomicin promote mistranslation at concentrations which do not inhibit polyphenylalanine synthesis. On the

TABLE 5. Effects of aminoglycoside antibiotics on enzymic translocation and peptide bond formation by yeast polysomes.

	% Control			
	Enzymic translocation		Peptide bond formation	
	$4x10^{-4}M$	$10^{-3}M$	$4x10^{-4}M$	$10^{-3}M$
Hygromycin B	15	15	100	100
Gentamicin C_1	45	-12	103	48
Gentamicin C_{1A}	50	22	118	107
Sisomicin	76	57	135	111
Paromomycin	9	0	89	64
Neomycin	29	-32	53	34
Kanamycin A	118	101	194	229
Kanamycin B	114	116	236	171
Tobramycin	131	158	260	282

other hand in the presence of neomycin, kanamycins A and B, tobramycin and gentamicin C_1 misreading of poly(U) and inhibition of polyphenylalanine synthesis in yeast cell-free systems have parallel dependences on antibiotic concentrations (ref. 52, and E. Zaera personal communication). However an excess of elongation factor EF-2 may effectively eliminate the inhibitory effect of hygromycin B on both polypeptide synthesis and translocation (53) but has no effect on the drug-promoted misreading in a reticulocyte cell-free system (A. González, personal communication) suggesting these might not be a relationship between misreading and inhibition of translocation by the antibiotic. Nevertheless it might be possible that some ribosomal components involved in translocation or aminoacyl-tRNA binding are differentially affected by the action of the antibiotics. It has been reported that hygromycin B stabilizes both the enzymic and the nomenzymic binding of Phe-tRNA to the 80S ribosomal acceptor site and therefore blocks translocation (53). A similar effect on the binding of N-Ac-Phe-tRNA to 80S ribosomes is also observed for kanamycin B, tobramycin, gentamicin C_{1A}, sisomicin, neomycin and paromomycin, all of which except neomycin are highly active in promoting misreading. Gentamicin C_1 which also induces misreading does not stabilize the binding of N-Ac-Phe-tRNA to 80S ribosomes. Therefore stabilization of peptidyl-tRNA in the acceptor site, that could result in a stronger interaction of both cognate and non-cognate aminoacyl-tRNA, would neither explain by itself the misreading effect nor the inhibition of translocation by all the aminoglycoside antibiotics.

There is genetic and biochemical evidence that cryptopleurine, tylophorine, tylocrebrine, emetine and tubulosine act on the 40S ribosomal subunit at identical or overlapping sites. All of these compounds inhibit EF-2- and GTP-dependent translocation but this effect can be overcome in cell-free systems by the addition of a large excess of EF-2. Cryptopleurine also inhibits peptide bond formation in cell-free systems at concentrations more than one hundred times higher than that required to block translocation (6 and references therein). Multiple binding sites for cryptopleurine on yeast ribosomes and their 60S and 40S subunit have been reported by spectro-fluorometric measurements (60), the existance of which are difficult to reconcil with the finding of one-step nuclear mutations in yeast conferring resistance to the drug (6, review). We have recently purified (^{3}H)cryptopleurine prepared by the NaB(^{3}H)$_4$-mediated reduction of the perchlorate salt of 9, 11, 12, 13, 14, 15 hexahydro-Δ^{10-14a} dehydro-2, 3, 6 trimethoxy-phenanthro (9, 10-b) quinolizidinium (M. Sõlhuber, A. Jiménez and D. Vázquez, unpublished). We have observed that there is a leveling off of the binding when 2 molecules of drug interact per 80S ribosome, while the onset of binding to the 60S and 40S ribosomal subunits takes place at about 1 molecule of (^{3}H)cryptopleurine per particle. Drug concentrations higher than $5x10^{-7}$M promote multiple binding sites, thought to be non specific. Scatchard plots of these data give values of $Kd^{80} = 8.46x10^{-8}$(n=2.15), $Kd^{60} = 16.20x10^{-8}$(n=1) and $Kd^{40} = 3.65x10^{-8}$(n= 0.84) for the 80S, 60S and 40S particles (H. Dölz, D. Vázquez and A. Jiménez, unpublished results). In addition there are multiple binding sites of very low affinities on 80S, 60S and 40S ribosomal particles as well as on E. coli 70S ribosomes. The two molecules of (^{3}H) cryptopleurine bound per 80S ribosome could account for the dual effect of the inhibitor on translocation at low concentrations and peptide bond formation at higher concentrations. In agreement with this proposal we have observed that one single molecule of (^{3}H)cryptopleurine interact with high affinity with one 40S ribosomal subunit. This interaction might promote the inhibition of translocation; this is in agreement with the existance of one step cell mutants resistant to cryptopleurine that express the mutation in the 40S subunit, at the level of enzymic translocation (6, review) and the finding of an altered 40S ribosomal protein (S20) in a CHO cell mutant resistant to emetine (61, 62). The single molecule of (^{3}H)cryptopleurine bound to each 60S ribosomal subunit, with an affinity 4 times lower than that for the 40S one, might account for the inhibition of peptide bond formation at higher concentrations of the drug. This might explain the observation that resistance to cryptopleurine both *in vivo* and *in vitro* is never complete, and a small increase in drug concentration has always a slowing down effect either on growth or on protein synthesis in cry^r cells or cell-free systems, respectively (63).

As indicated above, emetine and tubulosine are thought to share a common mechanism of action with cryptopleurine on translocation, although they do not affect peptide bond formation even at high concentrations. Accordingly, emetine and tubulosine inhibit (^{3}H) cryptopleurine interaction with its target site(s) to either 80S ribosomes or polysomes from yeasts. The results would suggest that the presence of peptidyl-tRNA on the ribosome does not alter the binding parameters of emetine and tubulosine to their target site(s). 50% inhibition of the binding of (^{3}H)cryptopleurine to 80S ribosomes requires 100 times higher concentrations of tubulosine and emetine than those of unlabeled cryptopleurine. These results might suggest lower binding affinities for emetine and tubulosine than for cryptopleurine, and an identical or overlapping binding site for the three drugs on the 40S ribosomal subunit.

Pederine and antibiotics of the glutarimide group are included within the inhibitors of translocation that interact with the 60S ribosomal subunit. The best known antibiotic of the glutarimide group is cycloheximide. Both pederine and cycloheximide block not only the EF-2-dependent translocation but also the non-enzymic translocation that takes place in the presence of rather high concentration of either K^+ or NH_4^+. However pederine is active on cycloheximide resistant ribosomes suggesting that both do not interact precisely with the same binding site on the ribosome.

INHIBITORS OF TERMINATION

The termination phase of translation begins when a chain-terminating codon of the mRNA(nonsense codon) (either UAA, UAG or UGA) is recognized, thus determining the interaction of the release factors with the ribosome. A single release factor (RF) has been isolated from eukaryotic systems, GTP being required in the termination step (g_1). After recognition of the release factor(s) the peptidyl transferase of the larger ribosomal subunit exercises its peptidyl hydrolase activity and cleaves the bond between the peptidyl and the tRNA moieties of peptidyl-tRNA so that the peptide chain is released (step (g_2)). A further reaction is required for termination to be completed in which the uncharged tRNA is released (step (h)). This step is not yet resolved in eukaryotic systems (6, review).

Protein synthesis inhibitors blocking translation specifically in the termination phase are not known. A number of the inhibitors described above, however inhibit termination as consequence of their interaction with the ribosome, besides inhibiting other steps of translation also. Thus it is most likely that tetracyclines and some other inhibitors of step (d) are also inhibitors of termination, with the selectivity indicated in Tables 1 and 2. (6, review).

Peptidyl transferase is involved not only in peptide bond formation (step (e)) but also in the peptidyl-tRNA hydrolisis reaction (step(g_2)). In fact all the peptidyl transferase inhibitors which have been tested block termination with the selectivity indicated in Tables 1 and 2. The only exception is anisomycin that does not inhibit this step in certain model systems (6, review).

As indicated in Figure 1 EF-2 and GTP appear to be required in step (h). It is therefore likely that inhibitors that interfere with EF-2 activity and certain analogues of GTP also inhibit step (h) (6, review).

REFERENCES

1. Gale, E.F., Cundliffe, E., Reynolds, P.E., Richmond, M.H. and Waring, M.J. (1972) The Molecular Basis of Antibiotic Action, pp. 278-379. London, John Wiley e Sons.
2. Kaji, A. (1973) Progress in Molecular and Subcellular Biology (Hahn, F. E., ed.), Vol 3, pp. 85-158, Berlin-Heildelberg-New York; Springer-Verlag.
3. Vázquez, D. (1974) Inhibitors of protein synthesis. FEBS Letters 40 (Suppl.), 563-584.
4. Pestka, S. (1977) Molecular Mechanisms of Protein Biosynthesis (Weisbach, H. and Pestka, S., eds.), pp. 447-553. Academic Press: New York-San Francisco-London.
5. Vázquez, D. (1978) International Review of Biochemistry, Vol. 18, Amino Acid and Protein Biosynthesis II. (Arnstein, H.R.V., ed.), pp. 169-232. University Park Press: Baltimore.
6. Vázquez, D. (1979) Inhibitors of protein biosynthesis. Springer-Verlag: Berlin-Heidelberg-New York.
7. Weisbach, H. and Pestka, S. (1977) Molecular Mechanisms of Protein Synthesis. Academic Press: New York-San Francisco-London.
8. Safer, B. and Anderson, W.F. (1978) CRC Critical Rev. Biochem. pp. 261-290.
9. Haro, C. de and Ochoa, S. (1979) Proc. Nat. Acad. Sci. 76, 1741-1745.
10. Haro, C. de and Ochoa, S. (1979) Proc. Nat. Acad. Sci. 76, 2163-2164.
11. Falcoff, R., Falcoff, E., Sancéau, J. and Lewis, J.A. (1978) Virology 86, 507-515.
12. Kerr, I.M. and Brown, R.E. (1978) Proc. Nat. Acad. Sci. 75, 256-260.
13. Ball, L.A. and White, C.N. (1978) Proc. Nat. Acad. Sci. 75, 1167-1171.
14. Clemens, M.J. and Williams, B.R.G. (1978) Cell 13, 565-572.
15. Zilberstein, A., Kimchi, A., Schmidt, A. and Revel, M. (1978) Proc. Nat. Acad. Sci. 75, 4734-4738.
16. Farrell, P.J., Sen. G.C., Dubois, M.F., Ratner, L., Slattery, E. and Lengyel (1978) Proc. Nat. Acad. Sci. 75, 5893-5897.
17. Aujean, O., Sancéau, J., Falcoff, E. and Falcoff, R. (1979) Virology 92, 583-586.
18. Sammuel, C.E. (1979) Virology 93, 281-285.

19. Gross, M. (1979) J. Biol. Chem. 254, 2378-2383.

20. Lewis, J.A., Falcoff, E. and Falcoff, R. (1978) Eur. J. Biochem. 86, 497-509.

21. Baglioni, C., Minks, M.A. and Maroney, P.A. (1978) Nature 273, 684-686.

22. Schmidt, A., Zilberstein, A., Shulman, L., Federman, P., Berissi, H. and Revel, M. (1978) FEBS Letters 95, 257-264.

23. Torrence, P.F. and Friedman, R. M. (1979) J. Biol. Chem. 254, 1259-1267.

24. Fresno, M., Carrasco, L. and Vázquez, D. (1976) Eur. J. Biochem. 68, 355-364.

25. Hunter, A., Jackson, R. and Hunt, T. (1977) Eur. J. Biochem. 75, 159-170.

26. Santon, J.B. and Stanley, W.M. (1978) Biochem. Biophys. Res. Commun. 84, 985-992.

27. Kozak, M. (1979) J. Biol. Chem. 254, 4731-4738.

28. Odom, O.W., Kramer, G., Henderson, A.B. Pinphanichakarn, P. and Hardesty, B. (1978) J. Biol. Chem. 253, 1807-1816.

29. Fresno, M. and Vázquez, D. (1978) Eur. J. Biochem. 83, 169-178.

30. Kaemfer, R., Rosen, H. and Israeli, R. (1978) Proc. Nat. Acad. Sci. 75, 650-654.

31. Weber, L.A., Hickey, D.A. and Baglioni, C. (1978) J. Biol. Chem. 253, 178-183.

32. Seal, S.N., Schmidt, A., Tomaszewski, M. and Marcus, A. (1978) Biochem. Biophys. Res. Commun. 82, 553-559.

33. Bergmann, J.E. and Lodish, H.F. (1979) J. Biol. Chem. 254, 459-468.

34. Fresno, M. and Vázquez, D. (1979) Eur. J. Biochem. (In press).

35. Tait, R.C. and Boyer, H.W. (1978) Cell 13, 73-81.

36. Chopa, I. and Ecles, S.J. (1978) Biochem. Biophys. Res. Commun. 33, 550-557.

37. Sompolinsky, D., Hammerman, I., Assaf, O. and Wojdani, A. FEMS Microbiology Letters 4, 23-26.

38. Conde, F.P., Fernandez-Puentes, C., Montero, M.T.V. and Vázquez, D. (1978) FEMS Microbiology Letters 4, 349-355.

39. Olsnes, S. and Refsnes, K. (1978) Eur. J. Biochem. 88, 7-15.

40. Cawley, D.B., Hedblom, M.L. and Houston, L.L. (1978) Arch. Biochem. Biophys. 190, 744-755.

41. Olsnes, S. and Abraham, A.K. (1979) Eur. J. Biochem. 93, 447-452.

42. Montanaro, L., Sperti, S., Zamboni, M.C., Denaro, M., Testoni, G., Gasperi-Campani, A. and Stirpe, F. (1979) Biochem. J. 176, 371-379.

43. Sperti, S. and Montanaro, L. (1979) Biochem. J. 178, 233-236.

44. González, A., Vázquez, D. and Jiménez, A. (1979) Biochim. Biophys. Acta 561, 403-409.

45. Wall, M.E., Wani, M.C. and Taylor, H. (1976) Cancer Chemother.

Reports 60, 1011-1030.

46. Chung, D.W., and Collier, R.J. (1977) Infection and Immunity 16, 832-841.

47. Bayliss, F.T. and Ingraham, J.L. (1974) J. Bacteriol. 118, 319.

48. Wilhelm, J.M., Pettit, S.E. and Jessop, J.J. (1978) Biochemistry 17, 1143-1149.

49. Wilhelm, J.M., Jessop, J.J. and Pettit, S.E. (1978) Biochemistry, 17, 1149-1153.

50. Palmer, E. and Wilhelm, J.M. (1978) Cell, 13, 329-334.

51. Singh, A., Ursic, D. and Davies, J. (1979) Nature 277, 146-148.

52. Palmer, E., Wilhelm, J.M. and Sherman, F. (1979) 277, 148-149.

53. González, A., Jiménez, A., Vázquez, D., Davies, J.E. and Schindler, D. (1978) Biochim. Biophys. Acta 521, 459-469.

54. Cabañas, M.J., Vázquez, D. and Modolell, J. (1978) Eur. J. Biochem. 87, 21-27.

55. Davies, J. and Davis, B.D. (1968) J. Biol. Chem. 243, 3312-3316.

56. Misumi, M., Tanaka, N. and Shiba, T. (1978) Biochem. Biophys. Res. Comm. 82, 971-976.

57. Misumi, M., Nishimura, T., Komai, T. and Tanaka, N. (1978) Biochem. Biophys. Res. Comm. 84, 358-365.

58. Campuzano, S., Vázquez, D. and Modolell, J. (1979) Biochem. Biophys Res. Comm. 87, 960-966.

59. Choi, E.C., Misumi, M., Nishimura, T., Tanaka, N., Nomoto, S., Teshima, T. and Shiba, T. (1979) Biochem. Biophys. Res. Comm. 79, 904-910.

60. Bucher, K. and Skogerson, L. (1976) Biochemistry 15, 4755-4759.

61. Boersma, D., Mc Gill, S. M., Mollenkamp, J. W. and Roufa, D.J. (1979) Proc. Nat. Acad. Sci. (USA) 76, 415-419.

62. Boersma, D., Mc Gill, S.M., Mollenkamp, J.W. and Roufa, D. J. (1979) J. Biol. Chem. 254, 559-567.

63. Grant, P., Sánchez, L. and Jiménez, A. (1974) J. Bacteriol. 120, 1308-1314.

ORGANIZATION OF THE RIBOSOMAL RNA GENES IN EUKARYOTES

R. J. PLANTA and J. H. MEYERINK

Biochemisch Laboratorium,
Vrije Universiteit,
Amsterdam, The Netherlands

INTRODUCTION

Ribosomes from eukaryotic cells are composed of 4 different RNA molecules (with sedimentation values of 5S, 5.8S, 17-18S and 25-28S) and about 70 different proteins. Therefore, the formation of ribosomes is a very complex process which requires: a) the synchronized production of all ribosomal constituents and thus a highly coordinated expression of a large number of genes; b) the assembly of the various components into preribosomal particles in an ordered fashion; and c) subsequent processing of the preribosomes into mature ribosomal subunits in a very accurate way (see e.g. Refs 1,2).

Regulation of the synthesis of ribosomes occurs both at the level of transcription and at the level of assembly and processing. The latter type of regulation is particularly important in situations where the growth of cells is slowing down or where there is an unbalanced production of the various ribosomal constituents. In case of an insufficient supply of one or more of the ribosomal constituents, irrespective which, the excess of the other ribosomal constituents is eliminated (3-6). Apparently, ribosomal (precursor) RNA and ribosomal proteins are degraded if they are not promptly assembled and processed into mature ribosomes. Under balanced growth conditions, however, there is little if any degradation of any of the ribosomal constituents, at least in so far as this has been examined. Therefore there must be an almost perfect balanced synthesis of all the ribosomal components, most likely by a coordinate control of transcription of the ribosomal genes (see Ref. 4). This coordinate synthesis of the ribosomal components in eukaryotes is the more remarkable as three different kinds of RNA polymerases are involved in the transcription of the ribosomal genes. The ribosomal protein genes are transcribed by RNA polymerase B, the genes coding for 17-18S,

5.8S and 25-28S rRNA form a transcriptional unit which is transcribed by RNA polymerase A, and 5S rRNA is synthesized separately by RNA polymerase C.

To understand the regulation of the ribosomal genes in eukaryotes we need information about the organization and sequence of those genes and their flanking DNA regions. In the past ten years a lot of information has become available about the organization of the rRNA genes. In eukaryotes the genes coding for the rRNA components are present in clusters of hundreds or even thousands of copies. For instance, the part of the yeast genome coding for rRNAs consists of more than a hundred tandemly repeated units. Each repeating unit codes for a large common precursor rRNA molecule, containing the sequences of 17S, 5.8S and 26S rRNA (in that order of transcription) and in addition non-ribosomal sequences that are removed during ribosome maturation (2,7). In contrast, 5S rRNA is synthesized as a primary transcript from genes that in eukaryotes are normally found at different chromosomal locations than the other rRNA genes (7,8). Much less is known about the genes coding for ribosomal proteins in eukaryotes, although in this field now a rapid progress is being made using recombinant DNA methodology. In the recent past suggestive evidence was obtained that in yeast, mRNAs for ribosomal proteins are transcribed from transcriptional units which are up to 12 times as long as can be expected (9). This can be explained by assuming that the yeast ribosomal protein genes are to some extent clustered in a number of transcriptional units. Such a genetic organization of the ribosomal protein genes in yeast can explain the observed coordination in the synthesis of ribosomal proteins (4). On the other hand, there are also some indications that the genes for ribosomal proteins in yeast are widely scattered, suggesting that the coordinate regulation of ribosomal protein synthesis derives from some common regulatory site at each of the ribosomal protein genes (4,10).

This review concentrates on some aspects of the organization and transcription of the rRNA genes in eukaryotes, using the yeast _Saccharomyces carlsbergensis_ as a reference.

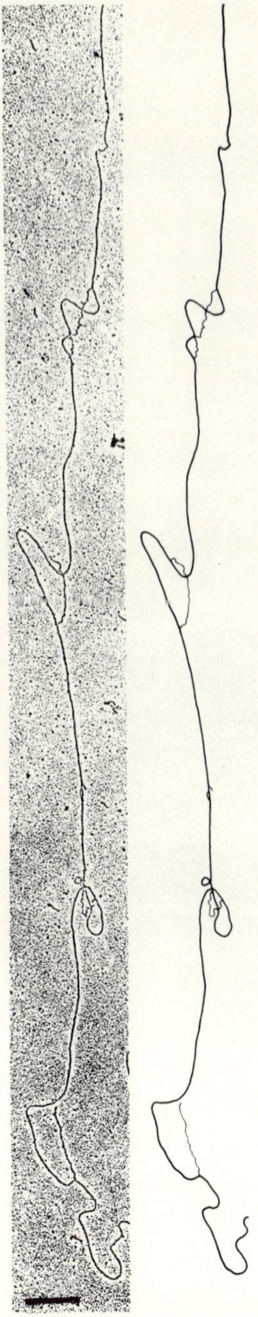

ORGANIZATION OF THE RIBOSOMAL REPEATING UNITS IN YEAST

The part of the yeast genome containing the rRNA genes, called ribosomal DNA (rDNA), can be easily separated from the rest of the nuclear DNA by centrifugation in Hg^{2+}/Cs_2SO_4 density-gradients (11). The arrangement of the ribosomal repeating units on this rDNA was studied by electron microscopy of purified rDNA hybridized to 26S rRNA using the R-loop technique (12). By measuring the distances between the midpoints of two consecutive R-loops (see Fig. 1) and comparing this distance to the size of the ribosomal repeating unit it could be concluded that the rRNA genes in yeast are arranged in multiple clusters made up of 4 to 5 repeating units and containing little or no spacer DNA between the units. This organization is quite similar to that proposed by Cramer *et al*. (13) for rDNA of *S. cerevisiae* on the basis of restriction endonuclease analysis. It is, however, quite different from the arrangement of the ribosomal repeating units in higher eukaryotes (see below), where individual repeating units are separated by long stretches of non-transcribed spacer DNA.

In a recent paper Petes presented genetic evidence that during the mitotic cell cycle of the yeast *S. cerevisiae* all, or almost all, of the rRNA genes are located on chromosome XII (14)

Fig. 1. Electron micrograph of R-loop structures obtained with yeast rDNA and 26S rRNA (the bar represents 0.5 μm). The electron microscopy was performed by Dr. A. van der Ende en Dr. E.F.J. van Bruggen (The University, Groningen, The Netherlands). For experimental details see Ref. 12.

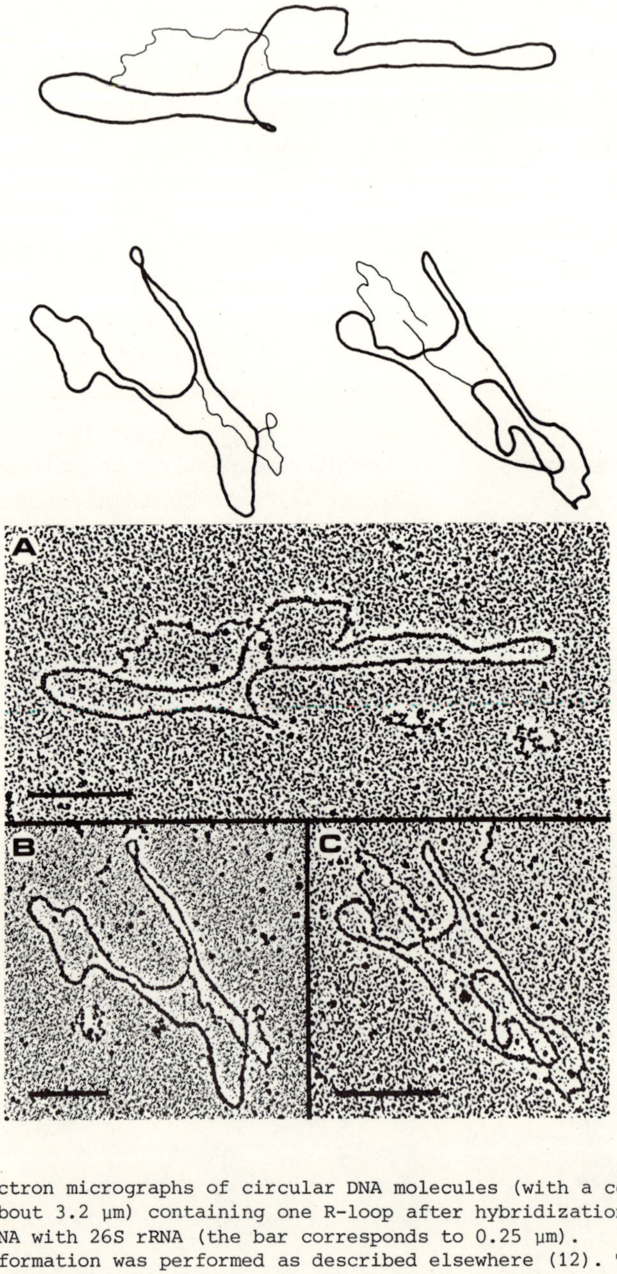

Fig. 2. Electron micrographs of circular DNA molecules (with a contour length of about 3.2 μm) containing one R-loop after hybridization of yeast ribosomal DNA with 26S rRNA (the bar corresponds to 0.25 μm).
The R-loop formation was performed as described elsewhere (12). The electron microscopic analysis was carried out by Dr. A. van der Ende and Dr. E.F.J. van Bruggen (The University, Groningen, The Netherlands).

in contrast to earlier reports in which was suggested that many of the rRNA genes in yeast are located on chromosome I. Petes could not rule out the presence of some extrachromosomal copies of rDNA in the yeast strain used,but he suggested that in yeast most, and possibly all, of the rRNA genes are chromosomal, both in mitosis and meiosis.

However, careful electron microscopic analysis of purified ribosomal DNA from S. carlsbergensis revealed a small number of circular DNA molecules with a contour length of about 3.2 µm. This length is identical to the length reported earlier (12) for the ribosomal repeating unit on yeast chromosomes. A few molecules having two or three times this length were observed also. These circular DNA molecules appear to be extrachromosomal copies of the rRNA genes, since they hybridize with homologous 26S rRNA under conditions favoring the formation of R-loops (see Fig. 2). The average length of the R-loops (1.1 µm) in these molecules is virtually identical to the length found for the corresponding R-loops in the chromosomal genes. The circular rDNA molecules are apparently replicated by a rolling circle mechanism, since occasionally rolling circle-type molecules (with one R-loop in the circular part of the molecule and one R-loop in the free DNA tail) were observed. If the extrachromosomal copies of rDNA in yeast indeed can replicate autonomically then an origin of replication must be present on the ribosomal repeating unit in yeast.

The methods we used do not allow us to determine accurately the number of circular rDNA molecules per yeast cell. They do comprise only a minor fraction (about 1%) of the total number of rRNA genes in yeast, however. Therefore, these circular rDNA molecules seem to have a rather transient function. For instance, one could envisage a mechanism whereby extrachromosomal copies of rDNA are involved in the regulation of the total number of rRNA genes. Amplification of rRNA genes in yeast has been described by Øyen et al. (15) and Kaback and Halvorson (16). It is conceivable that the vegetative yeast cell controls its number of chromosomal rRNA genes by excision and integration of the 3 µm circular ribosomal units. Excised molecules could either be degraded or replicated by a rolling circle mechanism followed by integration into the chromosome. The presence of a small amount of circular rDNA in Xenopus tissue cells and blood cells (17) suggests that the mechanism postulated for amplification of yeast rRNA genes may also apply to other eukaryotes.

Extrachromosomal rDNA has also been observed in other lower eukaryotes. Two basically different types of gene organization have been described. First, the macronuclear rDNA of *Paramecium* exists as relatively small molecules with a tandemly repeated basic unit, in both linear and circular forms (18). On the other hand, the free rRNA genes in *Physarum*, *Dictyostelium* and *Tetrahymena* exists as giant palindromes (Refs 19, 20,21; see also Fig. 6). In *Tetrahymena* the free rDNA copies found in the macronucleus represent a case of gene amplification, as only a single integrated rDNA gene occurs in the micronuclear genome (22).

Finally, a very large number of extrachromosomal copies of rDNA are present in the oocytes of many animals. Some of these extrachromosomal molecules are circular, and their sizes are integral multiples of a basic repeating unit (23).

STRUCTURE OF THE RIBOSOMAL REPEATING UNIT OF YEAST

In previous papers we have described in detail the physical and genetic map of the ribosomal repeating unit of the yeast *S. carlsbergensis* (see *e.g.* Refs 2,7). Digestion of yeast rDNA with the restriction endonuclease Hind III produces only two discrete fragments with molecular weights of 1.7×10^6 D (appr. 2600 base pairs) and 4.2×10^6 D (appr. 6500 base pairs). The first fragment hybridizes only to 5S rRNA and to 37S precursor rRNA, whereas the largest fragment hybridizes to 17S, 5.8S and 26S rRNA. These fragments are therefore generated by digestion of the repeating ribosomal units.

For a more detailed structural study of the repeating unit the two Hind III-generated fragments were cloned, as well as the whole repeating unit (5.9×10^6 D or appr. 9100 base pairs) which was obtained by partial digestion of yeast rDNA with Hind III. As a cloning vehicle plasmid pBR 322 was used. With the aid of these recombinants we studied the fine structure of the ribosomal unit by restriction endonuclease digestion, by hybridization of purified rRNAs to restriction fragments, by R-loop analysis and DNA sequence analysis and by *in vitro* transcription. Part of the results obtained so far

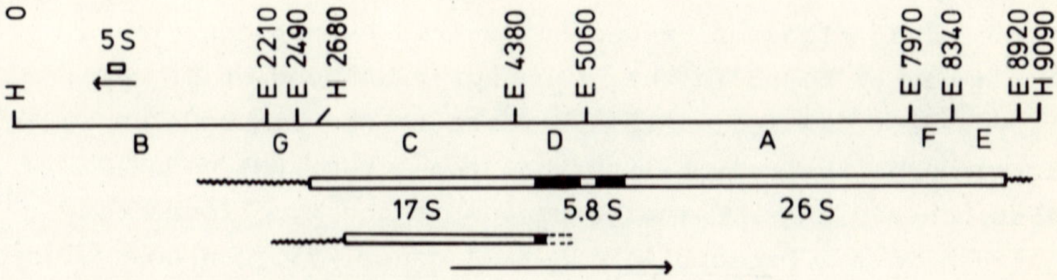

Fig. 3. Map of the ribosomal repeating unit of the yeast S. carlsbergensis. This map shows the cleavage sites for the restriction endonucleases EcoRI (E) and Hind III (H) of which the positions are indicated in numbers of base pairs relative to the first Hind III cleavage site. In addition, the location of the four rRNA genes is shown and their primary transcripts. The ribosomal sequences in the transcripts are represented by open boxes, and the non-ribosomal sequences by zigzag lines (terminal sequences) and black boxes (internal sequences).

is presented in Fig. 3. Recently, similar results have been described for the ribosomal repeating unit of the related yeast strain S. cerevisiae (24,25,26).

The ribosomal repeating unit of yeast contains two different transcription units: one is coding for a common precursor RNA for 17S, 5.8S and 26S rRNA, and the other unit is coding for mature 5S RNA. Thus, the 5S RNA genes and those coding for the precursor of the other three rRNA components are alternating in yeast. Such an arrangement differs from that found in higher eukaryotes where the genes for 5S RNA are located at different chromosomal positions than the other rRNA genes. However, in another lower eukaryote, Dictyostelium discoideum, 5S rRNA genes were also found to be physically linked to the other rRNA genes (20). Therefore, this type of arrangement may be a common feature of the organization of rRNA genes in lower eukaryotes. The linkage of 5S rRNA genes to the other rRNA genes in yeast is principally different from that found in bacteria. In yeast, 5S rRNA is synthesized separately by RNA polymerase C with opposite polarity as compared with the precursor rRNA, which is synthesized by RNA polymerase A.

The data presented in Fig. 3 also reconfirm that the polarity of the large precursor rRNA is: 5'-17S, 5.8S, 26S-3' proposed by us previously (2,8). This polarity of the high molecular weight precursor rRNA is now generally accepted for all eukaryotes.

The yeast ribosomal repeating unit does not contain any tRNA genes as found in the precursor rRNA gene of *E. coli*, since yeast tRNA failed to hybridize to the ribosomal unit. Furthermore, the possibility that the large rRNA genes in yeast contain intervening sequences of the type found for instance in the *Drosophila* 28S rRNA gene (27), can be excluded from the restriction endonuclease digestion pattern and from the R-loop analysis.

Large parts of the yeast ribosomal repeating unit have now been sequenced, but in spite of this knowledge the exact site of initiation and termination of the precursor rRNA gene are still not known. Recently, we were able to show that 37S RNA, the largest precursor rRNA which can be detected by pulse-labelling in wild-type yeast cells, can be considered as a primary transcription product, since it contains a 5'-polyphosphate terminus (pppA-Up-) in a close-to-one molar yield (28). The 37S precursor rRNA has an estimated molecular weight of about 2.8×10^6 D, corresponding to about 8100 nucleotides. The transcription of a precursor RNA of that size would start at a position located at about 1000 base pairs on the map of Fig. 3, *i.e.* very close to the promotor site of the 5S rRNA gene on the middle of EcoRI-fragment B. However, ^{32}P-labelled 37S pre-rRNA does not hybridize to this fragment B on Southern strips, at least not to a significant extent. On the other hand, *in vitro* transcription of total yeast nuclear DNA with purified RNA polymerase A produces ribosomal transcripts with molecular weights up to about 3×10^6 D, a size corresponding to an almost complete transcript of the repeating unit (29). In addition it was found that RNA polymerase A, before reading into the 17S RNA sequences transcribes a stretch of DNA with a length of about 2300 base pairs, which suggests a start of transcription at about 400 base pairs on the map of Fig. 3. With the available DNA sequence data this problem could not be solved, since we still lack information about 5'-terminal sequences of sufficient length of the 37S pre-rRNA and the *in vitro* transcripts. General structural relationships which could help determine transcription initiation points in eukaryotes are as yet difficult to perceive.

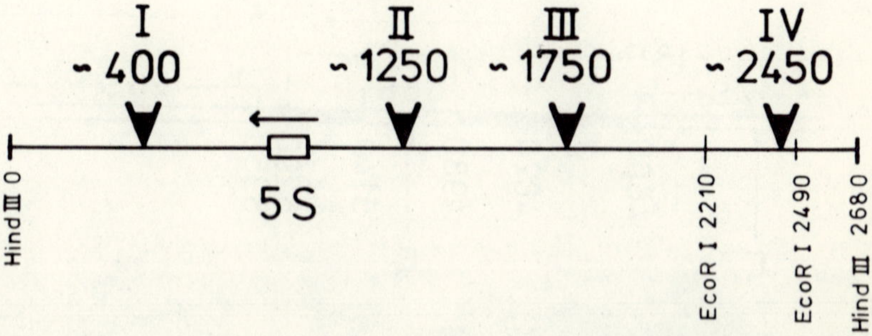

Fig. 4. Binding sites (I, II, III and IV) of RNA polymerase A in the initiating region of the yeast ribosomal repeating unit. Recombinant DNA, composed of pBR 322 and the 1.7 x 10^6 D-fragment of the ribosomal unit, generated by digestion with Hind III, was incubated with polymerase A at a molar ratio of 1:10. Incubation was for 20 min at 25°C in a buffer containing 70 mM Tris (pH 8.0), 0.2 mM EDTA, 7 mM NaCl, 50 mM $(NH_4)_2SO_4$, 10 mM $MgCl_2$ and 1 mM each of ATP, CTP and UTP. Fixation of the DNA-enzyme complexes was achieved in 0.1% glutaraldehyde at 25°C for 20 min. Fixed complexes were mounted on loaded carbon film and rotary shadowed with Pt/Pd at an angle of 5°. Preparations were examined with a Philips E.M. 200. Photography and length determinations were carried as described previously (12).

In an attempt to locate the binding site of RNA polymerase A we performed experiments in which complexes between polymerase A and a recombinant DNA containing the small Hind III-fragment (the first 2600 base pairs of the ribosomal unit; see Fig. 3) were formed under appropriate conditions. The complexes were examined under the electron microscope after fixation with glutaraldehyde. The positions of the polymerase molecules bound to the rDNA-fragment were determined for a large number of complexes and the results of these studies are summarized in Fig. 4.

As indicated in Fig. 4 there are four main binding sites in the initiating region of the ribosomal repeat, located at about 400 (± 100), 1250 (± 100), 1750 (± 100) and 2450 (± 100) base pairs, respectively, and designated as I, II, III and IV. Binding site I could be identical to the _in vitro_ initiation site for polymerase A in the production of the large transcripts of about 3 x 10^6 D molecular weight. The other three binding sites are all equally probable candidates for the initiation site in the transcription of 37S precursor rRNA.

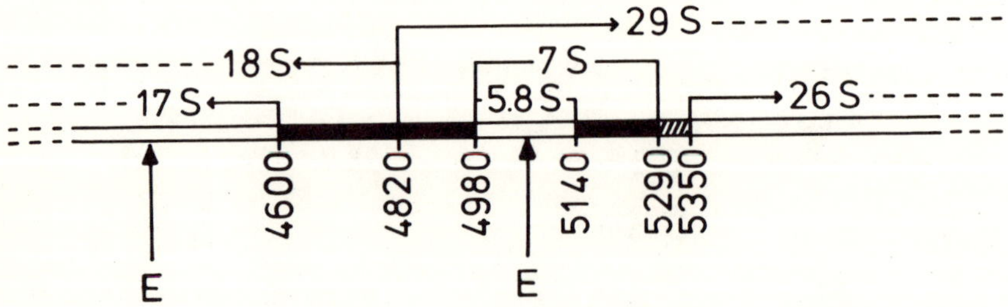

Fig. 5. Fine map of the intergenic region of the ribosomal repeating unit. In this region at least 6 processing sites are present, involved in the formation of 18S and 29S precursor rRNA, and the mature 17S, 5.8S and 26S rRNA. Ribosomal sequences are indicated by open boxes, nonribosomal sequences by black boxes.

Taking into account the estimated size of the 37S RNA (roughly 8,000 nucleotides) the sites II or III should be preferred as the initiation site. However, on the basis of the lack of hybridization of the EcoRI-fragment B with 37S pre-rRNA site, IV should be the choice for the real initiation site. Obviously, it is not yet possible to draw a solid conclusion from the present set of experimental results. More extensive sequence data at the 5'-end of the primary transcript are required to settle this point.

There is also still some uncertainty about the termination site of 37S precursor rRNA. The previously reported 3'-terminal sequence of 37S RNA, viz. U_{6-8}-A-N_{OH}, is found twice at the DNA level, once immediately adjacent to the 3'-end of 26S rRNA and another time about 200 base pairs downstream from the 3'-end of 26S rRNA.

The transcribed spacers between the 17S and 26S rRNA genes have been sequenced completely in a search for sequence characteristics of RNA processing sites (for a fine map of this region see Fig. 5). Generally, these processing regions are extremely A+T-rich. For instance, the nonconserved region of 40 nucleotides joining the 3'-end of 17S rRNA contains nearly 90% A+T-residues. In addition, sequence repeats frequently are observed around the processing sites. Whether duplex structures are functionally involved in the RNA processing, cannot yet be concluded from our data.

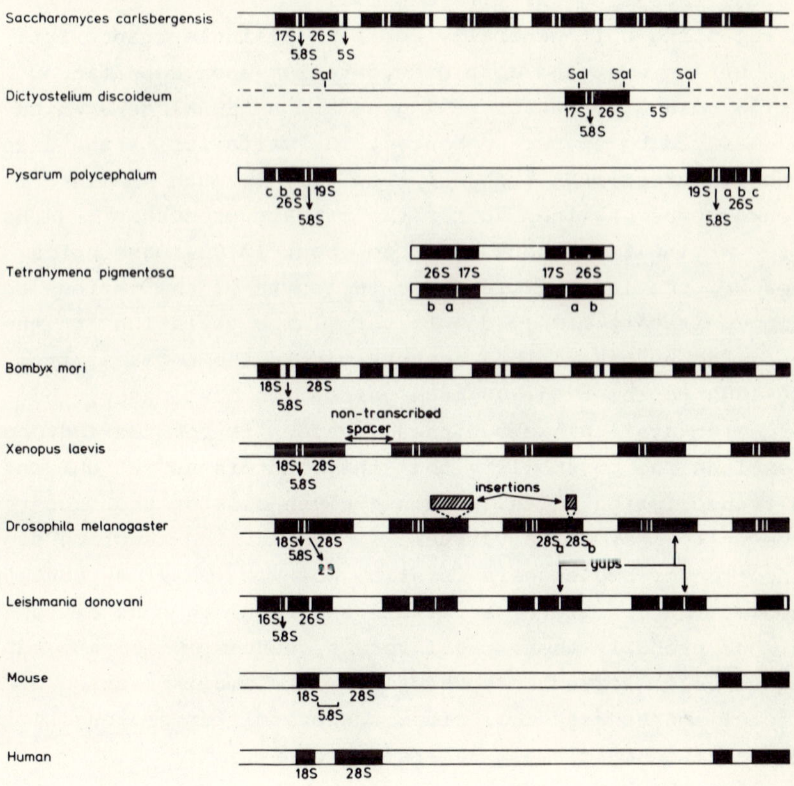

Fig. 6. Sequence arrangement of the ribosomal DNA from various eukaryotes. All the maps are drawn to the same scale. Sequences encoding mature rRNAs are represented by black boxes; spacers or nonribosomal sequences are represented by open boxes.
Data were taken for yeast from Refs 2,7,13,24,25,26; for Dictyostelium discoideum from Refs 20,34,35; for Physarum polycephalum from Refs 36, 37,38; for Tetrahymena from Refs 21,39,40; for Bombyx mori from Ref. 41; for Xenopus laevis from Refs 42,43,44,45,46; for Drosophila melanogaster from Refs 27,30,31,32,33,47; for Leishmania donovani from Ref. 48; for mouse from Ref. 49 and for man from Ref. 50.

ORGANIZATION OF RIBOSOMAL DNA IN VARIOUS EUKARYOTES

In Fig. 6 a schematic representation is given of the repeating unit structure in rDNA of a number of eukaryotes. All the maps are drawn to the same scale, and even a cursory survey of these maps reveals that there are large differences in the length of the ribosomal repeating unit between the various eukaryotes. The significance of this difference in size of the ribosomal repeating units is unknown.

A repeating unit generally contains a single transcription unit, but in yeast and also Dictyostelium each repeating unit contains both a precursor rRNA gene and a 5S RNA gene which are separated by spacer sequences. The variation in the size of the precursor rRNA transcription unit is much smaller: in all eukaryotes examined so far the primary precursor rRNA has a size varying from about 8,000 to about 13,000 base pairs. Therefore, the large differences in length of the various repeating units are mainly due to very large variations in the size of the nontranscribed spacers within the repeats (from about 1000 to about 35,000 base pairs). In most cases, the information available about these spacers is limited and does not exclude the possibility that they contain as yet unidentified transcription units. Within a given cluster the spacers can be uniform or quite variable in length. The function of the nontranscribed spacers is still obscure. They may contain sequences that initiate, terminate and regulate gene expression, but probably these regulatory sequences (which are supposed to be conserved) account for only a small fraction of the total spacer sequence, since spacers differ profoundly in sequence, even in closely related species.

In contrast, the rRNA genes are rather conserved and virtually homogeneous within an organism. Their organization and polarity in all eukaryotic organisms examined so far is very similar (5'-18S-5.8S-28S-3'), but in some organisms the large rRNA genes contain intervening sequences that are present in the primary transcript but not in the mature rRNA (for instance in Drosophila melanogaster (see Refs 27,30-33) and Physarum polycephalum (37,38)).

A very large ribosomal repeating unit was found in Dictyostelium discoideum (20,34,35). The DNA which codes for 17S and 25S rRNAs is located within a stretch of repetitive DNA at least 40,000 base pairs long, whereas the transcribed region of the unit, coding for the 36S precursor rRNA, is only about 8000 base pairs long. This repeated unit includes the 5S rRNA gene linked to the 25S rRNA gene, an organization comparable with the one in yeast. Very recently (51), it could be shown that Dictyostelium rDNA consists of free palindromic dimers of the repeating unit(shown in Fig. 6)of the type also found in *Physarum* and *Tetrahymena*.

The extrachromosomal copies of the rDNA in *Physarum polycephalum* (36,37,38) and various strains of *Tetrahymena* exist as giant palindromes with lengths of 57,500 and 21,000 base pairs, respectively. The significance of the rotational symmetry axis in the rDNA of these organisms is unknown. The two transcription units on these palindromic DNAs are apparently transcribed in opposite directions with the usual polarity from initiation sites located in the central part of the molecules. Recently it was found that the 26S rRNA gene on the rDNA of *Physarum polycephalum* consists of three coding regions interrupted by two relatively large intervening sequences (37, 38). The 2 intervening sequences are present in more than 85% of the ribosomal genes from active plasmodia, suggesting that genes containing these inserts are likely to be transcribed into pre-rRNA molecules. This notion is supported by the recent finding that in a certain strain of *Tetrahymena pigmentosa* all the 26S rRNA genes on the macronuclear rDNA contain an intervening sequence of about 400 base pairs, whereas the 26S rRNA genes in another, closely related strain lack the intervening sequence completely (40). Since both strains can survive and grow equally well, the presence or absence of the intervening sequence has no obvious functional importance. This observation makes it also unlikely that the intervening sequence has an important controlling or regulatory function. Intervening sequences have also been found in the genes for the large rRNAs on the mitochondrial DNA of yeast (52) and the chloroplast DNA of *Chlamydomonas* (53). In both cases, all gene copies contain the intervening sequences but nevertheless they are transcriptionally active.

The multiple genes for rRNA in *Bombyx mori* (41), *Xenopus laevis* (42), *Drosophila melanogaster* (27,30-33) and *Leishmania donovani* (48) are arranged in tandemly repeating units, like in yeast. Each repeat consists of a transcribed region containing the 18S, 5.8S and 28S rRNA genes (in that order of transcription) and a nontranscribed spacer. Thus the rDNA of these eukaryotes consists of alternating transcribed regions and nontranscribed spacers. The nontranscribed spacers in the four higher eukaryotes have, in contrast to yeast, a considerable length. In *Bombyx mori* the rDNA repeats have a striking length homogeneity (41); thus, there is neither variation in the length of the nontranscribed spacer nor the presence of intervening sequences in a detectable portion of the rDNA, as is the case in e.g. Xenopus laevis and Drosophila melanogaster, respectively.

The transcribed regions in Xenopus laevis are constant in length and exactly or virtually the same in sequence, whereas the nontranscribed spacers differ strongly in length (from about 2700 to about 9000 base pairs) in different repeating units (42,43). The pattern of spacer lengths is significantly different from one animal to another so that it can be followed as a genetic marker. Despite this difference in length of the nontranscribed spacer, spacers differ relatively little in sequence. This is because the nontranscribed spacers are internally repetitive with a periodicity of about 15 nucleotides, although also higher order periodicities, some approaching 750 base pairs, are apparent (43).

The tandemly repeated ribosomal units in Drosophila melanogaster are also producing their share of surprises (27,30-33). The transcription unit contains the sequences encoding the 18S, 5.8S, 28S and an additional 2S rRNA (of 30 nucleotides) associated with the mature 28S rRNA, in the order of transcription 5'-18S,5.8S,2S,28S-3'. R-loop analysis have revealed two types of repeating units in the rDNA of this fly, the first of which resembles the repeating unit in X. laevis. The second type of repeating unit has the same basic organization, but contains intervening sequences of various length within the 28S rRNA gene (27,30-32). There are at least two major classes of intervening sequences, designated type 1 (with a size range of 500-6000 base pairs) and type 2 (with a size range of 1500-4000 base pairs). Each class consists of discrete subclasses differing in length and sequence (47). Almost 60% of the repeating units comprising the nucleolar organizer on the X chromosome contain the intervening sequences of type 1, but such insertions are either rare or not present at all in the rDNA of the Y chromosome. Type 2 inserts occur in about 16% of the repeating units in rDNA from both X and Y chromosomes. Repeating units with and without intervening sequences are interspersed in the rDNA of the X chromosome, but are clustered on the Y chromosome (31,47). Although intervening sequences constitute the major source of length heterogeneity in the rDNA of D. melanogaster, there is also some length heterogeneity in the nontranscribed spacer sequence which has an average length of about 3500 base pairs.

Hybrids of 28S rRNA and rDNA of D. melanogaster are also interrupted by a short gap of about 120 nucleotides. This gap is due to the fact that Drosophila 28S rRNA has an internal break, most probably as a result of a post-transcriptional elimination of an internal segment of the rRNA. A similar, though larger, gap was recently found in the 26S rRNA of the pathogenic haemoflagellate Leishmania donovani (48).

Very large ribosomal repeating units of about 44,000 base pairs have been found in two mammalians, mouse (49) and man (50). The repeats in these two species are rather similar. In both cases the nontranscribed spacers are unusually long with a gene-to-spacer ratio of about 1:2.5. The evolutionary significance of this is obscure, the more so because long spacers occur also in the rDNA of Dictyostelium and Physarum (see Fig. 6). The nontranscribed spacers in human and mouse rDNA share with the spacers of some other organisms the property of length and sequence heterogeneity, although this heterogeneity appears to be more limited than in Xenopus.

In conclusion, although the basic structure of the precursor rRNA gene seems to be universal, there are large differences in the over-all organization of the ribosomal repeating units in eukaryotes. These differences are mainly found in the nontranscribed spacers and in the presence or absence of intervening sequences in the large rRNA gene, but they are as yet difficult to explain in functional terms.

ACKNOWLEDGEMENTS

The authors are grateful to Drs. J. Klootwijk and G. M. Veldman for critical comments and helpful discussions during the preparation of this manuscript.

REFERENCES

1. Perry, R.J. (1976) Ann. Rev. Biochem. 45, 605-629.
2. Planta, R.J., Retèl, J., Klootwijk, J., Meyerink, J.H., De Jonge, P., Van Keulen, H. and Brand, R.C. (1977) Biochem. Soc. Trans. 5, 462-466.
3. Warner, J.R. (1974) in Ribosomes, 461-488 (Nomura, M., Tissières, A. and Lengyel, P., eds), Cold Spring Harbor Press, New York.

4. Gorenstein, C. and Warner, J.R. (1976) Proc. Nat. Acad. Sci. USA 73, 1547-1551.

5. Tsurugi, K. and Ogata, K. (1977) Biochem. Biophys. Res. Commun. 75, 525-531.

6. Warner, J.R. (1977) J. Mol. Biol. 115, 315-333.

7. Planta, R.J., Meyerink, J.H. and Klootwijk, J. (1979) in Gene Function, 401-411, Pergamon Press, Oxford and New York.

8. Trapman, J. and Planta, R.J. (1975) Biochim. Biophys. Acta 414, 115-125.

9. Mager, W.H., Retèl, J., Planta, R.J., Bollen, G.H.P.M., De Regt, V.C.H.F. and Hoving, H. (1977) Europ. J. Biochem. 78, 575-583.

10. Warner, J.R. (1978) Personal communication.

11. Retèl, J. and Planta, R.J. (1972) Biochim. Biophys. Acta 281, 299-309.

12. Meyerink, J.H., Retèl, J., Raué, H.A., Planta, R.J., Van der Ende, A. and Van Bruggen, E.F.J. (1978) Nucl. Acids Res. 5, 2801-2808.

13. Cramer, J.H., Farelly, F.W. and Rownd, R.H. (1976) Molec. Gen. Genet. 148, 233-241.

14. Petes, T.D. (1979) Proc. Nat. Acad. Sci. USA 76, 410-414.

15. Øyen, T.B., Saelid, G. and Skuladottir, G.V. (1978) Biochim. Biophys. Acta 520, 88-102.

16. Kaback, D.B. and Halvorson, H.O. (1978) J. Bact. 134, 237-245

17. Bird, A.P. (1978) Cold Spring Harbor Symp. Quant. Biol. 42, 1179-1183.

18. Findley, R.C. and Gall, J.G. (1978) Proc. Nat. Acad. Sci. USA 75, 3312-3316.

19. Vogt, V.M. and Braun, R. (1976) J. Mol. Biol. 106, 567-587.

20. Maizels, N. (1976) Cell 9, 431-438.

21. Engberg, J., Andersson, P., Leick, V. and Collins, J. (1976) J. Mol. Biol. 104, 455-470.

22. Yao, M.-C. and Gall, J.G. (1977) Cell 12, 121-132.

23. Tobler, H. (1975) in The Biochemistry of Animal Development, Vol. 3, 91-143 (Weber, R., ed.), Academic Press, New York.

24. Bell, G.I., DeGennaro, L.J., Gelfand, D.H., Bishop, R.J., Valenzuela, P. and Rutter, W.J. (1977) J. Biol. Chem. 252, 8118-8125.

25. Valenzuela, P., Bell, G.I., Venegas, A., Sewell, E.T., Masiarz, F.R., DeGennaro, L.J., Weinberg, F. and Rutter, W.J. (1977) J. Biol. Chem. 252, 8126-8135.

26. Kramer, R.A., Philippsen, P. and Davis, R.W. (1978) J. Mol. Biol. 123, 405-416.

27. White, R.L. and Hogness, D.S. (1977) Cell 10, 177-192.

28. Klootwijk, J., De Jonge, P. and Planta, R.J. (1979) Nucl. Acids Res. 6, 27-39.

29. Van Keulen, H. and Retèl, J. (1977) Europ. J. Biochem. 79, 579-588.

30. Glover, D.M. and Hogness, D.S. (1977) Cell 10, 167-176.

31. Wellauer, P.K. and Dawid, I.B. (1977) Cell 10, 193-212.

32. Pellegrini, M., Manning, J. and Davidson, N. (1977) Cell 10, 213-224.

33. Jordan, B.R. and Glover, D.M. (1977) FEBS Letters 78, 271-274.

34. Cockburn, A.F., Newkirk, M.J. and Firtel, R.A. (1976) Cell 9, 605-613.

35. Frankel, G., Cockburn, A.F., Kindle, K.L. and Firtel, R.A. (1977)J. Mol. Biol. 109, 539-558.

36. Molgaard, H.V., Matthews, H.R. and Bradbury, E.M. (1976) Europ. J. Biochem. 68, 541-549.

37. Gubler, U., Wyler, T. and Braun, R. (1979) FEBS Letters 100, 347-350.

38. Campbell, G.R., Littau, V.C., Melera, P.W., Allfrey, V.G. and Johnson, E.M. (1979) Nucl. Acids Res. 6, 1433-1447.

39. Karrer, K.M. and Gall, J.G. (1976) J. Mol. Biol. 104, 421-453.

40. Wild, M.A. and Gall, J.G. (1979) Cell 16, 565-573.

41. Manning, R.F., Samols, D.R. and Gage, L.P. (1978) Gene 4, 153-166.

42. Wellauer, P.K., Reeder, R.H., Dawid, I.B. and Brown, D.D. (1976) J. Mol. Biol. 105, 487-505.

43. Botchan, P., Reeder, R.H. and Dawid, I.B. (1977) Cell 11, 599-607.

44. Buongiorno-Nardelli, M., Amaldi, F., Beccari, E. and Junakovic, N. (1977) J. Mol. Biol. 110, 105-117.

45. Boseley, P.G., Tuyns, A. and Birnstiel, M.L. (1978) Nucl. Acids Res. 5, 1121-1137.

46. Junakovic, N., Poretti, A., Amaldi, F. and Buongiorno-Nardelli, M. (1978) Nucl. Acids Res. 5, 1335-1343.

47. Wellauer, P.K., Dawid, I.B. and Tartof, K.D. (1978) Cell 14, 269-278.

48. Leon, W., Fouts, D.L. and Manning, J. (1978) Nucl. Acids Res. 5, 491-504.

49. Cory, S. and Adams, J.M. (1977) Cell 11, 795-805.

50. Wellauer, P.K. and Dawid, I.B. (1979) J. Mol. Biol. 128. 289-303.

51. Cockburn, A., Taylor, W. and Firtel, R. (1979) Chromosoma, in press.

52. Bos, J.L., Heyting, C., Borst, P., Arnberg, A.C. and Van Bruggen, E.F.J. (1978) Nature 275, 336-338.

53. Rochaix, J.D. and Malnoe, P. (1978) Cell 15, 661-670.

COORDINATION OF RNA AND PROTEINS IN EUKARYOTIC RIBOSOME PRODUCTION

JONATHAN R. WARNER, ROBERT J. TUSHINSKI, and PETER J. WEJKSNORA

Departments of Biochemistry and Cell Biology,
Albert Einstein College of Medicine,
Bronx, NY 10461

INTRODUCTION

The process of ribosome synthesis in eukaryotes is diagrammed in Figure 1. It can be considered as two concurrent sets of reactions: 1) the transcription of ribosomal precursor RNA, its assembly with ribosomal proteins and its processing to mature species, which occurs almost exclusively within the nucleus, and 2) the transcription and processing of mRNA molecules for 70 ribosomal proteins, again in the nucleus, and their translation on cytoplasmic polyribosomes into 70 ribosomal proteins.

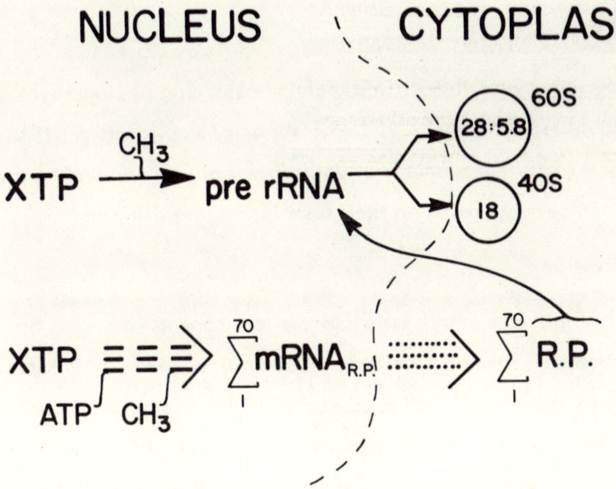

Figure 1. Schematic representation of the reactions involved in ribosome biosynthesis in a eukaryotic cell.

The regulation of ribosome biosynthesis, then, consists in the regulation of each set of reactions. We are concerned with the way in which each reaction is regulated, not only at the level of transcription but also at each of the subsequent steps. In particular we are concerned with the degree and mechanisms of intercourse between the two reactions. Is there a direct effect of one on the other, or does all regulation occur through a central "black box"? Also interesting is the possibility of understanding some aspects of nuclear-cytoplasmic interaction, since ribosome biosynthesis involves massive transfer of material across the nuclear envelope: ribosomal proteins from cytoplasm to nucleus, and both mRNA and finished ribosomal subunits from the nucleus to the cytoplasm.

This review is a subjective discussion of aspects of the regulation of ribosome synthesis that have emerged since we last reviewed this area (1). The major technical advance since that time has been the development of methods for measuring directly the synthesis of individual ribosomal proteins, isolated by two dimensional gel electrophoresis from total cell extracts (2,3,4). This permits independent analysis of the two sets of reactions outlined in Figure 1.

For convenience, we have segregated the discussion into simple eukaryotes and complex eukaryotes. It remains to be seen whether both share the same mechanisms of regulation. Other chapters of this volume examine more directly the structure and the genetics of eukaryotic ribosomal components.

SIMPLE EUKARYOTES

The yeast, Saccharomyces, has proved useful for studying ribosome synthesis, since it is amenable to both genetic and environmental manipulation. It is now possible to study directly the synthesis of both ribosomal RNA (5) and ribosomal proteins (3,4), without depending on their association into ribosomes.

Amino Acid Starvation

A key to analyzing the regulation of ribosomal components in E. coli has been the "stringent response", in which cells deprived of an amino acid shut down the synthesis of both ribosomal RNA and messenger RNA for ribosomal proteins (6). The isolation of relaxed mutants and the observation of unusual nucleotides, such as "magic spot", have provided the experimental tools to study the coordination of rRNA and ribosomal proteins.

It now seems clear that a similar "stringent response" occurs in yeast. It has been known for some time that the synthesis of ribosomal RNA declines sharply in cells deprived of an essential amino acid (7,8,9), reaching 10-20% of normal in two hours. Whether this is related to amino acid starvation, per se (7,9) or to the level of protein synthesis (8) is in dispute, perhaps because deprivation of an amino acid drastically reduces the uptake of labeled nucleotide precursors (7,8). Interestingly, however, deprivation of an amino acid has little effect on the synthesis of tRNA (7,8,9) and mRNA (8,10). Nevertheless, by assaying the template activity of the mRNA isolated from cells deprived of tyrosine, we found that there was a progressive decline of mRNA for ribosomal proteins, reaching about thirty percent of normal in three hours (10). The template activity for other proteins remains normal. The kinetics of decline of the synthesis of ribosomal precursor RNA and of mRNA for ribosomal proteins are closely parallel, suggesting that they are coordinately regulated and that yeast have a full "stringent response". The mechanism for this is yet to be understood. While "magic spot" has been identified in yeast (11), it appears to be involved in mitochondrial function. No unusual nucleotides or other potential signal molecules have been identified, nor have any relaxed mutants been isolated, in spite of substantial efforts by several laboratories.

Growth Conditions

In E. coli the concentration of ribosomes is proportional to the growth rate over a ten fold range (12). Thus any change in the growth rate has an exaggerated effect on the synthesis of ribosomes, coordinately affecting both ribosomal RNA and ribosomal proteins (13).

In simple eukaryotes, such as yeast (14) or Tetrahymena (15), the ribosome content appears to depend relatively little on the growth rate. Nevertheless, substantial changes in growth conditions have substantial effects on the synthesis of ribosomal components, although in ways that are not uniform from one organism to another.

When Tetrahymena are suspended in starvation medium, both ribosomal RNA and ribosomal protein synthesis are reduced ten fold from the rate in growing cells (16). Within two hours of refeeding the synthesis of ribosomal proteins increases 80 fold, to nearly twice the rate in growing cells, while the increase in ribosomal RNA synthesis is much more sluggish (17). In this case there is only rather loose coordination of RNA and protein synthesis.

Saccharomyces offers a somewhat different picture. The addition of glucose to a culture growing on ethanol as a carbon source leads within 15 minutes to a 2.5 fold increase in the transcription of ribosomal RNA, and nearly a 4 fold increase in synthesis of a large number of ribosomal proteins (Kief and Warner, unpublished). Conversely, when diploid yeast cells are deprived of a nitrogen source, and enter the sporulation cycle, within two hours there is a 70 to 80% decline in synthesis of both ribosomal RNA and ribosomal proteins (Pearson, N. J., personal communication). Thus, under several physiological conditions, Saccharomyces appears to coordinate effectively the synthesis of ribosomal RNA and ribosomal proteins.

How such coordination occurs remains a mystery. Is there a regulatory element which independently affects each of the reactions of Figure 1, or is one of them the primary target and the other dependent on the first? UV mapping experiments may shed some light on the problem. We find (Gorenstein and Warner, unpublished) that the synthesis of nearly all ribosomal proteins is equally affected by UV irradiation, that they are affected three-fold more than larger non-ribosomal proteins, and that they are affected to exactly the same degree as ribosomal RNA. Since the "target size" of rRNA is expected to be substantially greater than that of rprotein, these results suggest that synthesis of mRNA for ribosomal protein is dependent on the synthesis, or processing, of ribosomal precursor RNA.

Temperature Shifts

There are other situations in Saccharomyces, however, where there is a total lack of coordination between ribosomal RNA and ribosomal protein synthesis. For example, when a culture is shifted from 23° to 36°, there is a transient but severe inhibition of the synthesis of ribosomal proteins (3). We find that the synthesis of forty individual ribosomal proteins declines with a half life of about fifteen minutes, reaching a minimum of 25% of normal after 30 minutes, and returns to normal by 60 to 90 minutes. Such kinetics, as well as measurements of template activity (18), suggest that the temperature shift causes an immediate shut-off of the synthesis of mRNA for ribosomal proteins which lasts for 15 to 30 minutes. Nevertheless, the transcription of ribosomal precursor RNA is almost unaffected (19). This phenomenon occurs in all Saccharomyces strains that we have tested. Its magnitude depends on the magnitude of the temperature shift. No analogous phenomenon has been observed on a temperature down-shift.

There are a number of temperature sensitive mutants in which the synthesis of mRNA for ribosomal proteins is permanently turned off at the non-permissive temperature (3,18). The synthesis of ribosomal proteins declines to 10% of normal after one hour. At that time the transcription of ribosomal precursor RNA is continuing at 85% of its normal rate (20). Therefore, its transcription is not dependent on concurrent synthesis of ribosomal proteins or their mRNAs. Thus, the inhibition of transcription of rpreRNA which occurs when cells are deprived of an amino acid is caused not by a lack of ribosomal proteins but by some factor related to protein synthesis in general.

Whatever the regulation of its transcription, the processing of ribosomal precursor RNA is absolutely dependent on concurrent synthesis of ribosomal proteins. After a temperature shift in wild type cells, the rate of processing of rpreRNA is slowed temporarily, with kinetics almost identical to those of rprotein synthesis (19). Any ribosomal precursor RNA transcribed in the absence of ribosomal protein synthesis is processed not at all; it is rapidly degraded (19,21).

The temperature upshift phenomenon is remarkably widespread. Not only all strains of yeast tested but also bacteria (22) and mammalian cells (23) turn down the synthesis of components of the protein synthetic apparatus in response to a temperature upshift. It is perhaps a universal protective mechanism. Its importance in yeast may be judged by the fact that mutations in at least ten genes can cause a temperature sensitivity which now appears to be due to an inability to recover from the effects of a temperature upshift (24). It is interesting, however, that the effects of a temperature shift, and of these mutants, are found only in vegetative cells. The low but easily measurable synthesis of ribosomal proteins in sporulating yeast cells continues unaffected when either wild type or temperature sensitive strains are shifted from 23° to 36° (25).

Coordination Among Ribosomal Proteins

In the experiments described above, it has been possible to measure the synthesis of 40 to 45 proteins. Of these, all but six respond in an almost identical way to the various experimental manipulations (3). Those six proteins, however, are clearly different. Their synthesis is not affected by a shift-up in temperature in either wild type or mutant cells. The mRNA activity for some but not others is inhibited by deprivation for an amino acid. Yet these six proteins appear to be true ribosomal proteins, in that they are found in good yield on well washed ribosomal subunits. A final explanation of the regulation of ribosomal proteins must include a consideration of these proteins.

We know relatively little of the genetics of ribosomal proteins in any eukaryote. In yeast, three antibiotic markers, thought to be ribosomal proteins, are on different chromosomes (26,27,28). Four cloned ribosomal protein genes are unlinked (29). UV mapping data (Gorenstein and Warner, unpublished) and analysis of the synthesis of ribosomal proteins (30) suggest that neither the transcriptional nor translational units are polycistronic, although conflicting data has been presented (31). If, in fact, the genes for ribosomal proteins are individual units scattered all over the genome, their coordinate regulation must involve some common feature of these genes. Whether this is a promotor region, or a common site for processing of precursors to mRNA, or something quite unexpected, its elucidation should provide considerable insight into the way in which a simple eukaryote governs its economy. The clones of yeast ribosomal protein genes recently isolated by Woolford, et al. (29) may provide some answers.

COMPLEX EUKARYOTES

The regulation of ribosome biosynthesis is more difficult to study in complex eukaryotes, for both theoretical and practical reasons. The theoretical reason is that animal cells must be capable of growing rapidly, during embryogenesis and occasionally at other times, but spend most of their lives in a non-growing though metabolically active state. The practical reason is that since it is frequently difficult, if not impossible, to study ribosome synthesis in animals, cultured cells are used. But to what degree do cultured cells behave as they would in the organism, particularly if they are somewhat transformed?

Shift Up

The regulation of ribosome biosynthesis in animal cells has most frequently been studied in one form or another or a step-up, which may take the form of adding serum to cells deprived ot it (32), stimulating lymphocytes with phytohemagglutinin, (33), or partial hepatectomy of a rat (34).

In most cases, a shift-up is accompanied by an increase in synthesis of ribosomal RNA. Perhaps the clearest example is mouse fibroblast cells, where Mauck and Green (35) found that the number of growing ribosomal RNA chains was 2 fold greater within 2-3 hours of adding serum to starved monolayer cells. Our results with the same system suggest that there is a parallel stimulation of the synthesis of ribosomal protein (Tushinski and Warner, unpublished). Within an hour of serum addition there is a more than 50% increase in the synthesis of thirty individual ribosomal proteins as a proportion of total protein synthesis. Since total protein synthesis also increases during this time, the total rate of synthesis of individual ribosomal proteins is increased about three fold. It has yet to be determined whether this is due to the synthesis of new mRNA, to the mobilization of unused mRNA, or to the more efficient translation of mRNA already on polysomes. Nevertheless, in this case there does appear to be coordination between rRNA and rprotein production.

In comparing such experiments to analogous ones in bacteria or unicellular eukaryotes, it is important to keep in mind the differences. Stationary mammalian cells are truly stationary, i.e. there is no net increase of ribosomes, but not inert, i.e. there is a continuous synthesis of ribosomes, which is balanced by degradation. Thus, where a shift-up in bacteria or unicellular eukaryotes usually involves going from one exponential growth rate to another, the shift-up in animal cells involves a shift from emphasis on degradation to emphasis on synthesis. Perhaps the most dramatic example of this occurs on removal of one kidney from the mouse (36). The remaining kidney increases its content of ribosomes 20 to 40% within two days without any change in transcription. The half life of ribosomes is simply increased.

In cultured cells, whose growth can be manipulated more readily, ribosomes appear to be stable in cells that are growing exponentially. However, when growth is slowed due to serum starvation or approach to confluence, turnover commences (37,38). In transformed cells under analagous conditions, ribosome turnover does not occur (39). Such cells continue to accumulate ribosomes, to grow, and to die. To a partisan, it appears that the accumulation of ribosomes may be a key to understanding the transformed state. It would certainly be of great interest to understand the way in which the turnover itself is regulated. One careful study of the mechanism of turnover (40) suggests that it is a catastrophic event, for an individual ribosome, with the RNAs and each of the proteins being destroyed together.

Interspecific Cell Hybrids

Some beginnings have been made in studying the regulation of ribosomal components in interspecific hybrids. These offer the (dis?) advantage of two separate genetic and regulatory systems, one of which may be dispensable. The close similarity of ribosomal RNA and ribosomal proteins of genetically compatible species creates major difficulties in this approach. Nevertheless it has been shown that hybrid cells which lose preferentially the chromosomes of one parent generally express only one type of 28S rRNA (41) while hybrids of two rodent species generally express the RNA (42) and at least one protein (43) of both species. In certain mouse-human hybrids, which retain ribosomal DNA of both parents, only the mouse nucleolar organizer is active (41).

In recent work from this laboratory (44), we have fused mouse cells with CHO cells resistant to emetine, (45) an antibiotic affecting the 40S subunit. Emetine resistance, recessive in heterozygotes, is associated with alterations in a single 40S protein (46). Although hybrid lines were phenotypically sensitive, it was possible to isolate sublines in which either the mouse emetine sensitive gene was active, or the CHO emetine resistant gene was active, or both genes were active. In all three cases both 28S and 18S RNA were predominantly, but not completely, mouse. On the other hand, for those ribosomal proteins which could be distinguished, both mouse and hamster types were present in equal amounts. We conclude that hybrid cells can form hybrid ribosomes, i.e. ribosomes containing mouse RNA and at least some hamster proteins, and that within a single cell the amounts of ribosomal RNA and ribosomal protein of each species are not closely coupled (44).

Ribosomal RNA and Ribosomal Protein are not Always Coupled

One of the most elegant systems for analyzing the synthesis of ribosomal components is the developing embryo of the anucleolate Xenopus. Since this lacks all genes for 28S and 18S ribosomal RNA, it makes none (47), although 5S RNA synthesis occurs. Nevertheless, Steel and Hallberg (48) found that the synthesis of ribosomal proteins commenced on schedule in such anucleolate embryos. In this case de novo synthesis of ribosomal proteins occurs without concurrent synthesis of ribosomal RNA, unless of course 5S RNA plays an unexpected role.

The L6E9 line of rat myoblasts can be induced to differentiate, leading to the synthesis of muscle-specific proteins, to cell fusion, and to the eventual formation of myotubes (49). Under such conditions the transcription of ribosomal RNA declines rapidly to reach ten percent of that in growing cells by six days (50). Surprisingly, however, the synthesis of ribosomal proteins continues unimpaired during this time (50). These ribosomal proteins do not accumulate, but are degraded.

HeLa cells treated with a low concentration of actinomycin (10 ng/ml) are unable to synthesize detectable amounts of ribosomal precursor RNA. Nevertheless, ribosomal proteins are synthesized at a normal rate for thirty hours after the transcription of ribosomal precursor RNA has ceased (51,60). Either the mRNA for ribosomal proteins is exceptionally stable, or the synthesis of mRNA for ribosomal proteins is independent of concurrent synthesis of rpreRNA. Once again, these proteins do not accumulate, but are degraded with half lives of 20 to 120 minutes, depending on the individual protein (51).

It is striking that these experimental systems show a complete lack of regulation of the synthesis of ribosomal proteins, compared with ribosomal RNA. This is not observed in simple eukaryotes. Does the massive amount of protein turnover that occurs normally in animal cells obviate the need for careful control of protein synthesis? Do animal cells share a fundamentally different approach to the regulation of ribosomal components, depending largely on the degradation of excess RNA or protein rather than on a finely tuned coordination of the synthesis of ribosomal RNA and protein? This is certainly not entirely true. In HeLa cells deprived of an essential amino acid so that gross protein synthesis is reduced to 20 to 30% of normal,and net protein synthesis to zero, the synthesis of ribosomal precursor RNA as well as its maturation is significantly slowed (52).

Efficiency of Processing

Ribosome synthesis can also be regulated at the level of the efficiency of processing, i.e. the proportion of transcripts which eventually enter mature ribosomes. Such "wastage" (33) of unprocessed transcripts reaches 50% in some situations, such as quail myoblasts undergoing differentiation to myotubes (53). A careful study suggests that there is little if any wastage in growing WI-38 fibroblasts, but that as such cells reach senescence, after 50 doublings, wastage increases substantially(54).

The processing of ribosomal precursor RNA requires ribosomal proteins (19,31). To what degree is wastage a manifestation of deficient synthesis of one or more ribosomal proteins? No such measurements have yet been reported, but insofar as processing and degradation compete for rpreRNA transcripts, the abundance of ribosomal proteins may play a critical role.

Assembly of Ribosomes

It has been known for some time that newly formed ribosomal proteins migrate to the nucleolus where they are assembled onto ribosomal precursor RNA (55,56).

Kumar and Subramanian (57) have used two dimensional gel analysis to demonstrate that the 55S nucleolar precursor to the large ribosomal subunit, which contains 32S precursor RNA, also contains at least 30 of the large ribosomal subunit proteins. These are rapidly labeled, confirming their precursor role. The particle also contains more than 30 non-ribosomal proteins, which appear to be repeatedly used as scaffolding for ribosome assembly in the nucleolus (58). Extension of this approach to the ribonucleoprotein particle carrying the original transcript (59) suggests that the addition of many 40S proteins may occur relatively late in the assembly process.

Little is known about the way in which ribosomal proteins migrate to the nucleus. While the nuclear envelope is studded with polyribosomes, it seems clear that they are not enriched in mRNA for ribosomal proteins (60). Furthermore, the complex structure of the nuclear envelope, with a double layer of membranes, a proteinaceous lamina, and tightly packed chromatin (61), makes unlikely a direct extrusion of newly formed proteins into the nucleus. In attempts to do more precise bookkeeping on the location of newly formed ribosomal proteins we found (62) that all but a few were concentrated 10 to 50 fold in the nucleolus and two to five fold in the nucleoplasm. Nevertheless, substantial amounts were found in the cytoplasm. There was considerable variation in compartmentalization among the ribosomal proteins however, and several were not concentrated in the nucleus at all, suggesting that they may be assembled onto ribosomes in the cytoplasm. Pretreatment of cells with actinomycin D to deplete the nucleolar pool of ribosomal precursor RNA had no effect on the concentration of newly formed ribosomal proteins in the nucleus, but did lead to an increased amount in the nucleoplasm at the expense of the nucleolus. These results suggest that ribosomal proteins concentrate in the nucleus independent of the presence of ribosomal precursor RNA. Subsequent binding to RNA, however, may account for the very high concentration found in the nucleolus. This notion is supported by the finding (63) that most ribosomal proteins are found in equimolar amounts in the nucleolus. It should be emphasized that there is no direct evidence to determine whether the passage of proteins into the nucleus occurs by passive diffusion or by facilitated transport. *In vitro* studies of the migration of ribosomal proteins in the nucleus (64) may shed some light on the question.

In regard to the migration of newly formed ribosomal subunits from the nucleus to the cytoplasm, little has happened to alter my conclusion that it is a complex process (1).

SPECULATIONS

Autoregulation

The synthesis of each of the ribosomal proteins appears to be within about ten percent of the amount utilized (3,51,65), although more careful data must be obtained. This is remarkable if one considers the steps involved in determining the level of synthesis of a protein: transcription of nuclear RNA, its processing, its migration to the cytoplasm, the multiple initiations of protein synthesis, and finally the degradation of the mRNA. Because we find it difficult to envision a central regulatory system which could exert such fine tuning, we suggest that each ribosomal protein can limit its own synthesis to some degree. But at what level?

RNA Fragments

If the genes for ribosomal proteins are scattered over seventy different loci, what kind of controlling element can regulate their synthesis? We and others (Johnson, G. and Singer, R., personal communication) have considered that the fragments of RNA generated during the processing of rpreRNA might play such a role. They are generated in ample numbers, but their existence is so fleeting that they have never been identified.

CONCLUSIONS

One could conclude with a sense of disappointment. The ability to measure independently the synthesis of ribosomal RNA and of ribosomal proteins has broadened our knowledge of ribosome biosynthesis but has not sharpened our insight. In contemplating Figure 1, we are forced to realize that almost all combinations have been observed: coordination between RNA and protein, lack of coordination; RNA without proteins, proteins without RNA. No clear picture emerges.

Nevertheless, the prospects are in fact, bright. Broader and more accurate analysis of ribosomal protein synthesis should reveal whether its tight coordination with RNA synthesis is apparent or real. Regulatory mutants isolated from yeast, and the use of mammalian cell hybrids, may provide a genetic basis for the biochemical analysis. Recombinant DNA clones containing genes for ribosomal proteins, isolated already from yeast (29) and soon from mammalian cells, will make possible accurate measurements of nuclear RNA and messenger RNA molecules corresponding to ribosomal proteins. Eventually they may yield regulatory sequences offering a clue to the coordinate synthesis of the multitude of ribosomal proteins.

In the long run these contradictions and uncertainties may be turned to advantage. The ribosome has been highly conserved during evolution and is the prime example of a "housekeeping enzyme". Therefore a comparison of the regulation of ribosome biosynthesis in prokaryotes, in simple eukaryotes and in complex eukaryotes may provide fundamental insight into the way regulation, itself, has evolved.

ACKNOWLEDGEMENTS

Research from this laboratory was supported by grants from the NIH Nos. GM 25532, CA 13330, GM 19100 and the ACS No. NP 72 I. R.J.T. and P.J.W. are supported by postdoctoral fellowships from the NIH No. F32 GM 05667 and No. F32 GM 06137 respectively.

REFERENCES

1. Warner, J. R. (1974) Ribosomes, pp. 461-488, Cold Spring Harbor.
2. Dennis, P. P. (1974) J. Mol. Biol. 89, 223-232.
3. Gorenstein, C. and Warner, J. R. (1976) Proc. Nat. Acad. Sci. USA 73, 1547-1551.
4. Mager, W. H. and Planta, R. J. (1976) Eur. J. Biochem. 62, 193-197.
5. Udem, S. A. and Warner, J. R. (1972) J. Mol. Biol. 65, 227-242.
6. Gallant, J. and Lazzarini, R. A. (1976) Protein Synthesis, A Series of Advances, pp. 309-359, Marcel Dekker, New York.
7. Gross, K. J. and Pogo, A. O. (1974) J. Biol. Chem. 249, 568-576.
8. Shulman, R. W., Sripati, C. E. and Warner, J. R. (1977) J. Biol. Chem. 252, 1344-1349.
9. Oliver, S. G. and McLaughlin, C. S. (1977) Molec. Gen. Genet. 154, 145-153.
10. Warner, J. R. and Gorenstein, C. (1978) Nature 275, 338-339.
11. Pao, C. C., Paietta, J. and Gallant, J. A. (1977) Biochem. Biophys. Res. Commun. 74, 314-322.
12. Maaloe, O. and Kjeldgoard, N. O. (1966) Control of Macromolecular Synthesis, pp. 72-96, W. A. Benjamin, Inc., New York.

13. Gausing, K. (1977) J. Mol. Biol. 115, 335-355.

14. Boehlke, K. W. and Friesen, J. D. (1975) J. Bacteriol. 121, 429-433.

15. Leick, V. (1973) The Formation of Ribosomes in Tetrahymena pyriformis, Thesis, University of Copenhagen.

16. Hallberg, R. L. and Bruns, P. J. (1976) J. Cell Biol. 71, 383-394.

17. Sutton, C. A., Sylvan, P. and Hallberg, R. L. (1979) J. Cell Physiol., in press.

18. Warner, J. R. and Gorenstein (1977) Cell 11, 201-208.

19. Warner, J. R. and Udem, S. A. (1972) J. Mol. Biol. 65, 243-257.

20. Shulman, R. and Warner, J. R. (1978) Molec. Gen. Genet. 161, 221-223.

21. deKloet, S. R. (1966) Biochem. J. 99, 566-574.

22. Lemaux, P. G., Herendeen, S. L., Bloch, P. L. and Neidhardt, F. C. (1978) Cell 13, 427-434.

23. Kelley, P. M. and Schlessinger, M. J. (1978) Cell 15, 1277-1286.

24. Hartwell, L., McLaughlin, C. and Warner, J. R. (1970) Molec. Gen. Genet. 109, 42-56.

25. Pearson, N. J. and Haber, J. E. (1977) Molec. Gen. Genet. 158, 81-91.

26. Mortimer, R. K. and Hawthorne, D. C. (1973) Genetics 74, 33-54.

27. Skogerson, L., McLaughlin, C. and Wakatama, E. (1973) J. Bacteriol. 116, 818-822.

28. Schindler, D., Grant, P. and Davies, J. (1974) Nature (London) 248, 535-536.

29. Woolford, J. L. and Rosbash, M. (1979) Nucleic Acids Res., in press.

30. Gorenstein, C. and Warner, J. R. (1979) Molec. Gen. Genet., in press.

31. Mager, W. H., Retel, J., Planta, R. J., Bollen, H. P. M. G., DeRegt, C. H. F. V. and Hoving, H. (1977) Eur. J. Biochem. 78, 575-583.

32. Rudland, P., Weil, S. and Hunter, A. R. (1975) J. Mol. Biol. 96, 745-766.

33. Cooper, H. L. and Gibson, E. M. (1971) J. Biol. Chem. 246, 5059-5064.

34. Chaudhuri, S. and Lieberman, I. (1968) J. Biol. Chem. 243, 29-33.

35. Mauck, J. C. and Green, H. (1973) Proc. Nat. Acad. Sci. USA 70, 2819-2822.

36. Melvin, W. T., Kumar, A. and Malt, R. A. (1976) J. Cell Biol. 69, 548-556.

37. Abelson, H. T., Johnson, J. F., Penman, S. and Green, H. (1974) Cell 1, 161-165.

38. Melvin, W. T. and Keir, H. M. (1978) Biochem. J. 176, 933-941.

39. Liebhaber, S. A., Wolf, S. and Schlessinger, D. (1978) Cell 13, 121-127.

40. Tsurugi, K., Morita, T. and Ogata, K. (1974) Eur. J. Biochem. 45, 119-126.

41. Perry, R. P., Kelley, D. E., Schibler, U., Huebner, K. and Croce, C. (1979) J. Cell Physiol. 98, 553-560.

42. Elicieri, G. L. (1972) J. Cell Biol. 53, 177-184.

43. Kuter, D. J. and Rogers, A. (1975) Exp. Cell Res. 91, 317-325.

44. Wejksnora, P. J. and Warner, J. R. (1979) Proc. Nat. Acad. Sci. USA, in press.

45. Gupta, R. S. and Siminovitch, L. (1976) Cell 9, 213-219.

46. Boerama, D., McGill, S., Mallenkamp, J. and Roufa, D. S. (1979) J. Biol. Chem. 254, 559-567.

47. Brown, D. D. and Gurdon, J. B. (1964) Proc. Nat. Acad. Sci. USA 51, 139-146.

48. Steel, L. F. and Hallberg, R. L. (1979), personal communication.

49. Nadal-Ginard, B. (1978) Cell 15, 855-862.

50. Krauter, K. S., Soeiro, R. and Nadal-Ginard, B. (1979) J. Mol. Biol., in press.

51. Warner, J. R. (1977) J. Mol. Biol. 115, 315-333.

52. Maden, B. E. H., Vaughan, M. H., Warner, J. R. and Darnell, J. E. (1969) J. Mol. Biol. 45, 265-275.

53. Bowman, L. H. and Emerson, C. P. (1977) Cell 10, 587-595.

54. Wolf, S., Sameshima, M., Liebhaber, S. A. and Schlessinger, D. (1979) J. Cell Biol., in press.

55. Warner, J. R. and Soeiro, R. (1967) Proc. Nat. Acad. Sci. USA 58, 1984-1990.

56. Liau, M. C. and Perry, R. P. (1969) J. Cell Biol. 42, 272-280.

57. Kumar, A. and Subramanian, A. R. (1975) J. Mol. Biol. 94, 409-423.

58. Kumar, A. and Warner, J. R. (1972) J. Mol. Biol. 63, 233-244.

59. Fujisawa, T., Imai, K., Tanaka, Y. and Ogata, K. (1979) J. Biochem. 85, 277-286.

60. Craig, N. C. and Perry, R. P. (1971) Nature New Biol. 229, 75-87.

61. Kirschner, R. H., Rusli, M. and Martin, T. E. (1977) J. Cell Biol. 72, 118-132.

62. Warner, J. R. (1979) J. Cell Biol. 80, 767-772.

63. Phillips, W. F. and McConkey, E. H. (1976) J. Biol. Chem. 251, 2876-2881.

64. Roth, H., Bolla, R., Cox, G. S., Redfield, B., Weissbach, H. and Best, N. (1976) Biochem. Biophys. Res. Commun. 69, 608-612.

65. Gorenstein, C. and Warner, J. R. (1977) Molec. Gen. Genet. 157, 327-332.

BIOGENESIS OF CHLOROPLAST AND MITOCHONDRIAL RIBOSOMES

JOHN E. BOYNTON,* NICHOLAS W. GILLHAM,*
and ALAN M. LAMBOWITZ†

*Departments of Botany and Zoology,
Duke University,
Durham, NC 27706
†Department of Biochemistry,
St. Louis University School of Medicine,
St. Louis, MI 63104

INTRODUCTION

Organelle protein synthesis accounts for a small but essential fraction of the protein synthesis of eukaryotic cells. In the absence of mitochondrial protein synthesis, certain mitochondrial inner membrane proteins required for electron transport and ATP formation are not made (1-4). Similarly, a deficiency in chloroplast protein synthesis blocks photosynthetic electron transport, photophosphorylation and CO_2 fixation because certain polypeptides required for each of these processes are made on chloroplast ribosomes (1,5,6). Protein synthesis in both chloroplasts and mitochondria makes use of a unique set of tRNAs coded by the organelle genomes (cf. 1,3,7-9). Initiation involves F-met-$tRNA_F$ in both chloroplasts and mitochondria (1,3,10) as is the case in bacteria. Organelle ribosomes are distinct in virtually all respects from the ribosomes in the cytoplasm of eukaryote cells. Chloroplast ribosomes are similar in many ways to the 70S ribosomes of prokaryotes. In contrast, mitochondrial ribosomes form a very diverse class with respect to other kinds of ribosomes and even each other. Our purpose in this chapter is to discuss the properties of chloroplast and mitochondrial ribosomes with particular emphasis on findings that are new since this topic was last reviewed by Chua and Luck (12) in the book entitled Ribosomes (11). The structure and function of organelle ribosomes have also been considered in recent reviews by Buetow and Wood (3), Ellis and Hartley (14), O'Brien and Matthews (13), Stutz and Boschetti (10) and in the book by Gillham (1).

CHLOROPLAST RIBOSOMES

Structural Characteristics

Only a few general remarks will be made here concerning the overall structural features of chloroplast ribosomes since these have been comprehensively reviewed recently (1,10,12,14). Chloroplast ribosomes sediment at 70S like bacterial ribosomes and consist of two subunits of about 50S (46-54S) and about 30S (29-41S). Recently, Ledoigt *et al*. (15) compared the electrophoretic mobility and size of negatively stained chloroplast and cytoplasmic ribosomes from *Euglena* and spinach with prokaryotic ribosomes. In comparison to cytoplasmic ribosomes, chloroplast ribosomes from *Euglena* and spinach have a higher electrophoretic mobility and smaller dimensions. They are similar in electrophoretic mobility to prokaryotic ribosomes and are of similar size.

In *Chlamydomonas*, sucrose gradient analysis indicates that chloroplast ribosomes constitute between 29% and 35% of the ribosomes of the cell depending on growth conditions (16). Cytoplasmic ribosomes constitute the majority of the cell's ribosomes and mitochondrial ribosomes are negligible in quantity. Similar estimates are also obtained by electron microscopy and quantitative determination of rRNA. Comparable analyses in higher plants yield ratios for chloroplast:cytoplasmic ribosomes of 1:2 and 1:3 (10). Chloroplast ribosomes change shape and therefore sedimentation velocity as a function of the monovalent-divalent cation concentration (10) and can be dissociated into subunits under ionic conditions where the 80S cytoplasmic ribosomes remain intact (17).

Chloroplast rRNAs

The large subunit of the chloroplast ribosome contains 23S rRNA ($\underline{M}_r$ 1.1 x 10^6) while the small subunit contains 16S rRNA ($\underline{M}_r$ 0.56 x 10^6) (cf. 1,10,14). In addition the large subunit contains two low molecular weight rRNAs whose properties have been reviewed by Dyer *et al*. (18). One of these, chloroplast 5S rRNA ($\underline{M}_r$ 3.93 x 10^4, about 122 nucleotides), is distinct from the 5S rRNA of plant cytoplasmic (118 nucleotides) or mitochondrial (about 120 nucleotides) ribosomes. The second species, 4.5S rRNA ($\underline{M}_r$ 2.5 or 3.3 x 10^4), which is present in equimolar amounts with chloroplast 5S rRNA, varies in size in different flowering plants, and was discovered independently by three groups of workers (18,19,20). In *Chlamydomonas*, Rochaix and Malnoë (21) have reported the existence of three low molecular weight rRNAs in the large subunit: 7S, 5S, and 3S. Neither the relationship of the 7S or 3S rRNAs to 4.5S rRNA nor the possible homology of these rRNAs to the 5.8S rRNA of cytoplasmic ribosomes is known.

The GC contents of chloroplast rRNAs for several higher plants and for *Chlorella* and *Euglena* have been reviewed by Ellis and Hartley (14) and by Stutz and Boschetti (10). The GC content of 23+16S rRNA for six Angiosperm species was quite constant varying between 55.2 and 55.9%. In the cases of spinach and beet, the 23S rRNA has a slightly higher GC content than 16S rRNA. The overall GC content for Chlorella and Euglena chloroplast rRNA is about 51.5% and the GC contents of 23S and 16S rRNA in Chlorella are nearly identical. For comparison, E. coli 23S rRNA is 53.5% GC and 16S rRNA is 54.5% GC, and in the blue green alga Anacystis nidulans the overall GC content of 23+16S rRNA is 52.7%. As Ellis and Hartley (14) remark, base compositions of rRNA from the chloroplasts of other lower plants and algae must be determined before any significance can be attached to these figures with respect to putative prokaryotic ancestry.

Buetow and colleagues (22,23) have examined the relationship of chloroplast rRNA and prokaryotic rRNA by comparing the T_1 ribonuclease digests of 16S chloroplast rRNA from Euglena and the red alga Porphyridium, 18S cytoplasmic rRNA from yeast and L cells, and 16S rRNA from Anacystis and several bacteria. The two chloroplast rRNAs had the most similar fingerprints and were typically prokaryotic in terms of their capacity for base modification. At least 12 of the oligomers referred to as prokaryotic "universals," because they occur in prokaryotes but not eukaryotes (24), are found in Euglena 16S rRNA. Comparison of the fingerprints of chloroplast 16S rRNA to that of Anacystis suggests at least a distant relationship which may also extend to the Bacillaceae.

The similarity between the T_1 fingerprints of chloroplast 16S rRNA from Euglena and Porphyridium suggests that chloroplast rRNA may be highly conserved. Thomas and Tewari (25) have found homology among higher plant chloroplast rRNA genes based on saturation hybridization between rRNA from pea and chloroplast DNA from pea, lettuce, spinach, corn, bean, and oats. Heterologous as well as homologous hybrids showed a sharp T_m 20° above the hybridization temperature which indicates the existence of specific base pairing in such hybrids. The specificity of these hybrids was also confirmed by competition experiments where formation of heterologous DNA·rRNA hybrids was challenged with homologous or heterologous rRNA. In all cases, the observed competition was of the same order as that obtained with homologous systems.

The GC content of chloroplast rRNA is considerably higher than the GC content of total chloroplast DNA which is in the range of 37-39% in higher plants and Chlamydomonas (cf. 14). The difference is even more extreme in Euglena where the average GC content of chloroplast DNA is 25% while the chloroplast rRNA has an overall GC content of 56.1% (14). Gray and Hallick (26) have shown recently that the existence of GC rich rRNA genes in Euglena chloroplast DNA accounts at least in part for the heterogeneity that has been reported in this DNA on the basis of thermal denaturation profiles (27,28) and the observation of a GC rich satellite in CsCl gradients of partially sheared chloroplast DNA from Euglena (29-31). The GC rich segment contains the three tandemly repeated sets of rRNA genes, representing 13% of the chloroplast genome, and having an average mole percent GC content of 43-44%.

The 23S rRNA from chloroplasts is often cleaved during isolation. Ingle and collaborators (32,33) showed that the 23S rRNA from several species of higher plants and from Chlamydomonas broke apart during isolation, yielding a specific pattern of RNA fragments upon gel electrophoresis, while the 16S rRNA remained intact. The use of Mg^{2+} or Ca^{2+} in the isolation buffer stabilized the 23S rRNA. The extent to which these "hidden breaks" in 23S rRNA are present in vivo and their significance remains to be established.

Chloroplast Ribosomal Proteins

As pointed out by Stutz and Boschetti (10) none of the chloroplast ribosomal proteins has been obtained in sufficient amounts to allow for detailed analysis of its amino acid sequence or for producing specific antibodies. However, the amino acid composition of total chloroplast ribosomal proteins has been established in Euglena (34) and antibodies have been prepared against total ribosomal subunit proteins in several experimental systems (35,36).

As gel technology has improved the number of chloroplast ribosomal proteins resolved has increased. Stutz and Boschetti (10) have reviewed earlier estimates of the numbers of chloroplast ribosomal proteins. More recent estimates are to be found in papers by Bourque (37,38) for tobacco, Freysinnet (34) for Euglena, Gillham et al. (39) for Chlamydomonas, and Pitel and Durzan (40) for jack pine. Bourque, using the two-dimensional system of Mets and Bogorad (41), reported that the small subunit of the tobacco chloroplast ribosome contains 20-24 proteins while the large subunit contains 34-38 proteins. Freyssinet (34, 42), who employed a modification of the Kaltschmidt-Wittmann two-dimensional system, found that the small subunit of the Euglena chloroplast ribosome contained 22-24 proteins while the large subunit contained 30-34 proteins. Pitel and Durzan (40), using a one-dimensional urea gel system (43) separated 21 proteins from whole chloroplast ribosomes from jack pine, which is clearly a gross underestimate of the total number of chloroplast ribosomal proteins. In *Chlamydomonas* Hanson *et al*. (44), using the Mets-Bogorad system, found that the small subunit contained 22 proteins while the large subunit included 26 proteins. Brügger and Boschetti (45), on the other hand, separated chloroplast ribosomal proteins from *Chlamydomonas* with the Kaltschmidt-Wittmann system and reported that the small subunit contained 25 proteins while the large subunit had 34 proteins. In one-dimensional SDS gels both Brügger and Boschetti (45) and Gillham *et al*. (39) resolved six to seven proteins of $M_r > 50{,}000$ associated with the small subunit. These proteins were not seen by Hanson *et al*. in the second dimension of the Mets-Bogorad system which also separates proteins on the basis of molecular weight. In addition Hanson *et al*. reported the presence of several other proteins in small subunit preparations which they attributed to contamination by cytoplasmic small subunits. If we assume that the additional high molecular weight proteins not resolved by Hanson *et al*. and the five attributed to cytoplasmic contamination are really chloroplast ribosomal proteins, the total number of small subunit proteins could be as high as 34. In the case of the large chloroplast ribosomal subunit Hanson *et al*. found four spots which were not reproducible in addition to the 26 they regarded as true large subunit proteins. These four could be among the 34 proteins that Brügger and Boschetti counted. Since neither Brügger and Boschetti nor Gillham *et al*. reported finding high molecular weight proteins in one-dimensional SDS gels of the large subunit, the discrepancy between the two estimates of numbers of proteins cannot be accounted for by high molecular weight proteins as in the case of the small subunit. This detailed discussion of *Chlamydomonas* chloroplast ribosomal proteins emphasizes that even in the best characterized plant system, different authors using different gel systems obtain different results. Therefore, the precise number of chloroplast ribosomal proteins remains to be established for any plant system. If the high estimates prove to be correct, the small subunit of the *Chlamydomonas* chloroplast ribosome would clearly contain more proteins than the comparable subunit of *E. coli* (34 versus 21) while the large subunit would contain the same number as in *E. coli*.

Two groups of investigators have compared the immunological cross-reactivity of antibodies made against whole ribosomes and ribosomal subunits in an attempt to estimate the relatedness of prokaryotic ribosomal proteins to chloroplast ribosomal proteins. Gualerzi _et al_. (35) found no crossreactivity between chloroplast ribosomal proteins from spinach and beans with antibodies prepared against ribosomes or ribosomal subunits from _E. coli_, _Bacillus subtilis_, and _Bacillus stearothermophilus_. However, they only found a crossreactivity of 20% between _E. coli_ and _B. subtilis_ ribosomes. Furthermore, antibodies prepared against chloroplast ribosomes from bean or spinach did not react with ribosomal proteins from the blue green alga _Anacystis_. In contrast, Schneeman and Surzycki (36) reported that antisera prepared against the small subunit of the chloroplast ribosome from _C. reinhardtii_ reacted with both _E. coli_ ribosomal subunits and with the large subunit of the _Chlamydomonas_ chloroplast ribosome as well. More extensive experiments, preferably with purified ribosomal proteins, will be required to determine whether the chloroplast ribosomal proteins of this green alga really are more closely related to bacterial ribosomes than those of higher plants.

Functional Properties

Protein synthesis on chloroplast ribosomes like protein synthesis on mitochondrial ribosomes bears many similarities to prokaryotic protein synthesis. Chloroplasts contain N-formylmethionyl tRNA and active transformylases (see 1, 10 for reviews). Highfield and Ellis (46) have studied the initiation process in isolated intact pea chloroplasts using the puromycin initiation assay and have demonstrated the formation of N-formyl (^{35}S)-methionyl puromycin when chloroplasts are incubated with L-(^{35}S) methionine and puromycin. Chloroplasts also contain unique elongation factors similar to those in prokaryotes and these have been characterized in detail in _Chlorella_ and spinach by Ciferri and his colleagues (47-52). In the case of _Chlorella_ synthesis of the elongation factor G (EF-G_{chl}) is inhibited by chloramphenicol suggesting that it is made on chloroplast ribosomes. From spinach chloroplasts Ciferri _et al_. have purified EF-G_{chl}, elongation factor TU (EF-TU_{chl}), and partially purified elongation factor T (EF-T_{chl}). Both EF-G_{chl} and EF-TU_{chl} are labelled in isolated chloroplasts indicating that they are products of chloroplast protein synthesis.

Chloroplast ribosomes are resistant to inhibitors of eukaryotic cytoplasmic protein synthesis such as cycloheximide and anisomycin, but sensitive to virtually all inhibitors of prokaryotic protein synthesis including chloramphenicol, the aminoglycosides (i.e. kanamycin, neamine, neomycin, spectinomycin, streptomycin), the macrolides (i.e. carbomycin, clindamycin, and erythromycin) and lincomycin (cf. 1). All of these antibiotics except chloramphenicol (53) have been shown to be effective inhibitors of protein synthesis *in vitro* in a system employing purified *Chlamydomonas* chloroplast ribosomes, supernatant factors from whole cells, synthetic poly(U) and poly(C) nucleotide templates, and radioactive amino acids (17, 54-58). Chloramphenicol exhibits only a low level of inhibition in the poly(U) phenylalanine system (53). All of these antibiotics except neomycin and chloramphenicol are lethal to *C. reinhardtii* at appropriate concentrations *in vivo*. Streptomycin causes misreading of isoleucine for phenylalanine in the poly(U) system just as it does in bacteria (56,57). Subunit exchange experiments performed with wild type chloroplast ribosomes and chloroplast ribosomes from antibiotic dependent and resistant mutants demonstrate that the antibiotics have the same subunit specificities in this system as with *E. coli* ribosomes (17,55,58). Lee and Evans (59) found that the small subunit of the *Euglena* chloroplast ribosome was active in the poly(U) phenylalanine system in combination with the large subunit from *E. coli* and *E. coli* supernatant factors. However, the reciprocal combination was inactive. Grivell and Walg (60) found that both possible hybrid combinations between spinach chloroplast and *E. coli* subunits were active in the poly(U)-phenylalanine system.

Both free and thylakoid membrane bound chloroplast polysomes have been isolated from algae and higher plants (61-75). Chloroplast polysomes are usually small (N=2-6) whether free or bound to membranes (cf. 72-74). Margulies and Michaels (70) calculated that in log phase cells of *Chlamydomonas* about 5% of the chloroplast ribosomes occurred as free polysomes in the stroma, while 25% were present as polysomes bound to thylakoid membranes. In higher plants estimates of the fraction of thylakoid membrane bound ribosomes range from 18% (75) to 50% (63). Chua *et al*. (72) showed that the ratio of free to membrane bound chloroplast ribosomes in *Chlamydomonas* depends strongly on growth conditions. In cells growing synchronously on a light-dark cycle there are virtually no membrane bound chloroplast ribosomes during the dark cycle while the fraction increases to 20-30% of the total in the light. Chua *et al*. (65,72) and Margulies and Michaels (66) also showed that isolation of membrane bound chloroplast polysomes depends critically on conditions. If cells of *Chlamydomonas* are cooled rapidly or inhibitors of elongation (*e.g*. chloramphenicol, erythromycin, spectinomycin) are added prior to isolation of chloroplast ribosomes, the ratio of thylakoid bound to free chloroplast ribosomes increases dramatically. Thus, the ratio of free to membrane bound chloroplast ribosomes depends both on growth conditions and the isolation procedure used.

Free chloroplast ribosomes are primarily engaged in synthesizing the large subunit of the CO_2 fixing enzyme ribulose bisphosphate carboxylase (M_r~55,000) and possibly other soluble proteins of the chloroplast stroma (5,73). The ribulose bisphosphate carboxylase enzyme (M_r~500,000) is an aggregate of eight copies each of small and large subunits and constitutes 50% or more of the total soluble protein of green plant leaves (cf. 1). The small subunit is made in the cytoplasm and has an M_r of ~ 12,000 (cf. 6). Gelvin and Howell (73) used antibodies against the carboxylase large subunit of *Chlamydomonas* to precipitate polysomes bearing nascent large subunit polypeptide chains, and found that these polysomes contained only two to five ribosomes.

Membrane-bound polysomes are presumed to be involved in synthesis of those hydrophobic polypeptides of the thylakoid membrane known to be made on chloroplast ribosomes (cf. 6). In *Chlamydomonas* approximately 40-50% of these ribosomes can be released from the membranes by high salt alone, but the remainder are released only by high salt plus puromycin treatment (72). Margulies *et al*. (71) reported that when isolated thylakoid membranes from *Chlamydomonas* bearing attached chloroplast polysomes were allowed to incorporate labelled amino acids, radioactivity was found associated with the thylakoid membranes following release of the ribosomes with puromycin. Margulies *et al*. interpreted their results as demonstrating the vectorial release of the nascent polypeptides synthesized on the membrane bound ribosomes into the thylakoids. Michaels and Margulies (69) further characterized the membrane polypeptides labelled in this *in vitro* incorporating system. Although a large portion of the membrane-bound protein occurred as nascent polypeptides, at least four products of discrete molecular weight were observed, with the major *in vitro* product being a polypeptide of M_r 23,000. Precisely how these *in vitro* products relate to thylakoid membrane polypeptides known to be made on chloroplast ribosomes in *Chlamydomonas* *in vivo* from inhibitor studies is not clear (6,

76). As Chua et al. (72) have pointed out, the molecular events involved in the specific interaction between polysomes and thylakoid membranes are largely unknown. This interaction could involve a "signal" as in the case of polypeptides produced for secretion (77), or could depend largely on the hydrophobicity of the membrane polypeptides, or could involve a preparatory step such as proteolysis (78).

Chloroplast ribosome biogenesis can be influenced by a variety of factors. Feierabend (79,80) has found that chloroplast ribosome formation in rye, barley, oats, peas, and wheat is prevented specifically by growth of the seedlings at $32^{\circ}C$ while at $22^{\circ}C$ chloroplast ribosome formation proceeds normally. Feierabend (81-83) has used the rye system to investigate the sites of synthesis of different chloroplast components, since at $32^{\circ}C$ those chloroplast constituents present must be derived entirely from protein synthesis on cytoplasmic ribosomes.

In algae light is known to be important in the formation of both chloroplast and cytoplasmic ribosomes. Cohen and Schiff (84) made a detailed study of the effect of blue and red light on ribosome formation during light-induced chloroplast development in Euglena. In the case of cytoplasmic ribosomes the net amount of cytoplasmic rRNA did not change during white light-induced chloroplast development of dark-grown wild type cells, but $^{32}PO_4$ was incorporated with a peak rate at 24-36 hours, indicating that turnover of cytoplasmic rRNA was occurring during development to the extent of 60% at 36 hours. Labelling of cytoplasmic rRNA was induced by blue, but not red or green light both in dark-grown wild type cells and cells of the aplastidic mutant W_3BUL. Although the extent of labelling was lower in the mutant, the results indicate that the blue light receptor present in W_3BUL is not connected with protochlorophyll(ide). Plastid rRNA increased from about 2 to 25% of the total cellular rRNA between 0 and 48 hours after chloroplast development was initiated and did not increase thereafter. Labelling of plastid rRNA was enhanced by both blue and red light, and the ratio of effectiveness of blue or red light to green light corresponded well with the relative absorption maxima of the protochlorophyllide holochrome from beans. Cohen and Schiff observed that these results are consistent with other data indicating that certain cellular changes which accompany chloroplast development (e.g. increase in cytoplasmic rRNA formation) are under the control of a blue light receptor located outside of the chloroplast while events mediated by the chloroplast itself (e.g.

chloroplast rRNA formation) are under the control of the protochlorophyll(ide) system which is localized in the chloroplast.

In cells of *Chlamydomonas reinhardtii* growing synchronously on a 12-hour-light-12-hour dark cycle, Cattolico *et al.* (85) and Wilson and Chiang (86) found that both chloroplast and cytoplasmic rRNAs are synthesized extensively during the light period. Wilson and Chiang calculated that only 10% of the chloroplast and cytoplasmic rRNAs were made during the dark period. These results are certainly consistent with an inductive effect of light on rRNA synthesis. Steup and Ssymank (87) have investigated the effect of blue and red light on rRNA formation in cells of *C. reinhardtii* growing synchronously under 14 hours light-12 hours dark. Steup and Ssymank found that blue light enhanced incorporation of $^{32}PO_4$ into cytoplasmic and chloroplast rRNA. However, red light in equal quantum flux resulted in the same incorporation as the dark control. Since the *Euglena* experiments showed that blue light is three times as effective as red light in the induction of chloroplast rRNA formation, these experiments clearly should be repeated with higher red light intensities. Nonetheless they do document the blue light effect on chloroplast and cytoplasmic rRNA formation in *Chlamydomonas*. Blue light induction of cytoplasmic rRNA synthesis has also been demonstrated in *Chlorella* by Steup *et al.* (88,89). Matsuda (91) examined synthesis of chloroplast rRNA during light-induced greening in cells of the *y-1* mutant of *Chlamydomonas*, which lacks the ability to convert protochlorophyll(ide) to chlorophyll(ide) in the dark (cf. 90). He found that most chloroplast rRNA synthesis occurred during the middle phase of the regreening process, and that chloroplast ribosomes increased from about 35% of the total in dark-grown cells to 47% of the total at the end of regreening. Matsuda observed increases of similar magnitude when dark-grown wild-type cells, which accumulate chlorophyll and contain a normal chloroplast (cf. 16,90), were compared to cells grown in the light. These observations confirm those made previously on wild-type and *y-1* cells by Bourque *et al.* (16).

Siersma and Chiang (92) and Martin *et al.* (93) have demonstrated that nitrogen deprivation in *C. reinhardtii*, which results in gametogenesis (cf. 1), also leads to a massive degradation of cytoplasmic and chloroplast ribosomes. According to Martin *et al.*, 50% of the cytoplasmic ribosomes and 80% of the chloroplast ribosomes were degraded during this period, but some new chloroplast and cytoplasmic ribosomes were also being synthesized. Turnover of both classes of newly synthesized ribosomes was observed, with the rate for chloroplast ribosomes being distinctly faster than for cytoplasmic ribosomes.

Chloroplast rRNAs: Genes and Processing

The chloroplast rRNA genes are localized in chloroplast DNA (cf. 1). Their organization has been studied in detail in Chlamydomonas (21,94, 95), Euglena (26,96-99), maize (100), and spinach (101-103). Except in the case of Euglena there are two sets of rRNA genes per chloroplast DNA molecule which are localized in two inverted repeat regions widely separated from each other on the chloroplast genome (Figure 1).

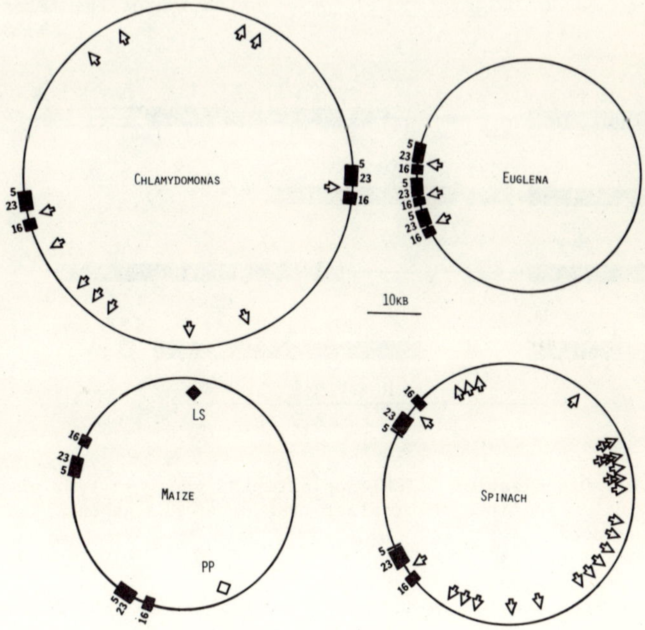

Figure 1. Location and orientation of rRNA genes in the chloroplast genome of four different plants. Arrows indicate the positions of 4S RNA genes in Chlamydomonas and Euglena and tRNA genes in spinach. In maize LS = gene for large subunit of ribulose bisphosphate carboxylase and PP = gene for the maize photoprotein. See text for references.

Bedbrook and Kolodner (104) have summarized these data recently pointing out that the size of the inverted repeat regions ranges from 19 kb in Chlamydomonas to 24.5 kb in maize and spinach. They point out that at most only 20% of the inverted sequence can code for chloroplast rRNA. In Euglena there are three sets of rRNA genes located very close together in the chloroplast genome arranged in straight tandem repeats (Figure 1). Since the organization of chloroplast rRNA genes has so far been reported in so few plants, the extent to which Euglena is exceptional remains to be seen. For example, Kolodner and Tewari (105) report the absence of the two inverted repeat regions in pea chloroplast DNA. This would imply a different organization for the rRNA genes in pea chloroplast DNA as well.

The organization of single sets of chloroplast rRNA genes is shown in Figure 2. In Chlamydomonas, the organization of the chloroplast rRNA

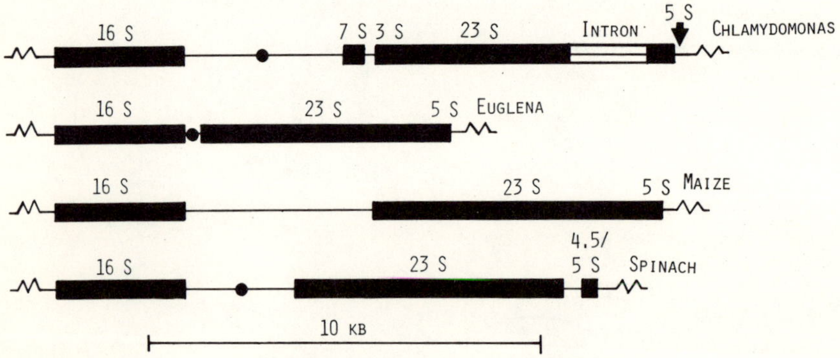

Figure 2. Organization of chloroplast rRNA genes in four plants. Small circles indicate chloroplast tRNA or 4S RNA genes. See text for references.

genes has been studied in detail by Rochaix and Malnoë (21) and the 4S RNA genes coded by the chloroplast genome have been mapped by Malnoë and Rochaix (95). In transcriptional mapping experiments, Surzycki and Rochaix (106) found that the direction of transcription was from 16S to 23S. Rochaix and Malnoë (21) demonstrated that the gene order in the rRNA region is 16S-7S-3S-23S-5S and that all rRNA genes are transcribed from the same strand. The 16S and 7S genes are separated by a spacer of 1.7 kb which Malnoë and Rochaix (95) have shown to contain at least one 4S RNA gene which is also transcribed from the same strand as the rRNA genes. The 7S rRNA is separated from the 3S and 23S rRNA by a short gap of 0.15 kb. Electron microscopy of duplexes formed between 23S rRNA and a cloned EcoRl fragment, R24, which contains the entire 23S rRNA sequence, reveals a single stranded loop 940 bp in length which is 0.27 kb from the 5' end of the 23S rRNA coding strand. R looping experiments confirmed that the single stranded loop was DNA since the DNA in this sequence was not displaced by mature 23S rRNA. Therefore, the 23S gene contains an intervening sequence. Distal to the 23S gene is the gene for chloroplast 5S rRNA. Whether there is a small spacer between the 23S gene and the 5S gene remains to be established.

Rochaix and Malnoë (21) point out that the organization of the chloroplast rRNA genes suggests that they may well be transcribed as a single precursor. So far, however, only small chloroplast rRNA precursors have been observed in *Chlamydomonas*. Miller and McMahon (107) identified a precursor of M_r 1.14 x 10^6 for 23S rRNA which has an M_r of 1.07 x 10^6 and two small precursors for 16S rRNA (M_r 0.54 x 10^6) of M_r 0.64 and 0.56 x 10^6 respectively. Clearly, at least one more precursor must exist since the 23S precursor containing the intervening sequence should have a minimum M_r of 1.38 x 10^6.

The organization of the chloroplast rRNA genes in *Euglena* has been studied independently by Gray and Hallick (96,99), Rawson *et al*. (97) and Jenni and Stutz (98) all of whom have come to similar conclusions (Figure 2). Rawson *et al*. (97) estimate the size of each of the three straight tandem repeat units at 5.4 kb and Gray and Hallick (96,99) at 5.6 kb. Gray and Hallick (99) have recently shown that the chloroplast 5S rRNA gene maps adjacent to the 23S gene and that there is a 4S RNA gene which maps in the small spacer between the 16S and 23S genes. Gray and Hallick (99) estimate that the 16S gene, the intergenic spacer and the 23S gene are contained completely within a segment of 4.8 kb. The size of the intergenic spacer, which contains the 4S RNA gene, is calculated to be 300 bp. The distance between each repeating unit was thought to be about 0.8 kb by Rawson *et al*. (97). Scott (108) found that the mature 16S and 23S rRNA species are made as precursors of M_r 0.64 and 1.16 x 10^6 respectively. Hallick *et al*. (109) report that the entire rRNA gene sequence may be transcribed initially into a primary transcript of $>$5000 nucleotides.

Bedbrook *et al*. (100) have shown in maize that the 16S, 23S, and 5S rRNA are transcribed from the same strand. There is no evidence of a spacer between the 5S and 23S genes, but the 16S and 23S genes are separated by a spacer of approximately 2.1 kb (Figure 2). Although Haff and Bogorad (110) have shown that maize chloroplast DNA contains 20-26 tRNA cistrons on the basis of DNA-RNA hybridization experiments, these have not so far been mapped with respect to the rRNA genes. Electron microscopy of rRNA-DNA hybrids has not given any evidence of loop structures so there is no reason to believe that the 16S and 23S genes in maize chloroplast DNA contains intervening sequences.

In the case of spinach Herrmann _et al_. (101) originally reported that the 16S and 23S rRNA genes were not located in the two inverted repeat regions of the chloroplast genome. However, Whitfeld _et al_. (102) and Crouse _et al_. (103) showed that the original conclusion was in error and that the rRNA genes are indeed localized in the two inverted repeats. Crouse _et al_. report that the order of the rRNA genes is: 16S-"spacer"-23S-4.5S-5S and that transcription is from 16S to 23S (Figure 2). A tRNA is located in the spacer region (111) which is 2.2 kb in length (104). Hartley and Ellis (112) and Bohnert _et al_. (113) have shown that isolated spinach chloroplasts synthesize a transcript of the rRNA genes of M_r 2.7 x 10^6 which is then processed in several steps to the mature rRNAs. This large precursor includes the spacer. The smaller precursors include a p23S (M_r 1.54 x 10^6) and a p16S (M_r 0.7 x 10^6) which have been characterized in detail by Mache _et al_. (114) under fully denaturing conditions. The fact that large rRNA precursors can be obtained from isolated spinach chloroplasts strongly suggests that processing may proceed more slowly in the isolated plastid system perhaps because of a deficiency of processing enzymes made on cytoplasmic ribosomes.

Sites of Synthesis of Chloroplast Ribosomal Proteins

The site of synthesis of specific organelle proteins has proved to be an accurate guide of where these proteins are coded (cf. 1,2,4). Only one detailed study of the sites of synthesis of individual chloroplast ribosomal proteins has so far been published. In _Euglena_, Freysinnet (42) has made use of specific inhibitors of chloroplast and cytoplasmic protein synthesis to show that at least nine chloroplast ribosomal proteins are made within the chloroplast while at least 12 are made on cytoplasmic ribosomes. The proteins were labelled with $^{35}SO_4$ following addition of either cycloheximide or lincomycin to inhibit cytoplasmic or chloroplast protein synthesis respectively. Lincomycin has the advantage that it blocks initiation of protein synthesis on chloroplast ribosomes rather than polypeptide chain elongation, permitting membrane bound chloroplast polysomes to run off normally and not become jammed onto the thylakoid membranes as they would if chloramphenicol were used (72).

In addition to those proteins whose synthesis was clearly inhibited either by lincomycin or cycloheximide, Freyssinet found a group of six proteins whose synthesis seemed to be inhibited by both antibiotics and 12 proteins whose synthesis was not blocked by either inhibitor. Those proteins affected by both inhibitors may either depend on a tightly coordinated assembly step requiring a second partner protein made elsewhere, or may require special processing or modification for assembly. For example, a given ribosomal protein might really be made in the chloroplast, but its integration into the chloroplast ribosome might depend on a cytoplasmically synthesized ribosomal protein or processing enzyme. Antibodies to specific proteins were not used in these experiments, but instead intact ribosomes were used as the source of ribosomal proteins analyzed. Absence of label in a protein could therefore mean either that it was not synthesized or that it was not integrated into the ribosome.

The group of 12 ribosomal proteins insensitive to either inhibitor is more difficult to explain. Freyssinet suggests that his inability to detect inhibition in these cases could be a consequence of experimental design or might be related to physiological effects (*e.g.* pool sizes). Despite these "ambiguous" proteins the results published by Freyssinet clearly show that, in contrast to mitochondria where only a single protein of the small subunit is made inside the organelle (see p. 34), a group of several ribosomal proteins is made within the chloroplast.

An earlier study by Ellis and Hartley (115) with greening pea apices showed that lincomycin blocked accumulation of chloroplast ribosomes, but not chloroplast rRNA synthesis. Similarly, Honeycutt and Margulies (116) observed that cycloheximide, but not chloramphenicol blocked ^{14}C arginine incorporation into *Chlamydomonas* chloroplast ribosomes. The results reported in these papers, discussed in more detail by Gillham *et al.* (6), very likely represent a gross oversimplification of the true situation, since neither study attempted to distinguish the sites of synthesis of individual chloroplast ribosomal proteins.

Genetic Analysis of Chloroplast Ribosomes

The genetic control of chloroplast ribosome biogenesis has been better studied in Chlamydomonas than in any other plant. A variety of chloroplast and nuclear mutations affecting chloroplast ribosome structure and function have been isolated, and their properties have been summarized in recent reviews (1,6,39,117).

Mutations affecting chloroplast ribosome function in Chlamydomonas characterized to date either cause defects in chloroplast ribosome assembly, or confer resistance to antibiotics which block protein synthesis on bacterial ribosomes (Table 1). These mutations are scattered between the chloroplast and nuclear genomes of the alga. By mapping and allele testing, Boynton et al. (118) and Harris et al. (57) have identified two recombinationally distinct loci for erythromycin resistance (er-u), one for neamine/spectinomycin resistance (nr/spr-u), and four for streptomycin resistance (sr-u) which map in the order: er-u-1a--er-u-37--nr/spr-u--sr-u-sm-3a--sr-u-2-60--sr-u-sm3--sr-u-sm2. Bolen et al. (119) have found that four chloroplast mutations to streptomycin dependence (sd-u) map in three recombinationally distinct loci whose genetic relationship to the seven loci previously mapped is currently under study. Chloroplast mutations to neamine dependence (nd-u) remain to be mapped. Nuclear mutations to streptomycin resistance (sr-1) map in a single gene (57) while nuclear mutations conferring resistance to erythromycin (ery-M1, ery-M2, ery-M3) map in three unlinked genes (120). Finally, mutations in four nuclear genes (es) suppress ery-M1 mutants (121).

Schlanger and Sager (55), in subunit exchange experiments using the poly (U) directed phenylalanine incorporating system, were the first to show that chloroplast mutations resistant to neamine, spectinomycin and streptomycin conferred resistance on the small subunit of the chloroplast ribosome while mutations resistant to carbomycin and cleocin (clindamycin) caused the large subunit to become resistant to these antibiotics. Subsequently, Conde et al. (56) improved this assay to include the poly (U)-isoleucine misreading reaction for streptomycin resistance and a poly (C)-proline assay for erythromycin resistance. Harris et al. (57) have used these in vitro assays to assess the drug resistance of chloroplast ribosomes from many nuclear and chloroplast mutations of C. reinhardtii while Bartlett et al. (17) have systemati-

Table 1. Summary of nuclear and chloroplast mutations in Chlamydomonas reinhardtii known to affect the function or assembly of chloroplast ribosomes.

					Effect of Mutation on Chloroplast Ribosomes		
					Levels of antibiotic resistance		
Type of alteration	Locus designation	Location of gene	Number of Alleles	Reference	In vitro	In vivo	Other phenotypic modifications
Antibiotic resistance							
Streptomycin	sr-1	N	3	57	Low(S)	-	---
	sr[1]	N	1	6	Low	-	Unnamed sr mutation reported to alter a ribosomal protein
	sr-u-2-23	C	13	17,57	Low to high(S)	Low to high	---
	sr-u-sm3	C	1	17,57	Intermediate(S)	-	---
	sr-u-2-60	C	2	17,57,279	High(S)	High	sr-u-2-60 affects assembly of the small subunit
	sr-u-sm2	C	2	17,57,124	Very high(S)	-	sr-u-sm2 alters a protein of the small subunit?
	sr-u-35[1]	C	1	17,45	High(S)	High	sr-u-35 reported to alter a protein of the large subunit?
	sr[1]	C	1	6	High	-	---
	sd-u-3-18[1]	C	1	119,279	High(S)	High	sd-u-3-18 affects assembly of the small subunit
	sd-u-ts-A8	C	1	119	High(S)	High	Requires streptomycin at 33°
	sd-u-2-24	C	2	119	High(S)	High	---
Spectinomycin or neamine	nr/spr-u[2]	C	7	17,57	Low to high(S)	Low to high	spr-u-1-27-3 alters a protein of the small subunit?
Erythromycin	ery-M-1	N	4	123,125	-	-	All 4 mutant alleles alter the same protein of the large subunit
	ery-M-2	N	4	123,125	-	-	ery-M-2d alters a protein of the large subunit?
	er-u-1a	C	2	17,57,123	High(L)	High	er-u-1a alters a protein of the large subunit?
	er-u-37[3]	C	5	17,57	High(L)	High	---
Carbomycin	car[1]	C	1	55	High(L)	-	---
Assembly defective							
Small subunit	cr-1	N	1	132,280	-	-	Accumulates large subunits, deficient in 70S monomers
	cr-2	N	1	132	-	-	Accumulates large subunits, deficient in 70S monomers
	cr-3	N	1	132	-	-	Accumulates large subunits, deficient in 70S monomers
	cr-5	N	1	39	-	-	Accumulates large subunits, deficient in 70S monomers
	ac-u-d	C	1	131	-	-	Accumulates large subunits, deficient in 70S monomers
	ac-u-h	C	1	131	-	-	Accumulates large subunits, deficient in 70S monomers
Both subunits	ac-20	N	1	132,280,281	-	-	Deficient in 70S monomers
	cr-4	N	1	132	-	-	Deficient in 70S monomers
	cr-6	N	1	39	-	-	Accumulates 40-60S material, deficient in 70S monomers
	cr-7	N	1	39	-	-	Accumulates 40-60S material, deficient in 70S monomers

N, gene located in nucleus; C, gene located in chloroplast; S, resistance in vitro localized to small subunit; L, resistance in vitro localized to large subunit; ?, data not entirely convincing; superscript 1, allelic relationship to other genes not determined; 2, neamine and spectinomycin resistant mutants appear to be allelic based on recombination analysis, 48 independently isolated spectinomycin resistant mutants all appear to be alleles at this locus in preliminary allele tests; 3, cle mutant studied by Sager (19) appears to be allele at this locus.

cally examined the chloroplast ribosomal subunits affected by antibiotic resistance mutations mapping at the seven chloroplast loci defined by Boynton et al. (118) and Harris et al. (57). The conclusion from all of these experiments is that chloroplast mutations to antibiotic resistance always confer resistance on chloroplast ribosomes with the subunit affected being the same as in E. coli. That is, mutations resistant to neamine, spectinomycin and streptomycin are small subunit mutations and mutations resistant to carbomycin, clindamycin and erythromycin are large subunit mutations. Finally, Bolen et al. (119) have recently reported that chloroplast mutations to streptomycin dependence confer resistance on the small subunit as they do in E. coli.

In several cases (45,122-124) specific chloroplast gene mutants have been reported to alter particular proteins of a given chloroplast ribosomal subunit, although each of these cases needs to be substantiated further. In one instance (45) the protein alteration is in the large subunit whereas resistance resides in the small subunit (17). To prove unequivocally that a specific gene codes for a given ribosomal protein, a minimal requirement is the demonstration that a series of allelic mutants at the same gene locus all affect that same protein. So far this has been done only by Davidson et al. (125) for the nuclear mutants at the ery-M1 locus (see below). Despite the fact that no chloroplast locus which mutates to antibiotic resistance has so far been rigorously shown to code for a chloroplast ribosomal protein, the likelihood that the resistance mutations arose in the rRNA genes seems small. Rochaix and Malnoë (21) showed that there are two sets of rRNA genes per chloroplast DNA molecule. Only if one of these sets were not expressed or if all of the mutations were semidominant or dominant would such mutations be expressed.

Davidson et al. (125) have demonstrated that the nuclear ery-M1 gene codes for a specific protein (LC6) of the large subunit. They found that three alleles at this locus alter the charge of LC6 while the fourth (ery-M1b) causes a deletion of 30% of this protein. Hanson and Bogorad (126) have shown that ery-M1 mutations are semidominant in terms of resistance level in diploids with wild type, and that the population of chloroplast ribosomes from such vegetative diploids contains both mutant and wild type forms of protein LC6. These results further strengthen the hypothesis that the ery-M1 gene codes for LC6 since if the ery-M1 gene were responsible for modifying enzymatically the structure of LC6, one would expect mutations in this gene to be recessive to wild type rather than being co-dominant as observed. Davidson et al. (120,121) have presented evidence that mutations in the ery-M1, ery-M2, and ery-M3 loci all reduce the ability of the large subunit of chloroplast ribosomes to bind ^{14}C labelled erythromycin while double mutants of several of the es mutants and ery-M1b enhance binding of ^{14}C erythromycin to chloroplast ribosomes compared to ery-M1b alone. So far the subunit responsible for suppression remains to be determined. Mets and Bogorad (123) presented evidence, in need of confirmation with other mutants, showing an alteration of a basic, low molecular weight protein of the large subunit in an ery-M2 mutant. Hanson and Bogorad (127) have shown that ery-M2 mutants are also cold sensitive and deficient in chloroplast ribosomes. The effect of specific antibiotic resistance mutations on antibiotic-induced binding of chloroplast polysomes to the thylakoid membranes has also been studied (56,128). The nuclear sr-1 mutations are known from subunit exchange experiments (17) to confer resistance on the small subunit.

To summarize, in *Chlamydomonas*, mutations conferring resistance on chloroplast ribosomes to a given antibiotic are scattered between two spatially separated genomes within the cell. Possibly, mutations conferring resistance to the same antibiotics affect proteins in the same ribosomal neighborhood. If so, there are at least two such neighborhoods, one involving streptomycin binding in the small subunit and the other involving erythromycin binding in the large subunit. One nuclear and four chloroplast loci can mutate to confer streptomycin resistance on the chloroplast ribosome (17,57,118) and three other chloroplast loci confer streptomycin dependence (119). In addition, nuclear mutations which raise (129) and lower (130) the resistance level conferred by streptomycin resistant mutations are known. If these also affect the ribosomal neighborhood which binds streptomycin, as many as seven chloroplast and three nuclear loci may specify components of this site or their processing and modification (Table 1). In the large subunit, an erythromycin resistance neighborhood has been identified (17,120,123). Mutations in two chloroplast and three nuclear loci confer resistance to erythromycin (17,120,123). Mutations in four nuclear genes also suppress the resistance of ery-M1 mutations (121). Potentially, then, as many as nine different genes may be involved in altering the properties of the erythromycin binding site.

Mutants deficient in chloroplast protein synthesis are viable in *C. reinhardtii* when provided with acetate as a carbon source (cf. 6). Such mutations have a characteristic, pleiotropic syndrome of photosynthetic defects as a result (cf. 6,16,117,131,132). So far all such mutants isolated have proved to have defects in chloroplast ribosome assembly and eight are nuclear gene mutants (6,39,117,132) while two are mutations in the chloroplast genome (131). The ten assembly defective mutants isolated thus far fall into three phenotypic classes, all of which are deficient in 70S chloroplast ribosome monomers (Table 1). Class I includes the nuclear mutants ac-20 and cr-4 which are deficient in both small and large ribosomal subunits. Class II is represented by the nuclear mutants cr-1, cr-2, cr-3, and cr-5 and the chloroplast mutants ac-u-d and ac-u-h, all of which accumulate 54S particles that appear to be large subunits of the chloroplast ribosome. Class III includes the nuclear mutants cr-6 and cr-7 which accumulate heterogeneously sedimenting material in the 40-60S region of sucrose gradients.

Genetic analysis indicates that the eight nuclear mutants deficient in chloroplast ribosomes complement one another in diploids, and, with the possible exception of ac-20 and cr-7, are nonallelic by recombination analysis (6,39,132). By comparing the photosynthetic capacity and chloroplast ribosome profiles of individual mutants, of their double mutants, and of their diploids (with each other and wild type), Harris et al. (132) have formulated a working model for the assembly of chloroplast ribosomes analogous to the formal schemes used to deduce enzymic pathways from the behavior of auxotrophic mutants. Shepherd et al. (131) have found that the two chloroplast mutants, ac-u-d and ac-u-h, also recombine and thus define two separate loci. Both the nuclear and chloroplast assembly defective mutations have a strong tendency to become suppressed by nuclear gene mutations (131,132).

Finally, Gyurgan et al. (133) have recently reported the isolation of a novel Chlamydomonas mutant called y-27 which has the pattern of inheritance characteristic of chloroplast genes and possesses two alterations in total chloroplast ribosomal proteins separated in two-dimensional gels. One high molecular weight protein present in wild type is virtually missing and is replaced by a second protein not found in wild type. Whether the two alterations are related remains to be seen.

Antibiotic resistant mutants have also been isolated in Euglena (134-136), Scenedesmus (137), and tobacco (138-141). Schiff and his colleagues (134,135) reported the isolation and characterization of a streptomycin resistant mutant of Euglena that was selected by its ability to grow photosynthetically at a streptomycin concentration (0.05%) that normally causes irreversible loss of plastids. The chloroplast ribosomes, unlike those of wild type, did not bind radioactively labelled dihydrostreptomycin. Freyssinet (136), on the basis of two-dimensional gel electrophoresis, reported that the mutant lacked four acidic proteins of the 30S subunit present in wild type. He suggested that these proteins might be part of the streptomycin binding site. This drastic alteration is surprising to say the least and Freyssinet points out that the proteins in question are present in low amounts even in wild type and stain faintly. The possibility that these proteins are lost during extraction of the mutant as opposed to wild type chloroplast ribosomes needs to be explored. The streptomycin resistant mutants in Euglena very likely arose in the chloroplast genome since the nuclear genome is probably at least diploid and nuclear mutants will not segregate during vegetative growth whereas plastid mutations will (see reference 1 for a discussion of these arguments).

Maliga et al. (138,139) isolated a streptomycin resistant mutant (SR1) from haploid callus cultures of tobacco and found, following regeneration, that resistance was transmitted maternally suggesting that the mutation may reside in the chloroplast genome. Bourque et al. (140), in an abstract, reported that 30S ribosomal subunits from the mutant bound one-tenth as much ^{3}H streptomycin as wild type 30S subunits and that at least two differences between wild type and mutant 30S subunit proteins could be detected in electrofocusing gels. A full account of these interesting results is awaited. Umiel and Goldner (141) have isolated other streptomycin resistant mutants in diploid callus culture of tobacco, but so far the chloroplast ribosomes have not been characterized for resistance. Bourque and Wildman (142) also reported two protein differences between the large subunits of chloroplast ribosomes of Nicotiana tabacum and N. glauca on one-dimensional gel electrophoresis. From genetic experiments, the authors concluded that both proteins were coded by nuclear genes. However, careful repetition of these experiments using improved gel techniques is desirable before these conclusions are taken for granted.

Plastid ribosome deficient mutants have been described in barley (143), maize (144,145), and Pelargonium (146-149). The first such mutant to be identified as being deficient in chloroplast ribosomes was the nonMendelian iojap mutant of maize. Shumway and Weier (144) showed by electron microscopy that plastids from green tissue in this mutant were normal and contained plastid ribosomes whereas the plastids from white tissue were abnormal and lacked plastid ribosomes. The non-Mendelian mutants of barley and Pelargonium, characterized by Börner, Hagemann, and their colleagues (143,146-149), are similar in that plastids from white tissue are abnormal and lack plastid ribosomes. In all these cases it is difficult to determine whether the primary defect is related directly to chloroplast ribosome formation or to some more general phenomenon (e.g. deletions in plastid DNA or failure to transcribe plastid DNA). These mutants have proved useful in determining which plastid components can be made in the absence of chloroplast ribosomes in higher plants and the results are to be compared to those of Feierabend (see p. 9) where loss of plastid ribosomes is induced in higher plants such as rye by growth at high temperature. Recently, Gyurgan et al. (145) have found that two Mendelian mutants of maize, one of which accumulates ζ-carotene and the other which accumulates lycopene, are deficient in chloroplast ribosomes. Both mutants also show a syndrome of other chloroplast defects which the authors attribute to a low level of chloroplast protein synthesis. In contrast to the other mutants just described, these two maize mutants may really be assembly defective in chloroplast ribosome formation since they do, in fact, contain some chloroplast ribosomes.

MITOCHONDRIAL RIBOSOMES

Structural Characteristics

The structural characteristics of mitochondrial ribosomes and their rRNAs have recently been reviewed by O'Brien and Matthews (13) and by Buetow and Wood (3). Because the previous reviews are comprehensive, we present here only a brief summary of these topics focusing on salient points. Mitochondrial ribosomes have been found to be a structurally diverse class. As a group, they contain higher protein to RNA ratios than *E. coli* ribosomes and mitochondrial rRNAs are characterized by an unusually low degree of secondary structure. Both properties are reflected in anomalous electrophoretic and sedimentation behaviors which have traditionally complicated molecular weight estimates. Table 2 shows that four classes of mitochondrial ribosomes can be distinguished by structural characteristics, and, as discussed by O'Brien (150), additional classes can be distinguished by taking into account the GC content of mitochondrial rRNAs. Mitochondrial ribosomes of animal cells are the same size or slightly larger than *E. coli* ribosomes in terms of molecular weight and volume (151). Their low sedimentation coefficient (55S) is due to low buoyant density (reflecting the high protein to RNA ratio) and possibly to larger volume (higher frictional coefficient) than *E. coli* ribosomes (152,153). Mitochondrial ribosomes of *Tetrahymena* and *Paramecium* have the unusual characteristic of dissociating at low Mg^{+2} concentrations into two subunits of equal size (55S) which appear indistinguishable by electron microscopy (154-157). Mitochondrial ribosomes from most organisms con-

TABLE 2. Sedimentation values of mitochondrial ribosomes and molecular weights of mitochondrial rRNAs.

	Monomer	Small subunit	Large subunit	rRNAs (Mdal)	
Animals	55-60S	30-35S	40-45S	0.3-0.5	0.5-0.8
Ascomycetes	70-75S	30-40S	50S	0.7	1.2-1.3
Ciliate Protozoans (*Tetrahymena*, *Paramecium*)	80S	55S	55S	0.5	0.9
Higher Plants	77-80S	40S	60S	0.65-0.8	1.1-1.3

For references see Borst and Grivell (159) and Buetow and Wood (3).

spicuously lack small (5S or 5.8S) RNA species (3,158,159). However, mitochondrial ribosomes from higher plants contain a unique 5S RNA distinguished from cytoplasmic 5S RNA by fingerprint analysis (160-162).

Mitochondrial rRNAs from different organisms are readily distinguished from cytoplasmic rRNAs by (a) low GC content (*e.g*. 25 to 40% in ascomycetes compared to 50 to 60% for cytoplasmic rRNAs, (b) a low degree of secondary structure and (c) distinctive patterns of methylated nucleotides (see 3,159). In yeast and *Neurospora*, the small rRNA appears to be completely unmethylated whereas the large RNA contains only two methylated ribose moieties (163-165). In addition, each of the yeast rRNAs has been found to contain (probably) one pseudouridine residue (163). In hamster cells, the small rRNA contains five methylated nucleotides including two residues of m_2^6Ap and one residue of the unusual methylated base m^4Cp, the natural occurrence of which had not been reported previously (166,167). The hamster cell large rRNA contains three methylated ribose moieties and the major methylated sequence is UmpGmpUp which is also characteristic of cytoplasmic 28S rRNA (166,168).

From indirect evidence, rRNA methylation has been inferred to play a role in RNA processing in eukaryotes (169) and in protein synthesis in prokaryotes (see references in Schlessinger, 170). Although the low level of methylation of the ascomycete mitochondrial rRNAs demonstrates that these processes can be carried out without substantial methylation, the few methyl groups remaining in fungal mitochondrial rRNAs may be those absolutely required for RNA processing or protein synthesis. If the mitochondrial system is in fact unembellished in this sense, it may prove useful in elucidating the role of rRNA methylation. In addition, fungal mitochondrial rRNAs may provide substrates to assay methylases from other cellular compartments or organisms.

As indicated previously, all mitochondrial ribosomes studied so far lack a 5.8S RNA and all but higher plant mitochondrial ribosomes lack a 5S RNA (3,158,159,161,162). Possible explanations for the absence of 5S RNA, first discussed by Lizardi and Luck (158) include: (a) that 5S RNA is lost during the isolation of mitochondria or mitochondrial ribosomes, (b) that 5S RNA is replaced by a 3S or 4S RNA which would be difficult to distinguish from mitochondrial tRNAs in preparations of whole mitochondrial RNA, and (c) that the functional equivalent of 5S RNA remains covalently attached to the 25S RNA. In prokaryotes, 5S RNA is required for reconstitution of 50S subunits and may function in protein synthesis by promoting the binding of tRNAs (171). Since mitochondrial ribosomes lacking 5S RNA carry out poly(U)-directed polyphenylalanine synthesis at rates comparable to those for E. coli ribosomes, it seems unlikely that 5S RNA is merely lost (172). Dubin et al. (173) have identified an unmethylated 3S RNA species in hamster cell mitochondria which is present in ca. 1:1 molar ratio with the rRNAs. However, the 3S RNA species could not be demonstrated in the mitochondrial large subunits. Lizardi and Luck (158) speculated that Neurospora 25S and 5S RNA might be part of the same transcription unit as they are in E. coli, but that mitochondria might lack enzymes required to cleave the two into separate molecules. In that case, characteristic 5S RNA sequences (see Monier, 171) should be present near the 3' end of 25S RNA. This possibility would now be relatively easy to examine by DNA sequencing of restriction enzyme fragments.

Datema et al. (174) have claimed that the conditions used for mitochondrial isolation can have a profound effect on ribosome structure in Neurospora and, by implication, in other organisms. Specifically, they suggest that addition of Mg^{+2} to the mitochondrial isolation medium makes it possible to isolate mitochondrial ribosomes which sediment at 80S and which contain a 5S RNA. Although Datema et al. (174) (see also, 175) claimed that contamination by cytoplasmic 80S ribosomes is negligible in their preparations as judged by analysis of rRNAs and ribosomal proteins, the data are in disagreement with those published by other laboratories. Lizardi and Luck (158) had previously studied the effect of Mg^{+2}-containing buffers during mitochondrial isolation and concluded that all of the 5S RNA in such preparations could be attributed to contamination by cytoplasmic 80S ribosomes. More recently, Michel *et al.* (176) have reached the same conclusion in a detailed appraisal of the work of Datema *et al.* (174).

Mitochondrial Ribosomal Proteins

Mitochondrial ribosomal proteins from many organisms have been analyzed by one- or two-dimensional gel electrophoresis (summarized by Buetow and Wood, 3). Yeast and *Neurospora* mitochondrial ribosomes contain 60 to 70 proteins and, as a group, these have higher molecular weights and are more acidic than cytoplasmic or *E. coli* ribosomal proteins (177-179). Mitochondrial ribosomes from animal cells have been found to contain up to 70 to 100 proteins (153,180), although the possibility remains that these include adventitiously associated nonribosomal proteins. In a particularly careful study, O'Brien *et al.* (181) examined the protein composition of bovine mitochondrial ribosomes as a function of K^+/Mg^{+2} ratio during centrifugation of ribosomal subunits. As expected, the number of proteins decreased with increasing K^+/Mg^{+2}. However, even under the most stringent washing conditions, just prior to losing peptidyl transferase activity, bovine mitochondrial ribosomes still contained about 90 proteins.

In all organisms studied, most mitochondrial ribosomal proteins appear to be totally distinct entities from cytoplasmic proteins as judged by electrophoretic mobility and antigenic sites (see, for example, Hallermayer and Neupert, 182). However, the published data are not sufficiently detailed or "clean" to exclude the possibility that a small number of proteins share antigenic determinants. Judged by amino acid composition, yeast mitochondrial ribosomal proteins appear more distant from prokaryotic than from eukaryotic ribosomal proteins (183). Immunochemical data comparing *Neurospora* and *E. coli* ribosomal proteins (182) do not rule out a limited number of proteins with common antigenic determinants.

Functional Properties

The protein synthetic activities of mitochondrial ribosomes are in some ways analogous to those of prokaryotic ribosomes. A species of F-met-$tRNA_F$ has been identified in mitochondria from a variety of organisms (184-186) and mitochondria have been shown to synthesize F-met-peptides *in vitro* (187,188) and *in vivo* (189). In a number of cases, mitochondrial and bacterial components are interchangeable. *Neurospora* mitochondrial ribosomes recognize, bind, and translocate E. coli F-met-$tRNA_F$ in response to AUG, and E. coli initiation factors can substitute for mitochondrial initiation factors in catalyzing the reaction (190). Also, purified initiator tRNA from Neurospora mitochondria can be aminoacylated and formylated by E. coli enzymes and is capable of initiating protein synthesis on E. coli ribosomes (191). Finally, mitochondrial elongation factors T and G have also been found to be interchangeable with those from E. coli but not with cytoplasmic elongation factors (3,192). However, in contrast to chloroplast ribosomes, yeast mitochondrial ribosomal subunits do not form active hybrid ribosomes with E. coli subunits (60). Also, the mitochondrial elongation factors appear to be nuclear gene products synthesized in the cytoplasm (193, 194).

Mitochondrial ribosomes are sensitive to most inhibitors of bacterial ribosomes, but in some cases, the degree of sensitivity is markedly different, and it is not clear that the inhibitors are acting by the classical mechanism. The peptidyl transferase activity of mitochondrial ribosomes is inhibited by the bacterial inhibitor chloramphenicol but not cycloheximide or anisomycin which inhibit eukaryotic cytoplasmic ribosomes. Studies using isolated ribosomes in vitro show that yeast mitochondrial ribosomes are as sensitive to lincomycin and carbomycin as bacterial ribosomes (195) whereas mammalian mitochondrial ribosomes are 100- to 1000-fold less sensitive (196-198). The study by Denslow and O'Brien (198) appears to be definitive for antibiotics affecting the peptidyl transferase of mammalian mitochondrial ribosomes. There seems to be no simple phylogenetic pattern since mitochondrial ribosomes from Neurospora are similar to mammalian rather than yeast ribosomes with respect to carbomycin sensitivity (196). To date, no detailed study of the effect of aminoglycosides on isolated mitochondrial ribosomes has been published. In yeast, neomycin and paromomycin have been found to be effective inhibitors of mitochondrial protein synthesis in whole cells and in isolated mitochondria (199). However, other aminoglycosides (neamine, kanamycin, streptomycin, viomycin) inhibit only at very high concentrations (100 to 2000 µM) at which they also inhibit cytoplasmic 80S ribosomes (199). Mitochondrial ribosomes have been found to be insensitive to fusidic acid (200, see Buetow and Wood, 3, for additional references), an inhibitor which interacts with elongation factor G from both prokaryotic and eukaryotic ribosomes (201,202).

Mitochondrial ribosomes from yeast and mammalian liver carry out poly(U)-directed polyphenylalanine synthesis at rates near 2000 pmoles per mg RNA per 30 min, comparable to rates for E. coli ribosomes (172, 203,204). However, mitochondrial ribosomes isolated from other organisms by similar procedures have rates of only 4 to 300 pmoles per mg RNA per 30 min (see Borst and Grivell, 159, for references). The extent to which the functional integrity of mitochondrial ribosomes is preserved by current methods of isolation, which necessarily involve the use of detergents for mitochondrial lysis, is not particularly clear. Although rates in the poly(U) assay are high in some cases, there have been no reports that mitochondrial ribosomes can function in in vitro systems to translate mitochondrial mRNAs into recognizable proteins.

Possibly related, it has also proven difficult to isolate mitochondrial polysomes, and, in particular, to make the distinction between true mitochondrial polysomes and aggregated monosomes. Putative mitochondrial polysomes have now been isolated from a number of organisms (see 3 for review). However, only the mitochondrial polysomes from Euglena sediment as a series of discrete peaks on sucrose gradients with higher order polysomes dissociated at low Mg^{+2} concentrations or by treatment with RNase (74,205). By contrast, both HeLa cell and yeast mitochondrial "polysomes" do not sediment as a series of discrete peaks on sucrose gradients (206,207) and those from HeLa cells possess atypical properties, e.g. resistance to dissociation by EDTA or RNase (206).

Although mitochondrial ribosomes synthesize certain hydrophobic membrane proteins, this does not constitute an a priori reason to believe that these ribosomes are membrane bound. In Neurospora, an extremely hydrophobic protein, the proteolipid subunit of the mitochondrial ATPase, is synthesized on cytoplasmic ribosomes (208,209). Apparently, therefore, there is no absolute requirement that intrinsic inner membrane proteins be synthesized at their site of incorporation into the membrane.

Linnane and coworkers postulated that mitochondrial ribosomes might be tightly bound to the inner mitochondrial membrane in order to account for several observations including differences between the in vivo and in vitro sensitivity to antibiotics which affect these ribosomes, cross-resistance to antibiotics in drug-resistant mutants, and the fact that mitochondrial protein synthesis shows breaks in Arrhenius plots corresponding to the phase transition temperatures for mitochondrial membrane lipids (210-212). Electron microscopic studies have shown mitochondrial ribosomes clustered near the periphery of the inner membrane _in vivo_, but the contact has not been considered as intimate as for membrane bound ribosomes of the endoplasmic reticulum (213-215).

Kuriyama and Luck (216) isolated membrane fractions from _Neurospora_ mitochondria from which mitochondrial ribosomes could be dissociated by treatment with high salt and puromycin. The best evidence that this fraction contains ribosomes active in protein synthesis was provided by studies with the [_poky_] mutant. This mutant is deficient in small subunits of the mitochondrial ribosome so that the mass ratio of small to large subunits is 1:10. However, the mass ratio of subunits for the membrane fraction is 1:2 as expected for active ribosomes whereas the mass ratio for the supernatant fraction is 1:50 (216). Since Kuriyama and Luck (216) did not present electron microscopic evidence showing _in vivo_ attachment of ribosomes to the inner membrane, the data do not completely exclude the possibility that active ribosomes become attached to the membrane after the mitochondria are disrupted, perhaps via their hydrophobic nascent chains.

Mitochondrial rRNAs: Genes and Processing

Mitochondrial DNAs from many organisms have been shown by quantitative hybridization to contain one copy each of the genes for the small and large subunit mitochondrial rRNAs (see Chua and Luck, 12, review). Figure 3 depicts the organization of the mitochondrial rRNA genes in several organisms. In all animal cell mitochondrial DNAs examined thus far, the genes encoding the small and large subunit RNAs are closely linked and separated by a short spacer region of 100 to 200 bp (217,218). By contrast, in yeast mitochondrial DNA, the genes encoding the small and large subunit rRNAs are separated by _ca._ 30 kb (219,220). In _Tetrahymena_ mitochondrial DNA, the genes for the small and large rRNAs are separated by at least 10 kb and, in addition, there is a duplication of the large but not the small rRNA gene (221,222). Initial studies suggested that the _Neurospora_ mitochondrial rRNA genes might be closely linked (223) but more recent work shows that they are in fact separated by 5 to 6 kb (224-227).

As shown in Fig. 3, the genes in yeast and _Neurospora_, encoding the large rRNA contain introns of 1.1 and 2.3 kb, respectively (224,225, 228). In both organisms, the intron is located about 500 bp from one

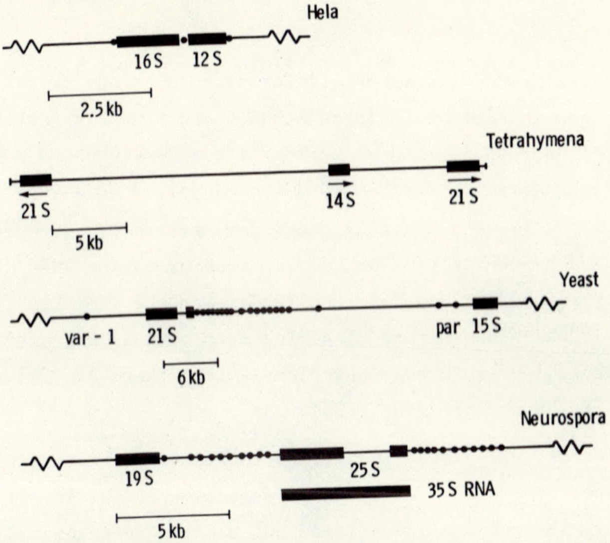

Figure 3. Organization of mitochondrial rRNA and mitochondrial tRNA (or 4S RNA) genes in several organisms. var-1 = variant one gene and par = paromomycin resistance gene in yeast. See text for references.

end of the gene, but it is not yet known whether this corresponds to the 5' or 3' end of the RNA transcript. In yeast the intron has been found only in S. cerevisiae strains carrying the ω^+ allele of the ω locus which is associated with polar recombination events (4,228). In Neurospora, however, the intron is present in both of the two different classes of wild-type strains (Lindegren and Abbott) which have been examined (224,225,227). Mitochondrial rRNA genes in animal cells do not seem to possess introns as judged by electron microscopy of RNA-DNA hybrids and by S_1-nuclease mapping (217,218,229).

Mitochondrial tRNA genes have been found associated with or clustered near the mitochondrial rRNA genes in several organisms (Fig. 3). In mitochondrial DNA of animal cells, a gene encoding a 4S RNA is present on the same DNA strand in the 160 bp spacer between the two rRNA genes; two additional 4S RNA genes are found flanking the 12S and 16S rRNA genes, but the remaining mitochondrial tRNA genes appear to be uniformly distributed around the genome (217,218,230). In yeast, mitochondrial tRNA genes are clustered in the 30 kb region separating the two rRNA genes (4,231,232). Also, in Neurospora, a cluster of tRNA genes has been found in the 6 kb region separating the two rRNA genes with an additional cluster located just outside of the gene encoding 25S rRNA (226,227). Recent evidence shows that most, if not all, of these tRNA genes are located on the same DNA strand as the rRNA genes (226).

Detailed studies on mitochondrial rRNA precursors and processing pathways are now in progress. These studies are probably most advanced in the case of *Neurospora*, and some older ideas have changed markedly. The 32S RNA considered by Kuriyama and Luck (233) to be a continuous transcript of the 19S and 25S rRNA genes now appears to have been an artifact caused by comigration on gels of true precursor RNAs and aggregates of mature 19S and 25S rRNA (234). At the same time, recent experiments in which the method of Alwine *et al.* (235) was used to identify potential precursors, show clearly that there are separate precursors for 19S and 25S rRNA (236).

Mannella *et al.* (225) have identified two temperature-sensitive, nuclear mutants of *Neurospora* which are defective in the splicing of the 25S rRNA. The mutants are allelic but can be distinguished in temperature shift-up experiments in which mitochondrial rRNA processing turns off more rapidly in one mutant than in the other. When grown at the nonpermissive temperature, the mutants accumulate a novel 35S RNA (see Fig. 3) which appears to be a continuous transcript of the 25S rRNA gene including the 2.3 kb intervening sequence. The transcript mapping experiments mentioned above suggest that 35S RNA is related to the normal precursor of 25S rRNA, although in wild-type strains, the intron appears to be removed rapidly, perhaps during transcription (236). Recently, 35S RNA from the mutants has been isolated as part of a ribonucleoprotein particle containing nearly the full complement of large subunit proteins (237). The latter result shows that binding of mitochondrial ribosomal proteins could precede the removal of the intron in the course of mitochondrial ribosome assembly.

In HeLa cells, both mitochondrial DNA strands are completely transcribed *in vivo* (238,239). Initial transcripts are apparently processed to yield the mature rRNA and mRNA species (240), but a detailed processing pathway for the rRNAs has not yet been described. Recent studies suggest that the region of mitochondrial DNA close to the origin of DNA replication, may contain one promoter for L strand transcription and one promoter for H strand transcription with the latter probably involved in rRNA synthesis (229). The direction of transcription is from the 12S to the 16S rRNA gene (241).

Yeast and Neurospora mitochondrial DNAs, having molecular weights of 50 and 40 Mdal, respectively, are considerably larger than mitochondrial DNA of animal cells (molecular weight, 10 Mdal). There is no indication that these larger mitochondrial DNAs are completely and symmetrically transcribed (see, e.g., Hendler et al., 242) and there certainly must be multiple promoters. Faye (243) has shown that yeast petite mutants in which one or the other mitochondrial rRNA genes is deleted continue to synthesize the other mitochondrial rRNA. In addition, the remaining mitochondrial rRNA is incorporated into ribonucleoprotein particles which sediment approximately like mature subunits. The results indicate that, at least in the case of yeast, the two rRNA genes can be transcribed independently and that the rRNAs can be separately processed and incorporated into subunit-like particles.

Sites of Synthesis of Mitochondrial Ribosomal Proteins

Almost all of the mitochondrial ribosomal proteins are nuclear gene products, synthesized in the cytoplasm and subsequently transported into the mitochondria (see review by Chua and Luck, 12). In Neurospora, however, one protein (S-5, M_r 52,000) associated with the small subunit of the mitochondrial ribosome is synthesized within the mitochondria (Lambowitz et al., 244). Two-dimensional gel analysis shows that S-5 is very basic (probably having a pI greater than 10) and has a high affinity for RNA, at least in the urea-containing gel buffers (170). Although a functional role for this protein has not been conclusively established, S-5 does satisfy several criteria for a ribosomal protein: (a) it is specifically associated with the small subunit of the mitochondrial ribosome, (b) it remains associated with this subunit after sedimentation through sucrose gradients containing 0.5 M KCl, and (c) its stoichiometry is comparable to that of other mitochondrial ribosomal proteins as judged by two-dimensional gel electrophoresis (179,244).

The role of S-5 in mitochondrial ribosome assembly has been investigated by examining the effect of chloramphenicol, a specific inhibitor of mitochondrial protein synthesis, on mitochondrial ribosome assembly in wild-type Neurospora (179,245). These studies have shown that chloramphenicol rapidly inhibits the assembly of mitochondrial small subunits and leads to the accumulation of incomplete small subunits (CAP-30S particles) which are enriched in a precursor of 19S rRNA and deficient in S-5 and several other proteins. The results support the idea that a mitochondrially-synthesized protein, presumably S-5, is required for the assembly of small subunits, specifically to facilitate processing of 19S rRNA and to stabilize the binding of additional small subunit proteins.

If S-5 or some other mitochondrially-synthesized protein is required for mitochondrial ribosome assembly, then any mutation which decreases the rate of mitochondrial protein synthesis may secondarily inhibit the assembly of small subunits. This prediction has been confirmed for four different mutants deficient in mitochondrial protein synthesis, all of which have been found to accumulate incompletely assembled small subunits, similar to CAP-30S particles (Collins et al., 246). In addition, correlation of mutant phenotypes with rates of mitochondrial protein synthesis show that inhibition of mitochondrial protein synthesis must be relatively severe (greater than 65%) before the assembly of mitochondrial small subunits is inhibited appreciably. In some mutants (i.e., those with ca. 50% of the rate of wild-type mitochondrial protein synthesis) mitochondrial ribosomes contain essentially normal levels of S-5 whereas the concentration of cytochrome b is decreased more than 70% and the concentration of cytochrome aa_3 is decreased more than 90%. The latter findings raise the possibility that translational control mechanisms operate so that S-5 is synthesized preferentially as the rate of mitochondrial protein synthesis decreases (246).

Pulse-labeling experiments suggest that S-5 has a relatively small free pool which could reflect a key regulatory role in mitochondrial ribosome assembly (244). However, if S-5 actually plays such a role, it must act at a relatively late stage in the assembly process since the CAP-30S particles mentioned above are almost fully assembled small subunits (179,245). In addition, recent results suggest that nearly all of the mitochondrial ribosomal proteins can bind to the particles in the absence of S-5, but that S-5 may play a role in stabilizing the binding of some proteins (179).

In yeast also, one mitochondrially-synthesized protein (var-1, M_r 43,000-47,000) has been found tightly bound to the mitochondrial small subunit (247). Even prior to this finding, the var-1 protein had been identified on SDS-polyacrylamide gels of whole mitochondrial protein (248) and the structural gene mapped on mitochondrial DNA taking advantage of the presence of polymorphic forms of the protein in different yeast strains (Perlman et al., 249, see Fig. 3). The structural alterations in the polymorphic forms of var-1 correspond to molecular weight differences of 10%; however, these alterations apparently do not affect the binding of var-1 to small subunits of the mitochondrial ribosome (4). On the basis of inhibitor experiments similar to those carried out for *Neurospora*, Butow *et al*. (250) have tentatively concluded that the var-1 polypeptide is required for the assembly of the small subunit.

Although the var-1 protein seems likely to be the cognate of *Neurospora* S-5 protein, there is really no evidence for this relationship at the present time (contrast Borst and Grivell, 4). The var-1 protein appears to be present at higher concentrations than the S-5 protein. The var-1 protein but not S-5 is readily detectable on SDS-gels of mitochondrial translation products from cells pulse-labeled with $^{35}SO_4$ in the presence of cycloheximide (248,251) possibly reflecting a large pool of nonribosomal var-1. In addition, yeast mutants resistant to paromomycin, an antibiotic assumed (but not yet demonstrated) to inhibit the small subunit of the mitochondrial ribosome, map to a mitochondrial DNA locus separate from var-1 or from 15S or 21S rRNA (4, see Fig. 3). The gene product of the *par* locus may be the cognate of *Neurospora* S-5 or there may conceivably be more than one mitochondrially-synthesized mitochondrial ribosomal protein in yeast.

Genetic Analysis of Mitochondrial Ribosomes

In yeast and *Neurospora*, a number of mutants have been isolated which are either defective in mitochondrial ribosome assembly or resistant to drugs which affect mitochondrial ribosome function. In principle, genetic and biochemical analysis of such mutants should give insight into the locations of the structural genes for mitochondrial ribosomal constituents as well as other genes which play a role in mitochondrial ribosome assembly. The genes encoding mitochondrial rRNAs have been mapped in several organisms (see above) and in yeast, the gene encoding var-1, presumably a mitochondrial ribosomal protein, has also been mapped (249). Aside from these cases, however, there has been relatively little progress in correlating mutations with primary defects in specific ribosomal components. None of the nuclear genes encoding mitochondrial ribosomal proteins has yet been identified in any organism.

In yeast, several laboratories have carried out detailed mapping studies of mitochondrial DNA mutations conferring resistance to chloroamphenicol (*cap*^{R}), erythromycin (*ery*^{R}) and spiramycin (*spi*^{R}). The mutations map to three main loci: *rib-1* (chloramphenicol), *rib-2* (erythromycin, spiramycin), and *rib-3* (erythromycin). All three loci may overlap the gene encoding 21S rRNA with *rib-1* and *rib-2* mapping in or near the short segment on the right of the ω^+ insert and *rib-3* mapping in the long segment on the left (see Fig. 3). Biochemical analysis indicates that drug resistance mutations at these loci confer resistance at the level of isolated mitochondrial ribosomes (195,252). Together, the results are considered suggestive evidence that at least some of the mutations affect the primary structure of 21S rRNA although rRNA alterations have not yet been directly demonstrated by RNA fingerprinting studies (cf. 4). At least two mutants with cold-sensitive defects in mitochondrial protein synthesis have been mapped in the same region of mitochondrial DNA near the *rib-2* and *rib-3* loci (253,254). These mutants show abnormal assembly of large subunits of the mitochondrial ribosome or non-functional large subunits, again consistent with an alteration in 21S mitochondrial rRNA. Chloramphenicol-resistance mutations mapping in additional loci have been identified between *rib-1*, *rib-2*, and *rib-3* (255,256). Despite attempts to squeeze all of the mutations inside the 21S rRNA gene, there have been persistent indications that some *cap*^{R} mutations map near, but to the right of this gene (Fig. 3; 257). Nevertheless, short of a direct demonstration, it seems almost certain that most of the mutations affect the primary structure of the 21S mitochondrial rRNA. The apparent ease of isolating such mutations is an important feature of the yeast system and should be useful in studying structure-function relationships of the rRNAs. In *E. coli* such rRNA mutants are probably masked by the redundancy of the rRNA genes.

As indicated above, yeast mutants conferring resistance to paromomycin (*par*^{R}) map to a locus separate from *var-1*, 15S or 21S rRNA (220,258,259, and see Fig. 3). Although questions have been raised about the mapping data (4), recent studies continue to support the idea that the *par* locus is separate from the 15S rRNA gene (257). The *par* locus could code for a ribosomal protein of the small subunit or for a factor involved in protein synthesis. Faye (260) examined mitochondrial ribosomal proteins from yeast strains carrying two different *par*^{R} alleles and found no alteration in the electrophoretic mobility of any ribosomal proteins on two-dimensional gels. These studies are not definitive, however, because small alterations, even small charge differences, could have escaped detection in the gel system employed (see discussion by Lambowitz in 261).

Waxman et al. (256) have isolated nuclear gene suppressors of mitochondrial cap^R and par^R mutations. Such suppressors potentially include strains with altered mitochondrial ribosomal proteins which can interact with the mutant component to modify the drug resistance phenotype. Two of the suppressors were found to discriminate among different chloramphenicol-resistance loci suggesting that specific RNA-protein or protein-protein interactions might be involved.

In Neurospora, a number of nuclear and nonMendelian mutants have been identified which have defects in the assembly of one or the other mitochondrial ribosomal subunits. Historically, the nonMendelian [poky] mutant has been the most intensively studied (262,263). Studies by Luck and coworkers showed that 19S rRNA is rapidly degraded in [poky] leading to a gross decrease in the ratio of small to large subunits (264,265). The apparently defective processing of 19S rRNA in [poky] could be accounted for by a defect in a mitochondrial rRNA or in a mitochondrial ribosomal protein. With respect to the first possibility, however, no alteration in base sequence or methylation of [poky] 19S or 25S rRNA could be detected in fingerprinting experiments (165). More recently, attention has been focused on the possibility that the mitochondrial ribosomal protein, S-5, might be the site of the primary defect in [poky]. Two-dimensional gel analysis has thus far failed to reveal an alteration in the electrophoretic mobility of S-5 in [poky], although small subunits of the mitochondrial ribosomes from [poky] do appear to be ca. 50% deficient in S-5 compared to other ribosomal proteins (179,244).

Recently, six additional nonMendelian mutants ([exn-1], [exn-2], [exn-4], [stp-B1], [SG-1] and [SG-3]) have been shown to be virtually identical to [poky] in terms of their defects in assembly of mitochondrial ribosomes. Each of the mutants is deficient in small subunits and, in each case, 19S rRNA appears to be synthesized but rapidly degraded (179,246,266). The mutants had previously been classified as genetically related on the basis of mitochondrial phenotype and response to the nuclear f-gene suppressor of [poky] (267). More recently, Bertrand and coworkers have isolated six, non-allelic nuclear suppressors of [poky], five of which at least partially alleviate the deficiency of small subunits of the mitochondrial ribosome (266,268). All of these suppressors have been found to suppress partially the remaining six nonMendelian mutants (266). Mitochondrial ribosomal proteins from each of the nonMendelian mutants have been analyzed by two-dimensional gel electrophoresis, but again no alteration could be detected in the electrophoretic mobility of S-5 (179). As discussed previously, the results do not rule out small changes in amino acid composition and this possibility will have to be addressed by polypeptide fingerprint analysis. Possibly the primary defects in the extra-nuclear mutants are in the mitochondrial rRNAs or in the unconserved regions of mitochondrial ribosomal precursor RNAs.

The nuclear suppressors of [poky] were isolated with the idea that they would include strains with structural alterations in mitochondrial ribosomal proteins capable of interacting with the defective component in [poky] so as to ameliorate the mitochondrial ribosome assembly defect (268). Although no alterations in the electrophoretic mobility of any of the ribosomal proteins have been found, in one case, su^{I}-5, the product of the suppressor gene appears to interact with the wild-type allele of the [poky] gene causing, for unknown reasons, a cold-sensitive defect in the assembly of large subunits of the mitochondrial ribosome (269).

A number of nuclear mutants with defects in assembly of mitochondrial ribosomes have also been described recently. These include two allelic mutations which affect mitochondrial rRNA splicing (225), three mutations which affect assembly of only the small subunit of the mitochondrial ribosome, one which affects assembly of only the large subunit and two which affect assembly of both subunits (246,270). To date, mutants resistant to drugs affecting mitochondrial ribosomes have not been isolated in *Neurospora*.

NonMendelian chloramphenicol- and erythromycin-resistant mutants have been isolated in many other organisms (see reviews 1, 271). In mammalian cells, nonMendelian chloramphenicol-resistant mutants have been isolated and resistance demonstrated *in vitro* for isolated mitochondria and detergent lysed mitochondria (272-275) but so far, not for isolated mitochondrial ribosomes. Ditta *et al*. (276) have described a nuclear mutant of CHO cells which is defective in mitochondrial protein synthesis. More recently, Wiseman and Attardi (277) have isolated nonMendelian mutants of human fibroblasts which are deficient in mitochondrial protein synthesis and found subsequently to be chloramphenicol-resistant. Biochemical analysis of the mammalian cell mutants has so far been limited by the difficulty of obtaining pure mitochondrial ribosomes and by the small amounts of material that can be obtained.

ACKNOWLEDGEMENTS

This work is supported by NIH grant GM-19427 to J.E.B. and N.W.G. and by NIH grant GM-23961 and by a Basil O'Connor grant from the National Foundation, March of Dimes to A.M.L. We thank Drs. Helmut Bertrand, Richard A. Collins, Robert J. La Polla, and T. W. O'Brien for their comments on those parts of the manuscript having to do with mitochondrial ribosomes.

NOTE ADDED

Recent studies by Spithill *et al*. (278) with isolated yeast mitochondrial ribosomes show that the aminoglycosides paromomycin and neomycin stimulate misreading of poly-(U). Furthermore, mitochondrial ribosomes isolated from a *par* resistant mutant were shown to be resistant to the misreading effect.

REFERENCES

1. Gillham N. W. (1978) *Organelle Heredity*, Academic Press, New York, 602 pp.
2. Schatz, G., and Mason, T. L. (1974) Annu. Rev. Biochem. 43, 51-87.
3. Buetow, D. E., and Wood, W. M. (1978) Subcell. Biochem. 5, 1-85.
4. Borst, P., and Grivell, L. A. (1978) Cell 15, 705-723.
5. Ellis, R. J. (1977) Biochim. Biophys. Acta 463, 185-215.
6. Gillham, N. W., Boynton, J. E., and Chua, N.-H. (1978) Curr. Topics in Bioenergetics 8, 211-260.
7. Weil, J. H., Burkard, G., Guillemaut, P., Jeannin, G., Martin, R., and Steinmetz, A. (1976) *Nucleic Acids and Protein Synthesis in Plants*, pp. 97-120, Plenum Press, New York.
8. Tewari, K. K., Kolodner, R., Chu, N. M., and Meeker, R. M. (1976) *Nucleic Acids and Protein Synthesis in Plants*, pp. 15-36, Plenum Press, New York.
9. Barnett, W. E., Schwartzbach, S. D., and Hecker, L. I. (1976) *Genetics and Biogenesis of Chloroplasts and Mitochondria*, pp. 661-666, Elsevier/North-Holland Biomedical Press, Amsterdam.
10. Stutz, E., and Boschetti, A. (1976) Handbook of Genetics 5, 425-450, Plenum Press, New York.
11. Nomura, M., Tissières, A., and Lengyel, P. ed. (1974) *Ribosomes*, Cold Spring Harbor Laboratory, New York, 930 pp.
12. Chua, N.-H., and Luck, D. J. L. (1974) *Ribosomes*, pp. 519-539, Cold Spring Harbor Laboratory, New York.
13. O'Brien, T. W., and Matthews, D. E. (1976) Handbook of Genetics 5, 535-580, Plenum Press, New York
14. Ellis, R. J., and Hartley, M. R. (1974) MTP International Review of Science 6, 323-348.
15. Ledoigt, G., Stevens, B. J., Curgy, J. J., and Andre, J. (1979) Exptl. Cell Res. 119, 221-229.
16. Bourque, D. P., Boynton, J. E., and Gillham, N. W. (1971) J. Cell Sci. 8, 155-183.

17. Bartlett, S. G., Harris, E. H., Grabowy, C. T., Gillham, N. W., and Boynton, J. E. (1979) Molec. gen. Genet. in press.

18. Dyer, T. A., Bowman, C. M., and Payne, P. I. (1976) Nucleic Acids and Protein Synthesis in Plants, pp. 121-133, Plenum Press, New York.

19. Bohnert, H. J., Driesel, A. J., and Herrmann, R. G. (1976) Biogenesis of Chloroplasts and Mitochondria, pp. 629-636, Elsevier/North-Holland Biomedical Press, Amsterdam.

20. Whitfeld, P. R., Leaver, C. J., Bottomley, W., and Atchison, B. A. (1977) Acides Nucléiques et Synthèse des Protéines chez les Végétaux, pp. 235-241, CNRS.

21. Rochaix, J. D., and Malnoë, P. (1978) Cell 15, 661-670.

22. Zablen, L., Kissil, M. S., Woese, C. R., and Buetow, D. E. (1975) Proc. Nat. Acad. Sci. U.S. 72, 2418-2422.

23. Buetow, D. E., Kissil, M. S., and Zablen, L. (1976) Biogenesis of Chloroplasts and Mitochondria, pp. 641-644, Elsevier/North-Holland Biomedical Press, Amsterdam.

24. Woese, C. R., Fox, G. E., Zablen, L., Uchida, T., Bonen, L., Pechman, K., Lewis, B. J., and Stahl, D. (1975) Nature (London) 254, 83-86.

25. Thomas, J. R., and Tewari, K. K. (1974) Proc. Nat. Acad. Sci. U.S. 71, 3147-3151.

26. Gray, P. W., and Hallick, R. B. (1979) Biochim. Biophys. Acta 561, 53-58.

27. Crouse, E., Vandrey, J., and Stutz, E. (1974) Proc. Third Int. Cong. on Photosynthesis, pp. 1775-1786, Elsevier, Amsterdam.

28. Slavik, N. S., and Hershberger, C. L. (1976) J. Mol. Biol. 103, 563-581.

29. Manning, J. E., and Richards, O. C. (1972) Biochim. Biophys. Acta 259, 285-296.

30. Stutz, E., and Vandrey, J. P. (1971) FEBS Lett. 17, 277-280.

31. Rawson, J. R. Y., and Haselkorn, R. (1973) J. Mol. Biol. 77, 125-132.

32. Ingle, J., Wells, R., Possingham, J. V., Leaver, C. J. and Loening, U. E. (1970) Control of Organelle Development, pp. 303-325, Cambridge University Press, Cambridge.

33. Leaver, C. J., and Ingle, J. (1971) Biochem. J. 123, 235-243.

34. Freyssinet, G. (1977) Biochimie 59, 597-610.

35. Gualerzi, C., Janda, H. G., Passow, H., and Stöffler, G. (1974) J. Biol. Chem. 249, 3347-3355.

36. Schneeman, R., and Surzycki, S. (1977) J. Cell Biol. 75, 304a.

37. Capel, M. S., and Bourque, D. P. (1977) Plant Physiol. 59, S-110.

38. Bourque, D. P. (1977) Acides Nucléiques et Synthèse des Protéines chez les Végétaux, pp. 285-289, CNRS.

39. Gillham, N. W., Boynton, J. E., Harris, E. H., Fox, S. B., and Bolen, P. L. (1976) Genetics and Biogenesis of Chloroplasts and Mitochondria, pp. 69-76, Elsevier/North-Holland Biomedical Press, Amsterdam.

40. Pitel, J. A., and Durzan, D. J. (1977) Phytochemistry 16, 1475-1480.

41. Mets, L. J., and Bogorad, L. (1974) Analyt. Biochem. 57, 200-210.

42. Freyssinet, G. (1978) Exptl. Cell Res. 115, 207-219.

43. Panyim, S., and Chalkley, R. (1969) Arch. Biochem. Biophys. 130, 337-346.

44. Hanson, M. R., Davidson, J. N., Mets, L. J., and Bogorad, L. (1974) Molec. gen. Genet. 132, 105-118.

45. Brügger, M., and Boschetti, A. (1975) Eur. J. Biochem. 58, 603-610.

46. Highfield, P. E., and Ellis, R. J. (1976) Biochim. Biophys. Acta 447, 20-27.

47. Ciferri, O., and Tiboni, O. (1973) Nature New Biology (London) 245, 209-210.

48. Tiboni, O., DiPasquale, G., and Ciferri, O. (1976) Plant Sci. Lett. 6, 419-429.

49. Ciferri, O., and Tiboni, O. (1976) Plant Sci. Lett. 7, 455-466.

50. Ciferri, O., Tiboni, O., Munoz-Calvo, M. L., and Camerino, G. (1976) Nucleic Acids and Protein Synthesis in Plants, pp. 155-166, Plenum Press, New York.

51. Tiboni, O., DiPasquale, G., and Ciferri, O. (1978) Chloroplast Development, pp. 675-678, Elsevier/North-Holland Biomedical Press, Amsterdam.

52. Tiboni, O., DiPasquale, G., and Ciferri, O. (1978) Eur. J. Biochem. 92, 471-477.

53. Chua, N.-H., Blobel, G., and Siekevitz, P. (1973) J. Cell Biol. 57, 798-814.

54. Schlanger, G., Sager, R., and Ramanis, Z. (1972) Proc. Nat. Acad. Sci. U.S. 69, 3551-3555.

55. Schlanger, G., and Sager, R. (1974) Proc. Nat. Acad. Sci. U.S. 71, 1715-1719.

56. Conde, M. F., Boynton, J. E., Gillham, N. W., Harris, E. H., Tingle, C. L., Wang, W. L. (1975) Molec. gen. Genet. 140, 183-220.

57. Harris, E. H., Boynton, J. E., Gillham, N. W., Tingle, C. L., and Fox, S. B. (1977) Molec. gen. Genet. 155, 249-265.

58. Fox, S. B., Grabowy, C. T., Harris, E. H., Gillham, N. W., and Boynton, J. E. (1977) J. Cell Biol. 75, 307a.

59. Lee, S. G., and Evans, W. R. (1971) Science 173, 241-242.

60. Grivell, L. A., and Walg, H. L. (1972) Biochem. Biophys. Res. Commun. 49, 1452-1458.

61. Stutz, E., and Noll, H. (1967) Proc. Nat. Acad. Sci. U.S. 57, 774-781.

62. Chen, J. L., and Wildman, S. G. (1967) Science 155, 1271-1273.

63. Chen, J. L., and Wildman, S. G. (1970) Biochim. Biophys. Acta 209, 207-219.

64. Philippovich, I. I., Bezsmertnaya, I. N., and Oparin, A. I. (1973) Exptl. Cell Res. 79, 159-168.

65. Chua. N.-H., Blobel, G., Siekevitz, P., and Palade, G. E. (1973) Proc. Nat. Acad. Sci. U.S. 70, 1554-1558.

66. Margulies, M. M., and Michaels, A. (1974) J. Cell Biol. 60, 65-77.

67. Baumgartel, D. M., and Howell, S. H. (1976) Biochim. Biophys. Acta 454, 338-348.

68. Baumgartel, D. M., and Howell, S. H. (1976) Biochim. Biophys. Acta 454, 349-361.

69. Michaels, A., and Margulies, M. M. (1975) Biochim. Biophys. Acta 390, 352-362.

70. Margulies, M. M., and Michaels, A. (1975) Biochim. Biophys. Acta 402, 297-308.

71. Margulies, M. M., Tiffany, H. L., and Michaels, A. (1975) Biochem. Biophys. Res. Commun. 64, 735-739.

72. Chua., N.-H., Blobel, G., Siekevitz, P., and Palade, G. E. (1976) J. Cell Biol. 71, 497-514.

73. Gelvin, S., and Howell, S. H. (1977) Plant Physiol. 59, 471-477.

74. Avadhani, N. G., and Buetow, D. E. (1972) Biochem. J. 128, 353-365.

75. Tao, K. L., and Jagendorf, A. T. (1973) Biochim. Biophys. Acta 324, 518-532.

76. Chua, N.-H., and Gillham, N. W. (1977) J. Cell Biol. 74, 441-452.

77. Blobel, G., and Dobberstein, B. (1975) J. Cell Biol. 67, 835-851.

78. Lodish, H. F., and Small, B. (1975) J. Cell Biol. 65, 51-64.

79. Feierabend, J., and Schrader-Reichardt, U. (1976) Planta (Berl.) 129, 133-145.

80. Feierabend, J. (1976) Genetics and Biogenesis of Chloroplasts and Mitochondria, pp. 99-102, Elsevier/North-Holland Biomedical Press, Amsterdam.

81. Feierabend, J. (1977) Planta (Berl.) 135, 83-88.

82. Rademacher, E., and J. Feierabend (1976) Planta (Berl.) 129, 147-153.

83. Schäfers, H.-A., and Feierabend, J. (1976) Cytobiologie 14, 75-90.

84. Cohen, D., and Schiff, J. A. (1976) Arch. Biochem. Biophys. 177, 201-216.

85. Cattolico, R. A., Senner, J. W., and Jones, R. F. (1973) Arch. Biochem. Biophys. 156, 58-65.

86. Wilson, R., and Chiang, K.-S. (1977) J. Cell Biol. 72, 470-481.

87. Steup, M., and Ssymank, V. (1978) Ber. Deutsch. Bot. Ges. 91, 243-250.

88. Steup, M. (1977) Arch. Microbiol. 112, 277-282.

89. Steup, M., Ssymank, V., Winkler, U., and Glock, H. (1977) Planta (Berl.) 137, 139-144.

90. Wang, W.-Y. (1978) Int. Rev. Cytol. Supp. 8, 335-354.

91. Matsuda, Y. (1974) Biochim. Biophys. Acta 366, 45-55.

92. Siersma, P., and Chiang, K.-S. (1971) J. Mol. Biol. 58, 167-185.

93. Martin, N. C., Chiang, K.-S., and Goodenough, U. W. (1976) Dev. Biol. 51, 190-201.

94. Rochaix, J.-D. (1978) J. Mol. Biol. 126, 597-617.

95. Malnoë, P., and Rochaix, J.-D. (1978) Molec. gen. Genet. 166, 269-275.

96. Gray, P. W., and Hallick, R. B. (1978) Biochemistry 17, 284-289.

97. Rawson, J. R. Y., Kushner, S. R., Vapnek, D., Alton, N. K., and Boerma, C. L. (1978) Gene 3, 191-209.

98. Jenni, B., and Stutz, E. (1978) Eur. J. Biochem. 88, 127-134.

99. Gray, P. W., and Hallick, R. B. (1979) Biochemistry 18, 1820-1825.

100. Bedbrook, J. R., Kolodner, R., and Bogorad, L. (1977) Cell 11, 739-749.

101. Herrmann, R. G., Bohnert, H.-J., Driesel, A., and Hobom, G. (1976) Genetics and Biogenesis of Chloroplasts and Mitochondria, pp. 351-359. Elsevier/North-Holland Biomedical Press, Amsterdam.

102. Whitfeld, P. R., Herrmann, R. G., and Bottomley, W. (1978) Nucleic Acids Res. 5, 1741-1751.

103. Crouse, E. J., Schmitt, J. M., Bohnert, H.-J., Gordon, K., Driesel, A. J., and Herrmann, R. G. (1978) Chloroplast Development, pp. 565 572, Elsevier/North-Holland Biomedical Press, Amsterdam.

104. Bedbrook, J. R., and Kolodner, R. (1979) Plant Physiol. 30, 593-620.

105. Kolodner, R., and Tewari, K. K. (1979) Proc. Nat. Acad. Sci. U.S. 76, 41-45.

106. Surzycki, S. J., and Rochaix, J.-D. (1971) J. Mol. Biol. 62, 89-109.

107. Miller, M. J., and McMahon, D. (1974) Biochim. Biophys. Acta 366, 35-44.

108. Scott, N. S. (1976) Phytochemistry 15, 1207-1213.

109. Hallick, R. B., Gray, P. W., Chelm, B. K., Rushlow, K. E., and Orozco, E. M., Jr. (1978) Chloroplast Development, pp. 619-622, Elsevier/North-Holland Biomedical Press, Amsterdam.

110. Haff, L. A., and Bogorad, L. (1976) Biochemistry 15, 4105-4109.

111. Driesel, A. J., Crouse, E. J., Gordon, K., Bohnert, H. J., Herrmann, R. G., Steinmetz, A., Mubumbila, M., Keller, M., Burkhard, M., and Weil, J. H., Gene, in press.

112. Hartley, M. R., and Ellis, R. J. (1973) Biochem. J. 134, 249-262.

113. Bohnert, H. J., Driesel, A. J., and Herrmann, R. G. (1976) Genetics and Biogenesis of Chloroplasts and Mitochondria, pp. 629-636, Elsevier/North-Holland Biomedical Press, Amsterdam.

114. Mache, R., Jalliffier-Verne, M., Rozier, C., and Loiseaux, S. (1978) Biochim. Biophys. Acta 517, 390-399.

115. Ellis, R. J., and Hartley, M. R. (1971) Nature (London), New Biol. 233, 193-196.

116. Honeycutt, R. C., and Margulies, M. M. (1973) J. Biol. Chem. 248, 6145-6153.

117. Harris, E. H., Boynton, J. E., and Gillham, N. W. (1976) The Genetics of Algae, pp. 119-144, University of California Press, Berkeley.

118. Boynton, J. E., Gillham, N. W., Harris, E. H., Tingle, C. L., Van Winkle-Swift, K., and Adams, G. M. W. (1976) Genetics and Biogenesis of Chloroplasts and Mitochondria, pp. 313-322, Elsevier/North-Holland Biomedical Press, Amsterdam.

119. Bolen, P. L., Boynton, J. E., and Gillham, N. W. (1979) Genetics, in press.

120. Davidson, J. N., Hanson, M. R., Bogorad, L. (1978) Genetics 89, 281-297.

121. Davidson, J. N., and Bogorad, L. (1977) Molec. gen. Genet. 157, 39-46.

122. Boynton, J. E., Burton, W. G., Gillham, N. W., Harris, E. H. (1973) Proc. Nat. Acad. Sci. U.S. 70, 3463-3467.

123. Mets, L., and Bogorad, L. (1972) Proc. Nat. Acad. Sci. U.S. 69, 3779-3783.

124. Ohta, N., Sager, R., and Inouye, M. (1975) J. Biol. Chem. 250, 3655-3659.

125. Davidson, J. N., Hanson, M. R., and Bogorad, L. (1974) Molec. gen. Genet. 132, 119-129.

126. Hanson, M. R., and Bogorad, L. (1977) Molec. gen. Genet. 153, 271-277.

127. Hanson, M. R., and Bogorad, L. (1978) J. Gen. Microbiol. 105, 253-262.

128. Hanson, M. R., and Bogorad, L. (1977) Biochim. Biophys. Acta 479, 279-289.

129. Sager, R., and Tsubo, Y. (1961) Z. Vererbungslehre 92, 430-438.

130. Lee, R. W., and Sapp, J. A. (1978) Genetics 88, 643-650.

131. Shepherd, H. S., Boynton, J. E. and Gillham, N. W. (1979) Proc. Nat. Acad. Sci. U.S. 76, 1353-1357.

132. Harris, E. H., Boynton, J. E., and Gillham, N. W. (1974) J. Cell Biol. 63, 160-179.

133. Gyurjan, I., Yurina, N. P., Thurischeva, M. S., and Odintsova, M. S. (1979) Molec. gen. Genet. 170, 203-211.

134. Diamond, J., and Schiff, J. A. (1974) Plant Sci. Lett. 3, 289-295.

135. Schwartzbach, S. D., and Schiff, J. A. (1974) J. Bacteriol. 120, 334-341.

136. Freyssinet, G. (1975) Plant Sci. Lett. 5, 305-311.

137. Mahanty, H. K. (1979) Nature (London) 277, 562-563.

138. Maliga, P., Sz-Breznovits, A., and Marton, L. (1973) Nature New Biology (London) 244, 29-30.

139. Maliga, P., Sz-Breznovits, A., Marton, L., and Joo, F. (1975) Nature (London) 255, 401-402.

140. Bourque, D. P., Horn, N. A., and Capel, M. S. (1977) Plant Physiol. 59, S-110.

141. Umiel, N., and Goldner, R. (1976) Protoplasma 89, 83-89.

142. Bourque, D. P., and Wildman, S. G. (1973) Biochem. Biophys. Res. Commun. 50, 532-537.

143. Börner, T., Schumann, B., and Hagemann, R. (1976) Genetics and Biogenesis of Chloroplasts and Mitochondria, pp. 41-48 Elsevier/North-Holland Biomedical Press, Amsterdam.

144. Shumway, L. K., and Weier, T. E. (1967) Amer. J. Bot. 54, 773-780.

145. Gyurjan, I., Nagy, A. H., and Rakovan, J.N. (1978) Biochem. Physiol. Pflanzen 173, 429-439.

146. Börner, T., Knoth, R., Herrmann, F., and Hagemann, R. (1972) Theoret. Appl. Genet. 42, 3-11.

147. Börner, T., Herrmann, F., and Hagemann, R. (1973) FEBS Lett., 37, 117-119.

148. Börner, T., Knoth, R., Herrmann, F. H., and Hagemann, R. (1974) Biochem. Physiol. Pflanzen 165, 429-432.

149. Hagemann, R., Herrmann, F. H., and Börner, T. (1975) Genetic Aspects of Photosynthesis, pp. 115-118, Junk, The Hague.

150. O'Brien, T. W. (1977) International Cell Biology, pp. 245-255, Rockefeller University Press, New York.

151. de Vries, H., and Van der Koogh-Schuuring, R. (1973) Biochem. Biophys. Res. Commun. 54, 308-314.

152. Hamilton, M. G., and O'Brien, T. W. (1974) Biochemistry 13, 5400-5403.

153. Leister, D. E., and Dawid, I. B. (1974) J. Biol. Chem. 249, 5108-5118.

154. Chi, J. C. H., and Suyama, Y. (1970) J. Mol. Biol. 53, 531-556.

155. Tait, A. (1972) FEBS Lett. 24, 117-120.

156. Curgy, J. J., Ledoigt, G., Stevens, B. J., and Andre, J. (1974) J. Cell Biol. 60, 628-640.

157. Tait, A., and Knowles, J. K. C. (1977) J. Cell Biol. 73, 139-148.

158. Lizardi, P. M., and Luck, D. J. L. (1971) Nature New Biology (London) 229, 140-142.

159. Borst, P., and Grivell, L. A. (1971) FEBS Lett. 13, 73-87.

160. Leaver, C. J., and Harmey, M. A. (1973) Biochem. Soc. Symp. 38, 175-193.

161. Leaver, C. J., and Harmey, M. A. (1976) Biochem. J. 157, 275-277.

162. Cunningham, R. S., Bonen, L., Doolittle, W. F., and Gray, M. W. (1976) FEBS Lett. 69, 116-122.

163. Klootwijk, J., Klein, I., and Grivell, L. A. (1975) J. Mol. Biol. 97, 337-350.

164. Lambowitz, A. M., and Luck, D. J. L. (1975) J. Mol. Biol. 96, 207-214.

165. Lambowitz, A. M., and Luck, D. J. L. (1976) J. Biol. Chem. 251, 3081-3095.

166. Dubin, D. T. (1974) J. Mol. Biol. 84, 257-273.

167. Dubin, D. T., Taylor, R. H., and Davenport, L. W. (1978) Nucleic Acids Res. 5, 4385-4397.

168. Dubin, D. T., and Taylor, R. H. (1978) J. Mol. Biol. 121, 523-540.

169. Maden, B. E. H., Salim, M., and Robertson, J. S. (1974) Ribosomes, pp. 829-839, Cold Spring Harbor Laboratory, New York.

170. Schlessinger, D. (1974) Ribosomes, pp. 393-416, Cold Spring Harbor Laboratory, New York.

171. Monier, R. (1974) Ribosomes, pp. 141-168, Cold Spring Harbor Laboratory, New York.

172. Grivell, L. A., Reijnders, L., and Borst, P. (1971) Biochim. Biophys. Acta 247, 91-103.

173. Dubin, D. T., Jones, T. H., and Cleaves, G. R. (1974) Biochem. Biophys. Res. Commun. 56, 401-406.

174. Datema, R., Agsteribbe, E., and Kroon, A. M. (1974) Biochim. Biophys. Acta 335, 386-395.

175. Van den Bogert, C., and deVries, H. (1976) Biochim. Biophys. Acta 442, 227-238.

176. Michel, R., Hallermayer, G., Harmey, M. A., Miller, F., and Neupert, W. (1977) Biochim. Biophys. Acta 478, 316-330.

177. Lizardi, P. M., and Luck, D. J. L. (1972) J. Cell Biol. 54, 56-74.

178. Faye, G., and Sor, F. (1977) Molec. gen. Genet. 155, 27-34.

179. Lambowitz, A. M., LaPolla, R. J., and Collins, R. A. (1979) J. Cell Biol. 82, 17-31.

180. O'Brien, T. W., Denslow, N. D., and Martin, G. R. (1974) The Biogenesis of Mitochondria, pp. 347-356, Academic Press, New York.

181. O'Brien, T. W., Matthews, D. E., and Denslow, N. D. (1976) Genetics and Biogenesis of Chloroplasts and Mitochondria, pp. 741-748, Elsevier/North-Holland Biomedical Press, Amsterdam.

182. Hallermeyer, G., and Neupert, W. (1974) FEBS Lett. 41, 264-268.

183. Faye, G., Sor, F., Glatigny, A., Lederer, F., and Lesquoy, E. (1979) Molec. gen. Genet. 171, 335-341.

184. Smith, A. E., and Marcker, K. A. (1968) J. Mol. Biol. 38, 241-243.

185. Galper, J. B., and Darnell, J. E. (1969) Biochem. Biophys. Res. Commun. 34, 205-214.

186. Epler, J. L., Shugart, L. R., and Barnett, W. E. (1970) Biochemistry 9, 3575-3579.

187. Bianchetti, R., Lucchini, G., and Sartirana, M. L. (1971) Biochem. Biophys. Res. Commun. 42, 97-102.

188. Bianchetti, R., Lucchini, G., Crosti, P., and Tortora, P. (1977) J. Biol. Chem. 252, 2519-2523.

189. Mahler, H. R., Dawidowicz, K., and Feldman, F. (1972) J. Biol. Chem. 247, 7439-7442.

190. Sala, F., and Küntzel, H. (1970) Eur. J. Biochem. 15, 280-286.

191. Heckman, J. E., Hecker, L. I., Schwartzbach, S. D., Barnett, W. E., Baumstark, B., and RajBhandary, U. L. (1978) Cell 13, 83-95.

192. Grandi, M., and Küntzel, H. (1970) FEBS Lett. 10, 25-28.

193. Parisi, B., and Cella, R. (1971) FEBS Lett. 14, 209-213.

194. Barath, Z., and Küntzel, H. (1972) Proc. Nat. Acad. Sci. U.S. 69, 1371-1374.

195. Grivell, L. A., Netter, P., Borst, P., and Slonimski, P. P. (1973) Biochim. Biophys. Acta 312, 358-367.

196. deVries, H., Arendzen, A. J., and Kroon, A. M. (1973) Biochim. Biophys. Acta 331, 264-275.

197. Denslow, N. D., and O'Brien, T. W. (1974) Biochem. Biophys. Res. Commun. 57, 9-16.

198. Denslow, N. D., and O'Brien, T. W. (1978) Eur. J. Biochem. 91, 441-448.

199. Davey, P. J., Haslam, J. M., and Linnane, A. W. (1970) Arch. Biochem. Biophys. 136, 54-64.

200. Grandi, M., Helms, A., and Küntzel, H. (1971) Biochem. Biophys. Res. Commun. 44, 864-871.

201. Tanaka, N., Kinoshita, T., and Masukawa, H. (1969) J. Biochem. 65, 459-464.

202. Bodley, J. W., and Lin, L. (1970) Nature (London) 227, 60-62.

203. Greco, M., Cantatore, P., Pepe, G., and Saccone, C. (1973) Eur. J. Biochem. 37, 171-177.

204. Ibrahim, N. G., and Beattie, D. S. (1973) FEBS Lett. 36, 102-104.

205. Avadhani, N. G., and Buetow, D. E. (1972) Biochem. Biophys. Res. Commun. 46, 773-778.

206. Ojala, D., and Attardi, G. (1972) J. Mol. Biol. 65, 273-289.

207. Mahler, H. R., and Dawidowicz, K. (1973) Proc. Nat. Acad. Sci. U.S. 70, 111-114.

208. Sebald, W., Graf, Th., and Wild, G. (1976) Genetics and Biogenesis of Chloroplasts and Mitochondria, pp. 167-174, Elsevier/North-Holland Biomedical Press, Amsterdam.

209. Wachter, E., Sebald, W., and Tzagoloff, A. (1977) Mitochondria 1977: Genetics and Biogenesis, pp. 441-449, Walter de Gruyter, Berlin.

210. Bunn, C. L., Mitchell, C. H., Lukins, H. B., and Linnane, A. W. (1970) Proc. Nat. Acad. Sci. U.S. 67, 1233-1240.

211. Towers, N. R., Kellerman, G. M., Raison, J. K., and Linnane, A. W. (1973) Biochim. Biophys. Acta 299, 153-161.

212. Spithill, T. W., Trembath, M. K., Lukins, H. B., and Linnane, A. W. (1978) Molec. gen. Genet. 164, 155-162.

213. Curgy, J. J. (1970) Exp. Cell Res. 62, 359-367.

214. Watson, K. (1972) J. Cell Biol. 55, 721-726.

215. Keyhani, E. (1973) J. Cell Biol. 58, 480-484.

216. Kuriyama, Y., and Luck, D. J. L. (1973) J. Cell Biol. 59, 776-784.

217. Wu, M., Davidson, N., Attardi, G., and Aloni, Y. (1972) J. Mol. Biol. 71, 81-93.

218. Dawid, I. B., Klukas, C. K., Ohi, S., Ramirez, J. L., and Upholt, W. B. (1976) The Genetic Function of Mitochondrial DNA, pp. 3-13, North-Holland, Amsterdam.

219. Sanders, J. P. M., Heyting, C., and Borst, P. (1975) Biochem. Biophys. Res. Commun. 65, 699-707.

220. Faye, G., Kujawa, C., Dujon, B., Bolotin-Fukuhara, M., Wolf, K., Fukuhara, H., and Slonimski, P. P. (1975) J. Mol. Biol. 99, 203-217.

221. Goldbach, R. W., Borst, P., Bollen-DeBoer, J. E., and Van Bruggen, E. F. J. (1978) Biochim. Biophys. Acta 521, 169-186.

222. Goldbach, R. W., Bollen-DeBoer, J. E., Van Bruggen, E. F. J., and Borst, P. (1978) Biochim. Biophys. Acta 521, 187-197.

223. Terpstra, P., Holtrop, M., and Kroon, A. M. (1977) Biochim. Biophys. Acta 478, 146-155.

224. Hahn, U., Lazarus, C. M., Lunsdorf, H. and Küntzel, H. (1979) Cell 17, 191-200.

225. Mannella, C. A., Collins, R. A., Green, M. R., and Lambowitz, A. M. (1979) Proc. Nat. Acad. Sci. U.S. 76, 2635-2639.

226. Heckman, J. E., and RajBhandary, U. L. (1979) Cell, in press.

227. deVries, H., de Jonge, J. C., Bakker, H., Meurs, H., and Kroon, A. (1979) Nucleic Acids Res. 6, 1791-1803.

228. Bos, J. L., Heyting, C., Borst, P., Arnberg, A. C., and van Bruggen, E. F. J. (1978) Nature 275, 336-338; also Faye, G., Dennebouy, N., Kujawa, C., and Jacq, C. (1979) Molec. gen. Genet. 168, 101-109.

229. Attardi, G., Cantatore, P., Ching, E., Crews, S., Gelfand, R., Merkel, C., and Ojala, D. (1979) Proceedings ICN-UCLA Symposium on Extrachromosomal DNA, in press. Academic Press, N.Y.

230. Angerer, L., Davidson, N., Murphy, W., Lynch, D., and Attardi, G. (1976) Cell 9, 81-90.

231. Fukuhara, H., Bolotin-Fukuhara, M., Hsu, H. J., and Rabinowitz, M. (1976) Mol. gen. Genet. 145, 7-17.

232. Van Ommen, G. J. B., Groot, G. S. P., and Borst, P. (1977) Mol. gen. Genet. 154, 255-262.

233. Kuriyama, Y., and Luck, D. J. L. (1973) J. Mol. Biol. 73, 425-437.

234. Grimm, M. F., and Lambowitz, A. M. (1979) J. Mol. Biol., in press.

235. Alwine, J. C., Kemp, D. J., and Stark, G. R. (1977) Proc. Nat. Acad. Sci. U.S. 74, 5350-5354.

236. Green, M. R., Goewert, R. R., Lambowitz, A. M., Heckman, J. E., and RajBhandary, U. L. manuscript in preparation.

237. LaPolla, R. J., and Lambowitz, A. M. (1979) J. Biol. Chem., submitted.

238. Aloni, Y., and Attardi, G. (1971) J. Mol. Biol. 55, 251-270.

239. Aloni, Y., and Attardi, G. (1971) Proc. Nat. Acad. Sci. U.S. 68, 1757-1761.

240. Amalric, F., Merkel, C., Gelfand, R., and Attardi, G. (1978) J. Mol. Biol. 118, 1-25.

241. Ojala, D., and Attardi, G. (1977) Plasmid 1, 78-105.

242. Hendler, F., Halbreich, A., Jakovcic, S., Patzer, J., Merten, S., and Rabinowitz, M. (1976) Genetics and Biogenesis of Chloroplasts and Mitochondria, pp. 679-684, Elsevier/North-Holland Biomedical Press, Amsterdam.

243. Faye, G. (1976) FEBS Lett. 69, 167-170.

244. Lambowitz, A. M., Chua, N.-H., and Luck, D. J. L. (1976) J. Mol. Biol. 107, 223-253.

245. LaPolla, R. J., and Lambowitz, A. M. (1977) J. Mol. Biol. 116, 189-205.

246. Collins, R. A., Bertrand, H., LaPolla, R. J., and Lambowitz, A. M. Molec. gen. Genet., in press.

247. Groot, G. S. P., Grivell, L. A., Van Harten-Loosbroek, N., Kreike, J., Moorman, A. F. M., and Van Ommen, G. J. B. (1977) Structure and Function of Energy Transducing Membranes, pp. 177-186, Elsevier, Amsterdam.

248. Douglas, M. G., and Butow, R. A. (1976) Proc. Nat. Acad. Sci. U.S. 73, 1083-1086.

249. Perlman, P. S., Douglas, M. G., Strausberg, R. L., and Butow, R. A. (1977) J. Mol. Biol. 115, 675-694.

250. Butow, R. A., Terpstra, P., and Strausberg, R. L. (1979) Proceedings ICN-UCLA Symposium on Extrachromosomal DNA, in press. Academic Press, N.Y.

251. Collins, R. A., Bertrand, H., LaPolla, R. J., and Lambowitz, A. M., to be submitted.

252. Grivell, L. A., Reijnders, L., and deVries, H. (1971) FEBS Lett. 16, 159-163.

253. Spithill, T. W., English, K. J., Nagley, P., and Linnane, A. W. (1978) Molec. Biol. Rep. 4, 83-86.

254. Singh, A., Mason, T. L., and Zimmerman, R. A. (1978) Molec. gen. Genet. 161, 143-151.

255. Knight, J. A. (1978) Abstr. 9th Intern. Conf. on Yeast Genetics and Molecular Biology, Rochester, New York, p. 26; Nr. 110.

256. Waxman, M. F., Knight, J. A., and Perlman, P. S. (1979) Molec. gen. Genet. 167, 243-250.

257. Morimoto, R., Merten, S., Lewin, A., Martin, N. C., and Rabinowitz, M. (1978) Molec. gen. Genet. 163, 241-255.

258. Sanders, J. P. M., Heyting, C., DiFranco, A., Borst, P., and Slonimski, P. P. (1976) The Genetic Function of Mitochondrial DNA, pp. 259-272. North-Holland, Amsterdam.

259. Linnane, A. W., and Nagley, P. (1978) Plasmid 1, 324-345.

260. Faye, G. (1977) Mitochondria 1977: Genetics and Biogenesis of Mitochondria, pp. 575-578, Walter de Gruyter, Berlin.

261. Lambowitz, A. M. (1979) Methods in Enzymology, Nucleic Acids and Protein Synthesis (Part G), pp. 421-433, Academic Press, New York.

262. Mitchell, M. B., and Mitchell, H. K. (1952) Proc. Nat. Acad. Sci. U.S. 38, 442-449.

263. Lambowitz, A. M. (1976) Genetics and Biogenesis of Chloroplasts and Mitochondria, pp. 713-720, Elsevier/North-Holland Biomedical Press, Amsterdam.

264. Rifkin, M. R., and Luck, D. J. L. (1971) Proc. Nat. Acad. Sci. U.S. 68, 287-290.

265. Kuriyama, Y., and Luck, D. J. L. (1974) J. Mol. Biol. 83, 253-266.

266. Collins, R. A., and Bertrand, H. (1978) Molec. gen. Genet. 161, 267-273.

267. Bertrand, H., and Pittenger, T. H. (1972) Genetics 71, 521-533.

268. Kohout, J., and Bertrand, H. (1976) Can. J. Genet. Cytol. 18, 311-324.

269. Collins, R. A., Bertrand, H., and Lambowitz, A. M., manuscript in preparation.

270. Nargang, F., Bertrand, H., and Collins, R. A. (1979) Curr. Genet. in press.

271. Birky, C. W., Jr. (1975) Genetics and Biogenesis of Mitochondria and Chloroplasts, pp. 182-224, Ohio State University Press, Columbus, Ohio.

272. Spolsky, C. M., and Eisenstadt, J. M. (1972) FEBS Lett. 25, 319-324.

273. Kislev, N., Spolsky, C. M., and Eisenstadt, J. M. (1973) J. Cell Biol. 57, 571-579.

274. Bunn, C. L., Wallace, D. C., and Eisenstadt, J. M. (1974) Proc. Nat. Acad. Sci. U.S. 71, 1681-1685.

275. Wallace, R. B., and Freeman, K. B. (1975) J. Cell Biol. 65, 492-498.

276. Ditta, G., Soderberg, K., and Scheffler, I. E. (1977) Nature (London) 268, 64-67.

277. Wiseman, A., and Attardi, G. (1979) Somatic Cell Genet. 5, 241-262.

278. Spithill, T. W., Nagley, P., and Linnane, A. W. (1979) Mol. gen. Genet. 173, 159-170.

279. Gillham, N. W., Boynton, J. E., and Burkholder, B. (1970) Proc. Nat. Acad. Sci. U.S. 67, 1026-1033.

280. Boynton, J. E., Gillham, N. W., and Chabot, J. F. (1972) J. Cell Sci. 10, 267-305.

281. Goodenough, U. W., Togasaki, R. K., Paszewski, A., and Levine, R. P. (1971) Autonomy and Biogenesis of Mitochondria and Chloroplasts, pp. 224-234, Elsevier/North-Holland, Amsterdam.

INDEX